PRODUCT AND PROCESS DESIGN
FOR
QUALITY, ECONOMY AND RELIABILITY

PRODUCT AND PROCESS DESIGN FOR QUALITY, ECONOMY AND RELIABILITY

(With Solution Manual)

RAO V. DUKKIPATI
Ph.D., P.E.
Fellow of ASME and CSME
Professor and Chair
Graduate Program Director
Department of Mechanical Engineering
Fairfield University
Fairfield, Connecticut, USA

PRADIP K. RAY
Ph.D.
Professor and Head
Department of Industrial Engineering
and Management
Indian Institute of Technology
Kharagpur, India

PUBLISHING FOR ONE WORLD

NEW AGE INTERNATIONAL (P) LIMITED, PUBLISHERS

New Delhi • Bangalore • Chennai • Cochin • Guwahati • Hyderabad
Jalandhar • Kolkata • Lucknow • Mumbai • Ranchi
Visit us at **www.newagepublishers.com**

Published by New Age International (P) Ltd., Publishers
First Edition: 2010

Branches:

- 37/10, 8th Cross (Near Hanuman Temple), Azad Nagar, Chamrajpet, **Bangalore**-560 018. Tel.: (080) 26756823, Telefax: 26756820, E-mail: bangalore@newagepublishers.com
- 26, Damodaran Street, T. Nagar, **Chennai**-600 017. Tel.: (044) 24353401, Telefax: 24351463 E-mail: chennai@newagepublishers.com
- CE-39/1016, Carrier Station Road, Ernakulam South, **Cochin**-682 016. Tel.: (0484) 2377004, Telefax: 4051303 E-mail: cochin@newagepublishers.com
- Hemsen Complex, Mohd. Shah Road, Paltan Bazar, Near Starline Hotel, **Guwahati**-781 008. Tel.: (0361) 2513881 Telefax: 2543669, E-mail: guwahati@newagepublishers.com
- No. 105, 1st Floor, Madhiray Kaveri Tower, 3-2-19, Azam Jahi Road, Nimboliadda, **Hyderabad**-500 027. Tel.: (040) 24652456, Telefax: 24652457, E-mail: hyderabad@newagepublishers.com
- RDB Chambers (Formerly Lotus Cinema) 106A, Ist Floor, S.N. Banerjee Road, **Kolkata**-700 014. Tel.: (033) 22273773, Telefax: 22275247, E-mail: kolkata@newagepublishers.com
- 16-A, Jopling Road, **Lucknow**-226 001. Tel.: (0522) 2209578, 4045297, Telefax: 2204098 E-mail: lucknow@newagepublishers.com
- 142C, Victor House, Ground Floor, N.M. Joshi Marg, Lower Parel, **Mumbai**-400 013. Tel.: (022) 24927869 Telefax: 24915415, E-mail: mumbai@newagepublishers.com
- 22, Golden House, Daryaganj, **New Delhi**-110 002. Tel.: (011) 23262370, 23262368, Telefax: 43551305 E-mail: sales@newagepublishers.com

ISBN: 978-81-224-2661-8

Rs. 895.00

C-10-03-4360

Printed in India at Glorious Printers, Delhi.
Typeset at Goswami Associates, Delhi.

PUBLISHING FOR ONE WORLD

NEW AGE INTERNATIONAL (P) LIMITED, PUBLISHERS
4835/24, Ansari Road, Daryaganj, New Delhi - 110002
Visit us at **www.newagepublishers.com**

Preface

Introduction

Design is the central activity of the engineering profession. It is an interdisciplinary activity and involves decision-making, optimization, engineering economy, planning, and applied statistics. Design engineers and managers of technology need to have the knowledge of these subjects. Considerations of quality and reliability permit a product to achieve a desired performance throughout its service life, thereby satisfying those who have purchased it. Reliability is achieved through careful thought and design. Furthermore, economy of manufacture is a natural outcome of that same careful thought and design. This book presents the prospective manager of technology in the considerations leading to creation of cost-effective products of quality, and presents instruction in the following: (1) An introduction to the book, (2) Basic statistics and Probability, (3) Product and Process Design, (4) Quality Principles in Design, (5) Engineering Economic Analysis, (6) Cost Evaluation, (7) Design for Reliability, (8) Quality by Experimental Design, and (9) Robust Design and Taguchi Method.

Audience

This book is a comprehensive text on product and process design for quality, economy and reliability. It is self-contained and the subject matter is presented in an organized and systematic manner. No previous knowledge of quality, basic statistics and probability, economy or reliability is assumed. This book is quite appropriate for several groups of people including:

- Graduate students taking the course "Product and Process Design for Quality, Economy and Reliability".
- The book can be adapted for a short professional course on the subject matter.
- Design and research engineers will be able to draw upon the book in selecting and developing mathematical models for analytical and design purposes.
- Practicing engineers and managers who want to learn about the basic principles and concepts involved in design for quality, economy and reliability, and how it can be applied at their own work place concerns.

Because the book is aimed at a wider audience, the level of mathematics is kept intentionally low. All the principles presented in the book are illustrated by numerous worked examples. The book draws a balance between theory and practice.

Contents

Each chapter in this book consists of a concise but thorough fundamental statement of the theory, principles and methods, followed by a selected number of illustrative worked examples. A number of unsolved sample problems for student's practice, to amplify and extend the theory are also included. Numerous references provided at the end of chapters serve as helpful sources for further study and research by interested readers. The distinctive features of the book are as follows: Chapter 1 reviews the necessary introductory background and principles underlying the product and process design for quality, economic and reliability considerations. Chapter 2 deals with the basic concepts in probability and applied statistics. Chapter 3 outlines the necessary background and principles underlying the Product and Process Design including the design methodologies, the process of considering concurrently the requirements of assembly and manufacturing with design requirements in order to reduce unit cost of production, improves quality, and reduces total lead-time. It involves managing the mutual dependences among all phases of the product life cycle. This chapter also details the various tools and techniques that are used in the manufacturing phase of product development. Chapter 4 presents quality principles in design and management, quality considerations in all the phases of product development from marketing to disposal, and the methods of the idea of quality evaluation throughout the system's entire life cycle. It demonstrates the importance of quality control and quality improvement in quality systems of modern organizations including the issues of quality of conformance and quality of performance to ensure customer satisfaction. This chapter also discusses the concepts of quality engineering and design robustness of products and processes as proposed by Taguchi. Chapter 5 details the engineering economic analysis or the decision making process. This chapter isolates those problems that are commonly faced by engineers and develop the tools to properly grasp, analyze, and solve them. Engineering decision-making process deals with the solution of substantial engineering problems where economic aspects dominate and economic efficiency is the criterion for selecting from possible alternatives. Some of the aspects of engineering decision making include: the cost-accounting systems, the various consequences- costs and benefits- of an alternative, the economic criteria for judging alternatives, selection among the alternatives and the post audit of results. Chapter 6 deals with the basics of cost evaluation in order to produce high-functioning, low-cost designs starting from the conceptual design through embodiment and detail design stage. Chapter 7 covers the fundamental concepts, models, and analysis techniques necessary to perform reliability engineering. Included are the basic reliability models, covariate models, and hazard rate functions, including the exponential, Weibull, normal, and lognormal; systems reliability including redundant, standby, and load sharing systems; reliability allocation and fault tree analysis. Chapter 8 treats the design of quality by experimental design. Chapters 9 and 10 deal with the two aspects of quality engineering, namely, design optimization and evaluation for quality. Design optimization involves the application of orthogonal arrays, linear graphs and signal-to-noise ratio in the design of experiments. Taguchi's methodology permits engineers to evaluate a product design with respect to robustness against noise, and the cost involved. Evaluation of quality, on the other hand, involves the application of loss function and signal-to-noise ratio. The main goal is the optimization of product/process design for minimal sensitivity to noise.

In addition, glossary of terms, symbols and appendices used in the book are also summarized towards the end of each chapter. Answers to selected end- of-chapter problems are given at the end of the book. All end-of chapter problems are fully solved in the Solution Manual.

—Authors

Acknowledgements

Our sincere acknowledgement is extended to all those who have had a direct impact on this work. Many people working in the general areas of design, quality, economy and reliability engineering have influenced the format of this book. The authors are indebted to their colleagues and to numerous authors who have made contributions to the literature in this field. In addition, we greatly owe our indebtedness to all the authors of the articles listed at the end of each Chapter in this book. The authors would like to thank the reviewers for their efforts and for the comments and suggestions, which have well served to compile the best possible book for the intent and targeted audience. Finally, we would very much like to acknowledge the encouragement, patience, and support provided by our family members: Sudha, Ravi, Madhavi, Anand, Ashwin, Raghav, Vishwa, and Bela and Poushali who have also a share in all the pain, frustration, and fun of producing a manuscript.

We would appreciate being informed of errors, or receiving other comments about the book. Please write to the authors' e-mail to us at Rdukkipati@mail.fairfield.edu or pkr@hijli.iitkgp.ernet.in.

Rao V. Dukkipati

Pradip K. Ray

Contents

CHAPTER 1

Introduction

This introductory chapter provides an overview of quality and reliability engineering techniques, the relationship between reliability and cost, and reliability, design, and assessment. This chapter also presents the scope and organization of the book.

1.1 INTRODUCTION

Design involves innovation and requires knowledge from the science and engineering fields. It includes the selection of materials, parts, and tentative parameter values in the product design stage and the selection of production equipment and tentative values for process factors in the process design stage.

The term *quality* has been defined in different ways by various authors. Crosby (1979) defines quality as, *quality is conformance to requirements or specifications.* Juran (1974) proposes the definition for quality as *quality is fitness for use.* A more general definition for quality is given as follows: *the quality of a product or service is the fitness of that product or service for meeting or exceeding its intended use as required by the customer.* The term quality *implies* different levels of expectations for different groups of consumers or customers. Quality should be designed into a product.

There are three aspects generally associated with the definition of quality. They are:

(*a*) quality of design

(*b*) quality of conformance

(*c*) quality of performance

Quality of Design: Quality of design focuses on the stringent conditions that the product must minimally possess to satisfy the requirements of a customer. In other words, the product must be designed to meet at least the minimal needs of the customer. Quality of design is generally influenced by factors including the type of product, cost, profit policy or criteria, product demand, materials and parts availability, and product safety. In general, the effect of an increase in the designed quality level will increase the cost at an exponential rate. On the other hand, the value of the product increases at a decreasing rate. The rate of increase approaches zero beyond a certain designed quality level.

Quality of Conformance: Quality of conformance requires that the manufactured product must meet the standards selected in the design phase. This would involve defect prevention, defect finding, and defect analysis and rectification. Defect prevention is accomplished using statistical process control techniques. Defects are usually located using inspection, test, and statistical analysis of data from the process. Once the sources or causes for the presence of defects are determined, then corrective measures or actions are taken accordingly.

Quality of Performance: Quality of performance focuses on how well the product performs when put to use. Quality of performance is a function of both quality of design and the quality of conformance. If a product does not meet the specified expectations, then adjustments are to be made in the design or conformance phase.

Reliability is described as *quality in the time dimension*. Reliability of an item or a product is defined as *the probability that the item or the product will perform satisfactorily for a specified period of time under a stated set of use conditions*. An item must also operate satisfactorily for an acceptable period of time in the field application for which the item is intended. This definition of reliability stresses the following four elements: probability, performance requirements, time, and use conditions. Probability is the likelihood of an event's occurrence or nonoccurrence. The value of probability is between 0 and 1. Performance requirements clearly define or describe what is considered to be satisfactory operation. Time is the period during which satisfactory performance is expected, and use conditions to describe the environmental conditions under which the item is expected to function.

To ensure customer satisfaction in the performance phase, measures to improve reliability in the design stage need to be addressed. In order to determine the reliability of a component, product or process, improvement is stressed to understand its design functionality. Invariably, the functionality relating to the concept of *failure rate* needs to be understood. A failure rate is a measure of the number of malfunctions occurring per unit of time. Most products, involve many components and require calculating the system reliability. Once designed and produced most components or products go through three distinct phases from product inception to wear-out. They are the *debugging phase*, the *chance-failure phase,* and the *wear-out phase* as shown in Fig. 1.1. This curve is often referred to as the *bathtub curve.* Figure 1.1 is a typical *life-cycle curve* or the *life-characteristic curve* showing the failure rate as a function of increasing age (hours/cycles). The *debugging phase* is also known as the *early-mortality phase* or the *burn-in period.* This phase shows a rapid decrease in the failure rate due to the fact that the initial problems found during the course of prototype testing are fixed. This phase stabilizes at an

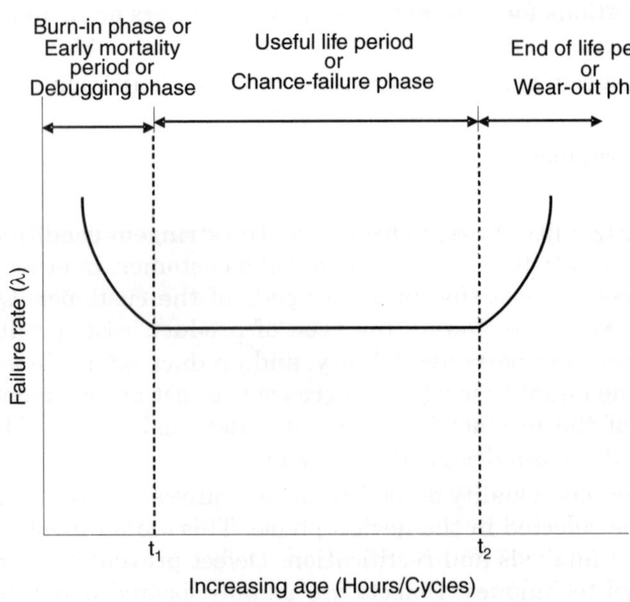

Fig. 1.1 A Typical Life-cycle Curve (Bathtub Curve).

approximate value of time t_1 when the weak units have died out as shown in Fig. 1.1. This may be the result of design compromises or flaws due to faulty workmanship, transportation damage or installation or setup errors. This initial failure rate is usually found in new components, products or systems. Examples of these failures are: poor welds, seals, solder joints, contamination on surfaces or in materials, voids, cracks in insulation or coatings, and chemical impurities in metals etc. These early failures can be avoided by improving the manufacturing processes or in design phase itself. These early failures are discovered during in-process and final tests, life tests, environmental tests and process audits.

During the *chance-failure phase* or the *useful-life period*, failures occur randomly and independently. The failure rate is constant and corresponds to the useful life of the component or the product. This period is characterized mainly by the occurrence of stress related failures. The exponential failure distribution is commonly used to approximate the time period. This period depends on the component, product or process and is considered the most significant period for reliability prediction and assessment activities. Humidity, vibration, shock and altitude contribute to the failure of components, products or systems. Point t_2 in Fig. 1.1 indicates the end of useful life or the start of wear out. Beyond this point, the failure rate increases rapidly.

In the *wear out phase* or the *end-of-life period*, an increase in the failure rate is noticed. In this situation, components or products or systems age and wear out at the end of their useful life. When the components, products or systems failure rate due to wear out becomes unacceptably high, replacement, or repair of that component or product or system should be made. Wear out failures are mainly due to deterioration of the design strength of the item as a consequence of operation and exposure to environmental fluctuations. Some of the chemical or physical phenomena causing the deterioration of the design strength of the product or component are corrosion or oxidation, insulation breakdown or leakage, frictional wear or fatigue, shrinkage and cracking in plastic materials and ironic migration of metals in vacuum or on surfaces.

Reliability is an inherent attribute of a system. The reliability level is established at the design phase. The designer in order to engineer reliability into products or processes during its basic design stage requires design data, guidelines for safety, mission, maintenance and cost factors together with basic elements of concepts of reliability, engineering, system engineering and cost effectiveness.

1.2 RELATIONSHIP BETWEEN RELIABILITY DESIGN AND COST

Design to cost represents a constraint added to the task of designing a system or component, which will meet performance, reliability, maintainability and cost goals. This means that during design stage one must aim at a balanced design, which will (1) maximize performance within unit cost goals and (2) minimize support cost in order to minimize life-cycle costs. A design that minimizes support cost involves the application of reliability principles during the design phase.

Life cycle cost represents all costs incurred from the point at which a decision is made to acquire a system through operational life and eventual disposal of the system. A total life cycle cost is represented by costs collected in two categories: (1) system acquisition costs, and (2) logistics and support costs.

The total life cycle cost can be expressed as

$$LCC = AC + LSC \qquad ...(1.1)$$

where, LCC = life cycle cost
AC = acquisition cost
LSC = logistic support cost

Acquisition cost elements include design and development (basic engineering, test and evaluation, experimental tooling, system management), manufacturing and quality engineering, fabrication, production tooling, quality control, test equipment, facilities, initial spares, and training. The major elements comprising the logistic and support cost include spares, personnel and training, overhaul and lower echelon maintenance facilities, and logistic factors.

1.3 RELIABILITY ASSESSMENT

There are three steps used in the assessment of the reliability of a system. These steps are as follows:

1. Development of a reliability model based on system (design) functionality

The developed model should be comprehensive enough to represent the main operational and other features of the system. The validity of the assumptions should be checked carefully. The developed model represents subsystems or subunits, with individual models and structural diagrams interconnected to emulate the overall behavior of the complete system under study. It is recommended to model the system into several blocks that are functionally independent of one another if possible.

2. Analyze the model

The analysis of the model includes the determination of the reliability performance measures from the block diagrams and the corresponding reliability parameters and functions. Generally, probabilistic and combinatorial aspects of the system reliability and approximate methods are used in this step.

3. Evaluation and interpretation of results

If the results obtained are not accurate and precise enough, the statistical techniques are used to establish their confidence intervals or significance levels. The validation of the model results can be accomplished by comparisons with data from field tests for similar systems or the same system. The determination of the system reliability also depends on a variety of economic and other considerations.

1.4 SCOPE AND ORGANIZATION OF THE BOOK

Noting that the role of management is to plan and organize activities that effectively apply and control the use of necessary resources, in order to achieve desired outcomes, the objectives of this text are established so that readers may learn methods by which this may be accomplished, in context with the activities of design and manufacturing. Particular emphasis is placed upon the creation of reliable and cost-effective products. Specific objectives for learning are as follows:

1. To understand the techniques for planning projects and the methods by which the critical functions of design and manufacturing may be unified in order to achieve a design activity that smoothly evolves into an efficient manufacturing process in which critical considerations may be concurrently addressed to the full capability of all necessary resources. The expected learning outcomes are:

(*a*) How to plan project activities ?

(*b*) How to plan cross-functional activities to achieve an efficient concurrent engineering process ?

(*c*) To understand the process and methodology for functional and conceptual design.

2. To learn methods for assessing the progress of on-going activities and instituting actions to control them. The expected learning outcomes are:

(*a*) When and how to apply the design review process ?

(*b*) How to initiate and control cross-functional design activities ?

3. To learn and apply appropriate analytical and experimental methods leading to the creation of reliable and robust products. The expected learning outcomes are:

(*a*) How to apply effective methods of concept formulations ?

(*b*) How to apply Quality Function Deployment (QFD) ?

(*c*) How to access and improve manufacturing process capability ?

(*d*) How to affect Robust Design using orthogonal arrays in *Design of Experiment* techniques ?

4. To learn and apply appropriate procedures for assuring the cost effectiveness of new products. The expected learning outcomes are:

(*a*) How to treat cost as explicit design criteria ?

(*b*) How cost estimation is accomplished ?

(*c*) How to apply the Loss Function to assure that quality is not compromised ?

Chapter 1: Introduction

This introductory chapter provides an overview of quality and reliability engineering techniques, design, the relationship between reliability and cost, reliability assessment, and the scope and organization of the book.

Chapter 2: Basic Probability and Statistics

This chapter introduces the reader to the basic concepts and definitions, population and sample, types of variables, graphical representation of qualitative and quantitative data, numerical summary measures, measures of central tendency for ungrouped data (mean, median and mode), measures of dispersion for ungrouped data (range, variance and standard deviation), mean, variance and standard deviation for grouped data, measures of position including quartiles, percentiles and box plot.

The basics on probability, events (simple and compound), Venn diagram, tree diagram, approaches to probability (classical, relative, frequency concept of probability, subjective probability, marginal probability and conditional probability) are presented. Special multiplication rule and Baye's Formula are also briefly presented. Random variables (both discontinuous and continuous) and their mean and standard deviation are discussed. Permutations and combinations are defined briefly. Discrete probability distributions (hyper-geometric, binomial and Poisson distributions) and their mean and standard deviations are derived. Continuous probability distribution, namely, the normal distribution, its properties, mean and variance and the standard normal distribution are presented. Approximate probability distributions are also presented. They include binomial approximation to the hyper-geometric, Poisson approximate to the binomial and normal approximation to the binomial are briefly discussed. Chebyshev's theorem and empirical rule for distribution are presented.

Chapter 3: Product and Process Design

This chapter introduces a number of topics, such as Important Approaches in Product Development, Stages in Product Development Process, Tools and Techniques used in Product Development, Technical Risk Assessment in Product Development, Analysis of Customer Needs and Product Development, Common Approaches for Customer Needs Analysis, Conceptual Design Stage, Detailed Design including Design Analysis, Modeling and Simulation, Test and Evaluation, and Manufacturing.

This chapter discusses in detail the various aspects of product and process design. In view of critical importance of product development and its contribution to a company's ability to succeed in a fiercely competitive market of a global economy, it is essential that the designers use a number of approaches at different stages of product development. Several tools and techniques commonly used in this context are described. Comparisons between the traditional serial engineering and simultaneous or concurrent engineering are clearly indicated. Technical risk assessment, a critical issue at product and process design, is emphasized in the approaches mentioned, and common approaches for customer needs analysis are described in detail. The reasons for forming design teams when the design is complex requiring inter-disciplinary approach are mentioned.

Various stages in product development are mentioned and explained in detail. Highlighting the important aspects and difficulties, and problems in the conceptual design stage, the activities in detailed design, such as modeling and simulation as well as test and evaluation are elaborately discussed. Different manufacturing tools and techniques as practised in modern organizations are briefly presented in this chapter.

Chapter 4: Quality Principles in Design

This chapter discusses the various principles that have been introduced and practised by organizations worldwide to design, produce, and maintain quality of products and services. Emphasizing the importance of quality in engineering design, the evolution of quality over the years in modern organizations is discussed, and the concept of 'quality loop' and its elements, from marketing through production to disposal, are explained with identification of quality-related activities at all elements. Comparing an inspection-based quality control with a prevention-based quality control, and elaborating the systems approach for quality problem solving at all three aspects–quality of design, of conformance, and of performance, several quality principles for product development, such as extended process, customer-driven improvement cycle, constancy and consistency of purpose for quality improvement, rules for inspection, principles of supplier selection, process improvement cycle, variability reduction including loss function, and principles of 'zero defects' are discussed in this chapter. Basic issues related to Taguchi's concept of quality engineering, Juran's quality trilogy process, and Crosby's 'absolutes' of quality management, that are useful for creation and management of both product and process quality are presented in the context of quality control and improvement in modern business and manufacturing organizations.

Chapter 5: Engineering Economic Analysis

In this chapter we consider the subject of engineering economic analysis, with particular emphasis on economic analysis in design and manufacturing. Economic analysis involves techniques for comparing and making decisions between alternatives on the basis of monetary and economic objectives. With the ever increasing complexity of our manufacturing and industrial technology, economic decision-making is becoming more critical and essential. In addition, investment decisions are important in determining the success or failure and profitability of a firm.

Engineering economic decisions between alternatives can be made by the engineer or manager of technology considering options in the design of components, systems, facilities or man-machine systems. However, considering a number of investment opportunities and alternatives within each opportunity, best makes these decisions.

Our objectives are the following:

1. The fundamental concepts involving the *time value of money*.
2. Various economic measures for determining a projects' worth. The measures considered are the annual worth, future worth, present worth, rates of return, and payback period.
3. Comparing investment alternatives, sensitivity analysis and replacing capital assets.
4. Incorporation of depreciation allowances and income taxes in economic analyses.
5. Benefit cost analysis for the economic analysis of projects.
6. Profitability of investments.
7. Sensitivity and break-even analysis.
8. Capitalized cost analysis.
9. Uncertainty and risk in economic analysis.

Chapter 6: Cost Evaluation

The American Association of Cost Engineers (AACE) defines cost engineering as *that area of engineering practice where engineering judgment and experience are utilized in the application of scientific principles and techniques to the problems of cost estimation, cost control and profitability*. Cost estimation is a complex process and requires knowledge of principles of engineering, materials, manufacturing processes, quality control and inspection methods, product and environmental safety, repair and servicing and principles of costing. Cost estimation is an important aspect of the product evaluation process. It is advisable to generate a cost estimate early in the design process in order to compare the product design with original objectives. The design of new components or products requires cost estimates from scratch while redesigns preliminary cost estimates will serve the purpose as the current costs are known. The design engineer shares the responsibility of generating the cost of manufactured products or components and selling prices of the products or components along with the cost-estimating department. Cost is next to performance in importance. Lowest cost design stands a good chance to be successful in a free market place.

In this chapter, we consider the subject of economy, with particular emphasis on cost evaluation in design and manufacturing. The identification of the elements of costs, general concepts of cost evaluation and some of the more generally accepted cost evaluation methods and models are described.

Chapter 7: Design for Reliability Predictions and Models

Although it is possible in some cases to develop failure models based on the mechanics of failure, the procedure is rather complex and needs a considerable amount of study and analysis. Four reliability functions are: the failure density function $f(t)$, hazard function (t), failure distribution function $Q(t)$ and reliability function $R(t)$ were defined. They were developed as limiting cases of the respective *p*-linear functions constructed using catastrophic failure data. The mean time to failure (MTTF), median time to failure and average failure rate over time interval were introduced and their determination using reliability functions was presented. Both constant failure rate (CFR) models and time-dependent failure models were presented and discussed. Complex systems are decomposed into functional entities composed of components, units, and subsystems for the purpose of reliability analysis. Combinational aspects of reliability analysis are used to connect the components in series, parallel, series-parallel or meshed structures, or in any combination of these. Probability concepts are applied to determine the reliability of systems in terms of the reliabilities of its subsystems.

Reliability allocation methods including the Aeronautical Radio Inc. (ARINC) method and the Advisory Group on Reliability of Electronic Equipment (AGREE) methods were presented and discussed. Fault trees, which are used to analyze complex systems, are presented.

Chapter 8: Quality by Experimental Design

This chapter discusses the fundamental concepts of experimental design in the context of ever-increasing demand of improved quality in products and services by the customers. Highlighting the important features and terminologies in experimental design and its uses and advantages, this chapter explains in detail the traditional experimental designs, such as completely randomized design, randomized block design, and Latin square design. The procedure of measuring the main and interaction effects on the response or output variable in an experiment is clearly described. The evaluation process using ANOVA procedures in this context is illustrated with examples.

The procedures and significance of conducting factorial experiment are presented in this chapter. While a full factorial experiment represents all possible treatment combinations of the factors considered, it may not be desirable because of time and resource constraints in many situations. A fractional factorial experiment, which is a subset of the full factorial design, selects the treatment combinations in such a way that the desirable effects can be estimated. For factorial experiments, the role of contrasts including orthogonal contrasts in estimating the treatment effects is explained in detail. The procedure of blocking by using the principle of confounding to increase the precision of the experiment is illustrated with a number of examples. The procedures for selecting the fractional replicates of the full factorial design to reduce the size of the experiment are also presented in this chapter. Overall, the experimental designs, as they are used in achieving the objective of improving and sustaining the quality of products and processes with reduced cost, are discussed at length in this chapter.

Chapter 9: Robust Design and Taguchi Methods

Taguchi has made a significant contribution in quality improvement and statistical methods. Taguchi's quality improvement activities which include product and process design provide a system to develop functional specifications, incorporate those specifications into a product and/or process and optimize the system to surpass the desired specifications. *Quality engineering* has the objective of designing quality into every product and corresponding process. Quality engineering directs quality improvement efforts from the manufacturing process to the design phase and hence is often called as an *off-line quality control* method. The techniques of statistical process control that are discussed in Chapter 8 are referred to as *on-line quality control* methods. Off-line methods reduce product development and lifetime costs and improve product manufacturability. Taguchi method seeks to minimize the effects of noise and to determine the optimal level of the important controllable factors based on the concept of *robustness*. In

Taguchi's method, quality is measured by the deviation of a characteristic from its target value. Uncontrollable factors also called as *noise* cause these deviations and thereby lead to a loss. A *loss function* is defined for this deviation. Thus, Taguchi's method seeks to create a product/process design that is insensitive to all possible combinations of the uncontrollable noise factors and at the same time is effective and cost efficient as a result of setting the key controllable factors at certain levels. Cost is a key consideration in Taguchi's method, with the objective of obtaining the best possible design at the lowest possible cost.

1.5 FURTHER READING

This book is intended as an introduction to the vast subject of design for quality, economy and reliability. Many excellent books and articles are available that deal with individual topics or branches of quality, economy, reliability, design, applications and related fields at varying levels of treatment. The list of references at the end of each chapter of this book is extensive, although not necessarily exhaustive, and will serve the reader well to gain further knowledge about and insight into the subject matter.

CHAPTER 2

Basic Probability and Statistics

This chapter introduces the reader the basic concepts and definitions, population and sample, types of variables, graphical representation of qualitative and quantitative data, numerical summary measures, measure of central tendency for ungrouped data (mean, median and mode), measures of dispersion for ungrouped data (range, variance and standard deviation), mean, variance and standard deviation for grouped data, and measures of position including quartiles, and percentiles. The basics on probability, events (simple and compound), Venn diagram, tree diagram, approaches to probability (classical, relative frequency, concept of probability, subjective probability, marginal probability and conditional probability) are presented. Special multiplication rule and Baye's formula are also briefly presented. Random variables (both discrete and continuous) and their mean and standard deviation are discussed. Permutations and combinations are defined briefly. Discrete probability distributions (Hypergeometric, binomial and Poisson distributions) and their mean and standard deviations are derived. Continuous probability distribution, namely, the normal distribution, its properties, mean and variance and the standard normal distribution are presented. Approximating probability distributions including binomial approximation to the hypergeometric, Poisson approximate to the binomial and normal approximation to the binomial are briefly discussed. Chebyshev's theorem and empirical rule for distribution are presented.

2.1 INTRODUCTION

Statistics refers to the field or discipline of study. *Statistics* is a group of methods that are used to collect, analyze, present, and interpret data and to make decisions. Decisions made by using statistical methods are called *educated guesses*. *Statistics* has two areas: *theoretical* or *mathematical* and *applied statistics*. *Theoretical* or *mathematical statistics* deals with the development, derivation, and proof of statistical theorems, formula, rules and laws. *Applied statistics* concerns the applications of those theorems, formulas, rules and laws which are used to solve real-world problems.

Applied statistics is divided into two branches: *descriptive statistics* and *inferential statistics*. *Descriptive statistics* involves methods for organizing, displaying and describing data by using tables, graphs and summary measures. The collection of all elements of interest is called a *population* in statistics. The selection of a few elements from this population is called a *sample*. Statistics that deals with decision-making procedures is called *inferential statistics*. This branch of statistics is also known as *inductive reasoning or inductive statistics*. *Inferential statistics* consists of methods that employ sample results to help make decisions or predictions about a population.

Probability, which measures the likelihood that an event will occur, is an important part of statistics. It is the basis of inferential statistics, where decisions are made under conditions of uncertainty. Probability theory is used to evaluate the uncertainty involved in those decisions. Probability statements are about occurrence or non-occurrence of a certain event under certain conditions.

2.1.1 Population and Sample

A *population* consists of all elements—individuals, items, or objects, whose characteristics are being studied. The population that is being studied is also called the *target population*. The collection of a few elements selected from a population is called *a sample*. The collection of information from the elements of a population or a sample is called a *survey*. A *survey* that includes each and every element of the population is called a *census*. A survey conducted on a sample is called a *sample survey*. A sample that represents the characteristics of the population as closely as possible is called a *representative sample*.

A sample drawn in such a way that each element of the population has same chance of being selected is called a *random sample*. If the chance of being selected is the same for each element of the population, it is called a *simple random sample*. An *element* or *member* of a sample or population is a specific subject or object about which the information is collected. A *variable* is a characteristic under study that assumes different values for different elements. In relation to a variable, the value of a *constant* is fixed. The value of a variable for an element is called an *observation or measurement*. A *data set* is a collection of observations on one or more variables.

2.1.2 Types of Variables

A variable that can be measured numerically is called a *quantitative variable*. The data collected on a quantitative variable are known as *quantitative data*. Quantitative variable is classified as either *discrete variable* or *continuous variable*. A variable whose values are continuous is called a *continuous variable*. Thus, a *discrete variable* can assume only certain values with no intermediate values. On the other hand, a variable that can assume any numerical value over a certain interval or intervals is called a *continuous variable*. The time taken to complete a trip to the next city is an example of a continuous random variable because it can assume any value, say, between one and two hours. A variable that cannot assume a numerical value but can be classified into two or more nonnumeric categories is called a *qualitative* or categorical variable. The data collected on such a variable are known as *qualitative data*.

Quantitative

Continuous: Weight, Height, Line width measurements

Discrete: Number of students, Number of accidents

Qualitative

Nominal: Black, Green, Yellow, etc.

Ordinal: High, Medium, Low, etc.

Ranked: 1, 2, 3, etc.

Data collected on different elements at the same point in time or for the same period of time are called *cross-section data*.

2.1.3 Organizing Data

Data recorded in the sequence in which they are collected and before they are processed or ranked are called *raw data*. To have meaningful information from the raw data, the obtained data are to be organized in several ways. The methods to be employed for this depends on the types of data.

2.1.3.1 Qualitative data

A *frequency distribution* for qualitative data lists all categories and the number of elements that belong to each of the categories. A *relative frequency distribution* lists the relative frequencies for all categories.

$$\text{Relative frequency of a category} = \frac{\text{Frequency of that category}}{\text{Sum of all frequencies}} \qquad ...(2.1)$$

A *percentage distribution* lists the percentages for all categories.

$$\text{Percentage} = \text{Relative frequency} \times 100 \qquad ...(2.2)$$

2.1.3.2 Graphical representation of qualitative data

The *bar graph* and the *pie chart* are two types of graphs used to represent or display qualitative data. A graph made of bars whose heights represent the frequencies of representative categories is called a *bar graph*. A *pie chart* is more commonly used to display percentages, although it can be used to display frequencies or relative frequencies. The pie chart represents the total sample or population, and is divided into different portions that represent the percentages of the population or sample belonging to different categories.

2.1.3.3 Graphical representation of quantitative data

A *frequency distribution* for quantitative data lists all the classes and the number of values that belong to each class. Data presented in the form of a frequency distribution are called *grouped data*. The *class boundary* is the midpoint of the upper limit of one class and the lower limit of the next class, while class width (also known as class size) = upper boundary – lower boundary.

$$\text{The } \textit{class} \text{ midpoint or } \textit{mark} = \frac{\text{Lower limit} + \text{upper limit}}{2} \qquad ...(2.3)$$

$$\text{Approximate class width} = \frac{\text{Largest value} - \text{smallest value}}{\text{Number of classes}} \qquad ...(2.4)$$

The relative frequencies and percentages for a quantitative data set are calculated as follows:

$$\text{Relative frequency of a class} = \frac{\text{Frequency of that class}}{\text{Sum of all frequencies}} = \frac{f}{\sum f} \qquad ...(2.5)$$

$$\text{Percentage} = \text{Relative frequency} \times 100$$

A *histogram* is a graph in which classes are marked on the horizontal axis and the frequencies, relative frequencies, or percentages are marked on the vertical axis. The frequencies, relative frequencies, or percentages are represented by the heights of the bars.

The most common shapes of histograms are: *symmetric, skewed, and uniform* or *rectangular*. A *symmetric histogram* is identical on both sides of its central point. A *skewed histogram* is nonsymmetrical. For a skewed histogram, the tail on one side is longer than the tail on the other side. A *skewed-to-the-right* histogram has a longer tail on the right side while a skewed-to-the-left histogram has a longer tail on the left side. A *uniform* or *rectangular histogram* has the same frequency for each class.

A *polygon* is another method that can be used to represent quantitative data in graphical form. A graph formed by joining the midpoints of the tops of successive bars in a histogram with straight lines is called a *polygon.* A polygon with relative frequencies marked on the vertical axis is called a *relative frequency polygon.* In a similar manner, a polygon with percentages marked on the vertical axis is called a *percentage polygon.*

A *cumulative frequency distribution* gives the total number of values that fall below the upper boundary of each class.

$$\text{Cumulative relative frequency} = \frac{\text{Cumulative frequency}}{\text{Total observations in the data set}} \quad ...(2.6)$$

$$\text{Cumulative frequency} = \text{Cumulative relative frequency} \times 100 \quad ...(2.7)$$

When plotted on a graph, the cumulative frequencies give a curve that is called an *ogive.* Thus, an *ogive* is a curve drawn for the cumulative frequency distribution by joining the straight lines with the dots marked above the upper boundaries of classes at height equal to the cumulative frequencies of respective classes.

Another technique that is used to represent quantitative data in condensed form is the *stem-and-leaf-display*. In a *stem-and-leaf display* of quantitative data, each value is divided into two portions: a stem and a leaf. The leaves for each stem are shown separately in a display.

Example E2.1. *The frequency distribution of ages of all 60 part-time graduate students in a Master of Science in Management of Technology degree program of a university are given below.*

Age (years)	Number of students
21–25	*22*
26–30	*13*
31–35	*11*
36–40	*8*
41–45	*6*

(*a*) *Prepare a cumulative frequency distribution table.*

(*b*) *Calculate the cumulative relative frequencies and cumulative percentages for all classes.*

(*c*) *Find the percentage of students who are 31 years of age or older.*

(*d*) *Using the ogive, determine the percentage of students who are 35 years or younger.*

Solution:

(*a*) and (*b*): The cumulative frequency, cumulative relative frequency and percentages are shown in Table E2.1.

Table E2.1

Age (years)	Frequency	Cumulate frequency	Cumulative relative frequency	Cumulative percentage
21–25	22	22	0.366	36.6
26–30	13	35	0.583	58.3
31–35	11	46	0.766	76.6
36–40	8	54	0.90	90.0
41–45	6	60	1.00	100.0

(*c*) Percentage of students who are 31 years of age or older = 100 – 58.3 = 41.7%

(*d*) The percentage of students who are 35 years or younger are about 70%

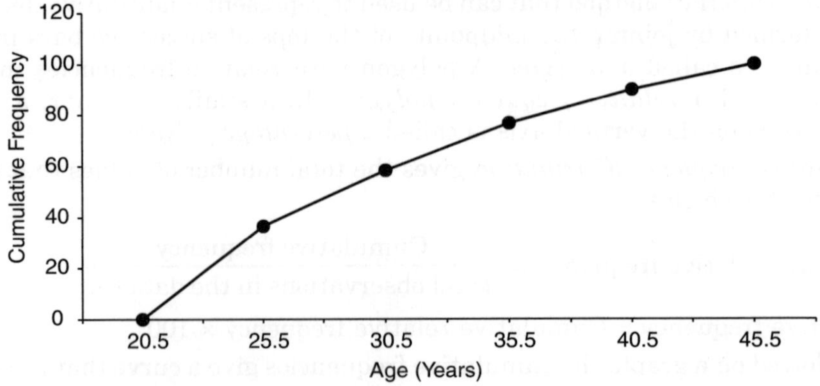

Fig. E2.1 Ogive Diagram for the Cumulative Percentage Distribution.

2.2 NUMERICAL SUMMARY MEASURES

There are three basic numerical summary (descriptive) measures used to organize and display large data sets. They are

1. Measures of central tendency
2. Measures of dispersion or scatter
3. Measures of position

2.2.1 Measures of Central Tendency for Ungrouped Data

A data set is generally represented by numerical summary measures called the *typical values*. A measure of central tendency gives the center of a histogram or a frequency distribution curve. There are three measures of central tendency: the mean, the median, and the mode.

2.2.1.1 Mean for ungrouped data

The mean, also called the *arithmetic mean* or *average*, for ungrouped data is obtained by dividing the sum of all values by the number of values in the data set.

Thus

Mean for population data, $$\mu = \frac{\sum X}{N} \qquad ...(2.8)$$

Mean for sample data, $$\bar{X} = \frac{\sum X}{n} \qquad ...(2.9)$$

Where μ is the population mean, ΣX the sum of all values, N the population size, $\bar{X}$ the sample mean, and n the sample size. Sometimes the data may contain a few very small or a few very large values. Such values are called *outliers* or *extreme values*.

2.2.1.2 Median

The median is the value of the middle term in a data set that has been ranked in either increasing or decreasing order. Median for ungrouped data = the value of the $\left(\frac{n+1}{2}\right)^{\text{th}}$ term in a ranked data set. If the given data set represents a population, replace n by N. If the number of observations in a data set is odd, then the median is given by the value of the middle term in the ranked data. Similarly, if the number of observations is even, then the median is the average of the values of the two middle terms.

2.2.1.3 Mode

The mode for an ungrouped data is the value that occurs with the highest frequency in a data set. A data set with only one value occurring with highest frequency has only one mode and such a data set is called *unimodal.* A data set with two values occurring with the same (highest) frequency has two modes. The distribution in this case is called *bimodal.* If more than two values in a data set occur with the same (highest) frequency, then the data set contains more than two modes and it is said to be *multi-modal.* It should be noted here that a data set with each value occurring only once has no mode.

2.2.2 Measures of Dispersion for Ungrouped Data

The mean, median and mode are not sufficient measures to reveal the shape of the distribution of a data set. The measures that show the spread of a data set are called the *measures of dispersion*. There are three measures of dispersion: range, variance and standard deviation.

2.2.2.1 Range

The range is the difference between the largest and smallest values in the data.

The range for ungrouped data = Largest value – Smallest value.

The range is not a very satisfactory measure of dispersion.

2.2.2.2 Variance and standard deviation

The standard deviation is the most used measure of dispersion. The formulas for calculating the variance and standard deviation are as follows:

Variance for ungrouped data

$$\sigma^2 = \frac{\sum (x-\mu)^2}{N} = \frac{\sum x^2 - \frac{\left(\sum x\right)^2}{N}}{N} \qquad \text{...(2.10)}$$

$$s^2 = \frac{\sum (x-\overline{X})^2}{n-1} = \frac{\sum x^2 - \frac{\left(\sum x\right)^2}{n}}{n-1} \qquad \text{...(2.11)}$$

where, σ^2 is the population variance, s^2 the sample variance, $(x - \mu)$ or $(x - \overline{X})$ the deviation of x value from the mean, μ the population mean, $\overline{X}$ the sample mean, N the population size, and n the sample size.

The population standard deviation $\sigma = \sqrt{\text{Population variance}} = \sqrt{\sigma^2}$ ...(2.12)

The sample standard deviation $s = \sqrt{\text{Sample variance}} = \sqrt{s^2}$ ...(2.13)

A numerical value such as the mean, median, mode, range, variance, or standard deviation computed for a population data set is called a *population parameter*, or simply a *parameter*. A summary measure calculated for a sample data set is called a *sample statistic* or simply a *statistic*.

Example E2.2. *The following data refer to the pay (in thousand of dollars) of 12 employees at a manufacturing company.*

75 69 65 49 21 21 18 18 17 16 16 15

(a) Calculate the mean, median, and mode for this data

(b) Find the range, variance and standard deviation.

Solution:

(*a*) $$Mean = \frac{\sum x}{n} = \frac{400 \times 100}{12} = \$33{,}333$$

Median is $\left[\frac{n+1}{2}\right]^{th}$ term in a ranked data set. Therefore,

$$\frac{n+1}{2} = \frac{12+1}{2} = 6.5$$

$$\text{Median} = \frac{(21+18) \times 1000}{2} = \$19{,}500$$

Mode = \$21,000, \$18,000, and \$16,000

(*b*) Range = Largest pay – Smallest pay = \$75,000 – \$15,000 = \$60,000

$$\text{Variance} = s^2 = \frac{\sum x^2 - \frac{(\sum x)^2}{n}}{n-1} = \left(\frac{19{,}528 - \frac{400^2}{12}}{12-1}\right) \times 1000 = \$563151.5$$

$$\text{Standard deviation} = \sigma = \sqrt{s^2} = \sqrt{563151.5} = \$750.43$$

2.2.3 Mean, Variance, and Standard Deviation for Grouped Data

2.2.3.1 Mean for grouped data

$$\text{Mean for population data: } \mu = \frac{\sum mf}{N} \qquad ...(2.14)$$

$$\text{Mean for sample data: } \overline{X} = \frac{\sum mf}{n} \qquad ...(2.15)$$

Where, m is the midpoint and f is the frequency of a class.

2.2.3.2 Variance and standard deviation for grouped data

Population variance

$$\sigma^2 = \frac{\sum f(m-\mu)^2}{N} = \frac{\sum m^2 f - \frac{\left(\sum mf\right)^2}{N}}{N} \qquad ...(2.16)$$

Sample variance

$$s^2 = \frac{\sum f(m-\bar{X})^2}{n-1} = \frac{\sum m^2 f - \frac{\left(\sum mf\right)^2}{n}}{n-1} \quad \text{...(2.17)}$$

where σ^2 is the population variance, s^2 the sample variance, m the midpoint of a class, N the population size, n the sample size, μ the population mean, $\bar{X}$ is the sample mean, and f the class frequency.

Population standard deviation $\sigma = \sqrt{\sigma^2}$ ...(2.18)

Sample standard deviation $s = \sqrt{s^2}$ ...(2.19)

Example E2.3. *Calculate the mean, variance and standard derivation for the following population data :*

x	*0–3*	*4–7*	*8–11*	*12–15*	*16–19*	*20–23*
f	*7*	*4*	*18*	*12*	*7*	*6*

Solution:

Classes	f	m	mf	m² f
0–3	7	1.5	10.5	15.75
4–7	4	5.5	22	121
8–11	18	9.5	171	1624.5
12–15	12	13.5	162	2187
16–19	7	17.5	122.5	2143.75
20–23	6	21.5	129	2773.50
Total	54		617	8865.50

Population mean $\mu = \frac{\sum mf}{N} = \frac{617}{54} = 11.426$

Variance $\sigma^2 = \frac{\sum m^2 f - \frac{\left(\sum mf\right)^2}{N}}{N} = \frac{8865.50 - \frac{617^2}{54}}{54} = \frac{8865.50 - 7049.796}{54} = 33.624$

Standard deviation $\sigma = \sqrt{\sigma^2} = \sqrt{33.624} = 5.798$

2.2.4 Measures of Position

A measure of position determines the position of a single value in relation to other values in a sample or a population data set. The measures of position considered here are the quartiles, percentiles, and percentile ranks.

2.2.4.1 Quartiles and interquartile range

The most commonly used percentiles are quartiles. Let n denote the number of observations and if we arrange the data in increasing order, then

– the first quartile is at position $\frac{n+1}{4}$

– the second quartile is the median, which is at position $\frac{n+1}{2}$

– the third quartile is at position $3\left(\frac{n+1}{4}\right)$.

Thus, a data set has three quartiles, and is denoted by Q_1, Q_2 and Q_3. Q_1 is the number that divides the bottom (25%) of the data from the top 75%; the second quartile, Q_2 or the median is the number that divides the bottom 50% of the data from the top 50%; the third quartile, Q_3, is the number that divides the bottom 75% of the data from the top 25%. The difference between the third quartile and the first quartile of data set is called the *interquartile range* ($IQR = Q_3 - Q_1$). Interquartile range is a preferred measure of variation when the median is used as the measure of the center. *IQR* gives the range of the 50% of the observations.

2.2.4.2 Percentiles

Percentiles are the summary measures that divide ranked data set into 100 equal parts. Each ranked data set has 99 percentiles that divide it into 100 equal parts. The kth percentile is denoted by P_k, where k is an integer in the range 1 to 99.

Hence, the kth percentile, P_k, can be described as a value in a data set such that about k% of the measurements are smaller than the value of P_k and about $(100 - k)$% of the measurements are greater than the value of P_k. The value of the kth percentile is given by

$$P_k = \text{The value of the } \left(\frac{kn}{100}\right)^{th} \text{ term in a ranked data set}$$

where k denotes the number of the percentile and n represents the sample size.

The percentile rank for a particular value x_i of a data set is given by

$$\text{Percentile rank of } x_i = \frac{\text{Number of values less than } x_i}{\text{Total number of values in the data set}} \times 100$$

The percentile rank of x_i gives the percentage of values in the data set that are smaller than x_i.

Deciles: Deciles are also useful in statistics. The deciles of a data set divide it into tenths, or 10 equal parts. A typical data set has nine deciles. If we denote it by $D_1, D_2 \ldots D_9$, then the first decile, D_1 is the number that divides the bottom 10% of the data from the top 90%; the second decile, D_2 is the number that divides the bottom 20% of the data from the top 80%; and so on. Hence, the first decile is the 10th percentile, the second decile is the 20th percentile, etc.

2.3 PROBABILITY

Inferential statistics deals with the information obtained from part of a population (a sample) to draw conclusions about the entire population. There is an uncertainty about whether our conclusions are correct or not. Hence, there is a greater need to be familiar with uncertainty in order to understand, develop, and apply the methods of inferential statistics. The science of uncertainty is known as the probability theory. Probability theory helps us to assess and control the likelihood that a statistical inference is correct and provides the analytical basis for inferential statistics. The term probability is used to describe the uncertainty, and *probability* measures the likelihood that an event will occur. Probability theory is used extensively in analysing and decision-making situations that involve risk or incomplete information.

2.3.1 Experiment, Outcome and Sample Space

An *experiment* is a process that, when performed, results in one and only one of many observations. These observations are known as the *outcomes* of the experiment. The collection of all outcomes for an experiment is called a *sample space*. The elements of a sample space are called *sample points*. The *sample space* for an experiment can be described by drawing either a *Venn diagram* or a *tree diagram*. A *Venn diagram* is a picture that depicts all the possible outcomes for an experiment. In a *tree diagram*, a branch of the tree represents each outcome.

2.3.2 Simple and Compound Events

An *event* is a collection of one or more of the outcomes of an experiment. An event may be simple event or a compound event. A simple event is also called an *elementary event*, and a compound event is also called a *composite event*. An event that includes one and only one of the final outcomes for an experiment is called a *simple event* and is generally denoted by E_i. A compound event is a collection of more than one outcome of an experiment.

The probability that a simple event E_i will occur is denoted by $P(E_i)$, and the probability that a compound event A will occur is denoted by $P(A)$.

2.3.3 Properties of Probability

The basic properties of probabilities are as follows:

1. The probability of an event always lies in the range of 0 and 1. That is,

$0 \le P(E_i) \le 1$ Simple event

$0 \le P(A) \le 1$ Compound event

2. The sum of the probabilities of all simple events or final outcomes for an experiment, denoted by $\Sigma P(E_i)$ is always equal to 1.

Hence,

$$\Sigma P(E_i) = P(E_1) + P(E_2) + \ldots = 1 \qquad \ldots(2.20)$$

3. The probability of an event that cannot occur is 0. (An event that cannot occur is called an *impossible event.*)

4. The probability of an event that must occur is 1. (An event that must occur is called a *certain event.*)

2.3.4 Venn Diagram

Graphical display of events are helpful for explaining and in understanding probability. Venn diagrams, named after English logician John Venn (1834-1923), are considered one of the excellent ways to visually display events and relationships among events. The sample space is shown as a rectangle and the various events are drawn as circles (or other geometric shapes) inside the rectangle. Venn diagram for one event is shown in Figure 2.1.

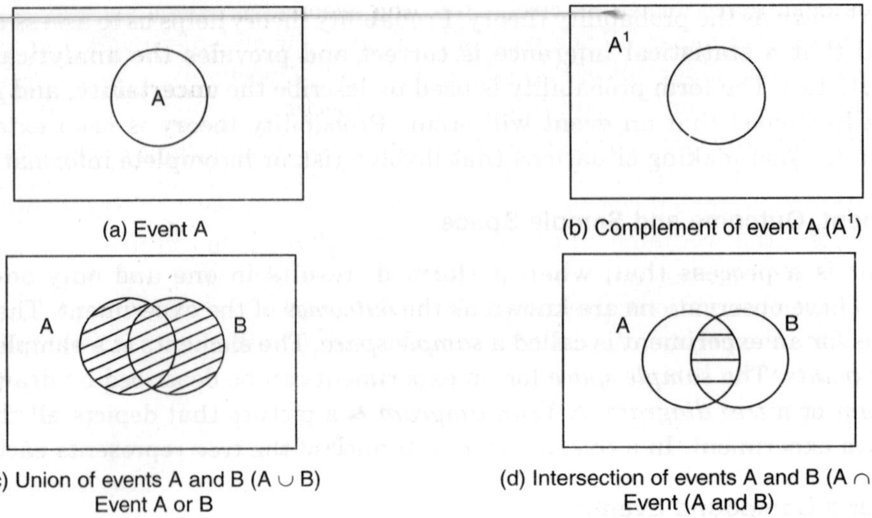

(a) Event A

(b) Complement of event A (A^1)

(c) Union of events A and B ($A \cup B$)
Event A or B

(d) Intersection of events A and B ($A \cap B$)
Event (A and B)

Fig. 2.1 Venn Diagrams.

2.3.5 Probability Tree or Tree Diagram

In a tree-diagram, each outcome is represented by a branch of the tree. Venn and tree diagrams help us to understand probability concepts by presenting them visually.

Example E2.4. *In Figure E2.4, H is the event that an employee has health insurance and D is the event that the employee has disability insurance.*

(*a*) *Express in words what events are represented by regions 1, 2, 3, and 4.*

What events are represented by

(*b*) *Regions 1 and 2 together,*

(*c*) *Regions 2 and 4 together,*

(*d*) *Regions 1, 2 and 3 together,*

(*e*) *Regions 2, 3 and 4 together ?*

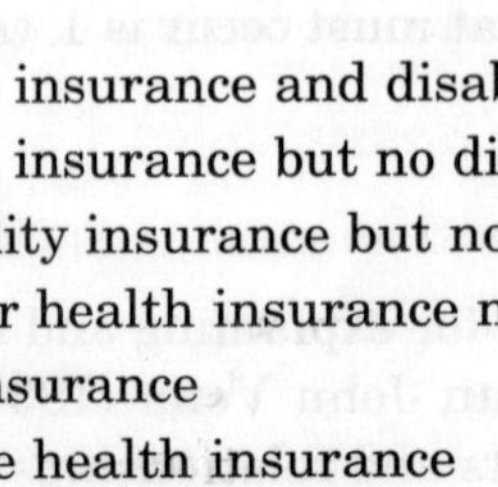

Fig. E2.4 Venn Diagram.

Solution:

(*a*) 1: the employee has health insurance and disability insurance

2: the employee has health insurance but no disability insurance

3: the employee has disability insurance but no health insurance

4: the employee has neither health insurance nor disability insurance.

(*b*) the employee has health insurance

(*c*) the employee does not have health insurance

(*d*) the employee has either health or disability insurance, but not both

(*e*) the employee does not have both kinds of insurance.

Various events are drawn as circles (or other geometric shape such as a rectangle, a square, or a circle) that depict all the possible outcomes for an experiment. For instance, the regions of the four Venn diagrams of Figure 2.1 represent respectively, event A, the complement of event A, the union of events A and B, and the intersection of events A and B.

If A and B are any two subsets of a sample space S, their union $A \cup B$ is the subset of S that contains all the elements that are either in A, in B, or in both; their intersection $A \cap B$ is the subset. S that contains all the elements that are in both A and B; or the complement A' of A is the subset of S that contains all the elements of S that are not in A.

Figure 2.2(a) indicates that events A and B are mutually exclusive, that is, the two sets have no elements in common (or the two events both cannot occur). This is written as $A \notin B = \phi$, where ϕ denotes the *empty set*, which has no elements at all. Figure 2.2(b) shows that A is contained in B, we write this as $A \subset B$.

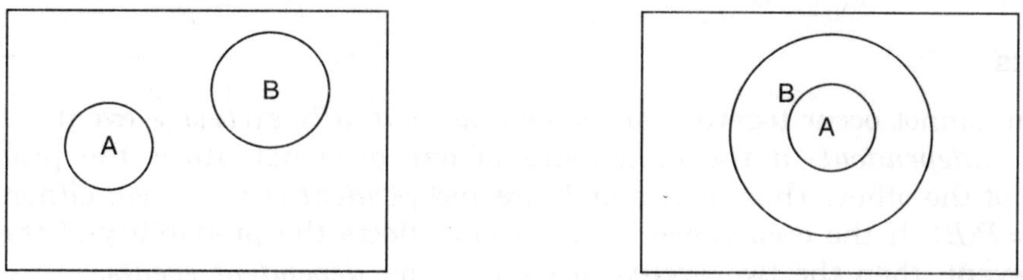

(a) Events A and B are Mutually Exclusive (b) Event A is contained in Event B

Fig. 2.2 Venn Diagram Sharing Special Relationships among Events A and B.

2.3.6 Approaches to Probability

There are three conceptual approaches to probability. They are referred to as

(1) classical probability,

(2) the relative frequency concept of probability, and

(3) the subjective probability concept.

2.3.6.1 Classical probability

Two or more outcomes or events that have the same probability of occurrence are said to be *equally likely outcomes or events*. If E is an event, then $P(E)$ stands for the probability that event E occurs.

$$P(E_i) = \frac{1}{\text{Total number of outcomes for the experiment}} \quad ...(2.21)$$

$$P(A) = \frac{\text{Number of outcomes favorable to } A}{\text{Total number of outcomes for the experiment}} \quad ...(2.22)$$

2.3.6.2 Relative frequency concept of probability

If an experiment is repeated n times and an event A is observed f times, then

$$P(A) = \frac{f}{n}$$

The *law of large numbers* states that if an experiment is repeated again and again, the probability of an event obtained from the relative frequency approaches the actual or theoretical probability.

2.3.6.3 Subjective probability

Subjective probability is assigned arbitrarily. It is the probability assigned to an event based on subjective judgment, experience, information and belief.

2.3.6.4 Marginal probability

Marginal probability is the probability of a single event without consideration of any other event. Marginal probability is also known as *simple probability*.

2.3.6.5 Conditional probability

It is the probability that an event will occur given that another event has already occurred. If A and B are two events, then the conditional probability of A and B is written as $P(A/B)$ and read as *the probability of A given that B has already occurred*.

2.3.7 Events

Events that cannot occur together are known as *mutually exclusive* events. Two events are said to be *independent*, if the occurrence of one does not affect the probability of the occurrence of the other. Thus, if A and B are *independent events*, then either $P(A/B) = P(A)$ or $P(B/A) = P(B)$. If the occurrence of one event affects the probability of the occurrence of the other event, then the two events are said to be *dependent events*.

Hence, *two* events are dependent if either $P(A/B) \neq P(A)$ or $P(B/A) \neq P(B)$. Two events are either *mutually exclusive* or *independent*. That is, mutually exclusive events are always dependent, and dependent and independent events are never mutually exclusive. Similarly, dependent events may or may not be mutually exclusive.

Two mutually exclusive events that are taken together include all the outcomes for an experiment are called *complementary events*. Thus, the complement of event A, denoted by $\bar{A}$, is the event that includes all the outcomes for an experiment that are not in A. It is clear that

$$P(A) + P(\bar{A}) = 1 \qquad ...(2.23)$$

Also $P(A) = 1 - P(\bar{A})$ and $P(\bar{A}) = 1 - P(A)$

The *intersection* of two events is given by the outcomes that are common to both events. If A and B are two events defined in a sample space, then the intersection of A and B represent the collection of all outcomes that are common to both A and B and is denoted by A and B or $A \cap B$ or AB.

2.3.7.1 Special addition rule

If event A and event B are mutually exclusive, then

$$P(A \text{ or } B) = P(A) + P(B) \qquad ...(2.24)$$

Generalizing, if events $A, B, C, \ldots$ are mutually exclusive, then

$$P(A \text{ or } B \text{ or } C \text{ or } \ldots) = P(A) + P(B) + P(C) + \ldots \qquad ...(2.25)$$

Therefore, for mutually exclusive events, the probability that one or another of the events occurs equals to the sum of the individual probabilities.

2.3.7.2 The general addition rule

For events that are not mutually exclusive, the general addition rule is applied. Referring to Figure 2.3, we see that

$$P(A \text{ or } B) = P(A) + P(B) - P(A \text{ and } B) \qquad ...(2.26)$$

For any two events, the probability that one or the other occurs equals the sum of the individual probabilities less (minus) the probability that both occur.

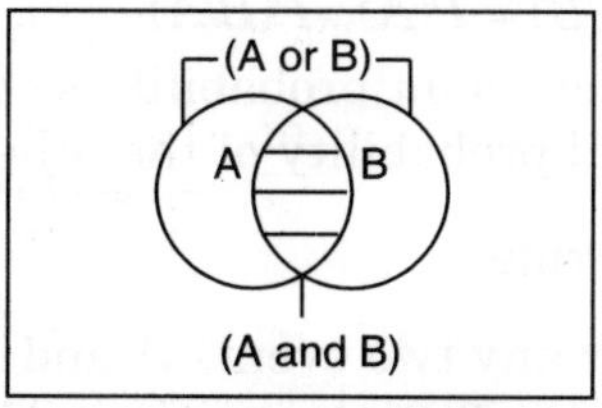

Fig. 2.3 Non-mutually Exclusive Events.

Example E2.5. *In a particular corporation, 56% of the employees are white, 95% are male, and 53% are white male. For a randomly selected employee, let*

W = Event that the employee selected is white

M = Event that the employee selected is male

(a) Find P(W), P(M), and P(W and M).

(b) Determine P(W or M) and express the answer in percentages.

(c) Find the probability that a randomly selected employee is female.

Solution:

(*a*) $P(W) = 0.56$

$P(M) = 0.95$

$P(W \text{ or } M) = 0.53$

(*b*) $P(W \text{ or } M) = 0.56 + 0.95 - 0.53 = 0.98$

98% of the employees are either white or male.

(*c*) $P(F) = 1 - 0.95 = 0.05$

2.3.7.3 Conditional probability

The conditional probability of an event is the probability that the event occurs under the assumption that another event has occurred. The probability that event *B* occurs given that event *A* has occur is called a *conditional probability*. It is denoted by the symbol *P*(*B*/*A*), which is read as the *probability of B given A*. *A* is called the *given event*.

2.3.7.4 Joint and marginal probabilities

Data obtained by observing values of one variable of a population are known as *univariate data*. Data obtained by observing two variables of a population are called *bivariate data*, and a frequency distribution for bivariate data is called a *contingency table* or *two-way table*.

2.3.7.5 Conditional probability rule

If *A* and *B* are any two events, then

$$P(B/A) = \frac{P(A \text{ and } B)}{P(A)} \qquad ...(2.27)$$

That is, for any two events *A* and *B*, the conditional probability that one event occurs given that the other event has occurred equals the joint probability of the two events divided by the probability of the given event.

2.3.7.6 The multiplication rule

The conditional probability rule is used to compute conditional probabilities in terms of unconditional probabilities:

$$P(B/A) = \frac{P(A \text{ and } B)}{P(A)} \quad ...(2.28)$$

or

$$P(A \text{ and } B) = P(A) \,.\, P(B/A) \quad ...(2.29)$$

Thus, for any two events, their joint probabilities equals the probability that one of the events occurs times the conditional probability of the other event, if given that event.

2.3.7.7 The special multiplication rule

The general multiplication rule for any two events A and B is given by

$$P(A \text{ and } B) = P(A) \,.\, P(B/A) \quad ...(2.30)$$

If A and B are independent events, then

$$P(B/A) = P(B) \quad ...(2.31)$$

Hence, for the special case of independent events, we replace the term $P(B/A)$ in the general multiplication rule by the term $P(B)$.

If A and B are independent events, then

$$P(A \text{ and } B) = P(A) \,.\, P(B) \quad ...(2.32)$$

and conversely, if

$P(A \text{ and } B) = P(A)\, P(B)$, then A and B are independent events.

Hence, two events are independent if and only if their joint probability equals the product of their marginal probabilities. Similarly, if events $A, B, C, \ldots$ are independent, then

$$P(A \text{ and } B \text{ and } C \text{ and } \ldots) = P(A) \,.\, P(B).\, P(C) \ldots \quad ...(2.33)$$

Mutually exclusive events are those that cannot occur simultaneously. *Independent events* are those for which the occurrence of some does not affect the probabilities of the others occurring. Two or more (non-impossible) events cannot be both mutually exclusive and independent.

2.3.7.8 Total probability rule

Events $A_1, A_2, A_3, \ldots, A_k$ are said to be *exhaustive*, if one or more of them must occur. Suppose events $A_1, A_2, A_3, \ldots, A_k$ are mutually exclusive and exhaustive; that is exactly one of the events must occur. Then for any event B

$$P(B) = \sum_{j=1}^{k} P(A_j) \,.\, P(B/A_j) \quad ...(2.34)$$

Example E2.6. *The two-way classification of data of 2000 randomly selected employees of a manufacturing company from a city based on gender and commuting time from residence to work place is given in Table E2.6.*

Table E2.6

Gender	Commuting time from residence to workplace			Total
	Less than 30 min	30 min to 1 hour	More than 1 hour	
Men	*525*	*450*	*225*	*1200*
Women	*410*	*250*	*140*	*800*
Total	*935*	*700*	*365*	*2000*

(a) If one employee is selected at random from these 2000 employees, find the probability that this employee

(i) commutes for more than 1 hour

(ii) commutes for less than 30 minutes

(iii) is a man commuting for 30 min to 1 hour

(iv) is a woman commuting for more than 1 hour.

(b) Determine if the following events are mutually exclusive

(i) 'men' and 'commuting for more than 1 hour'

(ii) 'less than 30 min' 'more than 1 hour'

(iii) 'woman' and 'commutes for 30 minutes to 1 hour'. Are these events independent?

Solution:

(a) *(i)* P (commutes for more than 1 hour) $= \dfrac{365}{2000} = 0.1825$

(ii) P (commutes for less than 30 min) $= \dfrac{935}{2000} = 0.4675$

(iii) P (commutes for 30 min to 1 hour) $= \dfrac{450}{700} = 0.6428$

(iv) P (commutes for more than 1 hour/woman) $= \dfrac{140}{800} = 0.175$

(b) *(i)* The term 'man' and 'commutes for more than 1 hour' are not mutually exclusive. The two events can occur together.

(ii) The events 'less than 30 minutes' and 'more than 1 hour' are mutually exclusive because they cannot occur together.

(iii) P (woman) $= \dfrac{800}{2000} = 0.4$

P (woman/commutes for 30 min to 1 hour) $= \dfrac{250}{700} = 0.3571$

Since these two probabilities are not equal, the events 'woman' and 'commute for 30 minutes to 1 hour' are not independent.

2.3.8 Bayes' Formula

Bayes' formula is quite useful in modifying a probability estimate as additional information becomes available. We have seen that a conditional probability is one in which the probability of the event depends upon whether the other event has occurred.

$$P(A/B) = \frac{P(AB)}{P(B)} \qquad \text{...(2.35)}$$

or

$$P(B/A) = \frac{P(AB)}{P(A)} \qquad \text{...(2.36)}$$

From Eqs. (2.35) and (2.36), we have

$$P(AB) = P(A/B)\, P(B) = P(B/A)\, P(A)$$

since

$$AB = BA$$

and

$$P(A/B) = \frac{P(B/A)\, P(A)}{P(B)} \qquad \text{...(2.37)}$$

Since, $P(A) + P(\bar{A}) = 1$, it follows that event B occurs jointly with either A or $\bar{A}$.

$$P(B) = P(AB) + P(\bar{A}B)$$

and from Eq. (2.36)

$$P(B) = P(A)\,P(B/A) + P(\bar{A})\,P(B/\bar{A}) \qquad \text{...(2.38)}$$

Substituting Eq. (2.38) into Eq. (2.39), we get

$$P(A/B) = \frac{P(A)\,P(B/A)}{P(A)\,P(B/A) + P(\bar{A})\,P(B/\bar{A})} \qquad \text{...(2.39)}$$

If event A has more than two available alternatives, we can write Eq. (2.39) as

$$P(A/B) = \frac{P(A_i)\,P(B/A_i)}{\sum_i P(A_i)\,P(B/A_i)} \qquad \text{...(2.40)}$$

Example E2.7. *The average acceptability of three automobile transmissions from three suppliers is given below:*

Supplier	Relative probability of supply	Average acceptability
1	*0.56*	*0.97*
2	*0.32*	*0.89*
3	*0.12*	*0.77*

(*a*) *Find the probability that any transmission received will perform acceptably.*

(*b*) *Find the probability that a particular transmission delivered by supplier 2 when it is known to have performed acceptably.*

Solution:

(*a*) $P(A) = (0.56)(0.97) + (0.32)(0.89) + (0.12)(0.77) = 0.5432 + 0.2848 + 0.0924 = 0.9204$

(*b*) $P(T_2/A) = \dfrac{P(T_2)\,P(A/T_2)}{\Sigma\, P(T_i)\,P(A/T_i)} = \dfrac{(0.32)(0.89)}{(0.56)(0.97) + (0.32)(0.89) + (0.12)(0.77)} = 0.3094$

2.4 RANDOM VARIABLES

A *random variable* is a variable whose value is determined by the outcome of a random experiment. Random variables are numerical-valued whose observed values are governed by the laws of probability. A random variable can be *discrete or continuous*. A *random variable* whose values are countable is called a *discrete random variable*. Some examples of this are the number of students in a class, toss of a coin, the number of customers visiting a bank during a given hour, the number of cars sold at a dealership in a given month, color of a ball drawn from a collection of balls, etc. On the other hand, a random variable that can assume any value in one or more intervals is called a *continuous random variable*. Examples of this are the height of a person, diameter of a manufactured shaft, time to failure of a component, repair time, the price of an automobile, the time taken to complete a medical examination, duration of a snow storm, etc.

2.4.1 Discrete Random Variables

Let X is a random variable. If the number of possible values of X is finite or countably infinite, X is called a *discrete random variable*. Each possible outcome is x_i. A number, $p(X_i) = p(X = x_i)$, gives the probability that the random variable X equals a value x_i. The numbers, $p(x_i)$, for $i = 1, 2, \ldots$ must satisfy the following two conditions

1. $p(x_i) \geq 0$ for all i

2. $\sum_{i=1}^{\infty} p(x_i) = 1$

The collection of pairs $[x_I, p(x_i)]$, $i = 1,2,\ldots$ is called the *probability distribution* of X, and $p(x_i)$ is called the *probability mass function* (or pmf) of X. The probability distribution for the variable X is characterized in Fig. 2.4.

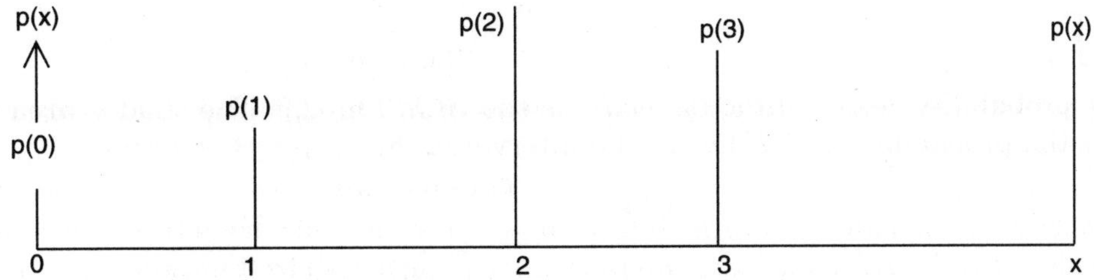

Fig. 2.4 Probability Distribution of a Discrete Random Variable x_i.

The *cumulative distribution function* of a discrete random variable is defined as

$$P(X = x_i) = \sum_{x_i \leq x} p(x_i) \qquad \ldots(2.41)$$

The function is a step function that is constant over every interval not containing any of the points x_i, as shown in Fig. 2.5.

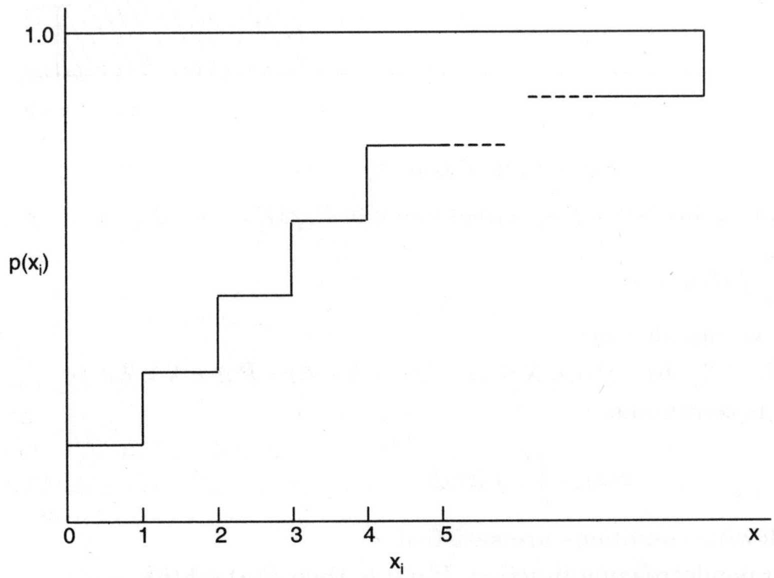

Fig. 2.5 Cumulative Distribution Function (CDF) of a Discrete Random Variable x_i.

2.4.2 Continuous Random Variables

If the range space of the random variable X is an interval or a collection of intervals, X is called a *continuous random variable*. If a and b are real numbers and $a < b$, then the probability that X lies in the interval $[a, b]$ is defined as

$$P[a \le X \le b] = \int_a^b f(x)\,dx \quad ...(2.42)$$

The function $f(x)$ is known as the *probability density function* (or *pdf*) of the random variable X. The *pdf* must satisfy the following conditions:

1. $f(x) \ge 0 \quad -\infty < x < \infty$

2. $\int_{-\infty}^{+\infty} f(x)\,dx = 1$ ...(2.43)

The probability density function (*pdf*) is shown in Fig. 2.6. The shaded area in Fig. 2.6 represents the probability that X lies in the interval $[a, b]$.

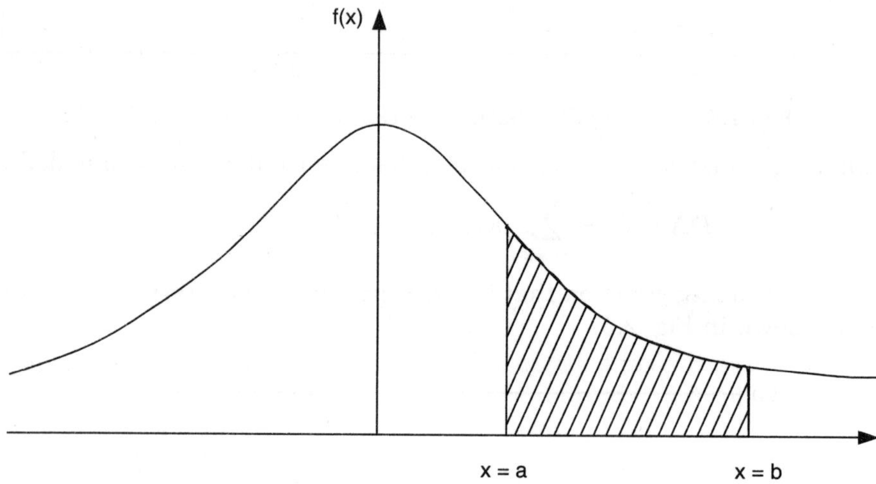

Fig. 2.6 The Probability Density Function (pdf).

From Eq. (2.43), for any specified value of X, $p(X = x) = 0$

Since $\int_x^x f(t)\,dt = 0$ ...(2.44)

Hence, we can also write

$$P(a \le X \le b) = P(a < X \le b) = P(a \le X < b) = P(a < X < b) \quad ...(2.45)$$

Since x is continuous

$$F(x) = \int_{-\infty}^{x} f(t)\,dt \quad ...(2.46)$$

The following conditions are satisfied:

1. F is a nondecreasing function. If $a < b$, then $F(a) \le F(b)$

2. $\lim_{x \to \infty} F(x) = 1$

3. $\lim_{x \to \infty} F(x) = 0$ ...(2.47)

All probability questions about X can be expressed in terms of the continuous density function (cdf).

For instance,

$$P(a < X \leq b) = F(b) - F(a) \text{ for } a < b \qquad ...(2.48)$$

2.4.2.1 Mean and standard deviation of a discrete random variable

The mean of a discrete random variable x is the value that is expected to occur per repetition on average, if an experiment is repeated for a large number of times. It is given by

$$\mu = \sum x\, P(x) \qquad (2.49)$$

where μ is the mean of a discrete random variable. The mean of a discrete random variable x is also called its *expected value E (x)* and is given by

$$E(x) = \sum x\, P(x) \qquad ...(2.50)$$

The standard deviation of a discrete random variable x measures the spread of its probability distribution and is given by

$$\sigma = \sqrt{\sum (x - \mu)^2\, P(x)} = \sqrt{\sum x^2\, P(x) - \mu^2} \qquad ...(2.51)$$

2.4.3 Expectation

If X is a random variable, the expected value of X, denoted by $E(X)$, for discrete and continuous variables is defined as

$$E(X) = \sum_{\text{all } i} x_i\, p(x_i) \text{ if } X \text{ is discrete} \qquad ...(2.52)$$

and

$$E(X) = \int_{-\infty}^{\infty} x\, f(x)\, dx \text{ if } X \text{ is continuous} \qquad ...(2.53)$$

The expected value $E(X)$ of a random variable x is also referred to as the *mean* of X, μ_X (or μ), or *the first moment* of X. The quantity $E(X^n)$, $n \geq 1$, is called the *nth moment of X* and is defined as

$$E(X^n) = \sum_{\text{all } i} x_i^n\, p(x_i) \quad \text{if } X \text{ is discrete} \qquad ...(2.54)$$

and

$$E(X^n) = \int_{-\infty}^{\infty} x^n\, f(x)\, dx \quad \text{if } X \text{ is continuous} \qquad ...(2.55)$$

The variance of a random variable, X, denoted by $V(X)$, Var(X) or σ^2, is defined by

$$V(X) = E([X - E(X)]^2) \qquad ...(2.56)$$

A useful identity in computing $V(X)$ is given by

$$V(X) = E(X)^2 - [E(X)]^2 \qquad ...(2.57)$$

The mean, $E(X)$, is a measure of the central tendency of a random variable. The variance, $V(X)$, is a measure of the spread or dispersion of the possible values of X around the mean $E(X)$. The standard deviation, σ, is the positive square root of the variance.

Example E2.8. *The following table lists the probability distribution of x, where x is the number of defects contained in a randomly selected manufactured product. Find the mean and standard deviation of x.*

x	*0*	*1*	*2*	*3*	*4*
P(x)	*0.41*	*0.12*	*0.05*	*0.03*	*0.01*

Solution:

x	P(x)	x P(x)	x² P(x)
0	0.41	0	0
1	0.12	0.12	0.12
2	0.05	0.10	0.20
3	0.03	0.09	0.27
4	0.01	0.04	0.16
Total		0.35	0.75

Mean $\mu = \sum x\, P(x) = 0.35$ defects

Standard deviation $\sigma = \sqrt{x^2\, P(x) - \mu^2} = \sqrt{0.75 - 0.35^2} = 0.79$ defects

Example E2.9. *The life of an electronic component is given by X, a continuous random variable assuming all values in the range $x \geq 0$. The pdf of the lifetime, in years, is given as*

$$f(x) = \begin{cases} \frac{1}{4} e^{-\frac{1}{4}} & x \geq 0 \\ 0 & \textit{otherwise} \end{cases}$$

(*a*) *Determine the probability that the life of this component is between 2 and 4 years.*

(*b*) *Determine cdf of the component.*

(*c*) *Determine the probability that this electronic component will last for 4 years.*

(*d*) *Find the mean and variance of the life of the component.*

Solution:

$$(a)\ P(2 \leq X \leq 4) = \frac{1}{4}\int_2^4 e^{-\frac{x}{4}} dx = -e^{-1} + e^{-\frac{x}{4}} = 0.2386$$

The probability that the component will last between 2 and 4 years is also obtained from

$$P(2 \leq X \leq 4) = F(4) - F(2)$$
$$= (1 - e^{-4/4}) - (1 - e^{-2/4}) = e^{-1/2} - e^{-1} = 0.6065 - 0.3679 = 0.2386$$

(*b*) The cdf of this component is given by

$$F(x) = \frac{1}{4}\int_0^x e^{-\frac{1}{4}}\, dx = 1 - e^{-\frac{x}{4}}$$

(*c*) The probability that the component will last for 4 years or less is given by

$$P(0 \leq X \leq 4) = F(4) - F(0) - F(4) = 1 - e^{4/4} = 0.6321$$

(*d*) The mean and variance of the life of the electronic component are determined as follows:

$$E(X) = \frac{1}{4}\int_0^\infty x e^{-\frac{x}{4}} dx = x e^{-\frac{x}{4}}\Big|_0^\infty + \int_0^\infty x e^{-\frac{x}{4}} dx = 0 + \frac{1}{1/4} e^{-\frac{x}{4}}\Big|_0^\infty = 4 \text{ years}$$

To compute $V(X)$ using Eq. (2.57), first compute $E(X^2)$ from Eq. (2.55) as follows:

$$E(X^2) = \frac{1}{4}\int_0^\infty x^2 e^{-\frac{x}{4}} dx = -x^2 e^{-\frac{x}{4}}\Big|_0^\infty + 8\int_0^\infty x e^{-\frac{x}{4}} dx = 32$$

Hence,

Variance is given by $V(X) = 32 - 4^2 = 16$ years and the standard deviation is given by

$$\sigma = \sqrt{16} = 4 \text{ years}$$

2.5 PERMUTATIONS AND COMBINATIONS

2.5.1 Permutations

Ordered arrangements of a set of objects are called *permutations*. When the order in which they are arranged is disregarded, then the arrangements are known as *combinations*. The number of permutations of n different items is the number of different arrangements in which these items can be placed. A permutation of n different objects taken r at a time is an arrangement of r out of the n objects with attention given to the order of arrangement. The number of permutations of n objects taken r at a time is denoted by $n\,P_r$, $P(n, r)$ or $P_{n,\,r}$, and is written as

$$n\,P_r = n(n-1)\,(n-2) \ldots (n-r+1) = \frac{n!}{(n-r)!} \qquad \ldots(2.58)$$

in which n is the symbol for factorial $n! = n(n-1) \ldots 3 \times 2 \times 1$, when $r = n$, $nP_n = n!$ (note by definition $0! = 1$). Tables of the values of factorial n are given in Appendix 2A.

The permutation of n items taken all at a time, when the n items consist of r_1 alike, r_2 alike, ..., r_k alike, so that by definition $r_1 + r_2 + \ldots + r_k = n$.

The number of permutation is then

$$P = \frac{n!}{r_1!\,r_2!\ldots r_k!} \qquad \ldots(2.59)$$

2.5.2 Combinations

The number of combinations of n different objects taken r at a time is a selection of r out of the n objects with no attention given to the order of arrangement. The number of combinations of n objects taken r at a time is denoted by nC_r, $C(n, r)$, $C_{n,\,r}$ or $\left(\frac{n}{r}\right)$ is $r\,!$ times smaller than the number of permutations.

Therefore

$$nC_r = \frac{nP_r}{r!} = \frac{n!}{(n-r)!\,r!} \qquad \ldots(2.60)$$

It should be noted from symmetry that the identity is valid.

$$nC_r = nC_{n-r}$$

Tables of binomial coefficients $\left\{\frac{n}{r}\right\}$ are given in Appendix 2B. ...(2.61)

Example E2.10. *Out of 5 men and 7 women, a committee consisting of 2 men and 3 women is to be formed. In how many ways can this be done if*

(*a*) *any man and any woman can be included*

(*b*) *one particular woman must be included on the committee* (*c*) *two particular men cannot be included on the committee*

Solution:

(*a*) 2 men out of 5 can be selected in $5C_2$ ways

3 women out of 7 can be selected in $7C_3$ ways

Hence, the number of possible selections = $(5C_2)(7C_3) = (10)(35) = 350$

(*b*) 2 men out of 5 can be selected in $5C_2$ ways

2 additional women out of 6 can be selected in $6C_2$ ways

Hence, the number of possible selections = $(5C_2)(6C_2) = (10)(15) = 150$

(*c*) 2 men out of 3 can be selected in $3C_2$ ways

3 women out of 7 can be selected in $7C_3$ ways

Hence, the number of possible selections = $(3C_2)(7C_3) = (3)(35) = 105$

2.6 DISCRETE DISTRIBUTIONS

Discrete random variables are used to describe random phenomena in which only integer values can occur.

2.6.1 Hypergeometric Distribution

Suppose there is a population of finite size N that contains a specified number of nonconforming components D. Then, the probability that a sample of size n containing x nonconforming components is given by

$$p(x) = \frac{\binom{D}{x}\binom{N-D}{n-x}}{\binom{N}{n}} \quad x = 0, 1, 2, ..., \min(n, D) \qquad ...(2.62)$$

where,

$$\binom{a}{b} = \frac{a!}{b!(a-b)!} \qquad ...(2.63)$$

is the number of a parts taken b at a time. The hypergeometric distribution is the appropriate probability model for sampling without replacement. Other distributions are often employed when the ratio of n/N becomes small, say 0.10 or less.

Example E2.11. *A lot of 75 gaskets contains five in which the variability in thickness around the circumference of the gasket is unacceptable. A sample of 10 gaskets is selected at random without replacement. What is the probability that*

(*a*) *None of the unacceptable gaskets is in the sample*

(*b*) *At least one unacceptable gasket is in the sample*

(*c*) *Exactly one unacceptable gasket is in the sample*

(*d*) *The mean number of unacceptable gaskets in the sample.*

Solution:

Let X denote the number of unacceptable gaskets in the sample of 10

$$(a)\quad P(X=0) = \frac{\binom{5}{0}\binom{70}{10}}{\binom{75}{10}} = \frac{\dfrac{70!}{10!\,60!}}{\dfrac{75!}{10!\,65!}} = \frac{65\times64\times63\times62\times61}{75\times74\times73\times72\times71} = 0.4786$$

$$(b)\quad P(X \geq 1) = 1 - P(X=0) = 0.5214$$

$$(c)\quad P(X=1) = \frac{\binom{5}{1}\binom{70}{9}}{\binom{75}{10}} = \frac{\dfrac{5!\,70!}{9!\,6!}}{\dfrac{75!}{10!\,65!}} = \frac{5\times65\times64\times63\times62\times10}{75\times74\times73\times72\times71} = 0.3923$$

$$(d)\quad E(X) = 10\left(\frac{5}{75}\right) = \frac{2}{3}$$

2.6.2 The Binomial Probability Distribution

The binomial probability distribution is one of the most widely used discrete probability distributions. It is used to find the probability that an outcome will occur x times in n performances of an experiment. The binomial distribution is applied to experiments that satisfy the four conditions of a *binomial experiment*. Each repetition of a binomial experiment is called a *trial* or a *Bernoulli trial* (named after Jacob Bernoulli).

2.6.3 The Binomial Experiment

An experiment that satisfies the following four conditions is called a *binomial experiment.*

1. There are n identical trials. In other words, the given experiment is repeated n times. All these repetitions are performed under identical conditions.
2. Each trial has two and only two outcomes. These outcomes are usually called a *success* and a *failure.*
3. The probability of success is denoted by p and that of failure by q, and $p + q = 1$. The probabilities p and q remain constant for each trial.
4. The trials are independent. In other words, the outcome of one trial does not affect the outcome of another trial.

One of the two outcomes of a trial is called a *success* and the other a *failure*.

2.6.4 The Binomial Formula

2.6.4.1 Binomial theorem

The nth power of $(p + q)$ can be expressed in terms of binomial coefficients as

$$(p+q)^n = p^n + np^{n-1}q + \frac{n(n-1)}{2}\,p^{n-2}q^2 + \ldots$$

$$\ldots\ldots + \frac{n!}{r!\,(n-r)!}\,p^{n-r}q^r + \ldots + q^n = \sum_{r=0}^{n}\binom{n}{r}p^{n-r}q^r = \sum_{r=0}^{n}{}_nC_r\,p^{n-r}q^r \quad \ldots(2.64)$$

Since $(p + q) = (q + p)$, we can also write

$$(p+q)^n = \sum_{r=0}^{n} {}_nC_n\, p^r\, q^{n-r} \qquad \text{...(2.65)}$$

Consider n trials in each of which the probability of success is p. Then the probability of failure is $1 - p = q$. To find the probability of r successes, we observe that:

the probability of 1 success in 1 try is p,

the probability of 2 successes in 2 tries is $p \times p$ or p^2,

the probability of 3 successes in 3 tries is $p \times p \times p$ or p^3,

the probability of r successes in r tries is p^r, and the probability of subsequent $(n - r)$ failures in $(n - r)$ tries is $(1 - p)^{n-r} = q^{n-r}$. The probability of r successes followed by $(n - r)$ failures is $p^r\,(1 - p)^{n-r}$. Here, we have considered only one particular group or combination of r events; *i.e.*, we have started with r successes and finished with $(n - r)$ failures; every other possible ordering of r successes and $(n - r)$ failures will also have the same probability.

The number of possible orderings or the number of selections for r successes and $(n - r)$ failures in n trials is $\dfrac{n!}{[r!(n-r)!]}$.

Therefore, the probability P_r of an event succeeding r times is

$$P_r = \frac{n!}{r!(n-r)!}\, p^r\,(1-p)^{n-r} \qquad \text{...(2.66)}$$

or

$$P_r = {}_nC_r\, p^r\, q^{n-r} \qquad \text{...(2.67)}$$

This term is similar to the rth term of the binomial expansion $(q + p)^n$ [see Eq. (2.65)], which can be written

$$(q+p)^n = \sum_{r=0}^{n} {}_nC_r\, p^r\, q^{n-r} \qquad \text{...(2.68)}$$

The successive terms of the expansion give the probability p_r of an event succeeding r times in n trials for values of r varying in steps of one from 0 to n.

2.6.4.2 Cumulative terms for binomial distribution

The probability P_r of an event succeeding *exactly* r times in n trials is given by Eq. (2.68).

$$P_r = {}_nC_r\, p^r\, q^{n-r}$$

The probability of an event succeeding at least r' times in n trials is given by

$$\sum_{r=r'}^{r=n} P_r = \sum_{r=r'}^{r=n} {}_nC_r\, p^r\, q^{n-r} \qquad \text{...(2.69)}$$

The values of the summation of Eq. (2.68) are given in Appendix 2C ranging from 0.05 to 0.50 with n between 2 and 20, and r' between 1 and 20.

For $p > 0.5$, we can utilize the fact that the probability is

$$P_r = 1 - \sum_{r=n-r'+1}^{r=n} {}_nC_r\, p^r\, q^{n-r} \qquad \text{...(2.70)}$$

2.6.4.3 Mean and standard deviation of binomial distribution

If p is the proportion of successes in the population, then the mean number of successes in n trials is

$$\mu = np \qquad \text{...(2.71)}$$

This is clear, as the mean number of successes in n trials is equal to the probability of success in one trial times the number of trials.

The standard deviation for a binomial frequency distribution is

$$\sigma = \sqrt{npq} \quad \text{or} \quad \sqrt{np(1-p)} \qquad \text{...(2.72)}$$

q is not independent but is equal to $(1-p)$. The binomial distribution can be expressed in terms of two parameters n and p.

Equations (2.71) and (2.72) will now be derived using the definition of expectation for the mean and variance. We have

$$E(r) = \mu = \sum_{r=0}^{n} rP_r = \sum_{r=0}^{n} \frac{r\,n!}{r!\,(n-r)!} p^r (1-p)^{n-r}$$

$$= np \sum_{r=1}^{n} \frac{(n-1)!}{(r-1)!\,(n-r)!} p^{r-1} (1-p)^{n-r}$$

$$= np[p+q]^{n-1} = np \qquad \text{...(2.73)}$$

For the variance σ^2

$$E(r-\mu)^2 = \sigma^2 = \sum_{r=0}^{n} (r-\mu)^2 P_r$$

$$= \sum_{r=0}^{n} r^2 P_2 - 2\sum_{r=0}^{n} rP_r + \mu^2 \sum_{r=0}^{n} P_r$$

$$= \sum_{r=1}^{n} r^2 P_r - 2\mu(\mu) + \mu^2$$

since $\displaystyle\sum_{r=0}^{n} P_r = 1$ and $\displaystyle\sum_{r=0}^{n} rP_r = \mu = np$

Thus, $$\sigma^2 = \sum_{r=0}^{n} r^2 P_r - (np)^2$$

$$= \sum_{r=0}^{n} [r(r-1) + r]P_r - (np)^2$$

$$= \sum_{r=2}^{n} r(r-1) \frac{n!}{r!\,(n-r)!} p^r (1-p)^{n-r} + \sum_{r=0}^{n} rp_r - (np)^2$$

$$= n(n-1)\,p^2 \sum_{r=2}^{n} \frac{(n-2)!}{(r-2)!(n-r)!}\,p^{r-2}(1-p)^{n-r} + np - (np)^2$$

$$= n(n-1)p^2\,[p+q]^{n-2}\,np - (np)^2$$

or
$$\sigma^2 = n(n-1)p^2 + np - (np)^2 = np\,(1-p) = npq$$

Hence,

$$\sigma = \sqrt{npq} \qquad \text{...(2.74)}$$

A binomial distribution with $n = 20$ and $p = 0.10$ is shown in Fig. 2.7.

Cumulative binomial distribution are given Table III in Appendix 2C.

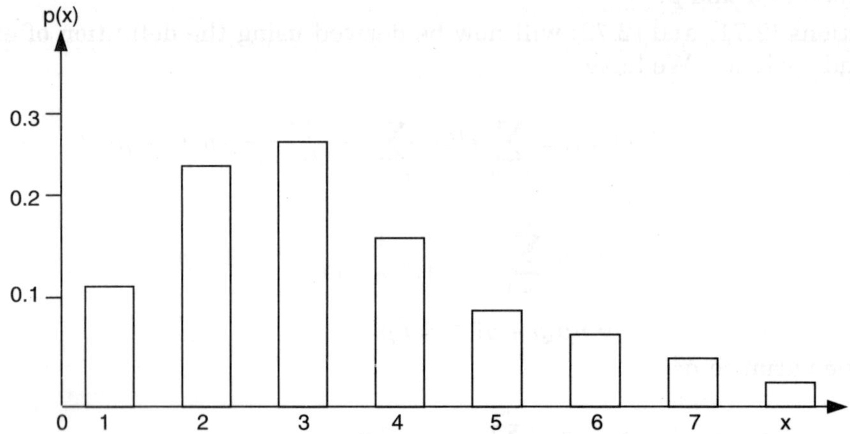

Fig. 2.7 Binomial pmf for $n = 20$, $p = 0.10$.

Example E2.12. *Let X denote the number of mechanical components that are defective in a testing process and assume that X is a binomial random variable with p = 0.001. If 1000 of these components are tested, find the following:*

(*a*) $P(X = 1)$

(*b*) $P(X \geq 1)$

(*c*) $P(X \leq 2)$

(*d*) *mean and variance of X*

Solution:

(*a*) $P(X = 1) = \binom{1000}{1}(0.001)^1(0.999)^{999} = 0.368$

(*b*) $P(X \geq 1) = 1 - P(X = 0) = 1 - \binom{1000}{1}(0.001)^1(0.999)^{999} = 0.632$

(*c*) $P(X \leq 2) = \binom{1000}{1}(0.001)^0(0.999)^{1000} + \binom{1000}{1}(0.001)^1(0.999)^{999}$

$$+ \binom{1000}{2}(0.001)^2(0.999)^{998} = 0.920$$

(*d*) $E(X) = 1000\,(0.001) = 1$

$V(X) = 1000\,(0.001)\,(0.999) = 0.999$

2.6.5 Poisson Distribution

The Poisson distribution, named after the French mathematician Simeon D. Poisson, is another important probability distribution of a discrete random variable that has many applications. The Poisson distribution is applied to experiments with random and independent occurrences. Independence of occurrences means that one occurrences (or nonoccurrence) of an event does not influence the successive occurrences or nonoccurrences of that event. The occurrences are always considered with respect to an interval. The interval may be a time interval, a space interval, or a volume interval. The actual number of occurrence within an interval is random and independent. If the average number of occurrences for a given interval is known, then by using the Poisson probability distribution we can determine the probability of a certain number of occurrences, x, in that interval. Note that the number of actual occurrences in an interval is denoted by x.

The following three conditions must be satisfied to apply the Poisson probability distribution:

1. X is a discrete random variable,
2. the occurrences are random,
3. the occurrences are independent.

The following are a few examples of discrete random variables for which the occurrences are random and independent, and follow Poisson distribution as:

1. flaws in castings,
2. number of vehicles on a highway,
3. number of customers visiting a bank,
4. number of accidents that occur on a given highway during a period of time, and
5. number of telephone calls

The Poisson distribution is made up of a series of terms

$$e^{-\mu}, e^{-\mu}\frac{\mu}{1!}, e^{-\mu}\frac{\mu^2}{2!}, e^{-\mu}\frac{\mu^3}{3!}, e^{-\mu}\frac{\mu^4}{4!}, \ldots$$

representing, respectively, the probability of the occurrence of 0, 1, 2, 3, 4, etc. events, where e is the base of natural logarithms and μ is the mean frequency of occurrence. The sum of all terms of the series is unity, as must be the case with a sum of all probabilities.

If r is the number of occurrences whose probability we require, we can write the general term of the series

$$e^{-\mu}, e^{-\mu}\frac{\mu}{1!}, e^{-\mu}\frac{\mu^2}{2!}, e^{-\mu}\frac{\mu^3}{3!}, e^{-\mu}\frac{\mu^4}{4!}, \ldots \qquad \ldots(2.75)$$

2.6.5.1 Derivation from binomial distribution

The Poisson distribution can also be deduced from the binomial distribution, provided that n is large ($\to \infty$), p is very small ($\to 0$), and np is finite. It has been shown earlier that in n trials the probability of an event succeeding r times is

$$P_r = \frac{n!}{r!\,(n-r)!}\, p^r\, q^{n-r}$$

When n is large compared to r,

$$\frac{n!}{(n-r)!} = n(n-1)(n-2)\ldots(n-r+1) = n^r$$

Therefore, the probability of r successes becomes

$$P_r = \frac{n!}{r!} p^r q^{n-r} \qquad \text{...(2.76)}$$

Now if p is very small and r is not large, then

$$q^r = (1-p)^r \cong 1$$

and $$q^{n-r} \cong q^n = (1-p)^n$$

Hence

$$P_r = \frac{(np)^r}{r!}(1-p)^n = \frac{(np)^r}{r!}\left[1 - np + \frac{n(n-1)(-p)^2}{2!} + \frac{n(n-1)(n-2)(-p)^3}{3!} + \ldots\right]$$

$$= \frac{(np)^r}{r!}\left[1 - np + \frac{(np)^2}{2!} - \frac{(np)^3}{3!} + \ldots\right]$$

Thus, $$P_r = \frac{(np)^r}{r!} e^{-np} \qquad \text{...(2.77)}$$

This, then, is the probability of r successes in n trials.

2.6.5.2 Mean and standard deviation

The mean number of occurrences of an event per unit of time (or space) is

$$\mu = np \qquad \text{...(2.78)}$$

and the standard deviation of the number of events is

$$\sigma = \sqrt{np} \qquad \text{...(2.79)}$$

Thus the mean and variance are equal to one another:

$$\mu = \sigma^2 = np \qquad \text{...(2.80)}$$

Equations (2.78) and (2.80) will now be derived.

$$\text{Mean} = E(r) = \sum_{r=0}^{\infty} rP_r = \sum_{r=0}^{\infty} \frac{re^{-\mu}\mu^r}{r!}$$

$$= 0 + \mu e^{-\mu} + 2\frac{\mu^2 e^{-\mu}}{2!} + 3\frac{\mu^3 e^{-\mu}}{3!} + \ldots$$

$$= \mu e^{-\mu}\left[1 + \mu + \frac{\mu^2}{2!} + \frac{\mu^3}{3!} + \ldots\right]$$

$$= \mu e^{-\mu} e^{\mu} = \mu \qquad \text{...(2.81)}$$

The variance σ^2 is given by

$$\sigma^2 = E(r-\mu)^2 = \sum_{0}^{\infty} (r-\mu)^2 P_r$$

$$\sigma^2 = \sum_{r=0}^{\infty} r^2 P_r - 2\mu \sum_{r=0}^{\infty} rP_r + \mu^2 \sum_{r=0}^{\infty} P_r$$

Now $$-2\mu \sum_{r=0}^{\infty} rP_r = -2\mu(\mu) = -2\mu^2$$

$$\mu^2 \sum_{r=0}^{\infty} P_r = \mu^2(1) = \mu^2$$

and $$\sum_{r=0}^{\infty} r^2P_r = \sum_{r=0}^{\infty} [r(r-1)+r]P_r$$

$$\sum_{r=0}^{\infty} r(r-1)P_r + \mu = \left[0+0+2\frac{\mu^2 e^{-\mu}}{2!}+6\frac{\mu^3 e^{-\mu}}{3!}+\ldots\right]+\mu$$

$$= \mu^2 e^{-\mu}\left[1+\mu+\frac{\mu^2}{2!}+\ldots\right]+\mu$$

$$= \mu^2 e^{-\mu}(e^{\mu}) + \mu = \mu^2 + \mu$$

Thus, collecting the terms, we have

$$\sigma^2 = \mu^2 + \mu - 2\mu^2 + \mu^2$$

or $$\sigma^2 = \mu \qquad \ldots(2.82)$$

Note that the Poisson distribution contains only one parameter, np, the mean occurrence of an event, and we do not know the value of n. In the binomial distribution we know the number of times an event occurs and the number of times an event does not occur.

Cumulative Poisson probabilities are given in Appendix 2D. A plot of the Poisson distribution for $\lambda = 3$ is shown in Fig. 2.8. This Poisson distribution has a very long tail to the right (the distribution is skewed).

2.7 THE NORMAL DISTRIBUTION

The normal distribution is in many respects the cornerstone of statistics. A random variable X is said to have a normal distribution with mean μ $(-\infty < \mu < \infty)$ and variance $\sigma^2 > 0$, if it has the density function

$$f(x) = \frac{1}{\sigma\sqrt{2\pi}} e^{-(1/2)[(x-\mu)\sigma]^2}, \; -\infty < X < \infty \qquad \ldots(2.83)$$

The distribution is illustrated graphically in Fig. 2.9. The normal distribution is used so extensively that the shorthand notation $X \sim N(\mu, \sigma^2)$ is often used to indicate that the random variable X is normally distributed with mean μ and variance σ^2.

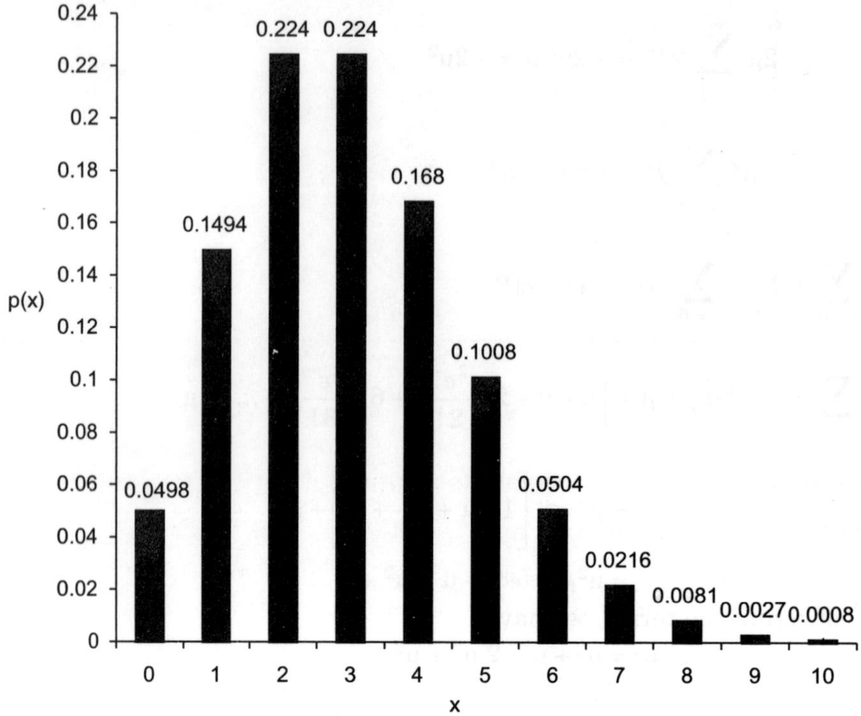

Fig. 2.8 Poisson Distribution for $\lambda = 3$.

2.7.1 Properties of the Normal Distribution

The normal distribution has several important properties:

1. $\int_{-\infty}^{\infty} f(x)\, dx = 1$
2. $f(x) \geq 0$ for all x [required of all density functions]
3. $\lim_{x \to \infty} f(x) = 0$ and $\lim_{x \to -\infty} f(x) = 0$
4. $f[(x + \mu)] = f[-(x - \mu)]$. The density is symmetric about μ
5. The maximum value of f occurs at $x = \mu$
6. The points of inflection of f are at $x = \mu \pm \sigma$

Property 1 may be demonstrated as follows:
Let $y = (x - \mu)/\sigma$ in Eq. (2.83) and denotes the integral as I.
That is,

$$I = \frac{1}{\sqrt{2\pi}} \int_{-\infty}^{\infty} e^{-(1/2)y^2} dy \frac{1}{\sqrt{2\pi}} \int_{-\infty}^{\infty} e^{-(1/2)x^2} dx$$

On changing to polar coordinates with the transformation of variables $y = r \sin \theta$ and $z = r \cos \theta$, the integral becomes

$$I^2 = \frac{1}{2\pi} \int_0^{\infty} \int_0^{2\pi} re^{-(1/2)r^2} d\theta\, dr = \int_0^{\infty} re^{-(1/2)r^2} dr = 1$$

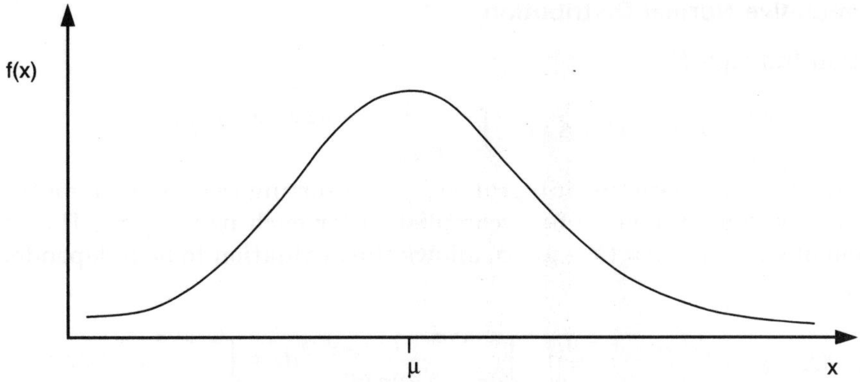

Fig. 2.9 The Normal Distribution.

2.7.2 Mean and Variance of the Normal Distribution

The mean of the normal distribution may be determined as follows:

Since $$E(X) = \int_{-\infty}^{\infty} \frac{x}{\sigma\sqrt{2\pi}} e^{-(1/2)[(X-\mu)/\sigma^2]} dx$$

and if we let $z = (x - \mu)/\sigma$, we obtain

$$E(X) = \int_{-\infty}^{\infty} \frac{1}{\sqrt{2\pi}} (\mu + \sigma z)\, e^{-z^2/2} dz$$

$$= \mu \int_{-\infty}^{\infty} \frac{1}{\sqrt{2\pi}} dz + \sigma + \int_{-\infty}^{\infty} \frac{1}{\sqrt{2\pi}} z e^{-z^2/2} dz$$

Since the integrand of the first integral is that of a normal density with $\mu = 0$ and $\sigma^2 = 1$, the value of the first integral is one. The second integral has value zero, that is

$$\int_{-\infty}^{\infty} \frac{1}{\sqrt{2\pi}} z e^{-z^2/2}\, dz = -\frac{1}{\sqrt{2\pi}} z e^{-z^2/2} \Big|_{-\infty}^{\infty} = 0$$

and thus $$E(X) = \mu\,[1] = \sigma\,[0] = \mu \qquad \text{...(2.84)}$$

To find the variance we must evaluate

$$V(X) = E[(X - \mu)^2] = \int_{-\infty}^{\infty} (x - \mu)^2 \frac{1}{\sigma\sqrt{2\pi}} e^{-(1/2)[(x-\mu)/\sigma]^2} dx$$

And letting $z = (x - \mu) / \sigma$, we obtain

$$V(X) = \int_{-\infty}^{\infty} \sigma^2 z^2 \frac{1}{\sqrt{2\pi}} e^{-z^2/2} dz = \sigma^2 \left[\int_{-\infty}^{\infty} \frac{z^2}{\sqrt{2\pi}} e^{-z^2/2} dz \right]$$

$$= \sigma^2 \left[\frac{-ze^{-z^2/2}}{\sqrt{2\pi}} \Big|_{-\infty}^{\infty} \right] + \int_{-\infty}^{\infty} \frac{1}{\sqrt{2\pi}} e^{-z^2/2} dz = \sigma^2\,(0 + 1)$$

so that $$V(X) = \sigma^2 \qquad \text{...(2.85)}$$

In summary, the mean and variance of the normal density given in Eq. (2.83) are μ and σ^2, respectively.

2.7.3 The Cumulative Normal Distribution

The distribution function F is

$$F(x) = P(X \le x) = \int_{-\infty}^{\infty} \frac{1}{\sigma\sqrt{2\pi}} e^{-(1/2)[(u-\mu)/\sigma]^2} du \qquad ...(2.86)$$

It is difficult to evaluate this integral without resorting to numerical methods, and even then the evaluation would have to be accomplished for each pair (μ, σ^2). However, a simple transformation of variables, $z = (X - \mu) / \sigma$, allows the evaluation to be independent of μ and σ.

That is

$$F(X) = P(X \le x) = P\left(Z \le \frac{x-\mu}{\sigma}\right) = \int_{-\infty}^{(x-\mu)/\sigma} \frac{1}{\sqrt{2\pi}} e^{-z^2/2} dz = \int_{-\infty}^{(x-\mu)/\sigma} \varphi(z)\, dz = \phi\left(\frac{x-\mu}{\sigma}\right) \qquad ...(2.87)$$

2.7.4 The Standard Normal Distribution

The probability distribution in Eq. (2.87) above, namely

$$\varphi(z) = \frac{1}{\sqrt{2\pi}} e^{-z^2/2} \qquad -\infty < z < \infty$$

is a normal distribution with mean 0 and variance 1; that is, $Z \sim N(0, 1)$ and we say that Z has a *standard normal distribution*. A graph of the probability density function is shown in Fig. 2.10.

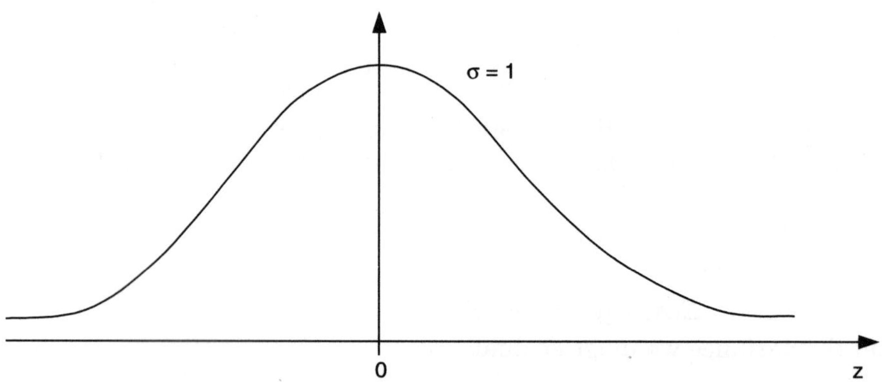

Fig. 2.10 The Standard Normal Distribution.

The corresponding distribution function is ϕ, where

$$\varphi(z) = \int_{-\infty}^{\infty} \frac{1}{\sqrt{2\pi}} e^{-u^2/2} du \qquad ...(2.88)$$

and this function has been well tabulated. A table of the integral in Eq. (2.88) has been provided in Tabular form in Appendix 2E.

2.7.4.1 Problem-solving procedure

The procedure of solving problems involves in calculating the cumulative normal probabilities is simple. For example, suppose that $X \sim N(50, 4)$, and we wish to find the probability that x is less than or equal to 54; that is $P(x \le 54) = F(54)$. Since that standard normal random variable is

$$z = \frac{x - \mu}{\sigma}$$

we can *standardize* the point of interest $x = 54$ to obtain

$$z = \frac{x - \mu}{\sigma} = \frac{54 - 50}{2} = 2$$

Now the probability that the *standard* normal random variable z is less than or equal to 2 is equal to the probability that the *original* normal random variable x is less than or equal to 54. Expressed mathematically, we have

$$F(x) = \phi = \left(\frac{x - \mu}{\sigma}\right) = \phi(z)$$

$$F(54) = \phi(2)$$

Appendix 2E contains cumulative standard normal probabilities for various values of z. From this table, we can read

$$\phi(2) = 0.9772$$

Note that in the relationship $z = (x - \mu)/\sigma$, the variable z measures the departure of x from the mean μ in standard deviation (σ) units. In our example $F(54) = \phi(2)$ indicates that 54 is *two* standard deviations ($\sigma = 2$) above the mean. In general, $x = \mu + \sigma z$. In solving problems, we sometimes need to use the symmetric property of φ in addition to the tables. It is useful to make a sketch in finding exactly which probabilities are required, since the area under the curve and over the interval of interest is the probability that the random variable will lie on the interval. Table 2.1 gives a summary of probability distributions.

Table 2.1. Summary of Probability Distributions

Name	Probability density function	Mean	Variance
Discrete			
Hypergeometric	$\dfrac{\binom{k}{x}\binom{N-k}{n-X}}{\binom{N}{n}}$ $x = \max(0, n - N + K), 1, \ldots \min(K, n), K \le N, n \le N$	np where $p = \dfrac{K}{N}$	$np(1-p)\left(\dfrac{N.m}{N-1}\right)$
Binomial	$\binom{n}{X} p^x (1-p)^{n-x}$, $x = 0, 1, \ldots, n, 0 \le p \le 1$	np	$np(1-p)$
Poisson	$\dfrac{e^{-\lambda}\lambda^x}{x!}, x = 0, 1, 2, \ldots, \lambda > 0$	λ	λ
Continuous			
Normal	$\dfrac{1}{\sqrt{2\pi}\sigma} e^{-\frac{(x-\mu)^2}{2\sigma^2}}$ $-\infty < x < \infty, -\infty < \mu < \infty, \sigma > 0$	μ	σ^2

Example E2.13. *The diameter of a metal shaft for a precision instrument is assumed to be normally distributed with a mean of 0.5 mm and a standard deviation of 0.025 mm.*

(a) What is the probability that a shaft diameter is greater than 0.31 mm?

(b) What is the probability that a shaft diameter is between 0.235 and 0.315 mm?

(c) The diameter of 90% of samples is below what value?

Solution:

(*a*)
$$P(X > 0.31) = P\left(Z > \frac{0.315 - 0.25}{0.025}\right) = P(Z > 3) = 0.00135$$

(*b*)
$$P(0.235 < X < 0.315) = P(-0.6 < Z < 2.6) = P(Z < 2.6) - P(Z < -0.6)$$
$$= 0.99534 - 0.27425 = 0.72109$$

(*c*)
$$P(X < x) = P\left(Z < \frac{X - 0.25}{0.025}\right) = 0.90$$

Hence
$$\frac{x - 0.25}{0.025} = 1.28$$

and
$$x = 0.282$$

Example E2.14. *The weight of an electronic component is normally distributed with a mean of 6 ounces and a standard deviation of 0.25 ounce.*

(a) Find the probability that the electronic component weighs more than 6.5 ounces.

(b) What must the standard deviation of weight be in order for the company to state that 99% of its electronic components are less than 6.5 ounces?

(c) If the standard deviation stays at 0.25 ounce, what must the mean weight be in order for the company to state that 99.9% of its electronic components are less than 6.5 ounces?

Solution:

(*a*) $P(X > 6.5) = P\left(Z > \frac{6.5 - 6}{0.25}\right) = P(Z > 2) = 0.02275$

(*b*) If $P(X < 13) = 0.999$, then $P\left(Z < \frac{6.5 - 6}{\sigma}\right) = 0.999$

Hence,
$$\frac{0.5}{\sigma} = 3.09 \quad \text{or } \sigma = 0.1618$$

(*c*) If $P(X < 6.5) = 0.999$, then $\text{P}\left(Z < \frac{6.5 - \mu}{\sigma 0.25}\right) = 0.999$

Therefore,
$$\frac{6.5 - \mu}{0.25} = 3.09 \text{ and } \mu = 5.7275$$

2.8 APPROXIMATING PROBABILITY DISTRIBUTIONS

In some engineering quality control applications, it is quite useful to approximate one probability distribution by another. Figure 2.11 shows the approximation guide, that is, the conditions under which one distribution may be approximated by another.

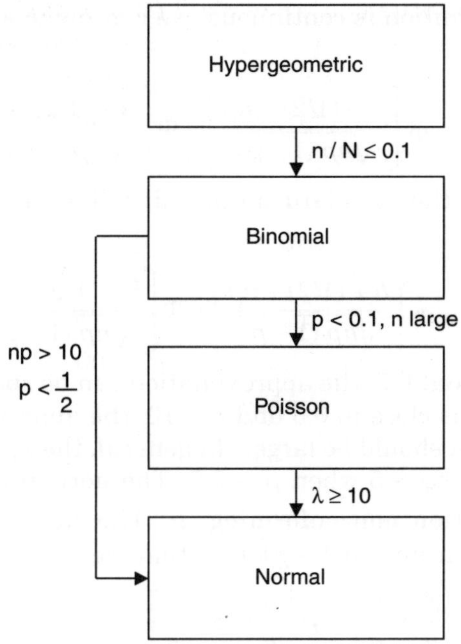

Fig. 2.11 Approximation Guide.

2.8.1 Binomial Approximation to the Hypergeometric

When the ratio $\frac{n}{N}$ is small, that is, say $\frac{n}{N} \leq 0.1$, then the binomial distribution parameter $p = \frac{D}{N}$ can be used as a good approximation to the hypergeometric distribution. The smaller the fraction $\frac{n}{N}$, the better the approximation.

2.8.2 Poisson Approximation to the Binomial

The Poisson distribution can be used as an approximation to the binomial distribution when the sample size n is large and the probability of success p is small (the same applies when q is small, p and q being of course interchangeable), *i.e.,* when the binomial distribution is highly skewed. As a guide, a good approximation is obtained when $n \geq 20$ and $p \leq 0.05$, and the approximation improves with a decrease in p.

The advantage of using the Poisson distribution to approximate the binomial distribution is that Poisson distribution, having only one parameter, is well tabulated.

When the parameter p is small, say, $p < 0.1$, and n is large, with $\lambda = np$ a constant, the Poisson distribution is used as an approximation to the binomial distribution. The larger the value of n and the smaller the value of p, the better the approximation.

2.8.3 Normal Approximation to the Binomial

If n is large, then one can justify the approximation of the binomial distribution by the normal distribution with mean np and variance $np(1 - p)$. Noting that the binomial distribution is discrete and the normal distribution is continuous, we can make a continuity correction such that

$$P(X = a) = \Phi\left[\frac{a + (1/2) - np}{\sqrt{np\,(1-p)}}\right] - \Phi\left[\frac{a - (1/2) - np}{\sqrt{np\,(1-p)}}\right] \quad ...(2.89)$$

where $\Phi(.)$ denotes the cdf of the standard normal distribution. We also write probability statements as

$$P(a \le X \le b) = \Phi\left[\frac{b + (1/2) - np}{\sqrt{np\,(1-p)}}\right] - \Phi\left[\frac{b - (1/2) - np}{\sqrt{np\,(1-p)}}\right] \quad ...(2.90)$$

Although p should be about 0.5, the approximation can be sued without excessive loss of accuracy for $0.1 \le p \le 0.9$. If p is close to 0.5 and $n > 10$, the approximation is fairly good. For other values of p , the value of n should be larger. In general, the approximation is good as long as $np > 5$ for $p \le 0.5$ or when $nq > 5$ when $p > 0.5$. The normal distribution is also used to approximate the sample fraction nonconforming $\hat{p}$. The random variable $\hat{p}$ is normally distributed with mean p and variance $p(1 - p)/n$, so that we have

$$P\left[\frac{a}{n} \le \frac{X}{n} \le \frac{b}{n}\right] = P\left[\frac{a}{n} \le \hat{p} \le \frac{b}{n}\right]$$

from which we obtain

$$P\left[\frac{a}{n} \le \hat{p} \le \frac{b}{n}\right] = \Phi\left[\frac{\dfrac{b + (1/2)}{n} - p}{\sqrt{\dfrac{p\,(1-p)}{n}}}\right] - \Phi\left[\frac{\dfrac{a - (1/2)}{n} - p}{\sqrt{\dfrac{p\,(1-p)}{n}}}\right] \quad ...(2.91)$$

2.8.4 Normal Approximation to the Poisson

The normal distribution with mean $\mu = \sigma^2 = \lambda$ can be used as an approximation to the Poisson distribution if the mean λ of the distribution is large, say 10 or more. The continuity correction used is given by

$$P(a \le X \le b) = \Phi\left[\frac{b + (1/2) - \lambda}{\sqrt{\lambda}}\right] - \Phi\left[\frac{a - (1/2) - \lambda}{\sqrt{\lambda}}\right] \quad ...(2.92)$$

Example E2.15. *Fifteen per cent of the parts produced in a manufacturing process fail a standardized quality test.*

(*a*) *Using the Poisson formula, find the probability that in a random sample of 100 parts which went through this test exactly 20 will fail.*

(*b*) *Using the Poisson probabilities table, find the probability that the number of parts which fail this test in a randomly selected 100 parts is*

(*i*) *At most 9*

(*ii*) *10 to 16*

(*iii*) *At least 20*

Solution:

Let X denote the number of parts in a random sample of 100, which fail the test. Since, on average, 15% of the parts fail the test, $\lambda = 0.15 \times 100 = 15$

(a) P (exactly 20 fail) $= P(x = 20) = \dfrac{\lambda^x e^{-x}}{x!} = \dfrac{(15)^{20} e^{-15}}{20!} = 0.0418$

(b) (i) P (at most 9 fail) $= P(x \le 9)$

$= P(0) + P(1) + P(2) + P(3) + P(4) + P(5) + P(6) + P(7) + P(8) + P(9)$

$= 0.0 + 0.0 + 0.0 + 0.0002 + 0.0006 + 0.0019 + 0.0048 + 0.0104 + 0.0194 + 0.0324$

$= 0.0697$

(ii) P (10 to 16 fail) $= P(10 \le x \le 16)$

$= P(10) + P(11) + P(12) + P(13) + P(14) + P(15) + P(16)$

$= 0.0486 + 0.0663 + 0.0829 + 0.0956 + 0.1024 + 0.1024 + 0.0960$

$= 0.5942$

(iii) P (at least 20) $= P(x \ge 20)$

$= P(20) + P(21) + P(22) + \ldots$

$= 0.0418 + 0.0299 + 0.0204 + 0.0133 + 0.0083 + 0.0050 + 0.0029$
$+ 0.0016 + 0.0009 + 0.0004 + 0.0002 + 0.0001 + 0.0001$

$= 0.1249$

2.9 CHEBYSHEV'S THEOREM

Chebyshev's theorem gives a lower bound for the area under a curve between two points that are on opposite sides of the mean and at the same distance from the mean. Chebyshev's theorem is stated as follows: For any number k greater than 1, at least $\left(1 - \dfrac{1}{k^2}\right)$ of the data values lie within k standard deviations of the mean. Chebyshev's theorem is illustrated in Fig. 2.12.

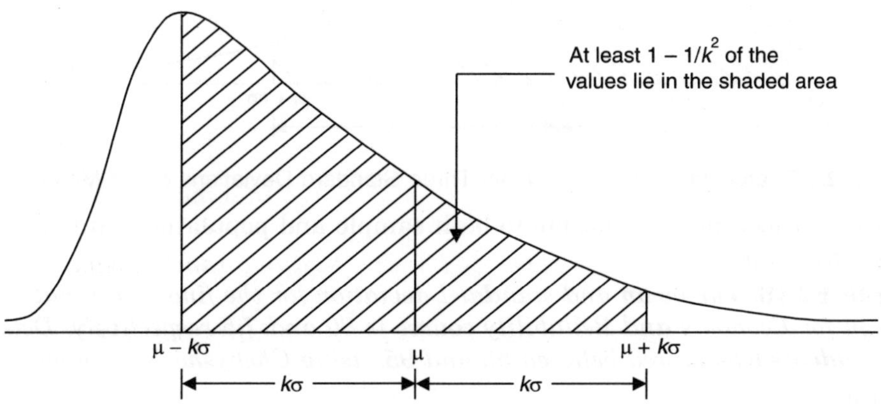

Fig. 2.12 Chebyshev's Theorem.

For $k = 2$

$$1 - \frac{1}{k^2} = 1 - \frac{1}{2^2} = 0.75 \text{ or } 75\%.$$

That is, at least 75% of the values of a data set lie within two standard deviations of the mean as shown in Fig. 2.13.

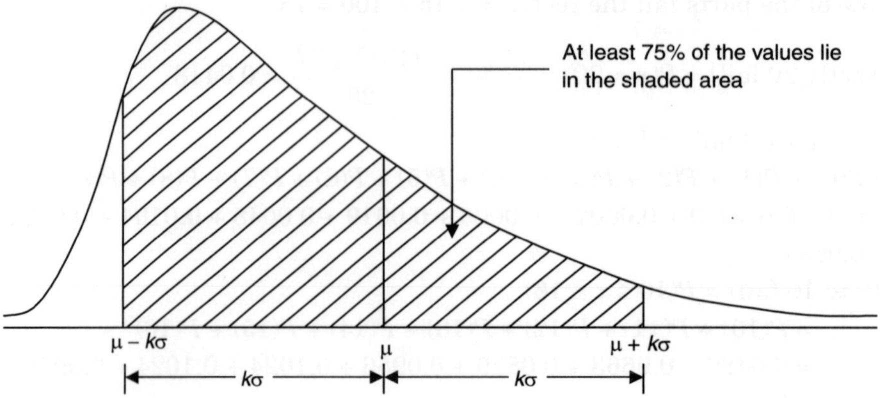

Fig. 2.13 Percentage of Values within Two Standard Deviations of the Mean.

For $k = 3$

$$1 - \frac{1}{k^2} = 1 - \frac{1}{3^2} = 0.89 \text{ or } 89\%.$$

Hence, 89% of the values fall within three standard deviations of the mean as shown in Fig. 2.14.

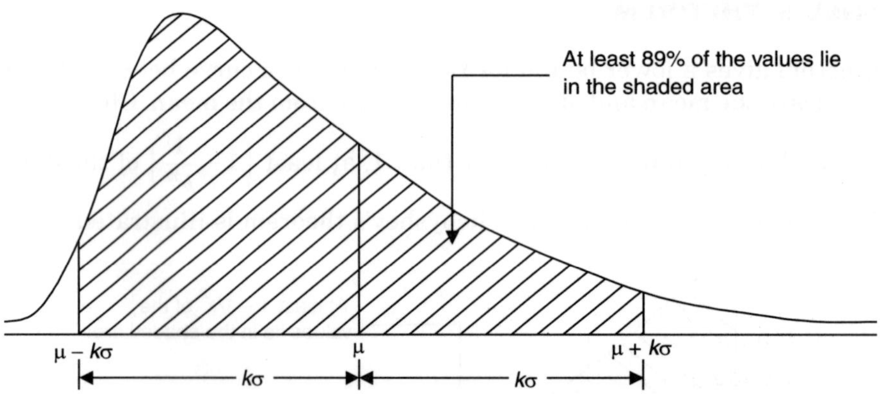

Fig. 2.14 Percentage of Values within Three Standard Deviations of the Mean.

Chebyshev's theorem is applicable to both sample and population data as well as to a distribution of any shape.

Example E2.16. *The mean and standard deviation for the final examination scores of DM 420: Design for Economy and Reliability course is 80 and 7.5 respectively. Determine the percentage of students who scored between 65 and 95, using Chebyshev's theorem.*

Solution:

From the given data

Mean $\mu = 80$ and standard deviation $\sigma = 7.5$

Each of the two points, 60 and 95, is 15 units away from the mean.

Hence, $$k = \frac{15}{\sigma} = \frac{15}{7.5} = 2$$

$$1 - \frac{1}{k^2} = 1 - \frac{1}{2^2} = 0.75 \text{ or } 75\%.$$

That is, at least 75% of the students scored between 60 and 95 as shown in Fig. E2.16.

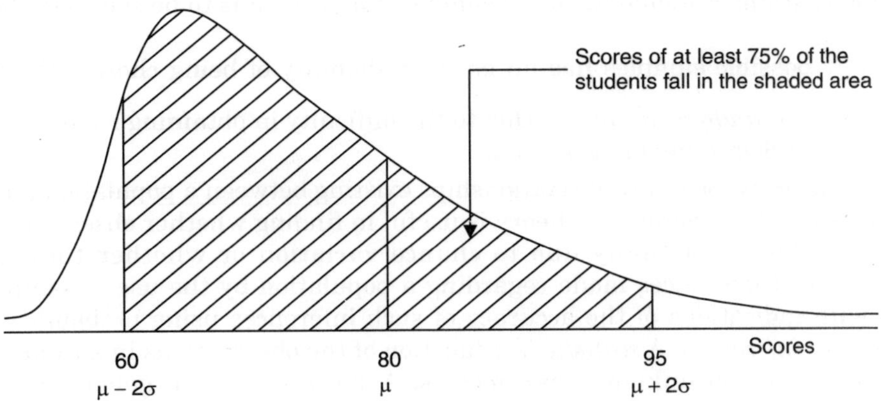

Fig. E2.16 Percentage of Students Who Scored Between 60 and 95.

2.10 EMPIRICAL RULE

If the distribution of the data is approximately bell-shaped, then empirical rule applies. The empirical rule is stated as follows:

For a bell-shaped distribution, approximately

1. 68% of the observations lie within one standard deviation of the mean
2. 95% of the observations lie within two standard deviations of the mean
3. 99.7% of the observations lie within three standard deviations of the mean

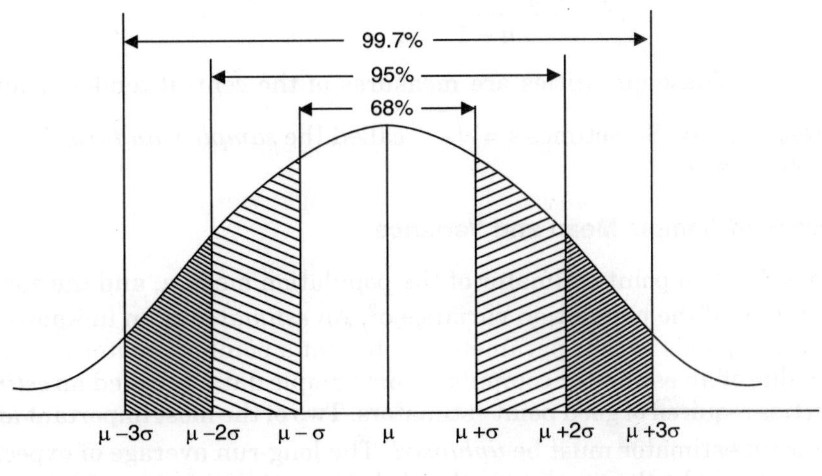

Fig. 2.15 The Empirical Rule.

The empirical rule is illustrated in Fig. 2.15. The bell-shaped distribution, also known as the normal distribution has been described earlier. The empirical rule is applicable to population data as well as sample data.

2.11 SAMPLING AND SAMPLING DISTRIBUTIONS

If the population contains N elements and a sample of n of them is to be selected, then if each of the $\frac{N!}{(N-n)!\,n!}$ possible samples has an equal probability of being chosen, the procedure employed is known as *random sampling*. Due to the difficulty in obtaining random samples in practice, tables of random numbers are used.

Sampling theory is the study of relationships existing between a population and samples drawn from the population. Sampling theory is useful in finding whether observed differences between two samples are actually due to chance variation or whether they are really significant. A study of inferences made regarding a population by the use of samples drawn from it, along with indications of the accuracy of such inferences using probability theory is known as *statistical inferences.* A *statistic* is a function of the observations in a random sample, which is not dependent on unknown parameters. A *parameter* is in general an unknown constant. For example, the parameters of a normal distribution are μ and σ^2, whereas $\overline{X}$ and σ^2 are statistics. The behavior of sample statistics is needed in order to draw conclusions about a sample. The behavior of the sample statistics is described by a *sampling distribution.* Suppose that $y_1, y_2, \ldots, y_n$ represents a sample. Then the *sample mean*

$$\bar{y} = \frac{\sum_{i=1}^{n} y_i}{n} \qquad \ldots(2.93)$$

and the *sample variance*

$$s^2 = \frac{\sum_{i=1}^{n} (y_i - \bar{y})^2}{n-1} \qquad \ldots(2.94)$$

are both statistics. These quantities are measures of the central tendency and dispersion of the sample, respectively. Sometimes $s = \sqrt{s^2}$, called the *sample standard deviation*, is used as a measure of dispersion.

2.11.1 Properties of Sample Mean and Variance

The sample mean $\bar{y}$ is a point estimator of the population mean μ, and the sample variance s^2 is a point estimator of the population variance σ^2. An *estimator* of an unknown parameter is a statistic that corresponds to that parameter. Note that a point estimator is a random variable. A numerical value of an estimator computed from sample data, is called an *estimate.* There are several properties required of good point estimators. Two of the most important are the following:

1. The point estimator must be *unbiased*. The long-run average of expected value of the point estimator should be the parameter that is being estimated. An unbiasedness is desirable property and this alone does not always make an estimator a good one.

2. An unbiased estimator must have *minimum variance*. The minimum variance point estimator has a variance that is smaller than the variance of any other estimator of that parameter. Here, we show that $\bar{y}$ and s^2 are unbiased estimators of μ and σ^2 respectively. Using the properties of expectation,

$$E(\bar{y}) = \left(\frac{\sum_{i=1}^{n} y_i}{n} \right) = \frac{1}{n} E\left(\sum_{i=1}^{n} y_i \right)$$

$$= \frac{1}{n} \sum_{i=1}^{n} E(y_i) = \frac{1}{n} \sum_{i=1}^{n} \mu = \mu \qquad \text{...(2.95)}$$

since the expected value of each observation y_i is μ. Thus, $\bar{y}$ is an unbiased estimator of μ. Considering the sample variances s^2, we get

$$E(S^2) = E\left(\frac{\sum_{i=1}^{n} (y_i - \bar{y})^2}{n-1} \right) = \frac{1}{n-1} E\left[\sum_{i=1}^{n} (y_i - \bar{y})^2 \right] = \frac{1}{n-1} E(SS) \qquad \text{...(2.96)}$$

where $SS = \sum_{i=1}^{n} (y_i - \bar{y})^2$ is the *corrected sum of squares* of the observation y_i. Now

$$E(SS) = E\left[\sum_{i=1}^{n} (y_i - \bar{y})^2 \right]$$

$$= E\left[\sum_{i=1}^{n} y_i^2 - n\bar{y}^2 \right] = \sum_{i=1}^{n} (\mu^2 + \sigma^2) - n(\mu^2 + \sigma^2/n) = (n-1)\,\sigma^2 \qquad \text{...(2.97)}$$

Hence, $$E(S^2) = \frac{1}{n-1} E(SS) = \sigma^2 \qquad \text{...(2.98)}$$

Therefore, s^2 is an unbiased estimator of σ^2. The quantity $n - 1$ in Eq. (2.97) is called the *number of degrees of freedom* of the sum of squares SS. If y is a random variable with variance σ^2 and $SS = \Sigma(y_i - \bar{y})^2$ has degrees of freedom, then

$$E\left(\frac{SS}{v} \right) = \sigma^2 \qquad \text{...(2.99)}$$

The number of degrees of freedom of a sum of squares is equal to the number of independent elements in that sum of squares. For example, $SS = \sum_{i=1}^{n} (y_i - \bar{y})^2$ in Eq. (2.97) consists of the sum of squares of the n elements $y_1 - \bar{y}, y_2 - \bar{y}, ..., y_n - \bar{y}$. These elements are not all independent since $\sum_{i=1}^{n} (y_i - \bar{y}) = 0$; in fact, only $n - 1$ of them are independent, implying that SS has $n - 1$ degrees of freedom.

The *standard error* of a statistic is the standard deviation of its sampling distribution. If the standard error involves unknown parameters whose values can be estimated, substitution of these estimates into the standard error results in an *estimated standard error*. Suppose, we are sampling from a normal distribution with mean μ and variance σ^2. Now the distribution is normal with mean μ and variance σ^2/n, so the *standard error* of $\overline{X}$ is

$$\frac{\sigma}{\sqrt{n}} \qquad ...(2.100)$$

The *estimated standard error* of $\overline{X}$ is

$$\frac{s}{\sqrt{n}} \qquad ...(2.101)$$

2.11.2 The Central Limit Theorem

If $y_1, y_2, ..., y_n$ is a sequence of n independent and identically distributed random variables with $E(y_1) = \mu$ and $V(y_i) = \sigma^2$ (both finite) and $x = y_1 + y_2 + ... + y_n$, then

$$z_n = \frac{x - n\mu}{\sqrt{n\sigma^2}} \qquad ...(2.102)$$

has an approximate $N(0,1)$ distribution in the sense that, if $F_n(z)$ is the distribution function of z_n and $\Phi(z)$ is the distribution function of the $N(0,1)$ random variable, then

$$\lim_{n \to \infty} [F_n(z)/\Phi(z)] = 1.$$

That is, the sum of n independent and identically distributed random variables is approximately normally distributed. This approximation is good for very small n, say $n < 10$, whereas in other cases large n is required, say $n > 100$.

2.11.3 The Chi-Square Distribution

Let $Z_1, Z_2, ..., Z_k$ be normally and independently distributed random variables, with mean $\mu = 0$ and variance $\sigma^2 = 1$. Then the random variable

$$\chi^2 = Z_1^2 + Z_2^2 + ... + Z_k^2$$

has the probability density function

$$f_{\chi^2}(u) = \frac{1}{2^{k/2}\Gamma\left(\frac{k}{2}\right)} u^{(k/2)} e^{-\mu/2}$$

$$= 0 \text{ otherwise} \qquad ...(2.103)$$

and is said to follow the chi-square distribution with k degrees of freedom abbreviated χ_k^2.

The mean and variance of the distribution are

$$\mu = k \qquad ...(2.104)$$

and

$$\sigma^2 = 2k \qquad ...(2.105)$$

Several chi-square distributions are shown in Fig. 2.16.

The chi-square random variable is non-negative, and the probability distribution is skewed to the right. As k increases, the distribution becomes more symmetric. As $k \to \infty$, the limiting form of the chi-square distribution is the normal distribution. The percentage points of the χ_k^2 distribution are given in Appendix 2F. Define $\chi_{\alpha,k}^2$ as the percentage point or value of the chi-square random variable with k degrees of freedom such that the probability that χ_k^2 exceed this value is α.

That is $$P\{\chi_k^2 \geq \chi_{\alpha,k}^2\} = \int_{\chi_{\alpha,k}^2}^{\infty} f_{\chi^2}(\mu)\, du = \alpha \qquad ...(2.106)$$

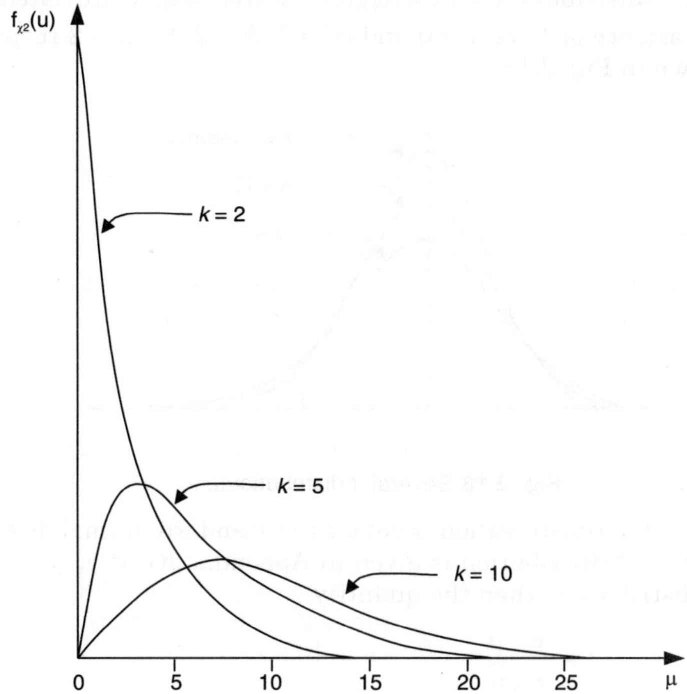

Fig. 2.16 Several χ^2 Distributions.

The probability in Eq. (2.106) is shown in Fig. 2.17.

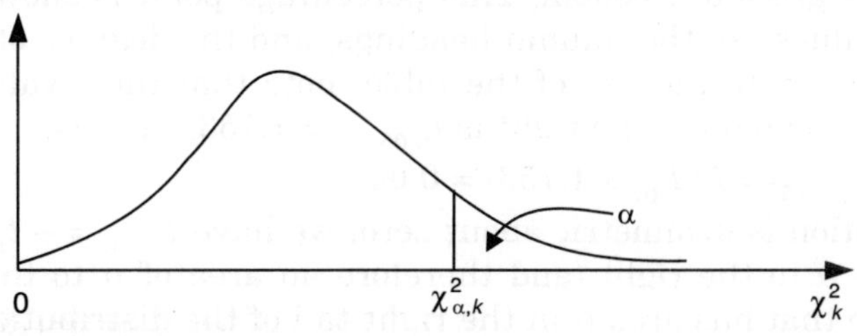

Fig. 2.17 Percentage Point $\chi_{\alpha,k}^2$ of the Chi-square Distribution.

2.11.4 The *t*-Distribution

Let $Z \sim N(0,1)$ and V be a chi-square random variable with k degrees of freedom. If Z and V are independent, then the random variable

$$T = \frac{Z}{\sqrt{V/k}} \qquad ...(2.107)$$

has the probability density function

$$f(t) = \frac{\Gamma\,[(k+1)/2]}{\sqrt{\pi k}\,\Gamma\,(k/2)} \times \frac{1}{[(t^2/k)+1]^{(k+1)/2}} \qquad -\infty < t < \infty \qquad \text{...(2.108)}$$

and is said to follow the t-distribution with k degrees of freedom, abbreviated t_k.

The mean and variance of t are $\mu = 0$ and $\sigma^2 = k/(k-2)$ for $k > 2$ respectively. Several t-distributions are shown in Fig. 2.18.

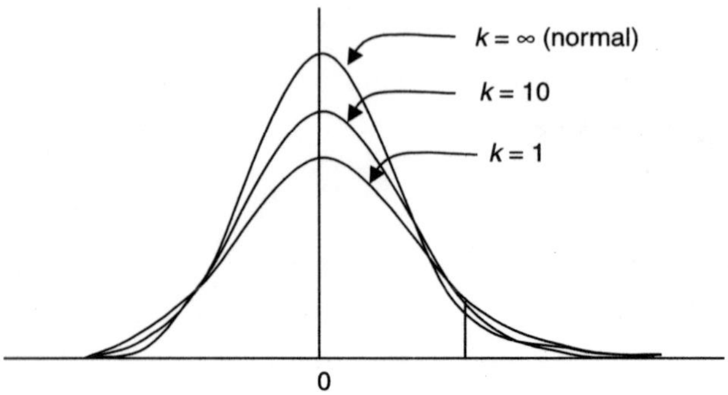

Fig. 2.18 Several t-distributions.

Note that if $k = \infty$, the t-distribution becomes the standard normal distribution. A table of percentage points of the t-distribution is given in Appendix 2G. If $y_1, y_2, \ldots, y_n$ is a random sample from $N(\mu, \sigma^2)$ distribution, then the quantity

$$t = \frac{\bar{y} - \mu}{s/\sqrt{n}} \qquad \text{...(2.109)}$$

is distributed as t with $n - 1$ degrees of freedom.

We will let $t_{\alpha,\,k}$ be the value of the random variable T with k degrees of freedom above which we find an area (or probability) α. Thus, $t_{\alpha,k}$ is an upper-tail 100α percentage point of the t-distribution with k degrees of freedom. This percentage point is shown in Fig. 2.19. In the Appendix 2G, the α values are the column headings, and the degrees of freedom are listed in the left column. To illustrate the use of the table, note that the t-value with 15 degrees of freedom having an area of 0.05 to the right is $t_{0.05,\,15} = 1.753$. That is,

$$P(T_{15} > t_{0.05,\,15}) = P(T_{15} > 1.753) = 0.05$$

Since t-distribution is symmetric about zero, we have $t_{1-\alpha} = -\,t_{\alpha}$; that is, the t-value having an area of $1 - \alpha$ to the right (and therefore an area of α to the left) is equal to the negative of the t-value that has area α in the right tail of the distribution.

Therefore, $t_{0.95,\,15} = -\,t_{0.05,\,15} = -\,1.753$.

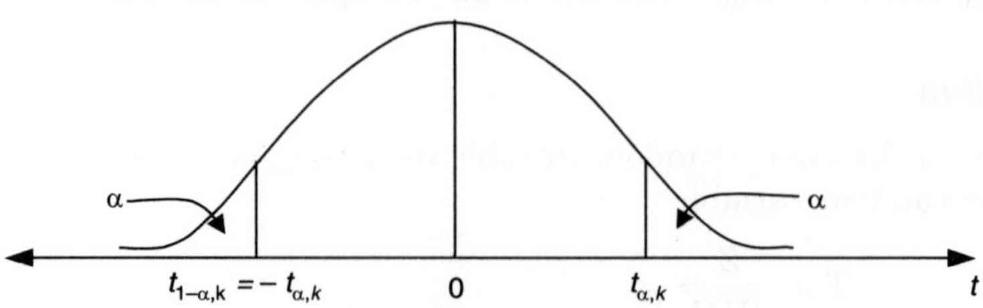

Fig. 2.19 Percentage Points of the t-distributions.

2.11.5 The *F*-Distribution

The random variable F is defined to be the ratio of two independent chi-square random variables, each divided by its number of degrees of freedom. That is,

$$F = \frac{W/u}{Y/v} \qquad ...(2.110)$$

where W and Y are independent chi-square random variables with u and v degrees of freedom, respectively.

The ratio

$$F = \frac{W/u}{Y/v} \qquad ...(2.111)$$

has the probability density function

$$f(x) = \frac{\Gamma\left(\frac{u+v}{2}\right)\left(\frac{u}{v}\right)^{u/2} x^{(u/2)-1}}{\Gamma\left(\frac{u}{2}\right)\Gamma\left(\frac{v}{2}\right)\left[\left(\frac{u}{v}\right)x+1\right]^{(u/2)}} \qquad 0 < x < \infty \qquad ...(2.112)$$

and is said to follow the F-distribution with u degrees of freedom in the numerator and v degrees of freedom in the denominator. It is denoted as $F_{u,v}$.

The mean and variance of F-distribution are $\mu = v/(v-2)$ for $v > 2$, and

$$\sigma^2 = \frac{2v^2\,(u+v-2)}{u\,(v-2)^2\,(v-4)}, \; v > 4$$

The F random variable is non-negative, and the distribution is skewed to the right. The F-distribution looks very similar to the chi-square distribution as shown in Fig. 2.20; however the two parameters u and v provide extra flexibility regarding shape.

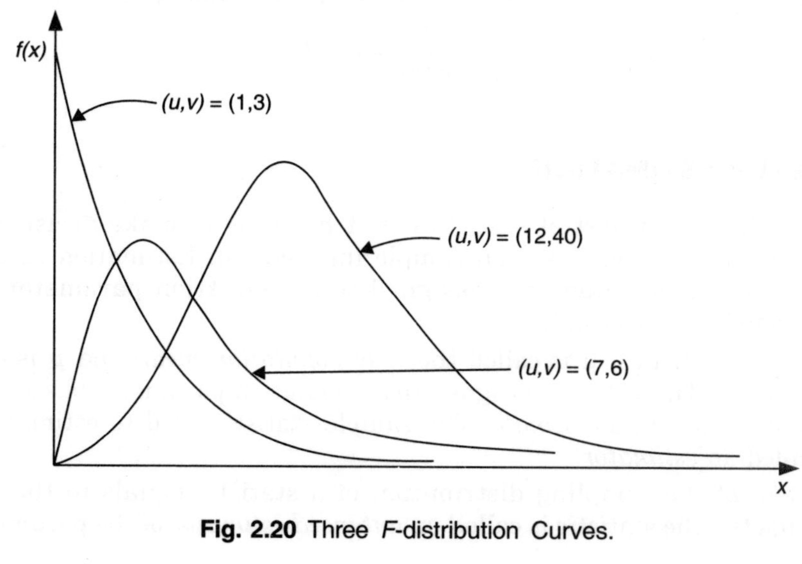

Fig. 2.20 Three *F*-distribution Curves.

The percentage points of the F-distribution are given in Appendix 2H. Let $f_{\alpha, u, v}$ be the percentage point of the F-distribution, with the numerator degrees of freedom u and the denominator degrees of freedom v such that the probability that the random variable F exceeds this value is

$$P(F > f_{\alpha, u, v}) = \int_{f_{\alpha,u,v}}^{\infty} f(x)\, dx = \alpha \quad ...(2.113)$$

This is illustrated in Fig. 2.21.

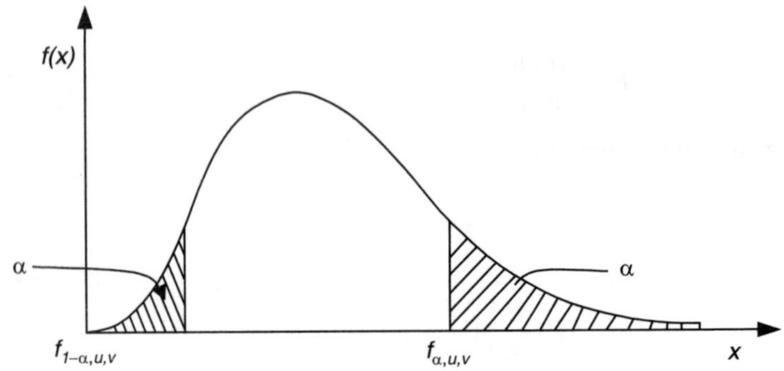

Fig. 2.21 Upper and Lower Percentage Points of the F-distribution.

For example, if $u = 6$ and $v = 10$, we find from Appendix 2H, that

$$P(F > f_{0.05, 6, 10}) = P(F_{6, 10} > 3.22) = 0.05$$

That is, the upper 5 percentage point of $F_{6, 10}$ is $f_{0.05, 6,10} = 3.22$. Appendix 2H contains only upper-tail percentage points (for selected values of $f_{\alpha, u, v}$ for $\alpha \le 0.25$) of the F-distribution. The lower-tail percentage point $f_{1-\alpha,u,v}$ can be found as follows:

$$f_{1-\alpha, u, v} = \frac{1}{f_{\alpha,u,v}} \quad ...(2.114)$$

For example, to find the lower-tail percentage point of $f_{0.95, 6, 10}$, note that

$$f_{0.95, 6, 10} = \frac{1}{f_{0.05,10,6}} = \frac{1}{4.06} = 0.246$$

2.12 PARAMETER ESTIMATION

Inferential statistics is the part of statistics that helps us to make decisions about some characteristics of a population based on sample information. Estimation is a procedure by which numerical value or values are assigned to a population parameter based on the information collected from a sample.

In inferential statistics, μ is called the *true population mean* and p is called the *true population proportion*. The value(s) assigned to a population parameter based on the value of a sample statistic is called an *estimate*. The sample statistic used to estimate a population parameter is called an *estimator*.

If the mean of the sampling distribution of a statistic equals to the corresponding population parameter, the statistic is called an *unbiased estimator* of the parameter, otherwise it is called a *biased estimator*. The corresponding values of such statistics are called *unbiased* or *biased estimates* respectively. If the sampling distributions of two statistics have the same mean (or expectation), the statistic with the smaller variance is called an *efficient estimator* of the mean while the other statistic is called an *inefficient estimator*. The corresponding values of the statistics are called *efficient* or *inefficient estimates* respectively.

Statistical inference is divided into two major categories: (1) Parameter estimation and (2) Hypothesis testing (see section 2.13). Parameter estimation is further classified into two major types: (1) Point estimation and (2) Interval estimation.

2.12.1 Point Estimation

An estimate of a population parameter given by a single number is called *point estimate* of parameter. An estimate of a population parameter is given by two numbers between which the parameter may be considered to lie is called an *interval estimate* of the parameter. The sample statistic used to estimate a population parameter is called an *estimator*. Hence, the sample mean $\overline{X}$ is an estimator of the population mean, μ and the sample proportion $\hat{p}$ is an estimator of the population proportion, p.

The desired estimators depend on the distribution used. These estimators are not always given by $\overline{X}$ and s^2. The common distributions and their estimators are given below:

Distribution	**Parameter**	**Suggested Estimator(s)**
Poisson	λ	$\hat{\lambda} = \overline{X}$
Binomial	p	$\hat{p} = \overline{X}$
Normal	μ, σ^2	$\hat{\mu} = \overline{X}$
		$\hat{\sigma}^2 = s^2$ (denominator is $n - 1$)

After a sample is taken and the data are analyzed, the result is an estimate.

2.12.2 Interval Estimation

Interval estimates indicate the precision or accuracy of an estimate and are therefore preferable to point estimates. A statement of the error or precision of an estimate is often called its *reliability*.

An interval estimate consists of two sample statistics, L and U. The probability that the interval formed by L and U contains the true value of a parameter $1 - \alpha$. For instance, to construct an interval estimate on the parameter θ, we find two statistics, L and U, such that

$$P(L \le \theta \le U) = 1 - \alpha \qquad ...(2.115)$$

The resulting interval $[L \le \theta \le U]$ is called a *100 (1 – α)% confidence interval* on unknown parameter θ. If a large number of independent samples were taken from the population being considered, approximately 100 $(1 - \alpha)$% of the intervals formed would be expected to contain the true value of θ. Confidence intervals that contain both an L, or lower side, and a U, or upper side, are called *two-sided confidence intervals*. Confidence intervals that contain only a lower or an upper side are called *one-sided confidence intervals*. A one-sided 100 $(1 - \alpha)$% confidence interval on θ is given by

$$L \le \theta$$

with the probability property

$$P(L \le \theta) = 1 - \alpha \qquad ...(2.116)$$

Similarly, the one-sided upper 100 (1 – α)% confidence interval on θ is given by

$$\theta \leq U$$

with the probability property

$$P(\theta \leq U) = 1 - \alpha \qquad (2.117)$$

The length of the observed confidence interval is an important measure of the quality of the information obtained from the sample. The half-interval length $\theta - L$ or $U - \theta$ is called *the accuracy of estimator*. The longer the confidence interval, the more confident we are that interval actually contains the true value of θ, and the less information we have about the true value of θ.

2.12.3 Confidence Interval on Mean, Variance Known

Let X be a random variable with unknown mean μ and known variance σ^2, and suppose that a random sample of size n, $X_1, X_2, ..., X_n$ is taken. A 100(1 – α)% confidence interval on μ can be obtained by considering the sampling distribution of the sample mean $\overline{X}$. The mean of $\overline{X}$ is μ and the variance σ^2/n. Hence, the distribution of the statistic

$$Z = \frac{\overline{X} - \mu}{\sigma/\sqrt{n}}$$

is taken to be a standard normal distribution.

The distribution of $Z = (\overline{X} - \mu)/(\sigma/\sqrt{n})$ is shown in Fig. 2.22. From Fig. 2.22, we observe that

$$P\{-Z_{\alpha/2} \leq Z \leq Z_{\alpha/2}\} = 1 - \alpha$$

or

$$P\left\{-Z_{\alpha/2} \leq \frac{\overline{X} - \mu}{\sigma/\sqrt{n}} \leq Z_{\alpha/2}\right\} = 1 - \alpha$$

Rearranging the above equation, we get

$$P\{\overline{X} - Z_{\alpha/2}\, \sigma/\sqrt{n} \leq \mu \leq \overline{X} + Z_{\alpha/2}\, \sigma/\sqrt{n}\} = 1 - \alpha \qquad ...(2.118)$$

From Eqs. (2.115) and (2.118), we note that the 100(1 – α)% two-sided confidence interval on μ is

$$\overline{X} - Z_{\alpha/2}\, \sigma/\sqrt{n} \leq \mu \leq \overline{X} + Z_{\alpha/2}\, \sigma/\sqrt{n} \qquad ...(2.119)$$

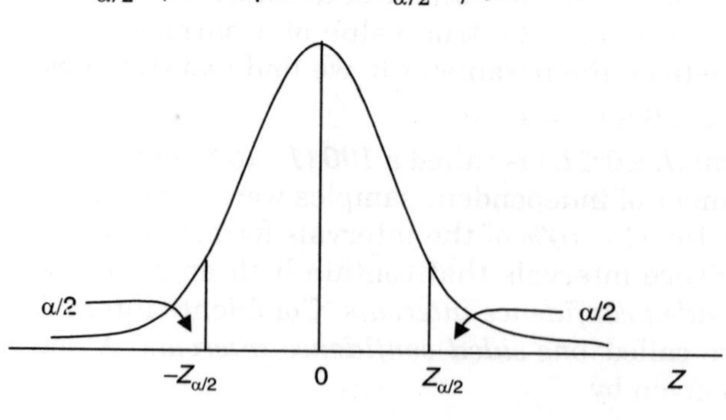

Fig. 2.22 The Distribution of *Z*.

Example E2.17. *It has been established that in a certain packaging process, the packages are of an average weight of μ but that μ changes over time as the process adjustment changes through wear. It was also found that the variance of the package weight is a constant at 16 lbs even though the mean changes overtime. A sample of 36 packages has revealed their mean weight $\overline{X}$ of 100 lbs. Assuming that the weight of the individual packages are normally distributed around μ, (a) find the 90% confidence limits for μ. (b) find the 95% confidence limits for μ.*

Solution:

(*a*) Here $1 - \alpha = 0.90$ or $\alpha = 0.10$ and $\alpha/2 = 0.05$.

$$Z_{0.05} = 1.645, \sigma = 16 = 4 \text{ lbs.}$$

By computation,

$$L = \overline{X} - \frac{Z_{\alpha/2}\sigma}{\sqrt{n}} = 100 - \frac{1.645\,(4)}{\sqrt{36}} = 100 - 1.0967 = 98.9033$$

$$U = \overline{X} + \frac{Z_{\alpha/2}\sigma}{\sqrt{n}} = 100 + \frac{1.645\,(4)}{\sqrt{36}} = 100 + 1.0967 = 101.0967$$

Hence, we are 90% confident that μ lies between 98.9033 and 101.0967 lbs.

(*b*) Here $Z_{\alpha/2} = 1.96$

$$L = \overline{X} - \frac{Z_{\alpha/2}\sigma}{\sqrt{n}} = 100 - \frac{1.96\,(4)}{\sqrt{36}} = 100 - 1.8067 = 98.6933$$

$$U = \overline{X} + \frac{Z_{\alpha/2}\sigma}{\sqrt{n}} = 100 + \frac{1.96\,(4)}{\sqrt{36}} = 100 + 1.8067 = 101.3067$$

Therefore, we are 95% confident that μ lies between 98.6933 and 101.3067 lbs.

One-Sided Confidence Intervals

One-sided confidence intervals for μ are obtained by setting either $L = -\infty$ or $U = \infty$ and replacing $Z_{\alpha/2}$ by Z_α. The $100(1 - \alpha)\%$ upper-confidence interval for μ is

$$\mu \le \overline{X} + Z_\alpha\, \sigma/\sqrt{n} \qquad \text{...(2.120)}$$

and the $100(1 - \alpha)\%$ lower-confidence interval for μ is

$$\overline{X} + Z_\alpha\, \sigma/\sqrt{n} \le \mu \qquad \text{...(2.121)}$$

2.12.4 Confidence Interval on the Difference in Two Means, Variances Known

Consider two independent random variables X_1 with unknown mean μ_1 and known variance σ_1^2 and X_2 with unknown mean μ_2 and variance σ_2^2. We like to find a $100(1 - \alpha)\%$ confidence interval on the difference in means $\mu_1 - \mu_2$. Let $X_{11}, X_{12}, \ldots$ be a random sample of n_1 observations from X_1, and $X_{21}, X_{22}, \ldots, X_{2n_2}$ be a random sample of n_2 observations from X_2. If $\overline{X}_1$ and $\overline{X}_2$ are the sample means, the statistic

$$Z = \frac{\overline{X}_1 - \overline{X}_2 - (\mu_1 - \mu_2)}{\sqrt{\dfrac{\sigma_1^2}{n_1} + \dfrac{\sigma_2^2}{n_2}}}$$

is standard normal if X_1 and X_2 are normal or approximately standard normal if the conditions of the central limit theorem apply, respectively. From Fig. 2.22 it is clear that

$$P\{-Z_{\alpha/2} \le Z \le Z_{\alpha/2}\} = 1 - \alpha$$

or $$P\left\{ -Z_{\alpha/2} \le \frac{\bar{X}_1 - \bar{X}_2 - (\mu_1 - \mu_2)}{\sqrt{\dfrac{\sigma_1^2}{n_1} + \dfrac{\sigma_2^2}{n_2}}} \le Z_{\alpha/2} \right\} = 1 - \alpha$$

Rearranging, we have

$$P\left\{\bar{X}_1 - \bar{X}_2 - Z_{\alpha/2}\sqrt{\frac{\sigma_1^2}{n_1} + \frac{\sigma_2^2}{n_2}} \le (\mu_1 - \mu_2) \le \bar{X}_1 - \bar{X}_2 - Z_{\alpha/2}\sqrt{\frac{\sigma_1^2}{n_1} + \frac{\sigma_2^2}{n_2}}\right\} = 1 - \alpha \quad \text{...(2.122)}$$

Comparing Eqs. (2.118) and (2.122), we note that 100(1 – α)% confidence interval for $\mu_1 - \mu_2$ is

$$\bar{X}_1 - \bar{X}_2 - Z_{\alpha/2}\sqrt{\frac{\sigma_1^2}{n_1} + \frac{\sigma_2^2}{n_2}} \le (\mu_1 - \mu_2) \le \bar{X}_1 - \bar{X}_2 + Z_{\alpha/2}\sqrt{\frac{\sigma_1^2}{n_1} + \frac{\sigma_2^2}{n_2}} \quad \text{...(2.123)}$$

One-sided confidence intervals on $\mu_1 - \mu_2$ may also be obtained. A 100(1 – α)% upper-confidence interval on $\mu_1 - \mu_2$ is

$$\mu_1 - \mu_2 \le \bar{X}_1 - \bar{X}_2 + Z_{\alpha}\sqrt{\frac{\sigma_1^2}{n_1} + \frac{\sigma_2^2}{n_2}} \quad \text{...(2.124)}$$

and a 100(1 – α)% lower-confidence interval is

$$\bar{X}_1 - \bar{X}_2 - Z_{\alpha}\sqrt{\frac{\sigma_1^2}{n_1} + \frac{\sigma_2^2}{n_2}} \le \mu_1 - \mu_2 \quad \text{...(2.125)}$$

Example E2.18. *The mean hourly wage for male workers was \$15 and \$13 for female workers in a particular manufacturing sector in the year 1997. These two estimates were based on random samples of 1000 and 1200 workers taken, respectively from two independent population. The standard deviations of the two population are known to be \$2 and \$1.50 respectively. Construct a 95% confidence interval for the difference between the mean hourly wages of the two population.*

Solution:

For 95% confidence level $Z_{\alpha/2} = 1.96$.

The 95% confidence interval for $\mu_1 - \mu_2$ is given by

$$\bar{X}_1 - \bar{X}_2 \pm Z_{\alpha/2}\sqrt{\frac{\sigma_1^2}{n_1} + \frac{\sigma_2^2}{n_2}}$$

$$= 15 - 13 \pm 1.96\sqrt{\frac{2^2}{1000} + \frac{1.5^2}{1200}} = 2 \pm 1.96(0.2046) = 2 \pm 0.4011 = \$1.60 \text{ to } \$2.40$$

2.12.5 Confidence Interval on the Mean of a Normal Distribution, Variance Unknown

Here we find a confidence interval on the mean of a distribution when the variance is unknown.

Suppose a random sample of size n, $X_1, X_2, \ldots, X_n$ is available, and $\bar{X}$ and s^2 are the sample mean and sample variance, respectively. In order to use a valid confidence interval when the sample size is small, we make an assumption that the underlying population is normally distributed. This leads to confidence intervals based on the t-distribution. Therefore, let $X_1, X_2, \ldots, X_n$ be a random sample from a normal distribution with unknown mean μ and unknown variance σ^2. We know the sampling distribution of the statistic

$$t = \frac{\overline{X} - \mu}{\sigma/\sqrt{n}}$$

is the t-distribution with $n - 1$ degrees of freedom.

The distribution of t is shown in Fig. 2.23. Letting $t_{\alpha/2,\, n-1}$ be the upper $\alpha/2$ percentage point of the t-distribution with $n - 1$ degrees of freedom, we note from Fig. 2.23, that

$$P\{-t_{\alpha/2,\, n-1} \le t \le t_{\alpha/2,\, n-1}\} = 1 - \alpha$$

or $$P\left\{-t_{\alpha/2,n-1} \le \frac{\overline{X} - \mu}{s/\sqrt{n}} \le t_{\alpha/2,n-1}\right\} \qquad \ldots(2.126)$$

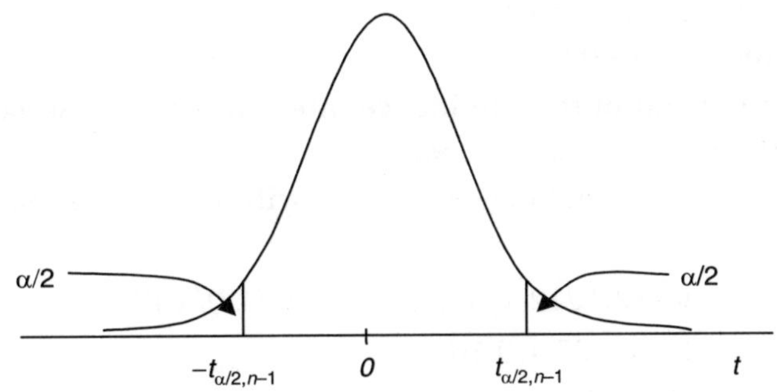

Fig. 2.23 The t-distribution.

Rearranging Eq. (2.125)

$$P\{\overline{X} - t_{\alpha/2,\, n-1}\, s/\sqrt{n} \le \mu \le \overline{X} + t_{\alpha/2,\, n-1}\, s/\sqrt{n}\} \qquad \ldots(2.127)$$

Comparing Eqs. (2.118) and (2.127), we see that a 100(1 – α)% two-sided confidence interval on μ is

$$\overline{X} - t_{\alpha/2,\, n-1}\, s/\sqrt{n} \le \mu \le \overline{X} + t_{\alpha/2,\, n-1}\, s/\sqrt{n} \qquad \ldots(2.128)$$

A 100(1 – α)% lower-confidence interval on μ is

$$\overline{X} - t_{\alpha,\, n-1}\, s/\sqrt{n} \le \mu \qquad \ldots(2.129)$$

and a 100(1 – α)% upper-confidence interval on μ is

$$\mu \le \overline{X} - t_{\alpha,\, n-1}\, s/\sqrt{n} \qquad \ldots(2.130)$$

The above method assumes that the samplings are from a normal population.

Example E2.19. *Find the following confidence intervals for* μ_d *assuming that the populations of paired differences are normally distributed.*

(*a*) *n = 9,* $\overline{X}$ *= 25, s = 13, confidence level 99%*

(*b*) *n = 26,* $\overline{X}$ *= 13, s = 5, confidence level 95%*

(*c*) *n = 12,* $\overline{X}$ *= 35, s = 12, confidence level 90%*

Solution:

(*a*) The *t*-value for $v = n - 1 = 9 - 1 = 8$ d.o.f and 0.005 area in the right tail is 3.355. The 99% confidence interval with $t_{0.005,\,8} = 3.355$ is

$$\bar{X} - t_{\alpha/2,\,\nu}\, s/\sqrt{n} \le \mu \le \bar{X} + t_{\alpha/2,\,\nu}\, s/\sqrt{n}$$

$$25 - 3.355(13)/\sqrt{9} \le \mu \le 25 + 3.355(13)/\sqrt{9}$$
$$25 - 14.538 \le \mu \le 25 + 14.538$$
$$10.462 \le \mu \le 39.538$$

The complete statement of the confidence interval, with the associated probability is

$$P(10.462 \le \mu \le 39.538) = 0.99$$

(*b*) Using the data given, a 95% confidence interval with $t_{0.025,\,25} = 2.060$ from Appendix 2G, is given

$$13 - 2.060(5)/\sqrt{26} \le \mu \le 13 + 2.060(5)/\sqrt{26}$$
$$13 - 2.02 \le \mu \le 13 + 2.02$$
$$10.98 \le \mu \le 15.02$$

The complete statement of the confidence interval with the associated probability is

$$P(10.98 \le \mu \le 15.02) = 0.95$$

(*c*) Using the data given, a 90% confidence interval with $t_{0.05,\,11} = 1.796$ from Appendix 2G, is given by

$$35 - 1.796(12)/\sqrt{12} \le \mu \le 35 + 1.796(12)/\sqrt{12}$$
$$35 - 0.16 \le \mu \le 35 + 0.16$$
$$34.84 \le \mu \le 35.16$$

The complete statement of the confidence interval with the associated probability is

$$P(34.84 \le \mu \le 35.16) = 0.90$$

2.12.6 Confidence Interval on the Difference in Means of Two Normal Distributions, Variances Unknown

Consider two independent normal random variables, say X_1 with mean μ_1 and variance σ_1^2, and X_2 with mean μ_2 and variance σ_2^2. Both the means μ_1 and μ_2 and the variances σ_1^2 and σ_2^2 are unknown. Assume that both variances are equal, that is, $\sigma_1^2 = \sigma_2^2 = \sigma^2$. We wish to find a $100(1 - \alpha)\%$ confidence interval on the difference in means $\mu_1 - \mu_2$.

Random samples of size n_1 and n_2 are taken on X_1 and X_2 respectively. Let the sample means be denoted by $\bar{X}_1$ and $\bar{X}_2$, and the sample variances be denoted by s_1^2 and s_2^2. Since both s_1^2 and s_2^2 are estimates of the common variance σ^2, we may obtain a combined (or *pooled*) estimator of σ^2 as

$$s_p^2 = \frac{(n_1 - 1)\, s_1^2 + (n_2 - 1)\, s_2^2}{n_1 + n_2 - 2} \qquad \text{...(2.131)}$$

To find the confidence interval for $\mu_1 - \mu_2$, note that the distribution of the statistic

$$t = \frac{\bar{X}_1 - \bar{X}_2 - (\mu_1 - \mu_2)}{s_p \sqrt{\dfrac{1}{n_1} + \dfrac{1}{n_2}}}$$

is the t-distribution with $n_1 + n_2 - 2$ degrees of freedom. Hence,

$$P\left\{-t_{\alpha/2,\, n_1+n_2-2} \le t \le t_{\alpha/2,\, n_1+n_2-2}\right\} = 1 - \alpha$$

or
$$P\left\{ -t_{\alpha/2,n_1+n_2-2} \le \frac{\bar{X}_1 - \bar{X}_2 - (\mu_1 - \mu_2)}{s_p\sqrt{\frac{1}{n_1} + \frac{1}{n_2}}} \le t_{\alpha/2,n_1+n_2-2} \right\} = 1 - \alpha$$

Rearranging, we obtain

$$P\left\{\bar{X}_1 - \bar{X}_2 - t_{\alpha/2,n_1+n_2-2}\, s_p\sqrt{\frac{1}{n_1} + \frac{1}{n_2}} \le \mu_1 - \mu_2 \le \bar{X}_1 - \bar{X}_2 + t_{\alpha/2,n_1+n_2-2}\, s_p\sqrt{\frac{1}{n_1} + \frac{1}{n_2}}\right\} = 1 - \alpha \quad ...(2.132)$$

Hence, a $100(1 - \alpha)\%$ two-sided confidence interval on the difference in means $\mu_1 - \mu_2$ is

$$\bar{X}_1 - \bar{X}_2 - t_{\alpha/2,n_1+n_2-2}\, s_p\sqrt{\frac{1}{n_1} + \frac{1}{n_2}} \le \mu_1 - \mu_2 \le \bar{X}_1 - \bar{X}_2 + t_{\alpha/2,n_1+n_2-2}\, s_p\sqrt{\frac{1}{n_1} + \frac{1}{n_2}} \quad ...(2.133)$$

A one-sided $100(1 - \alpha)\%$ lower-confidence interval on $\mu_1 - \mu_2$ is

$$\bar{X}_1 - \bar{X}_2 - t_{\alpha,n_1+n_2-2}\, s_p\sqrt{\frac{1}{n_1} + \frac{1}{n_2}} \le \mu_1 - \mu_2 \quad ...(2.134)$$

and a one-sided $100(1 - \alpha)\%$ upper-confidence interval on $\mu_1 - \mu_2$ is

$$\mu_1 - \mu_2 \le \bar{X}_1 - \bar{X}_2 + t_{\alpha,n_1+n_2-2}\, s_p\sqrt{\frac{1}{n_1} + \frac{1}{n_2}} \quad ...(2.135)$$

Example E2.20. *To compare the tensile strength of two materials, 12 test specimens of each material were tested. The order was decided at random. The higher the test value, the better the strength of these materials. The data obtained are summarized in Table E2.20:*

Table E2.20

Material A	*12.5*	*11.7*	*9.70*	*9.60*	*10.5*	*9.60*	*9.40*	*11.5*	*8.70*	*11.5*	*10.6*	*9.7*
Material B	*9.40*	*8.40*	*11.6*	*7.20*	*9.70*	*7.00*	*10.4*	*8.20*	*7.30*	*9.20*	*6.90*	*12.7*

Solution:

For both the materials

Material A

$\sum X_{1i} = 124.8$

$\bar{X}_1 = 10.4$

$\sum X_{1i}^2 = 1312$

$\left(\frac{1}{n_1}\right)\left(\sum X_{1i}\right)^2 = 1297.92$

$\sum (X_{1i} - \bar{X}_1)^2 = 14.08$

Material B

$\sum X_{2i} = 108$

$\bar{X}_2 = 9$

$\sum X_{2i}^2 = 1010.64$

$\left(\frac{1}{n_2}\right)\left(\sum X_{2i}\right)^2 = 972$

$\sum (X_{2i} - \bar{X}_2)^2$

Hence, the summary table is

Material	n	Degrees of freedom	$\overline{X}$	Sum of squares
A	12	11	10.4	14.08
B	12	11	9	38.64
				Total 52.72

Therefore $d = \overline{X}_1 - \overline{X}_2 = 1.4$

The pooled estimate of variance is $s^2 = 52.72 / 22 = 2.4$

The estimate variance of the difference between the means is

$$s_d^2 = s^2 \left(\frac{1}{n} + \frac{1}{n}\right) = 2.4 \left(\frac{2}{12}\right) = 0.4$$

The estimated standard deviation of d is

$$s_d = \sqrt{0.4} = 0.63$$

Hence, if we desire a 95% confidence interval, we must obtain $t_{0.025,22} = 2.074$ (Appendix 2G). The 95% confidence interval now be formed using the relationship

$$L = d - t_{\alpha/2,\, n_1+n_2-2}\, s_d = 1.40 - (2.074)\,(0.63) = 0.09$$

$$U = d + t_{\alpha/2,\, n_1+n_2-2}\, s_d = 1.40 + (2.074)\,(0.63) = 2.71$$

We can be 95% confident that the difference between the actual means is between 0.09 and 2.71.

When we cannot assume $\sigma_1^2 = \sigma_2^2$, we can find a $100(1 - \alpha)\%$ confidence interval on $\mu_1 - \mu_2$ using the fact that the statistic

$$t^* = \frac{\overline{X}_1 - \overline{X}_2 - (\mu_1 - \mu_2)}{\sqrt{s_1^2/n_1 + s_2^2/n_2}}$$

distributed approximately as t with degrees of freedom given by

$$v = \frac{\left(s_1^2/n_1 + s_2^2/n_2\right)^2}{\dfrac{\left(s_1^2/n_1\right)^2}{n_1+1} + \dfrac{\left(s_2^2/n_2\right)^2}{n_2+1}} - 2 \qquad \text{...(2.136)}$$

Therefore, an approximate $100(1 - \alpha)\%$ two-sided confidence interval on $\mu_1 - \mu_2$, when $\sigma_1^2 \neq \sigma_2^2$ is

$$\overline{X}_1 - \overline{X}_2 - t_{\alpha/2,\, v}\sqrt{\frac{s_1^2}{n_1} + \frac{s_2^2}{n_2}} \le \mu_1 - \mu_2 \le \overline{X}_1 - \overline{X}_2 + t_{\alpha/2,\, v}\sqrt{\frac{s_1^2}{n_1} + \frac{s_2^2}{n_2}} \qquad \text{...(2.137)}$$

Upper (lower) one-sided confidence limits may be found by replacing the lower (upper) confidence limit with $-\infty$ (∞) and changing $\alpha/2$ to α.

2.12.7 Confidence Interval on $\mu_1 - \mu_2$ for Paired Observations

Suppose the data consists of n pairs $(X_{11}, X_{21}), (X_{12}, X_{22}), \ldots, (X_{1n}, X_{2n})$. Both X_1 and X_2 are assumed to be normally distributed with mean μ_1 and μ_2 respectively. The random variables within different pairs are independent. Since, there are two measurements on the same experimental unit, the two measurements within the same pair may not be independent. Consider the n differences $D_1 = X_{11} - X_{21}, D_2 = X_{12} - X_{22}, \ldots, D_n = X_{1n} - X_{2n}$. The mean of the differences D, say μ_D, is

$$\mu_D = E(D) = E(X_1 - X_2) = E(X_1) - E(X_2) = \mu_1 - \mu_2$$

since the expected value of $X_1 - X_2$ is the difference in expected values regardless of whether X_1 and X_2 are independent. Hence, we can construct a confidence interval on $\mu_1 - \mu_2$ by just finding a confidence interval on μ_D. The differences D_i are normally and independently distributed, we can apply the t-distribution procedure described in Section 2.12.5 to find the confidence interval on μ_D. By analogy with Eq. (2.128), the $100(1 - \alpha)\%$ confidence interval on $\mu_D = \mu_1 - \mu_2$ is

$$\bar{D} - t_{\alpha/2, n-1}\, s_D/\sqrt{n} \leq \mu_D \leq \bar{D} + t_{\alpha/2, n-1}\, s_D/\sqrt{n} \qquad \text{...(2.138)}$$

where $\bar{D}$ and s_D are the sample mean and sample standard deviation of the differences D_i, respectively. This confidence interval is valid for the case where $\sigma_1^2 \neq \sigma_2^2$, because s_D^2 estimates $\sigma_D^2 = V(X_1 - X_2)$ and for large samples ($n \geq 30$ pairs) and the assumption of normality is not necessary.

2.12.8 Confidence Interval on the Variance of a Normal Distribution

Assume that X is normally distributed with unknown mean μ and unknown variance σ^2. Let $X_1, X_2, \ldots, X_n$ be a random sample of size n and s^2 be the sample variance. We note that the sampling distribution of

$$\chi^2 = \frac{(n-1)\,s^2}{\sigma^2}$$

is chi-square with $n - 1$ degrees of freedom. This distribution is shown in Fig. 2.24.

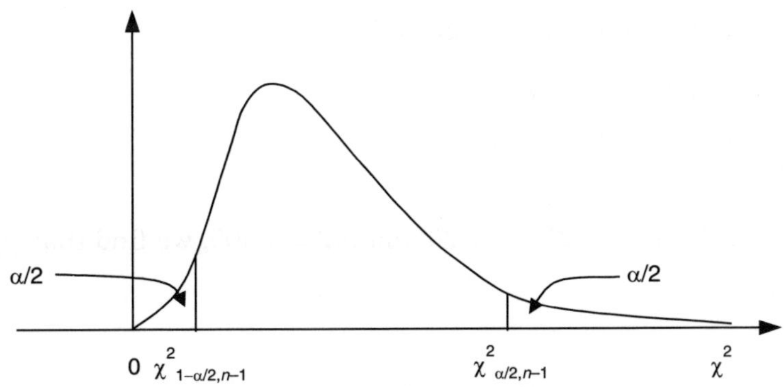

Fig. 2.24 The χ^2_{n-1} Distribution.

We observe from Fig. 2.24 that,

$$P\left\{\chi^2_{1-\alpha/2, n-1} \leq \chi^2 \leq \chi^2_{\alpha/2, n-1}\right\} = 1 - \alpha$$

or
$$P\left\{\chi^2_{1-\alpha/2,\,n-1} \le \frac{(n-1)\,s^2}{\sigma^2} \le \chi^2_{\alpha/2,\,n-1}\right\} = 1-\alpha \quad ...(2.139)$$

Equation (2.138) can be rearranged to give

$$P\left\{\frac{(n-1)\,s^2}{\chi^2_{\alpha/2,\,n-1}} \le \sigma^2 \le \frac{(n-1)\,s^2}{\chi^2_{1-\alpha/2,\,n-1}}\right\} = 1-\alpha \quad ...(2.140)$$

Comparing Eqs. (2.140) and (2.118), we see that a 100(1 – α)% two-sided confidence interval on σ^2 is

$$\frac{(n-1)\,s^2}{\chi^2_{\alpha/2,\,n-1}} \le \sigma^2 \le \frac{(n-1)\,s^2}{\chi^2_{1-\alpha/2,\,n-1}} \quad ...(2.141)$$

In order to find a 100(1 – α)% lower-confidence interval on σ^2, set $U = \infty$ and replace $\chi^2_{\alpha/2,\,n-1}$ with $\chi^2_{\alpha,\,n-1}$, giving

$$\frac{(n-1)\,s^2}{\chi^2_{\alpha,\,n-1}} \le \sigma^2 \quad ...(2.142)$$

The 100(1 – α)% upper-confidence interval is found by setting $L = 0$ and replacing $\chi^2_{1-\alpha/2,n-1}$ with $\chi^2_{1-\alpha,n-1}$, resulting in

$$\sigma^2 \le \frac{(n-1)\,s^2}{\chi^2_{1-\alpha,n-1}} \quad ...(2.143)$$

Example E2.21. *Data are collected on the driving time required to reach place A from place B. The driving time is assumed to follow a normal distribution. Twenty-one driving times are collected and the sample variance is calculated to be 2 hours. A 99% confidence interval on the true variance is desired.*

Solution:

The two-sided confidence interval is given by

$$\frac{(n-1)\,s^2}{\chi^2_{\alpha/2,\,v}} \le \sigma^2 \le \frac{(n-1)\,s^2}{\chi^2_{1-\alpha/2,\,v}}$$

where $v = n - 1$.

From Appendix 2F with $v = 21 - 1 = 20$ and $\alpha/2 = 0.005$, we find that $\chi^2_{0.005,\,20} = 39.997$ and $\chi^2_{0.995,\,20} = 7.434$.

Therefore

$$\frac{(21-1)\,(2^2)}{39.997} \le \sigma^2 \le \frac{(21-1)\,(2^2)}{7.434} \quad \text{or} \quad 2 \le \sigma^2 \le 10.761$$

The complete statement of the confidence interval with the associated probability is

$$P(2 \text{ hrs} \le \sigma^2 \le 10.761 \text{ hrs}) = 0.99$$

2.12.9 Confidence Interval on the Ratio of Variances of Two Normal Distributions

Let X_1 and X_2 are independent normal random variables with unknown means μ_1 and μ_2 and unknown variances σ_1^2 and σ_2^2, respectively. We wish to find a $100(1-\alpha)\%$ confidence interval on the ratio of σ_1^2/σ_2^2. Let two random samples of sizes n_1 and n_2 be taken on X_1 and X_2, and let s_1^2 and s_2^2 denote the sample variances. The sampling distribution of

$$F = \frac{s_2^2/\sigma_2^2}{s_1^2/\sigma_1^2}$$

is F with $n_2 - 1$ and $n_1 - 1$ degrees of freedom. This distribution is shown in Fig. 2.25.

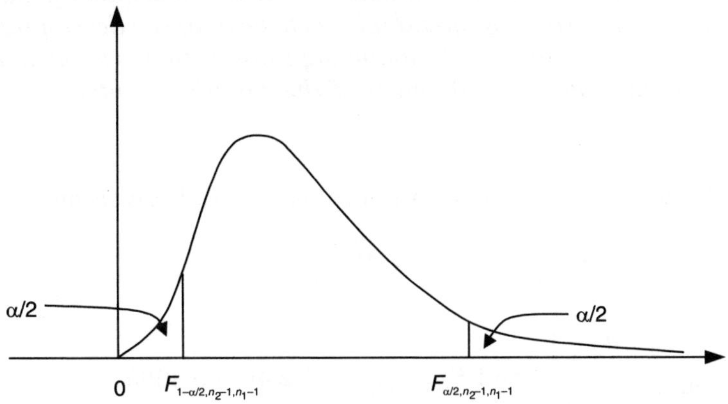

Fig. 2.25 The $F_{n_2-1,\, n_1-1}$ Distribution.

From Fig. 2.25, we observe that

$$P\left\{F_{1-\alpha/2, n_2-1, n_1-1} \le F \le F_{\alpha/2, n_2-1, n_1-1}\right\} = 1-\alpha$$

or

$$P\left\{F_{1-\alpha/2, n_2-1, n_1-1} \le \frac{s_2^2/\sigma_2^2}{s_1^2/\sigma_1^2} \le F_{\alpha/2, n_2-1, n_1-1}\right\} = 1-\alpha$$

Hence,

$$P\left\{\frac{s_1^2}{s_2^2} F_{1-\alpha/2, n_2-1, n_1-1} \le \frac{\sigma_1^2}{\sigma_2^2} \le \frac{s_1^2}{s_2^2} \le F_{\alpha/2, n_2-1, n_1-1}\right\} = 1-\alpha \qquad ...(2.144)$$

Comparing Eqs. (2.144) and (2.118), we note that a $100(1-\alpha)\%$ two-sided confidence interval on σ_1^2/σ_2^2 is

$$\frac{s_1^2}{s_2^2} F_{1-\alpha/2, n_2-1, n_1-1} \le \frac{\sigma_1^2}{\sigma_2^2} \le \frac{s_1^2}{s_2^2} \le F_{\alpha/2, n_2-1, n_1-1} \qquad ...(2.145)$$

where the lower $1-\alpha/2$ tail-point of the $F_{n_2-1,\, n_1-1}$ distribution is given by

$$F_{1-\alpha/2,\, n_2-1,\, n_1-1} = \frac{1}{F_{\alpha/2,\, n_1-1,\, n_2-1}} \qquad ...(2.146)$$

We can also construct one-sided confidence intervals. A 100(1 – α)% lower-confidence limit on σ_1^2/σ_2^2 is

$$\frac{s_1^2}{s_2^2} F_{1-\alpha, n_2-1, n_1-1} \le \frac{\sigma_1^2}{\sigma_2^2} \quad ...(2.147)$$

while a 100(1 – α)% upper-confidence interval on σ_1^2/σ_2^2 is

$$\frac{\sigma_1^2}{\sigma_2^2} \le \frac{s_1^2}{s_2^2} F_{\alpha, n_2-1, n_1-1} \quad ...(2.148)$$

Example E2.22. *Two samples are drawn from two independent populations, which are normally distributed. A comparison of the samples is to be made by developing a 99% confidence interval on the ratio of their variances. The following information is known: $n_1 = n_2 = 21$, $s_1^2 = 5$ and $s_2^2 = 8$. Determine the complete statement of the confidence interval, with the associated probability.*

Solution:

A 100(1 – α)% two-sided confidence interval on σ_1^2/σ_2^2 is given by

$$\frac{s_1^2}{s_2^2} F_{1-\alpha/2, \upsilon_2, \upsilon_1} \le \frac{\sigma_1^2}{\sigma_2^2} \le \frac{s_1^2}{s_2^2} F_{\alpha/2, \upsilon_2, \upsilon_1}$$

where $\upsilon_1 = n_1 - 1$ and $\upsilon_2 = n_2 - 1$.

Therefore, $F_{0.025,20,20} = 2.46$ and $F_{0.975,19,19} = 1/2.46 = 0.4065$

Confidence interval is given by

$$\frac{5}{8}(0.4065) \le \frac{\sigma_1^2}{\sigma_2^2} \le \frac{5}{8}(2.46)$$

or

$$0.254 \le \frac{\sigma_1^2}{\sigma_2^2} \le 1.538$$

The complete statement of the confidence interval with the associated probability is

$$P\left(0.254 \le \frac{\sigma_1^2}{\sigma_2^2} \le 1.538\right)$$

2.12.10 Confidence Interval on a Proportion

If a random sample size n has been taken from a large (possibly infinite) population, and $X (\le n)$ observations in this sample belong to a class of interest, then $\hat{p} = X/n$ is the point estimator of the proportion of the population belonging. n and p are the parameters of binomial distribution. The sampling distribution of $\hat{p}$ is approximately normal w.r.t. mean p and variance $p(1-p)/n$, if p is not too close to either 0 or 1, and if n is relatively large. Therefore, the distribution of

$$Z = \frac{\hat{p} - p}{\sqrt{\dfrac{p(1-p)}{n}}}$$

is approximately standard normal distribution.

We also observe that

$$P\{-Z_{\alpha/2} \le Z \le Z_{\alpha/2}\} = 1 - \alpha$$

or
$$P\left\{-Z_{\alpha/2} \le \frac{\hat{p} - p}{\sqrt{\dfrac{p(1-p)}{n}}} \le Z_{\alpha/2}\right\} = 1 - \alpha \quad \text{...(2.149)}$$

Rearranging Eq. (2.148), we get

$$P\left\{\hat{p} - Z_{\alpha/2}\sqrt{\frac{p(1-p)}{n}} \le p \le \hat{p} + Z_{\alpha/2}\sqrt{\frac{p(1-p)}{n}}\right\} = 1 - \alpha \quad \text{...(2.150)}$$

The quantity of $\sqrt{p(1-P)/n}$ as the standard error of the point estimator $\hat{p}$. Replacing p by $\hat{p}$ in the standard error, giving an estimated standard error.

$$P\left\{\hat{p} - Z_{\alpha/2}\sqrt{\frac{\hat{p}(1-\hat{p})}{n}} \le p \le \hat{p} + Z_{\alpha/2}\sqrt{\frac{\hat{p}(1-\hat{p})}{n}}\right\} = 1 - \alpha \quad \text{...(2.151)}$$

Example E2.23. *A sample of 300 of electronic components produced showed 36 non-conforming ones. Construct a 95% confidence interval for this. Assume binomial distribution.*

Solution:

A two-sided confidence interval on p is given by

$$\hat{p} - Z_{\alpha/2}\sqrt{\frac{\hat{p}(1-\hat{p})}{n}} \le p \le \hat{p} + Z_{\alpha/2}\sqrt{\frac{\hat{p}(1-\hat{p})}{n}}$$

Since n is large and the underlying distribution is binomial, $\hat{p}$ = 36/300 = 0.12.

A 95% confidence interval, with $Z_{0.025}$ = 1.96 (from Appendix 2E) is given by

$$0.12 - 1.96\sqrt{\frac{(0.12)\,(0.88)}{300}} \le p \le 0.12 + 1.96\sqrt{\frac{(0.12)\,(0.88)}{300}}$$

or
$$0.12 - 1.96\,(0.01876) \le p \le 0.12 + 1.96\,(0.01876)$$

$$0.12 - 0.0368 \le p \le 0.12 + 0.0368$$

$$0.0832 \le p \le 0.1568$$

The complete statement of the confidence interval with the associated probability is

$$P(0.0832 \le p \le 0.1568) = 0.95$$

2.12.11 Confidence Interval on the Difference in Two Proportions

If two independent samples of size n_1 and n_1 are taken from infinite populations so that X_1 and X_2 are independent, binomial random variables with parameters (n_1, p_1) and (n_2, p_2), respectively, where X_1 represents the number of sample observations from the first population that belongs to a class of interest and X_2 represents the number of sample observations from the second population that belongs to a class of interest, then $\hat{p}_1 = X_1/n_1$ and $\hat{p}_2 = X_2/n_2$ are independent estimators of p_1 and p_2, respectively. Considering the fact that the normal approximation to the binomial applies, the statistic

$$Z = \frac{\hat{p}_1 - \hat{p}_2 - (p_1 - p_2)}{\sqrt{\dfrac{p_1(1-p_1)}{n_1} + \dfrac{p_2(1-p_2)}{n_2}}}$$

is distributed approximately as standard normal. An approximate 100(1–α)% two-sided confidence interval for $p_1 - p_2$ is

$$\hat{p}_1 - \hat{p}_2 - Z_{\alpha/2}\sqrt{\frac{\hat{p}_1(1-\hat{p}_1)}{n_1} + \frac{\hat{p}_2(1-\hat{p}_2)}{n_2}}$$

or

$$\hat{p}_1 - \hat{p}_2 - Z_{\alpha/2}\sqrt{\frac{\hat{p}_1(1-\hat{p}_1)}{n_1} + \frac{\hat{p}_2(1-\hat{p}_2)}{n_2}} \le p_1 - p_2$$

$$\le \hat{p}_1 - \hat{p}_2 + Z_{\alpha/2}\sqrt{\frac{\hat{p}_1(1-\hat{p}_1)}{n_1} + \frac{\hat{p}_2(1-\hat{p}_2)}{n_2}} \quad \text{...(2.152)}$$

An approximate 100(1 – α)% lower-confidence interval for $p_1 - p_2$ is

$$\hat{p}_1 - \hat{p}_2 - Z_{\alpha/2}\sqrt{\frac{\hat{p}_1(1-\hat{p}_1)}{n_1} + \frac{\hat{p}_2(1-\hat{p}_2)}{n_2}} \le p_1 - p_2$$

and an approximate 100(1 – α)% upper-confidence interval for $p_1 - p_2$ is

$$p_1 - p_2 \le \hat{p}_1 - \hat{p}_2 + Z_{\alpha/2}\sqrt{\frac{\hat{p}_1(1-\hat{p}_1)}{n_1} + \frac{\hat{p}_2(1-\hat{p}_2)}{n_2}}$$

The two-sided confidence intervals are summarized in Table 2.2.

2.12.12 Sample Size Selection

Each level is constructed with regard to a given *confidence level* and is called a *confidence interval*. The confidence level associated with a confidence interval states how much confidence we have that this interval contains the true population parameter. The confidence level is denoted by 100(1 – α)%. When expressed as probability, it is called the *confidence coefficient* and is denoted by (1 – α). α is called the *significance level.*

The confidence interval for μ for large samples is given as follows:

The 100(1 – α)% confidence interval for μ is

$$\bar{x} \pm z\sigma_{\bar{x}} \quad \text{if } \sigma \text{ is known}$$

$$\bar{x} \pm z s_{\bar{x}} \quad \text{if } \sigma \text{ is unknown} \quad \text{...(2.153)}$$

where $\sigma_{\bar{x}} = \sigma/\sqrt{n}$ and $s_{\bar{x}} = s/\sqrt{n}$

The value of Z used here is obtained from standard normal distribution table for the given confidence level. The quantity $z\sigma_{\bar{x}}$ (or $zs_{\bar{x}}$ when σ is not known) in the confidence interval formula is called the *maximum error of estimate* and is denoted by E. Thus,

$$E = z\sigma_{\bar{x}}$$

or $$E = zs_{\bar{x}}$$

The standard deviation of the sample mean is $\sigma/\sqrt{n}$

$$E = z\frac{\sigma}{\sqrt{n}} \quad \text{...(2.154)}$$

Given the confidence level and the standard deviation of the population, the sample size that will produce a predetermined error E of the confidence interval estimate of μ is

$$n = \frac{z^2\sigma^2}{E^2} \quad \text{...(2.155)}$$

where n = required sample size.

Example E2.24. *At a 99% confidence level, the standard deviation σ for the population is 0.7. How large a sample should be selected if we want the estimate to be within 0.01 of the population mean?*

Solution:

The mean size is $\bar{x} \pm 0.01$

Therefore, the maximum size of the error of the estimate is to be 0.01, that is $E = 0.01$. The value of z for a 99% confidence level is 2.58. The value of σ is given to be 0.7. Hence, substituting the values and simplifying Eq. (2.154), we get

$$n = \frac{z^2\sigma^2}{E^2} = \frac{(2.58)^2\,(0.7)^2}{(0.01)^2} = 32{,}616$$

Hence, the required sample size is 32,616.

2.13 HYPOTHESIS TESTING

Decisions made about populations on the basis of sample information are called *statistical decisions*. A *hypothesis* is an assertion made about a population. The assertion concerns the numerical value of some parameter of the population. Procedures, which help us to decide whether to accept or reject hypothesis or to determine whether observed samples differ significantly from expected results, are called *test of hypotheses, test of significance, or results of decision.*

A *statistical hypothesis* is a statement about the parameters of a probability distribution. This may be stated as

$$H_0: \mu_1 = \mu_2$$
$$H_1: \mu_1 \neq \mu_2$$

where μ_1 and μ_2 are the means of two populations. The statement $H_0: \mu_1 = \mu_2$ is called the *null hypothesis* and $H_1: \mu_1 \neq \mu_2$ is called the *alternative hypothesis*.

Table 2.2. Summary of Confidence Intervals

Parameter	Assumptions	Point estimator	Two-sided 100(1 – α)% confidence interval
μ	Mean μ, variance σ^2 known, n large.	$\bar{X}$	$\bar{X} - Z_{\alpha/2}\, \sigma/\sqrt{n} \le \mu \le \bar{X} + Z_{\alpha/2}\, \sigma/\sqrt{n}$
$\mu_1 - \mu_2$	Difference in two means μ_1 and μ_2, variances ${\sigma_1}^2$ and ${\sigma_2}^2$ known, n large. Independent distributions.	$\bar{X}_1 - \bar{X}_2$	$\bar{X}_1 - \bar{X}_2 - Z_{\alpha/2}\sqrt{\dfrac{\sigma_1^2}{n_1} + \dfrac{\sigma_2^2}{n_2}} \le \mu_1 - \mu_2$ $\le \bar{X}_1 - \bar{X}_2 - Z_{\alpha/2}\sqrt{\dfrac{\sigma_1^2}{n_1} + \dfrac{\sigma_2^2}{n_2}}$
μ	Mean μ of a normal distribution, variance σ^2 unknown.	$\bar{X}$	$\bar{X} - t_{\alpha/2,n-1}\, s/\sqrt{n} \le \mu \le \bar{X} + t_{\alpha/2,n-1}\, s/\sqrt{n}$
$\mu_1 - \mu_2$	Difference in means of two normal distributions $\mu_1 - \mu_2$, variances ${\sigma_1}^2 - {\sigma_2}^2$ unknown. Independent distributions.	$\bar{X}_1 - \bar{X}_2$	$\bar{X}_1 - \bar{X}_2 - t_{\alpha/2,n_1+n_2-2}\, s_p\sqrt{\dfrac{1}{n_1} + \dfrac{1}{n_2}} \le \mu_1 - \mu_2$ $\le \bar{X}_1 - \bar{X}_2 + t_{\alpha/2,n_1+n_2-2}\, s_p\sqrt{\dfrac{1}{n_1} + \dfrac{1}{n_2}}$ where $s_p^2 = \dfrac{(n_1 - 1)\, s_1^2 + (n_2 - 1)\, s_2^2}{n_1 + n_2 - 2}$
D	Difference in means of two normal distributions for paired samples $\mu_0 = \mu_1 - \mu_2$.	$\bar{D}$	$\bar{D} - t_{\alpha/2,n-1}\, s_D/\sqrt{n} \le \mu_D$ $\le \bar{D} + t_{\alpha/2,n-1}\, s_D/\sqrt{n}$
σ^2	Variances σ^2 of a normal distribution.	s^2	$\dfrac{(n-1)\, s^2}{\chi^2_{\alpha/2,n-1}} \le \sigma^2 \le \dfrac{(n-1)\, s^2}{\chi^2_{1-\alpha/2,n-1}}$
$\dfrac{\sigma_1^2}{\sigma_2^2}$	Ratio of the variances ${\sigma_1}^2/{\sigma_2}^2$ of two normal distributions.	$\dfrac{s_1^2}{s_2^2}$	$\dfrac{s_1^2}{s_2^2} F_{1-\alpha/2,n_2-1,n_1-1} \le \dfrac{\sigma_1^2}{\sigma_2^2} \le \dfrac{s_1^2}{s_2^2} F_{\alpha/2,n_2-1,n_1-1}$
p	Proportion or parameter of a binomial distribution p, n large.	p	$\hat{p} - Z_{\alpha/2}\sqrt{\dfrac{\hat{p}(1-\hat{p})}{n}} \le p$ $\le \hat{p} + Z_{\alpha/2}\sqrt{\dfrac{\hat{p}(1-\hat{p})}{n}}$
$p_1 - p_2$	Difference in two proportions or two binomial parameters, $p_1 - p_2$.	$p_1 - p_2$	$\hat{p}_1 - \hat{p}_2 - Z_{\alpha/2}\sqrt{\dfrac{\hat{p}_1(1-\hat{p}_2)}{n_1} + \dfrac{\hat{p}_2(1-\hat{p}_2)}{n_2}}$ $\le p_1 - \hat{p}_2 \le \hat{p}_1 - \hat{p}_2 + Z_{\alpha/2}$ $\le \hat{p}_1 - \hat{p}_2 + Z_{\alpha/2}\sqrt{\dfrac{\hat{p}_1(1-\hat{p}_2)}{n_1} + \dfrac{\hat{p}_2(1-\hat{p}_2)}{n_2}}$

The alternative hypothesis here is known as a *two-sided alternative hypothesis* since it would be true if $\mu_1 < \mu_2$ or if $\mu_1 > \mu_2$.

A *null hypothesis* is a claim (or a statement) about a population parameter that is assumed to be true until it is declared false. An *alternative hypothesis* is a claim about a population parameter that will be true if the *null hypothesis* is false.

To test a hypothesis, we develop a procedure for taking a random sample, computing an appropriate *test statistic*, and then rejecting or failing to reject the null hypothesis H_0. Part of this procedure is specifying the set of values for the test statistic that leads to rejection of H_0. This set of values is called the *critical region* or *rejection region* for the test.

The steps involved in performing a test of hypothesis are:

1. State the null and alternative hypothesis
2. Select the distribution to use
3. Determine the rejection and non-rejection regions
4. Calculate the value of the test statistic
5. Make a decision

Part of this procedure is specifying the set of values for the test statistic.

2.13.1 Types of Sampling Error

In evaluating a hypothesis, two types of errors can be made:

1. **Type I Error:** If we reject a hypothesis when it should be accepted, we call that a *type I error* has been made. Hence, a type I error occurs when a true null hypothesis is rejected.

The value of α represents the probability of committing this type of error .

That is, the probability of this error is written as: $\alpha = P(\text{Type I Error}) = P(H_0/H_0 \text{ is true})$.

The value of α represents the *significance level* of the test.

2. **Type II Error:** A type II error occurs when a false null hypothesis is not rejected. The value of β represents the probability of making a type II error.

It represents the probability that H_0 is not rejected when actually H_0 is false. The value of $(1 - \beta)$ is called the *power of the test*. It represents the probability of not making a type II error.

That is, $\beta = P(\text{Type II Error}) = P(\text{fail to reject } H_0/H_0 \text{ is false})$.

$$\text{Power} = 1 - \beta = P(\text{reject } H_0/H_0 \text{ is false})$$

The two types of errors that occur in test of hypotheses depend on each other. The values of α and β cannot be lowered simultaneously for a test of hypothesis for a fixed sample size.

Lowering the value of one of them will raise the value of the other. On the other hand, one can decrease both α and β simultaneously by increasing the sample size.

The Type I and Type II errors are defined in terms of probability numbers and can be controlled to desired value. The results possible in testing a hypothesis are summarized in Table 2.3.

Table 2.3. Type I (α) Error and Type II (β) Error

If the decision on analysis is:	If H is:	
	True	**False**
Accept H	Correct decision $P = 1 - \alpha$	Wrong decision $P = \beta$
Reject H	Wrong decision $P = \alpha$	Correct decision $P = 1 - \beta$
	$\sum P = 1.0$	$\sum P = 1.0$

2.13.2 Level of Significance

In testing a given hypothesis, the maximum probability with which one would be willing or prepared to risk a Type I Error is called the *level of significance* of the test. This probability often denoted by α is generally specified before any samples are drawn, so that results obtained will not influence the choice made.

The general procedure in hypothesis testing is to specify a value of the probability of type I error α, often called the *significance level* of the test, and then design the test procedure so that the probability of type II error β has a suitably small value.

2.13.3 Tails of a Test

The rejection region for a hypothesis-testing problem can be on both sides with the non-rejection region in the middle, or it can be on the left side or on the right side of the non-rejection region. A test with two rejection regions is called a *two-tailed test*, and a test with one rejection region is called a *one-tailed test*. The one-tailed test is called a *left-tailed test* if the rejection region is on the left tail of the distribution curve, and it is called a *right-tailed test* if the rejection region is on the right tail of the distribution curve. In such cases, the critical region is a region to one side of the distribution with area equal to the level of significance.

Table 2.4 gives critical values of z for one-tailed and two-tailed tests at various levels of significance. If under a given hypothesis, the sampling distribution of a statistic s is a normal distribution with mean μ and standard deviation σ. Then, the distribution of a standardized variable (or z score) is given by $z = (s - \mu) / \sigma$, is the standardized normal distribution (mean 0, variance 1) as shown in Fig. 2.26. Table 2.5 summarizes the relationships between the signs in H_0 and H_1 and the tails of a test.

Table 2.4. Critical Value of z

Level of significance, z	**0.10**	**0.05**	**0.01**	**0.005**	**0.002**
Critical value of z for one-tailed test.	– 1.28 or 1.28	– 1.645 or 1.645	– 2.33 or 2.33	– 2.58 or 2.58	– 2.88 or 2.88
Critical value of z for two-tailed test.	– 1.645 or 1.645	– 1.96 or 1.96	– 2.58 or 2.58	– 2.81 or 2.81	– 3.08 or 3.08

Table 2.5. Summary of the Relationships Between the Signs in H_0 and H_1 and the Tails of a Test

	Two-tailed test	Left-tailed test	Right-tailed test
Sign in the null hypothesis H_0	=	= or ≥	= or ≤
Sign in the alternative hypothesis H_1	≠	<	>
Rejection region	In both tails	In the left tail	In the right tail

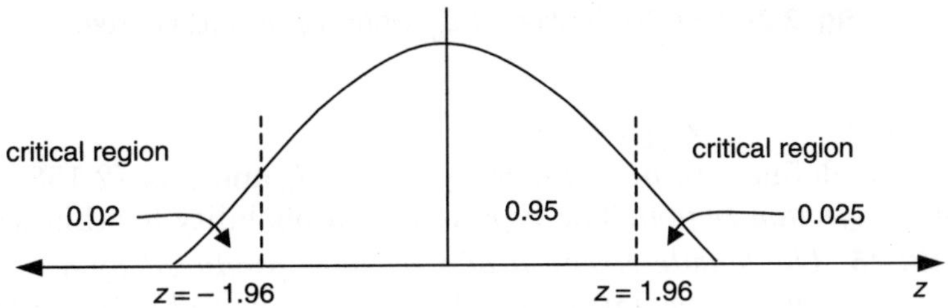

Fig. 2.26 Critical Region.

2.13.4 Hypothesis Test on the Mean, Variance Known

Suppose, the random variable X represents the population of interest. We assume that the distribution of X is either normal or that, if it is non-normal, the conditions of the central limit theorem hold.

Consider that the mean μ of X is unknown but that the variance σ^2 is known. Suppose, we are interested in testing the hypothesis

$$H_0: \mu = \mu_0$$
$$H_1: \mu \neq \mu_0 \quad \text{...(2.156)}$$

where μ_0 is a specified constant.

A random sample size of n, $X_1, X_2, \ldots X_n$ is available. Each observation in this sample has unknown mean μ and known variance σ^2. The test procedure for H_0: $\mu = \mu_0$ uses the test statistic

$$Z_0 = \frac{\overline{X} - \mu_0}{\sigma/\sqrt{n}} \quad \text{...(2.157)}$$

If the null hypothesis H_0: $\mu = \mu_0$ is true, then $E(\overline{X}) = \mu_0$, and it follows that the distribution of Z_0 is $N(0, 1)$. Hence, if H_0: $\mu = \mu_0$ is true, the probability is $1 - \alpha$ that a value of the test statistic Z_0 falls between $-Z_{\alpha/2}$ and $Z_{\alpha/2}$, where $Z_{\alpha/2}$ is the percentage point of the standard normal distribution such that $P\{Z \geq Z_{\alpha/2}\} = \alpha/2$ (*i.e.*, $Z_{\alpha/2}$ is the 100 $\alpha/2$ percentage point of the standard normal distribution). The situation is illustrated in Fig. 2.27.

Note that the probability is α that a value of the test statistic Z_0 would fall in the region $Z_0 > Z_{\alpha/2}$ or $Z_0 < -Z_{\alpha/2}$ when H_0: $\mu = \mu_0$ is true. We should reject H_0 if either

$$Z_0 > Z_{\alpha/2} \quad \text{...(2.158)}$$

or

$$Z_0 < -Z_{\alpha/2} \quad \text{...(2.159)}$$

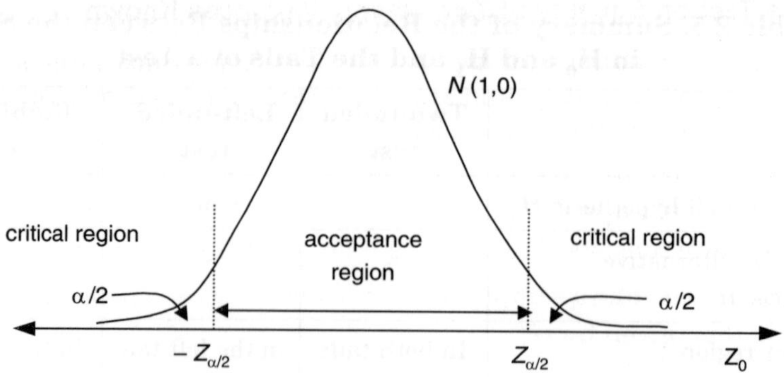

Fig. 2.27 The Distribution of Z_0, when H_0: $\mu = \mu_0$ is True.

and fail to reject H_0 if

$$-Z_{\alpha/2} \leq Z_0 \leq Z_{\alpha/2} \qquad ...(2.160)$$

Equation (2.160) defines the *acceptance region* for H_0 and Eqs. (2.158) and (2.159) define the *critical region* or *rejection region*. The type I error probability for this test procedure is α.

Example E2.25. *The tensile strength of steel bars produced by a manufacturer has a mean of 30 ksi and the standard deviation of 1.5 ksi. By a new technique in the manufacturing process, it is claimed that the tensile strength can be improved. To test this claim, a sample of 50 steel bars is tested and it is found that the mean tensile strength is 30.5 ksi. Can we support the claim at 0.01 level of significance?*

Solution:

A decision has to be made between the two hypothesis

H_0: $\mu = 30$ ksi, and there is really no change in tensile strength.

H_1: $\mu > 30$ ksi, and there is a change in tensile strength.

A one-tailed test should be used here. At a 0.01 level of significance, the decision rule is:

(1) If the z score observed is greater than 2.33, the results are significant at the 0.01 level and H_0 is rejected. The area between 0 and z_1 in Fig. E2.26 is 0.4900, and $z_1 = 2.33$.

(2) Otherwise, H_0 is accepted (or the decision is withheld). Under the hypothesis that H_0 is true, we find

$$Z_0 = \frac{\overline{X} - \mu_0}{\sigma/\sqrt{n}} = \frac{30.5 - 30}{1.5/\sqrt{50}} = 2.357$$

which is greater than 2.33. Hence, we can conclude that the results are *highly significant* and therefore the claim should be supported.

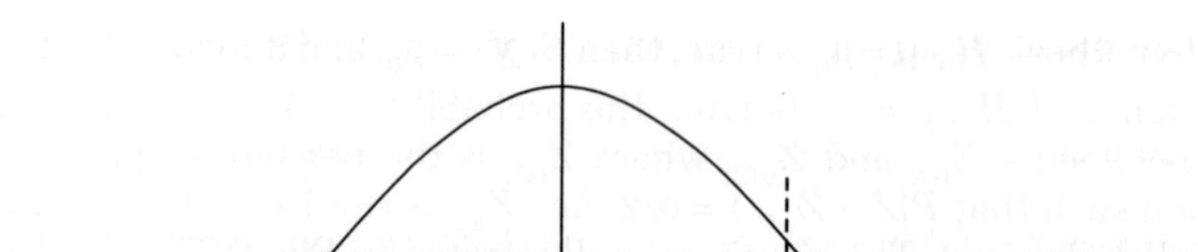

Fig. E2.25

2.13.5 Hypothesis Test on Equality of Two Means, Variances Known

There are two normal populations (or two populations from which large samples are taken) with unknown means, say μ_1 and μ_2, and known variances $\sigma_1{}^2$ and $\sigma_2{}^2$. To test the equality of the means, the following hypothesis are used:

$$H_0: \mu_1 = \mu_2$$
$$H_1: \mu_1 \neq \mu_2$$

The test procedure consists of taking a random sample of size n_1 from the first population and compute $\overline{X}_1$. Similarly take a sample of size n_2 from the second population and compute $\overline{X}_2$. The test statistic is then given by

$$Z_0 = \frac{\overline{X}_1 - \overline{X}_2}{\sqrt{\frac{\sigma_1^2}{n_1} + \frac{\sigma_2^2}{n_2}}} \qquad ...(2.161)$$

with critical region $| Z_0 | > Z_{\alpha/2}$.

The one-sided alternative hypothesis can also be used to test

$$H_0: \mu_1 = \mu_2$$
$$H_1: \mu_1 > \mu_2$$

The test statistic given by Eq. (2.161) is computed, and H_0 is rejected if $Z_0 > Z_\alpha$.

To test the other one-sided alternative hypothesis

$$H_0: \mu_1 = \mu_2$$
$$H_1: \mu_1 < \mu_2$$

The test statistic given by Eq. (2.161) is also used, but H_0 is rejected if $Z_0 < -Z_\alpha$.

Example E2.26. *Refer to example E2.18.*

(*a*) *Test at the 2.5 % significant level if the mean hourly wage for the male workers was higher than that of the female workers.*

(*b*) *What will be the conclusion in part (b) if the probability of making a Type I Error is zero and why?*

Solution:

(*a*) H_0: $\mu_1 - \mu_2 = 0$ and H_1: $\mu_1 - \mu_2 > 0$, for $\alpha = 0.025$, the critical value of $Z_{\alpha/2}$ is 1.96

$$Z_{\alpha/2} = \frac{(\overline{X}_1 - \overline{X}_2) - (\mu_1 - \mu_2)}{\sqrt{\frac{\sigma_1^2}{n_1} + \frac{\sigma_2^2}{n_2}}} = \frac{(15 - 13) - (2 - 1.5)}{\sqrt{\frac{2^2}{1000} + \frac{1.5^2}{1200}}} = \frac{1.5}{0.2046} = 7.331$$

Reject H_0. The mean hourly wage for male workers in 1997 was higher than that of female workers.

(*b*) If $\alpha = 0$, there is no rejection region, so we cannot reject H_0.

2.13.6 Tests of Hypothesis on the Mean of a Normal Distribution, Variance Unknown

Suppose that $X \sim N(\mu, \sigma^2)$ and wish to test H_0: $\mu = \mu_0$, and we do not know μ or σ^2. This test procedure depends on X being normally distributed, although slight departures from normally are not serious, especially when the sample size is large. A random sample of size n is taken, and $\overline{X}$ and s^2 are computed.

Now, to test the two-sided hypothesis

$$H_0: \mu = \mu_0$$
$$H_1: \mu \neq \mu_0$$

The test statistic formed is given by

$$t_0 = \frac{\overline{X} - \mu_0}{S/\sqrt{n}} \quad ...(2.162)$$

which follows the t-distribution with $n - 1$ degrees of freedom if H_0 is true. H_0 is rejected if $|t_0| > t_{\alpha/2, v}$ where $v = n - 1$. One-sided alternatives can also be informed in a similar manner.

Example E2.27. *A manufacturing company sent ten of its employees for a special training course to increase their productivity. The following table gives the one-week production of these employees before and after they attend this course:*

Before	*12*	*10*	*18*	*23*	*25*	*9*	*14*	*16*	*7*	*8*
After	*18*	*13*	*24*	*22*	*24*	*14*	*19*	*20*	*10*	*11*

Using the 1% significance level, can a conclusion be made saying that the mean weekly productions of all the employees increase as a result of attending this course?

Assume the populations of paired differences have a normal distribution.

Solution:

Since the data are for paired samples, we can test the hypothesis about paired differences mean μ_0 of population using the paired differences mean $\overline{X}$ of the sample.

Let X = weekly output before the course – weekly output after the course.

Before	After	Difference x	x^2
12	18	– 6	36
10	13	– 3	9
18	24	– 6	36
23	22	1	1
25	24	1	1
9	14	– 5	25
14	19	– 5	25
16	20	– 4	16
7	10	– 3	9
8	11	– 3	9

$$\sum x = -33 \quad \sum x^2 = 167$$

$$\overline{X} = \frac{\sum x}{n} = \frac{-33}{10} = -3.3$$

$$\sigma = \sqrt{\frac{\sum x^2 - \frac{\left(\sum x\right)^2}{n}}{n-1}} = \sqrt{\frac{167 - \frac{(-33)^2}{10}}{10-1}} = 2.541$$

Let $\dfrac{\sigma}{\sqrt{n}} = \dfrac{2.541}{\sqrt{10}} = 0.8035 \quad \mu_d = \mu_1 - \mu_2$

H_0: $\mu_d = 0$ ($\mu_1 - \mu_2 = 0$ or the mean output do not increase)

H_1: $\mu_d < 0$ ($\mu_1 - \mu_2 < 0$ or the mean output do increase)

Since the sample size is small (< 30), the population of paired differences is normal, and μ_d is unknown, we can use *t*-distribution to conduct the test. Degree of freedom = $n - 1$ = 10 –1 = 9. The critical value of *t* is obtained from Appendix 2G, $t_{0.01,\,9} = 2.821$. The test statistic *t* is

$$t = \frac{\overline{X} - \mu_0}{\sigma/\sqrt{n}} = \frac{-3.3 - 0}{0.8035} = -4.107$$

Since the test statistic $t = -4.107$ for $\overline{X}$ fall in the rejection region, we reject the null hypothesis. Therefore, we conclude that the mean weekly output for all employees increase as a result of the course. The critical value of *t* as obtained from the *t* distribution table is shown in Fig. E2.27.

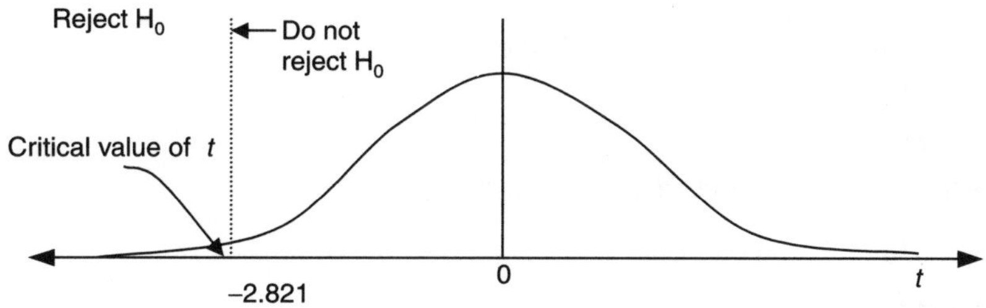

Fig. E2.27

2.13.7 Test of Hypothesis on the Means of Two Normal Distributions, Variances Unknown

We consider here the tests of hypotheses on the equality of the means μ_1 and μ_2 of two normal distributions where the variances σ_1^2 and σ_2^2 are unknown. A *t*-statistic is used to test the hypothesis.

2.13.7.1 Case 1: $\sigma_1^2 = \sigma_2^2 = \sigma^2$

Let X_1 and X_2 be two independent normal populations with unknown means μ_1 and μ_2, and unknown but equal variances, $\sigma_1^2 = \sigma_2^2 = \sigma^2$. We like to test

$$H_0: \mu_1 = \mu_2$$

$$H_1: \mu_1 \neq \mu_2 \qquad \text{...(2.163)}$$

Assume that $X_{11}, X_{12}, \ldots, X_{1n_1}$ is a random sample of n_1 observations from X_1, and $X_{21}, X_{22}, \ldots, X_{2n_2}$ is a random sample of n_2 observations from X_2.

Let $\overline{X}_1$, $\overline{X}_2$, s_1^2, and s_2^2 be the sample means and sample variances, respectively. s_1^2 and s_2^2 estimate the common variance σ^2 and we can combine them to yield a single estimate

$$s_p^2 = \frac{(n_1 - 1)\, s_1^2 + (n_2 - 1)\, s_2^2}{n_1 + n_2 - 2} \qquad \text{...(2.164)}$$

To test H_0: $\mu_1 = \mu_2$ in Eq. (2.163), we compute the test statistic

$$t_0 = \frac{\overline{X}_1 - \overline{X}_2}{s_p\sqrt{\dfrac{1}{n_1} + \dfrac{1}{n_2}}} \quad ...(2.165)$$

If H_0: $\mu_1 = \mu_2$ is true, t_0 is distributed as $t_{n_1+n_2-2}$. Therefore, if

$$t_0 > t_{\alpha/2, n_1+n_2-2} \quad ...(2.166)$$

or if

$$t_0 < -t_{\alpha/2, n_1+n_2-2} \quad ...(2.167)$$

we reject H_0: $\mu_1 = \mu_2$

The one-sided alternatives are treated similarly. To test

$$H_0: \mu_1 = \mu_2$$
$$H_1: \mu_1 > \mu_2 \quad ...(2.168)$$

Compute the test statistic t_0 in Eq. (2.165) and reject H_0: $\mu_1 = \mu_2$ if

$$t_0 > t_{\alpha,\, n_1+n_2-2} \quad ...(2.169)$$

For the other one-sided alternative, we have

$$H_0: \mu_1 = \mu_2$$
$$H_1: \mu_1 < \mu_2 \quad ...(2.170)$$

calculate the test statistic t_0 and reject H_0: $\mu_1 = \mu_2$, if

$$t_0 < -t_{\alpha,\, n_1+n_2-2} \quad ...(2.171)$$

2.13.7.2 Case 2: $\sigma_1^2 \neq \sigma_2^2$

In some cases, we cannot assume that the unknown variances σ_1^2 and σ_2^2 are equal. There is not an exact t-statistic available for testing H_0: $\mu_1 = \mu_2$ in this case. However, the statistic

$$t_0^* = \frac{\overline{X}_1 - \overline{X}_2}{\sqrt{\dfrac{s_1^2}{n_1} + \dfrac{s_2^2}{n_2}}} \quad ...(2.172)$$

is distributed approximately as t with degrees of freedom given by

$$v = \frac{\left(\dfrac{s_1^2}{n_1} + \dfrac{s_2^2}{n_2}\right)^2}{\dfrac{(s_1^2/n_1)^2}{n_1+1} + \dfrac{(s_2^2/n_2)^2}{n_2+1}} - 2 \quad ...(2.173)$$

if the null hypothesis H_0: $\mu_1 = \mu_2$ is true. Therefore, if $\sigma_1^2 \neq \sigma_2^2$, the hypothesis of Eqs. (2.163), (2.168), (2.170) are tested and t_0^* is used as the test statistic and $n_1 + n_2 - 2$ is replaced by v in determining the degrees of freedom for the test.

Example E2.28. *A company claims that its high strength steel, type A provides higher tensile strength than another company's steel, type B. A researcher tested both types of steel on two groups of randomly selected ones. The results of the test are given in the following table. The mean and standard deviation of tensile strength are in ksi.*

Type	Sample size	Mean of tensile strength	Standard deviation of tensile strength
A	*25*	*40*	*13*
B	*22*	*45*	*11*

(a) Construct a 99% confidence interval for the difference between the mean tensile strengths for the two types of steel.

(b) Test at a 1% significance level if the mean tensile strengths for type A are less than that for B.

Solution:

(*a*) Degrees of freedom $= \dfrac{\left[\dfrac{s_1^2}{n_1}+\dfrac{s_2^2}{n_2}\right]^2}{\dfrac{\left(\dfrac{s_1^2}{n_1}\right)^2}{n_1-1}+\dfrac{\left(\dfrac{s_2^2}{n_2}\right)^2}{n_2-1}}$

$$\text{d.o.f} = \frac{\left[\dfrac{13^2}{25}+\dfrac{11^2}{22}\right]^2}{\dfrac{\left(\dfrac{13^2}{25}\right)^2}{25-1}+\dfrac{\left(\dfrac{11^2}{22}\right)^2}{22-1}} = 44$$

The critical t-value from Appendix 2G, $t_{0.005,44} = 2.692$.

Hence,

$$(\overline{X}_1 - \overline{X}_2) \pm t\sqrt{\frac{13^2}{25}+\frac{11^2}{22}}$$

$$(40 - 45) \pm 2.692\ (3.5014)$$

99% confidence interval for $\mu_1 - \mu_2$ is

$$-5 \pm 9.43 \quad \text{or} \quad -14.42 \text{ to } 4.4179$$

Type A is between 14.42 ksi lesser and 4.42 ksi higher than type B.

(*b*) H_0: $\mu_1 - \mu_2 = 0$

H_1: $\mu_1 - \mu_2 < 0$

Degrees of freedom = 44

$$t_{0.01,\,44} = -2.414$$

$$t = \frac{(\overline{X}_1 - \overline{X}_2) - (\mu_1 - \mu_2)}{\sqrt{\dfrac{s_1^2}{n_1}+\dfrac{s_2^2}{n_2}}} = \frac{(40-45)-0}{3.5014} = -1.428$$

Do not reject H_0. The mean tensile strength for type A is not less than the mean tensile strength for type B.

2.13.8 Tests of Hypothesis on the Variance for a Normal Population

Here, we test the hypothesis that the variance of a normal distribution σ^2 equals a specified value of σ_0^2. Let $X \sim N(\mu, \sigma^2)$, where μ and σ^2 are unknown, and let $X_1, X_2, \ldots, X_n$ be a random sample of n observations from this population. To test

$$H_0: \sigma^2 = \sigma_0^2$$
$$H_1: \sigma^2 \neq \sigma_0^2 \quad \text{...(2.174)}$$

we use the test statistic

$$\chi_0^2 = \frac{(n-1)\, s^2}{\sigma_0^2} \quad \text{...(2.175)}$$

then s^2 is the sample variance. Now if H_0: $\sigma^2 = \sigma_0^2$ is true, then the test statistic χ_0^2 follow the chi-square distribution with $n - 1$ degrees of freedom.

Therefore $H_0: \sigma^2 = \sigma_0^2$

would be rejected if $\chi_0^2 > \chi^2_{\alpha/2,\, n-1}$ ...(2.176)

or if $\chi_0^2 < \chi^2_{1-\alpha/2,\, n-1}$ ...(2.177)

where $\chi^2_{\alpha/2,\, n-1}$ and $\chi^2_{1-\alpha/2,\, n-1}$ are the upper and lower $\alpha/2$ percentage point of chi-square distribution with $n - 1$ degrees of freedom.

For the one-sided hypothesis

$$H_0: \sigma^2 = \sigma_0^2$$
$$H_1: \sigma^2 > \sigma_0^2 \quad \text{...(2.178)}$$

We would reject H_0 if

$$\chi_0^2 > \chi^2_{\alpha,\, n-1} \quad \text{...(2.179)}$$

For the other one-sided hypothesis

$$H_0: \sigma^2 = \sigma_0^2$$
$$H_1: \sigma^2 < \sigma_0^2 \quad \text{...(2.180)}$$

we would reject H_0 if

$$\chi_0^2 < \chi^2_{1-\alpha,\, n-1} \quad \text{...(2.181)}$$

Example E2.29. *The copper content in an alloy material used to make a machine part is normally distributed, and the variance is $\sigma^2 = 20$. Test the hypothesis H_0: $\sigma^2 = 20$ versus $H_0 : \sigma^2 \neq 20$, if a random sample of $n = 10$ machine parts yields a sample standard deviation of $s = 4$. Use $\alpha = 0.05$.*

Solution:

$$\sigma^2 = 20,\ n = 10,\ s = 4$$
$$H_0: \sigma^2 = 20$$
$$H_0: \sigma^2 \neq 20$$
$$\alpha = 0.05$$

$$\chi_0^2 = \frac{(n-1)\, s^2}{\sigma_0^2} = \frac{(10-1)\,(4)^2}{20} = 7.2$$

$$\chi^2_{0.05/2,\, n-1} = \chi^2_{0.025,\, 9} = 19.022 \text{ from Appendix 2F}$$

$$\chi^2_{0.975,\,9} = 2.70$$

$$\chi^2_{0.975,\,9} \le \chi_0^2 \le \chi^2_{0.025,\,9}$$

Do not reject H_0, the data provide insufficient evidence to claim that the variance of the copper content of the machine parts is not 20 at $\alpha = 0.05$.

2.13.9 Hypothesis Test on the Equality of Two Variances

The two independent populations are $X_1 \sim N(\mu_1, \sigma_1^2)$ and $X_2 \sim N(\mu_2, \sigma_2^2)$ where μ_1, σ_1^2, μ_2 and σ_2^2 are unknown. We wish to test hypothesis about the equality of the two variances, say H_0: $\sigma_1^2 = \sigma_2^2$. Assume that two random samples of size n_1 from population 1 and of size n_2 from population 2 are available, and let s_1^2 and s_2^2 be the sample variances. To test the two-sided alternative

$$H_0: \sigma_1^2 = \sigma_2^2$$
$$H_1: \sigma_1^2 \neq \sigma_2^2 \qquad \text{...(2.182)}$$

The statistic

$$F_0 = \frac{s_1^2}{s_2^2} \qquad \text{...(2.183)}$$

is distributed as F, with $n_1 - 1$ and $n_2 - 1$ degrees of freedom, if the null hypothesis H_0: $\sigma_1^2 = \sigma_2^2$ is true. Hence, we would reject H_0 if

$$F_0 > F_{\alpha/2,\, n_1-1,\, n_2-1} \qquad \text{...(2.184)}$$

or if

$$F_0 < F_{1-\alpha/2,\, n_1-1,\, n_2-1} \qquad \text{...(2.185)}$$

where $F_{\alpha/2,\, n_1-1,\, n_2-1}$ and $F_{1-\alpha/2,\, n_1-1,\, n_2-1}$ are the upper and lower $\alpha/2$ percentage points of the F-distribution with $n_1 - 1$ and $n_2 - 1$ degrees of freedom. Appendix 2H gives only the upper tail points of F, so to find $F_{1-\alpha/2, n_1-1, n_2-1}$ we must use

$$F_{1-\alpha/2,\, n_1-1,\, n_2-1} = \frac{1}{F_{\alpha/2,\, n_1-1,\, n_2-1}} \qquad \text{...(2.186)}$$

To test one-sided alternative hypothesis, let X_1 denote the population that may have the largest variance, then the one-sided alternative hypothesis is

$$H_0: \sigma_1^2 = \sigma_2^2$$
$$H_1: \sigma_1^2 > \sigma_2^2 \qquad \text{...(2.187)}$$

If $F_0 > F_{\alpha,\, n_1-1,\, n_2-1}$ ...(2.188)

We would reject H_0: $\sigma_1^2 = \sigma_2^2$.

Example E2.30. *A company has two branch operations, A and B, producing a particular product. Branch A has15 employees while branch B has 25 employees on any given month, although there was no significant difference in mean production, branch A had a standard deviation of 8 while branch B had a standard deviation of 10. Can we conclude at the (a) 0.01, (b) 0.05 level of significance that the variability of branch B is greater than that of branch A?*

Solution:

$$n_1 = 15,\ n_2 = 25$$

(*a*) We have, on using subscripts 1 and 2 for branches *A* and *B* respectively,

$$s_1 = 8,\ s_2 = 10$$

so that

$$\hat{s}_1^2 = \frac{n_1}{n_1 - 1} s_1^2 = \frac{15}{(15 - 1)}(8^2) = 68.571$$

$$\hat{s}_2^2 = \frac{n_2}{n_2 - 1} s_2^2 = \frac{25}{(25 - 1)}(10^2) = 104.167$$

We have to decide between the hypothesis

H_0: $\sigma_1 = \sigma_2$, and any observed variability is due to chance

H_1: $\sigma_2 > \sigma_1$, and the variability of branch *B* is greater than that of *A*

The decision must therefore be based on a one-tailed test of the *F*-distribution. Hence,

$$F = \frac{\hat{s}_2^2}{\hat{s}_1^2} = \frac{104.167}{68.571} = 1.519$$

$v_2 = 25 - 1 = 24; v_1 = 15 - 1 = 14$. At 0.01 level for 24, 14 degrees of freedom, we have from Appendix 2H, $F_{0.99} = 3.43$

Since $F < F_{0.99}$, we cannot reject H_0 at the 0.01 level.

(*b*) $F_{0.95} = 2.35$ for 24, 14 degrees of freedom (see Appendix 2H).

We must note again that $F < F_{0.95}$. Thus, we cannot reject H_0 at the 0.05 level.

2.13.10 Hypothesis Test on a Proportion

Consider a binomial random variable to be tested

H_0: $p = p_0$

H_1: $p \neq p_0$

If $0.1 \leq p \leq 0.9$ and *n* is large, the procedure described is acceptable.

Pick a random sample of size *n* and compute $\hat{p} = X/n$, where X represents the number of observations that are nonconforming. If H_0 is true, then X ~ $N[p_0, p_0(1 - p_0)/n]$. To test $p = p_0$, calculate

$$Z_0 = \frac{\hat{p} - p_0}{\sqrt{\dfrac{p_0(1 - p_0)}{n}}} \qquad \text{...(2.189)}$$

and reject H_0 if $|Z_0| > Z_{\alpha/2}$. One-sided alternative could also be tested.

Example E2.31. *An Internet catalogue ordering company wants to deliver on time at least 90% of all the orders received from its customers. A sample of 100 orders taken for inspection showed that 83 of them were delivered on time.*

(*a*) *using the 2% significance level, can we conclude that the company's delivery policy is maintained?*

(*b*) *what will the decision be if the probability of making a type I error is zero?*

(*c*) *find the p-value for the test of hypothesis. Using this p-value, would the null hypotheses at $\alpha = 0.05$ and $\alpha = 0.01$ be rejected ?*

Solution:

(*a*) H_0: $p \geq 0.90$

H_1: $p < 0.90$

For $\alpha = 0.02$, the critical value of Z is -2.05.

$$\hat{p} = \frac{83}{100} = 0.83$$

$$Z = \frac{\hat{p} - p}{\sqrt{pq/n}} = \frac{0.83 - 0.90}{\sqrt{(.90)\,(0.10)/100}} = -2.333$$

Reject H_0. The conclusion is that the company's policy is not maintained.

(*b*) If $\alpha = 0$, there is no rejection region. Hence, we cannot reject H_0 and cannot conclude that the company's policy is maintained.

(*c*) $H_0: p \geq 0.90$

$H_1: p < 0.90$

$$\hat{p} = \frac{83}{100} = 0.83$$

$$Z = \frac{\hat{p} - p}{\sqrt{pq/n}} = \frac{0.83 - 0.90}{\sqrt{(0.90)\,(0.10)/100}} = -2.333$$

The area to the left of $Z = -2.33$ under the normal curve is $0.5 - 0.4901 = 0.0099$.

Hence, p-value = 0.0099.

If $\alpha = 0.05$, reject H_0 since p-value < 0.05

If $\alpha = 0.01$, reject H_0 since p-value < 0.01.

2.13.11 Tests of Hypothesis on Two Proportions

The two binomial parameters are p_1 and p_2 and we wish to test that they are equal. That is, $H_0: p_1 = p_2$

Suppose that the two random samples of sizes n_1 and n_2 are taken from two populations, and let X_1 and X_2 represent the number of observations that belong to the class of interest in sample 1 and 2, respectively. Assume that the normal approximation to the binomial applies to each population, so that the estimators of the population proportions $\hat{p}_1 = X_1/n_1$ and $\hat{p}_2 = X_2/n_2$ have approximate normal distributions. Now, if the null hypothesis $H_0: p_1 = p_2$ is true, then considering $p_1 = p_2 = p$, the random variable is

$$Z = \frac{\hat{p}_1 - \hat{p}_2}{\sqrt{p(1-p)\left[\frac{1}{n_1} + \frac{1}{n_2}\right]}}$$

is distributed approximately $N(0,1)$. An estimate of the common parameter p is

$$\hat{p} = \frac{X_1 + X_2}{n_1 + n_2}$$

The test statistic for $H_0: p_1 = p_2$ is then

$$Z_0 = \frac{\hat{p}_1 - \hat{p}_2}{\sqrt{\hat{p}(1-\hat{p})\left[\frac{1}{n_1} + \frac{1}{n_2}\right]}} \qquad \text{...(2.190)}$$

If $Z_0 > Z_{\alpha/2}$ or $Z_0 < -Z_{\alpha/2}$ ...(2.191)

the null hypothesis is rejected.

Procedures for testing hypothesis on means and variances are summarizes in Table 2.6.

Table 2.6. Summary of Hypothesis Tests

	Null hypothesis	Alternative hypothesis	Test statistic	Criteria for rejection
1	H_0: $\mu = \mu_0$ σ^2 known	H_1: $\mu \neq \mu_0$ H_1: $\mu > \mu_0$ H_1: $\mu < \mu_0$	$Z_0 = \dfrac{\overline{X} - \mu_0}{\sigma/\sqrt{n}}$	$Z_0 > Z_{\alpha/2}$ $Z_0 > Z_\alpha$ $Z_0 < -Z_\alpha$
2	H_0: $\mu = \mu_0$ σ^2 unknown	H_1: $\mu \neq \mu_0$ H_1: $\mu > \mu_0$ H_1: $\mu < \mu_0$	$t_0 = \dfrac{\overline{X} - \mu_0}{s/\sqrt{n}}$	$\lvert t_0 \rvert > t_{\alpha/2,\, n-1}$ $t_0 > t_{\alpha,\, n-1}$ $t_0 > -t_{\alpha,\, n-1}$
3	H_0: $\mu_1 = \mu_2$ σ_1^2 and σ_2^2 known and independent	H_1: $\mu_1 \neq \mu_2$ H_1: $\mu_1 > \mu_2$ H_1: $\mu_1 < \mu_2$	$Z_0 = \dfrac{\overline{X}_1 - \overline{X}_2}{\sqrt{\dfrac{\sigma_1^2}{n_1} + \dfrac{\sigma_2^2}{n_2}}}$	$Z_0 > Z_{\alpha/2}$ $Z_0 > Z_\alpha$ $Z_0 < -Z_\alpha$
4	H_0: $\mu_1 = \mu_2$ $\sigma_1^2 = \sigma_2^2 = \sigma^2$ unknown and independent	H_1: $\mu_1 \neq \mu_2$ H_1: $\mu_1 > \mu_2$ H_1: $\mu_1 < \mu_2$	$t_0 = \dfrac{\overline{X}_1 - \overline{X}_2}{s_p\sqrt{\dfrac{1}{n_1} + \dfrac{1}{n_2}}}$	$\lvert t_0 \rvert > t_{\alpha/2, n_1+n_2-2}$ $t_0 > t_{\alpha, n_1+n_2-2}$ $t_0 < -t_{\alpha, n_1+n_2-2}$
5	H_0: $\mu_1 = \mu_2$ $\sigma_1^2 \neq \sigma_2^2$ unknown σ^2 unknown and independent	H_1: $\mu_1 \neq \mu_2$ H_1: $\mu_1 > \mu_2$ H_1: $\mu_1 < \mu_2$	$t_0 = \dfrac{\overline{X}_1 - \overline{X}_2}{\sqrt{\dfrac{s_1^2}{n_1} + \dfrac{s_2^2}{n_2}}}$ $v = \dfrac{\left(\dfrac{s_1^2}{n_1} + \dfrac{s_2^2}{n_2}\right)^2}{\dfrac{(s_1^2/n_1)^2}{n_1+1} + \dfrac{(s_2^2/n_2)^2}{n_2+1}} - 2$	$\lvert t_0 \rvert > t_{\alpha/2,\, v}$ $t_0 > t_{\alpha,\, v}$ $t_0 > -t_{\alpha,\, v}$
6	H_0: $\sigma^2 = \sigma_0^2$	H_1: $\sigma^2 \neq \sigma_0^2$ H_1: $\sigma^2 > \sigma_0^2$ H_1: $\sigma^2 < \sigma_0^2$	$\chi_0^2 = \dfrac{(n-1)\, s^2}{\sigma_0^2}$	$\chi_0^2 > \chi^2_{\alpha/2,\, n-1}$ or $\chi_0^2 < \chi^2_{1-\alpha/2,\, n-1}$ $\chi_0^2 > \chi^2_{\alpha,\, n-1}$ $\chi_0^2 > \chi^2_{1-\alpha,\, n-1}$
7	H_0: $\sigma_1^2 = \sigma_2^2$	H_1: $\sigma_1^2 \neq \sigma_2^2$ H_1: $s_1^2 > s_2^2$	$F_0 = \dfrac{s_1^2}{s_2^2}$	$F_0 > F_{\alpha/2,\, n_1-1,\, n_2-1}$ $F_0 < F_{1-\alpha/2,\, n_1-1,\, n_2-1}$ $F_0 > F_{\alpha,\, n_1-1,\, n_2-1}$
8	H_0: $p = p_0$ $0.1 \le p \le 0.9$ n large	H_1: $p \neq p_0$ H_1: $p > p_0$ H_1: $p < p_0$	$Z_0 = \dfrac{\hat{p} - p_0}{\sqrt{\dfrac{p_0\,(1-p_0)}{n}}}$	$\lvert Z_0 \rvert > Z_{\alpha/2}$ $Z_0 > Z_\alpha$ $Z_0 < -Z_\alpha$

2.14 SUMMARY

In this chapter, the basic terms of statistics are explained. These terms will bridge the understanding of the concepts and techniques presented in subsequent chapters. The organization and display of data using tables and graphs, frequency distribution tables for qualitative and quantitative data, construction of bar graphs, and histograms are described. The numerical summary measures for measure of central tendency, measure of dispersion and measure of position are discussed. The basic concepts of probability and the rules for computing probability are presented. Random variables and types of random variables are explained. The concept of a probability distribution and its mean and standard deviation are discussed. Special probability distributions for a discrete random variable are presented. They are the Hypergeometric distribution, Binomial distribution and the Poisson distribution. Continuous random variables can possess one of many probability distributions. The normal probability distribution and the standard normal distribution are presented.

Approximating probability distributions are briefly discussed. They are the binomial approximation to the hypergeometric, Poisson approximation to the binomial, normal approximation to the binomial and the normal approximation to the Poisson distribution. Chebyshev's theorem and the empirical rule for distributions are briefly discussed.

The concept of random sampling and sampling distributions were introduced. The chi-square, student-t, and *F*-distributions useful for describing sampling variation are presented. The point and interval estimation of unknown parameters were introduced. Confidence intervals were derived for a variety of parameter estimation problems. The two-sided confidence intervals developed are summarized in a tabular form. Procedures for testing hypotheses on means and variances are presented.

PROBLEMS

2.1 The following table gives the frequency distribution of the amounts of long distance telephone bills for the month of August 2000, for a sample of 60 families in a city.

Amount of long distance telephone bill (dollars)	Number of families
25 to less than 50	15
50 to less than 75	12
75 to less than 100	20
100 to less than 125	8
125 to less than 150	5

Calculate the mean, variance and standard deviation.

2.2 In Fig. P2.2, *E*, *T*, and *N* are the events that an automobile brought to a service garage needs an engine oil-change, tune-up, or new tyres. Express in words the events represented by

(*a*) region 1

(*b*) region 3

(*c*) region 7

(*d*) regions 1 and 4 together

(*e*) regions 2 and 5 together

(*f*) regions 3, 5, 6, and 8 together.

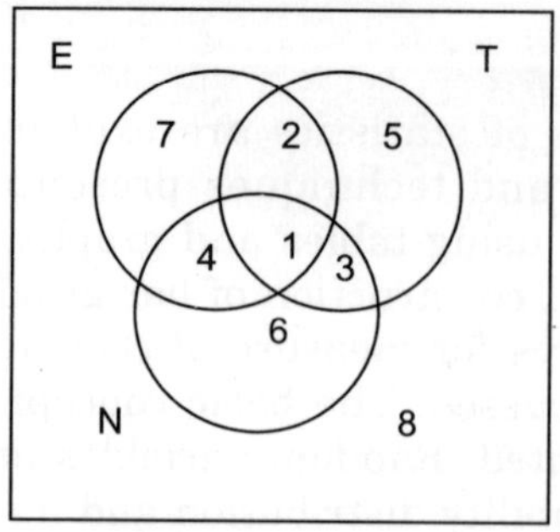

Fig. P2.2 Venn Diagram.

2.3 In Problem 2.2, list the region or combinations of regions representing the events that an automobile brought to the garage needs

(*a*) tune-up, but neither engine oil-change nor new tyres

(*b*) an engine oil-change and tune-up

(*c*) tune-up or new tires, but not an engine oil-change

(*d*) new tyres.

2.4 In a particular company, 50% of the adult employees are female, 10% are divorced and 5% are divorced females. For an employee selected at random, let

F = event the person is female

D = event the person is divorced

(*a*) Find $P(F)$, $P(D)$, and $P(F \text{ and } D)$

(*b*) Determine $P(F \text{ or } D)$ and interpret your answer in terms of percentages.

(*c*) Find the probability that a randomly selected adult is male.

2.5 In the research and development department of a company, the number of males and female employees are shown in the frequency distribution as given below in a tabular form

Sex	Frequency
Male	30
Female	20
Total	50

Two employees selected at random from this department. The first employee obtained is not returned to the department for possible reselection; that is, the sampling is without replacement.

(*a*) Determine the probability that the first employee obtained is female and the second is male.

(*b*) Draw the tree diagram.

2.6 In a random sample of 1000 companies surveyed showed that 35 of them have no design engineer, 210 have one design engineer each, 379 have two design engineers, 270 have three design engineers and 106 have four or more design engineers.

(*a*) Prepare a frequency distribution table and calculate the relative frequencies for all categories.

(*b*) If one company is selected randomly from these 1000 companies, find the probability that this company has two design engineers.

(*c*) Find the probability that the company selected has four or more design engineers.

2.7 The probability that an employee at a university is a female is 0.3. The probability that an employee is a female and married is 0.2. Find the conditional probability that a randomly selected female employee from this university is a married person.

2.8 Consider the following events for one roll of a die

A = an odd number is observed (1,3,5)

B = an even number is observed (2,4,6)

C = a number less than 5 is observed (1,2,3,4)

Are events A and B mutually exclusive? Are events B and C mutually exclusive?

2.9 Out of a collection of 100 gaskets the following are defective as indicated:

Defective type	Number of gaskets
Type I defect	10
Type II defect	8
Type III defect	3
Type I and II defect	5
Type I, II and III defect	4
No defects	70

(*a*) If a gasket picked out of the collection has a type II defect, what is the probability that it also has a type I defect?

(*b*) If the gasket picked has a type I defect, what is the probability that it also has a type II defect?

2.10 A mechanical system assembly plant purchases a certain component from three independent vendors; A, B, and C, who supply 20%, 30% and 50% respectively of the total number of components needed by the assembly plant. On an average, the percentage of defective components supplied by vendors A, B, and C are 5%, 3%, and 2% respectively.

(*a*) if a component is selected at random, find the probability that it is defective.

(*b*) if the selected component is defective, what is the probability that vendor C supplied it? By vendor A? By vendor B?

2.11 The frequency distribution of the number of orders received per day by a distribution company during the past 3 month (90 days) are given below:

Number of orders received per day	2	3	4	5	6
Number of days	11	20	23	22	14

(*a*) Construct a probability distribution table for the number of orders received per day.

(*b*) Are the probabilities in (*a*) exact or approximate?

(*c*) If x denotes the number of orders received on any given day, calculate the following probabilities.

(*i*) $P(x = 3)$

(*ii*) $P(x \geq 3)$

(*iii*) $P(2 \leq x \leq 4)$

(*iv*) $P(x < 4)$

2.12 Find the mean μ for the probability density function, $\rho(x)$, of the life of a projector bulb, random variable x, is given as:

$\rho(x) = 0$ for $x < 0$

and $\rho(x) = 1/900\ e^{-x/900}$ for $x \geq 0$

2.13 The daily consumption of an electric power of a certain machine (in units of power) is a random variable whose probability density is given by

$$f(x) = \begin{cases} 1/9xe^{-x/3} & \text{for } x > 0 \\ 0 & \text{for } x \leq 0 \end{cases}$$

Determine the probabilities that on a given day

(*a*) The consumption of this machine is no more than 6 units.

(*b*) The power supply is inadequate in the daily capacity if the supply is 9 units.

2.14 A lot contains 140 electronic components and 20 are selected without replacement for quality testing.

(*a*) If 20 components are defective, what is the probability that at least one defective component is in the sample?

(*b*) If 5 components are defective, what is the probability that at least one defective component appears in the sample?

2.15 A quality test engineer claims that 1 in 10 of certain manufactured parts is due to material defects. Using binomial distribution and rounding to four decimals, find the probability that at least 3 out of 5 of the tested parts are due to material defects.

2.16 A photocopying machine in an office breaks down an average of three times per month. Using the Poisson probability distribution formula, find the probability that during the next month this machine will have

(*a*) exactly two breakdowns

(*b*) at most one breakdown.

2.17 The diameter of a machine shaft produced in a manufacturing company is normally distributed with a mean diameter of 0.001 inches and a standard deviation of 0.0002 inches.

(*a*) what is the probability that the diameter of the shaft exceeds 0.0013 inches?

(*b*) what is the probability that a diameter of the shaft is between 0.0007 and 0.0013 inches?

(*c*) what standard deviation of diameters is needed so that the probability in part (*b*) is 0.995?

2.18 Out of a batch of 2800 electronic components received 25% of them were defective or non-conforming. In a sample of 50 randomly selected, the probability that between 12 and 14 are given by the binomial distribution. Use normal distribution to this binomial distribution.

2.19 Two suppliers supply machine systems for a manufacturing company. Machines from supplier 1 have a standard deviation of 25 units and supplier 2 have a standard deviation of 30 units. A random sample of 10 machines from supplier 1 has an average of 155 units and a similar sample of supplier 2 machines have an average of 137.4 units. Construct a 90% confidence interval on the true mean difference in the means of two machines.

2.20 The breaking strength of a machine part is required to be at least 75 psi. Based on previous test data, the standard deviation of breaking strength is 6 psi. A random sample of 16 specimens is tested and the average value is found to be 73 psi. Construct a 95% confidence interval on the mean breaking strength.

2.21 The diameters of copper shafts produced by a certain manufacturing process should have a mean diameter of 0.51 mm. The diameter is known to have a standard deviation of 0.0002 mm. A random sample of 40 shafts has an average diameter of 0.509 mm. Construct a 95% confidence interval on the mean shaft diameter.

2.22 Two slightly different steel specimen types of bar samples are being tested for tensile strength. Ten samples of each specimen type were selected one at random from ten different lots. The tensile tests on these bars resulted in

$\overline{X}_1 = 25$ ksi, $s_1^2 = 2.2$ (ksi)^2,

$\overline{X}_2 = 25.5$ ksi, and $s_2^2 = 2.8$ (ksi)^2. Assume the tensile strengths are normally distributed.

(*a*) Find a 99% confidence interval on the difference in means.

(*b*) If only 5 samples were selected for the second specimen type, how does this affect the confidence interval? Assume the value of $\overline{X}_2$ and s_2^2 remain unchanged.

2.23 The outputs of two manufacturing firms over one year period with random days picked gave the following data:

$n_1 = 20 \quad \overline{X}_1 = 250 \quad s_1^2 = 35$

$n_2 = 25 \quad \overline{X}_2 = 200 \quad s_2^2 = 40$

Assuming the two outputs are normally distributed with the same variance, find a 95% two-sided confidence interval on the difference in their means.

2.24 Random sample of weights were taken from two packages in a packaging process. The first sample consisted of $n_1 = 150$ and the second consisted of $n_2 = 250$ packages. The two sample means and standard deviations are as follows:

$\overline{X}_1 = 10$ kg $\quad \overline{X}_2 = 9.8$ kg

$s_1 = 0.25$ kg $\quad s_2 = 0.35$ kg

Determine the interval within which we can be 90% sure that $\mu_1 - \mu_2$ lies.

2.25 The length of steel bars manufactured on two different machines is being investigated. Two random samples of sizes $n_1 = 16$ and $n_2 = 18$ are selected, and the sample means and sample variances are $\overline{X}_1 = 10$, $S_1^2 = 0.4$, $\overline{X}_2 = 9.6$, and $S_2^2 = 0.45$, respectively. Assuming that $\sigma_1^2 = \sigma_2^2$, construct a 95% two-sided confidence interval on ratio of the population variances σ_1^2 / σ_2^2. Is it reasonable to conclude that the two variances are equal?

2.26 A study has been made to compare the aluminum contents of two brands of machine parts. Ten parts of brand A had an average aluminum content of 3 grams with a standard deviation of 0.4 grams, while 8 parts of brand *B* had an average aluminum content of 0.3 grams with a standard deviation of 0.5 grams. Assuming that the two sets of data are independent random samples from normal population, with equal variance find a 98% confidence interval for σ_1^2 / σ_2^2.

2.27 The manufacturer of a new line of automotive transmission is estimating the fraction of nonconforming units produced. Find a 95% two-sided confidence interval for the following:

(*a*) 150 transmissions in the sample with 15 nonconforming

(*b*) 1500 transmissions in the sample with 150 nonconforming

(*c*) 15,000 transmissions in the sample with 1500 nonconforming.

2.28 With a random sample of 1000 people, 400 are women who turn up to see a baseball game. Find a 95% confidence interval for the percentage of women at the game.

2.29 A sample of 600 observations selected from a population produced a sample proportion equal to 0.7.

(*a*) make a 90% confidence interval for *p*

(*b*) construct a 95% confidence interval for *p*

(*c*) make a 99% confidence interval for *p*

(*d*) Does the width of the confidence intervals constructed in part (*a*) through (*c*) increase as the confidence level increases?

2.30 (*a*) The standard deviation of the lifetimes of a particular brand of light bulbs is 120 hours. How large a sample must be so that we are 95% confident that the error in the estimated mean life of the lights bulbs is less than 8 hours?

(*b*) If the standard deviation of the lifetimes of the light bulbs is not known, how large the samples must be tested to be 95% confident that the error in the estimated mean life of all the light bulbs is less than 0.1σ.

2.31 The mean height of 45 students who showed an above-average participation in college sports was 68.1 inches with a standard deviation of 2.42 inches, while 45 other students who showed no interest in such sports participation had a mean height of 67.67 inches with a standard deviation of 2.76 inches.

(*a*) Test the hypothesis that the students who participate in sports activities are taller that the other students.

(*b*) by how much the sample size of each of the two groups be increased in order that the observed difference of 0.75 inches in the mean heights be significant at the level of significance (*i*) 0.05 and (*ii*) 0.01.

2.32 A garage has a line of electronic tune-up systems. Another line has been installed and the manager would like to find out if the output of the new line is greater than that of the old one. Twelve days of data was gathered at random from the old line and 10 days of data are gathered at random from the new line, with $\overline{X}_1$ =1,150 cases and $\overline{X}_2$ = 1,175 cases. It is known that $\sigma_1^2 = 50$ and $\sigma_2^2 = 60$. Test the appropriate hypotheses at $\alpha = 0.05$ assuming that the outputs are normally distributed.

2.33 The time required to change flat tyres in a garage is a normal distributed random variable. The tyre changing times for 15 automobiles selected at random are as follows:

5 9 7 6 24 11 4 13 10 9 20 8 19 17 25

(*a*) at $\alpha = 0.05$, does it seem reasonable that the mean tyre changing time is 10 hours?

(*b*) does it seem reasonable that the mean tyre changing time is less than 10 hours?

(*c*) does it seem reasonable that the mean tyre changing time is greater than 10 hours?

2.34 The results of tensile tests on parts made of steel specimens are as follows. The load at which specimen failure in Mpa are: the sample mean $\overline{X}$ = 12, and the sample standard deviation is s = 3. The number of samples are 25. Do the data suggest that the mean load at failure exceeds 10 Mpa? Assume that the load at failure has normal distribution, and use $\alpha = 0.05$.

2.35 The maximum baking temperature for a certain component is supposed to be 90°C. The oven temperatures have a standard deviation of 10°C at the maximum baking temperature. A random sample of 228 readings is taken and the average baking temperature is found to be 94°C. Do the data suggest that the mean temperature at baking process exceeds 90°C? Assume that the baking process has a normal distribution and use $\alpha = 0.05$.

2.36 A manufacturing company was testing two machines for production automation purpose. The first machine was run for 8 hours and it produced an average of 136 items per hour with a standard deviation of 9 items. The second machine was run for 10 hours and produced an average of 127 items per hour with a standard deviation of 6 items. Assume that the production rates of both machines are normally distributed.

(*a*) make a 95% confidence interval for the difference between the two population means.

(*b*) using a 2.5% significance level, can it be concluded that the mean number of items produced per hour by the first machine is higher than that of the second machine?

2.37 The percentage of manganese in an alloy used in metal castings is measured in 50 randomly selected parts. The sample standard deviation is $s = 0.4$. Test the hypothesis H_0: $\sigma = 0.25$ versus H_1: $\sigma \neq 0.25$ using $\alpha = 0.05$.

2.38 The thickness of a machine part used in a system is its critical dimension and that measurements of the thickness of a random sample of 18 such parts have a variance of $s^2 = 0.6$, where the measurements are in thousandths of an inch. The process is considered good if the variation of the thickness is given by a variance not greater than 0.3. Assuming that the measurements contribute a random sample from a normal population, test the null hypotheses $\sigma^2 = 0.3$ against the alternative hypotheses $\sigma^2 > 0.3$ at the 0.05 level of significance.

2.39 In a random sample, the weights of 24-machined parts of a certain process have a standard deviation of 3 grams. Assume that the weights constitute a random sample from a normal population, test the null hypothesis $\sigma = 4$ grams against the two-sided alternative $\sigma \neq 4$ grams at the 0.01 level of significance.

2.40 A manufacturing company has two production lines that turn parts to a predetermined length. From a sample of 21 items taken from production line 1 and 30 items taken from production line 2, it was found that $s_1 = 15$ and $s_2 = 12$. Test the hypothesis that the two production lines have the same variance from the predetermined length. Take $\alpha = 0.05$.

2.41 A new voltage-regulating device is installed in an electromechanical system. Before its installation, a random sample yielded the following information about the percentage of breakdown of the system: $\overline{X}_1 = 8$, $s_1^2 = 36$, and $n_1 = 8$. After installation, a random sample yielded $\overline{X}_2 = 7$, $s_2^2 = 25$, and $n_2 = 9$.

(*a*) can you conclude that the variances are equal? Use $\alpha = 0.05$

(*b*) has the new voltage-regulating device reduced the machine system breakdowns significantly? Use $\alpha = 0.05$.

2.42 In bolted joints using gaskets the variability in the thickness of the gasket material is a critical characteristic of the bolted joint and a low variability is desirable for better performance of the bolted joint. Two different materials are being considered to determine whether one is superior in reducing the variability of the thickness. Twenty gaskets are selected. The sample standard deviation of gasket thickness are $s_1 = 2.96$ mm and $s_2 = 3.05$ mm, respectively. Is there any evidence to indicate that either gasket material is preferable? Use $\alpha = 0.05$.

2.43 A manufacturing firm promises a distributor that not more than 0.10 of those delivered toys will be blemished. A random sample of 500 toys delivered over the past 6 months yields 75 blemished toys. Should the toy manufacturer be declared in default at $\alpha = 0.05$?

2.44 In a survey of 1,000 random samples of machined parts, 700 were found to be non-defective. The manufacturing company wants at least 60% of all parts it produces are non-defective. Using $\alpha = 0.05$ significance level, can we conclude that the company's manufacturing policy is maintained?

2.45 A political analyst knows that a certain district voted 70% Republican in the last Congressional election. He also knows, however, that drastic changes in this district have occurred since the last election. He believes that these changes may have changed the percentage of voters who will vote for the Republican candidate for Congress, but doesn't know which way it will be changed. He polls 2,000 voters in the district and finds that 960 of these plan to vote for the Democratic candidate. Is his belief supported at 0.01 level of significance?

2.46 Two different polishing materials are being evaluated for use in machining operations in a manufacturing company. One hundred machine parts were polished using polishing material *A*, and of this number 87 had no polishing-induced defects. Another 100 machine parts were polished using polishing material *B* and 75 parts were satisfactory upon completion. Is there any reason to believe that the two polishing materials differ? Use $\alpha = 0.01$.

2.47 Two different types of grinding machines are used to finish metal parts. A part is considered defective if it has excessive surface roughness or variation. Two random samples, each of size 100 are selected and 5 defective parts are found in the sample from machine 1, while 3 defective parts are found in the sample from machine 2. Is it reasonable to conclude that both machines produce the same fraction of defective parts, using $\alpha = 0.05$?

2.48 A sample of 500 observations taken from the first population gave $\overline{X}_1 = 400$. Another sample of 600 observations taken from the second population gave $\overline{X}_2 = 420$.

(*a*) Find the point estimate of $\hat{p}_1 - \hat{p}_2$

(*b*) Make a 97% confidence interval for $\hat{p}_1 - \hat{p}_2$

(*c*) Show the rejection and non-rejection regions on the sampling distribution of $\hat{p}_1 - \hat{p}_2$ for $H_0: p_1 = p_2$ versus $H_1: p_1 > p_2$. Use a significance level of 2.5%

(*d*) Find the value of the test statistic z for the test of part (*c*)

(*e*) Do you reject the null hypothesis mentioned in part (*c*) at a significance level of 2.5%?

REFERENCES

Bowker, A.H. and Lieberman, G. J., *Engineering Statistics,* 2nd ed., Prentice-Hall, Englewood Cliffs, NJ, 1972.

Chambers, J., Cleveland, W., Kleiner, B. and Toukai, P., *Graphical Methods for Data Analysis*, Wadsworth and Brooks/Cole, Pacific Grove, CA, 1983.

Conover, W.J., *Practical Nonparametric Statistics*, 2nd ed., Wiley, New York, 1980.

Derman, C., Gleser, L. and Olkin, I., *Probability Models and Applications*, Macmillan, New York, 1980.

Devore, J.L., *Probability and Statistics for Engineering and the Sciences*, 4th ed., Wadsworth and Brooks/Cole, Pacific Grove, CA, 1995.

Efron, B. and Tibshirani, R., *An Introduction to the Bootstrap*, Chapman and Hall, New York, 1993.

Freedman, D., Pisani, R., Purves, R. and Adbikari, A., *Statistics*, 2nd ed., Norton, New York, 1991.

Guttman, I. and Wilks, S. S., *Introductory Engineering Statistics,* Wiley, New York, 1965.

Helsel, D.R. and Hirsch, R.M., *Statistical Methods in Water Resources,* Elsevier, New York, 1992.

Hines, W.W. and Montgomery, D.C., *Probability and Statistics in Engineering and Management Science*, 3rd ed., Wiley, New York, 1990.

Hoaglin, D.C., Mosteller, F. and Tukey, J.W., (Eds.) *Understanding Robust and Exploratory Data Analysis*, Wiley, New York, 1983.

Hoel, P.G., *Introduction to Mathematical Statistics*, 5th ed., Wiley, New York, 1984.

Hoel, P.G., Port, S.C. and Stone, C.J., *Introduction to Probability Theory*, Houghton Mifflin, Boston, 1971.

Hogg, R. and Craig, A., *Introduction to Mathematical Statistics*, 4th ed., Macmillan, New York, 1978.

Iman, R.L. *A Data-Based Approach to Statistics,* Duxbury Press, Belmont, CA, 1994.

Kennedy, J.B. and Neville, A.M., *Basic Statistical Methods for Engineers and Scientists*, 2nd ed., Harper and Row, New York, 1976.

Larsen, R. and Marx, M., *Introduction to Mathematical Statistics*, 2nd ed. Prentice-Hall, Englewood Cliffs, New Jersey, 1986.

Larson, H.J., *Introduction to Probability Theory and Statistical Inference*, 3rd ed., Wiley, New York, 1982.

Mann, P. S., *Introductory Statistics*, 3rd ed., Wiley, 1998.

Mason, R.L., Gunst, R.F. and Hess, J.L., *Statistical Design and Analysis of Experiments*, Wiley, New York, 1989.

Microsoft Excel User's Guide Version 5.0, *Microsoft,* Redmond, Washington, 1994.

Minitab Reference Manual, Release 11, *Minitab,* State College, PA, 1996.

Mosteller, F., Rourke, R. and Thomas, G., *Probability with Statistical Applications*, 2nd ed., Addison-Wesley, Reading, MA, 1970.

Ross, S., *A First Course in Probability*, 3rd ed., Macmillan, New York, 1985.

Ross, S., *Introduction to Probability and Statistics for Engineers and Scientists*, Wiley, New York, 1987.

Ryan, B.F., Joiner, B.L., and Ryan, T.A., *Minitab Handbook*, 2nd ed., PWS-Kent, Boston, 1985.

Tanur, J. et al. (eds.), *Statistics: A Guide to the Unknown*, 3rd ed., Wadsworth and Brooks/Cole, Pacific Grove, CA, 1989.

Tufte, E. R., *The Visual Display of Quantitative Information,* Graphics Press, Cheshire, Connecticut, 1983.

Tukey, J., *Exploratory Data Analysis*, Addison-Wesley, Reading, MA, 1977.

Walpole, R.E., Myers, R.H. and Myers, S.L., *Probability and Statistics for Engineers and Scientists*, 6th ed., Prentice-Hall, Upper Saddle River, New Jersey, 1998.

APPENDICES

Statistical and Numerical Tables

Appendix–2A
Values for *n* Factorial

n	$n!$	n	$n!$	n	$n!$
0	1 (by definition)	40	8.15915×10^{47}	80	7.15695×10^{118}
1	1	41	3.34525×10^{49}	81	5.79713×10^{120}
2	2	42	1.40501×10^{51}	82	4.75364×10^{122}
3	6	43	6.04153×10^{52}	83	3.94552×10^{124}
4	24	44	2.65827×10^{54}	84	3.31424×10^{126}
5	120	45	1.19622×10^{56}	85	2.81710×10^{128}
6	720	46	5.50262×10^{57}	86	2.42271×10^{130}
7	5040	47	2.58623×10^{59}	87	2.10776×10^{132}
8	40,320	48	1.24139×10^{61}	88	1.85483×10^{134}
9	362,880	49	6.08282×10^{62}	89	1.65080×10^{136}
10	3,628,800	50	3.04141×10^{64}	90	1.48572×10^{138}
11	39,916,800	51	1.55112×10^{65}	91	1.35200×10^{140}
12	479,001,600	52	8.06582×10^{67}	92	1.24384×10^{142}
13	6,227,020,800	53	4.27488×10^{69}	93	1.15677×10^{144}
14	87,178,291,200	54	2.30844×10^{71}	94	1.08737×10^{146}
15	1,307,674,368,000	55	1.26964×10^{73}	95	1.03300×10^{148}
16	20,922,789,888,000	56	7.10999×10^{74}	96	9.91678×10^{149}
17	355,687,428,096,000	57	4.05269×10^{76}	97	9.61928×10^{151}
18	6,402,373,705,728,000	58	2.35056×10^{78}	98	9.42689×10^{153}
19	121,645,100,408,832,000	59	1.38683×10^{80}	99	9.33262×10^{155}
20	2,432,902,008,176,640,000	60	8.32099×10^{81}	100	9.33262×10^{157}
21	51,090,942,171,709,440,000	61	5.07580×10^{83}		
22	1,124,000,727,777,607,680,000	62	3.14700×10^{85}		
23	25,852,016,738,884,976,640,000	63	1.98261×10^{87}		
24	620,448,401,733,239,439,360,000	64	1.26887×10^{89}		
25	5,511,210,043,330,985,984,000,000	65	8.24765×10^{90}		
26	403,291,461,126,605,635,584,000,000	66	5.44345×10^{92}		
27	10,888,869,450,418,352,160,768,000,000	67	3.64711×10^{94}		
28	304,888,344,611,713,860,501,504,000,000	68	2.48004×10^{96}		
29	8,841,761,993,739,701,954,543,616,000,000	69	1.71122×10^{98}		
30	265,252,859,812,191,058,636,308,480,000,000				
31	8.22284×10^{33}	70	1.19786×10^{100}		
32	2.63131×10^{35}	71	8.50479×10^{101}		
33	8.68332×10^{36}	72	6.12345×10^{103}		
34	2.95233×10^{38}	73	4.47012×10^{105}		
35	1.03331×10^{40}	74	3.30789×10^{107}		
36	3.71993×10^{41}	75	2.48091×10^{109}		
37	1.37638×10^{43}	76	1.88549×10^{111}		
38	5.23023×10^{44}	77	1.45183×10^{113}		
39	2.03979×10^{46}	78	1.13243×10^{115}		
		79	8.94618×10^{116}		

Appendix–2 B
Binomial Coefficients

$$\binom{n}{k} = \frac{n!}{k!(n-k)!} = \frac{n(n-1)\ldots(n-k+1)}{k!} = \binom{n}{n-k},\ 0! = 1$$

For $k > 20$, use the fact that $\binom{n}{k} = \binom{n}{n-k}$

n \ x	0	1	2	3	4	5	6	7	8	9	10
1	1	1									
2	1	2	1								
3	1	3	3	1							
4	1	4	6	4	1						
5	1	5	10	10	5	1					
6	1	6	15	20	15	6	1				
7	1	7	21	35	35	21	7	1			
8	1	8	28	56	70	56	28	8	1		
9	1	9	36	84	126	126	84	36	9	1	
10	1	10	45	120	210	252	210	120	45	10	1
11	1	11	55	165	330	462	462	330	165	55	11
12	1	12	66	220	495	792	924	792	495	220	66
13	1	13	78	286	715	1,287	1,716	1,716	1,287	715	286
14	1	14	91	364	1,001	2,002	3,003	3,432	3,003	2,002	1,001
15	1	15	105	455	1,365	3,003	5,005	6,435	6,435	5,005	3,003
16	1	16	120	560	1,820	4,368	8,008	11,440	12,870	11,440	8,008
17	1	17	136	680	2,380	6,188	12,376	19,448	24,310	24,310	19,448
18	1	18	153	816	3,060	8,568	18,564	31,824	43,758	48,620	43,758
19	1	19	171	969	3,876	11,628	27,132	50,388	75,582	92,378	92,378
20	1	20	190	1,140	4,485	15,504	38,760	77,520	125,970	167,960	184,756
21	1	21	210	1.330	5.985	20,349	54,264	116,280	203,490	293,930	352,716
22	1	22	231	1,540	7,315	26,334	74,613	170,544	319,770	497,420	646,646
23	1	23	253	1,771	8,855	33,649	100,947	245,157	490,314	817,190	1,144,066
24	1	24	276	2,024	10,626	42,504	134,596	346,104	735,471	1,307,504	1,961,256
25	1	25	300	2,300	12,650	53,330	177,100	480,700	1,081,575	2,042,975	3,268,760

Appendix–2 B (Continued)
Binomial Coefficients

n \ x	11	12	13	14	15	16	17	18	19	20
1										
2										
3										
4										
5										
6										
7										
8										
9										
10										
11	1									
12	12	1								
13	78	13	1							
14	364	91	14	1						
15	1,365	455	105	15	1					
16	4,368	1,820	560	120	16	1				
17	12,376	6,188	2,380	680	136	17	1			
18	31,824	18,564	8,568	3,060	816	153	18	1		
19	75,582	50,388	27,132	11,628	3,876	969	171	19	1	
20	167,960	125,970	77,520	38,760	15,504	4,845	1,140	190	20	1
21	352,716	293,930	203,490	116,280	54,264	20,349	5,985	1,330	210	21
22	705,432	646,646	497,420	319,770	170,544	74,613	26,334	7,315	1,540	231
23	1,352,078	1,352,078	1,144,066	817,190	490,314	245,157	100,947	33,649	8,855	1,771
24	2,496,144	2,704,156	2,496,144	1,961,256	1,307,504	735,471	346,104	134,596	42,504	10,626
25	4,457,400	5,200,300	5,200,300	4,457,400	3,268,760	2,042,975	1,081,575	480,700	177,100	53,130

Appendix–2C
Cumulative Binomial Probabilities

n	*x*	p										
		.05	.10	.20	.30	.40	.50	.60	.70	.80	.90	.95
1	0	.9500	.9000	.8000	.7000	.6000	.5000	.4000	.3000	.2000	.1000	.0500
	1	.0500	.1000	.2000	.3000	.4000	.5000	.6000	.7000	.8000	.9000	.9500
2	0	.9025	.8100	.6400	.4900	.3600	.2500	.1600	.0900	.0400	.0100	.0025
	1	.0950	.1800	.3200	.4200	.4800	.5000	.4800	.4200	.3200	.1800	.0950
	2	.0025	.0100	.0400	.0900	.1600	.2500	.2600	.4900	.6400	.8100	.9025
3	0	.8574	.7290	.5120	.3430	.2160	.1250	.0640	.0270	.0080	.0010	.0001
	1	.1354	.2430	.3840	.4410	.4320	.3750	.2880	.1890	.0960	.0270	.0071
	2	.0071	.0270	.0960	.1890	.2880	.3750	.4320	.4410	.3840	.2430	.1354
	3	.0001	.0010	.0080	.0270	.0640	.1250	.2160	.3430	.5120	.7290	.8574
4	0	.8145	.6561	.4096	.2401	.1296	.0625	.0256	.0081	.0016	.0001	.0000
	1	.1715	.2916	.4096	.4116	.3456	.2500	.1536	.0756	.0256	.0036	.0005
	2	.0135	.0486	.1536	.2646	.3456	.3750	.3456	.2646	.1536	.0486	.0135
	3	.0005	.0036	.0256	.0756	.1536	.2500	.3456	.4116	.4096	.2916	.1715
	4	.0000	.0001	.0016	.0081	.0256	.0625	.1296	.2401	.4096	.6561	.8145
5	0	.7738	.5905	.3277	.1681	.0778	.0312	.0102	.0024	.0003	.0000	.0000
	1	.2036	.3280	.4096	.3602	.2592	.1562	.0768	.0284	.0064	.0005	.0000
	2	.0214	.0729	.2048	.3087	.3456	.3125	.2304	.1323	.0512	.0081	.0011
	3	.0011	.0081	.0512	.1323	.2304	.3125	.3456	.3087	.2048	.0729	.0214
	4	.0000	.0004	.0064	.0283	.0768	.1562	.2592	.3601	.4096	.3281	.2036
	5	.0000	.0000	.0003	.0024	.0102	.0312	.0778	.1681	.3277	.5905	.7738
6	0	.7351	.5314	.2621	.1176	.0467	.0156	.0041	.0007	.0001	.0000	.0000
	1	.2321	.3543	.3932	.3025	.1866	.0937	.0369	.0102	.0015	.0001	.0000
	2	.0305	.0984	.2458	.3241	.3110	.2344	.1382	.0595	.0154	.0012	.0001
	3	.0021	.0146	.0819	.1852	.2765	.3125	.2765	.1852	.0819	.0146	.0021
	4	.0001	.0012	.0154	0.595	.1382	.2344	.3110	.3241	.2458	.0984	.0305
	5	.0000	.0001	.0015	.0102	.0369	.0937	.1866	.3025	.3932	.3543	.2321
	6	.0000	.0000	.0001	.0007	.0041	.0156	.0467	.1176	.2621	.5314	.7351
7	0	.6983	.4783	.2097	.0824	.0280	.0078	.0016	.0002	.0000	.0000	.0000
	1	.2573	.3720	.3670	.2171	.1306	.0547	.0172	.0036	.0004	.0000	.0000
	2	.0406	.1240	.2670	.0177	.2613	.1641	.0774	.0250	.0043	.0002	.0000
	3	.0036	.0230	.1147	.2269	.2903	.2734	.1935	.0972	.0287	.0026	.0002
	4	.0002	.0026	.0287	.0972	.1935	.2734	.2903	.2269	.1147	.0230	.0036
	5	.0000	.0002	.0043	.0250	.0774	.1641	.2613	.3177	.2753	.1240	.0406
	6	.0000	.0000	.0004	.0086	.0172	.0547	.1306	.2471	.3670	.3720	.2573
	7	.0000	.0000	.0000	.0002	.0016	.0078	.0280	.0824	.2097	.4783	.6983

...*(Contd.)*

Appendix–2C (Continued)
Cumulative Binomial Probabilities

n	x	p .05	.10	.20	.30	.40	.50	.60	.70	.80	.90	.95
8	0	.6634	.4305	.1678	.0576	.0168	.0039	.0007	.0001	.0000	.0000	.0000
	1	.2793	.3826	.3355	.1977	.0896	.0312	.0079	.0012	.0001	.0000	.0000
	2	.0515	.1488	.2936	.2965	2090	.1094	.0413	.0100	.0011	.0000	.0000
	3	.0054	.0331	.1468	.2541	.2787	.2187	.1239	.0467	.0092	.0004	.0000
	4	.0004	.0046	.0459	.1361	.2322	.2734	.2322	.1361	.0459	.0046	.0004
	5	.0000	.0004	.0092	.0467	.1239	.2187	.2787	.2541	.1468	.0331	.0054
	6	.0000	.0000	.0011	.0100	.0413	.1094	.2090	.2965	.2936	.1488	.0515
	7	.0000	.0000	.0001	.0012	.0079	.0312	.0896	.1977	.3355	.3826	.2793
	8	.0000	.0000	.0000	.0001	.0007	.0039	.0168	.0576	.1678	.4305	.6634
9	0	.6302	.3874	.1342	.0404	.0101	.0020	.0003	.0000	.0000	.0000	.0000
	1	.2985	.3874	.3020	.1556	.0605	.0176	.0035	.0004	.0000	.0000	.0000
	2	.0629	.1722	.3020	.2668	.1612	.0703	.0212	.0039	.0003	.0000	.0000
	3	.0077	.0446	.1762	.2668	.2508	.1641	.0743	.0210	.0028	.0001	.0000
	4	.0006	.0074	.0661	.1715	.2508	.2461	.1672	.0735	.0165	.0008	.0000
	5	.0000	.0008	.0165	.0735	.1672	.2461	.2508	.1715	.0661	.0074	.0006
	6	.0000	.0001	.0028	.0210	.0743	.1641	.2508	.2668	.1762	.0446	.0077
	7	.0000	.0000	.0003	.0039	.0212	.0703	.1612	.2668	.3020	.1722	.0629
	8	.0000	.0000	.0000	.0004	.0035	.0176	.0605	.1556	.3020	.3874	.2985
	9	.0000	.0000	.0000	.0000	.0003	.0020	.0101	.0404	.1342	.3874	.6302
10	0	.5987	.3487	.1074	.0282	.0060	.0010	.0001	.0000	.0000	.0000	.0000
	1	.3151	.3874	.2684	.1211	.0403	.0098	.0016	.0001	.0000	.0000	.0000
	2	.0746	.1937	.3020	.2335	.1209	.0439	.0106	.0014	.0001	.0000	.0000
	3	.0105	.0574	.2013	.2668	.2150	.1172	.0425	.0090	.0008	.0000	.0000
	4	.0010	.0112	.0881	.2001	.2508	.2051	.1115	.0368	.0055	.0001	.0000
	5	.0001	.0015	.0264	.1029	.2007	.2461	.2007	.1029	.0264	.0015	.0001
	6	.0000	.0001	.0055	.0368	.1115	.2051	.2508	.2001	.0881	.0112	.0010
	7	.0000	.0000	.0008	.0090	.0425	.1172	.2150	.2668	.2013	.0574	.0105
	8	.0000	.0000	.0001	.0014	.0106	.0439	.1209	.2335	.3020	.1937	.0746
	9	.0000	.0000	.0000	.0001	.0016	.0098	.0403	.1211	.2684	.3874	.3151
	10	.0000	.0000	.0000	.0000	.0001	.0010	.0060	.0282	.1074	.3487	.5987
11	0	.5688	.3138	.0859	.0198	.0036	.0005	.0000	.0000	.0000	.0000	.0000
	1	.3293	.3835	.2362	.0932	.0266	.0054	.0007	.0000	.0000	.0000	.0000
	2	.0867	.2131	.2953	.1998	.0887	.0269	.0052	.0005	.0000	.0000	.0000
	3	.0137	.0710	.2215	.2568	.1774	.0806	.0234	.0037	.0002	.0000	.0000
	4	.0014	.0158	.1107	.2201	.2365	.1611	.0701	.0173	.0017	.0000	.0000
	5	.0001	.0025	.0388	.1321	.2207	.2256	.1471	.0566	.0097	.0003	.0000
	6	.0000	.0003	.0097	.0566	.1471	.2256	.2207	.1321	.0388	.0025	.0001
	7	.0000	.0000	.0017	.0173	.0701	.1611	.2365	.2201	.1107	.0158	.0014
	8	.0000	.0000	.0002	.0037	.0234	.0806	.1774	.2568	.2215	.0710	.0137
	9	.0000	.0000	.0000	.0005	.0052	.0269	.0887	.1998	.2953	.2131	.0867
	10	.0000	.0000	.0000	.0000	.0007	.0054	.0266	.0932	.2362	.3835	.3293
	11	.0000	.0000	.0000	.0000	.0000	.0005	.0036	.0198	.0859	.3138	.5688

...(Contd.)

Appendix–2C (Continued)
Cumulative Binomial Probabilities

n	x	p .05	.10	.20	.30	.40	.50	.60	.70	.80	.90	.95
12	0	.5404	.2824	.0687	.0138	.0022	.0002	.0000	.0000	.0000	.0000	.0000
	1	.3413	.3766	.2062	.0712	.0174	.0029	.0003	.0000	.0000	.0000	.0000
	2	.0988	.2301	.2835	.1678	.0639	.0161	.0025	.0002	.0000	.0000	.0000
	3	.0173	.0852	.2362	.2397	.1419	.0537	.0125	.0015	.0001	.0000	.0000
	4	.0021	.0213	.1329	.2311	.2128	.1208	.0420	.0078	.0005	.0000	.0000
	5	.0002	.0038	.0532	.1585	.2270	.1934	.1009	.0291	.0033	.0000	.0000
	6	.0000	.0005	.0155	.0792	.1766	.2256	.1766	.0792	.0155	.0005	.0000
	7	.0000	.0000	.0033	.0291	.1009	.1934	.2270	.1585	.0532	.0038	.0002
	8	.0000	.0000	.0005	.0078	.0420	.1208	.2128	.2311	.1329	.0213	.0021
	9	.0000	.0000	.0001	.0015	.0125	.0537	.1419	.2397	.2362	.0852	.0173
	10	.0000	.0000	.0000	.0002	.0025	.0161	.0639	.1678	.2835	.2301	.0988
	11	.0000	.0000	.0000	.0000	.0003	.0029	.0174	.0712	.2062	.3766	.3413
	12	.0000	.0000	.0000	.0000	.0000	.0002	.0022	.0138	.0687	.2824	.5404
13	0	.5133	.2542	.0550	.0097	.0013	.0001	.0000	.0000	.0000	.0000	.0000
	1	.3512	.3672	.1787	.0540	.0113	.0016	.0001	.0000	.0000	.0000	.0000
	2	.1109	.2448	.2680	.1388	.0453	.0095	.0012	.0001	.0000	.0000	.0000
	3	.0214	.0997	.2457	.2181	.1107	.0349	.0065	.0006	.0000	.0000	.0000
	4	.0028	.0277	.1535	.2337	.1845	.0873	.0243	.0034	.0001	.0000	.0000
	5	.0003	.0055	.0691	.1803	.2214	.1571	.0656	.0142	.0011	.0000	.0000
	6	.0000	.0008	.0230	.1030	.1968	.2095	.1312	.0442	.0058	.0001	.0000
	7	.0000	.0001	.0058	.0442	.1312	.2095	.1968	.1030	.0230	.0008	.0000
	8	.0000	.0000	.0011	.0142	.0656	.1571	.2214	.1803	.0691	.0055	.0003
	9	.0000	.0000	.0001	.0034	.0243	.0873	.1845	.2337	.1535	.0277	.0028
	10	.0000	.0000	.0000	.0006	.0065	.0349	.1107	.2181	.2457	.0997	.0214
	11	.0000	.0000	.0000	.0001	.0012	.0095	.0453	.1388	.2680	.2448	.1109
	12	.0000	.0000	.0000	.0000	.0001	.0016	.0113	.0540	.1787	.3672	.3512
	13	.0000	.0000	.0000	.0000	.0000	.0001	.0013	.0097	.0550	.2542	.5133
14	0	.4877	.2288	.0440	.0068	.0008	.0001	.0000	.0000	.0000	.0000	.0000
	1	.3593	.3559	.1539	.0407	.0073	.0009	.0001	.0000	.0000	.0000	.0000
	2	.1229	.2570	.2501	.1134	.0317	.0056	.0005	.0000	.0000	.0000	.0000
	3	.0259	.1142	.2501	.1943	.0845	.0222	.0033	.0002	.0000	.0000	.0000
	4	.0037	.0349	.1720	.2290	.1549	.0611	.0136	.0014	.0000	.0000	.0000
	5	.0004	.0078	.0860	.1963	.2066	.1222	.0408	.0066	.0003	.0000	.0000
	6	.0000	.0013	.0322	.1262	.2066	.1833	.0918	.0232	.0020	.0000	.0000
	7	.0000	.0002	.0092	.0618	.1574	.2095	.1574	.0618	.0092	.0002	.0000
	8	.0000	.0000	.0020	.0232	.0918	.1833	.2066	.1262	.0322	.0013	.0000
	9	.0000	.0000	.0003	.0066	.0408	.1222	.2066	.1963	.0860	.0078	.0004
	10	.0000	.0000	.0000	.0014	.0136	.0611	.1549	.2290	.1720	.0349	.0037
	11	.0000	.0000	.0000	.0002	.0033	.0222	.0845	.1943	.2501	.1142	.0259
	12	.0000	.0000	.0000	.0000	.0005	.0056	.0317	.1134	.2501	.2570	.1229
	13	.0000	.0000	.0000	.0000	.0001	.0009	.0073	.0407	.1539	.3559	.3593
	14	.0000	.0000	.0000	.0000	.0000	.0001	.0008	.0068	.0440	.2288	.4877

...(Contd.)

Appendix–2C (Continued)
Cumulative Binomial Probabilities

n	*x*	p										
		.05	**.10**	**.20**	**.30**	**.40**	**.50**	**.60**	**.70**	**.80**	**.90**	**.95**
15	0	.4633	.2059	.0352	.0047	.0005	.0000	.0000	.0000	.0000	.0000	.0000
	1	.3658	.3432	.1319	.0305	.0047	.0005	.0000	.0000	.0000	.0000	.0000
	2	.1348	.2669	.2309	.0916	.0219	.0032	.0003	.0000	.0000	.0000	.0000
	3	.0307	.1285	.2501	.1700	.0634	.0139	.0016	.0001	.0000	.0000	.0000
	4	.0049	.0428	.1876	.2186	.1268	.0417	.0074	.0006	.0000	.0000	.0000
	5	.0006	.0105	.1032	.2061	.1859	.0916	.0245	.0030	.0001	.0000	.0000
	6	.0000	.0019	.0430	.1472	.2066	.1527	.0612	.0116	.0007	.0000	.0000
	7	.0000	.0003	.0138	.0811	.1771	.1964	.1181	.0348	.0035	.0000	.0000
	8	.0000	.0000	.0035	.0348	.1181	.1964	.1771	.0811	.0138	.0003	.0000
	9	.0000	.0000	.0007	.0116	.0612	.1527	.2066	.1472	.0430	.0019	.0000
	10	.0000	.0000	.0001	.0030	.0245	.0916	.1859	.2061	.1032	.0105	.0006
	11	.0000	.0000	.0000	.0006	.0074	.0417	.1268	.2186	.1876	.0428	.0049
	12	.0000	.0000	.0000	.0001	.0016	.0139	.0634	.1700	.2501	.1285	.0307
	13	.0000	.0000	.0000	.0000	.0003	.0032	.0219	.0916	.2309	.2669	.1348
	14	.0000	.0000	.0000	.0000	.0000	.0005	.0047	.0305	.1319	.3432	.3658
	15	.0000	.0000	.0000	.0000	.0000	.0000	.0005	.0047	.0352	.2059	.4633
16	0	.4401	.1853	.0281	.0033	.0003	.0000	.0000	.0000	.0000	.0000	.0000
	1	.3706	.3294	.1126	.0228	.0030	.0002	.0000	.0000	.0000	.0000	.0000
	2	.1463	.2745	.2111	.0732	.0150	.0018	.0001	.0000	.0000	.0000	.0000
	3	.0359	.1423	.2463	.1465	.0468	.0085	.0008	.0000	.0000	.0000	.0000
	4	.0061	.0514	.2001	.2040	.1014	.0278	.0040	.0002	.0000	.0000	.0000
	5	.0008	.0137	.1201	.2099	.1623	.0667	.0142	.0013	.0000	.0000	.0000
	6	.0001	.0028	.0550	.1649	.1983	.1222	.0392	.0056	.0002	.0000	.0000
	7	.0000	.0004	.0197	.1010	.1889	.1746	.0840	.0185	.0012	.0000	.0000
	8	.0000	.0001	.0055	.0487	.1417	.1964	.1417	.0487	.0055	.0001	.0000
	9	.0000	.0000	.0012	.0185	.0840	.1746	.1889	.1010	.0197	.0004	.0000
	10	.0000	.0000	.0002	.0056	.0392	.1222	.1983	.1649	.0550	.0028	.0001
	11	.0000	.0000	.0000	.0013	.0142	.0666	.1623	.2099	.1201	.0137	.0008
	12	.0000	.0000	.0000	.0002	.0040	.0278	.1014	.2040	.2001	.0514	.0061
	13	.0000	.0000	.0000	.0000	.0008	.0085	.0468	.1465	.2463	.1423	.0359
	14	.0000	.0000	.0000	.0000	.0001	.0018	.0150	.0732	.2111	.2745	.1463
	15	.0000	.0000	.0000	.0000	.0000	.0002	.0030	.0228	.1126	.3294	.3706
	16	.0000	.0000	.0000	.0000	.0000	.0000	.0003	.0033	.0281	.1853	.4401

...(*Contd.*)

Appendix-2C (Continued)
Cumulative Binomial Probabilities

n	x	p .05	.10	.20	.30	.40	.50	.60	.70	.80	.90	.95
17	0	.4181	.1668	.0225	.0023	.0002	.0000	.0000	.0000	.0000	.0000	.0000
	1	.3741	.3150	.0957	.0169	.0019	.0001	.0000	.0000	.0000	.0000	.0000
	2	.1575	.2800	.1914	.0581	.0102	.0010	.0001	.0000	.0000	.0000	.0000
	3	.0415	.1556	.2393	.1245	.0341	.0052	.0004	.0000	.0000	.0000	.0000
	4	.0076	.0605	.2093	.1868	.0796	.0182	.0021	.0001	.0000	.0000	.0000
	5	.0010	.0175	.1361	.2081	.1379	.0472	.0081	.0006	.0000	.0000	.0000
	6	.0001	.0039	.0680	.1784	.1839	.0944	.0242	.0026	.0001	.0000	.0000
	7	.0000	.0007	.0267	.1201	.1927	.1484	.0571	.0095	.0004	.0000	.0000
	8	.0000	.0001	.0084	.0644	.1606	.1855	.1070	.0276	.0021	.0000	.0000
	9	.0000	.0000	.0021	.0276	.1070	.1855	.1606	.0644	.0084	.0001	.0000
	10	.0000	.0000	.0004	.0095	.0571	.1484	.1927	.1201	.0267	.0007	.0000
	11	.0000	.0000	.0001	.0026	.0242	.0944	.1839	.1784	.0680	.0039	.0001
	12	.0000	.0000	.0000	.0006	.0081	.0472	.1379	.2081	.1361	.0175	.0010
	13	.0000	.0000	.0000	.0001	.0021	.0182	.0796	.1868	.2093	.0605	.0076
	14	.0000	.0000	.0000	.0000	.0004	.0052	.0341	.1245	.2393	.1556	.0415
	15	.0000	.0000	.0000	.0000	.0001	.0010	.0102	.0581	.1914	.2800	.1575
	16	.0000	.0000	.0000	.0000	.0000	.0001	.0019	.0169	.0957	.3150	.3741
	17	.0000	.0000	.0000	.0000	.0000	.0000	.0002	.0023	.0225	.1668	.4181
18	0	.3972	.1501	.0180	.0016	.0001	.0000	.0000	.0000	.0000	.0000	.0000
	1	.3763	.3002	.0811	.0126	.0012	.0001	.0000	.0000	.0000	.0000	.0000
	2	.1683	.2835	.1723	.0458	.0069	.0006	.0000	.0000	.0000	.0000	.0000
	3	.0473	.1680	.2297	.1046	.0246	.0031	.0002	.0000	.0000	.0000	.0000
	4	.0093	.0700	.2153	.1681	.0614	.0117	.0011	.0000	.0000	.0000	.0000
	5	.0014	.0218	.1507	.2017	.1146	.0327	.0045	.0002	.0000	.0000	.0000
	6	.0002	.0052	.0816	.1873	.1655	.0708	.0145	.0012	.0000	.0000	.0000
	7	.0000	.0010	.0350	.1376	.1892	.1214	.0374	.0046	.0001	.0000	.0000
	8	.0000	.0002	.0120	.0811	.1734	.1669	.0771	.0149	.0008	.0000	.0000
	9	.0000	.0000	.0033	.0386	.1284	.1855	.1284	.0386	.0033	.0000	.0000
	10	.0000	.0000	.0008	.0149	.0771	.1669	.1734	.0811	.0120	.0002	.0000
	11	.0000	.0000	.0001	.0046	.0374	.1214	.1892	.1376	.0350	.0010	.0000
	12	.0000	.0000	.0000	.0012	.0145	.0708	.1655	.1873	.0816	.0052	.0002
	13	.0000	.0000	.0000	.0002	.0045	.0327	.1146	.2017	.1507	.0218	.0014
	14	.0000	.0000	.0000	.0000	.0011	.0117	.0614	.1681	.2153	.0700	.0093
	15	.0000	.0000	.0000	.0000	.0002	.0031	.0246	.1046	.2297	.1680	.0473
	16	.0000	.0000	.0000	.0000	.0000	.0006	.0069	.0458	.1723	.2835	.1683
	17	.0000	.0000	.0000	.0000	.0000	.0001	.0012	.0126	.0811	.3002	.3763
	18	.0000	.0000	.0000	.0000	.0000	.0000	.0001	.0016	.0180	.1501	.3972

...(Contd.)

Appendix–2C (Continued)
Cumulative Binomial Probabilities

n	*x*	p										
		.05	.10	.20	.30	.40	.50	.60	.70	.80	.90	.95
19	0	.3774	.1351	.0144	.0011	.0001	.0000	.0000	.0000	.0000	.0000	.0000
	1	.3774	.2852	.0685	.0093	.0008	.0000	.0000	.0000	.0000	.0000	.0000
	2	.1787	.2852	.1540	.0358	.0046	.0003	.0000	.0000	.0000	.0000	.0000
	3	.0533	.1796	.2182	.0869	.0175	.0018	.0001	.0000	.0000	.0000	.0000
	4	.0112	.0798	.2182	.1491	.0467	.0074	.0005	.0000	.0000	.0000	.0000
	5	.0018	.0266	.1636	.1916	.0933	.0222	.0024	.0001	.0000	.0000	.0000
	6	.0002	.0069	.0955	.1916	.1451	.0518	.0085	.0005	.0000	.0000	.0000
	7	.0000	.0014	.0443	.1525	.1797	.0961	.0237	.0022	.0000	.0000	.0000
	8	.0000	.0002	.0166	.0981	.1797	.1442	.0532	.0077	.0003	.0000	.0000
	9	.0000	.0000	.0051	.0514	.1464	.1762	.0976	.0220	.0013	.0000	.0000
	10	.0000	.0000	.0013	.0220	.0976	.1762	.1464	.0514	.0051	.0000	.0000
	11	.0000	.0000	.0003	.0077	.0532	.1442	.1797	.0981	.0166	.0002	.0000
	12	.0000	.0000	.0000	.0022	.0237	.0961	.1797	.1525	.0443	.0014	.0000
	13	.0000	.0000	.0000	.0005	.0085	.0518	.1451	.1916	.0955	.0069	.0002
	14	.0000	.0000	.0000	.0001	.0024	.0222	.0933	.1916	.1636	.0266	.0018
	15	.0000	.0000	.0000	.0000	.0005	.0074	.0467	.1491	.2182	.0798	.0112
	16	.0000	.0000	.0000	.0000	.0001	.0018	.0175	.0869	.2182	.1796	.0533
	17	.0000	.0000	.0000	.0000	.0000	.0003	.0046	.0358	.1540	.2852	.1787
	18	.0000	.0000	.0000	.0000	.0000	.0000	.0008	.0093	.0685	.2852	.3774
	19	.0000	.0000	.0000	.0000	.0000	.0000	.0001	.0011	.0144	.1351	.3774
20	0	.3585	.1216	.0115	.0008	.0000	.0000	.0000	.0000	.0000	.0000	.0000
	1	.3774	.2702	.0576	.0068	.0005	.0000	.0000	.0000	.0000	.0000	.0000
	2	.1887	.2852	.1369	.0278	.0031	.0002	.0000	.0000	.0000	.0000	.0000
	3	.0596	.1901	.2054	.0716	.0123	.0011	.0000	.0000	.0000	.0000	.0000
	4	.0133	.0898	.2182	.1304	.0350	.0046	.0003	.0000	.0000	.0000	.0000
	5	.0022	.0319	.1746	.1789	.0746	.0148	.0013	.0000	.0000	.0000	.0000
	6	.0003	.0089	.1091	.1916	.1244	.0370	.0049	.0002	.0000	.0000	.0000
	7	.0000	.0020	.0545	.1643	.1659	.0739	.0146	.0010	.0000	.0000	.0000
	8	.0000	.0004	.0222	.1144	.1797	.1201	.0355	.0039	.0001	.0000	.0000
	9	.0000	.0001	.0074	.0654	.1597	.1602	.0710	.0120	.0005	.0000	.0000
	10	.0000	.0000	.0020	.0308	.1171	.1762	1171	.0308	.0020	.0000	.0000
	11	.0000	.0000	.0005	.0120	.0710	.1602	.1597	.0654	.0074	.0001	.0000
	12	.0000	.0000	.0001	.0039	.0355	.1201	.1797	.1144	.0222	.0004	.0000
	13	.0000	.0000	.0000	.0010	.0146	.0739	.1659	.1643	.0545	.0020	.0000
	14	.0000	.0000	.0000	.0002	.0049	.0370	.1244	.1916	.1091	.0089	.0003
	15	.0000	.0000	.0000	.0000	.0013	.0148	.0746	.1789	.1746	.0319	.0022
	16	.0000	.0000	.0000	.0000	.0003	.0046	.0350	.1304	.2182	.0898	.0133
	17	.0000	.0000	.0000	.0000	.0000	.0011	.0123	.0716	.2054	.1901	.0596
	18	.0000	.0000	.0000	.0000	.0000	.0002	.0031	.0278	.1369	.2852	.1887
	19	.0000	.0000	.0000	.0000	.0000	.0000	.0005	.0068	.0576	.2702	.3774
	20	.0000	.0000	.0000	.0000	.0000	.0000	.0000	.0008	.0115	.1216	.3585

...(*Contd.*)

Appendix-2C (Continued)
Cumulative Binomial Probabilities

n	x	p .05	.10	.20	.30	.40	.50	.60	.70	.80	.90	.95
21	0	.3406	.1094	.0092	.0006	.0000	.0000	.0000	.0000	.0000	.0000	.0000
	1	.3764	.2553	.0484	.0050	.0003	.0000	.0000	.0000	.0000	.0000	.0000
	2	.1981	.2837	.1211	.0215	.0020	.0001	.0000	.0000	.0000	.0000	.0000
	3	.0660	.1996	.1917	.0585	.0086	.0006	.0000	.0000	.0000	.0000	.0000
	4	.0156	.0998	.2156	.1128	.0259	.0029	.0001	.0000	.0000	.0000	.0000
	5	.0028	.0377	.1833	.1643	.0588	.0097	.0007	.0000	.0000	.0000	.0000
	6	.0004	.0112	.1222	.1878	.1045	.0259	.0027	.0001	.0000	.0000	.0000
	7	.0000	.0027	.0655	.1725	.1493	.0554	.0087	.0005	.0000	.0000	.0000
	8	.0000	.0005	.0286	.1294	.1742	.0970	.0229	.0019	.0000	.0000	.0000
	9	.0000	.0001	.0103	.0801	.1677	.1402	.0497	.0063	.0002	.0000	.0000
	10	.0000	.0000	.0031	.0412	.1342	.1682	.0895	.0176	.0008	.0000	.0000
	11	.0000	.0000	.0008	.0176	.0895	.1682	.1342	.0412	.0031	.0000	.0000
	12	.0000	.0000	.0002	.0063	.0497	.1402	.1677	.0801	.0103	.0001	.0000
	13	.0000	.0000	.0000	.0019	.0229	.0970	.1742	.1294	.0286	.0005	.0000
	14	.0000	.0000	.0000	.0005	.0087	.0554	.1493	.1725	.0655	.0027	.0000
	15	.0000	.0000	.0000	.0001	.0027	.0259	.1045	.1878	.1222	.0112	.0004
	16	.0000	.0000	.0000	.0000	.0007	.0097	.0588	.1143	.1833	.0377	.0028
	17	.0000	.0000	.0000	.0000	.0001	.0029	.0259	.1128	.2156	.0998	.0156
	18	.0000	.0000	.0000	.0000	.0000	.0006	.0086	.0585	.1917	.1996	.0660
	19	.0000	.0000	.0000	.0000	.0000	.0001	.0020	.0215	.1211	.2837	.1981
	20	.0000	.0000	.0000	.0000	.0000	.0000	.0003	.0050	.0484	.2553	.3764
	21	.0000	.0000	.0000	.0000	.0000	.0000	.0000	.0006	.0092	.1094	.3406
22	0	.3235	.0985	.0074	.0004	.0000	.0000	.0000	.0000	.0000	.0000	.0000
	1	.3746	.2407	.0406	.0037	.0002	.0000	.0000	.0000	.0000	.0000	.0000
	2	.2070	.2808	.1065	.0166	.0014	.0001	.0000	.0000	.0000	.0000	.0000
	3	.0726	.2080	.1775	.0474	.0060	.0004	.0000	.0000	.0000	.0000	.0000
	4	.0182	.1098	.2108	.0965	.0190	.0017	.0001	.0000	.0000	.0000	.0000
	5	.0034	.0439	.1898	.1489	.0456	.0063	.0004	.0000	.0000	.0000	.0000
	6	.0005	.0138	.1344	.1808	.0862	.0178	.0015	.0000	.0000	.0000	.0000
	7	.0001	.0035	.0768	.1771	.1314	.0407	.0051	.0002	.0000	.0000	.0000
	8	.0000	.0007	.0360	.1423	.1642	.0762	.0144	.0009	.0000	.0000	.0000
	9	.0000	.0001	.0140	.0949	.1703	.1186	.0336	.0032	.0001	.0000	.0000
	10	.0000	.0000	.0046	.0529	.1476	.1542	.0656	.0097	.0003	.0000	.0000
	11	.0000	.0000	.0012	.0247	.1073	.1682	.1073	.0247	.0012	.0000	.0000
	12	.0000	.0000	.0003	.0097	.0656	.1542	.1476	.0529	.0046	.0000	.0000
	13	.0000	.0000	.0001	.0032	.0336	.1186	.1703	.0949	.0140	.0001	.0000
	14	.0000	.0000	.0000	.0009	.0144	.0762	.1642	.1423	.0360	.0007	.0000
	15	.0000	.0000	.0000	.0002	.0051	.0407	.1314	.1771	.0768	.0035	.0001
	16	.0000	.0000	.0000	.0000	.0015	.0178	.0862	.1808	.1344	.0138	.0005
	17	.0000	.0000	.0000	.0000	.0004	.0063	.0456	.1489	.1898	.0439	.0034
	18	.0000	.0000	.0000	.0000	.0001	.0017	.0190	.0965	.2108	.1098	.0182
	19	.0000	.0000	.0000	.0000	.0000	.0004	.0060	.0474	.1775	.2080	.0726
	20	.0000	.0000	.0000	.0000	.0000	.0001	.0014	.0166	.1065	.2808	.2070
	21	.0000	.0000	.0000	.0000	.0000	.0000	.0002	.0037	.0406	.2407	.3746
	22	.0000	.0000	.0000	.0000	.0000	.0000	.0000	.0004	.0074	.0985	.3235

...(*Contd.*)

Appendix–2C (Continued)
Cumulative Binomial Probabilities

n	*x*	p										
		.05	.10	.20	.30	.40	.50	.60	.70	.80	.90	.95
23	0	.3074	.0886	.0059	.0003	.0000	.0000	.0000	.0000	.0000	.0000	.0000
	1	.3721	.2265	.0339	.0027	.0001	.0000	.0000	.0000	.0000	.0000	.0000
	2	.2154	.2768	.0933	.0127	.0009	.0000	.0000	.0000	.0000	.0000	.0000
	3	.0794	.2153	.1633	.0382	.0041	.0002	.0000	.0000	.0000	.0000	.0000
	4	.0209	.1196	.2042	.0818	.0138	.0011	.0000	.0000	.0000	.0000	.0000
	5	.0042	.0505	.1940	.1332	.0350	.0040	.0002	.0000	.0000	.0000	.0000
	6	.0007	.0168	.1455	.1712	.0700	.0120	.0008	.0000	.0000	.0000	.0000
	7	.0001	.0045	.0883	.1782	.1133	.0292	.0029	.0001	.0000	.0000	.0000
	8	.0000	.0010	.0442	.1527	.1511	.0584	.0088	.0004	.0000	.0000	.0000
	9	.0000	.0002	.0184	.1091	.1679	.0974	.0221	.0016	.0000	.0000	.0000
	10	.0000	.0000	.0064	.0655	.1567	.1364	.0464	.0052	.0001	.0000	.0000
	11	.0000	.0000	.0019	.0332	.1234	.1612	.0823	.0142	.0005	.0000	.0000
	12	.0000	.0000	.0005	.0142	.0823	.1612	.1234	.0332	.0019	.0000	.0000
	13	.0000	.0000	.0001	.0052	.0464	.1364	.1567	.0655	.0064	.0000	.0000
	14	.0000	.0000	.0000	.0016	.0221	.0974	.1679	.1091	.0184	.0002	.0000
	15	.0000	.0000	.0000	.0004	.0088	.0584	.1511	.1527	.0442	.0010	.0000
	16	.0000	.0000	.0000	.0001	.0029	.0292	.1133	.1782	.0883	.0045	.0001
	17	.0000	.0000	.0000	.0000	.0008	.0120	.0700	.1712	.1455	.0168	.0007
	18	.0000	.0000	.0000	.0000	.0002	.0040	.0350	.1332	.1940	.0505	.0042
	19	.0000	.0000	.0000	.0000	.0000	.0011	.0138	.0818	.2042	.1196	.0209
	20	.0000	.0000	.0000	.0000	.0000	.0002	.0041	.0382	.1633	.2153	.0794
	21	.0000	.0000	.0000	.0000	.0000	.0000	.0009	.0127	.0933	.2768	.2154
	22	.0000	.0000	.0000	.0000	.0000	.0000	.0001	.0027	.0339	.2265	.3721
	23	.0000	.0000	.0000	.0000	.0000	.0000	.0000	.0003	.0059	.0886	.3074
24	0	.2920	.0798	.0047	.0002	.0000	.0000	.0000	.0000	.0000	.0000	.0000
	1	.3688	.2127	.0283	.0020	.0001	.0000	.0000	.0000	.0000	.0000	.0000
	2	.2232	.2718	.0815	.0097	.0006	.0000	.0000	.0000	.0000	.0000	.0000
	3	.0862	.2215	.1493	.0305	.0028	.0001	.0000	.0000	.0000	.0000	.0000
	4	.0238	.1292	.1960	.0687	.0099	.0006	.0000	.0000	.0000	.0000	.0000
	5	.0050	.0574	.1960	.1177	.0265	.0025	.0001	.0000	.0000	.0000	.0000
	6	.0008	.0202	.1552	.1598	.0560	.0080	.0004	.0000	.0000	.0000	.0000
	7	.0001	.0058	.0998	.1761	.0960	.0206	.0017	.0000	.0000	.0000	.0000
	8	.0000	.0014	.0530	.1604	.1360	.0438	.0053	.0002	.0000	.0000	.0000
	9	.0000	.0003	.0236	.1222	.1612	.0779	.0141	.0008	.0000	.0000	.0000
	10	.0000	.0000	.0088	.0785	.1612	.1169	.0318	.0026	.0000	.0000	.0000
	11	.0000	.0000	.0028	.0428	.1367	.1488	.0608	.0079	.0002	.0000	.0000
	12	.0000	.0000	.0008	.0199	.0988	.1612	.0988	.0199	.0008	.0000	.0000
	13	.0000	.0000	.0002	.0079	.0608	.1488	.1367	.0428	.0028	.0000	.0000
	14	.0000	.0000	.0000	.0026	.0318	.1169	.1612	.0785	.0088	.0000	.0000
	15	.0000	.0000	.0000	.0008	.0141	.0779	.1612	.1222	.0236	.0003	.0000
	16	.0000	.0000	.0000	.0002	.0053	.0438	.1360	.1604	.0530	.0014	.0000
	17	.0000	.0000	.0000	.0000	.0017	.0206	.0960	.1761	.0998	.0058	.0001
	18	.0000	.0000	.0000	.0000	.0004	.0080	.0560	.1598	.1552	.0202	.0008
	19	.0000	.0000	.0000	.0000	.0001	.0025	.0265	.1177	.1960	.0574	.0050
	20	.0000	.0000	.0000	.0000	.0000	.0006	.0099	.0687	.1960	.1292	.0238
	21	.0000	.0000	.0000	.0000	.0000	.0001	.0028	.0305	.1493	.2215	.0862
	22	.0000	.0000	.0000	.0000	.0000	.0000	.0006	.0097	.0815	.2718	.2232
	23	.0000	.0000	.0000	.0000	.0000	.0000	.0001	.0020	.0283	.2127	.3688
	24	.0000	.0000	.0000	.0000	.0000	.0000	.0000	.0002	.0047	.0798	.2920

...(*Contd.*)

Appendix–2C (Continued)
Cumulative Binomial Probabilities

n	*x*	p: .05	.10	.20	.30	.40	.50	.60	.70	.80	.90	.95
25	0	.2774	.0718	.0038	.0001	.0000	.0000	.0000	.0000	.0000	.0000	.0000
	1	.3650	.1994	.0236	.0014	.0000	.0000	.0000	.0000	.0000	.0000	.0000
	2	.2305	.2659	.0708	.0074	.0004	.0000	.0000	.0000	.0000	.0000	.0000
	3	.0930	.2265	.1358	.0243	.0019	.0001	.0000	.0000	.0000	.0000	.0000
	4	.0269	.1384	.1867	.0572	.0071	.0004	.0000	.0000	.0000	.0000	.0000
	5	.0060	.0646	.1960	.1030	.0199	.0016	.0000	.0000	.0000	.0000	.0000
	6	.0010	.0239	.1633	.1472	.0442	.0053	.0002	.0000	.0000	.0000	.0000
	7	.0001	.0072	.1108	.1712	.0800	.0143	.0009	.0000	.0000	.0000	.0000
	8	.0000	.0018	.0623	.1651	.1200	.0322	.0031	.0001	.0000	.0000	.0000
	9	.0000	.0004	.0294	.1336	.1511	.0609	.0088	.0004	.0000	.0000	.0000
	10	.0000	.0001	.0118	.0916	.1612	.0974	.0212	.0013	.0000	.0000	.0000
	11	.0000	.0000	.0040	.0536	.1465	.1328	.0434	.0042	.0001	.0000	.0000
	12	.0000	.0000	.0012	.0268	.1140	.1550	.0760	.0115	.0003	.0000	.0000
	13	.0000	.0000	.0003	.0115	.0760	.1550	.1140	.0268	.0012	.0000	.0000
	14	.0000	.0000	.0001	.0042	.0434	.1328	.1465	.0536	.0040	.0000	.0000
	15	.0000	.0000	.0000	.0013	.0212	.0974	.1612	.0916	.0118	.0001	.0000
	16	.0000	.0000	.0000	.0004	.0088	.0609	.1511	.1336	.0294	.0004	.0000
	17	.0000	.0000	.0000	.0001	.0031	.0322	.1200	.1651	.0623	.0018	.0000
	18	.0000	.0000	.0000	.0000	.0009	.0143	.0800	.1712	.1108	.0072	.0001
	19	.0000	.0000	.0000	.0000	.0002	.0053	.0442	.1472	.1633	.0239	.0010
	20	.0000	.0000	.0000	.0000	.0000	.0016	.0199	.1030	.1960	.0646	.0060
	21	.0000	.0000	.0000	.0000	.0000	.0004	.0071	.0572	.1867	.1384	.0269
	22	.0000	.0000	.0000	.0000	.0000	.0001	.0019	.0243	.1358	.2265	.0930
	23	.0000	.0000	.0000	.0000	.0000	.0000	.0004	.0074	.0708	.2659	.2305
	24	.0000	.0000	.0000	.0000	.0000	.0000	.0000	.0014	.0236	.1994	.3650
	25	.0000	.0000	.0000	.0000	.0000	.0000	.0000	.0001	.0038	.0718	.2774

Appendix–2D
Cumulative Poisson Distribution

	λ									
x	0.1	0.2	0.3	0.4	0.5	0.6	0.7	0.8	0.9	1.0
0	.9048	.8187	.7408	.6703	.6065	.5488	.4966	.4493	.4066	.3679
1	.0905	.1637	.2222	.2681	.3033	.3293	.3476	.3595	.3659	.3679
2	.0045	.0164	.0333	.0536	.0758	.0988	.1217	.1438	.1647	.1839
3	.0002	.0011	.0033	.0072	.0126	.0198	.0284	.0383	.0494	.0613
4	.0000	.0001	.0003	.0007	.0016	.0030	.0050	.0077	.0111	.0153
5	.0000	.0000	.0000	.0001	.0002	.0004	.0007	.0012	.0020	.0031
6	.0000	.0000	.0000	.0000	.0000	.0000	.0001	.0002	.0003	.0005
7	.0000	.0000	.0000	.0000	.0000	.0000	.0000	.0000	.0000	.0001

	λ									
x	1.1	1.2	1.3	1.4	1.5	1.6	1.7	1.8	1.9	2.0
0	.3329	.3012	.2725	.2466	.2231	.2019	.1827	.1653	.1496	.1353
1	.3662	.3614	.3543	.3452	.3347	.3230	.3106	.2975	.2842	.2707
2	.2014	.2169	.2303	.2417	.2510	.2584	.2640	.2678	.2700	.2707
3	.0738	.0867	.0998	.1128	.1255	.1378	.1496	.1607	.1710	.1804
4	.0203	.0260	.0324	.0395	.0471	.0551	.0636	.0723	.0812	.0902
5	.0045	.0062	.0084	.0111	.0141	.0176	.0216	.0260	.0309	.0361
6	.0008	.0012	.0018	.0026	.0035	.0047	.0061	.0078	.0098	.0120
7	.0001	.0002	.0003	.0005	.0008	.0011	.0015	.0020	.0027	.0034
8	.0000	.0000	.0001	.0001	.0001	.0002	.0003	.0005	.0006	.0009
9	.0000	.0000	.0000	.0000	.0000	.0000	.0001	.0001	.0001	.0002

	λ									
x	2.1	2.2	2.3	2.4	2.5	2.6	2.7	2.8	2.9	3.0
0	.1225	.1108	.1003	.0907	.0821	.0743	.0672	.0608	0550	.0498
1	.2572	.2438	.2306	.2177	.2052	.1931	.1815	.1703	.1596	.1494
2	.2700	.2681	.2652	.2613	.2565	.2510	.2450	.2384	.2314	.2240
3	.1890	.1966	.2033	.2090	.2138	.2176	.2205	.2225	.2237	.2240
4	.0992	.1082	.1169	.1254	.1336	.1414	.1488	.1557	.1622	.1680
5	.0417	.0476	.0538	.0602	.0668	.0735	.0804	.0872	.0940	.1008
6	.0146	.0174	.0206	.0241	.0278	.0319	.0362	.0407	.0455	.0504
7	.0044	.0055	.0068	.0083	.0099	.0118	.0139	.0163	.0188	.0216
8	.0011	.0015	.0019	.0025	.0031	.0038	.0047	.0057	.0068	.0081
9	.0003	.0004	.0005	.0007	.0009	.0011	.0014	.0018	.0022	.0027
10	.0001	.0001	.0001	.0002	.0002	.0003	.0004	.0005	.0006	.0008
11	.0000	.0000	.0000	.0000	.0000	.0001	.0001	.0001	.0002	.0002
12	.0000	.0000	.0000	.0000	.0000	.0000	.0000	.0000	.0000	.0001

...(*Contd.*)

Appendix–2D (Continued)
Cumulative Poisson Distribution

x	λ = 3.1	3.2	3.3	3.4	3.5	3.6	3.7	3.8	3.9	4.0
0	.0450	.0408	.0369	.0334	.0302	.0273	.0247	.0224	.0202	.0183
1	.1397	.1304	.1217	.1135	.1057	.0984	.0915	.0850	.0789	.0733
2	.2165	.2087	.2008	.1929	.1850	.1771	.1692	.1615	.1539	.1465
3	.2237	.2226	.2209	.2186	.2158	.2125	.2087	.2046	.2001	.1954
4	.1733	.1781	.1823	.1858	.1888	.1912	.1931	.1944	.1951	.1954
5	.1075	.1140	.1203	.1264	.1322	.1377	.1429	.1477	.1522	.1563
6	.0555	.0608	.0662	.0716	.0771	.0826	.0881	.0936	.0989	.1042
7	.0246	.0278	.0312	.0348	.0385	.0425	.0466	.0508	.0551	.0595
8	.0095	.0111	.0129	.0148	.0169	.0191	.0215	.0241	.0269	.0298
9	.0033	.0040	.0047	.0056	.0066	.0076	.0089	.0102	.0116	.0132
10	.0010	.0013	.0016	.0019	.0023	.0028	.0033	.0039	.0045	.0053
11	.0003	.0004	.0005	.0006	.0007	.0009	.0011	.0013	.0016	.0019
12	.0001	.0001	.0001	.0002	.0002	.0003	.0003	.0004	.0005	.0006
13	.0000	.0000	.0000	.0000	.0001	.0001	.0001	.0001	.0002	.0002
14	.0000	.0000	.0000	.0000	.0000	.0000	.0000	.0000	.0000	.0001

x	λ = 4.1	4.2	4.3	4.4	4.5	4.6	4.7	4.8	4.9	5.0
0	.0166	.0150	.0136	.0123	.0111	.0101	.0091	.0082	.0074	.0067
1	.0679	.0630	.0583	.0540	.0500	.0462	.0427	.0395	.0365	.0337
2	.1393	.1323	.1254	.1188	.1125	.1063	.1005	.0948	.0894	.0842
3	.1904	.1852	.1798	.1743	.1687	.1631	.1574	.1517	.1460	.1404
4	.1951	.1944	.1933	.1917	.1898	.1875	.1849	.1820	.1789	.1755
5	.1600	.1633	.1662	.1687	.1708	.1725	.1738	.1747	.1753	.1755
6	.1039	.1143	.1191	.1237	.1281	.1323	.1362	.1398	.1432	.1462
7	.0640	.086	.0732	.0778	.0824	.0869	.0914	.0959	.1002	.1044
8	.0328	.0360	.0393	.0428	.0463	.0500	.0537	.0575	.0614	.0653
9	.0150	.0168	.0188	.0209	.0232	.0255	.0281	.0307	.0334	.0363
10	.0061	.0071	.0081	.0092	.0104	.0118	.0132	.0147	.0164	.0181
11	.0023	.0027	.0032	.0037	.0043	.0049	.0056	.0064	.0073	.0082
12	.0008	.0009	.0011	.0014	.0016	.0019	.0022	.0026	.0030	.0034
13	.0002	.0003	.0004	.0005	.0006	.0007	.0008	.0009	.0011	.0013
14	.0001	.0001	.0001	.0001	.0002	.0002	.0003	.0003	.0004	.0005
15	.0000	.0000	.0000	.0000	.0001	.0001	.0001	.0001	.0001	.0002

...(*Contd.*)

Appendix–2D (Continued)
Cumulative Poisson Distribution

	λ									
x	**5.1**	**5.2**	**5.3**	**5.4**	**5.5**	**5.6**	**5.7**	**5.8**	**5.9**	**6.0**
0	.0061	.0055	.0050	.0045	.0041	.0037	.0033	.0030	.0027	.0025
1	.0311	.0287	.0265	.0244	.0225	.0207	.0191	.0176	.0162	.0149
2	.0793	.0746	.0701	.0659	.0618	.0580	.0544	.0509	.0477	.0446
3	.1348	.1293	.1239	.1185	.1133	.1082	.1033	.0985	.0938	.0892
4	.1719	.1681	.1641	.1600	.1558	.1515	.1472	.1428	.1383	.1339
5	.1753	.1748	.1740	.1728	.1714	.1697	.1678	.1656	.1632	.1606
6	.1490	.1515	.1537	.1555	.1571	.1584	.1594	.1601	.1605	.1606
7	.1086	.1125	.1163	.1200	.1234	.1267	.1298	.1326	.1353	.1377
8	.0692	.0731	.0771	.0810	.0849	.0887	.0925	.0962	.0998	.1033
9	.0392	.0423	.0454	.0486	.0519	.0552	0586	.0620	0.654	.0688
10	.0200	.0220	.0241	.0262	.0285	.0309	.0334	.0359	.0386	.0413
11	.0093	.0104	.0116	.0129	.0143	.0157	.0173	.0190	.0207	.0225
12	.0039	.0045	.0051	.0058	.0065	.0073	.0082	.0092	.0102	.0113
13	.0015	.0018	.0021	.0024	.0028	.0032	.0036	.0041	.0046	.0052
14	.0006	.0007	.0008	.0009	.0011	.0013	.0015	.0017	.0019	.0022
15	.0002	.0002	.0003	.0003	.0004	.0005	.0006	.0007	.0008	.0009
16	.0001	.0001	.0001	.0001	.0001	.0002	.0002	.0002	.0003	.0003
17	.0000	.0000	.0000	.0000	.0000	.0001	.0001	.0001	.0001	.0001

	λ									
x	**6.1**	**6.2**	**6.3**	**6.4**	**6.5**	**6.6**	**6.7**	**6.8**	**6.9**	**7.0**
0	.0022	.0020	.0018	.0017	.0015	.0014	.0012	.0011	.0010	.0009
1	.0137	.0126	.0116	.0106	.0098	.0090	.0082	.0076	.0070	.0064
2	.0417	.0390	.0364	.0340	.0318	.0296	.0276	.0258	.0240	.0223
3	.0848	.0806	.0765	.0726	.0688	.0652	.0617	.0584	.0552	.0521
4	.1294	.1249	.1205	.1162	.1118	.1076	.1034	.0992	.0952	.0912
5	.1579	.1549	.1519	.1487	.1454	.1420	.1385	.1349	.1314	.1277
6	.1605	.1601	.1595	.1586	.1575	.1562	.1546	.1529	.1511	.1490
7	.1399	.1418	.1435	.1450	.1462	.1472	.1480	.1486	.1489	.1490
8	.1066	.1099	.1130	.1160	.1188	.1215	.1240	.1263	.1284	.1304
9	.0723	.0757	.0791	.0825	.0858	.0891	.0923	.0954	.0985	.1014
10	.0441	.0469	.0498	.0528	.0558	.0588	.0618	.069	.0679	.0710
11	.0244	.0265	.0285	.0307	.0330	.0353	.0377	.0401	.0426	.0452
12	.0124	.0137	.0150	.0164	.0179	.0194	.0210	.0227	.0245	.0263
13	.0058	.0065	.0073	.0081	.0089	.0099	.0108	.0119	.0130	.0142
14	.0025	.0029	.0033	.0037	.0041	.0046	.0052	.0058	.0064	.0071
15	.0010	.0012	.0014	.0016	.0018	.0020	.0023	.0026	.0029	.0033
16	.0004	.0005	.0005	.0006	.0007	.0008	.0010	.0011	.0013	.0014
17	.0001	.0002	.0002	.0002	.0003	.0003	.0004	.0004	.0005	.0006
18	.0000	.0001	.0001	.0001	.0001	.0001	.0001	.0002	.0002	.0002
19	.0000	.0000	.0000	.0000	.0000	.0000	.0001	.0001	.0001	.0001

...(Contd.)

Appendix–2D (Continued)
Cumulative Poisson Distribution

	λ									
x	**7.1**	**7.2**	**7.3**	**7.4**	**7.5**	**7.6**	**7.7**	**7.8**	**7.9**	**8.0**
0	.0008	.0007	.0007	.0006	.0006	.0005	.0005	.0004	.0004	.0003
1	.0059	.0054	.0049	.0045	.0041	.0038	.0035	.0032	.0029	.0027
2	.0208	.0194	.0180	.0167	.0156	.0145	.0134	.0125	.0116	.0107
3	.0492	.0464	.0438	.0413	.0389	.0366	.0345	.0324	.0305	.0286
4	.0874	.0836	.0799	.0764	.0729	.0696	.0663	.0632	.0602	.0573
5	.1241	.1204	.1167	.1130	.1094	.1057	.1021	.0986	.0951	.0916
6	.1468	.1445	.1420	.1394	.1367	.1339	.1311	.1282	.1252	.1221
7	.1489	.1486	.1481	.1474	.1465	.1454	.1442	.1428	.1413	.1396
8	.1321	.1337	.1351	.1363	.1373	.1381	.1388	.1392	.1395	.1396
9	.1042	.1070	.1096	.1121	.1144	.1167	.1187	.1207	.1224	.1241
10	.0740	.0770	.0800	.0829	.0858	.0887	.0914	.0941	.0967	.0993
11	.0478	.0504	.0531	.0558	.0585	.0613	.0640	.0667	.0695	.0722
12	.0283	.0303	.0323	.0344	.0366	.0388	.0411	.0434	.0457	.0481
13	.0154	.0168	.0181	.0196	.0211	.0227	.0243	.0260	.0278	.0296
14	.0078	.0086	.0095	.0104	.0113	.0123	.0134	.0145	.0157	.0169
15	.0037	.0041	.0046	.0051	.0057	.0062	.0069	.0075	.0083	.0090
16	.0016	.0019	.0021	.0024	.0026	.0030	.0033	.0037	.0041	.0045
17	.0007	.0008	.0009	.0010	.0012	.0013	.0015	.0017	.0019	.0021
18	.0003	.0003	.0004	.0004	.0005	.0006	.0006	.0007	.0008	.0009
19	.0001	.0001	.0001	.0002	.0002	.0002	.0003	.0003	.0003	.0004
20	.0000	.0000	.0001	.0001	.0001	.0001	.0001	.0001	.0001	.0002
21	.0000	.0000	.0000	.0000	.0000	.0000	.0000	.0000	.0001	.0001

...(*Contd.*)

Appendix–2D (Continued)
Cumulative Poisson Distribution

	λ									
x	**8.1**	**8.2**	**8.3**	**8.4**	**8.5**	**8.6**	**8.7**	**8.8**	**8.9**	**9.0**
0	.0003	.0003	.0002	.0002	.0002	.0002	.0002	.0002	.0001	.0001
1	.0025	.0023	.0021	.0019	.0017	.0016	.0014	.0013	.0012	.0011
2	.0100	.0092	.0086	.0079	.0074	.0068	.0063	.0058	.0054	.0050
3	.0269	.0252	.0237	.0222	.0208	.0195	.0183	.0171	.0160	.0150
4	.0544	.0517	.0491	.0466	.0443	.0420	.0398	.0377	.0357	.0337
5	.0882	.0849	.0816	.0784	.0752	.0722	.0692	.0663	.0635	.0607
6	.1191	.1160	.1128	.1097	.1066	.1034	.1003	.0972	.0941	.0911
7	.1378	.1258	.1338	.1317	.1294	.1271	.1247	.1222	.1197	.1171
8	.1395	.1392	.1388	.1382	.1375	.1366	.1356	.1344	.1332	.1318
9	.1255	.1269	.1280	.1290	.1299	.1306	.1311	.1315	.1317	.1318
10	.1017	.1040	.1063	.1084	.1104	.1123	.1140	.1157	.1172	.1186
11	.0749	.0775	.0802	.0828	.0853	.0878	.0902	.0925	.0948	.0970
12	.0505	.0530	.0555	.0579	.0604	.0629	.0654	.0679	.0703	.0728
13	.0315	.0334	.0354	.0374	.0395	.0416	.0438	.0459	.0481	.0504
14	.0182	.0196	.0210	.0225	.0240	.0256	.0272	.0289	.0306	.0324
15	.0098	.0107	.0116	.0126	.0136	.0147	.0158	.0169	.0182	.0194
16	.0050	.0055	.0060	.0066	.0072	.0079	.0086	.0093	.0101	.0109
17	.0024	.0026	.0029	.0033	.0036	.0040	.0044	.0048	.0053	.0058
18	.0011	.0012	.0014	.0015	.0017	.0019	.0021	.0024	.0026	.0029
19	.0005	.0005	.0006	.0007	.0008	.0009	.0010	.0011	.0012	.0014
20	.0002	.0002	.0002	.0003	.0003	.0004	.0004	.0005	.0005	.0006
21	.0001	.0001	.0001	.0001	.0001	.0002	.0002	.0002	.0002	.0003
22	.0000	.0000	.0000	.0000	.0001	.0001	.0001	.0001	.0001	.0001

...(Contd.)

Appendix–2D (Continued)
Cumulative Poisson Distribution

	λ									
x	**9.1**	**9.2**	**9.3**	**9.4**	**9.5**	**9.6**	**9.7**	**9.8**	**9.9**	**10**
0	.0001	.0001	.0001	.0001	.0001	.0001	.0001	.0001	.0001	.0000
1	.0010	.0009	.0009	.0008	.0007	.0007	.0006	.0005	.0005	.0005
2	.0046	.0043	.0040	.0037	.0034	.0031	.0029	.0027	.0025	.0023
3	.0140	.0131	.0123	.0115	.0107	.0100	.0093	.0087	.0081	.0076
4	.0319	.0302	.0285	.0269	.0254	.0240	.0226	.0213	.0201	.0189
5	.0581	.0555	.0530	.0506	.0483	.0460	.0439	.0418	.0398	.0378
6	.0881	.0851	.0822	.0793	.0764	.0736	.0709	.0682	.0656	.0631
7	.1145	.1118	.1091	.1064	.1037	.1010	.0982	.0955	.0928	.0901
8	.1302	.1286	.1269	.1251	.1232	.1212	.1191	.1170	.1148	.1126
9	.1317	.1315	.1311	.1306	.1300	.1293	.1284	.1274	.1263	.1251
10	.1198	.1209	.1219	.1228	.1235	.1241	.1245	.1249	.1250	.1251
11	.0991	.1012	.1031	.1049	.1067	.1083	.1098	.1112	.1125	.1137
12	.0752	.0776	.0799	.0822	.0844	.0866	.0888	.0908	.0928	.0948
13	.0526	.0549	.0572	.0594	.0617	.0640	.0662	.0685	.0707	.0729
14	.0342	.0361	.0380	.0399	.0419	.0439	.0459	.0479	.0500	.0521
15	.0208	.0221	.0235	.0250	.0265	.0281	.0297	.0313	.0330	.0347
16	.0118	.0127	.0137	.0147	.0157	.0168	.0180	.0192	.0204	.0217
17	.0063	.0069	.0075	.0081	.0088	.0095	.0103	.0111	.0119	.0128
18	.0032	.0035	.0039	.0042	.0046	.0051	.0055	.0060	.0065	.0071
19	.0015	.0017	.0019	.0021	.0023	.0026	.0028	.0031	.0034	.0037
20	.0007	.0008	.0009	.0010	.0011	.0012	.0014	.0015	.0017	.0019
21	.0003	.0003	.0004	.0004	.0005	.0006	.0006	.0007	.0008	.0009
22	.0001	.0001	.0002	.0002	.0002	.0002	.0003	.0003	.0004	.0004
23	.0000	.0001	.0001	.0001	.0001	.0001	.0001	.0001	.0002	.0002
24	.0000	.0000	.0000	.0000	.0000	.0000	.0000	.0001	.0001	.0001

...(Contd.)

Appendix–2D (Continued)
Cumulative Poisson Distribution

	λ									
x	**11**	**12**	**13**	**14**	**15**	**16**	**17**	**18**	**19**	**20**
0	.0000	.0000	.0000	.0000	.0000	.0000	.0000	.0000	.0000	.0000
1	.0002	.0001	.0000	.0000	.0000	.0000	.0000	.0000	.0000	.0000
2	.0010	.0004	.0002	.0001	.0000	.0000	.0000	.0000	.0000	.0000
3	.0037	.0018	.0008	.0004	.0002	.0001	.0000	.0000	.0000	.0000
4	.0102	.0053	.0027	.0013	.0006	.0003	.0001	.0001	.0000	.0000
5	.0224	.0127	.0070	.0037	.0019	.0010	.0005	.0002	.0001	.0001
6	.0411	.0255	.0152	.0087	.0048	.0026	.0014	.0007	.0004	.0002
7	.0646	.0437	.0281	.0174	.0104	.0060	.0034	.0019	.0010	.0005
8	.0888	.0655	.0457	.0304	.0194	.0120	.0072	.0042	.0024	.0013
9	.1085	.0874	.0661	.0473	.0324	.0213	.0135	.0083	.0050	.0029
10	.1194	.1048	.0859	.0663	.0486	.0341	.0230	.0150	.0095	.0058
11	.1194	.1144	.1015	.0844	.0663	.0496	.0355	.0245	.0164	.0106
12	.1094	.1144	.1099	.0984	.0829	.0661	.0504	.0368	.0259	.0176
13	.0926	.1056	.1099	.1060	.0956	.0814	.0658	.0509	.0378	.0271
14	.0728	.0905	.1021	.1060	.1024	.0930	.0800	.0655	.0514	.0387
15	.0534	.0724	.0885	.0989	.1024	.0992	.0906	.0786	.0650	.0516
16	.0367	.0543	.0719	.0866	.0960	.0992	.0963	.0884	.0772	.0646
17	.0237	.0383	.0550	.0713	.0847	.0934	.0963	.0936	.0863	.0760
18	.0145	.0255	.0397	.0554	.0706	.0830	.0909	.0936	.0911	.0844
19	.0084	.0161	.0272	.0409	.0557	.0699	.0814	.0887	.0911	.0888
20	.0046	.0097	.0177	.0286	.0418	.0559	.0692	.0798	.0866	.0888
21	.0024	.0055	.0109	.0191	.0299	.0426	.0560	.0684	.0783	.0846
22	.0012	.0030	.0065	.0121	.0204	.0310	.0433	.0560	.0676	.0769
23	.0006	.0016	.0037	.0074	.0133	.0216	.0320	.0438	.0559	.0669
24	.0003	.0008	.0020	.0043	.0083	.0144	.0226	.0328	.0442	.0557
25	.0001	.0004	.0010	.0024	.0050	.0092	.0154	.0237	.0336	.0446
26	.0000	.0002	.0005	.0013	.0029	.0057	.0101	.0164	.0246	.0343
27	.0000	.0001	.0002	.0007	.0016	.0034	.0063	.0109	.0173	.0254
28	.0000	.0000	.0001	.0003	.0009	.0019	.0038	.0070	.0117	.0181
29	.0000	.0000	.0001	.0002	.0004	.0011	.0023	.0044	.0077	.0125
30	.0000	.0000	.0000	.0001	.0002	.0006	.0013	.0026	.0049	.0083
31	.0000	.0000	.0000	.0000	.0001	.0003	.0007	.0015	.0030	.0054
32	.0000	.0000	.0000	.0000	.0001	.0001	.0004	.0009	.0018	.0034
33	.0000	.0000	.0000	.0000	.0000	.0001	.0002	.0005	.0010	.0020
34	.0000	.0000	.0000	.0000	.0000	.0000	.0001	.0002	.0006	.0012
35	.0000	.0000	.0000	.0000	.0000	.0000	.0000	.0001	.0003	.0007
36	.0000	.0000	.0000	.0000	.0000	.0000	.0000	.0001	.0002	.0004
37	.0000	.0000	.0000	.0000	.0000	.0000	.0000	.0000	.0001	.0002
38	.0000	.0000	.0000	.0000	.0000	.0000	.0000	.0000	.0000	.0001
39	.0000	.0000	.0000	.0000	.0000	.0000	.0000	.0000	.0000	.0001

Appendix–2E
Cumulative Standard Normal Distribution Table

Z	– 0.09	– 0.08	– 0.07	– 0.06	– 0.05
– 3.9	0.000033	0.000034	0.000036	0.000037	0.000039
– 3.8	0.000050	0.000052	0.000054	0.000057	0.000059
– 3.7	0.000075	0.000078	0.000082	0.000085	0.000088
– 3.6	0.000112	0.000117	0.000121	0.000126	0.000131
– 3.5	0.000165	0.000172	0.000179	0.000185	0.000193
– 3.4	0.000242	0.000251	0.000260	0.000270	0.000280
– 3.3	0.000350	0.000362	0.000376	0.000390	0.000404
– 3.2	0.000501	0.000519	0.000538	0.000557	0.000577
– 3.1	0.000711	0.000736	0.000762	0.000789	0.000816
– 3.0	0.001001	0.001035	0.001070	0.001107	0.001144
– 2.9	0.001395	0.001441	0.001489	0.001538	0.001589
– 2.8	0.001926	0.001988	0.002052	0.002118	0.002186
– 2.7	0.002635	0.002718	0.002803	0.002890	0.002980
– 2.6	0.003573	0.003681	0.003793	0.003907	0.004025
– 2.5	0.004799	0.004940	0.005085	0.005234	0.005386
– 2.4	0.006387	0.006569	0.006756	0.006947	0.007143
– 2.3	0.008424	0.008656	0.008894	0.009137	0.009387
– 2.2	0.011011	0.011304	0.011604	0.011911	0.012224
– 2.1	0.014262	0.014629	0.015003	0.015386	0.015778
– 2.0	0.018309	0.018763	0.019226	0.019699	0.020182
– 1.9	0.023295	0.023852	0.024419	0.024998	0.025588
– 1.8	0.029379	0.030054	0.030742	0.031443	0.032157
– 1.7	0.036727	0.037538	0.038364	0.039204	0.040059
– 1.6	0.045514	0.046479	0.047460	0.048457	0.049471
– 1.5	0.055917	0.057053	0.058208	0.059380	0.060571
– 1.4	0.068112	0.069437	0.070781	0.072146	0.073529
– 1.3	0.082264	0.083793	0.085343	0.086915	0.088508
– 1.2	0.098525	0.100273	0.102042	0.103835	0.105650
– 1.1	0.117023	0.119000	0.121001	0.123024	0.125072
– 1.0	0.137857	0.140071	0.142310	0.144572	0.146859
– 0.9	0.161087	0.163543	0.166023	0.168528	0.171056
– 0.8	0.186733	0.189430	0.192150	0.194894	0.197662
– 0.7	0.214764	0.217695	0.220650	0.223627	0.226627
– 0.6	0.245097	0.248252	0.251429	0.254627	0.257846
– 0.5	0.277595	0.280957	0.284339	0.287740	0.291160
– 0.4	0.312067	0.315614	0.319178	0.322758	0.326355
– 0.3	0.348268	0.351973	0.355691	0.359424	0.363169
– 0.2	0.385908	0.389739	0.393580	0.397432	0.401294
– 0.1	0.424655	0.428576	0.432505	0.436441	0.440382
0.0	0.464144	0.468119	0.472097	0.476078	0.480061

...(*Contd.*)

Appendix–2E (Continued)
Cumulative Standard Normal Distribution Table

Z	– 0.04	– 0.03	– 0.02	– 0.01	– 0.00
– 3.9	0.000041	0.000042	0.000044	0.000046	0.000048
– 3.8	0.000062	0.000064	0.000067	0.000069	0.000072
– 3.7	0.000092	0.000096	0.000100	0.000104	0.000108
– 3.6	0.000136	0.000142	0.000147	0.000153	0.000159
– 3.5	0.000200	0.000208	0.000216	0.000224	0.000233
– 3.4	0.000291	0.000302	0.000313	0.000325	0.000337
– 3.3	0.000419	0.000434	0.000450	0.000467	0.000483
– 3.2	0.000598	0.000619	0.000641	0.000664	0.000687
– 3.1	0.000845	0.000874	0.000904	0.000935	0.000968
– 3.0	0.001183	0.001223	0.001264	0.001306	0.001350
– 2.9	0.001641	0.001695	0.001750	0.001807	0.001866
– 2.8	0.002256	0.002327	0.002401	0.002477	0.002555
– 2.7	0.003072	0.003167	0.003264	0.003364	0.003467
– 2.6	0.004145	0.004269	0.004396	0.004527	0.004661
– 2.5	0.005543	0.005703	0.005868	0.006037	0.006210
– 2.4	0.007344	0.007549	0.007760	0.007976	0.008198
– 2.3	0.009642	0.009903	0.010170	0.010444	0.010724
– 2.2	0.012545	0.012874	0.013209	0.013553	0.013903
– 2.1	0.016177	0.016586	0.017003	0.017429	0.017864
– 2.0	0.020675	0.021178	0.021692	0.022216	0.022750
– 1.9	0.026190	0.026803	0.027429	0.028067	0.028717
– 1.8	0.032884	0.033625	0.034379	0.035148	0.035930
– 1.7	0.040929	0.041815	0.042716	0.043633	0.044565
– 1.6	0.050503	0.050551	0.052616	0.053699	0.054799
– 1.5	0.061780	0.063008	0.064256	0.065522	0.066807
– 1.4	0.074934	0.076359	0.077804	0.079270	0.080757
– 1.3	0.090123	0.091759	0.093418	0.095098	0.096801
– 1.2	0.107488	0.109349	0.111233	0.113140	0.115070
– 1.1	0.127143	0.129238	0.131357	0.133500	0.135666
– 1.0	0.149170	0.151505	0.153864	0.156248	0.158655
– 0.9	0.173609	0.176185	0.178786	0.181411	0.184060
– 0.8	0.200454	0.203269	0.206108	0.208970	0.211855
– 0.7	0.229650	0.232695	0.235762	0.238852	0.241964
– 0.6	0.261086	0.264347	0.267629	0.270931	0.274253
– 0.5	0.294599	0.298056	0.301532	0.305026	0.308538
– 0.4	0.329969	0.333598	0.337243	0.340903	0.344578
– 0.3	0.366928	0.370700	0.374484	0.378281	0.382089
– 0.2	0.405165	0.409046	0.412936	0.416834	0.420740
– 0.1	0.444330	0.448283	0.452242	0.456205	0.460172
0.0	0.484047	0.488033	0.492022	0.496011	0.500000

...*(Contd.)*

Appendix-2E (Continued)
Cumulative Standard Normal Distribution Table

Z	0.00	0.01	0.02	0.03	0.04
0.0	0.500000	0.503989	0.507978	0.511967	0.515953
0.1	0.539828	0.543795	0.547758	0.551717	0.555760
0.2	0.579260	0.583166	0.587064	0.590954	0.594835
0.3	0.617911	0.621719	0.625516	0.629300	0.633072
0.4	0.655422	0.659097	0.662757	0.666402	0.670031
0.5	0.691462	0.694974	0.698468	0.701944	0.705401
0.6	0.725747	0.729069	0.732371	0.735653	0.738914
0.7	0.758036	0.761148	0.764238	0.767305	0.770350
0.8	0.788145	0.791030	0.793892	0.796731	0.799546
0.9	0.815940	0.818589	0.821214	0.823815	0.826391
1.0	0.841345	0.843752	0.846136	0.848495	0.850830
1.1	0.864334	0.866500	0.868643	0.870762	0.872857
1.2	0.884930	0.886860	0.888767	0.890651	0.892512
1.3	0.903199	0.904902	0.906582	0.908241	0.909877
1.4	0.919243	0.920730	0.922196	0.923641	0.925066
1.5	0.933193	0.934478	0.935744	0.936992	0.938220
1.6	0.945201	0.946301	0.947384	0.948449	0.949497
1.7	0.955435	0.956367	0.957284	0.958185	0.959071
1.8	0.964070	0.964852	0.965621	0.966375	0.967116
1.9	0.971283	0.971933	0.972571	0.973197	0.973810
2.0	0.977250	0.977784	0.978308	0.978822	0.979325
2.1	0.982136	0.982571	0.982997	0.983414	0.983823
2.2	0.986097	0.986447	0.986791	0.987126	0.987455
2.3	0.989276	0.989556	0.989830	0.990097	0.990358
2.4	0.991802	0.992024	0.992240	0.992451	0.992656
2.5	0.993790	0.993963	0.994132	0.994297	0.994457
2.6	0.995339	0.995473	0.995604	0.995731	0.995855
2.7	0.996533	0.996636	0.996736	0.996833	0.996928
2.8	0.997445	0.997523	0.997599	0.997673	0.997744
2.9	0.998134	0.998193	0.998250	0.998305	0.998359
3.0	0.998650	0.998694	0.998736	0.998777	0.998817
3.1	0.999032	0.999065	0.999096	0.999126	0.999155
3.2	0.999313	0.999336	0.999359	0.999381	0.999402
3.3	0.999517	0.999533	0.999550	0.999566	0.999581
3.4	0.999663	0.999675	0.999687	0.999698	0.999709
3.5	0.999767	0.999776	0.999784	0.999792	0.999800
3.6	0.999841	0.999847	0.999853	0.999858	0.999864
3.7	0.999892	0.999896	0.999900	0.999904	0.999908
3.8	0.999928	0.999931	0.999933	0.999936	0.999938
3.9	0.999952	0.999954	0.999956	0.999958	0.999959

...(*Contd.*)

Appendix–2E (Continued)
Cumulative Standard Normal Distribution Table

Z	0.05	0.06	0.07	0.08	0.09
0.0	0.519939	0.523922	0.527903	0.531881	0.535856
0.1	0.559618	0.563559	0.567495	0.571424	0.575345
0.2	0.598706	0.602568	0.606420	0.610261	0.614092
0.3	0.636831	0.640576	0.644309	0.648027	0.651732
0.4	0.673645	0.677242	0.680822	0.684386	0.687933
0.5	0.708840	0.712260	0.715662	0.719043	0.722405
0.6	0.742154	0.745373	0.748571	0.751748	0.754903
0.7	0.773373	0.776373	0.779350	0.782305	0.785236
0.8	0.802338	0.805106	0.807850	0.810570	0.813267
0.9	0.828944	0.831472	0.833977	0.836457	0.838913
1.0	0.853141	0.855428	0.857690	0.859929	0.862143
1.1	0.874928	0.876976	0.878999	0.881000	0.882977
1.2	0.894350	0.896165	0.897958	0.899727	0.901475
1.3	0.911492	0.913085	0.914657	0.916207	0.917736
1.4	0.926471	0.927855	0.929219	0.930563	0.931888
1.5	0.939429	0.940620	0.941792	0.942947	0.944083
1.6	0.950529	0.951543	0.952540	0.953521	0.954486
1.7	0.959941	0.960796	0.961636	0.962462	0.963273
1.8	0.967843	0.968557	0.969258	0.969946	0.970621
1.9	0.974412	0.975002	0.975581	0.976148	0.976705
2.0	0.979818	0.980301	0.980774	0.981237	0.981691
2.1	0.984222	0.984614	0.984997	0.985371	0.985738
2.2	0.987776	0.988089	0.988396	0.988696	0.988989
2.3	0.990613	0.990863	0.991106	0.991344	0.991576
2.4	0.992857	0.993053	0.993244	0.993431	0.993613
2.5	0.994614	0.994766	0.994915	0.995060	0.995201
2.6	0.995975	0.996093	0.996207	0.996319	0.996427
2.7	0.997020	0.997110	0.997197	0.997282	0.997365
2.8	0.997814	0.997882	0.997948	0.998012	0.998074
2.9	0.998411	0.998462	0.998511	0.998559	0.998605
3.0	0.998856	0.998893	0.998930	0.998965	0.998999
3.1	0.999184	0.999211	0.999238	0.999264	0.999289
3.2	0.999423	0.999443	0.999462	0.999481	0.999499
3.3	0.999596	0.999610	0.999624	0.999638	0.999650
3.4	0.999720	0.999730	0.999740	0.999749	0.999758
3.5	0.999807	0.999815	0.999821	0.999828	0.999835
3.6	0.999869	0.999874	0.999879	0.999883	0.999888
3.7	0.999912	0.999915	0.999918	0.999922	0.999925
3.8	0.999941	0.999943	0.999946	0.999948	0.999950
3.9	0.999961	0.999963	0.999964	0.999966	0.999967

Appendix–2F
Chi-Square Distribution Table

df	Area in the Right Tail under the Chi-square Distribution Curve				
	.995	**.990**	**.975**	**.950**	**.900**
1	0.000	0.000	0.001	0.004	0.016
2	0.010	0.020	0.051	0.103	0.211
3	0.072	0.115	0.216	0.352	0.584
4	0.207	0.297	0.484	0.711	1.064
5	0.412	0.554	0.831	1.145	1.610
6	0.676	0.872	1.237	1.635	2.204
7	0.989	1.239	1.690	2.167	2.833
8	1.344	1.646	2.180	2.733	3.490
9	1.735	2.088	2.700	3.325	4.168
10	2.156	2.558	3.247	3.940	4.865
11	2.603	3.053	3.816	4.575	5.578
12	3.074	3.571	4.404	5.226	6.304
13	3.565	4.107	5.009	5.892	7.042
14	4.075	4.660	5.629	6.571	7.790
15	4.601	5.229	6.262	7.261	8.547
16	5.142	5.812	6.908	7.962	9.312
17	5.697	6.408	7.564	8.672	10.085
18	6.265	7.015	8.231	9.390	10.865
19	5.844	7.633	8.907	10.117	11.651
20	7.434	8.260	9.591	10.851	12.443
21	8.034	8.897	10.283	11.591	13.240
22	8.643	9.542	10.982	12.338	14.041
23	9.260	10.196	11.689	13.091	14.848
24	9.886	10.856	12.401	13.848	15.659
25	10.520	11.524	13.120	14.611	16.473
26	11.160	12.198	13.844	15.379	17.292
27	11.808	12.879	14.573	16.151	18.114
28	12.461	13.565	15.308	16.928	18.939
29	13.121	14.256	16.047	17.708	19.768
30	13.787	14.953	16.791	18.493	20.599
40	20.707	22.164	24.433	26.509	29.051
50	27.991	29.707	32.357	34.764	37.689
60	35.534	37.485	40.482	43.188	46.459
70	43.275	45.442	48.758	51.739	55.329
80	51.172	53.540	57.153	60.391	64.278
90	59.196	61.754	65.647	69.126	73.291
100	67.328	70.065	74.222	77.929	82.358

...(Contd.)

Appendix–2F (Continued)
Chi-Square Distribution Table

df	Area in the Right Tail under the Chi-square Distribution Curve				
	.100	**.050**	**.025**	**.010**	**.005**
1	2.706	3.841	5.024	6.635	7.879
2	4.605	5.991	7.378	9.210	10.597
3	6.251	7.815	9.348	11.345	12.838
4	7.779	9.488	11.143	13.277	14.860
5	9.236	11.070	12.833	15.086	16.750
6	10.645	12.592	14.449	16.812	18.548
7	12.017	14.067	16.013	18.475	20.278
8	13.362	15.507	17.535	20.090	21.955
9	14.684	16.919	19.023	21.666	23.589
10	15.987	18.307	20.483	23.209	25.188
11	17.275	19.675	21.920	24.725	26.757
12	18.549	21.026	23.337	26.217	28.300
13	19.812	22.362	24.736	27.688	29.819
14	21.064	23.685	26.119	29.141	31.319
15	22.307	24.996	27.488	30.578	32.801
16	23.542	26.296	28.845	32.000	34.267
17	24.769	27.587	30.191	33.409	35.718
18	25.989	28.869	31.526	34.805	37.156
19	27.204	20.144	32.852	36.191	38.582
20	28.412	31.410	34.170	37.566	39.997
21	29.615	32.671	35.479	38.932	41.401
22	30.813	33.924	36.781	40.289	42.796
23	32.007	35.172	38.076	41.638	44.181
24	33.196	36.415	39.364	42.980	45.559
25	34.382	37.652	40.646	44.314	46.928
26	35.563	38.885	41.923	45.642	48.290
27	36.741	40.113	43.195	46.963	49.645
28	37.916	41.337	44.461	48.278	50.993
29	39.087	42.557	45.722	49.588	52.336
30	40.256	43.773	46.979	50.892	53.672
40	51.805	55.758	59.342	63.691	66.766
50	63.167	67.505	71.420	76.154	79.490
60	74.397	79.082	83.298	88.379	91.952
70	85.527	90.531	95.023	100.425	104.215
80	96.578	101.879	106.629	112.329	116.321
90	107.565	113.145	118.136	124.116	128.299
100	118.498	124.342	129.561	135.807	140.169

Appendix–2G
The *t*-Distribution Table

df	Area in the Right Tail under the *t*-distribution Curve					
	0.1	0.05	0.025	0.01	0.005	0.001
1	3.078	6.314	12.706	31.821	63.657	318.309
2	1.886	2.920	4.303	6.965	9.925	22.327
3	1.638	2.353	3.182	4.541	5.841	10.215
4	1.533	2.132	2.776	3.747	4.604	7.173
5	1.476	2.015	2.571	3.365	4.032	5.893
6	1.440	1.943	2.447	3.143	3.707	5.208
7	1.415	1.895	2.365	2.998	3.499	4.785
8	1.397	1.860	2.306	2.896	3.355	4.501
9	1.383	1.833	2.262	2.821	3.250	4.297
10	1.372	1.812	2.228	2.764	3.169	4.144
11	1.363	1.796	2.201	2.718	3.106	4.025
12	1.356	1.782	2.179	2.681	3.055	3.930
13	1.350	1.771	2.160	2.650	3.012	3.852
14	1.345	1.761	2.145	2.624	2.977	3.787
15	1.341	1.753	2.131	2.602	2.947	3.733
16	1.337	1.746	2.120	2.583	2.921	3.686
17	1.333	1.740	2.110	2.567	2.898	3.646
18	1.330	1.734	2.101	2.552	2.878	3.610
19	1.328	1.729	2.093	2.539	2.861	3.579
20	1.325	1.725	2.086	2.528	2.845	3.552
21	1.323	1.721	2.080	2.518	2.831	3.527
22	1.321	1.717	2.074	2.508	2.819	3.505
23	1.319	1.714	2.069	2.500	2.807	3.485
24	1.318	1.711	2.064	2.492	2.797	3.467
25	1.316	1.708	2.060	2.485	2.787	3.450
26	1.315	1.706	2.056	2.479	2.779	3.435
27	1.314	1.703	2.052	2.473	2.771	3.421
28	1.313	1.701	2.048	2.467	2.763	3.408
29	1.311	1.699	2.045	2.462	2.756	3.396
30	1.310	1.697	2.042	2.457	2.750	3.385
31	1.309	1.696	2.040	2.453	2.744	3.375
32	1.309	1.694	2.037	2.449	2.738	3.365
33	1.308	1.692	2.035	2.445	2.733	3.356
34	1.307	1.691	2.032	2.441	2.728	3.348
35	1.306	1.690	2.030	2.438	2.724	3.340
36	1.306	1.688	2.028	2.434	2.719	3.333
37	1.305	1.687	2.026	2.431	2.715	3.326
38	1.304	1.686	2.024	2.429	2.712	3.319
39	1.304	1.685	2.023	2.426	2.708	3.313
40	1.303	1.684	2.021	2.423	2.704	3.307

...(*Contd.*)

Appendix–2G (Continued)
The *t*-Distribution Table

df	Area in the Right Tail under the *t*-distribution Curve					
	0.1	**0.05**	**0.025**	**0.01**	**0.005**	**0.001**
41	1.303	1.683	2.020	2.421	2.701	3.301
42	1.302	1.682	2.018	2.418	2.698	3.296
43	1.302	1.681	2.017	2.416	2.695	3.291
44	1.301	1.680	2.015	2.414	2.692	3.286
45	1.301	1.679	2.014	2.412	2.690	3.281
46	1.300	1.679	2.013	2.410	2.687	3.277
47	1.300	1.678	2.012	2.408	2.685	3.273
48	1.299	1.677	2.011	2.407	2.682	3.269
49	1.299	1.677	2.010	2.405	2.680	3.265
50	1.299	1.676	2.009	2.403	2.678	3.261
51	1.298	1.675	2.008	2.402	2.676	3.258
52	1.298	1.675	2.007	2.400	2.674	3.255
53	1.298	1.674	2.006	2.399	2.672	3.251
54	1.297	1.674	2.005	2.397	2.670	3.248
55	1.297	1.673	2.004	2.396	2.668	3.245
56	1.297	1.673	2.003	2.395	2.667	3.242
57	1.297	1.672	2.002	2.394	2.665	3.239
58	1.296	1.672	2.002	2.392	2.663	3.237
59	1.296	1.671	2.001	2.391	2.662	3.234
60	1.296	1.671	2.000	2.390	2.660	3.232
61	1.296	1.670	2.000	2.389	2.659	3.229
62	1.295	1.670	1.999	2.388	2.657	3.227
63	1.295	1.669	1.998	2.387	2.656	3.225
64	1.295	1.669	1.998	2.386	2.655	3.223
65	1.295	1.669	1.997	2.385	2.654	3.220
66	1.295	1.668	1.997	2.384	2.652	3.218
67	1.294	1.668	1.996	2.383	2.651	3.216
68	1.294	1.668	1.995	2.382	2.650	3.214
69	1.294	1.667	1.995	2.382	2.649	3.213
70	1.294	1.667	1.994	2.381	2.648	3.211
71	1.294	1.667	1.994	2.380	2.647	3.209
72	1.293	1.666	1.993	2.379	2.646	3.207
73	1.293	1.666	1.993	2.379	2.645	3.206
74	1.293	1.666	1.993	2.378	2.644	3.204
75	1.293	1.665	1.992	2.377	2.643	3.202
infinite	1.282	1.645	1.960	2.326	2.576	3.090

Appendix–2H
The *F*-Distribution Table
Area in the Right Tail under the *F*-Distribution Curve = .01

	df	Degrees of Freedom for the Numerator						
		1	**2**	**3**	**4**	**5**	**6**	**7**
Degrees of Freedom for the Denominator	**1**	4052.00	5000.00	5403.00	5625.00	5764.00	5859.00	5928.00
	2	98.50	99.00	99.17	99.25	99.30	99.33	99.36
	3	34.12	30.82	29.46	28.71	28.24	27.91	27.67
	4	21.20	18.00	16.69	15.98	15.52	15.21	14.98
	5	16.26	13.27	12.06	11.39	10.97	10.67	10.46
	6	13.75	10.92	9.78	9.15	8.75	8.47	8.26
	7	12.25	9.55	8.45	7.85	7.46	7.19	6.99
	8	11.26	8.65	7.59	7.01	6.63	6.37	6.18
	9	10.56	8.02	6.99	6.42	6.06	5.80	5.61
	10	10.04	7.56	6.55	5.99	5.64	5.39	5.20
	11	9.65	7.21	6.22	5.67	5.32	5.07	4.89
	12	9.33	6.93	5.95	5.41	5.06	4.82	4.64
	13	9.07	6.70	5.74	5.21	4.86	4.62	4.44
	14	8.86	6.51	5.56	5.04	4.69	4.46	4.28
	15	8.68	6.36	5.42	4.89	4.56	4.32	4.14
	16	8.53	6.23	5.29	4.77	4.44	4.20	4.03
	17	8.40	6.11	5.18	4.67	4.34	4.10	3.93
	18	8.29	6.01	5.09	4.58	4.25	4.01	3.84
	19	8.18	5.93	5.01	4.50	4.17	3.94	3.77
	20	8.10	5.85	4.94	4.43	4.10	3.87	3.70
	21	8.02	5.78	4.87	4.37	4.04	3.81	3.64
	22	7.95	5.72	4.82	4.31	3.99	3.76	3.59
	23	7.88	5.66	4.76	4.26	3.94	3.71	3.54
	24	7.82	5.61	4.72	4.22	3.90	3.67	3.50
	25	7.77	5.57	4.68	4.18	3.85	3.63	3.46
	30	7.56	5.39	4.51	4.02	3.70	3.47	3.30
	40	7.31	5.18	4.31	3.83	3.51	3.29	3.12
	50	7.17	5.06	4.20	3.72	3.41	3.19	3.02
	100	6.90	4.82	3.98	3.51	3.21	2.99	2.82

...(*Contd.*)

Appendix–2H (Continued)
The *F*-Distribution Table
Area in the Right Tail under the *F*-Distribution Curve = .01

	df	Degrees of Freedom for the Numerator					
		8	9	10	11	12	15
Degrees of Freedom for the Denominator	1	5981.00	6022.00	6056.00	6083.00	6106.00	6157.00
	2	99.37	99.39	99.40	99.41	99.42	99.43
	3	27.49	27.35	27.23	27.13	27.05	26.87
	4	14.80	14.66	14.55	14.45	14.37	14.20
	5	10.29	10.16	10.05	9.96	9.89	9.72
	6	8.10	7.98	7.87	7.79	7.72	7.56
	7	6.84	6.72	6.62	6.54	6.47	6.31
	8	6.03	5.91	5.81	5.73	5.67	5.52
	9	5.47	5.35	5.26	5.18	5.11	4.96
	10	5.06	4.94	4.85	4.77	4.71	4.56
	11	4.74	4.63	4.54	4.46	4.40	4.25
	12	4.50	4.39	4.30	4.22	4.16	4.01
	13	4.30	4.19	4.10	4.02	3.96	3.82
	14	4.14	4.03	3.94	3.86	3.80	3.66
	15	4.00	3.89	3.80	3.73	3.67	3.52
	16	3.89	3.78	3.69	3.62	3.55	3.41
	17	3.79	3.68	3.59	3.52	3.46	3.31
	18	3.71	3.60	3.51	3.43	3.37	3.23
	19	3.63	3.52	3.43	3.36	3.30	3.15
	20	3.56	3.46	3.37	3.29	3.23	3.09
	21	3.51	3.40	3.31	3.24	3.17	3.03
	22	3.45	3.35	3.26	3.18	3.12	2.98
	23	3.41	3.30	3.21	3.14	3.07	2.93
	24	3.36	3.26	3.17	3.09	3.03	2.89
	25	3.32	3.22	3.13	3.06	2.99	2.85
	30	3.17	3.07	2.98	2.91	2.84	2.70
	40	2.99	2.89	2.80	2.73	2.66	2.52
	50	2.89	2.78	2.70	2.63	2.56	2.42
	100	2.69	2.59	2.50	2.43	2.37	2.22

...(*Contd.*)

Appendix–2H (Continued)
The *F*-Distribution Table
Area in the Right Tail under the *F*-Distribution Curve = .01

	df	Degrees of Freedom for the Numerator					
		20	25	30	40	50	100
Degrees of Freedom for the Denominator	1	6209.00	6240.00	6261.00	6287.00	6303.00	6334.00
	2	99.45	99.46	99.47	99.47	99.48	99.49
	3	26.69	26.58	26.50	26.41	26.35	26.24
	4	14.02	13.91	13.84	13.75	13.69	13.58
	5	9.55	9.45	9.38	9.29	9.24	9.13
	6	7.40	7.30	7.23	7.14	7.09	6.99
	7	6.16	6.06	5.99	5.91	5.86	5.75
	8	5.36	5.26	5.20	5.12	5.07	4.96
	9	4.81	4.71	4.65	4.57	4.52	4.41
	10	4.41	4.31	4.25	4.17	4.12	4.01
	11	4.10	4.01	3.94	3.86	3.81	3.71
	12	3.86	3.76	3.70	3.62	3.57	3.47
	13	3.66	3.57	3.51	3.43	3.38	3.27
	14	3.51	3.41	3.35	3.27	3.22	3.11
	15	3.37	3.28	3.21	3.13	3.08	2.98
	16	3.26	3.16	3.10	3.02	2.97	2.86
	17	3.16	3.07	3.00	2.92	2.87	2.76
	18	3.08	2.98	2.92	2.84	2.78	2.68
	19	3.00	2.91	2.84	2.76	2.71	2.60
	20	2.94	2.84	2.78	2.69	2.64	2.54
	21	2.88	2.79	2.72	2.64	2.58	2.48
	22	2.83	2.73	2.67	2.58	2.53	2.42
	23	2.78	2.69	2.62	2.54	2.48	2.37
	24	2.74	2.64	2.58	2.49	2.44	2.33
	25	2.70	2.60	2.54	2.45	2.40	2.29
	30	2.55	2.45	2.39	2.30	2.25	2.13
	40	2.37	2.27	2.20	2.11	2.06	1.94
	50	2.27	2.17	2.10	2.01	1.95	1.82
	100	2.07	1.97	1.89	1.80	1.74	1.60

Appendix–2H

The *F*-Distribution Table

Area in the Right Tail under the *F*-Distribution Curve = .025

	df	Degrees of Freedom for the Numerator						
		1	2	3	4	5	6	7
Degrees of Freedom for the Denominator	1	647.80	799.50	864.20	899.60	921.80	937.10	948.20
	2	38.51	39.00	39.17	39.25	39.30	39.33	39.36
	3	17.44	16.04	15.44	15.10	14.88	14.73	14.62
	4	12.22	10.65	9.98	9.61	9.36	9.20	9.07
	5	10.01	8.43	7.76	7.39	7.15	6.98	6.85
	6	8.81	7.26	6.60	6.23	5.99	5.82	5.70
	7	8.07	6.54	5.89	5.52	5.29	5.12	4.99
	8	7.57	6.06	5.42	5.05	4.82	4.65	4.53
	9	7.21	5.72	5.08	4.72	4.48	4.32	4.20
	10	6.94	5.46	4.83	4.47	4.24	4.07	3.95
	11	6.72	5.26	4.63	4.28	4.04	3.88	3.76
	12	6.55	5.10	4.47	4.12	3.89	3.73	3.61
	13	6.41	4.97	4.35	4.00	3.77	3.60	3.48
	14	6.30	4.86	4.24	3.89	3.66	3.50	3.38
	15	6.20	4.77	4.15	3.80	3.58	3.41	3.29
	16	6.12	4.69	4.08	3.73	3.50	3.34	3.22
	17	6.04	4.62	4.01	3.66	3.44	3.28	3.16
	18	5.98	4.56	3.95	3.61	3.38	3.22	3.10
	19	5.92	4.51	3.90	3.56	3.33	3.17	3.05
	20	5.87	4.46	3.86	3.51	3.29	3.13	3.01
	21	5.83	4.42	3.82	3.48	3.25	3.09	2.97
	22	5.79	4.38	3.78	3.44	3.22	3.05	2.93
	23	5.75	4.35	3.75	3.41	3.18	3.02	2.90
	24	5.72	4.32	3.72	3.38	3.15	2.99	2.87
	25	5.69	4.29	3.69	3.35	3.13	2.97	2.85
	30	5.57	4.18	3.59	3.25	3.03	2.87	2.75
	40	5.42	4.05	3.46	3.13	2.90	2.74	2.62
	50	5.34	3.97	3.39	3.05	2.83	2.67	2.55
	100	5.18	3.83	3.25	2.92	2.70	2.54	2.42

...(*Contd.*)

Appendix–2H (Continued)
The *F*-Distribution Table
Area in the Right Tail under the *F*-Distribution Curve = .025

	df	Degrees of Freedom for the Numerator					
		8	9	10	11	12	15
Degrees of Freedom for the Denominator	1	956.70	963.30	968.60	973.00	976.70	984.90
	2	39.37	39.39	39.40	39.41	39.41	39.43
	3	14.54	14.47	14.42	14.37	14.34	14.25
	4	8.98	8.90	8.84	8.79	8.75	8.66
	5	6.76	6.68	6.62	6.57	6.52	6.43
	6	5.60	5.52	5.46	5.41	5.37	5.27
	7	4.90	4.82	4.76	4.71	4.67	4.57
	8	4.43	4.36	4.30	4.24	4.20	4.10
	9	4.10	4.03	3.96	3.91	3.87	3.77
	10	3.85	3.78	3.72	3.66	3.62	3.52
	11	3.66	3.59	3.53	3.47	3.43	3.33
	12	3.51	3.44	3.37	3.32	3.28	3.18
	13	3.39	3.31	3.25	3.20	3.15	3.05
	14	3.29	3.21	3.15	3.09	3.05	2.95
	15	3.20	3.12	3.06	3.01	2.96	2.86
	16	3.12	3.05	2.99	2.93	2.89	2.79
	17	3.06	2.98	2.92	2.87	2.82	2.72
	18	3.01	2.93	2.87	2.81	2.77	2.67
	19	2.96	2.88	2.82	2.76	2.72	2.62
	20	2.91	2.84	2.77	2.72	2.68	2.57
	21	2.87	2.80	2.73	2.68	2.64	2.53
	22	2.84	2.76	2.70	2.65	2.60	2.50
	23	2.81	2.73	2.67	2.62	2.57	2.47
	24	2.78	2.70	2.64	2.59	2.54	2.44
	25	2.75	2.68	2.61	2.56	2.51	2.41
	30	2.65	2.57	2.51	2.46	2.41	2.31
	40	2.53	2.45	2.39	2.33	2.29	2.18
	50	2.46	2.38	2.32	2.26	2.22	2.11
	100	2.32	2.24	2.18	2.12	2.08	1.97

...(*Contd.*)

Appendix–2H (Continued)
The *F*-Distribution Table
Area in the Right Tail under the *F*-Distribution Curve = .025

	df	Degrees of Freedom for the Numerator					
		20	25	30	40	50	100
Degrees of Freedom for the Denominator	1	993.10	998.10	1001.00	1006.00	1008.00	1013.00
	2	39.45	39.46	39.46	39.47	39.48	39.49
	3	14.17	14.12	14.08	14.04	14.01	13.96
	4	8.56	8.50	8.46	8.41	8.38	8.32
	5	6.33	6.27	6.23	6.18	6.14	6.08
	6	5.17	5.11	5.07	5.01	4.98	4.92
	7	4.47	4.40	4.36	4.31	4.28	4.21
	8	4.00	3.94	3.89	3.84	3.81	3.74
	9	3.67	3.60	3.56	3.51	3.47	3.40
	10	3.42	3.35	3.31	3.26	3.22	3.15
	11	3.23	3.16	3.12	3.06	3.03	2.96
	12	3.07	3.01	2.96	2.91	2.87	2.80
	13	2.95	2.88	2.84	2.78	2.74	2.67
	14	2.84	2.78	2.73	2.67	2.64	2.56
	15	2.76	2.69	2.64	2.59	2.55	2.47
	16	2.68	2.61	2.57	2.51	2.47	2.40
	17	2.62	2.55	2.50	2.44	2.41	2.33
	18	2.56	2.49	2.44	2.38	2.35	2.27
	19	2.51	2.44	2.39	2.33	2.30	2.22
	20	2.46	2.40	2.35	2.29	2.25	2.17
	21	2.42	2.36	2.31	2.25	2.21	2.13
	22	2.39	2.32	2.27	2.21	2.17	2.09
	23	2.36	2.29	2.24	2.18	2.14	2.06
	24	2.33	2.26	2.21	2.15	2.11	2.02
	25	2.30	2.23	2.18	2.12	2.08	2.00
	30	2.20	2.12	2.07	2.01	1.97	1.88
	40	2.07	1.99	1.94	1.88	1.83	1.74
	50	1.99	1.92	1.87	1.80	1.75	1.66
	100	1.85	1.77	1.71	1.64	1.59	1.48

...(*Contd.*)

Appendix–2H
The *F*-Distribution Table
Area in the Right Tail under the *F*-Distribution Curve = .05

	df	Degrees of Freedom for the Numerator						
		1	2	3	4	5	6	7
Degrees of Freedom for the Denominator	1	161.50	199.50	215.70	224.60	230.20	234.00	236.80
	2	18.51	19.00	19.16	19.25	19.30	19.33	19.35
	3	10.13	9.55	9.28	9.12	9.01	8.94	8.89
	4	7.71	6.94	6.59	6.39	6.26	6.16	6.09
	5	6.61	5.79	5.41	5.19	5.05	4.95	4.88
	6	5.99	5.14	4.76	4.53	4.39	4.28	4.21
	7	5.59	4.74	4.35	4.12	3.97	3.87	3.79
	8	5.32	4.46	4.07	3.84	3.69	3.58	3.50
	9	5.12	4.26	3.86	3.63	3.48	3.37	3.29
	10	4.96	4.10	3.71	3.48	3.33	3.22	3.14
	11	4.84	3.98	3.59	3.36	3.20	3.09	3.01
	12	4.75	3.89	3.49	3.26	3.11	3.00	2.91
	13	4.67	3.81	3.41	3.18	3.03	2.92	2.83
	14	4.60	3.74	3.34	3.11	2.96	2.85	2.76
	15	4.54	3.68	3.29	3.06	2.90	2.79	2.71
	16	4.49	3.63	3.24	3.01	2.85	2.74	2.66
	17	4.45	3.59	3.20	2.96	2.81	2.70	2.61
	18	4.41	3.55	3.16	2.93	2.77	2.66	2.58
	19	4.38	3.52	3.13	2.90	2.74	2.63	2.54
	20	4.35	3.49	3.10	2.87	2.71	2.60	2.51
	21	4.32	3.47	3.07	2.84	2.68	2.57	2.49
	22	4.30	3.44	3.05	2.82	2.66	2.55	2.46
	23	4.28	3.42	3.03	2.80	2.64	2.53	2.44
	24	4.26	3.40	3.01	2.78	2.62	2.51	2.42
	25	4.24	3.39	2.99	2.76	2.60	2.49	2.40
	30	4.17	3.32	2.92	2.69	2.53	2.42	2.33
	40	4.08	3.23	2.84	2.61	2.45	2.34	2.25
	50	4.03	3.18	2.79	2.56	2.40	2.29	2.20
	100	3.94	3.09	2.70	2.46	2.31	2.19	2.10

...(*Contd.*)

Appendix–2H (Continued)
The *F*-Distribution Table
Area in the Right Tail under the *F*-Distribution Curve = .05

	df	Degrees of Freedom for the Numerator					
		8	9	10	11	12	15
Degrees of Freedom for the Denominator	1	238.90	240.50	241.90	243.00	243.90	246.00
	2	19.37	19.38	19.40	19.40	19.41	19.43
	3	8.85	8.81	8.79	8.76	8.74	8.70
	4	6.04	6.00	5.96	5.94	5.91	5.86
	5	4.82	4.77	4.74	4.70	4.68	4.62
	6	4.15	4.10	4.06	4.03	4.00	3.94
	7	3.73	3.68	3.64	3.60	3.57	3.51
	8	3.44	3.39	3.35	3.31	3.28	3.22
	9	3.23	3.18	3.14	3.10	3.07	3.01
	10	3.07	3.02	2.98	2.94	2.91	2.85
	11	2.95	2.90	2.85	2.82	2.79	2.72
	12	2.85	2.80	2.75	2.72	2.69	2.62
	13	2.77	2.71	2.67	2.63	2.60	2.53
	14	2.70	2.65	2.60	2.57	2.53	2.46
	15	2.64	2.59	2.54	2.51	2.48	2.40
	16	2.59	2.54	2.49	2.46	2.42	2.35
	17	2.55	2.49	2.45	2.41	2.38	2.31
	18	2.51	2.46	2.41	2.37	2.34	2.27
	19	2.48	2.42	2.38	2.34	2.31	2.23
	20	2.45	2.39	2.35	2.31	2.28	2.20
	21	2.42	2.37	2.32	2.28	2.25	2.18
	22	2.40	2.34	2.30	2.26	2.23	2.15
	23	2.37	2.32	2.27	2.24	2.20	2.13
	24	2.36	2.30	2.25	2.22	2.18	2.16
	25	2.34	2.28	2.24	2.20	2.16	2.09
	30	2.27	2.21	2.16	2.13	2.09	2.01
	40	2.18	2.12	2.08	2.04	2.00	1.92
	50	2.13	2.07	2.03	1.99	1.95	1.87
	100	2.03	1.97	1.93	1.89	1.85	1.77

...(*Contd.*)

Appendix–2H (Continued)
The *F*-Distribution Table
Area in the Right Tail under the *F*-Distribution Curve = .05

	df	Degrees of Freedom for the Numerator					
		20	**25**	**30**	**40**	**50**	**100**
Degrees of Freedom for the Denominator	**1**	248.00	249.30	250.10	251.10	251.80	253.00
	2	19.45	19.46	19.46	19.47	19.48	19.49
	3	8.66	8.63	8.62	8.59	8.58	8.55
	4	5.80	5.77	5.75	5.72	5.70	5.66
	5	4.56	4.52	4.50	4.46	4.44	4.41
	6	3.87	3.83	3.81	3.77	3.75	3.71
	7	3.44	3.40	3.38	3.34	3.32	3.27
	8	3.15	3.11	3.08	3.04	3.02	2.97
	9	2.94	2.89	2.86	2.83	2.80	2.76
	10	2.77	2.73	2.70	2.66	2.64	2.59
	11	2.65	2.60	2.57	2.53	2.51	2.46
	12	2.54	2.50	2.47	2.43	2.40	2.35
	13	2.46	2.41	2.38	2.34	2.31	2.26
	14	2.39	2.34	2.31	2.27	2.24	2.19
	15	2.33	2.28	2.25	2.20	2.18	2.12
	16	2.28	2.23	2.19	2.15	2.12	2.07
	17	2.23	2.18	2.15	2.10	2.08	2.02
	18	2.19	2.14	2.11	2.06	2.04	1.98
	19	2.16	2.11	2.07	2.03	2.00	1.94
	20	2.12	2.07	2.04	1.99	1.97	1.91
	21	2.10	2.05	2.01	1.96	1.94	1.88
	22	2.07	2.02	1.97	1.94	1.91	1.85
	23	2.05	2.00	1.96	1.91	1.88	1.82
	24	2.03	1.97	1.94	1.89	1.86	1.80
	25	2.01	1.96	1.92	1.87	1.84	1.78
	30	1.93	1.88	1.84	1.79	1.76	1.70
	40	1.84	1.78	1.74	1.69	1.66	1.59
	50	1.78	1.73	1.69	1.63	1.60	1.52
	100	1.68	1.62	1.57	1.52	1.48	1.39

...(*Contd.*)

Appendix–2H
The *F*-Distribution Table
Area in the Right Tail under the *F*-Distribution Curve = .10

	df	Degrees of Freedom for the Numerator						
		1	2	3	4	5	6	7
Degrees of Freedom for the Denominator	1	39.86	49.50	53.59	55.83	57.24	58.20	58.91
	2	8.53	9.00	9.16	9.24	9.29	9.33	9.35
	3	5.54	5.46	5.39	5.34	5.31	5.28	5.27
	4	4.54	4.32	4.19	4.11	4.05	4.01	3.98
	5	4.06	3.78	3.62	3.52	3.45	3.40	3.37
	6	3.78	3.46	3.29	3.18	3.11	3.05	3.01
	7	3.59	3.26	3.07	2.96	2.88	2.83	2.78
	8	3.46	3.11	2.92	2.81	2.73	2.67	2.62
	9	3.36	3.01	2.81	2.69	2.61	2.55	2.51
	10	3.29	2.92	2.73	2.61	2.52	2.46	2.41
	11	3.23	2.86	2.66	2.54	2.45	2.39	2.34
	12	3.18	2.81	2.61	2.48	2.39	2.33	2.28
	13	3.14	2.76	2.56	2.43	2.35	2.28	2.23
	14	3.10	2.73	2.52	2.39	2.31	2.24	2.19
	15	3.07	2.70	2.49	2.36	2.27	2.21	2.16
	16	3.05	2.67	2.46	2.33	2.24	2.18	2.13
	17	3.03	2.64	2.44	2.31	2.22	2.15	2.10
	18	3.01	2.62	2.42	2.29	2.20	2.13	2.08
	19	2.99	2.61	2.40	2.27	2.18	2.11	2.06
	20	2.97	2.59	2.38	2.25	2.16	2.09	2.04
	21	2.96	2.57	2.36	2.23	2.14	2.08	2.02
	22	2.95	2.56	2.35	2.22	2.13	2.06	2.01
	23	2.94	2.55	2.34	2.21	2.11	2.05	1.99
	24	2.93	2.54	2.33	2.19	2.10	2.04	1.98
	25	2.92	2.53	2.32	2.18	2.09	2.02	1.97
	30	2.88	2.49	2.28	2.14	2.05	1.98	1.93
	40	2.84	2.44	2.23	2.09	2.00	1.93	1.87
	50	2.81	2.41	2.20	2.06	1.97	1.90	1.84
	100	2.76	2.36	2.14	2.00	1.91	1.83	1.78

...(Contd.)

Appendix–2H (Continued)
The *F*-Distribution Table
Area in the Right Tail under the *F*-Distribution Curve = .10

	df	Degrees of Freedom for the Numerator					
		8	**9**	**10**	**11**	**12**	**15**
Degrees of Freedom for the Denominator	**1**	59.44	59.86	60.19	60.47	60.71	61.22
	2	9.37	9.38	9.39	9.40	9.41	9.42
	3	5.25	5.24	5.23	5.22	5.22	5.20
	4	3.95	3.94	3.92	3.91	3.90	3.87
	5	3.34	3.32	3.30	3.28	3.27	3.24
	6	2.98	2.96	2.94	2.92	2.90	2.87
	7	2.75	2.72	2.70	2.68	2.67	2.63
	8	2.59	2.56	2.54	2.52	2.50	2.46
	9	2.47	2.44	2.42	2.40	2.38	2.34
	10	2.38	2.35	2.32	2.30	2.28	2.24
	11	2.30	2.27	2.25	2.23	2.21	2.17
	12	2.24	2.21	2.19	2.17	2.15	2.10
	13	2.20	2.16	2.14	2.12	2.10	2.05
	14	2.15	2.12	2.10	2.07	2.05	2.01
	15	2.12	2.09	2.06	2.04	2.02	1.97
	16	2.09	2.06	2.03	2.01	1.99	1.94
	17	2.06	2.03	2.00	1.98	1.96	1.91
	18	2.04	2.00	1.98	1.95	1.93	1.89
	19	2.02	1.98	1.96	1.93	1.91	1.86
	20	2.00	1.96	1.94	1.91	1.89	1.84
	21	1.98	1.95	1.92	1.90	1.87	1.83
	22	1.97	1.93	1.90	1.88	1.86	1.81
	23	1.95	1.92	1.89	1.87	1.84	1.80
	24	1.94	1.91	1.88	1.85	1.83	1.78
	25	1.93	1.89	1.87	1.84	1.82	1.77
	30	1.88	1.85	1.82	1.79	1.77	1.72
	40	1.83	1.79	1.76	1.74	1.71	1.66
	50	1.80	1.76	1.73	1.70	1.68	1.63
	100	1.73	1.69	1.66	1.64	1.61	1.56

...(Contd.)

Appendix–2H (Continued)
The *F*-Distribution Table
Area in the Right Tail under the *F*-Distribution Curve = .10

	df	Degrees of Freedom for the Numerator					
		20	**25**	**30**	**40**	**50**	**100**
Degrees of Freedom for the Denominator	**1**	61.74	62.05	62.26	62.53	62.69	63.01
	2	9.44	9.45	9.46	9.47	9.47	9.48
	3	5.18	5.17	5.17	5.16	5.15	5.14
	4	3.84	3.83	3.82	3.80	3.80	3.78
	5	3.21	3.19	3.17	3.16	3.15	3.13
	6	2.84	2.81	2.80	2.78	2.77	2.75
	7	2.59	2.57	2.56	2.54	2.52	2.50
	8	2.42	2.40	2.38	2.36	2.35	2.32
	9	2.30	2.27	2.25	2.23	2.22	2.19
	10	2.20	2.17	2.16	2.13	2.12	2.09
	11	2.12	2.10	2.08	2.05	2.04	2.01
	12	2.06	2.03	2.01	1.99	1.97	1.94
	13	2.01	1.98	1.96	1.93	1.92	1.88
	14	1.96	1.93	1.91	1.89	1.87	1.83
	15	1.92	1.89	1.87	1.85	1.83	1.79
	16	1.89	1.86	1.84	1.81	1.79	1.76
	17	1.86	1.83	1.81	1.78	1.76	1.73
	18	1.84	1.80	1.78	1.75	1.74	1.70
	19	1.81	1.78	1.76	1.73	1.71	1.67
	20	1.79	1.76	1.74	1.71	1.69	1.65
	21	1.78	1.74	1.72	1.69	1.67	1.63
	22	1.76	1.73	1.70	1.67	1.65	1.61
	23	1.74	1.71	1.69	1.66	1.64	1.59
	24	1.73	1.70	1.67	1.64	1.62	1.58
	25	1.72	1.68	1.66	1.63	1.61	1.56
	30	1.67	1.63	1.61	1.57	1.55	1.51
	40	1.61	1.57	1.54	1.51	1.48	1.43
	50	1.57	1.53	1.50	1.46	1.44	1.39
	100	1.49	1.45	1.42	1.38	1.35	1.29

Chapter 3

Product and Process Design

3.1 INTRODUCTION

In contrast to what the manufacturing and business world had been in the past, there has been a radical change in the perspectives in which an organization has to function in the present times. In almost all aspects of organizational systems, the effect of technological advancement has become so significant and pronounced that it becomes very difficult to keep pace and stay competitive for any company. Wherever its plants and production units may be located, they have to face the challenges of modern times – fast changing customer requirements and aspirations, keeping pace with the rapid change in the technology of products and processes, and continuously evolving worksystems of extremely complex nature requiring long-sustaining solutions. In this context, the critical issues related to the design of products and processes assume significance.

Out of many factors responsible for this change in business and manufacturing perspectives, the product development plays a crucial role. In order to meet the continuously changing expectations and standard of living of the users and the customers, more and more number of newer products in different models must be designed and developed within a short and limited time span by the companies. They also should do so with minimum and acceptable investment level by selecting the most appropriate technologies for each and every activity or task they have to work on. This has posed a tremendous challenge to the management of almost all companies these days. As the whole product development entails mainly two important aspects, viz. product design and process design, it becomes imperative that design of products and processes should be explained in the light of new opportunities and constraints in the present business world.

In the present era of the so-called 'technology push', although many companies, particularly in the economically developed countries, apparently emphasize less on the importance of manufacturing activities having outsourcing many critical activities, including design, to less economically developed nations because of availability of cheaper labour and capital, and concentrating mainly on the assembly of parts and components in their own plants, the relevance of manufacturing cannot be undermined from a global perspective. Whatever may be the business policy of a company, product and process design remains its core issue. There are many instances of companies going 'bankrupt' or 'sick' simply because of their consistent neglect of and giving no importance to issues related to new product development.

A number of advantages a company may have if it emphasizes on product and process design. Some of these advantages are as follows:

(*a*) **Increased market demand for the products:** Because of rapid and large-scale changes and fluctuations in demand levels for most of the products, the companies must not wait for long to know customer responses regarding expectations from the products. Remaining reactive to customers' responses may not be a desirable approach. There may not be enough time to offer new products to satisfy a customer's current needs. Instead, companies must anticipate the effects of future and upcoming technologies on customer expectations and position themselves, well in advance, in the form of designing new products to satisfy current needs of the prospective customers.

(*b*) **Advantages of partnership:** Today, in almost all corners of the world, the business environment has become highly competitive, particularly with the onset of globalization. This environment demands agility and a commitment to excellence. To be successful, the companies should offer products with the best-possible quality comparable with the so-called 'world-class' standards. A typical 'world-class' company is capable of providing products and/or services that can successfully compete in an open global or international market. Extended enterprises, partnerships, and alliances are normally used in different forms to create a true synergy leading to a competitive advantage in most cases. Successful businesses will be truly global in their organizational structures and value-creating processes. Through the partnerships, the company is able to perform excellently in R & D, design, and manufacturing. The strength of a company is derived from its continuing enhanced performance in these activities. A collaborative effort with other companies will eventually create a growing international market for its products.

(*c*) **Improved customer service:** Well-established, renowned, and customer-friendly organizations consider satisfaction of customers needs and wants as the highest priority. For these organizations, the idea of being close to its customers is very important to compete against global competition. In a global economy, customer requests may come in many ways, and the means of communication may be verbal, written, and other forms, such as e-mail, fax and telephone call. In this context, customer response management (CRM) is an important approach as a means for improved customer service. The basic objective of this approach is to meet the service requirements of customers efficiently and effectively with minimum cost.

(*d*) **Reduced lead time to produce new products:** In today's market, there is tremendous pressure on management to produce products with much shorter lead times. Often referred to as 'speed' or 'lag time', a reduced lead time is a goal for both existing and new products. New products must be designed and manufactured quickly in order to get into the market before the competition sets in during the growth phase of the product life cycle. If a product is introduced late in the market, sales and market share are invariably reduced, or may be lost. The lead time from order to delivery of existing products must be constantly reviewed and reduced, and must be considered as the top priority for an effective product development strategy for a company.

In today's business environment, it is imperative that an established and leading company has the potential and capability to respond to various kinds of changes vis-à-vis product development. In this context, the factors that affect the product development process and the product life cycle are numerous. Some of the main factors are as follows:

(*i*) Rate of innovation,

(*ii*) Use of software tools, rapid prototyping, and virtual reality,

(*iii*) Mass customization and customized production,

(*iv*) Core competency, partnerships, and outsourcing,

(*v*) Modern telecommunication means and internet,

(*vi*) Electronic commerce (E-commerce)

(*vii*) Flexibility and agility, and

(*viii*) Global manufacturing

Important issues related to each of the above-mentioned factors are briefly described below.

(*i*) **Rate of innovation:** Innovation, as an important factor in the design of new products, represents the basic strength and the core competency of a company. Only an innovative company is able to design and produce new products with quality characteristics as desired by the customers. In most cases, innovativeness results in development of a unique product with which the company is branded. The advances of new technologies have been primarily responsible for creating opportunities for increasing the number of new products. New types of materials and processes including 'virtual reality' help design new and exciting products. For many companies, software is also becoming the focused area for innovation that differentiates their products from its competitors.

Significant pressures exist to develop new products much faster since the first innovative product is often the competitive winner. Many product development projects are now started before some key technologies are even ready. This so-called 'co-development method' increases the level of technical risk. In the future, there will be lesser time available to develop innovative and new products. Time constraints will become an even greater factor in the development and production of new innovative products.

(*ii*) **Use of software tools, rapid prototyping, and virtual reality:** Most businesses are driven by three essential activities, *viz.* design, manufacturing, and service. Competitive advantage is gained in the design process when software products or tools improve or shorten product development time and enhance service support. Available software tools, such as computer-aided design/engineering (CAD/CAE), expert systems, virtual reality, and prototyping tools may assist the design team in several ways, such as customer needs assessment, product layout, selection of components and materials, simulating the manufacturing processes and analysis, and assessment of usefulness to the perspective customers. Intelligent CAE systems are able to quickly access large knowledge bases, and provide answers to questions that might arise during the design process. Technical developments allow companies to have visualization and simulation capabilities that are more complete than a conventional CAD system. Virtual reality utilizes interactive graphics software to create computer-generated prototypes and simulations that are so close to reality that users believe they are participating in a real-world situation with a real final product. Similarly, prototyping consists of building an experimental product rapidly and inexpensively for users to evaluate. By interacting with a prototype, the users and business personnel may have a better idea of final manufacturing and functional requirements. The prototype approved by final users can then be used as a template to develop the final product requirements and the corresponding manufacturing operations. The future scenario may be in most likely in the form of design team defining customer requirements (as mentioned in 'product brief') in specific terms, and the software tools developing the possible design alternatives performing engineering analyses including testing of simulated and prototypes to evaluate, purchase of parts and components, and sending all the necessary information for manufacturing the product to the manufacturing units.

(*iii*) **Mass customization and customized production:** Today, in almost all over the world, the customers want products that are tailored to meet their unique and individual needs. The so-called 'customized' product which they want may be different from the 'generic' product in shape, size, and even functional characteristics. In this context, the mass customization may be an ideal and desirable strategy. Mass customization is the ability of a company to offer specific product or service solutions while still realizing the benefits of large-scale operations. The strategy may require building a generic product by the company's design team, with modifications being done by the concerned customers. This is actually a 'build-to-order' solution which requires manufacture of a customer-specified product only after it is ordered. This requires manufacturing to build many more versions or models of the same products incorporating the concept of product flexibility.

(*iv*) **Core competency, partnerships, and outsourcing:** Among many factors, the cost of staying competitive and developing new technologies is forcing many companies to focus internal development on a few core competencies that the company considers strategic to being competitive. Core competencies may be developed in any area of design, marketing, manufacturing, or service. These are the critical areas that make a company unique or better than the others in the market. Sometimes, companies may buy or merge with other companies that have needed core competency. Areas that are not a part of the core (i.e. not critical) but are still needed, are purchased or developed for the company from outside vendors and partners. The result of companies focusing on core competencies is a rising trend called 'outsourcing'. Historically, companies have tried to manufacture almost all or a part of a product's parts and components within their production units. This 'in-house' focus requires a company to be good in many different areas of manufacturing and software development. For many companies, the cost of doing everything is indeed very high. Outsourcing parts, components, sub-assemblies, assemblies, and software modules may save substantial resources and money. Successful outsourcing companies have developed into world-class manufacturers with high efficiency and lower costs. Companies who develop consumer products usually specialize in the design process and protecting their brand names and then leave manufacturing and distribution to other companies. For obvious reasons, an increased level in outsourcing places more importance on vendor selection and supply chain management.

(*v*) **Modern telecommunication means and internet:** Many types of telecommunication means, in their present form, have significantly helped increase the level of collaboration that is required and sometimes very much essential for new product development. The internet technology has created a new universal technology platform upon which to build a variety of new products or services. Its potential for reshaping the way products are developed is enormous. By eliminating geographic barriers obstructing the global flow of information, the internet and telecommunications are paving the way for increasing collaboration in new product development.

(*vi*) **Electronic commerce (E-commerce):** E-commerce is a paperless exchange, via computer networks of engineering and business information using e-mail and Electronic Data Interchange (EDI). Its introduction has dramatically changed the procurement process, eventually affecting all kinds of purchase and supply. For example, internet file transfer of design CAD data has become a part of the printed circuit board (PCB) production cycle at many companies. The factory electronically receives the data and then immediately manufactures the PCB as per the given specifications. Similarly, quotations from the prospective suppliers may be very quickly obtained and processed using e-commerce. The EDI may also be used to combine electronic transactions with production and inventory information to have parts directly delivered to the factory floor. The file transfer methods are expected to improve in near future for the performance of transaction processing and delivery of goods with least lead time and high effectiveness. E-commerce as a technique has a bearing on several aspects, such as marketing (customers and suppliers providing up-to-date information to the designers), purchasing and inventory management, maintenance of re-order schedules (real-time sales data and inventory information from the production shops are sent to the suppliers online), supply chain management (in which companies and suppliers working closely together as partners), and design of products and services where interdisciplinary and group approach is a necessity.

Through e-commerce, the design team is in a position to offer a product which may be state-of-the-art type having characteristics satisfying the current needs of the customers. This may invariably help the company increase the market share of the product in future.

(*vii*) **Flexibility and agility:** The requirements of reduced lead time, quick changeovers to new products, and mass customization have made a modern-day manufacturing a flexible one. A company's manufacturing system must now be able to produce different types of products quickly without modifying existing production lines or adding new ones. To meet the constantly changing requirements of the customers as well as the markets, there must be flexibility in productmix, design, volume of production, and delivery schedules. Although a truly flexible production line may be expensive to install, it is still desirable, because model-wise changeover in a rigid production line is prohibitively costly even if it is technically possible. However, flexibility pays off when a manufacturing system is able to change its production line by reprogramming the existing equipment rather than by replacing it. For this, non-conventional manufacturing using computer and numerical control based technology, such as Numerical Control (NC), Computer Numerical Control (CNC), Direct Numerical Control (DNC), and Flexible Manufacturing System (FMS) is recommended in most cases.

(*viii*) **Global manufacturing:** Globalization is instrumental in changing product development and manufacturing scenario. Products are now being collaboratively developed in many different locations all over the world. For manufacturing, this has resulted in worldwide production facilities and the use of many overseas suppliers. The same products may be manufactured in many different locations in several different countries. The different manufacturing facilities can vary in terms of process capabilities, labour skills, quality, cost, and many other factors. Critical parts that are purchased from another country must then be shipped to the various facilities. This causes many communication and logistics problems. Effective communication is difficult due to the differences in languages, cultures, time zones, education, and government regulations. Logistic costs that include shipping and handling costs become a significant part of the product cost. Scheduling, ordering and shipping of parts and products all over the globe also becomes very complex because of these reasons.

In addition to the above mentioned factors that define the path of the product development strategies for the companies, two more factors, *viz.* automation and environmental issues, have a far-reaching effect on the new product design and manufacturing.

Although automation of manual tasks has been a major strategy for many years, the level of automation in a manufacturing system of many companies is dependent on several factors, such as the type of product technologies to be adopted or required, and specified product requirements as stipulated by government regulations. For example, semiconductors and optoelectronics require automated processes and integrated manufacturing, planning and control.

Automation may also improve the quality level of a product. Because of automation, the processes may produce a product consistently. However, automation may not necessarily guarantee higher quality level in products in all cases. In combination with an effective quality improvement and control approach, automation may result in consistent quality in products. Thus, by solving quality problems in an automated process, the highest levels of product quality can be obtained.

In recent years, most of the companies have become aware about the consideration of environmental issues while adopting a product development strategy. Although the companies produce 'products', they also produce 'wastes' in different forms and levels. Hence, the companies have been active in minimizing process wastes and energy use, developing manufacturing process alternatives using environment friendly materials, and integrating pollution control and abatement into their facilities. In some cases, companies may invest in a particular technology only to discover a few years later that changes in some environmental factors make the investment absolutely unattractive and futile. For example, car companies are somewhat uncertain about investing in technology to manufacture electric cars, because they are not sure about the future of this technology and its impact on environment in the long run. Thus, the future ramifications of the environment on product development may be enormous.

Rapid changes caused by new technologies and globalization in recent times have altogether changed the scenario for product development. Implementing new technologies for new product development may be a high-risk venture for many companies. The key is to identify, reduce, and control various kinds of risks without limiting innovation, creativity, simplicity, and flexibility in the product development process. Most of the problems found in current product development may be traced back to inadequate design or design management. The focus needs to be on the fundamentals of successful product development, sometimes referred to as "best practices" as followed by successful companies in product development. They are developed from lessons learned that are examples of the best and worst methods that have been found on previous projects and at different companies. A successful company's experience with new product development may be worthwhile in several ways. It basically enriches the company with knowledge about the progress of product development process and the technical risk associated with new and upcoming technologies. It also gains knowledge in the recommendations that have been found to be successful in other products.

Using best industry practices have been shown to be a very effective method for product development. This approach identifies the "best" practices for a particular application from both inside and outside of the company. It ensures that the design team does not repeat the same mistakes from previous projects.

A successful design strategy requires a multidisciplinary and collaborative approach focusing an appropriate methodology to be used for product development. The strategy identifies tasks and activities that are most likely to produce competitive products with reduced technical risk. However, the selection of the best practice to be used depends on the particular company, business environment, and the product itself.

3.2 IMPORTANT APPROACHES IN PRODUCT DEVELOPMENT

The product development process in a company is one of the important activities that is primarily responsible for its long-term survival and sustainability in the market. The entire process of product development consists of several steps to be followed in sequence in traditional approach. This process is shown in Fig. 3.1. This process typically requires that the persons involved may be able to communicate freely and effectively among themselves regarding matters such as data collection, problem solving, and suggestions for improvement in products, processes, and manufacturing. The main objective is to improve communication between the manufacturing, the top management group, the designers, the suppliers, and the customers on a continuous basis. To become successful in product development in the long term, the company is required to establish a collaborative and multidisciplinary approach for product development.

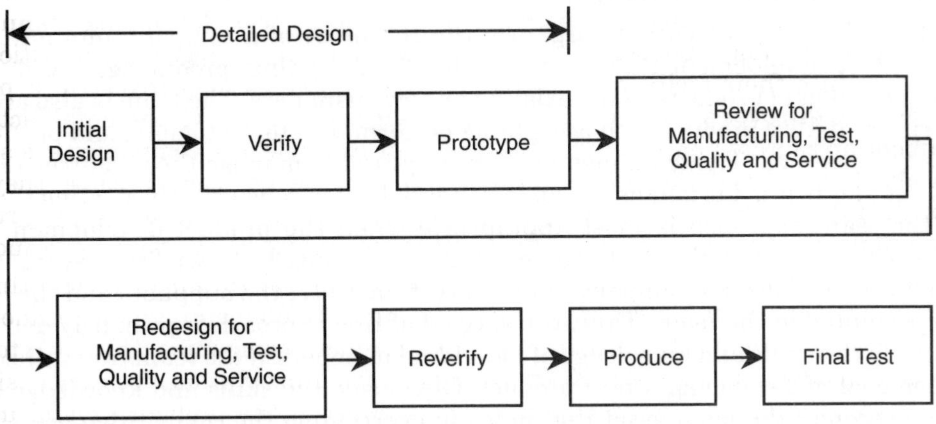

Fig. 3.1 Traditional Product Development Process.

A collaborative approach may be defined as "the process of people working in teams to pursue design innovation" in which information, ideas, and problem solving approaches are actively shared among the team members. This approach may be either synchronous or asynchronous. In a synchronous approach, the team members meet together either face-to-face in one location, or through audio or video conferencing tools when they are located at different places to discuss and decide on various aspects of product development jointly at the same time. On the other hand, in an asynchronous approach team members separately contribute to the respective aspects of product development at different stages and times, and the coordinator of the team becomes responsible for decision making at different stages of the product development process.

The following aspects are usually taken care of in a typical collaborative approach:

(*i*) the design objectives

(*ii*) existing capabilities of manufacturing systems (machine tools, quality standards, etc.),

(*iii*) changes in the existing designs of the products,

(*iv*) action steps for improvement in existing designs and initiation for new designs, and

(*v*) motivational issues related to active involvement of the team members.

To make the collaborative approach effective and result-oriented, the coordinator or the team leader plays a significant role. He or she has to ensure that (*i*) the state-of-the-art technologies and methods used are as far as practicable, (*ii*) the team members are adequately trained in their respective areas of expertise, (*iii*) the team members are encouraged to take up responsibilities for design and development tasks independently, and (*iv*) appropriate strategies and action steps are followed in order to have best support from the individual members.

In today's business environment, a number of tools, techniques, and methodologies may be quite useful in design and development of products. Product Development Team (PDT), Concurrent or Simultaneous Engineering (C/SE), and Computer-aided Design/Manufacturing/Engineering (CAD/CAM/CAE) and Virtual Reality (VR) are a few important methodologies and tools that are being widely used by many companies. In the following sections, the important characteristics of PDT and C/SE are briefly discussed.

Product Development Team (PDT)

The objective of this approach is to organize the resources of a company into an integrated team for product development that is responsible for designing, producing, and delivering a product or a number of products to the existing or new customers. The team is also accountable for delivering a quality product with acceptable performance to customers, and profitability to the company concerned. The team members are required to manage the resources under their command in order to provide maximum customer satisfaction concurrently meeting the business goal. As the team approach is most appropriate when the product development requires knowledge in different disciplines, the team members are usually drawn from several functions, departments, or sections of a company. Even in certain cases, the suppliers and the customers may be represented in the team. Thus, the success of this approach is primarily dependent on how effectively the skills and knowledge of individual members of the teams are utilized toward the common goal of developing a new product. Obviously, the skills and knowledge embodied in the team becomes the main asset that helps in overcoming the competitive pressure in the market. In specific terms, the team approach has the following advantages:

(*i*) It ensures that the product design is compatible with manufacturing or service capabilities, and life cycle requirements are most likely marketable with a competitive price.

(*ii*) It helps speed up the whole design process including its delivery and service as an integrated approach through team work is vital for success in a global economy.

(*iii*) It enables communication and collaboration between personnel in different organizations, functional areas, disciplines, and locations by using state-of-the-art tools and techniques at different phases of the product development.

(*iv*) By working collaboratively with suppliers, customers, and other partners on a continuous basis, there may be significant control on the external product cost.

Irrespective of the type of the company, a multifunctional approach becomes essential to have the maximum benefits in the team approach for designing and developing products. This approach is based on integrating product development with product life cycle issues to collectively achieve flexibility, high profit margins, and customer satisfaction.

In the current business environment, technologies offer an excellent opportunity to resolve the problems that might have stood in the way of streamlining the product development process in the past. It is now possible to organize product development projects globally and work locally. Information technologies such as e-mail, internet, and videoconferencing allow greater coordination between geographically dispersed workers and managers. Collaborative work across different countries spread over a large area and distance has become a reality as designers work on a new product together.

Concurrent/Simultaneous Engineering

The main objective of Concurrent or Simultaneous Engineering (C/SE) is to improve the process of developing a product. This approach is defined as "a systematic approach to the integrated and concurrent design of products and their related processes, including manufacture and support. This approach is intended to cause the developers, from the outset, to consider all elements of the product life cycle from conception through disposal, including quality, cost, schedule, and user requirements." In plain and simple terms, this means that the C/SE concept helps develop a product through simultaneous working on several aspects, issues, and functions as assigned at the same time such that the product development time gets shortened. With the use of CE concepts, benefits in the form of reduced direct labour costs, life cycle time, inventory, scrap, rework, and engineering changes are realized. Many of these changes are significant. With the application of this approach, simplification of the development process as well as specific tasks involved is also possible. The C/SE based product development process is shown in Fig. 3.2.

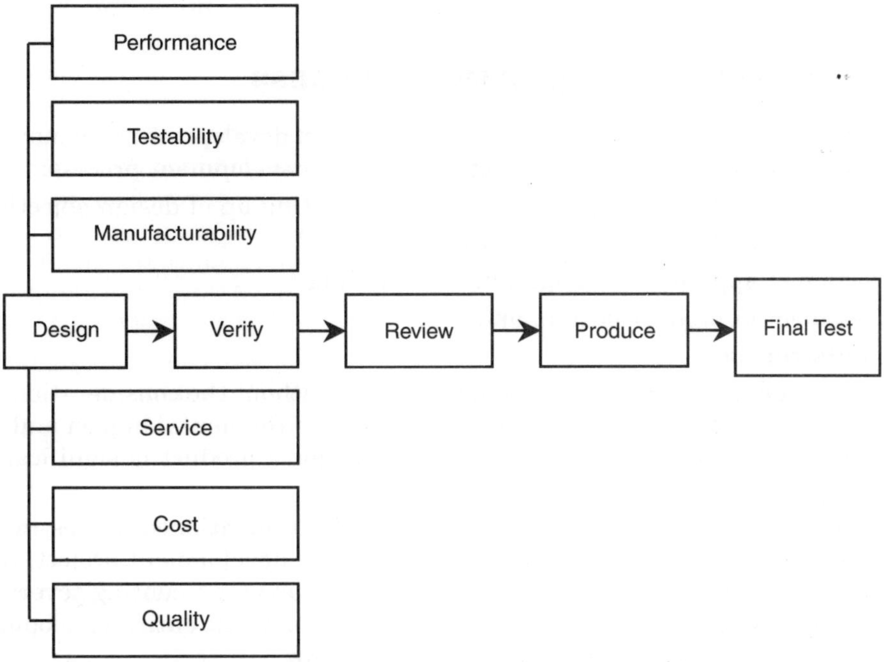

Fig. 3.2 Concurrent/Simultaneous Engineering-based Product Development Process.

In concurrent engineering, different phases of product design and development process are overlapped to reduce the time needed to develop a product. It requires simultaneous, interactive, and inter-disciplinary involvement of design, manufacturing, and service engineers to assure acceptable and improved levels of performance in design activities, product support service, technical support, and reliability of the product considered. Obviously, this requires involvement of several functions that justify to strive for improved quality, reduced cost, and shorter development time for the product. In many instances, team approach becomes effective and successful if SE is applied during the entire product development process. However, in order to apply SE concept, the individual members in the team must be aware of the different aspects of product development. Involvement in all these aspects and phases of product development by all the members of the team leads to a positive outcome if the conflicts between the members on any aspect, if any, are resolved. Hence, to have the best-possible synergy in the whole process, an effective CE approach requires the following conditions to be met:

(*i*) The decision models representing the process by which a team gets involved should be flexible in nature.

(*ii*) The kinds of knowledge to be developed for facilitating design should require integration of several functions.

(*iii*) Appropriate communication system must be in place.

(*iv*) A number of measurement and assessment tools and techniques (both qualitative as well as quantitative) should be used so that the evaluation of current and new design alternatives can be carried out and made known to all the members of the team.

Over the years, the C/SE approach has proved to be a valuable tool for designing competitive products in a changing and expanding world market. It is reported that the Japanese companies take almost half of the time that a typical US company takes to launch new products in the market. This success is due to the fact that C/SE has contributed to a significant reduction in the product development time.

3.3 STAGES IN PRODUCT DEVELOPMENT PROCESS

Traditionally, the interrelated activities for a typical product development task may be explained in terms of the following stages constituting the product development process:

(*i*) Identification of customer requirements and setting up of design objectives,

(*ii*) Identification of design alternatives,

(*iii*) Selection of appropriate design and its detailing,

(*iv*) Assessment of product design, and

(*v*) Manufacturing

The product development process is not complete without the consideration of a number of pertinent issues, such as logistics, supply chains, and environmental impact and assessment. In fact, in many cases the overall cost of development for a product is significantly affected mainly by these issues.

However, it should be kept in mind that the above mentioned stages in the product development process are not generic in nature. For a large development project, there may be other stages to be considered, where a large number of groups representing several disciplines and functions may be working together and engaged to complete the entire development process. All the above mentioned stages are briefly explained below.

(*i*) **Identification of customer requirements and setting up of design objective:** The first phase of the design process is to identify the overall needs of the customers and to define the design objectives for the product to be developed. It is basically the process of identifying, defining, and documenting specific needs for the development of a new product. The main objective at this stage is to identify, consolidate, and document all the features that the system may have in feasible, realistic, and complete specifications of product requirements. During this phase, a large number of new ideas for the product to be developed are converted into practical requirements and specifications. Determination of the specifications of the product is what is required as the outcome of the tasks carried out during this phase.

(*ii*) **Identification of design alternatives:** At this stage, identification of several design approaches or alternatives meeting the defined requirements, prior to identifying the best design to be used, is the main purpose. This process begins as and when a need for a new product is defined and explained in concrete terms, and ends only when the best design approach is recommended in order to meet the requirements from the new product. For this, responsibility should be shared among the members of the design team. Each member should be aware of the specific design goals and requirements that he/she is entrusted with for their inclusion in the total design of the product. Usually, several types of tools and techniques, such as mathematical models, simulation, and trade-off studies are used at this stage to determine the optimum or the 'best' design for the product. It is during this phase that the initial producibility, quality, and reliability requirements of the design are documented, and guidelines and action plans are provided.

(*iii*) **Selection of appropriate design and its detailing:** It is the process of finalizing a product's design which meets the requirements and design approach defined in the early phases. Critical feedback takes place as the design team develops an initial design, conducts analyses, and uses feedback from the design analyses to improve the design. Typical design analysis uses scientific methods, usually mathematical, to examine design parameters and their interaction with the environment. This is a continuous process until the various analyses indicate that the design is ready for testing. Usually, many types of design analyses are required and performed, such as stress analysis, failure modes, producibility, and reliability. These analyses require the support of the personnel having specialized knowledge of various disciplines. During this stage, the product development team may construct prototypes or laboratory working models of the design for testing and evaluation to verify performance results.

At this phase, the interactions among many disciplines and design professionals take place, and these interactions should be effective. For effective interaction, coordination among the members of the team becomes crucial during the evaluation and analysis of all possible design parameters. This ensures commitment and responsibility for product development of the team involved.

(*iv*) **Assessment of product design:** Assessment of the improved design is done through a series of test and evaluation of the prescribed design resulting in a set of quality features and improved performance meeting the requirements. In most cases, particularly when the product is complex in its requirements, the initial design may not meet all the requirements and hence, may not be ready for production. Test and evaluation is a "designer's tool" for identifying and correcting problems, and reducing technical risks. In this context, a mature design is defined "as one that has been tested, evaluated, and verified prior to production to meet 'all' requirements including producibility." Unless this maturity is adequately verified through design reviews, design verification, and testing, problems may occur because of unforeseen design deficiencies, manufacturing defects, and environmental conditions. The main goal of this assessment is to identify areas for design improvement for the product, keeping the issues like improved manufacturability and reliability in mind. The acceptable levels of the design parameters must also ensure minimum technical risk for the product.

(*v*) **Manufacturing:** At this stage, the main objective is to identify the problems (of different kinds) that may be encountered during manufacture of the product as per the specifications of the improved or most appropriate design, and to take corrective and preventive measures to minimize the cost of production and to improve quality of the product. The design team is required to perform data analysis using appropriate tools and techniques at this stage.

The issues of logistics, supply chains and environmental impact and assessment become relevant during post-production phase. These issues are concerned with delivery, distribution, logistics, technical support, after-sales services, and disposal procedures after the expiry of the product's life. The environmental issues are also taken into consideration at this stage. For many companies, logistics costs may be significantly more than the total direct costs. The supply chain refers to the 'flow' from suppliers to manufacturers to distributors that also include all information flow, processes and transactions between suppliers and customers. During delivery of a product, the packaging becomes crucial providing necessary protection to the product during shipping and necessary information ensuring minimum environmental impact and maximum safety for the users and others involved. During distribution and storage, the packaging is responsible for providing adequate maintenance and reliability of the product. Operational problems may result due to alternative maintenance techniques, customer handling, and other field environmental issues that are usually not readily known during the design stage. Design and management personnel of the company should evaluate how well the product performs at the users' sites. Alternate maintenance techniques and changes in certain aspects of the design may be required on the basis of actual results and feedback information obtained during operation of the product at the field.

As different kinds of problems including environmental factors may be encountered at this stage, and rules and regulations as set by the regulating authorities and government may be very stringent in this respect, it may be difficult to have perfect balance between economic product development and environment-friendly product development. The concept, called 'design for environment (DFE)' calls for consideration and estimation of the total cost to the environment at the design stages for the entire life cycle of the product. For this, a number of analyses and approaches may be required to be used in today's business environment.

3.4 TOOLS AND TECHNIQUES USED IN PRODUCT DEVELOPMENT

There are a number of tools and techniques that are quite useful for data analysis during the stages of product development. These are widely used by the design teams. Some of these tools and techniques are briefly discussed below:

(*i*) **Work-Breakdown Structure.** This structure lists and defines all the tasks and assignments in a hierarchical form that are required to be carried out for product development. Breaking the overall design programme down into several tasks and assignments at different levels is the key feature of this structure, often referred to as 'Work-Breakdown Structure (WBS).' It is desired that each task is appropriately coded (with a unique identification number), and the financial and technical performance for each task is monitored over time. A progress report on maintenance schedules for the task is also made periodically. It is better if the tasks as required are described in a clearly-understood written form so that both the team members and the customers are aware of the types of tasks and their influence on the product's performance and scope.

Thus WBS provides a framework and tool for detailed resource and financial planning during product development. Appropriate human resources hiring policy may be introduced and implemented at different levels based on the WBS. Action plans regarding development of work standards, objectives, goals, and work schedules are also documented, and are an outcome during WBS implementation.

(*ii*) **Network Diagram.** The activities to be carried out for product development need to broken down into a number of tasks each requiring smaller amount of resources and time to complete so as to have better control on the design project as the time passes. As there may be a large number of these tasks to be considered in a typical design project, and many of them are interrelated, the schedules of starting and completion times of these tasks are required to be known for continual monitoring of the design project. A network diagram is drawn for a design project principally to serve this purpose. In a typical network diagram, the following information is readily available:

(*a*) how many series and parallel activities are there in the project,

(*b*) against a particular task, how many tasks need to be completed before this task is taken up (predecessor tasks), and

(*c*) against a particular task, how many tasks need to be completed after this task is completed (follower tasks).

While using this diagram, it is the responsibility of the design team members (*i*) to estimate the task completion times for all the tasks considered, and (*ii*) to allocate resources (materials, equipment, and human resources) for each task.

A number of project scheduling techniques, such as Critical Path Method (CPM), Programme Evaluation and Review Technique (PERT), and Graphical Evaluation and Review Technique (GERT) may be used for following up and control of the activities of a design project so that the product development process is completed within a stipulated completion or due date. The critical path of the network diagram is determined for this purpose. However, two main difficulties that one may come across in using these techniques are as follows:

(*a*) **estimates of task completion time:** this time is probabilistic in nature, as uncertainty exists regarding the outcome of a task. It is particularly true for a new product development-related task. Moreover, the time estimates must be constantly updated and revised based on experience gained and knowledge required over time by the team members.

(*b*) **number of tasks to be considered:** it is most probable that the actual number of tasks to be considered and included in a product development project may not be known precisely because of inherent 'unpredictability' about the outcome of a task involved. Hence, the number of tasks is known only tentatively, and depending on the nature of the outcome actually obtained, the list of tasks needs to be updated on a continuous basis.

(*iii*) **Gantt Chart.** This chart, which is also known as a 'bar chart', shows all the tasks, in a sequence, of a product development project and their completion times (expected or estimated). For each task considered, the expected starting and ending times on a time scale are specifically marked and shown. As the project starts, control and monitoring of the project tasks is done by noting the actual starting and completion times of each task, comparing them with their estimated values, and taking corrective actions as deemed fit for the tasks to be completed for product development. This chart is a very simple but useful tool for project control and monitoring. However, the interrelationships that may exist between the project tasks are not shown in this chart.

(*iv*) **Budgets and Schedules.** The time estimates for each project task are required to be converted into resource requirement estimates and their budgeted expenditures. The budgets for the resources should be based on realistic assessment of the desired performance on the tasks involved under desirable conditions. For this, to have a realistic budget, continuous adjustments and review is the primary condition that should be met by the management of product development projects. This adjustment and review takes place on a periodic basis keeping in mind the variations in constraints and restrictions on the resources imposed over time. The main objective of preparation and review of budgets is to complete the project within the planned and accepted monetary resource and time period.

(*v*) **Technical Control Tools and Documentation.** During product development, the process tasks and elements need to be controlled. However, this is not a simple assignment. In majority of the cases, control of technical aspects of a project is not easy. It requires a properly and adequately documented system to support control of technical aspects. In this context, technical documentation with specifications, block and interface diagrams, drawings, design guidelines, purchased parts information, and process capabilities is a general necessity for any product development process. Specifications are usually required for the system, and they address the issues related to the performance, reliability, producibility and quality requirements, and environmental aspects of the product. The specifications at the higher level of a product form the basis of preparing lower-level written requirements so as to help the team members design the functions at the part or component level of the product. The documents must be significantly comprehensive to control the design task. Also, the written task definitions and specifications are a must for any kind of design task. The design team must take the initiative to find out the most appropriate specifications related to a design task including the requirements and implications. The types of specifications as required for a product development project may be varied in nature. Generally, they are of the following types:

(*a*) system specifications stating the identity of a product along with its functional characteristics,

(*b*) subsystem specifications stating the same aspects for an identified subsystem of the product,

(*c*) supplier and unit specifications stating the specifications of the externally supplied components as well as the parts or components that are to be internally manufactured,

(*d*) governmental regulations and professional standards as applicable for the product under consideration, and

(*e*) design guidelines, process specifications, and production and manufacturing standards.

A typical block diagram of a system or a product shows in block form the functional and physical partitioning of the elements of the system, and indicates the major input-output flow. A well-prepared block diagram is a very useful working document for communicating essential interface information and functional operations. Lower level block diagrams complement the system diagram by expanding the level of detail for the system-level blocks. For example, system interface drawing shows cable interconnections between system units. Detailed signal characteristics, cable type, connector type, and pin assignment information are defined in the system interface specifications.

The design guidelines and rules are prepared to encourage the use of proven design practices. They promote consistency of design for a specific product development project. Typically, the design guidelines may include the following aspects:

(*a*) a summary of common electrical, mechanical, and software design requirements for a particular project indicative of general design guidelines,

(*b*) specialized guidelines and rules, usually prepared by design, manufacturing, quality, reliability, packaging, repair, and safety groups containing requirements, helpful information, and best practices, and

(*c*) design policy specifying acceptable data gathering techniques, practices for common databases, and standards for reporting the results as well the detailed information about the participants in the project.

(*vi*) **Control and Assessment of Costs and Schedules.** The budgets and work and time schedules are established at the start of a product development project for each element of the WBS. Expected expenditures on resources to be used are estimated and entered into a system with which status on costs and schedules may be known at any point in time, and review is done accordingly. Periodic updates are made available for review of trends and corrective action. The Gantt charts can then be used effectively to indicate the progress of each activity against a time scale. The end points of the task are clearly defined, but judgment is required to measure progress between the end points. In setting up the lower level activities, the scope of each activity and its time span should be broken down into small units of time to provide correct understanding of progress. Review on the progress of the work done can be both formal as well as informal. A number of computer software programmes may be made available and subsequently used for this purpose.

(*vii*) **Design Review and Design Audit.** This is considered to be a crucial activity in the majority of new product development initiatives. The main objective of design review and audit is to measure and assess technical progress related to product development. In order to be aware of this progress by the design team members, they are to identify the potential problems and to suggest steps for design improvements. This will ensure identification of specific requirements in manufacturing, maintenance, tests, and logistic support systems for the product. In short, this is a systematic method to ensure that all aspects of the design are thoroughly looked into and assessed prior to production of the designed product. Successful completion of this task results in better design of the product and its acceptance to the users or customers in the long run.

(viii) **Design Release.** On completion of product development process, updating and improving the initial design of a product must lead to its final design. This design should not be further modified or improved, and hence, frozen before it is released to the production or manufacturing unit of the company. This activity in the developmental process is called 'design release'. Scheduling a design release is closely related to the status of other design activities, such as design review, production design, test results, and configuration control. Care must be taken to prevent the release of a design that is incomplete, inaccurate, or premature so that problems do not occur later. The release of the final design may be worked out following the backward scheduling from the manufacturing schedules. Quite often, however, the design team may be required to meet a time schedule that may not be realistic. Similarly, deviation from the standard procedures for design may result a 'not-so-good' product for the end users or customers. A realistic schedule for releasing the final design of the product can be made only when the details of the technical requirements and manufacturing capabilities are known and considered.

The design of the product should be validated in stages, using experienced personnel from technical and production units of the company, to ensure that the design is producible, the documentation is complete, and the final design is released on schedule. Before the design is ready for its release to the production or manufacturing unit, several activities should be undertaken. A few such activities are as follows:

(a) All engineering and business analyses, process capability studies, and vendor verification tests must be completed and corrective actions incorporated.

(b) Final technical reviews of hardware and software must be completed and corrective actions incorporated in the design, if needed.

(c) Manufacturing and service plans and procedures must be revised to reflect the latest design changes.

3.5 TECHNICAL RISK MANAGEMENT IN PRODUCT DEVELOPMENT

During product development, many kinds of unforeseen events may take place. Many of the variables and factors affecting the product development process may have an 'outcome' that is not at all predictable, and hence, unexpected. The occurrence of such 'outcome' may result in significant risk in initiating and completing tasks that are required for product development. In this context, management of 'technical risk' in any product development project is a serious issue, and needs adequate attention from the design team.

Technical risk is "a measure of the level of uncertainty for all the technical aspects of the product development process", and technical risk management attempts to control this uncertainty found in product development. Management of this risk is essential for identifying and resolving potential problems to ensure that the proposed product performs as intended in its design satisfying the customers' requirements. Obviously, management of technical risk calls for systematically following a number of steps in sequence. These steps are as follows:

(i) identify areas of potential technical risk,

(ii) determine the level of risk for each area as identified,

(*iii*) search for and implement solutions that eliminate or reduce the risk, and

(*iv*) monitor and measure progress on minimizing the risk.

For development of a new product, one must ensure that the product possesses a few unique features that make it an innovative one. This requires the application of the 'state-of-the-art' technologies at both product and process stages. However, even if a technology may seem to be innovative in nature under present condition, it may not pass the test of time. This means that many innovative products may simply be a 'novelty' item without any commercial success. Thus, for development of new products and technologies, it is imperative that the designer, as individual capacity or the design team as a group must have the mental intent to take risks. The technological risk while developing a new and innovative product remains at all stages of product development. Hence, the design team should identify all potential technical risks, monitor and control these risks as much as possible, and have the so-called 'contingency plans' when nothing tangibles may come out of the product development project. The responsibility of the management of a company in this context lies in its ability to develop systems for monitoring and evaluating technical progress throughout the development process. Usually analysis of past performance may lead to identification of types of risks that may occur in the future. Experienced designers and managers know that there are conditions, requirements, and situations that almost always create problems. When problems do occur as they most certainly will, the design team should take specific actions, known as contingency plans that are likely to have positive outcomes in future.

Technical risk assessment is, therefore, treated as a management control tool for quantifying design and technical progress during product development. It results in making decisions of three kinds:

(*i*) if a completely new design is needed and feasible,

(*ii*) if evaluation of alternate technologies is required, and

(*iii*) if a particular technology is required to be developed 'in-house' or to be purchased from an 'outside' source.

Through technical risk assessment, the management of a company may be in a position to identify quickly and at an early stage of product development any kind of serious production or reliability problems that may affect the product performance or its function. A systematic approach for technical risk assessment allows the management to regularly assess the states of product development, and to assign additional resources as and when problems are identified.

In this context, assessment of technical progress is essential for resolving problems, ensuring that the proposed product performs as intended, and is really producible when it reaches the production or manufacturing phase. Early detection of this aspect is critical because it is easier and cheaper to make changes early in a design. The members of the design team should realize that technical risk management is effective, if it

(*i*) develops a formal system for identifying and measuring important technical risks,

(*ii*) measures and monitors the progress from the very beginning with the help of appropriate indicators,

(*iii*) assesses the technical status of a product development project with early problem identification, and on 'real-time' basis as far as practicable,

(*iv*) develops and executes plans as developed during contingency situations with acute problems, and

(*v*) creates a database for crisis identification and trade-off analysis.

3.6 CUSTOMER NEEDS AND PRODUCT DEVELOPMENT

The basic process at the early design stage of product development requires identification of functions of the product under consideration. These identified functions should be able to meet the customer needs to the maximum extent. However, accomplishment of the task is not simple. The product development team should always attempt to offer a product in the market which is quite different (in its functions and technology) from other similar products already existing in the market. This calls for an attempt to design and manufacture an innovative product that meets the current requirements of the prospective customers. A number of approaches may be used to identify and quantify customer needs in this context. These approaches are as follows:

(*i*) **Analysis of the customer needs:** This is needed to ensure determination of correct user needs or predict new opportunities for product development.

(*ii*) **Analysis of product use and users:** This is needed to have realistic operational requirements that include scenarios, task analysis, and functional timeliness in all types of environments to determine the maximum performance limits.

(*iii*) **Analysis and forecast of technological capability:** Both the short and long term capabilities are required to be determined to ensure that the design is up-to-date (*i.e.,* not obsolete) when the product is ready for manufacture.

(*iv*) **Benchmarking studies:** These are usually performed to evaluate the company's capabilities and the level of competition, to identify best practices, and to finalize design and manufacturing requirements.

(*v*) **Use of tools and techniques such as Prototyping, Virtual Reality, and House of Quality:** These tools and techniques are used to systematically identify consumer needs and develop product requirements.

It may be summarised that all the above-mentioned approaches are primarily required to be used to determine the up-to-date product requirements. A set of product requirements become the concrete outcome of the early design process. A systematic approach for product requirement analysis is an important step in the total product development cycle.

A few important aspects of the requirement analysis and the problems commonly encountered in this regard are briefly explained below:

(*i*) **Requirement Analysis.** The reasons for a new product development are manifold. A new product may need to be developed to utilize the potential of an innovative 'breakthrough', for new use of an existing technology, for meeting the needs of new customers or markets, and/or for improving the performance of an existing product. Hence, before the product development cycle starts, requirement analysis is a must, and has to be pursued vigorously by any design team. It is the beginning of the product development cycle.

The main objective of this analysis is to systematically and comprehensively, as far as possible, identify and document all the features and characteristics that need to be built into a new product. The objective is also to develop an innovative, feasible, and realistic design of the product with the identified features and characteristics. As all the approaches as mentioned are dependent with one another, this analysis is carried out iteratively in almost all cases. The design specifications of the product are the outcome of this analysis. As new technology and/or customer requirements evolve through time, updates to the original specifications are also required to be developed following the same iterative process of requirement analysis.

At this stage, the design team may invariably encounter a typical problem. Even if the approach taken by the design team for requirement analysis is most elaborate and comprehensive, the product, that is considered to be the 'best' in design, may not be able to meet all the needs of the customers mainly because that the customer needs keep on changing all the time for most of the products, and the design team may not be able to engage 'unlimited' resources for product development. The product development in most cases takes place under constrained conditions and resources, in the form of a limited number of competent persons and technological capabilities of the organization as a whole or the design team or the department responsible for product design. Again, as the customer needs are influenced by various cultural, social, and economic factors, different customers do have different needs, and a single product is most likely unable to meet the requirements of everyone.

This situation calls for an appropriate 'trade-off' decisions, balancing a set of most up-to-date customer needs with the capability and resources to be made available and engaged by the design team. The trade-off analysis results in a design of a product, which is 'optimal' at best, and 'sub-optimal' at worst.

(ii) **Problems Commonly Encountered in Requirement Analysis.** As has been mentioned elsewhere, conversion of customer needs into the requirements for the product, in many cases, in a difficult and challenging task. In the conversion process, there may be several types of problems the design team may face. A number of such problems are discussed below.

In a particular case, the product requirements may be identified before a detailed design stage is completed. The 'state-of-the-art' technology may be too tempting to ignore, and hence, the requirements of the product under consideration may be prematurely determined, which, when tested for their feasibility, may not be realizable in the long run.

The design team may set a highly unrealistic goal while determining the expected performance from a new product. This situation may arise in order to make the product as innovative as possible. However, to achieve this goal, many types of risks may exist in the development process, and the product may not at all see the light of the day.

Many a time, the product requirements may be stated only in general terms. Obviously, it may be very difficult, if not impossible, for the design team to measure and monitor the performance of the product during the development stage as it is most likely that actual product parameters are not expressed or defined in quantifiable form if the requirements are stated in general terms. It is to be remembered in this context that the design effort calls for determining and measuring the parameters related to varied requirements from a product in concrete or measurable terms.

In many cases, while translating the customer needs into product requirements, it may be assumed by the design team members that the customers' views or opinions are sacrosanct, and whatever they opine it is to be looked into. We have to be very careful in the analysis of the customer needs in this respect. Bias, if any, in making judgment by the customers should be filtered out, and only the relevant information should be taken into consideration. However, there must not be neglect in the evaluation of the information that is obtained from the customers on a regular basis.

Another critical problem that the design team may face arises out of constantly changing product requirements. This results in constant design changes with the management periodically redirecting and changing its technical effort and design strategies over time. This may be, however, very common where the market conditions do constantly change actually for certain products, and there is no other alternatives but to change the requirements in the product. This is however an exceptional case.

The next critical problem is also common. In order to satisfy the needs of all types of customers, and incorporating them in a single product, the requirements may become varied and complex in nature. This may result in high probability of failure of the product. The key concept in this context is simple design. In order to meet all types of needs, products may be offered in different models or versions.

Another critical problem that has been occurring in the present era is related to the strategy wherein attempt is made to develop only one set of requirements for all types of customers. Obviously, this concept is not compatible with the strategy of 'mass customization.' The product requirements need to be established with the consideration of several factors, such as cultural differences among the prospective customers, delivery of spare items and parts to the customers on a timely basis, change in climatic condition and its effect on product performance, and varied standards in varied situations. These issues have become particularly relevant for the products for which an international market is sought.

3.7 COMMON APPROACHES FOR CUSTOMER NEEDS ANALYSIS

Defining the customer's needs may be an extremely complex process resulting in many different and conflicting types of information. Many a time, the customers themselves may not be in a position to identify and specify their exact requirements. The customers may not also realize the implications of a new or innovative technology to be used in a new or upcoming product. However, there are several approaches that may be used for identifying and determining customer needs. The design team should use several of these approaches so as to be sure that the final requirements are representative of the customers under consideration. There are several methods for capturing and documenting customer needs. A few important methods are as follows:

(*i*) **Interviews of customers** including techniques such as surveys and focus groups.

(*ii*) **Design partnerships or alliances** where the existing or prospective customers participate in the design process.

(*iii*) **Computer databases and data mining** of actual customer sales and preferences including internet activity.

(*iv*) **Consultants or experts, mainly customer research organizations,** who specialize in identifying what the customers want.

(*v*) **Brainstorming sessions** solicit inputs that encourages the generation of innovative ideas that may be totally different from the ordinary ideas, and thus help to change the features and functions of a new product.

(*vi*) **Personal and company experience** including previous successes and failures.

(*vii*) **Published information** such as articles in magazines, papers in journals, patents, and internet websites.

(*viii*) **Technology capability forecasting** of the future based on historical analysis of product markets and technologies.

(*ix*) **Market and competitor benchmark analysis** identifies best in class and innovative ideas.

(*x*) **Prototyping and virtual reality** are used to study customer responses and design team recommendations.

(*xi*) **House of quality or quality function deployment** is used to identify needs and determine product requirements.

There are several tools and techniques that may be used during the early design stage of a product. These are also used for determining the exact requirements of the customers. The important tools and techniques are as follows:

(*i*) **Scenarios and Use Cases.** Scenarios and use cases are a step-by-step description of how a product is used for a particular application or task. They are usually formatted as a list or a flow diagram. A product may have several scenarios depending on the number of product uses, features, and different users. They provide a focus for communication between users, experts, developers, and vendors. Usually, one scenario is developed to illustrate the desired functionality and problem solving focus for each application. The initial scenario helps the team members to learn system requirements and provide the basis for early prototyping. As each scenario becomes more detailed, they are used for developing design and test specifications. Each scenario can later be decomposed to define task responsibilities and constraints.

Scenarios are elicited from the users and the experts in the broad domain of the product, and are validated by independent experts. Scenario analysis identifies both typical (common) and atypical (uncommon) process flows within the system. The results of the analysis are used to define task responsibilities. They often involve time-dependent sequencing of these tasks. There are many types of documentation techniques used for scenario generation. The important documentation techniques are task lists, timeless topologies or "maps", event traces, scenario matrices and models, flowcharts, process diagrams, and activity state diagrams. Scenarios are generated using one or more of the following approaches:

(*a*) Group techniques,

(*b*) Interactive observation,

(*c*) Structured interviews,

(*d*) Demonstrations, and

(*e*) Focus groups

It is important to develop both the "as is" and the "to be" views of the scenario. The descriptions should include responsibilities, capabilities, and views. The scenario perspective focuses on each user's viewpoints on the tasks or activities to be performed.

Scenarios define important aspects of the domain *i.e.*, how the product will be used. They are initially made at the highest abstraction level, and are then broken down into several smaller sub-scenarios. In systems development with multiple users, small scenarios are developed for each user.

(*ii*) **Task Analysis and User Profile.** Task analysis is a design technique that evaluates specific task requirements for an operator with respect to his or her capability. The analysis lists each human activity or task required, and compares the demand of the tasks with human capabilities and the resources available. Tasks are derived from the previously defined scenarios and continuously updated during the design process. The level of detail for the task analysis is based on the design information required at each phase of the design and the importance of each task. In this context, user profiles are descriptions of the users and support personnel's capabilities. This includes descriptions of their physical, educational, training and motivational as well as mental ability.

During early design, task analysis and user profiles are used to ensure that the product's requirements are compatible with the user's capabilities. The design team must look at the product critically to see that the task requirements do not exceed human capabilities. In addition to the operator's tasks, task analysis should also be performed for maintenance and support personnel. It starts at a very simple, limited level of detail and progressively becomes more detailed during the design phase. Constraints are used to identify potential problems and enumerate reference requirements. Studying the results of task analysis can identify potential product and domain problems.

A typical task analysis results in breaking down each task into minute details and requirements. A detailed task analysis may address and describe the task in terms of its relevance and/or working with reference to several factors. A comprehensive list of such factors may include the following aspects:

(*i*) Expertise and training,
(*ii*) Organizational aspects,
(*iii*) Tools and equipment to be used,
(*iv*) Objectives and goals,
(*v*) Quality features and other dimensions,
(*vi*) Types of hazards and their implications,
(*vii*) Results and outcomes,
(*viii*) Types and quality of resources (physical and human) to be used,
(*ix*) Types of decisions and decision making process, and
(*x*) Environmental conditions.

(*iii*) **Environmental and Functional Conditions for Products.** It is imperative that both functional as well as environmental conditions for the product under consideration should be taken into account. However, in many cases, only the operational use of the product may be emphasized, ignoring the implications of physical environments in which the product is required to be used. For example, products such as automobiles, washing machines and computers can experience harsher physical environments during their shipping and transport, and even during their actual use. An environmental profile, in this context, shows, on a time scale, the significant environmental parameters, including their levels and duration that are expected to occur during the life of the product. It defines the total "environmental envelope" in which the product must perform, including conditions of storage, handling, transportation, and operational use. Similarly, a "functional profile" shows, on a time scale, the functions or tasks that will be utilized by the system to accomplish its intended use, including maintenance, storage, and transportation.

Obviously, a functional profile describes, in detail how a system or a product must perform in every potential situation in the total envelope of environment. It thus helps the design team describe the system completely at each level of detail. Examples of functional analysis tools include 'flow block diagrams' and 'time line analyses'. The time line analyses show sequences, overlay, and concurrence of function of a product (including time-critical functions that affect reaction time and availability).

(*iv*) **Forecasting of Technological Capability.** The analysis relating to prediction of future technology and its potential to be capable in product or system design is an important aspect of product development process. It predicts whether a new technology will be ready in time to be used in the design, its expected performance levels, and the manufacturing capabilities that are expected to be available when the product is ready for manufacturing. For many products, identification of its use and determination of the expected performance of the product where it is used are two major decisions. Product innovation and creativity are often designed around using new technology. If the technology is not ready and on schedule, the entire product development venture or project becomes risky. For many products, the design team must also anticipate the technology's performance level that will be available in the future when the product is manufactured and not the expected performance level only when the product is being designed. This requires the management and the designer of a company to predict the key parameters of future parts and software. The designers can often use the vendors' advanced information and the so-called pre-specifications. These specifications are sent out before the product hits the market. Internal funding may be required to insure that the technology is ready.

For determining producibility of a designed product, prediction of future process capability is an important responsibility. Predicting future process capabilities involves two technologies: (*i*) product technologies, and (*ii*) manufacturing process technologies. Another important aspect is the prediction of expected process capability in future. To address this aspect, the design team or the designer, as an individual, must be aware of the upcoming product as well as the manufacturing process technology. It is, therefore, essential that production personnel participate in its development to assure production compatibility. As the technology evolves, the requisite process capability can be predicted or determined. In many cases, experiments are required to be conducted with respect to selected process parameters or manufacturing criteria for a process involving a technology. This provides insight into the expected process performance and related capability requirements.

The typical activities required for forecasting the future technology and its possible impact on the product design include (*i*) identifying critical technologies, (*ii*) studying their historical trends, and (*iii*) getting the opinions of the experts (in most cases, in a structured way) to project or predict their expected levels in future for the design of the product under consideration. Usually, the relevant information in this respect may be made available from several sources, such as working groups in industries, trade journals and exhibitions, researchers in universities, research centers and laboratories and supplier source. A number of methods can be used for technological forecasting. The trend extrapolation (extrapolating the trend in the time series or historical data with respect to technologies of a particular product), the Delphi technique (a structured group communication process among a select group of experts making predictions about the future technologies and their uses through consensus), and Scenario (identifying the future situations and possible breakthroughs that may occur for consideration in the product design), generation are three widely-used techniques.

(*v*) **Benchmarking for Products and Processes.** In recent times, bench marking which is basically a process for studying a company's quality and design characteristics of products, services, and practices against their best-possible standards as achieved by the leading companies and taking effective steps or actions to minimize the gaps, if any, between them, has assumed significance. The purpose is mainly two fold: to determine product requirements, and to identify innovative ideas. It is the systematic process of identifying the "best" practices in industry, setting product and manufacturing goals based on results of what the competition has achieved or will achieve in the future, and identifying the best suppliers. Benchmarking in product development is crucial in current business environment for a number of reasons, such as

(*a*) Companies involved in product development should be aware of customer needs, preferences and values, and update their knowledge in these respects on a continuous basis.

(*b*) Each company must be aware of its strengths and weaknesses, vis-à-vis other competitor companies.

(*c*) Each company should be aware of its competitors' products, processes, technologies, and supplier base.

(*d*) Each company must constantly strive to incorporate more number of features and dimensions of quality in its products, and knowledge about competitors' products in these respects help it achieve this goal.

(*e*) Benchmarking identifies the best suppliers who can be used for product development to the advantage of the company.

(*f*) Benchmarking can be used to identify and evaluate "best" product and process parameters, and in turn help develop requirements for a successful product.

Benchmarking is carried out in a systematic manner. The steps involved in a typical benchmarking exercise are as follows:

1. **Analyze all aspects of the competition and other successful companies**.
2. **Determine the standing or ranking of the company** relative to competitors and leaders on key features, parameters, and processes both on today's products and on predictions of future products.
3. **Establish** *features and parameters of the 'best' products.*
4. **Set product, manufacturing, and supportability requirements** based on benchmarking information.
5. **Implement a practice of "innovative imitation"** (*i.e.*, improve on the best ideas and methods that were identified).

For example, a simple benchmarking study may include determining some manufacturing design requirements for a computer circuit board assembly. The company purchased several competitors' products. Each product was disassembled and studied. Note that the "best-in-class" competitor assembly time for the printed circuit board (PCB) is the lowest due to the smaller board size, fewer number of screws, and no ground plane requirement. As a result, the design team should attempt to eliminate screws, reduce board size, and eliminate the ground plane.

(*vi*) **Prototyping and Virtual Reality.** Prototyping and virtual reality in the early design phases use hardware and software conceptual prototypes to study customer responses to new products and to identify areas for improvements for both requirement definition and conceptual design. The basic philosophy behind the use of prototyping is to provide a communication platform for studying consumer needs and refining product use scenarios and task analysis. Evolutionary prototyping can be used for incrementing and refining the design. This can be especially helpful when the product is very different and innovative when compared to competing products. Prototyping helps the design team by encouraging the consideration of many product development issues during the early phases and identifying potential problems. The steps involved in a prototyping process are as follows:

1. Produce prototypes that provide the reviewer a realistic view or feel of the proposed design.
2. Develop as many prototypes as economically possible.
3. Continuously produce prototypes throughout the product development process.
4. Show the prototypes to everyone concerned.

The future of virtual reality and prototyping is to merge virtual reality software and CAD/CAE tools into a new kind of software that will allow consumers and the design team to "evaluate a product in a virtual environment." The designer may shape or manipulate a new concept by hand and give it to the customers and other engineers for analysis. The customer and product designer may also experience a design concept, walk into the simulation space and examine details or properties of the conceptual design on computer screen not even visible in the real world.

3.8 CONCEPTUAL DESIGN STAGE

During the conceptual design stage of a product, a number of activities are carried out. These activities are as follows: (*i*) identification of all possible design alternatives or options for the product, (*ii*) selection of the 'best' or 'most appropriate' design under the given constraints, based on 'trade-off' analysis, and (*iii*) transforming the product requirements into the design requirements at the lower level for the 'best' or 'most appropriate' design. This process of conceptual design continues till the final design for the product is selected. On completion of this process, the requirements at all levels are identified and properly documented. To cope up with increasing demand of knowledge and effective and formal communication between the design team members, the size of the design team may grow at this stage, and hence, effective teamwork is a must to have the best-possible outcome from the process. The most relevant aspects of the three main activities of this process as mentioned are briefly explained below.

3.8.1 Identification of Possible Design Alternatives

(*A*) **Identification of All Possible Design Alternatives.** This activity starts with the identification of all possible design solutions. A concerted effort on the part of the design team is required to identify them. A collaborative effort among the persons in the design team is a prerequisite before the team is able to offer a number of design options. For this, creativity and innovation must be encouraged not only for design but also for manufacturing, logistics and other areas in the whole product development process. It is recommended that 'brainstorming' session be conducted by the team leader for generating new and innovative ideas for design as well as for identifying design alternatives for the product. In most cases, the design team may be aware of new and upcoming technologies and 'styling trends' for the product and the methods of translating them into a good-looking, useful, and easy-to-use product. The team may also anticipate styling and social changes and identify new customer needs and desires as a result of those changes.

At this stage, it is always better that all the possible potential solutions are considered and examined as early as possible. This will ensure that the opportunity is optimized to take advantage of recent advances in technology and new styling trends to avoid outdated or preconceived solutions. Typically, the technologies should range from the 'mature', that are usually lower in technical risk and cost, but also have lower payoff potentials, to the 'leading edge' or newly emerging technologies, that have higher technical risks and costs, but also have the potential for a more significant impact on product success. The quality features, the functional characteristics, and the appearance value of the designed product must conform to the performance level as per the state-of-the-art technology. In this context, appropriate decisions are required to be taken by the design team on the types of software to be developed by the designers or to be purchased from the outside suppliers. This is called a "make-or-buy" decision to the potential problems of designing for the new technologies, developing new manufacturing processes, or using new vendors. At this stage, this decision should be based on consideration of a number of factors, such as cost, quality, capability, and schedules.

(*B*) **Trade-Off Analysis.** The next important activity in the conceptual design phase is to evaluate each of the identified design approaches for the product. In this context, trade-off analysis is carried out. This analysis examines all possible design alternatives and different parameters with the purpose of optimizing the overall performance of the product and minimizing technical risk. Both new or innovative and traditional approaches may be used for this purpose. Essentially, it is a formal decision-making method that can be used to solve many complex problems. A typical tradeoff analysis is used to rank user or customer needs in order of importance or preference, to develop cost models, and to identify realistic configurations meeting customer needs. This information covers many aspects of configurations, such as producibility, testability, and maintainability with quality, cost, and reliability at the required levels.

Trade-off analyses are directed at finding a proper balance between the many demands on a design, and should include all the relevant parameters, such as cost, schedule, technical risk, reliability, producibility, quality, and serviceability. In order to determine the ranking of each design alternative, the value against each parameter needs to be assumed. Usually, 'utility function' of a parameter may be established, and it may be used to quantify the different alternatives.

The following steps may be followed to carry out a typical trade-off study:

(*i*) **Form a 'cross-functional' team**: The team members may be drawn from multiple and diverse disciplines and background. A number of functional experts should act as an independent group or entity.

(*ii*) **Encourage customer involvement and innovation**: The design team should formally establish close contact with the customers in order to get first-hand knowledge about their creative ideas and requirements.

(*iii*) **Define the objectives of the trade-off analysis**: The design team should clearly spell out the specific objectives of the study. This also should focus on the scope and usefulness of the study.

(*iv*) **Determine the approach and resources required**: Here the basic framework of carrying out a quantitative assessment of each alternative is established. The parameters to be considered in the framework should include the constraints on the resources to be used.

(*v*) **Evaluate and select the preferred alternative**: The ranking of each alternative is known, and the alternative having the best or maximum ranking is to be selected.

(*vi*) **Validate the study results through experimentation and/or simulation**: The validation of the preferred design may be done either by carrying out required experiments on the prototype (physical modeling) or simulating the virtual.

(*vii*) **Iterate more detailed trade studies throughout the design process:** By changing settings of the design parameters, the experiments can be conducted at different phases of product development as the design becomes more refined. This will ensure the best design parameter settings for the product.

(*viii*) **Document the study and results:** This is an important step. An appropriately-designed database needs to be developed for future reference.

(*C*) **Transformation of Product Requirements into Design Requirements for the 'best' alternative.** Upon identification of the "best" or "most appropriate" design approach, the next step in conceptual design is to translate the high-level product requirements into lower level design requirements. Since these design requirements provide the performance baseline for each design team member, it is to be ensured that the design requirements are easy to understand, realistic, and measurable or tangible for the selected design approach.

The set of design requirements being the foundation for the design effort, it acts as the basic inputs to develop guidelines for part or component selection, producibility, and reliability, partners and suppliers selection, budget and investment analysis, and detailed process design. The design requirements at different levels should be established keeping in mind their following uses:

(*a*) they communicate essential requirements to all the members of the design team including suppliers.

(*b*) they permit complete technical control of the design process in all aspects of the product development,

(*c*) they provide a quantified baseline for design trade-offs, design reviews, and measurement of technical progress, and

(*d*) they provide quantified and testable requirements for test and evaluation.

Even when the design requirements are well-defined and documented, the design team must ensure that the stated requirements are reasonable and appropriate for the end user and within the limits of existing technology, or that the technology could be developed.

Once the higher-level design requirements are defined and known, they are further defined for and allocated to lower level subsystems. However, it should be remembered that, effectively establishing lower level design requirements from the higher-level product requirements is a difficult task. This whole process is often called requirements allocation or 'partitioning'. In specific terms, requirement allocation identifies how to decompose and allocate the system-level requirements to the levels of subsystems and components in terms of hardware, software, personnel, technical manuals, and facilities. Subsystems and components receive technical requirement budgets that together add up to the total system requirement. A total of all subsystem requirements should be equal or be less than the system requirement. This allows each designer to have assigned goals and responsibilities for a particular item of the design. This allocation should identify and compensate for areas of unusual technical problems or risk.

All design requirements should be expressed in measurable and "testable" terms. Figure 3.3 describes the requirements determination and their evaluation at different levels. As has been shown, each requirement at every level should be defined to allow the evaluation as to

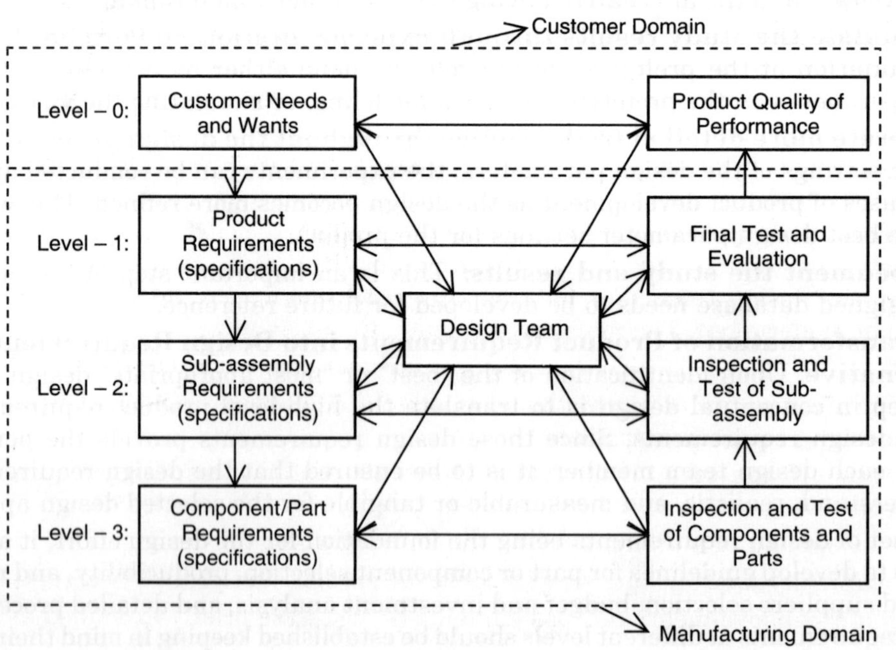

Fig. 3.3 Level wise Requirements, Test and Evaluation, and their Inter-relationships.

whether the design is satisfying the requirement. The basis condition that must be fulfilled for this process to be applicable is that every requirement at a level as mentioned should be both measurable and testable.

At the early design or conceptual design stage, three important aspects to be considered are as follows:

(*i*) Functional Task Allocation.

(*ii*) Producibility Assessment.

(*iii*) Reliability.

These three aspects are briefly explained below:

(*i*) **Functional Task Allocation.** Functional task allocation is the process of apportioning system performance functions among the user, the product, or some combination of the two. This will lead to identification of tasks that are to be performed by the product, by the user, and by both the user and the product. It is usually performed early in the design process during requirement definition or conceptual design. The necessary information needs to be provided to the user when a functional task is assigned to him/her, and this may not be possible without properly carrying out functional task allocation.

The major activities involved in functional task allocation are as follows:

(*a*) determine performance objectives of the product,

(*b*) determine functional requirements necessary to meet product objectives.

(*c*) allocate these functional tasks to the user, the product, the software, and/or some combination based on analysis,

(*d*) identify alternative design configurations, and

(*e*) verify that the user and the product are able to perform the assigned functions satisfying all design requirements, for each design alternative.

A critical but common mistake occurs when the design team attempts to automate every function or task that can be automated. Only functions that cannot be automated through hardware or software are then left for the human operator. This design approach results in excessive complexity, schedule delays, and assignment of tasks in a manner that the human may not be able to perform properly. Appropriate functions for humans and products are occasionally provided in handbooks, but they are only guidelines for evaluating specific functions in real-world design.

(*ii*) **Producibility Assessment.** To develop an effective design and related manufacturing approach prior to a detailed design of the product, the producibility criterion needs to be assessed by the design team. For making this assessment, a number of assignments are required to be undertaken. The main assignments are as follows:

(*a*) Identification of main characteristics and manufacturing resources,

(*b*) Identification of a design methodology that minimizes cost, and identification of manufacturing requirements goals,

(*c*) Trade-off analyses on alternative product and process designs including rapid prototyping,

(*d*) Development of a preliminary manufacturing plan and strategy, and

(*e*) Complexity analysis and simplification of product and process design.

Each of the above-mentioned assignments or tasks are briefly explained as follows:

Assignment – *(i)*

The objective of this task is to identify and control the design features that have the greatest impact on manufacturing time and cost, and product performance. Usually referred to as critical characteristics, they significantly affect the product's performance as well as its manufacturing. The design team is required to identify these characteristics not only at the final product level, but at all levels of the product development and manufacturing hierarchy. Completion of this task results in identifying and focusing the key characteristics of the product.

Assignment – *(ii)*

The design methodology that minimizes unit production cost for the product under consideration is known as 'Design-To-Cost' or DTC methodology. It reduces unit product cost through a rigorous approach of identifying and implementing cost-reducing design and manufacturing methods and approaches. For this, a DTC goal of providing a measure of the level of producibility for a product design is to be established at the requirements analysis phase. Obviously, setting appropriate and realistic cost targets is critical to successfully controlling product cost.

Assignment – *(iii)*

Relevant criteria are required to be considered for trade-off analysis. These variables may often have conflicting goals, making trade-off imperative. A host of factors, such as producibility measurements can be related to cost, schedule, quality, complexity, and risk. To make the trade-off analysis a reliable one, data on the variables related to the criteria should be properly documented.

Assignment – *(iv)*

A manufacturing plan identifies the required processes and suppliers to be used to create a product. The design team provides information on the critical features of the product design, resources, technologies, and the kinds of processes required. The manufacturing plan can help identify and highlight risk areas throughout product development. It is created early during conceptual design and updated frequently.

Assignment – *(v)*

In today's scenario, the majority of the product design, particularly in manufacturing sector, is complex in terms of the number of quality characteristics, technology, performance characteristics, methods of manufacturing, and materials quality to be considered. In this context, complexity analysis of product and process alternatives should be carried out to assess and reduce the levels of manufacturing risk and cost. Many a time, complex quality characteristics or design attributes or features may require the acquisition of new technologies, processes, equipment, or personnel of required skills and ability. A typical complexity analysis involves the following steps:

(a) Determine complexity metrics to be considered and analyzed (design attributes or features, part count, process required, schedule, cost, tooling, etc.),

(b) Analyze the proposed design against complexity metrics, and

(c) Incorporate suggestions into a modified design and manufacturing plan.

(iii) **Reliability.** Reliability is a major design parameter and depends on actions taken during design efforts, manufacturing, and testing. Developing and tailoring requirements and guidelines that achieve the user's reliability needs and at the same time provide adequate guidelines for designers is a difficult task. A major element of this effort is to base reliability requirements on realistic and accurate profiles of the product's use in the operational, maintenance, and storage environment.

It is also very difficult for designers to translate reliability requirements into everyday design decisions. Reliability goals, requirements, and guidelines must be translated into design-related parameters. The parameters should be easily understood, provide a basis for design decisions, and have the ability to be continuously measured by designers and management at all stages of the product development. Tests to measure and verify reliability should be specified in terms of the items to be measured, methods of measurement, failure definitions, testing environment, and test procedures.

Design guidelines on several aspects, such as part or component selection, derating criteria, materials and process selection may help establish norms for improved reliability. A number of information sources may be used to establish the design guidelines. Parts or components lists, military standards, and available design guides of research organizations may be very useful documents for this purpose.

After the total failure rate for a product has been established, failure rates are then apportioned or distributed among all identifiable lower level elements of the product. Apportioning the failure rate allows the design team to have a quantitative design requirement that can be measured for technical progress, similar to other design parameters.

The predicated failure rates of different design modules, piece parts, or software modules can then be compared to their actual value to ensure that the reliability requirements are continually met as the design progresses. Conversely, if any of the subsystem failure rate allocations are exceeded, the designer is alerted that corrections must be made. For improving reliability of the design, a number of options may be used. Some of these options are as follows:

(*a*) Reduce the number of parts and components in the design so that the design is simple.

(*b*) Use more reliable parts, materials, and software, wherever feasible and practicable.

(*c*) Reduce stress levels for the product.

(*d*) Design for variability and increase robustness of the design so that the effect of uncontrollable noise factor on the performance of the product is minimum.

(*e*) Add redundancy, self-corrective, and self maintaining features in the design of the product.

(*f*) Use standard parts and components in the assembly of the product as far as possible.

The product development team must consider the uncertainty of many of the assumptions and parameters used in their design analyses and in manufacturing the product. To compensate for the uncertainty of these assumptions, design margins, derating criteria, and safety margins are used. This requires the design to meet stress requirements that are greater than expected to compensate for uncertainty. Design and safety margins compensate for uncertainty by forcing the design to meet requirements higher than expected. They are usually expressed as a multiplier such as 200% or 2X. This means that the design must be more than twice the required stress. Derating is used in electronics where the maximum operating envelope stresses on the part can be no greater than some percentage of the maximum rating as determined by applicable inflection points. The purpose of setting derating criteria, safety factors, and design margins is to provide a design with the extra capability to compensate for any unforeseen problems or stresses. This "safety net" or "over design", as established by the customer, the management, or the designer, increases the strength and precision of the parts or components. This criterion must be carefully developed since over-ambitious derating and design criteria can significantly increase design costs, schedule, and risk. Because of this problem, criteria should be developed only after thorough analyses of the project's goals, product environment, and other applicable guidelines.

3.9 DETAILED DESIGN

During this stage, a certain group of tasks are required to be carried out to finalize the design of a product, as well as the design approach. On many aspects, both technical and non-technical, decisions have to be made by the design team. For this, a detailed and thorough analysis of the design should be made in order to ascertain the adequacy of the design as well as its correctness with respect to the assumptions made. For almost all types of products, three types of techniques, *viz.*, design analysis, modeling, and simulation, are used during this stage by the design team. Use of these techniques ensures development of an appropriate design serving the purpose as well as minimizing the technical risk.

Design analysis is the use of scientific methods, usually mathematical, to examine design parameters and their interactions with the environment. The purpose of this analysis is to gather enough information and adequate knowledge base of a situation so to make better decisions with minimum technical risk. Since the design team usually makes many assumptions regarding the design parameters and the design approach, the design process is often considered as an iterative procedure involving design, analysis, and test of the product utilizing the knowledge base obtained at a given time. Examples of knowledge base include rules of thumb, published standards textbooks, databases, and results obtained from actual analysis, modeling, simulation, and experimentation.

Under modeling and simulation, there exist a number of tools and techniques for evaluating and optimizing designs of products. The basic purpose of modeling and simulation is to assist the design team in the development of a product. They constitute a process in which models simulate one or more elements of either the product or the environment or interfaces between the product, environment, and users.

The above-mentioned tasks for detailed design are briefly discussed below.

3.9.1 Design Analysis

A multi-disciplinary approach is usually used for design analysis of a product. This has assumed importance in view of requirement of an enhanced functional and aesthetic value in a product design in the current competitive business scenario. For designing such products, knowledge in several disciplines, such as mechanical engineering, strength of materials, fluid dynamics, basic electronics, circuitry, digital communication, and computer science, is a primary requirement for design analysis. Moreover, knowledge in several other aspects of product development, such as manufacturability, testability, costing, supply chains and logistics, and reliability, should also be available with the design team. In many cases, effective coordination among all these disciplines and aspects of product development may be very difficult. Several computer-based designs, planning and manufacturing software and tools may be used by the design team for effective coordination. A comprehensive integrated database may contain information regarding drawings of assemblies, sub-assemblies, and parts, materials, suppliers data including performance statistics, design rules, specifications of the final design, guidelines on product use profiles, performance, producibility, reliability, supportability, and design-to-cost. Producibility criteria include capabilities of special and standard processes, testability, and estimated production quantity, and reliability criteria. This database should also contain results from prototype testing for the product.

Extensive use of CAD and CAE design automation systems at this stage help the design team create concept models easily and quickly using digital sketching or mathematical models, and evaluate many concepts in a short period of time. In future, automated design advisors and agent-based analysis technologies will allow product generation and evaluation to be completed almost instantaneously. The so-called 'paperless designs' automatically determine how the parts can be manufactured and assembled. Data from projects that have implemented computer-aided engineering tools indicate that design cycle time can be reduced as much as 60% while producing equal or superior product quality.

During design analysis, the built-in mechanism of the CAD and CAE software allows the design team to carry out trade-off analysis. This ensures optimization of design with respect to best overall performance from the product. Trade-off analyses are also directed at finding a proper balance between the many demands on a design. Hence, trade-off analysis is an integral part of the design analysis, and the detailed design process. The analysis that is carried out at this stage must be complete and comprehensive in the sense that the resulting final design of the product is expected to have no design problems in future. In order to be sure about the quality of performance when the product is put to use by the user, prototypes of the products are required to be made in most cases. Prototypes play a major role in all phases of development especially in detailed design. Physical models and software models (virtual reality) are used to gather information to reduce uncertainty in the product performance, and to optimize parameter settings of the design and to test the design. Prototyping provides valuable information regarding expected performance of the product, adequacy of the design features and parameters, types of problems and difficulties in process development and materials quality, and ways and means for improved performance of the product under consideration. It also confirms if the product design is 'robust', *i.e.*, to what extent the product performance is influenced by the uncontrollable 'noise' factors and how to estimate and then minimize or estimate this influence while the product is being used by the customers.

3.9.2 Modeling and Simulation

Modeling and simulation are the basic tools for evaluating and optimizing designs of products. They are used by the design team. In a typical modeling and simulation exercise, possible designs of a product are first modeled on a computer with the help of CAD-based software. These models then simulate one or more elements of either the product or the environment. In most situations, simulation and modeling can be least expensive and effective methods to gather and verify information regarding effectiveness of design and product performance. Modeling allows a designer to experiment with requirements, to optimize design decisions, and to verify product performance. There are several advantages of using simulation as a design tool. Some of these advantages are as follows:

(*i*) It increases the level of knowledge regarding design issues of product-environment interface.

(*ii*) It can assess the benefits, costs, and attributes of each design requirement

(*iii*) It can perform design trade-off analyses to optimize various design elements, such as performance, producibility or manufacturability, testability and reliability.

(*iv*) It may verify the completeness of a design in relation to the design's ability to meet the requirements as specified.

With the availability of different kinds of CAD-based software these days, almost any design can be simulated on the screen of a computer. Most design simulations involve mathematical models generated on computers. These models evaluate many kinds of system requirements, such as performance, reliability, cost, environment, and design trade-off. However, in many cases, a physical scale model of a product may be required to be made. For example, a scale model in a wind tunnel can simulate a full-sized airplane in flight. Sometimes, both computer-based simulation and physical scale models are required to be made.

Simulation is often used for experimenting with different design requirements to ensure that the requirements are feasible, and they lead to the development of the desired end product. In this use, performance standards are established and approaches are accordingly proposed to meet producibility and reliability goals. Sometimes design decisions made to enhance producibility or reliability result in losses in performance, weight, or power. Simulation allows the design team to perform trade-off studies before the product is manufactured so that the design can be optimized. As simulation may lead to increase the understanding of the product-environment interface, the designer may have a better appreciation of the benefits, costs, and attributes of each design requirement.

Development of the simulation has many steps that are analogous to actually building the system. For example, in the case of an electro-optical sensor that automatically detects objects, there are many parameters to be modeled even if the sensor is purchased from an outside source. Purchased sensors are advantageous in that the risks of design are removed and reliability, cost, weight, power, and size are precisely known. Modeling may still be required, however, to resolve such design questions as sensor look angles, optics diameter, field of view, software drivers and compatibility with other elements of the system.

Simulation is also used as a tool for design analysis, such as optimizing the choice of parts or components. Simulation can resolve questions about whether a part or a particular technology will improve performance, reliability or producibility and about its potential effects on other parameters. Simulation allows the product to be exercised more extensively than is possible in reality. Indeed, simulation of work environment to provide an analysis of external inputs and effects is one of the most important uses of simulation. In this context, it should be mentioned that simulation performs well if the following conditions and requirements are fulfilled:

(*i*) An effective simulation must be realistic and correct. An incorrect simulation is worse than no simulation at all. It leads to erroneous design decisions and causes users to lose confidence. Verifying the correctness of the simulation requires the same rigour used in the verification of software. Verification includes calculations, tests of limiting cases such as the response of the simulation to different kinds of inputs, and tests using observed responses to known inputs. Correctness of the simulation should be formally demonstrated and verified.

(*ii*) A simulation must be designed in such a way that it can solve real problems as identified and felt by the users. A simulation is usable if the designers understand how to use the simulation as a design tool. There are many cases where even very powerful simulation tools were proved to be useless as they were very complicated to use or their intended use was not clear. To make a simulation tool both useful and usable, one has to look into user's response and difficulty in its applicability. An effective simulation has to be accepted by the users.

(*iii*) Simulation requires extensive planning. Planning starts with initial identification of a need for simulation and continues through all the stages where the simulation is needed. Lack of planning and management is the principal reason cited as simulation failures.

The planning starts with the system evolution plan. The first step is a written simulation plan that guides the effort. The second step is generation of a written specification precisely defining the requirements for the simulation.

(*iv*) An effective simulation must be a cost-effective tool for design. It is to be ensured that it produces benefits in fair proportion to the cost. However, there may be situations where the simulation may not result in desired benefits, and it may cost more than it can save in design trade-offs. In this context, a 'capability versus cost analysis' may be relevant in defining what is worthwhile in increasing the scope of the simulation and what can be eliminated to cut costs. A simulation that is too expensive to be run will eventually not be run at all, and most of the work put into its development will be lost.

Often, two methods are prescribed to minimize the cost of a simulation. In the first method, the simulation is made with a modular concept, and usefulness of each module is evaluated relative to its cost. In the second method, the simulation is made problem-specific, and many cost-saving options are automatically incorporated. Often, the problem being solved requires only a limited amount of realism. For example, a physical model with 6 degrees of freedom may be very expensive to simulate and may require a lot of time and skills to make it work accurately. Although an actual aircraft experiences movement with 6 degrees of freedom, many aircraft components can be effectively simulated with fewer degrees of freedom. If the performance of the system being simulated is not sensitive to the vehicle motion, a simple motion model is sufficient. In addition, some systems are modeled for less cost by many small, special-purpose.

(*v*) A simulation should be flexible to accommodate changes in purpose, and should be able to adapt to different phases of product development and can accommodate new applications of the products. A top-down approach to the simulation design is helpful in building a control structure that accommodates changes. In this context, 'trans-portability' criterion of a simulation needs to be considered. Transportability refers to what extent a simulation can be used for new applications and design efforts without major changes. The types of computer and the programming languages used are the important factors that influence the transportability of a simulation. Among the options available for a simulation is the choice of digital, analog, or hybrid computers.

3.9.3 Types of Analysis

During detailed design phase, a number of critical aspects need to be looked into so as to make the product development process a cost-effective, flexible, and productive one. Related to these critical aspects, a number of analyses are required to be undertaken by the design team. These analyses are as follows:

(*i*) Uncertainty analysis,
(*ii*) Stress analysis,
(*iii*) Aging analyses,
(*iv*) Failure mode and effect analysis (FMEA), and
(*v*) Finite element analysis.

These analyses are briefly described below.

(*i*) **Uncertainty Analysis.** There are many types of changes that may occur in future. Unforeseen developments in many aspects of product development are very common in most product development projects. These developments cause variability in the design parameter values. The variations in the values of design parameters take place due to the following reasons:

(*a*) Variability in design, mainly change in precision, interaction, resolution, etc.
(*b*) Variability in manufacturing, mainly changes in specifications and tolerances of parts and assembly, processes, materials, quality, etc.

(*c*) Effects of stress and time, mainly fatigue, stress, failures, aging, etc.

(*d*) Variability due to changes in users as well as in the use of the product under consideration.

As has been the case in many product designs, if the product is more innovative in design, uncertainty increases during product development. More specifically, an altogether new product may significantly increase the level of uncertainty during manufacturing. For a new, unproven, or unique product, uncertainty can be caused by many factors, such as designs of parts and components, materials, software modules, features, technologies, options, vendors, logistic support methods, users and operators, manufacturing or software programming processed, design tools and testing methods used in development. The effects of this potential variation and its uncertainty must be considered.

To take into these variations and their impact on product development, three methods, *viz.*, worst-case analysis, parameter variation analysis, and statistical analysis, are used. All these methods determine the effect of variability over time on the performance of the product, and whether the performance is expected to be 'out of specifications'. In this context, a major decision in the product development process is to model this aspect for different design parameters to ensure an optimal design without "over designing."

As the name suggests, a 'worst-case analysis' is a rigorous evaluation of the ability of a design to meet requirements under the worst possible circumstances. This is done by determining the worst-case values of critical design parameters (expressed either by variables or by attributes), that may affect significantly performance, manufacturability or producibility, and reliability of the designed product. The design parameters are then evaluated for both the highest and lowest conditions. If the overall performance of each part or software module under these conditions remains within specified limits, the design is considered to be reliable in the worst possible conditions. To perform a worst-case analysis, the designer must identify variability limits including problems, such as manufacturing and supplier capabilities and degradation due to stress or time, and external problems due to environmental 'noise' factors or extreme conditions. Worst-case analysis is a traditional and conservative or negative approach which may assume that "everything will go against the product in future." The products developed with this approach usually may have increased manufacturing and software cost.

The 'parameter variation analysis' is a method that determines allowable parameter variation before a design fails to function. Parameters, either one at a time or two at a time, are varied in steps from their maximum to their minimum limits, while other input parameters are held at their nominal values. Data are generated to develop acceptable 'operating envelopes' for the various parameters selected. These parameter envelopes can also be plotted. If each parameter or plot is kept within the safe operating limits, the design is expected to perform satisfactorily. This kind of analysis may generally result in a design that can meet successfully several types of conditions before a failure may occur. It generally produces a design with lower manufacturing or software costs.

There are three main uses of 'statistical analysis'. It is used (*i*) to determine allowable parameter variations, (*ii*) to prevent the development of a 'over designed' product, and (*iii*) to lower manufacturing and software costs of the product. In statistical analysis, each individual parameter may be described as an independent random variable and its distribution. The allowed distributions of the product quality characteristics depend on the kind of the process selected and the supplier performance. In many cases, the distribution may be assumed to be normal for an acceptable significance level (0.01 or 0.05). Once the distribution with its parameter values is known (from analysis of test data set), the design tolerance is known in terms of individual parameter tolerances. The statistical analysis often results in wider tolerances with lower manufacturing costs.

(*ii*) **Aging Analysis.** For any product, its performance is a function of its environmental conditions and time. The term 'aging' is defined as the change in the performance of a product over a long period of time. For any component, or a product as a whole, the effect of aging on the performance may be known in terms of average and variance of the change in performance over time. Sufficient margin in the design of a product must exist so that no combinations of stress and aging distributions cause its failure or significant performance degradation over time.

The effect of aging on a product can be taken care of and minimized by adopting a number of measures. In the first measure, it is be checked if the product can be used only within specified limits of performance. The second option is to design sufficient margins or flexibility into the design. The first option usually requires controls by the manufacturer and continued maintenance to replace parts as needed. The second option is to design the part into the system so that there is adequate flexibility in the design to tolerate the expected effects of aging. The designer needs knowledge of typical part variation, both initially and with time, and the effects of various levels of stress on these variations. If they are not taken into account, failures occur due to the drift in the parameter values. Mathematical models describing output variables in terms of several interrelated input variables and process parameters may be an appropriate tool suitable for aging or degradation analysis.

(*iii*) **Stress Analysis.** The stress analysis in design is a systematic method of examining a design to determine the magnitude of operating stresses on the individual parts or components and their effects on performance and reliability. Its purpose is to identify potential problems and their remedies early in the design cycle. Failures of the majority most of the hardware components of a product may occur due to stress during the product's useful life, and hence, stress analysis is an important design method for ensuring high reliability of the product. Stress can be of different types, and it is caused by a host of factors related to manufacturing, testing, shipping, environment, users, disposal, and other products or systems. Stress can be explained and modeled by a number of parameters, such as temperature, current, corrosion, weight, number of customers, and bandwidth, or general product parameters, such as maintenance procedures. Stress analysis calls for minimizing the effects of stress by design modification, if needed.

The stress analysis consists of the following steps:

(*a*) Determine and understand all possible stress potentials that may be caused by product use and environment,

(*b*) Eliminate or minimize the level of stress whenever possible,

(*c*) Design a system that can adequately withstand all levels of stress, and

(*d*) Use environmental and developmental testing to ensure that the design can adequately meet all areas of stress.

For hardware component of a product, usually the three major types of stress analysis, *viz.*, mechanical, thermal, and electrical, are done. Loads induced from the interaction of two physical interfaces often cause mechanical stresses. Forces studied in mechanical stress analysis include axial, bending, shear, and torsion. Common types of mechanical failures are deformation, buckling, creep, fatigue, wear, and corrosion. Thermal stresses are caused when the environmental temperature of a product (or its components) changes. Problems include creep, deformation, material instability, temperature cycling stresses, and burnout. Electrical stresses are induced in much the same way as mechanical stresses do. The level of stress is usually caused by the level of power, current, voltage, and power cycling.

If the stress on a design is reduced, the reliability of the product is increased and production cost gets reduced. There may be substantial savings in the operating costs of a product (e.g., an equipment), if the operating temperature of the product is reduced, as high temperature usually causes early or premature failure of the electronic and other similar components.

For software components, a typical stress analysis may focus on several parameters, such as bandwidth, user traffic volume, types of user traffic, and computing resources.

When the basic goal of a design is to increase reliability of the product, it becomes imperative that the level of stress as measured through stress analysis be reduced by taking any one or more of the following options:

(*i*) Use of suitable materials with increased strengths. This will definitely increase the capability of the design to resist stress.

(*ii*) Decrease the level of heat stress on the product through system modifications, such as packaging, fans, or heat sinks.

(*iii*) Decrease variations of stress and strength by limiting conditions of product use or improving manufacturing methods.

Now-a-days, a typical kind of stress analysis, *viz.*, 'environmental stress analysis', is carried out in order to understand how a product performs under different anticipated environmental conditions. This analysis has become very relevant as many products may be developed for an international market, and they are required to be used under different environmental and climatic conditions.

Temperature is the most common environmental stress factor that affects the quality and performance of electronic parts. Reliability of electronic components is improved with a reduction in operating temperatures. Temperature changes also have a profound effect because of the differing thermal cycles cause materials to fatigue, resulting in open solder joints, cracking of seams, loss of hermeticity, and broken bond wires. The addition of vibration to thermal cycling increases further these effects. Derating reduces these strains by ensuring that all parts are rated in excess of the expected environmental conditions and all materials are selected with thermal expansion coefficients matched as closely as possible.

Another concern for the designers is high-humidity environments, which are extremely corrosive. Problems caused by high-humidity environments include corrosion of metallic materials, galvanic dissolution when dissimilar metals are in contact, surface films that increase leakage currents and degrade insulation, moisture absorption that causes deterioration of dielectric materials and changes in volume conductivity and the dissipation factor in insulators. One method of designing for a high-humidity environment is to seal the component in a hermetic package. A hermetic package has a cavity where the component element resides, sealed off from the outside environment. Other design considerations include reducing the number of dissimilar metals using moisture-resistant films, sealing the assembly in a hermetic package, or controlling the humidity environment around the system.

The environmental factors pertinent to each design must be identified along with their intensity. Some CAD systems have the ability to simulate the effects of environmental extremes on a product. The main as well as secondary effects also need to be considered in such a **stage.** The synergistic effect of various environmental combinations must also be considered.

(*iv*) **Failure Mode and Effect Analysis (FMEA).** Regarding any product however well designed, nobody can say that it will never fail, and hence, product failures and manufacturing problems are common occurrence. What is required is that appropriate techniques are used to minimize the number of failures and their effects when they occur. The failures may affect several parameters and aspects of product design, such as manufacturing, reliability, logistics, supply chains, safety, and user satisfaction.

The failure mode and effects analysis (FMEA) is one such appropriate technique for evaluating and reducing the effects caused by potential failure modes. It can be used for analyzing the design, user interface, or its manufacturing processes. In the context of product development, two types of FMEA analysis become relevant. The design FMEA or DFMEA is thus a design analysis technique that documents the failure modes of each part, signal, or software module, and determines the effect of the failure mode on the product performance. The process FMEA or PFMEA on the other hand, focuses on the processes and suppliers' failure modes. For design, critical failure modes are eliminated through design improvements that can include component and vendor selection, redundant circuit or software paths, alternative modes of signal processing, and design for safety. For manufacturing process, design improvements can include new processes, preventative maintenance, mistake proofing, and operator training.

Failure mode analysis starts early in the design process to minimize the number and the cost of design or manufacturing changes. Owing to the complexity of many products and the large number of failure modes, this analysis may require considerable effort and cost. Early in the development cycle, cost and schedule trade-offs must be made to determine the level of detail at which the analysis should be performed.

A "bottom-up" approach is followed in FMEA. Knowledge of the failure modes of each item or process is used to determine the effect of each failure mode on the performance of the system. There are a number of merits of using FMEA. Three important merits are as follows:

(*a*) It is capable of identification of single-point failures,

(*b*) It is capable of early identification of problems and their severity, and

(*c*) Design trade-off analyses are possible with the information obtained through FMEA.

The analysis usually assumes that conditions are within specifications, all inputs are correct, and no multiple failures occur. For example, if a computer is being analyzed, it is assumed that the environmental conditions, such as humidity and temperature, are within specified limits. The design under analysis must be clearly defined, including block diagrams. The function of each block and the interface between blocks should be defined. The second important aspect of FMEA is to establish the baseline or ground rules for the analysis.

A typical design FMEA requires the following steps:

(*a*) Identify all critical functional areas of the design,

(*b*) Identify the major failure rate contributors by studying the history of failure rates of items, such as parts, software modules, users, and signals,

(*c*) Examine output responses to input fluctuations, and identify the critical input parameters producing critical problems, such as out-of-tolerance output,

(*d*) Examine the effect of anticipated user and environmental extremes on the performance of the product,

(*e*) Determine the levels of tolerance build-up and parameter variations, and

(*f*) Identify areas for improvement and modify the design.

A process FMEA is similar except that here analysis focuses on process failures, and the available data can be combined with the measure of the severity of the effect to quantify criticality. A criticality analysis is then performed. The criticality analysis provides a quantitative measurement of the criticality of each failure mode based upon the qualitative results of the FMEA. Criticality is the probability that a failure mode will create an adverse effect, whereas a reliability prediction states the probability that a failure will occur during a period of time. Each failure mode is uniquely coded and its code number entered. Considering predetermined analysis guidelines, failure modes falling into specified classifications are declared critical and must be corrected.

(*v*) **Finite Element Analysis.** The finite element analysis (FEA) is a well-proven design technique, which uses mathematical techniques for predicting the stress and its effect on the physical behavior of a system. A set of simultaneous equations is developed that mathematically describes the physical properties of a product. Solutions to the equations are then obtained. These situations approximate the behavior of the product being studied.

The concept underlying the FEA is to divide a complex problem into a number of small solvable problems called 'elements'. Equations are developed that define both the elements and the entire product's response. These equations are then numerically evaluated iteratively going from element to element. Several types of problems in the domains of stress analysis, heat transfer, fluid flow, vibration, and elasticity use FEA techniques. It is an important analysis when extensive testing of the product is very expensive or time consuming.

3.10 TEST AND EVALUATION

Test and evaluation refers to an integrated approach for continual evaluation of the design of a product leading to the improvement of its original design. These evaluations are done so as to improve the product quality and performance. All identified problems and detected failures result in review analysis of the problems and corresponding corrective actions as deemed necessary. A systematic evaluation procedure requires that all available test data are collected in a consistent format, and analyzed to determine reliability growth of the product and its level of technical risk.

The goal of every test and evaluation method is to identify areas for design improvement. A number of design related tasks are to be undertaken at this stage. The important tasks are as follows:

(*i*) 'Coordination of Tests and Evaluations' so that a mature design by effectively coordinating all the tests to verify a design's maturity for a product is developed cost-effectively.

(*ii*) 'Design Reviews' use a multidisciplinary total approach for evaluating and improving all parameters of a design including producibility or manufacturability, reliability, testability and other support areas.

(*iii*) 'Prototyping, modeling and simulation' are used to both validate and verify the design, to identify problems, and to solicit ideas for improvement.

(*iv*) 'Design for Test' is used to design the product for easy and effective test.

(*v*) 'Test, Analyze, and Fix' methodology is used to identify areas for design improvement to maximize the reliability growth process.

(*vi*) 'Environmental and Life Testing' of critical components is initiated early in the product development process.

(*vii*) 'Qualifying New Parts, Technologies, and Suppliers' is required for improving the product performance with reduced technical risk.

(*viii*) 'Production and Field Testing' considers all quality control tests including incoming testing and environmental stress testing.

All the above-mentioned critical activities are briefly described below.

Coordination of Tests and Evaluations

An effective test and evaluation method can definitely reduce the technical risk associated with product development, and actually can improve the performance, producibility, reliability, and other aspects of a product. Its purpose is to make the design "visible" so that problems can be identified and resolved. In the early stage of product development, test and evaluation techniques are used to validate customer requirements, to evaluate design approaches, and to select alternative design solutions for further development. As the design matures, the tests usually become more complex in nature. These tests verify that the design can meet the requirements. This provides confidence to the design team, and ensures that the product performs satisfactorily in the actual operational environment.

The test and evaluation programme is an integrated series of evaluations leading to a design improvement and qualification. This improvement comes only as a result of a dedicated, implemented, and effective test and evaluation programme. This type of programme is accomplished by coordinating a series of evaluations and tests through the direction of an integrated test plan. This test plan is initiated early in a design to ensure proper levels of testing while minimizing costs. The plan becomes progressively more comprehensive and detailed as the product matures.

Another objective of this test and evaluation is the evaluation and maturing of newly emerging technologies, manufacturing processes, and suppliers. Breakthroughs in science and technology continue to pressure the design team to incorporate new technologies, suppliers, and new ideas in product design as soon as possible. Failure to use a new technology or process can result in very negative consequences, such as lost sales, as other competitors may be able to introduce innovative products quickly. On the other hand, using a new technology, process, or supply before it is ready, may be a critical mistake that results in cost overruns, poor quality, and high warranty costs. Since using new and often "unknown" technologies is becoming an integral part of the development process, an evaluation process must be established for accessing and studying promising new technologies prior to their use in actual products. This process of evaluating new technologies, prior to their use, is often known as, 'qualification testing' or 'off-line maturing'.

All tests should be organized in such a way that the product design gets improved at the end. An effective feedback loop, if appropriately built into the system of evaluation, may ensure that all detected failures and unusual results require analysis and corrective action.

Design Reviews

Design reviews should begin early in the design process to verify that the appropriate design analyses including trade-off analysis are carried out continually. Early design reviews usually focus on design drawings and design approaches to be used. As prototypes and test results become available, they can be reviewed on a periodic or 'as in, when is' basis.

A successful design review concentrates on a detailed analysis of the technical aspects and risks of the design. The best way to do the review in many situations is to hire senior technical experts. These experts are expected to review the design most objectively and impartially with respect to these aspects? Obviously the technical competence of the reviewers must be of very high order to ensure positive contributions and most-appropriate actions for continuous improvement of the product design.

The design review process is especially critical for producibility to ensure timely identification of potential manufacturing problems and their solutions. It is an efficient way to evaluate the maturity of the design, to find areas needing correction, to assess technical risks, to measure producibility and design margins, and to evaluate the manufacturing and quality aspects of the design and potential suppliers.

Prototyping, Modeling, and Simulation

Prototyping, modeling, and simulation are valuable methods for evaluating and testing new ideas, technologies, products, services and support systems. For instance, prototyping reduces the need for many tests and evaluations. They also provide a communication medium for information, integration and collaboration of different tasks to be undertaken for product development. For many products, product development may use desktop solid-modeling based CAD software, rapid-prototyping tools, and virtual-reality environments to design and evaluate products, components, parts, and software user interfaces. Prototypes can be physical, electrical, software models, or simple full-scale physical mock-ups. These can be made of insulation foam or cardboard or computer simulations using solid modeling and animation. Rapid prototyping describes any technology that can produce prototypes in a very short time. They can shorten product development time by evaluating design requirements quickly and allowing more design iterations, thereby improving the reliability of the final designs of the products.

Testing of many products is very expensive and time consuming in most cases. To reduce these costs and to verify the product's design simultaneously, these days many companies in recent times are using design modeling, simulation, and virtual reality to compliment the test and evaluation programme. This is especially true in the semiconductor, automobile, aircraft, and automobile industries. Complex models are developed which simulate designs of a product under different conditions. Solid CAD systems greatly enhance the ability to verify a design as "solid" models are much easier to visualize than wire frame-drawings, surface models, or multiple blueprints. The designer can find out the design errors easily with solid models even for complex parts.

Design For Test

Testability is a design characteristic that measures how quickly, effectively, and cost efficiently the problems of different types can be identified, isolated, and diagnosed in a product. Unlike production testing, validation testing focuses on characterizing all outputs and specifications. The test information should be sufficient to allow effective troubleshooting and "what if" analyses that facilitate quick decision-making. Key design parameters for testability are accessibility, controllability, observability, and compatibility. The design should be developed so that existing resources can be used with few modifications. This will save both time and cost.

'Test, Analyze, and Fix' Methodology

The most effective approach for obtaining a highly reliable product is to ensure reliability growth and design maturity through a structured 'test, analyze and fix' or TAAF methodology. The TAAF methodology identifies failure sources in the design by subjecting hardware and software to increasing levels of stress. Each failure is then analyzed and the design is improved so that the failure does not recur. Simulated user environments are used to develop parameters for the accelerated tests. These tests are usually performed at the subsystem level, with emphasis on those subsystems with low predicted reliability. Improvements in the reliability of these subsystems then have a major impact on overall system reliability. Subsystems that cannot meet the reliability requirements are then subjected to more extensive reliability development testing or a major redesign effort. These tests should be integrated with other tests to avoid duplication and to make effective use of limited test resources.

Reliability growth testing typically begins with early engineering models, and is especially important for new products. The environment for the reliability growth test is often more stringent than that expected during normal use to ensure that any critical problems are identified as early as possible.

In planning the TAAF test programme the predicted reliability should be greater than the required level. Tracking of reliability growth begins with initial laboratory testing and extends through operational field testing. Measures for different levels of risk can also be established.

In addition to the results of the reliability development tests, other development and operational tests are used to assess system reliability. Sufficient hardware must be dedicated to this function to realize a fast growth curve.

Reliability growth is measured by an increase in a reliability index such as the MTBF, approximately proportional to the square root of the cumulative test time. When plotted on log-log paper, with MTBF as the ordinate and cumulative test time as the abscissa, reliability growth approximates a straight line. The slope of the line is a measure of the growth rate. Growth rates have typically varied from 0.1 to 0.6. The rate varies based on how aggressive, well-planned, and fully-executed the test programme is.

Reliability growth requires TAAF's iterative design improvement process. A product is tested to identify failure sources, and further design effort is required to correct the identified problems. The three-step procedure in this case is (*i*) to investigate all possible causes, (*ii*) to identify and correct the problem, and (*iii*) to prevent the reoccurrence of this or a similar problem. The rate at which reliability grows during this process is dependent on how rapidly the sources of failure are detected, and how well the redesign effort solves the identified problems without introducing new problems.

A technical parameter that can be determined from a reliability growth curve is the instantaneous reliability for the design. The instantaneous or current reliability is the system reliability that is to be obtained if the test is stopped at that point in time. Periodic reliability growth assessments can then be made and compared with the planned reliability growth values.

Environment and Life Testing

Each design should be proven correct by complete operational and environmental testing. This should include part, subsystem, and system testing. Maximum stress tests are performed when they are found practical. If this is done early, future failures can be identified and avoided. Testing includes worst-case operating conditions, including operation at its maximum and minimum specified limits. Although there may be some degradation of performance at these operating extremes, the design should remain within specified limits. Test stresses can include input signals, temperature, rain, shock, vibration, salt, and fog. All failures and unusual results are analyzed thoroughly and not credited to "random failures or circumstances." All failures have a cause, and this cause must be found before any corrective action, such as a design change, can be taken. Design changes can include redesign, selecting different parts, or selecting new suppliers.

Life tests determine the predicted life of a part or product and information on how the product will fail or wear out. It consists of taking a random sample of parts or products from a lot and conducting tests under particular use and environmental conditions. There are three types of testing: life, accelerated life, and highly accelerated (HALT) testing. Testing starts at the lower part level and progresses up to the module and then to the product level.

In traditional life testing, samples of parts or products are tested in a reasonable environment for the life of the product. When the parts or products are found to be very reliable over a long period of time, complete life testing may not be very feasible since many years of testing under actual operating conditions would be required. Owing to this length of time, the parts may be obsolete by the time their reliability has been measured.

An accelerated life test is usually used under this condition. Accelerated life tests predict a product's life by putting the product under environmental conditions that are far more severe than those normally encountered in practice. This reduces the time required for the test since the higher levels of environmental stress cause the parts to fail more quickly. To properly conduct and analyze an accelerated life test, one needs to understand the relationships between use and environmental conditions and their effects on the possibility of failure of the part or the product. Models are often used that have been previously derived from the knowledge of the physical and other properties of the materials that are used in the product through empirical studies. These include the 'power rule' model, the Arrhenius reaction rate model, the Eyring model for a single stress, and the generalized Eyring model. These models account for the relationships between the failure rate of a part and the stress levels under a specific range of values of the stress. Success of the procedure depends not only on a proper selection of the model, but also on how accurately the model emulates the actual part. For example, the power rule model is preferred for the capacitors, whereas the Arrhenius and Eyring models seem to be the most favoured for such electronic components as semiconductors.

There are two important assumptions made when conducting and analyzing accelerated life test results. The first assumption is that the severity of the stress levels does not change the type of lifetime distribution, but does have an influence on the values of the parameters associated with the lifetime distribution. The second assumption is that the relationship between the stress level and the design parameters resulting in failures of the part or the product is known and valid for certain ranges of the stress level.

The third test type is the highly accelerated stress test (HALT) where stresses are applied in steps well beyond the expected environmental limits until the part or the product fails. The purpose is to detect any inherent design or manufacturing defects. The process includes test, analyze, and fix (TAAF) to identify the root cause of each failure. With the failure information available, the design team can improve the design of the product or manufacturing process. The design team must identify which stresses are appropriate for testing, such as vibration, temperature, cycling, and current.

All the above-mentioned three tests are considered to be an effective engineering tool, especially when normal life tests are not feasible. The key to success however is corrective actions. Because of high levels of environmental stress, the design team must ensure that unrealistic failure modes are not introduced. The physical properties of the part or the product and an analysis of its failure modes should indicate whether new failure modes have been initiated by the test itself. For example, increasing temperature beyond a certain level may change a material's characteristics, which would negate any information secured from the test.

Qualifying Parts, Technologies, and Suppliers

The purpose of qualifying new parts, technologies, and suppliers is to verify that a part or a supplier selected for a design is suitable even for the most severe conditions encountered in normal use. Acceptance of the part or the supplier can be based on the results of a statistical sample to ensure that the design, workmanship and materials used conform to acceptable minimum standards.

During the early design and development phase, it is particularly important to evaluate whether the suppliers and subcontractors have the capability and qualification to manufacture parts meeting all requirements, such as performance, cost, quality, quality control procedures, reliability, schedule, packaging, and technical risks. It is also important that critical items are identified and test plans are developed for them even though the actual assemblies may not be available for test until later.

The number of items to be tested and the types of environment to be used depend on experience, technical judgment, cost, schedule, and history of the supplier. The use of HALT or stress-to-failure testing can be very cost effective in ensuring an achieved 'safety margin' or the so-called 'figure of merit' (FOM) for any part tested. This type of testing requires an accurate realistic environmental analysis of the expected operational environment for the equipment being tested.

Production and Field Testing

Production testing refers to all quality control tests including incoming testing and environmental stress testing. As the production cycle begins, quality of all incoming parts and components are required to be measured and evaluated to eliminate manufacturing defects and early part failures.

One special type of production test in this context is called environmental stress screening (ESS). The goal of ESS is to test parts under stress conditions in order to transform latent defects into part failures that can be detected in testing. For electronic products, screening and conditioning are usually performed utilizing temperature cycling and vibration conditions. Testing for most commercial and consumer products are usually performed at the supplier's site and under less stringent conditions, because their environment is less severe. For commercial and consumer products, screening and conditioning criteria is based on an overall knowledge of the product. Owing to the cost-competitive nature of the commercial and consumer markets, product screening and conditioning are continuously tailored to maximize effectiveness. Initially, all products or new parts may be 100% screened and conditioned. As the product matures, the testing requirements are reduced.

User testing is extremely important in determining how well the product will satisfy the customer and actually perform in the real-world environment. Customer opinions and perceptions are the final measure of success and quality. These are especially important in service-oriented products. Tests are conducted with typical users under actual use conditions. These tests can also evaluate parameters such as aesthetics and identify problems that may not have been identified in any other tests. Since the users may take very unexpected actions, the response to this action by the product or the service becomes critical.

3.11 MANUFACTURING

Manufacturing, in general, refers to the approach of carrying out tasks that are necessary for a successful transition from the design to manufacture or production of the product. Planning for manufacturing is a basic requirement for the establishment of all production processes and their specifications. As the design develops, the comprehensive process plans, in different forms, are prepared. In this context, the design team needs to measure the 'producibility' criterion. Producibility is a criterion or a factor which specifies if the design requirements are compatible with the capabilities and realities of manufacturing. More specifically, producibility is a measure of the relative ease of manufacturing a product in terms of several factors, such as cost, quality, lead time or throughput time, and technical risk. The concept, 'design for producibility' is also known as 'manufacturability', or 'design for manufacturing', or simply 'design for production'. Designing for producibility is basically referred to as a method of designing a product so that it can be produced in an extremely efficient and effective manner with the highest or best levels of quality. This is possible only when the design team is aware of the effects that a design decision may have on the production processes. One must know if the capability of a process or an equipment in product is compatible with the design requirements as established.

A number of important issues need to be considered at this stage. The important issues to be considered are as follows:

(*i*) Manufacturing Strategy

(*ii*) Manufacturing or Process Planning

(*iii*) Process Development

(*iv*) Manufacturing Process, Qualification and Verification

(*v*) Design Release and Production Readiness

(*vi*) Manufacturing Methodologies and Technologies

These issues are briefly discussed below:

(*i*) **Manufacturing Strategy.** Historically, most successful companies were initially formed around an innovative or superior product design. This trend resulted in a situation where many companies considered design and marketing the company's most important functions and, therefore, received the most attention and resources neglecting manufacturing function. Manufacturing was treated as a service organization and evaluated in the negative terms of poor quality and low productivity, with high wage rates. Manufacturing in all likelihood was not expected to make a positive contribution to a company's objectives or business planning. Japanese success in manufacturing higher quality and lower cost products showed for the first time probably the error in this judgement. Unfortunately, large capital investment alone could not immediately correct problems caused by years of neglect. Improving a company's manufacturing capabilities is a difficult long-term process that requires considerable reserves of both expertise and capital.

One of the most popular new strategies in manufacturing is called 'lean manufacturing'. Perfected by Taichi Ohno, this strategy focuses mainly on the elimination of wastes in all areas with a focus on inventory, work-in-process, material handling, cost of quality, labour costs, set-up time, lead-time and worker skills. There are several principles to lean thinking including eliminating waste, standardize work, produce zero defects, and institute one-piece flow. The method focuses on the so-called 'Value Stream' that is defined as the specific network of activities required designing, ordering, and providing a specific product, from concept to delivery to the customer. This flow should have no excess steps, stoppages, scraps or backflows. It also emphasizes the complete elimination of "muda" (wastes and unnecessary tasks) so that only activities that create value are in the value stream. Timely responses are critical for this strategy to be effective. This provides a way to support value-creating action in the best sequence. The lean manufacturing strategy provides a way to do more and more with less and less: inventory, human effort, equipment, time and space, while coming closer and closer to providing the customers with exactly what they want.

Manufacturing strategies are the vision and framework for accomplishing long-term corporate goals. The vision establishes the company's goals for manufacturing. The framework helps to focus efforts on meeting manufacturing goals by planning for integrating the necessary functions and resources into a coordinated effort. Communication of this strategy sets the right climate for the teamwork and long-term planning that are necessary to improve manufacturing capabilities. The strategy should be well known throughout the company with regularly scheduled reviews to monitor progress toward the goals.

Without a well-defined manufacturing strategy, companies can too often look for short-term solutions that may prove detrimental in the long run. Long-range strategic plans allow sufficient emphasis to be placed on identifying and anticipating manufacturing technologies of the future. In this manner, manufacturing is prepared for new technologies with the expertise and equipment early enough to stay ahead of competitors.

The manufacturing strategy of a company should strengthen its ability to function in a much better or effective way in the long run. Specifically, it deals with the following important issues in order to be successful in this effort:

(*a*) future manufacturing technologies and requirements and essential expertise acquired early in development efforts,

(*b*) compatibility of the manufacturing strategy with long-range corporate objectives and modernization initiatives,

(*c*) long-term commitment for continuous improvement of manufacturing and supplier capability,

(*d*) product and process designs and their effectiveness,

(*e*) long term partnerships with suppliers, and

(*f*) "make or buy" decision criteria or parameters for outsourcing.

(*ii*) **Manufacturing or Process Planning.** A typical manufacturing plan coordinates the various production planning elements, such as production readiness and qualification. Without the benefit of thorough manufacturing planning, major problems may occur when a product is first produced. These problems include high rework and scrap rates, low quality, missed schedules, poor communication, cost overruns, and degraded product performance.

The manufacturing plan identifies the approach and details of all the tasks that are necessary for accomplishing manufacturing's strategies. This includes all critical paths between design and production and the tasks necessary to assure a successful transition from design to manufacturing. As the design develops, the plan becomes more comprehensive and thorough. Although no standard plan exists that is adequate for all products, all manufacturing plans are concerned with meeting the cost, schedule, quality, performance, and environmental goals established for the product.

An effective manufacturing plan should first be addressed during early product development and updated as the product design and customer needs change. This early involvement in manufacturing is essential to assure that the product design is compatible with the overall capability of the production system consisting of several types of fabrication, metal cutting operations, assembly, and tests that are needed to produce the product.

(*iii*) **Process Development.** The process development is involved typically in the following activities:

1. Identifying the company's manufacturing strategies and detailing their effects on product development process,
2. Studying the current business environment, competitors, and new technologies with sufficient detail, and incorporating this knowledge into the product development process,
3. Developing a detailed manufacturing plan that includes the supplier base and partners of the company,
4. Identifying manufacturing and supplier processes that need to be developed or improved, and
5. Establishing a comprehensive manufacturability program that includes design guidelines for manufacturing's capabilities, methodologies and processes and process qualification.

Effective communication and involvement between designers, suppliers, and manufacturing starts early in the design process. Often new processes must be developed or existing processes must be significantly improved. Just as design must continually use new technologies to stay competitive, manufacturing must also develop new processes and technologies to stay competitive and support new design technologies. Process development is similar to product development except that the "product" is a new manufacturing process, method, equipment, or technology. The methodologies of product development are used in process development.

(iv) **Manufacturing Process, Qualification and Verification.** The qualification of suppliers, methods, tools, software, and processes is an important task that is carried out to ensure readiness of manufacturing before the actual production begins. Each manufacturing process and supplier is reviewed and verified with respect to its adequacy to support the programme objectives. Prototypes are used to test each manufacturing process for verification of performance parameters, such as tolerances, cost, cycle time, and quality. Documentation such as production procedures and test requirements are verified. The process begins with the manufacturing plan. The process of qualification and verification requires the coordinated efforts of all members of the design team.

(v) **Design Release and Production Readiness.** One of the major milestones of any design programme is to know the time when the design of the product is stopped or frozen. The final design is released to manufacturing only when it is certified that the design is "mature" or "production ready". This is an important event for any company constantly involved in new product development. Many companies have their own guidelines that control the point in time when the design is released to manufacturing. The timing of design release can, however, vary from one company to another, depending on company's size and the design's complexity, technical risks, and schedule requirements among other factors.

One shortcoming in this respect arises when a design is released according to predetermined schedule requirements rather than its technical progress (*i.e.*, assessment of the design's maturity). Setting unrealistic completion dates may be a major cause of this problem. This situation may cause the concerned design team to deviate from the time-tested and standard procedures, thus increasing the probability of major problems.

Technical progress and the risks involved with initiating the manufacture of the product are the two main criteria that should determine the design's release date. When the programme involves the production of hardware or software that has been developed and manufactured before, a historical database can be used to schedule a design's release date. For an altogether new product, however, management must closely monitor the technical progress of the design and manufacturing process.

The technical risks involved in the design release allow management to schedule realistic delivery and design release dates. A standard checklist for evaluating the production readiness for the design may include many related factors, such as producibility or manufacturability of parts or components, completeness and comprehensiveness of design and test drawings, adequacy of tooling required, and documentation of such issues as environmental analysis of effectiveness and performance of purchased materials and parts, evaluation of supplier performance, involvement of experienced personnel in the design, and documentation effort.

(vi) **Manufacturing Methodologies and Technologies.** A successful product development team understands current manufacturing methodologies and technologies, how to design for them, and where the required expertise is available. The state-of-the-art manufacturing technology is changing so rapidly that everyone must continually stay abreast of the changes to insure the producibility of each design. There are several manufacturing

methodologies and technologies that are used worldwide. The common goal of all these methodologies and technologies is to evaluate and improve manufacturability or producibility of the manufacturing/or production system. How best a design can be manufactured and thus implemented is the major concern of these technologies and methodologies. A few important methodologies and technologies are as follows:

(*a*) Electronic commerce (e-commerce) and web purchasing

(*b*) Computer-aided design and manufacturing (CAD/CAM)

(*c*) Computer-aided process planning (CAPP)

(*d*) Ergonomics and human factors

(*e*) Material requirements and enterprise resource planning (MRP/ERP)

(*f*) Inventory control and just-in-time (JIT)-based manufacturing

(*g*) Group technology (GT) and cellular manufacturing system (CMS)

(*h*) Automation and robotics

(*i*) Supplier partnerships and supply chain management

A brief discussion on each of these technologies is given below:

Electronic commerce and web purchasing

Electronic Commerce (EC) is the paperless exchange of engineering and business information via computer networks using e-mail, Electronic Funds Transfer (EFT), and Electronic Data Interchange (EDI). Electronic commerce is dramatically changing the procurement process influencing significantly all suppliers and organizational purchase and supply. There are several advantages of e-commerce: reduced lead time, quick information transfer, no paperwork, better configuration control among multiple entities, and lower competitive prices for the purchases as many companies may bid for the work. The internet allows new and more worldwide suppliers to bid on contracts, as requests for bid can be sent to thousands of suppliers saving time and money. Many automotive and other companies have already developed and started using their own web supply systems.

Internet file transfer of design CAD data (electronic data interchange or EDI) has already become a part of the printed circuit board (PCB) production cycle at many companies. The factory electronically receives the data and then immediately manufactures the printed circuit board. Some companies solicit price quotes and bids from the suppliers. The electronic transactions may also be combined with manufacturing inventory and planning information to have parts directly delivered to the factory floor. As technology and the internet evolve, file transfer methods will continue to improve in future.

Computer-aided design and manufacturing (CAD/CAM)

The most common problem in manufacturing is ineffective communication between design and the various production departments. Previously, most communication was accomplished through drawings, design reviews, and company design guides. The computer-aided design (CAD) and computer network environments are beginning to provide a means of improving communication between production and design. This allows manufacturing to continuously communicate with design and evaluate design alternatives and design criteria. This is especially important when design and manufacturing are located in different areas.

However, a complete computerized approach for communicating between design databases and manufacturing databases is not currently viable, except in some applications, because it requires a greater level of software integration than that currently existing in industry. Future computer-aided drafting, design, analysis, and manufacturing tools will enable the electronic transfer of data for rapid development of problem solutions. Data standards will allow 'seamless' and efficient electronic interchange of business and technical information. Its future success is highly dependent on how well the company procedures and the design practices can be adapted to this concept.

In order to computerize producibility analyses, a complete understanding of the manufacturing process is necessary. Decision processes must be translated into logical representations and then developed into models that emulate the decision-making process. The software can make these decisions only when all the information required for consideration is available in a computer-readable format. This points need to have accurate design and manufacturing data. A challenge in future years will be for the manufacturing and product design functions to work together to define requirements and develop common integrated databases.

Computer-aided Process Planning (CAPP)

Process planning is the process with which product designs are transformed into a step-by-step manufacturing plan. A process plan includes a list of the manufacturing processes from the raw materials and vendor supplied parts to the finished product. Most companies depend on experienced manufacturing personnel to create process plans based on their specific manufacturing expertise. The process plans are either handwritten on paper or are simply keyed into a computer file. In recent years, computer-aided process planning has increased the overall productivity and quality of the process planning function. The full benefits of computer-aided process planning are realized when the manufacturing's capability database is comprehensive, accurate and updated continuously. Information should include cost, availability, precision, error rates and types, and functional properties.

Computer-aided process planning, as it exists today, utilizes either a group technology approach or a generative approach. The group technology approach involves designing a general process plan for a specific part family or redesigning an existing process plan to include a new addition to the part family. The generative approach utilizes decision logic processing to generate a unique process plan based on specific product design parameters.

When group technology in design is implemented, the designer retrieves an existing design that is most similar in manufacturing characteristics to the new part to be designed. The new design is then based, as much as possible, on the existing design's manufacturing processes. This approach will help to insure a maximum degree of standardization in product designs and manufacturing processes.

Ergonomics and Human Factors Engineering

Most manufacturing, assembly, inspection, and test operations are performed manually. Human component will continue to be the most important and dominant component in production in future, except in the cases of high-volume, simple-to-automate products having requirements that require automation. Manufacturing must ensure that the workplace is designed properly with adequate test capabilities. Also, the design of the workplace must ensure that proper controls are in place. The methodology for properly designing a worksystem or workstation for humans is called ergonomics, or human factors engineering. Consideration of ergonomics and human factors for designing and analyzing workplaces has assumed importance, because most of the workplaces are, complex in nature, the interfaces between humans, environment, and machine components are significant, and issues related to safety, comfort, and convenience of the humans involved at the workplaces have become a serious concern. There are a host of tools and techniques that one may have to apply for making the design of products, processes and workplaces as a whole an ergonomic one.

Material Requirements and Enterprise Resource Planning (MRP/ERP)

Planning, scheduling, producing, purchasing, and controlling all the parts and materials used in manufacturing including supplied parts and components is an extremely complex process. Due to the large number of parts and materials used by most companies, formal planning and control systems are utilized. The process is referred to as material requirements planning (MRP) and enterprise resource planning (ERP). It focuses on eliminating all types of wastes in production and manufacturing activities in terms of lead time, material, storage space, money, unnecessary processes, and inventories.

An enormous amount of design and manufacturing data are required for MRP. Data from each part is stored in databases. The first is the item master file or 'inventory status file', which contains information about each part, such as description, source (e.g., made or purchased), unit of measure, lot size, cost per unit and lead-time stock balance. The second is the 'product structure file', which contains the computerized version of the engineering drawing parts list with additional manufacturing information. The third is the 'master production schedule' (MPS), which provides the information of loading plans of the work to be considered during the planning horizon.

In MRP, the programmed logic evaluates the schedule and decides when parts will be needed. The program utilizes available inventory and offsets the release of orders by the necessary lead time. A total of seven parameters for each part number are used to develop the plans, and each must be continuously updated: Gross quantity, Gross need date, Available inventory, Open order quantity, Open order due date, Order quantity, and Order due date.

The enterprise resource planning or ERP expands this information to include all aspects of the entire enterprise, such as financial, human resources, and project management. The design team plays a major role in the success or failure of MRP and ERP. Since both are involved with the precision scheduling of thousands of different components and parts of the product, its success depends on the quality of the information in the system. Product delivery schedules should be correct and up-to-date if the resulting schedules are to be of any use. Engineering drawings must be released on time to support manufacturing lead times. Since the lowest levels of the product structure and long lead items are usually manufactured or purchased first, the design team should concentrate its efforts on designing and releasing these drawings first. Because of the large number of parts in the system, last-minute design changes or additions create major problems and usually result in schedule delays.

Inventory Control and Just In Time (JIT)-based Manufacturing

One of the major goals in any organization is to reduce its inventory of different types as excess/over required inventory is costly to maintain in all situation. Just-in-time (JIT) refers to several manufacturing techniques that are proved to be very successful to achieve this goal. Some of the techniques that are based on JIT philosophy are: mixed scheduling, pull dispatching rule and pull production system, 'Kanban' production control at the shop floor, and modular design.

A JIT-based inventory control system minimizes the inventory of parts and components, work-in-process, and finished products by manufacturing a product or ordering parts or components only when it is needed. An improved quality of the components and parts is the associated significant biggest benefit of this system. As work-in-process inventory levels are reduced, quality problems are identified and corrected quickly and immediately. This results in improved quality. The responsibility for quality is placed on each person at all stages of the supply chain. With reduced inventories between operations, the feedback on quality problems is instantaneous and results in a production line stoppage. Production problems quickly surface, and hence, must be solved.

However, implementing JIT-based production system is difficult. Companies that have experimented with JIT concepts have had various degrees of success. Some have tried to install the system into existing production operations, and some have tried to utilize Japanese-trained managers. The most successful have set up new operations with trained managers and production workers. The designer can greatly reduce a company's inventory costs and assist in implementing just-in-time based manufacturing systems.

Group Technology (GT) and Cellular Manufacturing System (CMS)

Group technology is a technique in which similar parts are identified, grouped together, and manufactured in a common production line environment. The purpose of group technology is to capitalize on similarities in manufacturing and design. These parts with common manufacturing characteristics are called part families. For example, a plant producing a large number of different products may be able to group the majority of these products into a few distinct families with common design and/or manufacturing process characteristics. Therefore, each part or component of a given part family can be produced similarly to every other part or component of the family. This results in higher production rates, greater manufacturing efficiencies, and lower manufacturing costs.

There are three general methods of grouping products into families:

1. **Visual inspection:** designers and engineers identify the families, but the method is least sophisticated and expensive method.
2. **Classification coding systems:** this is based on design and production data, however, this method is the most complicated and time consuming.
3. **Production flow analysis (PFA):** this method uses historical production information to identify similarities in production flow.

After the families are identified, special production lines, called cellular manufacturing system (CMS), are then developed to produce each family. Since the new cellular line treats each member of the family in the same manner, advantages of both job shop and flow or mass production system are possible to be achieved in CMS. The common production flow provides the basis for automation, quality improvement, and reduced levels of inventory.

The design plays a major role in the success or failure of GT and CMS. The designers must be familiar with the different part families and their production lines to ensure that the design parameters are compatible with group technology. Group technology generally affects the selection of the processes to be used, tolerances of the parts, materials, and part sizes. To ensure compatibility, many companies have instituted design reviews in their design process. Although certain design limitations may be imposed by group technology, the benefits of reduced lead times and improved quality should more than offset this inconvenience.

Automation and Robotics

Automation is a major method for improving productivity and quality. Low labour costs and high repeatability for quality improvement make automation a major goal for manufacturing. Identifying automation opportunities early in product design is important. Designing a product so that it is easy to automate is difficult, and must be integrated early in the design process. Moreover, automation equipment tends to require long periods of time for design and development, which are sometimes longer than developing the product itself.

Fixed automation occurs in high-volume products when the automation is specifically designed for a unique product or manufacturing function. This approach is used in industries with products that are not changed very often. These systems are usually hardware oriented, and require considerable cost and design effort when process changes must be made. When automation is designed for frequent modifications due to different products, the flexibility is usually provided by software. Robots are a special type of flexible automation that have become a major and essential component of several manufacturing systems, such as ship-building, painting booth, precision part manufacturing, welding, and such other hazardous production environments.

Supplier Partnerships and Supply Chain Management

Manufacturers usually rely upon many other companies (*i.e.*, outsourcing to suppliers) to provide many of the major and minor components of their products. Many electronic companies make fewer than 10% of the parts that go into their products. As a result, a company's success is highly dependent upon suppliers' performance. The suppliers are also dependent upon the company for their continued business and growth. The key to a successful relationship is communication and teamwork. The goal is to change supplier relationships from an adversarial, cutthroat win or lose battle to a cooperative, win-win situation. This is becoming increasingly harder with the use of more and more overseas suppliers. Although cost may be one reason for choosing an outside supplier, the primary reason is that in-house capability and technical expertise does not exist to produce the product.

Suppliers can be recognized as an extension of in-house capability, and as such be treated as partners. This allows a company to gain access to world-class capabilities, share risks, and release resources for other activities. Partnerships allow companies to share development costs, technical risk, logistics, and expertise. However, effective and shared communication is essential for an effective partnership relationship with a supplier, regardless of the location of the supplier. Information on product forecasts, sales, costs, and inventory, etc. should be shared. The amount and type of communication will vary depending on the location and historical relationship between the contractors. When this relationship is healthy and long-lasting the flow of information will be frequent and more informal. When the relationship becomes antagonistic, the communications will, by necessity, be formal and less frequent. Although the free flow of information can be very effective in producing results, the lack of formalized systems to document contractual agreements may later result in misunderstandings and confusion. The company should also keep the suppliers in confidence and be aware of long-range business projections to allow them to prepare for significant increase or decrease in production rate.

The initial selection and qualification of a supplier, as well as the decision to establish one source or multiple sources for a particular item, is dependent upon many considerations. As a minimum, suppliers should be committed to being the "best-in-class", willing to work in partnership, and have a working knowledge of statistical quality control. A strong positive correlation has been found between vendor partnering and effective product development.

The company's goal is to keep the number of suppliers to a minimum. This practice reduces the resource management needed to maintain control, and provides greater opportunity for cost improvement. Often, however, it is essential that additional suppliers be established for an item to assure sufficient production capacity, quality conformance, and pricing competition.

3.12 SUMMARY

This chapter discusses in detail the various aspects of product and process design. In view of critical importance of product development and its contribution to a company's ability to succeed in a fiercely competitive market of a global economy, it is essential that the designers use a number of approaches at different stages of product development. Several tools and techniques commonly used in this context are described. Comparisons between the traditional serial engineering and simultaneous or concurrent engineering are clearly indicated. Technical risk assessment, a critical issue at product and process design, is emphasized in the approaches mentioned, and common approaches for customer needs analysis are described in detail. The reasons for forming design teams when the design is complex requiring inter-disciplinary approach are mentioned.

Various stages in product development are mentioned and explained in detail. Highlighting the important aspects and difficulties and problems in the conceptual design stage, the activities in detailed design, such as modeling and simulation as well as test and evaluation are elaborately discussed. Different manufacturing tools and techniques as practiced in modern organizations are briefly presented in this chapter.

QUESTIONS

3.1 Identify and explain, in brief, the factors that affect the product development process. Why is this process crucial for a company's survival?

3.2 In which way product development is related to globalization of markets and rapid change in technology? Describe with suitable examples.

3.3 Identify and describe four new technologies that are likely to impact the way a manufacturing company develops new products.

3.4 What is 'product development life cycle'? Discuss the stages of this cycle.

3.5 What is 'collaborative design team approach' for product development? How is this approach different from the traditional approach for product development?

3.6 Identify the types of technical risks that a business organization may face during its endeavour for new product development.

3.7 Describe, in brief, the key steps for the requirement definition process in product development. Why is this process important for new product development?

3.8 Consider the following items or products. Prepare a set of product requirements for each in relation to its manufacturing requirements.

(*a*) a fountain pen

(*b*) a bike

(*c*) a washing machine

(*d*) a tape recorder

3.9 Identify and describe different types of design methodologies that may be used for identifying and evaluating user needs.

3.10 For each of the products mentioned in question no. (8), identify the types of technical risks that should be evaluated in product development.

3.11 What is Design-to-Cost (DTC) approach? Why is it recommended for product development?

3.12 What are the different considerations at the early design stage that a designer should be aware of in order to minimize manufacturing cost of the new product? What are the measurable parameters that you may need to identify in this context?

3.13 Why is 'trade-off analysis' essential for product development? How is it done? Describe with two suitable examples.

3.14 Make a library search of three practical cases of 'trade-off analysis', and identify the uniqueness and characteristic features of each of these cases.

3.15 There are different types of life cycle cost models for different products. Make a library search of these models, and write a report on merits, demerits, applicability, and limitations for each.

3.16 How are functional allocation, producibility and reliability aspects considered during early design phase of product development? What are the typical problems a design team may come across for this? How are they overcome?

3.17 "Overall, there may be two kinds of approaches you may adopt for product development – serial engineering (SE) and concurrent engineering (CE)." Describe the steps of both these approaches, and make a comparison between these two.

3.18 What is 'prototyping'? Why is it used during product development?

3.19 "Prototypes play a major role in all phases of development especially in detailed design." Discuss citing suitable examples.

3.20 In which situation, physical models of the product under development should be made, and why?

3.21 In which situation, software model or "virtual reality" of the product under development should be used, and why? Describe the future role of computer design tools in product development.

3.22 Why is simulation approach recommended for detailed design in product development? What are the human resources considerations for this approach to be fruitful? Discuss in detail.

3.23 Identify the basic elements of manufacturing process development, and discuss each.

3.24 List the different manufacturing methods and technologies that need to be considered for product development. Why is computer-aided process planning to be used in this context?

3.25 What are the elements of a typical manufacturing plan? Discuss each.

3.26 For the complex 'human-product' interface design, the concepts of ergonomics need to be applied for an acceptable product design. Make a library search, and make a report on the tools and techniques that may be used for ergonomic design of a product.

REFERENCES

A Comparative Analysis on New Product Programmes: European Vs. North American Companies, European Journal of Marketing; Bradford; 1994, vol. 28(7); pp. 5–29.

Alyssa S. Dver, *Software Product Management Essentials.*

Baldwn C.Y. and K.B. Clark, *Design Rules: The Power of Modularity*, MIT Press, Cambridge.

Crawfort C. Merle, *New Products Management*, 7th Edition; McGraw-Hill Higher Education Publishers.

Dan Conde, *Software Product Management: Managing Software Development from Idea to Product to Marketing to Sales*, 2002.

Darian W. Unger, *Product Development Process Design: Improving Development Response to Market Technical and Regulatory Risks*, Thesis, MIT, 2003.

David B. Sieger and Adedeji B. Badiru et al., *A Metric for Agility Measurement in Product Development*, IEEE Transactions, vol. 32(7), pp. 637, 2000.

David L. Deeds and Charles W. L. Hill, *Strategic Alliances and The Rate of New Product Development: An Empirical Study of Entrepreneurial Biotechnology Firms*, Journal of Business Venturing, vol.11, pp.41–55, 1996.

David L. Deeds, Dona Decarolis and Joseph Coombs, *Dynamic Capabilities and New Product Development in High Technology Ventures: An Empirical Analysis of New Biotechnology Firms*, Journal of Business Venturing, vol.15, pp.211–229, 1999.

David N. Ford and Durward K. Sobek II, *Adapting Real Options to New Product Development by Modeling the Second Tayota Paradox*, IEEE Transactions on Engineering and Management, October 21, 2004.

Edward B. Magrab, *Integrated Product and Process Design and Development*; CRC Press, 1997.

Frank M. Butrick, *How to Develop New Products for Sale to Industry*, Prentice Hall, 1971.

George M. Scott, *Critical Technology Management Issues of New Product Development in High-Tech Companies,* Journal of Product Innovation Management, vol.17, pp. 57–77, 2000.

Gianmario Verona, *A Resource-Based View of Product Development*, Academy of Management Review, vol. 24(1), pp. 132–143, 1999.

Griffin Abbie, *PDMA Research on New Product Development Practices: Updating Trends and Benchmarking Best Practices*, The Journal of Product Innovation Management, vol. 14(6), pp. 429–459, 1997.

Griffin Abbie, *The Effects of Project and Process Characteristics on Product Development Cycle Time*, Journal of Marketing Research, vol. 34(1), pp. 24–36, 1997.

Hearn Buck C., *Integrated Process Design & Development*, Pergamon Press.

Hilton, P., *Handbook of New Product Development*, Prentice Hall.

Hilton, P., *Keeping Old Products New*, Prentice Hall.

Ho, T. and Tang, C., *Managing Product Variety*, Kluwer Academic Publishers, 1998.

James T. Simpson and Christine Kollmannsberger et al., *New Product Development in German and US Technology Firms*, European Journal of Innovation Management, vol. 5(4), pp. 194–208, 2002.

Jerry Wind and Vijay Mahajan, *Issues and Opportunities in New Product Development: An Introduction to the Special Issue*, Journal of Marketing Research, vol. 34, pp. 1–12, 1997.

John M. Usher, Utpal Roy and Hamid and Parsael, *Integrated Product and Process Development, Methods, Tools & Technologies,* John Wiley & Sons, 1998.

John W. Priest and Jose M. Sanchez, *Product Development & Design for Manufacturing—A Collaborative Approach to Producibility & Reliability,* 2003.

John, E., Ettlie and Henry W. Stolleds, *Managing Design in Manufacturing*, McGraw-Hill, 1990.

Jonny Poolton and Ian Barclay, *New Product Development from Past Research to Future Applications,* Industrial Marketing Management, vol. 27, pp. 197–212, 1998.

Keah Choon Tan, *A Structural Equation Model of New Product Design and Development*, Decision Sciences, vol. 32(2), pp. 195–227, 2001.

Kim E. Schatzel, Roger J. Calantone and Cornelia Droge, *Beyond the Firm's Initial Declaration: Are Preannouncements of New Product Introductions and Withdrawals Alike?* The Journal of Product Innovation Management, vol. 18, pp. 82–95, 2001.

Kleinschmidt, E., Michael Song, J. X., William E. Souder and Barbara Dyer, *A Causal Model of the Impact of Skills, Synergy and Design Sensitivity on New Product Performance*, Journal of Production Innovation Management, vol. 14, pp. 88–101, 1997.

Kostas Axarloglou and Demetrios Tsapralis, *New Product Introductions and Price Makeups*, Eastern Economic Journal, vol. 30(2), pp. 223–236, 2004.

Meyer, M.H. and Lehnerd A.P., *The Power of Product Platforms,* Free Press, 1997.

Michael E. McGrath, *Product Strategy in High Technology Companies*, McGraw-Hill Publications, 2001.

Milton D. Rosenau, *The PDMA Handbook of New Product Development*, John Wiley & Sons, 1999.

Morgan Swink, *Completing Projects On-time: How Project Acceleration Affects New Product Development*, Journal of Engineering Technology Management, vol. 20, pp. 319–344, 2003.

Muanmmer Ozer, *A Survey of New Product Evaluation Models*, Journal of Production Innovation Management, vol. 16, pp. 77–94, 1999.

Nevins James L. and Daniel E. Whitney, *Concurrent Design of Products and Processes*, McGraw-Hill, 1989.

Papalambros, P.J. and Wilde D.J., *Principles of Optimal Design,* Cambridge University Press, 1988.

Paul Belliveau, Abbie Griffin and Stephen Somermeyer, *The PDMA Tool Book for New Product Development*, John Wiley & Sons, 2002.

Preston G. Smith and Donald G. Reinertsen, *Developing Products in Half the Time: New Rules, New Tools*, John Wiley & Sons, 1997.

R. Jeffrey Thieme, Michael Song and Roger J. Calantone, *Artificial Neural Network Decision Support Systems for New Product Development Project Selection,* Journal of Marketing Research, vol. 37(4), pp. 499–508, 2000.

Raymond M., Hainer, Sherman Kingsbury and David B. Gleicher, *Uncertainty in Research, Management and New Product Development*, Reinhold Publishing Corporation, New York, 1967.

Reinertsen, D. and Smith, P., *Developing Products in Half the Time*, Van Nostrand Publishers, New York, 1991.

Robert G. Cooper, *Product Development for Service Sector: Lessons from Marketing Leaders*, Perseus Publishers, 1999.

Robert G. Cooper, Scott J. Edgett and Elko J. Kleinschmidt, *Portfolio Management for New Products*, Perseus Publishers, 2001.

Robert G. Cooper, *Winning at New Products: Accelerating the Process from Idea to Launch,* Perseus Publishers, 2001.

Robin Cooper and Regine Slagmulder, *Target Costing for New-Product Development: Product-Level Target Costing*, Cost Management, vol. 16(4), pp. 5, 2002.

Roger J. Calantone and Shawnee K. Vickery et al., *Business Performance and Strategic New Product Development Activities: An Empirical Investigation*, Journal of Product Innovation Management, vol. 12, pp. 214–223, 1995.

Roger J. Calantone et al., *Principles of New Product Management: Exploring of Product Practitioners*, Journal of Product Innovation Management, vol.12, pp. 235–247, 1995.

Rudy K. Moenaert and William E. Souder, *An Information Transfer Model for Integrating Marketing and R&D Personnel in New Product Development Projects,* Journal of Product Innovation Management, vol. 7(2), pp. 91, 1990.

Sarah J. Marsh and Gregory N. Stock, *Building Dynamic Capabilities in New Product Development through Intertemporal Integration*, The Journal of Product Innovation Management, vol. 20(2), pp. 136, 2003.

Srikant Datar and Clark Jordan C. et al., *Advantages of Time-based New Product Development in a Fast Cycle Industry,* Journal of Marketing Research, vol. 34, pp. 36–49, 1997.

Stephen C. Armstrong, *Engineering & Product Development*, Cambridge University Press.

Stephen C. Wheelwright, *Revolutionizing Product Development: Quantum Leaps in Speed, Efficiency and Quality*, Free Press, 1992.

Stephen R. Rosenthal, *Effective Product Design and Development: How to Cut Leadtime and Increase Customer Satisfaction*, Business One Irvin, 1992.

Talley, W., *The Profitable Product: Its Planning, Launching and Management*, Prentice Hall.

Ulrich, K.T. and S.D. Eppinger, *Product Design and Development*, McGraw-Hill, 2000.

Urban Glen, L., John R. Hauser and Nikhilesh Dholakia, *Essentials of New Product Management*, Prentice Hall, Englewood Cliffs, N.J., 1987.

V. Krishnan and Karl T. Ulrich, *Product Development Decisions: A Review of the Literature*, Management Science, vol. 47(1), pp. 1–21, 2001.

Watton, H., *New Product Planning: A Practical Guide to Diversification*, Prentice Hall.

Wheelwright, S.C., *Managing New Product and Process Development: Text Cases*, Free Press, 1993.

CHAPTER 4

Quality Principles in Design

4.1 INTRODUCTION

Since time immemorial, quality has been a serious concern in all that the mankind has developed and used. Be it the Pyramids of Egypt, high roads and sculptures of the Roman Empire, or the finest paintings of Renaissance Europe, it had been a constant endeavour of any culture or society to design and develop the finest pieces of quality in all walks of life.

However, the quality as a formal discipline has developed only during the twentieth century. The modern quality has evolved through a number of stages. These stages are: inspection, quality control, quality assurance and total quality control/management (which are also referred to as Company-wide Quality Management or CWQM). In today's organizations, the issue of quality is given utmost importance, and as such managed quite differently that it used to be even thirty years ago. In this context, the basic characteristics or features of a modern quality system are manifold. These characteristics are discussed below.

First of all, the modern quality system is *customer focused*. Products are designed and produced and services provided to meet customer requirements. These customers may be of two types, *viz.*, internal and external customers to the company concerned. Everyone in the company should identify who are its customers and suppliers and what are their requirements. For example, managers are customers of their secretaries, but they are also suppliers to their secretaries. If a report needs to be prepared, it is the responsibility of the manager (supplier) to get the material to the secretary (customer) so that the report can be properly prepared and sent to the person who needs the report (another customer). The practice of each department optimizing its own operations without any consideration of the needs of other departments in the company is no longer acceptable.

The second characteristic of a modern quality system is that the quality improvement process is led by the top management. If the responsibility for quality is delegated to the QC department, everyone in the company understands that quality is not a key management concern. There must be active and visible involvement by top management in the quality effort. It is said that if quality is not on management's agenda, it isn't going to be on the workers. If workers perceive that management's only interest is in what they describe as the corporate bottom line, then their interest in turn will only be in their bottom line—wages and benefits.

The third characteristic is that everyone must understand their specific responsibilities for quality. Although it is true that "quality is everyone's job," it is also true that everyone has a different responsibility, depending on his or her position within the company. For example, the product development cell of a company is responsible for designing new products which meet customer requirements and which can be consistently and economically produced by manufacturing function of the company. Thus, manufacturing is a customer of product development. The purchasing department must not buy materials on price and delivery alone. The corporate management must show by word and action that quality is of utmost importance to the company's survival. By adopting a quality policy aimed at complete control of every activity, maximum quality, productivity, flexibility and minimum cost can be achieved.

The fourth characteristic of a modern quality system is that it is defect prevention oriented rather than being focused on defect detection. It has been found that quality through inspection is not enough, and may not be cost effective. Although it may be necessary to do some sort of inspection or audit of finished products, a company's quality efforts should be focused on preventing defects from occurring by "doing the right thing right the first time, and every time." A modern quality system works on controlling processes rather than focusing on finished products. If the process is right, the finished product will be right. It is the involvement of cost and quality in every function and process that matters. High quality may also reduce costs.

Finally, quality must be a way of life. Quality issues are discussed at every management meeting. There is no consideration given to shipping an unacceptable product in order to meet production schedules. Old technology is replaced by new and better one only when it is proven to be dependable and cost-effective. All employees are given training in modern quality concepts and methods. Everyone in the company is allowed to participate in the quality improvement effort. Quality does not necessarily come from simply having a large quality assurance staff responsible to remove defective products or services before they get to the customer. It is achieved by making the improvement process a permanent part of a company's culture.

In this context, the modern quality system consists of three interrelated components: Quality of Design, Quality of Conformance, and Quality of Performance.

Quality of Design (QOD) refers to those activities which assure that new or modified products and services are designed to meet customer needs and expectations, and are economically achievable. QOD is primarily the responsibility of Research and Development (*R* and *D*), Process Engineering, Market Research, and related groups of the organization.

Quality of Conformance (QOC) refers to manufacturing products or providing services which meet previously determined and clearly defined specifications of the products. In a manufacturing organization, Production, Scheduling, Purchasing, and Shipping and Dispatch have the primary responsibility for quality of conformance. The top management has the responsibility to see that people have the training, tools, and other resources to do their jobs, and have the opportunity to become involved in the quality improvement process.

Quality of Performance (QOP) refers to the performance of the products or services that is actually observed when the products or the services are put to use. This is primarily the responsibility of Product Test, Sales and Distribution, and Technical Services and Maintenance. For the quality of the product as designed to be accepted by the users or the customers, this is the final or ultimate test of quality. An effective communication channel with the customers, and objective analysis of customer response on product or service performance are the primary requirements to ensure an increased quality of performance of the product under consideration. These three aspects of quality are interrelated. The feedback information flows between the COF and COD, as well as between COP and COD play an important role in assessing the existing designed quality of a product and in taking effective steps and actions to improve the quality of design if needed. For continuous improvement of product quality, these information flows should be made most effective with the use of on-line and real-time data on different aspects and design parameters of the products. The interrelationships between COD, QOC, and QOP are schematically shown in Fig. 4.1.

These three aspects of quality will be explained in some more detail in a later section of this chapter.

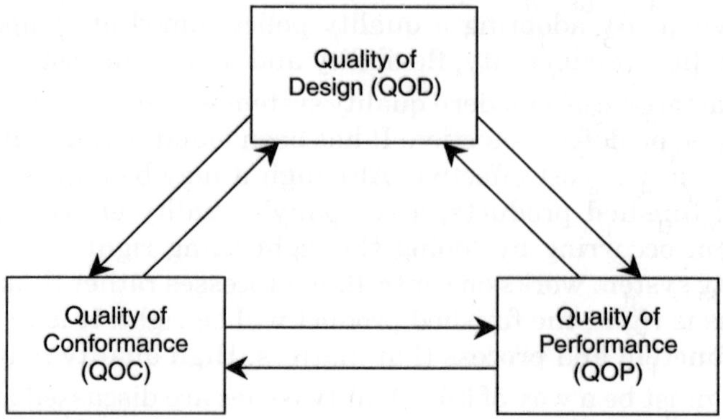

Fig. 4.1 Three Interrelated Aspects of Quality Systems.

4.2 DEFINITIONS OF QUALITY

Over the years, the concept of quality has been changing. It is a dynamic issue that is adapting to customer needs and expectations continuously. Several important trends in the international business and market are also responsible for this change.

In this context, the quality of a product or a service may be defined in several ways. A particular definition of quality is dependent on the perspective or viewpoint of the user of the product or the service. Broadly speaking, there are four definitions of quality.

Quality is conformance to applicable specifications and standards. Every organization, whether profit, non-profit, manufacturing, service, private or public, has specifications and standards for its products and services. Organizations develop these specifications and standards to measure performance and to correct deviations from expected levels of performance. For example, in a manufacturing operation, specifications detail dimensional limits or physical attributes of a quality characteristic of a part. In a service operation, standards dictate approved methods of behavior or service by the service personnel during transaction of an activity for the customer.

Quality is fitness for use. This is a market-or customer-based definition. A product or service is fit for use if it satisfies customer needs and requirements. An interesting point in this context is that a product may be fit for use in terms of satisfying the customer, but may not conform to the specifications. A surface finish specification is developed for a consumer product. The condition of the surface finish is important because it enhances the product's appearance and hence its marketability. The specification is written to include all surfaces, both external and internal. However, even if the inside product surface is blemished, it cannot be seen by a customer. So long it does not adversely influence any 'buy' decision for this product, the nonconformance is accepted. Hence, a product with a blemish may be fit for use if the blemish does not affect performance, safety, or marketability.

Quality is satisfaction of customer wants, needs, and expectations at a competitive price. The definition says that product or service quality is the producer's ability to satisfy customer needs while still being able to realize a profit. This definition has both a customer and a manufacturer orientation. While the customer is the reason for the organization's existence, the product manufacturer and service provider must still make a profit. This definition also focuses on satisfying the customer at a competitive price. Many customers will not purchase a product or service unless it is reasonably priced.

Quality is the "loss imported to the society after the product is shipped to the customers." Defined by Dr. Genichi Taguchi, an eminent authority on quality engineering, quality of a product is defined as the loss incurred due to bad or unacceptable performances of the product as a result of deviations from the target specifications or performance level. If this loss decreases, the quality of the product increases. This definition is different from all other definitions of quality. In order to measure this loss, the loss function is to be known, and, expressed in monetary terms. Taguchi proposes the concept of 'robustness' in design based on this definition.

4.3 QUALITY IN ENGINEERING DESIGN

For the quality system of an organization to be effective, the goal of any organization is not only to control and maintain the present quality level of the products or services, but also to improve their quality levels continuously. However, the traditional inspection-based quality control cannot improve the quality level. Through inspection, quality cannot be built into the product. In order to improve the quality levels of products and services, quite a large number of tools, techniques, and methodologies have been developed. The following six methods are used by the product development teams of organizations:

(*i*) Concurrent engineering,
(*ii*) Quality function deployment (QFD)
(*iii*) Reliability analysis,
(*iv*) Failure mode and effects analysis (FMEA),
(*v*) Taguchi methods, and
(*vi*) Statistically designed experiments.

A brief description of each of them is given below:

Concurrent engineering refers to the integrated and simultaneous development of products, their manufacturing processes, and support systems. In spite of holding Research and Development (*R* and *D*) section of a company to design a new product, the new design of the product is a responsibility of manufacturing engineers to develop manufacturing procedures, and to turn the new processes over to manufacturing, manufacturing and support personnel work together under concurrent engineering design. The result is reduced development and manufacturing costs, and a product that meets customer expectations. Concurrent engineering is more an engineering philosophy than a method.

Quality function deployment (QFD) was developed at the Kobe Shipyard in Japan in the early 1970s. It has been used successfully by many Japanese firms, most notably Toyota. QFD is a highly structured technique for assuring that the "voice of the customer" is not lost in the "noise of product development." There are four basic components in the QFD method:

(*i*) **Planning matrix:** The rows of this matrix represent customer requirements in customer-defined terms, and the columns list final product control characteristics. Marks made within the matrix represent relationships between customer requirements and product control characteristics.

(*ii*) **Deployment matrix:** Customer requirements and product control characteristics are listed in greater detail in this matrix, so that critical component characteristics can be identified.

(*iii*) **Process plan and quality control charts:** The process plan chart shows the relationship between manufacturing processes and critical component characteristics. It aids in identifying where monitoring points for the process are most needed. The quality control chart includes a schematic of the process flow and describes the necessary control points by location, type, and frequency of use.

(*iv*) **Operation instructions:** This document describes operations which the production personnel must perform to assure that critical product requirements are met.

QFD is a well-proven and time-tested technique which has been adopted by many leading companies of the world because of its ability to increase customer satisfaction with significant savings in product development cost.

Reliability analysis addresses the quality of a product over time. Reliability is one of important dimensions of quality. Through reliability analysis, the designer knows if the product operates satisfactory during the required or anticipated time period. Statistical analysis of failure data, life testing, and repair time data is the basic requirement for reliability analysis.

Failure mode and effects analysis (FMEA) is a tool used by design engineers and design review committees to protect against potential problems in processes and products. FMEA is a systematic approach to identify potential product or process failures (failure modes), their effects in terms of functionality and safety, and what steps need to be taken to protect against these failures. FMEA should be done during product development, and again when unanticipated failures occur or there are design changes. Timely use of FMEA can avoid expensive modifications by revealing potential design deficiencies or hazards prior to actual production. Usually an FMEA is carried out from the bottom up, starting at the component level and working up through subsystems to the complete system. FMEA should examine performance under extreme environmental conditions and should try to anticipate possible misuse or misapplication by customers. FMEA was originally used in the aerospace industry to prevent potentially catastrophic equipment failures. The technique is now applied to several types of production processes and service operations as well.

Taguchi methods refer to techniques of quality engineering developed by Dr. Genichi Taguchi. During the 1950s and 1960s, Dr. Taguchi developed a comprehensive approach to quality which touches every aspect of a product's design, manufacture and use. These methods are based on the concept of 'robustness' in design, and 'loss function' with respect to both products and process considered.

Statistically designed experiments are the heart of quality engineering. The days of experimentation by changing one variable at a time are over. Modern experimental designs make it possible to obtain an amazing amount of information about a process or product using a limited number of experimental runs.

4.4 EVOLUTION OF QUALITY PRINCIPLES AND MANAGEMENT

Many types of quality control and improvement techniques were first used in ancient times. Four thousand years ago, the Egyptians measured the rocks used in their pyramids. Then the Greeks and Romans measured buildings and aqueducts to ensure they conformed to requirements. Later, craft guilds in Renaissance Europe used to specify, measure, and control, the quality of paintings, sculpture, and architecture. To assure uniformity in quality of products, guild students used to go through exhaustive apprenticeship and other training programmes overseen by accomplished masters.

The quality function in modern organizations has evolved through four main stages: inspection, quality control, quality assurance, and company-wide quality management.

(*i*) **Inspection:** The concept of quality as a formal discipline was introduced in industries for the first time during 1920s. The first quality groups were inspection departments. During production, inspectors used to measure products against specifications. Inspection departments were not independent as the inspectors usually used to report to the manufacturing department whose efforts they were inspecting. This presented a conflict of interest. If the inspection department rejected a batch of nonconforming products and the manufacturing department wanted to push this batch of products out of the factory regardless of quality, the manufacturing department always got its way. Thus, the basic production philosophy was essentially based on "production at any cost" principle instead of a "quality is job # 1" principle. Product quality may only improve slowly, or may not improve at all if inspection is the only means of quality maintenance and control.

(*ii*) **Quality Control:** In the 1940s, inspection groups evolved into quality control (QC) departments. The start of World War II required that military products be defect-free. Product quality was crucial to winning the war and could only be ensured if the inspection department could control production processes. Quality defined as conformance to specification, was controlled during production instead of being inspected into products. Responsibility for quality was transferred to an independent QC department, which was now considered the "guardian" of quality. Also, the QC department was now separated from manufacturing to give its autonomy and independence. During this phase of development many organizations introduced the concept of statistical quality control (SQC) or statistical process control (SPC) in their processes. Control charting based on SQC or SPC concepts were proposed and implemented by Walter A. Stewhart in 1920s. A host of other tools and techniques, notably in the areas of process capability analysis and acceptance sampling for components and products were also made popular and worthwhile to pursue for quality control.

(*iii*) **Quality Assurance:** Quality control subsequently evolved into quality assurance (QA). The QA department focuses on assuring both process and product quality through executing operational and quality audits, importing skills-enhancement training programmes for operators, performing technical analysis, and identifying operational areas for quality improvement. The QA department, in close touch with other departments and functions, develops and implements QA programmes.

The QC department does exist in many organizations where QA has not yet evolved. It is considered to be a functional area, which is responsible for inspecting products calibrating instruments, testing products, and inspecting incoming material.

(*iv*) **Company-Wide Quality Management:** As the issue of quality becomes more prominent, QA is evolving into a company-wide quality management (CWQM) function. CWQM is also called total quality management (TQM), total quality control (TQC) or simply total quality (TQ). The quality organization is the prime facilitator and consultant in this effort. Corporate quality groups are small with more authority but less direct responsibility for quality. For example, the quality organization has authority to stop defective material from leaving the manufacturing area, while the responsibility for the control of quality rests with the manufacturing department operator.

In implementing total quality concept in an organization, its chief executive officer (CEO) often leads the quality management function. As the quality message permeates at all levels of an organization, more people become involved, and slowly a quality ethic and culture develop. The focus of the total quality programme is company-wide, customer-oriented, and competitively driven.

Quality no longer resides in one department. It is a company-wide issue essential to the organization's survival. To produce a quality product or deliver a quality service requires the attention and commitment of everyone in the organization. It is the responsibility of the person doing the work, whether it is the security guard controlling arrival of persons and transport, the manager supervising employees, the operator working on the machine, or the person delivering flowers. Every element in the organization, from the executive committee that establishes business and quality policy to the receptionist at the front desk, contributes to the quality effort. The top management usually defines a realistic policy, line management establishes operational goals and objectives, engineers design attractive, reliable, and functional products, and operators produce defect-free products in the total quality management initiative.

Another important aspect of TQM needs to be emphasized. This is customer orientation in any TQM programme. This is essential in any TQM programme because the customers' needs change and the organizations must adapt to changing needs. Designing aesthetic products, producing defect-free products, and delivering products on time, at a profit are some of the issues to be considered in adapting to this new philosophy. Most importantly, an organization must design, produce, and deliver what the customer wants, not what the organization thinks the customer wants.

4.5 CONCEPT OF 'QUALITY LOOP' AND ITS ELEMENTS

The American Society for Quality Control (ASQC) has developed quality standards that are basically a set of guidelines for implementing a quality management programme for any company. This series, standards (ANSI/ASQC Q90-94), embodies most elements of the different approaches to quality management. This standard refers to the concept of 'quality loop', and identifies its elements from market identification to disposal. All these elements may not be relevant for all types of products, particularly for the so-called 'consumer' products. However, for all 'industrial' products, all the elements of the quality loop are relevant. The elements of the quality loop are shown in Fig. 4.2, and briefly explained below.

(*i*) **Marketing and Market Research:** Marketing's responsibility is to identify market, identify customer requirements, develop a 'product brief', and establish feedback system for control and improvement purposes. The typical activities and responsibilities of this element are as follows:

***Identify a market*:** Marketing takes the lead in identifying a market, its customers, and its requirements. A market can be an entirely new market, an established market, or a niche in an established market. A company may develop an entirely new product for a new market. This product is usually high priced in order to have an acceptable return on the research and development investment within a limited payback period. This new product initially will not have any competition, so if customers want the product, they must pay its price.

For an established market, a company can produce an enhanced version of the product that is sufficiently different from competing products, or it can develop a similar product and price it lower. This is a mass marketing strategy based on lower priced products. Or a company can develop a specialized product for a niche in an established market. This product satisfies the needs of a specific target group of customers.

***Identify customer requirements*:** Marketing identifies customer needs, wants, and expectations in both products and services. Almost any product has a service component, which may be as important as the product itself. For example, people do not simply buy an IBM computer, they buy the service that is associated with the IBM name. To identify the customer requirements, survey and research methodologies need to be used. This is the most important activity of marketing. In a constantly changing customer requirements and business scenario, keeping marketing departments up-to-date and current requirements is a challenging assignment for a marketing executive.

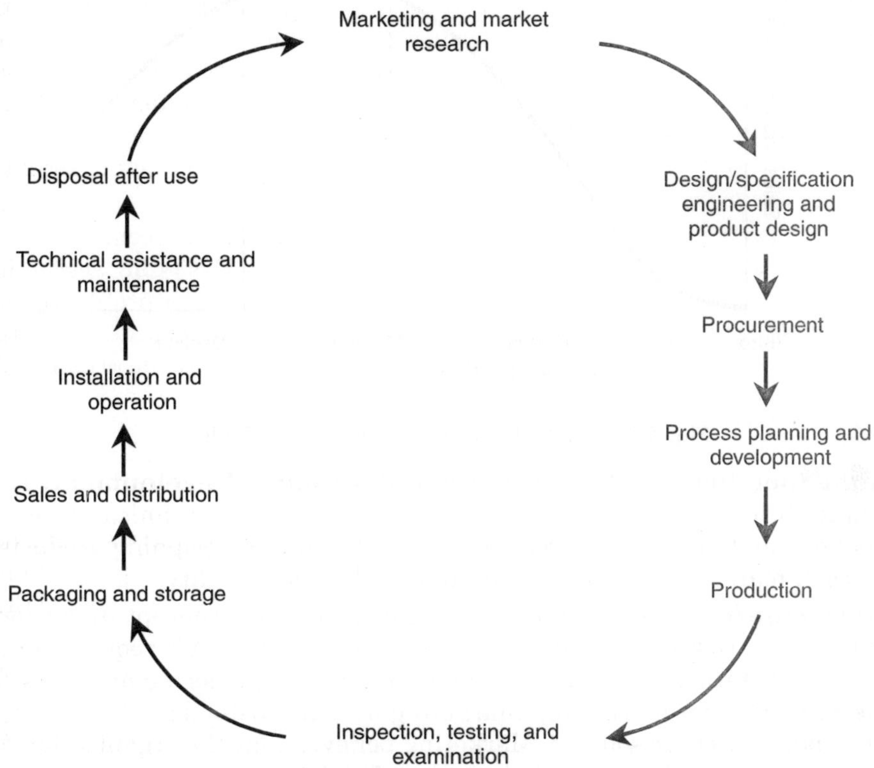

Fig. 4.2 Elements of 'Quality Loop'.

Develop a product brief: Once customers' needs are identified, these are communicated to the organization in terms of a set of requirements distilled into a 'product brief'. Customer requirements evolve into product and service specifications. A product brief can address many important features and aspects of product in simple and plain language that is easily understood by customers, such as performance characteristics, aesthetics, serviceability, packaging, price, and regulatory requirements.

Establish a feedback system: Finally, marketing establishes an information monitoring, and feedback system. Customer needs change, and an organization must continually accommodate to changes throughout a product's life cycle. Otherwise, products or services will age and not be able to satisfy current needs. Most products follow a life cycle consisting of four stages: introduction, growth, maturity, and decline. These four stages may show the sales pattern of a product. When a product is introduced in the marketplace, sales may be low because people are not aware of the product, or do not know what benefits will accrue from its use, or the product may be priced too high. In its growth stage, through advertising people become aware of its potential benefits, or of its ability to satisfy needs. Sales increase. In the mature stage, competitors develop enhanced products or similar lower priced products. Sales flatten out. Finally, in the decline stage, competition forces the company to develop new products or lower the price of the existing product, and sales decrease. Figure 4.3 illustrates the concept of product life cycle.

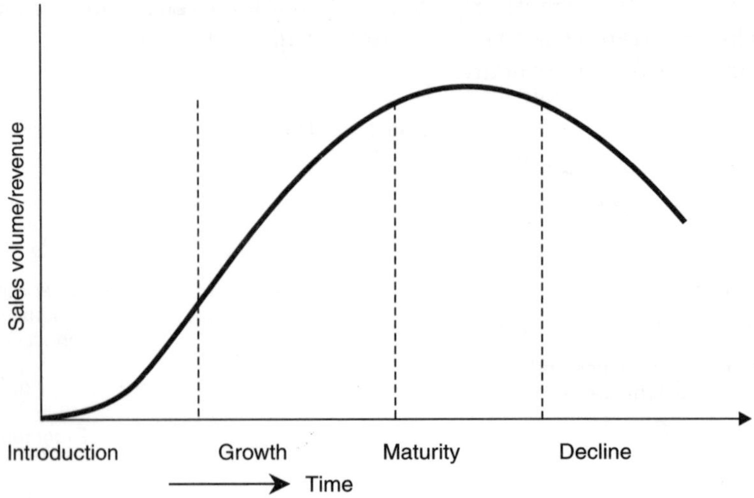

Fig. 4.3 Stages of a Typical Product Life Cycle.

(*ii*) **Design/Specification Engineering and Product Development:** Engineering, using the product brief, translates customer requirements into technical specifications for materials, products, and processes. It considers the following in designing products: customer requirements, cost, manufacturability, testability, and design quality.

***Customer requirements*:** Engineering first develops a concept or an idea and the knowledge of what the customer wants and expects. Often customer requirements, whether needs, wants, or expectations, are vaguely expressed, and engineering only gets an abstract idea about the product from marketing. Marketing obtains customer information from focus groups, surveys, polls, or other sources. Marketing believes that the organization can develop a product or service to satisfy these requirements. Engineering then determines whether it can develop the product within time and budget constraints.

***Cost*:** While always considering customer satisfaction, product and delivery costs are never far from the designer's mind. If a product is highly priced and has a quality image, it will sell to a narrow, affluent market, but usually not to a mass market.

***Manufacturability or Producibility*:** Sometimes engineers design products on a computer terminal and do not solicit inputs or advice with regard to manufacturability from manufacturing or purchasing departments. A product designed with tight tolerances is sometimes inoperable or not manufacturable. Tight tolerances might result in mating parts interfering with each other causing premature failure. A product designed with loose tolerances might result in unnecessary movement, again causing premature failure.

Testability: The ease and speed with which a product is accessible for testing is called testability. If the product is not positioned properly on a production line, it is difficult to test or even assemble. For example, in a cassette tape-recorder manufacturing facility, recorders travel on a conveyor belt where at set points, they are assembled and tested. If the production operator has to physically move the product or perform any unnecessary movements, this represents wasted effort, which adds cost to the product.

***Design quality*:** Quality in design is essential for a final product to be defect-free, safe, and reliable. Poor design can cause a product to fail prematurely. Also, if design errors are not eliminated, they are replicated in each manufactured product.

Good design also enhances safety and health. If a nuclear power plant is not designed and constructed with safety in mind, an accident can release radioactive and other kinds of waste in the atmosphere resulting in loss of life. The Bhopal gas tragedy, and the Chernobyl and Three Mile Island accidents are the cases of accidents in this respect.

While designing a product, one must also consider its **reliability** which may be defined as the long-term effects of quality or quality in the long-term. There are many instances where testability is not considered during design of the product. The resulting consequence may be very costly or catastrophic. Asbestos was originally installed in buildings because it was considered a good insulator, but its reliability was not fully understood by the researchers. Many years later it is found that asbestos is carcinogenic to those producing it, installing it, and living with it. The health risk of asbestos was not known initially when engineers developed the product. For many kinds of pharmaceutical drugs, knowing the long-term impacts on the human health and the harmful side effects is a critical issue, which must not be ignored by the designers of the products.

(iii) **Procurement:** Depending on the types of products and industries, many manufacturers procure as much as 40 to 90% (in terms of the manufacturing volumes) of their materials from outside suppliers. Hence, the quality of the final product or the assembly is significantly dependent on the quality of the procured components. Purchasing department plays a vital role in procuring the best quality materials and components from the 'right' suppliers. The responsibilities of the purchase department are to communicate requirements to suppliers, to select suppliers, and to measure and monitor the performance of suppliers.

A manufacturer may be thought of as being a customer to its suppliers. The customer must communicate requirements to suppliers explicitly through specifications, drawings, purchase orders, and contracts. In majority of cases, requirements include quality characteristics, service characteristics, operating environment, aesthetics and quality grade, inspection instructions, and specifications for the items to be purchased.

Supplier selection is a formal process that evaluates suppliers based on their ability to produce a defect-free product, which is competitively priced, delivered on time, and serviced properly.

Performance of the suppliers are measured and monitored continually, at periodic interval, to ensure that they are providing products that conform to specifications. Monitoring techniques include on-site assessment or audit of the supplier's quality system, product testing, and product improvement projects.

(iv) **Process Planning and Development:** Production processes, whether inspection, assembly, or fabrication, should be planned so that they run smoothly under controlled conditions. In a typical production system there are different kinds of fabrication and production equipment, and measuring instruments, and they are required to be monitored, calibrated, and controlled within prescribed operating limits so that defect-free products are produced. In this context, dimensional as well as process dispersion caused by operators, materials, methods, and machines should be kept at a minimum level. To achieve these, the details of the production activities, in terms of process planning are to be prescribed. The process plans are mode, and production department is to produce as per the process plan. Maintenance and measurement instructions are also included in the process plans. In modern production systems, particularly in repetitive manufacturing, where a long varieties of parts and components need to be produced with varieties of machines and equipments, computer-aided process planning (CAPP) techniques are widely used.

(*v*) **Production:** Production is responsible for creating reality out of engineering specifications and drawings. The term 'production' often used loosely. Actually, it refers to three types of activities: assembly, fabrication, and online inspection.

The responsibility for quality rests with the production supervisors and the workers. The production supervisor must communicate the importance of quality to the persons on the line. This person will only be committed if the organization is honestly dedicated to the pursuit of excellence and of never-ending improvement. If the organization's commitment to quality is not genuine, the workers may not be at all concerned about the quality they are producing. In this respect, production department's commitment to quality at all time is a pre-condition for creating an excellent system for production.

Another important issue in this context should also not be overlooked. This is the issue of process capability. The process capability is the ability of a process to manufacture a part or a component as per its given specifications. Thus, necessary tools and techniques are to be used to measure and assess the process capability on a continuous basis. Again, once a process is in operation, process should be 'in-control', and because of any reasons, the process is in 'out-of-control' state, corrective or remedial measures are to be taken to bring it back to 'in-control' state. Statistical process control techniques are used to achieve this.

(*vi*) **Inspection, Testing and Examination:** Inspection, testing, and examination are conducted on processes and products. The level of testing or inspection depends on the product, risks to the consumer, risk to the producer, cost, and regulations. For example, products involving public health or safety have associated risks to the producers and the consumers if the product fails during its use. This type of product may be governed by government regulations which mandate extensive testing and inspection. The standards and regulatory bodies in the government govern the testing, manufacturing, and distribution of such products. For example, the Nuclear Regulatory Commission (NRC) may dictate the construction, maintenance, operation, testing, and commissioning of nuclear power plants.

It is said that the quality of a product is as good as the measurement instrument used to check its quality characteristics. This implies that any measurement instrument should be accurate and precise to provide management with sufficient confidence in the decisions and actions based on the measurements. The gauges, instruments, and automatic test equipment therefore must be selected to meet or exceed the requirements of the customers and they should also be calibrated at regular intervals.

Product quality can be verified through incoming, in-process, and final inspection. SPC tools can be used at the incoming and in-process stages of a production system. Final inspection occurs just before the product is sent to inventory or to the customer. However, the degree and the frequency of inspection depend on the importance of the quality characteristic and the capability of the process. If a quality characteristic is essential to product performance, the component or the product may be 100% inspected, *i.e.*, all the units produced are to be inspected.

(*vii*) **Packing and Storage:** Quality can only be maintained if products are packaged, stored, handled, and transported properly. Proper packaging protects contents against damage due to vibration, shock, heat, abrasion, and corrosion. Some products require special storage, handling, and packaging. Electronic components may have to be packaged in antistatic containers. Perishable foods are transported in refrigerated containers.

(*viii*) **Sales and Distribution:** Once a product is sold, it must be delivered to the customer intact and in an expedient and courteous manner. If a product is damaged in transit, the problem must be handled effectively and efficiently with the customer. Also, discourteous delivery people can create a harmful impression of the product even before it is used. It is important to monitor and correct this situation because it may adversely affect future sales.

(*ix*) **Installation and Operation:** Complex industrial products may require specialized tools, equipment, methods, and trained personnel to install and to operate. The manufacturers of the complex machinery and products provide the necessary manuals in sufficient detail so that safe installation and operation procedures are followed by the customers. Installation and operating documentation includes instructions for assembly, repair, installation, and operation, spare parts lists, and product servicing information.

(*x*) **Technical Assistance and Maintenance:** Once a product is sold, the buyer of a technical or complex product might require technical assistance to maintain its operation. Technical assistance might be nothing more than answering questions on the phone. After-sales service for complex equipment might mean sending a trained person to repair, replace, or maintain a sophisticated piece of equipment. After-sales service creates product or brand loyalty, which enhances repeat product sales.

(*xi*) **Disposal After Use:** After use, a product has to be disposed of properly and safely. If the product no longer has a useful life, it is scrapped. However, a product can have its life extended if it is repaired. If it must be scrapped, this is done in such a way that it does not jeopardize safety, health, or environment. For example, environmental laws may regulate the disposal of hazardous chemicals.

After a product has gone through the quality loop, the loop will start again with another product, whether it is a new or an enhanced product. There are very few products or services that do not follow the quality loop because products must adapt and respond effectively to marketplace changes in order to survive in the long run.

4.6 INSPECTION- AND PREVENTION-BASED QUALITY CONTROL SYSTEMS

Before the quality-related principles are identified prior to their application in operations and production systems, two types of quality control systems, *viz*. inspection based and prevention-based, need to be explained.

The traditional method of quality control is inspection-based. This is illustrated in Fig. 4.4 (*a*). The inputs are of five types: machine, man (operator), material, methods and management. Each is a potential cause of dispersion or variation of dimensions of quality characteristics. Fabrication or assembly operations process inputs by adding value to the product in each operational step.

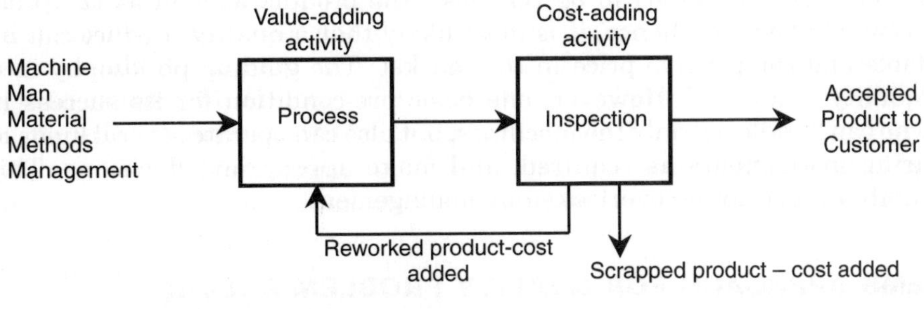

(*a*) *Inspection-based Quality Control*

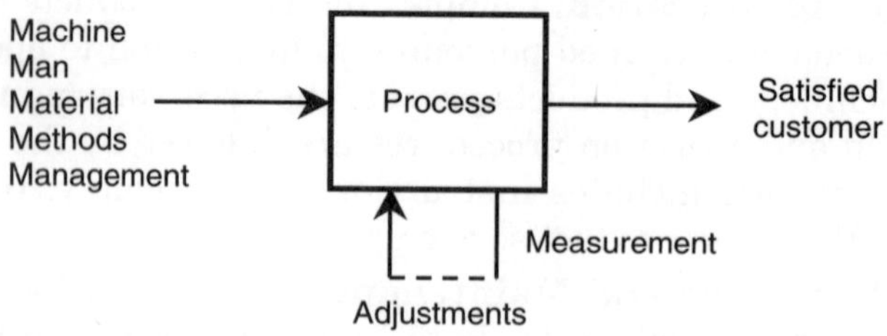

(*b*) *Prevention-based Quality Control*

Fig. 4.4 Comparison of Inspection- and Prevention-based Quality Control.

A process can be one step or 100 steps. At the end of the first step or the hundredth step, an operator may inspect a product to check its conformance with an acceptable specification or standard, and on acceptance, products are shipped to the customer. If rejected, products are scrapped or sent back through the process for rework. By the time product has been fully processed, value has been added to the product in terms of direct labour, raw materials, equipment, and training costs. At this point, if a product is scrapped or reworked, value has been lost. In this case, the numbers of cost-adding elements in comparison with value-adding activities are significantly more making the production process as well as the product more expensive. This type of quality control is also referred to as 'detection-based', and the guiding philosophy behind this mode of quality control has been 'production at any cost' which may not be a desirable feature of a modern production or manufacturing system where design of a best quality product with affordable price is always emphasized.

In order to achieve this objective, the quality control must be based on a philosophy which believes in the prevention of the occurrence of defects or nonconformance in the process itself. In the prevention-based quality control, the inputs are the same. Again, a process can be one step or 100 steps and value is added at each step. In the prevention mode, the process is controlled. The important difference between inspection and prevention modes is that during processing of parts, components, or raw materials, the operator continually measures the quality of these parts, and components or raw materials adjusts the process if dimensions deviate from calculated limits. Thus the operator is preventing defects from occurring by controlling the output of the operation. The objective is to send defect-free products to the customer. The prevention-based quality control is illustrated in Fig. 4.4 (*b*). In this case, as the number of cost-adding elements is expected to be very less, the product as well as the production cost remains reasonably low, and hence, it is most likely that a quality product can be developed with affordable and competitive price in the market. The guiding philosophy in this quality control is 'quality is job # 1'. However, one basic pre-condition for its success is to have a workforce who can handle not only the machines, but also can operate several kinds of inspection and measuring instruments as required, and make appropriate decisions. This aspect of operations ability must not be overlooked by management.

4.7 SYSTEMS APPROACH FOR QUALITY PROBLEM SOLVING

For solving a quality-related problem, a logical and systematic method is to be followed. Solving problems by a systems approach is a must for this. The value of this method lies in its ability to define a problem and to arrive at a solution through a logical process. The main objective of this approach is to identify the root cause of any problem and to take remedial actions to eliminate it. The methodology in systems approach is illustrated in Fig. 4.5. There are several elements in this methodology. Each element is described briefly.

(*i*) **Problem Definition:** The problem and its scope are first outlined. The problem is defined to ensure that it is solvable. If it is not solvable with available organizational resources, including knowledge, personnel, financial and political, additional resources need to be mobilized, otherwise a problem cannot be solved.

It is also important that a problem be defined to determine whether it is solvable within the required time period and with available resources. A problem might become a major and costly project that may be treated as tax burden to corporate resources and does not return the initial investment. In such a case, the company may be looser in financial terms.

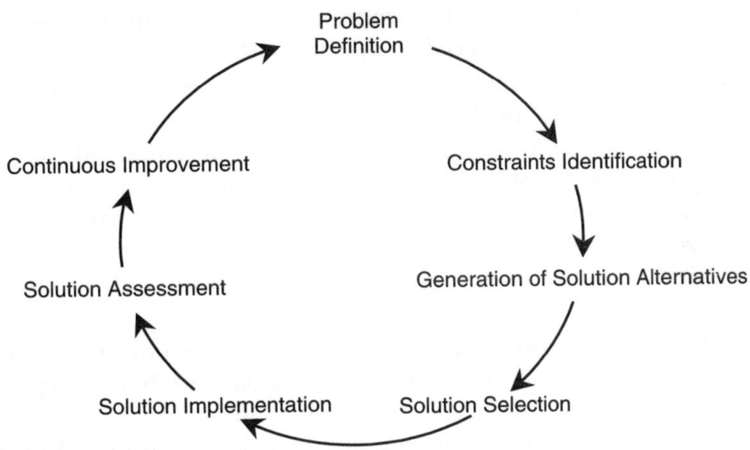

Fig. 4.5 System Approach to Design Problem Solution.

Most modern design problems are complex in nature. Usually one person does not have the skills or knowledge to solve them. So interdisciplinary design teams are formed to solve these problems. The design teams consist of representatives from the area where the problem is located as well as experts from peripheral areas who can supply information. For example, if a supplier is supplying a defective product, a team composed of personnel from purchasing, engineering, and manufacturing is formed to work with the supplier to improve product quality. A manufacturing process is suffering from chronic maintenance problem. A team consisting of experts or working personnel from engineering, production, and maintenance is formed to suggest solutions for improving availability of the process.

(*ii*) **Identification of Constraints:** At this stage, constraints to solving the problem or implementing solutions are identified. Constraints are factors that need to be anticipated and eliminated as problems arise. Common constraints are cost, time, size, rule, regulations, environment, permits, and culture. With any kind of quality-related decision, cost is always a major constraint. Any product or service has to be cost-effective. However good a product or service may be, it is only purchased if a competitive price is offered for it.

(*iii*) **Generation of Solution Alternatives:** The team brainstorms and process possible solutions on the basis of several ideas, however 'wild' they may seem to be initially. In a brainstorming session, organized methodically in free space and congenial environment, persons are not constrained, and are free to offer unconventional and novel ideas for problems even they may initially be thought of as unsolvable. Sometimes the most bizarre proposal turns out to be the most innovative solution. The design team invariably offers not only one but several competitive solutions to a problem. Thus, generation of a number of solution alternatives is a very common occurrence in most situations.

(*iv*) **Selection of Appropriate Solution:** Each proposed solution is evaluated in terms of its ability to overcome the constraints considered and to achieve corporate objectives depending on selection criteria. Proposals are ranked and quantified based on corporate requirements. The solution that is most cost-effective, easiest to implement, generates the most revenue, or saves the most money is selected.

If the problem is a defective product or a deficient machine, the selected solution should address the root cause of the problem. For example, if the identified cause of a failed pump is a defective seal, then replacing the seal would get the pump operating, but not prevent premature pump failure. The seal is not the root cause solution. A properly designed and implemented preventive maintenance programme may result in extending the life of the seal, and hence, this solution is considered to be the most appropriate.

(*v*) **Solution Implementation:** The solution is implemented so that problems do not recur. The affected personnel in the area where the problem is located are primarily responsible for its implementation. Most likely, they have a personal interest in its implementation and the elimination of the problem. They may also monitor implementation over time so that the problem does not recur. Schedules are established, and project milestones are identified and tracked. Network-based project scheduling approaches, such as CPM (Critical Path Method) and PERT (Programme Evaluation and Review Technique) are commonly used for this purpose.

(*vi*) **Solution Assessment:** The solution is finally monitored for effectiveness, cost, and reliability. If the solution does not solve the problem, further corrective action may be warranted. In corrective action, the team goes back to identifying the problem and repeating the whole cycle, always keeping in mind that the goal is continuous improvement. This forms a feedback loop, so that a solution is optimized and the problem does not recur.

(*vii*) **Continuous Improvement:** Continuous improvement in the quality of products manufactured or service has become a necessary element of an organization's survival in a global economy. In the past, if a firm had a quality or low-priced product, the firm probably used to dominate the market. There were several reasons for this. The number of firms producing the product was limited. There were also barriers for firms trying to enter the market, such as tariffs, cultural restrictions, cost, technical knowledge, or management skills.

Now there is no such thing as long-term ownership of a market. In a global economy, many of these constraints are gone, and if a market for a product or service evolves, an organization somewhere in the world may emulate or improve on the existing product. Hence, the company does not have any other alternative but to continuously improve the quality of the product. The continuous improvement philosophy is referred to as 'Kai'zen' in the Japanese firms.

4.8 THREE ASPECTS OF QUALITY

As has already been mentioned, there are three important aspects usually associated with the definition of quality: quality of design, quality of conformance, and quality of performance. These aspects assume importance in designing, controlling, and improving quality of products or services from the perspective of three stakeholders: designers, manufacturers, and customers. These three aspects are briefly discussed as follows:

(*i*) **Quality of Design.** Quality of design deals with the stringent conditions that the product or service must minimally possess to satisfy the requirements of the customer. It implies that the product or service must be designed to meet at least minimally the needs of the consumer. The objective of any design effort is to create a simple and least expensive product or service that meets the customers' expectations. This aspect should be taken into account while prescribing quality of design. Quality of design is influenced by many factors, such as the type of product, cost, demand for product, availability of parts and materials, government rules and regulations, and product safety. For example, suppose the quality level of the yield strength of steel cables desired by the customer is 100 kg/cm^2. When designing such a cable, the parameters that influence the yield strength are to be selected such that the requirement of yield strength is satisfied at least minimally. In practice, the product is typically over-designed so that the product may be able to withstand excessive loads in dynamic conditions. For this, concept of safety factor, k is used. Thus, to design a product with a 50 per cent stronger load characteristic over the specified weight, the value of k would be 1.50, and the product is designed for yield strength of $100 \times 1.50 = 150$ kg/cm^2.

In most situations, the effect of an increase in the designed quality level is to increase the cost at an exponential rate. The value of the product, however, increases at a decreasing rate, with the rate of increase approaching zero beyond a certain designed quality level. Figure 4.6 shows the impact of the designed quality level on the cost and value of the product or service. Sometimes, it might be of interest to choose a design quality level, b of a product which maximizes the difference between the value and the cost, given that the minimal customer requirements, a are met. This is done with the idea of maximizing the return on investment. It may however be observed for a particular designed equality level, the cost and the value may be equal. For any level above this the cost may exceed the value. This information is important for selection of an appropriate design level for a product or a service.

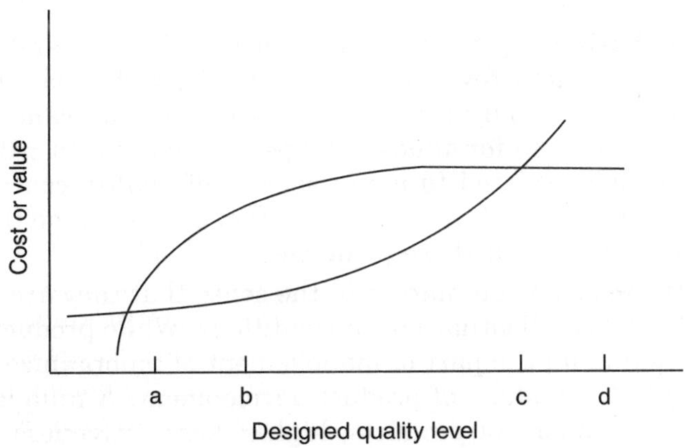

Fig. 4.6 Cost or Value Versus Designed Quality Level.

(*ii*) **Quality of Conformance.** When the manufactured product or the service rendered meets the standards selected in the design phase it is referred to as 'quality of conformance'. With respect to the manufacturing, this phase is concerned with the degree to which quality is controlled from the procurement of raw material to the shipment of finished goods. It consists of the three main areas, *viz.*, defect prevention, defect finding, and defect analysis and rectification. As the name suggests, defect prevention deals with the means to deter the occurrence of defects, and is usually achieved using statistical process control techniques.

Locating defects is conducted through inspection, test, and statistical analysis of data from the process. Finally, the causes behind the presence of defects are investigated, and remedial actions are taken.

The quality of design has an impact on the quality of conformance. Obviously, one must be able to produce what is designed. If the specification for the shaft diameter is 20 ± 0.2 mm, the question that must be addressed is how to design the tools, equipments, and operations such that the manufactured product meets the design specifications. If such a system of production is achieved, the conformance phase is capable of meeting the stringent requirements of the design phase. On the other hand, if such a production system cannot be installed (for instance, if the process is only capable of producing shaft with a specification of 20 ± 0.36 mm), the design phase is affected. This feedback may suggest that the product be redesigned because the current design cannot be produced using the existing capability. Therefore, there should be a constant exchange of information between the design and the manufacturing phases so that a feasible design can be achieved.

(iii) **Quality of Performance.** Quality of performance is concerned with how well the product functions or service performs when it is put to use. It measures the degree to which the product or service satisfies the customer. This is a function which is influenced and by both the quality of design and the quality of conformance. The final test of product or service acceptance always lies with the customers. Meeting their expectations is the major goal. If a product does not function well enough to meet these expectations, or if a service does not live up to customer standards, then adjustments need to be made in the design and/or conformance phase. The feedback from the performance to the design phase should help change in the design of the product or the service if so required.

4.9 QUALITY PRINCIPLES FOR PRODUCT DEVELOPMENT

It has always been emphasized, by the management practitioners and the researchers alike, that commitment of management towards fulfillment of quality goal at all stages of product development is key to organizational success. It has assumed significance in the current fiercely competitive business environment for almost all types of products. In order to fulfill the quality goals, a number of principles related to many aspects of quality control, improvement, and management need to be considered. It must be a constant endeavour on the part of management to take effective steps to implement these principles.

These principles are general in nature in the sense that they are applicable in all types of organizations and in almost all situations or conditions. While product development process is in progress, it is essential on the part of management of an organization to verify whether they are being followed at any stage of product development. A number of benefits, such as reduced cost, faster development of products, higher competitiveness of products, and long sustainability of products and processes, a company may have with the application of these principles. The principles, which are of immense importance in the current prevailing situations, are briefly discussed below.

4.9.1 Concept of Extended Process

The concept of 'extended process', as proposed by W.E. Deming, consists of not only the processes and employees of an organization, but also other components which are directly influencing the performance of the organization. These components are outside the traditional domain and control of the organization. A typical extended process includes the components, such as customers, suppliers, shareholders or investors, and the community and the environment. A typical extended process is schematically presented in Fig. 4.7. An organization consists of persons, machines, materials, methods, and capital or money. These elements are considered within the boundary of the organization. The extended process includes a number of key components that come from the external system to the organization. Among these components, customers or consumers are the most important. Satisfaction of customer is the primary goal of an organization whether it is of product or service producing type. Goals such as providing the investors with an acceptable rate of return are secondary. Achieving the primary goal—customer satisfaction—automatically causes secondary goals to be realized.

The community in which the organization operates should be considered in this extended process. This community includes consumers, employees, and anyone else who is influenced by the operations of the company, directly or indirectly.

Suppliers are another component of the extended process, and they play a very vital role. For most products, the majority of the components, parts, and raw materials required for a product need to be procured from outside suppliers. Many a time, as high as 80 per cent of parts and components are to be purchased from the suppliers. Hence, the quality of the product is very much dependent on the quality of supplied parts and components, and efforts must be made to ensure that suppliers supply quality parts and components all the time.

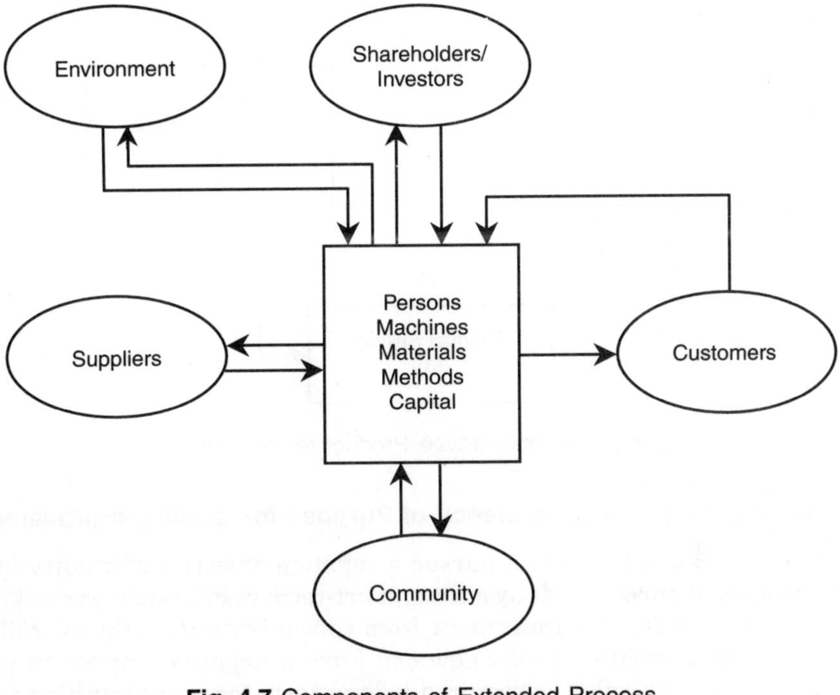

Fig. 4.7 Components of Extended Process.

A long-standing relationship between the suppliers and the organization is what is called for so that the organization has the maximum benefit in terms of lowered cost and improved service.

4.9.2 Customer-driven Improvement Cycle

In the present-day business and industrial environment, strategic goal of an organization should be based on the requirements and expectations of the customers out of its products and services being offered. The customers are the driving force in formulating and pursuing the strategic goal. This customer-driven approach ensures a better and faster product improvement cycle. This cycle consists of four main elements, shown in Fig. 4.8. In element (1), the product is designed based on current and anticipated future customer needs. Obviously an attempt to include these needs as far as possible will have a bearing on the product quality and cost. In element (2), the product is manufactured and tested in the laboratory in order to ascertain to what extent the actual quality of the product conforms to its designed quality, and if the product can be sold with the actual quality level. In element (3), the product needs to be sold with the expectation that the majority of the customers be satisfied with the present design. Once the product is sold, its ownership gets changed, and the customers start using it. In element (4), the quality of performance is judged through market research, and improvement in design, if needed, is achieved through incorporation of relevant changes or modifications in the design parameters for products and processes. This leads to improved customer satisfaction in the next cycle of product quality improvement.

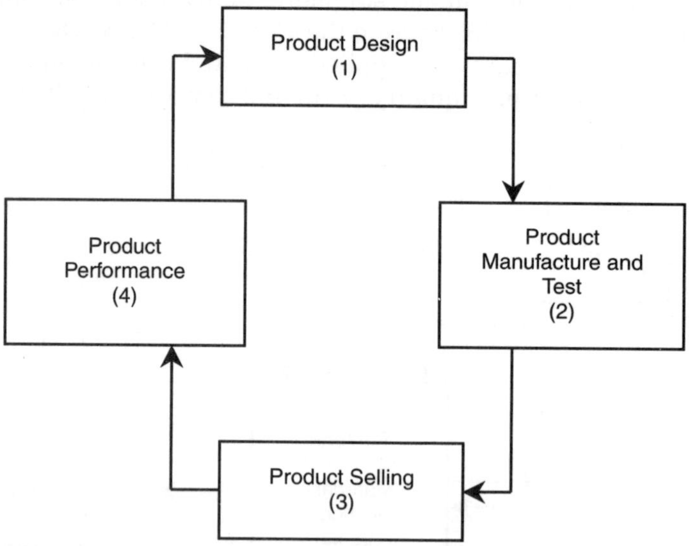

Fig. 4.8 Customer-driven Product Improvement.

4.9.3 Concept of Constancy and Consistency of Purpose for Quality Improvement

If all the departments of an organization pursue a common objective of quality improvement, it is known as 'constancy of purpose.' Many a time, short-term results such as weekly production reports or quarterly profits deter management from concentrating on the overall direction of the company. Actions that create a profit now can have a negative impact on profits in the future. As per this philosophy, what is required is long-term focus on anything to be done for quality improvement. Management should allocate resources for long-term planning, including consumer research and employee training and education that addresses not only work performance but also innovative approaches for quality improvement. They must ensure that there are resources available to cover the new costs of determining quality of performance and for new methods of production or changes in equipment. The priorities are research and education and a constant improvement in the product or service. Whatever actions management takes today must have a positive impact on the competitiveness of the product in the long run, say 5 to 10 years hence. Instead of focusing on short-term profit, management must strive for long-term survival, by taking strategic decisions and appropriate actions today. This will create a stable environment for the organization where improved performance can be sustained over a long period.

In addition to 'constancy of purpose', there should also be a 'consistency of purpose'. This means that all units of the company are working toward the common company goal to improve quality synergistically. Working for a departmental subgoal may not necessarily increase productivity or performance in other departments. Management should also accept the fact that variability exists and will continue to exist in any operation. They must try to determine ways in which this variation can be reduced. And this is a never ending process.

4.9.4 Rules for Inspection and Quality Improvement

It is stated that quality has to be designed into the product and it cannot be inspected into it. The concept of "design for manufacturability" emphasizes on producing the desired level of quality for a product. In this context, knowledge about the actual level of quality becomes the basis for assessment of quality of a product. Inspection is the means with which one may assess the level of quality of a product or a series. However, through inspection per se, the quality cannot be improved. Through inspection the unacceptable units of product are separated from the acceptable units. However, it does not address or may not be able to identify the actual or 'root' cause(s) of the quality deficiencies or problems. There are situations where inspection has to continue. For example, if a nonessential item is mass produced in large number and its standards are well established, sampling inspection may be used.

For all other cases, management should strive for achieving quality assurance in the system by implementing one or more of several quality improvement tools and techniques, such as Statistical Design of Experiments (SDOE) and Taguchi Method (TM). Sometimes, quality deficiency is directly influenced by excess process variability, which acts as the root cause in a given situation. The statistical process control (SPC) techniques may be used to minimize the process variability, thereby improving the quality of conformance for the product.

4.9.5 Principles of Supplier Selection

Selection of suppliers for its parts, components, and raw materials is a responsibility of management of an organization. Many a time, management is required to change the process through which the supplier selection is conducted at present. In this context, the gap between the supplier and the buyer must be narrowed down. They must work as a team to choose methods and materials that improve customer satisfaction.

In selecting a supplier, the total cost (which includes the first cost of purchase plus the cost to put the material into production) should be taken into account. As has been mentioned earlier, suppliers play an important role in the extended process. Knowing whether a product satisfies certain specifications is not enough. They must understand the precise problems encountered with the purchased material as it moves through the extended process of manufacture, assembly, and eventual shipment to the consumer. The buyers must be familiar with statistical methods so as to assess the quality of the supplier's plant. They must be able to determine the degree of customer satisfaction with the products, to specify what features are not liked by the customers, and to convey all related information to the supplier. Such information enable the suppliers to improve the quality of their products.

Another important principle involves reducing the number of suppliers. The goal is to move toward single-supplier items. Companies may have multiple suppliers for several reasons, including a fear of price increases, a supplier's inability to meet projected increase in demand,

and its lack of timeliness in meeting delivery schedules. However, there are several disadvantages to this policy. First, it promotes a feeling of mistrust between the buyer and the supplier and thereby creates a short-term, price-dependent relationship between them. Furthermore, suppliers have no motivation to change their processes to meet the buyer's specifications. Price, not quality, is the driving factor because another supplier may be selected if its price is lower. A long-term commitment cannot exist in such a situation.

Other disadvantages involve cost. Increased paperwork leads to increased order preparation costs. Travel costs of the supplier to purchaser's sites increase. Volume discounts do not kick in because order sizes are smaller when there is more than one supplier. Setup costs also go up because the buyer's process changes when the incoming supplier changes. Machine settings may have to be adjusted along with tooling. In continuous process industries, such as chemical companies producing sulfuric acid, differences in raw materials may require changes in the composition mix. Multiple setup periods mean idle production and therefore reduced capacity. Also, training the people who work with the suppliers cost more with multiple suppliers.

Many benefits are gained by moving to single-supplier items. As has been the experience of many well-known companies in this regard, the disadvantages as mentioned can be eliminated when a long-term relationship with a quality-conscious supplier is developed.

4.9.6 Process Improvement Cycle

Originally called the Shewhart cycle after its originator Walter A. Shewhart, or the Deming cycle when the Japanese started using the concepts of quality control and improvement on the advice and recommendation of W. Edwards Deming in 1950s, the process improvement cycle consists of four main stages: Plan(P), Do(D), Check(C), and Act(A). Hence, this cycle is also known as P-D-C-A cycle. The process improvement cycle is schematically shown in Fig. 4.9.

At the 'plan' stage, a framework is developed which describes the relationships between quality of performance as determined by the customers, and the process performance. The difference between them should be minimized, and for this, the effect of controllable process variables on process performance should be known and measured.

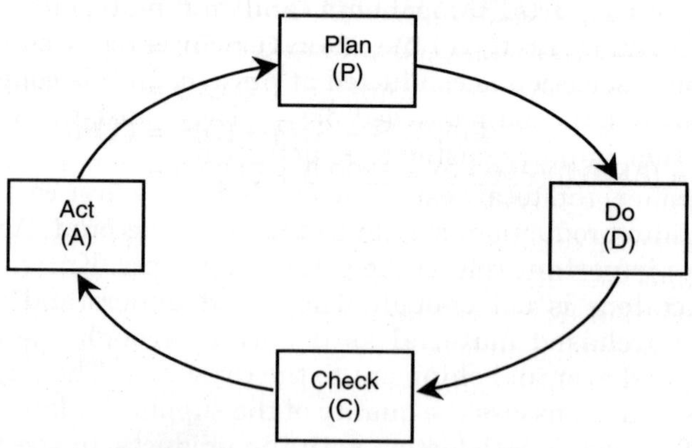

Fig. 4.9 Stages in Process Improvement Cycle.

For example, for given product, the customers are not satisfied with the present level of rejections. The process engineer finds that the main reason of rejection is non-conformance of a given quality characteristic, say diameter, of the product with the given specification, and the process performance needs to be improved. A study is planned to determine the appropriate process parameter settings so that the variability in the process is significantly reduced. The rejection proportion is also expected to be substantially very low once the process settings are maintained.

At the 'do' stage, the theory regarding the course of action developed in the plan stage is put into action. Experiments are conducted on the process to determine the required parameter settings ensuring minimum or no variability in the measurements. Feedback is obtained from the concerned customers and from the process. At this stage, the company may test the proposed plan on a small scale.

At the next 'check' stage, the results are analyzed. It is to be verified if the difference between the customer needs and the process performance is reduced by the proposed action. The potential drawbacks relating to other quality characteristics that are important to the customer are also to be looked into at this stage. Statistical analysis is required for this purpose. The performance of the process at the recommended parameter settings should be explained in terms of mean and variance of the output quality characteristics for a representative sample size. The performance must be indicative of positive improvement in the product design.

At the 'act' stage, a decision is made regarding implementation. If the results of the analysis conducted in the check stage are positive, the proposed plan is adopted. Customer and process feedback will again be obtained after full-scale implementation. Such information will provide a true measure of the plan's success. If the results of the check stage show no significant improvement, alternative plans should be proposed and developed, and the cycle continues. In this context, it should be remembered that customers' needs are not constant. They change with time, competition, societal outlook, and other factors. Therefore, the proposed plan is continuously verified to see if it reflects or represents the current needs of the customers. If the current needs are significantly different from the anticipated, the improvement cycle starts again from the plan stage. Continuous improvement in the process performance vis-à-vis fulfillment of current customer needs is the main objective of working as per the Shewhart or Deming cycle of process improvement.

4.9.7 Variability Reduction and Concept of Loss Function

Deming's philosophy calls for abandoning the idea that everything is fine if specifications are met. The idea behind this traditional and conservative attitude is that it is assumed that there is no loss associated with producing items that are off-target but within specifications. Of course, just the opposite is true, which is the reason for striving for continual process improvement. Reducing process variability is an objective that ensures minimization of loss that is due to deviation from the target value.

In pursuing the concept of quality improvement as proposed by Deming, Genichi Taguchi formalized certain loss functions in 1960. He based his approach on the belief that economic loss accrues with any deviation from the target value. Achieving the target value wins high praise from the customer and yields no loss. Small deviations yield small losses. However, the losses increase in a nonlinear relationship (say, quadratic) with larger deviations from the target.

Figure 4.7 demonstrates the difference between the loss function with the traditional approach and that as per Taguchi's view point. The new loss function implies that companies must strive for continual variability reduction and only then losses will be reduced. Any deviation from the customer target will not yield the fullest possible customer satisfaction. Losses may arise because of such problems as lost opportunities, warranty costs, customer complaint costs, and other tangible and intangible costs. There is even a loss associated with the customers not praising the product even though they are not unhappy with it. It is important to get people satisfied with the performance of the product or service because this affects public perception of the product and hence of the company. As is evident from Fig. 4.10, Taguchi loss function of a product is quadratic nature, *i.e.*, the loss to the customer is directly proportional to the square of the deviation of the actual value of the quality characteristic under consideration from its target value, as determined by the design team. Taguchi suggests that the loss should be measured in monetary terms so that everyone is capable of understanding the implications of the loss. A number of factors, both objective and subjective, are responsible for occurrence of this loss.

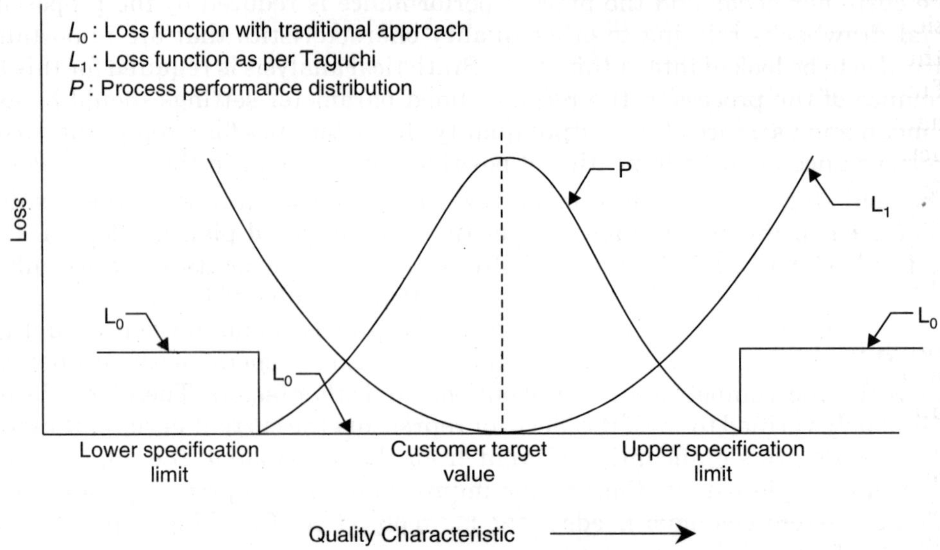

Fig. 4.10 Comparison of the Traditional Loss Function and Taguchi Loss Function.

4.9.8 Optimization of Product Development Performance

As has been mentioned, responsibility for product development rests principally with the design team. In order to be successful in this effort, the design team should work on several aspects of the product development process simultaneously. In this context, the design team must take an integrated approach for that ensures optimal performance of the functional components involved in product development. For example, the design team may consider simultaneously the product design, process design, and manufacturing aspects at the initial phase of the product development. This requires acceptable performances from all relevant functions involved such as marketing, engineering, procurement, and process planning. The principle of optimization is relevant to be followed in the sense that the performance of all these functions taken together should be optimized. In other words, the design team must not pursue for achieving goals of a specific function. It must try to achieve the goals of all the functions and departments at a time. This requires use of inputs from all related work units, such as sales and marketing, product design, products planning, purchasing, process planning, and manufacturing among others. Two conditions management must attempt to fulfill for this purpose. First, it creates open lines of communication at all levels and among different departments and functions. Second, management must take effective steps to remove 'organizational barriers' that may exist within the organization (between, say, workers and supervisors, design engineer and production in-charge), between the organization and others in the extended process (between the design team and the suppliers, plant managers and customers).

4.9.9 Principle of 'Zero Defects' in Product Design

Now-a-days, the principle of 'zero defects' (ZD) is known worldwide. For all types of products that organizations design, the ZD philosophy should be established across all functions and work-units involved in product development. The basic of any ZD programme is to assure a error-free best performance from a designed and developed product. Both quality of design and quality of conformance must guarantee the error-free performance of the product.

Originally introduced and implemented for the Pershing Missile Project in the USA in early sixties of the twentieth century, typical ZD programme calls for removal of errors of all types in the working of a product. There are three possible causes of error: (*i*) inferior or bad quality of materials, parts, or components, (*ii*) inferior or bad quality of equipment and facilities used in manufacturing, and (*iii*) human errors occurring during manufacturing or while using the product. Out of these three factors, the errors due to (*i*) and (*ii*) can be eliminated by using the quality materials, parts, and components from the right and reliable sources of supply, and by incorporating the best practices in selection and maintenance of equipment, and facilities, respectively. In ZD programmes, elimination of human errors is primarily emphasized. The kinds of human errors that may occur may be of two types: intentional and unintentional. Whereas unintentional human errors, which are basically a function of individual aptitude, can be minimized or eliminated by instituting proper training and skills enhancement programmes for the workers, the intentional error is a serious problem, which is primarily due to the human attitude. It can only be removed by changing the attitude or mindset of the persons considered, and by incorporating appropriate design parameters in the design of the product and also, by consideration of ergonomics of 'human-machine' interface. Achieving ZD remains a challenge to management in most cases, and only through removal of rot causes of a problem, ZD can be obtained. In order to sustain the benefits of ZD over the time, companies must strive for excelient use of all kinds of resources at their command, both physical as well as human.

4.9.10 Four 'Absolutes' of Quality Management

In order to achieve quality goals during product development, four aspects of quality management, *viz*. definition of quality, system for achievement of quality, performance standard, and measurement should be emphasized. These four aspects are referred to as 'absolutes' of quality management by Crosby. Whereas appropriate definition of quality, such as 'conformance to applicable specifications' should be focused for the given product, the most rational approach for achievement of quality is prevention of defects or defectives. The performance standard with which the existing quality level of a product is to be judged is Zero Defects (ZD). For this, appropriate quality goals need to be established at each stage of product development. The quality measurement system needs to be based on costs of quality. In order to measure the total cost of quality, four kinds of cost elements are to be considered and their measurement schemes need to be prescribed in a given situation. These costs are prevention, appraisal, internal, and external failure cost.

Prevention costs are those elements of cost that are incurred in planning, implementing, and maintaining a quality system. They include salaries and wages paid to the employees, product design cost, process design cost, costs of equipment design, information system design, and SQC techniques applications, and all other costs associated with manufacturing a product with the required quality levels.

Appraisal costs are those cost elements that are associated with measuring, evaluating, and assessing products, components, raw materials, and quality control systems in an organization to determine their degree of conformance to the specified standards. Different types of inspection costs and costs of calibrating and maintaining measuring instruments and equipment are included in this cost category. These costs are basically associated with managing the outcome of a manufacturing system, conceived as a quality management system.

Internal failure costs are incurred when products and service fail to meet quality requirements prior to their transfer of ownership to the customers through selling. Scrap, rework, labour, and overhead costs are some of the important cost elements in this category.

External failure costs are incurred when the product or the service does not perform satisfactorily after it is sold to the customer. If there are no nonconforming products or services, there exists no such cost. Costs due to customer complaints, and handling, repair, and replacement of nonconforming output are the main cost elements in this cost category.

4.9.11 'Quality Trilogy' Process

The iterative process of designing and developing a new product is based on the premise that the quality levels of a product are subject to changes or modifications over time, and these changes may be dependent on changes in process design and manufacturing. In this context, the quality improvement process assumes importance because of its impact on product development. As mentioned by Juran, the quality improvement process consists of three interacting phases, *viz*. quality planning, quality control, and quality improvement.

Introduced by Joseph Juran, the 'quality trilogy' process starts with quality planning at various levels of the organization, each of which has a distinct goal. At the upper management level, planning is termed 'strategic quality management'. Broad quality goals are established at this level. A structured approach is selected in which management chooses a plan of action and allocates resources to achieve the goals. Planning at the middle management level is termed 'operational quality management'. Departmental goals consistent with the strategic goals are established. At the workforce level, planning involves a clear assignment of work and responsibility to each worker. Each worker is made aware of how his or her individual goal contributes to departmental goals.

After the planning phase, quality control takes over. Here, the goal is to run the process effectively such that the plans are executed effectively. If there are deficiencies in the planning process, the process may operate at a high level of 'chronic waste'. Quality control will try to prevent the waste from getting worse. If unusual symptoms are sporadically detected, quality control will attempt to identify the possible causes behind this abnormal variation. Upon identifying the cause, remedial actions are taken to bring the process back to control. Figure 4.11 shows the three phases of planning, control, and improvement and the cost trends associated with poor quality. The objectives of the control phase are to eliminate the causes associated with the very high values, referred to as sporadic spikes and to bring the process output within the zone of quality control.

The next phase of the trilogy process is quality improvement, which deals with the continuous improvement of the product and the process. This phase is also called the 'quality breakthrough sequence'. Such improvements usually require an action on the part of upper and middle management. They deal with actions such as creating a new design, changing methods or procedures of manufacturing, and investing in new equipment. As can be seen from Fig. 4.11, quality improvement will usually cause a reduction in the cost of poor quality. The chronic waste drops to a lower level due to quality improvement. Repeating the whole cycle again result in further drops in the chronic wastes.

A number of activities need to be carried out in a sequence during quality planning phase. These activities are: identify the customer (both internal and external), determine customer needs, develop product features that respond to customer needs, establish quality goals that meet the needs of customers and suppliers at a minimum cost, develop a process that produce the required product features, and establish process capability.

During quality improvement sequence, actions, such as installing a new machine with improved or state-of-the-art technology, reengineering the existing method of manufacturing, and investment in new processes, are required and taken by the top management. These actions result in fundamental change in the process performance. The specific steps to be followed at this stage are: justify improvement requirements, identify and organize specific projects for improvement, identify the root causes of the problems, and take remedial actions. Once the remedial actions for improvement are known, their actual effectiveness should be tested. For this, actions need to be implemented, and the benefits assessed. Once the benefits in the form of process improvement are assured, there must be a control system to sustain the improved level of performance. Appropriate auditing tools may be used for this purpose. They provide a basis for further process improvement for better design of the product.

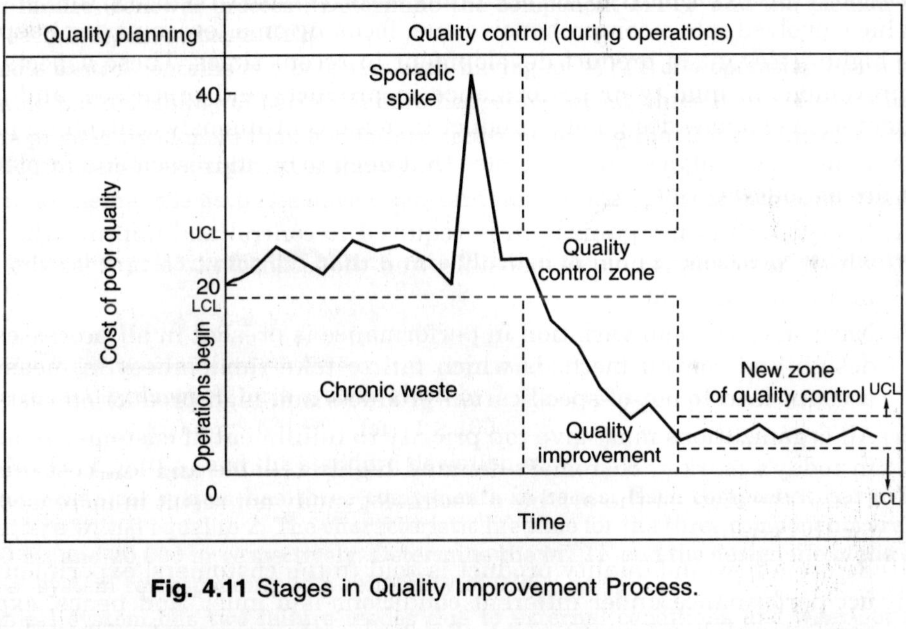

Fig. 4.11 Stages in Quality Improvement Process.

4.9.12 Removal of Problem Causes

There are two types of causes: common or natural and special or assignable causes for the problems that one may encounter in processes. While special causes refer to problems that arise because something unusual has occurred, the common causes refer to problems that are inherent to the system. Examples of special causes are problems due to poor quality from an unqualified vendor or use of an improper tool. With common causes, the system itself is the problem. Examples of common causes are inherent machine variability or worker capability. These problems are controllable only by management. Both Deming and Juran have claimed that about 85% of problems have common causes. Hence, only appropriate actions on the part of management can eliminate them. It is up to management to provide the necessary authority and tools to the workers so that the common causes can be removed.

There are numerous statistical techniques such as control charting that identifies special causes and common causes in processes. Variations outside the control limits are attributed to special causes. These variations are worker-controllable, and the workers are responsible for eliminating these causes. On the other hand, variations within the control limits are viewed as the result of common causes. These variations require management action. Juran says that special causes create 'sporadic problems' and common causes create 'chronic problems', and provides detailed guidelines for identifying sporadic problems.

Once the process performance is explained in terms of these causes, improvement actions may be implemented, and these tasks become the main responsibility of the design team during product development. The design team needs to work in close collaboration with quality improvement team so that benefits outweigh significantly the cost, and the product development cost is held at a minimum level.

4.10 QUALITY IMPROVEMENT PRINCIPLES

Since the time application of quality control and improvement tools and techniques in organizations started formally in the beginning of the twentieth century, the concepts of quality control have evolved into a comprehensive one, focusing on a few important aspects that have become highly relevant in product development in recent times. These aspects are related to (A) improvement in quality or performance in products and processes, and (B) systematic implementation of a new design of a product through continuously controlling process quality.

A number of fundamental principles that need to be addressed and implemented in this context are as follows:

(*i*) Everybody in an organization is required to control and improve the process(es) for which he or she is held responsible, and those closest to a process should participate in its management.

(*ii*) Natural or random variation in performance is present in all processes. Engineering design and control methods which fail to take randomness in measurements into account lead to out-of-specification products and high production costs.

(*iii*) All organizations must give top priority to fulfillment of customer requirements, and in today's market, customers demand highly reliable and low-cost quality products. Use of the state-of-the-art 'hard' technology may not result in increased market share for the product.

(*iv*) Before a new and quality product is sold to the customers, experimentation of product performance under different conditions is a must, and hence, experimental design is a valid and most sought-after tool in the process of product development.

(*v*) Use of statistical process control ensures prevention of occurrence of defects and defectives in processes, and the only way to increase market share and profits of an organization is by designing quality into all products and processes.

To put these principles into practice, management should take initiatives and provide right kind of leadership. The specific actions to be taken by management in this respect should be effective in meeting the customer expectations of the product. A product produced cannot compete in world markets if they do not meet customer expectations for quality, cost, and performance. These expectations cannot be met merely depending on quality control departments for product and process quality.

The main actions required in this regard are as follows:

(*i*) Management must reevaluate the way in which they design products and processes. Radically new and innovative ways of designing and developing a product becomes the 'hallmark' of any progressive management. The management must be empowered to provide leadership in the quality improvement effort.

(*ii*) Processes must be standardized, measured, and evaluated to ascertain their capability, and they must be stable and under statistical control. Statistically designed experiments need to be used to determine appropriate process settings for improved process performance.

(*iii*) Every employee in an organization should be thoroughly trained to be able to work in a group. For this, they should be skilled enough to form a team. This approach ensures a vibrant and effective quality improvement programme that is focused on designing quality and performance into products and processes.

The next set of principles is related to implementation of an improved product design through continuous quality improvement in processes. The following principles in this regard are worth mentioning:

(*i*) Improvement effort should be directed to an existing process in the beginning, where quality problems are very common. It is also to be seen that these processes can be controlled by the operators as far as possible.

(*ii*) Characteristics of the processes and the products that are required to be measured and monitored should be identified. The state of the process (stable or unstable, in-control or out-of-control) should be known. The level of process variability should be known, and reducing process variability should be a top priority for product development team.

(*iii*) Possible causes of process variability should be known and listed. The uncontrollable risk factors influencing significantly the product quality should be identified. Experimental design is to be used to known the impact of critical factors on the levels and variance of quality characteristics under consideration. Based on the experimental results, the key process-related factors to be controlled or adjusted should be selected.

(*iv*) A comprehensive plan to implement changes in the process based on the results of experiments conducted should be prepared. The specific changes required to be implemented in a process should be informed to the operators, supervisors, and line managers responsible for the process.

(*v*) The effectiveness of the change made in the process should be known. For this, the performance of the changed process needs to be monitored. If there is any adverse impacts or harmful side effects of the changes made, they should be immediately looked into as soon as they are seen or reported.

(*vi*) If the process changes produce the desired results, the changes should be standardized through training and documentation to make sure that the improvements become a part of the process.

(*vii*) The process performance needs to be reviewed at periodic intervals. Important issues, such as sustainability of the improvements, new opportunities for improvement, and effectiveness of experimental design techniques for new product and process development, should be considered, and decisions regarding the type of improvement methodology to be adopted needs to be taken during review of the improvement plan.

4.11 TAGUCHI'S CONCEPT OF QUALITY ENGINEERING

Genichi Taguchi, the well-known Japanese quality guru, has focused on an aspect of product quality that has remained as an important one for product development, but has been ignored or not emphasized by others in the past. He says that "quality is the loss a product causes to society after being shipped, other than any losses caused by its intrinsic functions." He also says that quality can be manufactured, or engineered into a product keeping in focus the performance of the product as the primary objective of creating its acceptable design. There are two types of 'loss' in this context:

(*i*) loss caused by variability of function, and

(*ii*) loss caused by harmful side effects.

There are several examples of these two types of losses. The breakdown of a machine at the workplace and non-starting of a car on the highway are two specific examples of the loss of first type. Because of some 'bad' settings, the machine cannot run at the optimal speed and the corresponding loss in production may be treated as a loss of the first type. Similarly, the variability in the quality of services provided by the hospital staff may result in unsatisfied patients, increased waiting time, and increased treatment cost. Asbestos is an excellent insulating material. However, the persons using it may be prone to cancer, and hence, this serious effect is considered to be a loss due to harmful side effects. This loss is present in most medicines used by patients for cure of a specific disease among humans. An unacceptable product which is scrapped or reworked prior to shipment is viewed by Taguchi as a cost to the company as it reflects production inefficiency, but not a loss to the society. Losses caused when a product performs as intended are not considered quality losses by Taguchi. For example, the effect on the health of a person consuming alcoholic beverages, designed and developed by a company, may not be treated as a loss at all. His reasoning here is that such situations reflect cultural or legal problems, not engineering problems. Another example of this case is an automobile accident caused by someone who has a radar detector and is driving too fast. The extent to which the proper functioning of the radar detector caused a loss to society is not treated as a quality loss.

Taguchi coins two terms, *viz.*, 'off-line quality control' and 'on-line quality control', in order to classify the 'quality systems' of organizations. There are four specific objectives of off-line quality control:

(*i*) to identify customer needs and expectations most accurately,

(*ii*) to design a product that meets customer needs and expectations,

(*iii*) to design a product which can be consistently and economically manufactured, and

(*iv*) to develop clear and adequate specifications, standards, procedures, and equipment for manufacture.

Activities relating to off-line quality control may be grouped under two categories: (*i*) product design, and (*ii*) process design. During the product design, a new product is developed or an existing product is modified. The goal here is to design a product which is manufacturable and meets customer requirements. During the process design, production and process engineers develop manufacturing processes to produce the specifications developed during the product design stage. Objectives (*i*), (*ii*) and (*iii*) as mentioned above are a part of the product design, and objectives (*iv*) a part of process design.

Taguchi proposes a three-step approach for assuring quality at each of product and process design. These steps are: system design, parameter design, and tolerance design.

The activities relating to on-line quality control are concerned with manufacturing products within the specifications established during product design using the procedures developed during process design. Also, product and process designs may be revised if feedback from customers reveals opportunities for improvement. Taguchi identifies two stages in on-line quality control. At the first stage, appropriate production quality control methods are developed and used for assurance of quality. At the second stage, the aspect of customer relations is looked into systematically.

For quality assurance, at product design stage, Taguchi's three-step approach addresses the following tasks:

(*i*) ***System design*:** Engineering and scientific knowledge is applied to develop a 'prototype' design which meets customer requirements. Selections of parts, components, and materials, as well as manufacturing technology are made. The emphasis here is on using the best available technology to meet customer requirements at lowest cost. In order to be a successful product designer, Taguchi recommends that time-tested and well-proven technology and low-cost parts should be used, and customer requirements need to be focused.

(*ii*) ***Parameter design*:** Optimal settings of product parameters are determined at this step. The main objective is to minimize manufacturing and product lifetime costs by minimizing performance variation. This involves making the product design robust, *i.e.*, making the product performance insensitive to noise factors. A noise factor is an uncontrollable source of variation in the functional characteristics of a product. Taguchi identifies three types of noise factors: external noise, or variation in environmental conditions, such as dust, temperature, humidity, or supply voltages; internal noise, or deterioration, such as product wear, material aging, or other changes in components or materials with time or use; and unit-to-unit noise, which is differences in product units built to the same specifications caused by variability in materials, manufacturing equipment, and assembly processes. The parameter design step involves use of experimental designs to determine the impact of both controllable and uncontrollable (noise) factors on product characteristics. The goal is to set controllable factors at levels which will make the product robust with respect to noise factors. Examples of robust design include an automobile part which can withstand shock and vibration (external noise), or a food product with a long shelf life (internal noise), or a replacement part that will fit properly (unit-to-unit noise). Parameter design is a critical step in building quality engineering, and lies at the heart of Taguchi's quality engineering concept.

(*iii*) ***Tolerance design*:** Tolerances around the target or nominal values of the parameters established during parameter design are determined at this step. The goal is to set tolerances as wide as possible so that the manufacturing costs are held at a minimum level, while keeping the product's functional characteristics within specified bounds.

For quality assurance at the process design stage, Taguchi's three-step approach requires completion of the following tasks:

(*i*) ***System design*:** The manufacturing process on the basis of knowledge of the product and current manufacturing technology is selected.

(*ii*) ***Parameter design*:** Appropriate levels for controllable production process parameters are determined at this step. The goal here is to make the process robust, *i.e.*, to minimize the effects of uncontrollable noise factors on the performance of production process and finished product. Many types of experimental designs are usually used during this step.

(*iii*) ***Tolerance design*:** Usually tolerances for the identified critical process parameters are determined. If the product or process parameter design steps are poorly done, it may be necessary here to tighten tolerances or specify higher-cost materials or better equipment. This may however increase the manufacturing costs.

During on-line quality control, Taguchi identifies three forms of production quality control methods. These are as follows:

(*i*) ***Process diagnosis and adjustment*:** The process is monitored at regular intervals and adjustments and corrections are made as needed.

(*ii*) ***Prediction and correction:*** A quantitative process parameter is measured at regular intervals. The data are used to project trends in the process. Whenever the process is found to be too far off-target, the process is adjusted to correct the situation. This method is also called 'feedback' or 'feed forward' control.

A number of activities need to be carried out during on-line customer relations building up. This may involve repair or replacement of defective products, compensation for losses, and complaint handling. Information on types of complaints and failures, and customer perceptions of products should be collected at periodic intervals and promptly fed back to relevant functions within the organization for corrective action.

4.12 SUMMARY

This chapter discusses the various principles that have been introduced and practiced by organizations worldwide to design, produce, and maintain quality of products and services. Emphasizing the importance of quality in engineering design, the evolution of quality over the years in modern organizations is discussed, and the concept of 'quality loop' and its elements, from marketing through production to disposal, are explained with identification of quality-related activities at all elements. Comparing an inspection-based quality control with a prevention-based quality control, and elaborating the systems approach for quality problem solving at all three aspects of quality of design, of conformance, and of performance, several quality principles for product development, such as extended process, customer-driven improvement cycle, constancy and consistency of purpose for quality improvement, rules for inspection, principles of supplier selection, process improvement cycle, variability reduction including loss function, and principles of 'zero defects' are discussed in this chapter. Basic issues related to Taguchi's concept of quality engineering, Juran's quality trilogy process, and Crosby's 'absolutes' of quality management, that are useful for creation and management of both product and process quality are presented in the context of quality control and improvement in modern business and manufacturing organizations.

QUESTIONS

4.1 State a few important definitions of quality, and discuss their significance and use in product development.

4.2 Taguchi's definition of quality mentions about minimization of 'quality loss'. How is this principle applied at the design stage of a product?

4.3 What are the important trends in quality? Discuss each.

4.4 Distinguish between quality control, quality assurance, and total quality management. Are they related? Discuss with suitable examples.

4.5 What is 'quality loop'? Discuss each element of this loop.

4.6 In designing a product, discuss how the following aspects may be considered and why:

(*i*) customer requirements

(*ii*) cost

(*iii*) manufacturability or producibility

(*iv*) testability

(*v*) reliability

4.7 Compare and contrast inspection-based and prevention-based quality control approaches. What are the design principles you may apply for each?

4.8 Make a literary search to make a report on the types of quality principles that a company needs to follow for new product development.

4.9 What are the steps you need to follow to improve the design of a product? Discuss significance of each step, and mention also the problems that one may face at each step.

4.10 Distinguish between product design and process design in the light of the specific quality principles to be applied for each. Discuss with suitable examples.

4.11 Explain the concept of 'extended process'. Why is this concept relevant for product development in today's business environment?

4.12 Describe the P-D-C-A cycle. Why and how does this cycle play a crucial role in quality improvement?

4.13 Consider a company of your choice at your locality. Collect data on its supplier selection and evaluation system by making a series of visits to the company. Identify 'good' and 'bad' points of that system, and make specific recommendation for improvement.

4.14 Why is 'procurement' function of an organization important for product development? How is quality of purchased parts and components maintained and improved?

4.15 How are the principles of 'continuous quality improvement', and improved customer satisfaction taken care of at different stages of product development process? Explain with examples.

4.16 How are the 'extended process' concept and goal of 'continuous quality improvement' related to each other?

4.17 Discuss, in brief, the concepts of 'quality of design', 'quality of conformance' and 'quality of performance'. How are they mapped on 'quality loop' elements?

4.18 Name a few critical organizational barriers that may slow down the product development process. How do they prevent a company from adopting the total quality philosophy?

4.19 Make a library/internet search on (*i*) major categories of quality costs, and (*ii*) various indices for measuring them.

4.20 Discuss the responsibilities of the different departments of an organization for (*i*) product development, and (*ii*) quality control improvement. Why is 'quality of performance' a matter of concern for the designers?

4.21 Select a manufacturing plant and identify its product development process. Analyze the strengths and weaknesses of 'manufacturing' in relation to its ability to support new product development.

4.22 Define and describe the 'four absolutes' of quality management as prescribed by Crosby. Can they be maintained always in product development? Why, or why not?

4.23 State and describe, in brief, the tasks that you need to carry out during (*i*) quality planning, (*ii*) quality control, and (*iii*) quality improvement.

4.24 It is said that 'control mechanisms' must be in place to hold the gains of quality improvement. State and explain a few control mechanisms that may be used for this purpose.

4.25 What are the major drawbacks of the traditional approach for product development from the viewpoint of quality improvement?

4.26 Even if product design is taken in right earnest, process design is neglected in the majority of Indian organizations. Make a report on the possible causes and their remedies based on a literary or internet search.

4.27 What is the difference between chance and assignable causes? Cite three examples of each. How do you eliminate (*i*) chance causes, and (*ii*) assignable causes from a system?

4.28 What are the possible actions you may take for product development if the existing 'manufacturing' system is found to be inadequate and incapable?

4.29 How is the quality of a product updated, and what sort of activities you may carry out for this? Discuss.

4.30 There are several dimensions of quality of a product. How are these dimensions be incorporated during product development process?

REFERENCES

Hutchins, G.B., *Introduction to Quality: Management, Assurance, and Control,* Macmillan Publishing Company, New York, 1991.

Banks, J., *Principles of Quality Control,* John Wiley & Sons, New York, 1989.

Besterfield, D.B., *Quality Control,* Prentice-Hall Inc., N.J., 3rd ed., 1990.

Garvin, D.A., *What Does Product Quality Really Mean?,* Sloan Management Review, 26(1), pp. 25–43, 1984.

Juran, J.M., *Quality Control Handbook,* McGraw-Hill Inc., New York, 3rd ed., 1974.

Juran, J.M. and Gryna, F.M., *Quality Planning and Analysis—From Product Development through Use,* McGraw-Hill Inc., New York, Int. ed., 1993.

Deming, W.E., *Quality, Productivity, and Competitive Position,* Center for Advanced Engineering Study, MIT, Cambridge, Mass, 1982.

Rich, A.B., *Continuous Improvement: The Key to Future Success,* Quality Progress, 30(6), pp.33–36, 1997.

Bogan, C.E. and English, M.J., *Benchmarking for Best Practices,* McGraw-Hill Inc., New York, 1994.

Mills, C.A., *The Quality Audit, Quality Process,* ASQC, Milwankee, Wis., 1989.

Montgomery, D.C., *Introduction to Statistical Quality Control,* John Wiley & Sons, New York, 3rd ed., 1996.

Gunter, B., *A Perspective on the Taguchi Methods,* Quality Progress, vol.20, No.6, pp.44–52,1987.

Halpern, S., *The Assurance Sciences: An Introduction to Quality Control and Reliability,* Prentice-Hall, Englewood Cliffs, N.J., 1978.

Halpin, J.F., *Zero Defects,* McGraw-Hill Inc., New York, 1966.

Ishikawa, K., *Guide to Quality Control,* UNIPUB, New York, 1976.

Juran, J.M., *Quality Problem, Remedies and Nostrums,* Industrial Quality Control, vol. 22, No.12, pp. 647-653.

Kivenko, K., *Quality Control for Management,* Prentice-Hall, Englewood Cliffs, N.J., 1984.

Simmons, D.A., *Practical Quality Control,* Addison-Wesley, Reading, Mass, 1970.

Mitra, A., *Fundamentals of Quality Control and Improvement,* Prentice-Hall, N.J., 1998.

Caplan, F., *The Quality System: A Sourcebook for Managers and Engineers,* Radnor, PA: Chilton, 1980.

Crosby, P.B., *Quality is free.* New York, McGraw-Hill, 1979.

Feigenbaum, A.V., *Total Quality Control,* 3rd ed., McGraw-Hill, New York, 1983.

Kackar, R.N., *Off-line Quality Control, Parameter Design and the Taguchi Method,* Journal of Quality Technology, vol.17, pp.176–188,1985.

Kackar, R.N., *Taguchi's Quality Philosophy: Analysis and Commentary,* Quality Progress, 19 (December), pp.21–29,1986.

Kume, H., *Statistical Methods for Quality Improvement*, 1987.

Phadke, M.S., *Quality Engineering using Robust Design,* Prentice-Hall, Englewood Cliffs, N.J., 1986.

Ross, P.J., *Taguchi Techniques for Quality Engineering,* McGraw-Hill, New York, 1988.

Taguchi, G., and Wu, Y., *Introduction to Off-Line Quality Control,* Central Japan Quality Control Association, Nagoya, Japan, 1980.

Taguchi, G., *Introduction to Quality Engineering,* Asian Productivity Organization, UNIPUB, White Plains, New York, 1986.

Ryan, T.P., *Statistical Methods for Quality Improvement,* John Wiley, New York, 1989.

CHAPTER 5

Engineering Economic Analysis

5.1 INTRODUCTION

In this chapter we consider the subject of engineering economic analysis, with particular emphasis on economic analysis in design and manufacturing. Economic analysis involves techniques for comparing and making decisions between alternatives on the basis of monetary and economic objectives. With the ever-increasing complexity of our manufacturing and industrial technology, economic decision-making is becoming more critical and essential. In addition, investment decisions are important in determining the success or failure and profitability of a firm.

Engineering economic decisions between alternatives can be made by the engineer or the manager of technology considering options in the design of components, systems facilities or man-machine systems. However, considering a number of investment opportunities and alternatives, within each opportunity best makes these decisions.

The objectives in the presentation are the following:

1. The fundamental concepts involving the *time value of money*
2. Various economic measures for determining a projects' worth, rates of return, and payback period
3. Comparing investment alternatives, sensitivity analysis, and replacing capital assets
4. Incorporation of depreciation allowances and income taxes in economic analyses
5. Benefit cost analysis for the economic analysis of projects
6. Profitability of investments
7. Sensitivity and break-even analysis
8. Capitalized cost analysis
9. Uncertainty and risk in economic analysis.

5.2 TIME VALUE OF MONEY

Money has a time value. The value of a given sum of money depends not only on the amount of money but when the money is received. In this section, we introduce the basic operations needed to conduct economic analysis incorporating time value of money concepts with an emphasis on cash flow profiles. A cash flow diagram represents a clear and concise description of the amount and timing of all cash flows associated with an economic analysis.

The term *interest* is used to designate rental amount charged by a financial institution for the use of money. Any economic gain that may occur from the use of the money gives money its time value. An interest rate (usually expressed as a percentage on a per year basis) is the rate of gain received from an investment.

5.2.1 Interest Calculations

When money is borrowed a sum of money called interest is paid for the use of money over a period of time. When the money is loaned, interest is received. The amount of a loan is called the *principal*. The future worth F_n of the money is greater than its present worth P. The type of interest used will dictate the relationship between F_n and P.

5.2.1.1 Simple interest

When simple interest is used F_n and P are related by the relationship

$$F_n = P(1 + in) \qquad ...(5.1)$$

where

F_n = accumulated value of P over n years or future value of P

P = current or present value of a single sum of money

i = annual interest rate (change in value for \$1 over a 1-year period)

n = number of years

The increase in the value of P over n years, I_n is given by

$$I_n = Pin \qquad ...(5.2)$$

I_n is referred to as the accumulated interest in borrowing and lending transactions and is a function of P, i and n. If we consider i as the *rate of change in the accumulated value of money*, then

$$I_n = i\,F_{n-1} \qquad ...(5.3)$$

and

$$F_n = F_{n-1}\,(1 + i) \qquad ...(5.4)$$

This approach is known as the *compound interest* approach. The effect of compound interest is illustrated in Table 5.1. The quantity $(1 + i)^n$ is tabulated in Appendix 5A for various values of i and n. Here i is the interest rate per interest period and n is the number of periods.

5.2.1.2 Single sums of money

The relationship between the future worth F and the present worth P is given as follows: (Refer to Table 5.1)

Table 5.1. Effect of Compound Interest

End of period	(A) Amount owed	(B) Interest for next period	(C) = (A) + (B) Amount owed for next period*
0	P	P_i	$P + Pi = P(1 + i)$
1	$P(1 + i)$	$P(1 + i)i$	$P(1 + i) + P(1 + i)i = P(1 + i)^2$
2	$P(1 + i)^2$	$P(1 + i)^2i$	$P(1 + i)^2 + P(1 + i)^2i = P(1 + i)^3$
3	$P(1 + i)^3$	$P(1 + i)^3i$	$P(1 + i)^3 + P(1 + i)^3i = P(1 + i)^4$
⋮	⋮	⋮	⋮
$n-1$	$P(1 + i)^{n-1}$	$P(1 + i)^{n-1}i$	$P(1 + i)^{n-1} + P(1 + i)^{n-1}i = P(1 + i)^n$
n	$P(1 + i)^n$		

$$F = P(1 + i)^n \quad ...(5.5)$$

where,

F = the equivalent value of an amount of money at time n or future worth

P = the equivalent value of an amount of money at time zero, or present worth

i = the interest rate per interest period (expressed as a decimal point)

Equation (5.5) can also be written as

$$F = \left(\frac{F}{P}, i, n\right) \quad ...(5.6)$$

where, i is expressed as a percentage amount. The cash flow diagram corresponding to Eqs. (5.5) or (5.6) showing the relationship between F and P is given in Fig. 5.1.

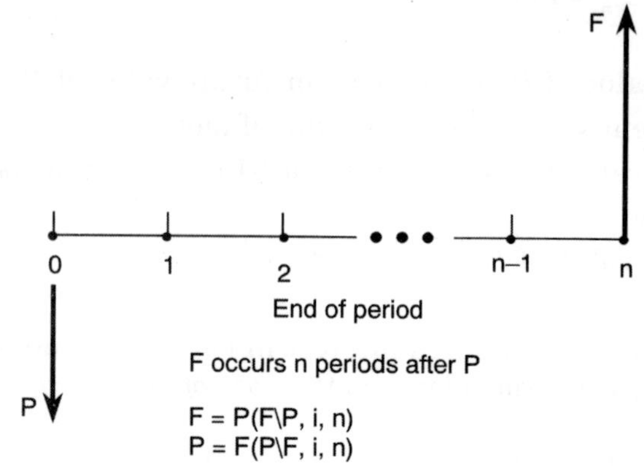

Fig. 5.1 Cash Flow Diagram Showing the Relationship between F and P.

Example E5.1. *A firm borrows a loan of 1 million dollars at 10% compounded annually. The loan is to be paid back after 6 years. What is the amount to be repaid?*

Solution: Refer to the compound interest tables in Appendix 5A for 10% and 6 periods, the value of (F/P, 10, 6) factor is read as 1.7716. Hence

$$F = P(F/P, 10, 6)$$
$$= \$\ 1,000,000(1 + 0.10)^6 = \$1,000,000(1.7716) = \$1,1,771,600$$

Thus, the amount to be repaid is $1,771,600.

The relationship between P and F is obtained by dividing both sides of Eq. (5.5) by $(1 + i)^n$

$$P = F(1 + i)^{-n} \quad ...(5.7)$$

Equation. (5.7) can be written as

$$P(F/P, i, n) \quad ...(5.8)$$

where

$(1 + i)^n$ = the single sum

and $P(F/P, i, n)$ = the present worth factor

Example E5.2. *Mr. Doe wishes to accumulate $20,000 in a savings account in 10 years. If the bank pays 6% compounded annually, how much should Mr. Doe deposit in the account today?*

Solution: This problem is of the form *find P when given F* and hence we can use Eq. (5.8).

$$P = F(P/F, i, n)$$
$$= \$20{,}000(P/F, 6, 10)$$

Referring to Appendix 5A for 6% and 10 periods, the value of $(P/F, 6, 10)$ factor is read as 0.5584.

Hence, $P = \$20{,}000(0.5584) = \$11{,}168.$

5.2.2 Series of Cash Flows

Consider the conversion of a series of cash flows to present worth and future worth equivalents.

Applying discrete compounding, the present worth equivalent for the cash flow series is equal to the sum of the present worth equivalent for the individual cash flows. Therefore, P can be written as

$$P = A_1(1 + i)^{-1} + A_2(1 + i)^{-2} + A_{n-1}(1 + i)^{n-1} + A_n(1 + i)^n \qquad ...(5.9)$$

where, A_t = the magnitude of a cash flow at the end of time period t.

Equation. (5.9) can be written as

$$P = \sum_{i=1}^{n} A_t(1+i)^{-1} \qquad ...(5.10)$$

or

$$P = \sum_{i=1}^{n} A_t(P/F, i, t) \qquad ...(5.11)$$

Example E5.3. *For the series of cash flows indicated by the cash flow diagram in Fig. E5.3 and using an interest rate of 7% per interest period, calculate the present worth equivalent.*

Solution:

$$P = \$300(P/F, 7, 2) - \$150(P/F, 7, 4) + \$150(P/F, 7, 5) + \$400(P/F, 7, 7) + \$300(P/F, 7, 9)$$

Referring to Tables in Appendix 5A, we have

$$P = \$300\ (0.8734) - \$150(0.7629) + \$150(0.7130) + \$400(0.6227) + \$300(0.5439) = \$754.13$$

Using discrete compounding, the future worth equivalent is equal to the sum of the future worth equivalents for the individual cash flows. Hence

$$F = A_1(1 + i)^{n-1} + A_2(1 + i)^{n-2} + \ldots + A_{n-1}(1 + i) + A_n \qquad ...(5.12)$$

or we can write

$$F = \sum_{i=1}^{n} A_t(1+i)^{n-T} \qquad ...(5.13)$$

$$P = \sum_{i=1}^{n} A_t(P/F, i, n-t) \qquad ...(5.14)$$

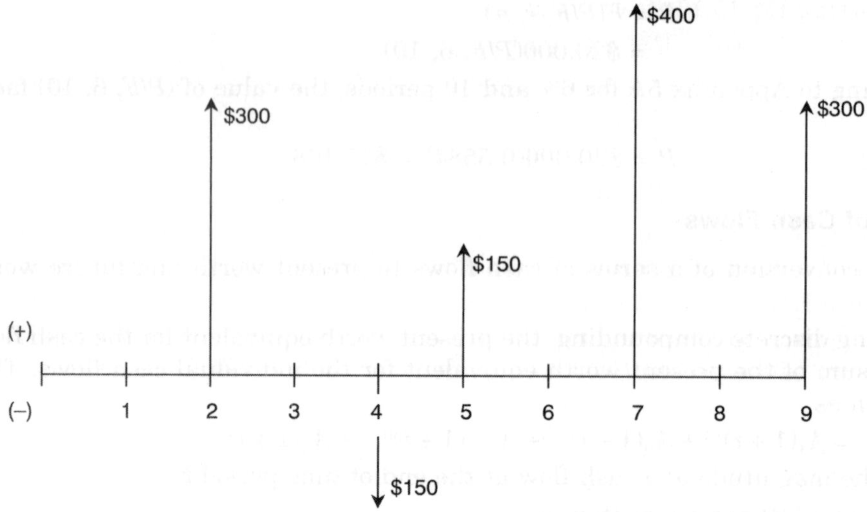

Fig. E5.3 Series of Cash Flows.

Example E5.4. *For the series of cash flows in Fig.E5.3, determine the future worth at the end of nth period using an interest rate of 7% interest period.*

Solution:

$$F = \$300(F/P, 7, 7) - \$150(F/P, 7, 5) + \$150(F/P, 7, 4) + \$400(F/P, 7, 2) + \$300$$

Referring to Table in Appendix 5A, we have

$$F = \$300(1.6058) - \$150(1.4026) + \$150(1.3108) + \$400(1.1449) + \$300 = \$1{,}226$$

From Example E5.3, the present worth is equal to \$ 754.13. Therefore,

$$F = P\,(F/P, 7, 9)$$
$$= \$754.13(1.6058) = 1226$$

5.2.3 Uniform Series of Cash Flows

A schematic model for money flow over time is shown in Fig. 5.2. A uniform series of cash flows exists when all the cash flows in a series are equal. The sum of the compound amount of this series may be obtained by reference to Fig. 5.2.

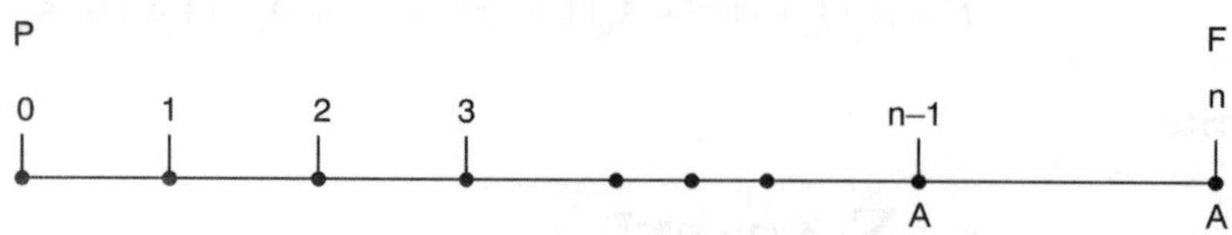

Fig. 5.2 The General Money Flow Model.

The sum of this series is

$$F = A(1) + A(1+i)^1 + A(1+i)^2 + \ldots + A(1+i)^{n-2} + A(1+i)^{n-1} \quad \ldots(5.15)$$

Multiplying both sides of Eq. (5.15) by $(1 + i)$ gives

$$F(1+i) = A\,[\,(1+i)^1 + (1+i)^2 + (1+i)^3 \ldots + (1+i)^{n-1} + (1+i)^n] \quad \ldots(5.16)$$

Subtracting Eq. (5.16) from (5.15) gives

$$F(1 + i) - F = A[(1 + i)^n - 1]$$

$$Fi = A[(1 + i)^n - 1]$$

or

$$F = A[\{(1 + i)^n - 1\}/i] \quad ...(5.17)$$

or equivalently

$$F = A[F/A, i, n]$$

where [*F*/*A*, *i*, *n*] is referred to as the *uniform series, future worth factor* or *equal payment-series compound amount factor* and is tabulated in Appendix 5A.

Example E5.5. *Joe wants to become a millionaire at retirement. How much would Joe have to deposit yearly in the fund to accumulate $1 million over a 36-year period at an annual compound interest rate of 7% ? Repeat the problem if the interest rate is 8%, 9% and 10%.*

Solution:

(*a*) A = \$1,000,000(*A*/*F*, 7, 36) = \$1,000,000(6.715×10^{-3}) = \$6,715

(*b*) A = \$1,000,000(*A*/*F*, 8, 36) = \$1,000,000(5.345×10^{-3}) = \$5,345

(*c*) A = \$1,000,000(*A*/*F*, 9, 36) = \$1,000,000(4.235×10^{-3}) = \$4,235

(*d*) A = \$1,000,000(*A*/*F*, 10, 36) = \$1,000,000(3.343×10^{-3}) = \$3,343

Example E5.6. *Six deposits of $1000 each are made at t =1,2,3,4,5 and 6 into a savings fund paying 7% compounded per year. How much money will be accumulated in the fund at (a) t = 6 and (b) t = 10?*

Solution:

(*a*) F = \$1,000(*F*/*A*, 7, 6) = \$1,000(7.1533) = \$7,153.30

(*b*) F = \$1,000(*F*/*A*, 7, 10) = \$1,000(13.8164) = \$ 13,816.40

The reciprocal relationship between *A* and *F* is obtained from Eq. (5.17).

$$A = F[(i/\{(1 + i)^n - 1\}] \quad ...(5.18)$$

or equivalently

$$A = F\left[\frac{A}{F}, i, n\right] \quad ...(5.19)$$

Hence, the expression $\left[\frac{A}{F}, i, n\right]$ is referred to as the *sinking fund factor* or the *equal-payment-series sinking fund factor* and tabulated in Appendix 5A.

Substituting $P[1 + i]^n$ for F in the equal-payment-series sinking fund formula of Eq. (5.18) gives

$$A = P(1 + i)^n \left[\frac{i}{(1+i)^n - 1}\right] \quad ...(5.20)$$

$$= P\left[\frac{i(1+i)^n}{(1+i)^n - 1}\right] \quad ...(5.21)$$

This can be written as

$$A = P\left[\frac{A}{P}, i, n\right] \quad ...(5.22)$$

where $\left[\frac{A}{P}, i, n\right]$ is called the *capital-recovery factor* or the *equal-payment-series capital recovery factor* and is tabulated in Appendix 5A.

The reciprocal relationship between A and P can be expressed as

$$P = A\left[\frac{(1+i)^n - 1}{i(1+i)^n}\right] \qquad ...(5.23)$$

or more commonly expressed as

$$P = A\left[\frac{P}{A}, i, n\right] \qquad ...(5.24)$$

where $\left[\frac{P}{A}, i, n\right]$ is referred to as the *uniform series, present worth factor* and is tabulated in Appendix 5A for the various values of i and n.

Example E5.7. *John borrows \$10,000 at 7% compounded annually. If he repays the loan in ten annual payments, what will be the size of the payments if the first payment is made 1 year after borrowing the money?*

Solution:

$$A = P\left[\frac{A}{P}, 7\%, 10\right]$$

From Appendix 5A (Table)

$$A = \$10{,}000[0.1424] = \$1424$$

Example E5.8. *Linda deposits a sum of money in an account that earns interest at a rate of 5% compounded annually. She plans to withdraw \$3000 a year from the account for 5 years, with the first withdrawal occurring 1 year after the deposit. How much money should she deposit so that her fifth withdrawal just depletes the account?*

Solution:

$$P = A\left[\frac{P}{A}, 5\%, 5\right]$$

From Appendix 5A (Table)

$$P = \$3{,}000[4.3295] = \$12{,}989$$

5.2.4 Uniform Gradient Series Formula

A gradient series of cash flows occurs when the value of a given cash flow is greater than the value of the previous cash flow by a constant amount, G. They may increase or decrease by a constant amount.

A uniformly increasing series of cash flows for n interest periods may be expressed as $A_1, A_1 + G, A_2 + 2G, \ldots A_1 + (n-1)G$, as schematically shown in Fig. 5.3. Here A represents the first year-end flow in the series and G is the annual change in the magnitude of the flows.

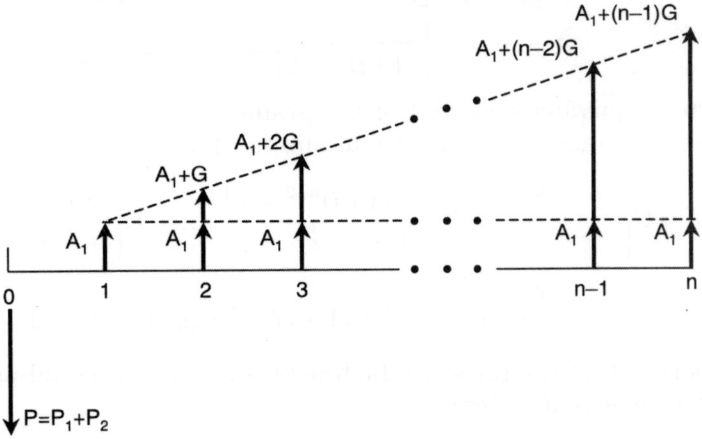

Fig. 5.3(a) A Uniformly Increasing Gradient Series.

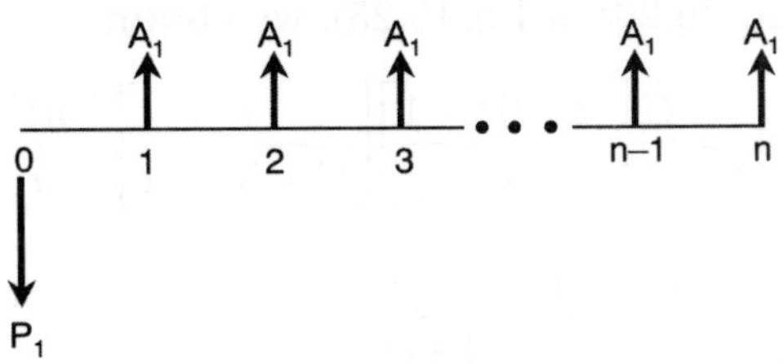

Uniform series

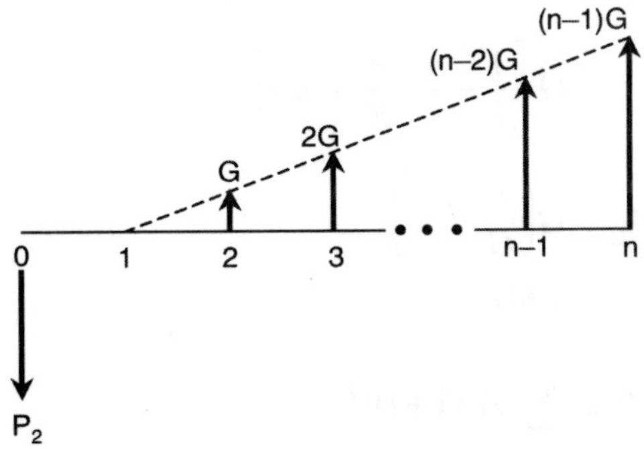

Gradient series

Fig. 5.3 (b) Cash Flow Diagram of a Combination of Uniform and Gradient Series.

Let n = *number of years*

A = *Annual equivalent equal amount*

A uniform-gradient series consists of two separate series:

(*a*) equal-flow series *base* A_1 and (*b*) a gradient series $O, G, 2G, \ldots (n-1)G$ at the end of each successive year. We can write each flow in the equal-flow series as

$$A = A_1 + F\left[\frac{i}{(1+i)^n - 1}\right] = A_1 + A_2 \qquad \text{...(5.25)}$$

where F is the future equivalent amount of the gradient series. The gradient series can be separated into $(n - 1)$ distinct series with equal annual flows of G. That is,

$$F = G\left[\frac{(1+i)^{n-1} - 1}{i}\right] + G\left[\frac{(1+i)^{n-2} - 1}{i}\right] + \ldots + G\left[\frac{(1+i)^2 - 1}{i}\right] + G\left[\frac{(1+i)^1 - 1}{i}\right]$$

$$\frac{G}{i}\,[(1 + i)^{n-1} + (1 + i)^{n-2} + \ldots\ (1 + i)^2 + (1 + i) + 1] - \frac{nG}{i}$$

It can be observed that the terms in the brackets denote the equal-payment series compound amount factor for n years. Hence,

$$F = \frac{G}{i}\left[\frac{(1+i)^n - 1}{i}\right] - \frac{nG}{i} \qquad \text{...(5.26)}$$

Substituting F from Eq. (5.26) in Eq. (5.25), we obtain

$$A_2 = \frac{G}{i}\left[\frac{(1+i)^n - 1}{i}\right]\left[\frac{i}{(1+i)^n - 1}\right] - \frac{nG}{i}\left[\frac{i}{(1+i)^n - 1}\right]$$

$$= \frac{G}{i} - \frac{nG}{i}\left[\frac{i}{(1+i)^n - 1}\right]$$

or

$$A_2 = \frac{G}{i} - \frac{nG}{i}\left(\frac{A}{F}, i, n\right)$$

$$= G\left[\frac{1}{i} - \frac{n}{i}\left(\frac{A}{F}, i, n\right)\right] \qquad \text{...(5.27)}$$

or

$$A_2 = G\left(\frac{A}{G}, i, n\right) \qquad \text{...(5.28)}$$

Recalling that (see Fig. 5.3(b)).

$$P = \sum_{t=1}^{n} A_t(1+i)^{-t} \qquad \text{...(5.29)}$$

and $A_t = (t - 1)G$ for $t = 1, 2, \ldots n$ ...(5.30)

Substituting A_t from Eq. (5.30) in Eq. (5.29), we obtain

$$P = \sum_{t=1}^{n} (t - 1)G(1+i)^{-t} \qquad \text{...(5.31)}$$

or

$$P = G\sum_{t=1}^{n} (t - 1)(1+i)^{-t} \qquad \text{...(5.32)}$$

Equation (5.32) can be reduced to

$$P = G\left[\frac{1-(1+n,i)\,(1+i)^{-n}}{i^2}\right] \qquad \text{...(5.33)}$$

Equation (5.33) in terms of interest factors reduces to

$$P = G\left[\frac{\left(\frac{P}{A}, i, n\right) - n\left(\frac{P}{F}, i, n\right)}{i}\right] \qquad \text{...(5.34)}$$

or

$$P = G\left(\frac{P}{G}, i, n\right) \qquad \text{...(5.35)}$$

where $\left(\frac{P}{G}, i, n\right)$ is referred to as the *gradient series, present worth factor* and is tabulated in Appendix 5A.

Example E5.9. *Linda borrows \$20,000 at 15% compounded annually; she pays off the loan over a 7-year period with annual payments. Each successive payment is \$1000 greater than the previous payment. Find the first payment she made.*

Solution:

$$\$20{,}000 = A\left(\frac{P}{A}, 15\%, 7\right) + \$1000\left(\frac{P}{G}, 15\%, 7\right)$$

Referring to Appendix 5A,

$$\$20{,}000 = A(4.1604) + \$1000\,(10.1924)$$

Hence, $A = \$2{,}357.37$

Example E5.10. *Solve example E5.9 for the case in which Linda's successive payments are \$1000 less than the previous payment. Find the first payment she made.*

Solution: $\$20{,}000 = A\left(\frac{P}{A}, 15\%, 7\right) - \$1000\left(\frac{P}{G}, 15\%, 7\right)$

From Appendix 5A,

$$\$20{,}000 = A(4.1604) - \$1000(10.1924)$$

Hence, $A = \$7{,}257.09$

5.2.5 Geometric Series of Cash Flows

The geometric series of cash flows is shown in Figure 5.4. For this series, the size of a cash flow increases or decreases by a fixed per cent from one time period to the next.

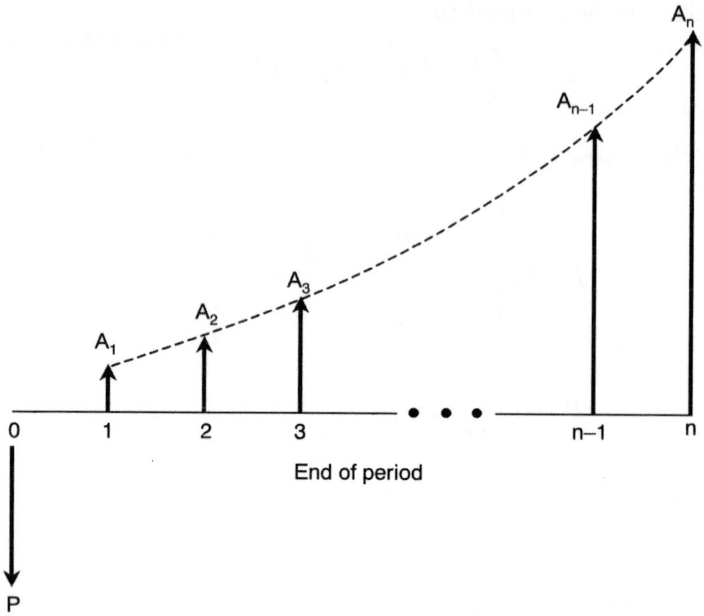

Fig. 5.4 Cash Flow Diagram of a Geometric Series of Cash Flows.

The size of the rth flow is given by

$$A_t = A_{t-1}(1 + j) \qquad \text{for } t = 2, 3, \ldots, n \qquad \ldots(5.36)$$

or

$$A_t = A_1(1 + j)^{t-1} \qquad \text{for } t = 1, 2, 3, \ldots, n \qquad \ldots(5.37)$$

The present worth equivalent of the cash flow series can be defined by substituting Eq. (5.37) in Eq. (5.10). That is,

$$P = \sum_{t=1}^{n} A_1(1 + j)^{t-1}(1 + i)^{-t} \qquad \ldots(5.38)$$

or

$$P = A_1(1 + j)^{-1} \sum_{t=1}^{n} \left(\frac{1 + j}{1 + i}\right)^t \qquad \ldots(5.39)$$

It can be shown that Eq. (5.39) reduces to

$$P = \begin{cases} A_1\left[\dfrac{1 - (1 + j)^n\,(1 + i)^{-n}}{i - j}\right] & i \neq j \\ \dfrac{nA_1}{1 + i} & i = j \end{cases} \qquad \ldots(5.40)$$

or equivalently,

$$P = A_1\left(\frac{P}{A_1}, i, j, n\right) \qquad \ldots(5.41)$$

where $\left(\frac{P}{A_1}, i, j, n\right)$ is referred to as the *geometric series present worth factor* and is tabulated in appendix 5A for various values of i, j, and n.

For the special case of $j \geq 0$ and $i \neq j$, the relationship between P and A can be written in terms of compound interest factors as

$$P = A_1\left[\frac{1-\left(\frac{F}{P}, j, n\right)\left(\frac{P}{F}, i, n\right)}{i-j}\right]_{j \geq 0}^{i \neq j} \quad \text{...(5.42)}$$

Example E5.11. *Solve example E5.9 for the case in which each successive payment is to be 10% greater than the previous payment. Find the first payment she made.*

Solution: $\$\ 20{,}000 = A_1\left(\frac{P}{A_1}, 15\%, 10\%, 7\right)$

From Appendix 5A,

$$\$20{,}000 = A_1(5.3481)$$

Hence,

$$A_1 = \$3{,}739.65$$

The future worth equivalent of the geometric series is determined by multiplying the value of the geometric series present worth factor and the $\left(\frac{F}{P}, i, n\right)$ factor.

The result is

$$F = \begin{cases} A_1\left[\dfrac{(1+i)^n - (1+j)^n}{i-j}\right] & i \neq j \\ nA_1(1+i)^{n-1} & i = j \end{cases} \quad \text{...(5.43)}$$

or equivalently,

$$F = A_1\left(\frac{F}{A_1}, i, j, n\right) \quad \text{...(5.44)}$$

where $\left(\frac{F}{A_1}, i, j, n\right)$ is referred to as the *geometric series, future worth factor* and is tabulated in Appendix 5A for various values of i, j, and n.

5.3 WORTH OF INVESTMENTS

In section 5.2 we dealt with the concept of equivalence involving the time value of money. The general goal of engineering economy studies is to minimize costs or maximize profits or savings. In this section we present a number of methods and criteria for measuring the worth of investments. The interest rate used is variously referred to as a minimum alternative rate of return (MARR), hurdle rate, required rate of return, return on investment, and discount rate.

If investment proves to be profitable after considering the MARR, then it is considered as a good candidate for investment.

There are several basic methods of measuring investment worth. All of these methods begin with the assumption that a minimum attractive rate of return has been defined as the criterion for judging whether or not the investment opportunity has merit. These methods are as follows:

1. Present worth method (PW)
2. Annual worth method (AW)
3. Future worth method (FW)
4. Internal rate of return method (IRR)
5. External rate of return method (ERR)
6. Savings/investment ratio method (SIR)
7. Payback period method (PBP)
8. Capitalized worth method (CW)
9. Capital recovery cost method (CRC)

These methods are briefly described in the following sections:

5.3.1 Present Worth Method (PW)

The term *present worth* means an amount at some beginning or base time, which is equivalent to a particular schedule of receipts and/or disbursements under consideration.

The present worth method (PW) uses the equivalent present value of all current and future cash flows to evaluate the investment opportunity. In this method all cash flows are converted to a single sum equivalent at time zero using i = MARR.

The present worth of investment alternative j can be written as

$$PW_j(i) = \sum_{t=0}^{n} A_{jt}(1+i)^{-t} \qquad ...(5.45)$$

where

$PW_j(i)$ = the present worth of alternative j using MARR of i %.

A_{jt} = the net cash flow for alternative j at the end of period t

i = MARR (%)

n = planning horizon

The decision criterion is that a project investment is desirable as long as its net present worth is greater than zero.

Example E5.12. *Refurbishing a mechanical system will cost \$ 20,000, lasts 10 years, and have no salvage value at that time. The net income will increase due to this by \$ 3,000 during each of the first 5 years and by \$ 5,000 during each of the remaining 5 years. Using a 12% MARR, determine the present worth for the investment.*

Solution: PW_1 (12%) = – \$ 20,000 + \$ 3,000 (*P/A*, 12, 5) + \$ 5,000 (*P/A*, 12, 5) (*P/F*, 12, 5)

= – \$ 20,000 + \$ 3,000 (3.6048) + \$ 5,000 (3.6048) (0.5674)

= – \$ 20,000 + \$ 10,814.40 + \$ 10,226.82

= \$ 1,041.22

Hence, a single sum of \$ 1,041.22 at time zero is equivalent to the cash flow profile for this investment.

5.3.2 Annual Worth Method (AW)

In this method, all cash flows are converted to an equivalent uniform annual series of cash flows over the planning horizon using i = MARR.

The annual worth of investment alternative j can be represented as

$$AW_j(i) = \left[\sum_{t=0}^{n} A_{jt}(P/F, i, t)\right](A/P, i, n)$$

or equivalently

$$AW_j(i) = PW_j(i) = (A/P, i, n) \qquad ...(5.46)$$

where

$AW_j(i)$ = the annual worth of alternative j using i = MARR

$PW_j(i)$ = the present worth of alternative j using MARR of i %

$(A/P, i, n)$ = the uniform series capital recovery factor

n = planning horizon

i = MARR

A_{jt} = the net cash flow for alternative j at the end of period t

It can be seen from Eq. (5.46) that the *annual worth method* (AW) and the *present worth method* (PW) are related by a constant multiplier $(A/P, i, n)$.

Example E5.13. *An earth moving machinery was purchased for $ 50,000, kept for 5 years, and sold for 4 $14,000. The annual operating and maintenance cost were $ 5,000. Using a 12 % minimum attractive rate of return (MARR), determine the annual worth for this machinery.*

Solution: Method 1:

AW_1 (12%) = –$ 50,000 (A/P, 12, 5) – $ 5,000 + $ 14,000 (A/F, 12,5)
= –$ 50,000 (0.1614) – $ 5,000 + $ 14,000 (0.0414)
= –$ 8,070 – $ 5,000 + $ 579.60
= –$ 12,490.40 / year.

Method 2:

PW_1 (12%) = –50,000 – $ 5,000 (P/A, 12, 5) + $ 14,000 (P/F, 12, 5)
= –$ 50,000 – $ 5,000 (6.1944) + $ 14,000 (0.2567)
= –$ 50,000 – $ 30, 972 + $ 3,593. 80
= –$ 77, 378.20

Using Eq. (5.46)

AW_1 (12%) = (–$ 77,378.20)(A/P, 12, 5)
= –$ 77,378.20 (0.1614)
= –$ 12,488.84

The slight difference in the solutions in method 1 and method 2 is due to round off error. The result shows that the expenditure of $ 12,490.40 at the end of each of the 5 years is the same as the cash flow profile for the machinery.

5.3.3 Future Worth Method (FW)

In this method, all cash flows are converted to a single sum equivalent at the end of the planning horizon using i = MARR. The future worth of investment alternative j can be computed as

$$\text{FW}_j\,(i) = \sum_{t=0}^{n} A_{jt}\,(1+i)^{\,n-t} \qquad \text{...(5.47)}$$

or equivalently

$$\text{FW}_j\,(i) = \text{PW}_j\,(i) = (F/P,\, i,\, n) \qquad \text{...(5.48)}$$

Equation (5.48) can further be written as

$$\text{FW}_j = \text{AW}_j\,(i)\left(\frac{F}{A}, i, n\right) \qquad \text{...(5.49)}$$

where

$\text{FW}_j\,(i)$ = the annual worth of investment j using a MARR of $i\%$
$\text{AW}_j\,(i)$ = the annual worth of alternative j using i = MARR
$(F/P, i, n)$ = single sum compound amount factor
$(F/A, i, n)$ = uniform series compound amount factor
$\text{PW}_j\,(i)$ = the present worth of alternative j using i = MARR
i = MARR
n = planning horizon
A_{jt} = the net cash flow for alternative j at the end of period t

It can be observed from Eqs. (5.47) and (5.48) that the *future worth method* (FW) is equivalent to the *present worth method* (PW) and the *annual worth method* (AW).

The decision criterion is that a project investment is desirable when the greatest future worth is obtained.

Example E5.14. Determine the future worth of the investment in Example E5.12.

Solution: AW_1 (12%) = PW_1 (12%) (A/P, 12, 10)
= \$ 1,041.22 (0.1770)
= \$ 184.22 / year

AW_1 (12%) = –\$ 20,000 (A/P, 12, 10) + \$ 3,000 + \$ 2,000 (F/A, 12, 5) (A/F, 12, 10)
= –\$20,000 (0.1770) + \$ 3,000 + \$ 2,000 (6.3528)(0.0570)
= –\$ 3,540 + \$ 3,000 + \$ 724.22
= \$ 184.22/year

FW_1 (12%) = –\$20,000 (F/P, 12, 10) + \$ 3,000 (F/A, 12, 10) + \$ 2,000 (F/A, 12, 5)
= –\$ 20,000 (3.1058) + \$ 3,000 (17.5487) + \$ 2,000 (6.3528)
= –62,116 + \$ 52,646.10 + \$ 12,705.60
= \$ 3, 235.70

or FW_1 (12%) = PW_1 (12%) (F/P, 12, 10)
= \$ 1,041.22 (3.1058)
= \$ 3,233.82

or FW_1 (12%) = AW_1 (12%) (F/A, 12, 10)
= \$ 184.22 (17.5487)
= \$ 3,232.82

The differences in answers are due to round-off errors.

5.3.4 Internal Rate of Return Method (IRR)

The *Internal rate of return* (IRR) or *the discounted cash flow rate of return* or *the true rate of return* for an alternative in engineering economic analysis is defined as the interest rate that equates the present worth or the future worth to zero. If we denote i_j^* as the rate of return for alternative j, then Eq. (5.46) can be written as

$$0 = \sum_{t=0}^{n} A_{jt} (1 + i_j^*)^{n-t} \quad ...(5.50)$$

The internal rate of return on an investment can be defined as the *rate of interest earned on the unrecovered balance of an investment*. Based on this definition, the following relationships can be written

$$\begin{aligned} E_0 &= A_0 \\ B_t &= E_{t-1} & t &= 1, 2, 3, ..., n \\ I_t &= B_t I & t &= 1, 2, 3, ..., n \\ E_t &= A_t + B_t + I_t & t &= 1, 2, 3, ..., n \end{aligned} \quad ...(5.51)$$

where

A_t = the cash flow at the end of period t

B_t = the unrecovered balance at the beginning of the period t

E_t = the unrecovered balance at the end of period t

I_t = the interest on the unrecovered balance during the period t

I = internal rate of return

We can therefore write

$$E_n = A_n + B_n + I_n$$

or

$$E_n = A_n + B_n (1 + i)$$

or

$$E_n = A_n + E_{n-1}(1 + i)$$

Hence

$$E_n = A_n + A_{n-1} (1+ i) + E_{n-2} (1 + i)^2$$

Generalizing the recursive relationship we have finally

$$E_n = A_n + A_{n-1} (1 + i) + A_{n-2} (1 + i)^2 + ... + A_0 (1 + i)^n \quad ...(5.52)$$

or equivalently

$$E_n = \text{FW } (i\%) \quad ...(5.53)$$

We can see from the above relationships that the time value of money operations involved in the internal rate of return method (IRR) are identical to considering that all monies received are reinvested and earn interest at a rate equal to the interval rate of return. That is,

$$\sum_{t=0}^{n} R_{jt} (1+ r_t)^{n-t} = \sum_{t=0}^{n} C_{jt} (1+ i')^{n-t} \quad ...(5.54)$$

where

A_{jt} = the net cash flow for investment j in period t

$$R_{jt} = \begin{cases} A_{ij} & \text{for } A_{jt} \geq 0 \\ 0 & \text{otherwise} \end{cases}$$

$$C_{jt} = \begin{cases} -A_{jt} & \text{for } A_{jt} < 0 \\ 0 & \text{otherwise} \end{cases}$$

r_t = the reinvestment rate for positive cash flows occurring in period t

i' = the rate of return for negative cash flows

R_{jt} = positive net cash flows for investment j during period t

C_j = negative net cash flows for investment j during period t

For a special case, when $r_t = i'$ then Eq. (5.54) can be written as

$$0 = \sum_{t=0}^{n} (R_{jt} - C_{jt})(1 + i')^{n-t} \quad \text{...(5.55)}$$

Comparing Eqs. (5.50) and (5.55) we notice that $A_{jt} = R_{jt} - C_{jt}$

Hence, the finding of the rate of return involves solving for i_j^*.

Example E5.15. *There exist two investment opportunities for a company. In the first case, an investment of $300,000 is expected to generate $90,000 in income after the first year. The income for the next four years is expected to be $150,000, $510,000, $0, and $120,000. In the second case, an investment of $310,000 can be made; two years following the investment, an income of $460,000 is expected; two years later another $400,000 is estimated. The cash flow profile for the difference in the two investments is given below.*

End of year	First alternative	Second alternative	Difference in cash flows
0	*– $310,000*	*– $300,000*	*– $10,000*
1	*90,000*	*0*	*90,000*
2	*150,000*	*460,000*	*– 310,000*
3	*510,000*	*0*	*510,000*
4	*0*	*400,000*	*– 400,000*
5	*120,000*	*0*	*120,000*

Determine the external rate of return when the MARR is (a) 0%, (b) 100%, and (c) 200%.

Solution:

EOY	CF(A)	CF(B)	CF(A – B)
0	– $310,000	– $300,000	– $10,000
1	90,000	0	90,000
2	150,000	460,000	– 310,000
3	510,000	0	510,000
4	0	400,000	– 400,000
5	120,000	0	120,000

We cannot use the MIRR function due to the multiple negative values in CF(A – B). The future value of the positive-valued cash flows is

$$\text{FW} = \$90{,}000(F/P, \text{MARR}, 4) + \$510{,}000(F/P, \text{MARR}, 2) + \$120{,}000$$

The future value of the negative-valued cash flows is

$$\text{FW} = \$10{,}000(F/P, \text{ERR}, 5) + \$310{,}000(F/P, \text{ERR}, 3) + 400{,}000(F/P, \text{ERR}, 1)$$

(*a*) MARR = 0% ERR = 0%

(*b*) MARR = 100% ERR = 100%

(*c*) MARR = 200% ERR = 200%

Example E5.16. *In example E5.15, determine the interval rate(s) of return for the difference in cash flows.*

Solution:

EOY	CF(A)	CF(B)	CF(A – B)
0	– $310,000	– 300,000	– 10,000
1	90,000	0	90,000
2	150,000	460,000	– 310,000
3	510,000	0	510,000
4	0	400,000	– 400,000
5	120,000	0	120,000

Multiple roots exist, i = 0%, i = 100%, and i = 200%

Hence IRR = 0%

IRR = 100%

IRR = 200%

Example E5.17. *A manufacturing company purchases a computer-controlled robot for $100,000 with a $50,000 loan from a bank at 15% compounded annually and the loan is to be paid back with equal annual payments over a 5-year period. The operating and maintenance costs are estimated to be $20,000/year over the 10-year period. If the company expects to earn 8% on its investment, determine the following:*

(a) present worth

(b) annual worth

(c) future worth

(d) internal rate of return

(e) external rate of return

(f) savings/investment ratio

Solution:

MARR = 0.08

PW = $96,380.00

AW = $14,363.46

FW = $208,077.20

IRR = 34.22%

ERR = 20.25%

SIR = 2.93

EOY	CF
0	– $50,000
1	$15,084
2	$15,084
3	$15,084
4	$15,084
5	$15,084
6	$30,000
7	$30,000
8	$30,000
9	$30,000
10	$40,000

Example E5.18. *An investment company places $100,000 in an investment fund. The company receives $15,000 at the end of first year and for the next 10 years. There is no income until 15 years after the initial investment, at which time $100,000 is received. Determine the following if MARR is 8% over the 15-year period.*

(*a*) *present worth* (*b*) *annual worth*

(*c*) *future worth* (*d*) *internal rate of return*

(*e*) *external rate of return* (*f*) *savings / investment ratio*

(*g*) *payback period with and without considering the time value of money.*

Solution:

MARR = 0.08 Payback Period Calculations

EOY	CF	Cum CF	Cum PW
0	– $100,000	– 100,000	– 100,000.00
1	15,000	– 85,000	– 86,111.11
2	15,000	– 70,000	– 73,251.03
3	15,000	– 55,000	– 61,343.55
4	15,000	– 40,000	– 50,318.10
5	15,000	– 25,000	– 40,109.35
6	15,000	– 10,000	– 30,656.81
7	15,000	5,000	– 21,904.45
8	15,000	20,000	– 13,800.42
9	15,000	35,000	– 6,296.68
10	15,000	50,000	651.22
11	0	50,000	651.22
12	0	50,000	651.22
13	0	50,000	651.22
14	0	50,000	651.22
15	100,000	150,000	32,175.39

PW = $32,175.39

AW = $3,759.04

FW = $102,065.78

IRR = 12.52%

ERR = 10.03%

SIR = 1.32

PBP = 7 yrs

DPBP = 15 yrs

5.3.5 External Rate of Return Method (ERR)

The external rate of return (ERR) method involves the determination of the value of the *external rate of return*, i' that satisfies Eq. (5.54) that yields a future worth of zero by explicitly assuming the reinvestment recovered funds at the MARR. A unique value of i is obtained by the ERR method by solving the following relationship.

$$\sum_{i=0}^{n} R_{jt} (1 + r_t)\, n^{-t} = \sum_{t=0}^{n} C_{jt} (1 + i')^{n-t} \qquad ...(5.56)$$

where

R_{jt} = positive net cash flows for investment j during period t

C_{jt} = negative net cash flows for investment j during period t

r_t = the reinvestment rate for positive cash flows occurring in period t

i' = the rate of return for negative cash flows.

It can be shown that the value of the external rate of return always lies between the interval rate of return and the MARR.

Example E5.19. *A manufacturing company purchases an electrochemical system for \$30,000 by borrowing the money at an annual compound rate of 12%. The loan is to be repaid with four equal annual payments. The expected use of the system is for 9 years and then sold for \$2,000. The annual operating and maintenance expenses are estimated to be \$9,000/year. The new system saves the company of \$15,000/year over the old system. If the MARR is 12%, determine the following:*

(a) present worth *(b) annual worth*

(c) future worth *(d) internal rate of return*

(e) external rate of return *(f) savings/investment ratio*

(g) payback period with and without considering the time value of money

Solution:

MARR = 0.12

PW = \$2,690.72

AW = \$504.99

FW = \$7,461.58

IRR = 17.23%

ERR = 11.14%

SIR = 1.23

PBP = 7 yrs

DPBP = 9 yrs

EOY	CF	Cum CF	Cum PW
0	\$0	0	0
1	– 3,877	– 3,877	– 3,371
2	– 3,877	– 7,754	– 6,303
3	– 3,877	– 11,631	– 8,852
4	– 3,877	– 15,508	– 11,069
5	6,000	– 9,508	– 8,086
6	6,000	– 3,508	– 5,492
7	6,000	2,492	– 3,236
8	6,000	8,492	– 1,275
9	8,000	16,492	999
		7 years	9 years

5.3.6 Savings/Investment Ratio Method (SIR)

In this method, the ratio of the present worth of savings to the present worth of the investment is determined. The savings/investment ratio method can be represented by the following formulation:

$$\text{SIR}_{j(i)} = \frac{\sum_{t=0}^{n} R_{jt}(1+i)^{-t}}{\sum_{t=0}^{n} C_{jt}(1+i)^{-t}}$$

$$= \frac{\text{Present worth of net positive cash flows}}{\text{Present worth of net negative cash flows}} \qquad \text{...(5.57)}$$

where

$\text{SIR}_j(i)$ = savings/investment ratio for investment alternative j based on a MARR of $i\%$

A_{jt} = the net cash flow for investment j in period t

R_{jt} = the net positive net cash flows for investment j during period t

$$= \begin{cases} A_{jt}, & \text{if } A_{jt} \geq 0 \\ 0, & \text{otherwise} \end{cases}$$

C_{jt} = the net negative cash flows for investment j during period t.

For a project to be preferred for investment at the MARR, the value obtained from (5.57) for $\text{SIR}_j(i)$ must be greater than one.

Example E5.20. *If a company invested \$80,000 in a venture with the cash flow shown in Table E5.27, results and the MARR is 10%, determine the following:*

(*a*) *present worth*

(*b*) *annual worth*

(*c*) *future worth*

(*d*) *internal rate of return*

(*e*) *external rate of return*

(*f*) *savings/sinvestment ratio*

(*g*) *payback period with and without considering the time value of money*

Table E5.20

EOY	CF
0	*–\$80,000*
1	*10,000*
2	*15,000*
3	*20,000*
4	*27,000*
5	*27,000*
6	*20,000*
7	*15,000*
8	*10,000*

Solution:

EOY	CF
0	–$80,000
1	10,000
2	15,000
3	20,000
4	27,000
5	27,000
6	20,000
7	15,000
8	10,000

MARR = 0.1

EOY	Cum CF	Cum PW
0	– $80,000	– 80,000
1	– 70,000	– 70,909.09
2	– 55,000	– 58,512.40
3	– 35,000	– 43,486.10
4	– 8,000	– 25,044.74
5	19,000	– 8,279.86
6	39,000	3,009.62

PW = $15,372.06
AW = $2,881.40
FW = $32,951.38
IRR = 14.86%
ERR = 12.44%
SIR = 1.19
PBP = 5 yrs
DPBP = 6 yrs

5.3.7 Payback Period Method (PBM)

Payback period method involves in determining how long it will take to recover the initial investment. The *payback period method* is not an equivalent method. If C_{jo} denotes the initial cost of investing alternative j, and R_{jt} denotes the net revenue obtained from investing j during period t, and assuming that no other negative cash flows occur, then the smallest value of m_j is such that

$$\sum_{t=1}^{m_j} R_{jt} \geq C_{jo} \qquad ...(5.58)$$

define the payback period for investment j. Therefore, the investment alternative with the smallest payback period is the desired alternative using the above method.

Example E5.21. *A heavy maintenance facility plant for machinery has a construction cost at t = 0 of $3,000,000, an annual maintenance cost of $70,000, and a major repair at 5-year intervals projected to cost $300,000 with first such repair occurring at t = 5. If interest is 10% annually, determine the amount of money needed at t = 0 to provide for construction and perpetual upkeep.*

Solution:

End of year (EOY)	Cash flow
0	– $3,000,000
1	– 70,000
2	– 70,000
3	– 70,000
4	– 70,000
5	– 70,000

MARR = 0.10 or 10%

Capitalized cost = $3,000,000 for initial construction
+ ($70,000/0.10) for annual maintenance
+ [$300,000(*A*/*F*, 10%,5)/0.10] for major repairs every 5 years
= $3,000,000 + $700,000 + $300,000(0.1638)/0.10
= $4,191,400

Hence, Capitalized Cost = $4,191,400

5.3.8 Capitalized Worth Method (CW)

Capitalized worth method determines the single sum at time zero that is equivalent at i = MARR to a cash flow pattern that continues indefinitely. This method is based on the fact that if a present value P is deposited into a fund at an interest rate i per period such that a payment of size A can be withdrawn each and every period perpetually, then we have

$$Pi = A \qquad ...(5.59)$$

The present value P is called the *capitalized worth* which is the size of each of the perpetual payments.

Example E5.22. *The maintenance of a large electromechanical system is cyclic with the following costs occurring over a 5-year period: $6,000, $2,500, $8,000, $0, and $2,200. It is expected that this sequence will repeat itself every 5 years forever. Determine the capitalized cost of the maintenance costs based on a time value of money of 12%.*

Solution:

EOY	CF
0	$0
1	6,000
2	2,500
3	8,000
4	0
5	2,200

MARR = 0.12
PW = – \$1,090,284.12
AW = – \$204,623.31
Endowment = – \$1,705,194.27

5.3.9 Capital Recovery Cost Method (CRC)

Figure 5.5 depicts the cash flow diagram of an investment of \$ P in an asset having a life of n years and deposited for a salvage value of \$ F. The capital recovery cost is a uniform annual amount defined by the following relationship.

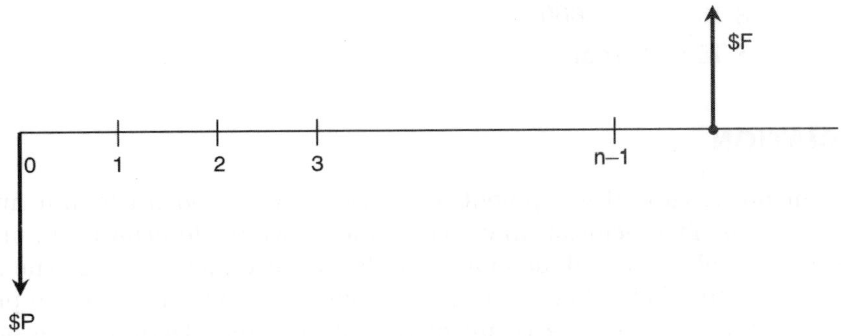

Fig. 5.5 Cash Flow Diagram of an Investment in an Asset.

$$CR = P(A/P, i, n) - F(A/F, i, n) \quad ...(5.60)$$

where

CR = Capital Recovery Cost

$(A/P, i, n)$ = the uniform series capital recovery factor

$(A/F, i, n)$ = the uniform series sinking factor

Also, we have
$$(A/P, i, n) = (A/F, i, n) + i \quad ...(5.61)$$
and
$$(A/F, i, n) = (A/P, i, n) - i \quad ...(5.62)$$

Substituting Eqs. (5.61) and (5.62) in (5.60) yields the following additional cost recovery formulas:

$$CR = (P - F)\,(A/F, i, n) + P\,i \quad ...(5.63)$$
and
$$CR = (P - F)\,(A/P, i, n) + F\,i \quad ...(5.64)$$

where

P = the equivalent value of an amount of money at time zero or present worth

F = the equivalent value of an amount of money at time n or future worth.

Example E5.23. *Determine the capital recovery cost of an electromechanical system purchased for P = \$ 100,000 with a service life of 10 years and a salvage value of \$ 5,000. Assume an interest rate of 12% per year.*

Solution: The capital recovery cost

CR = \$ 100,000 (A/P, 12, 10) – \$ 5,000 (A/F, 12, 10)
= \$ 100,000 (0.1614) – \$ 5,000 (0.0414)
= \$ 16,140 – \$ 207
= \$ 15,933 / year

Using Eq. (5.63), we have

$$\begin{aligned} CR &= (P-F)\,(A/F, i, n) + P\,i \\ &= (\$100{,}000 - \$5{,}000)\,(A/F, 12, 10) + \$\,100{,}000\,(0.12) \\ &= \$\,95{,}000\,(0.0414) + \$\,12{,}000 \\ &= \$\,3{,}933 + \$\,12{,}000 \\ &= \$\,15{,}933/\text{ year} \end{aligned}$$

Also using Eq. (5.64), we get

$$\begin{aligned} CR &= (P-F)\,(A/P, 12, 10) + F\,i \\ &= \$\,95{,}000\,(0.1614) + \$\,5{,}000\,(0.12) \\ &= \$\,15{,}333 + \$\,600 \\ &= \$\,15{,}933/\text{ year.} \end{aligned}$$

5.4 DEPRECIATION

Most property including capital equipment decreases in value with use and time and that is known as *depreciation*. Depreciation may occur due to wear, deterioration, or obsolescence. The allowance for loss of value is depreciation in depreciation accounting. The main objective of depreciation accounting is to allocate for the recovery of capital invested in property, which is expected to decline in value, as a result of time and/or use. Depreciation accounting also provides a means of placing a declared or unamortized value, called *book value* on property. Depreciation write-offs are not by themselves cash flows but they affect income taxes and hence affect cash flows.

We shall consider here four methods of spreading the depreciation over the recovery period n:

1. Straight-Line Depreciation
2. Declining Balance Depreciation
3. Sum of the Years' Digits Depreciation,
4. Modified Accelerated Cost Recovery System (MACRS) Depreciation Method

5.4.1 Straight-Line Depreciation

The straight-line depreciation method provides for uniform periodic depreciation charges over the write-off period.

Letting

P = cost of asset

F = salvage value

n = write-off period (in years)

D_t = depreciation deduction at the end of each year t

B_t = the unrecovered investment at the end of year t or book value.

Then

$$\text{Annual depreciation charge } D_t = \frac{(P-F)}{n} \qquad \text{...(5.65)}$$

$$\text{Depreciable investment} = (P-F) \qquad \text{...(5.66)}$$

and

$$B_t = P - \left(\frac{P-F}{n}\right) t \qquad \text{...(5.67)}$$

The straight-line depreciation is popular because of its simplicity. This method is required for depreciating an intangible property.

Example E5.24. *A computer system costs \$ 110,000 and has an estimated salvage value of \$ 10,000 at the end of 5 years useful life. Determine the depreciation schedule, and book value to the end of the useful life of the computer system by the straight-line depreciation method.*

Solution:

Annual depreciation charge

$$D_t = \frac{P - F}{n}$$

$$= \frac{110{,}000 - 10{,}000}{5} = \$\ 20{,}000 \text{ per year.}$$

Year	D_t	B_t
1	20,000	110,000 – 20,000 = 90,000
2	20,000	110,000 – 40,000 = 70,000
3	20,000	110,000 – 60,000 = 50,000
4	20,000	110,000 – 80,000 = 30,000
5	20,000	110,000 – 100,000 = 10,000
	Σ = 100,000	

5.4.2 Declining Balance Depreciation

The *declining balance depreciation method* provides for an accelerated write-off (depreciation) during the early years of the life of an asset and lower allowances throughout the rest of the property's useful life.

For this method, the depreciation charge, D_t, is equal to a fixed percentage of the unrecovered investment at the end of the previous year. That is,

$$D_t = p\, B_{t-1} \qquad \text{...(5.68)}$$

It can be shown that the book value at the end of the $(t-1)^{th}$ year is given by

$$B_{t-1} = P(1-p)^{t-1} \qquad \text{...(5.69)}$$

and

$$B_t = P(1-p)^t \qquad \text{...(5.70)}$$

Substituting Eq. (5.70) into Eq. (5.68) gives

$$D_t = p\, P(1-p)^{t-1} \qquad \text{...(5.71)}$$

In order for the book value to be equal to the estimated salvage value at the end of the write-off period, n years, p is calculated from

$$p = 1 - \sqrt[n]{\frac{F}{P}} \qquad \text{...(5.72)}$$

In the declining balance method of depreciation, the unrecovered investment must never fall below the salvage value. A special version of the declining balance depreciation is called the *double declining balance method* or *200% declining balance depreciation method.* For this method, p is calculated as $200\%/n$.

In this case, the book value at the end of the write-off period, n years, will not normally be equal to the salvage value. Due to this, the taxpayer is normally permitted to switch later to straight-line depreciation method for the remaining years.

The optimum switch occurs whenever the straight-line depreciation method on the unrecovered portion of an asset exceeds the declining balance allowance. Hence, a switch to a straight-line depreciation method occurs at the first year for which

$$\frac{B_{t-1} - F}{n - (t - 1)} > p\, B_{t-1} \qquad \text{...(5.73)}$$

Example E5.25. *Compute the depreciation schedule, and book value, to the end of the useful life of the computer system in Example E5.24 by the double Declining Balance Depreciation method.*

Solution: Double declining balance (DDB)

$$\text{DDB} = \frac{2}{n}[\text{P} - \Sigma\, D_t \text{ \{to date\}}]$$

Depreciated to end of useful life means to convert to straight line.

$$\text{3rd year : } D_t \text{ (straight line)} = \frac{39{,}600 - 10{,}000}{3} = 9{,}866.67 < 15{,}840$$

$$\text{4th year: } D_t \text{ (straight line)} = \frac{23{,}760 - 10{,}000}{2} = 6{,}880/\text{year}$$

Year	DDB	Σ DDB	B_t
1	$\frac{2}{5}(110{,}000 - 0) = 44{,}000$		110,000 – 44,000 = 66,000
2	$\frac{2}{5}(110{,}000 - 44{,}000) = 26{,}400$	70,400	110,000 – 70,400 = 39,600
3	$\frac{2}{5}(110{,}000 - 70{,}400) = 15{,}840$	86,240	110,000 – 86,240 = 23,760
4	$\frac{2}{5}(110{,}000 - 86{,}240) = 9{,}504$	95,744	110,000 – 95,744 = 14,256
5	$\frac{2}{5}(110{,}000 - 95{,}744) = 5{,}702.4$	101,446.4	110,000 – 101,446.4 = 8,553.6

Year	DDB	$B_t = P - \Sigma D_t$
1	44,000	110,000 – 44,000 = 66,000
2	26,400	110,000 – 70,400 = 39,600
3	15,840	110,000 – 86,240 = 23,760
4	6,880	110,000 – 93,120 = 16,880
5	6,880	110,000 – 100,000 = 10,000
	Σ = 100,000	

5.4.3 Sum of the Years' Digits Depreciation

The *sum of the years digits depreciation* method, like the *declining balance depreciation* method, is an alternative depreciation method for achieving accelerated write-off during the early years of life of an asset. In this method, the digits corresponding to the number of years of estimated life are added together which is used directly in the calculation of the allowable depreciation. This sum is obtained from

$$\text{Sum of the years' Digits} = \frac{n(n+1)}{2}$$

The depreciation deduction during any year t is given by

$$D_t = \frac{n-(t-1)}{n(n+1)/2}(P-F) \quad ...(5.74)$$

and the unrecovered investment at the end of each year t is given by

$$B_t = P - \sum_{j=1}^{t} \frac{n-(j-1)}{n(n+1)/2}(P-F) \quad ...(5.75)$$

or

$$B_t = (P-F)\,\frac{(n-t)\,(n-t+1)}{n(n+1)} + F \quad ...(5.76)$$

This method is allowed if it was elected for new pre-1981 tangible property having a useful life of 3 years or more.

Example E5.26. *Compute the depreciation schedule, and book value, to the end of the useful life of the computer system in Example E5.24 by the sum-of-years-digits (SOYD) method.*

Solution:

$$\text{SOYD} = \frac{n(n+1)}{2} = \frac{5}{2}\,(5+1) = 15$$

$$D_t = \frac{n}{\text{SOYD}}\,(P-F)$$

Year	D_t	B_t
1	$\frac{5}{15}$(110,000 – 10,000) = 33,333.33	110,000 – 33,333.33 = 76.666.67
2	$\frac{4}{15}$(100,000) = 26,666.66	110,000 – 60,000 = 50,000
3	$\frac{3}{15}$(100,000) = 20,000	110,000 – 80,000 = 30,000
4	$\frac{2}{15}$(100,000) = 13,333.33	110,000 – 93,333.33 = 16,666.67
5	$\frac{1}{15}$(100,000) = 6,666.66	110,000 – 100,000 = 10,000
	Σ = 100,000	

5.4.4 MACRS Depreciation Method

The *Modified Accelerated Cost Recovery System* (MACRS) depreciation method applies to most depreciable property placed in service after the year 1986. The U.S. Tax Reform Act of 1986 established the MACRS system. All property depreciated under MACRS is assigned a recovery period and deductions are calculated as specific percentages of the cost of the asset. The salvage value is not considered in this method. The MACRS system has pre-established recovery periods for most property. Most property is assigned to eight property classes. These property classes establish the number of years over which the cost basis of an item is recovered.

The MACRS deduction can be easily calculated by applying the appropriate percentages ($d_r \times 100\%$), for a specified property class and year, to the cost basis P. Table 5.2 gives the MACRS deduction rates. The allowable deduction in year t, D_t, is given by

$$D_t = d_t P \qquad ...(5.77)$$

In this method, the entire cost basis of a property can be depreciated over its recovery period, even if its useful life is much longer.

The unrecovered investment B_t at the end of year t is given by

Table 5.2. MACRS Deduction Rates

General depreciation system
Applicable depreciation method; 200 or 150 per cent
Declining balance switching to straight line
Applicable recovery periods; 3, 5, 7, 10, 15, 20 years
Applicable convention: Half-year

Recovery year	Recovery Period					
	3-year	5-year	7-year	10-year	15-year	20-year
	Depreciation rate					
1	33.33	20.00	14.29	10.00	5.00	3.750
2	44.45	32.00	24.49	18.00	9.50	7.219
3	14.81	19.20	17.49	14.40	8.55	6.677
4	7.41	11.52	12.49	11.52	7.70	6.177
5		11.52	8.93	9.22	6.93	5.713
6		5.76	8.92	7.37	6.23	5.285
7			8.93	6.55	5.90	4.888
8			4.46	6.55	5.90	4.522
9				6.56	5.91	4.462
10				6.55	5.90	4.461
11				3.28	5.91	4.462
12					5.90	4.461
13					5.91	4.462
14					5.90	4.461
15					5.91	4.462
16					2.95	4.461
17						4.462
18						4.461
19						4.462
20						4.461
21						2.231

Source: **U.S. Internal Revenue Service: Tax Information Depreciation, U.S. Department of the Treasury, Publication No. 534, published yearly.**

Note: 200 per cent declining balance applies to class lives of 10 years or less. 150 per cent applies to 15- and 20-year class lives.

$$B_t = P - \sum_{j=1}^{t} D_j = P\left(1 - \sum_{j=1}^{t} d_j\right) \qquad ...(5.78)$$

B_t in Eq. (5.78) represents the amount of the cost basis yet to be recovered. Under MACRS a one-half year convention generally applies. In other words, all acquisitions are assumed to be made in midyear. For the first tax year, the allowable deduction is one-half the allowance for subsequent years.

The declining-balance method of depreciation calculation is used to calculate MACRS allowances until the optimum year is reached for a switch to the straight-line method. The optimum year is the first year for which the straight-line method will give an equal or greater deduction than does the declining-balance method.

An option exists whereby most property may be depreciated over its class life using straight-line depreciation. The one-half year convention, however, continues to apply for the first year.

Example E5.27. *A company is purchasing a high-speed computer system. The first cost including installation will be \$ 150,000 with an estimated salvage value of \$ 10,000 at the end of a projected useful service life of 7 years. This is a 5-year property and MACRS–GDS 5-year recovery period applies. Calculate the depreciation deduction and the resulting unrecovered investment during each year of the assets' life.*

Solution: Referring to Table 5.2 for MACRS–GDS percentages for 5-year property for calculation of depreciation–the allowable MACRS deduction and the resulting unrecovered investment for each year are given in Table E5.27.

Table E5.27. MACRS Deduction and Unrecovered Investment

End of year t	MACRS-GDS deduction D_t	Unrecovered investment B_t
0		\$ 150,000
1	\$ 30,000	120,000
2	48,000	72,000
3	28,800	43,200
4	17,280	25,920
5	17,280	8,640
6	8,640	0

Example E5.28. *Reconsider the high-speed computer system in Example E5.27. Calculate the depreciation deduction and the resulting unrecovered investment during each year of the assets life. Use*

(a) straight line depreciation

(b) sum of years digits depreciation

(c) use MACRS–GDS allowances.

Solution:

(*a*) Straight-line depreciation (See Table E5.28(a)).

$$D_t = \frac{P - F}{n} = \frac{150{,}000 - 10{,}000}{7} = \$\ 20{,}000$$

(*b*) Sum of the years digits depreciation. Refer to Table E5.28(*b*).

(*c*) Refer to Example E5.27

Table E5.28(a). Straight-line Depreciation and Unrecovered Investment

End of year	Straight line deduction D_t	Unrecovered investment B_r
0		\$150,000
1	\$20,000	130,000
2	20,000	110,000
3	20,000	90,000
4	20,000	70,000
5	20,000	50,000
6	20,000	30,000
7	20,000	10,000

Table E5.28 (b). Sum of the Years Digits Depreciation and Unrecovered Investment

End of year t	Value of [n – (t – 1)]/[n(n + 1)/2]	SOYD	Unrecovered investment B_t
0			150,000
1	7/28	35,000	115,000
2	6/28	30,000	85,000
3	5/28	25,000	60,000
4	4/28	20,000	40,000
5	3/28	15,000	25,000
6	2/28	10,000	15,000
7	1/28	5,000	10,000

5.5 TAXES

Taxes make a significant difference in engineering economic analysis. Tax dollars are cash flows, and hence it is important to consider them. The Federal tax is based on the income after all costs have been deducted. All depreciation costs reduce the tax base and are a tax savings or credit in any year. This tax saving on credit is equal to tD, where t is the decimal tax rate and D is the depreciation dollars for the year. The knowledge and application of the tax laws can make the difference between accepting and rejecting a project and between profit and loss for a corporation. A corporation, in this context includes associations, business trusts, insurance companies, joint stock companies, trusts and partnerships that operate and function as associations or corporations.

This section provides a brief introduction to main tax considerations and a method for including the effect of income taxes in engineering economic studies.

5.5.1 Types of Taxes

There are numerous types of taxes important in engineering economic analysis. These include property (ad valorem), sales, excise (a tax on the manufacture, sale, or consumption of a commodity) and income taxes.

1. *Property taxes*: Property taxes are based on the value of the property owned by the corporation (land, buildings, equipment, inventory, etc.) and the established tax rates. They do not change with profits and are usually much lower than the income taxes.
2. *Sales taxes*: Sales taxes are imposed on the sales of products, usually at the retail level. They generally are not relevant in engineering economic analysis.
3. *Excise taxes*: Excise taxes are imposed on the manufacture of certain products such as alcohol and tobacco. These taxes are normally passed on to the consumer and not significant in engineering economic analysis.
4. *Income taxes*: Income taxes are taxes imposed on corporate profits or personal income, or on gains on the disposal of capital property. These taxes are most significant and relevant in engineering economic analysis.

5.5.2 Income Tax

A corporate includes business trusts, association, joint stock companies, insurance companies, and partnerships and trusts that operate as associations or corporations. The income taxes are greatly influenced by politics and economic conditions. The United States has a corporate graduated tax schedule as shown in Table 5.3. Great many states, most cities and countries in the US also have an income tax. The state taxes are deductible from the federal taxes. The tax rates are applied to the taxable income in corporate income tax rates for tax years beginning january 1, 1993 and beyond.

For this purpose, gross income less any monies specifically exempt from tax liability. Taxable income is the gross income less allowable deductions. The deductions for corporations include salaries, wages, repairs, rents, bad debts, taxes (excluding income taxes), charitable contributions (both cash and non cash), casualty, loses, interest on borrowed money and depreciation allowances. Taxable income is shown pictorially in Figure 5.6.

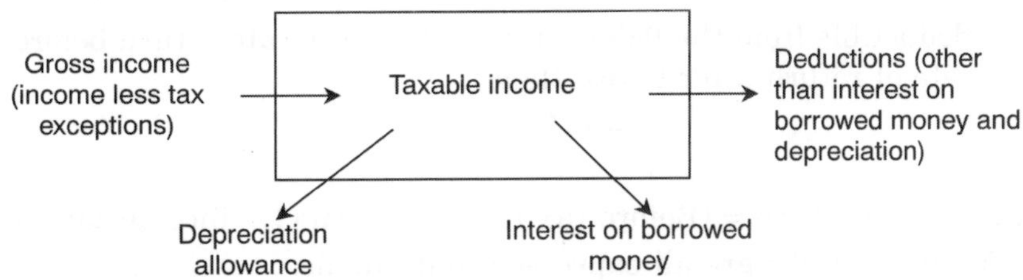

Fig. 5.6 Pictorial Representation of Taxable Income.

5.5.3 Cash Flow

Before tax cash flow is the gross income less deductions, not including depreciation. After-tax cash flow is the gross income less income taxes and deductions, interest on borrowed money but excluding depreciation allowance. The after-tax cash flow is represented pictorially in Figure 5.7. Before-tax and loan cash flow is the gross income less deductions, not including either depreciation or principal or interest on the loan.

Table 5.3. Corporate Income Tax Rates for Tax Years Beginning January 1, 1993 and Beyond

Taxable income (TI), in $	Tax rate (t)	Income tax (T)
	0.15	0.15(TI)
0 < TI ≤ 50,000	0.25	7500 + 0.25(TI – 50,000)
50,000 < TI ≤ 75,000	0.34	13,500 + 0.34(TI – 75,000)
75,000 < TI ≤ 100,000	0.39(0.34 + 0.05)	22,250 + 0.39(TI – 100,000)
100,000 < TI ≤ 335,000	0.34	113,900 + 0.34(TI – 335,000)
335,000 < TI ≤ 10,000,000	0.35	3,400,000 + 0.35(TI – 10,000,000)
10,000,000 < TI ≤ 15,000,000	0.38(0.35 + 0.03)	5,150,000 + 0.38(TI – 15,000,000)
15,000,000 < TI ≤ 18,333,333	0.35	0.35(TI)
18,333,333 < TI		

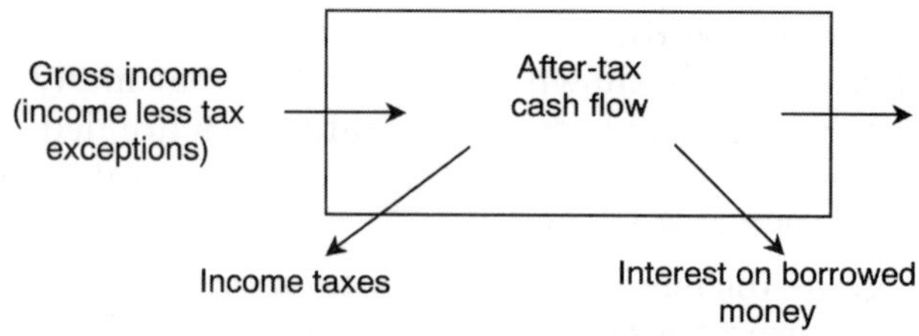

Fig. 5.7 Pictorial Representation of After-Tax Cash Flow.

5.5.4 Present Value after Taxes

State taxes are deductible from the federal taxes. If i is the rate return before taxes, t the tax rate, and r the rate of return after taxes, then

$$r = i\,(1 - t) \qquad ...(5.79)$$

That is

After-tax rate of return = (Before tax rate of return)(1 – income tax rate)

In Eq. (5.79), r, i and t are all expressed in decimals.

Also it should be noted that

Effective tax rate = State tax rate + (1 – State tax rate) (Federal rate) ...(5.80)

The effect introduced by the corporate income taxes is a reduction of the rate of return on a project or venture. Equation. (5.79) is valid only when there are no depreciable assets. The significance of depreciation in reducing taxes is depicted in Fig. 5.8. The gross profit is noticeably reduced by the depreciation charges and hence the taxes.

If an article lasts n years with initial cost C_i and salvage value C_{sal}, then the depreciable cost C_d is given by

$$C_d = C_i - C_{sal} \qquad ...(5.81)$$

There is no tax consequence at the time the article is purchased. But at the first-year end, a depreciation equal to $D_{f1}C_d$ is available as an expense recognized for tax purchases, where

$$D_{f1} = \text{the fractional depreciation.}$$

Hence, the taxable base reduction is $D_{f1}\, C_d$ and a savings or reduction in taxes equal to $D_{f1}C_d t$ will be realized. The time-cost diagram is depicted in Fig. 5.9.

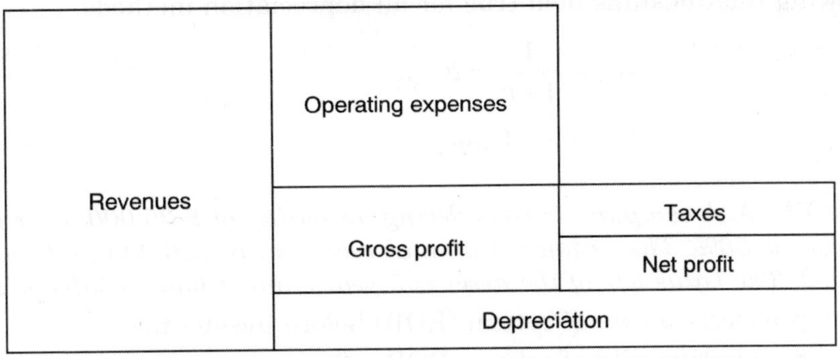

Fig. 5.8 Distribution of Corporate Revenues.

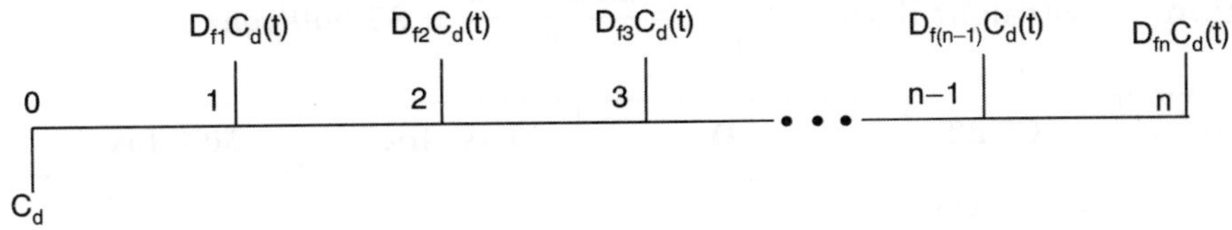

Fig. 5.9 The Time-cost Diagram.

Therefore, the present value can be written as

$$P = C_d - C_d\, t\left[\frac{D_{f1}}{1+r} + \frac{D_{f2}}{(1+r)^2} + \ldots.. + \frac{D_{fn}}{(1+r)^n}\right] \quad ...(5.82)$$

Equation. (5.82) can be rewritten as

$$P = C_d - C_d\, t\,\psi = C_d\,(1 - t\,\psi) \quad ...(5.83)$$

where

$$\psi = \sum_{1}^{n} \frac{D_{fm}}{(1+r)^m} \quad ...(5.84)$$

The factor ψ depends on r, n and the method selected for depreciation

For instance, for sum-of-the-years-digits and straight-line depreciation methods, we have

$$\psi_{SD}, r, n = F_{SDP,\, r,\, n} \quad ...(5.85)$$

$$\psi_{SL}, r, n = F_{SLP,\, r,\, n}$$

where

F_{SDP} = converts sum-of-the-years-digits depreciation to a present value

F_{SLP} = converts straight-line depreciation to a present value

These factors are generally found in tabular form

Hence, Eq. (5.83) can now be written as, for sum-of-the-years-digits depreciation.

$$P = C_d\,(1 - t\, F_{SDP,\, r,\, n}) \quad ...(5.86)$$

and for straight-line depreciation

$$P = C_d\,(1 - t\,F_{SLP,\,r,\,n}) \qquad ...(5.87)$$

The following relationships hold true for all depreciation methods.

$$\psi_{1r} = \frac{1}{1+r} = F_{SDP,\,r,\,1} \qquad ...(5.88)$$

and

$$\psi_{\alpha r} = 0 = F_{SDP,\,r,\,\alpha} \qquad ...(5.89)$$

Example E5.29. *A company is considering an outlay of $ 35,000 for a new asset. The company's tax rate is 50%. The estimated annual receipts are $20,000 and annual disbursements are $10,000. The value life of the asset is 5 years and it has no salvage value.*

(*a*) find the prospective rate of return (ROR) before income tax.

(*b*) find the prospective rate of return (ROR) after taxes, assuming straight-line depreciation for writing off asset for tax purpose.

Solution: D_t (Straight line) $= \dfrac{P-F}{n} = \dfrac{35{,}000-0}{5}$ = $7,000/year

Year	CFBT	D_t	Tax. Inc.	50% tax	CFAT
0	– 50,000				– 50,000
1	10,000	7,000	3,000	– 1,500	85,000
2	10,000	7,000	3,000	– 1,500	85,000
3	10,000	7,000	3,000	– 1,500	85,000
4	10,000	7,000	3,000	– 1,500	85,000
5	10,000	7,000	3,000	– 1,500	85,000
		Σ = 35,000			

(*a*) ROR (BT) PW (B) = PW (C)

$$A(P/A,\,i,\,n) = c\ 12\% = 3.605\ \frac{X}{3} = \frac{0.105}{0.253}$$

$$(P/A,\,i,\,5) = \frac{35{,}000}{10{,}000} = 3.5. \text{ Therefore, } X = 3.5$$

15% = 3.352 or X = 1.25

(ROR) i = 12% + 1.25% = 13.25%

(*b*) ROR (AT) $6\% = 4.212 = \dfrac{X}{1} = \dfrac{0.0944}{0.102}$

Since

$$(P/A,\,i,\,5) = \frac{35{,}000}{8{,}500} = 4.1176$$

X = 4.1176

7% = 4.11

X = 0.9255

(ROR)i = 6% + 0.9% = 6.9%

Example E5.30. *The data for a project undertaken by a company is summarized below. The company uses straight-line depreciation and pays an incremental tax rate of 30% and requires an after-tax rate of return of 9%. Investment tax credit is ignored.*

Item	Amount
Life	*25 years*
Investment	*$110,000*
Salvage value	*$10,000*
Revenues	*$25,000/year*
Operating expenses	*$10,000/year*

(*a*) *using Net Present Worth, determine whether the project should be undertaken.*

(*b*) *if the company used sum of years digits depreciation method, would the decision change?*

Solution:

$$(a)\ \text{Depreciation} = \frac{110{,}000 - 10{,}000}{25} = \$\ 4{,}000/\text{year}$$

Year	BTCF	Depreciation	Taxable income	Taxes@ 30%	ATCF
0	– $110,000	–	–	–	– 110,000
1–25	15,000	4,000	11,000	3,300	11,700
S	10,000	–	–		10,000

$$\begin{aligned} \text{NPW} &= \$11{,}700\ (P/A,\ 9\%,\ 25) + \$10{,}000(P/F,\ 9\%,\ 25) - \$110{,}000 \\ &= \$11{,}700\ (9.8226) + \$10{,}000(0.1166) - \$110{,}000 \\ &= \$114{,}924.42 + \$1.166 - \$110{,}000 \\ &= \$6{,}090.42 \end{aligned}$$

Since NPW is + $6,090.42 which is > 0, the project should be undertaken

(*b*) No. Although total depreciation is the same, SOYD is larger in the early years when it is worth more. Therefore, the NPW would increase with SOYD making the project even more desirable.

Example E5.31. *A company expects to receive $ 30,000 each year for 15 years from the sale of a project. The initial investment required is $120,000. The expected expenses of manufacturing and selling the project will be $7,000 per year. Assuming straight-line depreciation, a 15-year useful life and no salvage value, determine the projected after-tax rate of return. Assume a 46% income tax rate and ignore the investment tax credit.*

Solution: Straight-line depreciation

$$D_t = \frac{P - F}{n} = \frac{1{,}20{,}0000}{15} = \$8{,}000 \text{ per year}$$

Year	BTCF	Straight line depreciation	Taxable income	46% Income tax	ATCF
0	– 120,000				– 120,000
1–15	+ 23,000	8,000	15,000	6,900	+ 16,100

Take the after-tax cash flow and compute the rate of return at which PW of costs equals PW of benefits \$120,000 = \$16,100 (*P*/*A*, *i* %, 15)

$$(P/A, i\ \%, 15) = \frac{1,20,000}{16,100} = 7.4534$$

From interest tables, $i = 11\%$ (approx).

Example E5.32. *A large company in the 48% income tax bracket purchased an asset for \$100,000. The asset will be depreciated over 10 years using 175% declining balance depreciation with a \$75,000 salvage value assumed. The asset is expected to generate benefits of \$15,000 per year. Calculate the after-tax cash flow for years 1, 2, and 3.*

Solution:

Year	BTCF	175% DB	Taxable income	48% tax	ATCF
0	– \$100,000	–	–	–	– 100,000
1	+ 15,000	17,500	– 2,500	+ 1,200	16,200
2	+ 15,000	7,500	+ 7,500	– 3,600	11,400
3	+ 15,000	0	+ 15,000	– 7,200	+ 7,800

Book value at the end of year 1 = \$100,000 – (1.75/10) (\$100,000)

Book value = \$100,000 – \$17,500 = \$ 82,500

Book value at the end of year 2 = (1.75/10) (\$82,500) = \$14,437.50

Book value = \$100,000 – \$17,500 – \$14,437.50 = \$68,062.50

Since this value is less than the salvage value, the asset may only be depreciated by \$82,500 – \$75,000 (salvage value).

5.6 PROFITABILITY OF INVESTMENTS

Profitability is a measure of the total income for a project compared to the total outlay. Profitability can be applied to year by year or to the complete life of a project. Profitability is applicable to a single cost center, a complete project or to an entire organization. Legal, political or social consideration or issues of a corporate image often outweigh financial considerations in making decisions on which project to pursue. Four methods of evaluating profitability are commonly used. They are:

1. *Payout time*: The criterion is expressed as time that is time to recover the investment.
   (*a*) Without interest
   (*b*) With interest
   (*c*) Equivalent maximum investment period
   (*d*) Positive equivalent recovery period
2. *Return on investment (ROI)*: Uses rate of return but ignores the timing of cash flows.
   (*a*) Return on originals investment (ROI)
   (*b*) Return on average investment (RAI)
3. *Present value (PV)*: Based on present value recognizes the timing of cash flows.
   (*a*) Net present value (NPV)
   (*b*) Excess present-value index
4. *Discounted-cash-flow rate of return (DCFRR)*: Uses rate of return as a criterion and recognizes the timing of cash flows.

Of these four methods, the net present value (NPV) and the discounted-cash-flow rate of return (DCFRR) methods are the most commonly used methods for profitability since both these methods consider the time value of money.

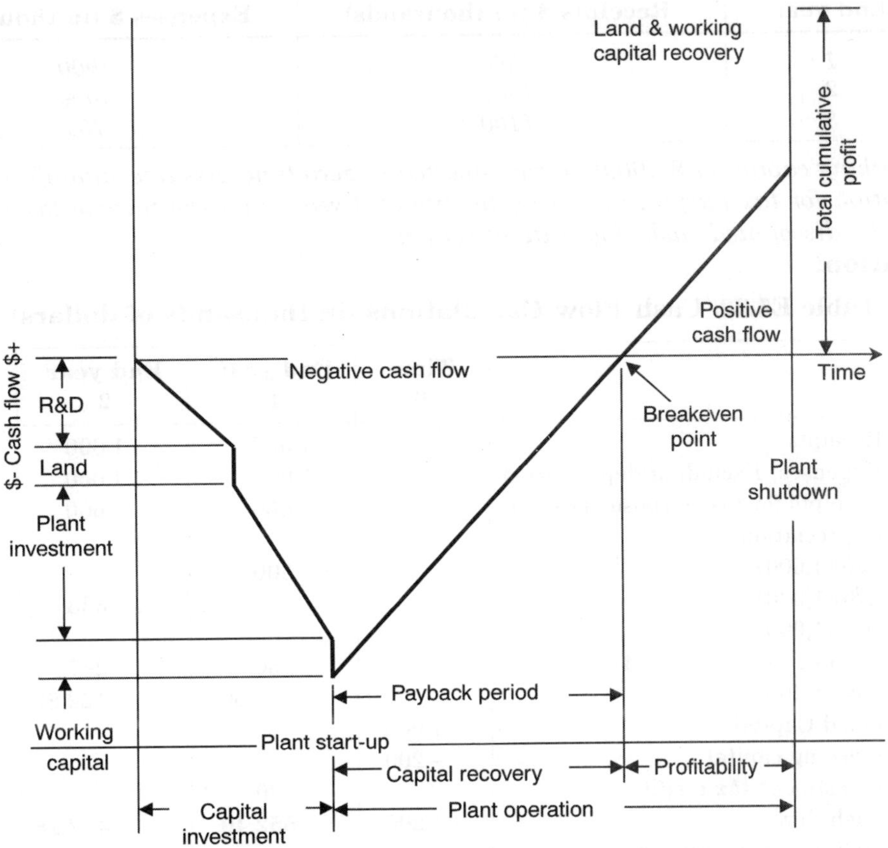

Fig. 5.10 Typical Costs in the Cycle of a Plant Investment.

Cash flow is the clue to profitability analyses. Cash flow measures the flow of cash into or out of a project. The cash flow can be defined as

Cash flow = net annual cash income + depreciation.

Cash flow can be positive or a negative quantity. Cash flowing in constitute positive cash flow while cash flowing out is negative cash flow. Figure 5.10 shows the cash flow diagram for a typical plant. Depreciation is an expense for tax purposes, even though it is not an out-of-pocket outlay at the time it is taken. Depletion also reduces the tax base with no actual expenditure being made for the period. The salvage value is not depreciable, but it does reduce the depreciable cost of the plant.

The *investment tax credit* (ITC) is meant to stimulate investment by providing reduced taxation in the year in which an asset is placed in service. The ITC permits taxpayers to claim a credit, not just a deduction, against income taxes for qualified investment in certain new or used depreciable property. The regular credit allowable is 10% of the eligible investment. However, if the 10% ITC was claimed, the cost basis of the asset was to be reduced immediately by 5%.

Example E5.33. *A particular venture by a company requires $1million now, with an investment tax credit of $20,000 at the end of the first year and no salvage value. The details of receipts and expenses, excluding depreciation are given below.*

End year	Receipts $ (in thousands)	Expenses $ (in thousands)
1	*1560*	*1000*
2	*1600*	*1000*
3	*1100*	*700*

A working capital of $ 200,000 was invested at zero time. Assume sum-of-the-years-digits depreciation for tax purposes for a useful life of 3 years and the income tax rate of 46%. Determine the discounted cash flow rate of return.

Solution:

Table E5.33. Cash Flow Calculations (in thousands of dollars)

	Time 0	End year 1	End year 2	End year 3
1. Receipts		1,560	1,600	1,100
2. Expenses, Excluding depreciation		– 1,000	– 1,000	– 700
3. Receipts for tax purpose (1) + (2)		560	600	400
4. Depreciation				
(3/6)(1,000)		500		
(2/6)(1,000)			333	
(1/6)(1,000)				167
5. Taxable income (3) – (4)		60	267	273
6. Tax at 46%		27.60	122.82	125.58
7. Fixed Capital	– 1,000			0
8. Working capital	– 200			200
9. Investment tax credit		20		
10. Cash flow (3) – (6) + (7) + (8) + (9)	– 1,200	552.40	477.18	474.42

Table E5.33 shows the cash flow calculations

For the present value of the cash flow to be zero

$$P = 0 = -\$1{,}200 + \frac{552.40}{(1+r)} + \frac{477.18}{(1+r)^2} + \frac{474.42}{(1+r)^3}$$

$$P = \begin{cases} -\$46.8967 \text{ for } r & = 0.15 \\ +\$11.3019 \text{ for } r & = 0.12 \\ -\$8.6516 \text{ for } r & = 0.13 \end{cases}$$

or by interpolation, $P = 0$ for $r = 12.82$.

Hence, the discounted cash flow rate of return is 12.82%.

5.6.1 Payout Time

5.6.1.1 Payout time without interest

The payout time is also known as payback time. Payback period without interest is defined as the length of time required to recover the first cost of an investment from the net cash flow produced by that investment for an interest rate of zero. It is the time required to reduce the investment to zero.

Payback period is defined as the smallest value of n that satisfies the equation

$$\sum_{t=0}^{n} F_t \geq 0 \qquad ...(5.90)$$

where,

F_0 = first cost of the investment

F_t = the net cash flow in period t.

The deficiency of the payback period is that it fails to consider the time value of money. Also the method does not consider the consequences of the investment following the payback period, including the magnitude and timing of the cash flows and the expected life of the investment. Payout time without interest does not consider working capital.

Example E5.34. *A sample project A with cash flows is given in Table E5.34. The cash flow here is the sum of the profit after taxes plus depreciation. Determine the payout time.*

Solution: The cumulative cash flows are shown in Table E5.34 (*a*).

The investment will be reduced to zero between 2 and 3 years, or approximately 2.334 years by interpolation. The values are also plotted in Fig. E5.34.

Table E5.34
Cash Flow for Project A

Time end year	After-tax profit, k$	Depreciation, k$	Cash flow, k$
0	– 1,000	0	– 1,000
1	200	200	400
2	300	200	500
3	100	200	300
4	50	200	250
5	0	200	200

Table E5.34(*a*)
Cumulative Cash Flow, k$

Time end year	Cumulative cash flow, k$
0	– 1,000
1	– 600
2	– 100
3	200
4	450
5	650

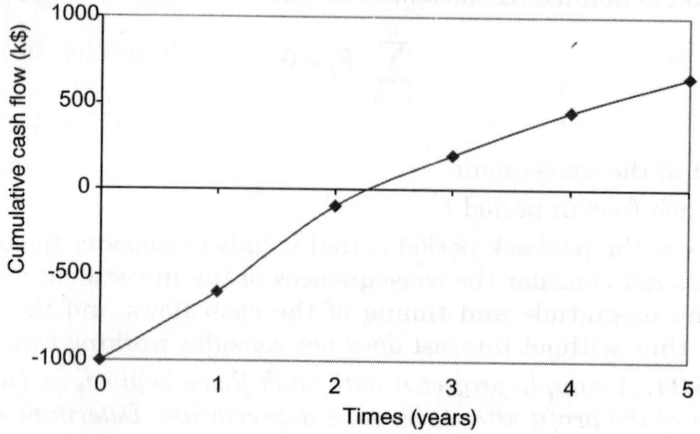

Fig. E5.34 Cumulative Cash Flow.

5.6.1.2 Payout time with interest

Payback with interest determines the length of time required until the investment's equivalent receipts exceed the equivalent capital outlays. The discounted payback period is the smallest value, n^1 that satisfies the relationship

$$\sum_{t=0}^{n^1} F_t (1+i)^{-1} \geq 0 \qquad \text{...(5.91)}$$

where,

F_t = the net cash flow in period t

i = interest rate (% per year)

n^1 = payback period (years)

Example E5.35. *Determine the payback time with interest on the remaining investment at 10% per year in Example E5.33.*

Solution: Table E5.35 shows the payout time is between 2 and 3 years or, 2.989 years by interpolation. The investment for each year is the investment for the previous year less the cash flow after the investment charge for the previous year.

Table E5.35. Cumulative Net Cash Flow

End year (1)	Investment for year (2)	Charge on investment for year (3) 10% of (2)	Cash flow (4)	Cash flow after investment charge (5) = (4) – (3)	Cumulative net cash flow (6) = (5) – (2)
0			– 1,000		– 1,000
1	1,000	100	400	300	– 700
2	700	70	500	430	– 270
3	270	27	300	273	+ 3

5.6.1.3 Equivalent Maximum investment period

The equivalent maximum investment period is a criterion for disclosing how fast the investment is reduced.

Example E5.36. *Determine the equivalent maximum period for the data in Table E5.33.*

Solution: Referring to Table E5.34(*a*).

The average investment for the 1st year is

$$\frac{1000 + 600}{2} = \$800$$

For 0 to 2.334 years, the cumulative value for the product of investment multiplied by time is

$$\left(\frac{1000 + 525}{2}\right)(1) = \$762.5$$

$$\left(\frac{600 + 100}{2}\right)(1) = \$350$$

$$\left(\frac{100 + 0}{2}\right)(0.334) = \$16.70$$

or a total of 1129.20 k\$/year.

The initial investment of 1000k\$ is therefore on the average $\frac{1129.20}{1000} = 1.129$ years.

Hence, the average time of 1.129 years for the initial investment supplements the 2.334 years required to retrieve the initial investment.

5.6.1.4 Positive equivalent recovery period

The positive equivalent recovery period addresses the period after the payout time. Referring to period 8.2 the data can be extended beyond 5 years by extrapolation. By trial, it is found that at 5.25 years the same area exists above the zero axis and the curve form 2.334 to 5.25 years. Hence, at 5.25 years the positive area of the curve between 2.334 and 5.25 years is the same as the negative area of the curve between 0 and 2.334 years.

5.6.2 Return on Investment

5.6.2.1 Return on original investment (ROI)

The rate of return on the investment (ROI) is the simplest measure of profitability. Again, it does not consider the time value of money. Return on original investment, also called the *Dupont or engineers method* is the percentage relationship of the average annual profit of the original investment, including such items as working capital. Hence,

$$\text{ROI} = \frac{\text{Average yearly profit during earning life}}{\text{Original fixed investment + working capital}} \times 100 \quad \text{...(5.92)}$$

There are many ways to assess the rate of return on the capital investment. ROI can be based on net annual profit before taxes, net annual profit after taxes, annual cash income before taxes and annual cash income after taxes. These ratios may be computed for each year or over the life of the project.

Example E5.37. *Determine the return on original investment for the data in Table E5.35.*

Solution: The average profit is

$$\frac{200 + 300 + 100 + 50 + 0}{5} = 130 \text{ k\$ /year}$$

Applying Eq. (5.92), the return on original investment is

$$\frac{130}{1000 + 0}(100) = 13 \text{ per cent per year.}$$

Note that the time value of money is not considered since only the average profit was used and not its timing.

5.6.2.2 Return on average investment (RAI)

The return on average investment is very similar to the return on original investment. The denominator in Eq.(5.92) is the average outstanding investment for this case.

Example E5.38. *Determine the return on average investment for the data in Table E5.34.*

Solution:

The average investment calculations are summarized below.

Year	Investment, k$
1	1,000
2	1,000 – 200 = 800
3	800 – 200 = 600
4	600 – 200 = 400
5	400 – 200 = 200
	Average = 600

Hence,

$$\frac{130}{600}(100) = 21.667 \text{ per cent per year.}$$

The return on average investment at 21.667% is higher than the return on original investment at 13%.

5.6.3 Present Value

5.6.3.1 Net present value

Net present value (NPV) method includes the time value of money since all cash flows are referred to zero time before an evaluation is made. This method also used to be known as *venture worth* method. The minimum acceptable rate of return (MARR) on capital is used for the reduction to present value discounting at a fixed rate. Net present values = present value of all inflows–present value of all outflows.

By this technique the expected net present value is a single amount referred to zero time (both +ve and –ve) through the life of the project. The project with greatest positive value of NPV is preferred. NPV depends upon the project life and therefore the NPVs of two projects should be compared if the projects have the same service life. Low interest rate values will tend to make NPV more positive for a given set of cash flows, and large values of interest will push NPV in a negative direction.

The value of interest rate for which the sum of the discounted cash flows equals zero, NPV = 0 is called the discounted cash flow rate of return.

Example E5.39. *Determine the net present value at 9%, 18% and 20% per year for the data in Table E5.35.*

Solution: At 9% per year rate of interest, the net present value (NPV) is

$$= \frac{-1{,}000}{(1.09)^0} + \frac{400}{(1.09)^1} + \frac{500}{(1.09)^2} + \frac{300}{(1.09)^3} + \frac{250}{(1.09)^4} + \frac{200}{(1.09)^5}$$

$$= -1000 + 366.973 + 420.840 + 231.655 + 177.106 + 129.986$$

$$= 326.560$$

At 18% per year, the net present value (NPV) is

$$= \frac{-1{,}000}{(1.18)^0} + \frac{400}{(1.18)^1} + \frac{500}{(1.18)^2} + \frac{300}{(1.18)^3} + \frac{250}{(1.18)^4} + \frac{200}{(1.18)^5}$$

$$= -1000 + 338.983 + 359.092 + 182.589 + 128.947 + 87.422$$

$$= 97.033$$

At 20% per year, the net present value (NPV) is

$$= \frac{-1{,}000}{(1.20)^0} + \frac{400}{(1.20)^1} + \frac{500}{(1.20)^2} + \frac{300}{(1.20)^3} + \frac{250}{(1.20)^4} + \frac{200}{(1.20)^5}$$

$$= -1000 + 333.334 + 347.222 + 241.048 + 120.564 + 80.376$$

$$= -122.544$$

From the above calculations, we notice that the project has a positive values or premium net present values (NPV) at 9% and 18% per year and a negative or deficiency in NPV value at 20% per year. The negative present value at 20% per year indicates that the project does not pay off at 20% per year.

5.6.3.2 Excess present value index

The excess present value index is defined as the ratio of the present value of receipts and the present value of outlays. Thus the net present value can be corrected for size using this approach.

The excess present value index for the Example E5.39 at 18% per year is

$$\frac{1097.033}{1000} = 1.097$$

If the projects have different service lives, the net present value should be corrected to annual cost or a capitalized cost so that both will have built-in factors for different service lives.

Example E5.40. *Compare project A with an initial investment of 1000 k\$, a 5-year service life, and a net present value (NPV) of 326.56 k\$, with project B, having an initial investment of 1600 k\$, a 7-year service life, and a net present value of 750 k\$. Both present values are at 9% per year. Also, if the initial investment is 1 k\$ for both projects, make a correction for the difference in size and duration by changing to a premium unicost.*

Solution: Project A:

$$(A/P, 9\%, 5) = 0.3265\ (0.2571) = 0.0839$$

Project B:

$$(A/P, 9\%, 7) = 0.4688\ (0.1987) = 0.0932$$

From the above, we notice that project A has premium of 0.0839 k\$/year per 1 k\$ of original investment against 0.0932 for project B. Hence, project B is more profitable than project A at 9% per year. The profitability changes with the interest rate used.

5.6.4 Discounted-Cash Flow Rate of Return (DCFRR)

When compound interest calculations are used, the measures of investment worth are referred to as *discounted cash flow (DCF) measures* and the methods used to compute the values of the investment worth are called *discounted cash flow methods.* Discounted cash flow (DCF) is the rate of return for which the net present worth equals zero. That is

$$\text{PW of benefits} - \text{PW of costs} = 0$$

or

$$\frac{\text{PW of benefits}}{\text{PW of costs}} = 1$$

and PW of benefits = PW of costs

Depreciation is considered implicitly in NPW and DCF calculations through the definition of cash flow. The discounted cash flow rate of return (DCFRR) is a criterion used for profitability that corrects for the differences in the duration and size of projects, including the time value of money and timing of all cash flows. This method is also known as the *interest-rate-of-return method, profitability index (PI), Interval rate of return,* and *investors method*. It includes all cash flows over the entire service life of the project.

Generally, a trial and error calculation is needed to find the compound interest rate at which the sum of the time-adjusted cash outflows equals the sum of the time-adjusted cash inflows. That is

$$\sum_{n=1}^{n} \frac{\text{cash flow for year } n}{(1+r)^n} = 0 \qquad ...(5.93)$$

Equation. (5.93) states that the present value of all cash flows is zero at the discounted cash flow rate of return.

Example E5.41. *Determine the discounted cash flow rate of return for data in Table E5.34.*

Solution: Referring to Eq. (5.93), the present value of all cash flows must be zero. Therefore,

$$\text{NPV} = 0 = \frac{-1000}{(1+r)^0} + \frac{400}{(1+r)^1} + \frac{500}{(1+r)^2} + \frac{300}{(1+r)^3} + \frac{250}{(1+r)^4} + \frac{200}{(1+r)^5}$$

From Example E5.35, we have

$$\text{NPV} = 97.033 \qquad \text{for } r = 0.18$$
$$= -122.544 \qquad \text{for } r = 0.20$$

By interpolation, NPV = 0 for r = 18.884.

Hence, the discounted cash flow rate of return is 18.884% per year.

5.7 INFLATION

Inflation exists when the prices of goods and services are increasing. Inflation is characterized by a decrease in the purchasing power of money caused by an increase in general price levels of goods and services without an increase in the value of goods and services. The price level increases are referred to as inflation and the price level decreases as deflation. When the purchasing power of money increases rather than decreases as time passes, the result is *deflation*. Although deflation is very rare, but nevertheless, it potentially exists. Deflation makes future dollars more worth or valuable than the current dollar. Under deflationary conditions, the deflation rate can be accommodated in engineering economic analysis as a negative inflation rate. Inflation plays an important role in today's engineering economic planning. That is, in order to select the most economical from among many alternatives, the inflationary loss in purchasing power of future income and costs must be accounted for and real, incremented rates of return calculated to determine the true anticipated benefits of proposed projects.

Interest rates and inflation are directly related. The basic interest rate is about 2 to 3 per cent higher than the inflation rate. In a period of high inflation, the purchasing power of dollar reduces and the cost of borrowing money rises.

5.7.1 Sources of Inflation

The following factors are considered to influence inflation either individually or jointly.

(*a*) *Money Supply:* If there is over supply of money in the system against the goods and services to purchase with that money, then it tends to decrease the value of dollars. The Federal Reserve regulates the increase in volume of money in the system at the same rate that the economy is growing.

(*b*) *Exchange Rates:* The profitability of the international corporations in the world markets is greatly influenced by the strength of the U.S. dollar. If the corporations' profits are weakened due to exchange rates in some markets, then prices are raised in other markets in order to compensate.

(*c*) *Cost-Push Inflation:* This situation arises whenever the producers of goods and services *push* their climbing operating costs (such as fabrication/manufacturing, marketing, and sales, etc.) to the customers through higher prices.

(*d*) *Demand-Pull Inflation*: This effect is caused by the consumers who are on *Spending Spree* at the expense of their savings. The prices of those goods and services will rise due to the demand exceeding the supply.

5.7.2 Definitions

To measure and illustrate how inflation and price-level changes affect two quantities: interest rates and cash flows. It is necessary to calculate a price index. A *price index* is the ratio of the historical price of some commodity or service at some point in time to the price at some earlier point.

It is desirable for interest rates and rates of returns to reflect both the time value of money and the effects of inflation. If we denote the time value of money (or desired returns) in the absence of inflation the real interest rate by d, the combined interest rate by i, and the inflation rate by j, then these three factors are related by

$$1 + i = (1 + d)(1 + j) \qquad \text{...(5.94)}$$

or

$$I = d + j + dj$$

Example E5.42. *If the combined interest rate is 25% per year and the real time value of money is 20% per year, determine the inflation rate.*

Solution:

$$(1 + i) = (1 + j)(1 + d)$$

$$1 + 0.025 = (1 + j)(1 + 0.20)$$

$$(1 + j) = \frac{1.25}{1.20}$$

or

$$j = 1.04167 - 1.0 = 0.04167 = 4.167\%$$

Price indexes are tracked for specific commodities as well as for bundles of commodities. As a result, in estimating the value of the inflation rate, data are collected and analyzed by the Department of Commerce (Bureau of Economic Analysis) and the Department of Labor (Bureau of Labor Statistics). A number of composite indexes are used to forecast the inflation rate. These composite indexes include:

(*a*) Consumer Price Index (CPI)

(*b*) Producer Price Indexes (PPI), also known as the Wholesale Price Indexes (WPI).

The Consumer Price Index represents the change in retail price for a selected *market basket*, including food, housing, clothing, transportation and utilities. This index measures the changes in retail prices needed in order to maintain a standard (or fixed) of living for an average consumer. Figures 5.11 and 5.12 show the Consumer Price Index and the Producer Price Index, the inflation rates for the year 1967 to 2000 are given in Fig. 5.11 in which the year 1967 was taken as the base year. The variations in these annual rates from 1967 through 2000 are shown in Figure 5.13.

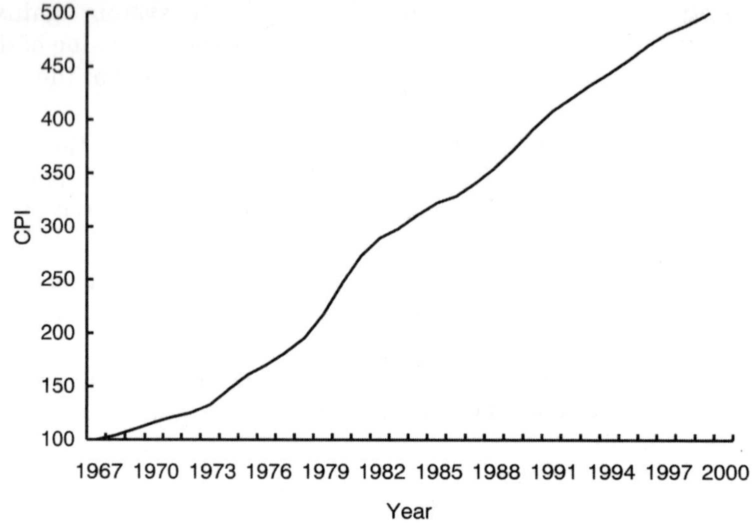

Fig. 5.11 Consumer Price Index (1967 = 100).
Source: **Statistical Abstract of the United States.**

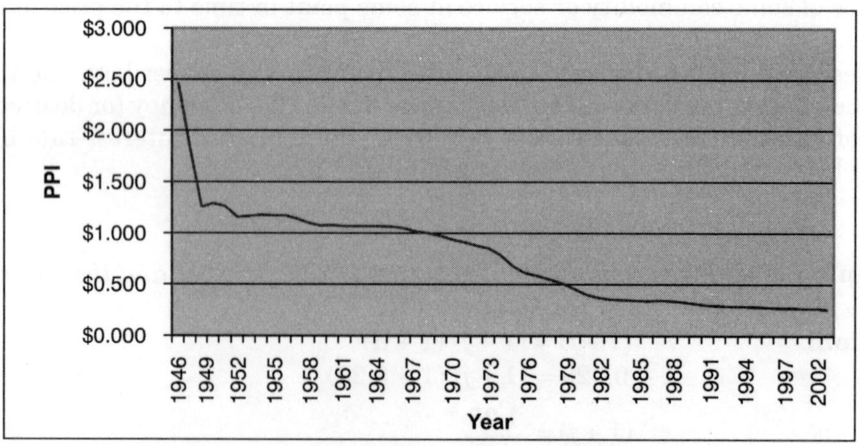

Fig. 5.12 Producer Price Index – All Commodities (1967 = 100).
Source: **Statistical Abstract of the United States.**

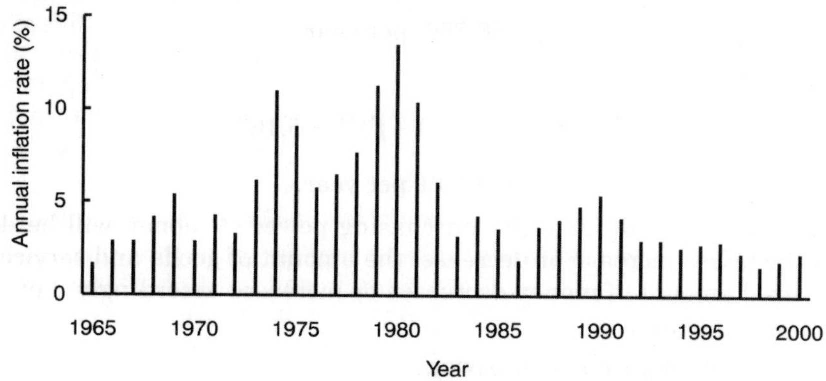

Fig. 5.13 Annual Inflation Rates, 1965–2000.

In addition to consumer prices, producer prices and whole prices, the Department of Commerce and the Department of Labor publish a wide range of economic data for a variety of industry groups. Depending on the specific industry and its needs, inflation estimates are more accurately developed using one or more such sources.

The annual inflation rates can be calculated using the CPI values given in Fig. 5.11.

Annual inflation rate for year

$$t + 1 = \frac{\text{CPI}_{t+1} - \text{CPI}_t}{\text{CPI}_t} \quad ...(5.95)$$

where,

CPI_t = Consumer price index at the end of year t.

CPI_{t+1} = Consumer price index at the end of year $t + 1$.

The average inflation rate from the end of the year t to the end of year $t + n$ is calculated as follows.

$$\text{CPI}_t(i + \bar{f})^n = \text{CPI}_{t+n} \quad ...(5.96)$$

where

$\bar{f}$ = the average annual inflation rate

n = the number of years $[(t + n) - t]$

CPI_t = Consumer price index at the end of year t.

CPI_{t+n} = Consumer price index at the end of year $t + n$.

Example E5.43. *Calculate the average inflation rate from the end of 1967 to the end of 1985 (15 years) and from the end of 1965 through 1998 (10 years).*

Solution: (1967 to 1985)

Set $\bar{f}$ = the average annual inflation rate

Then

$$100(1+\bar{f})^{18} = 322.2$$

$$(1+\bar{f})^{18} = 3.222$$

or

$$(1+\bar{f}) = (3.222)^{-18}$$

$$\bar{f} = 6.72\% \text{ per year}$$

1965 through 1998:

$$94.5(1+\bar{f})^{33} = 488.3 \quad \text{or} \quad (1+\bar{f})^{33} = 5.1672$$

or

$$\bar{f} = 5.10\% \text{ per year.}$$

Under inflationary conditions, the purchasing power of money will be decreasing. In other words, as the prices increase or decrease, the amount of goods and services that can be purchased for a fixed amount of money decreases or increases accordingly. Let

f = rate of price increase

k = the rate of loss of purchasing power

$\bar{f}$ = the rate of increase in price per year

$\bar{k}$ = the rate of loss in purchasing power per year

n = number of years

Then from Eq. (5.96)

$$\text{CPI}_t (1+\bar{f})^n = \text{CPI}_{t+n} \quad \text{...(5.97)}$$

But by definition,

$$\text{Purchasing Power}_t (1-\bar{k})^n = \text{Purchasing Power}_{t+n}$$

or

$$\frac{\text{CPI}_{\text{base year}}(1-\bar{k})^n}{\text{CPI}_t} = \frac{\text{CPI}_{\text{base year}}}{\text{CPI}_{t+n}}$$

That is

$$\text{CPI}_t = (1-\bar{k})^n \,\text{CPI}_{t+n} \quad \text{...(5.98)}$$

Hence

$$(1+\bar{f})^n = \frac{1}{(1-\bar{k})^n} \quad \text{...(5.99)}$$

Example E5.44. *Based on the CPI values in Fig. 5.11, compute the average inflation rate and the rate of loss of purchasing power from the end of year 1991 through 1999 years.*

Solution: From Fig. 5.11

$$408.0(1+\bar{f})^8 = 497.6$$

or

$$(1+\bar{f})^8 = 0.201$$

$$(1+\bar{f})^8 = \frac{497.6}{408.0} = 1.2196$$

or

$$\bar{f} = 0.0251$$

From Fig. 5.11

$$0.245(1-\bar{k})^8 = 0.201$$

$$(1-\bar{k})^8 = \frac{0.201}{0.245} = 0.8204$$

or

$$\bar{k} = 0.0244$$

Thus, a $\bar{f} = 0.0251$ per year increase in prices has the same effect as a $\bar{k} = 0.0244$ per year loss in purchasing power.

The following definitions are required to distinguish the market interest rate, the inflation-free rate, and the inflation rate and to illustrate how inflation and price changes affect two quantities: interest rate and cash flows.

Inflation Rate (*f*): Inflation rate is the annual percentage of increase in prices of goods and services. It captures the effect of goods and services costing more or a decrease in the purchasing power of dollars. Inflation rate is also known by other names: escalation rate, rate of increase in cost of living.

Inflation-Free Rate (*i′*): Inflation free interest rate represents the earning power of money with the effects of inflation removed. Inflation-free interest is also called the *real interest rate* and *constant-dollar interest rate*.

Market Interest Rate (*i*): The market interest rate is the rate of interest one obtains in the market place and represents the opportunity to earn as reflected by the actual rates of interest available in finance and business. When a corporation rates its minimum attractive rate of return (MARR), it is usually a market rate of interest. This rate is sometimes called the *combined interest rate, current dollar interest rate, actual interest rate* and *inflated interest rate*. Market interest rate has in it the *real growth of money* and the *effect of inflation*.

The relationship between the inflation, real and market interest rates is given by:

$$i = i' + f + i'f$$

where,

i = market interest rate

i' = inflation free or real interest rate

f = inflation rate

Example E5.45. *How much insurance should a person buy if he wants to leave enough money to his family, so they get \$50,000 per year in interest, of constant year 0 value dollars? The anticipated interest rate from bank is 10% while the inflation rate is expected to be 3% per year.*

Solution: The actual (effective) rate that the family will be getting is

$$i' = \frac{i-f}{i+f} = \frac{0.10-0.03}{1.03} = 0.06796 \quad \text{or} \quad 6.796\%$$

To calculate P, $n = \infty$ (capitalized cost)

$$P = \frac{A}{i'} = \frac{50{,}000}{0.06796} = \$735{,}727$$

Therefore, he needs to buy about \$736,000 of life insurance.

5.7.3 Cash Flows in Actual or Constant Dollars

Here we consider the effects of inflation on cash flows. Cash flows can be represented in terms of either *actual dollars* or *constant dollars*.

Actual Dollars (A\$): Actual dollars represent the out-of-pocket dollars received or disbursed at any point in time and is measured by summing the denominations of the money paid or received. Actual dollars are sometimes called *then-current dollars, current dollars,* *future dollars, escalated dollars* and *inflated dollars* because they carry any inflation that has reduced their growth.

Real Dollars (R$): Real dollars represent the hypothetical purchasing power of future receipts and disbursements in terms of the purchasing power of dollars at some base year. The base year is arbitrarily selected. Real dollars are also called *constant purchasing dollars*, or *constant dollars, deflated dollars, today's dollar, zero date dollars*, and *inflation-free dollars* because they do not carry the effects of inflation.

Figure 5.14 shows the relationship between *i*, *i'*, *f*, *A*$ and *R*$. Economic studies can be done in either actual dollars (*A*$) or real dollars (*R*$). The interest rate that is applicable depends on the domain in which the calculations are done and the principles are summarized in Table 5.4.

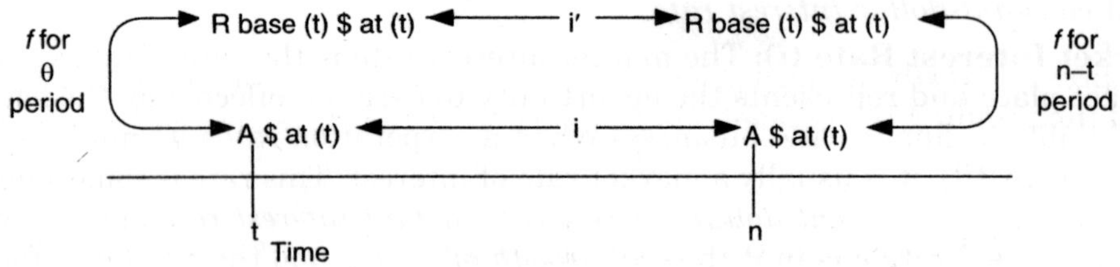

Fig. 5.14 Relationship between *i*, *f*, *i* ', *A*$ and *R*$.

Example E5.46. *A corporation has an option to purchase 5 years from now a complex electronic system that presently is priced at \$200,000. The option states that the price at that time will be adjusted for inflation, assuming inflation at the rate of 4% per year. The corporation had net earnings last year of \$800,000, which is assumed to have occurred at the present. The corporation's earnings are expected to increase at the rate of 25% per year, and the rate at which they can be invested is 10% compounded annually. What fixed percentage of net earnings must be saved at the end of each of the next 5 years so that the complex electronic system can be purchased 5 years from now?*

Solution: $$(\$800{,}000)(x)\left[\begin{array}{r}(1.25)(F/P,10\%,4)+(1.25)^2(F/P,10\%,3)\\ +(1.25)^3\,(F/P,10\%,2)+(1.25)^4\,(F/P,10\%1)+(1.25)^5\end{array}\right]$$

$$= \$200{,}000(F/P,\ 4\%,\ 5)\ (\$800{,}000)(x)\ [(1.25)(1.464 + (1.25)^2(1.331) + (1.25)^3(1.210) + (1.25)^4(1.1) + (1.125)^5$$

$$= \$200{,}000(1.611)\ (800{,}000)(x)[1.83 + 2.0797 + 2.3632 + 2.6855 + 3.0517]$$

$$= 322{,}200\ (8000{,}000)\ (x)\ (12.0101) = 322{,}200$$

or $$x = 0.03353 \quad \text{or} \quad 3.353\%$$

Therefore, the corporation must save 3.353% of the net earnings at the end of each of the next years so that the electronic system can be purchased 5 years from now.

5.7.4 Actual Dollar (A$) versus Constant-Dollar Analysis

Analyzing inflation in investment can be done in either actual dollars (*A*$) or constant dollars. Constant (real) dollar analysis requires the use of the inflation-free interest rate or the real interest rate (*i'*) for discounting while the actual dollar or the then current dollar analysis requires the market interest rate or the combined interest rate (*i*).

Example E5.47. *Calculate the present worth of a uniform series of payments of $2,000 per year to 5 years if the real interest rate is 10% per year and the inflation rate is 5% per year, assuming that the payments are in terms of (a) today's dollars, and (b) future dollars.*

Solution: (*a*) Since the dollars are today's dollars, the present worth is determined using the real i = 10% [Refer to Figure E5.47(*a*)].

$$PW = -\$2{,}000(P/A, 10\%, 5) = -\$2{,}000(3.7908)$$
$$= \$7{,}581.60$$

(*b*) Since the dollars are expressed in future dollars, use an inflated interest rate in the *P/A* factor.

$$i' = i + f + if = 0.10 + 0.05 + (0.10)(0.05) = 0.155 \quad \text{or} \quad 15.5\%$$
$$PW = -\$2{,}000(P/A, 15.5\%, 5) = -\$2{,}000(3.3133)$$
$$= \$6{,}626.50$$

The present worth can also be obtained by converting the future cash flows into today's dollars and then finding the present worth using the real interest rate of 10% [Refer to Figure E5.47(*b*)].

$$PW = -\$1904.76\left(\frac{P}{F},10\%,1\right) - \$1814.06\left(\frac{P}{F},10\%,2\right) - \$1727.68\left(\frac{P}{F},10\%,3\right)$$
$$- \$1645.40\left(\frac{P}{F},10\%,4\right) - \$1567.05\left(\frac{P}{F},10\%,5\right)$$
$$= -\$1904.76(0.9091) - \$1814.06(0.8264) - \$1727.68(0.7513)$$
$$- \$1645.40(0.6830) - \$1567.05(0.6209)$$
$$= -\$1731.62 - \$1499.14 - \$1298.00 - \$1123.81 - \$972.98$$
$$= -\$6{,}626.50$$

This is the same result obtained using i' as above.

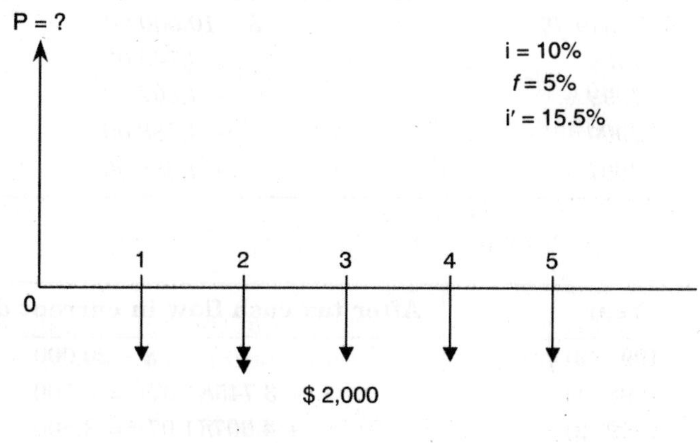

Fig. E5.47 (a) Cash Flow – Future Dollars.

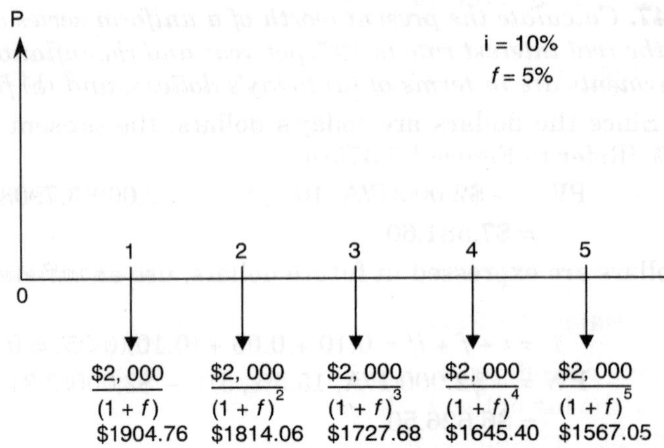

Fig. E5.47 (b) Cash Flow – Today's Dollars.

5.7.5 Different Inflation Rates for Cash Flow

In some situations, the inflation rates for the various cash flows for different elements contributing to the overall cash flow are found to inflate at different rates. Some material costs may be experiencing price changes at one rate while other material costs are changing at a different rate. Similarly, overhead and labor costs may be increasing or decreasing at their own particular rates. The procedure is to determine the component cash flows in actual dollars and then combine them to find the overall actual dollar cash flows.

Example E5.48. *Determine the internal rate of return based on constant (year 0) dollars for the followings after-tax cash flow given in current or actual dollars. Inflation is assumed to be 7% per year.*

Year	After tax cash flow in actual dollars
1997 (year 0)	*$ – 10,000.00*
1998	*+ 3,745.00*
1999	*+ 4,007.00*
2000	*+ 4,288.00*
2001	*+ 4,588.00*

Solution:

Year	After tax cash flow in current dollars
1997 (0)	$ – 20,000
1998 (1)	$+ 3,745/(1.07)^1 = 3,500$
1999 (2)	$+ 4,007/(1.07)^2 = 3,500$
2000 (3)	$+ 4,288/(1.07)^3 = 3,500$
2001 (4)	$+ 4,588/(1.07)^4 = 3,500$

$$\text{NPW} = 0 = -\ 10{,}000 + 3{,}500\ (P/A, i\%, 4)$$

$$(P/A, i\%, 4) = 10{,}000/3{,}500 = 2.857$$

Therefore, ROR = i = 15%.

Table 5.6. Principles for Applying Interest Rate

Cash flow domain	Interest rate to apply to find equivalences at different points in time
Actual dollars	
Convert between domains with inflation rate, f	Market interest rate, i
Constant dollars	Inflation-free interest rate, i

5.7.6 Different Inflation Rates Per Period

In Section 5.7.5 we have considered the case where the inflation rates for the various cash flows are changing over the life of the analysis period. In this section we address the situation where the interest rate for the same cash flow is changing over time. The procedure to follow for this situation is simply to apply the inflation rates in the years in which they are projected to occur or each of the individual cash flows for the entire period and then apply the measures of merit using the market interest rate. We can obtain the total marginal cost (in terms of actual dollars) by combining the individual yearly costs. We then calculate the equivalent uniform annual cost (EUAC) using the given market interest rate.

Example E5.49. *Determine the economic life of the following new mechanical system. The system costs $10,000 initially. First year maintenance cost is $1,000. Maintenance cost is forecast to increase $1,000 per year for each year after the first year. Assume the rate of interest as 13% and consider only the maintenance and capital recovery in the analysis.*

Solution:

Year	Maintenance cost	EUAC of capital recovery	EUAC of maintenance	Total EUAC
1	$1000	$11,300	$1,000	$12,300
2	2000	5,995	1,469	7,464
3	3000	4,235	1,919	6,154
4	4000	3,362	2,348	5,710
5	5000	2,843	2,757	5,600
6	6000	2,502	3,147	5,649
7	7000	2,261	3,517	5,778
8	8000	2,084	3,869	5,953

Economic life = 5 years (EUAC = minimum)

EUAC of Capital Recovery = $10,000 (A/P, 13%, n)

EUAC of Maintenance = $1,000 + 1,000 (A/G, 13%, n)

5.7.7 After-Tax Analysis under Inflation

When dealing with after-tax cash flows calculations under inflation, it should be realized that inflation affects some elements quite differently than the other elements. In Section 5.4 we have noted that depreciation charges are not affected by inflation, since depreciation charges represent the recover of an asset's original cost. Generally speaking, the revenues, operating and maintenance costs, and future salvage values are expressed in constant dollars, while repayment schedules associated in loans, and taxes paid are stated in actual dollars. The principle to follow is that when calculating after-tax cash flows from before-cash flows use actual dollar analysis.

Example E5.50. *A $15,000 investment will return annual benefits for six years with no salvage value at the end of six years. Assume a 46% income tax rate and straight-line depreciation. Solve the problem for both before and after-tax rates of return. The two situations for after-tax rate of return are:*

1. No inflation: the annual benefits are constant at $3,649 per year.

2. Inflation equal to 5%: the benefits from the investment increase at this same rate, such that they continue to be the equivalent of $3,500 in year-0 based dollars.

The benefit schedule for the two situations is given as follows:

Year	Annual benefit for both situations, in Year-0 based dollars	No inflation, actual dollars received	5% inflation factor	5% inflation, actual dollars received
1	$3649	$3649	$(1.05)^1$	$3831
2	$3649	$3649	$(1.05)^2$	$4023
3	$3649	$3649	$(1.05)^3$	$4224
4	$3649	$3649	$(1.05)^4$	$4435
5	$3649	$3649	$(1.05)^5$	$4657
6	$3649	$3649	$(1.05)^6$	$4890

*May be read from the 5% Compound Interest Table as (F/P, 5%, *n*).

Solution: Before-tax rate of return:

They have the same before-tax rate of return, since in both situations have annual benefit (no inflation and 5% inflation) stated in year-0 based dollars of $3500.

$$\text{PW of cost} = \text{PW of benefit}$$

$$\$15{,}000 = \$3469\ (P/A, i, 6)$$

or

$$(P/A, i, 6) = \$15000/\$3469 = 4.11$$

Hence, the before-tax rate of return equals to 12% (from interest tables of Appendix 5A, Section I) After-tax rate of return, no inflation:

Year	Before-tax cash flow	Straight line depreciation	Taxable income	46% income taxes	Actual dollars, and Year-0 based $ after-tax cash flow
0	– $15,000				$15,000
1–6	+ $3,649	$2500	$1149	– $529	+$3,120

$$\text{PW of cost} = \text{PW of benefit}$$

$$\$15{,}000 = \$3{,}120(P/A, i, 6)$$

$$(P/A, i, 6) = \$15{,}000/\$3120 = 4.81$$

Hence, after-tax rate of return equals 6.7%. (From Appendix 5A, Section I)

After-tax rate of return, 5% inflation:

Year	Before-tax cash flow	Straight-line depreciation	Taxable income	46% income taxes	Actual dollars, after-tax cash flow
0	–$15,000				–$15,000
1	+ $3,831	$2500	$1,331	– $612	+ $3,219
2	+ $4,023	$2500	$1,523	– $701	+ $3,322
3	+ $4,224	$2500	$1,724	– $793	+ $3,431
4	+ $4,435	$2500	$1,935	– $890	+ $3,545
5	+ $4,657	$2500	$2,157	– $992	+ $3,665
6	+ $4,890	$2500	$2,390	– $1,099	+ $3,791

Converting to year-0 based dollars and solving for the rate of return:

Year	Actual dollars, after-tax cash flow	Conversion factor	Year-0 based dollars, after-tax cash flow	Present worth at 5%	Present worth at 4%
0	– $15,000		–$15,000	–$15,000	–$15,000
1	+ $3,219	$x\ (1.05)^{-1} =$	$3,066	$2,920	+ $2,948
2	+ $3,322	$x\ (1.05)^{-2} =$	$3,013	$2,733	+ $2,786
3	+ $3,431	$x\ (1.05)^{-3} =$	$2,964	$2,560	+ $2,635
4	+ $3,545	$x\ (1.05)^{-4} =$	$2,916	$2,399	+ $2,493
5	+ $3,665	$x\ (1.05)^{-5} =$	$2,872	$2,250	+ $2,361
6	+ $3,791	$x\ (1.05)^{-6} =$	$2,829	$2,111	+ $2,236
Σ				–$27	+$459

Linear interpolation between 4% and 5%:

After-tax rate of return = 4% + 1%[459/(459+27)] = 4.94%

From the above example, we make the following observations:

1. The before-tax rate of return for both situations, namely, 5% inflation and no inflation is the same.

2. When future benefits fluctuate with changes in inflation or deflation, the effects do not alter the year-0 based dollar estimates.

3. After-tax calculations: With equal before-tax rates of return do not produce equal after-tax rates of return.

Situation	Before-tax rate of return	After-tax rate of return
No inflation	12%	6.70%
5% inflation	12%	4.94%

4. Even though the benefits increase at the same rate as the inflation, the 5% inflation results in a smaller after-tax rate of return.

5. Inflation gives increased taxable income and therefore larger income tax payments.

6. The after-tax cash flows in actual dollars are not large enough to offset both increased income taxes and inflation.

7. The year-0 based dollar after-tax cash flow with inflation is smaller than the year-0 based dollar after-tax cash flow without inflation.

5.8 BREAK-EVEN AND SENSITIVITY ANALYSIS

Break-even analysis is a technique of evaluating the effect of changes in the production output on costs, revenues and profits. Break-even analysis is a form of sensitivity analysis. The break-even point corresponds to the value for the parameter at which the project is marginally justified. This analysis is conducted whenever there is an uncertainty about one of the parameter in an economic analysis. Break-even analysis is used when a true estimate of a parameter is not known but certain judgements can be made as to whether or not the parameter's value is less than or greater than some break-even value. Hence, the term break-even is derived from the objective to determine the value of a given parameter for which there is neither a profit or nor a loss. *Break-even analysis* determines the conditions when the alternatives are equivalent and this is quite often depicted in the form of *break-even charts*. *Sensitivity analysis* is used to analyze the effects of making errors in estimating parameter values. *Sensitivity analysis* is used to determine how sensitive a decision is to estimates for the various parameters. Thus, sensitivity analysis is a thorough evaluation of a range of values for some parameters to assess their effect on a specific decision.

Break-even analysis can be applied for two situations:

1. *Profit analysis:* The break-even chart for this situation shows the effect of changes in output on costs and revenues. The break-even point corresponds to the output level at which the total costs equals revenues and the profit is zero.

2. *Production method cost analysis:* The break-even chart in this case shows the effect of changes in output level on the costs of different methods of production. The break-even point corresponds to the output level at which the costs of two production methods are the same.

Example E5.51. *Determine the break-even capital cost of Project B compared to project A if the interest rate is 9%.*

Year	A	B
0	*– 2000*	*?*
1–5	*+500/year*	*+300/year*

Solution: NPW of A = – \$2,000 + \$500(P/A, 9%, 5)

= –\$2,000 + \$500(3.8897) = \$55.15

NPW of A = NPW of B

\$55.15 = P_B + \$300(3.8897)

or P_B = –\$1,111.76

Example E5.52. *The cost of a new plant will cost \$10 million for the first stage now and \$12 million for the second stage at n years in the future. The cost to build the plant to full capacity now is \$14 million. All facilities will last until forty years from now regardless of when they are installed. The salvage value is zero. Assume the annual cost of operation and maintenance is the same for both two-stage construction, as well as full capacity construction. The annual interest rate is 10%. Use present worth calculations.*

1. *Plot the break-even chart*

2. *Determine the sensitivity of the decision to second stage construction 14 years or more in the future.*

Solution:

(*a*) Using present worth calculations, for two-stage construction

PW = \$10 m + \$12 m $(F_{A/P}, 8, n)$

n = 5 years: PW = 10 + 12(0.6806) = \$18.1672 m

n = 10 years: PW = 10 + 12(0.4632) = \$15.5584 m

n = 20 years: PW = 10 + 12(0.2145) = \$12.574 m

n = 30 years: PW = 10 + 12(0.0994) = \$11.1928 m

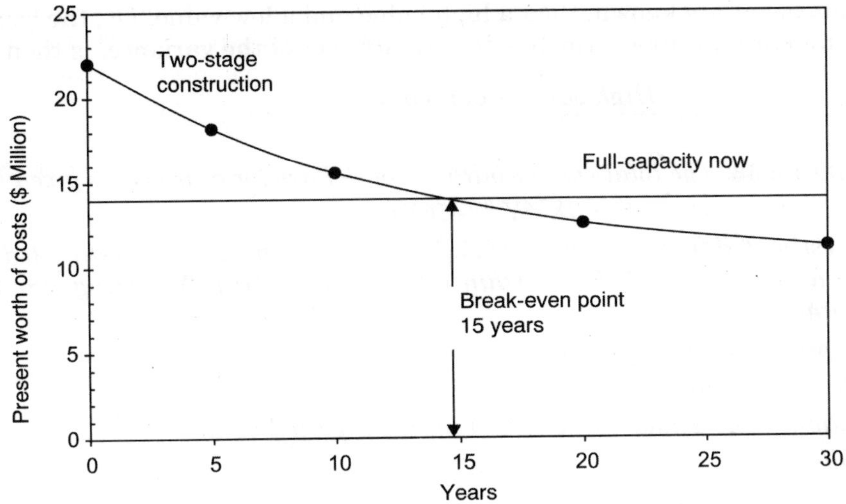

Fig. E5.52 Break-even Plot.

These results are shown in Figure E5.52. The break-even point (15 years) is the point at which both alternatives have the equivalent costs. If full capacity is needed before 15 years, then full capacity built now is the preferred action to be taken.

(*b*) The decision would be sensitive to the estimated value when the full capacity would be needed if the second stage construction were needed in 14 years or more.

5.9 UNCERTAINTY AND RISK ANALYSIS

In practice, decisions are based on future estimates of costs and benefits. Estimates are likely to contain uncertainties and may not be very precise. When a single estimate of a project is desired with an analysis of how the actual amount might vary from the expected value, such an analysis is known as *uncertainty analysis* and *risk analysis*. The procedure is to combine the uncertainty of the individual components to express uncertainty in the overall project. If the uncertainty in each component is assumed as normally distributed, then the variances are additive when they are weighted for their sensitivity to the overall effect.

If the estimate for a project is Z, then the variance for Z is given by:

$$\sigma_Z^2 = \left(\frac{\partial Z}{\partial A}\right)^2 \sigma_A^2 + \left(\frac{\partial Z}{\partial B}\right)^2 \sigma_B^2 + \left(\frac{\partial Z}{\partial C}\right)^2 \sigma_C^2 \quad \text{...(5.100)}$$

where $\frac{\partial Z}{\partial A}, \frac{\partial Z}{\partial B}$ and $\frac{\partial Z}{\partial C}$: the measures of $\frac{\Delta Z}{\Delta A}, \frac{\Delta Z}{\Delta B}$ and $\frac{\Delta Z}{\Delta C}$ which are the changes in Z for a unit charge in A, B and C respectively. They are the individual component sensitivities σ_A, σ_B, σ_C: the variances of A, B, and C respectively

σ_Z : variance of Z

$\left(\frac{\partial Z}{\partial A}\right)^2, \left(\frac{\partial Z}{\partial B}\right)^2$ *and* $\left(\frac{\partial Z}{\partial C}\right)^2$: *weighting factors for individual variances.* If only a few estimates for a component are known, then a high value and a low value, *i.e.*, the range can often be used. The standard duration, which is the square root of the variance, is then given by

$$\sigma = \frac{High\ value - Low\ value}{8} \qquad ...(5.101)$$

Example E5.53. *The total cost in millions of dollars for a project is given by:*

$$C = 3X^{1.4}Y^{-0.5}+16Y+F$$

where X is the estimated from five values: 2.5, 1.7, 1.4, 2.9, and 1.3. Y is estimated from a high value of 3 and a low value of 1. F is certain with a value of 0.4. Assuming standard normal distributions, find

(*a*) *the expected value for C*

(*b*) *sensitivity of C to X and Y*

(*c*) *variation of the actual value of C from the expected value*

(*d*) *variance of C*

Solution:

(*a*) The mean value for X is

$$(2.5 + 1.7 + 1.4 + 2.9 + 1.3)/5 = \frac{9.8}{5} = 1.96$$

The variance for X is given by

$$\sigma_X^2 = \frac{(2.5-1.96)^2 + (1.7-1.96)^2 + (1.4-1.96)^2 + (2.9-1.96)^2 + (1.3-1.96)^2}{(5-1)}$$

$$= \frac{(0.2916 + 0.0676 + 0.3136 + 0.8836 + 0.4356)}{4}$$

$$= 0.498$$

The mean for Y is $\frac{(3+1)}{2} = 2$ and the variance of Y is

$$\sigma_Y^2 = \left(\frac{3-1}{8}\right)^2 = 0.0625$$

The mean value of F is 0.4 and the variance is zero.

The expected value of C is given by:

$$C = 3(1.96)^{1.4}\ 2^{-0.5} + 16(2) + 0.4$$

$$= 5.4413 + 32 + 0.4 = 37.8413$$

(*b*) The sensitivity of C to X and Y is obtained from

$$\frac{\partial C}{\partial X} = 3(1.4)X^{0.4}\, Y^{-0.5} = 3(1.4)\,(1.96)^{0.4}\,(2)^{-0.5} = 3.8866$$

$$\frac{\partial C}{\partial Y} = 3X^{1.4}(-0.5)Y^{-1.5} + 16 = 3(1.96)^{1.4}\,(-0.5)\,(2)^{-1.5} + 16$$

$$= 3(2.5654)\,(-0.5)\,(0.3536) + 16$$

$$= 14.6395$$

Thus, C changes 3.8866 times as fast as X for a unit change in X and 14.6395 times as fast as Y for unit change in Y. A 1% change in $X(\Delta x = 0.0196)$ would cause a change in C equal to 3.8866(0.0196) = 0.07618 or (0.07618)(100)/37.8413 = 0.2013 per cent. Also, a 1% change in $Y(Y = 2)$ gives 14.6395(0.02) = 0.2928 or (0.2928) × (100)/37.8413 = 0.774 per cent error in C.

Hence, the total cost C is more sensitive to the estimate for Y than X.

(*c*) From Eq. (5.100)

$$\sigma^2 = (3.8866)^2\,(0.498) + (14.6395)^2\,(0.0625)$$

$$= 7.5226 + 13.3938$$

$$= 20.9164$$

or

$$\sigma = 4.5734$$

(*d*) Transforming the above information to a standard normal distribution, we have,

$$Z = \frac{C - 37.8413}{4.5734}$$

Assuming values for Z and from Table (Chapter 2), the following tabulation is obtained.

Z	Cumulative probability for Z or less	C
2.0	0.0228	28.6945
1.7	0.0446	30.0665
1.4	0.0808	31.4385
1.2	0.1151	32.3532
0.8	0.2119	34.1828
0.2	0.4207	36.9266
0.0	0.5000	37.8413
0.2	0.5793	38.0870
0.8	0.7881	41.5000
1.2	0.8849	43.3294
1.4	0.9192	44.2441
1.7	0.9554	45.6128
2.0	0.9773	46.9881

The plot of the cumulative probability for Z and C is shown in Fig. E5.53 from the plot, we notice that the probability that the project will cost \$34.1828 million or less is 0.2119 or 21.19%.

The probability of a cost of \$45.6128 million dollars or less is 0.9554 or 95.54%. The probability of a cost between \$34.1828 million dollars and \$45.6128 million dollars is

$$95.54 - 21.19 = 74.35\%$$

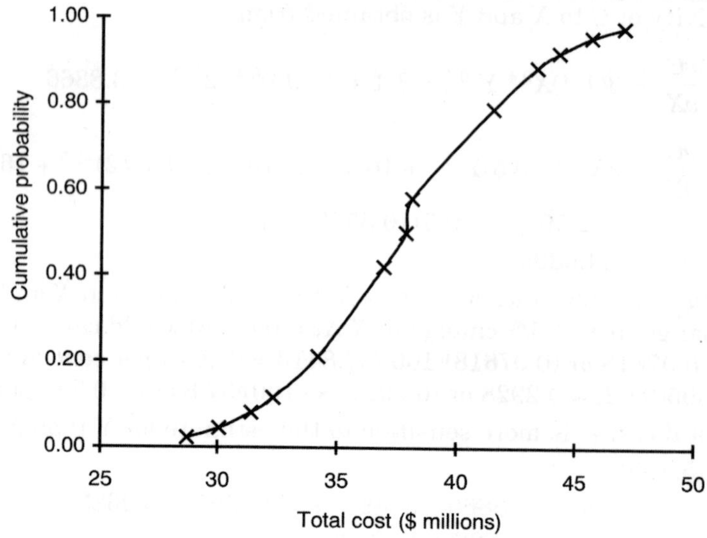

Fig. E5.53 Cumulative Probability versus Total Cost.

5.10 BENEFIT-COST ANALYSIS

At a given minimum attractive rate of return (MARR), one can consider an alternative acceptable provided that

$$\text{Benefit–Cost ratio, } (B/C) = \frac{\text{PW of benefit}}{\text{PW of costs}} = \frac{\text{EUAB}}{\text{EUAC}} \geq 1 \qquad ...(5.102)$$

This method gives an alternative way of solving problems than using the present worth or annual cash flow methods. The criteria for benefit-cost rates, *B*/C analysis are summarized in Table 5.5.

Table 5.5. Benefit-Cost Ratio Analysis

	Situation	Criterion
Fixed input	Amount of money or other input resources are fixed	Maximize *B/C*
Fixed output	Fixed task, benefit, or other output to be accomplished	Maximize *B/C*
Neither input nor output fixed	Neither amount of money or other inputs nor amount of benefits or other outputs are fixed	*Two alternatives:* Compute incremental benefit-cost ratio ("*B*/"*C*) on the increment of investment between the alternatives. If "B/"$C \geq 1$, choose higher-cost alternative; otherwise, choose lower-cost alternative. *Three or more alternatives:* Solve by benefit-cost ratio incremental analysis

Example E5.54. *A company is deciding which of the two computer systems to purchase. Using Benefit-Cost ratio analysis, determine which alternative should be selected if the interest rate is 10% per year.*

Computer system	Cost	Annual benefit	Salvage	Useful life
A	*$50,000*	*$15,000*	*0*	*5 years*
B	*$35,000*	*$12,000*	*0*	*5 years*

Solution:

Alternative *A*:

PW (*B*) = $15,000(*P*/*A*, 10%, 5) = $15,000(3.7908) = $56,862

PW(*C*) = $50,000

B/*C* = 56,862/50,000 = 1.1372

Alternative *B*:

PW (*B*) = 10,000(*P*/*A*, 10%, 5) = 12,000(3.7908) = $45,489

PW(*C*) = $35,000

B/*C* = $45,489/$35,000 = 1.2997

Therefore, select *B* since it gives largest *B*/*C* rates.

$$\frac{\Delta B}{\Delta C} = \frac{\$56,862 - \$45,489}{\$50,000 - \$35,000} = 11,373 = 0.7582$$

Since $\frac{\Delta B}{\Delta C}$ < 1.0, select smallest cost alternative *B*.

NPW = PW (*B*) – PW(*C*)

NPW (*A*) = $56,862 – $50,000 = $6,862

NPW (*B*) = $45,489 – $35,000 = 10,489

To maximize NPW, choose largest positive number, namely, alternative *B*.

Example E5.55. *A firm is considering selecting one alternative out of the given four mutually exclusive alternatives, each with a useful life of 25 years and no salvage value.*

	A	B	C	D
PW of Costs	*$4,000*	*$10,000*	*$7,000*	*$3,000*
PW of Benefits	*$7,000*	*$12,500*	*$5,500*	*$6,000*

Solution:

	A	B	C	D
B/*C* ratio:	1.75	1.25	0.76	2.0

Since *B*/*C* rates for alternative *C* is 0.786, which is less than 1.0, we can exclude alternative *C*.

Now rearranging the alternative *A*, *B*, and *D* as:

	D	A	B
PW (Benefits)	$6,000	$7,000	$12,500
PW (Costs)	$3,000	$4,000	$10,000

We have

	A – D	B – A
ΔB	1,000	5,500
ΔC	2,000	6,000
ΔB/ΔC	0.5	0.917

Since $\Delta B/\Delta C$ is ≤ 1.0, Choose D as smaller cost.

5.11 SUMMARY

In this chapter, cash flow tables, time value of money and equivalence are described briefly. Present worth analysis is desirable when we wish to know the present worth of future costs and benefits. In order to make valid comparisons, we need to analyze each alternative in a problem over the same analysis period or planning horizon. Rate of return analysis is the most frequently used method in practice, as the resulting rate of return is readily understood. The difficulties in selecting a suitable rate of interest in present worth and annual cash flow analysis are avoided. The different depreciation methods presented have an impact on the taxable income of a company in any year. For corporations, taxable income equals gross income minus all ordinary and necessary expenditures (except capital expenditures) and depreciation and depletion charges. Before-tax cash flows less income taxes equal the after tax cash flows. No simple relationship exists between before-tax and after-tax rate of return in the cases of investments involving depreciable assets.

The effect of inflation on the computed rate of return for an investment depends on how future benefits respond to the inflation. Estimates of future inflation or deflation may be important in assessing capital expenditure proposals. To incorporate risk and uncertainty about future events into a decision process, it is appropriate to utilize probability theory. A break-even analysis often is used when there is particular uncertainty about one of the factors in an economic study. The break-even point is the value for the factor at which the project is just marginally justified. Although, benefit-cost analysis methodology is used widely, it should be noted that it is a deterministic method that does not deal with uncertainty in a major way.

PROBLEMS

5.1 A bank loans Mr. Jones $10,000 with an interest rate of 8% compounded annually. How much will Mr. Jones pay the bank if he pays at the end of 6 years?

5.2 Linda deposits $10,000, $5,000 and $20,000 at $t = 1, 2$ and 3 years, respectively. If the fund pays 8% compounded per period, what sum will be accumulated in the fund at (*a*) $t = 3$ and (*b*) $t = 5$?

5.3 If a savings account pays 5% compounded annually, what single deposit now will accumulate $10,000 at the end of the tenth year. If the fund pays 6% compounded annually, what single deposit is required now in order to accumulate $5,000 at the end of the tenth year?

5.4 John decides to place his annual company bonus of $1,000 in a fund that pays 8% compounded annually. How much money will be in the fund immediately after the 30th deposit?

5.5 Find the annual payments needed to accumulate $10,000 over 10 years at a compound interest rate of 10%.

5.6 (*a*) A sum of $10,000 is deposited into an account that pays interest at a rate of 10% compounded annually. If ten equal annual withdrawals are made from this account with the first withdrawal occurring one year after the deposit, how much money can be withdrawn each year in order to deplete the fund with the last withdrawal?

(*b*) If the first withdrawal is delayed for two years, how much money can be withdrawn each of the 10 years?

5.7 (*a*) Mr. Jones wishes to deposit a single sum of money in a fund so that ten equal annual withdrawals of $1,000 can be made before depleting the fund. If the first withdrawal is to occur 1 year after the deposit and the fund pays interest at a rate of 10% compounded annually, how much money should Mr. Jones deposit?

(*b*) If the first withdrawal will not occur until 5 years after the deposit.

5.8 The maintenance expense for a heavy-duty earth moving machinery increases by $1,000 per year over the 10-year life of the equipment. The initial maintenance cost was $5,000. Determine the present worth equivalent for the maintenance costs at 7% interest rate compounded annually.

5.9 Seven annual deposits were made into a savings fund, which pays interest at a rate of 7% compounded annually. The first was $900; second, third, fourth, fifth, sixth and the seventh deposits made were $1,000, $900, $800, $700, $600, $500, and $400 respectively. Determine the accumulated amount in the savings fund right after the seventh deposit.

5.10 The maintenance costs of an ice-breaking machine are expected to be $2,000 at the end of the first year of the machine's life and increases 10% per year thereafter. The projected life of the machine is 20 years. If an endowment fund is set up which earns 10% per year compounded annually, determine the amount of money to be deposited initially in this endowment account.

5.11 A company invested $ 100,000 in a venture with the following cash flow data:

EOY	CF	EOY	CF	EOY	CF
0	– $100,000	3	20,000	6	21,000
1	15,000	4	25,000	7	20,000
2	20,000	5	25,000	8	15,000

If MARR is 10%, determine the following:

(*a*) present worth
(*b*) annual worth
(*c*) future worth
(*d*) internal rate of return
(*e*) external rate of return
(*f*) savings/investment ratio
(*g*) payback period with and without considering the time value of money

5.12 An investment of $25,000 was made on a computer system that will last 5 years and a zero salvage value at the end of service life. The operating, maintenance and software costs are estimated to be $ 10,000 for the first 3 years and $15,000, the last 3 years. If MARR is 10%, determine the following for this investment:

(*a*) present worth
(*b*) annual worth
(*c*) future worth

5.13 A drilling machine was purchased for $50,000, kept for 5 years, and sold for $ 8,000. The annual operating and maintenance costs were $5,000. Using a MARR of 12% per year, determine for this investment the following:

(*a*) present worth
(*b*) annual worth
(*c*) future worth

5.14 An investment of $50,000 for a new earth moving machinery is being considered. The estimated salvage value of this machinery is $10,000 at the end of the estimated life of 6 years. The annual income each year for 6 years is $10,000. Annual operating expenses are $3,000. If the money is worth 15% compounded annually, determine the following measures of investment worth in deciding whether this machinery should be purchased.

(*a*) present worth

(*b*) annual worth

(*c*) future worth

(*d*) internal rate of return

(*e*) external rate of return

(*f*) savings/investment ratio

5.15 A company purchases a welding machine for $15,000; the machine is used for 4 years and is then sold for $5,000. The annual operating, maintenance costs equals $5,000/year over the 4-year period. The welding machine replaced an older system that cost $ 12,000/year. If the MARR is 12%, determine if the decision to buy this welding machine was economically sound. Use the following measures of investment worth.

(*a*) present worth

(*b*) annual worth

(*c*) future worth

(*d*) internal rate of return

(*e*) external rate of return

(*f*) savings/investment ratio

5.16 If a fund pays 15% compounded annually, how much deposit is needed today such that $2,000 can be withdrawn every year forever?

5.17 If a fund pays 15% compounded annually, what deposit is required today such that $5,000 can be withdrawn every 4 years forever?

5.18 The maintenance costs of a numerically controlled milling machine are cyclic with the following costs occurring over a 5-year period: $5,000, $3,000, $4,000, $0, and $2,000. This sequence is expected to repeat itself every 5 years forever. Find the capitalized cost of the maintenance costs based on a time value of money of 12%.

5.19 A company can invest in a venture, which costs $200,000 and returns $80,000/year at the end of 4 years. This investment can be renewed perpetually every 4 years. If the company's MARR is 12%, find the capitalized worth of an infinite series of these investments.

5.20 Determine the capital recovery cost of a high-speed computer system costing $300,000, having a life of 10 years and a salvage value of $ 50,000. Use an interest rate of 12%.

5.21 Determine the capital recovery cost of a milling machine having a first cost of $70,000, a salvage value of $10,000, and a life of 10 years. Use an interest rate of 10%.

5.22 A molding machine for a manufacturing company is purchased at the beginning of the fiscal year for $30,000. The estimated salvage value after 10 years is $3,000. Calculate the depreciation deduction and the resulting unrecovered investment during each year of the molding machine's life using.

(*a*) straight-line depreciation method.

(*b*) double declining balance method

(c) sum of the year's digit depreciation method.

5.23 On January first a printing press was purchased for $50,000. The press has a 5-year useful life and a $5,000 expected salvage value. Compute yearly depreciation using:

(*a*) straight-line depreciation

(*b*) sum of year's digit depreciation

(*c*) double declining balance depreciation.

5.24 A front-end loader cost $80,000 and has an estimated salvage value of $20,000 at the end of 5 years useful life. Calculate the depreciation schedule, and book value, to the end of the useful life of the front-end loader by

(*a*) straight-line depreciation
(*b*) sum of years digit depreciation
(*c*) double declining balance depreciation.

5.25 Determine, which is best from the following two depreciation schedules assuming an interest rate of 6% per year.

Year	SOYD	Double digits/ Straight line
1	$16,000	19,000
2	14,000	14,000
3	12,000	10,000
4	10,000	8,000
5	8,000	6,000
6	6,000	5,000
7	4,000	5,000
8	2,000	5,000
Total	$72,000	$72,000

5.26 From the two depreciation schedules for a computer laser printer given in Table P5.26, which method of depreciation should be selected assuming an interest rate of 12% per year.

Year	DDB	SOYD
1	$300	$286
2	$200	$230
3	$150	$180
4	$100	$130
5	$100	$85
6	$100	$39

5.27 A company purchased a computer system for $ 50,000. It is expected to last 14 years and has an estimated salvage value of $6,000. Determine the depreciation charge on the computer system for the fourth year of its service life and the book value at the end of 8 years, using sum of digits depreciation method.

5.28 By purchasing a machine for $30,000, a large company in the 50% income tax bracket was able to save $5,000 during year 1, $4,000 during year 2, $3,000 during year 3, $2,000 during year 4 and $1,000 during year 5. Sum of year's digits depreciation method was used to depreciate the machine over its four-year life and a zero salvage value was assumed. The company would like to sell the machine at the end of year 5. Assuming that the capital gains/losses are taxed at a rate of 28% and the machine qualities for a 10% investment tax credit, determine the resale value that would yield a 10% after-tax rate of return for the company.

5.29 A company has a gross income in one year of $20 million. The operating expenses are $12 million. Depreciation is $3 million. In this year there is a depreciation recapture of $1 million realized when the company sold a machine that is no longer needed and sold for more than its book value.

(*a*) determine the company's federal income taxes
(*b*) determine the average federal tax rate
(*c*) if the state tax rate is 10 per cent, determine the total income taxes paid.

5.30 A project can be summarized by the data given in Table P5.30. The company uses straight-line depreciation, pays an incremental income tax rate of 30% and requires an after-tax rate of return of 10%. Ignore any investment tax credit.

Item	Amount
Life	30 years
Investment	$100,000
Salvage	$25,000
Revenues	$26,000/year
Operating expenses	$13,500/year

(*a*) using Net Present Worth, determine whether the project should be undertaken.

(*b*) if the company used Sum-of-Years-Digits depreciation, is it possible the decision would change? (No computations needed).

5.31 A company is considering an investment in computer controlled manufacturing system to reduce its production cost. The initial cost would be $15,000 and over a 10-year period it is estimated that the company would save $2,500 annually in production costs. The salvage value of the system is estimated at $3,000. Determine the after-tax rate of return.

5.32 A company expects to receive $60,000 each year for 15 years from the sale of a project. The initial investment will be $300,000. The expenses of manufacturing and selling the project will be $10,000 per year. Assume a straight-line depreciation, a 15-year useful life and no salvage value. Use a 48% income tax rate and ignore the investment tax credit. Determine the projected after-tax rate of return.

5.33 A venture by a firm requires $2 million now, with an investment tax credit of $40,000 at the end of the first year and no salvage value. The receipts and expenses (excluding depreciation) are:

End year	Receipts, $(in thousands)	Expenses, $(in thousands)
1	3,500	2,000
2	3,800	2,800
3	2,500	1,800

A working capital of $400,000 is invested at zero time and the income tax rate is 46%. Determine the discounted cash flow rate of return assuming sum-of-the-years-digits depreciation.

5.34 Compare project A with an initial investment of $2000K, a 5-year service life, and a net present value (NPV) of $500K, with project *B*, having an initial investment of $3500K a 9-year service life, and a net present value of $1500K. Both present values are at 10% per year. Also, if the initial investment is $1K for both projects, make a correction for the difference in size and duration by changing to a premium unit cost.

5.35 An initial capital investment is $400,000 and has a 7-year life and $ 100,000 salvage value. If the working capital is $ 50,000 and the total net profit after taxes over 7-years is estimated at $200,000, find the ROI.

5.36 Find the discounted cash flow rate of return for the data in Problem 5.33.

5.37 A machine has a first cost of $20,000 and a salvage value of $5,000 after a 5-year life. Annual benefits (savings) from its use are $8,000, and the annual cost of operation is $4,000. The tax rate is 50 per cent. Find the DCF rate of return.

5.38 A company in the 50% income tax bracket, is considering the purchase of a new machine that qualifies for a 10% investment tax credit. The new machine will yield benefits of $10,000 for year 1, $15,000 for year 2, $20,000 for year 3, $20,000 for year 4, and $20,000 for year 5.

The machine is to be depreciated using the Accelerated Cost Recovery System (ACRS) over three years. The percentages 33%, 45% and 22% respectively, for years 1,2 and 3. The company expects to sell the machinery at the end of five years of use for 25% of the original purchase price. Determine the maximum purchase cost the company can pay if it requires a 10% after-tax rate of return? The capital gains/losses are taxed/refunded at a rate of 28%.

5.39 A company is considering building a machine using two different materials *A* and *B*. The machine could be fabricated using material *A* with an initial cost of \$30,000 and will last 15 years. The maintenance expenses are \$1000 per year. If material *B* is used, the annual maintenance cost would be \$100 per year and its life also is 15 years. At the end of 15 years there is no salvage value for either machines. The company pays income tax at a rate of 48% and uses straight-line depreciation. There is no investment tax credit. If the minimum acceptable after-tax rate of return is 10, determine the maximum amount that should be spent on the machine fabricated using material *B*.

5.40 Calculate the average annual rate of inflation based on the CPI from the end of:

(*a*) 1990 through 1998
(*b*) 1985 through 1996
(*c*) 1970 through 1994
(*d*) 1969 through 1990

5.41 Using the data in Fig. 5.11, which gives the purchasing power of the dollar in various years, calculate the inflation rate based on consumer prices for the following years.

(*a*) 1953
(*b*) 1982
(*c*) 1997
(*d*) 1967
(*e*) 1991

5.42 For the time spans listed, find the average annual inflation rate from the producer price data given in Figs. 5.11 and 5.13.

(*a*) end of 1961 through 1991
(*b*) end of 1958 through 1968
(*c*) end of 1971 through 1989
(*d*) end of 1986 through 1999
(*e*) end of 1945 through 1986

5.43 Mr. Jones wants to invest some money in a bank for one year. Currently, the bank is paying a rate of 6% compounded annually. Assume that inflation is anticipated to be (*a*) 3% per year, and (*b*) 8% per year for the year Mr. Jones wants to invest. Determine i, f, and f' in each case.

5.44 A charitable establishment is considering replacing its present building with a new facility. When the present building was constructed in 1945, the total cost was 1 million dollars. Also, at that time, a wealthy trust gifted the establishment with 1 million dollars to be used for future replacement. The administrators are now considering building a new facility in the year 2005. Assume the following data:

Inflation is 5% per year from 1945 to 2005. In 1945, the charitable establishment invested the gift at a market interest rate of 10% per year.

(*a*) determine i, i', f, $a\$$ and $r\$$ for this problem.
(*b*) how many actual dollars in the year 2005 will the gift be worth?
(*c*) how much would the actual dollars in 2005 be in terms of 1945 purchasing power?

5.45 A series of four-end of the year payments of \$10,000 has been accrued. If over the next five years the market interest rate is 10% per year and the annual inflation rate is 4%, find the present equivalent of this series using:

(*a*) actual dollar analysis
(*b*) constant dollar analysis

5.46 A corporation is interested in evaluating a new manufacturing technology. Two competitors, companies *A* and *B* offered to deliver this technology at the end of a five-year period. The annual technology development cost for each of the two companies is:

Company *A*: Development costs will be $100,000 the first year and increase at a rate of 5% over a five-year period.

Company *B*: Development costs will be constant $100,000 per year in terms of today's dollars over the five-year period.

If the corporate MARR (investment market rate) is 30%, and the general price inflation is 4% per year over the next five years, which company should the corporation choose?

5.47 Determine the present worth of utility costs for the following data:

The company's minimum attractive rate of return (MARR) = 30%

The project's useful life: 10 years.

The electrical costs increase at 8% per year.

City water costs increase at 5% per year.

Natural gas costs increase at 6% per year.

The first year estimates of the utility costs (in today's dollars) are:

Electricity cost = $50,000

Water cost = $15,000

Natural gas cost = $35,000

5.48 The manufacturing manager of a company wants to know the equivalent annual cost of producing certain product over the five-year period. The company's before-tax MARR is 25%. The manager estimates that he will need an average of 500 lbs of each of material *A* and material *B* each day. The manufacturing company operates 7 days per week, 52 weeks per year. The current costs for material *A* are $50 per lb and $80 per lb for material *B*. Also, it was anticipated that the costs for material A will inflate 3% per year the next three years and then 5% the following two years. The costs for material *B* will inflate 8% per year the next two years and decrease 3% the following three years.

5.49 If the general price inflation is estimated to be 4% for the next five years, 7% for the three years after that, and 3% the following five years, at what market interest rate (i) would a company have to invest the money to maintain a real purchasing power growth rate (i^1) of 10% during those years.

5.50 A $10,000 investment will return annual benefits for six-years with no salvage value at the end of six years. Determine the before and after-tax rates of return assuming a straight-line depreciation and a 45% income tax rate. With no inflation, the annual benefits are $2,433 per year and with inflation equal to 6%, the benefits from the investment increase at this same rate, so that they continue to be the equivalent of $2,433 in year zero based dollars.

5.51 Consider a project that has the following cost series over a five-year period.

(*a*) if the rate of inflation is 3% and the market rate is 8.15%, what is the present equivalent of this series? What is the equal-annual series of payments equivalent to this series in actual dollars?

(*b*) if the rate of inflation is 5% and the market rate is 11.3%, what is the equivalent at the present of the series? What is the annual equivalent of this series in actual dollars?

End of year	1	2	3	4	5
Estimated future cost in constant dollars	$2,000	$2,000	$2,000	$2,000	$2,000

5.52 The minimal acceptable return for an investment of $30,000 with a 5-year life and salvage value of $5,000 is 9%. This investment results in an annual benefit of $10,000 at an operating cost of $2,000. Determine the break-even point in terms of life at which the project is marginally justifiable if there is an uncertainty as to whether the new machinery will survive 5 years of continuous use.

5.53 A new plant will cost \$100,000 for the first stage and \$150,000 for the second stage at n years in the future. If it is built to full capacity now, its cost is \$160,000. It is assumed that all facilities are expected to last 50 years and there is no salvage value. Conduct a break-even and sensitivity analysis. Assume an 8% interest rate.

5.54 An ice cream company has studied the possibility of marketing a new low-calorie ice cream in small region. The profits would depend upon the sales volume and there is some uncertainty on the sales-forecast figures. The estimated investment is \$200,000 and the profits are \$65,000 per year for the next 5 years. The company's MARR is 15%.

(*a*) Is the investment decision sensitive to the uncertainty of the sales forecast if the estimated profits will be reduced to \$40,000 per year?

(*b*) The minimum volume of sales for the investment projects to break-even, if there is a profit of \$ 7 per unit volume.

5.55 A conventional machine system will cost \$100,000 and has a service life of 10 years and an anticipated salvage value of \$10,000 at the end of life. The annual maintenance and repair cost for the machine system is estimated at \$3,000. The expected rate of return from the investment is 12%. The production rate for this machine system is 20 units per hour. The labor rate to operate this machine system is \$8.00 per hour. Assume the overhead rates on the machine system as 25% and 30% on labor.

Find the annual production output at which break-even will occur if the value added per piece of this operation is \$1.00.

5.56 Refer to Problem 5.55. Consider an alternative of an automated machine system to the machine system costing \$150,000 with an expected service life of 5 years with an anticipated service life of zero is proposed whose annual maintenance costs are estimated at \$8,000. Assume the same 12% rate of return and 25% overhead rates for this system. The production rate for this system will be 50 units per hour, and one-fourth of one operator's time at \$8.00 per hour will be needed to tend the machine. The overhead rate for labor is 30%.

Determine the annual production levels for which the automated system would be advantageous.

5.57 The total cost C for a project in millions of dollars is represented by

$C = 4\,X^{1.3}\,Y^{-0.5} + 12\,Y + F$

where X is estimated from five values: 2.6, 1.7, 1.6, 2.9, 1.5, and 2.0. Y is estimated from a high of 3 and a low of 1, F is certain with a value of 1.5. Assume normal distributions.

(*a*) determine the expected value for C, its sensitivity to X and Y.

(*b*) find variation of the actual value of C with the expected value.

(*c*) find the variance for C

5.58 A company is trying to decide which of the two subsystems to install in order to reduce costs in a particular machinery system. Both subsystems cost \$1500 and have the same useful life of 5 years and no salvage value. Subsystem A is expected to give \$400 in savings annually. Subsystem B will provide cost savings of \$500 in the first year but will decline \$50 annually. With an annual interest rate of 8%, determine which subsystem the company should purchase.

5.59 The data for two machines A and B which are being considered for purchase by a company is as given below. Assuming 10% interest rate per year, which machine the company should buy?

	Machine A	**Machine B**
Initial cost	\$1,000	\$3,000
Uniform annual benefit	90	200
End-of-useful life salvage value	150	800
Useful life in years	6	12

5.60 A section of a manufacturing company is under expansion. The initial investment and operating benefits and costs for alternatives A, B, and C are summarized as follows:

	Alternative A	Alternative B	Alternative C
Investment costs	$200,000	$150,000	$300,000
Annual operating costs	15,000	10,000	25,000
Annual benefits	60,000	35,000	85,000

If the company's MARR is 10%, which alternative should be selected for a ten-year useful life period using benefit/cost ratio analysis.

REFERENCES

Adler, M., "The True Rate of Return and the Reinvestment Rate", *The Engineering Economist*, 15 (3), pp. 185–187, 1970.

American Telephone and Telegraph Co., *Engineering Economy*, 3rd ed., McGraw-Hill, New York, NY, 1977.

Au, T. and Au, T.P., *Engineering Economics for Capital Investment Analysis*, Allyn and Bacon, Boston, MA, 1983.

Baldwin, R.H., "How to Assess Investment Proposals", *Harvard Business Review*, 37(3), pp. 98-99, 1959.

Barish, N.N. and Kaplan, S., Economic Analysis for Engineering and Managerial Decision-Making, 2nd ed., McGraw-Hill, New York, NY, 1978.

Beaujon, George J. and Singhal, V.R., "Understanding the Activity Costs in an Activity-Based Cost System", *Journal of Cost Management*, 4(1), pp. 51–72, September 1990.

Beaves, R.G., "Net Present Value and Rate of Return: Implicit and Explicit Reinvestment Assumptions", *The Engineering Economist*, 33(4), pp. 275–302, 1988.

Beenhakker, H.L., "Discounting Indices Proposed for Capital Investment Evaluation: A Further Examination", *The Engineering Economist*, 18(3), pp. 149–68, 1973.

Blank, L.T. and Tarquin, A.J., *Engineering Economy*, 4th ed., McGraw-Hill, New York, NY, 1998.

Bussey, L.E., *The Economic Analysis of Industrial Projects*, Prentice Hall, Upper Saddle River, NJ, 1978.

Campen, James T., *Benefit, Cost, and Beyond*, Ballenger Publishing Company, Cambridge, MA, 1986.

Canada, J.R., Sullivan, W.G. and White, J.A., *Capital Investment Decision Analysis for Management and Engineering*, 2nd ed., Prentice Hall, Englewood Cliffs, NJ, 1996.

Canada, J.R., *Intermediate Economic Analysis for Management and Engineering*, Prentice Hall, Englewood Cliffs, NJ, 1971.

Cassimatis, P.A., *Concise Introduction to Engineering Economics*, Unwin Hyman, Boston, 1988.

Collier, C.A. and Glagola, C.A., *Engineering Economic and Cost Analysis*, Addison Wesley, Reading, MA, 1998.

DeGarmo, E. P. Sullivan, W.G. and Boutadelli, J.A., *Engineering Economy*, 9th ed., Macmillan, New York, NY, 1993.

DeGarmo, E.P., Sullivan, W.G. and Canada, J.R., *Engineering Economy*, 7th ed. Macmillan, New York, NY, 1984.

Eschenbach, T., *Cases in Engineering Economy*, Wiley, New York, NY, 1989.

Ferguson, Earl J. and Shamblin, J.E., "Break-even Analysis", *The Journal of Industrial Engineering*, 18(8), August 1967.

Fleischer, G. A., *Engineering Economy—Capital Allocation Theory,* PWS Engineering, Boston, 1984.

Freidenfelds, J. and Kennedy, M., "Price Inflation and Long-Term Present-Worth Studies", *The Engineering Economist*, 24(3), pp. 143–160, 1979.

Grant, E.L., Ireson, W.G. and Leavenworth, R.S., *Principles of Engineering Economy,* 8th ed., Wiley, New York, NY, 1990.

Hertz, D.B., "Risk Analysis in Capital Investments", *Harvard Business Review*, 42 (1), January-February 1964.

Jones, B.W., *Inflation In Engineering Economy Analysis,* Wiley, New York, NY, 1982.

Kleinfeld, I., *Engineering and Managerial Economics,* Holt, Rinehart and Winston, New York, NY, 1986.

Kurtz, M. (ed)., *Handbook of Engineering Economics,* McGraw-Hill, New York, NY, 1984.

Newnan, D.G. and Johnson, R., *Engineering Economic Analysis,* 5th ed., Engineering Press, New York, NY, 1995.

Newnan, D.G. and Lavelle, J.P., *Engineering Economics Analysis,* 7th ed., Engineering Press, San Jose, CA, 1998.

Park, C.S., *Contemporary Engineering Economics,* 2nd ed., Addison Wesley, Reading, MA, 1996.

Park, C.S., *Modern Engineering Economic Analysis,* Addison Wesley, Reading, MA, 1990.

Park, C.S. and Sharp-Bette, G.P. *Advanced Engineering Economics,* Wiley, New York, NY, 1990.

Prest, A.R. and Turvey, R., "Cost-Benefit Analysis—A Survey", *Economic Journal,* 75, pp. 683–735, December 1965

Riggs, J.L., Bedworth, D. D. and Randhawa, S.U., *Engineering Economics,* 4th ed., McGraw-Hill, New York, NY, 1996.

Riggs, J.L. and West, T. M., *Engineering Economics,* 3rd ed., McGraw-Hill, New York, NY, 1986.

Swalm, R.O. "On Calculating the Rate of Return on An Investment", *The Journal of Industrial Engineering,* 9(2), pp. 99–103, 1958.

Swalm, R.O. and Lopez-Leautaud, J.L., *Engineering Economic Analysis: A Future Wealth Approach,* Wiley, New York, NY, 1984.

Tarquin, A.J. and Blank, L.T., *Engineering Economy,* 2nd ed., McGraw-Hill, New York, NY,1983.

Thuesen, G.J. and Fabrycky, W.J., *Engineering Economy,* 8th ed., Prentice Hall, Englewood Cliffs, NJ, 1993.

Thuesen, G.J., Fabrycky, W., and Thuessen, H.G., *Engineering Economy,* 5th ed., Prentice Hall, Englewood Cliff, NJ, 1977.

Waters, R.C. and Bullock, R.L., "Inflation and Replacement Decisions", *The Engineering Economist,* 21(4), pp. 249–258, 1976.

White, J.A., Case, K.E., Pratt, D.B., and Agee, M.H., *Principles of Economic Analysis,* 4th ed., Wiley, New York, NY, 1998.

Young, D., *Modern Engineering Economy,* Wiley, New York, NY, 1993.

Appendix–5A
Discrete Compounding

To Find	Given	Factor	Symbol	Name
P	F	$(1+i)^{-n}$	$(P/F, i, n)$	Present worth factor, Single payment
F	P	$(1+i)^{n}$	$(F/P, i, n)$	Compound amount factor, Single payment
P	A	$\dfrac{(1+i)^n - 1}{i(1+i)^n}$	$(P/A, i, n)$	Present worth factor, Uniform series
A	P	$\dfrac{i(1+i)^n}{(1+i)^n - 1}$	$(A/P, i, n)$	Capital recovery factor, Uniform series
F	A	$\dfrac{i(1+i)^n - 1}{i}$	$(F/A, i, n)$	Compound amount factor, Uniform series
A	F	$\dfrac{i}{(1+i)^n - 1}$	$(A/F, i, n)$	Sinking fund factor, Uniform series
P	G	$\dfrac{[1-(1+ni)(1+i)^{-n}]}{i^2}$	$(P/G, i, n)$	Present worth factor, Gradient series
A	G	$\dfrac{(1+i)^n - (1+ni)}{i(1+i)^n - 1}$	$(A/G, i, n)$	Uniform series factor, Gradient series
P	A_1, j	$\dfrac{1-(1+j)^n(1+i)^{-n}}{i-j}$, $i \neq j$	$(P/A_1, i, j, n)$	Present worth factor, Geometric series
F	A_1, j	$\dfrac{(1+i)^n - (1+j)^n}{i-j}$, $i \neq j$	$(F/A_1, i, j, n)$	Future worth factor, Geometric series

Values of Interest factors when n = ∞

Single Payment:

Uniform Payment Series:

$(P/F, i, \infty) = 0$ $\quad$ $(A/F, i, \infty) = 0$

$(F/P, i, \infty) = \infty$ $\quad$ $(A/P, i, \infty) = i$

Arithmetic Gradient Series:

$(A/G, i, \infty) = 1/i$ $\quad$ $(F/A, i, \infty) = \infty$

$(P/G, i, \infty) = 1/i^2$ $\quad$ $(P/A, i, \infty) = 1/i$

Appendix – 5A (Continued) Section – I
Time Value of Money Factors – Discrete Compounding (i = 0.5%)

n	Single Sums		Uniform Series	
	To find F Given P (F/P, i%, n)	To find P Given F (P/F, i%, n)	To find F Given A (F/A, i%, n)	To find A Given F (A/F, i%, n)
1	1.0050	0.9950	1.0000	1.0000
2	1.0100	0.9901	2.0050	0.4988
3	1.0151	0.9851	3.0150	0.3317
4	1.0202	0.9802	4.0301	0.2481
5	1.0253	0.9754	5.0503	0.1980
6	1.0304	0.9705	6.0755	0.1646
7	1.0355	0.9657	7.1059	0.1407
8	1.0407	0.9609	8.1414	0.1228
9	1.0459	0.9561	9.1821	0.1089
10	1.0511	0.9513	10.2280	0.0978
11	1.0564	0.9466	11.2792	0.0887
12	1.0617	0.9419	12.3356	0.0811
13	1.0670	0.9372	13.3972	0.0746
14	1.0723	0.9326	14.4642	0.0691
15	1.0777	0.9279	15.5365	0.0644
16	1.0831	0.9233	16.6142	0.0602
17	1.0885	0.9187	17.6973	0.0565
18	1.0939	0.9141	18.7858	0.0532
19	1.0994	0.9096	19.8797	0.0503
20	1.1049	0.9051	20.9791	0.0477
21	1.1104	0.9006	22.0840	0.0453
22	1.1160	0.8961	23.1944	0.0431
23	1.1216	0.8916	24.3104	0.0411
24	1.1272	0.8872	25.4320	0.0393
25	1.1328	0.8828	26.5591	0.0377
26	1.1385	0.8784	27.6919	0.0361
27	1.1442	0.8740	28.8304	0.0347
28	1.1499	0.8697	29.9745	0.0334
29	1.1556	0.8653	31.1244	0.0321
30	1.1614	0.8610	32.2800	0.0310
36	1.1967	0.8356	39.3361	0.0254
42	1.2330	0.8110	46.6065	0.0215
48	1.2705	0.7871	54.0978	0.0185
54	1.3091	0.7639	61.8167	0.0162
60	1.3489	0.7414	69.7700	0.0143
66	1.3898	0.7195	77.9650	0.0128
72	1.4320	0.6983	86.4089	0.0116
120	1.8194	0.5496	163.8793	6.102E – 03
180	2.4541	0.4075	290.8187	3.439E – 03
360	6.0226	0.1660	1.005E + 03	9.955E – 04

Appendix – 5A (Continued) Section – I (Continued)
Time Value of Money Factors – Discrete Compounding (i = 0.5%)

n	*Uniform Series*		*Gradient Series*	
	To find P Given A (P/A, i%, n)	*To find A Given* P (A/P, i%, n)	*To find P Given* G (P/G, i%, n)	*To find A Given* G (A/G, i%, n)
1	0.9950	1.0050	0.0000	0.0000
2	1.9851	0.5038	0.9901	0.4988
3	2.9702	0.3367	2.9604	0.9967
4	3.9505	0.2531	5.9011	1.4938
5	4.9259	0.2030	9.8026	1.9900
6	5.8964	0.1696	14.6552	2.4855
7	6.8621	0.1457	20.4493	2.9801
8	7.8230	0.1278	27.1755	3.4738
9	8.7791	0.1139	34.8244	3.9668
10	9.7304	0.1028	43.3865	4.4589
11	10.6770	0.0937	52.8526	4.9501
12	11.6189	0.0861	63.2136	5.4406
13	12.5562	0.0796	74.4602	5.9302
14	13.4887	0.0741	86.5835	6.4190
15	14.4166	0.0694	99.5743	6.9069
16	15.3399	0.0652	113.4238	7.3940
17	16.2586	0.0615	128.1231	7.8803
18	17.1728	0.0582	143.6634	8.3658
19	18.0824	0.0553	160.0360	8.8504
20	18.9874	0.527	177.2322	9.3342
21	19.8880	0.0503	195.2434	9.8172
22	20.7841	0.0481	214.0611	10.2993
23	21.6757	0.0461	233.6768	10.7806
24	22.5629	0.0443	254.0820	11.2611
25	23.4456	0.0427	275.2686	11.7407
26	24.3240	0.0411	297.2281	12.2195
27	25.1980	0.0397	319.9523	12.6975
28	26.0677	0.0384	343.4332	13.1747
29	26.9330	0.0371	367.6625	13.6510
30	27.7941	0.0360	392.6324	14.1265
36	32.8710	0.0304	557.5598	16.9621
42	37.7983	0.0265	747.1886	19.7678
48	42.5803	0.0235	959.9188	22.5437
54	47.2214	0.0212	1.194E+03	25.2899
60	51.7256	0.0193	1.449E+03	28.0064
66	56.0970	0.0178	1.722E+03	30.6932
72	60.3395	0.0166	2.012E+03	33.3504
120	90.0735	0.0111	4.824E+03	53.5508
180	118.5035	8.439E – 03	9.031E+03	76.2115
360	166.7916	5.996E – 03	2.140E+04	128.3236

Appendix – 5A (Continued) Section I (Continued)
Time Value of Money Factors – Discrete Compounding (i = 0.75%)

n	*Single Sums*		*Uniform Series*	
	To find F Given P (F/P, i%, n)	*To find P Given F (P/F, i%, n)*	*To find F Given A (F/A, i%, n)*	*To find A Given F (A/F, i%, n)*
1	1.0075	0.9926	1.0000	1.0000
2	1.0151	0.9852	2.0075	0.4981
3	0.0227	0.9778	3.0226	0.3308
4	1.0303	0.9706	4.0452	0.2472
5	1.0381	0.9633	5.0756	0.1970
6	1.0459	0.9562	6.1136	0.1636
7	1.0537	0.9490	7.1595	0.1397
8	1.0616	0.9420	8.2132	0.1218
9	1.0696	0.9350	9.2748	0.1078
10	1.0776	0.9280	10.3443	0.0967
11	1.0857	0.9211	11.4219	0.0876
12	1.0938	0.9142	12.5076	0.0800
13	1.1020	0.9074	13.6014	0.0735
14	1.1103	0.9007	14.7034	0.0680
15	1.1186	0.8940	15.8137	0.0632
16	1.1270	0.8873	16.9323	0.0591
17	1.1354	0.8807	18.0593	0.0554
18	1.1440	0.8742	19.1947	0.0521
19	1.1525	0.8676	20.3387	0.0492
20	1.1612	0.8612	21.4912	0.0465
21	1.1699	0.8548	22.6524	0.0441
22	1.1787	0.8484	23.8223	0.0420
23	1.1875	0.8421	25.0010	0.0400
24	1.1964	0.8358	26.1885	0.0382
25	1.2054	0.8296	27.3849	0.0365
26	1.2144	0.8234	28.5903	0.0350
27	1.2235	0.8173	29.8047	0.0336
28	1.2327	0.8112	31.0282	0.0322
29	1.2420	0.8052	32.2609	0.0310
30	1.2513	0.7992	33.5029	0.0298
36	1.3086	0.7641	41.1527	0.0243
42	1.3686	0.7306	49.1533	0.0203
48	1.4314	0.6986	57.5207	0.0174
54	1.4907	0.6680	66.2718	0.0151
60	1.5657	0.6387	75.4241	0.0133
66	1.6375	0.6107	84.9961	0.0118
72	1.7126	0.5839	95.0070	0.0105
120	2.4514	0.4079	193.5143	5.168E-03
180	3.8380	0.2605	378.4058	2.643E-03
360	14.7306	0.0679	1.831E+03	5.462E-04

Appendix – 5A (Continued) Section I (Continued)
Time Value of Money Factors – Discrete Compounding (i = 0.75%)

n	*Uniform Series*		*Gradient Series*	
	To find P Given A (P/A, i%, n)	*To find A Given P (A/P, i%, n)*	*To find P Given G (P/G, i%, n)*	*To find A Given G (A/G, i%, n)*
1	0.9926	1.0075	0.0000	0.0000
2	1.9777	0.5056	0.9852	0.4981
3	2.9556	0.3383	2.9408	0.9950
4	3.9261	0.2547	5.8525	1.4907
5	4.8894	0.2045	9.7058	1.9851
6	5.8458	0.1711	14.4866	2.4782
7	6.7946	0.1472	20.1808	2.9701
8	7.7366	0.1293	26.7747	3.4608
9	8.6716	0.1153	34.2544	3.9502
10	9.5996	0.1042	42.6064	4.4384
11	10.5207	0.0951	51.8174	4.9253
12	11.4349	0.0875	61.8740	5.4110
13	12.3423	0.0810	72.7632	5.8954
14	13.2430	0.0755	84.4720	6.3766
15	14.1370	0.0707	96.9876	6.8606
16	15.0243	0.0666	110.2973	7.3413
17	15.9050	0.0629	124.3887	7.8207
18	16.7792	0.0596	139.2494	8.2989
19	17.6468	0.0567	154.8671	8.7759
20	18.5080	0.0540	171.2297	9.2516
21	19.3628	0.0516	188.3253	9.7261
22	20.2112	0.0495	206.1420	10.1994
23	21.0533	0.0475	224.6682	10.6714
24	21.8891	0.0457	243.8923	11.1422
25	22.7188	0.0440	263.8029	11.6117
26	23.5422	0.0425	284.3888	12.0800
27	24.3595	0.0411	305.6387	12.5470
28	25.1707	0.0397	327.5416	13.0128
29	25.9759	0.0385	350.0867	13.4774
30	26.7751	0.0373	373.2631	13.9407
36	31.4468	0.0318	524.9924	16.6946
42	35.9137	0.0278	696.8709	19.4040
48	40.1848	0.0249	886.8404	22.0691
54	44.2686	0.0226	1.093E+03	24.6898
60	48.1734	0.0208	1.314E+03	27.2665
66	51.9070	0.0193	1.547E+03	29.7992
72	55.4768	0.0180	1.791E+03	32.2882
120	78.9417	0.0127	3.999E+03	50.6521
180	98.5934	0.0101	6.893E+03	69.9094
360	124.2819	8.046E-03	1.331E+04	107.1145

Appendix – 5A (Continued) Section I (Continued)
Time Value of Money Factors – Discrete Compounding (i=1%)

	Single Sums		*Uniform Series*	
n	*To find F Given* P (*F*/*P*, *i*%, *n*)	*To find P Given* F (*P*/*F*, *i*%, *n*)	*To find F Given* A (*F*/*A*, *i*%, *n*)	*To find A Given* F (*A*/*F*, *i*%, *n*)
1	1.0100	0.9901	1.0000	1.0000
2	1.0201	0.9803	2.0100	0.4975
3	1.0303	0.9706	3.0301	0.3300
4	1.0406	0.9610	4.0604	0.2463
5	1.0510	0.9515	5.1010	0.1960
6	1.0615	0.9420	6.1520	0.1625
7	1.0721	0.9327	7.2135	0.1386
8	1.0829	0.9235	8.2857	0.1207
9	1.0937	0.9143	9.3685	0.1067
10	1.1046	0.9053	10.4622	0.0956
11	1.1157	0.8963	11.5668	0.0865
12	1.1268	0.8874	12.6825	0.0788
13	1.1381	0.8787	13.8093	0.0724
14	1.1495	0.8700	14.9474	0.0669
15	1.1610	0.8613	16.0969	0.0621
16	1.1726	0.8528	17.2579	0.0579
17	1.1843	0.8444	18.4304	0.0543
18	1.1961	0.8360	19.6147	0.0510
19	1.2081	0.8277	20.8109	0.0481
20	1.2202	0.8195	22.0190	0.0454
21	1.2324	0.8114	23.2392	0.0430
22	1.2447	0.8034	24.4716	0.0409
23	1.2572	0.7954	25.7163	0.0389
24	1.2697	0.7876	26.9735	0.0371
25	1.2824	0.7798	28.2432	0.0354
26	1.2953	0.7720	29.5256	0.0339
27	1.3082	0.7644	30.8209	0.0324
28	1.3213	0.7568	32.1291	0.0311
29	1.3345	0.7493	33.4504	0.0299
30	1.3478	0.7419	34.7849	0.0287
36	1.4308	0.6989	43.0769	0.0232
42	1.5188	0.6584	51.8790	0.0193
48	1.6122	0.6203	61.2226	0.0163
54	1.7114	0.5843	71.1410	0.0141
60	1.8167	0.5504	81.6697	0.0122
66	1.9285	0.5185	92.8460	0.0108
72	2.0471	0.4885	104.7099	9.550E+03
120	3.3004	0.3030	230.0387	4.347E-03
180	5.9958	0.1668	499.5802	2.002E-03
360	35.9496	0.0278	3.495E+03	2.861E-04

Appendix – 5A (Continued) Section I (Continued)
Time Value of Money Factors – Discrete Compounding (i=1%)

n	*Uniform Series*		*Gradient Series*	
	To find P Given A (P/A, i%, n)	*To find A Given P (A/P, i%, n)*	*To find P Given G (P/G, i%, n)*	*To find A Given G (A/G, i%, n)*
1	0.9901	1.0100	0.0000	0.0000
2	1.9704	0.5075	0.9803	0.4975
3	2.9410	0.3400	2.9215	0.9934
4	3.9020	0.2563	5.8044	1.4876
5	4.8534	0.2060	9.6103	1.9801
6	5.7955	0.1725	14.3205	2.4710
7	6.7282	0.1486	19.9168	2.9602
8	7.6517	0.1307	26.3812	3.4478
9	8.5660	0.1167	33.6959	3.9337
10	9.4713	0.1056	41.8435	4.4179
11	10.3676	0.0965	50.8067	4.9005
12	11.2551	0.0888	60.5687	5.3815
13	12.1337	0.0824	71.1126	5.8607
14	13.0037	0.0769	82.4221	6.3384
15	13.8651	0.0721	94.4810	6.8143
16	14.7179	0.0679	107.2734	7.2886
17	15.5623	0.0643	120.7834	7.7613
18	16.3983	0.0610	134.9957	8.2323
19	17.2260	0.0581	149.8950	8.7017
20	18.0456	0.0554	165.4664	9.1694
21	18.8570	0.0530	181.6950	9.6354
22	19.6604	0.0509	198.5663	10.0998
23	20.4558	0.0489	216.0660	10.5626
24	21.2434	0.0471	234.1800	11.0237
25	22.0232	0.0454	252.8945	11.4831
26	22.7952	0.0439	272.1957	11.9409
27	23.5596	0.0424	292.0702	12.3971
28	24.3164	0.0411	312.5047	128516
29	25.0658	0.0399	333.4863	13.3044
30	25.8077	0.0387	355.0021	13.7557
36	30.1075	0.0332	494.6207	16.4285
42	34.1581	0.0293	650.4514	19.0424
48	37.9740	0.0263	820.1460	21.5976
54	41.5687	0.0241	1.002E+03	24.0945
60	44.9550	0.0222	1.193E+03	26.5333
66	48.1452	0.0208	1.392E+03	28.9146
72	51.1504	0.0196	1.598E+03	31.4389
120	69.7005	0.0143	3.334E+03	47.8349
180	83.3217	0.0120	5.330E+03	63.9697
360	97.2183	0.0103	8.720E+03	89.6995

Appendix – 5A (Continued) Section I (Continued)
Time Value of Money Factors – Discrete Compounding (i=1.5%)

n	Single Sums		Uniform Series	
	To find F Given P (F/P, i%, n)	To find P Given F (P/F, i%, n)	To find F Given A (F/A, i%, n)	To find A Given F (A/F, i%, n)
1	1.0150	0.9852	1.0000	1.0000
2	1.0302	0.9707	2.0150	0.4963
3	1.0457	0.9563	3.0452	0.3284
4	1.0614	0.9422	4.0909	0.2444
5	1.0773	0.9283	5.1523	0.1941
6	1.0934	0.9145	6.2296	0.1605
7	1.1098	0.9010	7.3230	0.1366
8	1.1265	0.8877	8.4328	0.1186
9	1.1434	0.8746	9.5593	0.1046
10	1.1605	0.8617	10.7027	0.0934
11	1.1779	0.8489	11.8633	0.0843
12	1.1956	0.8364	13.0412	0.0767
13	1.2136	0.8240	14.2368	0.0702
14	1.2318	0.8118	15.4504	0.0647
15	1.2502	0.7999	16.6821	0.0599
16	1.2690	0.7880	17.9324	0.0558
17	1.2880	0.7764	19.2014	0.0521
18	1.3073	0.7649	20.4894	0.0488
19	1.3270	0.7536	21.7967	0.0459
20	1.3469	0.7425	23.1237	0.0432
21	1.3671	0.7315	24.4705	0.0409
22	1.3876	0.7207	25.8376	0.0387
23	1.4084	0.7100	27.2251	0.0367
24	1.4295	0.6995	28.6335	0.0349
25	1.4509	0.6892	30.0630	0.0333
26	1.4727	0.6790	31.5140	0.0317
27	1.4949	0.6690	32.9867	0.0303
28	1.5172	0.6591	34.4815	0.0290
29	1.5400	0.6494	35.9987	0.0278
30	1.5631	0.6398	37.5387	0.0266
36	1.7091	0.5851	47.2760	0.0212
42	1.8688	0.5351	57.9231	0.0173
48	2.0435	0.4894	69.5652	0.0144
54	2.2344	0.4475	82.2952	0.0122
60	2.4432	0.4093	96.2147	0.0104
66	2.6715	0.3743	111.4348	8.974E-03
72	2.9212	0.3423	128.0772	7.808E-03
120	5.9693	0.1675	331.2882	3.019E-03
180	14.5844	0.0686	905.6245	1.104E-03
360	212.7038	0.0047	1.411E+04	7.085E-05

Appendix – 5A (Continued) Section I (Continued)
Time Value of Money Factors – Discrete Compounding (i=1.5%)

n	Uniform Series		Gradient Series	
	To find P Given A (P/A, i%, n)	*To find A Given* P (A/P, i%, n)	*To find P Given* G (P/G, i%, n)	*To find A Given* G (A/G, i%, n)
1	0.9852	1.0150	0.0000	0.0000
2	1.9559	0.5113	0.9707	0.4963
3	2.9122	0.3434	2.8833	0.9901
4	3.8544	0.2594	5.7098	1.4814
5	4.7826	0.2091	9.4229	1.9702
6	5.6972	0.1755	13.9956	2.4566
7	6.5982	0.1516	19.4018	2.9405
8	7.4859	0.1336	25.6157	3.4219
9	8.3605	0.1196	32.6125	3.9008
10	9.2222	0.1084	40.3675	4.3772
11	10.0711	0.0993	48.8568	4.8512
12	10.9075	0.0917	58.0571	5.3227
13	11.7315	0.0852	67.9454	5.7917
14	12.5434	0.0797	78.4994	6.2582
15	13.3432	0.0749	89.6974	6.7223
16	14.1313	0.0708	101.5178	7.1839
17	14.9076	0.0671	113.9400	7.6431
18	15.6726	0.0638	126.9435	8.0997
19	16.4262	0.0609	140.5084	8.5539
20	17.1686	0.0582	154.6154	9.0057
21	17.9001	0.0559	169.2453	9.4550
22	18.6208	0.0537	184.3798	9.9018
23	19.3309	0.0517	200.0006	10.3462
24	20.0304	0.0499	216.0901	10.7881
25	20.7196	0.0483	232.6310	11.2276
26	21.3986	0.0467	249.6065	11.6646
27	22.0676	0.0453	267.0002	12.0992
28	22.7267	0.0440	284.7958	12.5313
29	23.3761	0.0428	302.9779	12.9610
30	24.0158	0.0416	321.5310	13.3883
36	27.6607	0.0362	439.8303	15.9009
42	30.9941	0.0323	568.0201	18.3267
48	34.0426	0.0294	703.5463	20.6667
54	36.8305	0.0272	844.2184	22.9217
60	39.3803	0.0254	988.1674	25.0930
66	41.7121	0.0240	1.134E+03	27.1817
72	43.8447	0.0228	1.280E+03	29.1893
120	55.4985	0.0180	2.360E+03	42.5185
180	62.0956	0.0161	3.317E+03	53.4161
360	66.3532	0.0151	4.311E+03	64.9662

Appendix – 5A (Continued) Section I (Continued)
Time Value of Money Factors – Discrete Compounding (i=2%)

n	*Single Sums*		*Uniform Series*	
	To find F Given P (F/P, i%, n)	*To find P Given F (P/F, i%, n)*	*To find F Given A (F/A, i%, n)*	*To find A Given F (A/F, i%, n)*
1	1.0200	0.9804	1.0000	1.0000
2	1.0404	0.9612	2.0200	0.4950
3	1.0612	0.9423	3.0604	0.3268
4	1.0824	0.9238	4.1216	0.2426
5	1.1041	0.9057	5.2040	0.1922
6	1.1262	0.8880	6.3081	0.1585
7	1.1487	0.8706	7.4343	0.1345
8	1.1717	0.8535	8.5830	0.1165
9	1.1951	0.8368	9.7546	0.1025
10	1.2190	0.8203	10.9497	0.0913
11	1.2434	0.8043	12.1687	0.0822
12	1.2628	0.7885	13.4121	0.0746
13	1.2936	0.7730	14.6803	0.0681
14	1.3195	0.7579	15.9739	0.0626
15	1.3459	0.7430	17.2934	0.0578
16	1.3728	0.7284	18.6393	0.0537
17	1.4002	0.7142	20.0121	0.0500
18	1.4282	0.7002	21.4123	0.0467
19	1.4568	0.6864	22.8406	0.0438
20	1.4859	0.6730	24.2974	0.0412
21	1.5157	0.6598	25.7833	0.0388
22	1.5460	0.6468	27.2990	0.0366
23	1.5769	0.6342	28.8450	0.0347
24	1.6048	0.6217	30.4219	0.0329
25	1.6406	0.6095	32.0303	0.0312
26	1.6734	0.5976	33.6709	0.0297
27	1.7069	0.5859	35.3443	0.0283
28	1.7410	0.5744	37.0512	0.0270
29	1.7758	0.5631	38.7922	0.0258
30	1.8114	0.5521	40.5681	0.0246
36	2.0399	0.4902	51.9944	0.0192
42	2.2972	0.4353	64.8622	0.0154
48	2.5871	0.3865	79.3535	0.0126
54	2.9135	0.3432	95.6731	0.0105
60	3.2810	0.3048	114.0515	8.768E-03
66	3.6950	0.2706	134.7487	7.421E-03
72	4.1611	0.2403	158.0570	6.327E-03
120	10.7652	0.0929	488.2582	2.048E+03
180	35.3208	0.0283	1.716E+03	5.827E-04
360	1.248E+03	8.016E-04	6.233E+04	1.604E-05

Appendix – 5A (Continued) Section I (Continued)
Time Value of Money Factors – Discrete Compounding (i=2%)

n	*Uniform Series*		*Gradient Series*	
	To find P Given A (P/A, i%, n)	*To find A Given P (A/P, i%, n)*	*To find P Given G (P/G, i%, n)*	*To find A Given G (A/G, i%, n)*
1	0.9804	1.0200	0.0000	0.0000
2	1.9416	0.5150	0.9612	0.4950
3	2.8839	0.3468	2.8458	0.9868
4	3.8077	0.2626	5.6173	1.4752
5	4.7135	0.2122	9.2403	1.9604
6	5.6014	0.1785	13.6801	2.4423
7	6.4720	0.1545	18.9035	2.9208
8	7.3255	0.1365	24.8779	3.3961
9	8.1622	0.1225	31.5720	3.8681
10	8.9826	0.1113	38.9551	4.3367
11	9.7868	0.1022	46.9977	4.8021
12	10.5753	0.0946	55.6712	5.2642
13	11.3484	0.0881	64.9475	5.7231
14	12.1062	0.0826	74.7999	6.1786
15	12.8493	0.0778	85.2021	6.6309
16	13.5777	0.0737	96.1288	7.0799
17	14.2919	0.0700	107.5554	7.5256
18	14.9920	0.0667	119.4581	7.9681
19	15.6785	0.0638	131.8139	8.4073
20	16.3514	0.0612	144.6003	8.8433
21	17.0112	0.0588	157.7959	9.2760
22	17.6580	0.0566	171.3795	9.7055
23	18.2922	0.0547	185.3309	10.1317
24	18.9139	0.0529	199.6305	10.5547
25	19.5235	0.0512	214.2592	10.9745
26	20.1210	0.0497	229.1987	11.3910
27	20.7069	0.0483	244.4311	11.8043
28	21.2813	0.0470	259.9392	12.2145
29	21.8444	0.0458	275.7064	12.6214
30	22.3965	0.0446	291.7164	13.0251
36	25.4888	0.0392	392.0405	15.3809
42	28.2348	0.0354	497.6010	17.6237
48	30.6731	0.0326	605.9657	19.7556
54	32.8383	0.0305	715.1815	21.7789
60	34.7609	0.0288	823.6975	23.6961
66	36.4681	0.0274	930.3000	25.5100
72	37.9841	0.0263	1.034E+03	27.2234
120	45.3554	0.0220	1.710E+03	37.7114
180	48.5844	0.0206	2.174E+03	44.7554
360	49.9599	0.0200	2.484E+03	49.7112

Appendix – 5A (Continued) Section I (Continued)
Time Value of Money Factors – Discrete Compounding (i=3%)

n	Single Sums		Uniform Series	
	To find F Given P (F/P, i%, n)	*To find P Given* F (P/F, i%, n)	*To find F Given* A (F/A, i%, n)	*To find A Given* F (A/F, i%, n)
1	1.0300	0.9709	1.0000	1.0000
2	1.0609	0.9426	2.0300	0.4926
3	1.0927	0.9151	3.0909	0.3235
4	1.1255	0.8885	4.1836	0.2390
5	1.1593	0.8626	5.3091	0.1884
6	1.1941	0.8375	6.4684	0.1546
7	1.2299	0.8131	7.6625	0.1305
8	1.2668	0.7894	8.8923	0.1125
9	1.3048	0.7664	10.1591	0.0984
10	1.3439	0.7441	11.4639	0.0872
11	1.3842	0.7224	12.8078	0.0781
12	1.4258	0.7014	14.1920	0.0705
13	1.4685	0.6810	15.6178	0.0640
14	1.5126	0.6611	17.0863	0.0585
15	1.5580	0.6419	18.5989	0.0538
16	1.6047	0.6232	20.1569	0.0496
17	1.6528	0.6050	21.7616	0.0460
18	1.7024	0.5874	23.4144	0.0427
19	1.7535	0.5703	25.1169	0.0398
20	1.8061	0.5537	26.8704	0.0372
21	1.8603	0.5375	28.6765	0.0349
22	1.9161	0.5219	30.5368	0.0327
23	1.9736	0.5067	32.4529	0.0308
24	2.0328	0.4919	34.4265	0.0290
25	2.0938	0.4776	36.4593	0.0274
26	2.1566	0.4637	38.5530	0.0259
27	2.2213	0.4502	40.7096	0.0246
28	2.2879	0.4371	42.9309	0.0233
29	2.3566	0.4243	45.2189	0.0221
30	2.4273	0.4120	47.5754	0.0210
36	2.8983	0.3450	63.2759	0.0158
42	3.4607	0.2890	82.0232	0.0122
48	4.1323	0.2420	104.4084	9.578E-03
54	4.9341	0.2027	131.1375	7.626E-03
60	5.8916	0.1697	163.0534	6.133E-03
66	7.0349	0.1421	201.1627	4.971E-03
72	8.4000	0.1190	246.6672	4.054E-03
120	34.7110	0.0288	1.124E+03	8.899E-03
180	204.5034	4.890E-03	6.783E+03	1.474E-04
360	4.182E+04	2.391E-05	1.394E+06	7.173E-07

Appendix – 5A (Continued) Section I (Continued)
Time Value of Money Factors – Discrete Compounding (i=3%)

n	*Uniform Series*		*Gradient Series*	
	To find P Given A (*P/A, i%, n*)	*To find A Given* P (*A/P, i%, n*)	*To find P Given* G (*P/G, i%, n*)	*To find A Given* G (*A/G, i%, n*)
1	0.9709	1.0300	0.0000	0.0000
2	1.9135	0.5226	0.9426	0.4926
3	2.8286	0.3535	2.7729	0.9803
4	3.7171	0.2690	5.4383	1.4631
5	4.5797	0.2184	8.8888	1.9409
6	5.4172	0.1846	13.0762	2.4138
7	6.2303	0.1605	17.9547	2.8819
8	7.0197	0.1425	23.4806	3.3450
9	7.7861	0.1284	29.6119	3.8032
10	8.5302	0.1172	36.3088	4.2566
11	9.2526	0.1081	43.5330	4.7049
12	9.9540	0.1005	51.2482	5.1485
13	10.6350	0.0940	59.4196	5.5872
14	11.2961	0.0885	68.0141	6.0210
15	11.9379	0.0838	77.0002	6.4500
16	12.5611	0.0796	86.3477	6.8742
17	13.1661	0.0760	96.0280	7.2936
18	13.7535	0.0727	106.0137	7.7081
19	14.3238	0.0698	116.2788	8.1179
20	14.8775	0.0672	126.7987	8.5229
21	15.4150	0.0649	137.5496	8.9231
22	15.9369	0.0627	148.5094	9.3186
23	16.4436	0.0608	159.6566	9.7093
24	16.9355	0.0590	170.9711	10.0954
25	17.4131	0.0574	182.4336	10.4768
26	17.8768	0.0559	194.0260	10.8535
27	18.3270	0.0546	205.7309	11.2255
28	18.7641	0.0533	217.5320	11.5930
29	19.1885	0.0521	229.4137	11.9558
30	19.6004	0.0510	241.3613	12.3141
36	21.8323	0.0458	313.7028	14.3688
42	23.7014	0.0422	385.5024	16.2650
48	25.2667	0.0396	455.0255	18.0089
54	26.5777	0.0376	521.1157	19.6073
60	27.6756	0.0361	583.0526	21.0674
66	28.5950	0.0350	640.4407	22.3969
72	29.3651	0.0341	693.1226	23.6036
120	32.3730	0.0309	963.8635	29.7737
180	33.1703	0.0301	1.076E+03	32.4488
360	33.3325	0.0300	1.111E+03	33.3247

Appendix – 5A (Continued) Section I (Continued)
Time Value of Money Factors – Discrete Compounding (i=4%)

n	*Single Sums*		*Uniform Series*	
	To find F Given P (F/P, i%, n)	*To find P Given F (P/F, i%, n)*	*To find F Given A (F/A, i%, n)*	*To find A Given F (A/F, i%, n)*
1	1.0400	0.9615	1.0000	1.0000
2	1.0818	0.9246	2.0400	0.4902
3	1.1249	0.8890	3.1216	0.3203
4	1.1699	0.8548	4.2465	0.2355
5	1.2167	0.8219	5.4163	0.1846
6	1.2653	0.7903	6.6330	0.1508
7	1.3159	0.7599	7.8983	0.1266
8	1.3688	0.7307	9.2142	0.1085
9	1.4233	0.7026	10.5828	0.0945
10	1.4802	0.6756	12.0061	0.0833
11	1.5395	0.6496	13.4864	0.0741
12	1.6010	0.6246	15.0258	0.0666
13	1.6651	0.6006	16.6268	0.0601
14	1.7313	0.5775	18.2919	0.0547
15	1.8009	0.5553	20.0236	0.0499
16	1.8730	0.5339	21.8245	0.0458
17	1.9479	0.5134	23.6975	0.0422
18	2.0258	0.4936	25.6454	0.0390
19	2.1068	0.4746	27.6712	0.0361
20	2.1911	0.4564	29.7781	0.0336
21	2.2788	0.4388	31.9692	0.0313
22	2.3699	0.4220	34.2480	0.0292
23	2.4647	0.4057	36.6179	0.0273
24	2.5633	0.3901	39.0826	0.0256
25	2.6658	0.3751	41.6459	0.0240
26	2.7725	0.3607	44.3117	0.0226
27	2.8834	0.3468	47.0842	0.0212
28	2.9987	0.3335	49.9676	0.0200
29	3.1187	0.3207	52.9663	0.0189
30	3.2434	0.3083	56.0849	0.0178
36	4.1039	0.2437	77.5983	0.0129
42	5.1928	0.1926	104.8196	9.540E-03
48	6.5705	0.1522	139.2632	7.181E-03
54	8.3138	0.1203	182.8454	5.469E-03
60	10.5196	0.0951	237.9907	4.202E-03
66	13.3107	0.0751	307.7671	3.249E-03
72	16.8423	0.0594	396.0566	2.525E-03
120	110.6626	9.036E-03	2.742E+03	3.648E-04
180	1.164E+03	8.590E-04	2.908E+04	3.439E-05
360	1.355E+06	7.379E-07	3.388E+07	2.952E-08

Appendix – 5A (Continued) Section I (Continued)
Time Value of Money Factors – Discrete Compounding (i=4%)

n	*Uniform Series*		*Gradient Series*	
	To find P Given A (P/A, i%, n)	*To find A Given P (A/P, i%, n)*	*To find P Given G (P/G, i%, n)*	*To find A Given G (A/G, i%, n)*
1	0.9615	1.0400	0.0000	0.0000
2	1.8861	0.5302	0.9246	0.4902
3	2.7751	0.3603	2.7025	0.9739
4	3.6299	0.2755	5.2670	1.4510
5	4.4518	0.2246	8.5547	1.9216
6	5.2421	0.1908	12.5062	2.3857
7	6.0021	0.1666	17.0657	2.8433
8	6.7327	0.1485	22.1806	3.2944
9	7.4353	0.1345	27.8013	3.7391
10	8.1109	0.1233	33.8814	4.1773
11	8.7605	0.1141	40.3772	4.6090
12	9.3851	0.1066	47.2477	5.0343
13	9.9856	0.1001	54.4546	5.4533
14	10.5631	0.0947	61.9618	5.8659
15	11.1184	0.0899	69.7355	6.2721
16	11.6523	0.0858	77.7441	6.6720
17	12.1657	0.0822	85.9581	7.0656
18	12.6593	0.0790	94.3498	7.4530
19	13.1339	0.0761	102.8933	7.8342
20	13.5903	0.0736	111.5647	8.2091
21	14.0292	0.0713	120.3414	8.5779
22	14.4511	0.0692	129.2024	8.9407
23	14.8568	0.0673	138 1284	9.2973
24	15.2470	0.0656	147.1012	9.6479
25	15.6221	0.0640	156.1040	9.9925
26	15.9828	0.0626	165.1212	10.3312
27	16.3296	0.0612	174.1385	10.6640
28	16.6631	0.0600	183.1424	10.9909
29	16.9837	0.0589	192.1206	11.3120
30	17.2920	0.0578	201.0618	11.6274
36	18.9083	0.0529	253.4052	13.4018
42	20.1856	0.0495	302.4370	14.9828
48	21.1951	0.0472	347.2446	16.3832
54	21.9930	0.0455	387.4436	17.6167
60	22.6235	0.0442	422.9966	18.6972
66	23.1218	0.0432	454.0847	19.6388
72	23.5156	0.0425	481.0170	20.4552
120	24.7741	0.0404	592.2428	23.9057
180	24.9785	0.0400	620.5976	24.8452
360	25.0000	0.0400	624.9929	24.9997

Appendix – 5A (Continued) Section I (Continued)
Time Value of Money Factors – Discrete Compounding (i=5%)

n	Single Sums		Uniform Series	
	To find F Given P (F/P, i%, n)	*To find P Given F (P/F, i%, n)*	*To find F Given A (F/A, i%, n)*	*To find A Given F (A/F, i%, n)*
1	1.0500	0.9524	1.0000	1.0000
2	1.1025	0.9070	2.0500	0.4878
3	1.1576	0.8638	3.1525	0.3172
4	1.2155	0.8227	4.3101	0.2320
5	1.2763	0.7835	5.5256	0.1810
6	1.3401	0.7462	6.8019	0.1470
7	1.4071	0.7107	8.1420	0.1228
8	1.4775	0.6768	9.5491	0.1047
9	1.5513	0.6446	11.0266	0.0907
10	1.6289	0.6139	12.5779	0.0795
11	1.7103	0.5847	14.2068	0.0704
12	1.7959	0.5568	15.9171	0.0628
13	1.8856	0.5303	17.7130	0.0565
14	1.9799	0.5051	19.5986	0.0510
15	2.0789	0.4810	21.5786	0.0463
16	2.1829	0.4581	23.6575	0.0423
17	2.2920	0.4363	25.8404	0.0387
18	2.4066	0.4155	28.1324	0.0355
19	2.5270	0.3957	30.5390	0.0327
20	2.6533	0.3769	33.0660	0.0302
21	2.7860	0.3589	35.7193	0.0280
22	2.9253	0.3418	38.5052	0.0260
23	3.0715	0.3256	41.4305	0.0241
24	3.2251	0.3101	44.5020	0.0225
25	3.3864	0.2953	47.7271	0.0210
26	3.5557	0.2812	51.1135	0.0196
27	3.7335	0.2678	54.6691	0.0183
28	3.9201	0.2551	58.4026	0.0171
29	4.1161	0.2429	62.3227	0.0160
30	4.3219	0.2314	66.4388	0.0151
36	5.7918	0.1727	95.8363	0.0104
42	7.7616	0.1288	135.2318	1.395E-03
48	10.4013	0.0961	188.0254	5.318E-03
54	13.9387	0.0717	258.7739	3.864E-03
60	18.6792	0.0535	353.5837	2.828E-03
66	25.0319	0.0399	480.6379	2.081E-03
72	33.5451	0.0298	650.9027	1.536E-03
120	348.9120	2.866E-03	6.958E+03	1.437E-04
180	6.517E+03	1.534E-04	1.303E+04	7.673E-05
360	4.248E+07	2.354E-07	8.495E+07	1.177E-08

Appendix – 5A (Continued) Section I (Continued)
Time Value of Money Factors – Discrete Compounding (i=5%)

n	*Uniform Series*		*Gradient Series*	
	To find P Given A (P/A, i%, n)	*To find A Given P (A/P, i%, n)*	*To find P Given G (P/G, i%, n)*	*To find A Given G (A/G, i%, n)*
1	0.9524	1.0500	0.0000	0.0000
2	1.8594	0.5378	0.9070	0.4878
3	2.7232	0.3672	2.6347	0.9675
4	3.5460	0.2820	5.1028	1.4391
5	4.3295	0.2310	8.2369	1.9025
6	5.0757	0.1970	11.9680	2.3579
7	5.7864	0.1728	16.2321	2.8052
8	6.4632	0.1547	20.9700	3.2445
9	7.1078	0.1407	26.1268	3.6758
10	7.7217	0.1295	31.6520	4.0991
11	8.3064	0.1204	37.4988	4.5144
12	8.8633	0.1128	46.6241	4.9219
13	9.3936	0.1065	49.9879	5.3215
14	9.8986	0.1010	56.5538	5.7133
15	10.3797	0.0963	63.2880	6.0973
16	10.8378	0.0923	70.1597	6.4736
17	11.2741	0.0887	77.1405	6.8423
18	11.6896	0.0855	84.2043	7.2034
19	12.0853	0.0827	91.3275	7.5569
20	12.4622	0.0802	98.4884	7.9030
21	12.8212	0.0780	105.6673	8.2416
22	13.1630	0.0760	112.8461	8.5730
23	13.4886	0.0741	120.0087	8.8971
24	13.7986	0.0725	127.1402	9.2140
25	14.0939	0.0710	134.2275	9.5238
26	14.3752	0.0696	141.2585	9.8266
27	14.6430	0.0683	148.2226	10.1224
28	14.8981	0.0671	155.1101	10.4114
29	15.1411	0.0660	161.9126	10.6936
30	15.3725	0.0651	168.6226	10.9691
36	16.5469	0.0604	206.6237	12.4872
42	17.4232	0.0574	240.2389	13.7884
48	18.0772	0.0553	269.2467	14.8943
54	18.5651	0.0539	293.8208	15.8265
60	18.9293	0.0528	314.3432	16.6062
66	19.2010	0.0521	331.2877	17.2536
72	19.4038	0.0515	345.1485	17.7877
120	19.9427	0.0501	391.9751	19.6551
180	19.9969	0.0500	399.3863	19.9724
360	20.0000	0.0500	399.9998	20.0000

Appendix – 5A (Continued) Section I (Continued)
Time Value of Money Factors – Discrete Compounding (i=6%)

	Single Sums		*Uniform Series*	
n	*To find F Given P (F/P, i%, n)*	*To find P Given F (P/F, i%, n)*	*To find F Given A (F/A, i%, n)*	*To find A Given F (A/F, i%, n)*
1	1.0600	0.9434	1.0000	1.0000
2	1.1236	0.8900	2.0600	0.4854
3	1.1910	0.8396	3.1836	0.3141
4	1.2625	0.7921	4.3746	0.2286
5	1.3382	0.7473	5.6371	0.1774
6	1.4185	0.7050	6.9753	0.1434
7	1.5036	0.6651	8.3938	0.1191
8	1.5938	0.6274	9.8975	0.1010
9	1.6895	0.5919	11.4913	0.0870
10	1.7908	0.5584	13.1808	0.0759
11	1.8983	0.5268	14.9716	0.0668
12	2.0122	0.4970	16.8699	0.0593
13	2.1329	0.4688	18.8821	0.0530
14	2.2609	0.4423	21.0151	0.0476
15	2.3966	0.4173	23.2760	0.0430
16	2.5404	0.3936	25.6725	0.0390
17	2.6928	0.3714	28.2129	0.0354
18	2.8543	0.3503	30.9057	0.0324
19	3.0256	0.3305	33.7600	0.0296
20	3.2071	0.3118	36.7856	0.0272
21	3.3996	0.2942	39.9927	0.0250
22	3.6035	0.2775	43.3923	0.0230
23	3.8197	0.2618	46.9958	0.0213
24	4.0489	0.2470	50.8156	0.0197
25	4.2919	0.2330	54.8645	0.0182
26	4.5494	0.2198	59.1564	0.0169
27	4.8223	0.2074	63.7058	0.0157
28	5.1117	0.1956	68.5281	0.0146
29	5.4184	0.1846	73.6398	0.0136
30	5.7435	0.1741	79.0582	0.0126
36	8.1473	0.1227	119.1209	8.395E-03
42	11.5570	0.0865	175.9505	5.683E-03
48	16.3939	0.0610	256.5645	3.898E-03
54	23.2550	0.0430	370.9170	2.696E-03
60	32.9877	0.0303	533.1282	1.876E-03
66	46.7937	0.0214	763.2278	1.310E-03
72	66.3777	0.0151	1.090E+03	9.177E-04
120	1.088E+03	9.190E-4	1.812E+04	5.519E-05
180	3.590E+04	2.786E-05	5.983E+05	1.672E-06
360	1.289E+09	7.760E-10	2.148E+10	4.656E-11

Appendix – 5A (Continued) Section I (Continued)
Time Value of Money Factors – Discrete Compounding (i=6%)

n	*Uniform Series*		*Gradient Series*	
	To find P Given A (P/A, i%, n)	*To find A Given P (A/P, i%, n)*	*To find P Given G (P/G, i%, n)*	*To find A Given G (A/G, i%, n)*
1	0.9434	1.0600	0.0000	0.0000
2	1.8334	0.5454	0.8900	0.4854
3	2.6730	0.3741	2.5692	0.9612
4	3.4561	0.2886	4.9455	1.4272
5	4.2124	0.2374	7.9345	1.8836
6	4.9173	0.2034	11.4594	2.3304
7	5.5824	0.1791	15.4497	2.7676
8	6.2098	0.1610	19.8416	3.1952
9	6.8017	0.1470	24.5768	3.6133
10	7.3601	0.1359	29.6023	4.0220
11	7.8869	0.1268	34.8702	4.4213
12	8.3838	0.1193	40.3369	4.8113
13	8.8527	0.1130	45.9629	5.1920
14	9.2950	0.1076	51.7128	5.5635
15	9.7122	0.1030	57.5546	5.9260
16	10.1059	0.0990	63.4592	6.2794
17	10.4773	0.0954	69.4011	6.6240
18	10.8276	0.0924	75.3569	6.9575
19	11.1581	0.0896	81.3062	7.2867
20	11.4699	0.0872	87.2304	7.6051
21	11.7641	0.0850	93.1136	7.9151
22	12.0416	0.0830	98.9412	8.2166
23	12.3034	0.0813	104.7007	8.5099
24	12.5504	0.0797	110.3812	8.7951
25	12.7834	0.0782	115.9732	9.0722
26	13.0032	0.0769	121.4684	9.3414
27	13.2105	0.0757	126.8600	9.6029
28	13.4062	0.0746	132.1420	9.8568
29	13.5907	0.0736	137.3096	10.1032
30	13.7648	0.0726	142.3588	10.3402
36	14.6210	0.0684	170.0387	11.6298
42	15.2245	0.0657	193.1732	12.6883
48	15.6500	0.0639	212.0351	13.5485
54	15.9500	0.0627	227.1316	14.2402
60	16.1614	0.0619	239.0428	14.7909
66	16.3105	0.0613	248.3341	15.2254
72	16.4156	0.0609	255.5146	15.5654
120	16.6514	0.0601	275.6846	16.5563
180	16.6662	0.0600	277.6865	16.6617
360	16.6667	0.0600	277.7778	16.6667

Appendix – 5A (Continued) Section I (Continued)

Time Value of Money Factors – Discrete Compounding (i=7%)

n	*Single Sums*		*Uniform Series*	
	To find F Given P (F/P, i%, n)	*To find P Given F (P/F, i%, n)*	*To find F Given A (F/A, i%, n)*	*To find A Given F (A/F, i%, n)*
1	1.0700	0.9346	1.0000	1.0000
2	1.1449	0.8734	2.0700	0.4831
3	1.2250	0.8163	3.2149	0.3111
4	1.3108	0.7629	4.4399	0.2252
5	1.4026	0.7130	5.7507	0.1739
6	1.5007	0.6663	7.1533	0.1398
7	1.6058	0.6227	8.6540	0.1156
8	1.7182	0.5820	10.2598	0.0975
9	1.8385	0.5439	11.9780	0.0835
10	1.9672	0.5083	13.8164	0.0724
11	2.1049	0.4751	15.7836	0.0634
12	2.2522	0.4440	17.8885	0.0559
13	2.4098	0.4150	20.1406	0.0497
14	2.5785	0.3878	22.5505	0.0443
15	2.7590	0.3624	25.1290	0.0398
16	2.9522	0.3387	27.8881	0.0359
17	3.1588	0.3166	30.8402	0.0324
18	3.3799	0.2959	33.9990	0.0294
19	3.6165	0.2765	37.3790	0.0268
20	3.8697	0.2584	40.9955	0.0244
21	4.1406	0.2415	44.8652	0.0223
22	4.4304	0.2257	49.0057	0.0204
23	4.7405	0.2109	53.4361	0.0187
24	5.0724	0.1971	58.1767	0.0172
25	5.4274	0.1842	63.2490	0.0158
26	5.8074	0.1722	68.6765	0.0146
27	6.2139	0.1609	74.4838	0.0134
28	6.6488	0.1504	80.6977	0.0124
29	7.1143	0.1406	87.3465	0.0114
30	7.6123	0.1314	94.4608	0.0106
36	11.4239	0.0875	148.9135	6.715E-03
42	17.1443	0.0583	230.6322	4.336E-03
48	25.7289	0.0389	353.2701	2.831E-03
54	38.6122	0.02590	537.3164	1.861E-03
60	57.9464	0.0173	813.5204	1.229E-03
66	86.9620	0.0115	1.228E+03	8.143E-04
72	130.5065	7.662E-03	1.850E+03	5.405E-04
120	3.358E+03	2.978E-04	4.795E+04	2.085E-05
180	1.946E+05	5.139E-06	2.780E+06	3.598E-07
360	3.786E+10	2.641E-11	5.408E+11	1.849E-12

Appendix – 5A (Continued) Section I (Continued)
Time Value of Money Factors – Discrete Compounding (i=7%)

n	*Uniform Series*		*Gradient Series*	
	To find P Given A (*P/A, i%, n*)	*To find A Given* P (*A/P, i%, n*)	*To find P Given* G (*P/G, i%, n*)	*To find A Given* G (*A/G, i%, n*)
1	0.9346	1.0700	0.0000	0.0000
2	1.8080	0.5531	0.8734	0.4831
3	2.6243	0.3811	2.5060	0.9549
4	3.3872	0.2952	4.7947	1.4155
5	4.1002	0.2439	7.6467	1.8650
6	4.7665	0.2098	10.9784	2.3032
7	5.3893	0.1856	14.7149	2.7304
8	5.9713	0.1675	18.7889	3.1465
9	6.5152	0.1535	23.1404	3.5517
10	7.0236	0.1424	27.7156	3.9461
11	7.4987	0.1334	32.4665	4.3296
12	7.9427	0.1259	37.3506	4.7025
13	8.3577	0.1197	42.3302	5.0658
14	8.7455	0.1143	47.3718	5.4167
15	9.1079	0.1098	52.4461	5.7583
16	9.4466	0.1059	57.5271	6.0897
17	9.7632	0.1024	62.5923	6.4110
18	10.0591	0.0994	67.6219	6.7225
19	10.3356	0.0968	72.5991	7.0242
20	10.5940	0.0944	77.5091	7.3163
21	10.8355	0.0923	82.3393	7.5990
22	11.0612	0.0904	87.0793	7.8725
23	11.2722	0.0887	91.7201	8.1369
24	11.4693	0.0872	96.2545	8.3923
25	11.6536	0.0858	100.6765	8.6391
26	11.8258	0.0846	104.9814	8.8773
27	11.9867	0.0834	109.1656	9.1072
28	12.1371	0.0824	113.2264	9.3289
29	12.2777	0.0814	117.1622	9.5427
30	12.4090	0.0806	120.9718	9.7487
36	13.0352	0.0767	141.1990	10.8321
42	13.4524	0.0743	157.1807	11.6842
48	13.7305	0.0728	169.4981	12.3447
54	13.9157	0.0719	178.8173	12.8500
60	14.0392	0.0712	185.7677	13.2321
66	14.1214	0.0708	190.8927	13.5179
72	14.1763	0.0705	194.6365	13.7298
120	14.2815	0.0700	203.5103	14.2500
180	14.2856	0.0700	204.0674	14.2848
360	14.2857	0.0700	204.0816	14.2857

Appendix – 5A (Continued) Section I (Continued)
Time Value of Money Factors – Discrete Compounding (i=8%)

n	*Single Sums*		*Uniform Series*	
	To find F Given P (F/P, i%, n)	*To find P Given* F (P/F, i%, n)	*To find F Given* A (F/A, i%, n)	*To find A Given* F (A/F, i%, n)
1	1.0800	0.9259	1.0000	1.0000
2	1.1664	0.8573	2.0800	0.4808
3	1.2597	0.7938	3.2464	0.3080
4	1.3605	0.7350	4.5061	0.2219
5	1.4693	0.6806	5.8666	0.1705
6	1.5869	0.6302	7.3359	0.1363
7	1.7138	0.5835	8.9228	0.1121
8	1.8509	0.5403	10.6366	0.0940
9	1.9990	0.5002	12.4876	0.0801
10	2.1589	0.4632	14.4866	0.0690
11	2.3316	0.4289	16.6455	0.0601
12	2.5182	0.3971	18.9771	0.0527
13	2.7196	0.3677	21.4953	0.0465
14	2.9372	0.3405	24.2149	0.0413
15	3.1722	0.3152	27.1521	0.0368
16	3.4259	0.2919	30.3243	0.0330
17	3.7000	0.2703	33.7502	0.0296
18	3.9960	0.2502	37.4502	0.0267
19	4.3157	0.2317	41.4463	0.0241
20	4.6610	0.2145	45.7620	0.0219
21	5.0338	0.1987	50.4229	0.0198
22	5.4365	0.1839	55.4568	0.0180
23	5.8715	0.1703	60.8933	0.0164
24	6..3412	0.1577	66.7648	0.0150
25	6.8485	0.1460	73.1059	0.0137
26	7.3964	0.1352	79.9544	0.0125
27	7.9881	0.1252	87.3508	0.0114
28	8.6271	0.1159	95.3388	0.0105
29	9.3173	0.1073	103.9659	9.619E-03
30	10.0627	0.0994	113.2832	8.827E-03
36	15.9682	0.0626	187.1021	5.345E-03
42	25.3395	0.0395	304.2435	3.287E-03
48	40.2106	0.0249	490.1322	2.040E-03
54	63.8091	0.0157	785.1141	1.274E-03
60	101.2571	9.876-03	1.253E+03	7.979E-04
66	160.6822	6.223E-03	1.996E+03	5.010E-04
72	254.9825	3.922E-03	3.175E+03	3.150E-04
120	1.025E+04	9.753E-05	1.281E+05	7.803E-06
180	1.038E+06	9.632E-07	1.298E+07	7.706E-08
360	1.078E+12	9.278E-13	1.347E+13	7.422E-14

Appendix – 5A (Continued) Section I (Continued)
Time Value of Money Factors – Discrete Compounding (i=8%)

n	*Uniform Series*		*Gradient Series*	
	To find P Given A (P/A, i%, n)	*To find A Given P (A/P, i%, n)*	*To find P Given G (P/G, i%, n)*	*To find A Given G (A/G, i%, n)*
1	0.9259	1.0800	0.0000	0.0000
2	1.7833	0.5608	0.8573	0.4808
3	2.5771	0.3880	2.4450	0.9487
4	3.3121	0.3019	4.6501	1.4040
5	3.9927	0.2505	7.3724	1.8465
6	4.6229	0.2163	10.5233	2.2763
7	5.2064	0.1921	14.0242	2.6937
8	5.7466	0.1740	17.8061	3.0985
9	6.2469	0.1601	21.8081	3.4910
10	6.7101	0.1490	25.9768	3.8713
11	7.1390	0.1401	30.2657	4.2395
12	7.5361	0.1327	34.6339	4.5957
13	7.9038	0.1265	39.0463	4.9402
14	8.2442	0.1213	43.4723	5.2731
15	8.5595	0.1168	47.8857	5.5945
16	8.8514	0.1130	52.2640	5.9046
17	9.1216	0.1096	56.5883	6.3037
18	9.3719	0.1067	60.8426	6.4920
19	9.6036	0.1041	65.0134	6.7697
20	9.8181	0.1019	69.0898	7.0369
21	10.0168	0.0998	73.0629	7.2940
22	10.2007	0.0980	76.9257	7.5412
23	10.3711	0.0964	80.6726	7.7786
24	10.5288	0.0950	84.2997	8.0066
25	10.6748	0.0937	87.8041	8.2254
26	10.8100	0.0925	91.1842	8.4352
27	10.9352	0.0914	94.4390	8.6363
28	11.0511	0.0905	97.5687	8.8289
29	11.1584	0.0896	100.5738	9.0133
30	11.2578	0.0888	103.4558	9.1897
36	11.7172	0.0853	118.2839	10.0949
42	12.0067	0.0833	129.3651	10.7744
48	12.1891	0.0820	137.4428	11.2758
54	12.3041	0.0813	143.2229	11.6403
60	12.3766	0.0808	147.3000	11.9015
66	12.4222	0.0805	150.1432	12.0867
72	12.4510	0.0803	152.1076	12.2165
120	12.4988	0.0800	156.0885	12.4883
180	12.5000	0.0800	156.2477	12.4998
360	12.5000	0.0800	156.2500	12.5000

Appendix – 5A (Continued) Section I (Continued)
Time Value of Money Factors – Discrete Compounding (i = 9%)

	Single Sums		*Uniform Series*	
n	*To find F Given P (F/P, i%, n)*	*To find P Given F (P/F, i%, n)*	*To find F Given A (F/A, i%, n)*	*To find A Given F (A/F, i%, n)*
1	1.0900	0.9174	1.0000	1.0000
2	1.1881	0.8417	2.0900	0.4785
3	1.2950	0.7722	3.2781	0.3051
4	1.4116	0.7084	4.5731	0.2187
5	1.5386	0.6499	5.9847	0.1671
6	1.6771	0.5963	7.5233	0.1329
7	1.8280	0.5470	9.2004	0.1087
8	1.9926	0.5019	11.0285	0.0907
9	2.1719	0.4604	13.0210	0.0768
10	2.3674	0.4224	15.1929	0.0658
11	2.5804	0.3875	17.5603	0.0569
12	2.8127	0.3555	20.1407	0.0497
13	3.0658	0.3262	22.9534	0.0436
14	3.3417	0.2992	26.0192	0.0384
15	3.6425	0.2745	29.3609	0.0341
16	3.9703	0.2519	33.0034	0.0303
17	4.3276	0.2311	36.9737	0.0270
18	4.7171	0.2120	41.3013	0.0242
19	5.1417	0.1945	46.0185	0.0217
20	5.6044	0.1784	51.1601	0.0195
21	6.1088	0.1637	56.7645	0.0176
22	6.6586	0.1502	62.8733	0.0159
23	7.2579	0.1378	69.5319	0.0144
24	7.9111	0.1264	76.7898	0.0130
25	8.6231	0.1160	84.7009	0.0118
26	9.3992	0.1064	93.3240	0.0107
27	10.2451	0.0976	102.7231	9.735E-03
28	11.1671	0.0895	112.9682	8.852E-03
29	12.1722	0.0822	124.1354	8.056E-03
30	13.2677	0.0754	136.3075	7.336E-03
36	22.2512	0.0449	236.1247	4.235E-03
42	37.3175	0.0268	403.5281	2.478E-03
48	62.5852	0.0160	684.2804	1.461E-03
54	104.9617	9.527E-03	1.155E+03	8.657E-04
60	176.0313	5.681E-03	1.945E+03	5.142E-04
66	295.2221	3.387E-03	3.269E+03	3.059E-04
72	495.1170	2.020E-03	5.490E+03	1.821E-04
120	3.099E+04	3.227E-05	3.443E+05	2.905E-06
180	5.455E+06	1.833E-07	6.061E+07	1.650E-08
360	2.975E+13	3.361E-14	3.306E+14	3.025E-15

Appendix – 5A (Continued) Section I (Continued)
Time Value of Money Factors – Discrete Compounding (i = 9%)

n	*Uniform Series*		*Gradient Series*	
	To find P Given A (P/A, i%, n)	*To find A Given P (A/P, i%, n)*	*To find P Given G (P/G, i%, n)*	*To find A Given G (A/G, i%, n)*
1	0.9174	1.0900	0.0000	0.0000
2	1.7591	0.5685	0.8417	0.4785
3	2.5313	0.3951	2.3860	0.9426
4	3.2397	0.3087	4.5113	1.3925
5	3.8897	0.2571	7.1110	1.8282
6	4.4859	0.2229	10.0924	2.2498
7	5.0330	0.1987	13.3746	2.6574
8	5.5348	0.1807	16.8877	3.0512
9	5.9952	0.1668	20.5711	3.4312
10	6.4177	0.1558	24.3728	3.7978
11	6.8052	0.1469	28.2481	4.1510
12	7.1607	0.1397	32.1590	4.4910
13	7.4869	0.1336	36.0731	4.8182
14	7.7862	0.1284	39.9633	5.1326
15	8.0607	0.1241	43.8069	5.4346
16	8.3126	0.1203	47.5849	5.7245
17	8.5436	0.1170	51.2821	6.0024
18	8.7556	0.1142	54.8860	6.2687
19	8.9501	0.1117	58.3868	6.5236
20	9.1285	0.1095	61.7770	6.7674
21	9.2922	0.1076	65.0509	7.0006
22	9.4424	0.1059	68.2048	7.2232
23	9.5802	0.1044	71.2359	7.4357
24	9.7066	0.1030	74.1433	7.6384
25	9.8226	0.1018	76.9265	7.8316
26	9.9290	0.1007	79.5863	8.0156
27	10.0266	0.0997	82.1241	8.1906
28	10.1161	0.0989	84.5419	8.3571
29	10.1983	0.0981	86.8422	8.5154
30	10.2737	0.0973	89.0280	8.6657
36	10.6118	0.0942	99.9319	9.4171
42	10.8134	0.0925	107.6432	9.9546
48	10.9336	0.0915	112.9625	10.3317
54	11.0053	0.0909	116.5642	10.5917
60	11.0480	0.0905	118.9683	10.7683
66	11.0735	0.0903	120.5546	10.8868
72	11.0887	0.0902	121.5917	10.9654
120	11.1108	0.0900	123.4098	11.1072
180	11.1111	0.0900	123.4564	11.1111
360	11.1111	0.0900	123.4568	11.1111

Appendix – 5A (Continued) Section I (Continued)
Time Value of Money Factors – Discrete Compounding (i = 10%)

n	*Single Sums*		*Uniform Series*	
	To find F Given P (F/P, i%, n)	*To find P Given F (P/F, i%, n)*	*To find F Given A (F/A, i%, n)*	*To find A Given F (A/F, i%, n)*
1	1.1000	0.9091	1.0000	1.0000
2	1.2100	0.8264	2.1000	0.4762
3	1.3310	0.7513	3.3100	0.3021
4	1.4641	0.6830	4.6410	0.2155
5	1.6105	0.6209	6.1051	0.1638
6	1.7716	0.5645	7.7156	0.1296
7	1.9487	0.5132	9.4872	0.1054
8	2.1436	0.4665	11.4359	0.0874
9	2.3579	0.4241	13.5795	0.0736
10	2.5937	0.3855	15.9374	0.0627
11	2.8531	0.3505	18.5312	0.0540
12	3.1384	0.3186	21.3843	0.0468
13	3.4523	0.2897	24.5227	0.0408
14	3.7975	0.2633	27.9750	0.0357
15	4.1772	0.2394	31.7725	0.0315
16	4.5950	0.2176	35.9497	0.0278
17	5.0545	0.1978	40.5447	0.0247
18	5.5599	0.1799	45.5992	0.0219
19	6.1159	0.1635	51.1591	0.0195
20	6.7275	0.1486	57.2750	0.0175
21	7.4002	0.1351	64.0025	0.0156
22	8.1403	0.1228	71.4027	0.0140
23	8.9543	0.1117	79.5430	0.0126
24	9.8497	0.1015	88.4973	0.0113
25	10.8347	0.0923	98.3471	0.0102
26	11.9182	0.0839	109.1818	9.159E-03
27	13.1100	0.0763	121.0999	8.258E-03
28	14.4210	0.0693	134.2099	7.451E-03
29	15.8631	0.0630	148.6309	6.728E-03
30	17.4494	0.0573	164.4940	6.079E-03
36	30.9127	0.0323	299.1268	3.343E-03
42	54.7637	0.0183	537.6370	1.860E-03
48	97.0172	0.0103	960.1723	1.041E-03
54	171.8719	5.818E-03	1.709E+03	5.852E-04
60	304.4816	3.284E-03	3.035E+03	3.295E-04
66	539.4078	1.854E-03	5.384E+03	1.857E-04
72	955.5938	1.046E-03	9.546E+03	1.048E-04
120	9.271E+04	1.079E-05	9.271E+05	1.079E-06
180	2.823E+07	3.543E-08	2.823E+08	3.543E-09
360	7.968E+14	1.255E-15	7.968E+15	1.255E-16

Appendix – 5A (Continued) Section I (Continued)
Time Value of Money Factors – Discrete Compounding (i = 10%)

	Uniform Series		*Gradient Series*	
n	*To find P Given A (P/A, i%, n)*	*To find A Given P (A/P, i%, n)*	*To find P Given G (P/G, i%, n)*	*To find A Given G (A/G, i%, n)*
1	0.9091	1.1000	0.0000	0.0000
2	1.7355	0.5762	0.8264	0.4762
3	2.4869	0.4021	2.3291	0.9366
4	3.1699	0.3155	4.3781	1.3812
5	3.7908	0.2638	6.8618	1.8101
6	4.3553	0.2296	9.6842	2.2236
7	4.8684	0.2054	12.7631	2.6216
8	5.3349	0.1874	16.0287	3.0045
9	5.7590	0.1736	19.4215	3.3742
10	6.1446	0.1627	22.8913	3.7255
11	6.4951	0.1540	26.3963	4.0641
12	6.8137	0.1468	29.9012	4.3884
13	7.1034	0.1408	33.3772	4.6988
14	7.3667	0.1357	36.8005	4.9955
15	7.6061	0.1315	40.1520	5.2789
16	7.8237	0.1278	43.4164	5.5493
17	8.0216	0.1247	46.5819	5.8071
18	8.2014	0.1219	49.6395	6.0526
19	8.3649	0.1195	52.5827	6.2861
20	8.5136	0.1175	55.4069	6.5081
21	8.6487	0.1156	58.1095	6.7189
22	8.7715	0.1140	60.6893	6.9189
23	8.8832	0.1126	63.1462	7.1085
24	8.9847	0.1113	65.4813	7.2881
25	9.0770	0.1102	67.6964	7.4580
26	9.1609	0.1092	69.7940	7.6186
27	9.2372	0.1083	71.7773	7.7704
28	9.3066	0.1075	73.6495	7.9137
29	9.3696	0.1067	75.4146	8.0489
30	9.4269	0.1061	77.0766	8.1762
36	9.6765	0.1033	85.1194	8.7965
42	9.8174	0.1019	90.5047	9.2188
48	9.8969	0.1010	94.0217	9.5001
54	9.9418	0.1006	96.2763	9.6840
60	9.9672	0.1003	97.7010	9.8023
66	9.9815	0.1002	98.5910	9.8774
72	9.9895	0.1001	99.1419	9.9246
120	9.9999	0.1000	99.9860	9.9987
180	10.0000	0.1000	99.9999	10.0000
360	10.0000	0.1000	100.0008	10.0000

Appendix – 5A (Continued) Section I (Continued)
Time Value of Money Factors – Discrete Compounding (12%)

n	Single Sums		Uniform Series	
	To find F Given P (F/P, i%, n)	*To find P Given F (P/F, i%, n)*	*To find F Given A (F/A, i%, n)*	*To find A Given F (A/F, i%, n)*
1	1.1200	0.8929	1.0000	1.0000
2	1.2544	0.7972	2.1200	0.4717
3	1.4049	0.7118	3.3744	0.2963
4	1.5735	0.6355	4.7793	0.2092
5	1.7623	0.5674	6.3528	0.1574
6	1.9738	0.5066	8.1152	0.1232
7	2.2107	0.4523	10.0890	0.0991
8	2.4760	0.4039	12.2997	0.0813
9	2.7731	0.3606	14.7757	0.0677
10	3.1058	0.3220	17.5487	0.0570
11	3.4785	0.2875	20.6546	0.0484
12	3.8960	0.2567	24.1331	0.0414
13	4.3635	0.2292	28.0291	0.0357
14	4.8871	0.2046	32.3926	0.0309
15	5.4736	0.1827	32.3926	0.0268
16	6.1304	0.1631	42.7533	0.0234
17	6.8660	0.1456	48.8837	0.0205
18	7.6900	0.1300	55.7497	0.0179
19	8.6128	0.1161	63.4397	0.0158
20	9.6463	0.1037	72.0524	0.0139
21	10.8038	0.0926	81.6987	0.0122
22	12.1003	0.0826	92.5026	0.0108
23	13.5523	0.0738	104.6029	9.560E-03
24	15.1786	0.0659	118.1552	8.463E-03
25	17.0001	0.0588	133.3339	7.500E-03
26	19.0401	0.0525	150.3339	6.652E-03
27	21.3249	0.0469	169.3740	5.904E-03
28	23.8839	0.0419	190.6989	5.244E-03
29	26.7499	0.0374	214.5828	4.660E-03
30	29.9599	0.0334	241.3327	4.144E-03
36	59.1356	0.0619	484.4631	2.064E-03
42	116.7231	8.567E-03	964.3595	1.037E-03
48	230.3908	4.340E-03	1.912E+03	5.231E-04
54	454.7505	2.199E-03	3.781E+03	2.645E-04
60	897.5969	1.114E-03	7.472E+03	1.338E-04
66	1.772E+03	5.644E-04	1.476E+04	6.777E-05
72	3.497E+03	2.860E-04	2.913E+04	3.432E-05
120	8.057E+05	1.241E-06	6.714E+06	1.489E-07
180	7.232E+08	1.383E-09	6.026E+09	1.659E-10
360	5.230E+17	1.912E-18	4.358E+18	2.259E-19

Appendix – 5A (Continued) Section I (Continued)
Time Value of Money Factors – Discrete Compounding (12%)

n	*Uniform Series*		*Gradient Series*	
	To find P Given A (P/A, i%, n)	*To find A Given P (A/P, i%, n)*	*To find P Given G (P/G, i%, n)*	*To find A Given G (A/G, i%, n)*
1	0.8929	1.1200	0.0000	0.0000
2	1.6901	0.5917	0.7972	0.4717
3	2.4018	0.4163	2.2208	0.9246
4	3.0373	0.3292	4.1273	1.3589
5	3.6048	0.2774	6.3970	1.7746
6	4.1114	0.2432	8.9302	2.1720
7	4.5638	0.2191	11.6443	2.5515
8	4.9676	0.2013	14.4714	2.9131
9	5.3282	0.1877	17.3563	3.2574
10	5.6502	0.1770	20.2541	3.5847
11	5.9377	0.1684	23.1288	3.8953
12	6.1944	0.1614	25.9523	4.1897
13	6.4354	0.1557	28.7024	4.4683
14	6.6282	0.1509	31.3624	4.7317
15	6.8109	0.1468	33.9202	4.9803
16	6.9740	0.1434	36.3670	5.2147
17	7.1196	0.1405	38.6973	5.4353
18	7.2497	0.1379	40.9080	5.6427
19	7.3658	0.1358	42.9979	5.8375
20	7.4694	0.1339	44.9676	6.0202
21	7.5620	0.1322	46.8188	6.1913
22	7.6446	0.1308	48.5543	6.3514
23	7.7184	0.1296	50.1776	6.5010
24	7.7843	0.1285	51.6929	6.6406
25	7.8431	0.1275	53.1046	6.7708
26	7.8957	0.1267	54.4177	6.8921
27	7.9426	0.1259	55.6369	7.0049
28	7.9844	0.1252	56.7674	7.1098
29	8.0218	0.1247	57.8141	7.2071
30	8.0552	0.1241	58.7821	7.2974
36	8.1924	0.1221	63.1970	7.7141
42	8.2619	0.1210	65.8509	7.9704
48	8.2972	0.1205	67.4068	8.1241
54	8.3150	0.1203	68.3022	8.2143
60	8.3240	0.1201	68.8100	8.2664
66	8.3286	0.1201	69.0948	8.2961
72	8.3310	0.1200	69.2530	8.3127
120	8.3333	0.1200	69.4431	8.3332
180	8.3333	0.1200	69.4444	8.3333
360	8.3333	0.1200	69.4444	8.3333

Appendix – 5A (Continued) Section I (Continued)
Time Value of Money Factors – Discrete Compounding (15%)

n	*Single Sums*		*Uniform Series*	
	To find F Given P (F/P, i%, n)	*To find P Given* F (P/F, i%, n)	*To find F Given* A (F/A, i%, n)	*To find A Given* F (A/F, i%, n)
1	1.1500	0.8696	1.0000	1.0000
2	1.3225	0.7561	2.1500	0.4651
3	1.5209	0.6575	3.4725	0.2880
4	1.7490	0.5718	4.9934	0.2003
5	2.0114	0.4972	6.7424	0.1483
6	2.3131	0.4323	8.7537	0.1142
7	2.6600	0.3759	11.0668	0.0904
8	3.0590	0.3269	13.7268	0.0729
9	3.5179	0.2843	16.7858	0.0596
10	4.0456	0.2472	20.3037	0.0493
11	4.6524	0.2149	24.3493	0.0411
12	5.3503	0.1869	29.0017	0.0345
13	6.1528	0.1625	34.3519	0.0291
14	7.0757	0.1413	40.5047	0.0247
15	8.1371	0.1229	47.5804	0.0210
16	9.3576	0.1069	55.7175	0.0179
17	10.7613	0.0929	65.0751	0.0154
18	12.3755	0.0808	75.8364	0.0132
19	14.2318	0.0703	88.2118	0.0113
20	16.3665	0.0611	102.4436	9.761E-03
21	18.8215	0.0531	118.8101	8.417E-03
22	21.6447	0.0462	137.6316	7.266E-03
23	24.8915	0.0402	159.2764	6.278E-03
24	28.6252	0.0349	184.1678	5.430E-03
25	32.9190	0.0304	212.7930	4.699E-03
26	37.8568	0.0264	245.7120	4.070E-03
27	43.5353	0.0230	283.5688	3.526E-03
28	50.0656	0.0200	327.1041	3.057E-03
29	57.5755	0.0174	377.1697	2.651E-03
30	66.2118	0.0151	434.7451	2.300E-03
36	153.1519	6.529E-03	1.014E+03	9.859E-04
42	354.2495	2.823E-03	2.355E+03	4.246E-04
48	819.4007	1.220E-03	5.456E+03	1.833E-04
54	1.895E+03	5.276E-04	1.263E+04	7.918E-05
60	4.384E+03	2.281E-04	2.922E+04	3.422E-05
66	1.014E+04	9.861E-05	6.760E+04	1.479E-05
72	2.346E+04	4.263E-05	1.564E+05	6.395E-06
120	1.922E+07	5.203E-08	1.281E+08	7.805E-09
180	8.426E+10	1.187E-11	5.617E+11	1.780E-12
360	7.0999E+21	1.409E-22	4.733E+22	2.113E-23

Appendix – 5A (Continued) Section I (Continued)
Time Value of Money Factors – Discrete Compounding (15%)

n	*Uniform Series*		*Gradient Series*	
	To find P Given A (*P*/*A*, *i*%, *n*)	*To find A Given* P (*A*/*P*, *i*%, *n*)	*To find P Given* G (*P*/*G*, *i*%, *n*)	*To find A Given* G (*A*/*G*, *i*%, *n*)
1	0.8696	1.1500	0.0000	0.0000
2	1.6257	0.6151	0.7561	0.4651
3	2.2832	0.4380	2.0712	0.9071
4	2.8550	0.3503	3.7864	1.3263
5	3.3522	0.2983	5.7751	1.7228
6	3.7845	0.2462	7.9368	2.0972
7	4.1604	0.2404	10.1924	2.4498
8	4.4873	0.2229	12.4807	2.7813
9	4.7716	0.2096	14.7548	3.0922
10	5.0188	0.1993	16.9795	3.3832
11	5.2337	0.1911	19.1289	3.6549
12	5.4206	0.1845	21.1849	3.9082
13	5.5831	0.1791	23.1352	4.1438
14	5.7245	0.1747	24.9725	4.3624
15	5.8474	0.1710	26.6930	4.5650
16	5.9542	0.1679	28.2960	4.7522
17	6.0472	0.1654	29.7828	4.9251
18	6.1280	0.1632	31.1565	5.0843
19	6.1982	0.1613	32.4213	5.2307
20	6.2593	0.1598	33.5822	5.3651
21	6.3125	0.1584	34.6448	5.4883
22	6.3587	0.1573	35.6150	5.6010
23	6.3988	0.1563	36.4988	5.7040
24	6.4338	0.1554	37.3023	5.7979
25	6.4641	0.1547	38.0314	5.8834
26	6.4906	0.1541	38.6918	5.9612
27	6.5135	0.1535	39.2890	6.0319
28	6.5335	0.1531	39.8283	6.0960
29	6.5509	0.1527	40.3146	6.1541
30	6.5660	0.1523	40.7526	6.2066
36	6.6231	0.1510	42.5872	6.4301
42	6.6478	0.1504	43.5286	6.5478
48	6.6585	0.1502	43.9997	6.6080
54	6.6631	0.1501	44.2311	6.6382
60	6.6651	0.1500	44.3431	6.6530
66	6.6660	0.1500	44.3967	6.6602
72	6.6664	0.1500	44.4221	6.6636
120	6.6667	0.1500	44.4444	6.6667
180	6.6667	0.1500	44.4444	6.6667
360	6.6667	0.1500	44.4444	6.6667

Appendix – 5A (Continued) Section I (Continued)
Time Value of Money Factors – Discrete Compounding (i=18%)

n	*Single Sums*		*Uniform Series*	
	To find F Given P (F/P, i%, n)	*To find P Given F (P/F, i%, n)*	*To find F Given A (F/A, i%, n)*	*To find A Given F (A/F, i%, n)*
1	1.1800	0.8475	1.0000	1.0000
2	1.3924	0.7182	2.1800	0.4587
3	1.6430	0.6086	3.5724	0.2799
4	1.9388	0.5158	5.2154	0.1917
5	2.2878	0.4371	7.1542	0.1398
6	2.6996	0.3704	9.4420	0.1059
7	3.1855	0.3139	12.1415	0.0824
8	3.7589	0.2660	15.3270	0.0652
9	4.4355	0.2255	19.0859	0.0524
10	5.2338	0.1911	23.5213	0.0425
11	6.1759	0.1619	28.7551	0.0348
12	7.2876	0.1372	34.9311	0.0286
13	8.5994	0.1163	42.2187	0.0237
14	10.1472	0.0985	50.8180	0.0197
15	11.9737	0.0835	60.9653	0.0164
16	14.1290	0.0708	72.9390	0.0137
17	16.6722	0.0600	87.0680	0.0115
18	19.6733	0.0508	103.7403	9.639E-03
19	23.2144	0.0431	123.4135	8.103E-03
20	27.3930	0.0365	146.6280	6.820E-03
21	32.3238	0.0309	174.0210	5.746E-03
22	38.1421	0.0262	206.3448	4.846E-03
23	45.0076	0.0222	244.4868	4.090E-03
24	53.1090	0.0188	289.4945	3.454E-03
25	62.6686	0.160	342.6035	2.919E-03
26	73.9490	0.0135	405.2721	2.467E-03
27	87.2598	0.0115	479.2211	2.087E-03
28	102.0666	9.712E-03	566.4809	1.765E-03
29	121.5005	8.230E-03	669.4475	1.494E-03
30	143.3706	6.975E-03	790.9480	1.264E-03
36	387.0368	2.584E-03	2.145E+03	4.663E-04
42	1.045E+03	9.571E-04	5.799E+03	1.724E-04
48	2.821E+03	3.545E-04	1.566E+04	6.384E-05
54	7.614E+03	1.313E-04	4.230E+04	2.364E-05
60	2.056E+04	4.865E-05	1.142E+05	8.757E-06
66	5.549E+04	1.802E-05	3.083E+05	3.244E-06
72	1.498E+05	6.676E-06	8.322E+05	1.202E-06
120	4.225E+08	2.367E-09	2.347E+09	4.260E-10
180	8.685E+12	1.151E-13	4.825E+13	2.073E-14
360	7.543E+25	1.326E-26	4.190E+26	2.386E-27

Appendix – 5A (Continued) Section I (Continued)
Time Value of Money Factors – Discrete Compounding (i = 18%)

n	*Uniform Series*		*Gradient Series*	
	To find P Given A (*P/A, i%, n*)	*To find A Given* P (*A/P, i%, n*)	*To find P Given* G (*P/G, i%, n*)	*To find A Given* G (*A/G, i%, n*)
1	0.8475	1.1800	0.0000	0.0000
2	1.5656	0.6387	0.7182	0.4587
3	2.1743	0.4599	1.9354	0.8902
4	2.6901	0.3717	3.4828	1.2947
5	3.1272	0.3198	5.2312	1.6728
6	3.4976	0.2859	7.0834	2.0252
7	3.8115	0.2624	8.9670	2.3526
8	4.0776	0.2452	10.8292	2.6558
9	4.3030	0.2324	12.6329	2.9358
10	4.4941	0.2225	14.3525	3.1936
11	4.6560	0.2148	15.9716	3.4303
12	4.7932	0.2086	17.4811	3.6470
13	4.9095	0.2037	18.8765	3.8449
14	5.0081	0.1997	20.1576	4.0250
15	5.0916	0.1964	21.3269	4.1887
16	5.1624	0.1937	22.3885	4.3369
17	5.2223	0.1915	23.3482	4.4708
18	5.2732	0.1896	24.2123	4.5916
19	5.3162	0.1881	24.9877	4.7003
20	5.3527	0.1868	25.6813	4.7978
21	5.3837	0.1857	26.3000	4.8851
22	5.4099	0.1848	26.8506	4.9632
23	5.4321	0.1841	27.3394	5.0329
24	5.4509	0.1835	27.7725	5.0950
25	5.4669	0.1829	28.1555	5.1502
26	5.4804	0.1825	28.4935	5.1991
27	5.4919	0.1821	28.7915	5.2425
28	5.5016	0.1818	29.0537	5.2810
29	5.5098	0.1815	29.2842	5.3149
30	5.5168	0.1813	29.4864	5.3448
36	5.5412	0.1805	30.2677	5.4623
42	5.5502	0.1802	30.6113	5.5153
48	5.5536	0.1801	30.7587	5.5385
54	5.5548	0.1800	30.8207	5.5485
60	5.5553	0.1800	30.8465	5.5526
66	5.5555	0.1800	30.8570	5.5544
72	5.5555	0.1800	30.8613	5.5551
120	5.5556	0.1800	30.8642	5.5556
180	5.5556	0.1800	30.8642	5.5556
360	5.5556	0.1800	30.8642	5.5556

Appendix – 5A (Continued) Section I (Continued)

Time Value of Money Factors – Discrete Compounding (i = 20%)

	Single Sums		*Uniform Series*	
n	*To find F Given P (F/P, i%, n)*	*To find P Given F (P/F, i%, n)*	*To find F Given A (F/A, i%, n)*	*To find A Given F (A/F, i%, n)*
1	1.2000	0.8333	1.0000	1.0000
2	1.4400	0.6944	2.2000	0.4545
3	1.7280	0.5787	3.6400	0.2747
4	2.0736	0.4823	5.3680	0.1863
5	2.4883	0.4019	7.4416	0.1344
6	2.9860	0.3349	9.9299	0.1007
7	3.5832	0.2791	12.9159	0.0774
8	4.2998	0.2326	16.4991	0.0606
9	5.1598	0.1938	20.7989	0.0481
10	6.1917	0.1615	25.9587	0.0385
11	7.4301	0.1346	32.1504	0.0311
12	8.9161	0.1122	39.5805	0.0253
13	10.6993	0.0935	48.4966	0.0206
14	12.8392	0.0779	59.1959	0.0169
15	15.4070	0.0649	72.0351	0.0139
16	18.4884	0.0541	87.4421	0.0114
17	22.1861	0.0451	105.9306	9.440E-03
18	26.6233	0.0376	128.1167	7.805E-03
19	31.9480	0.0313	154.7400	6.462E-03
20	38.3376	0.0261	186.6880	5.357E-03
21	46.0051	0.0217	225.0256	4.444E-03
22	55.2061	0.0181	271.0307	3.690E-03
23	66.2474	0.0151	326.2369	3.065E-03
24	79.4968	0.0126	392.4842	2.548E-03
25	95.3962	0.0105	471.9811	2.119E-03
26	114.4855	8.735E-03	567.3773	1.762E-03
27	137.3706	7.280E-03	681.8528	1.467E-03
28	164.8447	6.066E-03	819.2233	1.221E-03
29	197.8136	5.055E-03	984.0680	1.016E-03
30	237.3763	4.213E-03	1.182E+03	8.461E-04
36	708.8019	1.411E-03	3.539E+03	2.826E-04
42	2.116E+03	4.725E-04	1.058E+04	9.454E-05
48	6.320E+03	1.582E-04	3.159E+04	3.165E-05
54	1.887E+04	5.299E-05	9.435E+04	1.060E-05
60	5.635E+04	1.775E-05	2.817E+05	3.549E-06
66	1.683E+05	5.943E-06	8.413E+05	1.189E-06
72	5.024E+05	1.990E-06	2.512E+06	3.981E-07
120	3.175E+09	3.150E-10	1.588E+10	6.299E-11
180	1.789E+14	5.590E-10	8.945E+14	1.118E-15
360	3.201E+28	3.124E-29	1.600E+29	6.249E-30

Appendix – 5A (Continued) Section I (Continued)
Time Value of Money Factors – Discrete Compounding (i=20%)

n	*Uniform Series*		*Gradient Series*	
	To find P Given A (P/A, i%, n)	*To find A Given P (A/P, i%, n)*	*To find P Given G (P/G, i%, n)*	*To find A Given G (A/G, i%, n)*
1	0.8333	1.2000	0.0000	0.0000
2	1.5278	0.6545	0.6944	0.4545
3	2.1065	0.4747	1.8519	0.8791
4	2.5887	0.3863	3.2986	1.2742
5	2.9906	0.3344	4.9061	1.6405
6	3.3255	0.3007	6.5806	1.9788
7	3.6046	0.2774	8.2551	2.2902
8	3.8372	0.2606	9.8831	2.5756
9	4.0310	0.2481	11.4335	2.8364
10	4.1925	0.2385	12.8871	3.0739
11	4.3271	0.2311	14.2330	3.2893
12	4.4392	0.2253	15.4667	3.4841
13	4.5327	0.2206	16.5883	3.6597
14	4.6106	0.2169	17.6008	3.8175
15	4.6755	0.2139	18.5095	3.9588
16	4.7296	0.2114	19.3208	4.0851
17	4.7746	0.2094	20.0419	4.1976
18	4.8122	0.2078	20.6805	4.2975
19	4.8435	0.2065	21.2439	4.3861
20	4.8696	0.2054	21.7395	4.4643
21	4.8913	0.2044	22.1742	4.5334
22	4.9094	0.2037	22.5546	4.5941
23	4.9245	0.2031	22.8867	4.6475
24	4.9371	0.2025	23.1760	4.6943
25	4.9476	0.2021	23.4276	4.7352
26	4.9563	0.2018	23.6460	4.7709
27	4.9636	0.2015	23.8353	4.8020
28	4.9697	0.2012	23.9991	4.8291
29	4.9747	0.2010	24.1406	4.8527
30	4.9789	0.2008	24.2628	4.8731
36	4.9920	0.2003	24.7108	4.9491
42	4.9976	0.2001	24.8890	4.9801
48	4.9992	0.2000	24.8581	4.9924
54	4.9997	0.2000	24.9844	4.9971
60	4.9999	0.2000	24.9942	4.9989
66	5.0000	0.2000	24.9979	4.9996
72	5.0000	0.2000	24.9992	4.9999
120	5.0000	0.2000	25.0000	5.0000
180	5.0000	0.2000	25.0000	5.0000
360	5.0000	0.2000	25.0000	5.0000

Appendix – 5A (Continued) Section I (Continued)
Time Value of Money Factors – Discrete Compounding (i=25%)

n	Single Sums		Uniform Series	
	To find F Given P (F/P, i%, n)	*To find P Given F (P/F, i%, n)*	*To find F Given A (F/A, i%, n)*	*To find A Given F (A/F, i%, n)*
1	1.2500	0.8000	1.0000	1.0000
2	1.5625	0.6400	2.2500	0.4444
3	1.9531	0.5120	3.8125	0.2623
4	2.4414	0.4096	5.7656	0.1734
5	3.0518	0.3277	8.2070	0.1218
6	3.8147	0.2621	11.2588	0.0888
7	4.7684	0.2097	15.0735	0.0663
8	5.9605	0.1678	19.8419	0.0504
9	7.4506	0.1342	25.8023	0.0388
10	9.3132	0.1074	33.2529	0.0301
11	11.6415	0.0859	42.5661	0.0235
12	14.5519	0.0687	54.2077	0.0184
13	18.1899	0.0550	68.7596	0.0145
14	22.7374	0.0440	86.9495	0.0115
15	28.4217	0.0352	109.6868	9.117E-03
16	35.5271	0.0281	138.1085	7.241E-03
17	44.4089	0.0225	176.6357	5.759E-03
18	55.5112	0.0180	218.0446	4.586E-03
19	69.3889	0.0144	273.5558	3.656E-03
20	86.7362	0.0115	342.9447	2.916E-03
21	108.4202	9.223E-03	429.6809	2.327E-03
22	135.5253	7.379E-03	538.1011	1.858E-03
23	169.4066	5.903E-03	673.6264	1.485E-03
24	211.7582	4.722E-03	843.0329	1.186E-03
25	264.6978	3.788E-03	1.055E+03	9.481E-04
26	330.8722	3.022E-03	1.319E+03	7.579E-03
27	413.5903	2.418E-03	1.650E+03	6.059E-04
28	516.9879	1.934E-03	2.064E+03	4.845E-04
29	646.2349	1.547E-03	2.581E+03	3.875E-04
30	807.7936	1.238E-03	3.227E+03	3.099E-04
36	3.081E+03	3.245E-04	1.232E+04	8.116E-05
42	1.175E+04	8.507E-05	4.702E+04	2.127E-05
48	4.484E+04	2.230E-05	1.794E+05	5.575E-06
54	1.711E+05	5.846E-06	6.842E+05	1.462E-06
60	6.525E+05	1.532E-06	2.610E+06	3.831E-07
66	2.489E+06	4.017E-07	9.957E+06	1.004E-07
72	9.496E+06	1.0523E-07	3.798E+07	2.633E-08
120	4.258E+11	2.349E-12	1.703E+12	5.871E-13
180	2.778E+17	3.599E-18	1.111E+18	8.998E-19
360	7.720E+34	1.295E-35	3.088E+35	3.238E-36

Appendix – 5A (Continued) Section I (Continued)
Time Value of Money Factors – Discrete Compounding (i=25%)

n	Uniform Series		Gradient Series	
	To find P Given A (P/A, i%, n)	*To find A Given* P (A/P, i%, n)	*To find P Given* G (P/G, i%, n)	*To find A Given* G (A/G, i%, n)
1	0.8000	1.2500	0.0000	0.0000
2	1.4400	0.6944	0.6400	0.4444
3	1.9520	0.5123	1.6640	0.8525
4	2.3616	0.4234	2.8928	1.2249
5	2.6893	0.3718	4.2035	1.5631
6	2.9514	0.3388	5.5142	1.8683
7	3.1611	0.3163	6.7725	2.1424
8	3.3289	0.3004	7.9469	2.3872
9	3.4631	0.2888	9.0207	2.6048
10	3.5705	0.2801	9.9870	2.7971
11	3.6564	0.2735	10.8460	2.9663
12	3.7251	0.2864	11.6020	3.1145
13	3.7801	0.2645	12.2617	3.2437
14	3.8241	0.2615	12.8334	3.3559
15	3.8593	0.2591	13.3260	3.4530
16	3.8874	0.2572	13.7482	3.5366
17	3.9099	0.2558	14.1085	3.6084
18	3.9279	0.2546	14.4147	3.6698
19	3.9424	0.2537	14.6741	3.7222
20	3.9539	0.2529	14.8932	3.7667
21	3.9631	0.2523	15.0777	3.8045
22	3.9705	0.2519	15.2326	3.8365
23	3.9764	0.2515	15.3625	3.8634
24	3.9811	0.2512	15.4711	3.8861
25	3.9849	0.2509	15.5618	3.9052
26	3.9879	0.2508	15.6373	3.9212
27	3.9903	0.2506	15.7002	3.9346
28	3.9923	0.2505	15.7524	3.9457
29	3.9938	0.2504	15.7957	3.9551
30	3.9950	0.2503	15.8316	3.9628
36	3.9987	0.2501	15.9481	3.9883
42	3.9997	0.2500	15.9843	3.9964
48	3.9999	0.2500	15.9954	3.9989
54	4.0000	0.2500	15.9986	3.9997
60	4.0000	0.2500	15.9996	3.9999
66	4.0000	0.2500	15.9999	4.0000
72	4.0000	0.2500	16.0000	4.0000
120	4.0000	0.2500	16.0000	4.0000
180	4.0000	0.2500	16.0000	4.0000
360	4.0000	0.2500	16.0000	4.0000

Appendix – 5A (Continued) Section I (Continued)
Time Value of Money Factors – Discrete Compounding (i=30%)

n	*Single Sums*		*Uniform Series*	
	To find F Given P (F/P, i%, n)	*To find P Given F (P/F, i%, n)*	*To find F Given A (F/A, i%, n)*	*To find A Given F (A/F, i%, n)*
1	1.3000	0.7692	1.0000	1.0000
2	1.6900	0.5917	2.3000	0.4348
3	2.1970	0.4552	3.9900	0.2506
4	2.8561	0.3501	6.1870	0.1616
5	3.7129	0.2693	9.0431	0.1106
6	4.8268	0.2072	12.7560	0.0784
7	6.2749	0.1594	17.5828	0.0569
8	8.1573	0.1226	23.8577	0.0419
9	10.6045	0.0943	32.0150	0.0312
10	13.7858	0.0725	42.6195	0.0235
11	17.9216	0.0558	56.4053	0.0177
12	23.2981	0.0429	73.3270	0.0135
13	30.2875	0.0330	97.6250	0.0102
14	39.3738	0.0254	127.9125	7.818E-03
15	51.1859	0.0195	167.2863	5.978E-03
16	66.5417	0.0150	218.4722	4.577E-03
17	86.5042	0.0116	285.0139	3.509E-03
18	112.4554	8.892E-03	371.5180	2.692E-03
19	146.1920	6.840E-03	483.9734	2.066E-03
20	190.0496	5.262E-03	630.1655	1.587E-03
21	247.0645	4.048E-03	820.2151	1.219E-03
22	321.1839	3.113E-03	1.067E+03	9.370E-04
23	417.5391	2.395E-03	1.388E+03	7.202E-04
24	542.8008	1.842E-03	1.806E+03	5.537E-04
25	705.6410	1.417E-03	2.349E+03	4.257E-04
26	917.3333	1.090E-03	3.274E+03	3.274E-04
27	1.193E+03	8.386E-04	3.972E+03	2.518E-04
28	1.550E+03	6.450E-04	5.164E+03	1.936E-04
29	2.105E+03	4.962E-04	6.715E+03	1.489E-03
30	2.620E+03	3.817E-04	8.730E+03	1.145E-04
36	1.265E+04	7.908E-05	4.215E+04	2.372E-05
42	6.104E+04	1.638E-05	2.035E+05	4.915E-06
48	2.946E+05	3.394E-06	9.821E+05	1.018E-06
54	1.422E+06	7.032E-07	4.740E+06	2.110E-07
60	6.864E+06	1.457E-07	2.288E+07	4.370E-08
66	3.313E+07	3.018E-08	1.104E+08	9.054E-09
72	1.599E+08	6.253E-09	5.331E+08	1.876E-09
120	4.712E+13	2.122E-14	1.571E+14	6.367E-15
180	3.234E+04	3.092E-21	1.078E+21	9.275E-22
360	1.046E+41	9.559E-42	3.487E+41	2.868E-42

Appendix – 5A (Continued) Section I (Continued)

Time Value of Money Factors – Discrete Compounding (i = 30%)

n	*Uniform Series*		*Gradient Series*	
	To find P Given A (P/A, i%, n)	*To find A Given* P (A/P, i%, n)	*To find P Given* G (P/G, i%, n)	*To find A Given* G (A/G, i%, n)
1	0.7692	1.3000	0.0000	0.0000
2	1.3609	0.7384	0.5917	0.4348
3	1.8161	0.5506	1.5020	0.8271
4	2.1662	0.4616	2.5524	1.1783
5	2.4356	0.4106	3.6297	1.4903
6	2.6427	0.3784	4.6656	1.7654
7	2.8021	0.3569	5.6218	2.0063
8	2.9247	0.3419	6.4800	2.2156
9	3.0190	0.3312	7.2343	2.3693
10	3.0915	0.3235	7.8872	2.5512
11	3.1473	0.3177	8.4452	2.6833
12	3.1903	0.3135	8.9173	2.7952
13	3.2233	0.3102	9.3135	2.8895
14	3.2487	0.3078	9.6437	2.9685
15	3.2682	0.3060	9.9172	3.0344
16	3.2832	0.3046	10.1426	3.0892
17	3.2948	0.3035	10.3276	3.1345
18	3.3037	0.3027	10.4788	3.1718
19	3.3105	0.3021	10.6019	3.2025
20	3.3158	0.3016	10.7019	3.2275
21	3.3198	0.3012	10.7828	3.2480
22	3.3230	0.3009	10.8482	3.2646
23	3.3254	0.3007	10.9009	3.2781
24	3.3272	0.3006	10.9433	3.2890
25	3.3286	0.3004	10.9773	3.2979
26	3.3297	0.3003	11.0045	3.3050
27	3.3305	0.3003	11.0263	3.3107
28	3.3312	0.3002	11.437	3.3153
29	3.3317	0.3001	11.0576	3.3189
30	3.3321	0.3001	11.0687	3.3219
36	3.3331	0.3000	11.1007	3.3305
42	3.3333	0.3000	11.1086	3.3326
48	3.3333	0.3000	11.1105	3.3332
54	3.3333	0.3000	11.1110	3.3333
60	3.3333	0.3000	11.1111	3.3333
66	3.3333	0.3000	11.1111	3.3333
72	3.3333	0.3000	11.1111	3.3333
120	3.3333	0.3000	11.1111	3.3333
180	3.3333	0.3000	11.1111	3.3333
360	3.3333	0.3000	11.1111	3.3333

Appendix – 5A (Continued) Section I (Continued)
Time Value of Money Factors – Discrete Compounding (i = 40%)

n	Single Sums		Uniform Series	
	To find F Given P (F/P, i%, n)	To find P Given F (P/F, i%, n)	To find F Given A (F/A, i%, n)	To find A Given F (A/F, i%, n)
1	1.4000	0.7143	1.0000	1.0000
2	1.9600	0.5102	2.4000	0.4167
3	2.7440	0.3644	4.3600	0.2294
4	3.8416	0.2603	7.1040	0.1408
5	5.3782	0.1859	10.9456	0.0914
6	7.5295	0.1328	16.3238	0.0613
7	10.5414	0.0949	28.8534	0.0419
8	14.7579	0.0678	34.3947	0.0291
9	20.6610	0.0484	49.1526	0.0203
10	28.9255	0.0346	69.8137	0.0143
11	40.4957	0.0247	98.7391	0.0101
12	56.6939	0.0176	139.2348	7.182E-03
13	79.3715	0.0126	195.9287	5.104 E-03
14	111.1201	8.999E-03	275.3002	3.632 E-03
15	155.5681	6.428E-03	386.4202	2.588 E-03
16	217.7953	4.591 E-03	541.9883	1.845 E-03
17	304.9135	3.280 E-03	759.7837	1.316 E-03
18	426.8789	2.343 E-03	1.065E+03	9.392 E-04
19	597.6304	1.673 E-03	1.492 E+03	6.704 E-04
20	836.6826	1.195 E-03	2.089 E+03	4.787 E-04
21	1.171E+03	8.537 E-04	2.926 E+03	3.418 E-04
22	1.640 E+03	6.098 E-04	4.097 E+03	2.441 E-04
23	2.296E +03	4.356 E-04	5.737 E+03	1.743 E-04
24	3.214E +03	3.111 E-04	8.033 E+04	1.245 E-04
25	4.500E+03	2.222 E-04	1.125 E+04	8.891 E-04
26	6.300E+03	1.587 E-04	1.575 E+04	6.350 E-05
27	8.820E+04	1.134 E-04	2.205 E+04	4.536 E-05
28	1.235 E+04	8.099 E-05	3.087 E+04	3.240 E-05
29	1.729 E+04	5.785 E-05	4.321 E+04	2.314 E-05
30	2.420 E+04	4.132 E-05	6.050 E+04	1.653 E-05
36	1.822 E+05	5.488 E-06	4.556 E+05	2.195E-06
42	1.372 E+06	7.288 E-07	3.430 E+06	2.915E-07
48	1.033 E+07	9.680 E-08	2.583 E+07	3.872E-08
54	7.779 E+07	1.268 E-08	1.954 E+08	5.142E-09
60	5.857 E+08	1.707 E-09	1.464 E+09	6.829E-10
66	4.410 E+09	2.268 E-10	1.103 E+10	9.070 E-11
72	3.321 E+10	3.011 E-11	8.302 E+10	1.205 E-11
120	3.431 E+17	2.915 E-18	8.576 E+17	1.166E-18
180	2.009 E+26	4.977 E-27	5.023 E+26	1.991E-27
360	4.037 E+52	2.477 E-53	1.009 E+53	9.908E-54

Appendix – 5A (Continued) Section I (Continued)
Time Value of Money Factors – Discrete Compounding (i=40%)

n	*Uniform Series*		*Gradient Series*	
	To find P Given A (*P/A, i%, n*)	*To find A Given* P (*A/P, i%, n*)	*To find P Given* G (*P/G, i%, n*)	*To find A Given* G (*A/G, i%, n*)
1	0.7143	1.4000	0.0000	0.0000
2	1.2245	0.8167	0.5102	0.4167
3	1.5889	0.6294	1.2391	0.7798
4	1.8492	0.5408	2.0200	1.0923
5	2.0352	0.4914	2.7637	1.3580
6	2.1680	0.4613	3.4278	1.5811
7	2.2628	0.4419	3.9970	1.7664
8	2.3306	0.4219	4.4713	1.9185
9	2.3790	0.4203	4.8585	2.0422
10	2.4136	0.4143	5.1696	2.1419
11	2.4383	0.4101	5.4166	2.2215
12	2.4559	0.4072	5.6106	2.2845
13	2.4685	0.4051	5.7618	2.3341
14	2.4775	0.4036	5.8788	2.3729
15	2.4389	0.4026	5.9688	2.4030
16	2.4885	0.4018	6.0376	2.4262
17	2.4918	0.4013	6.0901	2.4441
18	2.4941	0.4009	6.1299	2.4577
19	2.4958	0.4007	6.1601	2.4682
20	2.4970	0.4005	6.1828	2.4761
21	2.4979	0.4003	6.1998	2.4821
22	2.4985	0.4002	6.2127	2.4866
23	2.4989	0.4002	6.2222	2.4900
24	2.4992	0.4001	6.2294	2.4925
25	2.4994	0.4001	6.2347	2.4944
26	2.4996	0.4001	6.2387	2.4959
27	2.4997	0.4000	6.2416	2.4969
28	2.4998	0.4000	6.2438	2.4977
29	2.4999	0.4000	6.2454	2.4983
30	2.4999	0.4000	6.2466	2.4998
36	2.5000	0.4000	6.2495	2.4998
42	2.5000	0.4000	6.2499	2.5000
48	2.5000	0.4000	6.2500	2.5000
54	2.5000	0.4000	6.2500	2.5000
60	2.5000	0.4000	6.2500	2.5000
66	2.5000	0.4000	6.2500	2.5000
72	2.5000	0.4000	6.2500	2.5000
120	2.5000	0.4000	6.2500	2.5000
180	2.5000	0.4000	6.2500	2.5000
360	2.5000	0.4000	6.2500	2.5000

Appendix – 5A (Continued) Section-II
Geometric Series Factor Discrete Compounding (i=5%)

	Geometric series present worth factor				$(P/A_1\ i,j,n)$
n	*j=4%*	*j=6%*	*j=8%*	*j=10%*	*j=15%*
1	0.9524	0.9524	0.9524	0.9524	0.9524
2	1.8957	1.9138	1.9320	1.9501	1.9955
3	2.8300	2.8844	2.9396	2.9954	3.1379
4	3.7554	3.8643	3.9759	4.0904	4.3891
5	4.6721	4.8535	5.0419	5.2375	5.7595
6	5.5799	5.8521	6.1383	6.4393	7.2604
7	6.4792	6.8602	7.2661	7.6983	8.9043
8	7.3699	7.8779	8.4261	9.0173	10.7047
9	8.2521	8.9053	9.6192	10.3991	12.6765
10	9.1258	9.9425	10.8464	11.8467	14.8362
11	9.9913	10.9896	12.1087	13.3632	17.2016
12	10.8485	12.0466	13.4070	14.9519	19.7922
13	11.6976	13.1137	14.7425	16.6163	22.6295
14	12.5386	14.1910	16.1161	18.3599	25.7371
15	13.3715	15.2785	17.5289	20.1866	29.1407
16	14.1966	16.3764	18.9821	22.1002	32.8683
17	15.0137	17.4848	20.4769	24.1050	36.9510
18	15.8231	18.6037	22.0143	26.2052	41.4226
19	16.6248	19.7332	23.5956	28.4055	46.3200
20	17.4189	20.8736	25.2222	30.7105	51.6838
21	18.2054	22.0247	26.8952	33.1253	57.5584
22	18.9844	23.1869	28.6160	35.6550	63.9925
23	19.7559	24.3601	30.3860	38.3053	71.0394
24	20.5202	25.5445	32.2066	41.0817	78.7575
25	21.2771	26.7401	34.0791	43.9904	87.2106
26	22.0269	27.9472	36.0052	47.0375	96.4487
27	22.7695	29.7695	37.9863	50.2298	106.6086
28	23.5050	30.3959	40.0240	53.5741	117.7142
29	24.2335	31.6377	42.1199	57.0776	129.8774
30	24.9551	32.8914	44.2757	60.7480	143.1991
31	25.6698	34.1571	46.4931	64.5931	157.7895
32	26.3777	35.4348	48.7739	68.6213	173.7695
33	27.0789	36.7246	51.1198	72.8414	191.2713
34	27.7734	38.0267	53.5328	77.2624	210.4400
35	28.4612	39.3413	56.0146	81.8940	231.4343
36	29.1426	40.6683	58.5674	86.7461	254.4280
37	29.8174	42.0030	61.1932	91.8292	279.6116
38	30.4858	43.3605	63.8939	97.1544	307.1937
39	31.1478	44.7258	66.6719	102.7332	337.4026
40	31.8036	46.1042	69.5291	108.5776	370.4886
48	36.8296	57.6141	95.5310	166.5488	777.7891
50	38.0271	60.6306	102.9998	184.7384	934.9897
54	40.3544	66.8378	119.2613	226.6112	1349.7568
60	43.6826	76.6013	147.3617	306.0117	2236.9967
65	46.3138	85.1726	174.6930	391.0117	3688.7517
70	48.8221	94.1599	206.1585	499.1191	5819.0515
80	53.4925	113.4643	284.0874	806.6104	14467.1576
90	57.7367	134.6830	387.3737	1296.2390	35945.7801
100	61.5936	158.0219	524.2687	2075.8906	89290.5489
120	68.2835	211.8803	946.1871	5294.1808	550829.3461

Appendix – 5A (Continued) Section II (Continued)
Geometric Series Factor Discrete Compounding (i=5%)

	Geometric series present worth factor				$(F/A_1\ i,j,n)$
n	*j=4%*	*j=6%*	*j=8%*	*j=10%*	*j=15%*
1	1.0000	1.0000	1.0000	1.0000	1.0000
2	2.0900	2.1100	2.1300	2.1500	2.2000
3	3.2761	3.3391	3.4029	3.4675	3.6325
4	4.5648	4.6971	4.8328	4.9719	5.3350
5	5.9629	6.1944	6.4349	6.6846	7.3508
6	7.4777	7.8423	8.2260	8.6293	9.7297
7	9.1169	9.6530	10.2241	10.8323	12.5292
8	10.8886	11.6393	12.4492	13.3227	15.8157
9	12.8016	13.8151	14.9225	16.1324	19.6655
10	14.8650	16.1953	17.6677	19.2970	24.1666
11	17.0885	18.7959	20.7100	22.8556	29.4205
12	19.4824	21.6340	24.0771	26.8515	35.5439
13	22.0576	24.7279	27.7992	31.3325	42.6714
14	24.8255	28.0972	31.9087	36.3514	50.9577
15	27.7985	31.7630	36.4414	41.9664	60.5813
16	30.9893	35.7477	41.4356	48.2420	71.7475
17	34.4118	40.0754	46.9333	55.2491	84.6925
18	38.0803	44.7720	52.9800	63.0660	99.6883
19	42.0101	49.8649	59.6250	71.7792	117.0482
20	46.2175	55.3838	66.9220	81.4841	137.1324
21	50.7195	61.3601	74.9290	92.2858	160.3556
22	55.5342	67.8277	83.7093	104.3004	187.1948
23	60.6808	74.8226	93.3313	117.6557	218.1993
24	66.1796	82.3835	103.8694	132.4928	254.0008
25	72.0519	90.5516	115.4040	148.9672	295.3260
26	78.3203	69.3710	128.0227	167.2503	343.0112
27	85.0088	108.8890	141.8202	187.5310	398.0186
28	92.1426	119.1558	156.8992	210.0176	461.4548
29	99.7484	130.2252	173.3713	234.9395	534.5932
30	107.8545	142.1549	191.3572	262.5496	618.8983
31	116.4906	155.0061	210.9877	293.1266	716.0550
32	125.6883	168.8445	232.4047	326.9773	828.0013
33	135.4807	183.7401	255.7620	364.4400	956.9664
34	145.9032	199.7677	281.2262	405.8872	1105.5145
35	156.9926	217.0071	308.9776	451.7293	1276.5951
36	168.7884	235.5436	339.2119	502.4183	1473.6003
37	181.3317	255.4680	372.1406	558.4520	1700.4322
38	194.6664	276.8775	407.9933	620.3787	1961.5784
39	208.8385	299.8756	447.0182	688.8021	2262.2007
40	223.8968	324.5729	489.4844	764.3871	2608.2355
48	383.0741	599.2602	993.6434	1732.3243	8089.9942
50	436.0716	695.2754	1181.1404	2118.4755	10721.9001
54	562.4882	931.6324	1662.3477	3158.6756	18813.8487
60	815.9558	1430.8505	2752.5959	5716.0709	43653.1937
65	1104.1165	2030.5071	4164.6649	9330.6558	87939.4708
70	1485.4807	2864.9505	6272.6660	15186.4808	177052.9271
80	2651.1642	5623.4552	14079.779	39976.9935	717013.1369
90	4661.1032	10873.4146	831272.8241	104646.5051	2901915.7503
100	8099.6310	20780.0826	68941.9999	272984.1812	11741818.5907
120	23824.9425	73927.5762	330136.0319	1847219.6697	192190942.7662

Appendix – 5A (Continued) Section II (Continued)
Geometric Series Factor Discrete Compounding (i=8%)

	Geometric series present worth factor				$(P/A_1\ i,j,n)$
n	*j=4%*	*j=6%*	*j=8%*	*j=10%*	*j=15%*
1	0.9250	0.9259	0.9259	0.9259	0.9259
2	1.8176	1.8347	1.8519	1.8690	1.9119
3	2.6762	2.7267	2.7778	2.8295	2.9617
4	3.5030	3.6021	3.7037	3.8079	4.0796
5	4.2992	4.4613	4.6296	4.8043	5.2699
6	5.0659	5.3046	5.5556	5.8192	6.5374
7	5.8042	6.1323	6.4815	6.8529	7.8871
8	6.5151	6.9447	7.4074	7.9057	9.3242
9	7.1997	7.7420	8.3333	8.9780	10.8545
10	7.8590	8.5246	9.2593	10.0702	12.4839
11	8.4939	9.2926	10.1852	11.1826	14.2190
12	9.1052	10.0465	11.1111	12.3157	16.0665
13	9.6939	10.7863	12.0370	13.4696	18.0338
14	10.2608	11.5125	12.9630	14.6450	20.1286
15	10.8067	12.2252	13.8889	15.8421	22.3592
16	11.3324	12.9248	14.8148	17.0614	24.7343
17	11.8386	13.6114	15.7407	18.3033	27.2634
18	12.3260	14.2852	16.6667	19.5682	29.9564
19	12.7954	14.9466	17.5926	20.8565	32.8239
20	13.2475	15.5957	18.5185	22.1687	35.8773
21	13.6827	16.2329	19.4444	23.5051	39.1286
22	14.1019	16.8582	20.3704	24.8663	42.5906
23	14.5055	17.4719	21.2963	26.2527	46.2771
24	14.8942	18.0743	22.2222	27.6648	50.2024
25	15.2685	18.6655	23.1482	29.1031	54.3822
26	15.6289	19.2458	24.0741	30.5679	58.8329
27	15.9760	19.8153	25.0000	32.0599	63.5721
28	16.3102	20.3743	25.9259	33.5796	68.6184
29	16.6321	20.9229	26.8519	35.1273	73.9919
30	16.9420	21.4614	27.7778	36.7038	79.7136
31	17.2404	21.9899	28.7037	38.3094	85.8061
32	17.5278	22.5086	29.6296	39.9447	92.2935
33	17.8046	23.0177	30.5556	41.6104	99.2015
34	18.0711	23.5173	31.4815	43.3069	106.5571
35	18.3277	24.0078	32.4074	45.0348	114.3895
36	18.5749	24.4891	33.3333	46.7947	122.7296
37	18.8126	24.9615	34.2593	48.5872	131.6102
38	19.0419	25.4252	35.1852	50.4129	141.0664
39	19.2626	25.8803	36.1111	52.2724	151.1355
40	19.4751	26.3269	37.0370	54.1663	161.8573
48	20.9149	29.6150	44.4444	70.6365	276.8249
50	21.2119	30.3630	46.2963	75.1459	315.7844
54	21.7427	31.7777	50.0000	84.6766	410.0431
60	22.4027	33.7109	55.5556	100.3508	604.2248
65	22.8494	35.1643	60.1852	114.7974	832.3907
70	23.2192	36.4881	64.8148	130.6322	1144.7262
80	23.7790	38.7917	74.0741	167.0123	2157.5589
90	24.1629	40.7026	83.3333	210.7196	4055.4818
100	24.4260	42.2878	92.5926	263.2297	7611.9539
120	24.7302	44.6933	111.1111	402.1073	26764.5779

Appendix – 5A (Continued) Section II (Continued)
Geometric Series Factor Discrete Compounding (i=8%)

	Geometric series present worth factor				*(F/A$_1$ i,j,n)*
n	*j=4%*	*j=6%*	*j=8%*	*j=10%*	*j=15%*
1	1.0000	1.0000	1.0000	1.0000	1.0000
2	2.1200	2.1400	2.1000	2.1000	2.2300
3	3.3712	3.4348	3.4902	3.5644	3.7300
4	4.7658	4.9006	5.0388	5.1806	5.5502
5	6.3169	6.5551	6.8024	7.0591	7.7433
6	8.0389	8.4176	6.8160	9.2343	10.3741
7	9.9473	10.5097	11.1061	11.7446	13.5171
8	12.0590	12.8541	13.7106	14.6329	17.2585
9	14.3923	15.4763	16.6584	17.9472	21.6982
10	16.9670	18.4039	19.9900	21.7409	26.9519
11	19.8046	21.6670	23.7482	26.0739	33.1536
12	22.9284	25.2987	27.9797	31.0129	40.4583
13	26.3638	29.3348	32.7362	36.6324	49.0452
14	30.1379	33.8145	38.0747	43.0153	59.1216
15	34.2806	38.7805	44.0579	50.2540	70.9270
16	38.8240	44.2795	50.7547	58.4516	84.7383
17	43.8029	50.3623	58.2410	67.7227	100.8749
18	49.2551	57.0840	66.6003	78.1950	119.7062
19	55.2213	64.5051	75.9243	90.0105	141.6581
20	61.7459	72.6911	86.3139	103.3272	167.2226
21	68.8766	81.7135	97.8800	118.3209	196.9669
22	76.6655	91.6501	110.7442	135.1869	231.5458
23	85.1687	102.5857	125.0403	154.1421	271.7142
24	94.4469	114.6123	140.9150	175.4276	318.3428
25	104.5660	127.8302	158.5293	199.3116	372.4354
26	115.5971	142.3485	178.0601	226.0914	435.1492
27	127.6173	158.2858	199.7013	256.0970	507.8179
28	140.7101	175.7710	223.6654	289.6947	591.9786
29	154.9656	194.9443	250.1858	327.2914	689.4025
30	170.4815	215.9583	279.5179	369.3378	802.1302
31	187.3634	238.9784	311.9419	416.3343	932.5124
32	205.7256	264.1848	347.7649	468.8354	1083.2569
33	225.6917	291.7730	387.3232	527.4561	1257.4825
34	247.3954	321.9554	430.9850	592.8778	1458.7810
35	270.9814	354.9629	479.1539	665.8558	1691.2882
36	296.6060	391.0460	532.2715	747.2268	1959.7668
37	324.4384	430.4769	590.8214	835.9177	2269.7000
38	354.6616	473.5512	655.3327	938.9552	2627.4006
39	387.4733	520.5895	726.3845	1051.4760	3040.1360
40	423.0875	571.9402	804.6105	1176.7390	3516.2717
48	841.0011	1190.8351	1787.1327	2840.3400	1131.2874
50	994.8732	1424.0729	2171.3660	3524.4711	14810.7971
54	1387.3838	2027.7053	3190.4485	5403.1563	26164.4921
60	2268.4359	3413.4646	5625.3771	10161.2610	61182.0216
65	3399.5278	5231.7437	8954.3161	17079.6034	123842.9595
70	5075.8697	7976.5238	14168.8885	28557.1360	250244.4686
80	11222.6259	18307.9420	34959.4895	78822.6198	1018270.2942
90	24619.8939	41472.5289	84909.2410	214706.4761	4132191.3076
100	53731.4077	93022.9586	203680.6644	579045.9229	16744490.0582
120	253558.2595	458240.2597	1139215.1626	4122833.7485	274417003.9699

Appendix – 5A (Continued) Section II (Continued)
Geometric Series Factor Discrete Compounding (i=10%)

	Geometric series present worth factor				*(P/A₁ i,j,n)*
n	*j=4%*	*j=6%*	*j=8%*	*j=10%*	*j=15%*
1	0.9091	0.9091	0.9091	0.9091	0.9091
2	1.7686	1.7851	1.8017	1.8182	1.8596
3	2.5812	2.6293	2.6780	2.7273	2.8531
4	3.3495	3.4428	3.5384	3.6364	3.8919
5	4.0759	4.2267	4.3831	4.5455	4.9779
6	4.7627	4.9821	5.2125	5.4545	6.1123
7	5.4120	5.7100	6.0269	6.3636	7.3002
8	6.0259	6.4115	6.8264	7.2727	8.5411
9	6.6063	7.0874	7.6113	8.1818	9.8385
10	7.1550	7.7388	8.3820	9.0909	11.1948
11	7.6738	8.3664	9.1387	10.0000	12.6127
12	8.1644	8.9713	9.8817	10.9091	14.0951
13	8.6281	9.5542	10.6111	11.8182	15.6449
14	9.0666	10.1158	11.3273	12.7273	17.2651
15	9.4811	10.6571	12.0304	13.6364	18.9590
16	9.8731	11.1786	12.7208	14.5455	20.7296
17	10.2436	11.6812	13.3986	15.4545	22.5812
18	10.5940	12.1656	14.0640	16.3636	24.5167
19	10.9252	12.6323	14.7174	17.2727	26.5402
20	11.2384	13.0820	15.3589	18.1818	28.6556
21	11.5345	13.5154	15.9888	19.0909	30.8672
22	11.8144	13.9330	16.6071	20.0000	33.1794
23	12.0791	14.3354	17.2143	20.9091	35.5966
24	12.3293	14.7232	17.8104	21.8182	38.1238
25	12.5659	15.0969	18.3957	22.7273	40.7658
26	12.7896	15.4570	18.9703	23.6364	43.5278
27	13.0011	15.8041	19.5345	24.5455	46.4155
28	13.2010	16.1385	20.0884	25.4545	49.4343
30	13.5688	16.7712	21.1662	27.2727	55.8900
31	13.7377	17.0704	21.6904	28.1818	59.3396
32	13.8975	17.3588	22.2052	29.0909	62.9459
33	14.0485	17.6367	22.7105	30.0000	64.7162
34	14.1913	17.9044	23.2067	30.9091	70.6578
35	14.3264	18.1624	23.6938	31.8182	74.7786
36	14.4540	18.4111	24.1721	32.7273	79.0867
37	14.5747	18.6507	24.6417	33.6364	83.5907
38	14.6888	18.8816	25.1028	34.5455	88.2994
39	14.7967	19.1040	25.5555	35.4545	93.2221
40	14.8987	19.3184	25.9999	36.3636	98.3685
48	15.5379	20.7755	29.2766	43.6364	148.9196
50	15.6577	21.0772	30.0233	45.4545	164.6238
54	15.8605	21.6174	31.4370	49.0909	200.5507
60	16.0908	22.2915	33.3722	54.5455	267.9647
65	16.2317	22.7494	34.8299	50.0909	339.6376
70	16.3380	23.1290	36.1597	63.6364	429.1494
80	16.4791	23.7088	38.4799	72.7273	680.5553
90	16.5596	24.1085	40.4112	81.8182	1072.6824
100	16.6056	24.3845	42.0186	90.9091	1684.2979
120	16.6468	24.7066	44.4703	109.0909	4126.1845

Appendix – 5A (Continued) Section II (Continued)
Geometric Series Factor Discrete Compounding (i=10%)

	Geometric series present worth factor				*(F/A$_1$ i,j,n)*
n	*j=4%*	*j=6%*	*j=8%*	*j=10%*	*j=15%*
1	1.0000	1.0000	1.0000	1.0000	1.0000
2	2.1400	2.1600	2.1800	2.2000	2.2500
3	3.4356	3.4996	3.5644	3.6300	3.7975
4	4.9040	5.0406	5.1806	5.3240	5.6961
5	6.5643	6.8071	7.0591	7.3205	8.0169
6	8.4374	8.8260	9.2343	9.6631	10.3800
7	10.5464	11.1272	11.7446	12.4009	14.2261
8	12.9179	13.7435	14.6329	15.5897	18.3067
9	15.5773	16.7117	17.9471	19.2923	23.1986
10	18.5583	20.0724	21.7409	23.5795	29.0343
11	21.8944	23.8705	26.0739	28.5312	35.9856
12	25.6233	28.1558	31.0120	34.2374	44.2364
13	29.7886	32.9836	36.6324	40.7996	54.0103
14	34.4304	38.4149	43.0152	48.3318	65.5641
15	39.6051	44.5172	50.2539	54.9625	79.1963
16	45.3665	51.3655	58.4515	66.8360	95.2530
17	51.7762	59.0424	67.7226	76.1146	114.1359
18	58.9017	67.6395	78.1948	90.9805	136.3107
19	66.8177	77.2577	90.0103	105.6385	162.3173
20	75.6063	88.0091	103.3271	122.3183	192.7807
21	85.3580	100.0172	118.3207	141.2776	228.4254
22	96.1726	113.4184	135.1866	162.8057	270.0894
23	108.1598	128.3638	154.1418	187.2265	318.7431
24	121.4405	145.0200	175.4274	214.9035	375.5089
25	136.1478	163.5709	199.3113	246.2436	441.6849
26	152.4284	184.2198	226.0909	281.7027	518.7724
27	170.4438	207.1912	256.0964	321.7911	608.5064
28	190.3715	232.7327	289.6940	367.0803	712.8924
29	212.4074	261.1176	327.2905	418.2093	834.2472
30	236.7667	292.6478	369.3368	475.8934	975.2474
31	263.6868	327.6560	416.3331	540.9322	1138.9839
32	293.4286	366.5098	468.8341	614.2198	1329.0258
33	326.2796	409.6141	527.4545	696.7556	1549.4935
34	362.5559	457.4161	592.8760	789.6564	1805.1426
35	402.6058	510.4088	665.8537	894.1698	2101.4617
36	446.8125	569.1357	747.2244	1011.6893	2444.7834
37	495.5976	634.1965	837.9149	1143.7710	2842.4136
38	549.4255	706.2523	938.9520	1292.1522	3302.7795
39	608.8069	786.0318	1051.4724	1458.7719	3835.6008
40	657.3039	874.3384	1176.7348	1645.7940	4452.0857
48	1507.4451	2015.5841	2840.3278	4233.4883	14447.8602
50	1838.0695	2474.2675	3524.4553	5335.9507	19325.3312
54	2725.9689	3715.4232	5403.1300	8437.3705	34469.0327
60	4899.3669	6787.3487	10161.2061	16608.1339	81590.3302
65	7959.5332	11155.6438	17079.5033	28976.5362	166548.3266
70	12902.9223	18266.7757	28556.9556	50256.7821	338919.4468
80	33755.8403	48565.1055	78822.0497	148975.0958	1394049.5518
90	87981.7213	128088.9525	214704.7272	434703.6079	5699185.7115
100	228835.1232	336032.7564	579040.6774	1252788.5754	23210655.1891
120	1543306.7710	2290522.0267	4122788.8750	10113771.2878	382534696.2027

Appendix – 5A (Continued) Section II (Continued)
Geometric Series Factor Discrete Compounding (i=15%)

	Geometric series present worth factor				$(P/A_1\ i,j,n)$
n	*j=4%*	*j=6%*	*j=8%*	*j=10%*	*j=15%*
1	0.8696	0.8696	0.8696	0.8696	0.8696
2	1.6560	1.6711	1.6862	1.7013	1.7391
3	2.3671	2.4099	2.4531	2.4969	2.6087
4	3.0103	3.0908	3.1734	3.2579	3.4783
5	3.5919	3.7185	3.8498	3.9858	4.3478
6	4.1179	4.2971	4.4850	4.6821	5.2174
7	4.5936	4.8303	5.0616	5.3481	6.0870
8	5.0237	5.3219	5.6418	5.9851	6.9565
9	5.4128	5.7749	6.1680	6.5945	7.8261
10	5.7646	6.1926	6.6621	7.1773	8.6957
11	6.0828	6.5775	7.1261	7.7348	9.5652
12	6.3705	6.9323	7.5619	8.2681	10.4348
13	6.6307	7.2593	7.9712	8.7782	11.3043
14	6.8660	7.5608	8.3556	9.2661	12.1739
15	7.0789	7.8386	8.7165	9.7328	13.0435
16	7.2713	8.0947	9.0555	10.1792	13.9130
17	7.4454	8.3308	9.3739	10.6062	14.7826
18	7.6028	8.5484	9.6729	11.0146	15.6522
19	7.7451	8.7489	9.9537	11.4053	16.5217
20	7.8738	8.9338	10.2173	11.7790	17.3913
21	7.9903	9.1042	10.4650	12.1364	18.2609
22	8.0955	9.2613	10.6976	12.4783	19.1304
23	8.1907	9.4060	10.9160	12.8053	20.0000
24	8.2768	9.5395	11.1211	13.1181	20.8696
25	8.3547	9.6625	11.3137	13.4173	21.7391
26	8.4251	9.7759	11.4946	13.7035	22.6087
27	8.4888	9.8803	11.6645	13.9773	23.4783
28	8.5464	9.9767	11.8241	14.2392	24.3478
29	8.5985	10.0655	11.9739	14.4896	25.2174
30	8.6456	10.1473	11.1146	14.7292	26.0870
31	8.6882	10.2227	12.2468	14.9584	26.9565
32	8.7267	10.2922	12.3709	15.1776	27.8261
33	8.7615	10.3563	12.4874	15.3837	28.6956
34	8.7930	10.4154	12.5969	15.5878	29.5652
35	8.8215	10.4698	12.6997	15.7796	30.4348
36	8.8473	10.5200	12.7962	15.9631	31.3043
37	8.8706	10.5663	12.8869	16.1386	32.1739
38	8.8917	10.6089	12.9720	16.3065	33.0435
39	8.9107	10.6482	13.0520	16.4671	33.9130
40	8.9280	10.6845	13.1271	16.6207	34.7826
48	9.0180	10.8888	13.5847	17.6320	41.7391
50	9.0313	10.9222	13.6674	17.8334	43.4783
54	9.0510	10.9748	13.8048	18.1964	46.9565
60	9.0691	10.0275	13.9554	18.6109	52.1739
65	9.0777	10.0555	13.0447	18.8878	56.5217
70	9.0829	11.0741	14.1096	19.1094	60.8696
80	9.0880	11.0947	14.1917	19.4290	69.5652
90	9.0898	11.1039	14.2356	19.6339	78.2609
100	9.0905	11.1079	14.2590	19.7653	86.9565
120	9.0909	11.1105	14.2781	19.9035	104.3478

Appendix – 5A (Continued) Section II (Continued)
Geometric Series Factor Discrete Compounding (i=15%)

	Geometric series present worth factor				*$(F/A_1\ i,j,n)$*
n	*j=4%*	*j=6%*	*j=8%*	*j=10%*	*j=15%*
1	1.0000	1.0000	1.0000	1.0000	1.0000
2	2.1900	2.2100	2.2300	2.5000	2.3000
3	3.6001	3.6651	3.7300	3.7975	3.9675
4	5.2650	5.4059	5.5502	5.6981	6.0625
5	7.2246	7.4792	7.7433	8.0169	8.7450
6	9.5249	9.9394	10.3741	10.8300	12.0681
7	12.2190	12.8488	13.5171	14.2261	16.1914
8	15.3678	16.2797	17.2585	18.3087	21.2802
9	19.0415	20.3155	21.6982	23.1986	27.5312
10	23.3210	25.0523	26.9519	29.0363	36.1788
11	28.2994	30.6010	33.1536	35.9855	44.5011
12	34.0838	37.0895	40.4583	44.2365	55.8287
13	40.7974	44.6651	49.0452	54.0104	69.5532
14	48.5821	53.4978	59.1216	65.5642	96.1390
15	57.6011	63.7834	70.9270	79.1963	106.1356
16	68.0422	75.7474	84.7382	95.2530	130.1930
17	80.1215	89.6499	100.8749	114.1359	159.0795
18	94.0876	105.7902	119.7061	136.3108	193.7027
19	110.2266	124.5130	141.6581	162.3174	235.1336
20	128.8674	146.2156	167.2225	192.7809	284.6354
21	150.3886	171.3550	196.9668	228.4255	343.6972
22	175.2257	200.4579	231.5456	270.0896	414.0733
23	203.8795	234.1301	271.7140	318.7434	497.8291
24	236.9261	273.0694	318.3425	375.5092	597.3949
25	275.0283	318.0787	372.4351	441.6853	715.6293
26	318.9884	370.0824	435.1488	518.7729	855.8927
27	369.5631	430.1441	507.8175	608.5070	1022.1333
28	427.8810	499.4881	591.9781	712.8931	1218.9886
29	495.0618	579.5230	689.4019	834.2481	1451.9025
30	572.4398	671.8698	802.1295	975.2484	1727.2634
31	661.5491	778.3937	932.5115	1138.9851	2052.5646
32	764.1546	901.2409	1083.2559	1329.0273	2436.5929
33	882.2859	1042.8804	1257.4813	1549.4852	2889.6568
34	1018.2772	1206.1531	1458.7795	1805.1447	3423.7937
35	1174.8130	1394.3271	1691.2865	2101.4641	4053.1675
36	1354.9811	1611.1622	1959.7648	2444.7863	4794.3182
37	1562.3322	1860.9838	2269.6977	2842.4170	5666.6177
38	1800.9501	2142.7675	2627.3979	3302.7836	6692.7350
39	2075.5314	2480.2368	3040.1368	3835.6057	7899.1885
40	2391.4775	2861.9759	3516.2679	4452.0914	9316.9916
48	7389.3653	8922.2982	11131.2743	14447.6900	34201.0678
50	9786.8251	11835.9690	14810.7795	19325.3598	47115.5338
54	17154.6348	20800.7624	26164.4602	34469.0859	88897.7921
60	38758.9011	48344.5673	61181.9447	81590.4721	228730.3333
65	80045.2514	97484.9157	123842.8007	166548.6034	498396.5994
70	161092.2584	196407.1548	250244.1432	338920.0390	1079565.3949
80	652071.1782	796056.4823	1018268.9462	1394052.0739	4991364.6825
90	2638529.1443	3223142.8966	4132185.7913	5699196.5809	22716960.7653
100	10475117.6887	13044157.2069	16744457.6481	23210700.6830	102114195.2907
120	174721221.2670	213537297.9353	274416638.3034	382535463.5055	2005506937.5163

CHAPTER 6

Cost Evaluation

6.1 INTRODUCTION

The American Association of Cost Engineers (AACE) defines cost engineering as *that area of engineering practice where engineering judgment and experience are utilized in the application of scientific principles and techniques to the problems of cost estimation, cost control and profitability*. In this chapter, we shall be dealing, with cost estimation.

Cost estimation is a complex process and requires knowledge of principles of engineering, materials, manufacturing processes, quality control and inspection methods, product and environmental safety, repair and servicing, and principles of costing.

Cost estimation is an important aspect of the product evaluation process. It is advisable to generate a cost estimate early in the design process in order to compare the product design with original objectives. The design of new components or products requires cost estimates from scratch while for redesigns early cost estimates will serve the purpose as the current costs are known. The design engineer shares the responsibility of generating the cost of manufactured products or components and selling prices of the products or components along with the cost-estimating department. Cost is next to performance in importance. Lowest-cost stands a good chance to be successful in a free marketplace.

In this chapter, we consider the subject of economy, with particular emphasis on cost evaluation in design and manufacturing. The identification of the elements of costs, general concepts of cost evaluation and some of the more generally accepted cost evaluation methods and models would be described.

6.2 COST CONCEPTS

In chapter 5, we have seen that engineering economic analysis is mainly concerned with comparing alternative projects on the basis of an economic measure of effectiveness. This section provides an overview of terminology and concepts that are relevant and applicable to economic measures of effectiveness for comparing alternative projects. Concepts are crystallized thoughts that have withstood the test of time. The concepts are generally qualitative in nature and may not necessarily be universal in application. Cost estimating is needed for engineering economy. Current and future benefits should be estimated to provide the basis data to be analyzed. The cost information is generally obtained from variety of sources, such as the corporation production records, accounting records, manufacturer's catalogs, U.S. Government Printing Office (Publications), etc. Here, the emphasis is on the identification of the elements of costs and on some of the more generally accepted cost assessment methods. The cost terminology, cost factors, and cost concepts as used in practice are introduced here in order to make an effective comparison and intelligent recommendations and justifications.

6.2.1 Cost Terminology

Here, we discuss the cost definitions and cost concepts under the following six categories:

1. Life-cycle costs
2. Past and sunk costs
3. Future and opportunity costs
4. Direct, indirect and overhead costs
5. Fixed and variable costs, and
6. Average and marginal costs.

6.2.1.1 Life-cycle costs

The life-cycle cost of an item is the sum of all expenditures related to the item during its entire service life. Life cycle costs may also include engineering design and development costs, fabrication and testing costs, operating and maintenance costs, and disposal costs. The applications of life-cycle costing methods in system and product design and development is realized through the accomplishment of life cycle cost analyses. Life-cycle cost refers to all costs associated with the product or system as applied to the defined cycle. The life cycle tailored to the specific system forms the basis for life cycle costing. Throughout the system/product life cycle, many actions are required, both technical and non-technical. Life cycle costs may also be expressed as the summation of acquisition, operation, maintenance, and disposal costs. In general, life cycle costs fall into the following four categories based on organizational activity over the life cycle. They are:

1. *Research and development costs*: These costs include planning, market research, feasibility studies, product analysis, needs analysis, design, documentation, software, test and assessment of engineering models and management functions.

2. *Production and construction costs*: Production engineering analysis, operations analysis, manufacturing, facility development, process and production operations, quality control and the production of test and support equipment, and initial logistic support needs.

3. *Operation and maintenance costs*: Operation and maintenance costs are recurring costs that are necessary to operate and maintain a product or a system during its useful life. Operating and maintenance costs usually consist of consumer or user operations of the system or product in the field, product or system distribution, logistic support throughout the system or product life cycle, supply support, test and support equipment, transportation and handling, etc. Operating costs generally consist of labor, material, and overhead items. Typical overhead items are fuel, electrical power, insurance premiums, inventory charges, indirect labor, administrative and management expenses, etc. It is generally assumed that the operating and maintenance costs are annual costs, however, maintenance costs may not be on a recurring, annual basis. In most cases, the maintenance would consist of both preventive maintenance and maintenance on an *as needed* basis.

4. *Retirement and disposal costs*: These costs include the disposal of non-repairable system or products throughout the life cycle, system or product retirement, material recycling, and appropriate logistic support requirements. When the life cycle of a system/product has ended, disposal costs usually result. Disposal costs may include labor and material costs for removal of the system/product, shipping costs, or transportation costs. Generally, disposal costs are incurred at the end of the life cycle, but most systems/products have some momentary value at

the time of disposal. The value is known as the *market* or *trade-in value*. After deducting the cost of disposal, the net dollar worth at the time of disposal is called the *salvage* value. In general, the market value, the disposal costs, and the salvage value are not known with certainty and hence must be estimated. For a system/product that satisfies the US International Revenue Service (IRS) definition of a *capital asset* and that decreases in value over time through physical deterioration, the IRS has depreciation methods and guidelines that can serve to estimate the rate of deterioration and the consequent decrease in value of the asset. The value of the capital asset at the end of a given period during the asset's life is known as the *book value*. The value of the material from which the system/product is made is called the *scrap value*. The life cycle involves a time horizon and at the end of the system/product's life may be evaluated from either a functional or economic consideration. The economic life of a system/product is shorter than the functional life and is highly influenced by the company policy and the income tax considerations.

6.2.1.2 Past and sunk costs

Past costs are historical costs that have occurred for the system/product under consideration. *Sunk costs* are past costs for the system/product that are unrecoverable. Past costs and sunk costs provide valuable information that can help the accuracy of estimating future costs for similar systems/products.

6.2.1.3 Future and opportunity costs

Costs that may occur in the future are known as *future costs*. Future costs include the operating costs for labor and materials, maintenance costs, overhaul costs, and disposal costs. These costs are not readily available or known and must be estimated. The cost of foregoing the opportunity to earn interest, or a return on investment funds is called an *opportunity cost*. The concept of opportunity cost is important to the study of engineering economy and is a cost element that is included in all methodologies for comparing alternative projects. The cost of obtaining funds for financing project through debt obligations is termed as *cost of capital*. These funds are generally obtained from external sources by

(*a*) borrowing money from banks or other financial organizations

(*b*) issuing bonds.

The debt obligations are generally long term for the acquisition of supplies and raw materials.

The scheduled payments of interest are a cost of borrowed capital. Another way of financing engineering projects is through the use of *equity funds*. Equity funds are commonly obtained from one or both of the following sources:

(*a*) common or preferred stock authorized by the company and sold through brokers or investors, or

(*b*) earnings accumulated from previous years and retained by the company.

These two sources of funds incur an opportunity cost.

6.2.1.4 Direct, indirect, and overhead costs

A product's total cost to produce and market is given by

$$C_p = N_P (M + L + R) + T_0 + S + D \quad \text{...(6.1)}$$

where N_p = life time product volume (total number of units)

M = material cost/unit

L = direct labor for manufacturing and assembly/unit

R = production resource usage/unit

T_0 = tooling and capitalization costs (generally one-time costs)

S = system cost (overhead or indirect costs)

D = development costs.

The component of the total product cost given in Eq. (6.1), adapted from Ostwald [1974] is shown in Fig. 6.1.

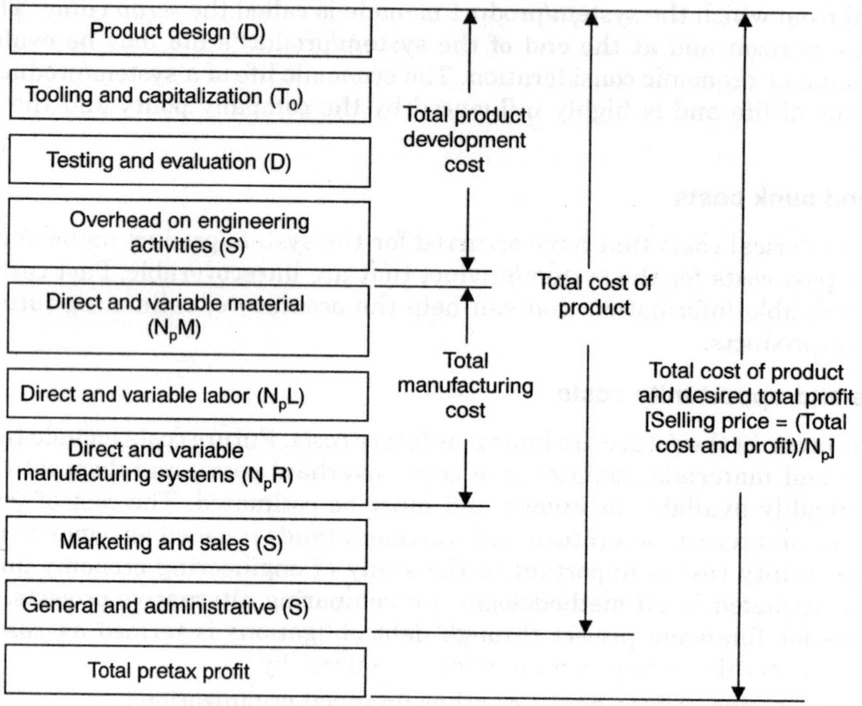

Fig. 6.1 General Cost Structure of a Product.

The total cost of product (Fig.6.1) is the total cost of manufacturing a product. An amount of profit is then added to this total cost of product to arrive at the total cost of product and desired total product or the selling price. That is,

Selling price = Total cost and profit.

Direct material and labor costs are the costs of material and labor that are easily measured and can be conveniently allocated to a specific operation, product or project. A cost is often called *a direct cost* when it can be directly assigned to a particular cost center, product line or part. *Indirect costs* cannot be directly assigned to a product but must be *spread* over an entire factory. *Indirect costs for both labor and material* are either very difficult or impossible to assign directly to a specific operation, product, or project.

Overhead costs or *burden* consists of all costs of manufacture other than direct material and direct labor. Typical specific items of cost included in the general category of overhead are: technical services (engineering), indirect materials, indirect labor, taxes, insurance premiums, rent, product design and development, maintenance and repairs, non technical services (office personnel, security, etc.), general supplies and rental equipment. Depreciation expenses are sometimes included in the general overhead costs, but may occasionally be considered a part of direct costs.

6.2.1.5 Fixed and variable costs

Fixed costs are independent of the rate of production of goods. That is, fixed costs do not vary in proportion to the quantity of output. Fixed costs may be fixed only over a given range of production; they may then change and be fixed for another range of production. Fixed costs include: 1) Indirect plant cost (*a*) Investment costs (depreciation on capital investment, interest on capital investment and inventory, property taxes and insurance) and (*b*) Overhead costs (burden) such as technical services (engineering), product design and development, non-technical services, general supplies and rental of equipment. 2) Management and administrative expenses (share of corporate executive staff, legal staff, share of corporate research and development staff and marketing staff and 3) Selling expenses such as sales force, delivery and warehouse costs, and technical service staff.

Sometimes, the fixed costs such as marketing and sales costs, legal costs, security costs, financial expenses, and general administrative costs are often combined into an overall category known as *general and administrative* expenses.

Variable costs change with production rate. That is, *variable costs* vary in proportion to quantity of output. These costs include: materials, direct labor, direct production supervision, maintenance costs, power and utilities, process control and quality control staff, royalty payments, packaging and storage costs, scrap losses and discard losses, etc. Variable costs are generally for direct material and direct labor.

Certain total costs (T) can be expressed as the sum of fixed (F) and variable costs (V). That is,

$$T(x) = F + V(x) \quad \text{...(6.2)}$$

This expression is linear in terms of output quantity, x as shown in Fig. 6.2.

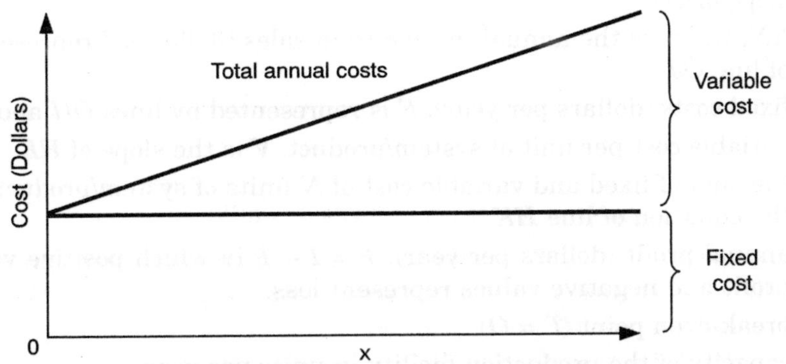

Fig. 6.2 Total Annual Costs as a Function of x.

6.2.1.5.1 Break-even analysis. There are two aspects to a manufacturing activity. The first one is the assembling facilities, labor, and materials for the production of a system product. The second one involves the marketing of goods or services, that is, the system/product produced. The success of the activity depends on the net difference between receipts for system/ products sold and the input necessary to produce and market them. Break-even analysis can be used to evaluate the situation.

Here, we present a linear break-even analysis, where the receipts and costs are assumed to be linear functions of the quantity of product produced, marketed and sold. The general pattern of income and costs is shown in Fig. 6.3. Line *HL* represents the fixed manufacturing

costs and line *HK* represents sum of variable production costs and marketing and sales costs. Line *OJ* represents the income from sales.

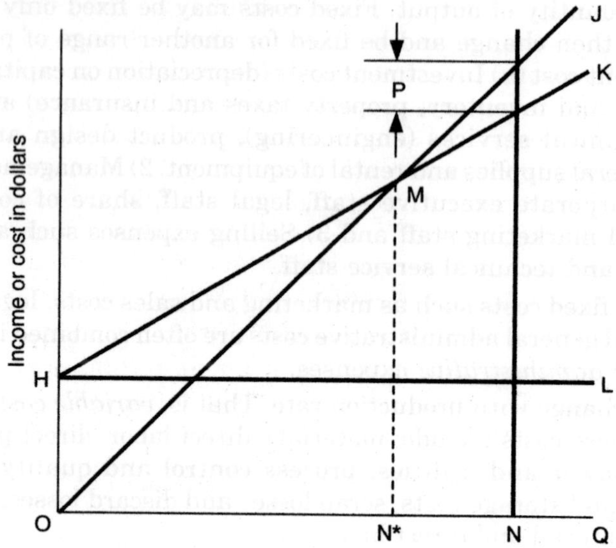

Fig. 6.3 General Graphic Representation for Income, Cost, Units of Output and Profit.

Let N = number of units of system/product made and sold per year

R = the amount in dollars received for an unit of system/product. It is equal to the slope of *OJ*

I = RN, which is the annual income from sales (dollars). I represent the equation of line *OJ*

F = fixed costs (dollars per year). F is represented by lines *OH* and *HL*

V = variable cost per unit of system/product. V is the slope of *HK*

T = the sum of fixed and variable cost of N units of system/product. $T = F + VN$ is the equation of line *HK*

P = annual profit (dollars per year). $P = I - T$ in which positive values represent profit and negative values represent loss.

N^* = break-even point ($P = O$)

Q = capacity of the production facility in units per year

The break-even point occurs when the total income equals the total cost. Referring to Fig.6.3, the intersection of lines *OJ* and *HK* is the break-even point. $I = T$ and $RN = F + VN$ at this point. The break-even point is given by

$$N^* = \frac{F}{R - V} \qquad \text{...(6.3)}$$

which gives the level of product output as shown in Fig. 6.3. The quantity ($R - V$) in the denominator of Eq. (6.3) is known as the *marginal contribution*.

Recalling,

$$I = RN \qquad \text{...(6.4)}$$

Substituting $N = F/(R - V)$ in Eq. (6.4), we have

$$I = R\left(\frac{F}{R-V}\right) \quad ...(6.5)$$

Similarly, $T = F + VN$ will give

$$T = F + \frac{VF}{(R-V)} \quad ...(6.6)$$

Now the annual profit

$$P = I - T \quad ...(6.7)$$

can be written as

$$P = RN - (F + VN)$$

or

$$P = (R - V)N - F \quad ...(6.8)$$

From Fig. 6.3, we observe that the profit amount P is represented by the difference between the lines OJ and HK. Equation (6.8) is displayed in Fig. 6.4.

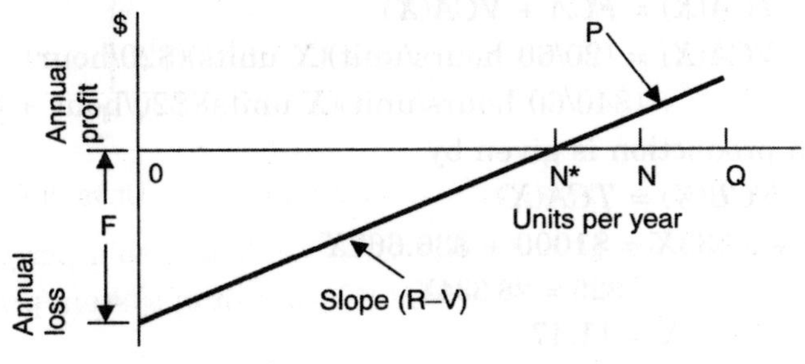

Fig. 6.4 Profit-volume Graph.

Example E6.1. *A small design and manufacturing company is converting from a manual inventory system to a computer based system. It is estimated that the incremental investment needed would be \$16,000, the salvage value will be zero at any time after the system is acquired, and the net annual savings of \$2,000 in operating costs when the new computer system is purchased. The time value of money is assumed to be zero.*

(a) determine the number of years of savings required to recover the incremental investment.

(b) the annual savings required to recover the incremental investment in five years.

Solution:

(*a*) Let x = number of years of savings required.

Then \$16,000 = \$2,000x

Hence x = 8 years.

(*b*) Let X = amount of annual savings required to recover the incremental investment in 5 years.

Then \$16,000 = 5X

or X = \$3,200.

Example E6.2. *Machine A or machine B can be used to manufacture a job order of part P. Machine B can perform all the required operations if the job order is produced using the machine A. The tooling set up time and machining time required on machine A to produce each part P are 20 minutes and 40 minutes respectively. The hourly rate for machinist B is \$20/hour* *and the overhead rate for each machining hour of machine A operation is $25. The tooling costs per lot for production on machine A are $1000 (estimated). For the job order to be produced using machine B, the machinist A requires five hours for the initial tooling set up for the job. Machinist B can operate the machine B once the tooling set up is done on machine B and no further tooling set up for the job is required. The hourly rate for machinist A is $25/hour. The machining time for part P is 10 minutes on machine B and the overhead rate for each machining hour is $30. The tooling costs per lot for production on machine B is $1200. Determine the job order size that would make the machine B economically preferred to machine A.*

Solution:

Let X = units of production.

The total cost for machine B is given by

$$TCB(X) = FCB + VCB(X)$$

$$FCB = \$1200 + (5 \text{ hours})(\$25/\text{hour}) = \$1325$$

$$VCB(X) = (10/60 \text{ hours/unit})(X \text{ units})(\$20/\text{hour} + \$30/\text{hour}) = \$8.333X$$

For the machine A, the total cost is given by

$$TCA(X) = FCA + VCA(X)$$

$$VCA(X) = (20/60 \text{ hours/unit})(X \text{ units})(\$20/\text{hour}) + (\$40/60 \text{ hours/unit})(X \text{ units})(\$20/\text{hour} + \$25/\text{hour}) = \$36.667X$$

The breakdown production is given by

$$TCB(X) = TCA(X)$$

$$\$1325 + 8.333X = \$1000 + \$36.667X$$

or $$325 = 28.334X$$

or $$X = 11.47$$

Hence, machine B is more economical for lot sizes of 12 units or greater.

Example E6.3. *A CNC machine system will have a total installed first cost of $60,000 including the training of the operator. The economic life of the machine is estimated to be 10 years with a salvage value of $20,000 at that time. The estimated decrease in asset value is $6,000/year and determined to be an annual fixed cost. The other fixed costs are $3,000/year. The raw components are purchased at $23.30/unit. The estimated variable unit costs are $2.30 for direct labor, $11 for direct material, and $5 for overhead (the depreciation of the machine of $4,000/year is excluded). Determine the annual sales volume (number of finished components from the raw components) is needed in order to breakeven on the CNC machine purchase assuming linearity.*

Solution:

Let X = number of finished components produced per year.

$TR(X)$ = total revenue for a production of X components/year.

$TC(X)$ = total cost for production of X components/year.

The breakeven is found when $TR(X) = TC(X)$.

Hence

$$\$23.30X = (\$4000 + \$3000) + (\$2.30 + \$11 + \$5)X$$

or $$23.30X = \$7000 + \$18.30X$$

or $$X = 1400$$

Hence, the total revenue exceeds the total costs if 1400 or more components are produced per year.

Example E6.4. *The manufacturer of certain electronic system has a capacity of 200,000 units annually. The fixed cost of production line is \$600,000 with a variable cost of \$6 and revenue of \$10 per unit.*

(a) determine the percent of present capacity that must be used to break-even

(b) if the manufacturer wants to use an equivalent of 100,000 units of the 200,000 unit capacity for another electronic product line, which will reduce the fixed cost to \$300,000, what variable cost per unit will make 200,000 units the new breakeven volume?

(c) compare this result with the current variable cost.

Solution:

(*a*) Using Eq.(6.3)

$$N^* = \frac{F}{R - V} = \frac{600{,}000}{10 - 6} = 150{,}000 \text{ units.}$$

This is 75% of the capacity of 200,000 units.

(*b*) Using Eq.(6.3) with $N^* = 100{,}000$ to compute V, we have

$$100{,}000 = \frac{300{,}000}{10 - V}$$

or

$$V = 10 - \frac{300{,}000}{100{,}000} = \$7 \text{ per unit.}$$

(*c*) Hence V can be \$1 higher per unit if N^* decreases from 150,000 to 100,000 units.

6.2.1.6 Average and marginal costs

The *average cost* of one unit of output (unit cost) is the ratio of total cost and quantity of output (production volume, etc.). That is,

$$A(x) = \frac{T(x)}{x} \qquad ...(6.9)$$

where $A(x)$ = average cost per unit of x

$T(x)$ = total cost for x units of output

x = output quantity

The average cost normally decreases with an increasing quantity of output.

Marginal cost is defined as the derivative of the total cost with respect to function x for a total cost function that is continuous in the output (or independent) variable x.

$$\text{Marginal cost} \qquad M(x) = \frac{dT(x)}{dx} \qquad ...(6.10)$$

If the total cost function is discontinuous and defined only for discrete values of x, then difference equations are used to compute the marginal costs. Tables 6.1 and 6.2 summarize the relationships between marginal cost (Table 6.1) and total cost, and marginal cost and average cost (Table 6.2).

Table 6.1. Relationship between Marginal Cost and Total Cost

If $M(x) > 0$, then $T(x + 1) > T(x)$
If $M(x) = 0$, then $T(x + 1) = T(x)$
If $M(x) < 0$, then $T(x + 1) < T(x)$

Table 6.2. Relationship between Marginal Cost and Average Cost

If $M(x) > A(x)$, then $A(x + 1) > A(x)$
If $M(x) = A(x)$, then $A(x + 1) = A(x)$
If $M(x) < A(x)$, then $A(x + 1) < A(x)$

Example E6.5. *The total cost per year (T) of transmitting a given quantity of chemical fluid through a pipe in a manufacturing plant over a fixed distance is given by*

$$T(A) = \$10 + \$20\,A + \frac{\$180}{A}$$

where A = cross-sectional area (square inches) of the pipe.

(a) *find the cross-sectional area of the pipe that will give minimum total cost per year*

(b) *determine the minimum total cost per year to operate the system with the cross-sectional area of the pipe in (a)*

(c) *for a 2-inch and 4-inch diameter pipes, find the total cost per year.*

Solution:

(*a*) $T\,(A) = \$10 + \$20A + \dfrac{\$180}{A}$

Differentiating with respect to A, we get

$$\frac{dT(A)}{dA} = 20 - 180\,A^{-2}$$

The cross-sectional area of the pipe that will give minimum total cost/year is obtained by setting $\dfrac{dT(A)}{dA} = 0.$

Therefore

$$0 = 20 - 180\,A^{-2}$$

or

$$A = 3 \text{ square inches.}$$

(*b*) $T(3) = \$10 + \$20(3) + \$\,\dfrac{180}{3} = \$130$

(*c*) For 2-inch diameter pipe

$$A = \frac{\pi}{4}\,(2)^2 = 3.1416$$

$$T\,(3.1416) = \$10 + \$20(3.1416) + \frac{180}{3.1416} = \$130.13$$

For 4-inch diameter pipe

$$A = \frac{\pi}{4}\,(4)^2 = 12.5664$$

$$T(12.5664) = \$10 + \$20(12.5664) + \frac{180}{12.5644} = \$275.65$$

6.2.2 Cost Estimation

Design is not complete until the cost required to build the design or manufacture of the product is determined. For most products and designs, cost is next to performance in importance. Here, we are mainly concerned with the estimation of specific factors relevant to comparing alternative engineering/investment projects and making a selection from these projects.

The National Estimating Society defines cost estimating as *the art of approximating the probable worth or cost of an activity based on information available at the time*. Estimating is not an exact science. Cost estimation is an approximation process that depends on the availability and relevancy of appropriate historical data, personal judgments based on experience, and the time period available for completing the cost-estimating activity. Cost estimates can be used to provide the following benefits:

1. the information can be used in establishing the selling price of a system/product or service
2. the information can be used to establish the most economical method, manufacturing process, or selecting materials for manufacturing the system/product
3. the information can be used as a basis for determining a cost-reduction program
4. to determine standards of manufacturing performances that can be used to control costs and establish the probability of a new product.

Cost estimating is needed for engineering economy. Future and current costs and benefits should be estimated to provide the basic data to be analyzed. Cost estimates can be developed using two approaches:

1. *The top-down or parametric approach*: This method uses historical data from prior work and establishes a new cost estimate based on the differences between the new system/product and the prior one.
2. *Bottom up or detailed approach*. This method needs estimates of costs for each element and subelement of a system/product and then applies appropriate pricing parameters or allowances for overhead and profit to determine a total cost estimate. The detailed approach method clearly gives more accurate estimates at the expense of more resources and time.

In general, during the early stages of decision-making process, the top-down or the parametric approach is used while in the final stages of decision-making process, the bottom-up or the detailed approach is used. As the level of detail increases, the higher the cost incurred in making the cost estimate. Ostwald (1992) gives the following relationship between the accuracy of an estimate and the cost of making the estimate.

$$C_T = C(M) + C(E) \qquad \text{...(6.11)}$$

Where, C_T = the total cost of making the estimate (dollars)

$C(M)$ = the functional cost of making the estimate (dollars)

$C(E)$ = the functional cost of errors in the estimate (dollars)

The relationship given by Eq. (6.11) is shown in Fig. 6.5. The total cost of making an estimate attains a minimum value, where the amount of detail reaches a value D_1. In general, the larger the scope of the system/product being estimated, the more difficult it is to make accurate estimates of the costs and revenues involved.

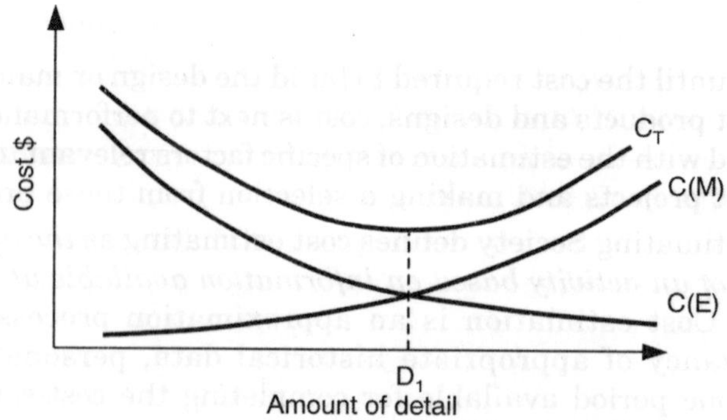

Fig. 6.5 Cost of Increasing Detail.

Example E6.6. *A small manufacturing firm compiled the following data in the first production run of a new product assuming no rejects or scrap.*

Direct material cost = \$7/unit
Direct labor cost = \$3.90/unit
Overhead cost = \$6.20/unit
Selling price = \$30/unit

Prior to shipping, the manufactured products are inspected only once. A scrap rate of 35% has occurred during the production run. Determine the maximum scrap rate permissible in order that the total costs equal to total revenues (breakdown).

Solution:

Let s = scrap rate.

In order to have X good products, a total of $\left(\dfrac{X}{1-s}\right)$ units must be produced.

The breakeven scrap rate is obtained when the total revenue (X) is equal to the total cost (X).

Hence $$\$30(X) = (\$7 + \$3.90 + \$6.20)\left[\frac{X}{1-s}\right]$$

or $$(1 - s)\,30(X) = 17.10X$$

or $$1 - s = 0.57$$

$$s = 0.43$$

Therefore, the maximum scrap rate permissible to breakeven is 43%.

Example E6.7. *The following two equations give the relationships that approximate the total cost and the total profit per week in dollars of producing certain types of products by a manufacturing company.*

Total cost, $TC(x) = 100{,}000 + 30x + 0.0005x^2$, *for* $x \geq 0$

$= 0$ *otherwise*

Total profit, $TP(x) = Sx - TC(x)$

where

x = *volume of products produced, products / week*

S = *sales price of product, dollars / product = \$50/product*

(a) determine the level of production in products/week is the cost/products minimum, and the minimum cost/product

(b) determine the maximum weekly profit that the company can make and the level of production the maximum weekly profit attainable

(c) determine the range of production for possible profit

(d) determine the range of production over which profit is possible if the company's production capacity is limited to 25,000/week.

Solution:

Let x = number of products manufactured (produced) per week.

S = sales price of product, \$/product = \$50/product.

$TC(x) = 100{,}000 + 30x + 0.0005x^2$

(*a*) Cost/product $= (100{,}000 + 30x + 0.0005x^2)/x$

$$= \frac{100{,}000}{x} + 30 + 0.0005x$$

$$\frac{dTC}{dx} = -100{,}000/x^2 + 0.0005 \qquad \text{...(E.1)}$$

Setting Eq. (E.1) to zero and solving, we get

$$0 = -100{,}000/x^2 + 0.0005$$

or $$x = 14{,}142.14 \text{ products/week}$$

$$\text{cost/product} = \frac{100{,}000}{x} + 30 + 0.0005x$$

$$= \frac{100{,}000}{14{,}142.14} + 30 + 0.0005(14{,}142.14)$$

$$= \$44.14/\text{product.}$$

(*b*) $TP(x)$ $= Sx - TC(x)$

$$= 50x - [100{,}000 + 30x + 0.0005x^2]$$

$$= -100{,}000 + 20x - 0.0005x^2$$

$$\frac{dTC}{dx} = 20 - 0.001x \qquad \text{...(E.2)}$$

Setting Eq.(E.2) to zero and solving, we obtain

$$0 = 20 - 0.001x$$

or $$x = 20{,}000 \text{ products/week}$$

$$TP(x) = -100{,}000 + 20x - 0.0005x^2$$

$$TP(20{,}000) = -100{,}000 + 20(20{,}000) - 0.0005(20{,}000)^2$$

$$= \$100{,}000$$

(*c*) From (*b*) above

$$TP(x) = -100{,}000 + 20x - 0.0005x^2 \qquad \text{...(E.3)}$$

Solving Eq.(E.3) for roots, we have

$$x = \frac{-20 \pm \sqrt{(20)^2 - 4(-100{,}000)(-0.0005)}}{2(-0.0005)}$$

The roots are $x = 5858$ and $x = 34{,}142$.

Hence, if the production is between 5858 and 34,142 products/week, then profit is obtained.

(*d*) Profit is obtained with production between 5858 and the cap of 25,000 products/week.

Example E6.8. ***A production worker is on an incentive system in a small manufacturing company and earns $1 for each part produced, and a guaranteed wage of $16/hour irrespective of production output. The product is sold at a price of (6 – 0.3 n) dollars/unit, where n is the number of units produced per hour.***

(a) assuming that all the produced units are sold, determine the rate of production at which the company can earn the most profit from this production worker's output

(b) determine the maximum profit per hour

(c) find the rate of production when the revenue be maximized and the corresponding profit per hour.

(d) determine the range of production for the company to make a profit.

Solution:

(*a*) $\text{Profit } (n) = (6 - 0.3\, n)n - [n + 16]$

$$= 6n - 0.3n^2 - n - 16$$

$$= -\,0.3n^2 + 5n - 16$$

$$\frac{dP}{dn} = -\,0.6n + 5 \qquad \text{...(E.1)}$$

Setting Eq. (E.1) to zero and solving, we get

$$0 = -\,0.6n + 5$$

or $n = 8.33$ units/hour

(*b*) $\text{Profit } (8.33) = -\,0.3(8.33)^2 + 5(8.33) - 16$

$$= -\,\$20.82 + \$41.65 - \$16 = \$4.83\text{/hour}$$

(*c*) $\text{Revenue } (n) = (6 - 0.3n)n = 6n - 0.3n^2$

$$\frac{dP}{dn} = 6 - 0.6n \qquad \text{...(E.2)}$$

Setting Eq.(E.2) to zero and solving, we obtain

$$0 = 6 - 0.6n$$

or $n = 10$ units/hour

$\text{Profit } (10) = -\,0.3n^2 + 5n - 16$

$$= -\,0.3(10)^2 + 5(10) - 16 = \$4\text{/hour}$$

(*d*) The points of zero profits are obtained by solving Profit (n) = 0

or $$-\,0.3n^2 + 5n - 16 = 0$$

or $$n = \frac{-\,5 \pm \sqrt{5^2 - 4(-\,0.3)(-\,16)}}{2(-\,0.3)}$$

$$= 8.333 \mp 4.014 = 4.32 \quad \text{or} \quad 12.348$$

Hence, the range is n = 4.32 (or 4) units/hour and n = 12.348 (or 12) units per hour.

Example E6.9. ***A heavy construction company owns and runs a fleet of four-wheel drive vehicles. The maintenance records over a period of seven years show that approximately 20% of the time, tires are replaced after 20,000 miles of use. Eighteen per cent of the tires are replaced***

after 18,000 miles, 21% are replaced after 25,000 miles, 32% are replaced after 30,000 miles, and 8% are replaced after 38,000 miles of use. Determine the weighted-average estimate of the tire life.

Solution:

Weighted-average estimate of the tire life

$$= (0.20)(20,000) + (0.18)(18,000) + (0.21)(25,000) + (0.32)(30,000) + (0.08)(38,000)$$
$$= 25,130 \text{ miles.}$$

6.3 GENERAL ACCOUNTING PRINCIPLES

The American Institute of Certified Public Accountants defines accounting as *the art of recording, classifying and summarizing in a significant manner and in terms of money, transactions and events which are, in part at least, of a financial character, and interpreting the results thereof.*

General accounting information is summarized in the basic financial reports: (1) the balance sheet, and (2) the income statement. These two statements have a close relationship.

1. *Balance sheet*: The balance sheet provides a statement of the financial condition of a company at a point in time by providing a summarized listing of the values of the assets, liabilities, and net worth of the company.

2. *Income statement*: The income statement provides a statement of the revenues and expenses incurred by a company during a period of time, generally a month, quarter, or year. The income statement gives a summary of the financial activities that occur during the period between the two balance sheets.

6.3.1 Balance Sheet

An example of a balance sheet for Vishwa Engineering Company Inc. is given in Table 6.3. There are basically three main groups of items listed on a balance sheet. They are:

1. *Assets*: Assets are properties owned by the company.

2. *Liabilities*: Liabilities are debts owed by the company against the assets.

3. *Net worth*: Net worth is the dollar difference between the assets and liabilities, and represents the investment made by the stockholders of the company plus any accumulated profits left in the company by the stockholders. Net worth is also known as *owners' equity*.

Hence

Net worth/Owners' equity = Assets – Liabilities

or

$$\text{Assets} = \text{Liabilities} + \text{Net worth/Owners' equity} \quad \text{...(6.12)}$$

Current assets include cash and other assets that can be readily converted into cash; an arbitrary period of 1 year is usually assumed as a criterion for conversion. Similarly, *current liabilities* are the debts that are due and payable within 1 year from the date of the balance sheet. *Fixed assets* are the properties owned by the firm that are not readily converted into cash within a 1-year period. *Fixed liabilities* are long-term debts due and payable after 1 year from the date of the balance sheet. Typical current-asset items are cash, accounts receivable, notes receivable, raw material inventory, work in process, finished goods inventory, and prepaid expenses. Fixed-asset items are land, buildings, equipment, furniture, and fixtures. Items that are typically listed under current liabilities are accounts payable, notes payable, interest payable, taxes payable, prepaid income, and dividends payable. Fixed liabilities may also be notes payable, bonds payable, mortgages payable, and so forth. Net worth items depend on whether the business is a sole proprietorship, a partnership, or a corporation. Items such as capital stock, retained earnings, capital surplus, or earned surplus appear under the net worth group.

The balance sheet in Table 6.3 is *balanced* since assets = liabilities + net worth. The depreciation reserve is an accumulated amount of funds held to repurchase the asset when its functionally useful life terminates. The first cost of the depreciable asset (building, in this case) minus the amount in the depreciation reserve equals the *book value*.

Income Statement

The income statement provides management with

(*a*) a summary of the revenues received

(*b*) a summary of the expenses incurred to obtain the revenues, and

(*c*) the profit or loss resulting from business operations.

The income statement begins with revenues and reflects the direct costs associated with generating the sales revenue. The same income statement for a manufacturing company and for a retail company is given in Tables 6.4 and 6.5 respectively. In both income statements, the gross profit is the difference between the cost of goods and the sales. The deduction of general and administrative expenses and the marketing expenses from the gross profit gives the net profit before taxes. Deducting income taxes gives the net profit (or loss) for the period.

Table 6.3. Sample Balance Sheet

(Figures in US $)

Vishwa Engineering Company, Inc.
Balance Sheet
December 31, 2004

Assets			
Current assets:			
Cash	50,000		
Accounts receivable	150,000		
Raw materials	10,000		
Work in process	10,000		
Finished goods inventory	5,000		
Small tool inventory	15,000		
Total current assets		240,000	
Fixed assets:			
Land	50,000		
Building	250,000		
Less: Depreciation Reserve	50,000	200,000	
Equipment	800,000		
Less: Depreciation Reserve	200,000	600,000	
Office equipment		20,000	
Total fixed assets			870,000
Total assets			1,110,000

Liabilities and Net Worth		
Current liabilities		
Accounts payable	50,000	
Taxes payable	20,000	
Total current liabilities		70,000
Fixed liabilities		
Mortgage loan payable	150,000	
Equipment loan payable	500,000	
Total fixed liabilities		650,000
Total liabilities		720,000
Common stock	250,000	
Retained earnings	100,000	
Earned surplus (for current year)	40,000	
Total equity		390,000
Total liabilities and equity		110,000

Table 6.4. Sample Income Statement

Vishwa Engineering Company, Inc.
Income Statement
Year Ended December 31, 2003 (Figures in US $)

Sales	1,500,000	
Less cost of goods manufactured:		
Direct labor	500,000	
Direct materials	300,000	
Indirect labor	100,000	
Depreciation	100,000	
Repairs and maintenance	60,000	
Utilities	40,000	1,100,000
Gross profit		400,000
Less other expenses:		
Administration	100,000	
Marketing	60,000	
Interest payments	40,000	200,000
Net profit before income taxes		200,000
Less income taxes		50,000
Net profit (posted to earned surplus)		150,000

Table 6.5. Sample Income Statement

(Figures in US $)

Vishwa Engineering Company, Inc.
Income Statement
Year Ended December 31, 2004

Sales		1,500,000
Less cost of goods sold:		
Inventory, January 1, 2004	350,000	
Plus purchases	550,000	
	900,000	
Less inventory, December 31, 2004	100,000	800,000
Gross profit		700,000
Less expenses:		
Direct labor	300,000	
Depreciation – building	20,000	
Depreciation – equipment	30,000	
Repairs and maintenance	40,000	
Indirect labor	195,000	
Utilities	10,000	
Supplies, tooling	5,000	600,000
Net profit before income taxes		200,000
Less income taxes		50,000
Net profit (posted to earned surplus)		150,000

Table 6.6. Vishwa Manufacturing Company Comparative Balance Sheets

(Figures in US $)

Vishwa Manufacturing
Comparative Balance Sheets
as of December 31, 2003 and 2004

Assets	**2004**	**2003**
Current assets		
Cash	300,000	200,000
Accounts Receivable (net)	250,000	150,000
Inventory	100,000	150,000
Prepaid Expenses	50,000	100,000
Total Current Assets	700,000	600,000
Fixed Assets		
Machinery	300,000	250,000
Furniture	150,000	100,000
Other	50,000	50,000
Total Assets	1,200,000	1,000,000

Liabilities and Capital		
Current Liabilities		
Notes Payable	300,000	150,000
Accounts Payable	200,000	250,000
Taxes Payable	100,000	150,000
Total Current Liabilities	600,000	550,000
Fixed Liabilities		
Loans	200,000	150,000
Total Liabilities	800,000	700,000
Capital		
Stock	200,000	150,000
Retained Earnings	125,000	100,000
Earned Surplus	75,000	50,000
Total Capital	400,000	300,000
Total Liabilities and Capital	1,200,000	1,000,000

Table 6.7. Vishwa Manufacturing Company Comparative Income Statements

Vishwa Manufacturing
Comparative Income Statements
For Years Ended December 31, 2003 and 2004 (Figures in US $)

	2004	**2003**
Net Sales	1,500,000	1,200,000
Cost of Goods Sold		
Beginning Inventory	100,000	50,000
Direct Materials	200,000	150,000
Direct Labor	800,000	600,000
Factory Overload	50,000	50,000
Total	1,150,000	850,000
Less: Ending Inventory	100,000	50,000
Cost of Goods Sold	1,050,000	800,000
Gross Profit	450,000	400,000
Other Operating Expenses		
Selling Expenses	50,000	50,000
General and Administrative	100,000	150,000
Total Other Operating Expenses	150,000	200,000
Net Operating Income	300,000	200,000
Less: Interest Expenses	30,000	20,000
Less: Income Taxes	70,000	30,000
Net Income	200,000	150,000

6.3.2 Ratio Analysis

Ratio analysis is used primarily to interpret a company's balance sheet and income statements to determine the attractiveness of a company as a healthy potential employer, or to determine the attractiveness of a company's stock as a potential investment. Ratio analysis examines the relationship between the values found in balance sheet and income statements. The analysis of financial statements of a company generally aims at the following three areas:

1. the earning power of the company
2. the short-term liability obligations and
3. the long-term liability obligations.

The comparative balance sheets for the years ended 2003 and 2004 for the Vishwa Manufacturing Company are summarized in Table 6.6 and the comparative income statements for the same company are contained in Table 6.7. The earning power, that is, using the following ratios as illustrated in Example E6.10 commonly assesses the viability of maintaining the financial health of a company.

Example E6.10. *Determine the earning power ratios for the data in Tables* 6.6 *and* 6.7.

$$\text{Return on assets employed} = \frac{\text{net income}}{\text{total average assets}} \quad \text{...(6.13)}$$

$$\text{Return on owner's equity} = \frac{\text{net income}}{\text{average owner's equity}} \quad \text{...(6.14)}$$

The calculation of earning power ratios for Vishwa Manufacturing Company for 2004 are given below:

Return on Assets Employed

$$\text{Return on assets employed} = \frac{\text{net income}}{\text{total average assets}}$$

$$= \frac{\$200{,}000}{0.5(\$1{,}000{,}000 + \$1{,}200{,}000)} = 0.1808 \text{ or } 18.18\%$$

Return on Owner's Equity

$$\text{Return on owner's equity} = \frac{\text{net income}}{\text{average owner's equity}}$$

$$= \frac{\$200{,}000}{0.5(\$300{,}000 + \$400{,}000)} = 0.5714 \text{ or } 57.14\%$$

The second major area of focus in ratio analysis is short-term liability obligations. This is referred to as *liquidity*. Liquidity ratios measure a company's ability to meet its current obligations. The following ratios are commonly used to measure liquidity.

$$\text{Current ratio} = \frac{\text{current assets}}{\text{current liabilities}} \quad \text{...(6.15)}$$

$$\text{Acid-test ratio} = \frac{\text{cash + receivables + short-term marketable securities}}{\text{current liabilities}} \quad \text{...(6.16)}$$

$$\text{Accounts receivable turnover} = \frac{\text{net sales}}{\text{average accounts receivable}} \quad \text{...(6.17)}$$

$$\text{Inventory turnover} = \frac{\text{cost of goods sold}}{\text{average inventory}} \quad \text{...(6.18)}$$

The calculation of liquidity ratios for Vishwa Manufacturing for 2004 is shown below.

(*a*) *Current Ratio*

$$\text{Current ratio} = \frac{\text{current assets}}{\text{current liabilities}} = \frac{\$700,000}{\$600,000} = 1.1667$$

(*b*) *Acid-Test Ratio*

$$\text{Acid-test ratio} = \frac{\text{cash} + \text{receivables} + \text{short-term marketable securities}}{\text{current liabilities}}$$

$$= \frac{\$300,000 + 250,000 + \$0}{\$600,000} = 0.9167$$

(*c*) *Accounts Receivable Turnover*

$$\text{Accounts receivable turnover} = \frac{\text{net sales}}{\text{average accounts receivable}}$$

$$= \frac{\$1,500,000}{0.5(150,000 + 250,000)} = 7.5$$

(*d*) *Inventory Turnover*

$$\text{Inventory turnover} = \frac{\text{cost of goods sold}}{\text{average inventory}} = \frac{\$1,050,000}{0.5(50,000 + 100,000)} = 14$$

The final major area of focus in ratio analysis is long-term liability obligations. This is referred to as *solvency*. Solvency ratios measure a company's ability to meet its long-term obligations based on its current debt structure. The following ratios are commonly used to measure solvency.

$$\text{Debt to equity ratio} = \frac{\text{total liabilities}}{\text{total capital worth}} \qquad \text{...(6.19)}$$

$$\text{Times interest earned ratio} = \frac{\text{net income before income taxes and interest}}{\text{interest charges}} \qquad \text{...(6.20)}$$

$$\text{Operating income to total assets ratio} = \frac{\text{net operating income}}{\text{total assets}} \qquad \text{...(6.21)}$$

The calculation of liquidity ratios for Vishwa Manufacturing for 2004 is shown below.

(*e*) *Debt to Equity Ratio*

$$\text{Debt to equity ratio} = \frac{\text{total liabilities}}{\text{total capital worth}} = \frac{\$800,000}{\$400,000} = 2.0$$

(*f*) *Times Interest Earned Ratio*

$$\text{Times interest earned ratio} = \frac{\text{net income before income taxes and interest}}{\text{interest charges}}$$

$$= \frac{\$300,000}{\$30,000} = 10$$

(*g*) *Operating Income to Total Assets*

$$\text{Operating income to total assets ratio} = \frac{\text{net operating income}}{\text{total assets}}$$

$$= \frac{\$300,000}{\$1,20,000} = 0.25 \text{ or } 25\%.$$

Ratio analysis provides a powerful means to express the relationships found in balance sheets and income statements. These ratios are generally useful when compared to a meaningful set of standards or expectations. The standards or expectations generally take the form of previous year's results, peer group comparisons, or industry averages.

6.3.3 Cost Allocation Methods

There are fundamental differences in cost accounting procedures for establishing manufacturing costs compared to determining the cost of providing a service. Similarly, there are differences in accounting principles if manufacturing job is on a job-shop or process basis. The total cost of producing any job order includes direct material, direct labor, and overhead costs. There are three methods generally employed for allocating overhead costs to manufacturing jobs. They are:

1. *Allocate Overhead Based on Direct Labor Hours*

(*i*) Determine the values for previous period direct labor hours and overhead cost for the manufacturing unit.

(*ii*) Compute the rate per direct labor hour:

$$\text{Rate} = \frac{\text{overhead cost}}{\text{direct labor hours}} \qquad ...(6.22)$$

(*iii*) Determine the number of direct labor hours required by the particular job for which overhead cost is being estimated.

(*iv*) Compute the overhead cost for the job.

Estimated overhead = rate × estimated direct labor hours

2. *Allocate Overhead Based on Direct Labor Dollars*

(*i*) Determine the values for previous period direct labor dollars and overhead cost for the manufacturing unit.

(*ii*) Compute the percentage ratio of overhead cost to direct labor dollars

$$\text{Ratio} = \frac{\text{overhead cost}}{\text{direct labor dollars}} \times 100\% \qquad ...(6.23)$$

(*iii*) Determine the direct labor dollars required by the particular job for which overhead cost for the job.

(*iv*) Compute the overhead cost for the job.

Estimated overhead = ratio × estimated direct labor dollars

3. *Allocate Overhead Based on Direct Labor Dollars and Direct Material Dollars*

(*i*) Determine the values for previous period direct labor dollars, direct material dollars, and overhead cost for the manufacturing unit.

(*ii*) Compute the percentage ratio of overhead cost to direct labor dollars plus direct material dollars.

$$\text{Ratio} = \frac{\text{overhead cost}}{\text{direct labor dollars} + \text{direct material dollars}} \times 100\% \quad ...(6.24)$$

(*iii*) Determine the direct labor dollars and direct material dollars required by the particular job for which overhead cost is being estimated.

(*iv*) Compute the overhead cost for the job.

$$\text{Estimated overhead} = \text{ratio} \times (\text{estimated direct labor dollars} + \text{estimated direct material dollars})$$

Example E6.11. *D. Ashwin purchased a small diner (retail restaurant business) and opened on July 1, 2004. The breakdown of $330,000 of equity funds invested is as follows: $15,000 in equipment, $10,000 in inventory items, and $5,000 in operating cash. The data in Table E6.11 summarizes the gross sales and expenses for the diner during the first 3-month period.*

Table E6.11

Gross sales	*$52,000*
Purchases	*22,500*
Salaries	*10,000*
Advertising expense	*2,000*
Rent expense	*2,800*
Expense for utilities	*1,900*
Expense for misc. supplies	*800*

For the purchases, $17,500 was paid in cash and $5,000 is still owed. At the close of the 3-month period, the end-of-period inventory is worth $7,500.

(a) prepare a balance sheet for the business as of July 1, 2004.

(b) prepare an income statement for the 3-month period covered.

(c) if at the end of the 3-month period on October 1, 2004, the net worth account is $30,000, find the amount of cash needed in order to balance the accounting equation as of October 1, 2004.

Solution:

(*a*)

D. Ashwin
Balance Sheet
July 1, 2004 (Figures in US $)

Assets		
Current Assets		
Cash	5,000	
Inventory	10,000	
Total Current Assets		15,000
Fixed Assets		
Equipment	15,000	
Total Fixed Assets		15,000
Total Assets		30,000

Liabilities and Net Worth		
Ownership	30,000	
Total Net Worth		30,000
Total Liabilities and Net Worth		30,000

(*b*)

D. Ashwin

Income Statement

July 1, 2004 – September 30, 2004 **(Figures in US $)**

Gross Sales		52,000
Less: Cost of Goods Sold:		
Inventory, July 1, 2004	10,000	
Plus: Purchases	22,500	
	32,500	
Less: Inventory, Sept. 30, 2004	7,500	25,000
Gross Profit		27,000
Less: Expenses		
Salaries	10,000	
Advertising	2,000	
Rent Expenses	2,800	
Utilities	1,900	
Miscellaneous	800	17,500
Net Profit Before Income Taxes		9,500

(*c*) Assets = Cash + Inventory + Equipment
= Cash + $7,500 + $15,000 = Cash + $22,500
Liabilities = Net Worth = Accounts Payable + Ownership
= $5,000 + $30,000 = $35,000
From Assets = Liabilities + Net Worth
Cash + $22,500 = $35,000
Cash = $12,500

Example E6.12. *The income statement for D. Anand company for the calendar year ending December 31, 2004 is given below:* **(Figures in US $)**

Income Statement

Gross income from sales	*360,000*	
Less: Cost of goods sold	*196,000*	
Net income from sales	*164,000*	
Operating expenses		
Rent	*13,000*	
Salaries	*62,000*	
Depreciation	*7,000*	
Advertising	*6,200*	
Insurance	*2,500*	*90,700*
Net profit before income taxes		*73,300*
Less: Income taxes		*35,200*
Net profit after income taxes		*38,100*

The Balance Sheet Accounts as of December 31, 2004 are:

Cash	*162,000*
Accounts receivable	*12,000*
Notes payable	*35,000*
Raw material inventory	*11,000*
Work-in-process inventory	*18,000*
Accounts payable	*12,000*
Declared dividends	*25,000*
Finished goods inventory	*20,100*
Land	*40,000*
Original cost of building	*90,000*
Building – Reserve for depreciation	*9,000*
Original cost of equipment	*43,000*
Equipment – Reserve for depreciation	*6,000*

(*a*) *determine the Earned Surplus for the year 2004*

(*b*) *prepare a Balance Sheet as of December 31, 2004.*

Solution:

(*a*) Earned surplus is $38,100 from the income statement.

(*b*)

D. Anand Company
Balance Sheet
December 31, 2004

(Figures in US $)

Current Assets		
Cash	162,000	
Accounts Receivable	12,000	
Raw Material Inventory	11,000	
WIP Inventory	18,000	
Finished Goods Inventory	20,100	
Total Current Assets		223,000
Fixed Assets		
Land		40,000
Building	90,000	
Less: Depreciation Reserve	9,000	81,000
Equipment	40,000	
Less: Depreciation Reserve	4,000	37,000
Total Assets		381,000

Liabilities and Net Worth

Liabilities	
Notes Payable	35,000
Accounts Payable	12,000

Declared Dividends	25,000
Total Liabilities	72,000
Ownership	
Capital Stock	270,900
Earned Surplus	38,100
Total Ownership	309,000
Total Liabilities and Net Worth	381,000

The Capital Stock account is established to balance the balance sheet.

Example E6.13. *D. Raghav Company manufactures a certain finished metal part for a machine system. Each unit of raw metal stock bar for the manufacturing process costs $50 and yields 25 finished parts. The manufacturing sequence, standard operation time per part, and the machine overhead rates are given in Table E6.13.*

Table E6.13

Manufacturing operation number	***Standard operation time per part***	***Machine overhead rate***
1	*3 minutes*	*$2.10/hour*
2	*6 minutes*	*$3.20/hour*
3	*4 minutes*	*$1.50/hour*
4	*2 minutes*	*1.80/hour*
Packaging	*0.3 minutes*	—

The tooling cost is $700, factory overhead costs is $25 per direct labor hour, and the average direct labor hour ra te (includes fringe benefits) is $20 per hour.

(a) determine the total estimated costs for the job order of 8000 parts

(b) what should be the unit selling price if profit is to be 35% of the total cost?

Solution:

(*a*) The various elements of total cost for an order of 8000 parts are

Raw material = [(8000 parts)/(25 parts/bar)]($50/bar) = $16,000

Tooling = $700

Factory overhead = (direct labor hours)($25/hour)

= [(3 + 6 + 4 + 2 + 0.3)minutes/(60 min./hr)](8000 parts)($25/hr)

= $51,000.

Direct labor and machine overhead:

Operation 1 = (3/60)hours (8000 parts)($20 + $2.10)/hour = $8,840

Operation 2 = (6/60)hours (8000 parts)($20 + $3.20)/hour = $18,560

Operation 3 = (4/60)hours (8000 parts)($20 + $1.50)/hour = $11,466.67

Operation 4 = (2/60)hours (8000 parts)($20 + $1.80)/hour = $29,067.39

Packaging = (0.3/60)hours (8000 parts)($20/hour) = $800

Total = $68,734.06

Therefore, the total cost is = raw material + tooling + factory overhead + total direct labor and machine overhead.

Total cost = $16,000 + $700 + $51,000 + $68,734.06 = $136,434.06

(*b*) Order selling price = 1.35(total cost) = 1.35($136434.06) = $184,185.98
Part selling price = $184,185.98/8000 = $23.02/part.

6.3.4 Activity-Based Costing

With the advent of computer controlled and automated manufacturing systems, overhead costs dominate the cost of producing items and direct labor is less significant in the cost pool. Activity-based costing is designed to address the changing cost mix by associating manufacturing costs with activities, which drive them. The procedure of activity-based costing is as follows:

1. Costs are identified by categories. Categories are associated with well-defined elements. Examples of such elements are material handling costs, energy costs, tooling costs and maintenance costs.

2. The activities, which are responsible for significant cost categories, are identified. These activities are then to be monitored and controlled under activity-based costing. Some examples of cost driving activities are machine hours for energy costs, trucking hours for material handling time (hours), and machine hours or production volume for tooling and maintenance costs.

3. The actual rate of activity for each of the cost drivers is used to predict the costs associated with each cost category. It is possible then to eliminate or weed out high cost activities. The analysis can also be used to improve process in manufacturing in order to reduce the costs. The activity-based costing system analyses can be used to redesign the process and products with cost reduction and profit improvements.

6.3.5 Standard Costs

The major objectives of cost accounting is to determine per-item or per-order costs in order to interpret financial data, by which the management can monitor and measure the production efficiency and the production performance. Establishing cost standards is of utmost importance in accomplishing these objectives. The standard-cost system includes the preparation of standard material rates, labor and overhead rates, and the application of these rates to the job-order. The cost-standards are more readily obtained for process companies compared to the job-shop companies.

A standard amount of material can be determined for each unit, and standard labor times and machine times can be established for each unit. By applying standard unit material costs and standard labor rates, standard unit costs for material, labor, and overhead are determined. The standard costs can serve as a basis for measuring production efficiency and performance over time. Deviations from standard costs are caused by several factors, such as raw material price variations and actual quantities of material and labor used versus the standard amount of these items.

6.3.6 Economic Value Added

In recent times, a management tool known as *economic value added* is being used by large number of companies in order to make investment decisions. It focuses on adding value for the shareholders. Economic value added is used to facilitate decisions involving major capital investment, acquisitions, and divestitures. It examines the difference between the net operating profit after taxes and the cost of capital, which includes the cost of both debt, and equity capital. Stern Stewart [1991] believes that there are ways to create value for the shareholder. They are:

1. Increase profitability without using additional capital, *e.g.*, increase profit margins and increase sales without using additional capital
2. Invest in projects that earn more than the cost of capital
3. Free-up capital that earns less than the cost of capital
4. Use debt to reduce the cost of capital.

Some examples include real estate, equipment, facilities, working capital, inventories, and capital investments in training and in research and development.

6.4 COST ESTIMATION METHODS

A cost estimate is an opinion based on analysis and judgment of the cost of a system, product, or structure. The methods used to develop cost estimates fall into three categories:

1. Estimating by engineering procedures
2. Estimating by analogy
3. Parametric estimating methods.

6.4.1 Estimating by Engineering Procedures

This method involves an assessment of separate segments at a low level of detail. The total cost is determined for a system or product by adding the costs that were assigned to each element at the lowest level of detail. The process is very time consuming and requires a lot of effort. Sketches, drawings, or descriptions for some items are considered for estimating. Some items like planning, rework, coordination, and testing are factored in as a percentage of the detailed estimates. Similarly, maintenance, inspection, and production control are factored in as a percentage of the production labor required. The overhead charges (fixed costs) often are many times greater than the direct labor costs. Significant variability that occurs in the fabrication of successive units can be a source of error in cost estimation by the method. There is a trend towards creating material and processing costs into a database and the computer is used to figure out the optimum processing sequence and compute the costs.

6.4.2 Estimating by Analogy

When a company is launching a new activity, estimating by analog is very effective. In this method, the future costs of a project, system, product, or design are based on past costs of a similar project, system, product or design, with due allowance for cost escalation and complexity differences. Estimating by analog needs a large database of experience or published cost data. One drawback of estimating by analogy is the high level of intuition and judgment required. This demands a great deal of experience and expertise on the part of the estimator. The method is best suited during the preliminary stage of development of a project, product, or system. The costs of estimating by analogy are relatively low and hence this method can be used as a check against other methods.

6.4.3 Parametric Estimating Methods

The parametric estimating methods typically use statistical techniques of varying complexity. The purpose of these techniques is to identify the functional relationship between changes in cost and the factor or factors upon which the cost depends. This method is not suitable when systematic historical database does not exist. Cost data is usually expressed in empirical form, which can be quite useful in trade-off studies. Parametric estimating method is considered to be sufficient in long range planning.

6.4.4 Developing Cost Data

In developing cost data, the cost estimator should investigate all possible data sources to find out what is readily available for direct use. It may be necessary to develop the cost breakdown structure and the cost model, whereby system parameters, estimating relationships and cost factors can be identified. This will lead to the identification of the input data needed to accomplish the cost analysis. The sources of cost data are many and diversified. There are five main sources of data for costing purposes. They are

1. Existing data banks: Historical data on existing systems similar to the one being developed can be used when applicable
2. Advanced product planning data
3. Individual cost estimates and predictions
4. Supplier documentation and data
5. Engineering test and field data.

6.4.5 Cost Indexes

Cost indexes are used to update past costs to present costs. The general expression for adjusting for different time periods is given by

$$C_{ti} = C_{tr}\left(\frac{I_{ti}}{I_{tr}}\right) \qquad ...(6.25)$$

where C_{ti} = cost at time of interest

C_{tr} = cost at time of reference

I_{ti} = cost index at time of interest for location of interest

I_{tr} = cost index at time of reference for location of interest

Published cost indexes correspond to national averages and may vary from one area to another. Location factors permit adjustment to reflect local condition. There are several sources available for cost indexes for general price indexes, construction cost indexes, industrial equipment and plant maintenance.

Table 6.8. Sources of Cost Indexes

Source of cost index	Type of cost
Bureau of Labor Statistics U.S. Department of Labor	General prices indexes – consumer (consumer price index) – producer (wholesale)
Engineering News Record Oil and Gas Journal Chemical Engineering	Construction cost indexes – General construction – Petroleum refinery – Chemical plants
Marshall and Swift Index	Industrial equipment
Factory	Plant maintenance

6.4.6 Cost-Capacity Factors

The cost of most capital equipment for facilities at different capacities or sizes is given by

$$C_d = C_r \left[\frac{Q_d}{Q_r}\right]^n \qquad ...(6.26)$$

where C_d = capital costs associated at the desired capacity or size

C_r = capital costs associated at reference capacity or size

Q_d = desired capacity or size

Q_r = reference capacity or size

n = exponent varies from 0.4 to 0.8 (for most process equipment items, n = 6).

In general, the value of the exponent n is approximately 0.6 and the maximum range is from 0.3 to 1.30.

Tables 6.9, 6.10 and 6.11 show some typical values of cost capacity or size factors for various types of equipment. The cost indexes and the capacity or size relationships (Eqs. 6.25 and 6.26) may also be combined to account for the timing and capacity or size as follows.

$$C(t_d, d) = C(t_r, r)\left[\frac{I_{ti}}{I_{tr}}\right]\left[\frac{Q_{di}}{Q_{rr}}\right]^n \qquad ...(6.27)$$

$C(t_d, d)$ = Cost at time of interest for desired capacity or size

$C(t_r, r)$ = Cost at time and capacity or size of reference

I_{ti} = Cost index at time of interest

I_{tr} = Cost index at time of reference

Q_{di} = Capacity or size at desired level

Q_{rr} = Capacity or size at reference level

n = Cost capacity or size factor

Table 6.9. Typical Values of Size Exponent for Equipment

Equipment	Size range	Capacity unit	Exponent x
Blower, single stage	1000–9000	ft^3/min	0.64
Centrifugal pumps, S/S	15–40	hp	0.78
Dust collector, cyclone	2–7000	ft^3/min	0.61
Heat exchanger, shell and tube S/S	50–100	ft^2	0.51
Motor, 440 V, fan-cooled	1–20	hp	0.59
Pressure vessel, unfired carbon steel	6000–30,000	lb	0.68
Tank, horizontal, carbon-steel	7000–16,000	lb	0.67
Transformer, 3-phase	9–45	kW	0.47

Source: R.H. Perry and C.H. Chilton, *Chemical Engineers Hand Book*, 5th (ed.), pp.25-18, McGraw-Hill, New York, 1973.

Table 6.10. Approximate Cost Capacity Factor for Various Types of Equipment

Equipment description	Equipment size range	Capacity units	Cost capacity value	Exponent range
Belt conveyor	30″ to 72″ wide 10 to 100 ft long	in-ft	1.00	
Buildings	Single story Multiple story	Floor area, rooms floor area, rooms	0.85	0.75–0.90
Vertical molding	360-1000 sq in.	Average mold volume	0.75	0.70–0.85
Machine			0.74	
Shredder			0.62	
Magnetic separator		tons/hr	0.27	
Aluminum separator		tons/hr	0.68	
General manufacturing equipment		tons/hr	0.60	0.50–0.95

Source: **From various sources.**

Table 6.11. Exponent Values for Typical Product Improvement Curves

Product improvement Percentage, P	Exponent
60	– 0.737
65	– 0.624
70	– 0.515
75	– 0.415
80	– 0.322
85	– 0.234
90	– 0.152
95	– 0.074

Example E6.14. *A company bought a molding machine with a capacity of producing 70 molds per hour in the year 1990 for $300,000.*

(a) what would be the estimate cost of a molding machine with 100 molds per hour capacity in the year 2010?

(b) if a machine capable of producing 100 molds per hour is desired in 2010, what would be the estimate cost? Assume a cost capacity factor of 0.6.

Solution:

$$(a)\ C_d = C_r\left[\frac{Q_d}{Q_i}\right]^n = \$300{,}000\left[\frac{100}{70}\right]^{0.6} = \$371{,}588.64$$

From Table E6.14

$$I_{ti} = 1038.9 \text{ and } I_{tr} = 330.4$$

Hence

$$C(t_d, d) = C(t_r, r)\left[\frac{I_{ti}}{I_{tr}}\right]\left[\frac{Q_{di}}{Q_{rr}}\right]^n$$

$$= \$300{,}000\left(\frac{1038.9}{330.4}\right)\left(\frac{100}{70}\right)^{0.6}$$

$$= \$1{,}168{,}412.34$$

Table E6.14. Cost Index Data for Selected Years for Construction Costs and Metal Working Equipment

Year	Construction cost index base 1913 = 100	Manufacturing metalworking equipment cost index base 1926 = 100
1960	824	242.4
1965	971	263.2
1970	1381	330.4
1975	2212	497.7
1980	3237	730.9
1985	4195	906.0
1986	4295	912.1
1987	4406	923.8
1988	4519	1011.2
1989	4615	1038.9
1990	4732	

Source: **From various references.**

6.4.7 Product Improvement Curves

It has been well established that the labor input necessary to complete a unit of production decreases by a constant percentage each time the production quantity is doubled. The amount and rate of time reduction are found to be directly influenced by a number of variables. These variables include the amount of pre-production planning, the production process, the stabilization of design, the time duration of production runs, and the management and worker attitudes. This time-reduction process on the improvement phenomenon has been referred to by several descriptive terms. They are *the learning curve, cost quantity relationship, start-up curve, manufacturing process improvement curve*, and *product improvement curve*.

A decaying experimental form expresses the product improvement curve

$$Y = KX^n \quad \text{...(6.28)}$$

where Y = production time expressed in either hours/unit or cumulative average hours/X units

K = the number of hours required to manufacture the first unit

X = the cumulative total number of units manufactured

n = a negative numerical value which determines the percentage by which Y decreases each time X is doubled.

The exponent n is a constant, which determines the improvement rate associated with an improvement curve. These rates are normally expressed as percentages. For instance, for a 60% improvement curve, consider two production quantities X_1 and X_2, where $X_2 = 2X_1$. The production times associated with these quantities X_1 and X_2 are Y_1 and Y_2 respectively. Due to 60% improvement rate, $Y_2 = 0.6Y_1$. That is, for a 60% curve, Y declines to 60% of its former value each time the production quantity is doubled. It may be noted that the lower the percentage associated with an improvement curve, the greater the resulting improvement rate.

If the value of an improvement curve percentage is known, then the corresponding value of n is found analytically as shown in the following example. Other values on n for different percentage product improvement curves are given in Table 6.11. Note that the product improvement percentage is $P = 2^n$.

Example E6.15. *Calculate the exponent n that corresponds to a 90% product improvement curve.*

Solution:

Consider two data points on the curve (Y_1, X_1) and (Y_2, X_2), where $X_2 = 2X_1$.

$$Y_1 = KX_1^n \quad \text{and} \quad Y_2 = KX_2^n$$

or

$$\frac{Y_1}{Y_2} = \frac{KX_1^n}{KX_2^n} = \frac{X_1^n}{X_2^n} = \left(\frac{X_1}{X_2}\right)^n$$

Since $X_2 = 2X_1$ and for a 90% product improvement curve, $Y_2 = 0.9Y_1$.

Then,

$$\frac{Y_1}{Y_2} = \frac{Y_1}{0.9Y_1} = \frac{1}{0.9} = 1.111$$

$$\frac{Y_1}{Y_2} = 1.111 = \left(\frac{X_1}{X_2}\right)^n = \left(\frac{X_1}{2X_1}\right)^n = \left(\frac{1}{2}\right)^n$$

Taking logarithms of the above equation, we have

$$\log(1.111) = \log\left(\frac{1}{2}\right)^n = n \log(0.5)$$

Solving for n,

$$n = \frac{\log(1.111)}{\log\left(\frac{1}{2}\right)} = \frac{0.0457}{-0.301} = -0.1518 \approx -0.152$$

Figure 6.6 shows the resulting improvement curves for two different assembly processes with equal amount of time required to manufacture the first unit. The first and second processes have an improvement percentage of 80% and 60% respectively. It can be seen from Fig. 6.6 that for a common starting value, lower curve percentages indicate higher rates of improvement as additional units are manufactured. There are two basic types of product improvement curves. They are the unit curves and the cumulative average curves.

(a) Unit Curves

Consider a product in which 100 hours are required to produce the first unit. If a unit curve with a percentage of 75% is applied, then the Y values are computed from $y = Kx^n$ = $100x - 0.415$. The Y values corresponding to a total of $x = 8$ units, the corresponding cumulative total hours, and cumulative average hourly values are given in Table 6.12. Both unit and cumulative average time values are plotted values of x are shown in Fig. 6.7. The cumulative average values consistently lie above the unit values. Also, while the unit values decreases by a factor of 0.75 each time the quantity is doubled, the cumulative average values do not.

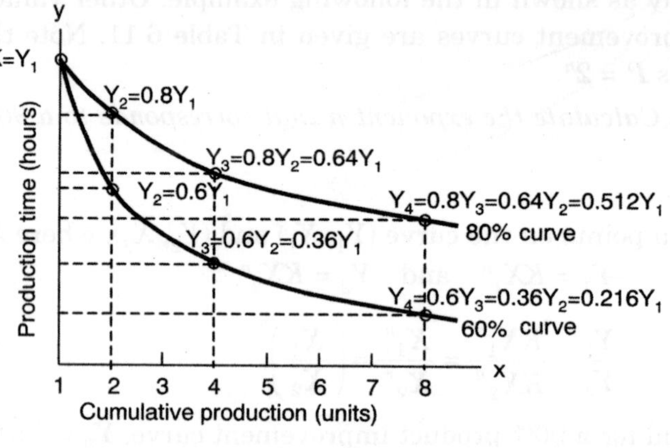

Fig. 6.6 Example of 80% and 60% Product Improvement Curves having a Common Starting Value.

Consider a quantity of consecutively manufactured units where the first unit produced is denoted by X_f. The last unit of the group that is built is denoted by X_e. The total hours T_t required to produce this group of units may be approximated by the expression.

$$T_t = \frac{K}{n+1}\left[(X_e + 0.5)^{n+1} - (X_f - 0.5)^{n+1}\right] \qquad \text{...(6.29)}$$

The total hours T required to build a cumulative total of X_e units may be found by substituting $X_f = 1$ into the above equation.

$$T = \frac{K}{n+1}\left[(X_e + 0.5)^{n+1} - (0.5)^{n+1}\right] \qquad \text{...(6.30)}$$

Equation (6.29) is usually accurate within 5% when the quantity $(X_e - X_f)$ is greater than 10. Similarly, Eq. (6.30) is usually accurate to within 5% when X_e is greater than 10. Finally, the cumulative average hours per unit required to manufacture a total of X_e units is given by

$$A = \frac{T}{X_e} \qquad \text{...(6.31)}$$

(b) Cumulative Average Curves

As before, assume that the time required to produce the first unit is 100 hours and that the improvement curve percentages is 75%. We have

$$Y = KX^n = 100\, X^{-0.415}$$

The value of Y now represents the cumulative average hours per unit needed to produce a total of X units. The cumulative total hours are obtained by multiplying the cumulative average hours per unit Y by the cumulative average hours/unit; Y by the cumulative total units produced X. The calculated values for the cumulative average hours/unit, cumulative total hours, plotted versus X are shown in Fig. 6.8. We note from Figure 6.8 that the cumulative average values consistently lie above the unit time values. The cumulative average values decreases by a factor of 0.75 each time the quantity is doubled, whereas the unit time values do not decrease by the same factor.

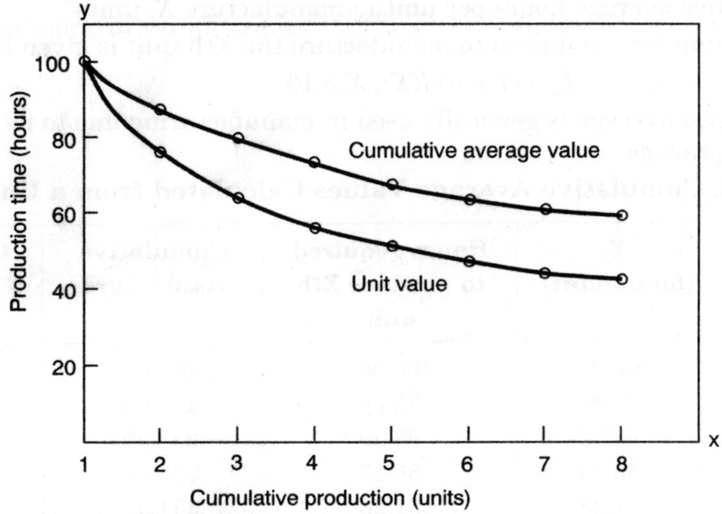

Fig. 6.7 Unit and Cumulative Values Starting with the Unit Curve.

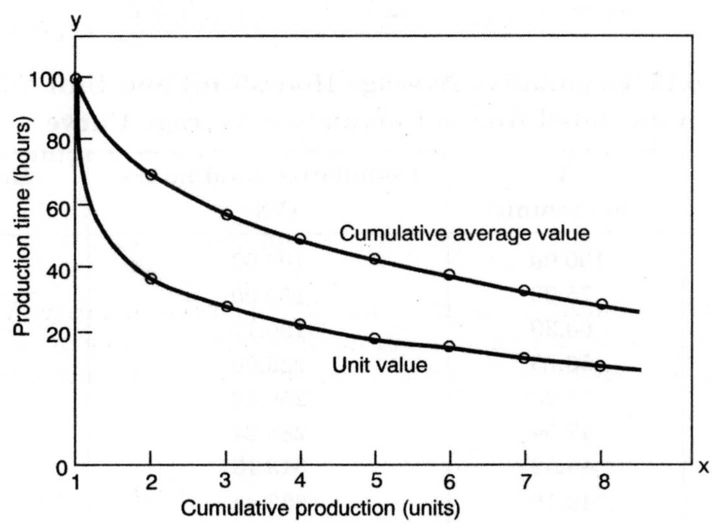

Fig. 6.8 Unit and Cumulative Average Values Starting with the Cumulative Average Curve.

Recalling that for cumulative average curves, Y represents the cumulative average hours per unit required to produce total of X units. The total hours T required to manufacture a cumulative total of X_e units is given by

$$T = YX_e \qquad ...(6.32)$$

Recalling that $y = KX^n$, Eq.(6.32) may be written as

$$T = KX_e^{-n} = K_e^{(1-n)} \quad ...(6.33)$$

The total hours required to manufacture a group of consecutively produced units where the first unit in the group is X_f and the last is X_e is

$$T_t = Y_e X_e - Y_f(X_f - 1) = KX_e^{-n} X_e - KX_f^{n}(X_f - 1) = KX_e^{1+n} - X_f^{n}(X_f - 1) \quad ...(6.34)$$

where Y_e = cumulative average hours per unit to manufacture X_e units

Y_f = cumulative average hours per unit to manufacture X_f units

Finally, the unit time required to manufacture the Xth unit is given by

$$U = (1 + n)\, KX^n,\ X \geq 10 \quad ...(6.35)$$

The cumulative average is generally used in manufacturing due to its wider utilization and acceptance in practice.

Table 6.12. Cumulative Average Values Calculated from a Unit Curve

X (units)	Y (hours/unit)	Hours required to produce Xth unit	Cumulative total hours	Cumulative average hours/unit
1	100.00	100.00	100.00	100.00
2	75.00	75.00	175.00	87.50
3	63.39	63.39	238.39	79.46
4	56.25	56.25	294.00	73.66
5	51.28	51.28	354.92	69.18
6	47.54	47.54	393.46	65.57
7	44.59	44.59	438.05	62.57
8	42.18	42.18	480.23	60.02

Table 6.13. Cumulative Average Hours/Unit and Unit Values Calculated from a Cumulative Average Curve

X (units)	Y (hours/unit)	Cumulative total hours (YX)	Hours per unit
1	100.00	100.00	100.00
2	75.00	150.00	50.00
3	63.39	190.17	40.17
4	56.25	225.00	34.83
5	51.28	256.40	31.40
6	47.54	285.24	28.84
7	44.59	313.13	26.89
8	42.18	337.44	25.31

(c) Multiple Improvements Curves

The product improvement curve or the learning curve during the start-up phases and there exists more than one product improvement curve. Consider two different product improvement curves expressed as

$$Y = K_1 X^{n1} \quad \text{for} \quad X = 1, 2, \ldots, X_1$$

and

$$Y = K_2 X^{n2} \quad \text{for} \quad X = X_1 + 1, X_1 + 2, \ldots, X_2$$

where K_1 = Intercept of unit improvement curve from 1 to X_1

K_2 = Intercept of unit improvement curve from X_1 + 1 to X_2

n_1 = Exponent for first improvement curve

n_2 = Exponent for second improvement curve

The total cumulative time to produce X units is

$$T_e = \left(\frac{K_1}{n_1+1}\right)[(X_1+0.5)^{n_1+1}] + \left(\frac{K_2}{n_2+1}\right)[(X_2+0.5)^{n_2+1}] \qquad \text{...(6.36)}$$

The Y values for the two curves must be equal where the improvement curves are the same. Hence

$$K_1 X_1^{n_1} = K_2 X_2^{n_2} \qquad \text{...(6.37)}$$

This relationship is useful to determine K_1 or K_2 when either K_1 or K_2 is known and both n_1 and n_2 are known. The time above the standard time is the time lost due to learning.

Example E6.16. *If the first unit takes 25 minutes, and the produce improvement for the process is 90% to unit 13 and 95% to unit, which is when the product improvement becomes zero.*

(a) determine the standard unit time at unit 60 when the product improvement curve is considered to end

(b) the total time for the 60 units

(c) time lost in the product improvement process.

Solution:

(a) From Table E6.11, n_1 = 0.152 and n_2 = – 0.074

From Eq.(6.37)

$$K_2 = K_1\,(X_1^{n_1}) = (25)(13)[-0.152 - (-0.074)] = 25.35 \text{ minutes}$$

Now $\quad Y = K_2\, X_2^{\,n} = 25.35\,(60^{-0.074}) = 18.724$ minutes

(b) From Eq.(6.36), the total time for the 60 units is

$$T_c = \left(\frac{K_1}{n_1+1}\right)[(X_1+0.5)^{n_1+1} - (0.5)^{n_1+1}]\left(\frac{K_2}{n_2+1}\right)[(X_2+0.5)^{n_2+1} - (0.5)^{n_2+1}]$$

Hence K_1 = 25, X_1 = 13, X_2 = 60, K_2 = 25.32, n_1 = – 0.152 and n_2 = – 0.074

Therefore,

$$T_c = \left(\frac{25}{-0.152+1}\right)[13.5^{(-0.152+1)} - (0.5)^{(-0.152+1)}]$$

$$+\left(\frac{25.35}{-0.074}\right)[60.5^{(-0.074+1)} - (0.5)^{(-0.074+1)}]$$

$$= 29.481[9.089 - 0.556] + 27.376[44.659 - 0.526]$$
$$= 251.561 + 1208.185 = 1459.746 \text{ minutes.}$$

(*c*) The total standard time for 60 units when each unit time is 18.724 minutes is 60(18.724) = 1123.44 minutes.

Time lost in the product improvement process for the first 60 units is

$$\text{Time lost} = T_c - 60(18.724) = 1159.746 - 1123.44 = 336.306 \text{ minutes.}$$

6.5 COST MODELS

The typical designer invariably has the necessary knowledge of the required manufacturing processes and the corresponding costs involved. Generally speaking, these costs are calculated after the fact. Designing to costs involves the designer to calculate both the physical properties such as strength, wear, fatigue, etc., and the costs. The cost of a product increases from conception, through design, development, utilization, and finally the scrapping or recycling. The technical and scientific knowledge of the designers is crucial in determining the product costs. The designer's objective is to determine the factory cost and the total life cycle cost.

Product costs are influenced at various stages of its life from design to marketplace. The most significant factors are

1. The concept, including physical effects, material type, number and type of active surfaces.
2. Size of the product, *i.e.*, dimensions and amount of material.
3. Number of parts, including standard, similar and same parts.

Quick cost estimation requires a simple method based on a defined cost entity which is at the designer's disposal at the given design stage. Cost models applicable to product design are briefly presented here. These cost estimation methods are based on operations, weight, material, throughput parameters, physical relationships, regression analysis and similarity laws.

6.5.1 Operation-based Cost Models

These models are used in the final design stage. Dewhurst and Boothroyd [1987] describe cost models for machining and injection molding operations.

6.5.1.1 Machining cost model

Neglecting non-productive costs, the cost of machining a feature in one component on one machine tool is given by

$$C_m = Mt_m + Q[(Mt_{ct} + C_t)t_m]/t \qquad ...(6.38)$$

where C_m = cost of machining a feature in one component on one machine tool

M = machine and operator rate

t_m = machining time (time to machine tool is operating)

Q = proportion of t_m for which a point on the tool cutting edge is contacting the work piece

t = tool life while the cutting edge is contacting the work piece

t_{ct} = tool changing time

C_t = cost of providing a new cutting edge

Two costs are associated with the second term in Eq.(6.38). They are

1. the cost of machine idle time (when the operator replaces the worn tool)
2. the cost of providing the new tool

By minimizing the sum of the tool replacement costs and the machining costs, the choice of best cutting speed is obtained. The machining time t_m is given by

$$t_m = k/v \qquad ...(6.39)$$

where k = a constant for a given operation

v = the cutting speed.

The tool life t is given by Taylor's tool life equation

$$vt^n = v_r t_r^{\ n} \qquad ...(6.40)$$

where v_r = the reference cutting speed

t_r = the reference tool life

n = the Taylor tool life index which depends on the tool material.

For high speed cutting steel tools, $n = 0.125$ and for carbide tools $n = 0.25$. Substituting Eqs.(6.39) and (6.40) in Eq.(6.38) and taking the first derivative of the resulting relationship, we obtain the cutting speed for minimum cost. The optimum or minimum cost of machining is then given by

$$C_{min} = M\, t_{mc}/(1 - n) \qquad ...(6.41)$$

where t_{mc} = the machining time corresponding to the optimum cutting speed.

The factor $1/(1 - n)$ is 1.14 for high-speed steel tools and 1.33 for carbide tools.

The machinery cost C_p for maximum power condition is given by

$$C_p = M\, t_{mp} + Q\,(M\, t_{ct} + c_t)\, t_{mp}/t_p \qquad ...(6.42)$$

where t_{mp} = the machining time

v_p = the cutting speed at maximum power condition

t_p = the tool life obtained under maximum power condition (obtained from Taylor's tool life equation).

That is

$$t_p = t_c \left(\frac{v_p}{v_c}\right)^{1/n} \qquad ...(6.43)$$

The tool life t_c under minimum cost conditions is given by

$$t_c \left[\frac{1-n}{n}\right] = Q(t_{ct} + c_t/M) \qquad ...(6.44)$$

From Eqs. (6.46), (6.49), (6.50) and (6.51) we obtain

$$C_p = M\, t_{mp} \left[1 + \left(\frac{n}{1-n}\right)\right]\left(\frac{t_{mc}}{t_{mp}}\right)^{1/n} \qquad ...(6.45)$$

Equation (6.45) is useful when the cutting speed is limited by the power available on the machine tool if the estimate of machining time t_m under optimum conditions, the rate for the machine and operator M, and the tool life index n for the tool material are known.

6.5.1.2 Injection molding machine cost model

A review of the capital costs of injection molding machines indicates that these can be associated directly with the clamping force. The linear relationship between machine cost and clamping force is given by

$$C_i = j_1 + n_1 F \qquad \text{...(6.46)}$$

where C_i = injection molding machine cost (dollars)

F = machine clamping force (kN)

j_1 and n_1 = coefficients whose values are 15,890 and 42.86 respectively.

The identification of the optimum machine size based on continuously variable size versus cost relationships can show the closest and best choice from a limited range of possible sizes. The area of the mold base for a given machine is determined from the assumption that efficient molding occurs when the full clamping force is used.

The die plate area A_c for a machine is given by

$$A_c = \frac{F}{P_j A_r} \qquad \text{...(6.47)}$$

where P_j = the recommended maximum polymer injection pressure for a part

F = the clamping force

A_r = the area ratio corresponding to the projected area of cavities and runners divided by the die plate area in the mold base.

For fixed cavity plate thickness, the cost of the mold base is found to depend on the mold base cavity plate area irrespective of its length to width ratio. For constant cavity plate thickness, mold base cost is given by

$$C_b = k_3 + n_3 A_c \qquad \text{...(6.48)}$$

where C_b = cost of mold base (dollars)

A_c = area of cavity plates (cm^2)

k_3 and n_3 = coefficients whose values are 890 and 0.97 respectively for 7 cm thick cavity plates. For the mold bases with combined cavity plate thickness equal to 30 cm, the corresponding values are k_3 = 1635 and m_3 = 1.49.

The required machine clamping force F is given by

$$F = n_c f \qquad \text{...(6.49)}$$

where n_c = number of cavities

f = the separating force acting on a single cavity.

The optimum number of cavities is obtained by generalizing the cost of a single cavity and associated cores to the cost of manufacturing multiple cavities/cores in a single mold base. This relationship is given by

$$\frac{C_n}{C_1} = (1 - k_5) + k_5 n_c^{\,m_5} \qquad \text{...(6.50)}$$

where C_n = cost of manufacturing n_c identical cavities/cores

C_1 = cost of manufacturing one unique cavity/core

n_c = number of identical cavities/cores in single mold base

k_5 and m_5 = coefficients whose values are 1.71 and 0.657 respectively.

In summary, the cost of manufacturing the total batch of components C_t is given by

$$C_t = \left(\frac{NT}{n_c}\right)C_r t_i + \left(\frac{NT}{n_c}\right) C_s\, t_i + C_n + C_b + NT\, C_p \qquad \text{...(6.51)}$$

where NT = the total number of components to be produced

C_r = machine rate (\$/component)

C_s = cost of machine supervision rate (\$/component)

C_n = cost of manufacture of n_c cavities in a single mold base (dollars)

C_b = cost of mold base (dollars)

C_p = cost of polymer per component (dollars).

The machine rate C_r is represented as

$$C_r = k_1 + m_1\, F$$

where k_1 and m_1 are same as in Eq. (6.46) by standard machine amortization calculations. For a payback period of PK years with SH shift working, these coefficients are

$$k_1 = \frac{j_1}{(PK)(SH)(5\times 10^6)} \text{ and } m_1 = \frac{m}{(PK)(SH)(5\times 10^6)} \qquad \text{...(6.52)}$$

where (5×10^6) is the number of seconds available in one shift year assuming approximately 70% plant efficiency.

Substituting Eqs.(6.48), (6.50) and (6.52) into Eq.(6.51) and n_c from Eq.(6.49) gives

$$C_t = NT\left[\left(\frac{k_1 f}{F}\right) + m_1 f\right] t_i + NT\left(\frac{f}{F}\right) C_s t_i + C_1\left[1 - k_5 + k_5\left(\frac{F}{f}\right)^{m_5}\right]$$

$$+ (k_3 + m_3 F) + (NT)\,(C_p) \qquad \text{...(6.53)}$$

where $m_3 = \dfrac{n_3}{P_j A_r}$ from Eqs. (6.47) and (6.48).

The machine size in terms of clamping force which will give the least cost manufacture is obtained by equating the first derivative of Eq.(6.53) to zero.

$$\frac{dC_t}{dF} = \frac{-(NTk_1 f)}{F^2 t_i} - \frac{(NT)(f)}{F^2 C_s t_i} + \left(\frac{1}{f}\right) C_1 k_2 m_5 \left(\frac{F}{f}\right)^{m_5 - 1} + m_3 = 0 \qquad \text{...(6.54)}$$

Also by substituting $F = n_c f$ in the resultant relationship we obtain an expression for least cost manufacture.

$$\left(\frac{NT}{n_c}\right)(k_1 + C_s)\,(t_i) = m_3\, n_c f + C_1\, k_5\, m_5 n_c^{\,m_5} \qquad \text{...(6.55)}$$

If the total production volume is known, then the cavity separating force f can be determined from the recommended polymer injection pressure for any given molded component. Equation (6.55) can be used to find the optimum number of cavities. Finally, Eq. (6.53) can then be used to estimate the minimum manufacturing cost. The major parameters affecting the optimum number of cavities are the part volume, wall thickness, geometry complexity and required production volume. Table 6.14 shows the number of identical cavities in a single core n_c and the ratio $\left(\dfrac{C_n}{C_1}\right)$.

Table 6.14. Number of Identical Cavities

Number of identical cavities in a single core n_c	Ratio $= \left(\frac{C_n}{C_1}\right) = \frac{\text{cost per cavity}}{\text{cost of single cavity}}$
1	1.00
2	0.99
4	0.89
8	0.75
16	0.62
32	0.50
64	0.40
128	

$$\frac{C_n}{C_1} = (1 - k_5) + k_5\, n_c^{\,m_5} \text{ with } k_5 = 1.71 \text{ and } m_5 = 0.657$$

Example E6.17. *The costs associated with a 100-sq.cm component, which has a molding cycle time of 20 seconds and of which a total of 1.2 million are to be produced. The part and production parameters are given in the following Table E6.17.*

Table E6.17

Description	*Symbol*	*Value*
Part parameters:		
Total production volume (1000's)	NT	*1200*
Projected part area (sq.cm.)	A_c	*100*
Max. molding depth (cm)	t_c	*5*
Max. section thickness (mm)	t_h	*3.3*
Cost of single cavity/core ($)	C_1	*2500*
Cavity area ratio		*0.5*
Max. injection pressure (MPa)	P_j	*125*
Production parameters:		
Amortization period (years)	PK	*6*
Number of working shifts	SH	*2*
Machine supervision ($/year)	SY	*28000*
Machine cycle time (sec.)	t	*20*

Solution:

To conduct preliminary tests of the optimum mold size and manufacturing cost equations presented here, a spreadsheet analysis was performed to solve Eq. (6.55) and then to tabulate the separate manufacturing costs defined in Eq. (6.54). For example, Table E6.17 (*a*) shows the costs associated with a 100 sq.cm. component, which has a molding cycle time of 20s. and of which a total of 1.2 million are to be produced. The parameters used in the calculations are included in the table together with two character names which correspond to those used in the equations.

It can be seen that a breakdown of the manufacturing cost is given for increasing numbers of cavities from 1 to 32. Column 2 indicates the appropriate machine cost and columns 3 and 4 give the cost of the mold base and cavity manufacturing cost respectively. The total mold cost (columns 3 plus 4) is divided by the total batch of components to give the mold cost per component in column 5. The total manufacturing cost, including machine and machine supervision time, total mold cost, but excluding material cost, is presented in column 6. Finally, column 7 is the percentage of the total manufacturing cost attributable to the cost of the mold.

The last row in the cost breakdown table corresponds to the optimum number of cavities predicted from Eq. (6.54) . In this case, a ten-cavity mold (the nearest integer value to the solution of 9.932) would represent 36.4% of the total manufacturing cost which has the minimum value of 4.55 cents.

Table E6.17 (*a*) Spreadsheet Results for Optimum Mold Cost

Number of cavities	Machine cost, $	Mold base cost, $	Cost of cav's, $	Mold cost/part, c	Cost/part, c	Cost ratio
1	2	3	4	5 ((3+4)/NT)	6	7 (5/6)
1	67322	1039	2500	0.295	12.74	2.1%
2	118754	1188	4966	0.513	8.09	6.3%
4	221618	1486	8854	0.862	5.51	15.6%
8	427346	2082	14985	1.422	4.60	30.9%
16	838802	3274	24652	2.327	4.77	48.7%
32	1661714	5658	39895	3.796	5.88	64.6%
9.932*	526701	2370	17544	1.659	4.55	36.4%
Part parameters:						
Total production volume (1000's)				NT	1200	
Projected part area (sq.cm.)				AC	100	
Maximum molding depth (cm)				TC	5	
Maximum section thickness (mm)				TH	3.3	
Cost of single cavity/core ($)				C1	2500	
Cavity area ratio				AR	0.5	
Max. injection pressure (MPa)				PJ	125	
Production parameters:						
Amortization period (years)				PK	6	
Number of working shifts				SH	2	
Machine supervision, $/year				SY	28000	
Machine cycle time, sec				T	20	

***Optimum value. **Source:* M.S. Hundal, "Rules and Models for Low-Cost Design", DE-Vol. 52, Design for Manufacturability, ASME 1993.**

Kiewert [1982] provides a formula for cost per part C of injection molded thermoplastic parts in quantities greater than 200,000 per machine as

$$C = \frac{t_1(C_r + C_s)}{n_c} + C_m + WC_{mt} \qquad \text{...(6.56)}$$

where t_i = mold cycle time

C_r = machine rate

C_s = machine supervision rate

n_c = cavities in mold base

C_m = machine rate

W = weight of part and

C_{mt} = material cost per unit.

6.5.2 Activity-based Costing

Keys et al. [1987] discuss electronic manufacturing process system cost modeling and simulation tools. The unit cost C of a printed wiring board assembly (PWBA) is given as:

$$C = (C_M + C_{MT} + C_A + C_T + C_{TSR} + C_{RM} + C_F)/[Y + Y_R(1 - Y)] + C_{FM} \qquad \text{...(6.57)}$$

where the costs are given by

C_M = materials

C_{MT} = materials testing

C_A = assembly

C_T = PWBA testing

C_{TSR} = troubleshoot and repair

C_{RM} = repair materials

C_F = fixed

C_{FM} = field maintenance.

Y and YR are the initial test yield and the repair yield, respectively. They give further breakdown of these individual costs in terms of operations.

6.5.3 Cost Model Based on Weight and Material

Weight-based calculation for estimating manufacturing costs is particularly suited for one-of-a-kind products and for all mass-produced items if only size changes are made. The production costs per unit weight of similar products are proportional. At the low end are simple, heavy machines such as pumps and agricultural machines, at \$5/lb. Complicated, high technology products can range upwards at \$100/lb. This method allows cost estimation for products for which material costs predominate and only minor changes in the product are involved.

Material costs constitute 40-50% of the total costs in the general machine industry. They are higher for mass-produced items, where rationalization has led to the lowering of other costs and lower for one-of-a-kind products. Materials cost include material direct costs and overhead costs. Direct costs are the product of material volume and the cost per unit volume. Methods are available for reducing each of these during embodiment. Material volume can be reduced by applying the basic principle of form design: aim for uniform stress. Two corollaries of this are: avoid using material where it serves no purpose, and avoid bending and torsion. In the latter cases specially shaped parts are needed to achieve uniform stress.

After material volume reduction the next steps to consider are those which reduce the cost per unit volume. This is achieved by using materials produced in large quantities, rather than specialty materials [See Hundal (1992)]. A convenient means of comparing materials is to generate values of relative costs, using a common material such as mild steel as the basis. Special lists of this type can be useful, such as for simple cast parts of different materials. A further means of reducing material costs is to reduce the material overhead costs, which can amount to 5-20% of the material costs. The means of achieving this is to use fewer types of material and fewer different purchased parts. This reduces storage requirements, handling costs and administrative expenses.

6.5.4 Cost Model Based on Physical Laws

6.5.4.1 Use of physical relationships

Costs can be estimated based upon physical laws applied in the design analysis of a product. For instance, stress equations can relate load for a structural member of pressure in a pressure vessel to the amount of material. Analyses can lead not only to costs but also their optimization. This is the most commonly used and obvious method of minimizing costs during embodiment. As a simple example, consider a pipe of constant thickness and a diameter d. Its annual cost is the sum of (a) Operating (pumping) costs which are proportional to $1/d^4$, and (b) Investment cost proportional to d. The total cost is given by Eq.(6.57), which, by its minimization, leads to an optimum diameter given by Eq. (6.58) as

$$C = C_1/d^4 + C_2 d; \; d_{opt} = (4C_1/C_2)^{0.2} \qquad ...(6.58)$$

6.5.4.2 Application of regression analysis

The dependence of costs on product characteristics such as throughput, size, weight, etc., can be found through statistical analyses. These analyses can be extensive and computationally intensive. Regression analysis shows this dependence in the form of coefficients and exponents in the regression equation. The degree to which such analyses may be carried is shown by Pacyna et al. [1982] for simple gray iron castings. The cost per unit C_{MTD} can be approximated by

$$C_{MTD} = K_p \, K_c \, n^{-0.0782} \, n^{-0.0782} \, v_s^{0.8179} \, r_e^{-0.1124} \, r_t^{0.1655} \, r_v^{0.1786} \, n_c^{0.0387} \, S_u^{0.2301} \qquad ...(6.59)$$

where, K_p = a proportionality factor, \$/unit

K_c = complexity factor (0.9 to 1.4)

v_s = volume of casting material

r_e = elongation ratio = d_c/d_s

r_t = wall thinness ratio = S_s/f_c

r_v = volume ratio = v_c/v_s

n_c = number of cores

S_u = tensile strength.

The terms needed for defining the ratios r_e, r_t and r_n are based upon considering a solid of similar overall shape and same volume as the net volume of the casting as shown in Fig. 6.9. The authors also provide cost models for patterns and for small forged parts.

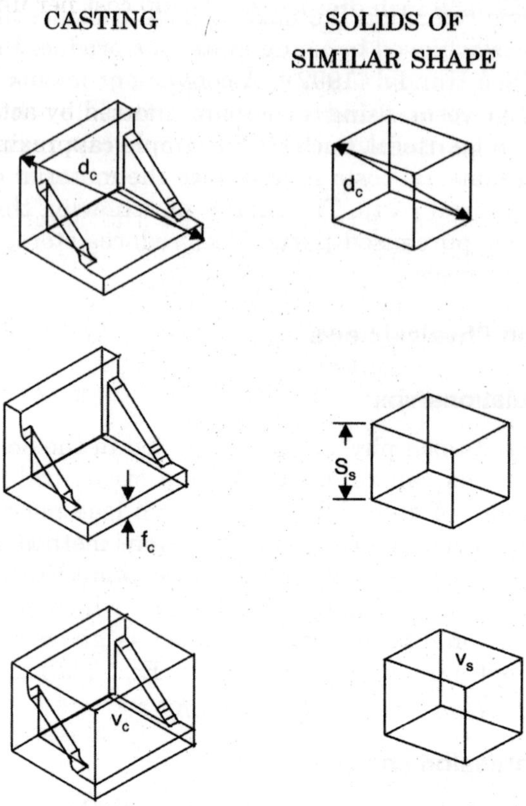

Fig. 6.9 Casting Geometry [Hundal (1993)].

6.5.5 Application of Similarity Principles

Similarity principles are used extensively in predicting costs of products made in size ranges. If a design has been built in one size, then the cost of designs in other sizes can be predicted by this method more accurately than using weight or material alone. The premise is that the linear dimension ratio of two products, $\phi_l = l_2/l_1$, is the primary parameter in relating their costs. For linear size ratio ϕ_l of two parts, their area ratio is ϕ_l^2 and the volume ratio ϕ_l^3.

The product costs can be broken down into components which change with ϕ_l^i, $i = 0, 1, 2, 3$, *i.e.*, fixed costs, those varying with linear size, with area and volume, respectively, Thus the manufacturing cost ratio, relating the cost of product 2 to the cost of product 1, is given by

$$\frac{C_{MFT2}}{C_{MFT1}} = \phi_{MFT} = \sum_{i=0}^{3} a_i \phi_l^{\,i} \qquad \text{...(6.60)}$$

The parameters a_i in general can be calculated by using regression analysis. Pahl and Rieg [1984] provide polynomials for obtaining relative costs of commonly purchased parts, *e.g.*, bearings, screws, etc. The exponents may also have non-integer values for certain types of parts as shown below.

In machining operations the similarity parameters for operation, idle and set-up times have been found to be of the form $\phi_t = \phi_l^n$. Set-up costs for machined parts have been found to be proportional to ϕ_l^s, where s (for gears) varies from 0.14 for 2–8 inch diameters, 0.56 for 8–40 inch diameters, to 1.8 for 40–60 inch diameters. Set-up cost per unit for product 2 in terms of cost for product 1 is given by

$$C_{SET2} = C_{SET1}\phi_l^s \qquad ...(6.61)$$

The production costs for machined parts are affected by actual cutting time, idle time, etc., and are found to be proportional to the surface area (approximately to ϕ_l^p), *i.e.*,

$$C_{PR2} = C_{PR1}\phi_l^p \qquad ...(6.62)$$

where the exponent p varies from 1.8 to 2.2. Likewise, the material costs for similar parts are proportional to volume and have been found to vary as

$$C_{MTL2} = C_{MTL1}\phi_l^m \qquad ...(6.63)$$

where the exponent m varies from 2.4 for small gears to 3 for large sizes.

A combination of above equations yields the combined effects of product size and production rates. The manufacturing cost per unit for product 2 in terms of costs for product 1 is given by

$$C_{MFT2} = \frac{C_{SET1}}{n}\phi_l^s + C_{PR1}\phi_l^p + C_{MTL1}\phi_l^m \qquad ...(6.64)$$

where the nominal values of the exponents are: $s = 0.5$, $p = 2$ and $m = 3$.

Example E6.18. *As an example, consider a machined part such as a gear or gear coupling. A company has produced a part in the size given below :*

Diameter, $d_1 = 8$ inch

Weight, $W_1 = 40$ lb

Manufacturing cost per unit, $C_{MFT1} = \$210$, made up of

Set-up cost, $C_{SET1} = \$110$

Production cost per unit, $C_{PR1} = \$70$

Material cost per unit, $C_{MTL1} = \$30$

It is required to estimate the costs for a similar part of diameter, $d_2 = 32$ inch. We find that the linear size ratio, $\phi_l = d_2/d_1 = 4$. Therefore, $\phi_1^{0.5} = 2$, $\phi_l^2 = 16$, and $\phi_l^3 = 64$. For the new part, the data are shown in Table E6.18.

Table E6.18. Weight and Costs for the New Part

Weight, W_2	$W_1\phi_l^3$	$= 40 \times 64 = 2560$ lb
Set-up cost, C_{SET2}	$(C_{SET1})(\phi_l^{0.5})$	$= 110 \times 2 = \$ 220$
Production cost, C_{PR2}	$(C_{PR1})(\phi_l^2)$	$= 70 \times 16 = \$1,120$
Material cost, C_{MTL2}	$(C_{MTL1})(\phi_l^3)$	$= 30 \times 64 = \$1,920$

Manufacturing costs per unit for the new part are 220 + 1,120 + 1,920 = \$3,260, if produced in single quantities. Thus for the larger part, material costs dominate.

For cylindrical objects in general, *e.g.*, gears, drums, bearings, etc., the ratio of diameter d and width b affect the costs. For spur gears the torque ratio ϕ_t is related to the diameter ratio ϕ_d and the width ratio ϕ_b, and thus to the linear size ratio ϕ_l by

$$\phi_t = \phi_d^2\,\phi_b = \phi_l^3 \qquad ...(6.65)$$

6.5.6 Accuracy and Errors During Cost Estimation

The total estimated costs of a unit (part, assembly, etc.) arise by summation of many separate items: costs of individual parts, production steps, material costs, etc. If the individual estimated costs c_i have errors E_i, which are uniformly distributed about the true cost C, the total estimated cost $c_{tot} \sum_{i=1}^{n} c_i$ will exhibit a smaller total error E_{tot} than do the individual costs c_i. This comes about due to the partial cancellation of positive and negative errors.

Kiewert [1982] has shown that the total relative error for a unit is given by

$$e_{tot} = \left[\sum_{i=1}^{n} (c_i e_i^{\,2})/c_{tot}^2 \right]^{1/2} \quad ...(6.66)$$

where $e_i = E_i/c_i$ is the relative error in c_i. If the individual costs are equal: $c_1 = c_2 = ... = c_n = c$, and have same relative error : $e_1 = e_2 = ... = e_n = e$, then Eq. (6.62) simplifies to

$$e_{tot} = e/n^{1/2} \quad ...(6.67)$$

i.e., the total error is $1/n^{1/2}$ of the individual error.

In order to determine the accuracy e_{al-i} to which the individual cost c_i must be estimated such that the total cost c_{tot} has the error e_{tot}, Kiewert [1982] gives the equation (all quantities as percentage)

$$e_{al\_i} = 10\, e_{tot}/c_i^{1/2} \quad ...(6.68)$$

as shown in Fig. E6.18. Thus if the allowable error in the total cost is to be e_{tot} = 10%, a cost which is c_i = 25% of the total cost, it must be estimated to within e_{al_i} = ± 20%. For a cost which is 1% of the total, the allowable error for it is ± 100%.

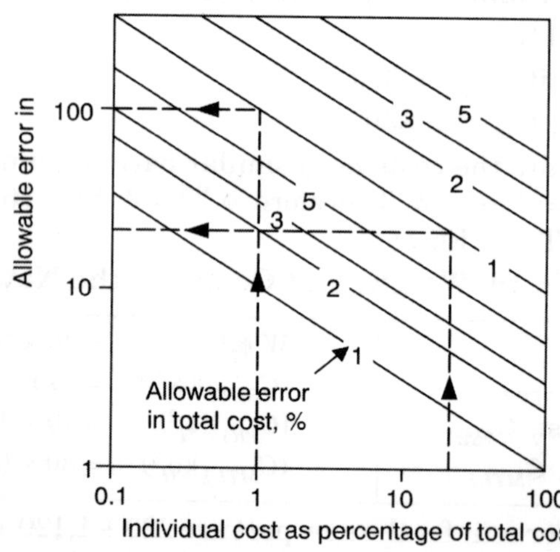

Fig. E6.18 Allowable Errors in Cost Estimation.

6.5.7 Designing to a Cost Goal

A cost goal for the product must be set by the company and provided to the design group. The designer then sets up cost structures according to parts or other criteria, based upon similar products. Only if cost calculation is carried out in parallel with design steps is a reasonable chance to hold to the goal. This is illustrated below by an example. In case of products for which no cost guidelines are available, a qualitative approach can be used in making decisions. By using relative costs for material, production process, assembly, etc., the optimum solution can be chosen.

Example E6.19. *As an example of designing to a preset cost goal we consider a centrifugal pump. The total manufacturing cost is to be \$10,000, within ± 5%. The first step is to set the cost goals of parts by using cost structures of similar pumps. Using MBC method, the parts are divided into categories A, B and C according to manufacturing cost. The calculations are shown in Table E6.19. On the basis of the percentage of total cost of each part (column 3), the maximum allowable deviation is found from Fig.E6.19 and shown in column 4. Column 5 shows the maximum cost of each part. By using models and methods described previously, the cost of each part in the initial design is calculated, shown in column 6. At this stage the cost of more expensive parts (category A) needed to be calculated more accurately, whereas category C parts require only a rough estimation. Comparing the figures in columns 5 and 6 we see that parts A2 and B1 exceed the limits and must therefore be redesigned. After redesign and calculation the final cost estimates are shown in column 8. The total cost of the redesigned pump is \$10,070 which is close to the preset goal.*

Table E6.19*

Calculations for Designing to a Preset Cost Goal

Cost goal: \$10,000 (C_{tot}) ±5% (e_{tot}); Maximum cost: \$10,500

Part	Part cost goal, \$	% total cost, c_i	% allowable deviation e_{al-i}	Maximum cost, \$	First design cost, \$	Decision	Redesigned cost, \$
(1)	(2)	(3)	(4)	(5)	(6)	(7)	(8)
A1 (housing)	4,500	45	7	4,835	4,400	acceptable	4,400
A2 (impeller)	3,500	35	8	3,796	4,000	redesign	3,600
B1 (shaft)	700	7	19	832	900	redesign	750
B2 (bearings)	600	6	20	722	600	acceptable	600
B3 (seals)	400	4	25	500	400	acceptable	400
B4 (wear rings)	200	2	35	271	220	acceptable	220
ΣC (key, bolts, gasket, etc.	100	1	50	150	100	acceptable	100
Total	10,000	100			10,630		10,070

***Source:* M.S. Hundal, "Rules and Models for Low-Cost Design", DE-Vol. 52, Design for Manufacturability, ASME 1993.**

In this section, a survey of cost models used for different types of products are presented. Examples of models based on operations, weight, material, throughput parameters, physical relationships, regression analysis and similarity laws have been given. Design decisions influence all of the subsequent costs of the product. Thus it is necessary for the designer to have as much of the cost information as possible available. Most of the cost tools today address the problem only in the final design stages, whereas the major portion of the product cost is set at much earlier. There is need to develop methods to forecast at the conceptual design stage. Function costing holds the promise of being such a tool.

6.6 SUMMARY

Cost is a primary factor in design and its buildup begins in conceptual design and continues through embodiment and detail design. Cost estimates can be done by engineering procedures, by analogy, or by statistical methods depending upon the phase of the project life cycle. For best results in estimating the final outcome of an undertaking, it will almost always be advantageous to begin with detailed estimates of income and costs. The tools available for estimating include statistics used for parametric forecasting, indexes of inflation and productivity, capacity functions, learning curves and factored estimates. In addition to the coverage of cost concepts, a brief introduction to cost estimation, ratio analysis, activity based costing, economic value added, various cost estimation methods and models, manufacturing cost estimation, and life cycle costing was presented. Life cycle costing attempts to capture all the costs associated with a system/product throughout its life cycle from design to retirement from service.

PROBLEMS

6.1 The Raghu Manufacturing Company produces metal brackets using three different machine tools $m1$, $m2$, or $m3$. An analysis of the bracket production resulted in the following data:

Machine tool	Fixed cost/order	Variable cost/unit
$m1$	\$400	\$10
$m2$	\$800	\$4
$m3$	\$600	\$6

(*a*) determine the most economical machine tool to be used for all order sizes between 1 and 250 units

(*b*) determine the machine tool to be used for an order of size 100

(*c*) in part (*b*), determine the total production cost.

6.2 The direct labor hours for the job are estimated to be 50 hours at a rate of \$15 per hour. Direct material costs are estimated at \$1000. The overhead calculations are to be based on the following previous period cost totals:

Total direct material dollars	\$58,000
Total direct labor dollars	\$520,000
Total direct material dollars	\$700,000
Total direct overhead dollars	\$400,000

Determine the overhead allocation based on

(*a*) direct labor hours

(*b*) direct labor dollars

(*c*) direct labor dollars plus direct material dollars.

6.3 Table PT6.3 summarizes the information collected for two departments of a company.

Table P6.3

	Department A	**Department B**	**Total**
Direct materials cost	\$750,000	\$250,000	\$1,000,000
Direct labor cost	\$250,000	\$150,000	\$400,000
Direct labor hours	25,000	15,000	40,000
Number of employees	16	9	25
First cost of equipment	\$250,000	\$200,000	\$450,000
Annual depreciation	\$25,000	\$20,000	\$45,000
Other factory overhead			\$150,000
General overhead			\$350,000

The company distributes depreciation overhead based on (1) the first cost of equipment in each department, (2) a zero salvage value of the equipment in 10 years, and (3) a constant annual (or straight-line) rate of depreciation. All overheads other than depreciation is first distributed to each department, and then an overhead rate per direct labor hour is computed for each department. Determine the selling price the company should quote on Job Order if the raw material costs are estimated as \$1000, estimated direct labor hours required in Departments *A* and *B* are 40 hours and 100 hours respectively, and profit is to be calculated as 35% of selling price.

6.4 A machine that is used in a manufacturing process to make a certain product has an initial cost of \$60,000 and it is expected to have a zero salvage value. The maximum capacity of this machine is 8,000 units per year, and its life is estimated to be 4 years. The unit variable cost (raw material, energy, etc.) is \$15, the total fixed annual cost of supervision and taxes is \$25,000. The company uses 15% interest compounded continuously. If the machine operates at (1) 50%, (2) 60%, (3) 70%, (4) 85% of maximum capacity, determine

(*a*) the average unit cost and annual total cost

(*b*) the incremental production cost.

6.5 A certain manufacturing process equipment costs \$12,000. The equipment has a life of 5 years and a salvage value of \$2,000. The maintenance, taxes, interests, and other fixed costs amount to \$500/year. The cost of power and supplies is \$5/hr, and the operator costs \$10/hr. The cycle time per unit of product is 1.2 minutes. If the MARR is 17%, determine the cost per unit of processing the product if the number of units of product made per week are 200, 400, 1200 and 2500. Assume the equipment operates 52 weeks per year.

6.6 A large distribution company is to contain 12 million square feet of floor space. The total cost of the building (*TC*) is given by

$$TC = (200 + 85n + 2n^2)A$$

where n = number of floors

A = area of floor in ft²/floor.

(*a*) determine the optimal number of floors for the building based on the *total cost*

(*b*) determine the optimal number of floors for the building based on the *marginal cost per floor*, $MC(n) = TC(n + 1) - TC(n)$.

6.7 The production of particular type of grain is a function of fertilizer used and is given by the following relationships:

$$P(t) = \text{grain production, tons/acre}$$
$$= 0.5f - 0.001f^2$$

where f = amount of fertilizer used, pounds/acre.

The selling price of the grain is \$300/ton. Assuming that 1 ton of grain weighs about 2000 lbs, the total cost of grain production is given by

$$TC(f) = \$(12200 + 2f) \text{ per acre.}$$

(*a*) determine the amount of fertilizer that need to be used for this grain per acre of land to maximize the annual profit

(*b*) determine the maximum annual profit per acre of land for this grain

(*c*) find the range of fertilizer application for possible profit.

6.8 An electronic component manufacturing company has historically operated at about 80% of line capacity with an output of $14,000 units per month. Due to the decreasing demand for this component, the production has been lowered to 9000 units per month for the foreseeable future. Given that

Fixed cost, FC = \$80,000 per month

Variable cost, v = \$3 per unit

Revenue, r = \$10 per unit.

(*a*) determine where the reduced level of 9000 units per month will place the production relative to the linear breakeven point and the profit

(*b*) determine the variable cost per part, v, needed to breakeven at the 9000 units level assuming the revenue per unit, r, and fixed costs remain constant.

6.9 Given that the total annual inventory cost for a particular item carried in inventory is a function of the order size Q and is given by

$$TC(Q) = \text{annual cost of ordering} + \text{annual cost of carrying inventory (\$/year)}$$

$$= P_c\left(\frac{A}{Q}\right) + (0.55Q + 160)$$

where P_c = cost of preparing a purchase order (constant value) = \$17 per purchase order

Q = the number of items ordered each time when a purchase order is placed

A = total annual demand for this item = 2200 units

A/Q = the number of orders placed each year.

(*a*) determine the value of Q that will minimize the total annual inventory cost (the economic order quantity Q^*).

(*b*) determine the corresponding total annual inventory cost, $TC(Q^*)$.

(*c*) determine the number of orders that will be placed each year for the economic order quantity Q^*.

(*d*) determine the effect on the economic order quantity if the cost of placing a purchase order is decreased by 50%.

6.10 The fixed costs for a particular exporting company are \$700,000 annually. Its mainline export is sold at a revenue of \$3 per unit and has \$2.50 variable costs.

(*a*) determine the annual breakeven quantity

(*b*) plot the revenue and total cost relationships and determine the annual profit if 1.6 million units are sold and if 2 million units are sold.

6.11 An electronic component manufacturing company has the following fixed and variable cost components for its product over a one year period.

Fixed costs, \$		**Variable costs, \$ per unit**	
Administrative	11,000	Materials	6
Lease cost	18,000	Labor	3
Insurance	8,000	Indirect labor	5
Utilities	4,000	Other overhead	18
Taxes	11,000		
Miscellaneous operations	55,000		

(*a*) determine the revenue per unit needed to breakeven if the domestic sales volume is estimated to be 6000 units

(*b*) if international sales of 4000 units can be added to the 6000-unit domestic sales, find the total profit assuming the revenue per unit obtained in part (*a*) is realized.

6.12 Two machines can be used for machining a metal part. Machine *A* costs $900 and is expected to last 3 years. Machine *B* will cost $1900 and last 5 years. An overhaul costing $400 will be required after 2500 operating hours for the machine *A*, while an overhaul costing $800 will be required for machine *B* after 10,000 hours. The operating cost of each machine is $1 per hour. Determine the number of hours per year must the machine be required to justify the purchase of the more expensive machine. Use an interest rate of 10% per year.

6.13 The total cost (*TC*) per year of transmitting a given quantity of current over a fixed distance by a copper conductor is given by

$$TC = \$25 + \$35A + \frac{\$300}{A}$$

where A = cross-sectional area of the copper conductor.

(*a*) determine the cross-sectional area of the conductor which will give the minimum total cost/year

(*b*) using the results in (*a*), determine the minimum total cost per year to operate the conductor system

(*c*) determine the total cost per year for 1-inch and 3-inch diameter copper conductors.

6.14 For a certain process in a manufacturing company, the fixed costs are $70,000, variable costs are $40 per unit of production. The total cost function is given by

$$TC(x) = 70{,}000 + 40x$$

Determine the marginal cost at x = 10, 15 and 20.

6.15 The financial data for two manufacturing companies with a 12% cost of capital is given below (in $ millions).

	Company A	**Company B**
Equity	$100	$200
Annual operating profit	$30	$40
Taxes (40%)	$12	$16
Net income	$18	$24

Determine the economic value added (EVA).

6.16 Determine the EVA for the data in Problem 6.15 using an alternative method (based on return on assets and cost of capital).

6.17 The purchase price of a 200 acre agricultural farm was $200,000. The purchaser paid 30% of the purchase price in cash and the remainder was financed at a 10% annual simple interest rate with a mortgage payable to a bank over 15 years. Also purchased the seeds for $30,000, paid $12,000 in cash and for the rest gave a promissory note to the seller for the balance. The promissory note carried an 8% annual simple interest rate and was to be paid off within 5 years. The farm reported the following revenues and expenses during the first full year of operations.

Produce sold (income)	$17,000
Timber sold (income)	6,000
Labor expenses	2,000
Expenses for machinery hired	3,500
Professional fees	1,000
Fertilizer expenses	2,000
Property taxes	1,700

Expenses for repairs	1,000
Interest expenses	12,300
Insurance expenses	1,600
Legal fees	1,200
Expenses for miscellaneous supplies	500

Prepare an income statement for the farming operation and determine the net profit (loss) before income taxes.

6.18 The Comparative Balance Sheet and Income Statements for D. Raghav Manufacturing for 2003 and 2004 are shown below. Based on these financial statements, determine the following ratios.

(*a*) return on assets employed

(*b*) current ratio

(*c*) acid-test ratio

(*d*) Accounts receivable turnover

(*e*) inventory turnover

(*f*) debt to equity ratio

(*g*) operating income to total assets ratio

(*h*) if the industry average for return on assets employed is 9.6%, assess D. Raghav's position relative to the industry on this measure

(*i*) if the industry average for current ratio is 2.05, assess D. Raghav's position relative to the industry on this measure

(*j*) if the industry average for debt to equity ratio is 82%, assess D. Raghav's position relative to the industry on this measure.

D. Raghav's Manufacturing
Comparative Balance Sheets
as of December 31, 2003 and 2004
(in thousands of $)

Assets	**2004**	**2003**
Current assets:		
Cash	130	
Temporary investments	–	410
Accounts receivable (net)	2,600	2,100
Inventory	2,000	2,200
Prepared expenses	70	50
Total current assets	4,800	5,060
Fixed assets net	2,200	1,300
Total assets	7,000	5,360
Liabilities and Capital		
Current liabilities		
Notes payable	130	60
Accounts payable	2,400	2,200
Taxes payable	140	130
Total current liabilities	2,670	2,390
Fixed liabilities		
Mortgage payable	150	90
Long term loans	300	–

Total liabilities	3,120	2,480
Capital		
Stock	2,330	1,360
Retained earnings	820	910
Earned surplus	730	610
Total capital	3,880	2,880
Total liabilities and capital	7,000	5,360

D. Raghav's Manufacturing
Comparative Income Statements
For years ended December 31, 2003 and 2004
(in thousands of $)

	2004	2003
Net sales	28,000	21,000
Cost of goods sold	23,000	17,000
Gross profit	5,000	4,000
Other operating expenses	4,000	3,200
Net operating income	1,000	800
Less: Interest expenses	50	10
Less: Income taxes	210	170
Net income	740	620

6.19 The soldering department of an equipment manufacturing company consists of four cost centers: *A*, *B*, *C* and *D*. The following data for the soldering department were compiled for the year 2004.

Cost center	Departmental expenses	Direct labor		Direct material
		Hour	Cost	
A	$22000	9800	$80000	$12,000
B	11000	3700	$17000	$11,800
C	5500	1400	$8000	$4,600
D	4700	2700	$10900	$2,900
		17600		

6.20 Refer to Problem 6.19. The additional data is given below:

Cost center	Square feet occupied	Cost of machinery	No. of direct labor employees
A	1000	$11000	6
B	500	$10000	3
C	700	$12000	2
D	600	$5000	2
Total	2800	$38000	13

Expenses other than for direct labor and materials chargeable to the soldering department last year were

Maintenance	\$7,000
Gas and electricity	20,000
Supervision and other indirect labor	32,000
Miscellaneous supplies	8,000
Equipment depreciation	5,000
Building depreciation	9,000

Determine an overhead rate per direct labor hour for each cost center if the soldering department expenses above are first allocated to each cost center as follows:

(*a*) maintenance expenses and equipment depreciation expenses are allocated according to the value of equipment (per cent of total) in each cost center

(*b*) building depreciation expenses are allocated according to the floor space occupied (per cent of total) by each cost center

(*c*) supervision and other indirect labor expenses are allocated according to the number of direct labor employees (per cent of total) of each cost center

(*d*) supplies, gas and electricity expenses are allocated according to the number of direct labor hours (per cent of total) for each cost center.

6.21 The costs of a plant built in the year 1985 with a two-storey building facility are as follows:

Cost of building = \$900,000
Cost of equipment = \$2,500,000
Building area = 70,000 square feet
Capacity of the plant = 8,000 tons per year

It is desired to build a two-storey building in the year 2005 with the following:

Desired building area = 100,000 square feet
Desired capacity = 10,000 tons per year

Determine

(*a*) the cost of the new building

(*b*) the cost of the new equipment

(*c*) the total estimated cost of the building and equipment.

6.22 Determine the product improvement rate curve exponents for 55%, 68% and 77% product improvement curves.

6.23 The first of a group of 100 machines takes 200 hours to build and assemble. Assuming a 70% product improvement curve, find the time needed to complete the fiftieth machine and the last hundredth machine.

6.24 The time taken for the first unit of production is 20 minutes and the total number of units produced is 8. Assuming a 95% product improvement curve, determine

(*a*) cumulative time required to produce the eight units

(*b*) the average time per unit required to manufacture a total of 8 units

(*c*) the time saved after the production of the first 8 units.

6.25 If the first unit takes 20 minutes and the production improvement for the process is 85% to unit 10 and 95% to unit 50, which is when the production improvement becomes zero.

(*a*) find the standard unit time at unit 50 when the product improvement curve is considered to terminate

(*b*) the total time for the 50 units

(*c*) time loss in the product improvement process.

REFERENCES

Activity-Based Cost System, *Journal of Cost Management,* 4, (1), September 1990, pp. 51–72.

Anonymous, "Engineering New Record", Vol. 142, No. 11, 1949.

Anonymous, *Mathematical Modeling of Material Removal Processes*, AFML-TR-154, September 1977.

Anonymous, MIL-HDBK 259, Life Cycle Costs in Navy Acquisition.

Apple, James M., *Material Handling Systems Design,* Ronald Press, 1972.

Armarego, E.J.A. and Brown, R.H., *The Machining of Metals*, Prentice Hall, Englewood Cliffs, NJ, 1969.

Baasel, W. D., *Preliminary Chemical Engineering Plant Design*, 2nd ed., Van Nostrand Reinhold, New York, 1990.

Barth, C.F., Blake, D.E. and Stelson, T.S., *Cost Analysis of Advanced Turbine Blade Manufacturing Processes*, NASA CR 135203, October 1977.

Blank, Leland T. and Tarquin, A.J., *Engineering Economy,* 3rd ed., McGraw-Hill, 1989.

Boothroyd, G., *Fundamentals of Metal Machining and Machine Tools*, McGraw-Hill, New York, 1975.

Bridgwater, A.V., "Chemical Engineering", November 5, 1979.

Brown, R.J. and Yanuck R.R., *Introduction of Life Cycle Costing*, Prentice Hall, Englewood Cliffs, NJ, 1985.

Chilton, C.H., "Chemical Engineering", Vol. 73, No. 9, 1966.

Chow, W.W.L., *Cost Reduction in Product Design*, Van Nostrand Reinhold, New York, 1978.

Clark, Forest D. and Lorenzoni, A.B., *Applied Cost Engineering,* Second Edition, Revised and Expanded, Marcel Dekker, 1985.

Collier, C.A. and Ledbetter, W.B., *Engineering Economic and Cost Analysis*, 2nd ed., Harper and Row, New York, 1988.

Cooper, R. and Kaplan, R.S., *Profit Priorities from Activity-Based Costing,* Harvard Business Review, May-June 1991, pp. 130–135.

Creese, C.R., "Unit Cost and Unit Production Standards for the Minimum Cost Metal Cutting Model", *AFS Transactions*, Vol. 92, p. 515 (1984).

Creese, C.R., Standards for the Minimum Cost and Maximum Production Metal Cutting Models, Turning, Shaping, Drilling, and Milling, *Trans. of the American Foundry men's Society*, Vol. 93, pp. 183–186, 1985.

Creese, R.C., Adithan, M. and Pabla, B.S., *Estimating and Costing for the Metals Manufacturing Industries,* Marcel Dekker, New York, 1992.

Dce Fazio, T.L. and Whitney, D.E., "Simplified Generation of all Mechanical Assembly Sequences", *IEEE Journal of Robotics and Automation*, Vol. RA-3, No. 6, December 1987.

De La Mare, R.F., *Manufacturing Systems Economics,* Holt, Rinehart & Winston, 1982.

DeGarmo, E. Paul, Sullivan, W.G., and Bontadein, J.A., *Engineering Economy,* 9th ed., Macmillan, 1993.

Dewhurst, P. and Boothroyd, G., "Early Cost Estimating in Design", *Journal of Manufacturing Systems,* Vol. 7 (3), pp. 183–191, 1987.

Dudick, T.S., *Manufacturing Cost Controls*, 2nd ed., Prentice Hall, Englewood Cliffs, NJ, 1992.

Fabrycky, W.J. and Blanchard, B.S., *Life Cycle Cost and Analysis*, Prentice Hall, Englewood Cliffs, NJ, 1991.

Ferguson, Earl J. and Shamblin, J.E., "Break Even Analysis," *The Journal of Industrial Engineering*, 18 (8), August 1967.

Ferririnha, P., "Computer Aided Pre Calculation in Mechanical Engineering for Designers and Work Planners with HKB "*Proc. ICED* 90, vol. 3, pp. 1346–1353, Zurich, Heurista, 1990.

Fowler, T.C., *Value Analysis in Design,* Van Nostrand Reinhold, New York, 1990.

French, M.J. and Widden, M.B., "Design for Manufacturability", *DE-Vol. 52, ASME*, New York, 1993.

Gilbert, W. W., "Economics of Machinery", *Machining – Theory and Practice*, American Society for Metals, p. 465–485, (1950).

Grant, E.L., Ireson, W.G. and Leavenworth, R.S., *Principles of Engineering Economy*, 6th ed., Wiley, New York, 1976.

Hartley, R.V., *Cost and Managerial Accounting*, Allyn and Bacon, New York, 1983.

Hendrickson, C. and Au, T., *Project Management for Construction,* Prentice Hall, Englewood Cliffs, NJ, 1989.

Hoult, D.P. and Meador, C.L., "Manufacturing Cost Estimating", *ASM Handbook*, Vol. 20, pp. 716–722, ASM International, Materials Park, OH, 1997.

Humphreys, Kenneth K. and Wellman, P., *Basic Cost Engineering,* 2nd ed., Marcel Dekker, 1987.

Hundal, M.S., "Systematic Designing for Cost", *Proceedings of ASME Manufacturing International Conference,* Dallas, Book No. G00657, pp. 393–405, March 29–April 1, 1992.

Hundal, M.S., "Design for Manufacturability", *ASME - DE*, Vol. 52, pp. 75–84, New York, 1993.

Hundal, M.S., "Rules and Models for Low Cost Design", *ASME - DE*-Vol. 52, Design for Manufacturability.

Hundal, M.S., *Systematic Mechanical Designing*, ASME Press, New York, 1997.

Jones, B.W., *Inflation in Engineering Economic Analysis*, Wiley, New York, 1982.

Kepner, Charles H., and Benjamin B. Tregoe, B.B., *The Rational Manager,* McGraw Hill, New York, 1965.

Keys, L., Balmar, J.L., and Creswell, R.A., "Electronic Manufacturing Process Systems Cost Modeling and Simulation Tools", IEEE Trans. Components Hybrids and Manufacturing Technology, CHMT-10(3). pp. 401–410, 1987.

Kiewert, A., "Short Form Calculation and Evaluation of its Accuracy", VDI-Z, Vol. 124, pp. 443–446, 1982.

Kiewert, A., *Systematic Working with Aids for Low Cost Design*, Dissertation, TU Munich, 1979.

Kleinfield, I., *Engineering and Managerial Economics,* Holt, Reinhart and Winston, New York, 1986.

Leech, D.J., and Turner, B.T., *Engineering for Design and Profit*, Ellis Horwood Ltd., Chichester, 1985.

Lindberg, R.A., *Processes and Materials of Manufacture,* 3rd ed., Boston, Ally and Bacon, Inc., 1983, pp. 138, 154–158.

Magrab, E.B., *Integrated Product and Process Design and Development*, CRC Press, Boça Raton, FL, 1997, pp. 55–63 Malstrom, E.M. (ed.), *Manufacturing Cost Engineering Handbook*, Marcel Dekker, New York, 1984.

Malstrom, E.M. and Shell, R.L., "Manufacturing Engineering", May 1979, pp. 70–75.

Michaels, J.V. and Wood, W.P., *Design to Cost*, Wiley, New York, 1989.

Mileham, A.R., Currie, G.C., Miles, A.W. and Bradford, D.T., *Journal of Engineering Design*, Vol. 4, No. 2, 1993, pp. 117–125.

Nanda, Ravinder and Alder, G.L., (eds.), *Learning Curve: Theory and Applications,* Institute of Industrial Engineers, 1982.

Nasr, N. and Varel, E.A ., "Total Product Life Cycle Analysis and Costing", *Proceedings of the* 1997 *Total Life Cycle Conference*, p. 310, pp. 9–15, Society of Automotive Engineers, Warrendale, PA, 1997.

Nelson, W. L., *Oil Gas Journal*, Vol. 65, May 15, 1967.

Nevins, J.L. and Whitney, D.L., *Concurrent Design of Products and Processes*, McGraw-Hill, New York, 1989.

Newman, D.G. and Lavelle, J.P., *Engineering Economic Analysis*, 7th ed., Engineering Press, Austin, Texas, 1998.

Niebel, B.W. and Freivalds, A., *Methods Standard Work Design*, WCB/McGraw-Hill, New York, 1999.

Ostwald, P.F., *Cost Estimating for Engineering and Management*, Prentice Hall, Englewood Cliffs, New Jersey, 1974.

Ostwald, P.F., *Engineering Cost Estimating,* 3rd ed., Prentice Hall, Englewood Cliffs, NJ, 1992.

Ott, H. H. and Hubka, V., "Pre-determination of Manufacturing Costs of Welded Designs", *Proc. ICED* 85, Vol. 1, pp. 478–487, Zurich: Heurista, 1985.

Pacyna, H., Hillebrand, A., and Rutz, A., "Early Cost Estimation for Castings", *VDI Berichte Nr.* 457, Designers lower manufacturing costs, Duesseldorf, VDI, Verlag, 1982.

Pal, G. and Baits, W., *Engineering Design: A Systematic Approach*, Springer-Verlag, New York, NY, 1988.

Pahl, G. and Rieg, F., "Relative Cost Diagrams for Purchased Parts: Approximation Polynomials Aid in Estimating Costs", *Konstruktion*, 36, pp. 1–6, 1984.

Park, W. R. and Jackson, A.E., *Cost Engineering Analysis*, 2nd ed., Wiley, New York, 1984.

Patrusco, A., *Construction Cost Engineering Handbook*, Marcel Dekker, New York, 1988.

Perry, R.H. and Chilton, C.H., *Chemical Engineers Handbook*, 5th ed., McGraw Hill, New York, 1973.

Peters, M.S. and Timmerhaus, K.D., *Plant Design and Economics for Engineers,* 4th ed., McGraw Hill, New York, 1992.

Pikulik, A. and Draz, H.E., *Chemical Engineering*, October 10, 1977.

Pilcher, R. *Appraisal and Control of Project Costs*, McGraw-Hill, New York, 1973.

Riggs, James L., Bedworth, D.S. and Randhawa S.U., *Engineering Economics,* 4th Edition, McGraw Hill, 1996.

Roberts, A.D. and Lapidge, S.C., *Manufacturing Processes*, McGraw-Hill, New York, pp. 436–437.

Sheldon, D.F., Huang, G.Q. and Perks, R., "Designing for Cost: Past Experience and Recent Developments", *Journal of Engineering Design*, 2(2), pp. 127–139, 1991.

Sims, E.R., *Precision Manufacturing Costing*, Marcel Dekker, New York, 1995.

Stewart, R.D., Wyskida, R.M., and Johannes, J.D., *Cost Estimator's Reference Manual,* 2nd ed., Wiley, New York, 1995.

Stewert, R.D., *Cost Estimating*, 2nd ed., Wiley, New York, 1991.

Tarquin, A. J. and Blank, L.T., *Engineering Economy*, McGraw-Hill, New York, 1976.

Taylor, R.W., *Art of Cutting Metals*, Vol. 28, pp. 31, 1906.

Thuesen, G.J., Fabrycky, W. and Thuessen, H.G., *Engineering Economy*, 5th ed., Prentice Hall, Englewood Cliff, NJ, 1977.

Tippet, Donald D. and Hoekstra, P., "Activity Based Costing: A Manufacturing Management Decision Making Aid," *Engineering Management Journal,* Volume 5, Number 2, June 1993, American Society for Engineering Management, pp. 37–42.

Trucks, H.E., *Designing for Economic Production*, 2nd ed., Society of Manufacturing Engineers, Dearborn, MI, 1987.

U.S. Air Force, *HQ USAF Annual Construction Pricing Guide for FY 90–94 Major Construction Program,* January 1988.

Ulrich, K.T. and Eppinger, S.D., *Product Design and Development*, McGraw-Hill, New York, 2000.

Vinton, I.R., *Realistic Cost Estimating for Manufacturing,* Society of Manufacturing Engineers, Dearborn, MI, 1968.

***Walker's Building Estimator's Reference Book*,** published periodically, Frank R. Walker Co.

Westney, R.E. (ed.), *The Engineer's Cost Hand Book,* Marcel Dekker, New York, 1997.

White, J.A., "Three Ways to Build Economic Value," *Modern Materials Handling.*

White, J. A., Agee, M.H. and Case, K.E., *Principles of Engineering Economic Analysis,* 3rd ed., Wiley, New York, 1989.

White, J.A., Case, K.E., Pratt, D.B. and Agee, M.H., *Principles of Economic Analysis*, 4th edition, John Wiley & Sons, New York, 1998.

Wilson, D.R., *An Exploratory Study of Complexity in Axiomatic Design*, Doctoral Thesis, MIT, 1980.

Winchell, W., *Realistic Cost Estimating for Manufacturing*, 2nd ed., Society of Manufacturing Engineers, Dearborn, MI, 1989.

CHAPTER 7

Design for Reliability

7.1 INTRODUCTION

In earlier chapters we dealt with product and process design, economic analysis, cost evaluation, quality principles in design and explored the notion of quality during the product design or during the manufacture of the product or the rendering of a service. The discussion was limited on a specific instant in time. In this chapter, we introduce the concept of reliability.

Reliability study deals with the random occurrences of undesirable events, failures, during the life of a physical system. In other words, reliability is a measure of the quality of the product or process over the long run. The concept of reliability is an extended period of time over which the expected operation of the product is considered. The product is expected to function according to certain expectations over a specified time period. The reliability of a product or a system is the probability that, when operating under stated environmental conditions, the product or system will perform its intended function adequately for a specified interval or period of time. In the broadest sense, reliability is a *measure of performance*. Reliability is applicable to various types of human activities as well as to the performance of physical systems. A variety of terms are used to denote the system or entity under study. The terms *unit*, *part*, *component*, *device*, *equipment*, *subsystem*, *element* and *system* are most commonly used interchangeably.

System reliability is the probability that the system will perform its intended or required function for a specified interval of time under stated operating conditions. It is the probability of a non-failure over time. Reliability is closely associated with the quality of a product and therefore can be considered as a subset of quality. Quality can be defined as the amount by which the product satisfies the customers' requirements. Product quality can be considered as a part function of design and conformance to design specifications. Quality also depends on the production or manufacturing system and on adherence to manufacturing procedures and tolerances. Reliability, on the other hand is concerned with how long the product continues to function once it becomes operational. A high-quality product will have a high reliability. Hence, reliability can be viewed as the quality of the products' operational performance over time, and therefore it extends quality into the time domain.

This chapter introduces the concept of reliability and methods for its computation. Reliability is required in the performance phase of quality, but plans for its achievement are actually addressed in the design phase. Many terms and measures used in reliability analysis are first defined, and then various distribution and hazard models are developed. Procedures for improving the reliability of a system by means of adding components in parallel (redundant systems) or through standby systems are discussed. A probabilistic design methodology for assessing the reliability of a product at the design stage is presented, and relationships between this probabilistic approach and conventional design approaches are also discussed briefly.

The life cycle of curve of a product is considered to identify the period of useful life of the product. Probability distributions that model different phases of the life cycle curve are presented. Methods of reliability allocation including the ARINC Research Corporation (ARINC) method and Advisory Group on Reliability of Electronic Equipment (AGREE) method are presented. Use of Fault tree diagrams to calculate the system, unreliability is also presented.

7.2 GENERAL RELIABILITY FUNCTION

If we represent the time to failure of a component or system by the continuous random variable T, then $T > 0$. Then, reliability is defined as

$$R(t) = \Pr(T \geq t) \quad ...(7.1)$$

In the above expression, $R(t) \leq 0$, $R(0) = 1$, and $\lim_{t \to \infty} R(t) = 0$. Hence, $R(t)$ represents the probability that the time to failure is t. Therefore, we can define the probability that a failure occurs before time t.

$$F(t) = 1 - R(t) = \Pr(T < t) \quad ...(7.2)$$

where $F(0) = 0$ and $\lim_{t \to \infty} F(t) = 1$

In equation (7.2), $R(t)$ is known as the *reliability function* and $F(t)$, the *cumulative distribution function* of the failure distribution.

Another function defined by

$$f(t) = \frac{dF(t)}{dt} = -\frac{dR(t)}{dt} \quad ...(7.3)$$

is known as the *probability density function*. $f(t)$ describes the shape of the failure distribution where

$$f(t) \geq 0 \quad ...(7.4)$$

and

$$\int_0^\infty f(t)\, dt = 1$$

If $f(t)$ is given, then

$$F(t) = \int_0^t f(t')\, dt' \quad ...(7.5)$$

and

$$R(t) = \int_t^\infty f(t')\, dt' \quad ...(7.6)$$

It should be noted that both $F(t)$ and $R(t)$ represent the areas under the curve defined by $f(t)$. Since the area under the curve is equal to unity, we have $0 \leq F(t) \leq 1$ and $0 \leq R(t) \leq 1$.

7.2.1 Mean Time to Failure (MTTF)

The *mean*, or the *expected value* $E(t)$ of the probability distribution function $f(t)$ of the continuous random variable time t, is given by

$$E(t) = \text{MTTF} = \int_0^\infty t\, f(t)\, dt \quad ...(7.7)$$

From Eq. (7.3)

$$\text{MTTF} = \int_0^\infty -\frac{dR(t)}{dt}\, t\, dt$$

Integration by parts gives

$$\text{MTTF} = -t\,R(t)\Big|_0^\infty + \int_0^\infty R(t)\, dt \qquad \text{...(7.8)}$$

$$\text{MTTF} = \int_0^\infty R(t)\, dt$$

Noting $\lim_{t\to\infty} t\,R(t) = \lim_{t\to\infty} t \exp\left[-\int_0^t \lambda(t')\,dt'\right] = 0$

and $0 \,.\, R(t) = 0$

Also, the mean time to failure is defined as

$$R(t_{med}) = 0.5 = \Pr(t \geq t_{med}) \qquad \text{...(7.9)}$$

and the mode, that is, the most likely observed failure time is given by

$$f(t_{mode}) = \max_{0 \leq t \leq \infty} f(t) \qquad \text{...(7.10)}$$

7.2.2 Variance

Variance is often used to describe a failure distribution. It is defined by

$$\sigma^2 = \int_0^\infty (t - \text{MTTF})^2 f(t)\, dt \qquad \text{...(7.11)}$$

Variance gives a measure of the spread or dispersion of the failure times about the mean. Equation (7.11) can be written as

$$\sigma^2 = \int_0^\infty [t^2 - 2t\,\text{MTTF} + (\text{MTTF})^2]\, f(t)\, dt$$

$$= \int_0^\infty t^2 f(t)\, dt - 2\text{MTTF}\int_0^\infty t\, f(t)\, dt + (\text{MTTF})^2 \int_0^\infty f(t)\, dt$$

$$= \int_0^\infty t^2 f(t)\, dt - 2(\text{MTTF})^2 + (\text{MTTF})^2$$

since $\int_0^\infty t\, f(t)\, dt = \text{MTTF}$

and $\int_0^\infty f(t)\, dt = 1$

Hence,

$$\sigma^2 = \int_0^\infty t^2 f(t)\, dt - (\text{MTTF})^2 \qquad \text{...(7.12)}$$

7.2.3 Instantaneous Failure or Hazard Rate Function

The probability of failure occurring in interval of time (a, b) is

$$\Pr(a \le T \le b) = F(b) - F(a) = R(a) - R(b) = \int_a^b f(t)\, dt$$

Therefore, $\Pr(t \le T \le t + \Delta t) = R(t) - R(t + \Delta t)$

The conditional probability of failure per unit of time or failure rate

$$\Pr(t \le T \le t + \Delta t \mid T \ge t) = \frac{R(t) - R(t + \Delta t)}{R(t)}$$

Let $$\lambda(t) = \lim_{\Delta t \to 0} \frac{-[R(t + \Delta t) - R(t)]}{\Delta t} \frac{1}{R(t)} = \frac{-dR(t)}{dt} \frac{1}{R(t)} = \frac{f(t)}{R(t)} \quad \text{...(7.13)}$$

Here, $\lambda(t)$ is known as the *instantaneous hazard rate or failure rate*.

From Eq. (7.13), we can write

$$\lambda(t)\, dt = \frac{-dR(t)}{R(t)}$$

Integrating on both sides, we have

$$\int_0^t \lambda(t')\, dt' = \int_1^{R(t)} \frac{-dR(t')}{R(t')} \qquad R(0) = 1$$

or

$$-\int_0^t \lambda(t')\, dt' = \ln R(t)$$

Hence

$$R(t) = \exp\left[-\int_0^t \lambda(t')\, dt'\right] \quad \text{...(7.14)}$$

Equation (7.14) is used to calculate the reliability function $R(t)$ when the hazard rate function $\lambda(t)$ is known or given.

7.2.4 Cumulative and Average Failure Rate

The cumulative failure rate (CFR) over a period of time t is given by

$$L(t) = \int_0^t \lambda(t')\, dt' \quad \text{...(7.15)}$$

The average failure rate (AFR), defined between two times t_1 and t_2:

$$\text{AFR}(t_1, t_2) = \frac{1}{t_2 - t_1} \int_{t_1}^{t_2} \lambda(t')\, dt' = \frac{\ln R(t_1) - \ln R(t_2)}{t_2 - t_1} \quad \text{...(7.16)}$$

Equation (7.16) can be simplified by noting $t_1 = 0$ and $t_2 = T$ and $\ln R(0) = \ln 1 = 0$. Hence,

$$\text{AFR}(t) = \frac{\ln R(0) - \ln R(t)}{t - 0} = -\frac{\ln R(t)}{t} = \frac{L(t)}{t}$$

If AFR(t) is a nondecreasing function, the failure distribution is characterized as having an increasing failure rate average (IFRA). If the function is nonincreasing, the distribution has a decreasing failure rate average (DFRA).

7.2.5 Bathtub Hazard Rate Curve

The hazard rate curve, shown in Figure 7.1, is a typical hazard rate curve, especially when representing the failure rate of electronic components. Mechanical components may not necessarily follow this type of failure pattern.

As shown in Figure 7.1, the decreasing hazard rate is also called the *burn-in period*. Other names for this period are the *debugging period*, *infant mortality period*, and *break-in period*. Failures during this period are normally attributed to design, manufacturing defects, welding flaws, cracks, defective parts, poor quality control, contamination, etc.

The constant part of the bathtub hazard rate is known as the *useful period*, which starts after the infant mortality period and ends just before the *wear-out period*. Environment, random loads, human error, etc. cause this.

The wear-out period starts when a component has aged or bypassed its useful operating life. The number of failures during this time begins to increase. Failures that occur during the useful life are called *random failures* since they occur randomly. This is caused by fatigue, corrosion, aging, friction, cyclic loadings, etc.

The hazard rate shown in Figure 7.1 can be represented as

$$\lambda(t) = k\lambda ct^{c-1} + (1-k)bt^{b-1}\ \beta e^{\beta t^b} \qquad \text{...(7.17)}$$

for $b, c, \beta, \lambda > 0$; $0 < k < 1$; $t > 0$ and $c = 0.5, b = 1$

where b, c = shape parameters

β, λ = scale parameters

t = time

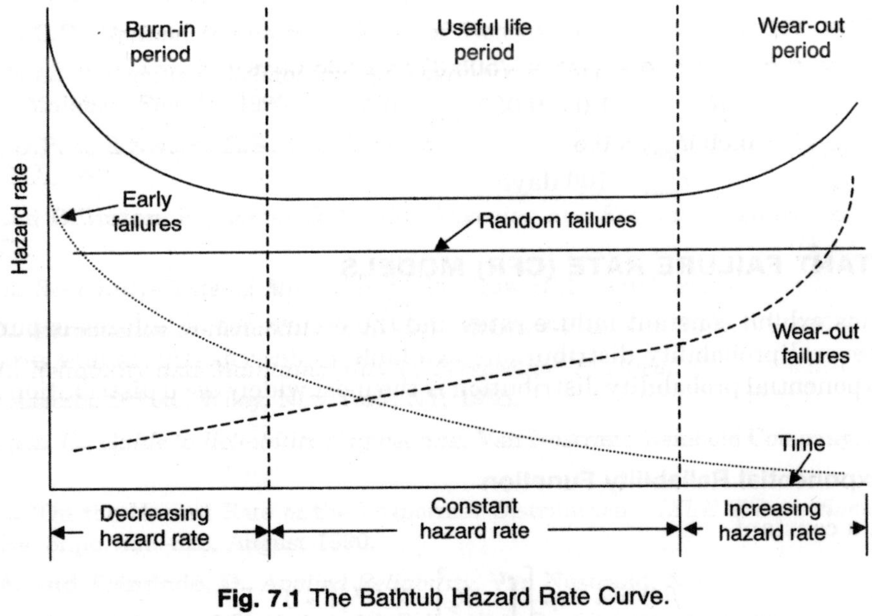

Fig. 7.1 The Bathtub Hazard Rate Curve.

Example E7.1. *The time-to-failure density function (PDF) for a system is given by*

$$f(t) = 0.005 \ ; 0 \le t \le 75 \text{ days}$$

Determine

(a) $R(t)$

(b) the hazard rate function
(c) the MTTF
(d) the standard deviation
(e) the median time to failure.

Solution:

(*a*) $$R(t) = \int_t^{75} f(t')\,dt' = \int_t^{75} 0.005\,dt' = 0.005 \int_t^{75} (75 - t)$$

$$= 0.375 - 0.005t\ ;\ 0 \le t \le 75 \text{ hours}$$

(*b*) $$\lambda(t) = \frac{-dR(t)}{dt}\frac{1}{R(t)} = \frac{f(t)}{R(t)} = \frac{0.005}{0.375 - 0.005t} \qquad 0 \le t \le 75$$

(*c*) $$\text{MTTF} = \int_0^{75} R(t)\,dt = \int_0^{75} (0.375 - 0.005t)\,dt$$

$$= 0.375t - 0.0025t^2 \Big|_0^{75} = 28.125 - 14.0625 = 14.0265 \text{ days}$$

(*d*) $$\sigma^2 = \int_0^{75} t^2 f(t)\,dt - (\text{MTTF})^2 = 0.005 \int_0^{75} t^2\,dt - (14.0625)^2$$

$$= \frac{0.005}{3}\left[t^3\right]_0^{75} - 14.0265^2 = 703.125 - 197.754 = 505.37\ (\text{days})^2$$

Hence,

$$\sigma = \sqrt{\sigma^2} = \sqrt{505.37} = 22.48 \text{ days}$$

(*e*) $$R(t_{med}) = 1.0 - 0.005t_{med} = 0.5$$

$$0.005t_{med} = 0.5$$

or $$t_{med} = 100 \text{ days}$$

7.3 CONSTANT FAILURE RATE (CFR) MODELS

Many systems exhibit constant failure rates and the occurrence of failures is purely random. There are several probability distributions available to describe the failure process. Among them, the exponential probability distribution is the most widely used distribution in reliability analysis.

7.3.1 The Exponential Reliability Function

Let $\lambda(t) = \lambda$, a constant ...(7.18)

Then $$f(t) = \lambda \exp\left[\int_0^t \lambda\,d\xi\right] = \lambda e^{-\lambda t}, t > 0 \qquad ...(7.19)$$

$$R(t) = \exp\left[-\int_0^t \lambda\,d\xi = e^{-\lambda t}\right] \qquad ...(7.20)$$

$$Q(t) = 1 - e^{-\lambda t} \qquad ...(7.21)$$

Figure 7.2 shows the four functions representing the exponential distribution. Expressions for $Q(t)$ and $R(t)$ can be derived using the failure rate λ as follows:

$$Q(t) = \int_0^t \lambda e^{-\lambda\xi}\, d\xi = \left[\frac{\lambda e^{-\lambda\xi}}{(-\lambda)}\right]_0^t = (1 - e^{-\lambda t}) \qquad ...(7.22)$$

$$R(t) = \int_t^\infty \lambda e^{-\lambda\xi}\, d\xi = \left[\frac{\lambda e^{-\lambda\xi}}{(-\lambda)}\right]_t^\lambda = e^{-\lambda t} \qquad ...(7.23)$$

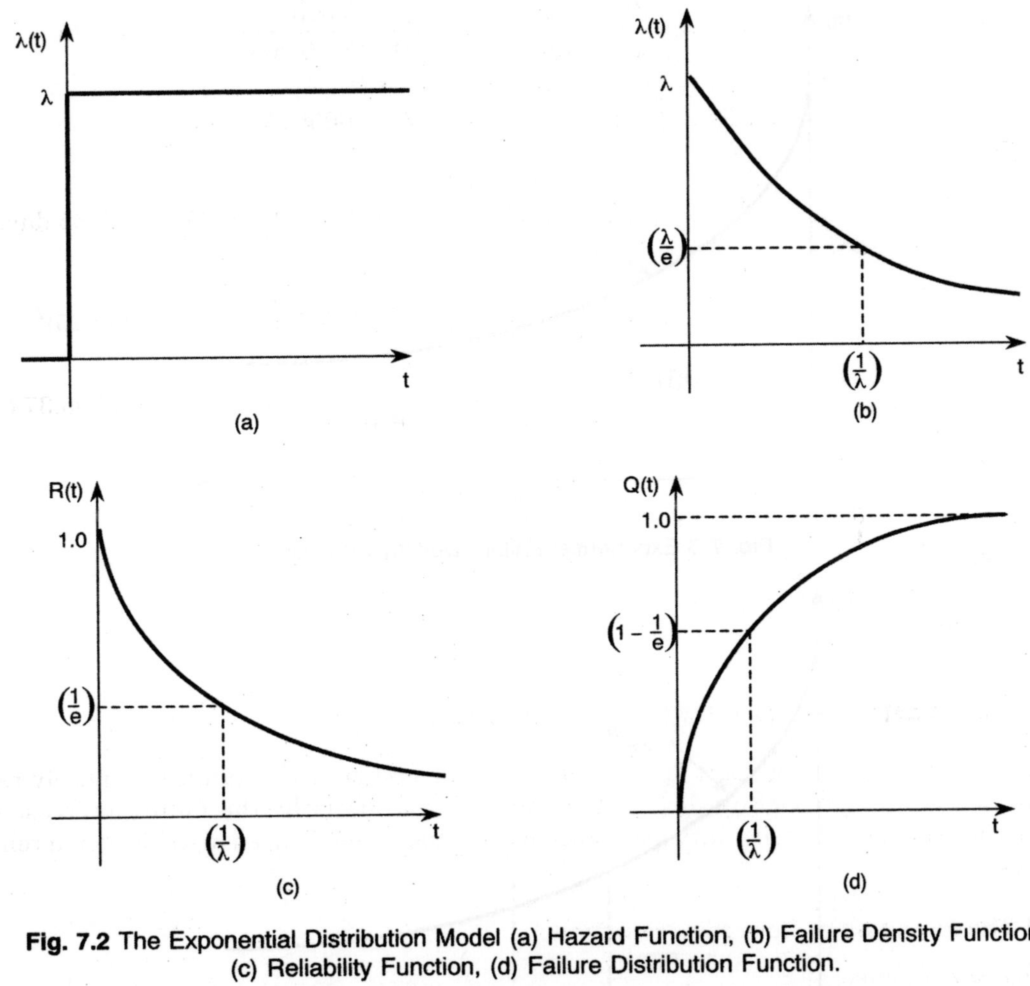

Fig. 7.2 The Exponential Distribution Model (a) Hazard Function, (b) Failure Density Function (c) Reliability Function, (d) Failure Distribution Function.

The sum of the equations (7.22) and (7.23) is always equal to unity, as illustrated in Figure 7.3.

The parameter λ is called the *failure rate* and it is equal to the number of failures per unit of time. We can derive the expressions for reliability and unreliability as follows:

P(no failure in the time interval 0 to T) = $R(T) = e^{-\lambda T}$ ...(7.24)

P(failure in the time interval 0 to T) = $Q(T) = 1 - e^{-\lambda T}$ ...(7.25)

Next, let us consider an interval of time $(T, T + t)$ as illustrated in Figure 7.4.

Let event A = failure during time t

event B = survival up to time T = no failure in the time interval (0, T)

Then

$$A \cap B = \text{survival up to } T \text{ and failure during } (T, T + t)$$

$$P(A \cap B) = \int_T^{T+t} \lambda e^{-\lambda\xi}\, d\xi = \left(\frac{\lambda e^{-\lambda\xi}}{-\lambda}\right)_T^{T+t} = e^{-\lambda t} - e^{-\lambda t(T+t)}$$

Also,

$$P(B) = \int_T^{\infty} \lambda e^{-\lambda\xi}\, d\xi = e^{-\lambda t}$$

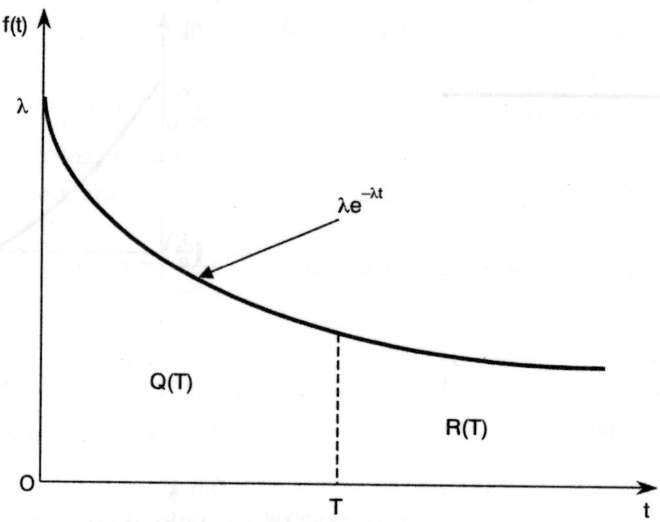

Fig. 7.3 Exponential Failure Density Function.

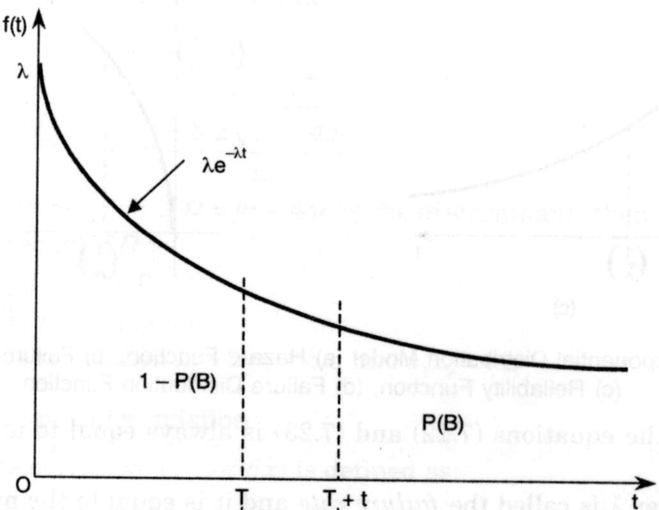

Fig. 7.4 A Posteriori Probability of Failure, $Q_c(t)$ = [$P(AB)$]/[$P(B)$].

The *a posteriori* failure probability $Q_c(t)$ is defined as the probability of failure during t, given survival up to T. Therefore,

$$Q_c(t) = P(A \mid B) = \frac{P(A \cap B)}{P(B)}$$

or
$$Q_c(t) = \frac{e^{-\lambda T} - e^{-\lambda(T+t)}}{e^{-\lambda T}} = 1 - e^{-\lambda t} \qquad ...(7.26)$$

$Q_c(t)$ is independent of the prior operating time T, and depends only on the duration t of the time interval. The component does not degrade in quality with the time of operation.

We note here that *a priori* failure probability of the component, which is the probability of failure between 0 and t, is

$$Q(t) = 1 - e^{-\lambda t} \qquad ...(7.27)$$

Hence, for the exponential distribution,

$$Q(t) = Q_c(t) = 1 - e^{-\lambda t} \qquad ...(7.28)$$

The probability of failure depends only on the duration of time considered, and not on previous history. In other words, the exponential distribution is a memoryless distribution.

If $\lambda t << 1$

$$Q(t) = Q_c(t) \approx \lambda t$$

and $R(t) \cong (1 - \lambda t)$

If λ is not constant, we can always find $Q_c(t)$ as

$$Q_c(t) = \frac{\int_T^{T+t} f(\xi)d\xi}{\int_T^{\infty} f(\xi)d\xi} \qquad ...(7.29)$$

7.3.1.1 Expected value and standard deviation

The expected value μ is obtained from

$$\mu = \int_0^{\infty} t\lambda e^{-\lambda t} dt = \int_0^{\infty} t d(-e^{-\lambda t})$$

Integrating, we get

$$\mu = \left[-te^{-\lambda t}\right]_0^{\infty} - \int_0^{\infty} (-e^{-\lambda t})dt$$

$$= 0 - \left[\frac{e^{-\lambda t}}{\lambda}\right]_0^{\infty} \qquad ...(7.30)$$

$$= \frac{1}{\lambda}$$

For the case of a failure density function, μ is called the *mean time to failure* (MTTF). The variance is given by

$$\text{var } t = \sigma^2 = E[t^2] - (E[t])^2$$

Since

$$E[t^2] = \int_0^{\infty} t^2 \lambda e^{-\lambda t} dt$$

$$= \int_0^{\infty} t^2 d\,(-e^{-\lambda t})$$

$$= (-t^2\, e^{-\lambda t})_0^{\infty} - \int_0^{\infty} -\, e^{-\lambda t}\; 2t\; dt$$

We obtain

$$= 0 + \frac{2}{\lambda} \int_0^{\infty} t\lambda e^{-\lambda t}\; dt = \frac{2}{\lambda}\frac{1}{\lambda} = \frac{2}{\lambda^2}$$

$$\sigma^2 = \frac{2}{\lambda^2} - \frac{1}{\lambda^2} = \frac{1}{\lambda^2}$$

and the standard deviation is therefore

$$\sigma = \frac{1}{\lambda} \qquad \text{...(7.31)}$$

For components with exponentially distributed failure times, the MTTF is equal to the reciprocal of the failure rate. It is possible for the MTTF to be greater then the useful lifetime of the component. The failure rate is equal to the reciprocal of the MTTF.

For repairable components, the term MTBF is used to indicate the mean time between failures:

MTBF = cycle time between failures
= mean time to failure + mean time to repair
= MTTF + MTTR

If MTTR < MTTF, then MTBF ≅ MTTF.

If a failed component is replaced by a new component, then

MTBF = MTTF + mean time to install a replacement
= MTTF + MTTI

Also, if MTTI << MTTF, then MTBF ≅ MTTF,

According to Eq. (7.31), the variability of failure time increases with increasing reliability (MTTF).

We also observe that

$$R(\text{MTTF}) = e^{-\text{MTTF/MTTF}} = 0.368$$

For a given reliability R, the design life of a component or system can be obtained from the inverse of the reliability function.

$$R(t_R) = e^{-\lambda t_R} = R$$

or

$$t_R = \frac{-1}{\lambda} \ln R \qquad \text{...(7.32)}$$

The median of the distribution (for $R = 0.5$) from Eq. (7.32) is

$$t_{\text{med}} = \frac{-1}{\lambda} \ln 0.5 = \frac{0.69315}{\lambda} = 0.69315 \text{ MTTF} \qquad \text{...(7.33)}$$

Equation (7.33) shows that the median is less than the mean.

$$R(t\,|\,T_0) = \Pr\{T > T_0 + t\,|\,T > T_0\} = \frac{\Pr\{T > T_0 + t\}}{\Pr\{T > T_0\}} = \frac{R(T_0 + t)}{R(T_0)}$$

$$= \frac{\exp\left[-\int_0^{T_0+1} \lambda(t')\,dt'\right]}{\exp\left[-\int_0^{T_0} \lambda(t')\,dt'\right]} = \exp\left[-\int_{T_0}^{T_0+1} \lambda(t')\,dt'\right] \quad ...(7.34)$$

$$R(t \mid T_0) = \frac{R(t+T_0)}{R(T_0)} = \frac{e^{-\lambda(t+T_0)}}{e^{-\lambda T_0}} = \frac{e^{-\lambda t}\,.\,e^{-\lambda T_0}}{e^{-\lambda T_0}} = e^{-\lambda t} = R(t)$$

Hence, the burn-in period T_0 has no subsequent effect on reliability and will not improve the component or system reliability. The time to failure depends only on the length of the observed operating time (t) and not on its current age (T_0).

Example E7.2. *Given* $R(t) = e^{-\sqrt{0.002t}}$; $t \geq 0$

(a) calculate the reliability for a 100-hr mission

(b) is the hazard rate increasing or decreasing?

(c) if the burn-in period is 15 hr, find the reliability for a 75-hr mission

(d) calculate the design life for a reliability of 0.96 given a 15-hr burn-in.

Solution:

$$(a)\; R(100) = e^{-\sqrt{0.002t}} = e^{-\sqrt{0.002(100)}} = e^{-\sqrt{0.4472}} = e^{-0.6687} = \frac{1}{1.1952} = 0.512$$

$$(b)\; \lambda(t) = \frac{-d\left(e^{-(0.002t)^{\frac{1}{2}}}\right)}{dt} \cdot \frac{1}{e^{-(0.002t)^{\frac{1}{2}}}} = \frac{1}{2}(0.002t)^{-\frac{1}{2}}(0.002)\, e^{-(0.002t)^{\frac{1}{2}}} \cdot \frac{1}{e^{-(0.002t)^{\frac{1}{2}}}}$$

$$= \frac{1}{2}(0.002 \times t)^{-\frac{1}{2}}(0.002) = \frac{0.001}{\sqrt{0.002t}}$$

From the above expression for $\lambda(t)$, we can see that $\lambda(t)$ goes to zero as t goes to infinity. Hence $\lambda(t)$ is decreasing.

$$(c)\; R\left(\frac{t}{T_0}\right) = \frac{R(T_0+t)}{R(T_0)}$$

$$R\left(\frac{75}{15}\right) = \frac{R(75+15)}{R(15)} = \frac{R(90)}{R(15)} = \frac{e^{-\sqrt{0.002(90)}}}{e^{-\sqrt{0.002(15)}}} = \frac{e^{-0.4243}}{e^{-0.1732}} = \frac{1.1891}{1.5285} = 0.778$$

$$(d)\; R\left(\frac{t}{15}\right) = \frac{R(t+15)}{R(15)} = \frac{e^{-\sqrt{0.002(t+15)}}}{e^{-\sqrt{0.002(15)}}} = 0.96$$

$$e^{-\sqrt{0.002(t+15)}} = 0.96e^{-\sqrt{0.002(15)}} = \frac{0.96}{1.1891} = 0.8073$$

$$t = \frac{[\ln 0.8073]^2}{0.002} - 15 = 22.91 - 15 = 7.91 \text{ hrs.}$$

Example E7.3. *The time-to-failure, in years, for a system is given by*

$$f(t) = \frac{400}{(t+14)^3} \qquad \text{for } t \geq 0$$

(a) find the reliability for the first year of operation
(b) calculate the MTTF
(c) determine the design life for a reliability of 0.96
(d) comment on the failure rate
(e) compute a one-year burn-in period and comment on the failure rate

Solution:

(a)
$$R(t) = \int_t^{\infty} f(t')dt' = 400 \int_0^{\infty} (t'+14)^{-3} dt' = \left[400 \frac{(t'+14)^{-2}}{-2} \right]_t^{\infty}$$

$$= 0 - \left[\frac{-200}{(t+14)^2} \right] = \frac{200}{(t+14)^2}$$

Hence,

$$R(1) = \frac{200}{(1+14)^2} = 0.889 \text{ years}$$

(b)
$$\text{MTTF} = \int_0^{\infty} R(t)dt = \int_0^{\infty} \frac{200}{(t+14)^2} dt$$

$$= 200 \int_0^{\infty} (t+14)^{-2} dt = 200 \left[\frac{(t+14)^{-1}}{-1} \right]_0^{\infty} = \left[\frac{-200}{t+14} \right]_0^{\infty}$$

$$= 0 - \left(\frac{-200}{14} \right) = 14.2857 \text{ years}$$

(c)
$$R(t) = \frac{200}{(t+14)^2} = 0.96$$

$$t + 14 = \sqrt{\frac{200}{0.96}} = 14.4337 \quad \text{or} \quad t = 0.4337 \text{ years}$$

(d)
$$\lambda(t) = \frac{f(t)}{R(t)} = \frac{\dfrac{400}{(t+14)^3}}{\dfrac{200}{(t+14)^2}} = \frac{2}{(t+14)}$$

Therefore,

$$\lambda(0) = 0.2 \quad \text{and} \quad \lambda(t \to \infty) = 0$$

Hence, the failure rate is decreasing (DFR).

(e)
$$R\left(\frac{2}{1}\right) = \frac{R(1+1)}{R(1)} = \frac{R(2)}{R(1)} = \frac{\dfrac{200}{(2+14)^2}}{\dfrac{200}{(1+14)^2}} = \frac{15^2}{16^2} = 0.879$$

A one year burn-in period will improve the reliability because the failure rate is decreasing.

Example E7.4. *The reliability of a mechanical component is given as*

$$R(t) = \left(1 - \frac{t}{2t_0}\right)^2 \quad ; \quad 0 \le t \le t_0$$

where t_0 is the maximum life of the mechanical component.

(*a*) *find out whether this mechanical component is experiencing wear out*

(*b*) *calculate the MTTF*

(*c*) *find the design life for a reliability of 0.95 if the maximum life of the component is given as 3000 operating hours.*

Solution:

(*a*) Wear out is indicated by an increasing failure rate

$$\lambda(t) = \frac{-dR(t)}{dt} \cdot \frac{1}{R(t)} = \frac{-d}{dt}\left(1 - \frac{t}{2t_0}\right) \cdot \frac{1}{\left(1 - \frac{t}{2t_0}\right)^2}$$

$$= \left[-2\left(1 - \frac{t}{2t_0}\right)\left(\frac{-1}{2t_0}\right)\right] \cdot \frac{1}{\left(1 - \frac{t}{2t_0}\right)^2}$$

$$= \frac{1}{t_0\left(1 - \frac{t}{2t_0}\right)} = \frac{1}{\left(t_0 - \frac{t_0 t}{2t_0}\right)} = \frac{1}{\left(t_0 - \frac{t}{2}\right)}$$

Now $\lambda(0) = \frac{1}{t_0}$

and $\lambda(t \to t_0)$ is at IFR

(*b*) $$\text{MTTF} = \int_0^{t_0} R(t)dt = \int_0^{t_0}\left(1 - \frac{t}{2t_0}\right)^2 dt = \left[\left(1 - \frac{t}{2t_0}\right)^3 \left(\frac{-2t_0}{3}\right)\right]_0^{t_0}$$

$$= \frac{-2t_0}{3}\left[\left(1 - \frac{t_0}{2t_0}\right)^3 - (1-0)^3\right]$$

$$= \frac{-2t_0}{3}\left[\left(1 - \frac{1}{2}\right)^3 - 1\right] = \frac{-2t_0}{3}\left[\frac{1}{8} - 1\right]$$

$$= \frac{-2t_0}{3}\left(\frac{-7}{8}\right) = \frac{7}{12}t_0$$

(c)
$$R(t) = \left[1 - \frac{t}{2(3000)}\right]^2 = 0.95$$

$$t = 6000\,(1 - \sqrt{0.95}) = 151.9234 \text{ hours}$$

Example E7.5. *Determine the following for a component experiencing chance failures (CFR) with an MTTF of 1200 hrs.*

(a) *reliability for a 300-hr mission*

(b) *design life for a 0.92 reliability*

(c) *the median time to failure*

(d) *the reliability for a 300-hr mission when a second redundant component is added.*

Solution:

(a)
$$R(t) = e^{-\lambda t} = e^{\frac{-t}{MTTF}} = e^{\frac{-t}{1200}}$$
$$R(200) = e^{-200/1200} = 0.8465$$

(b)
$$R(t_d) = e^{\frac{-t_d}{1200}} = 0.92$$
$$t_d = -\,1200 \ln (0.92) = 100.06 \text{ hrs.}$$

(c)
$$R(t_{med}) = e^{\frac{-t_{med}}{1200}} = 0.5$$
$$t_{med} = -\,1200 \ln (0.5) = 831.77 \text{ hours}$$

(d)
$$R(t) = 1 - \text{Prob(both components fail)}$$
$$= 1 - [1 - R_i(t)]^2 = 1 - \left[1 - e^{\frac{-t}{1200}}\right]^2$$

Since components are identical,

$$R(300) = 1 - \left[1 - e^{\frac{300}{1200}}\right]^2 = 0.9193$$

7.3.2 Failure Modes

Certain complex systems fail through various means resulting from different physical phenomena or different failure characteristics of individual components. The categories of failures are here referred to as failure modes.

If $R(t)$ is the reliability function for the ith failure mode, then, assuming independence among failure modes, the system reliability $R(t)$ is given as

$$R(t) = \prod_{i=1}^{n} R_i(t) \qquad \text{...(7.35)}$$

Where $R(t)$ is the probability that the ith failure mode does not occur before time t_0. Hence $R(t)$ is the product of these probabilities, the probability that none of the n failure modes occurs before time t. The system hazard rate function can be derived directly by summing the hazard rate functions of all failure modes.

If $\lambda_i(t)$ be the failure rate function for the ith failure mode, then

$$R_i(t) = \exp\left[-\int_0^t \lambda_i(t')dt'\right]$$

and

$$R(t) = \prod_{i=1}^{n} \exp\left[-\int_0^t \lambda_i(t')dt'\right]$$

$$= \exp\left[-\int_0^t \sum_{i=1}^{n} \lambda_i(t')dt'\right] = \exp\left[-\int_0^t \lambda(t')\right]dt'$$

where

$$\lambda(t) = \sum_{i=1}^{n} \lambda_i(t) \qquad ...(7.36)$$

Hence, the hazard rate function for the system is determined by summing the hazard rate functions of n independent failure modes.

7.3.3 Failure Modes with CFR Model

If a system consists of n independent, serially related components each having a constant failure rate , then from Eq. (7.36), we have

$$\lambda(t) = \lambda = \sum_{i=1}^{n} \lambda_i$$

and

$$R(t) = \exp\left[-\int_0^t \lambda dt'\right] = \exp[-\lambda t]$$

where

$$\text{MTTF} = \frac{1}{\lambda} = \frac{1}{\sum_{i=1}^{n} \lambda_i} = \frac{1}{\sum_{i=1}^{n} \frac{1}{\text{MTTF}_i}}; \qquad \text{MTTF} = \frac{1}{\lambda_i} \qquad ...(7.37)$$

That is, the system itself will have an exponential failure time (CFR model). If the components are also all identical (*i.e.*, $\lambda_i = \lambda_1$ for $i = 2,3,\ldots, n$) then

$$\lambda = n\lambda_i \quad \text{and} \quad \text{MTTF} = \frac{1}{n\lambda_i} \qquad ...(7.38)$$

Example E7.6. *The failure rate of an amplifier with an exponential time-to-failure distribution is 10% per 2000 hours. Find (a) the reliability of the amplifier at 6000 hours, (b) the mean time to failure.*

Solution:

(*a*) The constant failure rate is given as

$$\lambda = \frac{0.10}{200} = 0.00005/\text{hour}$$

The reliability at 6000 hours is

$$R(t) = e^{-\lambda t} = e^{-(0.00005)(6000)} = e^{-0.3} = 0.7408$$

(*b*) The mean time to failure is

$$\text{MTTF} = \frac{1}{\lambda} = \frac{1}{0.00005} = 20{,}000 \text{ hours}$$

Example E7.7. *Determine the highest failure rate for a component if it is to have a probability of survival (successful operation) of 98% at 5000 hours? It is given that the time to failure follows an exponential distribution.*

Solution:

The reliability at 5000 hours is 0.98. Let the constant failure rate is λ , then $R(t) = e^{-\lambda t}$. That is,

$$0.98 = e^{-\lambda(5000)}$$

or

$$0.98 = e^{-5000\lambda}$$

$$-5000\,\lambda \log_e e = \log_e (0.98)$$

$$-5000\,\lambda = \ln (0.98)$$

$$\lambda = \frac{\ln (0.98)}{-5000} = 4 \times 10^{-6}/\text{hr}$$

or

$$\lambda = 4 \times 10^{-6}$$

Hence, the highest failure rate is 247,491.58 hr for a reliability of 0.98 at 5000 hr.

7.3.4 Failures on Demand

Certain components that operate on a cyclical basis may fail on demand. Light bulbs, are more likely to fail when being turned on than while operating. If a constant failure rate can be assumed for each failure mode with a constant probability of failure on demand can be assumed, then the component will have an effective constant failure rate on a clock-hour basis.

Let λ_I = the average failure rate while idle (may be zero; e.g., failures per hour)
λ_O = the average failure rate while operating (e.g., failures per operating hour)
p = the probability of failure on demand
t_I = average length of the idle time period per cycle
t_O = average length of the operating time per cycle

Then

$$\lambda_{eff} = \frac{t_I}{t_I + t_O}\lambda_I + \frac{t_O}{t_I + t_O}\lambda_O + \frac{p}{t_I + t_O} \quad \text{...(7.39)}$$

is the failure rate in failures per clock unit of time.

A constant failure rate implies completely random and independent failures over time and hence results in a lack of memory. It was shown earlier that given the exponential distribution, the memoryless property followed.

$$R(t/T_0) = \frac{\exp\left[-\int_0^{t_0+t} \lambda(t')\,dt'\right]}{\exp\left[-\int_0^{T_0} \lambda(t')\,dt'\right]} = \exp\left[-\int_{T_0}^{T+t_0} \lambda(t')\,dt'\right]$$

$$= R(t) = \exp\left[-\int_0^t \lambda(t')\,dt'\right]$$

Then $$\int_{T_0}^{T_0+t} \lambda(t')dt' = \int_0^t \lambda(t')dt'$$

which requires $\lambda(t) = \lambda$, a constant. Hence, if failures are determined by completely random events not associated with the age of the system, a constant failure rate model will result.

7.3.5 Reliability Bounds

Let $0 < \lambda_L \le \lambda(t) \le \lambda_U$

Then $$\int_0^t \lambda_L dt' \le \int_0^t \lambda(t')dt' \le \int_0^t \lambda_U dt'$$

and $$\exp\left[-\int_0^t \lambda_L dt'\right] \ge \exp\left[-\int_0^t \lambda(t')dt'\right] \ge \exp\left[-\int_0^t \lambda_U dt'\right]$$

or $$e^{-\lambda_L t} \ge R(t) \ge e^{-\lambda_U t} \qquad ...(7.40)$$

Equation (7.40) implies that if the hazard rate is bounded, then the reliability function can also be bounded.

The exponential, because of its constant failure rate, provides a boundary between distributions having decreasing failure rates and those having increasing failure rates. In addition to knowing that the process has an increasing or decreasing failure rate, one needs to find the mean time to failure. For a decreasing failure rate process, we have

$$R(t) \le \begin{cases} e^{-t/\text{MTTF}}; & t \le \text{MTTF} \\ \text{MTTF}\dfrac{e^{-1}}{t}; & t > \text{MTTF} \end{cases} \qquad ...(7.41)$$

For an increasing failure rate process, we have

$$R(t) \ge \begin{cases} e^{-t/\text{MTTF}}; & t < \text{MTTF} \\ 0 \;; & t \ge \text{MTTF} \end{cases} \qquad ...(7.42)$$

The following upper bound is used when the hazard rate is increasing

$$R(t) \le \begin{cases} 1 \;; & t \le \text{MTTF} \\ e^{-\omega/\text{MTTF}}; & t > \text{MTTF} \end{cases} \qquad ...(7.43)$$

where

$$1 - \omega\text{MTTF} = e^{-\omega t}$$

A different ω must be found for each t.

7.3.6 The Two-Parameter Exponential Distribution

If a failure will never occur prior to some specified time t_0, then t_0 is a minimum, or threshold time of the *guaranteed lifetime*. The parameter t_0 is a location parameter that shifts the distribution of an amount equal to t_0 to the right on the time (horizontal) axis. This is same as rewriting the density function by replacing t with $t - t_0$, with the domain of the random variable. Now $t \ge t_0$. For the exponential distribution, the probability density function is given by

$$f(t) = \frac{-dR(t)}{dt} = \lambda e^{-\lambda(t-t_0)}; \quad 0 < t_0 \le t < \infty \qquad ...(7.44)$$

and the reliability function is

$$R(t) = e^{-\lambda(t-t_0)}; \qquad t \ge t_0 \qquad \text{...(7.45)}$$

From Eqs. (7.44) and (7.45), the failure rate is $\lambda(t) = f(t)/R(t) = \lambda$. The mean of the distribution is no longer $1/\lambda$ but is shifted a distance t_0 along the t-axis. Integrating, we get

$$\text{MTTF} = \int_{t_0}^{\infty} \lambda t e^{-\lambda(t-t_0)}\, dt = t_0 + \frac{1}{\lambda} \qquad \text{...(7.46)}$$

The median of the distribution is obtained by solving Eq. (7.45) for t_{med},

$$R(t_{med}) = e^{-\lambda(t_{med} - t_0)} = 0.5 \qquad \text{...(7.47)}$$

or

$$t_{med} = t_0 + \frac{\ln 0.5}{-\lambda} = t_0 + \frac{0.69315}{\lambda} \qquad \text{...(7.48)}$$

The design life t_R for a specified design reliability R is given by

$$t_R = t_0 + \frac{\ln R}{-\lambda} \qquad \text{...(7.49)}$$

The variance and standard deviation of the two-parameter exponential distribution unaffected by the location parameter. Hence $\sigma = 1/\lambda$ and the mode occurs at t_0.

7.3.7 Poisson Process

The probability of observing n failures in time t is given by the Poisson probability mass function as

$$\rho_n(t) = \frac{e^{-\lambda t}(\lambda t)^n}{n!}, \qquad n = 0, 1, 2,... \qquad \text{...(7.50)}$$

The Poisson distribution is discrete. The mean or expected number of failures over time t is given by λt, and the variance of the distribution is also λt.

7.3.8 Redundancy and the CFR Model

Consider the case of two independent and redundant components each having the same constant failure rate. A system failure will occur only when both components have failed. Since $(1 - e^{-\lambda t})^2$ is the probability that both components will fail by time t, the system reliability $R(t)$ is given by

$$R(t) = 1 - (1 - e^{-\lambda t})^2 \qquad \text{...(7.51)}$$

$$= 1 - (1 - 2e^{-\lambda t} + e^{-2\lambda t}) = 2e^{-\lambda t} - e^{-2\lambda t}$$

The hazard rate is given by

$$\lambda(t) = \frac{f(t)}{R(t)} = \frac{2\lambda e^{-\lambda t} - 2\lambda e^{-2\lambda t}}{2e^{-\lambda t} - e^{-2\lambda t}} \qquad \text{...(7.52)}$$

$$= \frac{\lambda(1-e^{-\lambda t})}{(1-0.5e^{-\lambda t})}$$

λt is clearly not a constant failure rate. Hence, the system does not have an exponential failure distribution even though both components are CFR. From Eq. (7.52), as $t \to \infty$, $\lambda(t) \to \lambda$, the redundancy system failure rate asymptotically approaches a constant failure rate.

The MTTF is given by

$$\text{MTTF} = \int_0^{\infty} R(t)dt = \int_0^{\infty} (2e^{-\lambda t} - e^{-2\lambda t})dt = \frac{2}{\lambda} - \frac{1}{2\lambda} = \frac{3}{2\lambda} = \frac{1.5}{\lambda} \qquad ...(7.53)$$

That is, the two-component redundant system will increase the MTTF by a factor of 1.5 over the single component MTTF.

Example E7.8. *The following data refer to a mechanical system with five components and has five constant failure modes:*

$$\lambda_1 = 0.002, \lambda_2 = 0.003, \lambda_3 = 0.005, \lambda_4 = 0.0015, \text{ and } \lambda_5 = 0.00002.$$

(a) determine the system MTTF and standard deviation

(b) determine the system design life for a reliability of 0.98.

Solution:

(*a*)
$$\lambda(t) = \sum_{\lambda=1}^{5} \lambda_i(t) = 0.002 + 0.003 + 0.0005 + 0.0015 + 0.00002 = 0.00702$$

$$\text{MTTF} = \frac{1}{\lambda} = \frac{1}{0.00702} = 138.89 \text{ days}$$

$$\sigma^2 = \frac{1}{\lambda^2} \quad \text{or} \quad \sigma = \frac{1}{\lambda} = 138.89 \text{ days}$$

(*b*)
$$t_{0.98} = -\frac{1}{\lambda} \ln(0.98) = -138.89 \ln(0.98) = 2.8 \text{ days}$$

Example E7.9. *If the system reliability consisting of two identical and CFR units placed in a redundant configuration is 0.97 at 2,500 operating hours, find the corresponding MTTF for each of the units.*

Solution:

$$R(2500) = 2e^{-\lambda(2500)} - e^{-2\lambda(2500)} = 0.97$$

By trial and error, we find = 0.000065.

Therefore

$$\text{MTTF} = \frac{1}{\lambda} = \frac{1}{0.000065} = 15{,}385 \text{ hours}$$

Example E7.10. *If two redundant components of a system have constant but different failure rates, find the following:*

(a) the reliability function and the MTTF

(b) the reliability at 1500 hours and MTTF of a two-component redundant system given that

$$\lambda_1(t) = 0.00025 \text{ per hour and } \lambda_2(t) = 0.0015 \text{ per hour}$$

Solution:

(*a*)
$$R(t) = 1 - (1 - e^{-\lambda_1 t})(1 - e^{-\lambda_2 t}) = e^{-\lambda_1 t} e^{-\lambda_2 t} - e^{-(\lambda_1 + \lambda_2)t}$$

$$\text{MTTF} = \int_0^{\infty} R(t)dt = \frac{1}{\lambda_1} + \frac{1}{\lambda_2} - \frac{1}{\lambda_1 + \lambda_2}$$

(*b*)
$$R(1500) = e^{-0.00025(1500)} + e^{-0.0015(1500)} - e^{-(0.00025 + 0.0015)(1500)}$$
$$= 0.68729 + 0.105399 + 0.072439 = 0.86513$$

$$\text{MTTF} = \frac{1}{0.00025} + \frac{1}{0.0015} - \frac{1}{(0.00025 + 0.0015)}$$
$$= 5238.1 \text{ hours}$$

Example E7.11. *(a) Derive a general expression for R(t) and the MTTF for a two-component system. Assume them as two identical and redundant CFR components.*

(b) If the guaranteed life is 3 months and a failure rate of 0.2 failures per year, find the system reliability for 15,000 hours of continuous operation.

Solution:

(*a*)
$$R(t) = \begin{cases} 1-[1-e^{-\lambda(t-t_0)}]^2 = 2e^{-\lambda(t-t_0)} - e^{-2\lambda(t-t_0)} & ; \; t > t_0 \\ 1 & ; \; t \le t_0 \end{cases}$$

$$= \int_0^{\infty} R(t)dt + \int_{t_0}^{\infty} 2e^{\lambda(t-t_0)} - e^{-2\lambda(t-t_0)} \, dt$$

$$\text{MTTF} = \int_0^{\infty} R(t)dt = \int_{t_0}^{\infty} 2e^{-\lambda(t-t_0)} - e^{-2\lambda(t-t_0)} \, dt = t_0 + 2e^{\lambda t} \int_{t_0}^{\infty} e^{-\lambda t} \, dt - e^{2\lambda t_0} \int_{t_0}^{\infty} e^{-2\lambda t} \, dt$$

$$= t_0 + 2e^{\lambda t_0} \left[\frac{e^{-\lambda t}}{-\lambda}\right]_{t_0}^{\infty} - e^{2\lambda t_0} \left[\frac{e^{-2\lambda t_0}}{-2\lambda}\right]_{t_0}^{\infty} = t_0 + \frac{2}{\lambda} - \frac{1}{2\lambda} = t_0 + \frac{3}{2\lambda}$$

(*b*)
$$\lambda_i = 0.2 \text{ failures/year}$$
$$t_0 = 3 \text{ months} = 1/4 \text{ year}$$
$$t = 15{,}000 \text{ hours} = 1.71 \text{ years}$$

$$R_i(t) = e^{-\lambda_i(t-t_0)}$$

$$R(t) = 1 - [1 - \mathrm{R}_i(t)]^2 = 1 - \left[1 - e^{-\lambda(t-t_0)}\right]^2$$

$$= 2^{-\lambda(t-t_0)} - e^{2\lambda(t-t_0)}$$

$$R(1.71) = 2e^{-0.2(1.71-1/4)} - e^{-2(0.2)(1.71-1/4)}$$
$$= 2(0.74677) - 0.55766 = 0.9359$$

7.4 TIME-DEPENDENT FAILURE MODELS

7.4.1 The Weibull Distribution

Weibull distribution is one of the most useful continuous probability distributions in reliability studies. The distribution can be used to model both increasing and decreasing failure rates.

The hazard rate function is given by

$$\lambda(t) = \frac{\beta t^{\beta-1}}{\alpha^{\beta}} \quad \text{...(7.54)}$$

where $\alpha > 0$, $\beta > 0$, and $t \ge 0$. The corresponding failure density function is given by

$$f(t) = \frac{\beta t^{\beta-1}}{\alpha^{\beta}} \exp\left[-\left(\frac{t}{\alpha}\right)^{\beta}\right] \quad ...(7.55)$$

The reliability function is

$$R(t) = \exp\left[-\left(\frac{t}{\alpha}\right)^{\beta}\right] \quad ...(7.56)$$

The failure distribution function is

$$Q(t) = 1 - \exp\left[-\left(\frac{t}{\alpha}\right)^{\beta}\right] \quad ...(7.57)$$

Special case of β = 1: If β = 1, the Weibull distribution reduces to the exponential distribution with a constant hazard rate of 1/α.

when $$\lambda(t) = \frac{1}{\alpha} \quad ...(7.58)$$

Hence,

$$f(t) = \left(\frac{1}{\alpha}\right) \exp\left(-\frac{t}{\alpha}\right) \quad ...(7.59)$$

and $$\text{MTTF} = \alpha. \quad ...(7.60)$$

Special case of β = 2: If β = 2, the Weibull distribution reduces to the Rayleigh distribution with $k = 2/\alpha^2$.

When $$\lambda(t) = \left(\frac{2}{\alpha^2}\right) t \quad ...(7.61)$$

we have

$$f(t) = \left(\frac{2}{\alpha^2}\right) t \exp\left[-\left(\frac{t}{\alpha}\right)^{2}\right] \quad ...(7.62)$$

In general, a value of β less than 1 represents a decreasing hazard rate, β greater than 1 represents an increasing hazard rate, and β equal to 1 represents a constant hazard rate (exponential distribution).

The expected value of the Weibull distribution is given by

$$\mu = E(t) = \int_0^{\infty} t f(t) dt \quad ...(7.63)$$

where *f(t)* is given by Eq. (7.55). Performing the integration by defining a new variable $y = (t/\alpha)^{\beta}$, we get

$$\mu = \alpha \Gamma \left(1 + \frac{1}{\beta}\right) \quad ...(7.64)$$

where Γ is the gamma function

$$\Gamma(x) = \int_0^{\infty} t^{x-1} e^{-t} dt \quad ...(7.65)$$

with $x > 0$.

The following relationships apply for integer values of x and n.

$$\Gamma(x) = (x - 1)! \qquad ...(7.66)$$

$$\Gamma(x + 1) = x\Gamma(x) \qquad ...(7.67)$$

$$\Gamma(n + 1) = n! \qquad n = 0, 1, 2 \qquad ...(7.68)$$

$$\Gamma(n) = \frac{\Gamma(n+1)}{n} \text{ for } n > 0 \qquad ...(7.69)$$

Table in Appendix 9B contains selected values of (x). With these tables, along with the fact that, for $x > 0$

$$\Gamma(x) = (x - 1)\Gamma(x - 1)$$

The variance is given by

$$\text{Var}[t] = \sigma^2 = \int_0^{\infty} t^2 f(t)dt - \mu^2 \qquad ...(7.70)$$

where $f(t)$ and μ are given by Eqs. (7.55) and (7.64) respectively.

Hence,

$$\sigma^2 = \alpha^2 \left[\Gamma\left(1+\frac{2}{\beta}\right) - \Gamma^2\left(1+\frac{1}{\beta}\right)\right] \qquad ...(7.71)$$

The conditional (*a posteriori*) probability of failure, $Q_c(t)$ is

$$Q_c(t) = 1 - \exp\left[\frac{T^{\beta} - (T+t)^{\beta}}{\alpha^{\beta}}\right] \qquad ...(7.72)$$

Figure 7.5 shows the functions associated with the Weibull distribution for $\beta = 0.5$, 2, and 3 and for $\alpha = 1$.

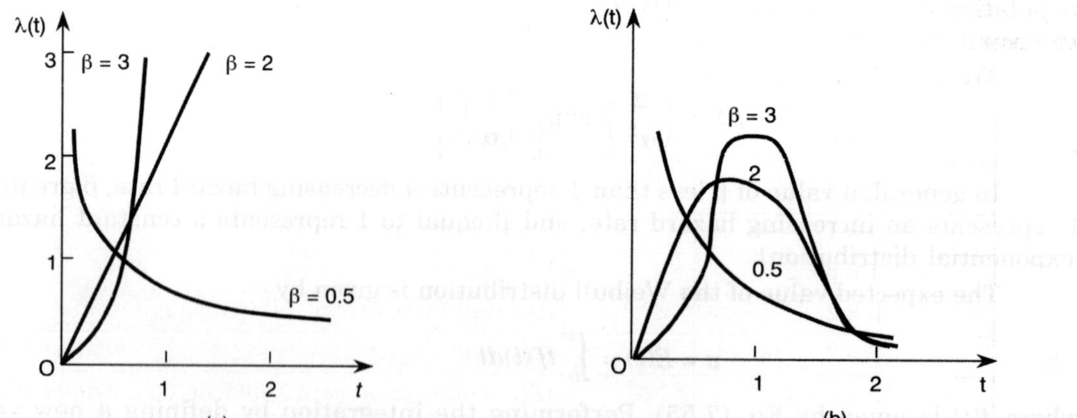

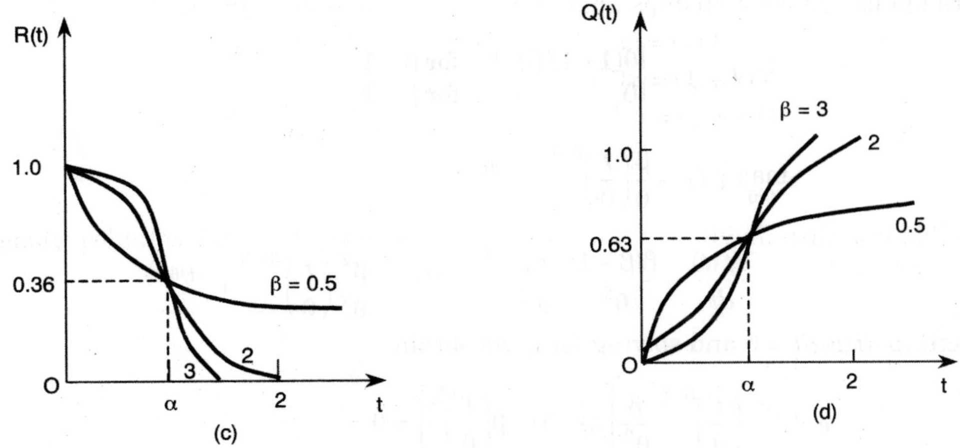

Fig. 7.5 The Weibull Model and the Associated Reliability Functions. (*a*) Hazard Function, (*b*) Failure Density Function, (*c*) Reliability Function, (*d*) Failure Distribution Function.

7.4.2 Design Life, Median, and Mode

Given a desired reliability $R(t)$

$$R(t) = e^{-(t/\theta)^{\beta}} = R$$

the design life is given by

$$t_R = \theta(-\ln R)^{1/\beta} \qquad ...(7.73)$$

When $R = 0.99$, $t_{0.99}$ is referred to as the B1 life, *i.e.*, the time at which 1 per cent of the population will have failed. Similarly, $t_{0.999}$ is called the B.1 life, the time when 0.1 per cent of the population will have failed.

Weibull Shape Parameter

Value	Property
$0 < \beta < 1$	Decreasing failure rate (DFR)
$\beta = 1$	Exponential distribution (CFR)
$1 < \beta < 2$	IFR, concave
$\beta = 2$	Rayleigh distribution (LFR)
$\beta > 2$	IFR, convex
$3 \le \beta \le 4$	IFR, Approaches normal distribution; symmetrical

If $R = 0.50$, then

$$t_{0.50} = t_{\text{med}} = \theta(-\ln 0.5)^{1/\beta} = \theta\,(0.69315)^{1/\beta} \qquad ...(7.74)$$

is the median to failure.

Since the median is a more representative measure of central tendency for highly skewed data, it may be preferred to the MTTF when β is small. The mode of the distribution can be found by solving for t^* such that

$$f(t^*) = \max_{t \ge 0} f(t)$$

which results in

$$t_{\text{mode}} = \begin{cases} \theta(1 - 1/\beta)^{1/\beta} & \text{for } \beta > 1 \\ 0 & \text{for } \beta \le 1 \end{cases} \qquad ...(7.75)$$

$$\max_{t \ge 0} f(t) = \frac{\beta}{\theta}\left(\frac{t}{\theta}\right)^{\beta-1} e^{-(t/\theta)^\beta}$$

$$\frac{df(t)}{dt} = \frac{\beta(\beta-1)}{\theta^2}\left(\frac{t}{\theta}\right)^{\beta-2} e^{-(t/\theta)^\beta} - \frac{\beta^2}{\theta^2}\left(\frac{t}{\theta}\right)^{2\beta-2} e^{-(t/\theta)^\beta}$$

Setting $df(t)/dt = 0$ and solving for t, we obtain

$$e^{-(t/\theta)^\beta}\left(\frac{t}{\theta}\right)^{\beta-2}\frac{\beta}{\theta^2}\left[(\beta-1) - \beta\left(\frac{t}{\theta}\right)^\beta\right] = 0$$

Hence,

$$(\beta - 1) - \beta\left(\frac{t}{\theta}\right)^\beta = 0$$

and

$$t_{\text{mode}} = \theta\left(\frac{\beta-1}{\beta}\right)^{1/\beta}$$

Since t_{mode} is not defined for $\beta < 1$, the mode occurs at the origin.

Example E7.12. *A mechanical system has demonstrated a Weibull failure pattern with a shape parameter of 1.4 and a scale parameter of 500 days. Determine the following:*

(a) R(150 days)

(b) the B_1 life

(c) MTTF

(d) the standard deviation σ

(e) t_{med} and t_{mode}

(f) the design life for a reliability of 0.95

Solution:

(a)
$$R(t) = e^{-\left(\frac{t}{\theta}\right)^\beta}$$

$$R(150) = e^{-\left(\frac{150}{500}\right)^{1.4}} = 0.8308$$

(b)
$$B_1 = R(t_{0.99}) = e^{-\left(\frac{t_{0.99}}{\theta}\right)^\beta} = 0.99$$

$$t_{0.99} = \theta\left[-\ln(0.99)\right]^{1/\beta} = 500\left[-\ln(0.99)\right]^{1/1.4}$$

$$= 500\left[-\ln(0.99)\right]^{0.71428} = 18.705 \text{ days}$$

(c)
$$\text{MTTF} = \theta \cdot \Gamma\left(1 + \frac{1}{\beta}\right) = 500 \,.\, \Gamma\left(1 + \frac{1}{1.4}\right)$$

$$= 500 \cdot \Gamma(1.71) = 500(0.910572)$$
$$= 455.286 \text{ days}$$

(*d*) $$\sigma^2 = \theta^2 \left[\Gamma\left(1 + \frac{2}{\beta}\right) - \left[\Gamma\left(1 + \frac{1}{\beta}\right) \right]^2 \right]$$

$$= 500^2 \left[\Gamma\left(1 + \frac{2}{1.4}\right) - \left[\Gamma\left(1 + \frac{1}{1.4}\right) \right]^2 \right]$$

$$= 500^2 \left[\Gamma(2.43) - [\Gamma(1.71)]^2 \right] = 500^2 \left[1.267032 - (0.910572)^2 \right]$$
$$= 109{,}472.66$$

Therefore, σ = 330.86 days

(*e*) $$t_{0.5} = t_{\text{med}} = \theta[-\ln(0.5)]^{1/\beta} = 500\,[-\ln(0.5)]^{1/1.4} = 384.836 \text{ days}$$

(*f*) $$t_{\text{mode}} = \theta\left[1 - \frac{1}{\beta}\right]^{1/\beta} = 500\left[1 - \frac{1}{1.4}\right]^{1/1.4} = 204.336 \text{ days}$$

(*g*) $$t_d = \theta(-\ln R)^{1/\beta} = 500(-\ln(0.95))^{1/1.4} = 59.92 \text{ days}$$

Example E7.13. *The following parameters correspond to a three-parameter Weibull distribution:*

$$\beta = 1.7 \text{ hours}, \ \theta = 7000 \text{ hours}, \ t_0 = 60 \text{ hours}$$

Determine the following:

(*a*) *R(200 hrs)*

(*b*) *MTTF*

(*c*) t_{med}

(*d*) *the standard deviation*

(*e*) *the design life for a reliability of 0.97*

Solution:

(*a*) $$R(t) = e^{-\left(\frac{t - t_0}{\theta}\right)^{\beta}}$$

$$R(200) = e^{-\left(\frac{200-60}{7000}\right)^{-1.7}} = e^{-(0.02)^{-1.7}} = 0.9987$$

(*b*) $$\text{MTTF} = t_0 + \theta \,.\, \Gamma\left(1 + \frac{1}{\beta}\right) = 60 + 7000 \,.\, \Gamma\left(1 + \frac{1}{1.7}\right)$$

$$= 650 + (7000)\Gamma(1.5882)$$
$$= 50 + 7000(0.8917) = 6291 \text{ hours}$$

(*c*) $$t_{\text{med}} = t_0 + \theta\,[-\ln(0.50)]^{1/\beta} = 60 + 7000\,[0.69315]^{1/1.7} = 5702.45 \text{ hrs.}$$

(*d*)
$$\sigma^2 = \theta^2 \left[\Gamma\left(1+\frac{2}{\beta}\right) - \left[\Gamma\left(1+\frac{1}{\beta}\right)\right]^2\right]$$

$$= 7000^2 \left[\Gamma\left(1+\frac{2}{1.7}\right) - (0.8917)^2\right]$$

$$= 7000^2 \left[\Gamma(2.1765) - 0.8409\right]$$

$$= 7000^2 \,[1.088 - 0.8409] = 12107900$$

Therefore,

$$\sigma = \sqrt{\sigma^2} = \sqrt{12107900} = 3479.64 \text{ hrs.}$$

(*e*)
$$t_d = t_0 + \theta[-\ln R]^{1/\beta}$$

$$t_{0.98} = 60 + 7000\left[-\ln 0.97\right]^{1/1.7} = 60 + 7000\;[-\ln 0.97]^{0.5882}$$

$$= 957.88 \text{ hrs.}$$

Example E7.14. *The time to failure in operating hours of a system has the hazard rate function given by:*

$$\lambda(t) = 0.002\left(\frac{T}{400}\right)^{0.5} \quad \text{for } t \geq 0$$

(*a*) *if the system must operate continuously for 70 hours, find the reliability*

(*b*) *if a reliability of 0.89 is desired, find the design life*

(*c*) *find the MTTF*

(*d*) *find the probability that the system will survive a second 70 hours of operation if the system already operated for the first 70 hours.*

(*e*) *if the system is also subjected to chance failure at a rate of 0.003 failure per operating hour, find the system's reliability for the first 70 hours of operation considering both failures.*

Solution:

(*a*) Failure time is Weibull with $\beta = 2$ and $\theta = 400$

$$R(70) = e^{-\left(\frac{70}{400}\right)^2} = e^{-0.030625} = 0.9938$$

(*b*)
$$t_d = \theta\left[-\ln R\right]^{1/\beta}$$

$$t_{0.89} = 400[-\ln (0.89)]^{1/2} = 136.548 \text{ hours}$$

(*c*)
$$\text{MTTF} = \theta \cdot \Gamma\left(1+\frac{1}{\beta}\right) = 400\,.\,\Gamma\left(1+\frac{1}{2}\right) = 400(0.88623) = 354.49 \text{ hours.}$$

(*d*)
$$R(70\,|\,70) = \frac{R(70+70)}{R(70)} = \frac{e^{-\left(\frac{140}{400}\right)^{0.5}}}{0.9938} = \frac{e^{-0.5916}}{0.9938} = 0.5569$$

(*e*) $$\lambda(t) = 0.002\left(\frac{t}{400}\right)^{0.5} + 0.003$$

$$R(t) = e^{-\left[\left(\frac{t}{400}\right)^2 + 0.003t\right]}$$

$$R(70) = e^{-\left[\left(\frac{70}{400}\right)^2 + 0.003(70)\right]} = e^{-[0.0306+0.21]} = 0.7861$$

Example E7.15. *The mechanical components in an electromechanical system have a time-to-failure distribution that can be modeled by the Weibull distribution with a scale parameter of 500 hours and a shape parameter of 2. Determine:*

(*a*) *the reliability of the components after 800 hr of operation.*

(*b*) *the mean time to failure.*

(*c*) *whether the failure rate is increasing or decreasing with time?*

Solution:

(*a*) The parameters of the Weibull distribution are $\theta = 500$ hr and $\beta = 0.3$. The reliability after 800 hr of operation is

$$R(t) = \exp\left[-\left(\frac{t}{\theta}\right)^{\beta}\right]$$

$$= \exp\left[-\left(\frac{800}{500}\right)^2\right] = \exp\,[-2.56] = 0.0773$$

(*b*) The mean time to failure is

$$\text{MTTF} = \theta\,\Gamma\left(\frac{1}{\beta}+1\right) = 500\,\Gamma\left(\frac{1}{2}+1\right) = 500\Gamma(1.5) = 500(0.88623) = 443.115$$

where $\Gamma(4.3333) = 0.88623$ from Appendix-7B, Table for Gamma functions.

(*c*) The failure rate-function is

$$r\,(t) = \frac{2\,t^{2-1}}{(500)^2} = 8(10^{-6})\,t$$

The failure rate function clearly increases with time, *t*.

7.4.3 The Lognormal Distribution

The lognormal distribution assumes that the natural logarithm of the random variable is normally distributed with a mean value of μ and a standard deviation of σ. Repair times of repairable components can be modeled using this distribution. The repair density function is given by

$$f(t) = \frac{1}{t\sigma\sqrt{2\pi}}\exp\left[-\frac{(\ln t-\mu)^2}{2\sigma^2}\right] \qquad \text{...(7.76)}$$

where t_0 is the repair time. The corresponding cumulative failure distribution function $Q(t)$ is given by

$$Q(t) = F(t) = \int_0^t \frac{1}{\xi\sigma\sqrt{2\pi}} \exp\left[-\frac{(\ln \xi - \mu)}{2\sigma^2}\right] d\xi \qquad ...(7.77)$$

where the failure is construed to be a failure to repair the component. Substitution of $z = (\ln \xi - \mu)/\sigma$ into equation (7.77), we obtain

$$Q(t) = \frac{1}{\sqrt{2\pi}} \int_{-\infty}^{\frac{\ln t - \mu}{\sigma}} \exp\left(-\frac{z^2}{2}\right) dz \qquad ...(7.78)$$

which indicates that $(\ln t)$ is normally distributed.

The expected value of t is given by

$$E(t) = \int_0^\infty t f(t) dt = \frac{1}{\sigma\sqrt{2\pi}} \int_0^\infty \exp\left[\frac{(\ln t - u)^2}{2\sigma^2} dt\right] \qquad ...(7.79)$$

With the change in variable $x = \ln t$, we get ...(7.80)

$$E(t) = \int_{-\infty}^{\infty} \frac{e^x}{\sigma\sqrt{2\pi}} \exp\left[-\frac{(x-\mu)^2}{2\sigma^2}\right] dx \qquad ...(7.81)$$

Now, let

$$(x - \mu) = y \qquad ...(7.82)$$

Then

$$E(t) = e^{\mu} \int_{-\infty}^{\infty} \frac{e^y}{\sigma\sqrt{2\pi}} \exp\left[-\frac{y^2}{2\sigma^2}\right] dy \qquad ...(7.83)$$

Since

$$e^y = 1 + y + \frac{y^2}{2!} + \frac{y^3}{3!} + \ldots \qquad ...(7.84)$$

We obtain

$$E(t) = e^{\mu} \left[E(y^0) + e(y^1) + \frac{E(y^2)}{2!} + \frac{E(y^3)}{3!} + \ldots\right] \qquad ...(7.85)$$

or

$$E(t) = e^{\mu} \left[1 + \frac{\sigma^2}{2} + \frac{\sigma^4}{8} + \frac{\sigma^6}{48} + \ldots\right]$$

$$= e^{\mu} \left[e^{\sigma^2/2}\right]$$

or

$$E(t) = \exp\left(\mu + \frac{\sigma^2}{2}\right) \qquad ...(7.86)$$

The variance is given by

$$\text{Var}[t] = \sigma^2 = \exp(2\mu + 2\sigma^2) - \exp(2\mu + \sigma^2) \qquad ...(7.87)$$

Figure 7.6 illustrates the nature of the functions involved.

The density function for the lognormal can be written as

$$f(t) = \frac{1}{\sqrt{2\pi}st}\exp\left[-\frac{1}{2s^2}\left(\ln\frac{t}{t_{med}}\right)^2\right]; \ t \geq 0 \qquad ...(7.88)$$

where the parameter s is a shape parameter and t_{med}, the location parameter, is the median time to failure. The distribution is defined for only positive values of t and is therefore more suitable than the normal as a failure distribution.

The mean, variance, and mode of the lognormal are given by

$$\text{MTTF} = t_{med}\exp(s^2/2) \qquad ...(7.89)$$

$$\sigma^2 = t^2_{med}\exp(s^2)\,[\exp(s^2)-1] \qquad ...(7.90)$$

$$t_{med} = \frac{t_{med}}{\exp(s^2)} \qquad ...(7.91)$$

To obtain the failure probabilities, the lognormal's relationship to the normal is used. Since the logarithm is a normally increasing function, we have

$$F(t) = \Pr\{T \leq t\} = \Pr\{\ln T \leq \ln t\} = \Pr\left\{\frac{\ln T - \ln t_{med}}{s} \leq \frac{\ln t - \ln t_{med}}{s}\right\}$$

$$= \Pr\left\{z \leq \frac{1}{s}\ln\frac{t}{t_{med}}\right\} = \Phi\left(\frac{1}{s}\ln\frac{t}{t_{med}}\right)$$

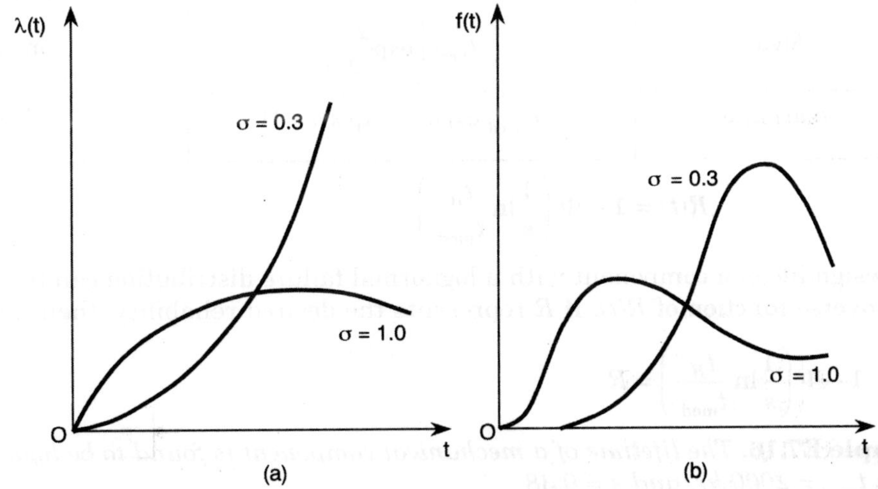

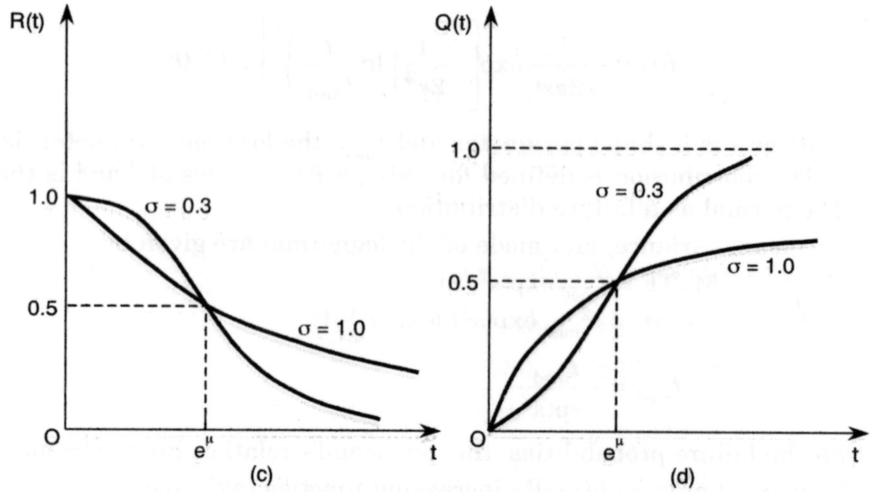

Fig. 7.6 The Lognormal Reliability Model
(*a*) Hazard Function, (*b*) Failure Density Function,
(*c*) Reliability Function, (*d*) Failure Distribution Function.

Relationship between Lognormal and Normal Distributions		
Distribution	**Lognormal**	**Normal**
Mean	$t_{med}\left(\exp\frac{x^2}{2}\right)$	$\ln t_{med}$
Variance	$t^2_{med}\exp(s^2)\,[\exp(s^2)-1]$	s^2

or $$R(t) = 1 - \Phi\left(\frac{1}{s}\ln\frac{t_R}{t_{med}}\right) \quad \text{...(7.92)}$$

The design life of a component with a lognormal failure distribution can be obtained by finding the inverse function of *R(t)*. If *R* represents the desired reliability, then we have

$$1 - \Phi\left(\frac{1}{s}\ln\frac{t_R}{t_{med}}\right) = R$$

Example E7.16. *The lifetime of a mechanical component is found to be lognormal after burn-in with t_{med} = 2000 hr. and s = 0.38*

(*a*) *find its design life for a reliability of 0.99*

(*b*) *if the component is required to have 6 weeks of continuous use, determine the probability of failure*

(*c*) *find the MTTF*

(*d*) *determine the standard deviation of time to failure and the mode*

Solution:

(*a*) $$t_R = t_{med}\, e^{sZ_{1-R}}$$

That is

$$t_{0.99} = 200e^{0.38(-2.33)} \; [\, Z_{1-R} \text{ is from Table A.1}]$$

for $\Phi(Z_{1-R}) = 1 - R = 1 - 0.99 = 0.01 = 825.098$

(*b*) $$R(t) = 1 - \Phi\left[\frac{1}{s}\ln\frac{t}{t_{\text{med}}}\right]$$

$$R(6 \text{ weeks}) = R(1008 \text{ hrs}) = 1 - \Phi\left[\frac{1}{0.38}\ln\frac{1008}{2000}\right] = 1 - \Phi(-1.803)$$

$$= 1 - 0.0358 = 0.9642$$

[From Table A.1, $\Phi(-1.803) = 0.035$]

(*c*) $$\text{MTTF} = t_{\text{med}}\, e^{s^2/2} = 2000e^{0.38^2/2} = 2149.74 \text{ hrs.}$$

(*d*) $$\sigma^2 = t^2_{\text{med}}\, e^{s^2}\,(e^{s^2} - 1) = 2000^2\, e^{0.38^2}\,(e^{0.38^2} - 1)$$

$$= 2000^2\, e^{0.38^2}\,(0.1553)$$

or $$\sigma^2 = (2000)^2(1.1553)(0.1553) = 717{,}672.36 = 847.155 \text{ hrs.}$$

$$t_{\text{mode}} = \frac{t_{\text{med}}}{e^{s^2}} = \frac{2000}{e^{0.38^2}} = \frac{2000}{1.1553} = 1731.15 \text{ hrs.}$$

7.4.4 The Normal Distribution

The density function for the normal distribution is given by

$$f(t) = \frac{1}{\sigma\sqrt{2\pi}} \exp\left[-\frac{(t-\mu)^2}{2\sigma^2}\right]; -\infty < t < +\infty \qquad \text{...(7.93)}$$

The corresponding distribution function, reliability function, and hazard function are given by

$$F(t) = \frac{1}{\sigma\sqrt{2\pi}} \int_{-\infty}^{t} \exp\left[-\frac{(\xi-\mu)^2}{2\sigma^2}\right] d\xi \qquad \text{...(7.94)}$$

$$= Q(t)$$

$$R(t) = 1 - Q(t) = 1 - F(t) = \int_t^{\infty} f(\xi)\,d\xi \qquad \text{...(7.95)}$$

and $$\lambda(t) = \frac{f(t)}{R(t)} \qquad \text{...(7.96)}$$

Typical density and distribution curves for the normal distribution are shown in Fig. 7.7. Larger values of σ result in low, broad curves, and smaller values result in narrow, high curves.

Also, by definition of the density function, we have

$$P(t_1 < t < t_2) = \int_{t_1}^{t_2} f(t)\,dt$$

This is shown in Fig. 7.8.

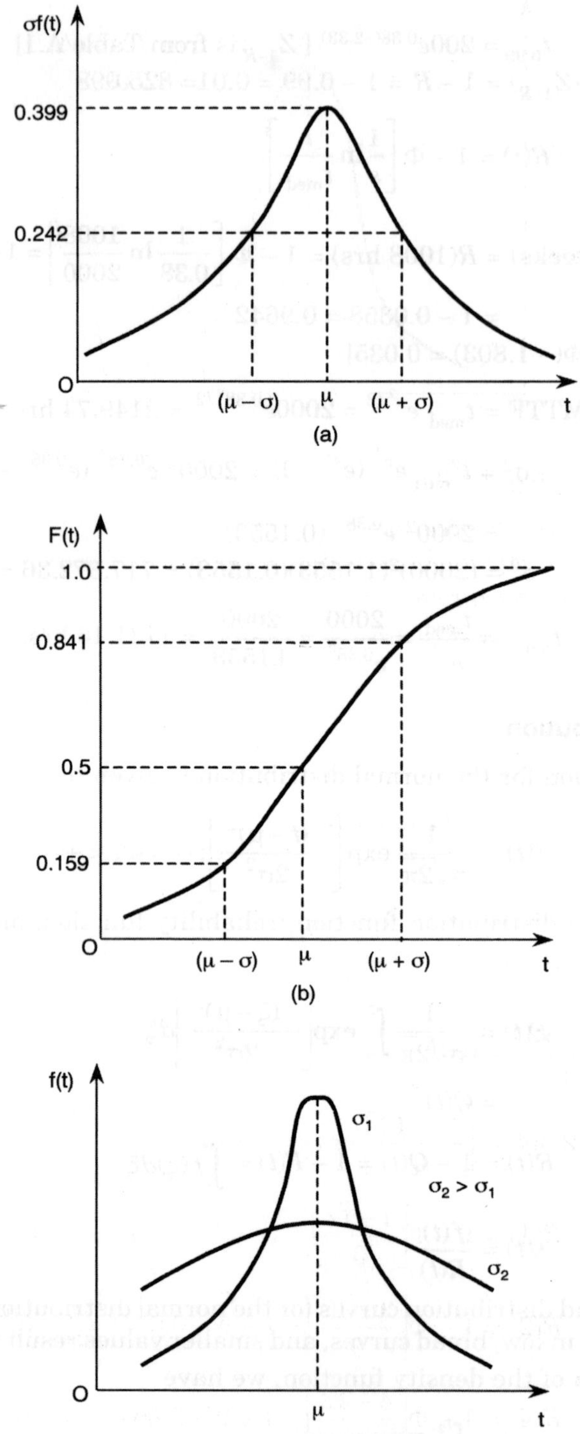

Fig. 7.7 The Normal Distribution
(*a*) Probability Density Function Multiplied by s, (*b*) Distribution Function, (*c*) Density Functions for Different σ Values.

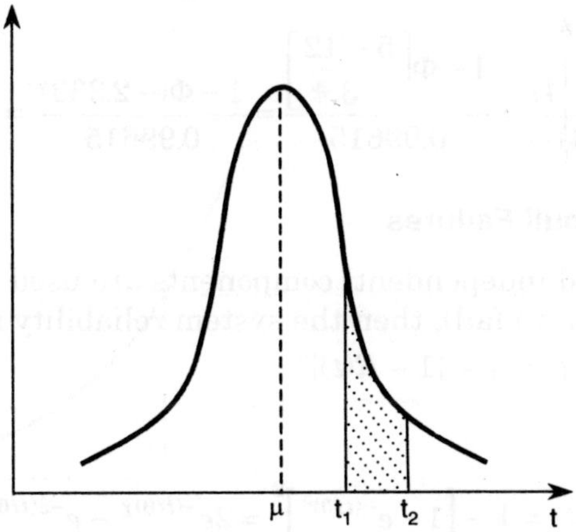

Fig. 7.8 $P(t_1 < t_2)$ for the Normal Distribution.

Example E7.17. *A mechanical tool wears out with a time to failure that is found to be normally distributed with a mean of 12 working days and a standard deviation of 3 days.*

(*a*) *find the design life for a reliability of 0.98*

(*b*) *determine its reliability if the tool is replaced*

(*i*) *every day*

(*ii*) *every two days*

(*iii*) *every four days*

(*iv*) *every seven days*

(*c*) *find the probability that the tool will function one more day when it has already been used for four days*

Solution:

(*a*) $$R(t_d) = 1 - \Phi\left[\frac{t_d - \mu}{\sigma}\right] = 1 - \Phi\left[\frac{t_d - 12}{3}\right] = 0.98$$

Therefore, $$\Phi\left[\frac{t_d - 12}{3}\right] = 0.02$$

or $$-2.055 = \left[\frac{t_d - 12}{3}\right] \text{ or } t_d = 5.835 \text{ days}$$

(*b*) (*i*) $$R(1) = 1 - \Phi\left[\frac{1-12}{3}\right] = 1 - \Phi(-3.667) = 1 - 0.000125 = 0.999875$$

(*ii*) $$R(2) = 1 - \Phi\left[\frac{2-12}{3}\right] = 1 - \Phi(-3.334) = 1 - 0.000425 = 0.999575$$

(*iii*) $$R(4) = 1 - \Phi\left[\frac{4-12}{3}\right] = 1 - \Phi(-2.667) = 1 - 0.00385 = 0.99615$$

(*iv*) $$R(7) = 1 - \Phi\left[\frac{7-12}{3}\right] = 1 - \Phi(-1.667) = 1 - 0.0478 = 0.9522$$

(c) $$R\left(\frac{1}{4}\right)=\frac{R(1+4)}{R(4)}=\frac{1-\Phi\left[\frac{5-12}{3}\right]}{0.99615}=\frac{1-\Phi(-2.333)}{0.99615}=\frac{1-0.0079}{0.99615}=0.99393$$

7.4.5 Redundancy with Weibull Failures

If two identical (and assumed independent) components are used to form a redundant system (both must fail for the system to fail), then the system reliability is given by

$$R_S(t) = 1 - [1 - R(t)]^2$$

If $R(t) = e^{-(t/\theta)^\beta}$

then $$R_S(t) = 1 - \left[1 - e^{-(t/\theta)^\beta}\right]^2 = 2e^{-(t/\theta)\chi} - e^{-2(t/\theta)^\beta} \quad ...(7.97)$$

The MTTF is obtained as

$$\text{MTTF} = \int_0^\infty R_S(t)dt$$

$$= \int_0^\infty \left[2e^{-(t/\theta)\chi} - e^{-2(t/\theta)^\beta}\right]dt = 2\int_0^\infty e^{-(t/\theta)^\psi}dt - \int_0^\infty e^{-2(t/\theta)^\beta}dt$$

The first integral value is given as $\theta\Gamma(1 + 1/\beta)$. The second integral is with parameters β and $\theta' = \theta/2^{1/\beta}$. Therefore, its value is given by

$$\frac{\theta}{2^{1/\beta}}\Gamma\left(1+\frac{1}{\beta}\right)$$

Hence,

$$\text{MTTF} = 2\theta\Gamma\left(1+\frac{1}{\beta}\right) - \frac{\theta}{2^{1/\beta}}\Gamma\left(1+\frac{1}{\beta}\right) \quad ...(7.98)$$

$$= \theta\Gamma\left(1+\frac{1}{\beta}\right)\left(2-2^{-1/\beta}\right)$$

The hazard rate for this system is derived as follows.

To find $\lambda_S(t)$ for this system,

$$\frac{dR_S(t)}{dt} = 2e^{-(t/\theta)^\beta}\left[-\frac{\beta}{\theta}\left(\frac{t}{\theta}\right)^{\beta-1}\right] - e^{-2(t/\theta)^\beta}\left[-\frac{2\beta}{\theta}\left(\frac{t}{\theta}\right)^{\beta-1}\right]$$

$$= -\frac{\beta}{\theta}\left(\frac{t}{\theta}\right)^{\beta-1} e^{-(t/\theta)^\beta}\left[2 - 2e^{-(t/\theta)^\beta}\right]$$

or $$\lambda_S(t) = -\frac{dR(t)}{dt}\frac{1}{R(t)} \quad ...(7.99)$$

$$\lambda_S(t) = \frac{\beta}{\theta}\left(\frac{t}{\theta}\right)^{\beta-1}\frac{2-2e^{-(t/\theta)^\beta}}{2-e^{-(t/\theta)^\beta}} \quad ...(7.100)$$

From Eq. (7.100) it is clear that the system formed from two redundant Weibull components is not itself Weibull. For larger values of t, $\exp[-(t/\theta)^\beta]$ is small and $\lambda_S(t)$ approximates the hazard rate function of a single Weibull component.

7.4.6 Identical Weibull Components

If a system of n serially related and independent components have identical hazard rate functions

$$\lambda(t) = \frac{\beta}{\theta}\left(\frac{t}{\theta}\right)^{\beta-1} \qquad ...(7.101)$$

then from Eq. (7.53), the system hazard rate is given by

$$\lambda(t) = \sum_{i=1}^{n}\frac{\beta}{\theta}\left(\frac{t}{\theta}\right)^{\beta-1} = \frac{n\beta}{\theta^\beta}(t)^{\beta-1}$$

and

$$R(t) = \exp\left[-n\left(\frac{t}{n}\right)^\beta\right] \qquad ...(7.102)$$

which n Weibull with shape parameter β and scale parameter $\theta/n^{1/\beta}$.

7.4.7 The Three-Parameter Weibull Distribution

Here it is assumed that no failures will take place prior to time t_0.
Hence,

$$R(t) = \exp\left[-\left(\frac{t-t_0}{\theta}\right)^\phi\right]; \qquad t \geq t_0 \qquad ...(7.103)$$

and

$$\lambda(t) = \frac{\beta}{\theta}\left(\frac{t-t_0}{\theta}\right)^{\beta-1}; \qquad t \geq t_0 \qquad ...(7.104)$$

The parameter t_0 is referred to as the *location parameter*. The MTTF is given by

$$\text{MTTF} = t_0 + \theta\Gamma\left(1+\frac{1}{\beta}\right) \qquad ...(7.105)$$

and

$$t_{med} = t_0 + \theta(0.69315)^{1/\beta} \qquad ...(7.106)$$

The design life t_d for a reliability of R is given by

$$t_d = t_0 + \theta(-\ln R)^{1/\beta} \qquad ...(7.107)$$

The three-parameter Weibull can be transformed into the two-parameter Weibull with the transformation $t' = t - t_0$.

7.4.8 Failure Modes

For a system with n serially related components and n independent failure modes, each with an independent Weibull failure distribution with shape parameter β and scale parameter θ_i, the system failure rate function is given by

$$\lambda(t) = \sum_{i=1}^{n}\frac{\beta}{\theta_i}\left(\frac{t}{\theta_i}\right)^{\beta-1} = \beta t^{\beta-1}\left[\sum_{i=1}^{n}\left(\frac{1}{\theta_i}\right)^\beta\right] \qquad ...(7.108)$$

Equation (7.108) is a hazard rate for a Weibull distribution having a shape parameter of β and a characteristic life given by

$$= \left[\sum_{i=1}^{n}\left(\frac{1}{\theta_i}\right)^{\beta}\right]^{-1/\beta}$$

Here, it is assumed that each component has the same shape parameter.

Example E7.18. *Determine the average failure rate of a Weibull distribution between two points in time, t_1 and t_2.*

Solution:

$$\text{Average failure rate} = \text{AFR} = \frac{-\ln R(t_1) - (-\ln R(t_2))}{t_2 - t_1}$$

$$= \frac{-\left(\dfrac{t_1}{\theta}\right)^{\beta} + \left(\dfrac{t_2}{\theta}\right)^{\beta}}{t_2 - t_1} = \frac{1}{\theta^{\beta}} \cdot \frac{(t_2{}^{\beta} - t_1{}^{\beta})}{(t_2 - t_1)}$$

Example E7.19. *The time to failure of a mechanical system in years has the probability density function $f(t) = 1/(t+1)^2$, $t \geq 0$*

(*a*) *if four of these systems are placed in parallel for a specific application, find the system reliability for the first one year of operation*

(*b*) *determine the system design life if reliability of 0.99 is required*

(*c*) *determine the system reliability for 9 months, if two out of three systems must function.*

Solution:

$$R(t) = \int_t^{\infty} t(t')dt' = \int_t^{\infty} \frac{1}{(t'+1)^2}dt' = \frac{1}{(t+1)}$$

(*a*)
$$R_S(t) = 1 - \left[1 - \frac{1}{t+1}\right]^4$$

$$R_S(1) = 1 - \left[1 - \frac{1}{2}\right]^4 = 0.9375$$

(*b*)
$$1 - \left[1 - \frac{1}{t+1}\right]^4 = 0.99$$

or
$$\left[1 - \frac{1}{t+1}\right] = (1 - 0.99)^{1/4} = 0.3162$$

or
$$\frac{1}{t+1} = 0.6838$$

$$t = \left[\frac{1}{0.6838} - 1\right] = 0.46247 \text{ year} = 168.80 \text{ days}$$

(c) $$R(0.75) = \frac{1}{t+1} = \frac{1}{0.75+1} = 0.5714$$

$$R_S(0.75) = \sum_{x=2}^{3} \binom{3}{x} (0.5714)^x (0.4286)^{3-x}$$
$$= 3\,(0.5714)^2\,(0.4286) + (0.5714)^3$$
$$= 0.4198 + 0.1866 = 0.6064$$

7.5 RELIABILITY OF SYSTEMS

Components, units, or subsystems within a complex can be related to one another in series, parallel, series-parallel, or meshed structures or in any combination of these. Probability concepts are applied to compute the reliability of the system in terms of the reliabilities of its sub units.

7.5.1 Series or Chain Structure

In series or chain structure, all components must function for the system to function. The block diagram of such a system is shown in Figure 7.9.

If the series system component failures are statistically independent, then the reliability R_S of a series system with nonidentical components is given by

$$R_S = \prod_{i=1}^{n} R_i \qquad ...(7.109)$$

where n is the number of components or subsystems and R_i is the reliability of the ith component or subsystem.

If the failure times of components are exponentially distributed (*i.e.*, if components have constant failure rates), then the ith component reliability is

$$R_i(t) = e^{-\lambda_i t} \qquad ...(7.110)$$

From Eqs. (7.109) and (7.110),

$$R_S(t) = e^{-\sum_{i=1}^{n} \lambda_i t} \qquad ...(7.111)$$

and MTTF is given by

$$\text{MTTF} = \int_0^{\infty} e^{-\sum_{i=1}^{n} \lambda_i t}\, dt = \frac{1}{\sum_{i=1}^{n} \lambda_i} \qquad ...(7.112)$$

Hence, the MTTF of a series system is the reciprocal of the sum of the series network component failure rates.

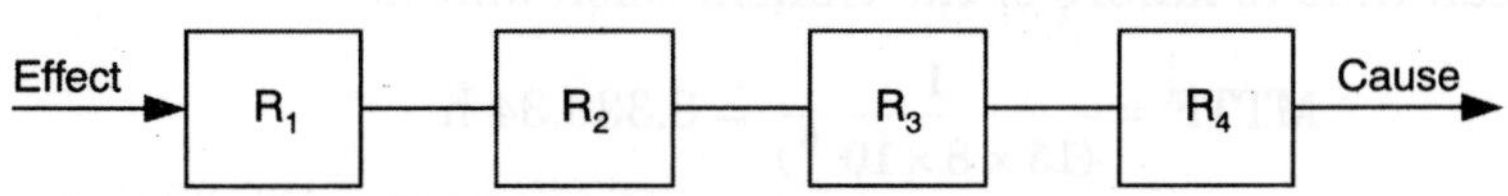

Fig. 7.9 A Series System Block Diagram.

Example E7.20. *Determine the minimum allowable component reliability to attain a system reliability of 0.95 for a series system consisting of 2, 3, 4, 5, 6, 7, 8, 9 and 10 identical components.*

Solution:

For n identical components in series, to attain a system reliability of 0.95, each component must have a reliability of $(0.95)^{1/n}$.

Hence, we obtain

n	2	3	4	5	6	7	8	9	10
Component Reliability	0.9747	0.9831	0.9873	0.9898	0.9915	0.9927	0.9936	0.9943	0.9949

Example E7.21. *A complex system has 600 components in series. The reliability of each individual component is 0.998.*

(*a*) *determine the reliability of the system*

(*b*) *if the number of components in series is reduced to 500, find the reliability of the resulting system.*

Solution:

(*a*) The system reliability is

$$R_S = (0.998)^{600} = 0.3008$$

The system reliability is only 30.08%.

(*b*) If the number of components in the series system is reduced to 500, the reliability of this system is $R_S = (0.998)^{500} = 0.3675$ or 36.75%.

Example E7.22. *The transmission unit of a mechanical system has 15 components in series. Each individual component has an exponential time-to-failure distribution with a constant failure rate of 0.04 per 5000 h.*

(*a*) *find the reliability of each individual component after 3000 h of operation*

(*b*) *find the reliability of the transmission unit for 3000 h of operation*

(*c*) *determine the mean time-to-failure.*

(*d*) *if the desired system reliability for the transmission on unit is 0.98 after 3000 h of operation, determine the mean time to failure of the individual components.*

Solution:

The failure rate of individual component is

$$\lambda = \frac{0.04}{5000\text{ h}} = 8 \times 10^{-6}\text{ h}^{-1}$$

(*a*) The reliability of each component after 3000 h of operation is

$$R = \exp[-(8 \times 10^{-6})\, 3000] = 0.9763$$

(*b*) The reliability of the transmission unit after 3000 h of operation is

$$R_S = \exp[-(15 \times 8 \times 10^{-6})\, 3000] = 0.6376$$

(*c*) The mean time to failure of the transmission unit is

$$\text{MTTF} = \frac{1}{(15 \times 8 \times 10^{-6})} = 8{,}333.34\text{ h}$$

(*d*) Let λ_s = failure rate of the transmission unit. Then,

$$\lambda_s = 15\,\lambda$$

where λ = the failure rate of individual component.

To accomplish the desired reliability of 0.98 after 3000 h of operation, we can find λ_s from

$$0.98 = \exp[-\lambda_s (3000)]$$

or

$$\lambda_s = 6.734 \times 10^{-6}/\text{h}$$

The failure rate of each component is

$$\lambda = \lambda_s/15$$

$$= \frac{(6.734 \times 10^{-6})}{15}$$

$$= 4.48 \times 10^{-7}/\text{h}$$

The mean time to failure for each component will be

$$\text{MTTF} = \frac{1}{\lambda} = \frac{1}{(4.48)(10^{-7})}$$

$$= 2{,}227{,}424.24\text{h}$$

7.5.2 Parallel Structure

This configuration is shown in Figure 7.10. This system will fail if and only if all the units in the system fail. The assumption is that not all system units are active and load sharing. It is assumed that the component failures are statistically independent. A parallel structure reliability R_P with nonidentical units or component reliability is given by

$$R_P = 1 - \prod_{i=1}^{n}(1 - R_i) \quad \text{...(7.113)}$$

where n is the number of units, R_i is the reliability of the ith component or system.

If the component failure rates are constant, then

$$R_P(t) = 1 - \prod_{i=1}^{n}\left[\left(1 - e^{-\lambda_i t}\right)\right] \quad \text{...(7.114)}$$

MTTF is obtained by integrating (7.114) over the interval [0, ∞],

$$\text{MTTF} = \int_0^{\infty} R_P(t)dt = \int_0^{\infty}\left\{1 - \prod_{i=1}^{n}\left[\left(1 - e^{-\lambda_i t}\right)\right]\right\}dt$$

$$= \left(\frac{1}{\lambda_i} + \frac{1}{\lambda_2} + \ldots + \frac{1}{\lambda_n}\right) - \left(\frac{1}{\lambda_1 + \lambda_2} + \frac{1}{\lambda_1 + \lambda_3} + \ldots\right)$$

$$+ \left(\frac{1}{\lambda_1 + \lambda_2 + \lambda_3} + \frac{1}{\lambda_1 + \lambda_2 + \lambda_4} + \ldots\right) + (-1)^{n+1}\frac{1}{\sum_{i=1}^{n}\lambda_i} \quad \text{...(7.115)}$$

For identical components, the above equation reduces to

$$\text{MTTF} = \frac{1}{\lambda}\sum_{i=1}^{n}\frac{1}{i} \quad \text{...(7.116)}$$

In general, parallel structure improves system reliability.

Example E7.23. *Determine the number of identical components required in parallel structure in order to achieve a system reliability of 0.97. The individual component reliability is given as 0.85. Determine if the desired level of system reliability is achieved if the same components are arranged as series structure.*

Solution:

(*a*) Here, we have

$$1 - (1 - 0.68)^n = 0.99$$

or
$$-(1 - 0.68)^n = -0.01$$
$$(0.32)^n = 0.01$$

which gives $n = 4$

(*b*) Series structure

System reliability $= R^n = (0.68)^n = (0.68)^4 = 0.2138$

Hence, the system reliability is not achieved with series structure.

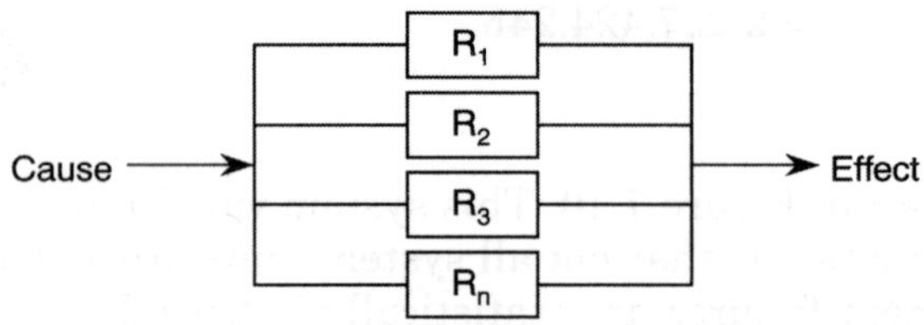

Fig. 7.10 A Parallel Network Block Diagram.

Example E7.24. (*a*) *For the parallel system shown in Figure E7.26, determine the system reliability if the reliabilities of R_1, R_2, and R_3 are 0.98, 0.96, and 0.94 respectively.*

(*b*) *Find the system reliability for 3000 h of operation.*

(*c*) *Find the mean time to failure for the system.*

(*d*) *If all the components in the system have an identical time-to-failure exponential distribution with a constant failure rate of 0.0003 / h, find the mean time to failure for the system is 5000 h, what is the mean time to failure for each component should be?*

Solution:

(*a*) The reliability of the system is

$$R_S = 1 - (1 - 0.98)(1 - 0.96)(1 - 0.94) = 0.999952$$

The system reliability is much higher than that of the individual component reliabilities.

(*b*) The failure rate of each component is $\lambda = 0.0003$/h for 3000 h of operation, the reliability of the system is

$$\begin{aligned} R_S &= 1 - [1 - \exp[-(0.0003)(3000)]]^3 \\ &= 1 - [1 - \exp(-0.9)]^3 \\ &= 1 - [1 - 0.40657]^3 \\ &= 1 - [0.59343]^3 \\ &= 1 - 0.20898 \\ &= 0.791018 \end{aligned}$$

(*c*) The mean time to failure for the system is

$$\text{MTTF} = \frac{1}{0.0003}\left(1 + \frac{1}{2} + \frac{1}{3}\right) = 6111.3334 \text{ h}$$

(*d*) The mean time to failure for each component is

$$\text{MTTF} = \frac{1}{\lambda} = \frac{1}{0.0003} = 3333.3334 \text{ h.}$$

Hence, when three identical components are in parallel, the system MTTF has been increased by

$$\left(\frac{6111.3334 - 3333.34}{3334.34}\right) = 83.33\%$$

For a desired system MTTF of 5000h, we calculate the required MTTF of the individual components.

Hence,

$$5000 = \frac{1}{\lambda}\left(1 + \frac{1}{2} + \frac{1}{3}\right) \qquad \text{...(E.1)}$$

where λ is the failure rate of each component. Solving for λ from Eq. (E.1), we obtain

$$\lambda = \frac{1.83334}{5000}$$

$$= 0.00036667$$

Hence, the MTTF for each component is

$$\text{MTTF} = \frac{1}{\lambda}$$

$$= \frac{1}{0.00036667}$$

$$= 2727.263 \text{ h}$$

7.5.3 Standby Redundancy

Standby redundancy represents a situation with one system operating and n units as standbys. The arrangement is shown in Figure 7.11. Unlike a parallel structure where all units in the configuration are active, the standby units are not active.

The system reliability of the $(n + 1)$ unit, in which one unit is operating and n units on the standby mission until the operating unit fails, is given by

$$R_{st}(t) = \sum_{i=0}^{n} \frac{(\lambda t)' e^{-\lambda t}}{i!} \qquad \text{...(7.117)}$$

The above equation is valid only if the following are valid:

1. The switching arrangement is perfect
2. The units are identical
3. The units failure rates are constant
4. The standby units are as good as new
5. The unit failures are statistically independent.

For $(n + 1)$, nonidentical units whose failure time density functions are different, the standby redundant system failure density becomes

$$f_{st}(t) = \int_{y_n=0}^{t} \int_{y_{n-1}=0}^{y_n} \cdots \int_{y_1=0}^{y_2} f_1(y_1) f_2(y_2 - y_1) \ldots f_{n+1}(t - y_n) \, dy_1 dy_2 \ldots dy_n \quad \ldots(7.118)$$

Hence, the system reliability can be obtained by integrating $f_{st}(t)$ over the interval [t,∞], or

$$R_{st}(t) = \int_t^{\infty} f_{st}(t)dt \quad \ldots(7.119)$$

7.5.4 k-out-of-n Configuration

This is another form of redundancy and is used where a specific number of units must be good for the system success. The series and parallel configuration are special cases of this configuration, that is, $k = n$ and $k = 1$, respectively.

By applying the binomial distribution, the system reliability for *k-out-of-n* number of independent and identical units can be written as

$$R_{k/n} = \sum_{i=k}^{n} \binom{n}{i} R^i (1-R)^{n-1} \quad \ldots(7.120)$$

For the constant unit failure rate λ, Eq. (7.120) becomes

$$R_{k/n}(t) = \sum_{i=k}^{n} \binom{n}{i} (e^{-\lambda t})^i \frac{\left(1-e^{-\lambda t}\right)^n}{\left(1-e^{-\lambda t}\right)^i} \quad \ldots(7.121)$$

7.5.5 Star and Delta Structures

A system of three components can be connected in a star or a delta configuration, as illustrated in Figure 7.12. For these configurations to be equivalent, the reliabilities computed across the terminals *AB*, *BC*, and *CA* must be equal for both the configurations.

For the reliabilities across the terminal *AB*, we have

$$R_A R_B = 1 - (1 - R_{AB})(1 - R_{BC} R_{CA}) \quad \ldots(7.122)$$

Similarly, the reliabilities across the terminals *BC* and *CA* are given by

$$R_B R_C = 1 - (1 - R_{BC})(1 - R_{CA} R_{AB}) \quad \ldots(7.123)$$

and

$$R_C R_A = 1 - (1 - R_{CA})(1 - R_{AB} R_{BC}) \quad \ldots(7.124)$$

The three equations (7.122), (7.123), and (7.124) can be solved simultaneously for the star quantities R_A, R_B, and R_C in terms of the delta quantities R_{AB}, R_{AC}, and R_{CA}. Hence

$$R_A = \sqrt{\frac{(R_A R_B)(R_C R_A)}{R_B R_C}} \quad \ldots(7.125)$$

$$= \sqrt{\frac{[1-(1-R_{AB})(1-R_{BC}R_{CA})][1-(1-R_{CA})(1-R_{AB}R_{BC})]}{1-(1-R_{BC})(1-R_{CA}R_{AB})}}$$

Similarly, R_B and R_C are given by

$$R_B = \sqrt{\frac{(R_A R_B)(R_B R_C)}{R_A R_C}} \quad \ldots(7.126)$$

and $$R_C = \sqrt{\frac{(R_B R_C)(R_C R_A)}{R_A R_B}} \quad ...(7.127)$$

For the special case of

$$R_{AB} = R_{BC} = R_{CA} = R_\Delta \quad ...(7.128)$$

and $$R_A = R_B = R_C = R_y \quad ...(7.129)$$

we get $$R_y = \sqrt{R^2{}_\Delta + R_\Delta - R^3{}_\Delta} \quad ...(7.130)$$

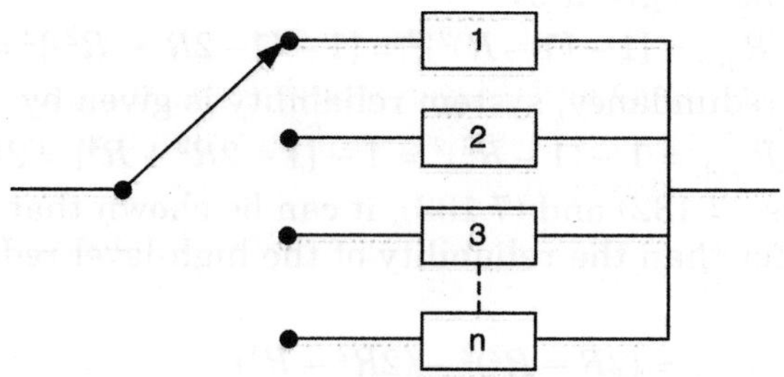

Fig. 7.11 A Standby Redundancy Model.

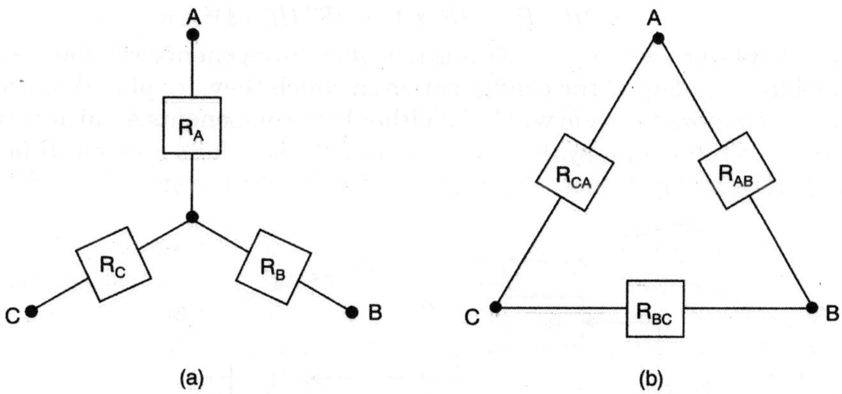

Fig. 7.12 (*a*) Star (Wye) Structure, (*b*) Delta (mesh) Structure.

7.5.6 Combined Series–Parallel Systems

These systems contain components in both serial and parallel relationships. In Figure 7.13, R_i represents the reliability of the ith component. To find the system reliability, the system may be split into serial or parallel subsystems. The reliability of each subsystem is computed first. The system reliability is then obtained on the basis of the relationship among the subsystems. For the network in Fig. 7.13 and the subsystems we have

$$R_A = [1 - (1 - R_1)(1 - R_2)]$$

$$R_B = R_A(R_3), \quad R_C = R_4(R_5)$$

Since R_B and R_C are in parallel with one another and in series with R_6, we have

$$R_S = [1 - (1 - R_B)(1 - R_C)]\,(R_6) \quad ...(7.131)$$

7.5.7 High-Level Versus Low-Level Redundancy

System redundancy may be obtained by two methods. Each component comprising the system may have one or more parallel components, or the entire system may be placed in parallel with one or more identical systems. The first method is known as *low-level redundancy*, and the second is called the *high-level redundancy.*

Consider a system comprised of two serial components, *A* and *B*. Figure 7.14 shows the system having low-level redundancy, and Figure 7.15 shows high-level redundancy. If we assume that both components have the same reliability *R*, then the system reliability for the case of low-level redundancy is given by

$$R_{low} = [1 - (1 - R)^2]^2 = [1 - (1 - 2R + R^2)]^2 = (2R - R^2)^2 \qquad ...(7.132)$$

For the high-level redundancy, system reliability is given by

$$R_{high} = 1 - (1 - R^2)^2 = 1 - [1 - 2R^2 + R^4] = 2R^2 - R^4 \qquad ...(7.133)$$

Comparing the Eqs. (7.132) and (7.133), it can be shown that the reliability of the low-level redundancy is greater than the reliability of the high-level redundancy.

Hence

$$\begin{aligned} R_{low} - R_{high} &= (2R - R^2)^2 - (2R^2 - R^4) \\ &= R^2 (2 - R)^2 - R^2 (2 - R^2) \\ &= R^2 (4 - 4R + R^2 - 2 + R^2) \\ &= 2R^2 (R^2 - 2R + 1) = 2R^2 (R - 1)^2 \geq 0 \end{aligned}$$

which the equality obtained when $R = 1$. This is true if the components' reliabilities are mutually independent and independent of the configuration in which they are placed. Both the low-level and the high-level redundant system will fail if either both components *A* fail or both components *B* fail. The high-level redundant system can also fail if one *A* fails and one *B* fails, as long as they fail on separate paths. Hence, the high-level redundant system has additional failure paths.

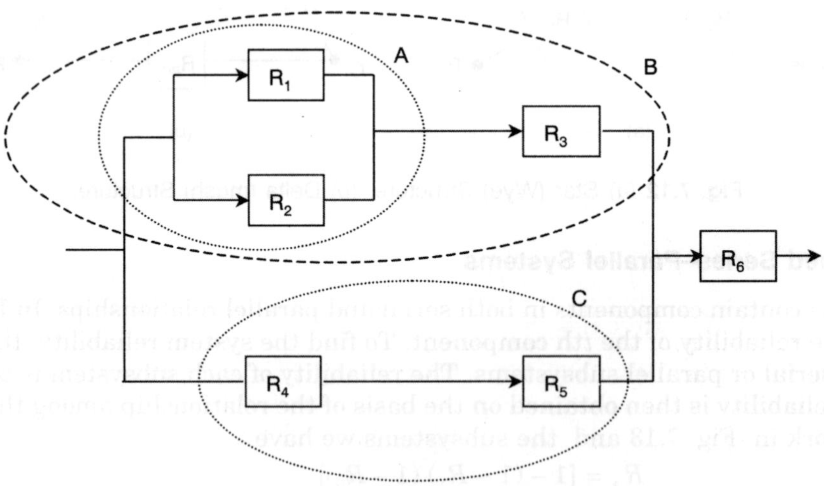

Fig. 7.13 A System Comprised of Components in a Combined Series and Parallel Relationship.

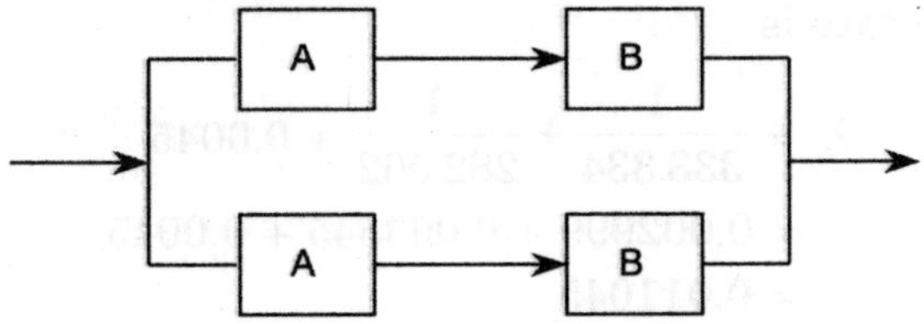

Fig. 7.14 Two Components in a High-Level Redundancy.

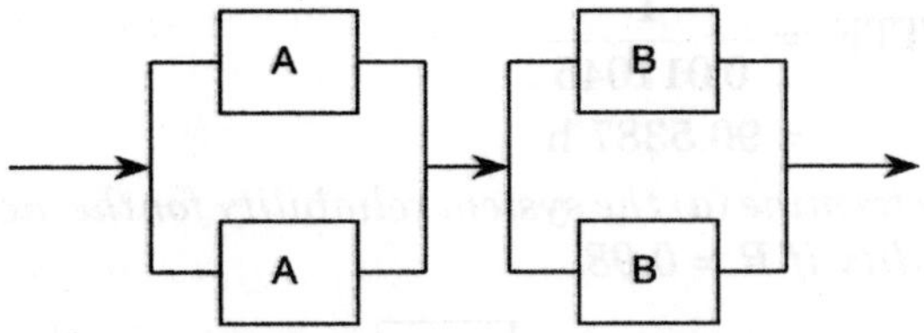

Fig. 7.15 Two Components in a Low-Level Redundancy.

Example E7.25. *(a) Determine the reliability of the system shown in Fig. P7.26. The reliabilities of the individual components are given as follows:*

$R_1 = 0.91, R_2 = 0.89, R_3 = 0.87, R_4 = 0.95, R_5 = 0.94, R_6 = 0.92, R_7 = 0.90$ *and* $R_8 = 0.95.$

(b) Determine the failure rate of the system in Fig.P7.7A time to failure. The individual component failure rates (number of units per hour) are given as follows:

$\lambda_{R1} = 0.001, \lambda_{R2} = 0.0035, \lambda_{R3} = 0.003, \lambda_{R4} = 0.0015,$

$\lambda_{R5} = 0.0065, \lambda_{R6} = 0.0065, \lambda_{R7} = 0.0065,$ *and* $\lambda_{R8} = 0.0045,$

Solution:

(a) The reliabilities of each subsystem are obtained as follows.

For the subsystem with components R_1, R_2, R_3, R_4, the reliability is

$$R_A = 1 - (1 - R_1R_2)(1 - R_3R_4)$$
$$= 1 - [1 - (0.91)(0.89)]\,[1 - (0.87)(0.95)]$$

Similarly, the reliability of the subsystem with components R_5, R_6, R_7, is

$$R_B = 1 - (1 - R_5)\,(1 - R_6)\,(1 - R_7)$$
$$= 1 - (1 - 0.94)(1 - 0.92)(1 - 0.90)$$
$$= 1 - (0.06)(0.08)(0.10)$$
$$= 0.99952$$

The system reliability is given by

$$R_S = R_A \times R_B \times R_8 = (0.9670)\,(0.99952)\,(0.95) = 0.9182$$

(b) The failure rates for the subsystems can be computed as follows:

The failure rate for the R_1/R_2 subsystem is $\lambda_{R_1} + \lambda_{R_2} = 0.001 + 0.0035 = 0.0045$

Similarly, the failure rate for the R_3/R_4 subsystem is $\lambda_{R3} + \lambda_{R4} = 0.003 + 0.0015 = 0.0045.$

The mean time to failure for the $R_1/R_2/R_3/R_4$ subsystem is

$$\text{MTTF}_1 = \frac{1}{0.0045}\left[1 + \frac{1}{2}\right] = 333.334 \text{ h}$$

The mean time to failure for the subsystem consisting of components R_5, R_6 and R_7 is

$$\text{MTTF}_2 = \frac{1}{0.0065}\left(1 + \frac{1}{2} + \frac{1}{3}\right)$$
$$= 282.062 \text{ h}$$

The system failure rate is

$$\lambda_s = \frac{1}{333.334} + \frac{1}{282.062} + 0.0045$$
$$= 0.002999 + 0.003545 + 0.0045$$
$$= 0.011045$$

The mean time to failure for the system is

$$\text{MTTF}_s = \frac{1}{0.011045}$$
$$= 90.5387 \text{ h}$$

Example E7.26. *Determine (a) the system reliability for the network shown in Fig. E7.26. (b) Find the system reliability if R = 0.98.*

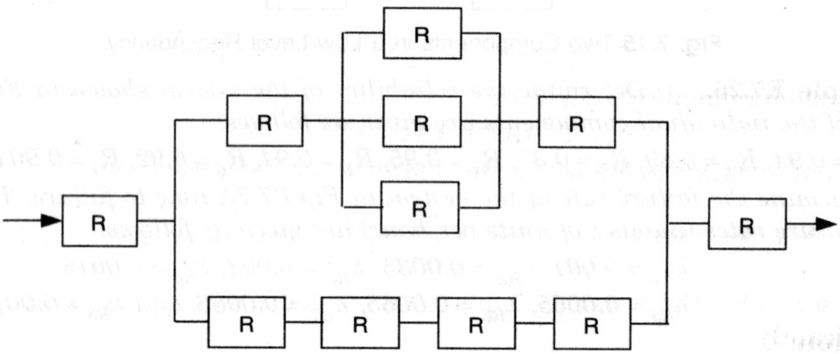

Fig. E7.26

Solution:

(*a*)

$$R_{\text{top}} = R[1 - (1 - R)^3]R = R^2[1 - (1 - R)^3]$$
$$R_{\text{bot}} = R^4$$
$$R_{\text{system}} = R[1 - (1 - R_{\text{top}})(1 - R_{\text{bot}})]\, R = R^2[1 - (1 - R_{\text{top}})(1 - R_{\text{bot}})]$$

(*b*) For $R = 0.98$

$$R_{\text{top}} = R^2[1 - (1 - R)^3] = 0.98^2[1 - (1 - 0.98)^3] = 0.9604$$
$$R_{\text{bot}} = R^4 = 0.98^4 = 0.9224$$
$$R_{\text{system}} = R^2\,[1 - (1 - R_{\text{top}})(1 - R_{\text{bot}})]$$
$$= 0.98^2\,[1 - (1 - 0.9604)(1 - 0.9224)] = 0.9574$$

Example E7.27. (*a*) *Determine the reliability of the standby system. Each component has an exponential time-to-failure distribution. The failure rate of each component is 0.003/h and the period of operation is 400 h.*

(*b*) *find the mean time to failure.*

(*c*) *if the three components are arranged in parallel, find the reliability of the system.*

(*d*) *find the mean time to failure for the parallel system in part (c).*

Solution:

(*a*) The failure rate λ for each component is 0.003/h. For 400 h of operation, the system reliability is

$$R_s = e^{-\lambda t}\left(1 + \lambda t + \frac{(\lambda t^2)}{2!}\right)$$

$$= e^{-0.003(400)}\left[1 + 0.003\,(400) + \frac{[(0.003)(400)]^2}{2}\right]$$

$$= e^{-12}[1 + 1.2 + 0.72] = 0.879$$

(*b*) The mean time to failure is

$$\text{MTTF} = \frac{n+1}{\lambda}$$

$$= \frac{3}{0.003}$$

$$= 1000 \text{ h}$$

(*c*) If the system has all three components in parallel, the probability of failure of each component is

$$F_1 = 1 - e^{-\lambda t} = 1 - e^{-0.003(400)} = 0.6988$$

The system reliability is given by

$$R_S = 1 - (0.6988)^3 = 0.659$$

(*d*) The mean time to failure for this parallel system is given by

$$\text{MTTF} = \frac{1}{\lambda}\left(1 + \frac{1}{2} + \frac{1}{3}\right) = \frac{1}{0.003}\left(1 + \frac{1}{2} + \frac{1}{3}\right) = 611.1107 \text{ h}.$$

We observe that the system reliability and the MTTF of the standby and parallel system differ significantly.

7.5.8 k-out-of-n Redundancy

A generalization of n parallel components occurs when a requirement exists for k out of n identical and independent components of the function, for the system to function. Here $k = n$. If $k = 1$, complete redundancy occurs, and if $k = n$, the n components are in series. The reliability is obtained from the application of the binomial probability distribution.

If each component is an independent trial with R as a probability of success, then we have

$$P(x) = \binom{n}{x} R^x\,(1 - R)^{n-x} \qquad \text{...(7.134)}$$

Equation (7.134) is the probability of exactly x components operating. Since

$$\binom{n}{x} = \frac{n!}{x!(n-x)!}$$

is the number of ways (arrangements) in which x successes (nonfailures) can be obtained from n components. $R^x\,(1 - R)^{n-x}$ is the probability of x successes and $n - x$ failures for a single arrangement of successes and failures.

Hence,

$$R_S = \sum_{x=k}^{n} P(x) \qquad \text{...(7.135)}$$

is the probability of k or more successes from among the n components.

Exponential failures

If the failure distribution is exponential, then

$$R_S(t) = \sum_{x=k}^{n} \binom{n}{x} e^{-\lambda x t} [1 - e^{-\lambda t}]^{n-x}$$

The MTTF is given by

$$\text{MTTF} = \int_0^\infty R_S(t)dt = \frac{1}{\lambda}\sum_{x=k}^{n}\frac{1}{x} \qquad \text{...(7.136)}$$

7.5.9 Series Structure

Consider a simple structure consisting of two switches in series. For the system to fail short, both of the switches must fail short. For the system to fail open, at least one switch must fail open. Hence we can write the events as

E_1 = the event that both switches fail short

E_2 = the event that at least one switch fails open

Let $Q = 1 - R$ equals the probability of a system failure. Then

$$Q = P(E_1 \cup E_2) = P(E_1) + P(E_2)$$

The events are mutually exclusive and let

q_{oi} = the probability that component i fails open

q_{si} = the probability that component i fails short

Then
$$Q = (q_{s1}q_{s2}) + (q_{o1} + q_{o2} - q_{o1}q_{o2})$$

and
$$R = 1 - [(q_{s1}q_{s2}) + (q_{o1} + q_{o2} - q_{o1}q_{o2})]$$
$$= (1 - q_{o1})(1 - q_{o2}) - q_{s1}q_{s2}$$

Therefore, the system reliability is the probability that neither component fails open minus the probability that both components fail short. Generalizing for n components in series, we have

$$R = \prod_{i=1}^{n}(1 - q_{oi}) - \prod_{i=1}^{n} q_{si} \qquad \text{...(7.137)}$$

System reliability is the probability that there are no opens minus the probability that all components short.

7.5.10 Parallel Structure

For this system to fail short, one or both of the switches must fail short. For the system to fail open, both switches must fail open. Hence, we can define two failure events as

E_1 = the event that both switches fail open

E_2 = the event that at least one switch fails short

Then
$$Q = P(E_1 \cup E_2) = P(E_1) + P(E_2)$$
$$= (q_{o1}q_{o2}) + (q_{s1} + q_{s2} - q_{s1}q_{s2})$$

and
$$R = 1 - [(q_{o1}q_{o2}) + (q_{s1} + q_{s2} - q_{s1}q_{s2})]$$
$$= (1 - q_{s1})(1 - q_{s2}) - q_{o1}q_{o2}$$

The system reliability is the probability that neither component fails short minus the probability that both components fail open. For n redundant components, we have

$$R = \prod_{i=1}^{n}(1 - q_{si}) - \prod_{i=1}^{n} q_{oi} \qquad ...(7.138)$$

Reliability is the probability that there are no shorts minus the probability that all components are open.

7.5.11 Low-Level Redundancy

Figure 7.16 shows a low-level redundant system having m serial components each having n redundant units. The system reliability is a generalization of the series configuration (Eq. 7.137) and is obtained by calculating the probability of at least one unopen path in each redundant string minus the probability of at least one short in each path. The reliability is given by

$$R_L = \prod_{i=1}^{m}(1 - q_{oi}^{n}) - \prod_{i=1}^{m}[1 - (1 - q_{si})^{n}] \qquad ...(7.139)$$

7.5.12 High-Level Redundancy

Figure 7.17 shows a high-level redundant system having n redundant units each having m serial components. The system reliability is a generalization of the parallel configuration (Eq. 7.138) and is obtained by calculating the probability that no path is completely shorted minus the probability of at least one open on each path. The reliability is given by

$$R_H = \left(1 - \prod_{i=1}^{m} q_{si}\right)^{n} - \left(1 - \prod_{i=1}^{m}(1 - q_{oi})\right)^{n} \qquad ...(7.140)$$

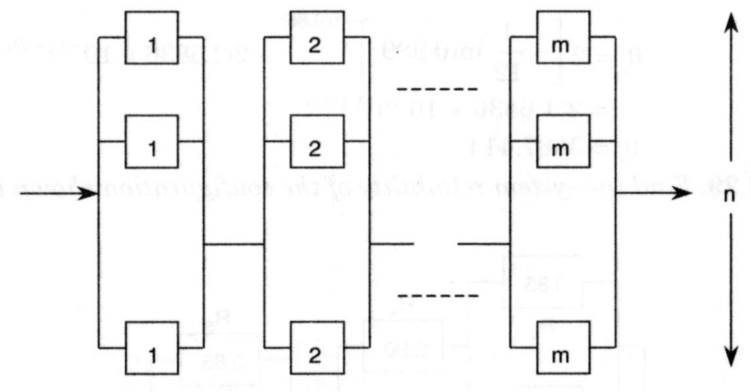

Fig. 7.16 Block Diagram of *m* Three-state Devices in a Low-level Redundant Configuration.

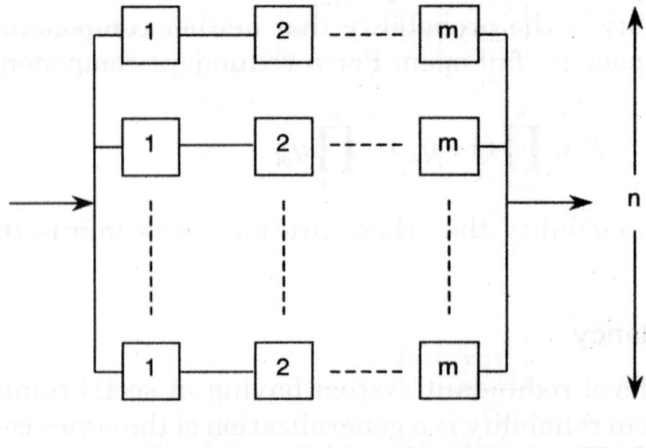

Fig. 7.17 Block Diagram of *m* Three-state Devices in a High-level Redundant Configuration.

Example E7.28. *If twelve Weibull components, each having a shape parameter of 0.85 are connected in series, find a common characteristic life in order that they may have a design life of 2 years with a reliability of 0.989.*

Solution:

$$R_S(t) = e^{-\sum_{i=1}^{12}\left(\frac{t}{\theta}\right)^{0.85}} = e^{-12\left(\frac{t}{\theta}\right)^{0.85}}$$

Therefore

$$R_S(2) = e^{-12\left(\frac{t}{\theta_i}\right)^{0.85}} = 0.989$$

$$-12\left(\frac{2}{\theta_i}\right)^{0.85} = \ln(0.989)$$

Hence

$$\theta_i = 2\left[-\frac{1}{12}\ln(0.989)\right]^{-1/0.85} = 2(1.6836 \times 10^{-3})^{-1/0.85}$$

$$= 2(1.6836 \times 10^{-3})^{-1.1765}$$

$$\theta_i = 3667.414$$

Example E7.29. *Find the system reliability of the configuration shown in Fig. E7.29 (a) and (b).*

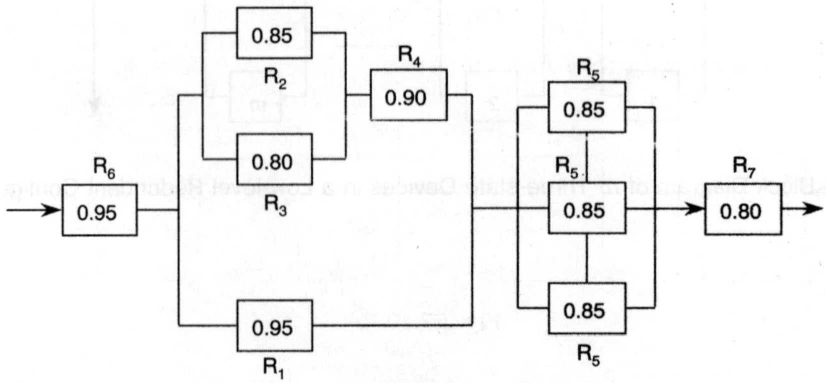

Fig. E7.29 (a)

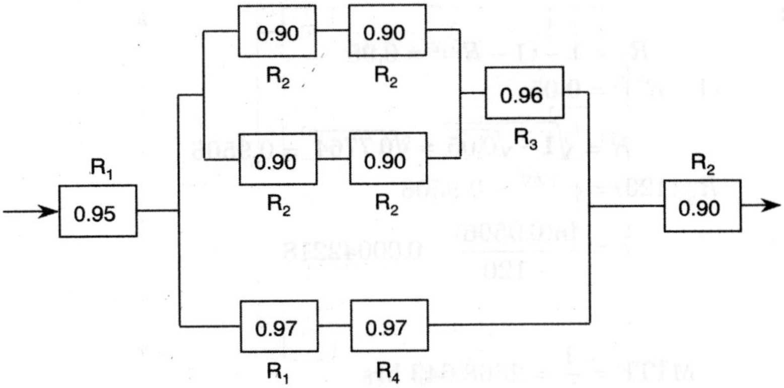

Fig. E7.29 (b)

Solution:

(a) $R_S = R_6 \{1 - (1 - R_1) \{1 - ((1 - (1 - R_2)(1 - R_3)) R_4\}\} \{1 - (1 - R_5)^3\} R_7$

$= 0.95 \{1 - (1 - 0.95) \{1 - (1 - (1 - 0.85) (1 - .80)(0.90\}\} \{1 - (1 - 0.85)^3\} 0.80$

$= 0.95 \{1 - 0.05\{1 - (1 - (0.15)(0.20))(0.90)\}\}\{1 - (0.15)^3\} 0.880$

$= 0.95 (0.99365)(0.9966)(0.80) = 0.7526$

(b) $R_S = R_1 \{1 - \{ 1 - ((1 - (1 - (0.9)(0.9))^2)(0.96)\} \{1 - (R_4)(R_4)\}\} R_2$

$= R_1 \{1 - \{1 - (1 - (1 - (R_2)(R_2))^2)R_3)\} \{1 - {R_4}^2\}\} R_2$

$= 0.95 \{1 - \{1 - (1 - (1 - (0.9)(0.9))^2)0.96\} \{1 - 0.80^2\}\} 0.90$

$= 0.95(0.9653)(0.96)(0.90) = 0.7923$

Example E7.30. *Find the component MTTF necessary to provide a system reliability of 0.95 after 120 hours of operation. Assume that the components have the same constant failure rate.*

(a) high-level redundancy (See Fig. 7.30(a)).

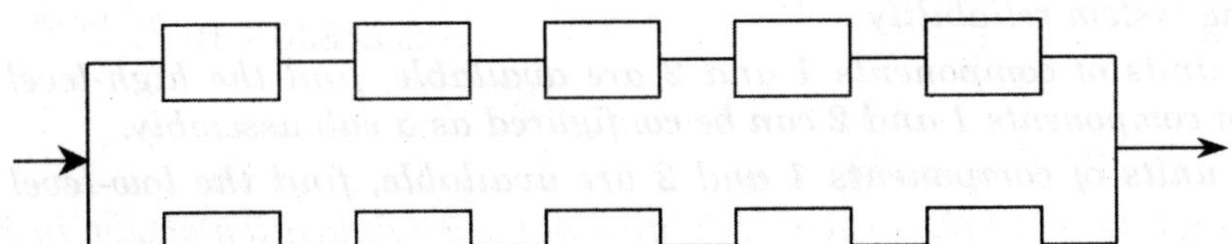

Fig. E7.30 (a)

(b) low-level redundancy (See Fig. 7.30(b)).

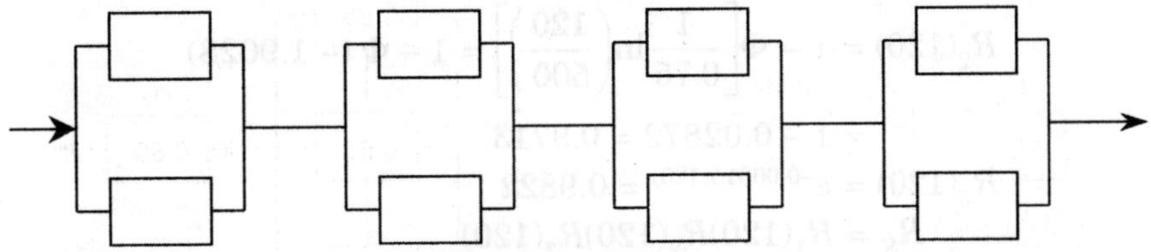

Fig. E7.30 (b)

Solution:

(*a*)
$$R_S = 1 - (1 - R^5)^2 = 0.95$$
$$(1 - R^5)^2 = 0.05$$
$$R = \sqrt[5]{1 - \sqrt{0.05}} = \sqrt[5]{0.7764} = 0.9506$$
$$R_S\,(120) = e^{-120\lambda} = 0.9506$$
$$\lambda = \frac{\ln(0.9506)}{-120} = 0.00042218$$

Hence,
$$\text{MTTF} = \frac{1}{\lambda} = 2368.643 \text{ hrs}$$

(*b*)
$$R_S = [1 - (1 - R)^2]^4 = 0.95$$
$$1 - (1 - R)^2 = \sqrt[4]{0.95}$$

or
$$R = 1 - \sqrt{1 - \sqrt[4]{0.95}} = 0.8871$$
$$R_S(120) = e^{-120\lambda} = 0.8871$$
$$\lambda = \frac{\ln(0.8871)}{-120} = 0.0009983$$

Therefore,
$$\text{MTTF} = \frac{1}{\lambda} = 1001.69 \text{ hrs.}$$

Example E7.31. *A mechanical system shown in Figure E7.31 is designed to operate for 120 days consists of three components in series. Their failure distributions are (1) Weibull with shape parameter 1.4 and scale parameter 1000 days; (2) lognormal with shape parameter (s) 0.75 and median 500 days; (3) constant failure rate of 0.00015.*

(*a*) *find the system reliability*

(*b*) *if two units of components 1 and 2 are available, find the high-level redundancy reliability, when components 1 and 2 can be configured as a sub-assembly.*

(*c*) *if two units of components 1 and 2 are available, find the low-level redundancy reliability.*

Solution:

$$R_1(120) = e^{-\left(\frac{120}{1000}\right)^{1.4}} = 0.9499$$
$$R_2(120) = 1 - \Phi\left[\frac{1}{0.75}\ln\left(\frac{120}{500}\right)\right] = 1 - \Phi\,(-\,1.9028)$$
$$= 1 - 0.02872 = 0.9713$$
$$R_3(120) = e^{-0.00015(120)} = 0.9822$$

(*a*)
$$R_S = R_1(120)R_2(120)R_3(120)$$
$$= (0.9499)(0.9713)(0.9822) = 0.9062$$

(*b*)
$$R_S = [1 - \{1 - (0.9499)(0.9713\}^2]\ 0.9822 = 0.9763$$

(*c*)
$$R_S = [1 - (1 - (0.9499)^2][1 - (1 - 0.9713)^2][0.9822]$$
$$= (0.9975)(0.9814) = 0.9789$$

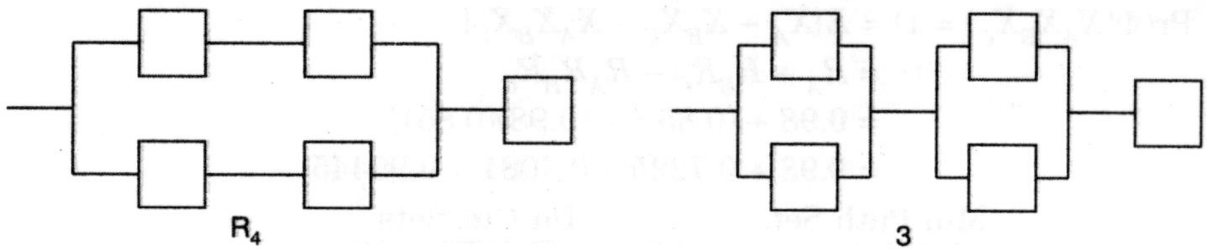

Fig. E7.31

7.5.13 Minimal Cut Set Method

Some physical systems have complex structures that are not purely series, parallel, or series-parallel. The minimal cut set method is a technique that forms the basis for the network evaluation methods for computing system reliability.

A *minimal cut set* is a set of system components such that if all components fail, system failure results, but if any one component has not failed, no system failure results. In other words, all the components of a minimal cut set must fail for the system to fail. If the cut sets of the system are denoted as $C_1, C_2, ..., C_n$, and $P(C_i)$ denotes the probability of all the components of C_i failing, then the system unreliability is

$$Q_S = P(C_1 + C_2 + ... + C_n) \quad ...(7.141)$$

where "+" denotes the logical sum (or union) of the operands. The system reliability is

$$R_S = 1 - Q_S \quad ...(7.142)$$

The expansion of the probability of the union of n nondisjoint events contains $(2^n - 1)$ terms. However, an upper bound for Q_S can be found by using the approximation

$$P(C_1 + C_2 + ... + C_n) \cong P(C_1) + P(C_2) + ... + P(C_n) \quad ...(7.143)$$

Hence, a lower bound for the system reliability is

$$R_S \geq 1 - [P(C_1) + P(C_2) + ... + P(C_n)] \quad ...(7.144)$$

This lower bound is a good approximation in the high-reliability region of components, the region with component reliabilities lie close to unity.

Example E7.32. *For the network shown in Figure E7.32.*

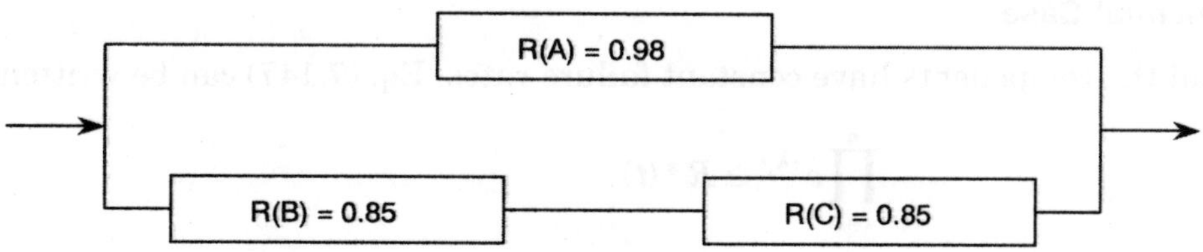

Fig. E7.32

(a) develop the structure function

(b) determine the system reliability

(c) determine the minimal path sets and cut sets

(d) determine the lower and upper bound reliabilities.

Solution:

(a)
$$\Phi(X_A X_B X_C) = 1 - (1 - X_A)(1 - X_B X_C)$$
$$= 1 - [1 - X_A - X_B X_C + X_A X_B X_C]$$
$$= X_A + X_B X_C - X_A X_B X_C$$

(b) $\Pr(\Phi(X_A X_B X_C) = 1) = E[X_A + X_B X_C - X_A X_B X_C]$

$$= R_A + R_B R_C - R_A R_B R_C$$
$$= 0.98 + (0.85)^2 - (0.98)(0.85)^2$$
$$= 0.98 + 0.7225 - 0.7081 = 0.99445$$

(c)

Min Path Sets	Min Cut Sets
A	A, B
B, C	A, C

(d) $R_L = [1 - (1 - 0.98)(1 - 0.85)]^2 = 0.9940$

$$R_U = 1 - (1 - 0.98)(1 - 0.85^2) = 0.99445$$

7.6 RELIABILITY ALLOCATION

When the system reliability goals have been defined and established, reliability must then be allocated to the components and subcomponents. Reliability block diagrams and the relationships developed earlier are useful tools in accomplishing this. In general, the following inequality must hold:

$$h(R_1(t), R_2(t), \ldots, R_n(t)) \geq R^*(t) \qquad \ldots(7.145)$$

where $R_i(t)$ is the reliability at time t of the ith component, $R^*(t)$ is the system reliability goal at time t, and h is a function that relates component reliabilities to system reliability. If MTTF* is a system reliability goal and component failures are exponential, then we can write

$$g(\text{MTTF}_1, \text{MTTF}_2, \ldots, \text{MTTF}_n) \geq \text{MTTF}^* \qquad \ldots(7.146)$$

where MTTF_i is the MTTF of component i and g is a function that provides the system MTTF. The form of the functions h and g depend on the component serial-parallel configuration. For instance, if all the components are serially related and their failures are independent of one another, then we have

$$\prod_{i=1}^{n} R_i(t) \geq R^*(t) \qquad \ldots(7.147)$$

7.6.1 Exponential Case

Assuming all the components have constant failure rates, Eq. (7.147) can be written as

$$\prod_{i=1}^{n} e^{-\lambda_i t} \geq R^*(t) \qquad \ldots(7.148)$$

or

$$\sum_{i=1}^{n} \lambda_i \leq \lambda_S \qquad \ldots(7.149)$$

where λ_S is the system failure rate goal.

7.6.2 Optimal Allocations

Ideally, reliability allocations should be accomplished in a least-cost manner. For example, assume Eq. (7.147) applies that R^* has been specified for some time t (we will drop the argument t as being understood). If each component has a current reliability R_i where $\pi R_i < R^*$, we would be interested in solving the following problem:

$$\min z = \sum_{i=1}^{n} C_i(x_i) \qquad ...(7.150)$$

subject to

$$\prod_{i=1}^{n} (R_i + x_i) \geq R^* \qquad ...(7.151)$$

$$0 < R_i + x_i \leq B_i < 1 \quad \text{for} \quad i = 1,2,..., n \qquad ...(7.152)$$

where x_i is the increase in reliability of the ith component, $C_i(x_i)$ is the corresponding cost of achieving this growth, and B_i is an upper bound on the attainable component reliability. Since specifying the cost functions may be quite difficult, we assume that the cost of component i is determined by $C_i x_i^2$ over the range of interest. A quadratic cost function is the simplest nonlinear function, which shows that the reliability growth cost increases at an increasing rate. Equation (7.150) can be written as

$$\min z = \sum_{i=1}^{n} C_i x_i^2 \qquad ...(7.153)$$

7.6.3 ARINC Method

There are many reliability allocation strategies available that do not require optimization. One such simplest one is the Aeronautical Radio Incorporated (ARINC) method. This method assumes that the components are in series, are independent, and have constant failure rates. If λ_i is the current estimated failure rate for the ith component and λ^* is the target system failure rate, then

$$\text{new } \lambda_i = w_i \lambda^*$$

where

$$w_i = \frac{\lambda_i}{\sum_{i=1}^{n} \lambda_i} \quad \text{for } i = 1, 2,..., n \qquad ...(7.154)$$

That is, allocated failure rates are proportional to the current failure rates. This method can also be used if failure rates are not constant by using an average failure rate.

7.6.4 AGREE Method

The AGREE (Advisory Group on Reliability of Electronic Equipment) method assumes that a system is comprised of n components each having n_i modules, or subcomponents. This method allows component operating times to be less than the system operating and it allows the inclusion of an importance index. Let

t = system operating time

$R^*(t)$ = system reliability goal at time t

n = number of components

n_i = a complexity number, *e.g.*, the number of module within component i

N = $*n_i$ = total number of modules in a system

t_i = operating time of the ith component (t_i t)

λ_i = failure rate of the ith component

w_i = probability that the system will fail given component i has failed (importance index).

The method is to allocate an equal share of the reliability to each module in the system. The ith component's contribution to system reliability is given by $[R^*(t)]^{n_i/N}$. The relationship for the ith component is given by

$$w_i\left(1-e^{-\lambda_i t_i}\right) = 1-\left[R^*(t)\right]^{n_i/N} \qquad \text{...(7.155)}$$

The left side of Eq. (7.155) is the joint probability that the ith component fails and results in a system failure. The right side of the equation is the failure probability allocated to the ith component. Solving Eq. (7.155) for λ_i results in

$$\lambda_i = -\frac{1}{t_i}\ln\left(1-\frac{1-R^*(t)^{n/N}}{w_i}\right),\quad i = 1, 2, ..., x_n \qquad \text{...(7.156)}$$

We note here that $\prod_{i=1}^{n} e^{-\lambda_i t_i} \leq R^*(t)$ since not all component failures result in system failures (if $w_i < 1$).

7.6.5 Redundancies

A general method to follow is in allocating reliability among the components is to consider only serially related components. Redundancy can then be applied to achieve the allocated component reliability as necessary. It is often required to account for the redundancy as part of the reliability allocation process. Consider the reliability block diagram shown in Figure 7.18. If R* is the system reliability goal, we can write $R^* = R_1 \times R' \times R_4$ and use one of the above allocation methods based on serially related components. Components 2 and 3 can be treated initially as a single component. If R' is the allocated reliability, then

$$R' = 1-(1-R_2)(1-R_3) = R_2 + R_3 - R_2R_3 \qquad \text{...(7.157)}$$

We can assign a reliability to one of the components, say component 3, such that $R_3 < R'$.

then
$$R_2 = \frac{R'-R_3}{1-R_3} \qquad \text{...(7.158)}$$

For two components with the same reliability R, we have $R' = 2R - R^2$, which gives the solution $R = 1-(1-R')^{0.5}$ from the quadratic formula. The system can be reduced initially to serially related components and we can further decompose it as required.

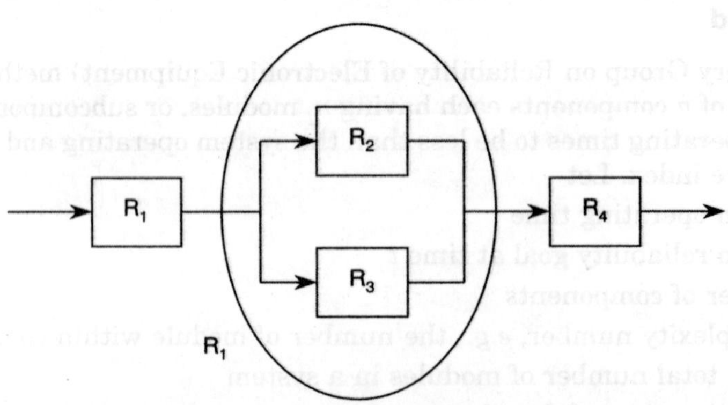

Fig. 7.18 Reliability Block Diagram for Reliability Allocation.

Example E7.33. *In the design of an electromechanical system, four major subsystems have been identified, each having a Weibull failure distribution with the parameter values as follows:*

Major subsystem	Scale parameter, θ, (yr)	Shape parameter, β
1	6.3	0.89
2	7.7	0.79
3	10.7	1.79
4	12.3	0.89

If the desired reliability of the system is 0.987 at the end of the first 2 years, find the percentage increase in reliability for each of the major subsystems needed in order to reach the system goal assuming all subsystems have equal reliability goals.

Solution:

$$R(t) = e^{-\left(\frac{t}{\theta}\right)^{\beta}}$$

$$R_1(1) = e^{-\left(\frac{2}{6.3}\right)^{0.89}} = 0.6976$$

$$R_2(1) = e^{-\left(\frac{2}{7.7}\right)^{0.79}} = 0.7084$$

$$R_3(1) = e^{-\left(\frac{2}{10.7}\right)^{1.79}} = 0.9515$$

$$R_4(1) = e^{-\left(\frac{2}{12.3}\right)^{0.89}} = 0.8199$$

$$R_{sys} = R_1R_2R_3R_4 = (0.6976)(0.7084)(0.9515)(0.8199) = 0.3855$$

$$\text{Reliability goal} = y = \sqrt[4]{0.987} = 0.996734$$

$$\%\ \text{subsystem 1} = 100\left[\frac{0.996734 - 0.6976}{0.6976}\right] = 0.4288 \times 100 = 42.88\%$$

$$\%\ \text{subsystem 2} = 100\left[\frac{0.996734 - 0.7084}{0.7084}\right] = 0.4082 \times 100 = 40.82\%$$

$$\%\ \text{subsystem 3} = 100\left[\frac{0.996734 - 0.9515}{0.9515}\right] = 0.0484 \times 100 = 4.84\%$$

$$\%\ \text{subsystem 4} = 100\left[\frac{0.996734 - 0.8199}{0.8199}\right] = 0.2167 \times 100 = 21.67\%$$

Example E7.34. *Assuming the redundancy is the only way of achieving further reliability growth for each of the subsystems in Example E7.33, find the minimum number of redundant units of each necessary in order to achieve the component reliability goals?*

Solution:

$$R = 1 - (1 - R_i)^n$$

or
$$n = \frac{\ln(1-R)}{\ln(1-R_i)}$$

Therefore,

$$n_1 = \frac{\ln(1-0.996734)}{\ln(1-0.6736)} = \frac{-5.72416}{-1.11963} = 5.1126 \approx 6 \text{ units}$$

$$n_2 = \frac{\ln(1-0.996734)}{\ln(1-0.7084)} = \frac{-5.72419}{-1.23237} = 4.6449 \approx 5 \text{ units}$$

$$n_3 = \frac{\ln(1-0.996734)}{\ln(1-0.9515)} = \frac{-5.72419}{-3.02619} = 1.8916 \approx 2 \text{ units}$$

$$n_4 = \frac{\ln(1-0.996734)}{\ln(1-0.8199)} = \frac{-5.72419}{-1.71424} = 3.3392 \approx 4 \text{ units}$$

Example E7.35. *On the basis of the current average failure rate of each subsystem over the first year, apply the ARINC method to determine an average failure rate goal for each subsystem of Example E7.33. If the reliability growth improves only the characteristic life but not the shape parameter, determine the characteristic life goal of each subsystem.*

Solution:

$$\lambda(t) = \frac{\beta}{\theta}\left(\frac{t}{\theta}\right)^{\beta-1} \quad \text{and AFR} = \frac{1}{(t_2 - t_1)}\int_{t_1}^{t_2} \lambda(t')\,dt'$$

$$\lambda = \text{AFR} = \frac{1}{1-0}\int_0^2 \frac{\beta}{\theta}\left(\frac{1}{\theta^{\beta-1}}\right) t'^{(\beta-1)}dt' = \left[\frac{\frac{\beta}{\theta}\left(\frac{1}{\theta^{\beta-1}}\right)t'^{\beta}}{\beta}\right]_0^2$$

$$= \frac{1}{\theta}\left(\frac{1}{\theta^{\beta-1}}\right)2^{\beta} = \frac{2^{\beta}}{\theta^{\beta}} = \left(\frac{2}{\theta}\right)^{\beta}$$

$$\lambda_1 = \text{AFR}_1 = \left(\frac{2}{6.3}\right)^{0.89} = 0.3602$$

$$\lambda_2 = \text{AFR}_2 = \left(\frac{2}{7.7}\right)^{0.79} = 0.3447$$

$$\lambda_3 = \text{AFR}_3 = \left(\frac{2}{10.7}\right)^{1.79} = 0.0497$$

$$\lambda_4 = AFR_4 = \left(\frac{2}{12.3}\right)^{0.89} = 0.1986$$

$$\lambda_{i\ \text{new}} = w_i\lambda^* = w_i\ [-\ln(0.987)] = w_i(0.013082)$$

$$w_i = \frac{\lambda_i}{\sum_{i=1}^{n}\lambda_i} = \frac{\lambda_i}{(0.3602) + (0.3447) + (0.0497) + (0.1986)} = \frac{\lambda_i}{0.9532}$$

$$\lambda_{1\ \text{new}} = \left(\frac{0.3602}{0.9532}\right)(0.013082) = 4.94349 \times 10^{-3}$$

$$\lambda_{2\ \text{new}} = \left(\frac{0.3447}{0.9532}\right)(0.013082) = 4.73066 \times 10^{-3}$$

$$\lambda_{3\ \text{new}} = \left(\frac{0.0497}{0.9532}\right)(0.013082) = 0.68208 \times 10^{-3}$$

$$\lambda_{4\ \text{new}} = \left(\frac{0.1986}{0.9532}\right)(0.013082) = 2.72559 \times 10^{-3}$$

$$\lambda_i = \left(\frac{2}{\theta_i}\right)^{\beta_i}$$

$$\left(\frac{\theta_i}{2}\right) = \left(\frac{1}{\lambda_i}\right)^{1/\beta_i}$$

$$\theta_i = 2\left(\frac{1}{\lambda_i}\right)^{1/\beta_i}$$

$$\theta_1 = 2\left(\frac{1}{4.94349 \times 10^{-3}}\right)^{1/0.89} = 779.836 \text{ years}$$

$$\theta_2 = 2\left(\frac{1}{4.73066 \times 10^{-3}}\right)^{1/0.79} = 1,754.555 \text{ years}$$

$$\theta_3 = 2\left(\frac{1}{0.68208 \times 10^{-3}}\right)^{1/1.79} = 117.446 \text{ years}$$

$$\theta_4 = 2\left(\frac{1}{2.72559 \times 10^{-3}}\right)^{1/0.89} = 1,552.4219 \text{ years}$$

Example E7.36. *Apply the AGREE allocation method to a mechanical system having the following components:*

Component	*No. of Parts*	*Importance Index*	*Operating Time, hr/yr*
1	*207*	*0.98*	*3013*
2	*63*	*0.95*	*1517*
3	*44*	*1.00*	*2864*

A reliability of 0.995 is required over the first year of its customer use.

Solution:

Given $R^* = 0.995$

Total number of parts in the system, $N = 207 + 63 + 44 = 314$

R_i = Reliability allocated to the ith component = $R^{n_i/N} = R^{*n_i/314}$

$$\lambda_i = -\frac{1}{t_i}\ln\left[1 - \frac{1 - R^{*n_i/314}}{w_i}\right]$$

$$\text{MTTF}_i = \frac{1}{\lambda_i}$$

R_{Imp} = Reliability accounting for importance = $e^{-\lambda_i t_i}$

The results are summarized below:

Component	R_i	λ_i	MTTF_i	R_{Imp}
1	0.9967	1.1195×10^{-6}	893,256	0.99663
2	0.9989	0.7637×10^{-6}	1,309,414	0.99884
3	0.9993	0.2449×10^{-6}	4,083,299	0.99929

Example E7.37. *Establish a criticality index for the components in Example E7.36 assuming that the conditional probability β_k is equal to the importance index. Assume that failure modes equate to components.*

Solution:

$$c_k = \alpha_{kp}\beta_k\lambda_t t$$

$$c_1 = \alpha_1\beta_1\lambda_1 t_1 = \alpha_1 w_1\lambda_1 t_1 = (1)(0.98)(1.1195 \times 10^{-6})(3013)$$

$$= 3.3056 \times 10^{-3}$$

$$c_2 = \alpha_2\beta_2\lambda_2 t_2 = \alpha_2 w_2\lambda_2 t_2 = (1)(0.95)(0.7637 \times 10^{-6})(1517)$$

$$= 1.1006 \times 10^{-3}$$

$$c_3 = \alpha_3\beta_3\lambda_3 t_3 = \alpha_3 w_3\lambda_3 t_3 = (1)(1.0)(0.2449 \times 10^{-6})(2864)$$

$$= 7.0139 \times 10^{-3}$$

7.7 FAULT TREE ANALYSIS

A *fault tree* is a diagrammatic representation of the relationship between a specific event and the causes that may lead to that event. Fault tree diagrams are used to calculate the system unreliability. System failure is shown at the top of the tree and the causes of failure progressively developed beneath, in the simplest form of diagram using fault tree symbols. The most commonly used fault tree symbols and their definitions are summarized in Table 7.1.

The purpose of a fault tree diagram are as follows:

(*a*) To find out which combination of events or situations may cause the undesirable event to occur

(*b*) To determine the probability of the event occuring

(*c*) To initiate and lead the corrective action to reduce the probability to an acceptable minimum level

The development of fault trees requires the following four steps for a systematic approach:

1. System Definition. Define the top event and the analysis boundaries beyond which the casual roots of the tree will not be developed. All the assumptions made on the system definition and fault tree are to be listed.

2. Fault Tree Construction. The objective of fault tree construction is to represent the system conditions which may cause the system to fail symbolically.

Fault tree analysis is aided by classifying failures into three groups:

(1) A *primary failure* is a failure that occurs while a component is functioning within the operating parameters for which it is designed.

(2) A *secondary failure* is a failure that is due to excessive environmental or operational stress placed on a component.

(3) *Command failure* is a failure that results from the proper operation of a component, but at the wrong time or place.

It is required to thoroughly understand the system and its intended use. The system description must be a part of the analysis documentation. Construct the fault tree by showing the top event at the top and below it is diverging multi-branch sequence of successively lower levels of events that may lead to the 'top-event'. The fault tree diagram is drawn using logic gate symbols of Table 7.1. The output from each gate is labeled.

1. **Qualitative Fault Tree Evaluation**

Evaluate the fault tree qualitatively using minimal cut sets (see Section 7.5.13)

2. **Quantitative Fault Tree Evaluation**

Collect statistical data regarding the total and modal failure rates of components and the frequency of all other events that constitute inputs to the tree. Evaluate the component failure parameters, critical path and the top event. Monte Carlo simulation technique is used to create combinations of all the tree inputs according to the appropriate failure rates and probabilities.

Example E7.38. (*a*) *Use Boolean algebra to conduct a qualitative analysis for the fault tree shown in Figure E7.38.*

(*b*) *Find the probability of the top event if the events are all independent and the probability of top event is 0.0045*

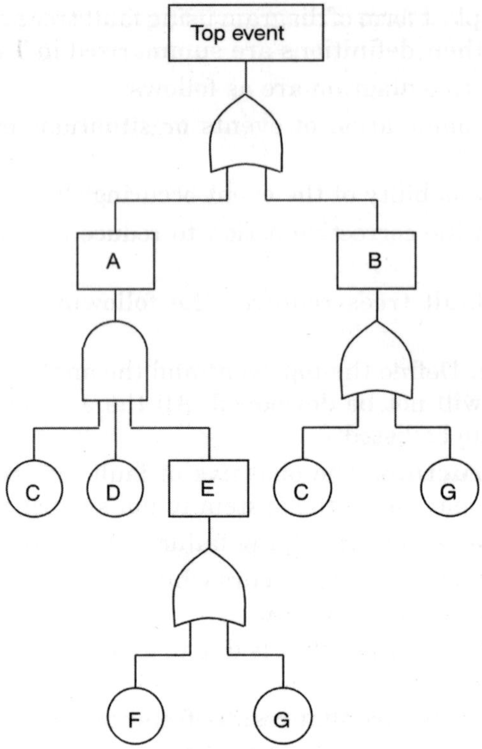

Fig. E7.38

Solution:

(*a*)
$$
\begin{aligned}
T &= A \cup B \\
&= [C \cap D \cap E] \cup [C \cup G] \\
&= [C \cap D \cap (F \cup G)] \cup [C \cup G] \\
&= [C \cap (D \cap F) \cup [D \cap G]] \cup C \cup G \\
&= [(C \cap D \cap F) \cup C] \cup [(D \cap G) \cup G] \\
&= C \cup G
\end{aligned}
$$

(*b*)
$$
\begin{aligned}
P(T) &= P(C) + P(G) - P(C)P(G) \\
&= 0.0045 + 0.0045 - (0.0045)(0.0045) = 0.008979
\end{aligned}
$$

Table 7.1. Fault Tree Symbols and Definitions

Symbol	Gate	Definition
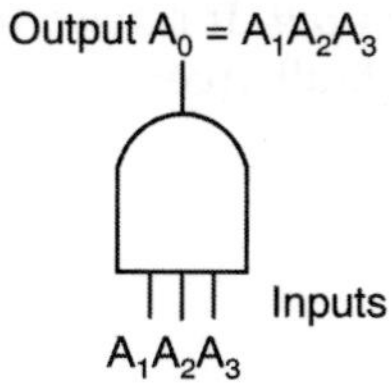	AND	An output event occurs if and only if all the input events occur. In Boolean algebra, the output is the intersection (∩) of the input events.
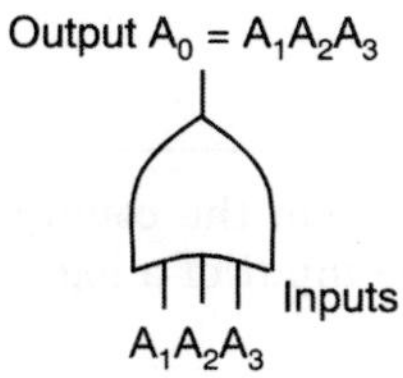	OR	Denotes that an output event occurs if and only if any one or more of the input events occur. In Boolean algebra the output is the union (∪) of the input events.
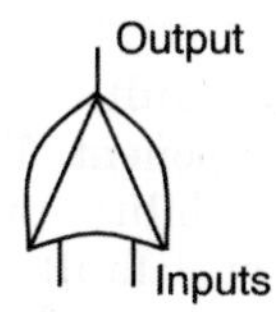	EXCLUSIVE OR	The output is an intermediate event. There is no output when one and only one of the input events occurs.
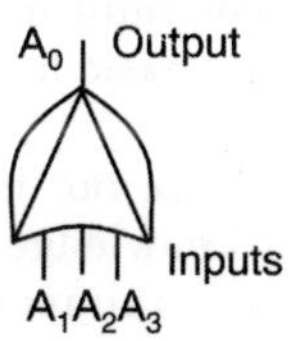	PRIORITY AND	Equivalent to an AND gate, except that the input events must occur in a specific order.Events A_1 and A_2 must occur before event A_3 in order for event A_0 to occur.
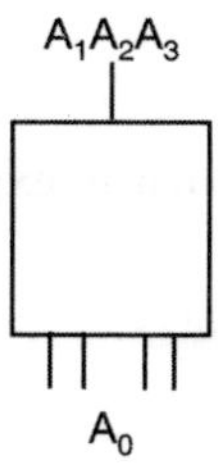	SPECIAL GATE OR RESULTANT-EVENT	Represents any other legitimate combination of the input events and usually an output to a logic gate.
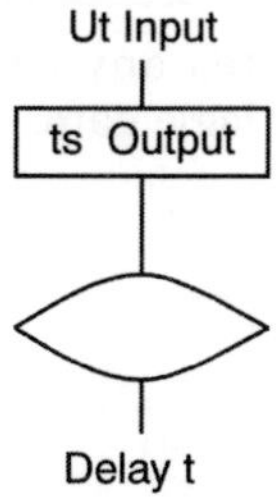	DELAY GATE	A gate whose output only occurs after a specified delay time has occurred.

Contd...

Table 7.1. Fault Tree Symbols and Definitions (continued)

Symbol	Gate	Definition
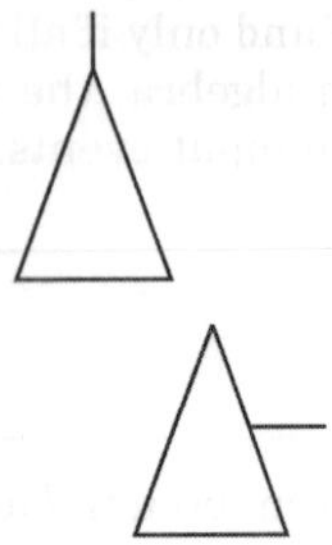	TRIANGLE IN AND TRANSFER OUT	Represents a transfer *in* or *out*. It is used to avoid repeating sections of the fault tree.
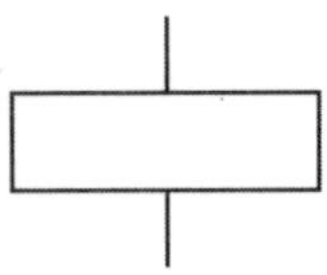	RESULTANT EVENT	An event which results from the combination of fault events through the input of a logic gate.
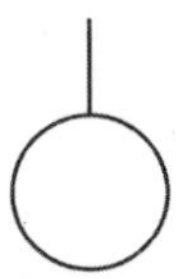	BASIC FAULT EVENT	A circle represents a basic fault event or the failure of an elementary component. The failure parameters such as unavailability, probability, failure, and repair rates of a fault event are obtained from the empirical data or other sources. The analysis ends with a basic fault event, that is, there are no events below a basic fault event.
	INCOMPLETE EVENT	A fault event whose causes have not been fully developed because of lack of knowledge concerning its cause or because of its unimportance in the analysis.
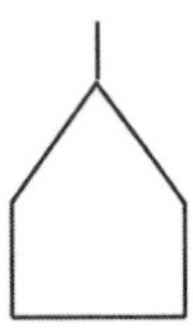	TRIGGEREVENT	Denotes that a fault event which is expected to occur.
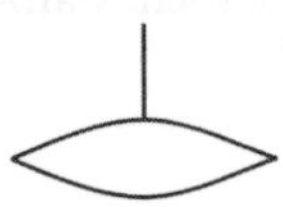	CONDITIONAL EVENT	An elliptical symbol denotes any condition or restriction that is tied to a logic gate.

Contd...

Table 7.1. Fault Tree Symbols and Definitions (continued)

Symbol	Gate	Definition
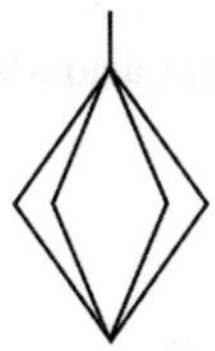	DOUBLE DIAMOND EVENT	Represents an undeveloped fault event that requires further development to occur.
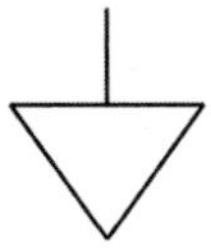	UPSIDE DOWN TRIANGLE	Denotes a similarity transfer, that is, the input is similar but not identical to the identified input.
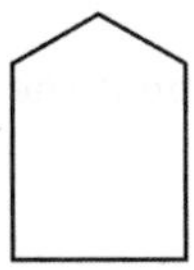	NORMAL EVENT	A normally occuring event that is not a fault.

7.8 SUMMARY

Although it is possible in some cases to develop failure models based on the physics of failure, the procedure is rather complex and needs a considerable amount of study and analysis. Four reliability functions: the failure density function $f(t)$, hazard function $\lambda(t)$, failure distribution function $Q(t)$, and reliability function $R(t)$ were defined. They were developed as limiting cases of the respective piecewise-linear functions conducted using catastrophic failure data. The mean time to failure (MTTF), median time to failure, and average failure rate over time interval were introduced and their determination using reliability functions was presented. Both constant failure rate (CFR) models and time-dependent failure models were presented and discussed. Complex systems are decomposed into functional entities composed of components, units, subsystems for the purpose of reliability analysis. Combinational aspects of reliability analysis are used to connect the components in series, parallel, series-parallel, meshed structures, or in any combination of these. Probability concepts are applied to determine the reliability of systems in terms of the reliabilities of its subunits. Reliability allocation methods including the ARINC and AGREE methods were presented and discussed. Fault trees which are used to analyze complex systems were presented.

PROBLEMS

7.1 The failure density function for a class of components is given by

$$f(t) = 0.20 - 0.02t$$

where t is in years. Determine the reliability functions $Q(t)$, $R(t)$, and $\lambda(t)$ and sketch them.

7.2 Refer to Problem 7.1. Determine

(*a*) the mean time to failure for the components

(*b*) the median time to failure for the components

7.3 The time to failure in hours of a system has the density function given as

$$f(t) = 0.002 \exp[-0.002t]$$

Find the probability that the system will fail in

(*a*) less than 100 hours

(*b*) less than 1,000 hours

7.4 The failure rate of a certain electronic component is $\lambda = 3 \times 10^{-9}$/hour

(*a*) determine $R(t)$ for t in hours

(*b*) find the 2-year reliability, MTTF in hours, and allowable operating time if the reliability is to be not less than 0.998.

7.5 The reliability of a class of components is approximated by

$$R(t) = (1 - bt)^n \quad \text{for} \quad 0 \le t \le 1/b \quad \text{and} \quad a < 0,\ n > 0$$

Determine

(*a*) the corresponding failure density function,

(*b*) distribution function,

(*c*) hazard function,

(*d*) MTTF, and

(*e*) the median time to failure.

7.6 The time to failure of a mechanical component follows the exponential distribution. The probability that the component would not survive for more than 60 days is 0.97. How often can one expect to replace this mechanical component?

7.7 The lifetime of an electromechanical component is estimated to be 50 hours. Find the probability that an electromechanical component of this type would survive for 2 weeks when subjected to a similar working environmental conditions.

7.8 The reliability function for a mechanical component is given by

$$R(t) = 1 - \frac{t^3}{b^3} \text{ for } 0 \le t \le b$$

where b is a constant parameter. Find $\lambda(t)$, MTTF, AFR and the average failure rate upto the MTTF.

7.9 The hazard rate function for a system is given by

$$\lambda(t) = \frac{0.3}{700}\left(\frac{t}{700}\right)^{-0.5}, \ t \text{ in years}$$

(*a*) calculate the design life for a reliability of 0.87

(*b*) if the burn-in period T_0 is 0.5 for a six-month burn-in period, calculate the design life for a reliability of 0.87.

7.10 The lifetime of a mechanical component is exponentially distributed with a mean value of 600 hours. Find the probability of this component operating successfully for at least 800 hours.

7.11 The time-to-failure density function (PDF) for a system is given by

$$f(t) = \frac{2}{a^2}\ ; \quad 0 \le t \le a$$

Determine $F(t)$, $R(t)$, $\lambda(t)$, MTTF, T_{med}, T_{mode} and σ.

7.12 A vacuum cleaner is guaranteed to have more than 15 years of life. If the time-to-density function (PDF) for this vacuum cleaner is given by

$$f(t) = 0.2(1 + 0.03t)^{-3} \quad \text{for} \quad t \ge 0$$

(*a*) find the reliability for the next 15 years if it has survived a 1-year warranty period.

(*b*) find the MTTF for this vacuum cleaner before the warranty period and after the warranty period.

7.13 Find the reliability of an electronic system for 150 hr operation, which has two independent failure modes, namely, a constant failure rate of 0.0002 and a linearly increasing failure rate given by

$$\lambda(t) = 3 \times 10^{-5}\, t.$$

7.14 Find the system MTTF and the median time to failure if the design life is 15 years with a reliability of 0.98 is desired. Assume that the system has 25 identical components and will be replaced on failure (renewal process). Determine the component MTTF and the median time to failure if each component has a CFR.

7.15 A TV remote control contains two batteries each having an MTTF of 4 operating hours (CFR).

(*a*) determine the probability of battery failure happening within three hours of operation.

(*b*) find the probability of more than one failure occurring during the first 4 hours of operation, if the failed batteries are replaced instantly.

(*c*) determine whether the batteries have a constant failure rate (CFR) or increasing failure rate (IFR).

7.16 Derive a general expression for the hazard rate function for an exponentially reliability model given by

$$R(t) = b^{-at}$$

where a and b are constant parameters.

7.17 Given the reliability function for a system given by

$$R(t) = e^{-0.002(t-100)} \quad \text{for} \quad t \ge 100$$

Find the MTTF, t_{med}, $t_{0.95}$, and the standard deviation σ.

7.18 An electromechanical system has three components in series. Each component has a Weibull failure pattern with β equal to 2. The characteristic life time for the three components are 10,000 hr, 15,000 hr, and 20,000 hr respectively. Determine the MTTF and the design life of the electromechanical system corresponding to a reliability of 0.95.

7.19 A mechanical system has two failure modes due to external conditions and wear-out respectively. The first one has a constant failure rate of 0.06 failures per year. The second has a Weibull distribution with a characteristic life of 15 years and a shape factor of 2. Find the system reliability for a design life of 2 years. Also calculate the median time to failure.

7.20 A Physician Specialist can serve four patients on any one day. The number of patients arriving has a Poisson distribution with an expected value of three per day. Find

(*a*) the probability of turning away patients on any given day.

(*b*) the expected number of patients served per day

(*c*) patients turned away per day

7.21 Extensive testing of prototypes of two designs of a mechanical component showed that the failure is a Weibull distribution in both cases. Based on the following information, which design should be selected if the component should have

(*i*) a 12-hour guaranteed life

(*ii*) a 24-hour guaranteed life

Design I: Costs 6,000 to build and b = 3 and α = 180 hours

Design II: Costs 5,000 to build and b = 4 and α = 120 hours

7.22 The current gain of a transistor is proportional to $\ln\beta$, where β is the ratio of output to input currents. The current gains of a certain transistor are normally distributed with a mean of 3 and a variance of 0.1. Find the probability that β is between 8 and 10.

7.23 The breakdown of a certain machine component is normally distributed with 90 per cent of the failures occurring between 200 and 270 hours of use.

(*a*) find the MTTF and the standard deviation of failure times.

(*b*) determine the reliability of the component to be used for 220 hours and then replaced.

(*c*) find the design life if no more that a 2 per cent probability of failure prior to replacement is to be tolerated.

(*d*) determine the conditional reliability for a 20-hour use if the part has been in operation for 200 hours.

7.24 Past experience has indicated that an average of 8% of the critical mechanical components supplied by a particular vendor are defective. In a recent purchase of a lot of 300 components, determine the probability of finding at least 35 defective components?

7.25 A mechanical component has a normally distributed life with a mean value of 5,000 hours. The design need of such a system using these components requires that 95% of them should last for at least 4,500 hours. Find the largest variance allowable for the lifetime of the components.

7.26 The lifetime of a certain component manufactured by a company are found to be normally distributed with a mean of 1,000 days and a standard deviation of 100 days. Taking one month as 30 days, for how many months can the manufacturer guarantee these components so that no more than 6% of them need replacement?

7.27 A mechanical system component has been found to have a Weibull distribution with a shape parameter of 2.2 and a characteristic life of 15,000 hours.

Determine:

(*a*) R(7000 hr)

(*b*) the B1 and B.1 life

(*c*) the MTTF and the standard deviation

(*d*) the median and the mode

(*e*) the probability of failure in the first year of continued use.

7.28 The reliability deterioration ratio for n identical and independent components in a series structure each with a reliability of R is given by

$$\text{Deterioration ratio} = \frac{\text{Reliability of single component}}{\text{Reliability of the series structure}}$$

Compute the deterioration ratio as a function of R (say R = 0.1 to 0.99) and n = 2 to 10.

7.29 The reliability of individual components in a parallel structure is 0.85

(*a*) determine the reliability of the system if two such components are used

(*b*) determine the reliability of the system if three such components are used

(*c*) find the number of n components in parallel needed to achieve an overall system reliability of 0.9999.

7.30 The time-dependent reliability of some components is given by a function of $e^{-\lambda t}$. Determine the reliability of a system consisting of two such components in parallel structure.

7.31 Determine the k-out-of-n system reliability for $n = 4$, $r = 2$, and $p = e^{-\lambda t}$.

7.32 Determine the system reliability of the series-parallel configurations shown in Fig. P7.32 (*a*) and (*b*).

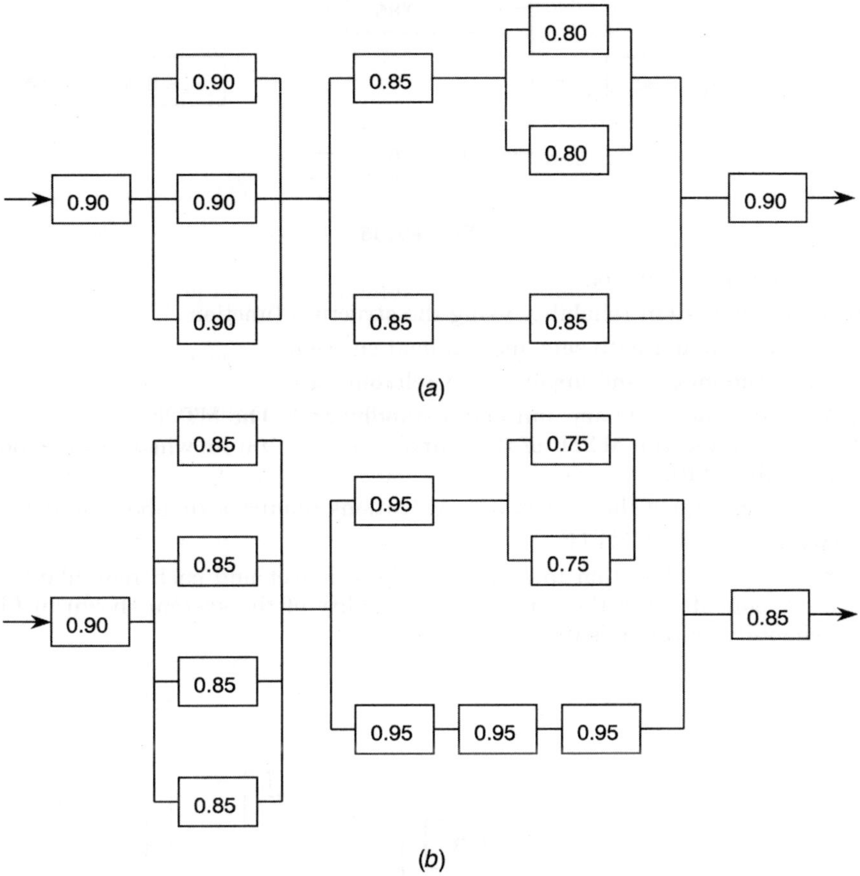

Fig. P7.32

7.33 Determine the system reliability if $R = 0.90$ and $R_1 = 0.98$ for the system shown in Fig. P7.33.

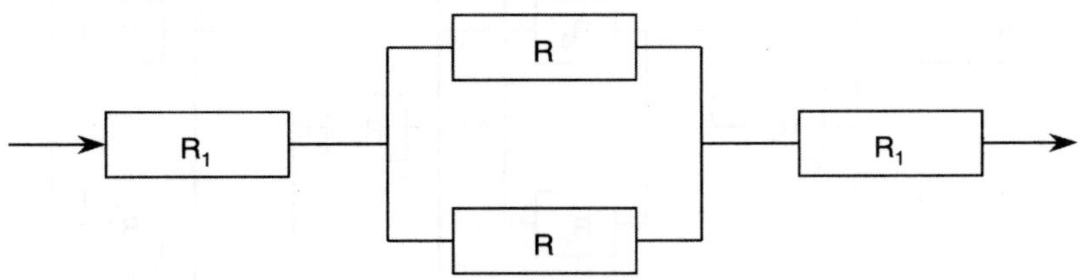

Fig. P7.33

7.34 Find the reliability of the system shown in Fig. P7.34.

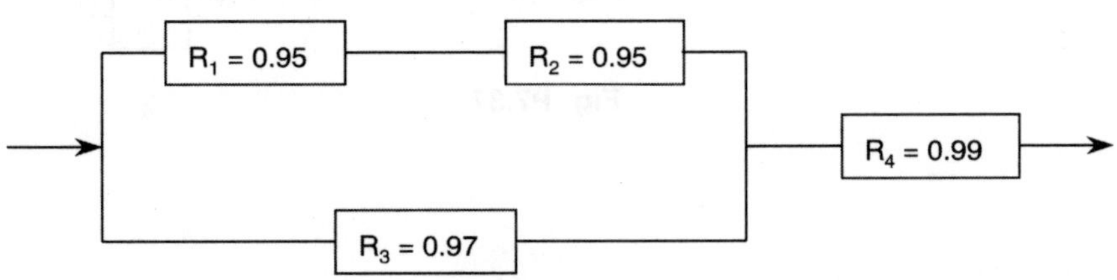

Fig. P7.34

7.35 For the network shown in Fig. P7.35.

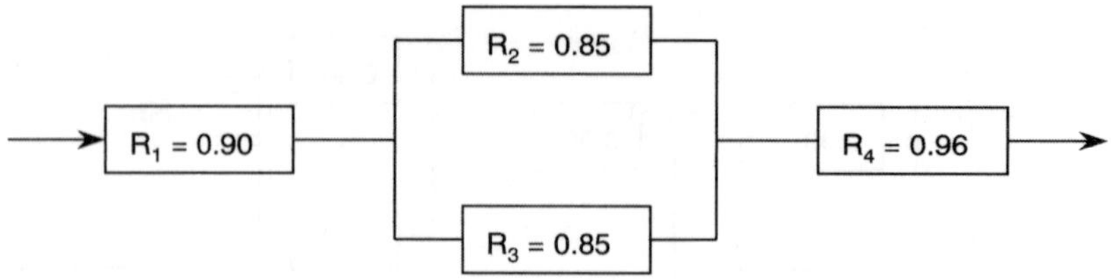

Fig. P7.35

(*a*) find the structure function

(*b*) determine the system reliability using the structure function

(*c*) identify the minimal path sets and minimal cut sets

(*d*) determine the lower- and upper-bound reliabilities

7.36 A system consists of a primary unit and a standby unit. The MTTF of the primary unit is 1200 operating hours, and the MTTF of the standby unit is 500 hr when in operation and 3000 hr while in standby status.

(*a*) find the design life of the system if specifications require a reliability of 0.95.

(*b*) determine the system MTTF

7.37 Assuming an equal allocation to each redundant subset and each redundant subset contains identical components, find the component reliability of the system shown in Fig. P7.37 if the system reliability objective is 0.97.

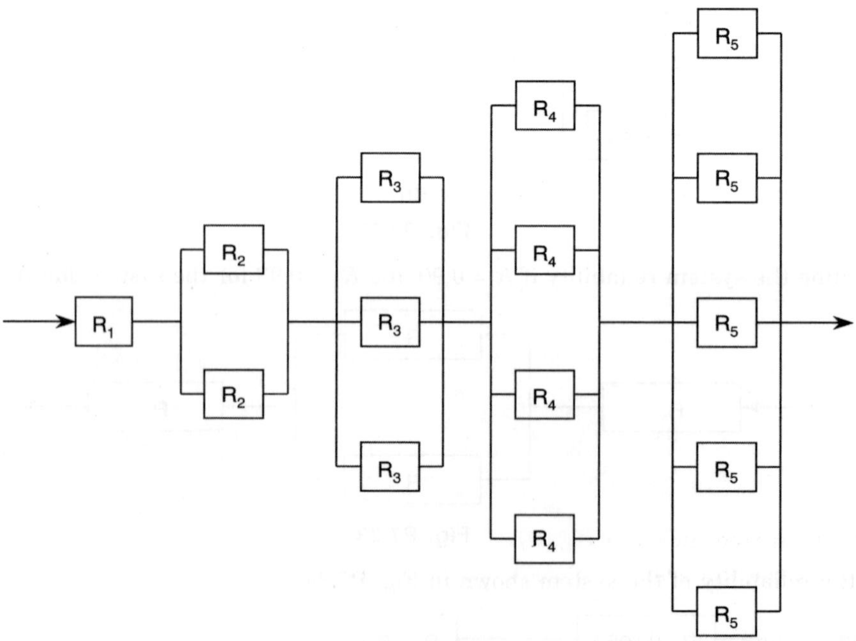

Fig. P7.37

7.38 The Weibull distribution parameters of a system with four components in series are given in Table P7.38.

Table P7.38

Component	Scale parameter, θ	Shape parameter, β
1	5,000	1.50
2	10,000	1.75
3	15,000	2.00
4	20,000	2.25

Assuming the cost of improving the current reliability is linear and equal to all the components, find the percentage of improvement needed for each component in order to achieve the system reliability goal of 0.989, at 1500 hours of operation.

7.39 Consider the fault tree shown in Fig. P7.39

(*a*) Express the top event in Boolean algebra terms consisting of basic events and incomplete events

(*b*) find the minimal cut sets

(*c*) estimate the probability of the top event given the reliability of each basic event is 0.015 and the probability of each incomplete event is 0.025.

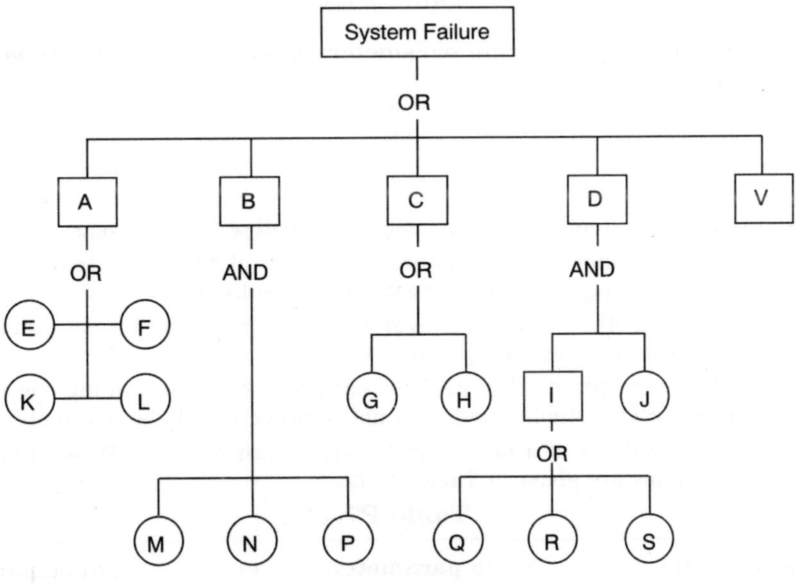

Fig. P7.39

7.40 For the fault tree shown in Fig. P7.40,

(*a*) express the top event in Boolean algebra terms consisting of basic events and incomplete events.

(*b*) find the minimal cut sets.

(*c*) Estimate the probability of the top event given the reliability of each basic event is 0.017 and the probability of each incomplete event is 0.025.

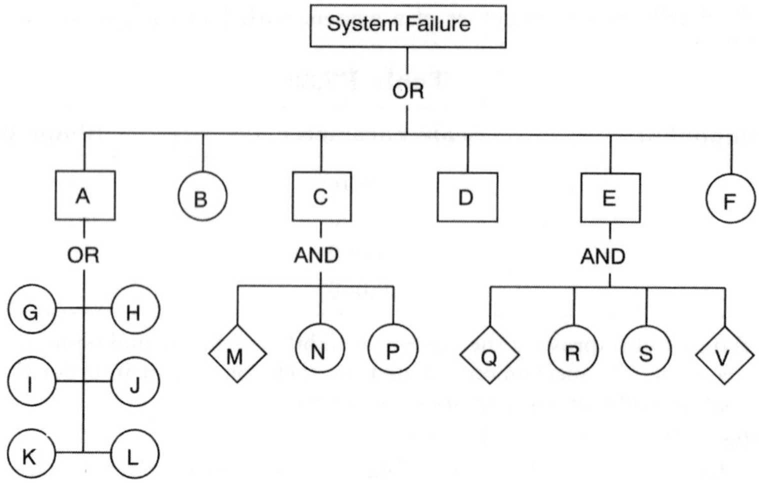

Fig. P7.40

7.41 The three major subsystems of an electromechanical system each having a Weibull failure distribution with parameter values is given in Table P7.41.

Table P7.41

Subsystem	Scale parameter, θ, (yr)	Shape parameter, β
A	3.2	0.92
B	4.1	0.88
C	5.3	1.28

The reliability of the system should be 0.998 at the end of the first year.

(*a*) find the percentage increase in reliability for each of the subsystems needed to meet the system goal. Assume equal reliability goals for all subsystems.

(*b*) apply the ARINC method to find an average failure rate goal for each subsystem assuming average failure rate for each subsystem.

(*c*) determine the characteristic life goal for each subsystem assuming the reliability growth improves the characteristic life and does not improve the shape parameter.

7.42 The four major subsystems of a mechanical system each having a Weibull failure distribution with parameter values are given in Table P7.42.

Table P7.42

Subsystem	Scale parameter, θ, (yr)	Shape parameter, β
A	3.5	0.94
B	4.0	0.87
C	5.5	1.75
D	6.5	1.20

The reliability of the system should be 0.997 at the end of the first year.

(*a*) find the percentage increase in reliability for each of the subsystems needed to meet the system goal. Assume equal reliability goals for all subsystems.

(*b*) apply the ARINC method to find an average failure rate goal for each subsystem assuming average failure rate for each subsystem.

(*c*) determine the characteristic life goal for each subsystem assuming the reliability growth improves the characteristic life and does not improve the shape parameter.

7.43 The five major subsystems of an electromechanical system each having a Weibull failure distribution with parameter values are given in Table P7.43.

Table P7.43

Subsystem	Scale parameter, θ, (yr)	Shape parameter, β
A	3.0	0.98
B	4.0	0.95
C	5.0	1.20
D	3.5	1.75
E	4.5	1.20

The reliability of the system should be 0.99 at the end of the first year.

(*a*) find the percentage increase in reliability for each of the subsystems needed to meet the system goal. Assume equal reliability goals for all subsystems.

(*b*) apply the ARINC method to find an average failure rate goal for each subsystem assuming average failure rate for each subsystem.

(*c*) determine the characteristic life goal for each subsystem assuming the reliability growth improves the characteristic life and does not improve the shape parameter.

7.44 For the fault tree shown in Fig.P7.44 performs a qualitative analysis for the top event in terms of non-redundant basic events using Boolean algebra if the probability of each basic event is 0.0005 and the events are independent, what is the probability of the top event?

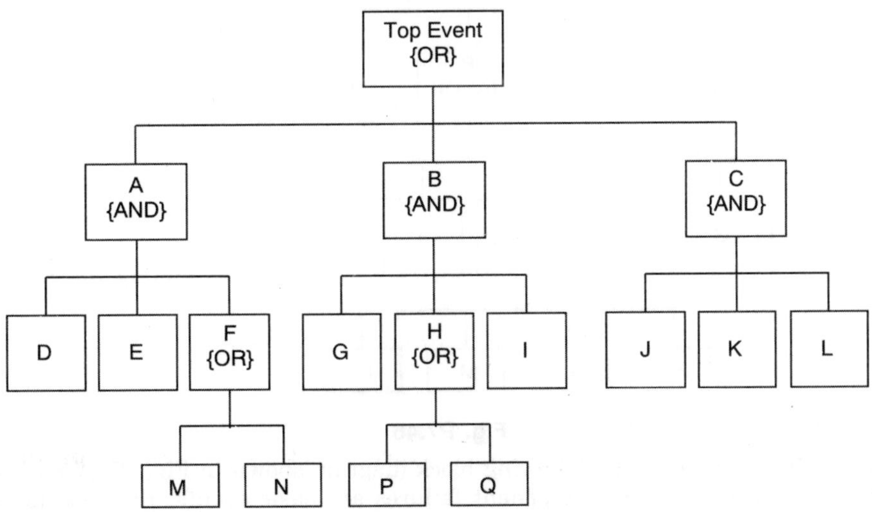

Fig. P7.44

7.45 For the fault tree shown in Figure P7.45 performs a qualitative analysis for the top event in terms of non-redundant basic events using Boolean algebra if the probability of each basic event is 0.0005 and the events are independent, what is the probability of the top event?

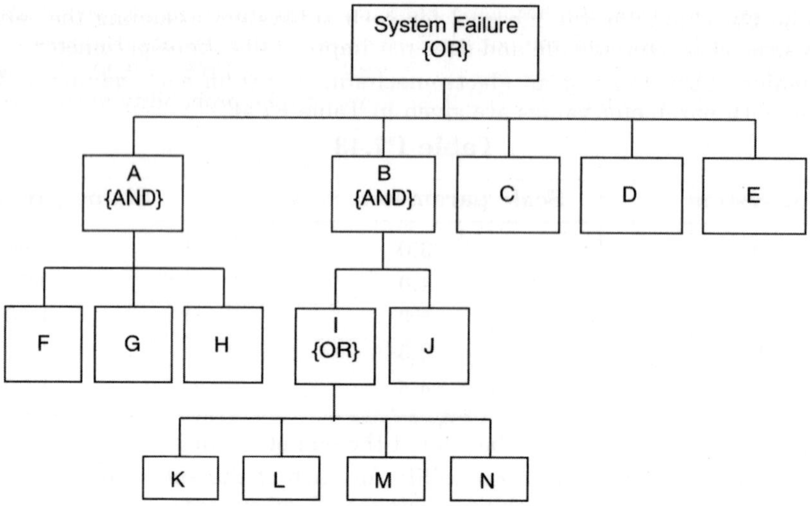

Fig. P7.45

7.46 For the fault tree shown in Fig. P7.46 performs a qualitative analysis for the top event in terms of non-redundant basic events using Boolean algebra if the probability of each basic event is 0.0005 and the events are independent, what is the probability of the top event?

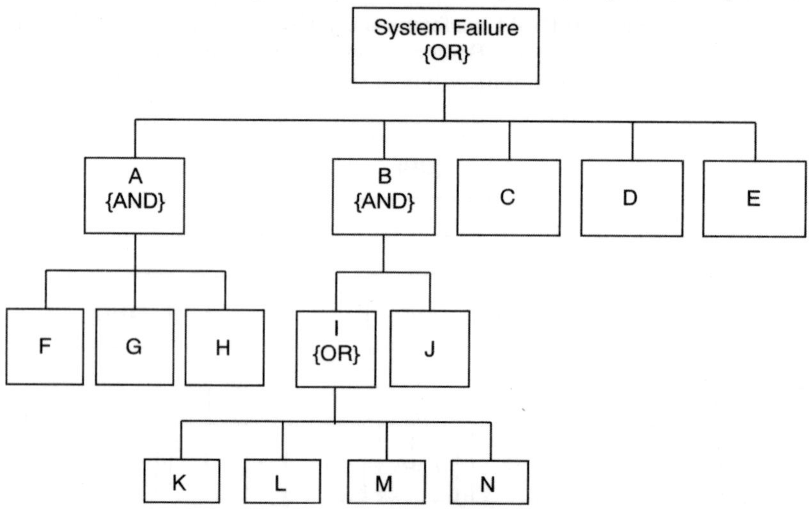

Fig. P7.46

7.47 Construct a fault tree from the following block diagram shown in Fig. P7.47 such that the top event is a system failure and component failures are basic events. Express the top event in Boolean algebra terms of basic events. If Pr{*A*} = Pr{*B*} = 0.92, Pr{*C*} = Pr{*D*} = 0.85, and Pr{*E*} = Pr{*F*} = 0.81, compute the probability of the top event.

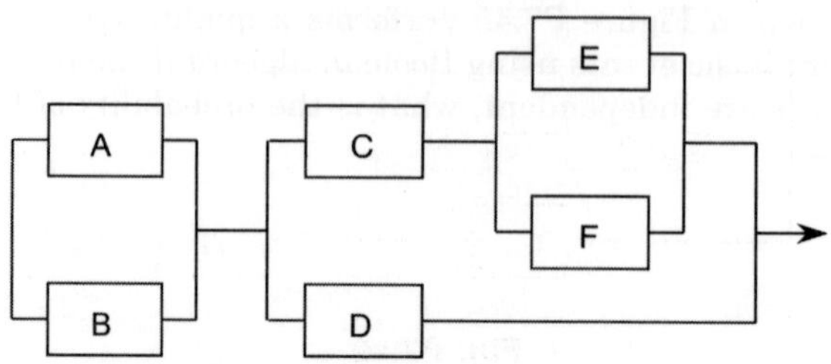

Fig. P7.47

7.48 Construct a fault tree from the following block diagram shown in Fig. P7.48 such that the top event is a system failure and component failures are basic events. Express the top event in Boolean algebra terms of the basic events. If Pr(A) = 0.90, Pr(B) = 0.90, Pr(C) = 0.90, Pr(D) = 0.90, Pr(E) = 0.95, Pr(F) = 0.97, Pr(G) = 0.95. Compute the probability of the top event.

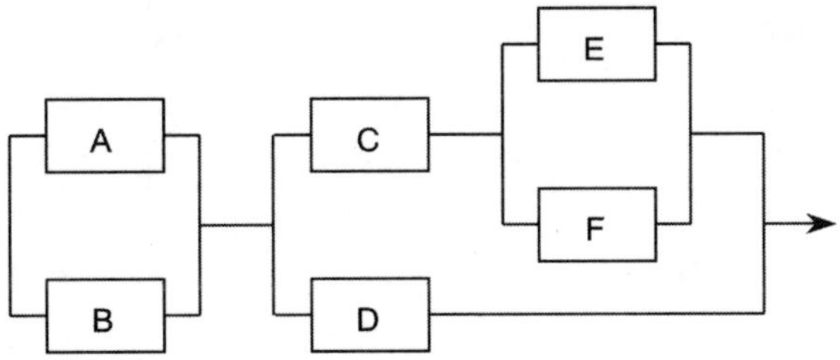

Fig. P7.48

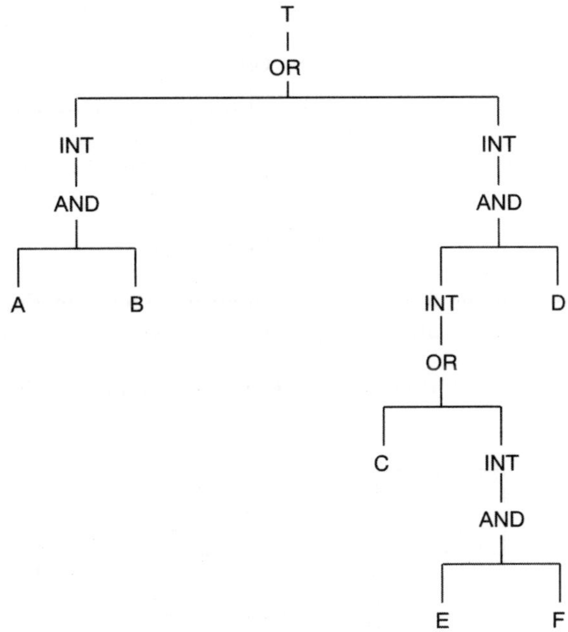

Fig. P7.48 (a)

7.49 Construct a fault tree from the following block diagram shown in Fig.P7.49 such that the top event is a system failure and component failures are basic events. Express the top event in Boolean algebra terms of the basic events. If Pr(A) = 0.90, Pr(B) = 0.90, Pr(C) = 0.80, Pr(D) = 0.80, Pr(E) = 0.75, Pr(F) = 0.75, Pr(G) = 0.75. Compute the probability of the top event.

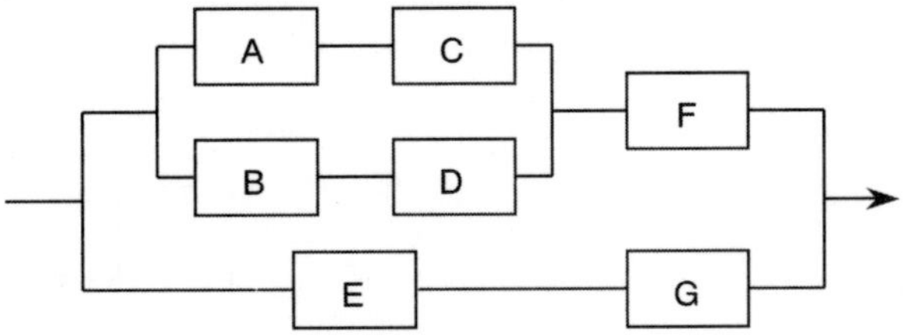

Fig. P7.49

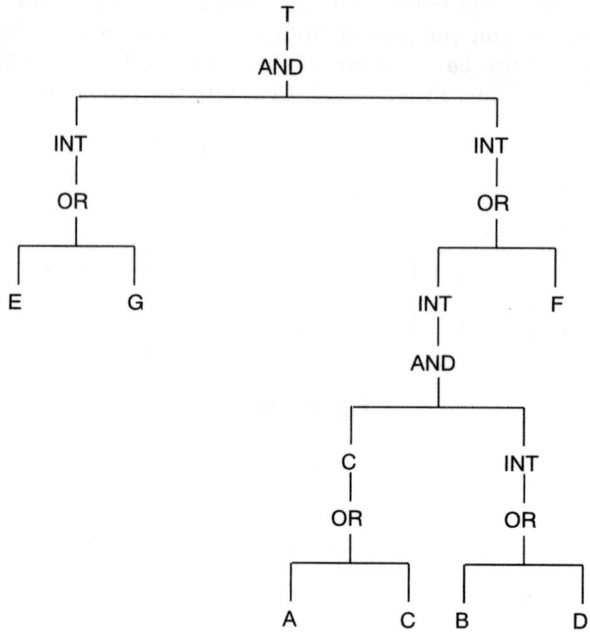

Fig. P7.49 (a)

7.50 Construct a fault tree from the following block diagram shown in Fig. P7.50 such that the top event is a system failure and component failures are basic events. Express the top event in Boolean algebra terms of the basic events. If Pr(*A*) = Pr(*B*) = Pr(*C*) = 0.80, Pr(*D*) = 0.80, Pr(*E*) = 0.90, Pr(*F*) = 0.70, Pr(*G*) = 0.70, Pr(*H*) = 0.90. Compute the probability of the top event.

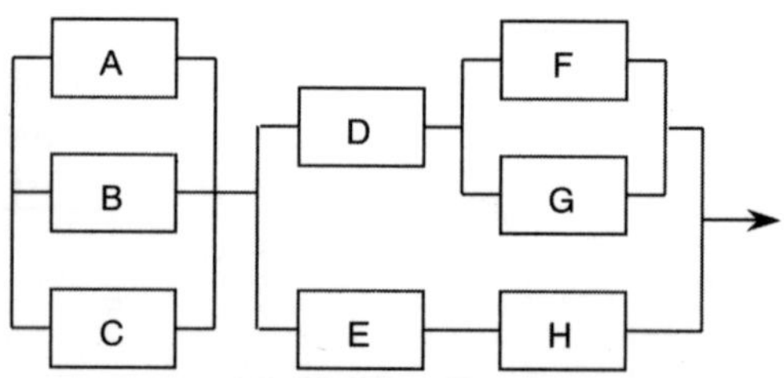

Fig. P7.50

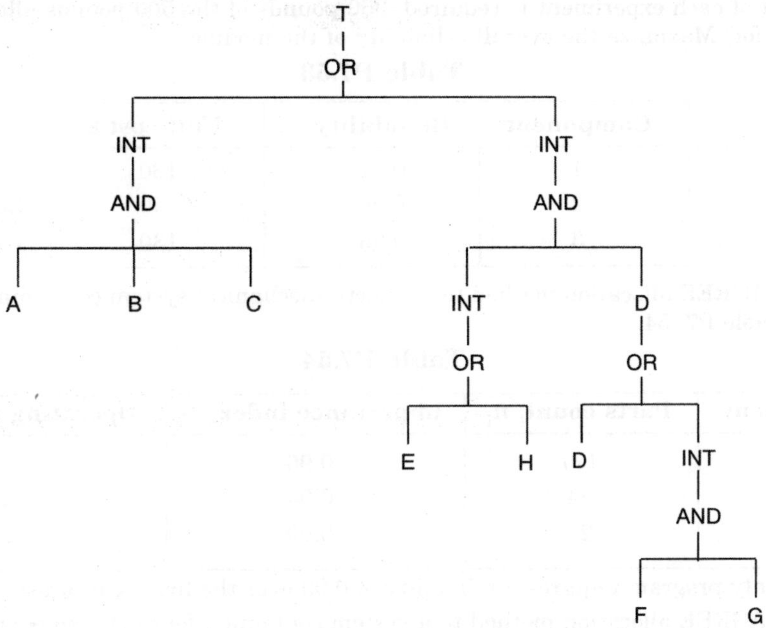

Fig. P7.50 (a)

7.51 An engineer has a $1,000 budget (per system) to be used to increase the reliability of a four-component series system. The reliability of each component at the desired system lifetime has been established through a reliability growth-testing program with the data given in Table P7.51.

Table P7.51

Component	Reliability	Unit cost
1	0.98	$100
2	0.92	170
3	0.96	80
4	0.94	350

Component = k
Reliability = R_i
Unit Cost = c_i

Determine using marginal analysis the optimum number of redundant units. Compute the achieved reliability.

7.52 Given a budget of $1,000 and the data given in Table P7.52 on three components that must operate in series, determine using marginal analysis, the optimum number of redundant units. Compute the achieved reliability.

Table P7.52

Component	Reliability	Unit cost
1	0.92	$250
2	0.95	150
3	0.98	100

Component = k
Reliability = R_i
Unit Cost = c_i

7.53 A certain space probe to be launched contains a module of three experiments. Testing has resulted in a (nonredundant) reliability estimate over the duration of the mission of 0.92, 0.96, and 0.98 for the respective experiments as shown in Table P7.53. There is a weight limitation of 500 pounds to be allocated to the module. How many redundant experiments of each type should be included in the module if their respective weights are 180, 50 and 130 pounds? Since a minimum of one unit of each experiment is required, 360 pounds of the 500 pounds allocated are already accounted for. Maximize the overall reliability of the module.

Table P7.53

Component	Reliability	Unit cost $
1	0.92	180
2	0.96	50
3	0.98	130

Component = k
Reliability = R_i
Weight = c_i

7.54 Apply the AGREE allocation method to an electromechanical system containing the components given in Table P7. 54.

Table P7.54

Component	Parts count, n_i	Importance index, ω_i	Operating time, hr/yr, t_i
1	177	0.96	2384
2	54	0.93	1762
3	28	0.99	2065

The warranty program requires a reliability of 0.99 over the first year of use.

7.55 Apply the AGREE allocation method to a system containing four (4) components given in Table P7. 55.

Table P7.55

Component	Parts count, n_i	Importance index, ω_i	Operating time, hr/yr, t_i
e	283	0.96	2000
2	154	0.97	1000
3	87	0.98	2000
4	46	1.00	1000

The warranty program requires a reliability of 0.999 over the first year of use.

7.56 Apply the AGREE allocation method to a system containing five (5) components given in Table P7. 56.

Table P7.56

Component	Parts count, n_i	Importance index, ω_i	Operating time, hr/yr, t_i
1	237	0.96	5000
2	184	0.95	3000
3	93	1.00	1000
4	64	0.97	2000
5	33	0.98	1500

The warranty program requires a reliability of 0.99 over the first year of use.

7.57 For the reliability block diagram for a mechanical system shown in Fig.P7.57, determine the overall system reliability.

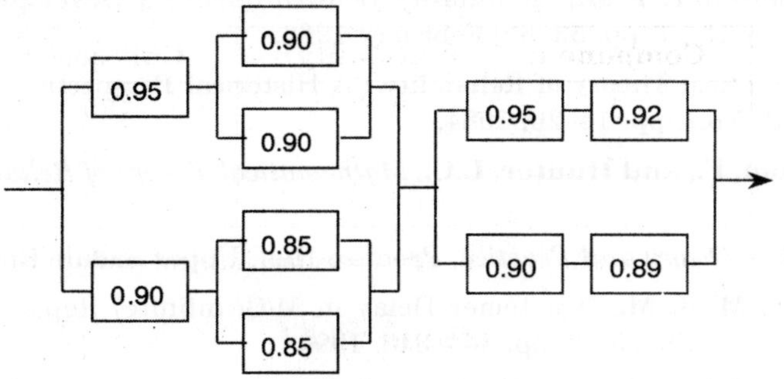

Fig. P7.57

7.58 An electronic component has an average life of 300 hours. Determine the probability of this component operating for at least 500 hours before failing. Assume an exponential failure distribution.

7.59 A group of components in a large electromechanical system follows a Weibull distribution. Determine the probability that one of the components will have a life of 30,000 hours. The parameters of the Weibull distribution are $\theta = 10^6$ hours, $m = 4.5$, and $t_0 = 0$.

7.60 A complex electromechanical system has 800 components in series configuration. Extensive testing of 200 components revealed that 3 failures occurred after 2000 hours.

(*a*) Assuming a constant failure rate, determine the reliability of the system to operate for 2000 hours.

(*b*) If the desired overall system reliability is 0.99 in 2000 hours, determine the individual failure rate of the components.

7.61 The mean time to failure of a single unit and standby unit of a complex mechanical system are 50,000 hours and 30,000 hours respectively. Determine the mean time between failures of a single unit without standby that will have the same reliability as the standby system for the system to operate for 20,000 hours. Assume constant failure rate for the single unit.

REFERENCES

Abernethy, R. et al., *Weibull Analysis Handbook,* (AFWAL-TR-83-2079), Aero Propulsion Laboratory, Air Force Wright Aeronautical Laboratories, Wright Patterson Air Force Base, Dayton, OH, 1983.

Abramowitz, M. and Stegun, I.A., (eds.), *Handbook of Mathematical Functions with Formulas, Graphs and Mathematical Tables*, National Bureau of Standards, Applied Mathematics Series, 55, Washington DC, 1966.

Alder, H.L. and Roessler, E.B., *Introduction to Probability and Statistics*, 4th ed., W.H. Freeman & Company, San Francisco, CA, 1968.

Arsenault, J.E. and Roberts, J.A., (eds.), *Reliability and Maintainability of Electronic Systems*, Maryland, Computer Science Press, Potomac, Maryland,1980.

Ascher, H. and Feingold, H., *Repairable Systems Reliability*, Marcel Dekker, New York, NY, 1984.

Aven, T., "Availability Formula for Standby Systems of Similar Units That Are Preventively Maintained", *IEEE Transactions on Reliability*, Vol. 39, No. 5, pp. 603–606, 1990.

Bain, L.J. and Engelhardt, M., *Statistical Analysis of Reliability and Life Testing Models*, 2nd ed., Marcel Dekker, New York, NY, 1991.

Banks, J., Carson, J.S. and Nelson, B.L., *Discrete Event System Simulation*, Prentice Hall, Upper Saddle River, NJ, 1996.

Barlow, R.E., and Scheuer, E.M., "Reliability Growth during a Development Testing Program, "*Technometrics*, Vol. 8, No. 1, pp. 53–60, February 1966.

Barlow, R.E., "Mathematical Theory of Reliability : A Historical Perspective", *IEEE Transactions on Reliability,* Vol. R-33, No.1, pp. 16–20, 1984.

Barlow, R.E., Proschan, F., and Hunter, L.C., *Mathematical Theory of Reliability*, Wiley, New York, NY, 1967.

Bazovsky, I., *Reliability Theory and Practice*, Prentice Hall, Upper Saddle River, NJ, 1961.

Berg, M. and Posner, M. J. M., "Customer Delay in M/G/(infinite) Repair Systems with Spares", *Operations Research*, Vol. 38, No .2, pp. 344–348, 1990.

Billinton, R. and Allan R.N. *Reliability Evaluation of Engineering Systems: Concepts and Techniques*, Plenum, New York, 1983.

Billinton, R., and Allan, R.N., *Reliability Evaluation of Power Systems*, Pitman/Plenum, Boston, MA, 1984.

Billinton, R., *Power System Reliability Evaluation*, Gordon and Breach, New York, NY, 1970.

Billinton, R., Ringlee, R.J., and Wood, A.J., *Power System Reliability Calculations*, MIT press, Cambridge, MA,1973.

Blanchard, B.S. and Lowery, E.E., *Maintainability*, McGraw-Hill, New York, NY, 1969.

Blanchard, B.S. and Fabrycky, W.J., *Systems Engineering and Analysis*, Prentice Hall, Upper Saddle River, NJ, 1990.

Blanchard, B.S., *Logistics Engineering and Management,* Prentice Hall, Upper Saddle River, NJ, 1992.

Blanks, H.S., *Reliability in Procurement and Use: From Specification to Replacement*, Wiley, New York, NY, 1992.

Bolz, R.E., and Tuve, G.L., (eds.), *CRC Handbook of Tables for Applied Engineering Science*, 2nd ed., CRC Press, Boca Raton, FL, 1973.

Bowles, J.B., "A Survey of Reliability-Prediction Procedures for Microelectronic Devices, *IEEE Transactions on Reliability*, Vol. 41, No.1, pp. 2–12, March 1992.

Breipohl, A.M., *Probabilistic Systems Analysis*, Wiley, New York, NY, 1970.

Bryant, J.L. and Murphy, R.A., "Uptime of Systems Subject to Repairable and Non-repairable Failures, *AIIIE Transactions*, Vol. 12, No.3, pp. 226–232, September 1980.

Bunday, B.D., *Statistical Methods in Reliability Theory and Practice*, Ellis Horwood, New York, NY, 1991.

Callister, W.D., *Materials Science and Engineering—An Introduction*, 3rd ed., Wiley, New York, NY, 1994.

Carrillo, M.J., "Extensions of Palm's Theorem: A Review", *Management Science*, Vol. 37, No.6, pp. 739–744, 1991.

Carter, A.D.S., *Mechanical Reliability*, Wiley, New York, NY, 1972.

Catuneanu, V.M. and Mihalache, A., "Some Aspects of the Relationship between Reliability and Efficiency", *Revue Roumaine des Sciences Techniques, Serie Electro-technique et Energetique,* 27(3), pp. 337–342, 1982.

Chang, N.E., "Application of Availability Analysis Techniques to Improve Power Plant Productivity", *IEEE paper No.82, JPGC 609-6*, 1982 IEEE/ASME/ASCE, Joint Power Generation Conference, Denver, CO, October 1982.

Close, C.M. and Frederick, D.K., *Modeling and Analysis of Dynamic Systems,* Houghton Mifflin Company, Boston, MA, 1978.

Clymer, J.R., *Systems Analysis Using Simulation and Markov Models*, Prentice Hall, Upper Saddle River, NJ, 1990.

Crow, L.H., "Methods for Assessing Reliability Growth Potential, "*Proceedings of the Annual Reliability and Maintainability Symposium*, IEEE, pp. 484–489, New York, NY, January 1984.

Crowder, M.J. et al., *Statistical Analysis of Reliability Data*, Chapman and Hall, New York, NY, 1991.

Dai, S., and Ming, W., *Reliability Analysis in Engineering Applications*, Van Nostrand Reinhold, New York, NY, 1992.

Deshmukh, R.G., and Ramakumar, R., "Reliability Analysis of Combined Wind Electric and Conventional Generation Systems", *Solar Energy,* 28(4), PP: 345–352, 1982.

Dhakar, T.S., Schmidt, C.P. and Miller, D.M., " Base Stock Level Determination for High Cost Low Demand Critical Repairable Spares Research", *Computers and Operations*, Vol. 21, No.4, April 1994, pp. 411–420.

Dhillon, B.S., *Reliability Engineering in Systems Design and Operations*, Van Nostrand Reinhold, New York, NY, 1985.

Dhillon, B.S. and Reiche, H., *Reliability and Maintainability Management*, Van Nostrand Reinhold, New York, NY, 1985.

Dhillon, B.S. and Singh, C., *Engineering Reliability : New Techniques and Applications*, Wiley, New York, NY, 1981, 1983.

Dhingra, A., "Optimal Apportionment of Reliability and Redundancy in Series Systems Under Multiple Objectives", *IEEE Transactions of Reliability*, Vol. 41, No. 4, pp. 576–582, December 1992.

Dieter, G.E., *Engineering Design*, 2nd ed., McGraw-Hill, New York, NY, 1991.

Doyle, E.A. Jr., "How Parts Fail", *IEEE Spectrum, (special issue on reliability),* 18(10), pp. 36–43, 1981.

Drake, A.W., *Fundamental of Applied Probability Theory*, McGraw-Hill, New York, 1967.

Duane, J.T., "Learning Curve Approach to Reliability Monitoring", *IEEE Transactions on Aerospace* 2, No. 2, pp. 563–566, April 1964.

Dudley, R.H., Hekimian, K.K., and Lattin, H., "The Economics of Reliability", *Annals of Reliability and Maintainability, Vol. 9: Assurance Technology Spinoffs*, Ninth Reliability and Maintainability Conference, Detroit, MI, pp. 451–470, 1970.

Ebeling, C.E., *An Introduction to Reliability and Maintainability Engineering*, McGraw-Hill, New York, NY, 1997.

Ebeling, C.E., "Optimal Stock Levels and Service Channel Allocations in a Multi-Item Repairable Asset Inventory System", *IIE Transactions*, Vol. 21, pp. 115–120, 1991.

Endrenyi, J., *Reliability Modeling in Electric Power Systems*, Wiley, New York, NY, 1978.

Engelhardt, M. and Bain, L.J., "Prediction Intervals for the Weibull Process", *Technometrics,* Vol. 20, No. 2, 167–169, May 1978.

Fabrycky, W.J. and Blanchard, B.S., *Life-Cycle Cost and Economic Analysis*, Prentice Hall, Upper Saddle River, NJ, 1991.

Fox, B., "Discrete Optimization Via Marginal Analysis", *Management Science*, Vol. 13, No. 3, pp. 210–216, 1996.

Fox, B.L. and Landi, D.M., " Searching for the Multiplier in One Constraint Optimization Problems", *Operations Research*, Vol. 18, pp. 253–262, 1970.

Gertman, D.I. and Blackman, H.S., *Human Reliability and Safety Analysis Data Handbook*, Wiley, New York, NY, 1994.

Gertsbakh, I.B., *Statistical Reliability Theory*, Marcel Dekker, New York, NY, 1989.

Gibra, I.N., *Probability and Statistical Inference for Scientists and Engineers*, Prentice Hall, Upper Saddle River, NJ, 1973.

Goldberg, H., *Extending the Limits of Reliability Theory*, Wiley, New York, NY, 1981.

Gottfried, P., "Comment On : On the Hazard Rate of the Lognormal Distribution", *IEEE Transactions on Reliability*, Vol. 39, No. 5, pp. 519, December 1990.

Graves, S.C., "A Multi-Echelon Inventory Model for a Repairable Item with One-for-one Replenishment", *Management Science*, Vol. 31, No.10, pp. 1247–1256, 1985.

Green, A.E. and Bourne, A.J., *Reliability Technology,* Wiley-Interscience, New York, NY, 1972.

Greenman, L.R, "Is Maintainability Keeping Up with Electronics", *Proceedings of the 1976 Reliability and Maintainability Symposium*, 1976.

Grosh, D.L., *A Primer of Reliability Theory*, Wiley, New York, 1989.

Gross, D., Miller, D.R. and Soland, R.M., "On Some Common Interests Among Reliability, Inventory, and Queuing, "*IEEE Transactions of Reliability*" Vol. R-34, No. 3, pp. 204–208, August 1985.

Gross, D., "On the Ample Service Assumptions of Palm's Theorem in Inventory Modeling", *Management Science*, Vol. 28, No.9, pp. 1065–1079, 1982.

Gross, D. and Ince, J.F., "The Machine Repair Problem with Heterogeneous Populations", *Operations Research*, Vol. 29, No. 3, pp. 532–549, 1981.

Gross, D. and Harris, C.M., *Fundamentals of Queuing Theory*, 2nd ed., Wiley, New York, 1985.

Guthrie, V.H., et al., "Guidelines for Integrating RAM Consideration into an Engineering Project", *IEEE Transactions on Reliability*, Vol. 39, No. 2, pp. 133–139, June 1990.

Gyllenhall, P.R. and Robinson, J.E ., "A Reliability-Cost Optimization Procedure", *Proceedings of the Fifth National Symposium on Reliability and Quality Control in Electronics*, Philadelphia, pp. 43–54, 1959.

Hall, F. and Clark, A., "ACIM: Availability Centered Inventory Model", *Proceedings of the Annual Reliability and Maintainability Symposium*, IEEE, New York, NY, pp. 247–252, January 1987.

Hammer, W., *Occupational Safety Management and Engineering*, 4th ed., Prentice Hall, Upper Saddle River, NJ, 1989.

Healy, J., "A Simple Procedure for Reliability Growth Modeling", *Proceedings of the Annual Reliability and Maintainability Symposium,* IEEE, New York, NY, pp. 171–175, January 1987.

Hecht, H., "Economic Formulation of Reliability Objectives", *Proceedings of the 1971 Annual Symposium on Reliability,* Washington, DC, pp. 280–284, 1971.

Henley, E.J. and Kumamoto, H., *Reliability Engineering and Risk Assessment*, Prentice Hall, Upper Saddle River, NJ, 1981.

Hicks, C.R., *Fundamental Concepts in the Design of Experiment*, 4th ed., Saunders College Publishing, New York, NY, 1993.

Hillier, F.S. and Lieberman, G.J., *Introduction to Stochastic Models in Operations Research*, McGraw-Hill, New York, NY, 1990.

Ireson, W.G. and Coombs, C.F., eds., *Handbook of Reliability Engineering and Management*, McGraw-Hill, New York, NY, 1988.

Jacobowitz, D.L., "A Software Tool for Designing Burn-In Programs", *Proceedings of the Annual Reliability and Maintainability Symposium* (January 1987), IEEE, New York, NY, pp. 302–305.

Jager, R. and Krause, G.S., "Generic Automated Model for Early MTTR Predictions", *Proceedings of the Annual Reliability and Maintainability Symposium*, IEEE, New York, NY, pp. 280–285, January 1987.

Jensen, F. and Peterson, N.E., *Burn-In*, Wiley, New York, NY, 1982.

Jumonville, P. and Lesso, W.G., "Determining the Mean Time to Failure for Certain Redundant Systems", *IIE Transactions*, Vol. 1, No.1, pp. 81–82, March 1969.

Kabak, I.W., "Some Aspects of Optimal Design", *IIE Transactions*, Vol. 1, No.4, pp. 371–374, December 1969.

Kapur, K.C. and Lamberson, L.R., *Reliability in Engineering Design*, Wiley, New York, 1977.

Kececioglu, D., *Reliability Engineering Handbook*, Vols. 1 and 2, Prentice Hall, Upper Saddle River, NJ, 1991.

Kececioglu, D., *Reliability and Life Testing Handbook*, Vol. 1, Prentice Hall, Upper Saddle River, NJ, 1993.

Kemeny, J.G. and Snell, J.L., *Finite Markov Chains*, Van Nostrand, Princeton, NJ, 1960.

Kivenson, G., *Durability and Reliability in Engineering Design*, Hayden Book Company, New York, 1971.

Kleinrock, L., *Queuing Systems*, Vol.1–Theory, Wiley, New York, NY, 1975.

Kolarik, W.J., *Creating Quality: Concepts, Systems, Strategies and Tools*, McGraw-Hill, New York, 1995.

Law, A.M. and Kelton, W.D., *Simulations Modeling and Analysis*, McGraw-Hill, New York, 1991.

Lawler, E.L. and Bell, M.D., "A Method for Solving Discrete Optimization Problems", *Operations Research*, Vol. 14, pp. 1098–1112, 1966.

Lawless, J.F., *Statistical Models and Methods for Lifetime Data*, Wiley, New York, 1982.

Leemis, L.M. and Beneke, M., "Burn-In Models and Methods", *IIE Transactions*, Vol. 22, No. 2, pp. 172–180, June 1990.

Leemis, L.M., *Reliability, Probabilistic Models and Statistical Methods*, Prentice Hall, Upper Saddle River, NJ, 1995.

Lewis, E.E., *Introduction to Reliability Engineering*, Wiley, New York, 1987.

Lie, C.H., Hwang, C.L. and Tillman, F.A., "Availability of Maintained Systems: A State-of-the-Art Survey", *AIIE Transactions*, Vol. 9, No. 3, pp. 247–259, November 1977.

Lilliefors, H.W., "On the Kolmogorov-Smirnov Test for Normality with Mean and Variance Unknown", *Journal of the American Statistical Association*, Vol. 62, pp. 399–402, 1967.

Lipschutz, S., *Theory and Problems on Probability*, Schaum's Outline Series, McGraw-Hill, New York, 1968.

Lloyd, D.K. and Lipow, M., *Reliability: Management, Methods and Mathematics*, Prentice Hall, Upper Saddle River, NJ, 1962.

Luenberger, D.G., *Introduction to Dynamic Systems Theory, Models and Applications*, Wiley, New York, NY, 1979.

Malik, M.A.K., "Reliable Preventive Maintenance Scheduling", *AIIE Transactions*, Vol. 11, No. 3, pp. 221–228, September 1979.

Mani, V. and Sarma,V.V.S., "Queuing Network Models for Aircraft Availability and Spares Management", *IEEE Transactions on Reliability*, Vol. R-33, No. 3, pp. 257–262, August 1984.

Mann, N.R., Schafer, R.E. and Singpurwalla, N.D., *Methods for Statistical Analysis of Reliability and Life Data*, Wiley, New York, 1974.

Melchers, R.E ., *Structural Reliability: Analysis and Prediction*, Ellis Horwood Limited, West Sussex, 1987.

Military Handbook: *Maintainability Prediction* (MIL-HDBK-472), Naval Publications and Forms Center Philadelphia, PA, 1984.

Military Handbook: *Reliability Growth Management* (MIL-HDBK-189), Naval Publications and Forms Center Philadelphia, PA, 1981.

Military Handbook: *Reliability Prediction of Electronic Equipment* (MIL-HDBK-217F), Rome Air Development Center, Griffiss Air Force Base, New York, 1991.

Military Standard: *Procedures for Performing a Failure Mode, Effects and Criticality Analysis* (MIL-STD-1629A), Naval Publications and Forms Center, Philadelphia, PA, 1980.

Military Standard: *Sampling Procedures and Tables for Inspection by Attributes* (MIL-STD-105), Naval Publications and Forms Center, Philadelphia, PA,1963.

Miller, I. and Freund, J.E., *Probability and Statistics for Engineers*, Prentice Hall, Englewood Cliffs, NJ, 1965.

Mitchell, R.A ., *Introduction to Weibull Analysis*, PWA 3001, Pratt & Whitney Aircraft, East Hartford, CT, January 1967.

Montgomery, D.C., *Design and Analysis of Experiments*, 3rd ed., Wiley, New York, 1991.

Moskowitz, F. and McLean, J.B., "Some Reliability Aspects of System Design", *IRE Transactions on Reliability and Quality Control*, RQC 87–35, 1956.

Muralidhar., K.H. and Zanakis, S., "A Simple Minimum-Bias Percentile Estimator of the Location Parameter for the Gamma, Weibull and Log Normal Distributions", *Decision Sciences*, Vol. 23, No. 4, pp. 862–879, August 1992.

Nelson, W., "Analysis of Accelerated Life Test Data–Part 1: The Arrhenius Model and Graphical Methods", *IEEE Transactions on Electrical Insulation* EI-6(4), 165–181, 1971.

Nelson, W., *Accelerated Testing*, Wiley, New York, 1990.

Nelson, W., *Applied Life Data Analysis*, Wiley, New York, 1982.

Neufelder, A.M., *Ensuring Software Reliability*, Marcel Dekker, New York, 1993.

O'Connor, P.D.T., *Practical Reliability Engineering*, 2nd ed., Wiley, New York, 1985.

Olkin, I., L.J. Gleser and Derman, C., *Probability Models and Applications*, Macmillan, New York, 1980.

Papoulis, A., *Probability, Random Variables, and Stochastic Processes,* McGraw-Hill, New York, 1965.

Pearson, K., (ed.), *Tables of The Incomplete T-Function*, Cambridge University Press and the Office of Biometrica, Cambridge, England, 1951.

Ploe, R.J. and Skewis, W.H., *Handbook of Reliability Prediction Procedures for Mechanical Equipment*, David Taylor Research Center, Bethesda, Maryland, May 1990.

Proll, L.G., "Marginal Analysis Revisited", *Operations Research Quarterly*, Vol. 27, No. 3, pp. 765–767, 1976.

Ramakumar, R., *Engineering Reliability: Fundamentals and Applications*, Prentice Hall, Upper Saddle River, NJ, 1993.

Rao, S.S., *Reliability-Based Design*, McGraw-Hill, New York, NY, 1992.

Rawicz, A.H., "Strongly Correlated Functions in Reliability Physical Approach", *International Journal of Reliability, Quality, and Safety Engineering,* Vol. 1, No. 1, pp. 63–69, 1994.

Raze, J.D. et al., "Reliability Models for Mechanical Equipment", *Proceedings of the Annual Reliability and Maintainability Symposium*, IEEE, New York, NY, pp. 130–133, January 1987.

Reklaitis, G.V., Ravindran, A. and Ragsdell, K.M., *Engineering Optimization: Methods and Applications*, Wiley, New York, 1983.

Roberts, N.H. et al., *Fault Tree Handbook*, (NUREG-0492), Office of Nuclear Regulatory Research, U.S. Nuclear Regulatory Commission, Washington DC, 1981.

Rome Air Development Center, *Reliability Growth Study* (RADC-TR-75-253), Hughes Aircraft Company, Air Force Systems Command, Griffis Air Force Base, New York, NY, 1975.

Rome Laboratory, *Rome Laboratory Reliability Engineers Tool Kit,* Air Force Material Command (AFMC), Griffis Air Force Base, New York, NY, 1993.

Rook, P., *Software Reliability Handbook,* Elsevier Science Publishing Company, New York, NY, 1990.

Ross, S.M., *Introduction to Probability and Statistics for Engineers and Scientists,* Wiley, New York, 1987.

Sampford, M.R. and Taylor, J., "Censored Observations in Randomized Block Experiments", *Journal of the Royal Statistical Society,* Vol. 21, pp. 214–237, 1959.

Sanders, M.S. and McCormick, E. J., *Human Factors in Engineering and Design,* 7th ed., McGraw-Hill, New York, NY, 1993.

Schlager, N., (ed)., *When Technology Fails*, Gale Research Inc., Detroit, MI, 1994.

Schneeweiss, W.G., "Computing Failure Frequency, MTBF & MTTR via Mixed Products of Availabilities and Unavailabilities", *IEEE Transactions on Reliability*, R-30 (4), 362–363, 1981.

Sherbrooke, C.C., "METRIC: A Multi-Echelon Technique for Recoverable Item Control", *Operations Research,* Vol. 16, No. 1, pp. 122–141, 1968.

Sherbrooke, C.C., *Optimal Inventory Modeling of Systems,* Wiley, New York, NY, 1992.

Shooman, M.L., *Probabilistic Reliability: An Engineering Approach.* 2nd ed., R.E. Krieger Publishing Company, Malabar, Florida, 1990.

Siewiorek, D.P. and Swarz, R.S., *The Theory and Practice of Reliable System Design*, Digital Press, Bedford, MA,1982.

Singh, C. and Billinton, R., *System Reliability Modeling and Evaluation*, Hutchinson, London, England, 1977.

Smith, A.M., *Reliability-Centered Maintenance,* McGraw-Hill, New York, NY, 1993.

Smith, C.O., *Introduction to Reliability in Design,* McGraw-Hill, New York, NY, 1976.

Smith, D.J., *Reliability and Maintainability in Perspective: Practical, Contractual, Commercial, and Software Aspects,* 2nd ed., Wiley, New York, NY, 1985.

Sundararajan, C., *Guide to Reliability Engineering,* Van Nostrand Reinhold Company, New York, NY, 1991.

Sweet, A.L., "On the Hazard Rate of the Lognormal Distribution", *IEEE Transactions of Reliability*, Vol. 39, No. 3, pp. 325–328, August 1990.

Tobias, P.A. and Trindade, D., *Applied Reliability,* Van Nostrand, New York, 1986.

Trivedi, K.S., *Probability and Statistics with Reliability, Queuing, and Computer Science Applications*, Englewood Cliffs, NJ, Prentice Hall, 1982.

Wald, A., *Sequential Analysis*, Wiley, New York, 1947.

Witherell, C.E., *Mechanical Failure Avoidance, Strategies and Techniques*, McGraw-Hill, New York, 1994.

Woodson, W.E., Tillman, B. and Tillman, P., *Human Factors Design Handbook*, 2nd ed., McGraw-Hill, New York, NY, 1992.

Xie, M., *Software Reliability Modeling,* World Scientific Publishing Company, Singapore, 1991.

Yurkowsky, W. et al., "Accelerated Testing Technology, Vol. 11, Handbook of Accelerated Life Testing Methods", *Technical Report No. RADC-TR-67-420,* Prepared for Rome Air Development Center by Hughes Aircraft Company, 1967.

Zacks, S., *Introduction to Reliability Analysis,* Springer-Verlag, New York, 1992.

Appendix–7A
Basic Mathematics

7A.1 EXPONENTS

In the following, m and n are real numbers, a and b are positive numbers.

$$a^m \times a^n = a^{m+n}$$
$$a^m / a^n = a^{m-n} = a^m a^{-n}$$
$$a^{m+n} = a^m a^n$$
$$a^m / a^n = 1/ a^{n-m}, a \neq 0$$
$$(a^m)^n = a^{mn}$$
$$(ab)^n = a^n b^n$$
$$(a/b)^n = a^n / b^n, b \neq 0$$
$$a^0 = 1, a \neq 0$$
$$a^{-n} = 1/ a^n, a \neq 0.$$

7A.1.1 Fractional exponents

In the following, a and b are positive numbers, and m and n are the positive integers.

$$a^{1/n} = \sqrt[n]{a}$$
$$a^{m/n} = \sqrt[n]{a^m} = (\sqrt[n]{a})^m$$

7A.1.2 Radicals

In the following, a and b are positive numbers, and m and n are the positive integers.

$$\sqrt[n]{a^n} = (\sqrt[n]{a})^n = a$$
$$\sqrt[n]{a}\,\sqrt[n]{b} = \sqrt[n]{ab}$$
$$\sqrt[m]{\sqrt[n]{a}} = \sqrt[mn]{a}$$
$$\frac{\sqrt[n]{a}}{\sqrt[n]{b}} = \sqrt[n]{\frac{a}{b}}, b \neq 0$$

7A.2 EXPONENTIAL AND LOGARITHMIC FUNCTIONS

Exponential function $y = b^x$, where $b \neq 0$ or 1, then $x = \log_b y$ is called the logarithm of y to the base b. The number $y = b^x$ is called the antilogarithm of x to the base b, written as $\text{antilog}_b\, x$.

Logarithmic form

$$x = \log_b y$$

Logarithmic function

$$y = \log_b x$$

Laws of exponents

$$b^u\, b^v = b^{u+v}$$

$$\frac{b^u}{b^v} = b^{u-v}$$

$$(b^u)^n = b^{un}$$

Properties of logarithms

$$\log_b xy = \log_b x + \log_b y$$

$$\log_b \left(\frac{x}{y}\right) = \log_b x - \log_b y$$

$$\log_b(x^n) = n \log_b x$$

$$\log_b 1 = 0$$

$$\log_b b = 1$$

$$\log_b (b^n) = n$$

Changing base of logarithms

$$\log_b x = \frac{\log_a x}{\log_a b}$$

$$\ln x = \frac{\log x}{\log e}$$

$$\log x = \frac{\ln x}{\ln 10}$$

7A.3 LOGARITHMIC FUNCTION WITH BASE b

For b > 0 and b ≠ 1, the logarithmic function with base b is defined as $y = \log_b x$ if and only if $b^y = x$ for $x > 0$. Hence, the equation $\log_b u = v$ is equivalent to $b^v = u$. The equation $\log_b u = v$ is said to be in *logarithmic form* and the equation $b^v = u$ is said to be in *exponential form.* There are two *logarithmic bases:* base 10 and base e. We refer to a base 10 logarithm ($\log_{10}$) as a *common logarithm* and a base e logarithm ($\log_{10} e$) as a *natural logarithm* or *Napier logarithm*. The common logarithm of a positive number x is usually written as log x and the natural logarithm of a positive number x is usually written as ln x.

Properties of logarithms:

$$\log_b 1 = 0 \text{ and } \log_b b = 1 \text{ for } b > 0 \text{ and } b \neq 1$$

$$\log_b a = \frac{\ln a}{\ln b}$$

$$\log_b PQ = \log_b P + \log_b Q$$

The log of a product is the sum of the logs of the factors.

$$\log_b (P/Q) = \log_b P - \log_b Q$$

The log of a quotient is the log of the numerator minus the log of the denominator.

$$\log_b (1/Q) = -\log_b Q$$

$$\log_b P^n = n \log_b P$$

$$\log_b x^r = r \log_b x \quad \text{for} \quad x > 0, b > 0, b \neq 1$$

$$\log_b \sqrt[n]{x} = \frac{1}{n} \log_b x \text{ for } \quad b > 0, b \neq 1$$

$$e^0 = 1, e^{-1} = \frac{1}{e} = 0.37$$

$$b^{\log_b u} = u \text{ for } u > 0$$

$$\log_b b^u = u$$

If $\log_b x$ is defined, then

$$\log_b x = \frac{\log_a x}{\log_a b} \quad \text{for} \quad a > 0 \text{ and } a \neq 1$$

$$\log_b a = \frac{1}{\log_a b}$$

or

$$\log_b a \,.\, \log_a b = 1$$

7A.4 QUADRATIC EQUATION AND FORMULA

Quadratic equation in x, where a, b, and c are constants is written as:

$$ax^2 + bx + c = 0$$

The roots of this quadratic equation are given by:

$$x = \frac{-b \pm \sqrt{b^2 - 4ac}}{2a}$$

If a, b and c are real and if $D = b^2 - 4ac$ is the discriminant, then the roots are

(*i*) real and unequal if $D > 0$

(*ii*) real and equal if $D = 0$

(*iii*) complex conjugate if $D < 0$

If x_1, x_2 are the roots, then $x_1 + x_2 = -b/a$ and $x_1x_2 = c/a$

7A.4.1 General Rules of Differentiation

Suppose $y = f(x)$. The derivative of y or $f(x)$ is defined as:

$$\frac{dy}{dx} = \lim_{h \to 0} \frac{f(x+h)}{h} = \lim_{h \to 0} \frac{f(x + \Delta x) - f(x)}{\Delta x}$$

where $h = \Delta x$. The derivative is also denoted by y', df/dx or $f'(x)$. The process of taking a derivative is called differentiation.

In the following u, v, w are functions of x; a, b, c, n are constants and $e = 2.71828...$ is the natural base of logarithms; $\ln u$ is the natural logarithm of u(*i.e.*, the logarithm to the base e), where it is assumed that $u > 0$ and all angles are in radians.

$$\frac{d}{dx}(c) = 0$$

$$\frac{d}{dx}(cx) = c$$

$$\frac{d}{dx}(cx^n) = ncx^{n-1}$$

$$\frac{d}{dx}(u \pm v \pm w \pm ......) = \frac{du}{dx} \pm \frac{dv}{dx} \pm \frac{dw}{dx} \pm ......$$

$$\frac{d}{dx}(cu) = c\frac{du}{dx}$$

$$\frac{d}{dx}(uv) = u\frac{dv}{dx} + v\frac{du}{dx}$$

$$\frac{d}{dx}(uvw) = uv\frac{dw}{dx} + uw\frac{dv}{dx} + vw\frac{du}{dx}$$

$$\frac{d}{dx}\left(\frac{u}{v}\right) = \frac{v(du/dx) - u(dv/dx)}{v^2}$$

$$\frac{d}{dx}(u^n) = nu^{n-1}\frac{du}{dx}$$

$$\frac{dy}{dx} = \frac{dy}{du}\frac{du}{dx} \qquad \text{(Chain rule)}$$

$$\frac{du}{dx} = \frac{1}{dx/du}$$

$$\frac{dy}{dx} = \frac{dy/du}{dx/du}$$

7A.4.2 Derivatives of Exponential and Logarithmic Functions

$$\frac{d}{dx}\log_a u = \frac{\log_a e}{u}\frac{du}{dx}, \ a \neq 0, 1$$

$$\frac{d}{dx}\log u = \frac{d}{dx}\log_e u = \frac{1}{u}\frac{du}{dx}$$

$$\frac{d}{dx}a^u = a^u \ln a \, . \frac{du}{dx}$$

$$\frac{d}{dx}e^u = e^u \frac{du}{dx}$$

$$\frac{d}{dx}u^v = \frac{d}{dx}e^{v \ln u} = e^{v \ln u}\frac{d}{dx}[v \ln u] = vu^{v-1}\frac{du}{dx} + u^v \ln u \frac{dv}{dx}$$

7A.4.3 Differentials

Let $y = f(x)$ and $\Delta y = f(x + \Delta x) - f(x)$. Then

$$\frac{\Delta y}{\Delta x} = \frac{f(x+\Delta x) - f(x)}{\Delta x} = f'(x) + \varepsilon = \frac{dy}{dx} + \varepsilon, \text{ where } \varepsilon \to 0 \text{ as } x \to 0.$$

Thus $\Delta y = f'(x)\Delta x + \varepsilon \Delta x$

If we call $\Delta x = dx$, the differential of x, then we define the differential of y to be

$$dy = f'(x)dx$$

7A.4.4 Rules for Differentials

The rules for differentials are exactly analogies to those for derivatives. As examples we observe that

$$d(u \pm v \pm w \pm ......) = du \pm dv \pm dw \pm ......$$

$$d(uv) = udv + vdu$$

$$d\left(\frac{u}{v}\right) = \frac{vdu - udv}{v^2}$$

$$d(u^n) = nu^{n-1}du$$

$$d(\sin u) = \cos u \, du$$

$$d(\cos u) = -\sin u \, du$$

7A.4.5 Partial Derivatives

Let $z = f(x, y)$ be a function of the two variables x and y. Then we define the partial derivative of z or $f(x, y)$ with respect to x, keeping y constant, to be

$$\frac{\partial f}{\partial x} = \lim_{\Delta x \to 0} \frac{f(x, y + \Delta y) - f(x, y)}{\Delta x}$$

This partial derivative is also denoted by $\frac{\partial z}{\partial x}$, f_x or z_x

Similarly the partial derivative of $z = f(x, y)$ with respect to y, keeping x constant, is defined to be:

$$\frac{\partial f}{\partial y} = \lim_{\Delta y \to 0} \frac{f(x, y + \Delta y) - f(x, y)}{\Delta y}$$

This partial derivative is also denoted by $\frac{\partial z}{\partial y}$, f_y or z_y

7A.5 INDEFINITE INTEGRALS

If $\frac{dy}{dx} = f(x)$, then y is the function whose derivative is $f(x)$ and is called the *anti-derivative* of $f(x)$ or the *indefinite integral* of $f(x)$, denoted by $\int f(x)\,dx$. Similarly if $y = \int f(u)\,du$, then $\frac{dy}{du} = f(u)$. Since the derivative of a constant is zero, all indefinite integrals differ by an arbitrary constant. The process of finding an integral is called *integration*.

7A.5.1 General Rules of Integration

In the following, u, v, w are functions of x; a, b, p, q, n are any constants, restricted if indicated; $e = 2.71828..$ is the natural base of logarithms; $\ln u$ denotes the natural logarithm of u where it is assumed that $u > 0$ [in general to extend formulas to cases where $u < 0$ as well, replace ln/u]; all angles are in radians; all constants of integration are omitted but implied.

$$\int a\,dx = ax$$

$$\int af(x)\,dx = a\int f(x)\,dx$$

$$\int (u \pm v \pm w \pm \ldots)\,dx = \int u\,dx \pm \int v\,dx \pm \int w\,dx + \ldots$$

$$\int u\,dv = uv - \int v\,du$$ [Integration by parts]

$$\int f(ax)\,dx = \frac{1}{a}\int f(u)\,du$$

$$\int F\{f(x)\}\,dx = \int F(u)\frac{dx}{du}\,du = \int \frac{F(u)}{f'(x)}\,du, \text{ where } u = f(x)$$

$$\int u^n\,du = \frac{u^{n+1}}{n+1}, n \neq -1$$

$$\int \frac{du}{u} = \ln u, \text{ if } u > 0 \text{ or } \ln(-u) \text{ if } u < 0$$
$$= \ln|u|$$

$$\int e^u\,du = e^u$$

$$\int a^u\,du = \int e^{u\ln a}\,du = \frac{e^{u\ln a}}{\ln a} = \frac{a^u}{\ln a}, a > 0, a \neq 1$$

7A.5.2 Integrals Involving e^{ax}

$$\int e^{ax}\,dux = \frac{e^{ax}}{a}$$

$$\int xe^{ax}dx = \frac{e^{ax}}{a}\left(x - \frac{1}{a}\right)$$

$$\int x^2 e^{ax}dx = \frac{e^{ax}}{a}\left(x^2 - \frac{2x}{a} + \frac{2}{a^2}\right)$$

$$\int x^n e^{ax}dx = \frac{x^n e^{ax}}{a} - \frac{n}{a}\int x^{n-1}e^{ax}dx$$

$$= \frac{e^{ax}}{a}\left(x^n - \frac{nx^{n-1}}{a} + \frac{n(n-1)x^{n-2}}{a^2} - \ldots + \frac{(-1)^n n!}{a^n}\right)$$

if n = positive integer

$$\int \frac{e^{ax}}{x}dx = \ln x + \frac{ax}{1.1!} + \frac{(ax)^2}{2.2!} + \frac{(ax)^3}{3.3!} + \ldots$$

$$\int \frac{e^{ax}}{x^n}dx = \frac{-e^{ax}}{(n-1)x^{n-1}} + \frac{a}{n-1}\int \frac{e^{ax}}{x^{n-1}}dx$$

$$\int \frac{dx}{p + qe^{ax}} = \frac{x}{p} - \frac{1}{ap}\ln(p + qe^{ax})$$

7A.5.3 Integrals Involving ln x

$$\int \ln\ x dx = x \ln x - x$$

$$\int x \ln x dx = \frac{x^2}{2}\left(\ln x - \frac{1}{2}\right)$$

$$\int x^m \ln x dx = \frac{x^{m+1}}{m+1}\left(\ln x - \frac{1}{m+1}\right)$$

$$\int \frac{\ln x}{x}dx = \frac{1}{2}\ln^2 x$$

$$\int \frac{\ln x}{x^2}dx = -\frac{\ln x}{x} - \frac{1}{x}$$

$$\int \ln^2 x dx = x\ln^2 x - 2x\ln x + 2x$$

$$\int \frac{\ln^n x}{x}dx = \frac{\ln^{n+1} x}{n+1} \text{ (for } n \neq -1\text{)}$$

$$\int \frac{1}{x \ln x}dx = \ln(\ln x)$$

7A.6 DEFINITE INTEGRALS

Let $f(x)$ be defined in an interval $a \le x \le b$. Divide the interval into n equal parts of length $\Delta x = (b - a)/n$. Then the definite integral of $f(x)$ between $x = a$ and $x = b$ is defined as

$$\int_a^b f(x)dx = \lim_{n\to\infty} \{f(a)\,\Delta x + f(a + \Delta x)\,\Delta x + f(a + 2\Delta x)\,\Delta x + \ldots$$

$$+ f(a + (n - 1)\Delta x)\,\Delta x\}$$

The limit will certainly exist if $f(x)$ is piecewise continuous.

If $f(x) = \frac{d}{dx}g(x)$, then by the fundamental theorem of the integral calculus the above definite integral can be evaluated by using the result

$$\int_a^b f(x)dx = \int_a^b \frac{d}{dx}g(x)dx = g(x)\Big|_a^b = g(b) - g(a)$$

If the interval is infinite or if $f(x)$ has a singularity at some point in the interval, the definite integral is called an *improper integral* and can be defined by using appropriate limiting procedures. For example,

$$\int_a^\infty f(x)dx = \lim_{b\to\infty} \int_a^b f(x)dx$$

$$\int_{-\infty}^\infty f(x)dx = \lim_{\substack{a\to-\infty \\ b\to\infty}} \int_a^b f(x)dx$$

$$\int_a^b f(x)dx = \lim_{\varepsilon\to 0} \int_a^{b-\varepsilon} f(x)dx, \text{ if } b \text{ is a singular point.}$$

$$\int_a^b f(x)dx = \lim_{\varepsilon\to 0} \int_{a+\varepsilon}^{b} f(x)dx, \text{ if } a \text{ is a singular point.}$$

7A.6.1 General Formulas Involving Definite Integrals

$$\int_a^b \{f(x) \pm g(x) \pm h(x) + \ldots.\}\ dx$$

$$= \int_a^b f(x)dx \pm \int_a^b g(x)dx \pm \int_a^b h(x)dx \pm \ldots$$

$$\int_a^b cf(x)dx = c\int_a^b f(x)dx, \text{ where } c \text{ is any constant.}$$

$$\int_a^a f(x)dx = 0$$

$$\int_a^b f(x)dx = -\int_b^a f(x)dx$$

$$\int_a^b f(x)dx = \int_a^c f(x)dx + \int_c^b f(x)dx$$

$$\int_a^b f(x)\,dx = (b-a)\,f(c), \text{ where } c \text{ is between } a \text{ and } b.$$

This is called the *mean-value theorem* for definite integrals and is valid if $f(x)$ is continuous in $a \le x \le b$.

7A.6.2 Definite Integrals Involving Exponential Functions

Some integrals contain Euler's constant $\gamma = 0.5772156$

$$\int_0^\infty e^{-ax}\cos bx\,dx = \frac{a}{a^2+b^2}$$

$$\int_0^\infty e^{-ax}\sin bx\,dx = \frac{b}{a^2+b^2}$$

$$\int_0^\infty \frac{e^{-ax}\sin bx}{x}\,dx = \tan^{-1}\frac{b}{a}$$

$$\int_0^\infty \frac{e^{-ax}-e^{-bx}}{x}\,dx = \ln\frac{b}{a}$$

$$\int_0^\infty e^{-ax^2}dx = \frac{1}{2}\sqrt{\frac{\pi}{a}}$$

$$\int_0^\infty e^{-ax^2}\cos bx\,dx = \frac{1}{2}\sqrt{\frac{\pi}{a}}\,e^{-b^2/4a}$$

$$\int_0^\infty e^{-(ax^2+bx+c)}\,dx = \frac{1}{2}\sqrt{\frac{\pi}{a}}\,e^{(b^2-4ac)/4a}\,erfc\frac{b}{2\sqrt{a}}, \text{ where } erfc(p) = \frac{2}{\sqrt{\pi}}\int_p^\infty e^{-x^2}\,dx$$

$$\int_{-\infty}^{\infty} e^{-(ax^2+bx+c)}\,dx = \sqrt{\frac{\pi}{a}}\,e^{(b^2-4ac)/4a}$$

$$\int_{0}^{\infty} x^n e^{-ax}\,dx = \frac{\Gamma(n+1)}{a^{n+1}}$$

$$\int_{0}^{\infty} x^m e^{-ax^2}\,dx = \frac{[\Gamma(m+1)/2]}{2a^{(m+1)/2}}$$

$$\int_{0}^{\infty} e^{-(ax^2+b/x^2)}\,dx = \frac{1}{2}\sqrt{\frac{\pi}{a}}\,e^{-2\sqrt{ab}}$$

$$\int_{0}^{\infty} \frac{x}{e^x-1}\,dx = \frac{1}{1^2}+\frac{1}{2^2}+\frac{1}{3^2}+\frac{1}{4^2}+\ldots. = \frac{\pi^2}{6}$$

$$\int_{0}^{\infty} \frac{x^{n-1}}{e^x-1}\,dx = \Gamma(n)\left(\frac{1}{1^n}+\frac{1}{2^n}+\frac{1}{3^n}+\ldots.\right)$$

$$\int_{0}^{\infty} \frac{x}{e^x+1}\,dx = \frac{1}{1^2}+\frac{1}{2^2}+\frac{1}{3^2}+\frac{1}{4^2}+\ldots. = \frac{\pi^2}{12}$$

$$\int_{0}^{\infty} \frac{x^{n-1}}{e^x+1}\,dx = \Gamma(n)\left(\frac{1}{1^n}-\frac{1}{2^n}+\frac{1}{3^n}-\ldots.\right)$$

For some positive integer values of n, the series can be summed

$$\int_{0}^{\infty} \frac{\sin mx}{e^{2\pi x-1}}\,dx = \frac{1}{4}\coth\frac{m}{2}-\frac{1}{2m}$$

$$\int_{0}^{\infty} \left(\frac{1}{x+1}-e^{-x}\right)\frac{dx}{x} = \gamma$$

$$\int_{0}^{\infty} \frac{e^{-x^2}-e^{-x}}{x}\,dx = \frac{1}{2}\gamma$$

$$\int_{0}^{\infty} \left(\frac{1}{e^x-1}-\frac{e^{-x}}{x}\right)dx = \gamma$$

$$\int_{0}^{\infty} \frac{e^{-ax}-e^{-bx}}{x\sec px}\,dx = \frac{1}{2}\ln\left(\frac{b^2+p^2}{a^2+p^2}\right)$$

$$\int_0^\infty \frac{e^{-ax} - e^{-bx}}{x \csc px}\, dx = \tan^{-1} \frac{b}{p} - \tan^{-1} \frac{a}{p}$$

$$\int_0^\infty \frac{e^{-ax}(1 - \cos x)}{x^2}\, dx = \cot^{-1} a - \frac{a}{2} \ln (a^2 + 1)$$

7A.6.3 Definite Integrals Involving Logarithmic Functions

$$\int_0^1 x^m (\ln x)^n \, dx = \frac{(-1)^n n!}{(m+1)^{n+1}}, \; m > -1, \; n = 0, 1, 2, \ldots$$

If $n \neq 0, 1, 2, \ldots$ replace $n!$ by $\Gamma(n + 1)$

$$\int_0^1 \frac{\ln x}{1+x}\, dx = -\frac{\pi^2}{12}$$

$$\int_0^1 \frac{\ln x}{1-x}\, dx = -\frac{\pi^2}{6}$$

$$\int_0^1 \frac{\ln (1+x)}{x}\, dx = \frac{\pi^2}{12}$$

$$\int_0^1 \frac{\ln (1-x)}{x}\, dx = -\frac{\pi^2}{6}$$

$$\int_0^1 \ln x \ln (1 + x)\, dx = 2 - 2 \ln 2 - \frac{\pi^2}{12}$$

$$\int_0^1 \ln x \ln (1 - x)\, dx = 2 - \frac{\pi^2}{6}$$

$$\int_0^\infty \frac{x^{p-1} \ln x}{1+x}\, dx = -\pi^2 \csc p\pi \cot p\pi, \; 0 < p < 1$$

$$\int_0^1 \frac{x^m - x^n}{\ln x}\, dx = \ln \frac{m+1}{n+1}$$

$$\int_0^\infty e^{-x} \ln x \, dx = -\gamma$$

$$\int_0^{\infty} e^{-x^2} \ln x \, dx = -\frac{\sqrt{\pi}}{4}(\gamma + 2\ln 2)$$

$$\int_0^{\infty} \ln \frac{e^x + 1}{e^x - 1} dx = \frac{\pi^2}{4}$$

7A.7 THE GAMMA FUNCTION

[De Morgan's law]

$$A + \bar{A}\,B = \text{a function } \Gamma(n) \text{ for } n > 0$$

$$\Gamma(n) = \int_0^{\infty} t^{n-1} \, e^{-t} \, dt, \; n > 0$$

Recursion formula

$$\Gamma(n + 1) = n \, \Gamma(n)$$

If $n = 0, 1, 2, \ldots$ a nonnegative integer, we have the following $(0\,! = 1)$

$$\Gamma(n + 1) = n\,!$$

The gamma function for $n \le 0$

If $n < 0$, the gamma function can be defined as

$$\Gamma(n) = \frac{\Gamma(n+1)}{n}$$

Special values for the Gamma Function

$$\Gamma\left(\frac{1}{2}\right) = \sqrt{\pi}$$

$$\Gamma\left(m + \frac{1}{2}\right) = \frac{1.3.5\ldots(2m-1)}{2^m}\sqrt{\pi}, \; m = 1, 2, 3, \ldots$$

$$\Gamma\left(-m + \frac{1}{2}\right) = \frac{(-1)^m \, 2^m \, \sqrt{\pi}}{1.3.5\ldots(2m-1)}, \; m = 1, 2, 3, \ldots$$

Example 7A.1. *Solve the following logarithmic equations.*

(*a*) $3 - \log x = \log 5$

(*b*) $\log_4 (x - 5) + 2 \log_4 x = 1 + \log_4 2x$

(*c*) $\log(y + a) - 2 \ln(x + b) = c$

(*d*) $e^x + 3e^{-x} = 4$

(*e*) $4^{5x+2} = 70$

Solution:

(*a*) $3 - \log x = \log 5$

Isolate the logarithms on one side, we have

$$3 = \log 5 + \log x$$

$$3 = \log 5x$$

Changing to exponential form

$$10^3 = 5x \quad \text{or} \quad x = \frac{1000}{5} = 200$$

(*b*) $\log_4 (x - 5) + 2 \log_4 x = 1 + \log_4 2x$

Isolate the logarithmic expression on one side,

$$\log_4(x - 5) + 2 \log_4 x - \log_4 2x = 1$$

$$\log_4 \frac{x^2 (x - 5)}{2x} = 1$$

or

$$\log_4 \frac{x (x - 5)}{2} = 1 \quad \text{for } x \neq 0$$

Changing to exponential form

$$4^1 = \frac{x (x - 5)}{2}$$

or

$$x^2 - 5x - 8 = 0$$

which gives

$$x = \frac{25 \pm \sqrt{25 - (-32)}}{2} = \frac{25 \pm 57}{2} = 41 \text{ or } - 16$$

(*c*) $\log(y + a) - 2 \ln(x + b) = c$

Applying the log properties and write in logarithmic form

$$\ln \frac{y + a}{(x + b)^2} = c$$

Changing it to exponential form

$$e^c = \frac{y + a}{(x + b)^2}$$

or

$$y = e^c(x + b)^2 - a$$

Letting $e^c = K$ = a constant

$$y = K(x + b)^2 - a, \text{ where } K = e^c$$

(*d*) $e^x + 3e^{-x} = 4$

Multiplying both sides by e^x

$$e^{2x} + 3 = 4e^x$$

or

$$e^{2x} - 4e^x + 3 = 0.$$

Let $u = e^x$, then the above equation becomes

$$u^2 - 4u + 3 = 0 \quad \text{or} \quad (u - 1)(u - 3) = 0$$

Hence,

$u = e^x = 1$ and or $u = e^x = 3$ or $x = \ln 1$ and $x = \ln 3$, which gives $x = 0$

or

$$x = 1.099.$$

(*e*) $4^{5x+1} = 70$

Taking the base ten logarithm on both sides of the given equation, we have

$$\log_{10} 4^{5x+1} = \log_{10} 70$$

$$(5x + 1) \log_{10} 4 = \log_{10} 70$$

$$5x + 1 = (\log_{10} 70) / (\log_{10} 4)$$

or
$$x = \frac{[(\log_{10} 70)/(\log_{10} 4)] - 1}{5}$$

which gives
$$x = 0.413$$

QUESTIONS

7A.1 Solve $2 \log_b t = \log_b 3 + \log_b(t + 5)$ for t

7A.2 Solve $\frac{1}{2} \log_3 t + \frac{1}{2} \log_3 (t - 3) = \log_3 6$ for t

7A.3 Simplify the following expressions.

(*a*) $3 \log_2 m + \frac{1}{3} \log_2 n - 3 \log_2 p$ (*b*) $\frac{1}{3} (\log_5 c + 3 \log_5 d - \log_5 g)$

7A.4 Simplify the following:

(*a*) $(27\, a^2\, b^4)^{1/3}$ (*b*) $a^{1/3}\, a^{2/5}$

7A.5 Simplify the following:

(*a*) $\left[\dfrac{8^{-3/2}\, x^{2/3}\, y^{-7/4}}{4^{3/2}\, x^{-1/3}\, y^{3/4}}\right]^{5/3}$ (*b*) $(4\, x^4)^{-1/2} - 5\, x^{-5}$

7A.6 Simplify the following:

(*a*) $\sqrt[3]{4^3}$ (*b*) $\sqrt[5]{2}\ \sqrt[5]{7}$ (*c*) $\sqrt[3]{\sqrt{7}}$ (*d*) $\dfrac{\sqrt{7}}{\sqrt{2}}$

7A.7 Simplify the following:

(*a*) $\sqrt{98}$ (*b*) $\sqrt{a^5\, b^4}$ (*c*) $\sqrt[3]{135}$ (*d*) $\sqrt[5]{32x^8\, y^{12}}$

7A.8 Simplify the following:

(*a*) $\dfrac{\sqrt[4]{25}}{\sqrt{5}}$ (*b*) $\dfrac{\sqrt[6]{125}}{\sqrt{3}}$ (*c*) $\sqrt[9]{125x^3\, y^{16}}$

7A.9 What value of A satisfies the expression $A^{-6/8} = 0.001$?

7A.10 What expression is equivalent to $\log(x/(y + z))$?

7A.11 What is the value of $\ln(7.3891)^{xy}$?

7A.12 If the $\log_{10}$ of 4 is $0.703x$, what is the $\log_{10}$ of $\frac{1}{4}$?

7A.13 Find $h'(t)$ if $h(t) = \exp(2t \sin 2t) = e^{2t \sin 2t}$

7A.14 Given $x = e^y$, find $\dfrac{dy}{dx}$

7A.15 Find $f'(t)$ if $f(t) = \ln(2t^2 - 5)$

7A.16 Find $\frac{dy}{dx}$ if $y = \frac{(2 + x^2)\sqrt{\sin 2x}}{(3x^3 - 2x - 1)^5}$

REFERENCES

Cohen, D., *College Algebra*, West Publishing Company, New York, NY, 1996.

Coughlin, R.F. and Zitarelli, D.E., *Brief Calculus with Applications*, 2nd ed., Saunders College Publishing, New York, 1993.

Johnston, E.H. and Mathews, J., *Calculus for Engineering and the Sciences*, Addison-Wesley, New York, NY, 1997.

Lial, M.L. and Miller, C.D., *Essential Calculus with Applications*, 2nd ed., Scott, Foresman and Company, London, 1980.

Nichols, E.D. Edwards, M.L. Henry Garland, E. Hoffman, S.A. Mamary, A. and Palmer, W.F., *Holt Algebra with Trigonometry*, Holt, Reinhart and Winston, Orlando, 1992.

Rothbart, H.A., *Mechanical Design and Systems Handbook*, McGraw-Hill, New York, NY, 1964.

Stevens, D.E., *College Algebra with Graphing Technology*, West Publishing Company, New York, 1996.

Washington, A.J., *Basic Technical Mathematics with Calculus*, 6th ed., Addison-Wesley, New York, 1995.

Appendix–7B
The Gamma Function

The complete *gamma function* also called the *factorial function*, is defined as

$$\Gamma(x) = \int_0^{\infty} t^{x-1}\, e^{-t}\, dt \qquad \text{...(7B.1)}$$

When x is an integer, we have

$$\Gamma(x + 1) = x! = x\Gamma(x) \qquad \text{...(7B.2)}$$

$$\Gamma(x) = (x - 1)! = (x - 1)\,\Gamma(x - 1)$$

For negative values of x

$$\Gamma(x) = \frac{\Gamma(x+1)}{x} \qquad \text{...(7B.3)}$$

For very large positive values of x

$$\Gamma(x + 1) = x\,! = x^x\, e^{-x}\, \sqrt{2\pi x}\left[1 + \frac{1}{12x} + \frac{1}{288x^2} + \ldots\right] \qquad \text{...(7B.4)}$$

or

$$\Gamma(x + 1) = x\,! = x^x\, e^{-x}\, \sqrt{2\pi x} \qquad \text{...(7B.5)}$$

This approximation is known as *Sterling's formula for large factorials*.

Special values of the gamma function are

$$\Gamma(1) = 1.0$$

$$\Gamma(2) = 1.0 \qquad \text{...(7B.6)}$$

$$\Gamma\left[\frac{1}{2}\right] = \int_0^{\infty} \frac{e^{-t}}{\sqrt{t}}\, dt = \sqrt{\pi}$$

Table 7B shows a record of calculated values of gamma function.

Appendix–7B
Gamma Function

x	$\Gamma(x)$	x	$\Gamma(x)$	x	$\Gamma(x)$	x	$\Gamma(x)$
1.01	.99433	1.51	.88659	2.01	1.00427	2.51	1.33857
1.02	.98884	1.52	.88704	2.02	1.00862	2.52	1.34830
1.03	.98355	1.53	.88757	2.03	1.01306	2.53	1.35798
1.04	.97844	1.54	.88818	2.04	1.01758	2.54	1.36779
1.05	.97350	1.55	.88887	2.05	1.02218	2.55	1.37775
1.06	.96874	1.56	.88964	2.06	1.02687	2.56	1.38784
1.07	.96415	1.57	.89049	2.07	1.03164	2.57	1.39807
1.08	.95973	1.58	.89142	2.08	1.03650	2.58	1.40844
1.09	.95546	1.59	.89243	2.09	1.04145	2.59	1.41896
1.10	.95135	1.60	.89352	2.10	1.04649	2.60	1.42962
1.11	.94740	1.61	.89468	2.11	1.05161	2.61	1.44044
1.12	.94359	1.62	.89592	2.12	1.05682	2.62	1.45140
1.13	.93993	1.63	.89724	2.13	1.06212	2.63	1.46251
1.14	.93642	1.64	.89864	2.14	1.06751	2.64	1.47377
1.15	.93304	1.65	.90012	2.15	1.07300	2.65	1.48519
1.16	.92980	1.66	.90167	2.16	1.07857	2.66	1.49677
1.17	.92670	1.67	.90330	2.17	1.08424	2.67	1.50851
1.18	.92373	1.68	.90500	2.18	1.09000	2.68	1.52040
1.19	.92089	1.69	.90678	2.19	1.09585	2.69	1.53246
1.20	.91817	1.70	.90864	2.20	1.10180	2.70	1.54469
1.21	.91558	1.71	.91057	2.21	1.10785	2.71	1.55708
1.22	.91311	1.72	.91258	2.22	1.11399	2.72	1.56964
1.23	.91075	1.73	.91467	2.23	1.12023	2.73	1.58237
1.24	.90852	1.74	.91683	2.24	1.12657	2.74	1.59528
1.25	.90640	1.75	.91906	2.25	1.13300	2.75	1.60836
1.26	.90440	1.76	.92137	2.26	1.13954	2.76	1.62162
1.27	.90250	1.77	.92376	2.27	1.14618	2.77	1.63506
1.28	.90072	1.78	.92623	2.28	1.15292	2.78	1.64868
1.29	.89904	1.79	.92877	2.29	1.15976	2.79	1.66249
1.30	.89747	1.80	.93138	2.30	1.16671	2.80	1.67649
1.31	.89600	1.81	.93408	2.31	1.17377	2.81	1.69068
1.32	.89464	1.82	.93685	2.32	1.18093	2.82	1.70506
1.33	.89338	1.83	.93969	2.33	1.18819	2.83	1.71963
1.34	.89222	1.84	.94261	2.34	1.19557	2.84	1.73441
1.35	.89115	1.85	.94561	2.35	1.20305	2.85	1.74938
1.36	.89018	1.86	.94869	2.36	1.21065	2.86	1.76456
1.37	.88931	1.87	.95184	2.37	1.21836	2.87	1.77994
1.38	.88854	1.88	.95507	2.38	1.22618	2.88	1.79553
1.39	.88785	1.89	.95838	2.39	1.23412	2.89	1.81134
1.40	.88726	1.90	.96177	2.40	1.24217	2.90	1.82736
1.41	.88676	1.91	.96523	2.41	1.25034	2.91	1.84359
1.42	.88636	1.92	.96877	2.42	1.25863	2.92	1.86005
1.43	.88604	1.93	.97240	2.43	1.26703	2.93	1.87673
1.44	.88581	1.94	.97610	2.44	1.27556	2.94	1.89363
1.45	.88566	1.95	.97988	2.45	1.28421	2.95	1.91077
1.46	.88560	1.96	.98374	2.46	1.29298	2.96	1.92814
1.47	.88563	1.97	.98769	2.47	1.30188	2.97	1.94574
1.48	.88575	1.98	.99171	2.48	1.31091	2.98	1.96358
1.49	.88595	1.99	.99581	2.49	1.32006	2.99	1.98167
1.50	.88623	2.00	1.00000	2.50	1.32934	3.00	2.00000

Appendix–7C
Boolean Algebra

Problems involving binary variable use Boolean algebra. A binary variable has only two values, denoted by 1 and 0, or ***A*** and $\overline{A}$, or ***true*** and ***false***, or ***high*** and ***low***, or ***switch closed*** and ***switch open,*** among other things. In algebra involving binary states, the plus sign, + is used to denote the ***or*** function, and the multiplication sign "**.**" is used to denote the ***and*** function. These two signs are called *logical sum* and *logical product,* respectively. The **+** and **–** signs used in this context will not follow conventional arithmetic rules. The following theorems are summarized here for easy reference:

$$1 \cdot 1 = 1$$
$$1 + 1 = 1$$
$$1 \cdot 0 = 0$$
$$1 + 0 = 1$$

Let *A*, *B*, and *C* be Boolean variables. Then

$$A \cdot 1 = A$$
$$A + A = A$$
$$A \cdot 0 = 0$$
$$A + 0 = A$$
$$A \cdot A = A$$
$$A + 1 = A$$
$$A + \overline{A} = 1$$
$$A \cdot \overline{A} = 0$$
$$A + AB = A$$
$$A(A + B) = A$$

$(A + B) + C = A + (B + C)$ [Associative law]

$(AB)C = A(BC)$ [Associative law]

$A + B = B + A$ [Commutative law]

$A \cdot B = B \cdot A$ [Commutative law]

$A(B + C) = AB + AC$ [Distributive law]

$A + BC = (A + B)(A + C)$ [Distributive law]

$\overline{\overline{A}} = A$

$\overline{A + B = \overline{A}\overline{B}}$ [De Morgan's law]

$\overline{AB} = \overline{A} + \overline{B}$ [De Morgan's law]

$$A + \overline{A}\,B = A + B$$
$$A(\overline{A} + B) = AB$$
$$(A + B)(\overline{A} + C) = AC + \overline{A}\,B$$
$$(AC + B\overline{C}) = (\overline{A}\,C + \overline{B}\,\overline{C})$$

CHAPTER 8

Quality by Experimental Design

8.1 INTRODUCTION

An experiment is defined as an analytic study designed to provide a basis for action. In case of planned experimentation, systematic procedure is employed to carry out experiments or test under specified conditions. There are different types of experimental designs one may opt for. The analysis of data from the planned experimentation is done keeping in mind the purpose of doing experiments. While designing and carrying out experiments, a number of terms associated with experiments need to defined. The important ones are as follows:

(*i*) *Response or dependent variables:* This is variable that is observed or measured in an experiment is also known as the output. This is the outcome of an experiment and is often a quality characteristic or a measure of performance of the process. An experiment may have one or multiple response variables.

(*ii*) *Factor:* A variable that is deliberately varied or changed in a controlled manner in an experiment to observe its impact on the response variable is known as a factor. It is also called an independent or causal variable. The factor can be either qualitative (for example, part 1, 2, or 3 *A, B,* or *C*) or quantitative (for example, a temperature of 90°, 100°, or 110°C).

(*iii*) *Noise variable:* A variable that potentially can affect a response variable in an experiment but is not an interest as a factor. This variable is considered to be uncontrollable in nature. Typical noise variable are lot, time, operator, cavity within a mold, and instrument, and in an experiment, they can be controlled by several means. In order to measure its effect the noise variable should be held at prespecified values during experimentation.

(*iv*) *Nuisance variable:* This variable can affect a response variable in an experiment. A nuisance variable is basically a noise variable that is unknown at the time the experiment is planned. A nuisance variable appears in an experiment as noise. The impact of nuisance variable can be minimized during experimentation.

(*v*) *Experimental unit:* Any combination of factors at which an experiment is conducted is known as experimental unit. Then, two units may have different combinations. Examples of experimental units are parts, batches, or an individual person.

(*vi*) *Blocks:* There are groups of experimental units treated similarly in an experimental design. Blocks are usually defined by noise variables. The variation of a response variable within a block is expected to be less than the variation within the entire experiment. For example, experimental units produced and tested at one time (a block defined by time) might be expected to vary less than experimental units produced at other time periods.

(*vii*) *Level:* This refers to a given value or specific setting of a quantitative factor or a specific option of a qualitative factor that is included in the experiment. The levels of a factor selected for study in the experiment may be fixed at certain values of interest, or they consist of a random selection from many possible values.

(*viii*) *Effect:* The change in the response variable that occurs as a factor or noise variable is changed from one level to another. The effect must be further described in terms of the context in which it is used, such as main effects and interaction effects.

The distinction between response variables, factors, and noise variables is especially important in planning an experiment. The factors and background variables can be thought of as causes and the response variables as effects. The objective of the experiment should clarify the distinction between response variables, factors, and background variables.

In specific terms, the experiments are conducted to study the performance of processes and systems. Such a process or a system may be represented by a model, the schematic diagram of which is shown in **Figure 8.1**.

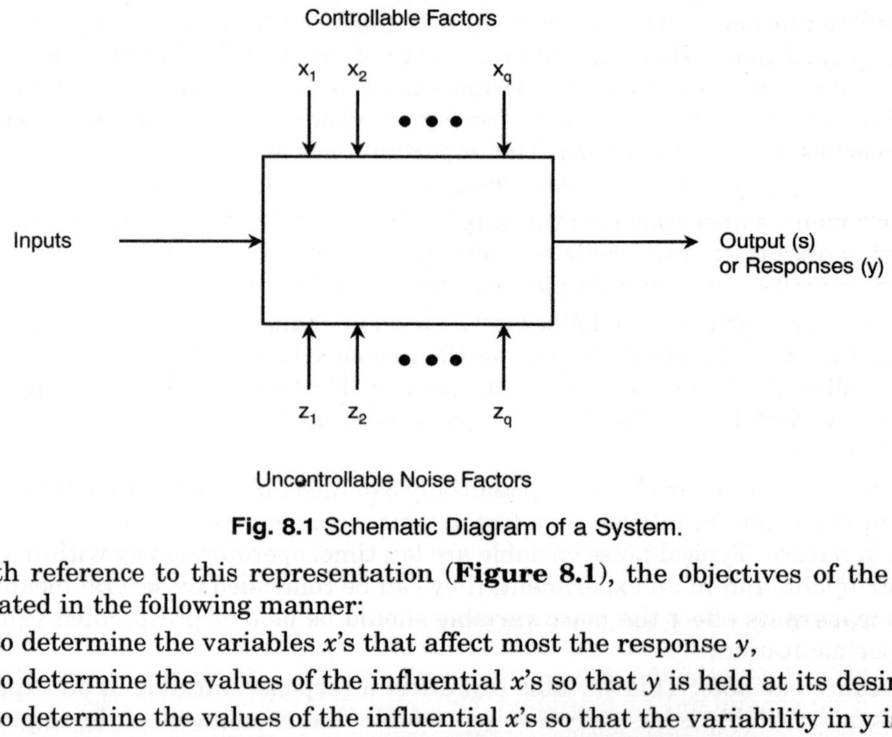

Fig. 8.1 Schematic Diagram of a System.

With reference to this representation (**Figure 8.1**), the objectives of the experiment may be stated in the following manner:

(*i*) to determine the variables x's that affect most the response y,

(*ii*) to determine the values of the influential x's so that y is held at its desired value,

(*iii*) to determine the values of the influential x's so that the variability in y is small, and

(*iv*) to determine the values of the influential x's so that the effects of z's are minimized.

8.2 IMPORTANT FEATURES OF EXPERIMENTAL DESIGN

Experiments on processes or products are very common in all types of industrial environments, particularly in production system. Experiments range from very informal investigations to well-planned studies involving a series of specific experiments. In order to provide an effective basis for action, an experimental design must have certain features. The British scientist, R.A. Fisher proposed the principles of experimentation in late forties of the twentieth century. Many other authors and researchers have contributed immensely in this area in the next sixty years. Among the many features that a good experiment would possess, the following are most important:

(*i*) An experiment must have a clearly defined objective that spells out planned action based on the results.

(*ii*) Experimentation should proceed sequentially with knowledge gained in previous experiments used to design new experiments.

(*iii*) The variations in the response variables in terms of factors and noise variables considered should be explained in an experiment. For this variation in the response variable are to be partitioned into components due to factors and noise variables.

(*iv*) Conclusions from the experiment should be made with an adequate degree of belief. The study should be conducted under a range of conditions to increase the degree of belief.

(*v*) The experimental design should be as simple as possible, while still satisfying the first four properties.

The above mentioned features are briefly discussed below.

Objectives of the Experiments

At the outset, the objectives of an experiment should be stated in clear and unambiguous terms. This will ensure efficient allocation of resources for designing experiment. The objective should clarify whether the experiment involves screening a large number of variables to find the most important ones, studying in depth a few variables, or confirming the results of past studies under new conditions. All experiments are conducted with a view to improving the existing state of the system with reference to some critical aspects. Thus, a statement of the result required for taking action may be considered in stating the objective. "Identify factors that can be used to improve productivity by 7%", "determine if two machines differ in their performance", and "identify the main factors that have the greatest effect on the variations in component quality characteristic", are some of the typical objective statements that refer to the improvement goal of a given system.

The objective of the experiment is also to make clear the analytic nature of the experiment and identify the various courses of action that can be taken based on the results of the study. Once the improvement goals are identified, it becomes easier to identify the response variables and appropriate factors that are to be considered for experimentation.

Sequential Approach in Experimentation

Experiments need to be planned in a sequence. The sequential nature of learning in analytic studies should be considered in planning. The quality improvement model emphasizes on continuous improvement in process performance. Through experimentation process knowledge is gained through interactive approach. Many iterations of the cycle will include the design of an experiment in the planning phase of the cycle. Studies in the initial phase of improvement of quality will be screening studies to focus on the key factors.

The general approaches in planning and conducting experiments are as follows:

1. Best-guess approach,
2. One-factor-at-a-time approach,
3. Factorial design approach, and
4. Fractional factorial design approach.

The first two approaches are often adopted in practice. But they have a number of shortcomings. In a best-guess approach, arbitrary combinations of factors are selected to test their effect on responses, and change the levels of one factor keeping other factors at their previous levels, combination of factors is made. This approach may continue for a long time without any guarantee of success, and one may be tempted to stop experimentation if the initial best-guess produces an acceptable result. In one-factor-at-a-time approach, a baseline set of values for each factor, is selected, and each factor is successively varied over its range with other factors held constant at the baseline level. However, this approach fails to consider any possible interactions between the factors, and moreover it is not very efficient approach.

A factorial design allows all the factors to vary simultaneously and thus gets over the disadvantages of the above-mentioned approaches. But this approach requires a large number of runs when the number of factors is high. A fractional factorial design allows only a fraction of the total number of runs required in the factorial design.

There are several applications of experimental design. It is applied for the development of new processes, having improved process yields, reduced variability and closer conformance to nominal (or target) requirements, reduced development time, and reduced overall costs. It is also a very effective tool in engineering design in the context of evaluation and comparison of basic design configurations, evaluation of material alternatives, selection of design parameters so that the product performs well under a wide variety of field conditions, and design of key product design parameters that impact product performance. Additionally, with the conduct of experiments, the working of a process or a product or a system is learned in depth.

As knowledge is gained, experiments will be repeated using new levels of factors previously studied and some new factors. If it is desired to study in depth the relationships between the factors, two to four factors can be studied in a factorial design. A fractional factorial design can be used for five factors. As new theories are developed, experiments will be confirmatory in order to increase the degree of belief from previous experiments.

Partitioning of Variation in Response Variable

Determination of important factors and estimation of the effects of these factors on the response variables are important in any experiment. The factors chosen for the experiment are usually those that the experimenter believes will have the greatest effect on the response variable. In many experiments, the variation due to noise variables may be as great as or greater than the variation due to the factors chosen. To help determine if the most significant factors have been studied, the experimental design must allow the variation in the response variable to be partitioned into components due to factors, due to noise variables, and due to other variables.

Degree of Belief

The wider the range of conditions included in the experiment, the more generally applicable will be the conclusions from the experiment. The degree of belief in the validity of the conclusions is increased by running the experiment using different machines, different operators, different days, different times of the year, different batches of raw materials, and so on.

The range of conditions selected for the study ultimately determines the degree of belief in the actions taken as a result of the experiment. The expert in the subject matter must determine what is an "adequate" degree of belief for taking action. This determination will depend on the magnitude of the change in the process being considered and the degree of extrapolation necessary. Changes in existing manufacturing processes may require a completely different degree of belief than changes in the process of development of a new product.

Simplicity of Experiment

The simplicity of the experiment should be considered as one of the most important properties of a planned study to improve a process. Simplicity is important in the design, the conduct, and the analysis of a planned experiment. An experimental design should be as simple as possible while still satisfying the other properties of a well-planned experiment. Simplicity allows all interested parties to be involved in all aspects of the study. Simplicity also allows the experimenter the flexibility to adjust for changes that are often required during the conduct of the study.

Simplicity requires that all the practical aspects of conducting an experiment be considered. Some important aspects include the degree of difficulty in changing levels of a factor, the ability to control background variables, and the ability to measure important response variables.

8.3 BASIC PRINCIPLES OF EXPERIMENTAL DESIGN

A properly-planned experiment must possess the above mentioned features. While planning an experiment, it is to be ensured that these features are included. A number of approaches may be used for this purpose. The approaches are as follows:

(*i*) *Experimental Plan:* The arrangement of factor levels and experimental units in the design.

(*ii*) *Grouping Methods:* Blocking of experimental units.

(*iii*) *Randomization:* The objective assignment of combinations of factor levels to experimental units.

(*iv*) *Replication:* Repetition of experiments, experimental units, measurements, treatments, and the other components as part of the planned experiment.

Each of these approaches is briefly discussed below.

Experimental Plan

It is defined as the schedule for conducting the experiment. Each test to be conducted or measurement to be made is identified in the plan. Usually, three aspects are considered in the plan. These aspects are grouping rules, randomization, and replication.

The selection of an appropriate plan is the primary tool for assuring the attainment of the objectives of the experiment. The test plan identifies the factor combinations and factor levels to be included in the study. An appropriate test plan results in establishment of several good features, such as

(*i*) an acceptable or high degree of belief for the results obtained from the study,

(*ii*) accurate estimation of important effects,

(*iii*) the analysis of data from the study, is simplified, and

(*iv*) costs incurred and other resources required to conduct the experiment are most likely in control.

There are many types of experimental plans. One of the most commonly used experimental plan to improve experimental quality is the factorial design. In a factorial design, tests are arranged in a pattern such that multiple factors are studied with each test.

In a nested or hierarchical design, levels of different factors are studied within a given level of another factor. In the fractional factorial design and the unbalanced or partially balanced nested design, only a subset of a full factorial or nested experimental pattern is considered.

In the composite designs, combination of factorial and nested designs are used. Designs for experiments involving mixtures are another important class.

Grouping Method

The method of grouping is an important tool for addressing noise variables in an experiment. There are two important decisions to make concerning noise variables:

(*i*) How to control the noise variables so that the effects of the factors are not distorted by them, and

(*ii*) How to use the noise variables to establish a wide range of conditions for the study to increase the degree of belief or to aid in designing a robust product or process.

Three methods of controlling noise variables are used:

(*i*) hold them constant in the study,

(*ii*) measure them and adjust for their effects in data analysis, and

(*iii*) use planned grouping to set up blocks.

The concept of planned grouping takes advantage of naturally occurring homogeneous experimental units in the conduct of an experiment. When planned grouping is used in developing the test plan, the design is called a block design.

Blocking means grouping of relatively homogeneous experimental conditions. It helps in reducing or eliminating the variability transmitted from *nuisance variables* (factors that may not be of any interest, but that influence the experimental response), and improving the precision with which factor comparisons can be made.

Examples of common noise variables used to form blocks are time of day, operator, machine, shift, batch of raw materials, and month. In studying tire wear for cars during normal use, an individual car is a natural blocking variable. The amount and type of driving for different cars could vary considerably. Different brands (or factor combinations) of tires could be tested on any particular car to minimize the car-to-car variation when comparing the brands. If the natural block size is four, four different brands could be "grouped" on each car. In studying the wear of soles of shoes, a natural block size of two would correspond to the two feet of each individual wearing the shoes.

In an analytic study, planned grouping plays a very important role in establishing the degree of belief for the results of the study. Should the study be conducted on one machine in the plant or on multiple machines? Including different machines will add additional variation but will also increase the degree of belief that the results can be extrapolated to all machines of that type in the plant. By treating the machines as blocks, the experiment can include multiple machines to increase the degree of belief in the results through analysis of interactions between the machines and the factors in the study.

Grouping important noise variables into blocks is an important part of designing an experiment. If the sizes of parts can possibly alter the effect of certain factors in the study, how can we incorporate part size in the experiment? One way is to form two groups of parts based on their measured sizes. One group would be parts at the low end of the size specification, and the other group would be parts at the high end of the size specification. Combinations of factors can be tested using parts from each of these planned groups or blocks.

Randomization

Randomization means random determination of the allocation of the experimental material and the order in which the test runs are performed. It helps in validating the assumption that observations and errors are independently distributed random variables, and averaging out the effects of extraneous effects, if any.

Randomization is the use of random numbers to determine the assignment of factor combinations to experimental units. Whereas an experimental design addresses the variables identified as factors in the experiment, and planned grouping is used to accommodate noise variables, randomization addresses the nuisance variables, that affect the response variables but that have not been identified by the experimenter. It is most likely that many variables that can affect the response variables may not be known to the experimenter. Typical nuisance variables are environmental effects such as temperature or humidity, drifts in measurement equipment, and batch-to-batch variation in raw materials.

Randomization helps prevent the variation due to nuisance variables from being confused with the variation due to the factors or the noise variables. If important unknown nuisance variables are present during an experiment, randomization can allow their effects to be separated from the effects of factors or noise variables. Diagnostic analyses to uncover nuisance variables are also facilitated by randomization. If randomization is not done, the experimenter may develop incorrect conclusions from the study results.

For this, random number tables should be used. Mechanical devices such as flipping a coin, rolling a pair of dice, and pulling numbers (after mixing) from a container that simulate a random process can also be used.

In most situations it is not desirable to completely randomize the conduct of an experiment. In an analytic study, randomization should be restricted as much as possible based on knowledge of background variables. Planned grouping is the most common restriction put on randomization. When blocking is employed, the randomization is done within each block rather than overall tests in the design. For example, an experiment might require 20 parts, 5 from each of four machines. Randomization would be used to assign combinations of factors to the 5 parts for each machine.

Sometimes, randomization may be expensive. For example, changing the levels of one particular factor might require a machine to be shut down for a day, whereas changing all the other factors could be done in a few minutes. Extra care is required in analysis and interpretation in these situations.

The decision of what type of level of randomization to use should be based on the objective of the experiment. The advantages of randomization should be evaluated against the additional resources required to randomize. In the following situations, in particular, randomization is especially important:

(*i*) the important response variables are not into a state of statistical control,

(*ii*) experiments need to be conducted by many different operators and on different resources and machines, and

(*iii*) the variation due to nuisance variables is large relative to the magnitude of effects of important factors.

Replication

Replication refers to repeating particular aspects of an experiment. It means doing an experiment a number of times with same set of factors and their levels. Replication plays an important role in attaining many of the properties of a good experiment. It is the primary tool for studying stability of effects and for increasing the degree of belief in the results. It helps in estimating the experimental error that in turn, helps in determining if the observed differences are statistically different, and permitting a more precise estimate of the effect of a factor. There are many different types of replication. A few of them are as follows:

(*i*) repeated measurements of experimental units,

(*ii*) multiple experimental units for each combination of factors,

(*iii*) partial replication of the experimental pattern,

(*iv*) complete replication of the experimental pattern.

Each of these types of replication requires different interpretations in the analysis of the results of an experiment. For example, for comparing differences between factor combinations, replications that include all sources of variation typical of a run of either combination are desirable.

Replication plays the primary role in providing a measure of the magnitude of variation in the experiment due to nuisance variables. Replication also aids in minimizing the impact of nuisance variables on factor affected by (*i*) possibly enabling nuisance variables to be averaged out, and (*ii*) allowing the study of interactions between factors and experimental conditions. Replication, in conjunction with planned grouping, can be used to expand the study to a wide range of conditions.

Since the amount of replication in an experiment directly affects the resources (mainly time, budget, and materials) required for the study, replication is often established by constraints on these resources. Usually it is desirable to obtain 5 to 10 comparisons for the different levels of each factor for the initial stages of the study.

The most important consideration in an experiment is not the number of experimental units per level of a factor, but the breadth of conditions under which the comparisons can be made. Replication over similar conditions provides little increase in the degree of belief. For example, in a high-speed manufacturing operation, a study based on 5 to 10 experimental units selected from different days' production and made from different material lots would be much preferred to a study based on 50 consecutively produced units.

8.4 EXPERIMENTAL DESIGN METHODS

As has been mentioned, an experiment is a planned method of inquiry conducted to support or refute a hypothesized belief or to discover new information about a product, process, or service, and in this method, the effects of the controllable factors on the responses are studied. The factors are also known as parameters. Examples of parameters are the type of materials, the spindle speed of a machine, the temperature setting of a furnace, type of coolant used, and the supplier selected for a process. For quantitative factors, the range of values, how they are to be measured, and the levels at which they are to be controlled during experimentation are required to be known. For example, the effect of temperature on the output of a product may be required to be studied through experimentation. Three selected levels of the factor may be 100, 200, and 300°C, and the raw material can be purchased from vendor *A* or vendor *B*. The treatments for this experiment are the various combinations of these two factor levels. The vendor *A* with 100°C oven temperature is one treatment. A total of six possible treatments can be considered in this case. While conducting an experiment, the effect on a response variable for a chosen treatment combination is measured. If this effect is measured for the whole batch of production, the experimental unit becomes the population. In most case, consideration of the whole unit may not be necessary. In this case, a sampling unit is a fraction of the experimental unit on which the treatment effect is measured.

The variation in experimental units that have been exposed to the same treatment is attributed to experimental error. This variability is due to uncontrollable noise factors.

In majority of the cases, more than one factor is considered in experimental analysis. In such cases, experiments do not usually involve changing one variable at a time. With two variables the experimental design is presented in **Figure 8.2(*a*)**. Each factor can be set at two levels, denoted by –1 and +1. The point (0, 0) may represent the current setting of the factor levels. In this figure, the first coordinate represents the level of factor *A*, and the second coordinate the level of factor *B*. *A* level of –1 represents a value below the current setting of 0, and a level of +1 represents a value above the current setting. Thus, four new design points, viz., (– 1, –1), (+1,–1), (–1, +1), and (+1, +1), can be investigated. The value of the response variable at these design points represents the results of experiments performed.

The response function value at the experimental design points is shown in **Figure 8.2(*b*)** and the response surface shown in **Figure 8.2(*c*)**. The objective is to maximize or minimize the response function. Accordingly, the direction at which the gradient is steepest is to be considered.

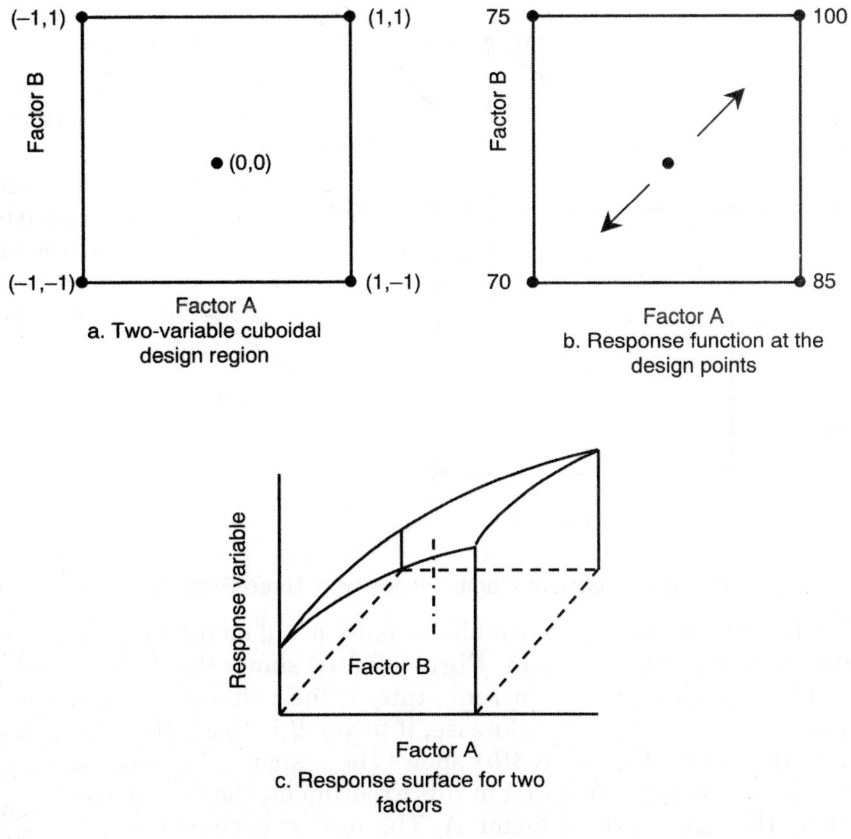

Fig. 8.2 Experimental Design with Two Variables.

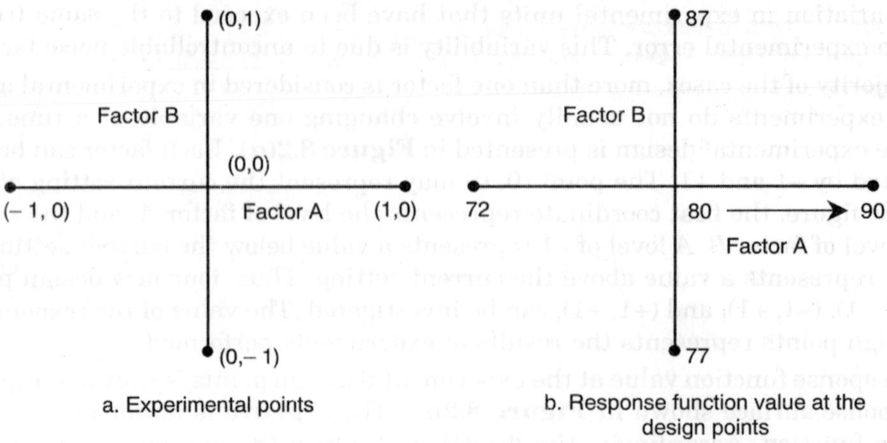

Fig. 8.3 Design Approach Involving Changes in One Variable at a Time.

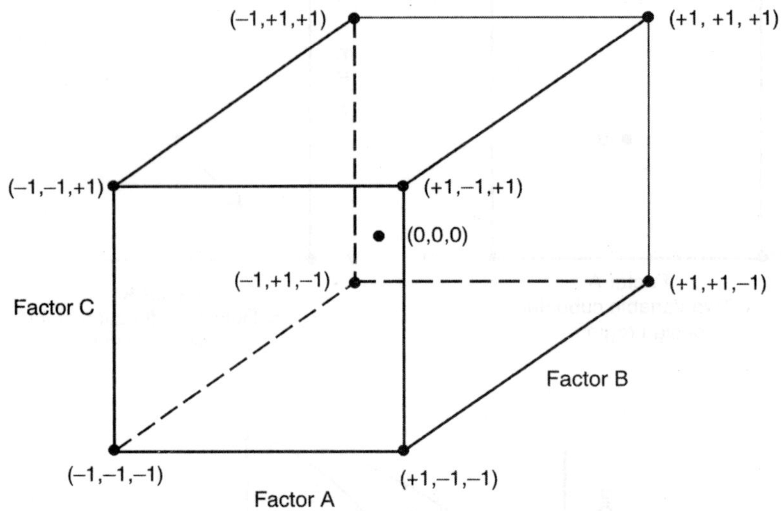

Fig. 8.4 Treatment Combinations with Three Factors.

Experiments with two or more variables usually avoid changing only one variable at a time keeping the other variables constant. **Figure 8.3(*a*)** shows the design points for such an approach with (0, 0) representing the current state. If the value of factor *A* is fixed, the two design points are (0, – 1) and (0, + 1). Likewise, if factor *B* is fixed, the two additional design points are (– 1,0) and (+1,0). **Figure 8.3(*b*)** shows the response function value at the design points. To maximize the response function in this experiment, the direction of the arrow shown should move along the major axis of factor *A*. The next experiment may be conducted in a region with the center at (2, 0) with respect to the current origin. When three variables need to be considered, the same approach as outlined for two variables may apply. As shown in **Figure 8.4**, there are nine treatment combinations with factors *A*, *B*, and *C*. To obtain the same amount of information given in this design using the one-variable-at-a-time approach, four sets of experiments, each with seven design points, need to be run. Thus, experiments with one-variable-at-a-time are not cost-effective. However, the main drawback of the one-variable-at-a-time approach is that it cannot detect interactions between two or more factors. When the relationship between the response variable and a certain factor is influenced by the levels of some other factors, there exists an interaction effect. This concept is illustrated graphically in **Figure 8.5.**

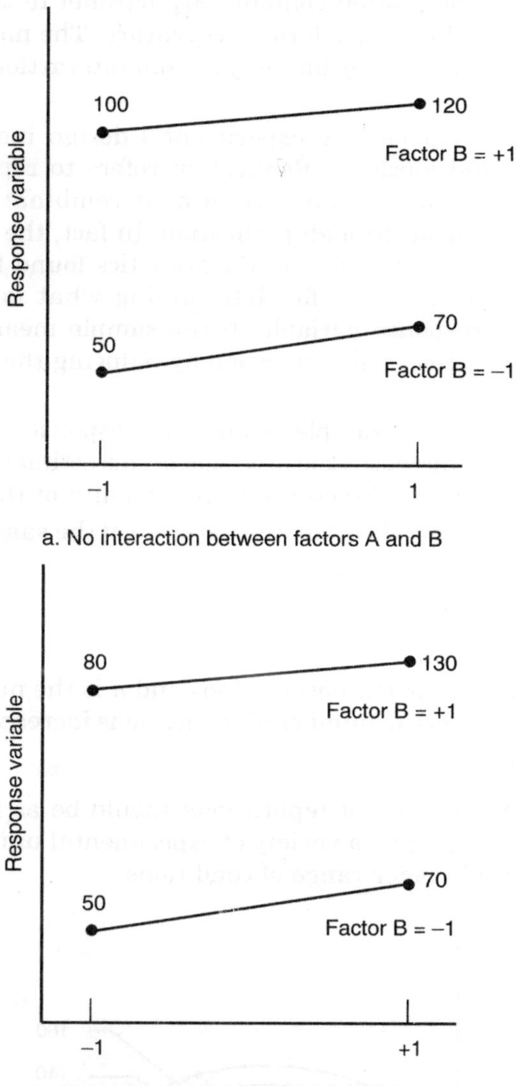

Fig. 8.5 Interaction between Factors.

Suppose, there are two factors, *A* and *B*, considered to be influencing a response variable. As shown in **Figure 8.5(*a*)**, the functional relationship between the response variable and factor *A* remains same regardless of the level of factor *B*. Hence, there is no interaction between the factors *A* and *B*. In **Figure 8.5(*b*)**, a different situation is depicted. The level of factor *B* influences the relationship between the response variable and factor *A*. Hence, there exists interaction between factors *A* and *B*.

Interactions show the joint relationship of factors on the response function. Such effects should be accounted when multiple factors are considered in experiments. For example, a new training programme for employees (factor *B*) may impact productivity (response variable) differently for different departments (factor *A*) within the organization. With the help of contour plots of the response function, the interaction effects can also be depicted. **Figure 8.6(*a*)** shows contour plots of the response function with factors *A* and *B* when there is no interaction between the factors. Here the axes of the plotted contours are parallel to the axes of the factors. In **Figure 8.6(*b*)**, the response surface has a form of curvature. The natural axes of the contours are rotated with respect to the factor coordinate axes, and interaction between factors *A* and *B* exists.

As has been mentioned earlier, the experimental design incorporates the features of replication, randomization, and blocking. Replication refers to repeating the conduct of an experiment a number of times at the same treatment combination, and estimate of the experimental error can be obtained through replication. In fact, the experimental error forms a basis for determining whether differences in the statistics found from the observations are significant, and serves as a benchmark for determining what factors or interactions are influencing significantly the response variable. If the sample mean is used to estimate the effect of a factor, replication increases the precision by reducing the standard deviation of the mean.

In an experimental design, the sample mean of the response variable is used to estimate the effect of a factor. Replication increases the precision of this estimate by reducing the standard deviation of this mean. This follows directly from the validity of the Central Limit Theorem during experimentation. As per this theorem, the variance of the sample mean, $\bar{y}$ of a response variable, y is given by

$$\sigma_{\bar{y}}^2 = \frac{\sigma^2}{n}$$

where, σ^2 represents the variance of the observations and n is the number of replications used to determine the sample mean. As the number of replications increases, the standard deviation of the mean ($\sigma_{\bar{y}}$) decreases.

For randomization the number of replications should be as high as possible. With an increase in the number of replications, a variety of experimental units can be used, permitting inferences to be drawn over a broader range of conditions.

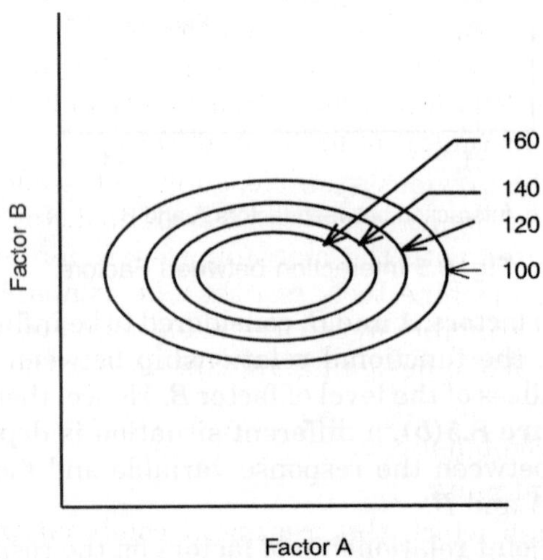

a. No interaction between factors A and B

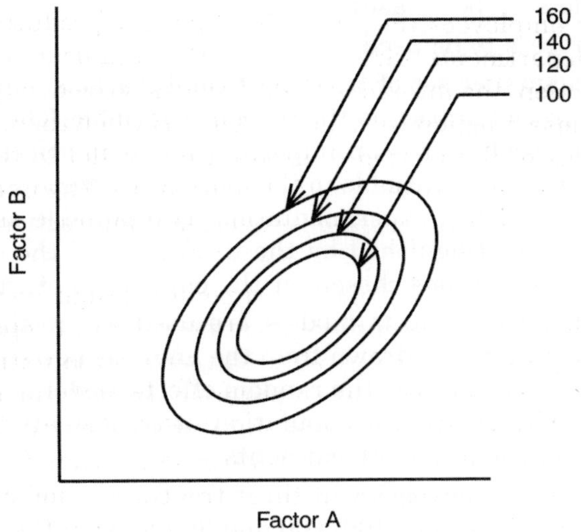

b. Interaction between factors A and B

Fig. 8.6 Contour Plots of the Response Function.

Randomization means that the treatments should be assigned to the experimental units in such a way that every unit has an equal chance of being assigned to any treatment. It thus eliminates bias and ensures that no particular treatment is given more importance or preference. Random assignments of the treatments and the order in which the experimental trials are run ensure that the observations and therefore the experimental errors may be independent of each other.

For example, the surface finish of a component is to be tested for three types of cutting tools. The factor is the type of cutting tool with three levels. The components which are processed with these tools should be randomized so that the effects of all other factors are averaged out. Any difference in the observed surface finish is thus attributed to differences in the types of cutting tools. When the principle of blocking is applied, similar experimental units are arranged into groups, or blocks. Variability of the response function within a block can then be attributed to the differences in the treatments as the impact of other extraneous variables has been minimized. Treatments are assigned at random to the units within each block. Variability between blocks is thus eliminated from the experimental error, leading to an increase in the precision of the experiment. Blocking allows treatments to be compared under nearly equal conditions because the blocks consist of uniform units.

As an extension of the previous example, suppose the effect of three different machine tools on the surface finish of component is to be evaluated. In this case, blocks consisting of different feed rates associated with the machining operation may be formed. The effect on surface finish of the three machine tools can then be compared within blocks. Differences between blocks due to the variability in the feed rates would be distinct from determining the effect of the machine tool. While designing an experiment, the following aspects need to be considered:

(*i*) set of treatment combinations,

(*ii*) set of experimental units,

(*iii*) procedures through which the treatment combinations are to be assigned to the experimental units, (*iv*) response variable(s) to be measured, and

(*v*) measurement process to be used.

In the beginning, a smaller set of treatment combinations is chosen from a larger set based on their relative degree of impact on the response variable. The cost associated with the application of a particular treatment is an important factor in selecting treatments. Ideally, the goal is to optimize the response variable in the most cost-effective manner. While selecting the treatment combinations during experimentation, two possible situations one may come across. These situations are distinguished by the usefulness of the experimentation for the sample of the treatment combinations chosen, or the population of treatment combinations.

In the first case, the fixed effects models are used where the treatment effects are considered to be fixed. The inferences drawn from the analysis pertain only to the treatments selected in the study. In the second case, the random effects model is used, where the selected treatments are randomly chosen from a population of treatments. This model is to make inferences about the whole population of treatments.

For example, suppose in an experiment three treatment combinations are the quantity of a chemical compound at three levels: 100, 200 and 300 *g*. An infinite number of treatments can be generated. From this population of treatments, the three treatments are randomly selected, and inferences are made on a population of treatments that is normally distributed. This is a random effects model.

8.5 COMPLETELY RANDOMIZED EXPERIMENTAL DESIGN

Among many types of designs recommended a completely randomized design is the most simple one. In this design, treatments are randomly assigned to the experimental units, and each unit has an equal chance of being assigned to any treatment. Assignment is made through the use of random number table, where the random numbers are generated from the uniform distribution.

Suppose there are four treatment combinations, denoted as *A*, *B*, *C*, and *D* and each treatment is replicated three times. Hence, there are a total of 12 experimental units. For assignment of these units randomly, 12 random numbers are required. Suppose the following three-digit random numbers are selected from the random number table: 984, 618, 240, 348, 655, 126, 952, 847, 076, 593, 744, and 351. Treatment *A* is assigned to first three numbers, treatment *B* the second three numbers, and so on. Once these numbers are ranked (from smallest to largest), these ranks correspond to the experimental unit numbers. Thus, the experimental unit 1 is assigned to treatment *C*, experimental unit 2 to treatment *B*, and so on. Thus, the time order of the data collection and experimentation is known.

A number of merits of this design need to be mentioned. First, any number of treatments or replications can be used in this design. Different treatments need not be replicated the same number of times making the design flexible. Sometimes it is not feasible to replicate treatments an equal number of times. Suppose, the treatments represent different machines. Machine-1, because of its capacity and resource advantages may run as many times as desired, whereas machine-2 and 3 cannot be run that frequently. Whereas treatments-1 can be replicated ten times each, treatments 1 and 2 can only be replicated only five times. This experimental design is an 'unbalanced' one. When all treatments are replicated an equal number of times, the design is known as a 'balanced' one. However, in this design, the precision of the experiment may be at a low level, if the experimental units are not uniform. The principle of blocking may be used to overcome this problem.

The statistical procedure used to analyze the experimental data is known as the analysis of variance or in short, ANOVA, which determines the effects of the treatments, as reflected by their means, through an analysis of their variability. In this technique, the total variability in the observations is partitioned into two components, viz., the variation among the treatment means (or treatment sum of squares) and the variation among the experimental units within treatments (or error sum of squares). Thus,

Total sum of squares (SS_T) = Treatment sum of squares (SS_{TR}) + Error sum of squares (SS_E)

The mean squares for treatments and for error are obtained by dividing the corresponding sum of squares by the appropriate number of degrees of freedom. This number is 1 less than the number of observations in each source of variation. For example, for a balanced design with p treatments, each with r replications, the total number of observations is rp. The total variability has (rp – 1) degrees of freedom. The number of degrees of freedom for the treatments is (p – 1). For each treatment, there are r observations. Hence, (r – 1) degrees of freedom apply toward the experimental error. The total number of degrees of freedom for the experimental error is, therefore, $p(r - 1)$.

The following notations are used for this design:

p = Number of treatments

r = Number of replications for each treatment

y_{ij} = Response variable value of the jth experimental unit that is assigned treatment i, (i = 1, 2,....., p, and j = 1, 2,....., r)

$y_{i\cdot}$ = Sum of the responses for the ith treatment; that is, $\sum_{j=1}^{r} y_{ij}$

$\bar{y}_{i\cdot}$ = Mean response of the ith treatment = $\frac{y_{i\cdot}}{r}$

$y_{\cdot\cdot}$ = Grand total of all observations = $\sum_{i=1}^{p}\sum_{j=1}^{r} y_{ij}$

$\bar{\bar{y}}$ = Grand mean of all observations = $\frac{y_{\cdot\cdot}}{rp}$

The data collection formats for p number of treatments with r number of replications for each along with the notations used for means are shown in Table 8.1.

With the given notations, the total sum of squares is computed as,

$$SS_T = \sum_{i=1}^{p}\sum_{j=1}^{r} y_{ij}^2 - C \qquad ...(8.1)$$

The treatment sum of squares is given as

$$SS_{TR} = \frac{\sum_{i=1}^{p} y_{i\cdot}^2}{r} = C \qquad ...(8.2)$$

where, C = a correction factor = $\frac{y_{\cdot\cdot}^2}{rp}$ ...(8.3)

The error sum of squares is given as

$$SS_E = SS_T + SS_{TR} \quad \text{...(8.4)}$$

The mean square for treatment is given by

$$SS_{TR} = \frac{SS_{TR}}{p-1} \quad \text{...(8.5)}$$

and the mean square error is given by

$$MS_E = \frac{SSE}{p(r-1)} \quad \text{...(8.6)}$$

In experimentations, for different treatments considered it may be desirable to know if the treatment means are a function of treatments, or in other words, if the values of treatment means as obtained are varying from one treatment to another. Hence, it is desirable to test the null hypothesis that the treatment means are equal against the alternative hypothesis that at least one treatment mean is different from the others. Thus, the hypotheses are as follows:

$H_0 : \mu_1 = \mu_2 = \ldots = \mu_p$

H_1 : at least one treatment mean (μ) is different from the others

In this case, the test statistic (F-statistic) is computed as

$$F = \frac{MS_{TR}}{MS_E} \quad \text{...(8.7)}$$

With $(p - 1)$ degrees of freedom in the numerator and $p(r - 1)$ degrees of freedom in the denominator.

Table 8.1. Data Collection Format in a Completely Balanced Randomized Design

Treatment	Replication				Sum	Mean
	1	2	...	r		
1	y_{11}	y_{12}	...	y_{1r}	$y_{1\cdot}$	$\bar{y}_1\ (\mu_1)$
2	y_{21}	y_{22}	...	y_{2r}	$y_{2\cdot}$	$\bar{y}_2\ (\mu_2)$
⋮	⋮	⋮	⋱	⋮	⋮	⋮
p	y_{p1}	y_{p2}	...	y_{pr}	$y_{p\cdot}$	$\bar{y}_p\ (\mu_p)$
					$y_{\cdot\cdot}$	$\bar{\bar{y}}$

For a chosen level of significance α, the critical value of F, (obtained from the table) is denoted by $F_{\alpha,\,(p-1),\,p(r-1)}$. If the computed test statistic $F > F_{\alpha,\,(p-1),\,p(r-1)}$, the null hypothesis is rejected, and we conclude that the treatment means are not all equal at the chosen level of significance. This computational procedure is known as analysis of variance (ANOVA). The typical format of an ANOVA table for this experiment is shown in Table 8.2.

If the null hypothesis is not rejected, there is no significant difference among the treatment means. It means that the response variable is not affected by any treatment, and hence, the most cost-effective treatment should be selected.

Table 8.2. ANOVA Table for a Completely Balanced Randomized Design

Source of variation	Degree of freedom	Sum of squares	Mean square	F-statistic
Treatments	$p-1$	SS_{TR}	$MS_{TR} = \dfrac{SS_{TR}}{(p-1)}$	$F = \dfrac{MS_{TR}}{MS_E}$
Error	$p(r-1)$	SS_E	$MS_E = \dfrac{SS_E}{[p(r-1)]}$	
Total	$rp-1$	SS_T		

If the null hypothesis is rejected, management might be interested in knowing which treatments are preferable. Each sample treatment mean is a point estimate of the population treatment mean. If interval estimates are desired, a 100(1 – α)% confidence interval for the treatment mean can be computed. For the *i*th treatment mean, μ_i, the confidence interval (CI) is given by

$$CI(\mu_i) = \bar{y}_i \pm t_{\alpha/2,\, p(r-1)}\sqrt{\frac{MS_E}{r}} \qquad \text{...(8.8)}$$

where, $t_{\alpha/2,\, p(r-1)}$ is the *t*-value for a right-tail area of α/2 and $p(r-1)$ degree of freedom. The quantity $\sqrt{MS_E/r}$ represents the standard deviation of the sample treatment mean $\bar{y}_i$.

In order to know if one treatment is preferable to the other, a 100(1 – α)% confidence interval for the difference of two treatment means, needs to be computed. The confidence interval for the difference between two treatment means, μ_{i1} and μ_{i2} is given by,

$$L(\mu_{i1} - \mu_{i2}) = (\bar{y}_{i1} - \bar{y}_{i2}) \pm t_{\alpha/2,\, p(r-1)}\sqrt{\frac{MS_E}{r}} \qquad \text{...(8.9)}$$

where, $t_{\alpha/2,\, p(r-1)}$ is the *t*-value for a right-tail area of α/2 and $p(r-1)$ degrees of freedom. The quantity $\sqrt{MS_E/r}$ represents the standard deviation of the difference between the sample treatment means, $(\bar{y}_{i1} - \bar{y}_{i2})$.

Fixed and Random Effects

The terms "fixed" and "random" are used in the context of analysis of variance (ANOVA) and regression models, and refer to a certain type of statistical model. Almost always, researchers use fixed effects regression or ANOVA and they are rarely faced with a situation involving random effects analyses. A fixed effect ANOVA refers to assumptions about the independent variables and their error distributions. An experimental design is the easiest example for illustrating this principle. Usually, the researcher is interested in only generalizing the results to experimental values used in the study. For instance, a study on the effect of firing temperature (specifically chosen by the experimenter) using 100°C, 125°C, and 150°C on density of bricks. This is a case where a fixed effect ANOVA may be appropriate. In this case, the extrapolation is to other studies or treatments that might use the same values of the temperature factor (*i.e.*, 100°C, 125°C, and 150°C). In this situation, the experimenter wishes to test the hypothesis about the treatment means, and his/her conclusion may apply only to the factor levels considered in the analysis. The conclusion cannot be extended to similar treatments that are not explicitly considered. Alternatively, if the researcher wants to make inferences (based on sample of treatments) beyond the particular values of the independent variables used in the study, or to all treatments in the population, a random effects model is used. In such case, different treatments (temperature levels) could be a random sample from a large population of treatments. The experimenter assumes that the population of factor levels is either of infinite size, or is large enough to be considered infinite. Random effects models are sometimes referred to as "components of variance models." In this context, analyses using both fixed and random effects are sometimes referred to as "mixed models".

Fisher's LSD Procedure: Comparing Pairs of Treatment Means

An approach suggested by statistician R.A. Fisher (called the "least significant difference method" or Fisher's *LSD*) is to first test the null hypothesis that all the population means are equal using analysis of variance. The null hypothesis is expressed as

$H_o : \mu_i = \mu_j$ and the alternate hypothesis is

$H_1 : \mu_i \neq \mu_j$

for all $i \neq j$. If the analysis of variance is not significant, then the null hypothesis is rejected, implying significant differences among the means. If the analysis of variance is significant, then the means are compared with one another using an *F* statistic expressed as

$$t_o = \frac{\bar{y}_{i.} - \bar{y}_{j.}}{\sqrt{MS_E\left(\frac{1}{n_i} + \frac{1}{n_j}\right)}}$$

Assuming a two-sided alternative, the pair of means μ_i and μ_j is declared significantly different if

$$|\bar{y}_{i.} - \bar{y}_{j.}| > t_{\alpha/2,\, N-\alpha}\sqrt{MS_E\left(\frac{1}{n_i} + \frac{1}{n_j}\right)},$$

where, *N* is the total number of observations, n_i and n_j are the number of observations for *i*-th and *j*-th treatment, respectively, and α is the significance level of the test.

The mathematical expression

$$LSD = t_{\alpha/2,\, N-\alpha}\sqrt{MS_E\left(\frac{1}{n_i} + \frac{1}{n_j}\right)},$$

is called the least significant difference or *LSD*. In Fisher *LSD* procedure, the experimenter compares the observed difference between each pair of averages to the corresponding *LSD*. If $|\bar{y}_{i.} - \bar{y}_{j.}| > LSD$, the population means μ_i and μ_j are significantly different.

Example E8.1. *A company has to deliver its goods to distributors throughout the country. Three transportation companies have offered their services for this purpose. To test the efficiency of these three transporters, the company randomly assigns its outgoing product shipments to them and determines the degree of lateness as a proportion of time allocated for delivery. Table E8.1 depicts the values of the degree of lateness of the transporters for five shipments.*

Table E8.1. Values of Degree of Lateness of the Transporters

Transporter	Values of degree of lateness				
1	*0.04*	*0.00*	*0.02*	*0.03*	*0.02*
2	*0.15*	*0.11*	*0.07*	*0.09*	*0.12*
3	*0.06*	*0.03*	*0.04*	*0.04*	*0.05*

(*i*) *Is there a difference in the mean degree of lateness of the three transporters? Test at 10% level of significance.*

(*ii*) *Find a 95% confidence interval for the mean degree of lateness of transporter 1.*

(*iii*) *Find a 90% confidence interval for the difference in the mean degree of lateness of transporters 2 and 3. Is there a difference in their means at 10% level of significance?*

Solution:

(*i*) The analysis of variance (ANOVA) table for the given problem is shown below.

Source of variation	Degree of freedom	Sum of squares	Mean square	F-statistic
Treatments	2	0.01996	0.00998	23.57
Error	12	0.00508	0.00042	
Total	14	0.02504		

For a level of significance, $\alpha = 0.10$, $F_{0.10,2,12} = 2.81$ which is less than $F = 23.57$, we conclude that there is a difference in the mean degree of lateness among the three transporters.

(*ii*) The mean degree of lateness for transporter 1 is denoted as μ_1. The 95% Confidence Interval for μ_1 is given as

$$0.022 \pm t_{0.025,12}\sqrt{0.00042/5} = 0.022 \pm 0.01997 = (0.00203, 0.04197)$$

(*iii*) The mean degree of lateness for transporters 2 and 3 are μ_2 and μ_3, respectively. The 90% Confidence Interval for the difference between the mean degree of lateness of transporters 2 and 3 is calculated as,

$$(0.108 - 0.044) \pm t_{0.05,12}\sqrt{(2 \times 0.00042)/5} = 0.064 \pm 0.0231 = (0.0409, 0.0871)$$

8.6 RANDOMIZED BLOCK DESIGN

In this design, the criterion with which blocking is made is identified first for experimentation. Successful blocking minimizes the variation among the experimental units *within* a block while maximizing variation *between* blocks. An advantage of blocking is that differences between blocks can be eliminated from the experimental error, thereby increasing the precision of the experiment. However, blocking is effective only if the variability within blocks is smaller than the variability over the entire set.

In a randomized block design, the experimental units are first divided into blocks. Within a block, the treatments are randomly assigned to the experimental units as in a completely randomized design. If there are *p* treatments and *r* blocks in a completely randomized block design, each of the *p* treatments is applied once in each block, resulting in a total of *rp* experimental units. Thus, *r* blocks, each consisting of *p* units, are chosen such that the units within blocks are homogeneous. In the next step, the p treatments are randomly assigned to the units within the blocks, with each treatment occurring only once within each block. The blocks serve as replications for the treatments.

Table 8.3 shows an experimental design in which there are four treatments, denoted as A, B, C, and D, and three blocks denoted as I, II, and III (hence, $p = 4$ and $r = 3$). Within each block, the four treatments are randomly applied to the experimental units. Each treatment appears only once in each block. The treatments can also be assigned as for the completely randomized design.

Table 8.3. Assignment of Treatments in a Randomized Block Design

Block I	Block II	Block III
B	C	C
A	A	B
C	D	D
D	B	A

Some of the merits of this design are as follows:

(*i*) blocking increases the precision by removing visibility between blocks from the experimental error,

(*ii*) any number of treatments and blocks, can be considered in this design, and

(*iii*) the statistical analysis is relatively simple.

However, the main shortcomings in this design is that the number of degrees of freedom for the experimental error is not as large as that for a completely randomized design. If there are r blocks, then there are $(r - 1)$ fewer degrees of freedom in the experimental error. The completely randomized design is more efficient than this design if the experimental units are homogeneous.

The following notations are used for this design:

p = Number of treatments

r = Number of blocks

y_{ij} = Response variable value of the ith treatment applied to the jth block, $(i = 1,2,\ldots\ldots, p; j = 1,2,\ldots\ldots, r)$

$$y_{i.} = \text{Sum of the responses for the } i\text{th treatment} = \sum_{j=1}^{r} y_{ij}$$

$$\bar{y}_{i.} = \text{Mean response of the } i\text{th treatment} = \frac{y_{i\cdot}}{r}$$

$$y_{.j} = \text{Sum of the response for the } j\text{th block} = \sum_{i=1}^{p} y_{ij}$$

$$\bar{y} = \text{Mean response of the } j\text{th block} = \frac{y_{.j}}{p}$$

$$y_{..} = \text{Grand total of all observations} = \sum_{i=1}^{p}\sum_{j=1}^{r} y_{ij}$$

$$\bar{\bar{y}} = \text{Grand mean of all observations} = \frac{y_{..}}{rp}$$

These notations are shown in Table 8.4.

Table 8.4. Data Collection Format for Randomized Block Design

	Replication					
Treatment	*1*	*2*	...	*r*	*Sum*	*Mean*
1	y_{11}	y_{12}	...	y_{1r}	$y_{1.}$	$\bar{y}_1$
2	y_{21}	y_{22}	...	y_{2r}	$y_{2.}$	$\bar{y}_2$
⋮	⋮	⋮	⋱	⋮	⋮	⋮
p	y_{p1}	y_{p2}	...	y_{pr}	$y_{p.}$	$\bar{y}_p$
Sum	$y_{.1}$	$y_{.2}$	...	$y_{.r}$	$y_{..}$	$\bar{\bar{y}}$
Mean	$\bar{y}_{.1}$	$\bar{y}_{.2}$	...	$\bar{y}_{.r}$		

Using the equations for the total sum of squares (equation 8.1) and the treatment sum of squares (equation 8.2), the block sum of squares for this design is given as

$$SS_B = \frac{\sum_{j=1}^{r} y_{.j}^2}{p} - C \qquad \text{...(8.10)}$$

and the error sum of squares is given as

$$SS_E = SS_T - SS_{TR} - SS_B \qquad \text{...(8.11)}$$

The number of degrees of freedom for the treatments is $(p - 1)$, and for the blocks is $(r - 1)$. The number of degrees of freedom for the error is $(p - 1)(r - 1)$, and the total number is $(rp - 1)$. The mean squares for the treatments, blocks, and experimental error are computed by dividing the respective sum of squares by the corresponding numbers of degrees of freedom.

In order to test the difference between the treatment means, the hypothesis is same as in the completely randomized design.

The calculated F-statistic is compared to the critical value of F for a chosen significance level α, with $(p - 1)$ degrees of freedom in the numerator and $(p - 1)(r - 1)$ in the denominator. If the computed test statistic $F > F_{\alpha, (p-1), (p-1)(r-1)}$, the null hypothesis is rejected, and hence, the treatment means are significantly different from one another at the chosen level of significance. The ANOVA table is shown in Table 8.4. In order to know if blocking is effective in reducing the experimental error, an F-statistic for the block is given as $F_B = \frac{MS_B}{MS_E}$ (where, MS_B = mean squares for blocks),

Source of variation	Degree of freedom	Sum of squares	Mean square	F-statistic
Treatments	$p - 1$	SS_{TR}	$MS_{TR} = \dfrac{SS_{TR}}{(p-1)}$	$F = \dfrac{MS_{TR}}{MS_E}$
Blocks	$r - 1$	SS_B	$MS_B = \dfrac{SS_B}{(r-1)}$	
Error	$(p - 1)(r - 1)$	SS_E	$MS_E = \dfrac{SS_E}{[(p-1)(r-1)]}$	
Total	$rp - 1$	SS_T		

The computed value of F_B is compared to $F_{\alpha,(r-1),(p-1)(r-1)}$ for a level of significance a. If $F_B > F_{\alpha,(r-1),(p-1)(r-1)}$, blocking is considered to be effective. On the other hand, if $F_B < F_{\alpha,(r-1),(p-1)(r-1)}$, blocking is not considered to be effective, and thus, the efficiency of the experiment has decreased relative to the completely randomized design due to blocking. In effect, $(r - 1)$ degree of freedom are lost from the error without a sufficiently compensating reduction in the mean square error. This means that although SS_E is decreased due to blocking, the MS_E is not decreased because the reduction in the number of degrees of freedom is not adequately compensated for.

A $100(1 - \alpha)\%$ confidence interval for μ_i is given by

$$CI(\mu_i) = \bar{y}_{i\cdot} \pm t_{\alpha/2,(p-1)(r-1)}\sqrt{\frac{MS_E}{r}} \qquad ...(8.12)$$

where, $t_{\alpha/2,\,(p-1)\,(r-1)}$ is the t-value for a right-tail area of $\alpha/2$ and $(p - 1)(r - 1)$ degrees of freedom.

A $100(1 - \alpha)$ % confidence interval for the difference between two treatment means, $\mu_{i1} - \mu_{i2}$ is given by

$$CI(\mu_{i1} - \mu_{i2}) = (\bar{y}_{i1} - \bar{y}_{i2}) \pm t_{\alpha/2,(p-1)(r-1)}\sqrt{\frac{2MS_E}{r}} \qquad ...(8.13)$$

where, $t_{\alpha/2,(p-1)(r-1)}$ is the t-value for a right-tail area of $\alpha/2$ and $(p - 1)(r - 1)$ degrees of freedom.

Example E8.2. *Three training programmes for the production executives are conducted by a company. The company has grouped its production executives into three classes based on their years of experience. Three executives are selected from each class and randomly assigned to the training programmes. The success of a training programme is measured on a rating scale of (0 to 100). Table E8.2 shows the scale values of ratings assigned to the executives on completion of training.*

Table E 8.2. Ratings of Training Programmes for Production Executives

Class	Training programme		
	1	2	3
1 (upto 5 years experience)	*70*	*65*	*81*
2 (5–10 years experience)	*75*	*80*	*87*
3 (above 10 years experience)	*87*	*82*	*94*

(*i*) Is there a difference in the training programmes based on their mean ratings? Test at 10% level of significance.

(*ii*) Find a 90% confidence interval for the difference in the mean ratings of training programmes 1 and 3. Is there a difference between these two programmes?

Solution:

(*i*) The analysis of variance (ANOVA) table is shown below.

Source of variation	Degree of freedom	Sum of squares	Mean square	F-statistic
Training Programmes	2	238.889	119.444	11.81
Classes (Blocks)	2	369.555	184.777	
Error	4	40.444	10.111	
Total	8	648.888		

At 10% level of significance, $F_{0.10,2,4} = 4.32$ which is less than the calculated, $F = 11.81$. Hence, the null hypothesis of equality of mean ratings associated with the training programmes is rejected, and it is concluded that there is a significant difference in the training programme.

(*ii*) The mean ratings for training programmes 1 and 3 are denoted as μ_1 and μ_3, respectively. The 90% confidence interval for their difference, $(\mu_1 - \mu_3)$ is computed as

$$\mu_1 = (70 + 75 + 87)/3 = 77.333,$$

$$\mu_3 = (81 + 87 + 94)/3 = 87.333,$$

and

$$CI = (77.333 - 87.333) \pm t_{0.05,4}\sqrt{(2 \times 10.111)/3}$$

$$= -10.00 \pm 5.535 = (-15.535 - 4.465)$$

Since the confidence interval does not include 'zero', there is a significant difference in the mean ratings of training programmes 1 and 3.

Example E8.3. *An experimenter wishes to determine if four different tips produce four different readings on a hardness testing machine as part of a gauge capability study. The machine presses a tip into a metal test coupon. The hardness of the coupon can be determined from the depth of the resulting depression. Four observations for each tip are to be taken in experimentation. It is known that metal coupons are taken from ingots that are produced in different heats, contributing to the variability in the hardness data. In order to minimize the experimental error because of variability, a randomized complete block design is recommended for the experiment by testing tips once on each of the four coupons. Thus, one coupon with four tips forms a block. Within a block, the order in which the four tips are tested is randomly determined. The observed values are in Rockwell C scale hardness are given in Table E8.3. Determine which treatment differs significantly from the others. Test at 5% level of significance.*

Table E8.3. Randomized Complete Block Design for the Hardness Testing Experiment

Type of tip	Coupon (Block)			
	1	2	3	4
1	*49.3*	*49.4*	*49.6*	*50.0*
2	*49.4*	*49.3*	*49.8*	*49.9*
3	*49.2*	*49.4*	*49.5*	*49.7*
4	*49.7*	*49.6*	*50.0*	*50.2*

Solution:

Subtracting 49.5 from each observation and multiplying by 10, the resulting coded data are shown in the Table E8.3 (*a*).

Table E8.3(a). Coded Data for the Hardness Testing Experiment

Type of tip	Coupon (Block)				$y_{i.}$
	1	2	3	4	
1	– 2	– 1	1	5	3
2	– 1	– 2	3	4	4
3	– 3	– 1	0	2	– 2
4	2	1	5	7	15
$y_{.j}$	– 4	– 3	9	18	$20 = y_{..}$

Here, the sums of the squares as obtained are as follows:

$$SS_T = 12.9;\ SS_{TR} = 38.5;\ SS_B = 82.5;\ SS_E = 8$$

From the ANOVA table shown below, it is found that the computed F-statistic = 14.44, which is greater than $F_{0.05,\,3,\,9}$ (= 3.86) indicating that the type of tool affects the mean hardness reading.

Source of variation	Sum of squares	Degrees of freedom	Mean square	F_0
Treatments (type of tip)	38.50	3	12.83	14.44
Blocks (coupons)	82.50	3	27.50	
Error	8.00	9	0.89	
Total	129.00	15		

To test which treatment differs substantially from the others, we apply the ***Fisher LSD procedure:***

$$LSD = t_{0.025,12}\sqrt{\frac{2MS_E}{b}} = 2.179\sqrt{\frac{(2)(0.89)}{4}} = 1.45$$

$$|\bar{y}_{1.} - \bar{y}_{2.}| = 0.25,\ |\bar{y}_{1.} - \bar{y}_{3.}| = 1.25,$$

$$|\bar{y}_{1.} - \bar{y}_{4.}| = 3,\ |\bar{y}_{2.} - \bar{y}_{3.}| = 1.5,$$

$$|\bar{y}_{3.} - \bar{y}_{4.}| = 4.25$$

From the above results, it is concluded that tip 4 produces significantly higher mean hardness than the other tips.

8.7 LATIN SQUARE DESIGN

This is a special design, where two sources of variation are controlled. In this design, blocking is done over two variables, and the number of blocks for each of the two variables is equal, which in turn is equal to the number of treatments. One of the blocking variables is denoted as the row variable and the other as the column variable. The treatments are assigned in such a way that each treatment appears once and only once in each row and column. The number of experimental units required is square of the number of treatments.

As an example of a Latin square design, let us consider a production system where performance of a machine varies from one worker to another and also from one day to another. It is to be known if the performance depends on types of incentive schemes. First, five machines (M1, M2, M3, M4 and M5) are randomly selected, and they comprise the first blocking or row variable. The five days of the week (1, 2, 3, 4 and 5) comprise the second blocking or column variable. The treatments are the types of incentive scheme (*A, B, C, D* and *E*). The assignment of treatments are shown in Table 8.5. In this design, treatments are always denoted by Latin letter (by convention). As the number of row and column classifications are equal, and they are also equal to the number of treatments, the design is known as a square design. Hence, this experimental design is referred to as a Latin square design. As shown in Table 8.5, each treatment occurs only once in each row and in each column. For five treatments, 25 experimental units are needed for the Latin square design

Table 8.5. A Latin Square Design for Five Treatments

Machine	Day of week				
	1	2	3	4	5
M1	A	B	C	D	E
M2	B	C	D	E	A
M3	C	D	E	A	B
M4	D	E	A	B	C
M5	E	A	B	C	D

The Latin square design is an example of an incomplete block design, in which not all treatments are necessarily assigned to each block. In this example there are actually 25 blocks, which represent the different combinations of machines and day of the week. For a randomized block design, in which each treatment appears once in each block, a total of 125 experimental units are needed. In the Latin square design there are only 25 experimental units. In general, if we have p treatments, we need p^2 experimental units and each of the two blocking variables should have p classes.

As there are two blocking variables used in this design, significant reduction in the variability of experimental error is obtained. This is the main advantage of the Latin square design. This design has a few limitations. First, since the number of classes of each blocking variable must equal the number of treatments, very few degrees of freedom are left for the experimental error if the number of treatments is small. When the number of treatments is large, the number of degrees of freedom for the experimental error may be larger than necessary. Second, in this design as the number of experimental units is equal to the square of the number of treatments, the number of treatments has to be limited (10 or less) for practical reasons. Third, it is assumed that there is no interaction between either the blocking variable and the treatment or between the blocking variables themselves. Hence, the design is restrictive.

For analysis of variance, the following notations are used in this design:

p = Number of treatments

$y_{ij(k)}$ = Response variable value of the experimental unit in the ith row and jth column with the kth treatment $i = 1, 2, \ldots, p$, and $j = 1, 2, \ldots, r$

$y_{i.} = \sum_{j=1}^{p} y_{ij}$ = Sum of the responses in the ith row

$y_{.j} = \sum_{j=1}^{p} y_{ij}$ = Sum of the responses in the jth column

$y_{..} = \sum_{i=1}^{p} \sum_{j=1}^{p} y_{ij}$ = Grand total of all observations

T_k = Sum of the responses for the kth treatment

$\bar{y}_k = \frac{T_k}{p}$ = Sample mean response for the kth treatment

The notations used for the Latin square design are shown in Table 8.6.

Table 8.6. Notation for the Latin Square Design

Row	Column				Sum
	1	**2**	**...**	**p**	
1	y_{11}	y_{12}	...	y_{1p}	$y_{1.}$
2	y_{21}	y_{22}	...	y_{2p}	$y_{2.}$
⋮	⋮	⋮	⋮	⋮	⋮
p	y_{p1}	y_{p2}	...	y_{pp}	$y_{p.}$
Sum	$y_{.1}$	$y_{.2}$	...	$y_{.p}$	$y_{..}$
				Treatment	
		1	*2*	...	*p*
Sum		T_1	T_2	...	T_p
Mean		$\bar{y}_1$	$\bar{y}_2$	...	$\bar{y}_p$

A correct factor C is computed as

$$C = \frac{y_{..}^2}{p^2}$$

The total sum of squares (SS_T), the sum of squares for the row blocking variables (SS_R), the sum of squares for the column blocking variables (SS_C), and the sum of squares for the treatments (SS_{TR}) are given as

$$SS_T = \sum_{i=1}^{p} \sum_{j=1}^{p} y_{ij}^2 - C \qquad \text{...(8.14)}$$

$$SS_R = \frac{\sum_{i=1}^{p} y_{.i}^2}{p} - C \qquad \text{...(8.15)}$$

$$SS_C = \frac{\sum_{j=1}^{p} y_j^2}{p} - C \qquad ...(8.16)$$

$$SS_{TR} = \frac{\sum_{k=1}^{p} T_k^2}{p} - C \qquad ...(8.17)$$

where, C = correction factor = $\frac{y_{..}^2}{p^2}$

The error sum of squares is given by

$$SS_E = SS_T - SS_R - SS_C - SS_{TR} \qquad ...(8.18)$$

The number of degrees of freedom for both the row blocking variable and the column blocking variables is $(p - 1)$. The number of degrees of freedom for the treatments is $(p - 1)$. The total number of degrees of freedom is $p^2 - 1$. The number of degrees of freedom for the experimental error is computed as follows:

Error degrees of freedom = $(p^2 - 1) - (p - 1) - (p - 1) - (p - 1) = (p - 1)(p - 2)$

As usual, the mean squares are found by dividing the sum of squares by the appropriate number of degrees of freedom. The format for the ANOVA Table is shown in Table 8.7. In order to test the difference among the treatment means, the hypothesis are same as in the completely randomized design.

Table 8.7. ANOVA Table for the Latin Square Design

Source of variation	Degrees of freedom	Sum of squares	Mean square	F-statistic
Rows (blocking variable I)	$p - 1$	SS_R	$MS_R = \frac{SS_R}{(p-1)}$	
Columns (blocking variable II)	$p - 1$	SS_C	$MS_C = \frac{SS_C}{(p-1)}$	
Treatments	$p - 1$	SS_{TR}	$MS_{TR} = \frac{SS_{TR}}{(p-1)}$	$F = \frac{MST}{(p-1)}$
Error	$(p - 1)(p - 1)$	SS_E	$MS_E = \frac{SS_E}{(p-1)(p-2)}$	
Total	$p^2 - 1$	SS_T		

If the calculated F-statistic $> F_{\alpha,(p-1),(p-1)(p-2)}$, for a right-tail area of α with $(p - 1)$ degrees of freedom in the numerator and $(p - 1)(p - 2)$ degree of freedom in the denominator, the null hypothesis is rejected, and hence, at least one of the treatment means is significantly different from the others at a chosen level of significance, α.

For determining the effectiveness of the row and column blocking variables in this design, the following two F-statistics are computed:

$$F_R = \frac{MS_R}{MS_E} \quad \text{for row blocking variable, and}$$

$$F_C = \frac{MS_C}{MS_E} \quad \text{for column blocking variable}$$

The computed values of these two statistics can be compared with $F_{\alpha,(p-1),(p-1)(p-2)}$ to make inferences.

A $100(1 - \alpha)\%$ confidence interval for a single treatment mean, μ_k is given by

$$CI(\mu_k) = \bar{y}_k \pm t_{\alpha/2,(p-1)(p-2)} \sqrt{\frac{2MS_E}{p}} \quad \text{...(8.19)}$$

where, $t_{\alpha/2,(p-1)(p-2)}$ is the t-value for a right-tail area of $\alpha/2$ and $(p - 1)(p - 2)$ degrees of freedom.

A $100(1 - \alpha)\%$ confidence interval for the difference between two treatment means, $\mu_{k1} - \mu_{k2}$, is given by

$$CI(\mu_{k1} - \mu_{k2}) = (\bar{y}_{k1} - \bar{y}_{k2}) \pm t_{\alpha/2,(p-1)(p-2)} \sqrt{\frac{2MS_E}{p}} \quad \text{...(8.20)}$$

where, $t_{\alpha/2,(p-1)(p-2)}$ is the t-value for a right-tail area of $\frac{\alpha}{2}$ and $(p - 1)(p - 2)$ degrees of freedom.

An interesting feature of this design should be mentioned. In this design, the number of error degrees of freedom is given by $(p - 1)(p - 2)$. If p is very small, the number of error degrees of freedom is small, which causes the mean square error to be large. Again, the number of treatments cannot be less than 3, because if $p = 2$, the number of error degrees of freedom is zero. As p becomes large, the number of experimental units required p^2 becomes very large, and may not be feasible from a cost and operational standpoint.

Example E8.4. *A firm wishes to evaluate the performance of four software packages as measured by their computational times. These softwares are designated as A, B, C, and D. It is desirable that two variables, viz., the type of the problem, and the operating system configuration used, are to be controlled. Four types of configurations and four types of problems are identified. Each package is used only once on each problem type, and only once on each configuration. The data are given in Table E8.4.*

Table E8.4. Computational Times for Software Packages in Controlled Conditions

Type of problem	Types of operating system configuration			
	I	**II**	**III**	**IV**
1	*B 14.2*	*A 37.3*	*D 24.3*	*C 22.6*
2	*D 18.1*	*C 18.1*	*B 39.4*	*A 39.6*
3	*A 36.1*	*B 16.8*	*C 13.2*	*D 20.3*
4	*C 17.5*	*D 24.6*	*A 36.7*	*B 19.5*

(i) Test at 5% level of significance for the equality of the mean computational times of the four software packages.

(ii) Find a 95% confidence interval for the mean computational time using software package C.

(iii) Is the difference in the mean computational times of software packages A and C significant? Find a 95% confidence interval for this difference.

(iv) Is blocking of the two variables considered in this problem effective?

Solution:

(*i*) The analysis of variance (ANOVA) is computed below:

Source of variation freedom	Degrees of squares	Sum of square	Mean statistic	F-statistic
Problem Type	3	39.7869	13.262	20.122
Operating System	3	40.0669	13.356	
Software Packages (Treatments)	3	970.8819	323.627	
Error	6	96.4987	16.083	
Total	15	1147.2344		

At 5% level of significance, $F_{0.05,3,6} = 4.76$ (table value). Since computed value of F-statistic is 20.112, which is greater than 4.76, we conclude that the mean computational times of the software packages are significantly different from one another.

(*ii*) The mean computational time of software package, *C* is denoted as μ_C. The 95% confidence interval for μ_C is given as

$$17.85 \pm t_{0.025,6}\sqrt{16.083/4} = 17.85 \pm 2.447 \times 2.005 = (12.9433, 22.7567)$$

(*iii*) The mean computational time of software package, *A* is denoted as μ_A. The 95% confidence interval for the difference $(\mu_A - \mu_C)$, is computed as

$$(37.425 - 17.85) \pm t_{0.025,6}\sqrt{(2 \times 16.083)/4} = 19.575 \pm 2.447 \times 2.8357 = (12.636, 26.514)$$

Since the confidence interval does not include 'zero', there is a significant difference in the mean computational times of softwares *A* and *C*.

(*iv*) To investigate the effectiveness of the blocking variable, problem type, we compute $F = 13.262/16.083 = 0.824$. At 5% level of significance, $F_{0.05,3,6} = 4.76$ (table value) which is greater than the computed F-value 0.824. Hence, the problem type as a blocking variable is not effective. Similarly, for the second blocking variable, operating system configuration, we compute, $F = 13.356/16.083 = 0.830$, which is also less than $F_{0.05,3,6} = 4.76$. Hence, the use of operating system configuration as a blocking variable is also not effective.

8.8 FACTORIAL EXPERIMENTS

In a factorial experiment, all possible treatment combinations of the factors are represented for each complete replication of the experiment. The number of treatments is equal to the product of the number of factor levels and can therefore become very large when many factors and/or levels of factors need to be considered in an experiment. In fact, the experiment size may increase exponentially with the increase in the number of the factors and/or the number of levels. This is a major disadvantage with factorial experiments. However, factorial experiments are widely used in numerous manufacturing and service situations when the number of factors and levels are restricted to a limited number.

Suppose, there are three factors to be considered in an experiment dealing with the quality level of a component being produced on a machine. If the first factor (quality of rough stock) has three levels, the second factor (experience of the operator) has four levels, and the third factor (age of the machine) has three levels, the factorial experiment requires $3 \times 4 \times 3 = 36$ treatment combinations for each replication. Treatments usually need to be replicated because the purpose of a factorial experiment is to ensure that any possible interaction effects between the factors can be estimated. This increases the total number of required trials.

The main advantage of this experiment is that it has the ability to estimate the interaction effects between factors. In the presence of two or more factors, knowledge of interaction effects is of prime importance to the experimenter or the researcher. If the effect of both factors on the response variable is to be determined, interaction effects cannot be ignored. If the interaction effects are present, testing for the main effects is not necessary. The main effects need to be estimated when the factors are independent of one another and interaction between the factors is not present. In such cases, fewer number of trials are required to estimate the main effects to the same degree of precision. Usually, single-factor experiments require more trials to achieve the same degree of precision that is obtained in typical factorial experiments.

Factorial experiments play an important role in exploratory studies to determine which factors are important. By conducting experiments for a factor at several levels of other factors, the inferences from a factorial experiment are valid over a wide range of conditions.

The factorial designs involve typically the study of effects of two or more factors, and all possible combinations of the levels of the factors in each complete trial or replication of the experiment. Thus, if factor A has $\boldsymbol{a}$ levels and factor B has $\boldsymbol{b}$ levels, each replicate or trial of the experiment contains all the $(\boldsymbol{ab})$ treatment combinations.

As an example, let us consider a two-factor factorial experiment, shown in **Figure 8.7**. Here, the factor levels are indicated as low (– 1) and high (+ 1), and the value of response (y) at the treatment combination are shown at the corners. The value of the responses (20, 30, 40, 52) are plotted in **Figure 8.8**.

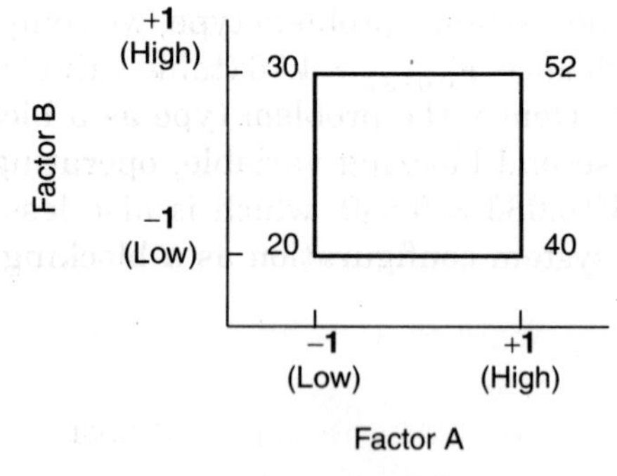

Fig. 8.7 A Two-Factor Factorial Experiment with the Response (y) Shown at the Corners.

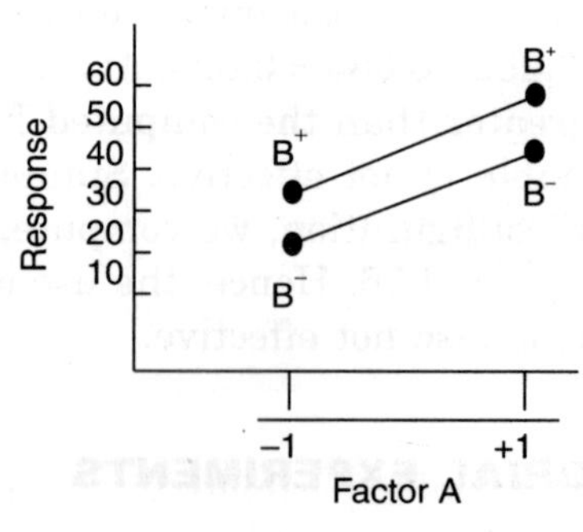

Fig. 8.8 A Factorial Experiment without Interaction.

With these experimental results the ***main*** and the ***interaction effects*** need to be computed. The procedure for computing these effects is as follows:

(*i*) **Main effect, A** = Average response at high value of A – Average response at low value of A.

$$= \frac{1}{2}(A^+B^- + A^+B^+) - \frac{1}{2}(A^-B^- + A^-B^+)$$

$$= \frac{40+52}{2} - \frac{20+30}{2} = 21$$

(*ii*) **Main Effect, B** = Average response at high value of B – Average response at low value of B.

$$= \frac{1}{2}(A^-B^+ + A^+B^+) - \frac{1}{2}(A^-B^- + A^+B^-)$$

$$= \frac{1}{2}(30 + 52) - \frac{1}{2}(20 + 40) = 11$$

(*iii*) **Interaction effect, AB** = Average A effect at B^+ – Average A effect at B^-

$$= \frac{1}{2}(A^+B^+ - A^-B^+) - \frac{1}{2}(A^+B^- - A^-B^-)$$

$$= \frac{1}{2}(52 - 30) - \frac{1}{2}(40 - 20) = 1$$

Let us consider another example of two-factor factorial experiment, as shown in **Figures 8.9** and **8.10.**

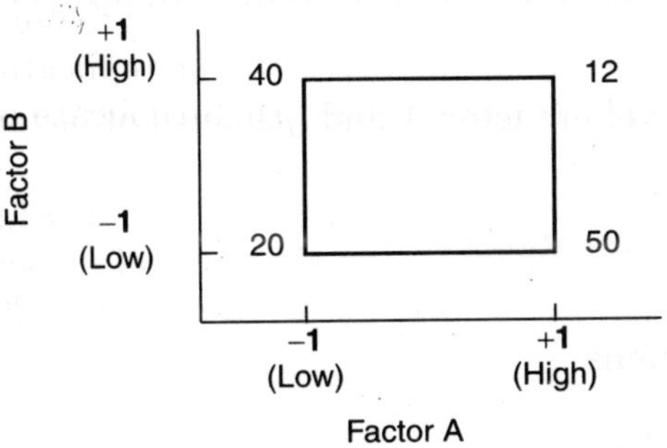

Fig. 8.9 A Two-factor Factorial Experiment with Interaction.

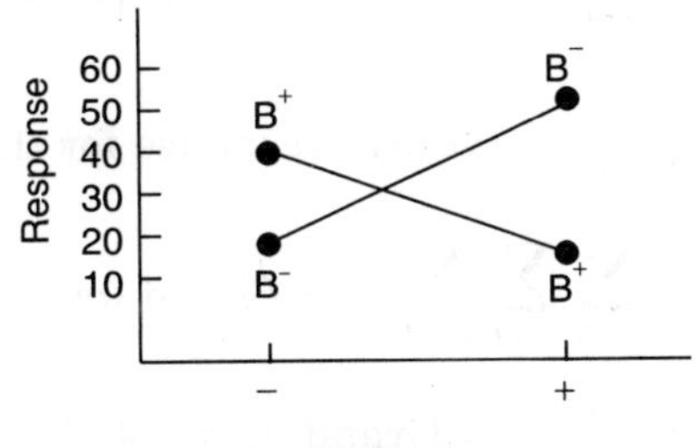

Fig. 8.10 A Factorial Experiment with Interaction.

Following the same principles as mentioned in the previous example, the main and interaction effects of the factors A and B are computed as

$$A = 30, B = -28, AB = -29.$$

In this case, the interaction effects are more significant than that in the previous example.

The factorial experiment can be used for both in completely randomized design as well as in the randomized block design. The factorial experiments under these two situations, when two factors are considered, are discussed below.

8.8.1 Two-Factor Experiment Using a Completely Randomized Design

Let two factors, denoted by A and B be considered in an experiment. There are a number of levels for factor A and b number of levels for factor B. Each of the total (ab) number of treatment is randomly assigned to the experimental units. Suppose, the experiment is replicated n times yielding a total of abn observations. The model is mathematically represented as

$$y_{ijk} = \mu + \alpha_i + \beta_j + (\alpha\beta)_{ij} + \varepsilon_{ijk} \quad \text{...(8.21)}$$

$$i = 1, 2, ...a \qquad j = 1, 2, ..., b \qquad k = 1, 2, ..., n$$

where, y_{ijk} is the response for the ith level of factor A, the jth level of factor B, and the kth replication, μ is the overall mean effect, α_i is the effect of the ith level of factor A, β_j is the effect of the jth level of factor B, $(\alpha\beta)_{ij}$ is the effect of the interaction between factors A and B, and is the random error component (assumed to be normally distributed with mean zero and constant variance, σ^2).

The following notations are used for the analysis of variance in this experiment:

$$y_{i..} = \sum_{j=1}^{b}\sum_{k=1}^{n} y_{ijk} = \text{Sum of the responses for the } i\text{th level of factor } A$$

$$y_{.j} = \sum_{i=1}^{a}\sum_{k=1}^{n} y_{ijk} = \text{Sum of the responses for the } j\text{th level of factor } B$$

$$\bar{y}_{i..} = \frac{y_i}{(bn)} = \text{Average response for the } i\text{th level of factor } A$$

$$\bar{y}_{.j..} = \frac{y_{.j.}}{(an)} = \text{Average response for the } j\text{th level of factor } B$$

$$y_{ij.} = \sum_{k=1}^{n} y_{ijk} = \text{Sum of the responses for the } i\text{th level of factor } A \text{ and } j\text{th level of factor } B$$

$$\bar{y}_{ij} \cdot = \frac{y_{ij}}{n} = \text{Average response for the } i\text{th level of factor } A \text{ and } j\text{th level of factor } B$$

$$\bar{y}_{...} = \sum_{i=1}^{a}\sum_{j=1}^{b}\sum_{k=1}^{n} y_{ijk} = \text{Grand total of all observations}$$

$$\bar{y}_{...} = \frac{y_{...}}{(abn)} = \text{Grand mean of all observations}$$

The above notations are shown in Table E8.8.

Table E8.8. Notations for Factorial Experiments Using a Completely Randomized Design

Factor A	Factor B					
	1	2	...	b	Sum	Average
1	$y_{111}, y_{112}, ..., y_1$ Sum = $y_{11\cdot}$	$y_{121}, y_{122}, ..., y_{12}$ Sum = $y_{12\cdot}$	...	$y_{1b1}, y_{1b2}, ..., y_{1l}$ Sum = $y_{1b.}$	$y_{1\cdots}$	$\bar{y}_{1..}$
2	$y_{211}, y_{212}, ..., y_2$ Sum = $y_{21\cdot}$	$y_{221}, y_{222}, ..., y_2$ Sum = $y_{22\cdot}$	... ...	$y_{2b1}, y_{2b2}, ..., y_2$ Sum = $y_{2b.}$	$y_{2...}$	$\bar{y}_{2..}$
⋮	⋮	⋮	⋮	⋮	⋮	⋮
a	y_{a11}, y_{a12}, y_a Sum = $y_{a1\cdot}$	$y_{a21}, y_{a22}, ..., y_a$ Sum = $y_{a2\cdot}$	...	$y_{ab1}, y_{ab2}, ..., y_a$ Sum = $y_{ab\cdot}$		
			...		$y_{a..}$	$\bar{y}_{a..}$
Sum	$y_{\cdot 1\cdot}$	$y_{\cdot 2\cdot}$	...	$y_{\cdot b\cdot}$	$y_{..}$	
Average	$\bar{y}_{\cdot 1\cdot}$	$\bar{y}_{\cdot 2\cdot}$	...	$\bar{y}_{\cdot b\cdot}$		$\bar{y}_{..}$

The total sum of squares (SS_T), the sum of squares for the main effects of factor A (SS_A), the sum of squares for the various effects of factor B (SS_B), and the sum of squares for the subtotal between the cell totals [$SS_{(subtotal)}$] are given respectively.

$$SST = \sum_{i=1}^{a} \sum_{j=1}^{b} \sum_{k=1}^{n} y_{ijk}^2 - C \qquad ...(8.22)$$

$$SSA = \frac{\sum_{i=1}^{a} y_{i..}^2}{(bn)} - C \qquad ...(8.23)$$

$$SSB = \frac{\sum_{j=1}^{b} y_{.j.}^2}{(an)} - C \qquad ...(8.24)$$

$$SS_{(subtotal)} = \frac{\sum_{i=1}^{a} \sum_{j=1}^{b} y_{ij.}^2}{n} - C \qquad ...(8.25)$$

where, C = correction factor = $\dfrac{y^2}{(abn)}$ ...(8.26)

The sum of squares due to the interaction between factors A and B is given by

$$SS_{AB} = SS_{(subtotal)} - SS_A - SS_B \qquad ...(8.27)$$

and the error sum of squares is

$$SS_E = SS_T - SS_{(subtotal)} \qquad ...(8.28)$$

There are $(a - 1)$ degrees of freedom for factor A and $(b - 1)$ degrees of freedom for factor B. The number of degrees of freedom for the interaction between factors A and B is the product of the number of degrees of freedom for each, *i.e.*, $(a - 1)(b - 1)$. Since the total number of degrees of freedom is $(abn - 1)$, the number of degrees of freedom for the experimental error is $ab(n - 1)$. The mean squares for each factor, interaction, and error are computed by dividing the respective sum of squares by the corresponding number of degrees of freedom. To test for the significance of the factors and the interaction, the F-statistic is calculated as usual. The ANOVA computations are shown in Table E8.9.

Table E8.9. ANOVA Table for the Two-Factor Factorial Experiment Using a Completely Randomized Design

Source of variation	Degrees of freedom	Sum of squares	Mean square	F-statistic
Factor A	$a - 1$	SS_A	$MS_A = \dfrac{SS_A}{(a-1)}$	$F_A = \dfrac{-MS_A}{MS_E}$
Factor B	$b - 1$	SS_B	$MS_B = \dfrac{SS_B}{(b-1)}$	$F_B = \dfrac{MS_B}{MS_E}$

Interaction between A and b	$(a-1)(b-1)$	SS_{AB}	$MS_{AB} = \dfrac{SS_{AB}}{(a-1(b-1)}$	$F_{AB} = \dfrac{MS_{AB}}{MS_E}$
Error	$ab(n-1)$	SS_E	$MS_E = \dfrac{SS_E}{ab(n-1)}$	
Total	$abn-1$	SS_T		

In order to test the significance of the interaction between factors A and B, the following hypothesis are used:

$H_0 : (\alpha\beta)_{ij}$ for all i,j; $i = 1,2,\ldots, a; j = 1,2,\ldots,b$

H_a : At least one $(\alpha\beta)_{ij}$ is different from zero

In this case the F-statistic is computed as

$$F_{AB} = \frac{MS_{AB}}{MS_E} \qquad \ldots(8.29)$$

If the calculated value of, $F_{AB} > F_{\alpha}$, $(a-1)(b-1)$, $ab(a-1)$, the null hypothesis is rejected, and the interactions between factors are assumed to be significant at the chosen level of significance, α.

As the presence of significant interaction effects may 'mask' the main effects of the factors, the main effects are usually not tested. When interaction effect is significant, the value of the response variable due to changes in factor A depends on the level of factor B. In such a situation, the mean for a level of factor A, averaged over all levels of factor B, does not have any practical significance. The treatment means along with their standard deviations may be required to be computed in this situation. Also, the preferable levels of the factors may be required to be identified.

The standard deviation of the treatment mean, which is the mean of any combination of factors A and B, is given as

$$S_{\bar{y}(AB)} = \sqrt{\frac{MS_E}{n}} \qquad \ldots(8.30)$$

A $100(1-\alpha)\%$ confidence interval for a treatment mean when factor A is at level i and factor B is at level j is given by

$$\bar{y}_{ij\cdot} \pm t_{\alpha/2,ab(n-1)}\sqrt{\frac{MS_E}{n}} \qquad \ldots(8.31)$$

where, $t_{\alpha/2,ab(n-1)}$ is the t-value for a right-tail area of $\alpha/2$ and $ab(n-1)$ degrees of freedom. The standard deviation of difference d between two treatment means is given as

$$S_{d(AB)}\sqrt{\frac{2MS_E}{n}} \qquad \ldots(8.32)$$

Let us consider the treatments for which factor A is at level i_1 and factor B is at level j_1, which yields a sample average of $\bar{y}_{i_1j_1\cdot}$. The other treatment has factor A at level i_2 and factor B at level j_2, the sample average being $\bar{y}_{i_2j_2\cdot}$. A $100(1-\alpha)$ % confidence interval for the difference between the two treatment means is given by

$$\left(\bar{y}_{i_1j_1\cdot} - \bar{y}_{i_1j_1\cdot}\right) \pm t_{\alpha/2,ab(n-1)}\sqrt{\frac{2MS_E}{n}} \qquad ...(8.33)$$

To test the significance of factor A, the F-statistic F_A is computed as the ratio of the mean squares due to factor A to the mean squares for error. Hence,

$$F_A = \frac{MS_A}{MS_E} \qquad ...(8.34)$$

The calculated value of F_A is compared to $F_{\alpha,a-1,ab(n-1)}$, the critical value of F for a right-tail area of α, $(a - 1)$ degrees of freedom for the numerator, and $ab\ (n - 1)$ degrees of freedom for the denominator. If $F_A > F_{\alpha,a-1,ab(n-1)}$, the factor means at the different levels of A are not assumed to be equal.

A $100(1 - \alpha)\%$ confidence interval for the mean of factor A when at level i is given by

$$\bar{y}_{i..} \pm t_{\alpha/2,ab(n-1)}\sqrt{\frac{2MS_E}{(bn)}} \qquad ...(8.35)$$

Also, a $100(1 - \alpha)\%$ confidence interval for the difference between the factor A means that are at levels i_1 and i_2 is found from

$$\left(\bar{y}_{i_1..} - \bar{y}_{i_2..}\right) \pm t_{\alpha/2,ab(n-1)}\sqrt{\frac{2MS_E}{(bn)}} \qquad ...(8.36)$$

For factor B, the similar computations can be made for making inferences about the factor B.

8.8.2 Two-Factor Experiment Using a Randomized Block Design

Suppose there are two factors, A and B. The factor A has a number of levels, and the factor B has b number of levels. Let the number of blocks be r with each block containing ab units. The ab treatments are randomly assigned to the ab units within each block. The units within each block are chosen to be homogeneous with respect to the blocking variable. For example, the quality characteristic of a certain component is dependent on both temperature and pressure. The number of levels for temperature be three and that for pressure be four. If the raw materials for this component are supplied from five different sources, each batch of production using raw materials from a particular supplier is treated as a block. Hence, five batches are selected, and within each batch (or block), 12 treatments are applied to selected units during experimentation.

The model for the two-factor factorial experiment using a randomized block design is mathematically represented as

$$y_{ijk} = \mu + \alpha_i + \beta_j + \rho_k + (\alpha\beta)_{ij} + \varepsilon_{ijk} \qquad ...(8.37)$$

$i = 1,2,...a;\ j = 1, 2,...b;\ k = 1, 2,..., r$

where, y_{ijk} is the response for the ith level of factor A, the jth level of factor B, and the kth block; μ is the overall mean effect; α_i is the effect of the ith level of factor A; β_j is the effect of the jth level of factor B; and $(\alpha\beta)_{ij}$ is the interaction effect between factors A and B. The quantity ρ_k represents the effect of block k, and represents the random error component, assumed to be distributed normally with mean zero and constant variance, σ^2.

For analysis of variance in this case, the following notations are used:

$$y_{i\cdot\cdot} = \sum_{j=1}^{b}\sum_{k=1}^{r} y_{ijk} = \text{Sum of the responses for the } i\text{th level of factor } A$$

$\bar{y}_{i...} = \dfrac{y_{i..}}{(br)}$ = Average response for the ith level of factor A

$y_{.j.} = \sum_{i=1}^{a} \sum_{k=1}^{r} y_{ijk}$ = Sum of the responses for the jth level of factor B

$y_{i.j.} = \dfrac{y_{.j.}}{(ar)}$ = Average response for the jth level of factor B

$y_{ij.} = \sum_{k=1}^{r} y_{ijk}$ = Sum of the responses for the treatment if factor A is at the ith level and factor B is at the jth level over the blocks.

$\bar{y}_{ij.} = \dfrac{y_{ij.}}{r}$ = Average response for the treatment consisting of the ith level of factor A and jth level of factor B over the blocks.

$y_{..k} = \sum_{i=1}^{a} \sum_{j=1}^{b} y_{ijk}$ = Sum of the responses for the kth block

$\bar{y}_{...} = \sum_{i=1}^{a} \sum_{j=1}^{b} \sum_{k=1}^{r} y_{ijk}$ = Grand total of all observations

The total sum of squares (SS_T), the sum of squares due to blocks (SS_{BL}), the sum of squares for the main effects of factor A (SS_A), and the sum of squares for the main effects of factor B (SS_B) are computed as

$$SS_T = \sum_{i=1}^{a} \sum_{j=1}^{b} \sum_{k=1}^{r} y_{ijk}^2 - C \qquad \text{...(8.38)}$$

$$SS_{BL} = \frac{\sum_{k=1}^{r} y_{i..k}^2}{(ab)} - C \qquad \text{...(8.39)}$$

$$SS_A = \frac{\sum_{i=1}^{a} y_{i..}^2}{(br)} - C \qquad \text{...(8.40)}$$

$$SS_B = \frac{\sum_{j=1}^{b} y_{.j.}^2}{(ar)} - C \qquad \text{...(8.41)}$$

where, C = correction factor $= \dfrac{y_{..}^2}{(abr)}$ ...(8.42)

The sum of squares for the interaction between factors A and B are computed as

$$SS_{AB} = \frac{\sum_{i=1}^{a}\sum_{j=1}^{b} y_{ij.}^2}{r} - C - SS_A - SS_B \qquad ...(8.43)$$

and the error sum of squares is

$$SS_E = SS_T - SS_{BL} - SS_A - SS_B - SS_{AB} \qquad ...(8.44)$$

The number of degrees of freedom is $(a - 1)$ for factor A, $(b - 1)$ for factor B, $(r - 1)$ for the blocks, and $(a - 1)(b - 1)$ for the interaction between factors A and B. Since the total number of degrees of freedom is $(abr - 1)$, the number of error degrees of freedom is $(ab - 1)(r - 1)$. The ANOVA computations are shown in Table E8.10.

Table E8.10. ANOVA Table for the Two-Factor Factorial Experiment Using a Randomized Block Design

Source of variation	Degrees of freedom	Sum of squares	Mean square	F-statistic
Factor A	$a - 1$	SS_A	$MS_A = \frac{SSA}{(a-1)}$	$F_A = \frac{MS_A}{MSE}$
Factor B	$b - 1$	SS_B	$MS_B = \frac{SS_B}{(b-1)}$	$F_B = \frac{MS_B}{MS_E}$
Interaction between factors A and B	$(a-1)(b-1)$	SS_{AB}	$MS_{AB} = \frac{SS_{AB}}{[(a-1)(b-1)]}$	$F_{AB} = \frac{MS_{AB}}{MS_{SE}}$
Blocks	$r - 1$	SS_{BL}	$MS_{BL} = \frac{SS_{BL}}{(r-1)}$	$F_{BL} = \frac{MS_B}{MS_E}$
Error	$(ab- 1)(r-1)$	SS_E	$MS_E = \frac{SS_E}{[(ab-1)(r-1)]}$	
Total	$abr - 1$	SS_T		

The mean squares for factors, block, interactions, and error are computed by dividing the respective sum of squares by the corresponding degrees of freedom. In order to test the significance of the interaction effects between factors A and B, the F-statistic is as usual computed as

$$F_{AB} = \frac{MS_{AB}}{MS_E}$$

If, $F_{AB} > F_{\alpha,(a-1),(b-1),(ab-1)(r-1)}$, which is the table value for a level of significance α and $(a - 1)(b - 1)$ degrees of freedom in the numerator and $(ab - 1)(r - 1)$ in the denominator, the interaction effects between factors A and B are assumed to be significant. A $100(1 - \alpha)\%$ confidence interval for the treatment mean is found from

$$\bar{y}_{ij.} \pm t_{\alpha/2,(ab-1)(r-1)}\sqrt{\frac{MS_E}{r}} \qquad ...(8.45)$$

where, $t_{\alpha/2,(ab-1)(r-1)}$ is the t-value for a right-tail area of $\alpha/2$ and $(ab - 1)(r - 1)$ degrees of freedom.

Inferences about the difference between two treatment means through confidence intervals with a level of confidence of $(1 - \alpha)$ may be found as follows:

$$(\bar{y}_{i_1 j_1 .} - \bar{y}_{i_2 j_2 .}) \pm t_{\alpha/2,(ab-1)(r-1)} \sqrt{\frac{2MS_E}{r}} \quad ...(8.46)$$

The significance of the means of factor A is tested by computing the F-statistics as

$$F_A = \frac{MS_A}{MS_E}$$

At a level of significance α, if $F_{AB} > F_{\alpha,(a-1),(ab-1)(r-1)}$, the means of factor A are not assumed to be all equal at different levels.

In making an inference about an individual mean for a certain level of factor A, a 100 $(1 - \alpha)$% confidence interval is given by

$$\bar{y}_{i..} \pm t_{\alpha/2,(ab-1)(r-1)} \sqrt{\frac{MS_E}{(br)}} \quad ...(8.47)$$

Similarly, a $100(1 - \alpha)$% confidence interval for the difference between the factor A means, at levels i_1 and i_2 is computed as

$$(\bar{y}_{i_1 .} - \bar{y}_{i_2 ..}) \pm t_{\alpha/2,(ab-1)(r-1)} \sqrt{\frac{2MS_E}{(br)}} \quad ...(8.48)$$

The above-mentioned procedure for factor A can be applied in a similar manner to factor B.

Example E8.5. *It is observed that two factors, temperature and pressure, are affecting the ductility of an alloy. It is decided for experimentation that each of these factors is kept at three levels (for temperature: 150, 250, and 300°C; and for pressure 50, 100, and 150 kg/cm²). Each of the nine treatment combinations is replicated five times. The data collected with experimentation are shown in Table E8.5.*

(i) Determine the significance of interaction effect between temperature and pressure on the ductility of the alloy at 5% level of significance.

(ii) Determine the desirable settings of the temperature and pressure if a ductile alloy is preferred.

(iii) Find a 90% confidence interval for the mean ductility when the process temperature is set at 150° C and the pressure at 150 kg/cm².

Table E8.5. Ductility Values at Selected Treatments for the Alloys

Process temperature	**Process pressure (kg/cm²)**		
	50	**100**	**150**
	50	*44*	*75*
	65	*49*	*80*
150°C	*70*	*52*	*81*
	73	*40*	*76*
	68	*45*	*82*
	40	*25*	*86*
	45	*30*	*88*
250°C	*56*	*20*	*82*

	50 52	33 34	81 76
300°C	72 60 75 73 70	62 71 65 68 72	60 62 55 52 50

Solution:

(*i*) The ANOVA table for the given problem is shown below.

Source of variation	Degrees of freedom	Sum of squares	Mean square	F-statistic
Temperature (Factor A)	2	1154.533	577.267	18.24
Pressure (Factor B)	2	4732.133	2366.067	74.77
Interaction (AB)	4	6064.133	1516.033	49.91
Error	36	1139.200	31.644	
Total	44	13090.000		

At 5% level of significance, $F_{0.05,4,36} = 2.642$ (table value), which is found to be less than the computed $F = 47.91$. Hence, it is concluded that the interaction effect (*AB*) on the ductility of the alloy is significant.

(*ii*) Since the interaction effect is significant, the desirable settings are those values of temperature and pressure for which the ductility value is maximum. From the given data set, the maximum average value of ductility is found as (86 + 88 + 82 + 81 + 76)/5 = 82.6 for which the settings are: temperature: 250°C, and pressure = 150 kg/cm^2. Hence, these are the desirable settings of the process.

(*iii*) The average value of the ductility at a pressure of 150 kg/cm^2 and process temperature of 150°C is (75 + 80 + 81 + 76 + 82)/5 = 78.8. Hence, the 90% confidence interval for the mean ductility at this setting is given as

$$78.8 \pm t_{0.05,36} \sqrt{31.644/5} = 78.8 \pm 1.6888 \times 2.5157 = (74.5515, 83.0465)$$

8.8.3 Concept of Contrast and Its Use in Factorial Experiments

In factorial experiments with multiple treatments, it may be necessary to test hypotheses on a linear combination of the treatment means or to partition the total sum of squares of all treatments into component sums of squares of certain desirable treatments. The concept of 'contrast' is useful in these cases. A contrast of treatment means, in this context, is defined as a linear combination of the means such that the sum of the coefficients of the linear combination equals zero. For example, if there are p numbers of treatments with means, $\mu_1, \mu_2, \ldots \mu_p$ a contrast of means, denoted as L is defined as

$$L = k_1\mu_1 + k_2\mu_2 + \ldots + k_p\mu_p = \sum_{i=1}^{p} k_i\mu_i \quad \ldots(8.49)$$

where, k_i are constants such that $\sum_{i=1}^{p} k_i = 0$

In many situations, it is desired that the difference between the treatment means is to be tested for its significance. For testing the hypothesis of equality of treatment means or otherwise, the sample mean and sample variance of the contrast need to be estimated to compute the test statistic. If $\bar{y}_i$ is the sample mean of treatment i, the sample estimate of L is computed as

$$\hat{L} = k_1\bar{y}_1 + k_2\bar{y}_2 + \ldots + k_p\bar{y}_p = \sum_{i=1}^{p} k_i\ \bar{y}_i \qquad \ldots(8.50)$$

If r_i is the number of observations from which the ith sample mean is computed and MSE is the mean square error for the design used in the experiment, the variance of $\hat{L}$ is computed as

$$\text{Var}(\hat{L}) = \left(\frac{k_1^2}{r_1} + \frac{k_2^2}{r_2} + \ldots + \frac{k_p^2}{r_p}\right) MS_E = \left(\sum_{i=1}^{p} \frac{k_i^2}{r_i}\right) MS_E \qquad \ldots(8.51)$$

The t-static is calculated as

$$t = \frac{\hat{L}}{\sqrt{\text{Var}(\hat{L})}} \qquad \ldots(8.52)$$

At a level of significance α, if $|\ t\ | > t_{\alpha/2,\upsilon}$ where υ is the number of degrees of freedom of the experimental error and $\alpha/2$ is the right-tail area of the t-distribution, the null hypothesis is rejected.

In case there is a significant difference between the treatment means, a $100(1 - \alpha)$ % confidence interval for L may be required to be computed, and is given as

$$CI = \hat{L} \pm t_{\alpha/2,\upsilon} \sqrt{Var\ (\hat{L})} \qquad \ldots(8.53)$$

where, $t_{\alpha/2,\ \upsilon}$ represents the t-value for a right-tail area of $\alpha/2$ and υ degrees of freedom.

Although equation (8.49) refers to one contrast formed with a linear combination of a number of treatment means, it is also possible to form more than one contrast for a given set of treatments. In this context, the concept of 'orthogonal' contrast becomes relevant. Let us define orthogonal contrast in this case.

If there are p number of treatments, there can be $(p - 1)$ number of contrasts formed that are statistically independent of one another. Two contrasts, say L_1 and L_2 are independent if they are orthogonal to each other. Suppose the following two contrasts are formed with p number of treatment means:

$$L_1 = k_{11}\mu_1 + k_{12}\mu_2 + \ldots + k_{1p}\mu_p$$
$$L_2 = k_{21}\mu_1 + k_{22}\mu_2 + \ldots + k_{2p}\mu_p$$

The contrasts L_1 and L_2 are orthogonal if and only if

$$\sum_{j=1}^{p} k_{1j}k_{2j} = 0$$

This means that the sum of the products of the corresponding coefficients associated with the treatment means is zero.

For example, the two contrasts involving three treatment means are as follows:

$$L_1 = 2\mu_1 - \mu_3$$
$$L_2 = -\mu_1 + \mu_2 - 2\mu_3$$

Here, L_1 is used to test the null hypothesis that the mean of treatment 3 is equal to twice of treatment mean 2, and L_2 is used to test the null hypothesis that the mean of treatment 1 and twice of treatment 3 equals the mean of treatment 2. The contrasts L_1 and L_2 are orthogonal, because the sum of the products of the corresponding coefficients of the means is

$$(2 \times -1) + (0 \times 1) + (-1)(-2) = 0$$

Contrasts of totals are also useful in analysis of variance. Such contrasts aid in partitioning the treatment sum of squares into components such that each component is associated with a contrast of interest. Let r_j represent the number of replications for the jth treatment (out of p treatments). The sum of the responses of the ith treatment is given as

$$T_i = y_{i.} = \sum_{j=1}^{r_j} y_{ij}$$

The contrast of totals is thus given as

$$\hat{L}_1 = k_1T_1 + k_2T_2 + \ldots + k_pT_p \qquad \ldots(8.54)$$

if and only if

$$r_1k_1 + r_2k_2 + \ldots + r_pk_p = 0$$

If the number of replications is given by r for each of the treatments, $\hat{L}$ is a contrast if

$$\sum_{i=1}^{p} k_i = 0$$

The variance of the contrast of totals is obtained from

$$\text{Var}(\hat{L}) = (r_1k_1^2 + r_2k_2^2 + \ldots + r_pk_p^2) = MS_E \qquad \ldots(8.55)$$

Two contrasts of totals is given by

$$\hat{L}_1 = k_{11}T_1 + k_{12}T_2 + \ldots + k_{1p}T_p$$
$$\hat{L}_2 = k_{21}T_1 + k_{22}T_2 + \ldots + k_{2p}T_p$$

are orthogonal if and only if

$$\sum_{i=1}^{p} k_{1j}k_{2j} = 0$$

This condition is similar to the condition for the contrasts of means.

Orthogonal contrasts for totals are important because they can be used to decompose the treatment sum of squares in analysis of variance, and thereby test hypotheses on the effectiveness of selected treatments. If $\hat{L}_1$ is a contrast of totals, then a single degree-of-freedom component for that contrast of the treatment sum of squares (SS_{TR}) is given as

$$S_1^2 = \frac{\hat{L}_1^2}{D_1} \qquad \ldots(8.56)$$

where $$D_1 = r_1k_{11}^2 + r_2k_{12}^2 + \ldots + r_pk_{1p}^2 = \sum_{j=1}^{p} r_jk_{1j}^2$$

Likewise, if $\hat{L}_1$ and $\hat{L}_2$ are orthogonal, then the sum of squares for the contrast $\hat{L}_2$ is given by

$$S_2^2 = \frac{\hat{L}_2^2}{D_2}$$

where $$D_2 = r_1k_{21}^2 + r_2k_{22}^2 + \ldots + r_pk_{2p}^2$$

Similarly, if $\hat{L}_1, \hat{L}_2, \ldots, \hat{L}_{p-1}$ are mutually orthogonal contrasts of totals, then

$$S_1^2 + S_2^2 + \ldots + S_{p-1}^2 = SS_{TR}$$

Example E8.6. *Refer to Example 8.1 concerning the effectiveness of the training programmes for the production executives.*

(i) Find a 90% confidence interval for the difference in the mean effectiveness of programme 1 and the average of that using programmes 2 and 3

(ii) Consider the following two contrasts of total: (a) difference between the totals for training programmes 1 and 3, (b) difference between the totals for the sum of training programmes 1 and 3 and twice of training programme 2. Find the sum of squares due to each of these two contrasts, and what null hypothesis is indicated by contrast (b).

Solution:

(*i*) The contrast, L is given as

$$L = 2\mu_1 - \mu_2 - \mu_3$$

An estimate of L is : $$\hat{L} = 2 \times 77.333 - 75.666 - 87.333 = -8.333$$

$$\text{Var}(\hat{L}) = (4/3 + 1/3 + 1/3) \times 10.111 = 20.222$$

Thus, the 90% confidence interval for $(2\mu_1 - \mu_2 - \mu_3)$ is computed as

$$-8.333 \pm t_{0.05,4}\sqrt{20.222} = -8.333 \pm 2.132 \times 4.4962 = (-17.920, 1.254)$$

(*ii*) The two contrasts are as follows:

$\hat{L}_1 = T_1 - T_3$, and $\hat{L}_2 = T_1 + T_3 - 2T_2$ (both are orthogonal contrasts). The estimates of $\hat{L}_1$ and $\hat{L}_2$ are:

$$\hat{L}_1 = 232 - 262 = -30$$

$$\hat{L}_2 = 232 + 262 - 2 \times 227 = 40$$

Now, $$D_1 = 3 + 3 = 6$$

$$D_2 = 3 + 3 + 4 \times 3 = 18$$

Hence, the sum of squares for the two contrasts are:

$$S_1^2 = \hat{L}_1^2/D_1 = (-30)^2/6 = 150.00$$

$$S_2^2 = \hat{L}_2^2/D_2 = (40)^2/18 = 88.89$$

The second contrast indicates the null hypothesis, $H_0 = \mu_1 + \mu_3 - 2\mu_2 = 0$

An estimate of the contrast of means = 77.333 + 87.333 – 2 × 75.667 = 13.333

The variance of the contrast of means = (1/3 + 1/3 + 4/3) × 10.111 = 20.222. Hence, the test statistic, $t = 13.333/\sqrt{20.222} = 2.965$.

At 5% level of significance, the table value of $t_{0.025,4} = 2.776$ which is less than the computed $t = 2.965$. Thus, we reject H_0 and conclude that there is significant difference between the average ratings of training programmes 1 and 3 and that of training programme 2.

8.9 2ᵏ FACTORIAL EXPERIMENT

A planned experimentation requires a number of steps to be followed. First, the set of factors affecting the response variable(s) is to be identified. On their identification, desirable levels of the selected factors should be known in the second stage. Finally, it may be of interest to find the relationship between the factor levels, the corresponding responses, and the physical and economic constraints that may have to be considered. In the literature on planned experimentation, a large number of experimental designs are developed and recommended. In deciding on a particular design, the objective of carrying out an experiment should be kept in mind. In most cases, in order to represent the reality, multifactor experiments are usually employed. Among the multifactor designs, 2^k factorial experiment is widely used. In this experiment, there are k numbers of factors, each factor at two desirable levels. The total number of treatment combinations is 2^k. For example, if there are three factors each of two levels, there are 2^3 or 8 treatments in each replication.

Usually, the level of each factor is thought of as either low or high. A treatment is represented by a series of lowercase letters; the presence of a particular letter indicates that the corresponding factor is at the high level. The absence of this letter indicates that the factor is at the low level. The notation (1) is used to indicate the treatment for which all factors are at the low level. In many instances, the low level signifies the absence of a factor and the high level its presence. Suppose, there are three factors A, B, and C, each at two levels. The set of treatments for this experiment is as follows:

1. (1)
2. a
3. b
4. ab
5. c
6. ac
7. bc
8. abc

Figure 8.11 shows the layout of a 2^3 experiment with the factors denoted as A, B, and C. In this layout high levels of the factors are denoted by + 1 and the low levels by – 1.

To estimate the main effects of the factors, and interaction effects between the factors, the concept of contrasts is used. For example, to determine the main effect of factor B, the average value of the observations for which B is at the high level is computed, and subtract from this value the average of the observations for which B is at the low level. If there are r replications for each treatment, the main effect of factor B is given by

$$B = \frac{b + ab + bc + abc}{4r} - \frac{(1) + a + c + ac}{4r}$$

$$= \frac{1}{4}\ [b + ab + bc + abc - (1) - a - c - ac] \qquad \text{...(8.57)}$$

In equation (8.57), the lowercase letters denote the sum of the responses for the corresponding treatments. The main effects of other factors can be similarly computed. As has been observed, the contrast for factor B are the orthogonal contrasts of totals.

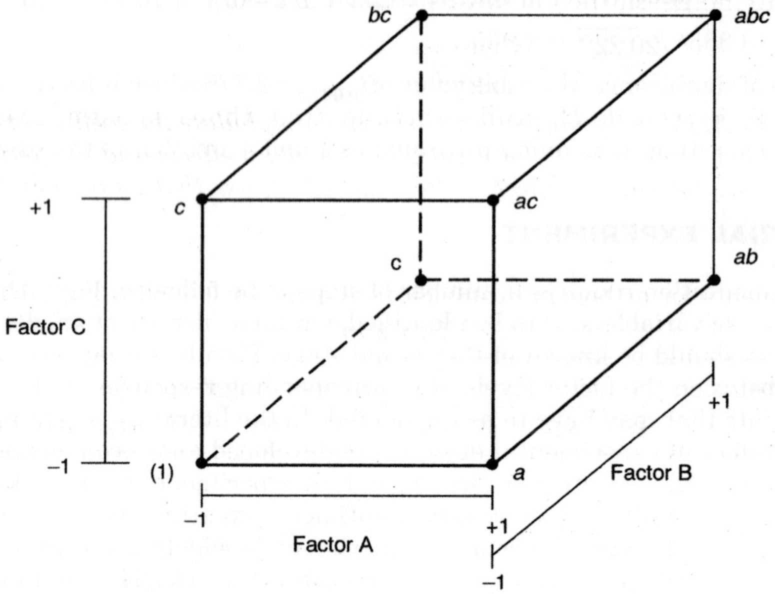

Fig. 8.11 Layout of a 2^3 Factorial Experiment.

Table 8.11 represents the coefficients for orthogonal contrasts in a 2^3 factorial experiment. In this table for each main effect contrast, a (+) sign is used for each treatment combination in which the letter for the effect occurs. All other combinations receive a (–) sign. The signs for the interaction contrasts are the product of the signs of the main effects whose letters appear in the interaction. For example, consider the contrast *AB*. Its sign under treatment column a is –, which is the product of the sign of *A* in the treatment column a (+) and the sign of *B* in the treatment column a (–). The contrast for any of the main effects or interactions can thus be obtained from this table.

Table 8.11: Coefficients for Orthogonal Contrasts in a 2^3 Factorial Experiment

	Treatment combination							
Contrast	(1)	*a*	*b*	*ab*	*c*	*ac*	*bc*	*Abc*
A	−	+	−	+	−	+	−	+
B	−	−	+	+	−	−	+	+
AB	+	−	−	+	+	−	−	+
C	−	−	−	−	+	+	+	+
AC	+	−	+	−	−	+	−	+
BC	+	+	−	−	−	−	+	+
ABC	−	+	+	−	+	−	−	+

The sum of squares for the set of orthogonal contrasts for the total is given as

$$SS = \frac{(\text{Contrast})^2}{r2^k} \qquad \ldots(8.58)$$

To estimate the effect, the following equation is used.

$$\text{Effect} = \frac{\text{Contrast}}{r2^{k-1}} \qquad \text{...(8.59)}$$

Example E8.7. *For a 2^4 full factorial experiment, tabulate the coefficients for orthogonal contrasts. Write down the contrasts for estimating the main effects and the four-factor interactions. If the four-way interaction effect is not significant, use that as a basis to confound the design into two blocks.*

Solution:

The coefficients for orthogonal contrasts are given in the following tables (*a*) and (*b*):

Table (a)

Treatment	Factor effects/Contrast											
	A	B	AB	C	AC	BC	ABC	D	AD	BD	ABD	CD
(1)	−	−	+	−	+	+	−	−	+	+	−	+
a	+	−	−	−	−	+	+	−	−	+	+	+
b	−	+	−	−	+	−	+	−	+	−	+	+
ab	+	+	+	−	−	−	−	−	−	−	−	+
c	−	−	+	+	−	−	+	−	+	+	−	−
ac	+	−	−	+	+	−	−	−	−	+	+	−
bc	−	+	−	+	−	+	−	−	+	−	+	−
abc	+	+	+	+	+	+	+	−	−	−	−	−
d	−	−	+	−	+	+	−	+	−	−	+	−
ad	+	−	−	−	−	+	+	+	+	−	−	−
bd	−	+	−	−	+	−	+	+	−	+	−	−
abd	+	+	+	−	−	−	−	+	+	+	+	−
cd	−	−	+	+	−	−	+	+	−	−	+	+
acd	+	−	−	+	+	−	−	+	+	−	−	+
bcd	−	+	−	+	−	+	−	+	−	+	−	+
abcd	+	+	+	+	+	+	+	+	+	+	+	+

Table (b)

Treatment	Factor effects/Contrast		
	ACD	*BCD*	*ABCD*
(1)	−	−	+
a	+	−	−
b	−	+	−
ab	+	+	+
c	+	+	−
ac	−	+	+
bc	+	−	+
abc	−	−	−
d	+	+	−
ad	−	+	+
bd	+	−	+
abd	−	−	−
cd	−	−	+
acd	+	−	−
bcd	−	+	−
abcd	+	+	+

(*i*) Contrast (*ABCD*) = $(1) - a - b + ab - c + ac + bc - abc - d + ad + bd - abd + cd - acd - bcd + abcd$

To confound the design into two blocks, the treatments with a positive sign in the contrast ABCD are put into one block, and the treatments with a negative sign are put into another block.

Hence, the treatments in the first block are: (1), *ab*, *ac*, *bc*, *ad*, *bd*, *cd* and *abcd*, and the treatments in the second block are: *a*, *b*, *c*, *abc*, *d*, *abd*, *acd* and *bcd*.

(*ii*) Considering *AB* as the confounding factor, divide the experiment into two blocks. Estimate the effect of factor A.

$$\text{Contrast }(AB) = (1) - a - b + ab + c - ac - bc + abc + d - ad - bd + abd + cd - acd - bcd + abcd$$

The whole experiment is divided in two blocks. In the first block, the treatments with a positive sign are included, whereas in the second block, the treatments with a negative sign are grouped. Thus, the first block contains the treatments: (1), *ab*, *c*, *abc*, *d*, *abd*, *cd*, and *abcd*, and the second block contains the treatments: *a*, *b*, *ac*, *bc*, *ad*, *bd*, *acd*, and *bcd*.

Contrast (*A*) is to be computed in order to estimate the effect of Factor *A*

$$\text{Contrast }(A) = a + ab + ac + abc + ad + abd + acd + abcd - 1 - b - c - bc - d - bd - cd - bcd.$$

An important aspect in factorial experiments, known as 'confounding', need to be considered. In multifactor factorial experiments, the number of treatments increases exponentially with the increase in the number of factors. For example, if 6 factors with 2 levels each are to be considered, the number of treatments is $2^6 = 64$. Thus, for a complete replication of the treatments in one block, 64 experimental units are needed. It is often difficult to obtain a large number of homogeneous units per block for complete replication of all treatments.

To increase the precision of an experiment by blocking, the block size must be kept as small as possible. One way to achieve this is through **confounding**, where the estimation of a certain treatment effect is 'confounded' (made not distinguishable) with the blocks. In confounding, not all treatments are required to be replicated in each block. The block contrast is then equivalent to the treatment contrast that is confounded. Thus, it is not possible to estimate the effects between blocks because estimating one treatment effect essentially matches it. Thus it may not be possible to know if any observed differences in the response variable between blocks are solely due to differences in the blocks. Such an effect cannot be isolated from that of one of the treatments, implying that differences between blocks could be due to this particular treatment (that is, confounded with the blocks), or the blocks themselves, or both. Although the treatment that it confounded with the blocks cannot be estimated, there is possibility of improving the precision of the experiment through reduced block sizes. A 2^k experiment may be confounded into blocks of size 2^p, where $p < k$.

For example, in a 2^3 factorial experiment 8 experimental units are needed, for a complete replication, which then influences the block size. Suppose, for one blocking variable, only 4 experimental units are made available instead of 8 experimental units for the block, because of some resource and time constraints. For dividing the total 8 treatment combinations for this design into two blocks of size 4 each, the procedure as outlined below should be employed.

Referring to the coefficients for orthogonal contrasts in a 2^3 factorial experiment shown in **Table 8.11**, the confounding contrast is typically selected to be a high-order interaction. The contrast for *ABC* is given by

$$\text{Contrast }(ABC) = a + b + c + abc - (1) - ab - ac - bc$$

The plus-sign terms are grouped in one group (*a, b, c, abc*) and the minus-sign terms in another group [(1), *ab, ac, bc*]. Two blocks of size 4 each are constructed. In block 1, the applied treatments are *a, b, c* and *abc*. In block 2, the applied treatments are (1), *ab, ac* and *bc*.

An estimate of the block effect is obtained by determining the difference in the average responses obtained from the two blocks. To estimate the effect of the treatment *ABC*, the same procedure is employed. Since the four positive coefficients in the contrast for *ABC* are assigned to block 1 and the four negative coefficients are assigned to block 2, the block effect and the *ABC* interaction effects are identical. In other words, the *ABC* interaction effects is confounded with the block. Thus, to estimate the main effect of factor *A*, the following contrast is used:

$$\text{Contrast }(A) = a + ab + ac + abc - (1) - b - c - bc$$

Hence, there are four runs from each block. Also, from each block there is one treatment whose effect is added and another treatment whose effect is subtracted. For example, from block 1, the response variable is added for treatments *a* and *abc,* and it is subtracted for treatments *b* and *c*. So, any difference between the two blocks cancels out, and estimating the main effect of factor *A* is not affected by blocking. The same is true for the remaining main effects and two-way interactions. The scheme for blocking as demonstrated can be used for any chosen contrast. Usually, the factors chosen to confound with blocks are the higher-order interactions, which are normally not as important as the main effects and the low-order interactions.

Example E8.8. *In order to develop a good quality synthetic fuel, experiments are conducted with three different factors, each controlled at two levels. Factor A is a concentration of corn extract at 5% and 10%, Factor B is a concentration of another compound at 15% and 25%, and Factor C is the distillation temperature at 120°C and 150°C. The degree of undesirable emission level at each treatment combination is shown in Table 8.11 for three replications.*

(*i*) *Find all the main and the interaction effects.*

(*ii*) *Find the sum of squares for each of the main and interaction effects.*

(*iii*) *At 5% level of significance, which effects are significant?*

Table 8.12. Values of Undesirable Emission Levels of the Fuel at Different Treatment Combinations

Treatment combination	Emission level		
	Replication 1	Replication 2	Replication 3
(1)	30	24	26
a	18	22	24
b	30	32	25
ab	43	47	41
c	28	24	22
ac	54	49	46
bc	58	48	50
abc	24	20	22

Solution:

(*i*) In this case, a 2^3 full factorial experiment is conducted, and the number of treatment combinations is 8 for each replication.

The main effect of factor A is computed as

$$A = \frac{1}{3 \times 4}[a + ab + ac + abc - (1) - b - c - bc]$$

$$= \frac{1}{12}[64 + 131 + 149 + 66 - 80 - 87 - 74 - 156] = 1.038$$

The other main effects and interaction effects are calculated in the same manner, and shown in the ANOVA table given below.

Source of variation	Effect	Degree of freedom	Sum of squares	Mean squares	F-statistic
A	1.083	1	7.0417	7.0417	0.57
B	6.083	1	222.0417	22.0417	18.00
C	6.917	1	287.0417	287.0417	23.27
AB	– 8.750	1	459.3750	459.3750	37.25
AC	– 3.583	1	77.0417	77.0417	6.25
BC	– 6.250	1	234.3750	234.3750	19.00
ABC	– 18.750	1	2109.3750	2109.3750	171.03
Error		16	197.3333	197.3333	
Total		23	3593.6250	3593.6250	

(*ii*) The sum of squares for factor A is computed as

$$(13^2)/(3 \times 8) = 7.0417$$

The sum of squares for all other main and interaction effects are calculated and shown in the ANOVA table. The total sum of squares (SS_T) is computed as

$$SS_T = [30^2 + 24^2 + 26^2 + 18^2 + \ldots\ldots + 20^2 + 22^2] - (807)^2/24$$

$$= 3593.625$$

(*iii*) At 5% level of significance, $F_{0.05,1,16} = 4.4967$ (table value). Comparing this value with computed F-statistic values as shown in the ANOVA table, it is observed that the interaction effects: AB, AC, BC, and ABC are all significant. The main effects, B and C are also having F-statistic values greater than the table value of F-statistic. However, since the interaction effects are significant, it is difficult to make any definite conclusion regarding the significance of the main effects.

8.10 FRACTIONAL FACTORIAL EXPERIMENTS

As has already been mentioned, the number of runs in a factorial experiment increases exponentially with increase in the number of factors. Many a time, due to cost and time considerations, it may not be possible or necessary to run all the experiments as required for a full factorial experiment. A properly designed fractional replication of the factorial experiments may serve the purpose of planned experimentation with reduced cost and time as well as achieving reasonable accuracy in the estimates. For example, for a 2^6 factorial experiment, 64 experimental runs are required for a full replication. There are 6 main effects, 15 two-way interactions (C_2^6), 20 three-way interactions, 15 four-way interactions, 6 five-way interactions, and 1 six-way interaction. Thus, there are 63 degrees of freedom for estimating different treatment effects. If only the main effects and two-way interactions (which require 21 degrees of freedom for estimation in total) are considered relevant in this experiment, it is better if a fractional factorial experiment is opted for. This is the basic purpose and objective of fractional replication in factorial experiment. Among the fractional factorial experiment, a half-fraction of 2^k experiment, denoted as 2^{k-1} factorial experiment, is a special case. Details of this experiment are given below.

A half-fraction of the 2^k experiment is denoted by 2^{k-1} factorial experiment. To determine the treatment combinations in this experiment, the defining contrast (also known as generator) is used. This usually represents a high-order interaction, which may be unimportant or negligible and may not be required to be estimated. For example, for a 2^3 full factorial design with factors A, B, and C, coefficients of orthogonal contrasts are shown in **Table 8.11**. The defining contrast, I associated with the three-way interaction ABC and, hence, $I = ABC$. For a 2^{3-1} fractional experiment, the treatment combinations associated with plus signs in the defining contrast are only selected. From **Table 8.11** these are represented as a, b, c, and abc, which involve only 4 runs of the experiment. Alternatively, the four other treatments with minus signs in the defining contrast can also be selected. In this case, the treatment combinations are (1), ab, ac, and bc, and the defining contrast is represented as $I = -ABC$. Conventionally, the fraction with the plus sign in the defining contrast is called the 'principal fraction', and the other is called the 'alternate fraction'.

In a fractional replication of a full factorial experiment, it is to be mentioned that certain effects cannot be separately estimated. This is a serious disadvantage of this design. Suppose, a 2^{3-1} experiment with generator $I = ABC$ is considered. With four funs (a, b, c and d), there will be 3 degrees of freedom to estimate the main effects. Using **Table 8.12** and noting that only four of the eight possible treatments are run, it seems that the estimates of the main effects are given by

$$A = \frac{1}{2}(a + abc - b - c)$$

$$B = \frac{1}{2}(b + abc - a - c)$$

$$C = \frac{1}{2}(c + abc - a - c)$$

where, the lowercase letters represent the sum of the responses in the corresponding treatments. However, in a fractional factorial experiment, these expressions, strictly speaking, represent the combined effect of the factor shown and its alias. Similarly, the estimates of the two-factor interactions are expressed as

$$AB = \frac{1}{2}(c + abc - a - b)$$

$$AC = \frac{1}{2}(b + abc - a - c)$$

$$BC = \frac{1}{2}(a + abc - b - c)$$

Again, these expressions also represent the combined effect of the factor shown and its alias.

The linear combination of the sum of the treatment responses that estimates the main effect A also estimates the interaction BC. In this case, A and BC are said to be **aliases.** If the effect of the contrast is significant, we cannot conclude whether it is due to the main effect of A, interaction effect BC, or a mixture of both. Similarly, B and AC are aliases, and so are C and AB.

To find the aliases of a given contrast, its interaction with the defining contrast is considered. For example, the alias of A, using the defining contrast as ABC, is found as follows:

Alias $(A) = A * ABC = BC$

Similarly, the aliases for B and C are

Alias $(B) = B * ABC = AC$

Alias $(C) = C * ABC = AB$

In determining an alias, the letters appearing in both the contrasts are combined, and any letter that appears twice is deleted. Since the contrasts A and BC are aliases, if only four treatments a, b, c, and abc are run, the *combined effect* of both contrasts is obtained from $1/2(a + abc - b - c)$. Therefore, using the principal fraction, an estimate of $(A + BC)$ is obtained from, an estimate of $(B + AC)$ is obtained from $1/2(b + abc - b - c)$; and that of $(C + AB)$ is obtained from $1/2(b + abc - a - b)$. If the two-factor interactions are significant, then by running the alternate fraction [that is the treatments (1), ab, ac and bc], the main and the interaction effects can be estimated by combining the results of the principal fraction and the alternate function.

If the defining contrast $I = -ABC$ is selected, the runs consist of the alternate fraction of treatment (1), ab, ac and bc. The alias of A is $A* -ABC = -BC$, the alias of B is $-AC$, and the alias of C is $-AB$. Thus, the estimate of $(A - BC)$ is obtained from $1/2[ab + ac - bc - (1)]$, the estimate of $(B - AC)$ is obtained from $1/2[ab + bc - ac - (1)]$, and that of $(C - AB)$ from $1/2[ac + bc - ab - (1)]$.

Upon combining the results of the principal and alternate fractions, the estimate of the main effect of factor A is expressed as

$$A = \frac{1}{2}(A + BC) + \frac{1}{2}(A - BC)$$

Similarly, the other two main effects are given by

$$B = \frac{1}{2}(B + AC) + \frac{1}{2}(B - AC)$$

$$C = \frac{1}{2}(C + AB) + \frac{1}{2}(C - AB)$$

The interaction effects can also be obtained by combining the results of the two fractions as follows:

$$AB = \frac{1}{2}(C + AB) - \frac{1}{2}(C - AB)$$

$$AC = \frac{1}{2}(B + AC) - \frac{1}{2}(B - AC)$$

$$BC = \frac{1}{2}(A + BC) - \frac{1}{2}(A - BC)$$

This process of combining the results of two fractional factorial experiments to isolate the estimates of main and interaction effects that ordinarily cannot be estimated from either fraction separately is a unique feature in a fractional factorial experiment. If a full factorial experiment cannot be conducted because of some genuine reasons, it may be possible to run sequences of small size fractional factorial experiments. The accumulated knowledge from the ongoing sequence of experiments can be used to update or modify the experiments that follow. For example, if a fractional replication suggests the non-significance of certain factors, they can be dropped in future experiments, which may lead to a full factorial experiment with the remaining factors.

In this context, the concept of 'design resolution' is relevant. In fractional factorial experiments, the main effects of the factors are not aliased with each other as the main effects should be estimated independently. The degree to which the main effects are aliased with the interaction terms (two-factor or higher-order) is represented by the 'resolution' of the corresponding design. In particular, designs of resolution III, IV and V are important. These three design resolutions are defined as follows:

Resolution III designs: Here, the main effects are not aliased with each other, but they are aliased with two-factor interactions. The two-factor interactions may be aliased with each other. The 2^{3-1} design with the defining contrast given by $I = ABC$ is of resolution III. In such a design, the alias structure is given by $A = BC, B = AC, and\ C = AB$. The main effects can be estimated only if the two-factor and higher interactions are not present.

Resolution IV designs: Here, main effects are not aliased with each other or with any two-factor interactions. Two-factor interactions are aliased with each other. Thus, the main effects can be estimated regardless of the significance of the two-factor interactions as long as three-factor and higher order interactions are not present. An example of a resolution IV design is a 2^{4-1} design with the defining contrast given by $I = ABCD$. In this case, the alias structure is $A = BCD, B = ACD, C = ABD, D = ABC, AB = CD, AC = BD$, and $AD = BC$.

Resolution V designs: Here, no main effects or two-factor interactions are aliased with any other main effects or two-factor interactions. Two-factor interactions, however, are aliased with three-factor interactions. An example of such a design is the 2^{5-1} design with a defining contrast given by $I = ABCDE$.

In general, the fractional factorial experiment is represented as 2^{k-p} factorial experiment. The 2^{k-p} experimental design is a $1/2^p$ fraction of the 2^k factorial experiment. For example, a 2^{k-2} experiment employs one-fourth of the total number of treatments. Although this is a smaller size experiment, it is possible for this experiment to provide sufficient information with significant reduction in experimental cost. Suppose, there are 6 factors, each at 2 levels and a factorial experiment is to be conducted. Suppose, also that a full factorial experiment is not cost effective. A 2^{6-1} fractional factorial experiment requires 32 experimental runs, which gives us 31 degrees of freedom to estimate effects. There are 6 main effects and 15 two-factor interactions. Thus, 22 runs are sufficient if only main effects and two-factor interactions are considered important, making the 2^{6-1} experiment inefficient because it requires 32 runs. However, a 2^{6-2} fractional factorial experiment need 16 runs and has 15 degrees of freedom to estimate effects. All the main effects and some (not more than 9) of the two-factor interactions can be estimated. If this design meets the specified needs the 2^{6-2} experiment can be treated as more cost efficient than the 2^{6-1} experiment.

To generate the treatment combinations for a 2^{k-p} fraction factorial experiment, p defining contrasts or generators are used. Using the first contrast, a 2^{k-1} fractional experiment is designed. Starting with the treatments selected in the 2^{k-1} experiment and using the second contrast, a half-replicate of the 2^{k-1} experiment yielding a 2^{k-2} fractional experiment is also created. This procedure continues until the pth contrast is used on the treatment combinations in the 2^{k-p+1} experiment to yield the 2^{k-p} fractional experiments. The aliases for the contrasts can be found using the previously explained procedure. In a fractional factorial experiment, the objective is to reduce the size of the experiment and is to reduce the number of treatment combinations. Depending on the degree of fractionalization, the contrasts may have aliases. Thus, it is not possible to estimate all of the factorial effects separately.

Example E8.9. *(i) Consider a 2^4 factorial experiment, using BC as the defining contrast. Find the treatment combinations in a 2^{4-1} fractional factorial experiment. Find also the alias of the contrasts, and estimate the effect of the contrast BC.*

(ii) Determine the treatment combinations in a 2^{4-2} fractional factorial experiment if AD is taken as the defining contrast. Find also the alias of the contrast AD.

Solution:

(i) Contrast $(BC) = (1) + a + bc + abc + d + ad + bcd + abcd$

$$- b - ab - c - ac - bd - abd - cd - acd.$$

The treatment combinations for 2^{4-1} fractional factorial experiment with BC as the defining contrast are: (1), *a*, *bc*, *abc*, *d*, *ad*, *bcd*, and *abcd*.

The alias structure (where $I = BC$):

$$A = ABC, B = C, D = BCD, AB = AC = ABC,$$
$$AD = ABCD, BD = CD, ABD = ACD.$$

To estimate the effect of the contrast *BC*, an experiment is to be conducted with the treatment combinations corresponding to the alternate fraction ($I = -BC$). The information from the two experiments is combined to estimate the effect of contrast *BC*.

(ii) The treatment combinations in a 2^{4-2} fractional factorial experiment using *AD* as the defining contrast are as follows:

(1), *bc*, *ad*, and *abcd*

The alias structure (where $I = BC = AD = ABCD$):

$$A = ABC = D = BCD, B = C = ABD = ACD, AB = AC = BD = CD.$$

8.11 SUMMARY

This chapter discusses the fundamental concepts of experimental design in the context of ever-increasing demand of improved quality in products and services by the customers. Highlighting the important features and terminologies in experimental design and its uses and advantages, this chapter explains in detail the traditional experimental designs, such as completely randomized design, randomized block design, and Latin square design. The procedure of measuring the main and interaction effects on the response or output variable in an experiment is clearly described. The evaluation process using ANOVA procedures in this context is illustrated with examples.

The procedures and significance of conducting factorial experiment are presented in this chapter. While a full factorial experiment represents all possible treatment combinations of the factors considered, it may not be desirable because of time and resource constraints in many situations. A fractional factorial experiment, which is a subset of the full factorial design, selects the treatment combinations in such a way that the desirable effects can be estimated. For factorial experiments, the role of contrasts including orthogonal contrasts in estimating the treatment effects is explained in detail. The procedure of blocking by using the principle of confounding to increase the precision of the experiment is illustrated with a number of examples. The procedures for selecting the fractional replicates of the full factorial design to reduce the size of the experiment are also presented in this chapter. Overall, the experimental designs, as they are used in achieving the objective of improving and sustaining the quality of products and processes with reduced cost, are discussed at length in this chapter.

QUESTIONS

8.1 Select a process of your choice. Identify one important factor for this process. Plan an experiment with this factor for the process that details objectives, problem definitions, experimental variables, randomization, replication, design of the experiment, analysis procedures, and resource requirements.

8.2 Select a product of your choice. For an important quality characteristic of this product, list the potential causes that affect the level (average) and the variability of the quality characteristic. You may draw a cause-and-effect diagram for dispersion analysis. Plan a factorial experiment to determine the effects of some of the potential causes.

8.3 List and explain the desirable features of an experimental design.

8.4 Mention the four important tools that need to be used for a planned experiment. Provide an example of the specific application for each.

8.5 Identify a process of your choice, and cite examples of the following terms in relation to it:

(*i*) Response variables (*ii*) Input variables
(*iii*) Process parameter (*iv*) Noise factors
(*v*) Experimental units (*vi*) Blocks.

8.6 Citing a few examples, discuss the significance of experimental design for quality control and improvement.

8.7 State the reasons of using the following principles in experimental design, with suitable examples,

(*i*) Randomization (*ii*) Replication
(*iii*) Blocking

8.8 Give an example of completely randomized design. Under what conditions this design is preferred to be used?

8.9 Give an example of randomized block design. Why is it used and when? Suppose you have to evaluate three alternative teaching methods that a professor may be using at your department. How do you use the blocking principle in this evaluation study?

8.10 Make an industrial visit to the manufacturing plants of different companies in your locality. Get the data on the numbers and types of planned experiments carried out by the design and process engineers during the past three years at each plant, and write a report on the status of 'planned experimentation' as a quality improvement strategy for each.

8.11 List the major shortcomings of using 'one-variable-at-a-time' approach in experimental design.

8.12 What are the relative merits and demerits of factorial experiments? Mention a few industrial or production situations where experimental design is not practicable.

8.13 Distinguish with examples between

(*i*) fixed effects and random effects models.
(*ii*) randomized block and Latin Square designs.

8.14 Define 'main effects' and 'interaction effects' with examples. In a factorial experiment, significance of main effects is not of much relevance if the interaction effects are found significant. Why?

8.15 Define the concepts of 'contrasts' and 'orthogonal contrasts' in experimental design. Explain how are they determined and why are they used, with one example.

8.16 Compare and contrast full factorial and fractional factorial experiments.

8.17 Select a group of companies. Study for each the effect of the existing approaches for product and process designs on the quality of the products it manufactures and sells in the market.

8.18 State the important features of (*i*) 2^k factorial experiment, and (*ii*) 2^{k-1} fractional factorial experiment.

8.19 What is 'confounding'? What is 'fractionalization'? Why are these principles used in experiments?

8.20 Define 'a defining contrast' in fractional factorial experiment. Why is it used?

8.21 Distinguish, with a suitable example, between 'principal' and 'alternate' fractions in fractional factorial experiments.

PROBLEMS

8.1 A manufacturer is interested in investigating the surface finish of a new disc that will be used for brake system in automobiles. She suspects that change in material influences surface finish, and also knows that acceptable materials are of four different types. She decides to test four specimens at each material type. Random test sequence is necessary to prevent the effect of nuisance variables. The random run sequence and results are provided in Table P8.1. Test the hypothesis that material type influences surface finish, using $\alpha = 0.05$.

Table P8.1

Test sequence (randomized)	Specimen material type	Surface finish (micron)
11	1	3.9
19	1	3.8
20	1	4
17	1	3.9
3	1	3.5
15	2	3.1
2	2	3.0
14	2	2.9
9	2	3.2
16	2	2.9
7	3	2.8
13	3	2.6
6	3	2.9
5	3	3.0
10	3	2.8
18	4	2.3
12	4	1.9
1	4	1.8
8	4	2.0
4	4	2.1

8.2 An engineer is interested in investigating the tensile strength of a composite material. He suspects that change in mixing technique type influences tensile strength, and also knows that available mixing are of four different types. He decides to test four specimens at each mixing technique type. Random test sequence is necessary to prevent the effect of nuisance variables. The test sequence and corresponding tensile strength is provided in the following Table P8.2. Test the hypothesis that mixing technique type influences tensile strength, using $\alpha = 0.05$.

Table P8.2

Test sequence (randomized)	Mixing technique	Tensile strength
10	A	3120
13	A	3001
9	A	2861
12	A	2880
8	B	3100
6	B	3310
11	B	2990
16	B	3120
7	C	2785
4	C	2901
3	C	2986
2	C	3054
5	D	2605
15	D	2750
1	D	2640
14	D	2755

8.3 An industrial engineer is interested to determine whether or not four different tips produce different reading on hardness testing machine. The engineer has decided to obtain four different observations for each tip into metal test material coupons. He is also interested to remove the variability between coupons from the experimental error. Therefore he performs a randomized complete block design. The scaled value of hardness as obtained during experimentation is provided in Table P8.3. Analyze the data and interpret the results.

Table P8.3

Type of tip	Coupon (block)	Hardness (scaled values)
1	1	– 2.2
1	2	– 1.3
1	3	0.9
1	4	4.8
2	1	– 1.1
2	2	– 2.3
2	3	3.1
2	4	4.2
3	1	– 3.1
3	2	– 0.8
3	3	0.1
3	4	2.2
4	1	2.1
4	2	1.1
4	3	5.3
4	4	7.4

8.4 Suppose the same industrial engineer is not aware of randomized block design, and uses four material coupons, randomly assigned the tips to each. The design resulted is same data as provided in Table P8.2 of Question 8.3. Analyze the data and interpret the results.

8.5 A chemical engineer is studying the effect of different formulation of engine oil used in automobile on the observed burning rate. Each formulation is mixed from a batch of raw material that is only enough for five formulations to be tested. Furthermore, the formulations are prepared by five different operators, and their may be substantial differences in the experience and skill of operators. Therefore, the engineer performs a Latin square design to eliminate two nuisance source of variability, *viz.* batch of raw material and operator. Analyze the experimental data as provided in Table P8.5 and interpret the results.

Table P8.5

Batch	Operator	Formulation	Burning rate
1	1	A	24.3
1	2	B	20.2
1	3	C	19.2
1	4	D	23.9
1	5	E	23.8
2	1	B	17.3
2	2	C	24.1
2	3	D	30.2
2	4	E	26.9
2	5	A	36.3
3	1	C	18.4
3	2	D	39
3	3	E	26.1
3	4	A	27.2
3	5	B	21.5
4	1	D	26.4
4	2	E	30.9
4	3	A	25.8
4	4	B	23.2
4	5	C	22.1
5	1	E	22.3
5	2	A	30.4
5	3	B	20.1
5	4	C	28.7
5	5	D	30.7

8.6 A process engineer is studying the effect of different tightening type (method) of engine cylinder head used in automobile on the observed tightening time. Four operators are selected for the study. Moreover, the tightening type can produce fatigue, regardless of the method. Therefore, the engineer performs a Latin square design to eliminate two nuisance source of variability, *viz.* order of fitting and operator. Analyze the experimental data ($\alpha = 0.05$), as provided in Table P8.6, and interpret the results.

Table P8.6

Order of tightening	Operator	Tightening type	Time
1	1	C	10.5
1	2	D	14.2
1	3	A	7.30
1	4	B	8.30
2	1	B	6.90
2	2	C	18.10
2	3	D	11.20
2	4	A	8.10
3	1	A	5.10
3	2	B	10.20
3	3	C	11.10
3	4	D	9.20
4	1	D	10.30
4	2	A	10.10
4	3	B	12.20
4	4	C	13.90

8.7 An engineer is interested into the change of concentration of reactant, and amount of catalyst on the growth of a typical virus. Assuming the high and low level of both the factor is represented by + 1 and – 1, respectively, a full factorial design (2^2) is performed. The full factorial (2^2) design and data of the experiment with replication is provided in the Table P8.7. Each experiment is replicated three times. Analyze the data and interpret the results.

Table P8.7

Std. order*	Concentration of reactant	Amount of catalyst	Growth of virus
7	1	– 1	37.0
4	1	1	29.5
8	1	1	29.1
2	– 1	1	18.2
6	– 1	1	22.9
10	– 1	1	19.4
11	1	– 1	32.3
5	– 1	– 1	25.0
12	1	1	32.0
1	– 1	– 1	27.0
9	– 1	– 1	27.5
3	1	– 1	32.1

8.8 Three factors are thought to influence chemical loss, namely nozzle type, pressure and temperature, respectively. Each experiment is replicated twice in the design. Therefore the experiment is based on 3 factors at 3 levels with 2 replicates. The standard run order and corresponding chemical loss is provided in the Table P8.8. Determine the influential factors and their setting based on full factorial design to minimize the chemical loss.

Table P8.8. Standard Run Order and Corresponding Chemical Loss

Std. order	Nozzle	Pressure	Temperature	Chemical loss (scaled data)
48	3	100	20	– 32.0
32	1	120	15	– 11.0
20	3	100	15	88.0
17	2	140	15	105.0
47	3	100	15	111.0
9	1	140	20	30.0
1	1	100	10	– 34.0
49	3	120	10	– 54.5
10	2	100	10	16.4
50	3	120	15	– 29.0
14	2	120	15	– 55.2
15	2	120	20	– 64.1
46	3	100	10	– 38.9
26	3	140	15	108.6
37	2	100	10	23.9
45	2	140	20	– 18.9
28	1	100	10	– 25.6
41	2	120	15	– 44.9
43	2	140	10	21.0
40	2	120	10	– 65.3
52	3	140	10	15.0
6	1	120	20	– 40.1
51	3	120	20	– 62.1
42	2	120	20	– 60.9
4	1	120	10	– 45.3
44	2	140	15	42.6
31	1	120	10	– 60.1
54	3	140	20	55.0
25	3	140	10	– 29.7
21	3	100	20	– 56.0
16	2	140	10	3.50
19	3	100	10	– 33.8
18	2	140	20	– 31.6

Table P8.8 (Contd.)

Std. order	Nozzle	Pressure	Temperature	Chemical loss (scaled data)
30	1	100	20	4.10
38	2	100	15	54.3
39	2	100	20	– 23.5
8	1	140	15	79.9
24	3	120	20	– 51.9
5	1	120	15	31.0
33	1	120	20	– 29.8
23	3	120	15	– 26.5
3	1	100	20	5.10
11	2	100	15	120.3
34	1	140	10	– 41.0
35	1	140	15	54.3
53	3	140	15	135.3
27	3	140	20	4.0
13	2	120	10	– 59.0
36	1	140	20	36.8
7	1	140	10	15.2
12	2	100	20	– 5.1
2	1	100	15	109.5
22	3	120	10	– 66.4
29	1	100	15	75.6

8.9 A process engineer wants to determine the effect of seven factors, namely *A, B, C, D, E, F* and *G*, on machining time. He suspects that only a few of these seven factors are of major importance. Two levels of each factor, – 1 and 1 respectively are considered. To screen these seven factors, he runs the treatment combination of 2_{III}^{7-4} design as provided in the Table P8.9, in random order. Analyze the fractional factorial design and interpret the results.

Table P8.9

Std. order	A	B	C	D=AB	E=AC	F=BC	G=ABC	Time
5	– 1	– 1	1	1	– 1	– 1	1	94.2
6	1	– 1	1	– 1	1	– 1	– 1	144
1	– 1	– 1	– 1	1	1	1	– 1	91.2
4	1	1	– 1	1	– 1	– 1	– 1	73.1
3	– 1	1	– 1	– 1	1	– 1	1	82.2
7	– 1	1	1	– 1	– 1	1	– 1	87.1
2	1	– 1	– 1	– 1	– 1	1	1	137.1
8	1	1	1	1	1	1	1	71.5

8.10 Determine the effect and final setting of six factors, namely *A, B, C, D, E* and *F* to minimize the shrinkage of a molding process, based on one-half fractional factorial design. All the factors are at two levels, – 1 and 1 respectively. Assume *ABC* and *BCD* as design generators. The experimental data, based on randomized standard run order is provided in Table P8.10.

Table P8.10

Std. order	A	B	C	D	E=ABC	F=BCD	Shrinkage
1	– 1	– 1	– 1	– 1	– 1	– 1	6.1
11	– 1	1	– 1	1	1	– 1	34.2
5	– 1	– 1	1	– 1	1	1	3.9
16	1	1	1	1	1	1	51.8
4	1	1	– 1	– 1	– 1	1	59.9
6	1	– 1	1	– 1	– 1	1	15.2
8	1	1	1	– 1	1	– 1	60.2
10	1	– 1	– 1	1	1	1	12.2
15	– 1	1	1	1	– 1	1	37.1
13	– 1	– 1	1	1	1	– 1	16.3
14	1	– 1	1	1	– 1	– 1	5.1
2	1	– 1	– 1	– 1	1	– 1	10.2
9	– 1	– 1	– 1	1	– 1	1	7.8
7	– 1	1	1	– 1	– 1	– 1	25.8
3	– 1	1	– 1	– 1	1	1	32.3
12	1	1	– 1	1	– 1	– 1	59.8

8.11 Determine the effect and final setting of five factors, namely *A, B, C, D* and *E* to maximize the chemical yield of a process, based on one-half fractional factorial design. All the factors are at two levels, –1 and 1 respectively. The experimental data, based on randomized standard run order is provided in the Table P8.11.

Table P8.11

Std. order	A	B	C	D	E	Yield
11	– 1	1	– 1	1	1	30.2
14	1	– 1	1	1	– 1	20.9
10	1	– 1	– 1	1	1	10.2
15	– 1	1	1	1	– 1	43.8
3	– 1	1	– 1	– 1	– 1	34.8
12	1	1	– 1	1	– 1	50.2
6	1	– 1	1	– 1	1	21.9
1	– 1	– 1	– 1	– 1	1	8.3
7	– 1	1	1	– 1	1	45.4
8	1	1	1	– 1	– 1	61
16	1	1	1	1	1	64.1
13	– 1	– 1	1	1	1	14.9
2	1	– 1	– 1	– 1	– 1	9.1
5	– 1	– 1	1	– 1	– 1	16.3
9	– 1	– 1	– 1	1	– 1	6.1
4	1	1	– 1	– 1	1	63.4

8.12 Consider a chemical process in which two factors are considered each at two level (–1 and 1), with 3 replication, to determine their effect on process yield. Suppose four experimental trial runs can be made from a single batch of raw material. Therefore, 3 batches of raw material are required to run 3 replications. The experimental data is provided in Table P8.12. Analyze the data, and interpret the results.

Table P8.12

Std. order	Blocks	A	B	Yield
12	3	1	1	29.1
11	3	– 1	1	23.2
10	3	1	– 1	31.9
9	3	– 1	– 1	27.1
8	2	1	1	29.8
5	2	– 1	– 1	25.2
7	2	– 1	1	19.2
6	2	1	– 1	31.9
4	1	1	1	32
3	1	– 1	1	18.1
2	1	1	– 1	36.4
1	1	– 1	– 1	28.3

8.13 Consider a chemical process in which four factors are considered each at two level (–1 and 1), with single replication, to determine their effect on process reaction time. Suppose eight experimental trial runs can be made from a single batch of raw material. Therefore, 2 batches of raw material are required to run the factorial design. The experimental data is provided in the Table P8.13. Analyze the data, and interpret the results.

Table P8.13

Std. order	Blocks	A	B	C	D	Reaction time
12	2	– 1	1	1	– 1	60.2
14	2	– 1	1	– 1	1	24.9
11	2	1	– 1	1	– 1	39.8
13	2	1	– 1	– 1	1	80.5
16	2	1	1	1	1	76.2
10	2	1	1	– 1	– 1	45.1
9	2	– 1	– 1	– 1	– 1	24.9
15	2	– 1	– 1	1	1	55
6	1	1	1	– 1	1	104.5
8	1	– 1	1	1	1	70.2
7	1	1	– 1	1	1	85.9
1	1	1	– 1	– 1	– 1	71.1
4	1	1	1	1	– 1	65.2
5	1	– 1	– 1	– 1	1	43.3
3	1	– 1	– 1	1	– 1	68.3
2	1	– 1	1	– 1	– 1	48.1

REFERENCES

Anscombe, F.J. and Tukey, J.W., *The Examination and Analysis of Residuals.* Technometrics, vol. 5, pp. 141–160, 1963.

Barker, T.B., *Quality by Experimental Design*, Dekker, 1985.

Bisgaard, S., *Industrial Use of Statistically Designed Experiments: Case Study References and Some Historical Anecdotes,* Quality Engineering, 4(4), pp. 547–562, 1992.

Box, G., Hunter, W. and Hunter, J.S., *Statistics for Experimenters*, John Wiley & Sons Inc., New York, 1978.

Box, G.E.P. and Draper, N.R., *An Analysis of Unreplicated Fractional Factorials.* Technometrics, vol. 28, pp. 11–18, 1986.

Box, G.E.P. and Hunter, J.S., *The 2^{k-p} fractional factorial designs.*, Technometrics, vol. 2, pp. 311–351 (Part I), pp. 449–458 (Part II), 1961.

Box, G.E.P., *Sequential Experimentation and Sequential Assembly of Designs.* Quality Engineering, vol. 5, No. 2, pp. 321–330, 1992–1993.

Cochran, W.G. and Cox, G.M., *Experimental Design*, John Wiley & Sons Inc., New York, 1957.

Coleman, D.E. and Montgomery, D.C., *A Systematic Approach to Planning for a Designed Industrial Experiment (with commentary),* Technometrics, 35(1): pp. 1–27, 1993.

Connor, W.S. and Zelen, M., *Fractional Factorial Experimental Designs for Factors at Three Levels.* National Bureau of Standards, Washington, D.C., Applied Mathematics Series, No. 54, 1959.

Cornell, J.A., *Experiments with Mixtures: Designs, Models, and the Analysis of Mixture Data.* 2nd Edition, Wiley, New York.

Cornell, J.A., *How to Run Mixture Experiments for Product Quality*, American Society for Quality Control, Milerankee, W.I., 1983.

Daniel, C., *Applications of Statistics to Industrial Experiments.* New York: Hafner (Macmillan), 1976

Fisher, R.A., *The Design of Experiments,*8th Edition, Hafner Publishing Company, New York, 1966.

Gunter, B., *Statistically Designed Experiments—Part 1: Quality Improvement, the Strategy of Experimentation, and the Road to Hell,* Quality Progress, vol. 22, pp. 63–64, 1989.

Gunter, B., *Statistically Designed Experiments—Part 2: The Universal Structure Underlying Experimentation,* Quality Progress, vol.23, pp.87–89, 1990a.

Gunter, B., *Statistically Designed Experiments—Part 3: Interaction,* Quality Progress, vol. 23, pp. 74–75, 1990b.

Hamada, M. and Wu, C.F.J., *Analysis of Designed Experiments with Complex Aliasing,* Journal of Quality Technology, vol. 24, No. 3, pp. 130–137, 1992.

Hunter, J.S., *Statistical Design Applied to Product Design,* Journal of Quality Technology, 17(4): pp. 210–221, 1985.

John, P.W.M., *Statistical Design and Analysis of Experiments.* Macmilan, New York, 1971.

Moen, R.D., Nolan, T.W. and Provost, L.P., *Improving Quality through Planned Experimentation*, McGraw-Hill, New York, Int. Edition, 1991.

Montgomery, D.C., *Design and Analysis of Experiments*, John Wiley & Sons, New York, 1997.

Nachtsheim, C.J., *Tools for Computer-Aided Design of Experiments.* Journal of Quality Technology, vol. 19, pp. 132–160, 1987.

Nelson, L.S., *Using Nested Designs I: Estimation of Standard Deviations*, Journal of Quality Technology, vol. 27, No. 2, pp. 169–171, 1995*a*.

Nelson, L.S., *Using Nested Designs II: Confidence Limits for Standard Deviations*, Journal of Quality Technology, vol. 27, No. 3, pp. 265–267, 1995*b*.

Peterson, R.G., *Design and Analysis of Experiments.* New York: Marcel Dekker, 1985.

Scheffe, H., *The Analysis of Variance,* Wiley, New York, 1959.

Smith, J.R. and Beverly, J.M., *The Use and Analysis of Staggered Nested Factorial Design,* Journal of Quality Technology, vol. 13, pp. 166–173, 1981.

Snee, R.D., *Graphical Analysis of Process Variation Studies*, Journal of Quality Technology, vol.15, No. 2, April, pp. 76–88, 1983.

Taguchi, G., *System of Experimental Design*, Unipub/ Krause International Publication, White Plains, N.Y., 1987.

Trout, R., *Design and Analysis of Experiments to Estimate Components of Variation—Two Case Studies*, Experiments in Industry, Chemicals and Process Industries Division, American Society for Quality Control, Quality Process, Milerankee, W.I., 1985.

Winner, B.J., *Statistical Principles in Experimental Design,* 2nd Edition. McGraw-Hill, New York, 1971.

Wordsworth, H.M. Stephens, K.S. and Godfrey, A.B., *Modern Methods for Quality Control and Improvement*, John Wiley & Sons, New York, 2004.

CHAPTER 9

Robust Design and Taguchi Method

9.1 INTRODUCTION

In this chapter we present the philosophy and robust experimental design principles originally developed by Taguchi. The objective of quality engineering is to design quality into every product and process. Robust design is the systematic approach to find the optimum values of design factors, which lead to economical designs with low variability.

Taguchi's methods are useful in identifying the key factors that have the greatest contribution to variation and to ascertain those settings or values that result in the least variability.

Robustness deals with product and process. Robustness can be further defined as *the ability of the product to perform consistently as designed with minimal effect from the changes in uncontrollable operating influences*. For a process point of view, robustness can be defined as *the ability of the process to produce consistently good product with minimal effect from changes in uncontrollable manufacturing influences.*

Taguchi's methods provide a way to increase the validity and reproducibility of experiments, while minimizing the time and cost associated with them. Taguchi's approach is to identify and control those factors that are significant, cause variation and to do so in a way that minimizes the effect of those variables that cannot be controlled or that are not practical to control.

Although engineers at Bell Labs and Xerox were the first in the United States (US) to experiment with Taguchi methods, Ford was the leader in changing company policy.

The new quality technology based on the Taguchi methods was first introduced to the US automotive industry in March 1982. In the beginning, this effort was fostered by a very aggressive Ford drive to improve quality and reduce cost. This has evolved fundamental changes in US engineering and quality control methods that have changed many US companies' operations.

9.2 TAGUCHI PHILOSOPHY

In order to improve an overall quality and productivity, company wide activities are needed. These activities include:

(*a*) Product planning
(*b*) Product design
(*c*) Process design
(*d*) Production
(*e*) Service after delivery

There are two types of undesirable and uncontrollable factors that cause deviation from the target values in a product's functional characteristics. They are known as (1) external noise factors, and (2) internal noise factors. *External noise factors* are those variables that are external to the product and affect its performance. Examples of external noise factors include changes in operating environmental variables such as temperature, humidity and dust, and variability in human operators of *internal noise factors* who are responsible for making the product. They are due to the settings of the product and process parameters.

Examples include:

(*a*) *Manufacturing variations:* Manufacturing variations or manufacturing process imperfections such as variations in machine settings, and

(*b*) *Product deterioration:* Product deterioration over time due to wear and tear, such as the wearing out of parts caused by friction and the loss of spring resilience. They create *product-to-product noise*, or *variability*. The main objective of the overall quality system is to produce a product that is *robust* with respect to all noise factors. By reducing the impact of the noise factors and by selecting the controllable factor levels that will ensure the desirable quality characteristic close to target values, a robust design of the product, and hence of the process, is accomplished. Taguchi recommends a three-stage design methodology to determine the target values and tolerances for the respective parameters in the product and the process namely,

1. System design
2. Parameter design
3. Tolerance design.

In order to achieve robustness, quality control efforts should start in the product design phase and be continued through production engineering and production operation phase.

1. *System design:* In this stage, scientific, economic, and engineering principles and experience are used to create a prototype of the product that will meet the functional requirements and also to create the process that will build it. For instance, in the research and development phase, system design involves the development of the prototype design and establishing the parts, components, materials, and assembly system. The determination of the manufacturing process is involved in the production-engineering phase.

2. *Parameter design:* In this stage, the optimal settings of the product and process parameters are determined in order to minimize to performance variability, that is, to minimize the effect of noise factors on the functional characteristics of the product. Taguchi defines a performance measure known as the *signal-to-noise* ratio (*s*/*N* ratio) so that the parameter levels are selected will maximize this ratio. The term *signal* here represents the square of the mean value of the quality characteristic and *noise* is a measure of the variability (as measured by the variance) of the characteristic. The noise factors should be selected to represent those that mostly impact the product's performance. The parameter design process is the most important step in the Taguchi method.

3. *Tolerance design:* In this stage, the tolerances are set, that is, a range of acceptable values around the target values of the control parameters identified in the parameter design stage. Narrow tolerances are specified for the deviations of design parameters in relation to the levels determined by the parameter design.

Taguchi and Wu [1979] summarize these activities for a typical manufacturing facility as shown in Table 9.1.

An asterisk (*) in column four indicates that the external noise can be controlled at that step in the product's life cycle. A plus (+) indicates that it is not desirable to control the external noise at that step. A double plus (++) indicates the impossibility of controlling the noise at a step.

Table 9.1. Functional Euality Engineering Activities

1 Quality control activities	2 Product phase	3 Steps	4 External noise	5 Internal noise	6 Variation among units (Tolerances)
Off-line	Product design	System design	*	*	*
		Parameter design	*	*	*
		Tolerance design	+	*	*
	Production engineering	System design	++	++	*
		Parameter design	+ +	+ +	*
		Tolerance design	+ +	+ +	*
On-line	Production operation	Process control	++	++	*
		Feedback	+ +	+ +	*
		Inspection, etc.	+ +	+ +	*

Key:

* : Possible

\+ : Not desirable

++ : Not possible

9.3 TAGUCHI METHODS

The term *Taguchi methods* refer to the parameter design, tolerance design, the quality loss function, on-line quality control, design of experiments using orthogonal arrays, and methodology applied to evaluate measuring systems. These methods were developed by Genichi Taguchi to simultaneously reduce cost and improve quality.

The new quality technology based on the Taguchi methods was first introduced to the US automotive industry in March, 1982 to improve quality, industrial productivity and reduce cost. What has evolved, are fundamental changes in US engineering and quality control methods that have changed many US companies' operations to further improve the competitive position.

Taguchi's approach can be summarized into two fundamental concepts:

1. Quality losses must be defined as deviations from target, not conformance to arbitrary specifications. Quality losses are to be measured by systems-wide costs –*loss to society* and not local costs at points of defect detection.

2. Product quality must be engineered or designed in. Quality cannot be achieved economically through inspection and product screening.

9.3.1 Quality Control Activities

The major quality control activities during the four stages of the life cycle of a product are summarized in Table 9.2.

Table 9.2. Quality Control Activities During Various Product Realization Steps

Product realization step	Quality control activity	Ability to reduce effect of noise factors			Comments
		External	Unit to unit	Deterioration	
Product design	(*a*) System design	Yes	Yes	Yes	Innovation activity to reduce sensitivity to all noise factors.
	(*b*) Parameter design	Yes	Yes	Yes	Most important step for reducing sensitivity to all noise factors. Uses Robust Design method.
	(*c*) Tolerance design	Yes	Yes	Yes	Method for selecting most economical grades of materials, components and manufacturing equipment, and operating environment for the product.
Manufacturing process design	(*a*) System design	No	Yes	No	Involves innovation to reduce unit-to-unit variation.
	(*b*) Parameter design	No	Yes	No	Important for reducing sensitivity of unit-to-unit variation to manufacturing variations.
	(*c*) Tolerance design	No	Yes	No	Method for determining tolerances on manufacturing process parameters.
Manufacturing	(*a*) Detection and correction	No	Yes	No	Method of detecting problems when they occur and correcting them.
	(*b*) Feed forward control	No	Yes	No	Method of compensating for known problems.
	(*c*) Screening	No	Yes	No	Last alternative, useful when process capability is poor.
Customer usage	Warranty and repair	No	No	No	

Source: **Adapted from G. Taguchi, *Off-line and On-line Quality Control System*, International Conference on Quality Control, Tokyo, Japan, 1978.**

The four stages of the life cycle of a product are (1) Product design, (2) Manufacturing process design, (3) Manufacturing, and (4) Customer usage. The robustness is defined from both—a product and a process related standpoint as follows:

Product: The ability of the product to perform consistently as designed with minimal effect from changes in uncontrollable operating influences.

Process: The ability of the process to produce consistently good product with minimal effect from changes in uncontrollable manufacturing influences. Taguchi's overall quality philosophy is called *quality engineering*.

Quality engineering has the objective of designing quality into every product and corresponding process. Quality engineering directs quality improvement efforts upstream from the manufacturing process to the design phase and is therefore called as *an off-line quality control* method. Off-line methods improve the product manufacturability and reduce product development and lifetime costs.

The quality control (*QC*) activities in product and process design are known as *off-line quality control*, and the quality control activities in manufacturing are called *on-line quality control* as shown in Figure 9.1.

On-line quality control involves the actual production of the product. The off-line quality control involves the optimization of product and process design in support of on-line quality control. Design of experiment is one such basic tool of *off-line quality control* and it gives (1) identification of the sources of variation and (2) determination of design and process optimization. There are three stages in optimizing a product or process as shown in Table 9.2. These are (1) system design, (2) parameter design and (3) tolerance design.

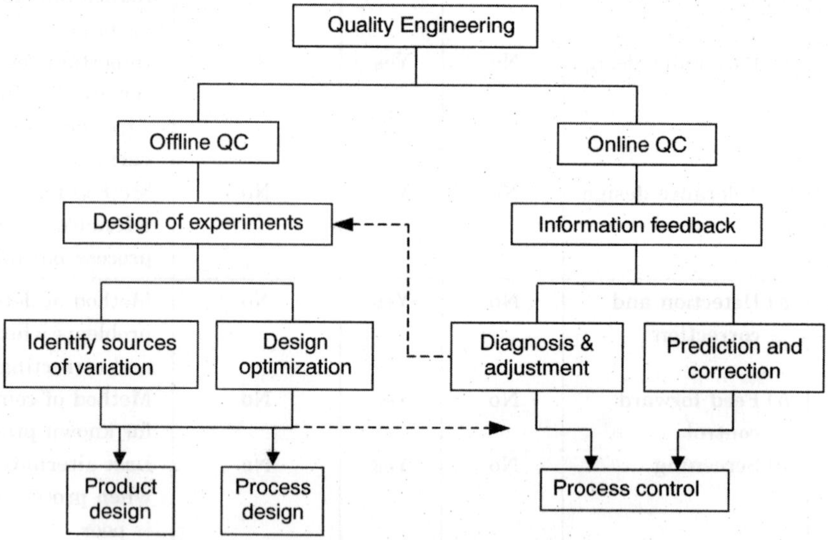

Fig. 9.1 Quality Control Activities.

In system design stage, the designer examines a variety of architectures and technologies for accomplishing the desired function of the product and selection of the most appropriate ones is made. The innovation, which has been proven to work in system design and improve it so that it will consistently function as desired. To achieve this objective, design of experiments are used in reducing variability without adding cost by determining the best settings or values of those factors that are least expensive to change or control. In tolerance design, the acceptable range of variability around the nominal settings is determined. Design of experiments is used to investigate the product or process and analysis of variance (ANOVA) gives the interpretation of the experimental data.

Taguchi classifies quality engineering into two major elements: (1) Design optimization, which involves the use of orthogonal arrays, linear graphs and signal-to-noise ratio in the design of experiments and (2) Evaluation of quality, which involves the loss function and signal-to-noise ratio. Taguchi methods are engineering tools, which do not require a high level of statistics and are cost-driven. The methods are highly efficient in their ability to extract large amounts of useful information from small-scale experiments. Taguchi methods are based on *orthogonal arrays* and *linear graphs*, which offer the flexibility and capability for assigning a number of variables to them.

The conclusions drawn from a full factorial experiment with potential interactions are highly reproducible. The full factorial experiment requires the testing of all combinations of the factor levels under investigation, which is often time consuming and expensive. Taguchi methods using orthogonal arrays and linear graphs produce smaller, less costly experiments with high rates of reproducibility.

To achieve a successful experiment and determine reproducible results requires a well planned and executed effort in the experimental design phase. The suggested steps in designing experiments are outlined below:

1. Define the problem: a clear and concise statement of the problem to be solved.
2. Planning the experiment
   (*a*) determine the objective
   (*b*) identify the quality characteristic (output characteristics, preferably measurable and with good additivity)
   (*c*) determine the measurement methods
   (*d*) identify independent variables
   (*e*) variable settings specification
   (*f*) identify potential interactions
   (*g*) identify control factors and noise factors
   (*h*) determine levels and values for factors
3. Design the experiment
   (*a*) determine the degrees of freedom
   (*b*) select the appropriate orthogonal arrays and control factors
   (*c*) assign control factors (and interactions) to orthogonal array columns
   (*d*) select an outer array for noise factors and assigns factors to columns
4. Conduct the experiment and collect data
   (*a*) develop test plan
   (*b*) coordination
   (*c*) experimental runs
   (*d*) testing and inspection
5. Analyze the data
   Regular analysis
   s/N analysis
   Average response tables and graphs
   s/N response tables and graphs
   s/N analysis of variance (ANOVA)

6. Interpretation of results
   (*a*) select the optimum levels of control factors
   (*b*) predict results for the optimal conditions
7. Confirmation experiment
   (*a*) verify the predicted results
   (*b*) conduct additional experiments as needed (if results obtained are not confirmed or unsatisfactory).

9.4 QUALITY LOSS FUNCTION

Taguchi defines the quality level of a product to be the total loss incurred by society due to failure of the product to deliver the target performance and due to harmful side effects of the product, including its operating costs.

The quality loss function is given as

$$L(y) = k(y - m)^2 \quad \text{...(9.1)}$$

where,

$L(y)$ = loss in dollars when the quality characteristics is equal to y. It is the average quality loss incurred by the customers.

y = the value of the quality characteristic (*i.e.*, length, width, concentration, surface finish, flatness, etc.)

m = target value of y

k = constant called *quality loss coefficient*

The quality loss function is plotted in Figures 9.2 and 9.3.

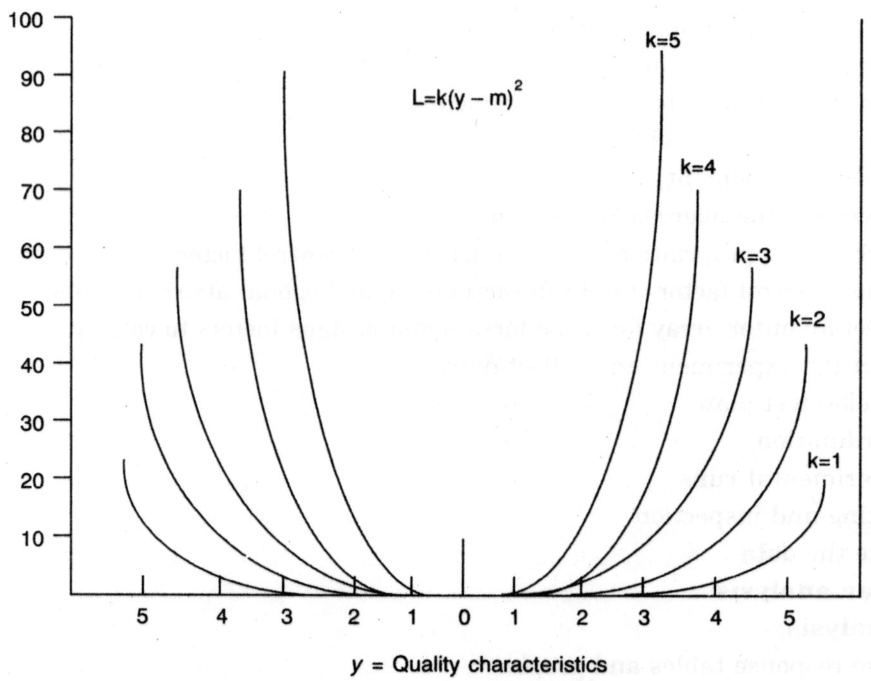

Fig. 9.2 Quality Loss as a Function of *k*.

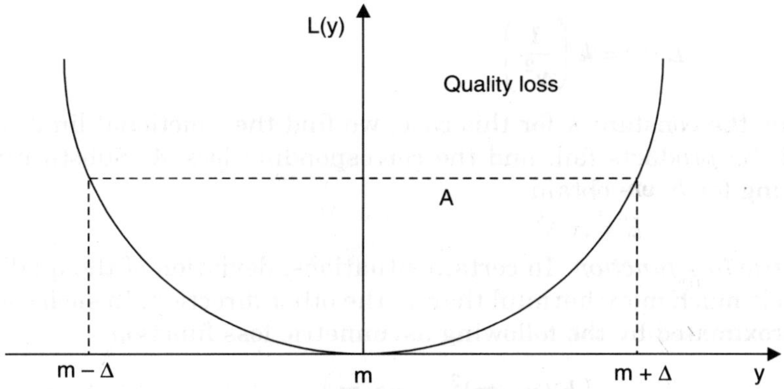

Fig. 9.3 Quality Loss Function.

A convenient way to determine k is to determine first the *functional limits* for the value of y. Functional limit is the value of y at which the product would fail in half of the applications. Let $m \pm \Delta$ be the functional limits. Suppose, the loss at $m \pm \Delta$ is A. Then by substitution in Eq. (9.1), we have

$$k = \frac{A}{\Delta^2} \qquad ...(9.2)$$

where A is the cost of repair or replacement of the product and includes the loss due to the unavailability of the product during the repair period.

Substituting Eq. (9.2) in Eq. (9.1), we get

$$L(y) = \frac{A}{\Delta^2} (y - m)^2 \qquad ...(9.3)$$

9.5 VARIATIONS OF THE QUADRATIC LOSS FUNCTION

The quadratic loss function given by Eq. (9.1) is applicable whether the quality characteristic y has a finite target value, usually non-zero and the quality loss is symmetric on either side of the target. Such quality characteristics are called *nominal-the-best type quality characteristics* and Eq. (9.1) is called the *nominal-the-best type quality loss function.*

Some variations of the quadratic loss function in Eq. (9.1) are given below:

(*a*) *Smaller-the-better type characteristic*: Some characteristics which have the ideal value is equal to zero, and has their value increase, the performance is poor. Such characteristics are called *smaller-the-better* type quality characteristics. The quality loss in such situations can be approximated by the following function, which is obtained from Eq.(9.1) by substituting $m = 0$.

$$L(y) = ky^2 \qquad ...(9.4)$$

This is a one-sided loss function since y cannot be negative. The quality loss coefficient k can be determined from the functional limit, Δ_0 and the quality loss A_0, can be determined at the functional limit by using Eq. (9.4).

(*b*) *Larger-the-better type characteristic*. Some characteristics do not take negative values and as their value becomes larger, the performance becomes better, that is, the quality loss becomes smaller. Their ideal value is infinity and at that point the loss is zero. Such characteristics are called *larger-the-better type* quality characteristics. The loss function for a larger-the-better type characteristics is obtained by substituting $1/y$ for y.

$$L(y) = k\left(\frac{1}{y^2}\right) \quad ...(9.5)$$

To determine the constant k for this case, we find the functional limit, Δ, below which more than half of the products fail, and the corresponding loss A. Substituting Δ and A in Eq. (9.5), and solving for k, we obtain

$$K = A\,\Delta^2 \quad ...(9.6)$$

(c) *Asymmetric loss function.* In certain situations, deviation of the quality characteristic in one direction is much more harmful than in the other direction. In such cases, the quality loss would be approximated by the following asymmetric loss function:

$$L(y) = \begin{pmatrix} k_1(y-m)^2, & y > m \\ k_2\,(y-m)^2, & y \le m \end{pmatrix} \quad ...(9.7)$$

The four different versions of the quadric loss function are shown in Figure 9.4.

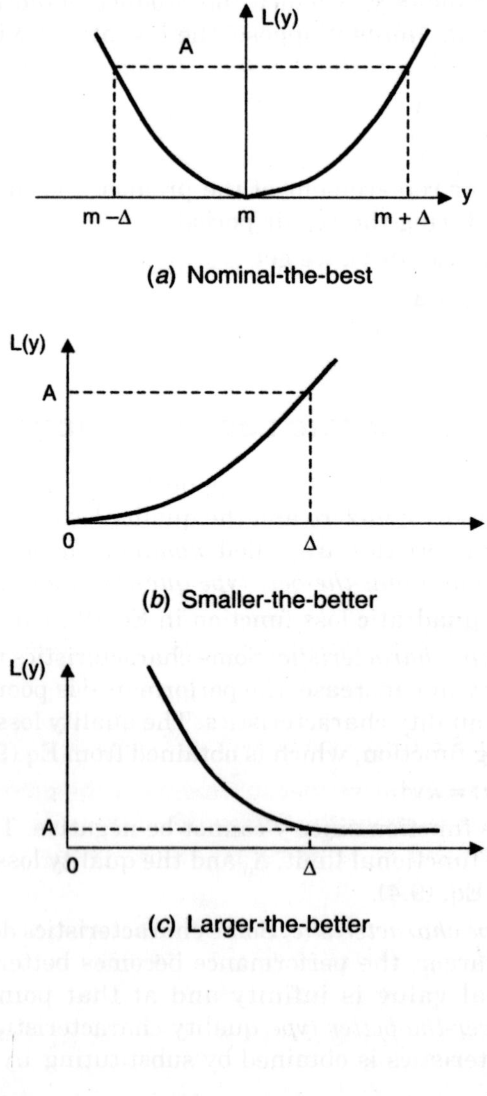

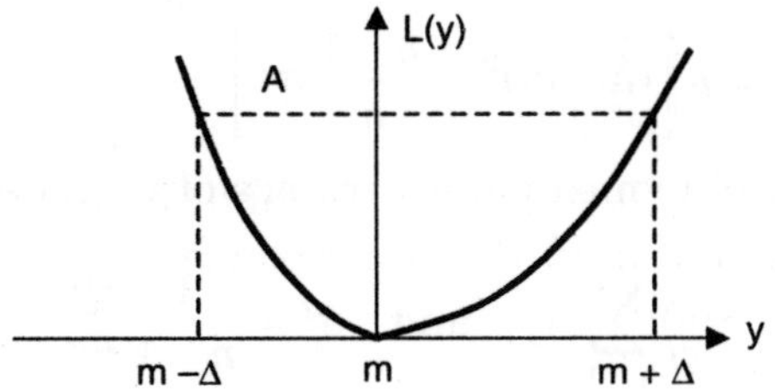

(*d*) Asymmetric

Fig. 9.4 Variation of the Quadratic Loss Function.

The performance of a product varies in the field due to causes known as *noise factors*. The noise factors are classified as (*a*) external, (*b*) unit-to-unit variation, and (*c*) deterioration. The environments in which a product is operated and the load to which it is subjected are the two major external sources of variation of product's performance. The variation that is inevitable in a manufacturing process leads to variation in the product parameters from unit-to-unit. When a product is sold, all its functional characteristics may be on target. As time passes, however, the values of individual components may change leading to deterioration in product's performance.

9.6 AVERAGE QUALITY LOSS

The quality characteristic *y* of a product varies from unit to unit, and from time to time during the usage of the product due to noise factors. Consider the distribution of *y* resulting from all sources of noise as shown in Figure 9.5.

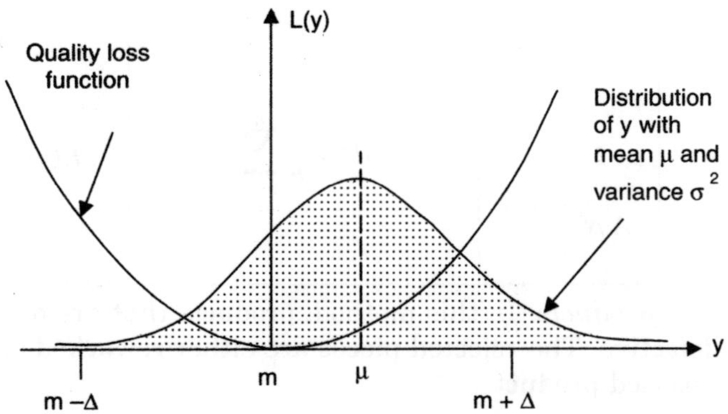

Fig. 9.5 Evaluation of Average Quality Loss.

Let $y_1, y_2, \ldots, y_n$ be *n* representative measurements of the quality characteristic *y* taken on a few representative units throughout the design life of the product. Let *y* be a nominal-the-best type quality characteristic and *m* be its target value. Then, the average quality loss, *Q*, resulting from this product is given by

$$Q = \frac{1}{n}\,[L(y_1) + L(y_2) + \ldots + L(y_n)]$$

$$= \frac{k}{n}[(y_1 - m)^2 + (y_2 - m)^2 + \ldots + (y_n - m)^2] \qquad \ldots(9.8)$$

$$= k\left[(\mu - m)^2 + \frac{n-1}{n}\sigma^2\right]$$

In Eq. (9.8), μ and σ^2 are the mean and variance of y, respectively. They are given by

$$\mu = \frac{1}{n}\sum_{i=1}^{n} y_i \quad \text{and} \quad \sigma^2 = \frac{1}{n-1}\sum_{i=1}^{n}(y_i - \mu)^2$$

When the number of measurements, n is large, Eq. (9.8) can be written as

$$Q = k[(\mu - m)^2 + \sigma^2] \qquad \text{...(9.9)}$$

Hence, the average quality loss has the following two components:

1. $k(\mu - m)^2$ resulting from the deviation of the average value of y from the target.
2. $k\sigma^2$ resulting from the mean squared deviation of y around its own mean.

The first component is easier to eliminate and the second component reduction needs decreasing the variance. There are three methods commonly used to reduce the variance. They are described as follows:

Table 9.3 Loss Functions

No.	Type	Loss function	Mean square deviation	Average loss function
1.	Nominal-the-best or target is best.	$L(y) = \frac{A}{\Delta^2}(y-m)^2$	$MSD = s^2 + (\bar{y} - m)^2$	$\bar{L}(y) = \frac{A}{\Delta^2}[s^2 + (\bar{y} - m)^2]$
2.	Smaller-the-best	$L(y) = \frac{A}{\Delta^2}y^2$	$MSD = s^2 + (\bar{y})^2$	$\bar{L}(y) = \frac{A}{\Delta^2}[s^2 + (\bar{y})^2]$
3.	Larger-the-better	$L(y) = A\Delta^2\left(\frac{1}{y^2}\right)$	$MSD = \frac{1}{n}\sum_{i=1}^{n}\frac{1}{y_i^2}$	$\bar{L}(y) = A\Delta^2\left[\frac{1}{n}\sum_{i=1}^{n}\frac{1}{y_i}\right]^2$

1. *Screening out bad products:* In this method, products that are outside certain limits, $M \pm \Delta'$, are rejected as defective. The rejected pieces are either reworked or scrapped and this leads to higher cost per passed product.

2. *Identifying the cause of malfunction and eliminating it*: If the tolerance on a particular component is identified as a major contributor to system performance, then a narrower tolerance is specified for that component.

3. *Application of the robust design method*: The method involves in making a product's performance is insensitive to noise factors.

The loss functions are summarized in Table 9.3.

9.7 CLASSIFICATION OF PARAMETERS

A schematic representation of a product/process is shown in Figure 9.6. The response is denoted by y and is called *quality characteristic*. The parameters that influence the quality characteristic are classified as follows:

1. *Signal factor (M)*: These are the parameters set by the user or operator of the product to express the intended value for the response of the product based on the knowledge of the product being developed. Sometimes two or more signal factors may also be used in combination to express the desired response.

2. *Noise factors* (x): Certain parameters are not controllable by the designer and are known as *noise factors*. Parameters whose settings (also called *levels*) are difficult to control in the field or whose levels are expensive to control are also considered *noise factors*. The levels of the noise factors change from one unit to another, from one environment to another, and from time to time. The noise factors cause the response y to deviate from the target specified by the signal factor M and lead to quality loss.

3. *Control factors* (z): These are parameters that can be specified by the designer. Control factor can take multiple values, called *levels*.

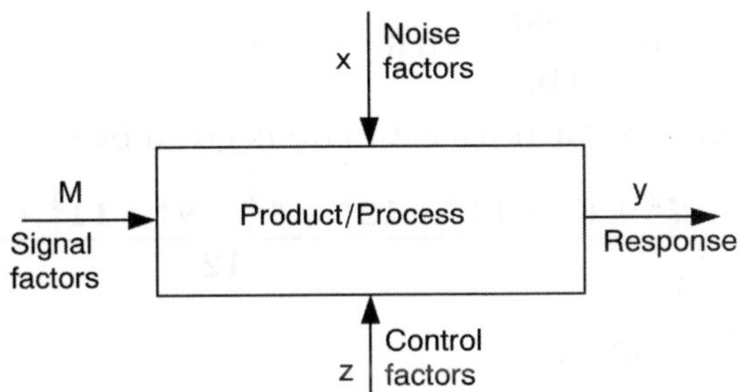

Fig. 9.6 Block Diagram of a Product/Process.

Example E9.1. *The critical dimension of a mechanical component made using a turning machine must be 0.45 ± 0.005 inches. The scrapping cost is \$100 per component. Samples taken from the turning machine had the following dimensions: 0.451, 0.447, 0.448, 0.452, 0.450, 0.453, 0.449, 0.447, 0.454, 0.456, 0.450, and 0.452.*

(a) find the Taguchi loss equation for this operation

(b) determine the average loss function for the parts made using this turning machine.

Solution:

$$(a)\ L(y) = k(y-m)^2 = \frac{A}{\Delta^2}(y-m)^2 = \frac{\$100}{(0.005)^2}(y-m)^2 = 4 \times 10^{-6}\,(y-m)^2$$

$$(b)\quad Q = k[(\mu - m)^2 + \sigma^2]$$

$$\mu = \bar{x} = \frac{\begin{bmatrix} 0.451 + 0.447 + 0.448 + 0.452 + 0.450 + 0.453 + 0.449 \\ + 0.447 + 0.454 + 0.456 + 0.450 + 0.452 \end{bmatrix}}{12} = \frac{5.409}{12} = 0.451$$

$$\sigma = \sqrt{\frac{1}{n-1}\sum (y_i - \mu)^2} = \sqrt{\frac{1}{12-1}} \times \sqrt{(0.000090)} = \sqrt{\frac{0.000090}{11}} = 0.00286$$

$$Q = 4 \times 10^6 [(0.451 - 0.450)^2 + (0.00286)^2] = 4 \times 10^6 [(0.001)^2 + (0.00286)^2]$$
$$= 4 \times 10^6 [(1 \times 10^{-6} + 8.180 \times 10^{-6}] = 36.72$$

Average quality loss = $ 36.72.

Example E9.2. *An auto service facility has identified two of the most important factors that help it attract and retain customers. They are (1) the price of the service item and (2) the time it had taken to serve. Based on the price for similar service items in other neighboring auto service facilities, it is estimated that the customer tolerance limit for price is $10, and the associated customer loss is $60. Similarly, the customer tolerance limit for the service time is 15 min and the associated customer loss is $50. A random sample of 12 customers gave the following values of price: 9, 7, 8, 11, 12, 6, 9, 11, 8, 9, 7 and 12. The sample service times (in minutes) are: 10, 8, 16, 9, 11, 13, 16, 14, 12, 7, 17 and 18.*

(a) determine the total expected loss per customer

(b) if the service facility expects 3000 customers monthly, determine the expected monthly loss.

Solution:

(*a*) The proportionality constant for the factor of price, where smaller is better is

$$k_1 = \frac{60}{(10)^2} = 0.6$$

The mean square deviation for price estimate is given by

$$\sum \frac{y_i^2}{12} = \frac{[9^2 + 7^2 + 8^2 + 11^2 + 12^2 + 6^2 + 9^2 + 11^2 + 8^2 + 9^2 + 7^2 + 12^2]}{12}$$

$$= \frac{1035}{12} = 86.25$$

Hence, the expected loss per customer due to the factor of price is = (0.6) (86.25) = $51.75

The proportionality constant for the factor of service time is

$$k_2 = \frac{50}{(15)^2} = 0.22222$$

The mean square deviation for service time is given by

$$\sum \frac{y_i^2}{12} = \frac{[10^2 + 8^2 + 16^2 + 9^2 + 11^2 + 13^2 + 16^2 + 14^2 + 12^2 + 7^2 + 17^2 + 18^2]}{12}$$

$$= \frac{2049}{12} = 170.75$$

Hence, the expected loss per customer due to the factor of service time

= (0.22222) (170.75) = $ 37.94

Therefore, the total expected loss per customer due to the both factors (price and service) is

$51.75 + $ 37.94 = $89.69

(*b*) The expected monthly loss = (3000) ($89.69) = $ 269,070

Example E9.3. *Refer to Example E 9.2. If the service facility employs more service personnel in order to cut down the service time, the estimated additional cost is $1 per customer. A random sample of 12 customers with the added personnel gave the following waiting times (in minutes): 9, 7, 10, 12, 14, 8, 13, 9, 13, 11, 6 and 7.*

(a) determine whether it is cost effective to add more service personnel

(b) determine the total expected monthly loss.

Solution:

(*a*) The mean square deviation for service time with added service personal is given by

$$\sum \frac{y_i^2}{12} = \frac{[9^2 + 7^2 + 10^2 + 12^2 + 14^2 + 8^2 + 13^2 + 9^2 + 13^2 + 9^2 + 13^2 + 11^2 + 6^2 + 7^2]}{12}$$

$$= \frac{1249}{12} = 104.916$$

The expected loss per customer due to the factor of service time is

= (0.22222) (104.916) = \$23.32

The total expected loss per customer due to the factors of price and service time as well as additional cost of personal is

(\$ 51.75 + \$ 23.32 + \$ 1) = \$ 76.10 < \$ 89.69

Hence, it is cost effective to add personal.

(*b*) The total expected monthly loss = (3000) (\$76.10) = \$228,300

9.8 QUALITY EVALUATIONS

The traditional loss function is shown in Figure 9.7. According to the traditional concept of loss function that as long as the product's quality characteristic is within the specification limits or tolerances, there are no losses incurred. The loss takes the form of a step function with a constant value A outside the specification limits or tolerances. From Fig. 9.7, we observe that the loss function $L(y)$ makes no distinction between products

(*a*) whose quality characteristic is exactly on target at m, or

(*b*) whose quality characteristic is just above lower specification limit (LSL), or

(*c*) whose quality characteristic is just below the upper specification limit (USL).

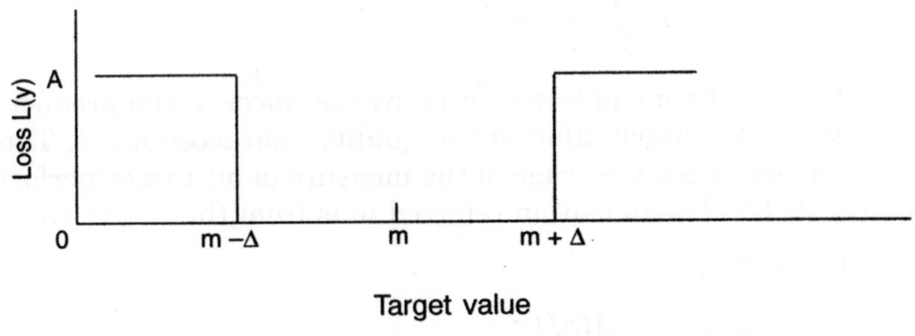

Fig. 9.7 Traditional Loss Function.

It is also doubtful that the loss function $L(y)$ remains constant at a value of A beyond the specification limits USL and LSL. Since the Taguchi loss function increases quadratically with increasing deviation from the target value, the Taguchi loss function can be applied to three different situations or cases, namely

(*a*) Target is best

(*b*) Smaller is better

(*c*) Larger is better.

9.8.1 Target is Best or the Nominal-the-Best Case

This function was shown in Fig. 9.4 (*a*). In the Target is the best or the nominal-the-best case the measured response *y*, always has a specific target value *m*. The quality loss is undesirable on either side of the target. The quality loss increases in a quadratic manner as the quality characteristic deviates from the target value. Examples of quality characteristics of the target is best or the nominal-the-best category include product's component measurements such as diameter, thickness, and length; and a product's quality characteristic such as the properties of a fluid etc. The nominal-the-best loss function can be applied to just one-part or product or to the average loss associated with more than one unit. Taguchi's loss function is given by

$$L(y) = k(y - m)^2 \quad \text{...(9.10)}$$

where k = a proportionality constant called *the quality loss coefficient*

m = the target value of y

y = the value of the quality characteristic

$L(y)$ = loss in dollars when the quality characteristic is equal to y

Note that $L(y) = 0$ when $y = m$.

Consider the quality loss function shown in Fig. 9.3. The functional tolerance range of the quality characteristic is $(m - \Delta)$ and $(m + \Delta)$ which represents the maximum permissible variation. If the consumer's average loss is *A* when the quality characteristic is at the limit of the functional tolerance, then *A* represents the customer cost for repair or replacement of the product. The proportionality constant *k* is obtained from

$$A = k\Delta^2 \quad \text{...(9.11)}$$

or

$$k = \frac{A}{\Delta^2} \quad \text{...(9.12)}$$

Then $L(y)$ in Eq. (9.10) can be written as

$$L(y) = \frac{A}{\Delta^2}(y - m)^2 \quad \text{...(9.13)}$$

The expected loss is the mean loss over many instances of the product. The expectation is taken with respect to the distribution of the quality characteristic *y*. The average quality loss is obtained by considering an average of the measure of off-target performance $(y - m)^2$ of the loss function Eq. (9.10). The deviation referred to is from the target value *m*. The average loss $\overline{L}(y)$ can be written as

$$\overline{L}(y) = k\,(MSD) \quad \text{...(9.14)}$$

where *MSD* represents the mean square deviation of *y* where *MSD* is estimated

$$MSD = \frac{1}{n}[(y_1 - m)^2 + (y_2 - m)^2 + (y_3 - m)^2 + ... + (y_n - m)^2] \quad \text{...(9.15)}$$

$$= \frac{1}{n}\sum_{i=1}^{n}(y_i - m)^2 = \frac{1}{n}\sum_{i=1}^{n}(y_i^2 - 2y_i m + m^2) \quad \text{...(9.16)}$$

Since the sample mean is $\bar{y} = \frac{1}{n}\sum_{i=1}^{n} y_i$, we can write Eq. (9.16) as

$$MSD = \frac{1}{n}\sum_{i=1}^{n}(y_i^2 - 2\bar{y}m + m^2) = \frac{1}{n}\sum_{i=1}^{n}(y_i^2 - \bar{y}^2 + \bar{y}^2 - 2\bar{y}m + m^2)$$

$$= \frac{1}{n}\sum_{i=1}^{n}\left((y_i - \bar{y})^2 + (\bar{y} - m)^2\right) \qquad ...(9.17)$$

or

$$MSD = \frac{1}{n}\sum_{i=1}^{n}(y_i - \bar{y})^2 + \frac{1}{n}\sum_{i=1}^{n}(\bar{y} - m)^2$$

= (Average square deviation from the mean)
+ (Square of the deviation of mean from the target value)
= (Population standard deviation) + (Squared bias of y)
= (Variance of y) + (Squared bias of y)

$$= \sigma^2 + (\bar{y} - m)^2 \qquad ...(9.18)$$

Also for $n > 30$, the sample variance s^2 = population variance σ^2

Hence, the loss function for n-units is

$$L(y) = k\,[s^2 + (\bar{y} - m)^2] \qquad ...(9.19)$$

To minimize the loss, the *MSD* must be minimized. This can be achieved by either minimizing s^2, *i.e.* reducing the variability that is causing the deviation of the data from $\bar{y}$ or by adjusting the average response $\bar{y}$ to coincide with the target value m, that is

$$(\bar{y} - m)^2 = 0$$

It should be noted here that the form of the distribution of the quality characteristics also has an influence on the expected loss. Two typical probability distributions of a quality characteristics are shown in Fig.9.8, where distribution A is a normal distribution and distribution B is a uniform distribution. Eq. (9.19) is applicable to both the distributions in Fig. 9.8. The expected loss for the product from distribution A will be less than that for distribution B, since most of the product is tightly clustered around the target value for distribution A compared to that of distribution B. The standard deviation of distribution is about 1/6 of the tolerance equals to 10.

In quality control, the index of tolerance divided by 6 standard deviations are known as the *process capability index*, denoted by C_p.

$$C_p = \frac{\text{Tolerance}}{6(\text{Standard deviation})} = 1.0 \qquad ...(9.20)$$

In Figure 9.8, $C_p = 1$ for distribution A.

Since the standard deviation of the uniform distribution is given by $\frac{1}{\sqrt{12}}$ of the tolerance, the process capability index for distribution B is given as follows:

$$C_p = \frac{\text{Tolerance}}{6(\text{Tolerance}/\sqrt{12})} = 0.577 \qquad ...(9.21)$$

The expected quality losses of products with similar functions can be evaluated collectively irrespective of their sizes or specifications using the above approach given by Eq. (9.19). Suppose that components of n different types are supplied to an assembly operation.

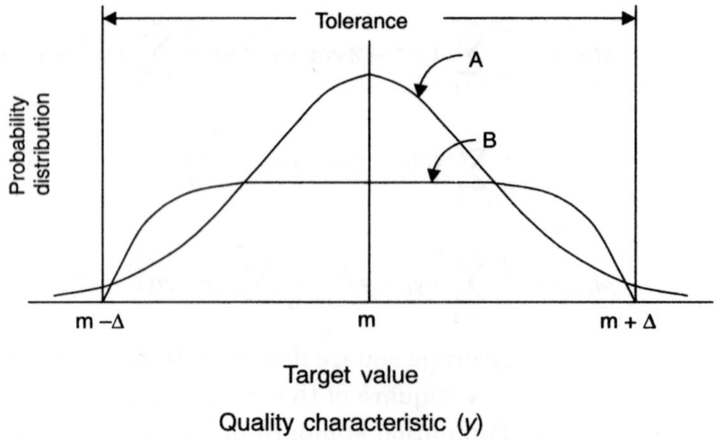

Fig. 9.8 Probability Distributions of the Quality Characteristic.

The quality of supplied components can be measured as follows:

Expected quality losses:

$$L(y) = k_1\,[\bar{y}_1 - m_1]^2 + k_2[\bar{y}_2 - m_2]^2 + \ldots + k_n\,[\bar{y}_n - m_n]^2 \qquad \ldots(9.22)$$

where,

$k_i = \dfrac{A_i}{\Delta_i^{\,2}}$ $(i = 1, 2, \ldots, n)$ =quality loss coefficient of component i

$2\Delta_i\,(i = 1, 2, \ldots, n)$ = tolerance limit of component i

$A_i\,(i = 1, 2, \ldots, n)$ = price (loss) per unit of component i

$[\bar{y}_i - m_i]\ (i = 1, 2, \ldots, n)$ = deviation of the component i from the target value m_i

$[\bar{y}_i - m_i]^2\ (i = 1, 2, \ldots, n)$ = the mean squared deviation of component i from the target value m_i

$m_i\,(i = 1, 2, \ldots, n)$ = the target value of component i

n = number of measurable functional quality characteristics of the product

Therefore, losses due to deviations are given by

$$L(y) = \sum_{i=1}^{n} \frac{A_i}{\Delta_i^{\,2}}(\bar{y}_i - m_i)^2 \qquad \ldots(9.23)$$

or

$$L(y) = \sum_{i=1}^{n} k_i(\bar{y}_i - m_i)^2 \qquad \ldots(9.24)$$

Eq. (9.24) gives an indication of the total quality of the product, since it considers all the measurable functional quality characteristics n of the product.

For many products, parts, elements and components when a nominal size or characteristics is preferred, we require to use the nominal-the-best or target is best type.

Bilateral tolerances (tolerances related to a basic dimensions given in two directions– plus and minus) are generally employed in this type of application.

Example E9.4. *Assume the cost of the repair of a failed DVD player in the factory is $50 per unit. Compare the losses caused by variations from the target value for two DVD players,* *one produced in factory A, which follows the normal distribution A and the other produced in a factory B, which follows the uniform distribution B as described earlier in Fig.9.8. The tolerance interval ranges from* $m - \Delta$ *to* $m + \Delta$*, where* $\Delta = 20$*.*

Solution:

The constant k is computed from Eq. (9.12) as

$$k = \frac{A}{\Delta^2} = \frac{50}{(20)^2} = 0.125$$

The expected losses caused by deviations in the production of the DVD player in case A is obtained from Eq. (9.13)

$$L(y) = \left(\frac{A}{\Delta^2}\right)(y - m)^2 = k(y - m)^2$$

$$= k \text{ (mean squared deviation from target value } m)$$

$$= k\left(\frac{40}{6}\right)^2 = \$5.56\text{/unit}$$

The loss caused by deviation in the production of the DVD player in case B is

$$L = k(y - m)^2 = k\left(\frac{40}{\sqrt{12}}\right)^2 = 0.125\left(\frac{40}{\sqrt{12}}\right)^2 = \$16.67\text{/unit}$$

The above results indicate that the losses caused by deviation for DVD players produced in case B are three times those for the same type of DVD players produced in case A.

Example E9.5. *Refer to Example E9.4. Assuming that the factory A has significantly improved the process of manufacturing so that a new standard deviation from the target value of* $\left(\frac{40}{8}\right)$ *is achieved. Determine the losses caused by deviations from the target value.*

Solution:

Here again $k = 0.125$ from Example E9.4.

$$L(y) = k(y - m)^2 = (0.125)\left(\frac{40}{8}\right)^2 = \$3.13\text{/unit}$$

The loss per unit of production would decrease from \$5.56 to \$3.13. Hence a saving of \$5.56 – \$3.13 = \$2.43/unit.

Example E9.6. *A manufacturer of circular washers required that the washers meet certain roundness standards for the diameter as well as having a specified flatness (surface roughness) standards for each washer. The losses incurred due to defective washers are as follows: the loss caused by unacceptable diameter is \$5 and the loss caused by unacceptable flatness is \$10. The specifications for a 2-in washer are given as*

Diameter 2 ± 0.00020

Flatness (Surface roughness) 0.00020 μm or less

The following twelve diameter measurements were taken:

2.000020	*1.999990*	*1.999998*	*1.999990*	*2.000020*	*2.000000*
1.999995	*2.000000*	*2.000020*	*2.000010*	*1.999995*	*2.000010*

The following twelve surface roughness measurements were taken:

0.00005	*0.00020*	*0.00015*	*0.00015*	*0.00015*	*0.00003*
0.00007	*0.00010*	*0.00010*	*0.00020*	*0.00010*	*0.00005*

Determine the expected total losses caused by the deviations.

Solution:

The mean squared deviation due to the diameter deviations is

$$MSD_1 = \frac{1}{12}\,[(2.000020 - 2)^2 + (1.999990 - 2)^2 + (1.999998 - 2)^2 + (1.999990 - 2)^2 + (2.000020 - 2)^2 + (2 - 2)^2 + (1.999995 - 2)^2 + (2 - 2)^2 + (2.000020 - 2)^2 + (2.000010 - 2)^2 + (1.999995 - 2)^2 + (2.000010 - 2)^2]$$

$$= \frac{10^{-10}}{12}\,[4 + 1 + 4 + 1 + 4 + 0 + 0.25 + 0 + 4 + 1 + 0.25] = 1.625 \times 10^{-10}$$

The mean squared deviation due to the flatness is

$$MSD_2 = \frac{1}{12}\,[0.00005^2 + 0.00020^2 + 0.00015^2 + 0.00015^2 + 0.00015^2 + 0.0003^2 + 0.00007^2 + 0.00010^2 + 0.00010^2 + 0.00020^2 + 0.00010^2 + 0.00005^2]$$

$$= \frac{10^{-8}}{12}\,[0.25 + 4 + 2.25 + 2.25 + 2.25 + 0.09 + 0.49 + 1 + 1 + 4 + 1 + 0.25]$$

$$= \frac{18.83(10^{-8})}{12} = 1.569(10^{-8})$$

$$\text{Expected total loss} = \frac{A_1}{\Delta_1^{\,2}}\,(MSD_1) + \frac{A_1}{\Delta_2^{\,2}}\,(MSD_2)$$

$$= \frac{5}{(0.00020)^2}\,(1.625)(10^{-10}) + \frac{10}{(0.00020)^2}\,(1.569)\,(10^{-8}) = \$3.95/\text{unit.}$$

Example E9.7. *The mechanical component of a machine system has a specification on width of 25 ± 5 mm. Three suppliers of the mechanical component produced the results as shown in Table E9.7. The replacement cost of the component is $50. Three suppliers had the following number of components out-of-spec (defectives) in deliveries of 300,000 parts.*

A : 0.25%; B : 0.15%; and C : 0.125%

Table E9.7

Supplier	Mean width	Variance	C_{pk}
A	*25*	*1.5*	*1.0*
B	*23*	*0.5*	*1.0*
C	*22*	*0.2*	*1.0*

(a) compare the three suppliers based on the loss function

(b) compare the three suppliers based on the cost of defective units.

Solution:

$$L(y) = k(y - m)^2 = \frac{A}{\Delta^2}(y - m)^2$$

Given A = \$50 and Δ = 5 mm

Therefore $$L(y) = \frac{50}{(5)^2}(y-m)^2 = 2(y-m)^2$$

(*a*) $Q_A = 2[(25 - 25)^2 + 1.5] = \$3$ per unit

$Q_B = 2[(23 - 25)^2 + 0.5] = \$9$ per unit

$Q_C = 2[(22 - 25)^2 + 0.2] = \$18.4$ per unit

(*b*) Cost based on 300,000 components is summarized in Table E9.7(*a*).

Table E9.7 (a)

Supplier	$\bar{x}$	s^2	C_{pk}	Defects %	Loss function	Out-of-spec costs
A	25	1.5	1.0	0.25	\$900,000	\$37,500
B	23	0.5	1.0	0.15	\$2,700,000	\$22,500
C	22	0.2	1.0	0.125	\$5,520,000	\$18,750

Out-of-spec costs = (Total production) (% Defects) (Average replacement cost)

From Table E9.7 (*a*), we observe that suppliers *B* and *C* are best based on cost of replacing the out-of-spec components. We also notice that supplier *A* consistently produces components on target and has the lowest cost to the customer. Hence, supplier *A* is the best quality supplier for long-term customer satisfaction.

Example E9.8. *A mechanical component has a strength specifications of 150 ± 10 MN with a target value of 150 MN. The standard deviation of the components coming out of the production line is 3 MN. The mean or average value of the strength is initially 145 MN, but it shifts to 153 MN and then to 156 MN without any change in the variability. Determine the* C_p *and* C_{pk}.

Solution:

$$C_p = \frac{USL - LSL}{6\sigma} = \frac{160 - 140}{6(3)} = 1.11$$

The process mean is not centered between the specification limits and it is given as

$$\mu = 145 \text{ MN}$$

Accordingly, the C_{pk} values are given by

$$C_{pk1} = \frac{USL - \mu}{3\sigma} = \frac{160 - 145}{3(3)} = \frac{15}{9} = 1.667$$

$$C_{pk2} = \frac{\mu - USL}{3\sigma} = \frac{145 - 140}{3(3)} = \frac{5}{9} = 0.556$$

The process C_{pk} is the lesser of these two values C_{pk1} and C_{pk2}. The values of C_{pk} for other values of μ are summarized in Table E9.8.

Table E9.8

Process mean, μ	C_p	C_{pk1}	C_{pk2}	C_{pk}
145	1.11	1.667	0.556	0.556
153	1.11	0.778	1.445	0.778
156	1.11	0.445	1.778	0.445

9.9 DETERMINATION OF MANUFACTURING TOLERANCES

There are four types of quality loss functions. They are

(1) target is best or the nominal-the-best case

(2) the smaller-the-better case

(3) the larger-the-better case

(4) the asymmetric nominal-the-best case

9.9.1 Target is Best or the Nominal-the-Best Case

There exists a breakeven point in terms of loss between the manufacturer and the customer. Figure 9.9 shows the manufacturing tolerance for the case in which target is best or the nominal-the-best case.

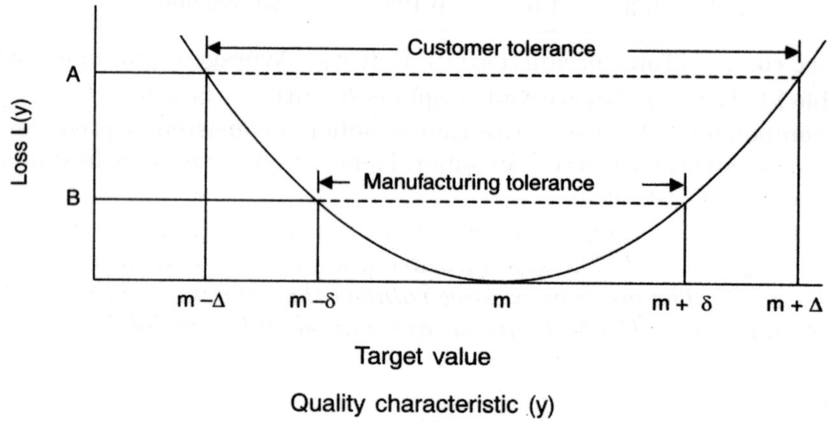

Fig. 9.9 Manufacturing Tolerance for a Situation in which Target is Best.

The following notation is used in Fig.9.9.

A = the associated loss

$m \pm \Delta$ = the customer tolerance

$m \pm \delta$ = the manufacturer tolerance

Let B represents the cost to the manufacturer for repairing or reworking on an item prior to shipping the item to the customer such that the item will exceed the customer's tolerance limits of $m \pm \Delta$.

The loss function is given by

$$L(y) = \frac{A}{\Delta^2}(y - m)^2$$

or

$$L(y) = k(y - m)^2 \quad \text{...(9.25)}$$

Now, for the manufacturing tolerance of $m \pm \delta$, the associated cost B is given by

$$B = k\delta^2 = \left(\frac{A}{\Delta^2}\right)\delta^2 \quad \text{...(9.26)}$$

or

$$\delta = \left(\frac{B}{A}\right)^{1/2}\Delta \quad \text{...(9.27)}$$

Hence, the manufacturer will not spend any extra money as long as the quality characteristic is within δ units from the target value of m.

9.9.2 The Smaller-the-Better Case

When the targeted response is ideally zero it is referred to as smaller-the-better case. Since $y \geq 0$ and $m = 0$, we have

$$L(y) = k(y - 0)^2 = ky^2 \qquad \text{...(9.28)}$$

We notice from Eq. (9.28) that as y drifts further away from zero, the loss starts to increase quadratically and the performance deteriorates.

Characteristics such as paper jam in a fax machine, automobile seat vibration, corrosion of a pipeline, road vehicle exhaust pollution, fuel consumption of a vehicle, an impurity in fresh water, customer waiting time in a restaurant, and the shrinkage of a gasket are a few examples of smaller-the-better case. The loss function for the case given by Eq. (9.28) is shown in Figure 9.4 (*b*). Denoting the loss caused by exceeding the customer's tolerance level Δ as *A*, the quality loss coefficient '*k*' is given by

$$k = \frac{A}{\Delta^2} \qquad \text{...(9.29)}$$

The expected or average loss over many produced items is given by

$$E\,[L(y)] = kE\,(y^2) = \left(\frac{A}{\Delta^2}\right) E\,(y^2) = \left(\frac{A}{\Delta^2}\right)$$

$$[\text{var}(y) + \mu^2] = \frac{A}{\Delta^2}\,[MSD] \qquad \text{...(9.30a)}$$

where, MSD = the mean square deviation $= \frac{1}{n}\sum_{i=1}^{n} y_i^{\,2}$ for a sample of n items.

y = the variance of y

μ = the mean value of y

Hence,

$$L(y) = k\left[\frac{1}{n}\sum_{i=1}^{n} y_i^{\,2}\right] = k\,[s^2 + y^2] = k\,[s^2 + \bar{y}^2] \qquad \text{...(9.30b)}$$

Since the target is zero in this case, $\bar{y}$ is the deviation of the mean from zero in Eq. (9.30*b*).

9.9.3 The-Large-the-Better Case

In this case, the quality characteristics never have negative values. As the performance value approaches infinity, the quality loss approaches zero. Also, no target value is predetermined. The loss function for this case is shown in Fig.9.4 (*c*). The loss function is simply the reciprocal of the smaller-the-better case.

$$L(y) = k\left(\frac{1}{y^2}\right) \qquad \text{...(9.31)}$$

To determine the value of the quality loss coefficient k, the reciprocal relationship must be employed. Hence, the loss A when the quality characteristic falls below the customer's tolerance level Δ, is given by

$$A = k\left[\frac{1}{\Delta^2}\right] \quad ...(9.32)$$

The quality loss coefficient, k is given by

$$k = A\,\Delta^2 \quad ...(9.33)$$

Examples of larger-the-better case include the soldering strength of a joint, the gas mileage of an automobile, weld strength of the beams, the customer acceptance rate of a product, the traction capacity of a tire of a vehicle, and the adhesive strength of glue, etc.

The mean square deviation [*MSD*] for the larger-the-better case is given by

$$MSD = \frac{1}{n}\sum_{i=1}^{n}\left(\frac{1}{y_i^2}\right) = \frac{1}{n}\left[\left(\frac{1}{y_1}\right)^2 + \left(\frac{1}{y_2}\right)^2 + ... + \left(\frac{1}{y_n}\right)^2\right] \quad ...(9.34)$$

The average loss function for the larger-the-better case is given by

$$L(y) = k\,(MSD) = k\left[\frac{1}{n}\sum_{i=1}^{n}\left(\frac{1}{y_i^2}\right)\right] \quad ...(9.35)$$

or

$$L(y) = A\Delta^2\left[\frac{1}{n}\sum_{i=1}^{n}\left(\frac{1}{y_i^2}\right)\right] \quad ...(9.36)$$

where the sample size is n.

9.9.4 The Asymmetric Nominal-the-Best Loss Function

There are two distinct quality loss coefficients needed for this case each corresponding to the direction away from the target value. The asymmetric loss function can be written as

$$L^+(y) = k^+(y - m)^2 \quad y > m \quad ...(9.37)$$

$$L^-(y) = k^-(y - m)^2 \quad y \le m \quad ...(9.38)$$

Examples of the asymmetric loss function include the amount of toner a printer uses to print one page and the variation of temperature in an ice cooler (deep freezer), etc.

Example E9.9. *The force needed to break apart parts joined by the adhesive determines the strength of an adhesive. Assume the functional limits for the strength of the adhesive are m ±0.2 kgf. Assume that the average cost for repairing or replacing an adhesive with unacceptable strength is $50. The target value of strength is 3 kgf. Determine*

(a) the quality loss to customers who purchased the adhesive with a strength of 2 kgf

(b) the average quality loss for a set of adhesives.

Assuming that there are two machines making the adhesives, one new machine and an older model that has a good deal of wear and tear on it.

New machine: 2.8, 2.9, 3.1, 2.7, 3.3, 3.05, 2.75, 2.85, 3.15, 3.25

Old machine : 3.5, 2.7, 3.4, 3.3, 2.9, 3.45, 3.35, 2.8, 2.75, 3.6

Solution:

(*a*) Here A = \$50, Δ = 0.2 kgf, and m = 3 kgf.

$$L(y) = \frac{\$50}{(0.2)^2}(y - 3)^2 = \$1250\,(y - 3)^2$$

Hence, the quality loss to customers who purchased an adhesive with an adhesive strength of 2 kgf is

$$L(2) = \$1250(y - 3)^2 = 1250(2 - 3)^2 = \$1250$$

(*b*) The average loss function is

$$L(y) = 1250[s^2 + (\bar{y} - m)^2]$$

$$m = 3 \text{ kgf}$$

Table E9.9

Machine	Data (y_i)	s^2	$\bar{y}$	$(\bar{y} - m)^2$	L(y)
New machine	2.8, 2.9, 3.1, 2.7, 3.3, 3.05, 2.75, 2.85, 3.15, 3.25	0.045583	2.985	0.000225	\$57.29
Old machine	3.5, 2.7, 3.4, 3.3, 2.9, 3.45, 3.35, 2.8, 2.75, 3.6	0.120139	3.175	0.030625	\$63.46

Note:

$$\bar{y} = \frac{1}{n}\sum_{i=1}^{n} y_i$$

and

$$s^2 = \frac{1}{n-1}\sum_{i=1}^{n} (y_i - \bar{y})^2$$

New machine:

$$s^2 = \frac{1}{(10-1)}[(2.8 - 2.985)^2 + (2.9 - 2.985)^2 + (3.1 - 2.985)^2$$
$$+ (2.7 - 2.985)^2 + (3.3 - 2.985)^2 + (3.05 - 2.985)^2 + (2.75 - 2.985)^2$$
$$+ (2.85 - 2.985)^2 + (3.15 - 2.985)^2 + (3.25 - 2.985)^2]$$
$$= 0.045583$$

Old machine:

$$s^2 = \frac{1}{(10-1)}[(3.5 - 3.175)^2 + (2.7 - 3.175)^2 + (3.4 - 3.175)^2$$
$$+ (3.3 - 3.175)^2 + (2.9 - 3.175)^2 + (3.45 - 3.175)^2 + (3.35 - 3.175)^2$$
$$+ (2.8 - 3.175)^2 + (2.75 - 3.175)^2 + (3.6 - 3.175)^2]$$
$$= 0.120139$$

$$L(y)_{\text{new machine}} = 1250[s^2 + (\bar{y} - m)^2] = 1250[0.045583 + 0.00025] = \$57.29$$

$$L(y)_{\text{old machine}} = 1250[s^2 + (\bar{y} - m)^2] = 1250[0.020139 + 0.030625] = \$63.45 \text{ or } = \$63.46.$$

Example E9.10. *In the copier industry, the amount of background toner particles that adhere to the portion of the copy that is intended to be white is one measure of acceptability of a copy. It was found that half of the customers will not accept a background level beyond a standard measure of 1 background unit. Any deviation from this level would cost \$250 for the service call plus copier downtime at \$200 per hour. The average copier downtime is 3 hours. For the two machines in a print shop the performance data collected is summarized in Table E9.10. Determine*

(a) the average quality loss for each machine for smaller-the-better case
(b) which machine is producing the smallest amount of loss.

Table E9.10

Machine	Performance data
1	0.7, 0.65, 0.63, 0.75, 0.8, 0.85, 0.73, 0.61, 0.55, 0.52
2	0.8, 0.55, 0.9, 0.65, 0.75, 0.85, 0.83, 0.53, 0.89, 0.61

Solution:
The average copier downtime is 3 hours. The customers loss is 3($200) = $600. Hence, the total cost is

$$A = \$250 \text{ (service call)} + \$600 \text{ (copier downtime)} = \$850$$

The quality loss coefficient

$$k = \frac{A}{\Delta^2} = \frac{\$850}{(1)^2} = \$850$$

The smaller-the-better loss function is given by

$$L(y) = ky^2 = \$850(y)^2$$

(*a*) The average loss for the smaller-the-better case is given by

$$L(y) = k\,[s^2 + \bar{y}^2]$$

Here, the target is zero and $\bar{y}$ is the deviation of the mean from zero.

$$k = \$850$$

The results are summarized in the Table E9.10 (*a*).

Table E9.10 (*a*)

Machine	Data	s^2	$\bar{y}$	$s^2 + \bar{y}^2$	$L(y) = 850(s^2 + \bar{y}^2)$
1	0.7, 0.65, 0.63, 0.75, 0.8, 0.85, 0.73, 0.61, 0.55, 0.52	0.011321	0.679	0.47237	401.51
2	0.8, 0.55, 0.9, 0.65, 0.75, 0.85, 0.83, 0.53, 0.89, 0.61	0.019671	0.736	0.56137	477.16

Machines 1 and 2:

$$\bar{y} = \frac{1}{n}\sum_{i=1}^{n} y_i \text{ and } s^2 = \frac{1}{n-1}\sum_{i=1}^{n} (y_i - \bar{y})^2$$

(*b*) From Table E9.10 (*a*), we note that machine 1 is producing the smallest amount of loss.

Example E9.11. *Compare the quality levels of two metals, M1 and M2 that cost $100 and $120 per unit weight respectively. The lower specification limit for tensile strength is 5 N/m². The out-of-specification units are discarded, resulting in a loss of $75 per unit. The annual production rate is 20,000 units. Twenty units were tested for each type of metal, and the following data for the tensile strength were obtained.*

Table E9.11

Metal type	Tensile strength N/m^2
M1	10.5, 7.8, 9.2, 15.5, 18.1, 17.5, 9.5, 12.5, 13.5, 14.8, 8.9, 11.2, 9.5, 10.8, 11.3, 13.4, 14.5, 15.2, 11.9, 12.3
M2	8.3, 10.3, 15.2, 14.1, 10.3, 8.7, 9.3, 14.1, 15.0, 11.2, 9.2, 8.8, 9.8, 10.5, 11.5, 12.3, 13.1,14.3, 15.1, 9.7

Solution:

In this problem, the larger the tensile strength, the better the metal. The average loss function for the larger-the-better is given by

$$L(y) = A\Delta^2 \left[\frac{1}{n} \sum_{i=1}^{n} \left(\frac{1}{{y_i}^2} \right) \right]$$

Here A = \$75, Δ = 5 N/m^2 and n = 20.

Metal 1: $\frac{1}{n} \sum_{i=1}^{n} \left(\frac{1}{{y_i}^2} \right) = 0.007587$, $L_{M1}(y) = (75)\,(5)^2\,(0.007587) = 14.23$

Metal 2: $\frac{1}{n} \sum_{i=1}^{n} \left(\frac{1}{{y_i}^2} \right) = 0.008453$, $L_{M2}(y) = (75)\,(5)^2\,(0.008453) = 15.85$

The cost per unit of Metal 1 (\$100) and Metal 2 (\$120), the savings that result from the use of Metal 1 instead of Metal 2 are

$$- [\$100 + L_{M1}(y)] + [\$120 + L_{M2}(y)] = \$\,21.62 \text{ per unit}$$

The annual savings are (20,000 units) (\$ 21.62 per unit) = \$ 432,400

9.10 TOLERANCE DESIGN AND TOLERANCING

It is important for the designer to establish the optimal values of a product's parameters and their tolerances besides a product's performance, appearance, and reliability. In this section, we examine the effect of tolerance design on the quality evaluation of a product and present methods for economic design of product tolerancing and tolerance. The nominal values and functional limits generally define the characteristics of a product. The tolerance of a product's characteristics has a significant effect on the total quality loss. Therefore, optimal tolerances are those that minimize the total quality loss. The three sources of variation of product characteristics from their nominal values are:

(1) Environmental factors

(2) Deterioration factors

(3) Imperfections in manufacturing processes.

In general, the product designer addresses the first two causes of variation by considering the parameter design and the tolerance design. The production and manufacturing people address on the third cause of variation, that is, variations due to manufacturing process imperfections.

There are three approaches for obtaining the optimal tolerances. They are

(1) The nominal-the-best

(2) The larger-the-better

(3) The smaller-the-better.

9.10.1 The Nominal-the-Best

In this type of tolerance, the nominal measurement of product characteristic such as dimensions and viscosity are considered and the variation is reduced to a minimum.

9.10.2 The Larger-the-Better

In larger-the-better is applicable to product's characteristic such as the hardness of materials, fuel efficiency, and thermal properties of materials. The product designer and the manufacturer aim for the larger value of the characteristic.

9.10.3 The Smaller-the-Better

In this case, the product designer and the manufacturer aim for the smaller value of the characteristic of a product such as wear and tear, deterioration and noise level, etc.

9.10.4 Multiple Components

The objective here is to confirm the output characteristic of the product consisting of k components to the functional limit. The tolerance design of each of these k components is required since the tolerances of the components have a direct effect on the quality characteristics of the product or process.

The desired characteristic of the product y, with functional limit Δ can be written as

$$y = m_0 + C(x - m) \quad \text{...(9.39)}$$

where m_0 = the nominal value of y (the target value)

C = a constant

x = characteristic

m = the nominal value.

The functional limits are $m_0 \pm \Delta$. Hence, the loss function is given by

$$L(y) = \frac{A}{\Delta^2}(y - m_0)^2 \quad \text{...(9.40)}$$

From Eqs. (9.39) and (9.40), we have

$$L(y) = \frac{A}{\Delta^2}[C(x - m)]^2 \quad \text{...(9.41)}$$

Now, the price of the component with the characteristic x is equal to $L(y)$.

Therefore

$$L(y) = A_0 = \frac{A}{\Delta^2}[C(x - m)]^2 \quad \text{...(9.42)}$$

The value of x from Eq. (9.42) is given by

$$x = m \pm \sqrt{\frac{A_0}{A}}\left(\frac{\Delta}{C}\right) \quad \text{...(9.43)}$$

which is the optimal tolerance specification for the component characteristic x.

For k components, the tolerances for each component are given by

$$\Delta_k = \sqrt{\frac{A_{0i}}{A}}\left(\frac{\Delta}{C_k}\right) \qquad ...(9.44)$$

when $x_i = m_i + \Delta_i$, $C_i \Delta_i$ represents the contribution to the displacement y caused by quality characteristic x_i. Hence,

$$\delta^2 = \sum_{i=1}^{k} (C_i \Delta_i)^2 \qquad i = 1, 2, ..., k \qquad ...(9.45)$$

$$= \left(\sqrt{\frac{A_{01}}{A}}\,\Delta\right)^2 + \left(\sqrt{\frac{A_{02}}{A}}\,\Delta\right)^2 + ... + \left(\sqrt{\frac{A_{0k}}{A}}\,\Delta\right)^2 \qquad ...(9.46)$$

$$= \left(\frac{A_{01} + A_{02} + ... + A_{0k}}{A}\right)\Delta^2 \qquad ...(9.47)$$

or

$$\delta = \sqrt{\frac{\sum_{i=1}^{k} A_{0i}}{A}}\,\Delta \qquad ...(9.48)$$

From Eq. (9.48), we observe that if $\sum_{i=1}^{k} A_{0i} << A$, then $\delta << \Delta$.

If $$\sum_{i=1}^{k} A_{0i} >> A, \text{ then } \delta >> \Delta \qquad ...(9.49)$$

and if $$\sum_{i=1}^{k} A_{0i} \approx A, \text{ then } \delta = \Delta$$

Eq.(9.43) can be used to determine the tolerance of each component.

9.10.5 Tolerances are Not Equal

In this case, the tolerances in general are set at unequal distances from the target value m. That is, they are set at $m^{-\Delta_1}_{+\Delta_2}$, where Δ_1 and Δ_2 are the lower and the upper limits of the tolerance respectively.

The loss due to the deviation of y from m is given by (Figure 9.10)

$$L(y) = \begin{cases} \dfrac{A_1}{\Delta_1^2}(y - m)^2 & \text{if } y \le m \\ \dfrac{A_2}{\Delta_2^2}(y - m)^2 & \text{if } y > m \end{cases} \qquad ...(9.50a)$$

where A_1 is the loss caused by y being below the lower limit of tolerance, and A_2, the loss caused by y being the upper limit of tolerance. When n observations are taken, the expected loss L is obtained as

$$L = \frac{1}{n}\left[\frac{A_1}{\Delta_1^2}\Sigma'(y-m)^2 + \frac{A_2}{\Delta_2^2}\Sigma''(y-m)^2\right] \quad ...(9.50b)$$

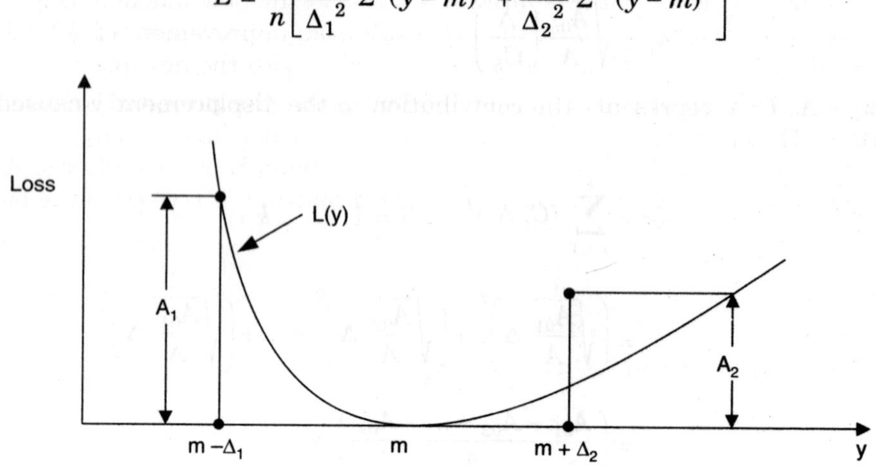

Fig. 9.10 Loss Due to Deviations from m.

Example E9.12. *A new surface roughness finishing machine is being contemplated to replace an old existing machine. The new machine will reduce the manufacturing cost by $2 per unit of production. The tolerance on the flatness (surface roughness) of the production unit produced by the new machine is 15 μm or less. The cost of defective products is $10 per defective unit. Compare the quality levels of the new and the old machine based on 50 units produced by the two machines under exactly the same conditions. The measurements of the flatness measured for 25 samples in μm is given in Table E9.12.*

Table E9.12. Flatness Measurements

Machine	Flatness (μm)
New machine	2, 5, 0, 8, 10, 1, 4, 6, 9, 11, 3, 7, 1, 3, 4, 5, 8, 7, 0, 1, 2, 3, 4, 5, 9
Old machine	11, 4, 12, 6, 8, 0, 9, 7, 13, 2, 5, 14, 9, 5, 13, 7, 9, 1, 10, 8, 14, 3, 6, 7, 11

Solution:

The expected average loss for smaller-the-better case is given by Eq.(9.30) for a sample of n items.

$$L(y) = \left[\frac{A}{\Delta^2}\right]\frac{1}{n}\sum_{i=1}^{n} y_i^2$$

Here, $A = \$10$, $\Delta = 15$ μm, and $n = 25$.

New machine:

$$L_N(y) = \frac{\Delta}{n\Delta^2}[2^2 + 5^2 + 0^2 + \ldots + 9^2] = \frac{10}{(15)^2(25)}(806) = \$1.43$$

Old machine:

$$L_O(y) = \frac{A}{n\Delta^2}[11^2 + 4^2 + 12^2 + \ldots + 11^2] = \frac{10(1886)}{(25)^2(15)} = \$3.35$$

Comparing $L_N(y)$ and $L_O(y)$, the quality loss of old machine exceeds that of the new machine by \$1.92. It is given that the production cost of the new machine is \$2 less than the production cost of old machine. Hence, the net result is an improvement of (\$2 + \$1.92) = \$3.92 per unit of production when we replace the old machine with the new machine.

Assuming the annual production is 500,000 units, the projected annual gain will be (500,000 units) (\$3.92) or \$1,960,000. Again, when a new machine replaces the old machine, then the new limit on the flatness of the products is determined as follows. Assuming the upper tolerance limit of products produced by the old machine is related to the factor of safety ϕ_0, we have

$$\phi_0 = \sqrt{\frac{A}{\$10}}$$

The new machine's factor of safety is given by

$$\phi = \sqrt{\frac{A}{8.08}}$$

Therefore, the new tolerance limit for the flatness = (existing tolerance limit) × $\left(\frac{\phi_0}{\phi}\right)$

$$= (15\ \mu\text{m})\sqrt{\frac{A/10}{A/8.08}} = 15\sqrt{\frac{8.08}{10}} = 13.48 \text{ or } 14\ \mu\text{m}.$$

Example E9.13. *Consider two types of metal rods R_1 and R_2 whose price and tensile strength are proportional to the rod's circular cross-sectional area. The prices are P_1 = \$500/mm^2 and P_2 = \$700/mm^2, and the tensile strength are $T_1 = 200\ N/mm^2$ and $T_2 = 250\ N/mm^2$ respectively. The lower tolerance limit of the rod's breaking strength is 20,000 N and the loss caused by failing below the tolerance limit is \$50,000. Determine the tolerance limits for the better rod by conducting the tolerance design.*

Solution:

For rod type R_1:

The total cost for each rod = price of the rod + quality loss

or $$C = P_1 x + \frac{A\Delta^2}{(T_1 x)^2} \tag{E.1}$$

where, A = \$50,000 and $\Delta = 20{,}000\ N$

C = the total cost

P_1 = the price of rod type R_1

T_1 = the tensile strength of rod R_1

x = the circular cross-sectional area of the rod.

Eq. (E.1) becomes

$$C = 500x + \frac{(50{,}000)(20{,}000)}{[(200)(x)]^2} \tag{E.2}$$

Differentiating Eq. (E.1) with respect to x, and equating to zero, we minimize the total cost C. Therefore,

$$\frac{dC}{dx} = P_1 - \left(\frac{2A\Delta^2}{T_1^2 x^3}\right) \tag{E.3}$$

or
$$x = \left(\frac{2A\Delta^2}{P_1 T_1^2}\right)^{1/3} = \left[\frac{2(50{,}000)(20{,}000)^2}{(500)(200)^2}\right]^{1/3} = 126 \text{ mm}^2$$

Therefore, the price of this rod is ($500/mm²) (126 mm²) = $63,000

For rod type R_2:

The cross-sectional area is

$$x = \left[\frac{2(50{,}000)(20{,}000)^2}{(700)(250)^2}\right]^{1/3} = 97.066 \text{ mm}^2$$

The price for rod R_2 is ($700) (97.066) = $679,042

The price of rod R_1 is less than the price of rod R_2. Therefore, we can select rod type R_1. The tolerance of this rod is obtained from Eq. (9.49)

$$\delta = \sqrt{\frac{A_{0i}}{A}}\,\Delta = \sqrt{\frac{50{,}000}{63{,}000}} \times (20{,}000) = 17{,}817.41 \text{ N}$$

Example E9.14. *A manufacturer of metal washers used in a machine system requires that the tolerances of the inside diameter and hardness of each washer be as follows:*

Tolerance of diameter = $d_t \pm 0.5\ \mu m$

Tolerance of hardness = $h_t \pm 2.0$ *(Brinell hardness)*

where d_t *and* h_t *are the target values of the inside diameter and Brinell hardness respectively. The cost of production is 0.50¢ per washer and the production rate is 100,000 washers per day. The following deviations from the inside diameter and the hardness target values were obtained from a sample size of 25.*

Deviations from the target inside diameter:

0.2, 0.1, 0, – 0.1, 0.2, 0.3, 0.1, – 0.1, 0.5, 0.3, 0.2, 0.4, – 0.1, 0.2, 0.0, – 0.4, 0.5, – 0.3, 0.4, 0.1, – 0.2, 0.1, 0.0, 0.1, 0.2

Deviations from the target hardness values:

1, – 1.0, – 0.5, 0.3, 0.5, 0.6, – 0.3, – 1.1, – 1.2, 0.6, – 0.3, – 0.5, 0.3, 0.6, 0.6, – 0.9, – 0.8, – 0.3, – 0.7, – 1.1, 1.5, – 1.2, – 1.4, 1.1, 1.4

Assuming that the defective washers cannot be reworked and are scrapped, determine the quality levels of the production process for the inside diameter and hardness attributes of the washers.

Solution:

The loss caused by deviations from the target value is given by Eq.(9.25)

$$L(y) = \frac{A}{\Delta^2} \sum_{i=1}^{n} (y_i - m)^2 / n \tag{E.1}$$

where A = cost of a defective product

2Δ = tolerance limit

n = the number of measurements or sample size

y_i = the value of measurement i.

Given that $A = 0.50'$ and $\Delta = 0.5$ μm. The quality level of the inside diameter of the washers is

$$L = \frac{0.50}{(0.5)^2}\left[\frac{1}{25}(0.2^2 + 0.1^2 + 0^2 + \ldots + 0.2^2)\right] = \$0.1256$$

or $\quad 12.56' \qquad$ (E.2)

Similarly, the quality level of the hardness of the washers is given by

$$A = \$0.50$$

$$\Delta = 2.0$$

$$L = \frac{0.50}{(2.0)^2}\left[\frac{1}{25}(1^2 + (-1.0)^2 + (-0.5)^2 + \ldots + (1.4)^2\right] = \$0.0963$$

or $\quad 9.63' \qquad$ ...(E.3)

Difference in quality levels

$$= \$(0.1256 - 0.0963)\,(225)\,(100{,}000)$$

$$= \$659{,}250 \text{ per year} \qquad \text{...(E.4)}$$

In Eq. (E.4), it was assumed that there are 225 working days per year.

Example E9.15. *A mechanical system manufacturer requires that shrink-fitting the shaft into the disk assemble the shaft and a disk, thereby attaining a highly desirable stress distribution. The specification of the diameter of the shaft is m ±10 μm and the loss resulting from a defective shaft is $50. The deviations observed for the shaft diameter from three different suppliers A, B and C for a sample size of 25 is given in Table E9.15.*

(a) determine the quality levels of the three suppliers

(b) assuming adjustments can be made, determine the quality levels after the adjustments.

Table E9.15 Supplier's Deviations from Target Data

Supplier	**Deviations from target value**	**Price per unit**
A	*– 4, 5, 6, 0, 2, – 3, 7, 9, – 2, – 5, – 4, – 8, 1, – 4, – 5, – 2, 0, – 6, 8, 7, 3, 2, 3, 4, 1*	*$60*
B	*– 6, – 2, – 5, – 4, – 3, – 2, – 1, – 6, – 3, – 2, – 3, – 2, – 1, – 4, – 2, – 3, – 3, – 4, – 6, – 5, – 2, – 3, – 5, – 2, – 1*	*$65*
C	*– 7, 3, 10, – 12, 13, 8, – 4, – 7, 9, 11, – 7, –8, – 3, – 4, – 9, 10, 14, 11, 4, 5, 6, – 3, – 6, 7, 8*	*$55*

Adjustments are made such that $\bar{y}_1 = \bar{y}_2 = \bar{y}_3 = m$. The *MSD* after the adjustments for the suppliers 1, 2 and 3 are given by

$$MSD_j = \frac{1}{(n-1)}\left[\sum_{i=1}^{n} y_i^2 - \frac{\left(\sum_{i=1}^{n} y_i\right)^2}{n}\right] \quad j = 1, 2, 3 \qquad \text{...(E.1)}$$

Substituting the values for y_i and n from Table E9.15 and solving Eq.(E.1), we obtain

$$MSD_1 = 22.52$$
$$MSD_2 = 12.64$$
$$MSD_3 = 66.92$$

Suppliers' losses before adjustments are summarized in Table E9.15 (*a*) and the suppliers' loss after adjustments are given in Table E9.15 (*b*).

Table E9.15 (*a*): Suppliers' Losses before Adjustments

Supplier	MSD	$L_i = k_i\ (MSD)_i$
1	22.52	\$1.351
2	12.64	\$0.822
3	66.92	\$3.681

where $k_i = \dfrac{A_i}{\Delta^2}$

in which A_i are given as \$60, \$65 and \$55 respectively and $\Delta = 10$.

Table E9.15 (*b*): Suppliers' Losses after Adjustments

Supplier	MSD	L = (0.05) (MSD)
1	22.52	\$1.126
2	12.64	\$0.632
3	66.92	\$3.346

The data in Tables E9.15 (*a*) and (*b*) suggest that the manufacturer should choose supplier 2 as the source if supplier 2 is able to keep the distribution of the sizes centered on the target value *m*.

Example E9.16. *An equipment manufacturer requires that the clearance between two components of a particular system be* 5^{-2}_{17} *μm. The loss due to a defective assembly of these two components is \$100, and the weekly production is 1000 units. Table E9.16 shows the deviations from the target value for the first two weeks of production for a sample size of 25.*

(*a*) *determine the quality levels during these two weeks*

(*b*) *is there any improvement of the quality level?*

Table E9.16

Week	Deviations				
1	– 2	– 1	0	1	– 2
	– 1	5	– 1	2	0
	0	6	– 2	5	– 1
	1	– 2	4	7	5
	3	7	5	– 1	6
2	3	5	7	6	4
	2	– 1	– 2	0	– 1
	0	1	2	3	4
	– 1	5	6	7	– 2
	– 2	– 1	0	4	5

Solution:

The quality level of the production during week-1 is determined from

$$L_1 = \frac{A_1}{{\Delta_1}^2}\left[\frac{1}{n}\sum_{i=1}^{n} {y_i}^2\right]$$

where, A_1 = \$100, $\Delta_1 = -2$ μm and $n = 25$.

Therefore

$$L_1 = \frac{100}{(-2)^2}\left[\frac{1}{25}\{(-2)^2 + (-1)^2 + (0)^2 + ... + 6^2\}\right] = \frac{100}{100}[322] = \$322$$

The loss during week-2, L_2, is

$$L_2 = \frac{A_2}{(\Delta_2)^2}\left[\frac{1}{n}\sum_{i=1}^{n} {y_i}^2\right]$$

where, A_2 = \$100, $\Delta_2 = 7$ μm and $n = 25$

$$L_2 = \frac{100}{7^2}\left[\frac{1}{25}\{3^2 + 5^2 + 7^2 + ... + 5^2\}\right] = \frac{100}{(49)(25)} \times 336$$

$$= \$27.43/\text{unit}$$

Hence, the losses resulting from the variation on the lower limit are greater than those of the upper limit in each week. The improvement of quality level per unit during week-2 is \$322 – \$27.43 = \$294.57 or an improvement of (\$294.57) (1,000 units) = \$294,570 per week.

Example E9.17. *An electromechanical system has four mechanical components, 1, 2, 3, and 4. Their prices are p_1 = \$25, p_2 = \$50, p_3 = \$100, and p_4 = \$150 respectively. The component characteristics have an approximately linear effect on a producer's characteristics around the nominal values of m_1, m_2, m_3 and m_4. The corresponding linear constants are 2.5, 3.5, 2 and 1.5 for components 1, 2, 3 and 4 respectively. The cost of defective component that does not conform to the specification is \$1000. The functional tolerance of the product characteristic is 5. Determine the tolerance of each component assuming the losses incurred by component failures are equal to the component prices given above.*

Solution:

Denoting y as the quality characteristic of the product, the relationships between y and the components are given by Eq. (9.40).

Component 1 : $y = m_0 + C_1(x_1 - m_1)$

Component 2 : $y = m_0 + C_2(x_2 - m_2)$

Component 3 : $y = m_0 + C_3(x_3 - m_3)$

Component 4 : $y = m_0 + C_4(x_4 - m_4)$

The total quality loss for the product is given by Eq. (9.42)

$$L(y) = \frac{A}{\Delta^2}\left[\sum_{i=1}^{n} C_i(x_i - m_i)\right]^2 \qquad ...(E.1)$$

Eq. (E.1) can be solved to obtain

$$x_i = m_i \pm \left[\sqrt{\left(\frac{A_{0i}}{A}\right)}\right]\left[\frac{\Delta}{C_i}\right] \quad i = 1, 2, 3, 4 \qquad \text{...(E.2)}$$

or

$$\Delta_i = \left[\sqrt{\left(\frac{A_{0i}}{A}\right)}\right]\left[\frac{\Delta}{C_i}\right] \quad i = 1, 2, 3, 4 \qquad \text{...(E.3)}$$

Hence, the tolerance limits for the four components are given by

$$\Delta_1 = \left(\sqrt{\frac{A_1}{A}}\right)\left(\frac{\Delta}{C_1}\right) = \left(\sqrt{\frac{25}{1000}}\right)\left(\frac{5}{2.5}\right) = 0.3162$$

$$\Delta_2 = \left(\sqrt{\frac{A_2}{A}}\right)\left(\frac{\Delta}{C_2}\right) = \left(\sqrt{\frac{50}{1000}}\right)\left(\frac{5}{3.5}\right) = 0.3194$$

$$\Delta_3 = \left(\sqrt{\frac{A_3}{A}}\right)\left(\frac{\Delta}{C_3}\right) = \left(\sqrt{\frac{100}{1000}}\right)\left(\frac{5}{2}\right) = 0.7906$$

$$\Delta_4 = \left(\sqrt{\frac{A_4}{A}}\right)\left(\frac{\Delta}{C_4}\right) = \left(\sqrt{\frac{150}{1000}}\right)\left(\frac{5}{1.5}\right) = 1.2910$$

Finally, the tolerances for the product characteristic, Δ, is obtained using Eq. (9.49)

$$\delta = \sqrt{\frac{\sum_{i=1}^{k} A_{0i}\Delta}{A}} = \sqrt{\frac{(25 + 50 + 100 + 150)5}{1000}} = \sqrt{1.625} = 1.2748$$

9.11 SIGNAL-TO-NOISE RATIO

The quality of a product or process is greatly affected by two classes of input parameters: (1) design factors, and (2) disturbance factors. The designer can specify the design factors or parameters rather freely while the disturbance factors are the parameters that are uncontrollable or not practical to control. The input and output parameter variability are classified into four categories. As mentioned earlier, the term *signal*, or the average value of the quality characteristic represents the desirable component, which will preferably close to a specified target value. *Noise* represents the undesirable component and it is a measure of the variability of the output characteristic, which will preferably be as small as possible. The four categories of the variability of the input and output parameters are:

Input Variability

1. *Tolerances:* Tolerances are defined as the limit at which some economically measurable action is taken. Tolerances are the normal variability in design factors. The customer tolerance corresponds to the point at which a significant number of customers take economic action because of off-target performance.

2. *Outer noise:* Outer noise corresponds to the variability of the disturbance factors that contribute to output variability. Environmentally related noise factors, which affect variation within a product (temperature, humidity, operators, etc.). The variables external to a product that affect the product performance are known as external noise factors. Examples include variations in temperature, humidity, and dust.

Output Variability

3. *Variational noise:* Variational noise is the short term unit-to-unit variation due to the manufacturing process.

4. *Inner noise:* Inner noise is a result of variation due to the deterioration of parts and material. Examples include wear out of parts due to friction, and increase in resistance of resistors with age. Thus, inner noise is a long term change in product characteristics over time due to deterioration and wear.

Taguchi has combined the two components signal and noise into one measure known as the *signal-to-noise (S/N) ratio*. The signal-to-noise metric is designed to be used to optimize the robustness of a product or process. The performance measure combines both the parameters, the mean and the variation around the mean. Hence, when the performance measure is maximized, the expected loss will be minimized. Taguchi uses the signal-to-noise ratio, *S/N*, as the objective function to be optimized. Here, we consider three situations:

1. Target is best or the nominal-the-best
2. Smaller is better
3. Larger is better

9.11.1 Target is Best or the Nominal-the-Best

The expected loss associated with the output characteristic is given by Eq.(9.30)

$$E[L(y)] = \frac{A}{\Delta^2}\,[\text{variance}[y] + (\mu - m)^2] = k[MSD]$$

$$= \frac{A}{\Delta^2}\,[\text{variance of } y + \text{squared bias of } y] \qquad ...(9.51)$$

where, A = the associated loss

Δ = the tolerance

m = the target value of y

y = the value of the quality characteristic

$k = \frac{A}{\Delta^2}$ = quality loss coefficient

MSD = the mean square deviation

μ = the mean value of y

It is desirable for the output characteristic to be on target while at the same time the variance to be as small as possible.

The mean square deviation [*MSD*] for the target is best or nominal the best can be written as

$$MSD = [s^2 + (\bar{y} - m)^2] \qquad ...(9.52)$$

There are two different nominal-the-best cases that require modifying the *MSD* in different ways to obtain the form most likely to be independent of the adjustment.

9.11.1.1 Variance independent of the mean

In this type, the response values are continuous and can take on either positive or negative values and the target can be zero. Typical examples of this type are temperature controller errors around the nominal value, and the offset voltage for a power supply.

In this situation, the performance measure, which reduces the variance of the output characteristic, is given by

$$\xi = -\log(\text{variance } y) \quad ...(9.53)$$

an estimate of which the performance statistic is given by

$$z = -\log(s^2) \quad ...(9.54)$$

where s^2 is the sample variance of y, which is given by

$$s^2 = \sum_{i=1}^{n} \frac{(y_i - \bar{y})^2}{n-1} \quad ...(9.55)$$

in which

n = sample size

$\bar{y}$ = the sample mean of y_i

From Eq. (9.55), we observe that the larger the performance statistic Z, the smaller the variability.

9.11.1.2 Variance related to the mean

In this type, the standard deviation of the output characteristic is assumed to be linearly proportional to the mean and the bias can be reduced independently of the coefficient of variation. The coefficient of variation being defined as the ratio of standard deviation to the mean. The response values in this type are continuous and non-negative, ranging from zero to infinity. Also, when the mean response is at zero, the target value is non-zero and the variance is zero. Some examples of this type of nominal-the-best case is the image density, and the dimension of an elastomer, that is, the variation due to temperature and humidity.

The method involves identifying the adjustment parameters and their ranges for which the *S/N* ratio is constant. The *s/N* ratio defined by Taguchi is given by

$$\xi = \frac{\mu^2}{\sigma^2} \quad ...(9.56)$$

where,

μ = the mean of the output characteristic y

σ = standard deviation of the output characteristic y

σ^2 = variance of the output characteristic y

y = output characteristic

The performance statistic for this type is given by

$$z = 10 \log\left(\frac{\bar{y}^2}{s^2}\right) \quad ...(9.57)$$

where,

$\bar{y}$ = the sample mean of the output characteristic y

s = the sample standard deviation of the output characteristic

9.11.2 Smaller is Better

The smaller-the-better *s/N* ratio is based on the smaller-the-better loss function. The loss function is given by Eq. (9.31).

$$L(y) = \frac{A}{\Delta^2}\,[s^2 + \bar{y}^2] = k[s^2 + \bar{y}^2] \qquad ...(9.58)$$

The expected loss or average loss is given by Eq.(9.30*a*)

$$E(L(y)) = \frac{A}{\Delta^2}\,[MSD] = \frac{A}{\Delta^2}\left(\frac{\sum_{i=1}^{n} y_i^2}{n}\right) \qquad ...(9.59)$$

The response values or quality characteristics for this case are continuous and non-negative and the desired value of the response is zero. The objective here is to minimize the mean and variance simultaneously.

The performance characteristic is given by

$$z = -\,10\log\left(\frac{\sum_{i=1}^{n} y_i^2}{n}\right) \qquad ...(9.60)$$

or

$$z = -\,10\log\,[s^2 + \bar{y}^2] \qquad ...(9.61)$$

9.11.3 Larger is Better

The larger-the-better case is the inverse of the smaller-the-better case. The response values are continuous non-negative numbers ranging from 0 to infinity and the desired value of the response is infinity or the largest number possible.

The loss function for this case is given by Eq. (9.31)

$$L(y) = k\left(\frac{1}{y^2}\right) \qquad ...(9.62)$$

The expected loss is given by Eq. (9.36)

$$E[L(y)] = A\Delta^2\left[\frac{1}{n}\sum_{i=1}^{n}\left(\frac{1}{y_i^2}\right)\right] \qquad ...(9.63)$$

or

$$= A\Delta^2[MSD] \qquad ...(9.64)$$

The performance characteristic is given by

$$z = -\,10\log\left(\frac{\sum_{i=1}^{n}\left(\frac{1}{y_i^2}\right)}{n}\right) \qquad ...(9.65)$$

The objective here is to find the parameter settings that will maximize the performance statistic given by Eq. (9.65).

In an experimental design, there are many design factors involved. Two of such design factors of importance are (1) control factors – which affect mainly the *s/N* ratio, but not the mean, and (2) signal factors – which affect mainly the mean response of the performance characteristic. The control factors are first used to minimize the output variability and then the signal factors are used to shift the mean to the desired target value. Taguchi method advocates the economical maximization of the output performance characteristics while simultaneously minimizing the effect of output variability.

Example E9.18. *In order to determine which injection molding machine has the best performance with respect to part shrinkage, a full factorial experiment based on four different machines and three noise factors (three different types of plastic materials) is designed and carried out. The control factors are the machines 1, 2, 3, and 4. The data in Table E9.18 gives the noise in inches of the four machines under each of the three noise conditions. The customer tolerance is given as 0.03 in and the customer loss is set at $35. Determine which machine gives the best performance.*

Table E9.18

Machine	Noise (inches)		
	Noise 1	**Noise 2**	**Noise 3**
1	*0.009*	*0.012*	*0.014*
2	*0.005*	*0.008*	*0.006*
3	*0.008*	*0.018*	*0.014*
4	*0.020*	*0.018*	*0.020*

Solution:

The most desirable condition is that the shrinkage should be zero. This example illustrates the smaller-the-better case and the use of the quality loss function and the *s/N* ratio to quantify the performance of several alternatives.

For smaller-the-better case, the *s/N* ratio is given by Eq. (9.61)

$$s/N = -10 \log [s^2 + \bar{y}^2] \quad \text{...(E.1)}$$

where

$$\bar{y} = \sum_{i=1}^{n} \frac{1}{n} y_i \quad \text{...(E.2)}$$

and

$$s = \sqrt{\frac{1}{n-1} \sum_{i=1}^{n} (y_i - \bar{y})^2} \quad \text{...(E.3)}$$

The average loss is given by Eq. (9.59)

$$E[L(y)] = \frac{A}{\Delta^2}\left[\frac{1}{n}\sum_{i=1}^{n} y_i^2\right] \quad \text{...(E.4)}$$

Eqs. (E.1) to (E.4) are used to complete the data in Table E9.18 (*a*).

Table E9.18 (a)

Machine	Noise 1	Noise 2	Noise 3	Mean	s	Loss	s/N
1	0.009	0.012	0.014	0.01166	3.56×10^{-6}	0.3150	38.6611
2	0.005	0.008	0.006	0.00633	8.89×10^{-6}	0.0972	43.9673
3	0.008	0.018	0.014	0.01333	1.42×10^{-6}	0.2489	37.5012
4	0.020	0.018	0.020	0.01933	2.20×10^{-7}	1.5555	34.2739

We observe from Table E9.19 (*a*) that machine 2 gives the lowest shrinkage and the best performance. The value of *s/N* ratio increases as the mean $\bar{y}$ increases. Similarly, *S/N* ratio increases as the variability decreases.

Example E9.19. *In order to determine which welding machine has the best performance with respect to weld strength, a full factorial experiment based on four different machines and three noise factors (welding machine operators) is designed and carried out. The control factors are the welding strength in N/m² of the machines under each of the three noise conditions. The customer tolerance is given as 1500 N/m² and the loss is set at \$1000. Determine which welding machine gives the best performance.*

Table E9.19

Machine	Noise (N/m^2)		
	Noise 1	Noise 2	Noise 3
1	8230	5126	7159
2	9567	8905	7986
3	2850	3400	4120
4	6925	6750	6850

Solution:

Ideally, the weld strength should be the largest. Hence, the example illustrates the larger-the-better case and the use of the quality loss function and the *s/N* ratio to quantify the performance of several alternatives. For the larger-the-better case, the *s/N* ratio is given by Eq. (9.65)

$$z = s/N = -10 \log \left[\frac{\sum_{i=1}^{n} \left(\frac{1}{y_i^2} \right)}{n} \right] \qquad \text{...(E.1)}$$

The expected loss is given by Eq. (9.64)

$$E[L(y)] = A\Delta^2[MSD] \qquad \text{...(E.2)}$$

where

$$MSD = \frac{\sum_{i=1}^{n} \left(\frac{1}{y_i^2} \right)}{n} \qquad \text{...(E.3)}$$

Eqs. (E.1) to (E.3) are used to complete the data in Table E9.19 (*a*).

Table E9.19 (*a*)

Machine	Noise 1	Noise 2	Noise 3	Mean	MSD	Loss	s/N
1	8230	5126	7159	6838.33	2.41×10^{-8}	54.249	76.1778
2	9567	8905	7986	8819.33	1.31×10^{-8}	29.412	78.8366
3	2850	3400	4120	3456.67	8.95×10^{-8}	201.399	70.4813
4	6925	6750	6850	6841.67	2.14×10^{-8}	48.084	76.7018

From Table E9.20 (*a*), we find that Machine 2 gives the highest weld strength and hence the best performance. Machine 3 gives the worst performance, which is also clear from the raw data in Table E9.20. Machine 1 and 4 are much closer in performance.

9.12 EXPERIMENTAL DESIGN

Experiments are performed to answer a question. An experiment can be defined as a test performed to discover some information about a particular process or system. If the design of the test or experiment is done correctly, then there is an excellent chance of obtaining the desired information. Experimental design may be defined as *a formal plan that details the specifics for conducting an experiment, such as which responses, factors, levels, blocks, treatments, and tools are to be used*. Designed experiments are the tools necessary for optimizing the processes, identifying interactions among the various process variables, and reducing the variation in process. Experimental design is used to determine the impact of factors on the response variable. With quantitative factors, which vary on a continuous manner, the information about the variable's behavior even for factor levels that were not experimentally determined can be obtained.

Quality must be designed into a product. Experimental design facilitates in selecting optimal product and process parameters. The quality control phase is followed by quality improvement, which will lead to the creation of better products.

In Chapter 8, a class of procedures called *off-line control methods*, which are used in the design phase of product or process were discussed. Taguchi method of experimental design is used during the initial design and prototype testing to determine optimal settings for the product and process parameters.

The product or process is designed to optimize any detrimental effects earlier in the cycle in order to avoid problems that might appear in the production phase. The proper selection in the design phase of parameter settings for the product or process reduces costs and improves the quality.

9.13 ORTHOGONAL ARRAYS

The basis for designing an experiment using Taguchi methodology is the *orthogonal array*. The *orthogonal array* is very efficient in achieving only a relatively small amount of data and drawing meaningful, valuable and verifiable conclusions from that data. Orthogonal means being balanced and not mixed. In the design of experiments context, orthogonality means statistically independent.

Design of experiments helps one in selecting combinations of the various factor levels in order to determine the output characteristic, which allows calculating the performance statistic.

The matrix that designates the settings of the controllable factors or design parameters for each experiment (run) is called an *inner array*. The matrix that designates the setting of the incontrollable or noise factors is called the *outer array*. Each experiment or run involves setting of the design parameters and an associated setting of the noise factors. The inner and outer arrays are designated as *design* and *noise* matrices respectively.

An important and major step in the *DOE* process is the determination of the combination of factors and levels, which will provide with the desired information one way of accomplishing. This is to use a fractional-factorial approach whenever several factors involved are using the aid of orthogonal arrays. Taguchi used orthogonal arrays for designing and analyzing experimental data. Orthogonal arrays are used in Taguchi methods to measure the effect of a factor on the average (mean) result as well as determine the variation from the mean.

For the design factors that are quantitative, three levels are needed to determine the quadratic (or non-linear) effect, if any. If only two levels of factors are tested, then only its linear effects on the response variable can be estimated. The principal advantage of orthogonal arrays is the relationship between the factors under study. For each level of any factor, all levels of the other factors occur an equal number of times. This constitutes a balanced experiment and the results of the findings of the experiment are reproducible. On pre-selecting the design and noise factors and the number of settings, a series of experiments can be run to establish the optimum setting of the design parameters. Taguchi's method relies on running only a portion of the total number of possible experiments instead of running experiments at all possible combinations of design and noise factors. Taguchi's method accomplishes this using the concept of orthogonal arrays. The objective here is to focus on the main few rather than the trivial many within the confines of engineering and cost constraints. An orthogonal array is a matrix of numbers. Each column represents a specific factor whose effects on the response variable is of interest and each, row represents the levels or states of the selected factors. Any two columns of an orthogonal array form a two-factor complete factorial design. Application of orthogonal arrays optimizes the number of runs while retraining the pair wise balance property. Taguchi recommends the use of orthogonal arrays for the inner array (design matrix) and the outer array (noise matrix). Orthogonal arrays are generalized versions of Latin square designs (see Chapter 8) where in a Latin square design, a blocking of two factors is accomplished.

To enhance the flexibility of the arrays, Taguchi uses linear graphs with the orthogonal arrays. By using these graphs, one can study the interactions between the experimental factors as well as the effects of the individual factors (main-effects) themselves. These linear graphs show the assignment of factors to the column of the orthogonal array and show the interactions between the factors. Linear graphs make this possible by providing a logical scheme for assigning interactions to the orthogonal array without confounding the effects of the interactions with the effects of the individual factors being investigated.

Table 9 A.1 to 9 A.8 in Appendix-9A show several orthogonal arrays and their associated linear graphs. The notation used in Appendix-9A is described as follows: The orthogonal array represented by the L_4 design (see Table 9A.1 in Appendix-9A) is a half-replicate of a 2^3 full factorial experiment, in which three factors, each at two levels, are considered. Four experiments are conducted in the L_4 design (one-half of 2^3, the full factorial experiment). Denoting the two levels of each factor by 1 and 2, each of the three factors is at level 1 in the first experiment. Factors 1 and 2 are at level 2 and factor 3 at level 1 in the fourth experiment. The linear graph for the L_4 table (Table 9A.1) has two nodes, labeled 1 and 2, and a line joining them labeled as 3. These numbers correspond to the columns of the orthogonal array.

As shown in Table 9A.2 in Appendix-9A, there are two linear graphs for the L_8 array. The graph to the right in Table 9A.2 is used if all the interactions of interest involve a common factor. If the design requires the study of 3 interactions involving same 3 factors $A \times B$, $B \times C$, and $A \times C$, the graph on the left in Table 9A.2 can be used. If the main effects of two of the factors are to be estimated, then they are placed in columns 1 and 2 of the orthogonal array. Column 3 then represents the interaction effect of the factors assigned to columns 1 and 2.

The graphs in Appendix-9A are constructed of interconnecting dots, with each dot representing a column within the array in which a factor (main effect) can be assigned. The connecting line represents the interaction between the two dots (factors) at each end of the line segment. The number accompanying the line segment represents the column within the array to which the interaction should be assigned.

Four different types of points represent the linear graphs shown in Appendix-9A. They are circles, double circles, dot with circles, and completely shaded circles. These points correspond to the degree of difficulty associated with the number of changes that should be made in the level of a factor assigned to the corresponding column.

The L8 (2^7) orthogonal array shows the factor assignment levels for seven factors, each at two levels. This represents a fractional replicate of the full factorial experiment that requires a total of 2^7 or 128 experiments. On the other hand, the L^8 orthogonal array requires only 8 experiments. From the two linear graphs for the L^8 orthogonal array, we can estimate the four main effects independently of the other if the assignment is made to columns 1, 2, 4 and 7. If the interaction effect between factors 1 and 2 is needed, then it should be assigned to column 3. The L_9 (3^4) orthogonal array shown in Table 9A.3 (Appendix-9A) represents the design for which four factors are varied at three levels each. It represents the fractional replicate of the full factorial experiment, which needs 3^4 or 81 experiments. The main effects of two factors can be estimated independently of the others when assigned to columns 1 and 2 in the linear graph for the L_9 orthogonal array. The main effects of the two other factors cannot be estimated independently of the effects of the remaining factors as they will be confounded by the interaction effects between the factors assigned to the first two columns.

Appendix-9A also lists the orthogonal arrays and the associated linear graphs for several other experiments. It should be noted here that all possible interactions between factors cannot be estimated from the orthogonal arrays. In order to accomplish this, a full factorial experiment is needed. According to Taguchi, interactions can be ignored during the initial experiments as it is difficult to identify the significant interactions without some prior knowledge or experience. The unassigned columns in the orthogonal arrays can be assigned to the possible interactions after assigning the factors whose main effects are desirable.

Taguchi's method is to identify and control those factors that are significant in cause variation and to do so in a way that minimizes the effect of those variables that cannot be controlled or that are not practical to control.

Example E9.20. *Determine the optimum values for factors based on a biggest are best for the array given in Table E9.20.*

Table E9.20

Experiment	A	B	C	y_i
1	1	1	1	6
2	1	2	2	4
3	2	1	2	8
4	2	2	1	5

Solution:

$$A_1 = 6 + 4 = 10$$
$$A_2 = 8 + 5 = 13$$
$$B_1 = 6 + 8 = 14$$
$$B_2 = 4 + 5 = 9$$
$$C_1 = 6 + 5 = 11$$
$$C_2 = 4 + 8 = 12$$

Level	A	B	C
Level 1	10	14	11
Level 2	13	9	12
Total	23	23	23

Hence, the optimum mix is A_2, B_1, C_2. The corresponding parameter values are known. The degree of improvement to be acquired by using the optimum mix is estimated as follows:

$$y_{\text{opt } n} = \frac{A_2 + B_1 + C_2}{4} = \frac{13 + 14 + 12}{4} = 9.75$$

Before the change

$$y_{\text{opt}} = \frac{23}{4} = 5.75$$

Hence the predicted improvement is

$$= \frac{y_{\text{opt } n} - y_{\text{opt}}}{y_{\text{opt}}} = \frac{9.75 - 5.75}{5.75} = 0.6957 \text{ or } 69.57\%$$

9.14 ESTIMATION OF EFFECTS

After selecting the experimental design, the factors are set at the levels as per the chosen design. Experiments are then conducted to obtain the response variable y. The noise factors are placed in an outer array and control factors are placed in an inner array of a selected orthogonal array. Table 9.4 shows the layout of the design. In Table 9.4, four control factors A, B, C and D, each at three levels are placed in an L_9 inner array, and three noise factors, E, F, and G each at two levels, are placed in an L_4 outer array. There are nine settings of the design factors, based on the L_9 orthogonal array. There are four replications corresponding to the setting of the three noise factors for each of these settings.

A quantitative measure of the mean and variability of the output characteristic can be computed for each setting of the design factors.

The *main effect* of a factor is obtained by calculating the average of the response variable over all replications for a given level of the factor. The main effect of a factor are obtained by assuming that the interaction effects between factors are insignificant. The mean response when factor A is at level 1 in Table 9.4 is given by

$$\overline{A_1} = \frac{\bar{y}_1 + \bar{y}_2 + \bar{y}_3}{3}$$

The mean responses when factor A is at level 2 and 3 are

$$\overline{A_2} = \frac{\bar{y}_4 + \bar{y}_5 + \bar{y}_6}{3}$$

and
$$\overline{A}_3 = \frac{\overline{y}_7 + \overline{y}_8 + \overline{y}_9}{3}$$

The procedure is extended to obtain the main effects of the remaining factors at each level.

The mean response when factor B is at level 3 is

$$\overline{B}_3 = \frac{\overline{y}_3 + \overline{y}_6 + \overline{y}_9}{3}$$

Table 9.4.
Experimental Layout Using an L_9 Inner Array and an L_4 Outer Array

	Noise factors (L_4 outer array)			E	1	1	2	2			
				F	1	2	1	2			
				G	1	2	2	1			
	Design factors (L_9 inner layer)										
Experiment	A	B	C	D						*Mean response,* $\overline{y}$	*s/N ratio, z*
1	1	1	1	1	y_{11}	y_{12}	y_{13}	y_{14}		$\overline{y}_1$	z_1
2	1	2	2	2	y_{21}	y_{22}	y_{23}	y_{24}		$\overline{y}_2$	z_2
3	1	3	3	3	.	.	.	.		.	.
4	2	1	2	3	.	.	.	.		.	.
5	2	2	3	1	.	.	.	.		.	.
6	2	3	1	2	.	.	.	.		.	.
7	3	1	3	2	.	.	.	.		.	.
8	3	2	1	3	.	.	.	.		.	.
9	3	3	2	1	y_{91}	y_{92}	y_{93}	y_{94}		$\overline{y}_9$	z_9

The significance or non-significance of the factor is determined by plotting the main effects. That is the values of $\overline{A}_1, \overline{A}_2$ and $\overline{A}_3$ are plotted versus the three levels of factor A. Figure 9.11 shows the different possible effects of factor A on the average response. Figure 9.11 (*a*) shows that factor A's main effect is not significant. Figure 9.11 (*b*) shows that factor A's main effect is a non-linear function. To create a design that is robust to the noise factors, level 2 for factor A can be selected. This will give the smallest variability in the average response. Figure 9.11 (*c*) indicates that the factor A has a linear relationship with the average response. An adjustment factor can be used to shift the average response toward a desired value without altering its variability.

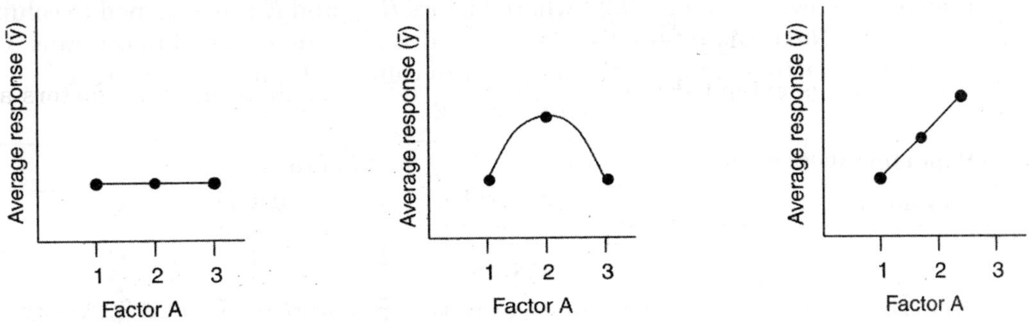

(*a*) Insignificant effect of factor *A* (*b*) Non-linear effect of factor *A* (*c*) Linear effect of factor *A*

Fig. 9.11 Effect of Factor on Response Variable.

Example E9.21. *A materials manufacturing company has identified four alloy elements (A, B, C and D), each at two levels, that are significant in their effect on the thermal properties of the material produced at the facility. The company also feels that the interaction effects A × C, A × B and B × C are important. Set up an experimental design that can estimate these effects using a minimal number of experiments.*

Solution:

We note that there is a total of seven factors of which three are the interaction effects $A \times C$, $A \times B$, and $B \times C$. Therefore, L_8 (2^7) orthogonal array can be selected. The corresponding factor assignments for all the eight experiments are summarized in Table E9.21.

Table E9.21

Experiment number	Factor						
	A	B	A × B	C	A × C	B × C	D
1	1	1	1	1	1	1	1
2	1	1	1	2	2	2	2
3	1	2	2	1	1	2	2
4	1	2	2	2	2	1	1
5	2	1	2	1	2	1	2
6	2	1	2	2	1	2	1
7	2	2	1	1	2	2	1
8	2	2	1	2	1	1	2

Example E9.22. *A quality control engineer believes that five factors (A, B, C, D and E), each at two levels, are significant in affecting the quality characteristic. The engineer also feels that the interactions B × C and B × E are important. Set up an experimental design using an orthogonal array that can estimate these effects.*

Solution:

We note that there are a total of seven factors of importance, which include the interaction effects $B \times C$, and $B \times E$. The linear graph for this design indicates the assignment of three interaction terms. Therefore, we need to estimate only two interaction terms.

For the interaction that is not utilized, the corresponding two column numbers can be removed and they can be considered as individual points assigned to the main effects. The final assignment is shown in Table E9.22 where factors B, C and E are assigned to columns 1, 2 and 4 respectively. Next, the interactions $B \times C$ and $B \times E$ are assigned to columns 3 and 5 respectively. Finally, factors A and D are assigned to columns 6 and 7 respectively.

Table E9.22

Experiment number	Factor						
	B	C	B × C	E	B × E	A	D
1	1	1	1	1	1	1	1
2	1	1	1	2	2	2	2
3	1	2	2	1	1	2	2
4	1	2	2	2	2	1	1
5	2	1	2	1	2	1	2
6	2	1	2	2	1	2	1
7	2	2	1	1	2	2	1
8	2	2	1	2	1	1	2

Example E9.23. *The management board of a large hotel chain in a very large metropolitan city is seeking ways to promote hotel occupancy. The board has identified five factors A, B, C, D and E that they feel have a great impact on typical hotel guest satisfaction. Factor C has four levels, and each of the other four factors has two levels. Set-up an experimental design that will estimate the factor effects using an orthogonal array.*

Solution:

It is given that factor C has four levels and hence has 3 degree of freedom. Similarly the other factors A, B, D and E has 1 degree of freedom each. Hence, the total degrees of freedom is 7 and we can select the $L_8(2^7)$ orthogonal array design with 7 degrees of freedom. Combining columns corresponding to factors A, B and D, a new column is created that is assigned to factor C as shown in Table E9.23. The column corresponding to the factor C has 4 unique levels. The final assignment including the column corresponding to the factor E as summarized in Table E9.23.

Table E9.23

Experiment number	Factor				
	C	A	A	D	E
1	1	1	1	1	1
2	1	2	2	2	2
3	2	1	1	2	2
4	2	2	2	1	1
5	3	1	2	1	2
6	3	2	1	2	1
7	4	1	2	2	1
8	4	2	1	1	2

Example E9.24. *In a manufacturing operation, four factors (A, B, C and D), each at three levels, are considered to be of importance in influencing the volume of parts produced. An L_9 orthogonal array is used with the factors A, B, C and D assigned to columns 1, 2, 3 and 4, respectively. The response variable showing the number of parts (in thousands) produced per day for each of the nine experiments as summarized in Table E9.24.*

Table E9.24

Experiment	1	2	3	4	5	6	7	8	9
Parts per day produced (in thousands)	*12.5*	*18.5*	*14*	*8.5*	*17.5*	*8.1*	*8.5*	*16.2*	*15.5*

(*a*) *set up the experimental design and the response variable for the corresponding experiments*

(*b*) *determine the main effects*

(*c*) *plot the average response curves*

(*d*) *determine the optimal settings of the design parameters.*

Solution:

(*a*) The experimental design is given in Table E9.24 (*a*)

Table E9.24 (*a*)

Experiment number	Factor				Parts per day (in thousands)
	A	B	C	D	
1	1	1	1	1	12.5
2	1	2	2	2	18.5
3	1	3	3	3	14.0
4	2	1	2	3	8.5
5	2	2	3	1	17.5
6	2	3	1	2	8.1
7	3	1	3	2	8.5
8	3	2	1	3	16.2
9	3	3	2	1	15.5

(*b*) The main effects of each factor are calculated as follows:

$$\bar{A}_1 = (12.5 + 18.5 + 14)/3 = 15$$
$$\bar{A}_2 = (8.5 + 17.5 + 8.1)/3 = 11.367$$
$$\bar{A}_3 = (8.5 + 16.2 + 15.5)/3 = 13.4$$
$$\bar{B}_1 = (12.5 + 8.5 + 8.5)/3 = 9.834$$
$$\bar{B}_2 = (18.5 + 17.5 + 16.2)/3 = 17.4$$
$$\bar{B}_3 = (14 + 18.1 + 15.5)/3 = 12.534$$
$$\bar{C}_1 = (12.5 + 8.1 + 16.2)/3 = 12.267$$
$$\bar{C}_2 = (18.5 + 8.5 + 15.5)/3 = 14.167$$
$$\bar{C}_3 = (14 + 17.5 + 8.5)/3 = 13.334$$
$$\bar{D}_1 = (12.5 + 17.5 + 15.5)/3 = 15.167$$
$$\bar{D}_2 = (18.5 + 8.1 + 8.5)/3 = 11.7$$
$$\bar{D}_3 = (14 + 8.5 + 16.2)/3 = 12.9$$

(*c*) The average effects are plotted in Figure E9.24.

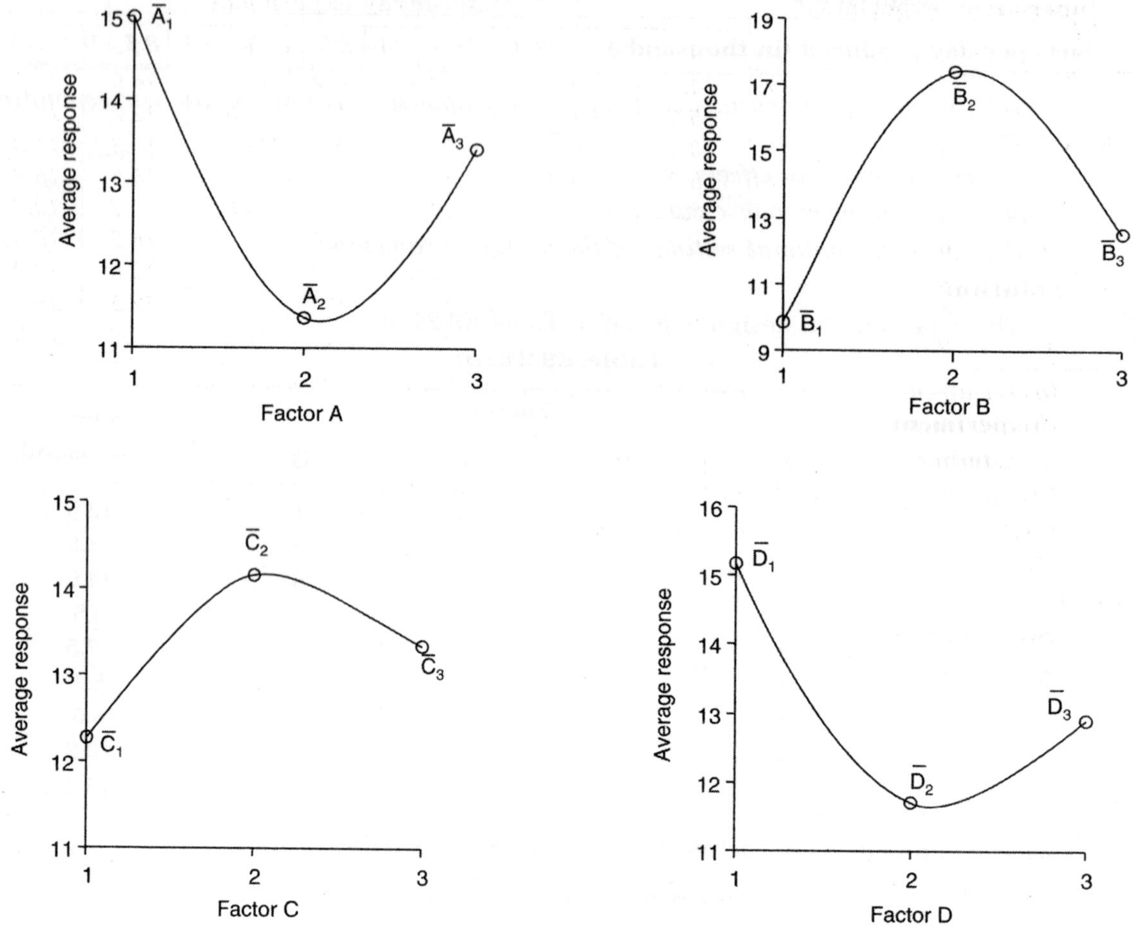

Fig. E9.24

(*d*) The response variable is of the type *larger-is-better*. The optimum settings of the design parameters, assuming the interaction effects are negligible or insignificant, are factor *A* at level 1, factor *B* at level 2, factor *C* at level 2, and factor *D* at level 1.

Example E9.25. *In a certain food processing plant, four design parameters, A, B, C, and D, each at three levels has been identified as having an effect on the moisture content in packaged product. These design factors, A, B, C, and D are assigned to an* L_9 *orthogonal inner array to columns 1, 2, 3, and 4 respectively. Also, in addition to the main effects of the noise factors E, F, and G, it is also found that the interaction effects E × F and E × G are also important ones to be investigated. An outer array selected is* L_8 *in the assignment of E to column 1, F to column 2, E × F to column 3, G to column 4 and E × G to column 5. The moisture content of the eight replications for each of the nine experiments in the inner array is given in Table E9.25. The last two columns in the* L_8 *outer array are assigned to the experimental error. In Table E9.25, the coded target value is 15.*

Table E9.25

Inner-array experiment	Outer-array experiment							
	1	*2*	*3*	*4*	*5*	*6*	*7*	*8*
1	*14.3*	*15.3*	*14.2*	*13.3*	*16.2*	*15.7*	*14.9*	*13.7*
2	*15.5*	*14.1*	*15.3*	*14.5*	*15.1*	*11.8*	*12.2*	*14.9*
3	*13.4*	*15.6*	*14.5*	*12.8*	*15.4*	*12.3*	*13.3*	*14.8*
4	*15.7*	*16.3*	*15.3*	*14.9*	*16.7*	*17.3*	*15.5*	*15.6*
5	*13.7*	*14.9*	*14.4*	*12.8*	*13.7*	*14.8*	*13.7*	*13.7*
6	*16.3*	*15.2*	*17.2*	*15.6*	*13.8*	*16.4*	*16.7*	*15.8*
7	*12.8*	*13.7*	*15.0*	*13.8*	*15.3*	*13.3*	*12.9*	*13.2*
8	*15.5*	*16.3*	*16.3*	*16.9*	*16.4*	*15.9*	*17.5*	*16.7*
9	*13.1*	*15.4*	*13.5*	*17.4*	*13.1*	*16.8*	*13.6*	*14.9*

(a) set up the complete parameter design layout using the Taguchi method

(b) determine the mean or the average response

(c) determine the appropriate s/N ratios

(d) plot the average s/N ratios and the average responses

(e) determine the existence of the interaction effects A × E, B × F, E × F, and E × G

(f) determine the optimal settings of the design parameters using the Taguchi's parameter design method.

Solution:

Equation (9.57) gives the signal-to-noise ratio for target is the best condition.

$$s/N = 10 \log \left(\frac{\bar{y}^2}{s^2} \right) \quad \text{...(E.1)}$$

where

$$s^2 = \sum_{i=1}^{n} \frac{(y_i - \bar{y})^2}{n-1} \quad \text{...(E.2)}$$

The calculations for $\bar{y}$ and s/N are summarized in Table E9.25 (*a*).

Table E9.25 (*a*)

Outer array (L_8)													Responses		
				E	1	1	1	1	2	2	2	2			
				F	1	1	2	2	1	1	2	2			
				E × *F*	1	1	2	2	2	2	1	1			
				G	1	2	1	2	1	2	1	2			
				E × *G*	1	2	1	2	2	1	2	1			
Inner array (L_9)															
Run	*A*	*B*	*C*	*D*									$\bar{y}$	*s*	*s/N*
1	1	1	1	1	14.3	15.3	14.2	13.3	16.2	15.7	14.9	13.7	14.700	1.00142	23.333
2	1	2	2	2	15.5	14.1	15.3	14.5	15.1	11.8	12.2	14.9	14.175	1.41698	20.003
3	1	3	3	3	13.4	15.6	14.5	12.8	15.4	12.3	13.3	14.8	14.013	1.22991	21.132
4	2	1	2	3	15.7	16.3	15.3	14.9	16.7	17.3	15.5	15.6	15.912	0.79361	26.042
5	2	2	3	1	13.7	14.9	14.4	12.8	13.7	14.8	13.7	13.7	13.962	0.69680	26.037
6	2	3	1	2	16.3	15.2	17.2	15.6	13.8	16.4	16.7	15.8	15.875	1.05119	23.580
7	3	1	3	2	12.8	13.7	15.0	13.8	15.3	13.3	12.9	13.2	13.750	0.93350	23.363
8	3	2	1	3	15.5	16.3	16.3	16.9	16.4	15.9	17.5	16.7	16.437	0.61164	28.586
9	3	3	2	1	13.1	15.4	13.5	17.4	13.1	16.8	13.6	14.9	14.725	1.69009	18.803

Note: $s/N = +10 \log \left(\frac{\bar{y}^2}{s^2} \right)$

We can now calculate the average *S/N* ratio and the average response for each factor level.

Factor *A* – Average *s/N*:

Level 1: (23.333 + 20.003 + 21.132)/3 = 21.489

Level 2: (26.042 + 26.037 + 23.580)/3 = 25.219

Level 3: (23.363 + 28.586 + 18.803)/3 = 23.584

Factor *A* – Average response:

Level 1: (14.700 + 14.175 + 14.013)/3 = 14.296

Level 2: (15.912 + 13.962 + 15.875)/3 = 15.249

Level 3: (13.750 + 16.437 + 14.725)/3 = 14.971

Factor *B* – Average *s/N*:

Level 1: (23.333 + 26.042 + 23.363)/3 = 24.246

Level 2: (20.003 + 26.037 + 28.586)/3 = 24.875

Level 3: (21.132 + 23.580 + 18.803)/3 = 21.172

Factor *B* – Average response:

Level 1: (14.700 + 15.912 + 13.750)/3 = 14.787

Level 2: (14.175 + 13.962 + 16.437)/3 = 14.858

Level 3: (14.013 + 15.875 + 14.725)/3 = 14.871

Factor *C* – Average *s/N*:

Level 1: (23.333 + 23.580 + 28.586)/3 = 25.166

Level 2: (20.003 + 26.042 + 18.803)/3 = 21.616

Level 3: (21.132 + 26.037 + 23.363)/3 = 23.511

Factor *C* – Average response:

Level 1: (14.700 + 15.875 + 16.437)/3 = 15.671

Level 2: (14.175 + 15.912 + 14.725)/3 = 14.937

Level 3: (14.013 + 13.962 + 13.750)/3 = 13.908

Factor *D* – Average *s*/N:

Level 1: (23.333 + 26.037 + 18.803)/3 = 17.057

Level 2: (20.003 + 23.580 + 23.363)/3 = 21.614

Level 3: (21.132 + 26.042 + 28.586)/3 = 25.253

Factor *D* – Average response:

Level 1: (14.700 + 13.962 + 14.725)/3 = 14.462

Level 2: (14.175 + 15.875 + 13.750)/3 = 14.600

Level 3: (14.013 + 15.912 + 16.437)/3 = 15.454

In maximizing *s/N* ratio, the selected factor levels would be: factor *A* – level 2, factor *B* – level 2, factor *C* – level 1, and factor *D* – level 3. In considering the average response, where the target value is 15, the selected factor levels would be: factor *A* – level 2, factor *B* – level 3, factor *C* – level 1, factor *D* – level 3. So, in considering both criteria, *i.e.*, maximizing *s/N* ratio and meeting the target value, factor *A* at level 2 and factor *D* at level 3 satisfy both criteria. For the other two factors *B* and *C*, in order to decide between the two chosen levels, other considerations such as cost could be taken into account.

The interaction effects of $A \times E$, $B \times F$, $E \times F$, $E \times G$ and $A \times F$ can be calculated.

9.15 ROBUST DESIGN

Robust design is the systematic method of determining the optimum values of design factors, which yield the economical designs with low or minimum variability. Taguchi proposes to achieve the goal by conducting the system design, parameter design and the tolerance design.

9.15.1 System Design

System design is the process of applying scientific and engineering knowledge to produce a basic functional prototype design. The prototype model defines the initial settings of product or process design characteristics.

9.15.2 Parameter Design

Parameter design is based on classifying the variables that affect a product's performance into two categories: design parameters and sources of noise. Design parameters are the product design characteristics whose nominal settings can be specified by the product designer. The actual values of design parameters in a manufactured product may deviate from the nominal settings. Sources of noise are all those variables that cause performance variation both during a product's life span and across different units of the product. Sources of noise can be classified into two categories: *External sources of noise* and *Internal sources of noise*. External sources of noise are variables external to a product that affect the product's performance. Examples are: variation in environmental variables such as temperature, humidity, dust and vibrations, and human variations in operating the product. Internal sources of noise are the deviations of the actual characteristics of a manufactured product from the corresponding nominal settings. The primary internal sources of noise are: manufacturing imperfections and product deterioration.

Parameter design is the process of determining the specification settings of the product design parameters or process variables in terms of nominal values so that the final product or the process will be less sensitive to the three common sources of variation caused by environmental factors, product deterioration, and manufacturing variations or imperfections.

Environmental factors: Environmental factors are conditions in the environment in which the customer will use the product. These factors include the human variations in operating the product.

Product deterioration: Product deterioration consists of variations in product parameters over time from wear and tear on the product during its life cycle.

Manufacturing variations or imperfections: Manufacturing variations or imperfections are manufacturing conditions that affect the product's production that deviates from its nominal values.

Once the variance has been reduced, then the mean response can easily be adjusted by using a suitable design parameter, called the *signal factor*. Parameter design makes use of statistically planned experiments. Two and three-level orthogonal arrays are most commonly used. These experiments can be either physical experiments or computer based simulation trials.

Noise factors are those sources of noise that can be included in a parameter design experiment.

Parameter design is possible because the effects of both internal and external noise factors can change with the settings of the design parameters.

A parameter design experiment consists of two parts: a *design matrix* and a *noise matrix*. The columns of a design matrix represents the design parameters, entries in the columns represent the test settings of the design parameters, and each row of the matrix (also called *a test run*) represents a product design. The columns of a noise matrix represent noise factors, and the rows of the matrix represents different combinations of the levels of the noise factors. A complete parameter design experiment consists of a combination of design and noise matrices as indicated in Fig.9.12. If the design matrix has m rows and the noise matrix has n rows, the total number of rows in the combined parameter design experiment is $m \times n$. For each of the m rows of the design matrix, the n rows of the noise matrix provide n or more repeat observations on the performance characteristic. The levels of the noise factors and the noise matrix are chosen so that these repeat observations are representative of the effects of all possible levels of the noise factors. The repeat observations on the performance characteristic from each test run in the design matrix are then used to compute a criterion called a *performance statistic*. The m values of the performance statistic associated with m test runs in the design matrix are then used to predict settings of design parameters that minimize the expected loss.

In Fig. 9.12, there are four parameters θ_1, θ_2, θ_3, and θ_4, each tested at three levels for nine test runs in the design parameter matrix [see also Table 9.4].

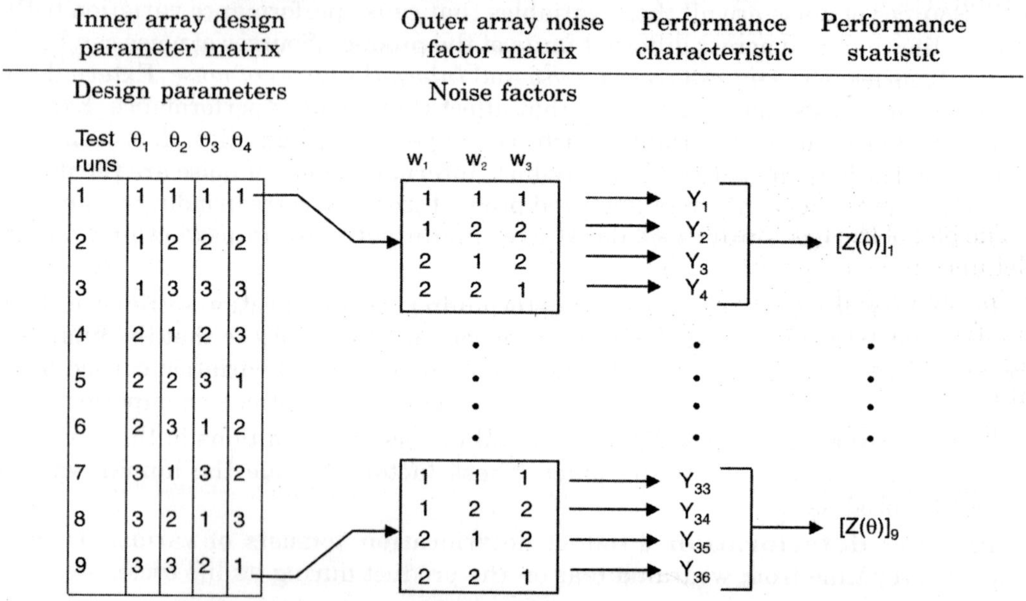

Fig. 9.12 A Typical Taguchi Design of Experiments for Parameters (robust) Design.

Taguchi recommends using a performance statistic called *a signal-to-noise ratio* as described in Sec. 9.11. Signal-to-noise ratios can similarly be defined for attribute-type performance characteristics. Consequently, statistically designed experiments using signal-to-noise ratio as a measure of product or process performance can be used to reduce variation by careful selection of parameter settings.

9.15.3 Tolerance Design

Tolerance design is a method for determining tolerances that minimize the sum of product manufacturing and lifetime costs. The final step in specifying product and process designs is to determine tolerances around the nominal settings identified by parameter design. It is still a common practice in industry to assign tolerances by convention rather than scientifically. Tolerances that are too narrow increase manufacturing cost, and tolerances that are too wide increase performance variation and, hence, a product's lifetime cost.

9.16 PARAMETER INTERACTION IN ORTHOGONAL ARRAYS

Parameter interaction occurs when one parameter modifies the condition of another parameter. Interaction is illustrated by the exclusive or *XOR* relationship. Consider the following two parameter interaction example.

No Interaction:

	A_1	A_2	Total
B_1	64	68	132
B_2	70	74	144
Total	134	142	276

Interaction:

	A_1	A_2	Total
B_1	65	67	132
B_2	69	75	144
Total	134	142	276

The plot of the two means indicates the existence of interactions between the parameters. Parallel lines indicate no interaction as shown in Figure 9.13. *XOR* is a logical relationship as shown in Table 9.5. In Table 9.5, *A* and *B* are the inputs and C is output. Array L_8, columns 1, 2, and 3 form such an *XOR* relationship (see Table 9A.2, $L_8(2^7)$ series in Appendix-9*A*). These parameters can be grouped into primary parameters and interaction parameters. Interaction indicates the effect of one parameter's variation on other parameters. If the interaction column in the array is assigned to a parameter, then the analysis for the effect of that parameter could be either negated or amplified by the interaction effect. This effect is known as *confounding*. In reality, the effect of interactions is generally much smaller than expected.

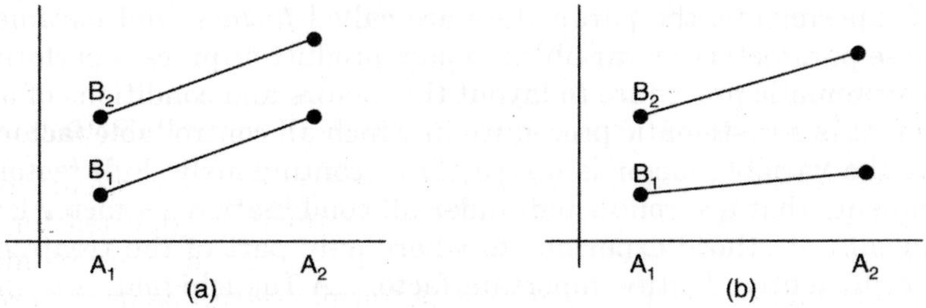

Fig. 9.13 (*a*) No Interaction. (*b*) Interaction of *A* × *B*.

Table 9.5

XOR		
A	B	C
1	1	1
1	2	2
2	1	2
2	2	1

The *interaction effects* between factors can be estimated based on the assumption that the interaction effect has been assigned an appropriate column using the linear graph and the orthogonal array. Consider the $L_9(3^4)$ orthogonal array and its associated graph shown in Figure 9A.3 (Appendix-9A). If we assume the factor C in Table 9.4 represents the interaction between factors A and B (that is $C = A \times B$) and assign to column 3 of the L_9 array, then it is in agreement with L_9's linear graph in Table 9A.3. The interaction effect between A and B is then obtained by treating C like any other factor.

From Table 9.4, the average response when $A \times B$ is at level 1 is given by

$$\overline{(A \times B)}_1 = \frac{\bar{y}_1 + \bar{y}_6 + \bar{y}_8}{3}$$

The average responses at other levels of $A \times B$ are given by

$$\overline{(A \times B)}_2 = \frac{\bar{y}_2 + \bar{y}_4 + \bar{y}_9}{3}$$

and

$$\overline{(A \times B)}_3 = \frac{\bar{y}_3 + \bar{y}_5 + \bar{y}_7}{3}$$

The above three averages can now be plotted against the levels of $A \times B$. If the plot obtained is a horizontal line, then the interaction effect between A and B is insignificant. In such a case, the main effects of A and B are investigated in order to determine the optimal settings of the factor levels. To determine the significant interaction effects, the optimal level of each factor has to be found based on the joint effect of both factors.

9.17 COMPARISON OF ROBUST AND CLASSICAL DESIGN OF EXPERIMENTS

In the design of experiments, the parameters are called *factors*, and parameter settings are called *levels*. These parameters or variables impact product or process performance. Design of experiment is a systematic procedure to layout the factors and conditions of an experiment. A factorial experiment is a systematic procedure in which all controllable factors expect one are held constant as the variable factor is discreetly or continuously. Full factorial experiments are those experiments that are conducted under all combinations of factor levels. Fractional-factorial experiments are those experiments where only part of the treatment conditions is tested to be more efficiently identify important factors. A Taguchi matrix is one such example.

The classical design of experiments is used to obtain the optimum settings of a product design characteristics or of a process through a two-stage process: (1) Screening and (2) Modeling.

1. *Screening:* Screening is a process in which a large number of possible factors are varied in order to establish which ones are to be included in further study.

2. *Modeling:* Modeling in classical design of experiments (*DOE*) is a process in which the factors that are identified to be important or significant in screening process are further refined by experimental techniques.

The main objective in conducting the classical design of experiments is to determine the optimum settings for the mean values of the factors to accomplish a given quality characteristic. On the other hand, robust design is always concerned with the mean and the variability of the design. Thus, robust design is an excellent technique for optimizing designs by shifting the mean characteristic of the design to target and reducing variability.

9.18 SOFTWARE PACKAGE FOR EXPERIMENTAL DESIGNS

Over the past two decades with the advent of the personal microcomputers a range of statistical software packages for forecasting statistical process control (SPC) and experimental design are available for product and process development and optimization. There are several different types of experimental designs in common use as well as software that are available to the engineer. Listing of the commercial statistical software packages of interest is given in Table 9.6. The ease of use and the number of features available with each package varies. The packages range in price from $100 to $5000 and in complexity also. Most of these software packages can be used to develop simplex, screening, and Response Surface Experiments. Other packages that include experimental design capability are NCSS, Turbo Spring-Stat, CSS, Sigstat, BMDP/PC, P-Stat, SAS and Systat. These packages contain a wide variety of other statistical features including factor analysis, multiple regression analysis, time series analysis, Statistical Process Control (SPC) routines and produce graphical output. Although the software packages in Table 9.6 assist the user in defining and analyzing experimental plans, it is strongly suggested that the researcher unfamiliar with statistical designs, to begin on a small scale and gradually build confidence with the programs and techniques.

Table 9.6. List of Statistical Software Packages

Software	Contact address
NCSS	NCSS, Inc., 865 E 400, North Kayville, UT, 84037
Experimental Design	Statistical Programs, 9941 Rowlett, Suite 6, Houston, TX, 77075
Simplex-V	Statistical Programs, 9941 Rowlett, Suite 6, Houston, TX, 77075
Screen	Statistical Programs, 9941 Rowlett, Suite 6, Houston, TX, 77075
C-Comp	Statistical Programs, 9941 Rowlett, Suite 6, Houston, TX, 77075
The Bass System	Bass Institute, P. O. Box 349, Chapel Hill, NC, 27514
Turbo Spring-Stat	Spring Systems, P. O. Box 10073, Chicago, IL, 60610
CSS	Stat Soft, Inc., 2325 E. 13th St., Tulsa, OK, 74104
Sigstat	Significant Statistics, 3336 N. Canyon Rd., Provo, UT, 84604
STATA	Computing Resource Center, 1080 National Blvd., Los Angels, CA, 90064
Statpac Gold	Walonick Associates, Inc., 6500 Nicollet Ave. S., Minneapolis, MN, 55423
BMDP/PC	BMDP Statistical Software, 1440 Sepulveda Blvd., # 316, Los Angeles, CA, 90025
P-Stat	P-STAT, Inc., P. O. Box AH, Princeton, NJ, 08542
SAS	SAS Institute, SAS Circle, Box 8000, Cary, NC, 27512-8000
SPSS/PC	SPSS, Inc., 444 N. Michigan Dr., Chicago, IL, 60611
Systat (With Sygraph)	SYSTAT, Inc., 1800 Sherman Ave, Evanston, IL, 60201
Statgraphics	STSC, Inc., 2115 E. Jefferson St., Rockville, MD, 20852
Prodas	Conceptual Software, Inc., P. O. Box 56627, Houston, TX, 77256-6627
RS/1	BBN Software Products Corp., 10 Fawcett St., Cambridge, MA, 02138

9.19 SUMMARY

Taguchi has made a significant contribution to quality improvement and statistical methods. Taguchi's quality improvement activities, which include the product or process design, provide a system to develop specification, design those specification into a product and/or process and produce products that continuously surpass the desired specifications. Fundamental concepts and procedures associated with the Taguchi method of design and analysis are presented in this chapter. Taguchi's philosophy has been described and procedures through which quality improvement through design were presented. The objective of Taguchi method is to achieve quality at a reduced cost. A measure of the deviation of an output characteristic from a target value known as the *Taguchi loss function* has been defined. The use of the loss function for evaluating the quality level of a product or process by quantifying the deviations from the target value is discussed. The use and application of loss function in the quality evaluations for three types of tolerances, namely, the nominal-the-best, the smaller-the-better, and the larger-the-better were presented. The concept of tolerance design, and the methods for tolerancing were introduced. The effect of tolerance on the quality of a product was discussed. Tolerance designs for the nominal-the-best, the smaller-the-better and the larger-the-better characteristics were illustrated with numerical examples. The three phases of the Taguchi method, namely, the systematic design, parameter design, and tolerance to the uncontrollable factors and achieve a desired target value with a minimum variability. To determine the effectiveness of a design using Taguchi method, performance measures known as signal-to-noise ratio was introduced. The term *noise* represents the undesirable component and is a measure of the variability of the output characteristic, which will preferably be as small as possible. The term *signal*, or average value of the characteristic, represents the desirable component, which will preferably be close to a specified target value. The expressions for signal/noise ratio are dependent on the three situations (target is best, smaller is better, and larger is better). The performance measure has the property that when it is maximized, the expected loss will be minimized. The expected loss is the mean loss over many instances of the product. Taguchi method aims at obtaining the best possible design at the lowest possible cost.

PROBLEMS

9.1 The functional limits for a television set for the color density are $m \pm 8$, which implies that about one-half of the customers would find the television set to be defective if the color density is $m \pm 8$. If the repair of a television set in the field cost is an average $A = \$128$. Compare the average quality loss incurred by the customers receiving sets with color density $(m + 4)$ and $(m + 2)$ respectively.

9.2 Determine the average loss associated with the output voltage value for a stereo system given that the target output voltage of the power supply circuit used for the stereo system is 110 volts. The stereo system fails in half of the situations and needs repair whenever the output voltage falls outside 100 ± 25 volts and the cost of repair is \$156.25.

9.3 A mechanical component has a specification length of 25 ± 5 mm. Three suppliers 1, 2, and 3 of the mechanical component produced the following results.

Supplier	Mean length $\overline{X}$	Variance σ^2	C_{pk}
1	25	1.8	1.0
2	23	0.5	1.0
3	22	0.3	1.0

The replacement cost of component which is out of specification is \$50. The number of components out of specification in delivery of 300,000 components for the three suppliers is:

Supplier 1 = 0.30%; Supplier 2 = 0.2% and Supplier 3 = 0.15%

(*a*) compare the three suppliers based on the loss function

(*b*) compare the three suppliers based on the cost of defective units.

9.4 Customer tolerances for the length of a machine component are 2 ± 0.01 m. For a product that exceeds these limits, the cost to the customer for getting it fixed is \$75. Ten products are randomly selected and yield the following lengths (in meters): 2.03, 1.99, 2.0, 1.99, 1.98, 2.02, 2.04, 2.03, 2.01, and 2.02. Determine the average loss per product item.

9.5 Refer to Prob. P9.1 regarding the length of the machine component. The manufacturing company is considering modifying the production process in order to reduce the variability in the output. The estimated additional cost for the modified production process is \$30/component. The estimated annual production of these machine components is 50,000 components. Eight components are randomly selected from the new production process, resulting in the following lengths of the component: 2.01, 2.0, 1.99, 2.02, 2.02, 2.0, 1.98, and 2.01.

(*a*) Determine whether the new production process is cost effective or not

(*b*) If the new production process is cost effective, determine the annual savings.

9.6 Refer to Problem P9.4 regarding the length of the machine component. If the manufacturing company decides to rework the length of the component prior to shipping the component at a cost of \$20 per component, determine the manufacturer's tolerance.

9.7 A manufacturer of copper wiring is concerned with reducing the variability of the diameter of the wiring. It is estimated that the loss to the customer is \$20 per reel if the diameter of the wire exceeds 2 ± 0.04 mm. Each reel has 100 m of wire. A random sample of 10 gave the following diameters (in mm): 1.98, 2.01, 2.03, 1.97, 2.01, 1.99, 2.02, 2.01, 2.02, and 2.04. Determine the average loss per reel.

9.8 Refer to Problem P9.7 regarding the diameter of the copper wire. In order to reduce the variability in the diameter of the wire, the manufacturer is considering a new and improved production process. It is estimated that the additional cost for this improved process is \$0.05 per linear meter. The annual production is 5,000 reels. Each reel has 100 m of wire. A random sample of size 10 from the new and improved production process yielded the following diameter (in millimeters): 2.01, 2.04, 1.98, 1.99, 2.02, 2.03, 2.01, 2.03, 1.97, and 1.98.

(*a*) Is the new and improved production process cost effective?

(*b*) Determine the annual savings or loss.

9.9 Refer Problem P9.7, if the manufacturer can rework the diameter of the wire prior to shipping the product at a cost of \$5 per reel, determine the manufacturer's tolerance for the produced wire.

9.10 Refer to Problem P9.7, the manufacturer is considering adopting a new process so that the average diameter of the wire produced is 2 mm, which is the target value. However, the manufacturer also estimated the standard deviation of the new process will be 0.01 mm and the cost of making this change is \$3 per reel. Determine

(*a*) if the new process will be cost effective compared to the original process

(*b*) the annual savings or loss if the annual production is 5,000 reels.

9.11 The critical dimension of a certain turned part using a lathe machine must be 0.5 ±0.003 inches and any part falls out of spec in the range of 0.497 and 0.503 will have to be scrapped at a cost of \$100. Fifteen samples were taken from the lathe operations of this part and their dimensions are as follows: 0.497, 0.499, 0.501, 0.502, 0.498, 0.501, 0.497, 0.500, 0.501, 0.497, 0.500, 0.498, 0.501, 0.501, and 0.497.

(*a*) Write the Taguchi loss equation for the turning operation

(*b*) Determine the average Taguchi loss function for the parts made using this lathe machine.

9.12 For the data in Problem P9.11, determine the process capability index C_p. Is the process capable of meeting the specifications?

9.13 The manufacturer of an electromechanical system wishes to improve the quality of the system by reducing deviations from the target value such that the new standard deviation will be 1.2. This improvement is technologically possible at an additional cost of $0.10 per unit of production. Assuming that no inspection is necessary or performed. Determine

(*a*) whether the manufacturer should improve the process.

(*b*) if the production rate of this manufacturer is 200,000 units per month, determine the expected savings per annum.

9.14 The tensile strength of steel cable wire of 30 samples collected for two consecutive months of production are given in Table P9.14. The lower specification limit is 10 N/m^2. Defect loss is $1 per linear meter. The monthly production is 500 meter. Determine the quality level during

Table P9.14

Month (30 work days)	Tensile strength data (N/m^2)								
1	12.3	12.4	13.5	14.6	12.9	13.1	11.5	12.3	15.1
	12.5	13.9	14.1	12.6	14.1	13.7	14.1	13.1	14.7
	12	15	13.8	13.1	12.8	15.1	15.1	14.1	13.8
2	15.1	14.2	12.5	14.3	15.1	14.7	13.1	14.1	14.2
	12.7	13.7	12.1	13.2	12.2	15.1	13.6	14.6	14.6
	13.5	14.2	12.7	11.8	10.1	14.7	13.4	14.1	13.2

(*a*) the first month of production

(*b*) the second month of production.

9.15 When the blower seal of a vacuum blower housing is a copier machine fails to operate, it costs $50 to replace, $25 in part costs and another $25 in installation labor costs. The machine downtime to the customer is $350 per hour. It takes approximately 30 minutes to replace the blower seal. The seal level at which the vacuum loss becomes unacceptable is 25 psi. Determine

(*a*) the larger-the-better loss function

(*b*) the expected quality loss for the vacuum blower for two separate sites. Each has ten machines operating in it. The performance data collected for the two sites is summarized in Table P9.15.

Table P9.15

Site	Data
1	18, 25, 30, 19, 35, 32, 26, 28, 23, 24
2	17, 24, 31, 18, 34, 31, 27, 22, 23, 28

9.16 The output voltage of a high-voltage transformer was found to alter due to the deterioration of transistors in the power circuit. When the output voltage exceeds the tolerance limit of ± 20 *V*, the transformer is unsuitable for use and the estimated loss when the transformer exceeds the limits is $500. The manufacturer can adjust the voltage by replacing the resistor at a cost of $2 before shipping the transformer to the customer. Determine the manufacturers specifications assuming the target voltage is 110 *V*.

9.17 The tolerance of the deviations of the flatness of a metal component is specified as ± 0.005 inches. The observations of the derivations recorded using the existing machining system for a sample of 30 are given in Table P9.17 (*a*).

Table P9.17 (a)

Derivations observed on a sample of 30 (inches)				
0.002	0.002	0.005	– 0.004	– 0.003
– 0.003	0.003	0.003	– 0.003	– 0.004
– 0.004	0.001	0.004	– 0.004	– 0.005
0.003	0.005	0.005	– 0.003	0.003
– 0.003	0.002	– 0.003	0.003	0.004
0.004	– 0.002	– 0.001	0.001	0.005

A new machining system is introduced to increase the quality of the flatness. Table P9.17 (*b*) gives the derivations observed after introducing the new system for a sample size of 30. Determine

Table P9.17 (b)

Derivations observed on a sample of 30 (inches)				
0.001	0.002	– 0.001	– 0.002	0.003
– 0.002	0.003	– 0.001	– 0.003	0.002
– 0.003	– 0.002	0.001	– 0.001	0.001
0.003	– 0.001	0.001	0.003	– 0.001
0.000	0.000	0.002	0.002	0.001
0.001	– 0.001	– 0.002	0.001	0.000

(*a*) the effect of the introduction of the new machining system

(*b*) the savings (if any) with the new system assuming the cost of a defective component is $200.

9.18 The nominal thickness of the corrosion resistance coating on a metal component is found to be an important variable in establishing the wear characteristics of the component. The loss caused by the wear of the coating is $25, when the coating thickness is ± 0.001 inch from its nominal value. A sample of twenty measurements of the coating taken from selected points along the component length gave the following deviations from the nominal thickness:

0.8, 1.0, 0, 0.4, 1.8, 1.3, 0.3, 0.7, 0.6, 1.1, 1.4, 1.5, 1.2, 1.7, 0.8, 0.9, 1.1, 1.3, 1.4, 0.5

A new coating process was found to reduce the loss by 50%. The following deviations were measured for the new process:

1.0, 1.0, 1.1, 0.2, 0.4, 0.5, 1.2, 0.9, 0.8, 0.7, 0.3, 0.2, 1.0, 1.1, 0.6, 1.0, 1.2, 0.3, 0.4, 0.5

(*a*) determine the tolerance of the coating thickness for the new process

(*b*) the loss for each unit for the old and new processes.

9.19 A high tech manufacturing company has established three factors *A*, *B*, and *C*, each at three levels that influence the satisfaction of their valued customers. The company also believes that the interaction effect $B \times C$ is important. Set up an appropriate experimental design using an orthogonal array.

9.20 A large electromechanical components manufacturing company has identified four factors, *A*, *B*, *C*, and *D*, each to be analyzed at two levels. The company also believes that the interactions $B \times C$, $B \times D$, and $C \times D$ are important. Set up an appropriate experimental design using an orthogonal array.

9.21 The quality engineering department of a large manufacturing company has identified five factors, *A*, *B*, *C*, *D*, and *E*, each to be investigated at two levels. The department also believes that the interactions $A \times C$ and $A \times D$ are important. Set up an appropriate experimental design for estimating the main effects and the two interactions.

9.22 A manufacturing company has identified five factors *A*, *B*, *C*, *D*, and *E* that have a great impact on its volume of production. There are four levels for factor *A*, and each of the other four factors is to be tested at two levels. Set up an appropriate experimental design for estimating the main effects and the interactions.

9.23 In a certain manufacturing operation, four factors *A*, *B*, *C*, and *D* each at three levels, are thought to be of importance in influencing the volume of parts produced. Using an L_9 orthogonal array, the factors *A*, *B*, *C*, and *D* are assigned to columns 1, 2, 3, and 4 respectively. The response variable showing the number of parts (in thousands) produced per day for each of the nine experiments is summarized below:

Experiment	1	2	3	4	5	6	7	8	9
Parts per day (in thousands)	6.2	15.2	11.5	6.5	16.5	4.8	6.7	13.2	9.5

(*a*) Setup the experimental design and response variable for the corresponding experiment

(*b*) Determine the main effects

(*c*) Plot the average response curves

(*d*) Determine the optimal settings of the design parameters.

9.24 In a certain meat processing plant, four design parameters, *A*, *B*, *C*, and *D*, each at three levels, have been identified as having an effect on the moisture content in packed meat. Three noise factors, *E*, *F*, and *G*, each at two levels, are also to be investigated in the experiment. An L_9 orthogonal array is used in the inner array with the factors *A*, *B*, *C*, and *D* assigned to column 1, 2, 3, and 4, respectively. An L_4 orthogonal array is used for the outer array with the noise factors *E*, *F*, and *G* assigned to columns 1, 2, and 3, respectively. The moisture content in a coded scale for four replications of each of the nine experiments in the inner array are given in Table P9.24. It is highly desirable that the moisture content be around a coded value of 15.

(*a*) Set up the complete parameter design layout using the Taguchi method

(*b*) Determine the mean response and the appropriate signal-to-noise ratio

(*c*) Plot the average *s*/*N* ratios and the average responses

(*d*) Determine the existence of possible interactions $B \times E$ and $C \times F$

(*e*) Determine the optimal settings of the design parameters using the Taguchi's parameter design approach.

Table P9.24

Inner-array experiment	**Outer-array experiment**			
	1	**2**	**3**	**4**
1	13.5	16.2	15.5	14.9
2	11.8	12.1	15.8	14.5
3	16.2	16.8	15.7	14.8
4	15.3	13.5	14.8	15.8
5	11.8	16.5	16.3	16.5
6	14.1	13.5	13.4	13.5
7	15.3	16.4	12.8	14.9
8	12.5	15.0	16.1	15.4
9	15.4	14.8	14.9	14.5

9.25 The tensile strength of a steel alloy has positive and negative effects depending on the amount of alloy used. Four independent factors have been identified that have an effect on the tensile strength of the steel alloy. Each factor can be tested at three levels. An L_9 orthogonal array is used in the inner array with the factors *A*, *B*, *C*, and *D* to columns 1, 2, 3, and 4 respectively. The response variable, measuring the impact of the alloy, is recorded on a coded scale for the nine experiments. The target value is zero, with the observed coded responses being positive and negative.

Table P9.25 shows the experimental design and the values of the coded response variables. For each of the settings of the design factors only one replication is conducted and the outer array of the noise factors is not shown.

(*a*) determine the main effects

(*b*) plot the average response curves

(*c*) determine the optimal settings of the design parameters.

Table P9.25

Experiment	Factor				Coded response
	A	B	C	D	
1	1	1	1	1	– 3.2
2	1	2	2	2	6.8
3	1	3	3	3	1.9
4	2	1	2	3	– 4.9
5	2	2	3	1	9.5
6	2	3	1	2	– 5.7
7	3	1	3	2	– 5.2
8	3	2	1	3	2.7
9	3	3	2	1	– 2.3

9.26 The monetary loss induced by premature failure of a product is defined by Taguchi's loss function. For an electrical power tool, the cost of repair to the tool merchant for a specific type of failure is estimated as $465/unit. Given the deviations from the target value are given for 15 units measured at the factors are: (0, 0.5, 0.3, 0.2, – 0.3, 0.4, 0, – 0.4, – 0.3, 0.6, 0.5, – 0.3, 0, – 0.4, 0.6). Determine

(*a*) the financial loss resulting from a machine

(*b*) the improvement in monetary loss that would result from the reduction of the standard deviation by two to one.

9.27 Determine the optimum combination of factor values for the orthogonal array shown below. Determine also the per cent improvement in results produced by using the optimum combination of factors. *Note:* Biggest is best.

Trial No.	A	B	C	D	E	F	G	Results
1	1	1	1	1	1	1	1	42.00
2	1	1	1	2	2	2	2	50.00
3	1	2	2	1	1	2	2	36.00
4	1	2	2	2	2	1	1	45.00
5	2	1	2	1	2	1	2	35.00
6	2	1	2	2	1	2	1	55.00
7	2	2	1	1	2	2	1	30.00
8	2	2	1	2	1	1	2	54.00

9.28 An experiment has been designed to study the influence of five factors, which were identified during a brain storming session as influencing the torque delivered by the valve-train of a newly developed small scale engine. Two levels were chosen for the experiment. The factor and their levels are given in the following:

Factors	Level 1	Level 2
Valve guide clearance	Low	High
Valve geometry	Type 1	Type 2
Seat concentricity	Quality 1	Quality 2
Lower guide length	Location 1	Location 2
Valve face run out	Run out type 1	Run out type 2

An L_8 orthogonal array was selected and torque levels were measured under appropriate conditions for the eight experiments. Interaction is expected between the seat concentricity and the valve face run out. Also, interaction is expected between the valve geometry and the valve face run out.

The eight measured torque levels are as follows:

$$y = \begin{bmatrix} 45 \\ 34 \\ 56 \\ 45 \\ 46 \\ 34 \\ 38 \\ 43 \end{bmatrix}$$

y is measured in units of inch pounds

(*a*) Set up the experimental array in the provided L_8 format as shown in Fig. 9.28 and Table P9.28

(*b*) Determine the optimum combination of factors to maximize torque

(*c*) Determine the contribution of each factor to this improvement

(*d*) Are either or both interactions significant?

(*e*) Estimate the overall improvement of the optimized system.

Table P9.28

Condition \ Column	$L_8(2^7)$ 1	2	3	4	5	6	7
1	1	1	1	1	1	1	1
2	1	1	1	2	2	2	2
3	1	2	2	1	1	2	2
4	1	2	2	2	2	1	1
5	2	1	2	1	2	1	2
6	2	1	2	2	1	2	1
7	2	2	1	1	2	2	1
8	2	2	1	2	1	1	2

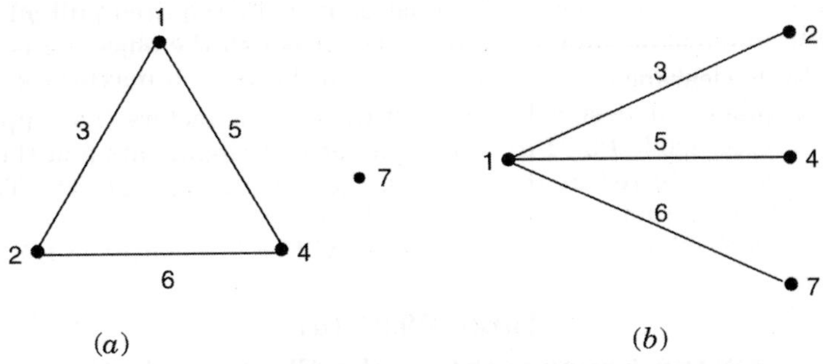

Fig. P9.28 Linear Graphs for L_8

9.29 In a certain manufacturing operation, four factors *A*, *B*, *C*, and *D*, each at three levels is thought to be a great importance (control factors) in influencing the volume of the parts produced. The environmental noise factors that could affect the performance of the part are recognized as 3 noise factors at 2 levels. They are given as *E*, *F*, and *G*.

Factor	Description	Level 1	Level 2
E	Time	25 hours	125 hours
F	Temperature	70 ºF	160 ºF
G	Humidity	30%	80%

(*a*) set up an orthogonal arrays for the inner and the outer array

(*b*) determine the number of runs required to complete the tests

(*c*) if the s/N ratios for the 9 experimental conditions of the control matrix are given in order as 24.03, 25.50, 25.31, 25.95, 26.92, 25.34, 25.75, 24.84, and 26.25

(*d*) discuss the results obtained from the above analysis.

9.30 Determine the most robust design of paper airplanes using robust design experiment. The control parameters and noise parameters are given in Tables P9.30 (*a*) and (*b*).

Table P9.30 (*a*)
Control Parameters

Parameter	Level 1	Level 2	Level 3
Weight of paper (A)	One sheet	Two sheets	Three sheets
Configuration (B)	Design 1	Design 2	Design 3
Width of paper (C)	4 in	6 in	8 in
Length of paper (D)	6 in	8 in	10 in

Table P9.30 (*b*)
Noise Parameters

Parameter	Level 1	Level 2
Launch height (X)	Standing on ground	Standing on chair
Launch angle (Y)	Horizontal to ground	45º above horizontal
Ground surface	Concrete	Polished tile

[**Source:** *Adopted from S. Sarin, Quality Progress, pp.102–106, May 1997*]

9.31 An elastomeric connector to a nylon tube must deliver the required pull-off performance so that it can be used in automobile engine components. Although the objective is to maximize pull-off force and minimize assembly effort, this experiment focuses on maximizing pull-off force.

Researchers identified four design factors and three noise factors that appear to affect the assembly's pull-off force. Table P9.31(a) shows the factor assignments and their associated levels. Each design factor was tested at three levels, the noise factors at two. The noise factors are uncontrollable during normal operations, but they were controlled over the two levels for the purpose of tests. Determine a robust design that will be insensitive to the noise factors during actual operation.

Table P9.31 (a)

Factors for the Experiment on the Elastomeric Connector

Design factors	**Levels**		
A. Interference	Low	Medium	High
B. Connector wall thickness	Thin	Medium	Thick
C. Insertion depth	Shallow	Medium	Deep
D. Per cent adhesive in connector pre-dip	Low	Medium	High
Noise factors	**Levels**		
E. Conditioning time	24 h	120 h	
F. Conditioning temperature	72 ºF	150 ºF	
G. Conditioning relative humidity	25%	75%	

Source: **Adapted from D.M. Byrne and S. Taguchi., "The Taguchi Approach to Parameter Design",** ***Quality Progress*****, Dec. 1987, pp. 19-26.**

REFERENCES

Anderson, V. L. and R. A. McLean, R. A., *Design of Experiments: A Realistic Approach*, Marcel Dekker, New York, NY, 1974.

Anscomble, F. J. and Tukey, J. W., "The Examination and Analysis of Residuals". *Technometrics,* Vol. 5, pp. 141–160, 1963.

Barker, T., *Engineering Quality by Design*, Marcel Dekker, New York, NY, 1990.

Barker, T., *Quality by Experimental Design*, Marcel Dekker, New York, NY, 1994.

Bisgaard, S., "Industrial Use of Statistically Designed Experiments: Case Study References and Some Historical Anecdotes", *Quality Engineering*, Vol. No. 4, pp. 547–562, 1992.

Bishop, L., Hill, W. J. and Lindsay, W.S., "Don't be fooled by the Measurement System", *Quality Progress*, pp. 35–38, December 1987.

Box, G.E.P. and Wilson, K.B., "On the Experimental Attainment of Optimum Conditions", *Journal of the Royal Statistical Society, Series B,* Vol. 13, No. 1, pp. 1–45, 1951.

Box, G.E.P., "Signal-to-Noise Ratios, Performance Criteria, and Transformation", *Technometrics*, Vol. 30, pp. 1–40, 1988.

Box, G.E.P., Bisgaard, S. and Fund, C.A., "An Explanation and Critique of Taguchi's Contributions to Quality Engineering", *Quality and Reliability Engineering International*, Vol. 4, pp. 123–131, 1988.

Box, G.E.P., Hunter, W.G. and Hunter, J.S., *Statistics for Experiments*, Wiley, New York, NY, 1978.

Box, G.E.P., *Studies in Quality Improvement: Signal to Noise Ratios, Performance Criteria and Statistical Analysis: Part I,* Report No. 11, Center for Quality and Productivity Improvement, Madison, WI, 1986.

Byrne, D.M. and Taguchi, S., "The Taguchi Approach to Parameter Design", *Quality Progress,* 20 (12), pp. 19–26, 1987.

Cochran, W.G., "Analysis of Covariance: Its Nature and Uses", Biometrics, Vol. 13, No. 3, pp. 262–281, 1957.

Cochran, W.G. and Cox, G.M., *Experimental Designs,* 2nd ed., Wiley, New York, NY, 1957.

Coleman, D.E. and Montgomery, D.C., "A Systematic Approach to Planning for a Designed Industrial Experiment", Technometrics, 35(1), pp. 1-27, 1993.

Daniel, C., *Applications of Statistics to Industrial Experimentation,* Wiley, New York, NY, 1976.

Davies, O.L., *Design and Analysis of Industrial Experiments,* 2nd ed., Hafner Publishing Company, New York, NY, 1956.

Dehand, K., *Quality Control, Robust Design and the Taguchi Method,* Wadsworth & Brooks/Cole, Pacific Grove, CA, 1989.

Diamond, W.J., *Practical Experiment Design for Engineers and Scientists,* Van Nostrand Reinhold, New York, NY, 1981.

Ealey, L., *Quality by Design: Taguchi Methods and the U.S. Industry,* ASI Press, Dearborn, MI, 1988.

Feigenbaum, A.V., *Total Quality Control,* McGraw Hill, New York, NY, 1983.

Fowlkes, W.Y. and Creveling, C.M., *Engineering Methods for Robust Product Design,* Addison Wesley, Reading, MA, 1995.

Gunter B., "Statistically Design Experiments-Part 1: Quality Improvement, the Strategy of Experimentation and the Road to Hell", *Quality Progress,* Vol. 22, pp. 63–64, 1989.

Gunter B., "Statistic Design Experiments-Part 2: The Universal Structure Underlying Experimentation", *Quality Progress,* Vol. 23, pp. 87–89, 1990.

Gunter B., "Statistically Design Experiments-Part-3: Interaction", *Quality Progress,* Vol. 23, pp. 74, 1990.

Gunter B., "Statistically Design Experiments-Part 5: Robust Process and Product Design and Related Matters", *Quality Progress,* Vol. 23, pp. 107–108, 1990.

Gunter, B., "A Perspective on the Taguchi Methods", *Quality Progress,* American Society of Quality Control, Milwaukee, WI, 1987.

Hendrix, C. D., "What Every Technologist Should Know About Experimental Design", *Chemical of Technology,* Vol. 9, No. 3, pp. 167–174, 1979.

Hill, W.J. and Delmer, W.R., "More on Planning Experiments to Increase Research Efficiency". *Industrial and Engineering Chemistry,* Vol. 62, No. 10, pp. 60–65, 1970.

Hoaglin, D., Mosteller, F. and Tukey, J., *Understanding Robust and Exploratory Data Analysis,* Wiley, New York, NY, 1983.

Hunter, J.S., "Statistical Design Applied to Product Design", *Journal of the Quality Technology,* Vol. 17, No. 4, pp. 210–221, 1985.

John, J.A. and Prescott, P., "Critical Values of a Test to Detect Outliers in Factorial Experiments", *Applied Statistics,* Vol. 24, pp. 56–59, 1975.

John, P.W.M., *Statistical Design and Analysis of Experiments,* Macmillan, New York, NY, 1971.

Kackar, R.N., "Off-line Quality Control, Parameter Design, and the Taguchi Method", *Journal of the Quality Technology,* Vol. 17, No. 4, pp.176–209, 1985.

Kackar, R.N., "Taguchi's Quality Philosophy: Analysis and Commentary", *Quality Progress*, Vol. 19, No. 12, pp. 21–29, 1986.

Kempthorne, O., *The Design and Analysis of Experiments*, Wiley, New York, NY, 1952.

Lawson, J.S. and Madrigal, J.L., "Robust Design Through Optimization Techniques", *Quality Engineering*, Vol. 6, No. 4, pp. 593–608, 1994.

Lochner, R.H. and Matar, J.E., *Designing for Quality: An Introduction to the Best of Taguchi and Western Methods of Statistical Experimental Design*, Chapman and Hall, London and New York, NY, 1990.

Mason, R.L., Gunst, R.F. and Hess, J.L., *Statistical Design and Analysis of Experiments*, Wiley, New York, NY, 1989.

Miller, A., Sitter, R.R., Wu, C.F.J. and Long, D., "Are Large Taguchi-Style Experiments Necessary? A Reanalysis of Gear and Pinion Data", *Quality Engineering*, Vol. 6, No. 1, pp. 21–37, 1993–1994.

Minitab, Inc., *Minitab Reference Manual*, Release 11, Minitab, State College, PA, 1996.

Mitra, A., *Fundamentals of Quality Control and Improvement*, Prentice Hall, Upper Saddle River, NJ, 1998.

Montgomery, D.C., *Introduction to Statistical Quality Control*, 3rd ed., Wiley, New York, NY, 1996.

Taguchi, G. and Konishi, S., "Orthogonal Arrays and Linear Graphs", *Tools for Quality Engineering*, American Supplier Institute, Dearborn, Michigan, 1987.

Taguchi, G., *Introduction to Quality Evaluation and Quality Control*, Japanese Standards Association, Tokyo, Japan, 1978.

Taguchi, G., "Off-line and On-line Quality Control Systems", *Proceedings of the International Conference on Quality Control*, Vol. B4 (1978), pp. 1–5.

Taguchi, G., *On-line Quality Control During Production*, Japanese Standards Association, Tokyo, Japan, 1981.

Taguchi, G., *Introduction to Quality Engineering: Designing Quality into Products and Processes*, Asian Productivity Organization, UNIPUB, White Plains, New York, NY, 1986.

Taguchi, G., *System of Experimental Design: Engineering Methods to Optimize Quality and Minimize Costs*, Vol. 1 and 2, American Supplier Institute, Dearborn, MI, 1987.

Taguchi, G. and Wu, Y., *Introduction to Off-line Quality Control Systems*, Central Japan Quality Control Association, Nagoya, Japan, 1980. Available from American Supplier Institute Dearborn, MI.

Taguchi, G., and Konishi, S., *Quality Engineering Series, Vol. 1, Taguchi Methods, Signal-to-Noise Ratio for Quality Evaluation*, ASI Press, Dearborn, MI, 1992.

Taguchi, G., and Konishi, S., *Quality Engineering Series, Vol. 3, Taguchi Methods, Research and Development*, ASI Press, Dearborn, MI, 1992.

Taguchi, G., Chowdhury, S. and Taguchi, S., *Robust Engineering*, McGraw-Hill, New York, NY, 2000.

Taguchi, G., Elsayed, E. and Hsiang, T., *Quality Engineering in Production Systems*, McGraw-Hill, New York, NY, 1989.

Taguchi, G., *Taguchi on Robust Technology Development*, ASME Press, New York, NY, 1993.

Taguchi, G., Yokohoma, Y. and Wu, Y., *Quality Engineering Series, Vol. 2, Taguchi Methods, On-line Production*, ASI Press, Dearborn, MI, 1994.

Taguchi, G., Yokohoma, Y. and Wu, Y., *Quality Engineering Series, Vol. 4, Taguchi Methods, Design of Experiments*, ASI Press, Dearborn, MI, 1994.

Tsyi, K.L., "An Overview of Taguchi Method and Newly Developed Statistical Methods for Robust Design", *IIE Transactions*, 24(5), pp. 44–57, 1992.

Wadsworth, H.M., Stephens, K.S. and Godfrey, A.B., *Modern Methods for Quality Control and Improvement*, Wiley, New York, NY, 1986.

Winer, B.J., *Statistical Principles in Experimental Design*, 2nd ed., McGraw-Hill, New York, NY, 1971.

Wu, Y. and Moore W.H., *Quality Engineering Product and Process Optimization*, American Supplier Institute, Dearborn, Michigan, 1986.

Wu, Y., *Orthogonal Arrays and Linear Graphs*, American Supplier Institute, Dearborn, Michigan, 1986.

Yates, F., *Design and Analysis of Factorial Experiments*, Tech. Comm. No. 35, Imperial Bureau of Soil Sciences, London, 1937.

Appendix–9A

Orthogonal Arrays and Associated Graphs

Source: G. Taguchi (1987), *System of Experimental Design-Engineering Methods to Optimize Quality and Minimize Costs*, Vol. 2, American Supplier Institute, Dearborn, MI, USA.

Legend for Linear Graphs

Symbol	Group	Degree of difficulty
○	1	Easy
◉	3	Fairly difficult
◎	2	Moderately easy
●	4	Difficult

Table 9A.1

L_4 (2^3) Series

Experiment	Factor		
	1	**2**	**3**
1	1	1	1
2	1	2	2
3	2	1	2
4	2	2	1
	Group 1	Group 2	

Linear Graph for L_4 Table

Table 9A.2

L_8 (2^7) Series

Experiment	Factor						
	1	**2**	**3**	**4**	**5**	**6**	**7**
1	1	1	1	1	1	1	1
2	1	1	1	2	2	2	2
3	1	2	2	1	1	2	2
4	1	2	2	2	2	1	1
5	2	1	2	1	2	1	2
6	2	1	2	2	1	2	1
7	2	2	1	1	2	2	1
8	2	2	1	2	1	1	2
	Group 1	Group 2		Group 3			

Linear Graphs for L_8 Table

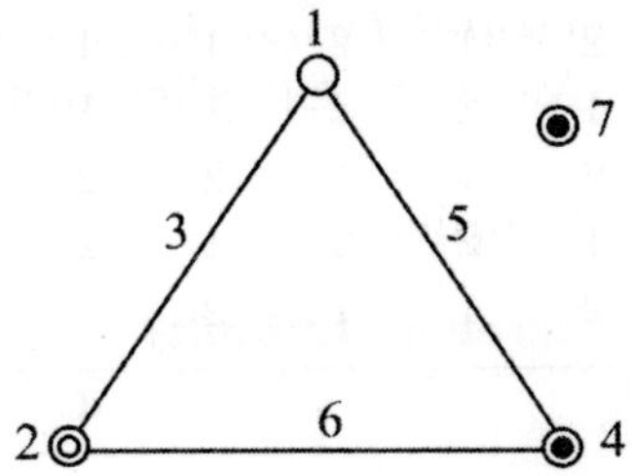

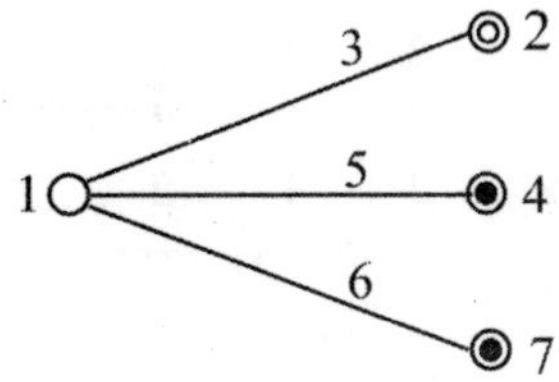

Table 9A.3

L_9 (3^4) Series

Experiment	Factor			
	1	2	3	4
1	1	1	1	1
2	1	2	2	2
3	1	3	3	3
4	2	1	2	3
5	2	2	3	1
6	2	3	1	2
7	3	1	3	2
8	3	2	1	3
9	3	3	2	1
	Group 1	Group 2		

Linear Graph for L_9 Table

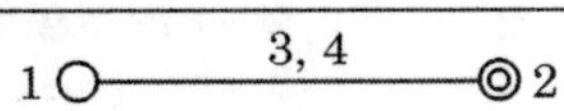

Table 9A.4

L_{16} (2^{15}) Series

Experiment	Factor														
	1	2	3	4	5	6	7	8	9	10	11	12	13	14	15
1	1	1	1	1	1	1	1	1	1	1	1	1	1	1	1
2	1	1	1	1	1	1	1	2	2	2	2	2	2	2	2
3	1	1	1	2	2	2	2	1	1	1	1	2	2	2	2
4	1	1	1	2	2	2	2	2	2	2	2	1	1	1	1
5	1	2	2	1	1	2	2	1	1	2	2	1	1	2	2
6	1	2	2	1	1	2	2	2	2	1	1	2	2	1	1
7	1	2	2	2	2	1	1	1	1	2	2	2	2	1	1
8	1	2	2	2	2	1	1	2	2	1	1	1	1	2	2
9	2	1	2	1	2	1	2	1	2	1	2	1	2	1	2

10	2	1	2	1	2	1	2	2	1	2	1	2	1	2	1
11	2	1	2	2	1	2	1	1	2	1	2	2	1	2	1
12	2	1	2	2	1	2	1	2	1	2	1	1	2	1	2
13	2	2	1	1	2	2	1	1	2	2	1	1	2	2	1
14	2	2	1	1	2	2	1	2	1	1	2	2	1	1	2
15	2	2	1	2	1	1	2	1	2	2	1	2	1	1	2
16	2	2	1	2	1	1	2	2	1	1	2	1	2	2	1
Groups	1	2		3				4							

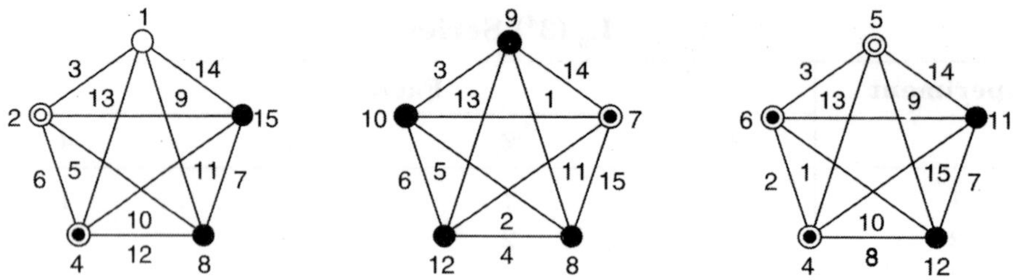

Table 9A.5
L_{18} ($2^1 \times 3^7$) Series

Experiment	Factor 1	2	3	4	5	6	7	8
1	1	1	1	1	1	1	1	1
2	1	1	2	2	2	2	2	2
3	1	1	3	3	3	3	3	3
4	1	2	1	1	2	2	3	3
5	1	2	2	2	3	3	1	1
6	1	2	3	3	1	1	2	2
7	1	3	1	2	1	3	2	3
8	1	3	2	3	2	1	3	1
9	1	3	3	1	3	2	1	2
10	2	1	1	3	3	2	2	1
11	2	1	2	1	1	3	3	2
12	2	1	3	2	2	1	1	3
13	2	2	1	2	3	1	3	2
14	2	2	2	3	1	2	1	3
15	2	2	3	1	2	3	2	1
16	2	3	1	3	2	3	1	2
17	2	3	2	1	3	1	2	3
18	2	3	3	2	1	2	3	1
Groups	1	2	3					

Linear Graph for L_{18} Table

1 ○————◎ 2

Table 9A.6

L_{27} (3^{13}) Series

Experiment	Factor 1	2	3	4	5	6	7	8	9	10	11	12	13
1	1	1	1	1	1	1	1	1	1	1	1	1	1
2	1	1	1	1	2	2	2	2	2	2	2	2	2
3	1	1	1	1	3	3	3	3	3	3	3	3	3
4	1	2	2	2	1	1	1	2	2	2	3	3	3
5	1	2	2	2	2	2	2	3	3	3	1	1	1
6	1	2	2	2	3	3	3	1	1	1	2	2	2
7	1	3	3	3	1	1	1	3	3	3	2	2	2
8	1	3	3	3	2	2	2	1	1	1	3	3	3
9	1	3	3	3	3	3	3	2	2	2	1	1	1
10	2	1	2	3	1	2	3	1	2	3	1	2	3
11	2	1	2	3	2	3	1	2	3	1	2	3	1
12	2	1	2	3	3	1	2	3	1	2	3	1	2
13	2	2	3	1	1	2	3	2	3	1	3	1	2
14	2	2	3	1	2	3	1	3	1	2	1	2	3
15	2	2	3	1	3	1	2	1	2	3	2	3	1
16	2	3	1	2	1	2	3	3	1	2	2	3	1
17	2	3	1	2	2	3	1	1	2	3	3	1	2
18	2	3	1	2	3	1	2	2	3	1	1	2	3
19	3	1	3	2	1	3	2	1	3	2	1	3	2
20	3	1	3	2	2	1	3	2	1	3	2	1	3
21	3	1	3	2	3	2	1	3	2	1	3	2	1
22	3	2	1	3	1	3	2	2	1	3	3	2	1
23	3	2	1	3	2	1	3	3	2	1	1	3	2
24	3	2	1	3	3	2	1	1	3	2	2	1	3
25	3	3	2	1	1	3	2	3	2	1	2	1	3
26	3	3	2	1	2	1	3	1	3	2	3	2	1
27	3	3	2	1	3	2	1	2	1	3	1	3	2
Groups	1	2			3								

Linear Graphs for L_{27} Table

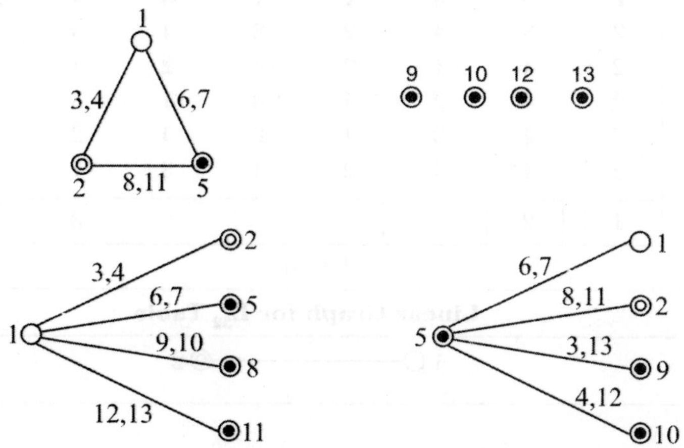

Table 9A.7
L_{32} ($2^1 \times 4^9$) Series

Experiment	Factor									
	1	**2**	**3**	**4**	**5**	**6**	**7**	**8**	**9**	**10**
1	1	1	1	1	1	1	1	1	1	1
2	1	1	2	2	2	2	2	2	2	2
3	1	1	3	3	3	3	3	3	3	3
4	1	1	4	4	4	4	4	4	4	4
5	1	2	1	1	2	2	3	3	4	4
6	1	2	2	2	1	1	4	4	3	3
7	1	2	3	3	4	4	1	1	2	2
8	1	2	4	4	3	3	2	2	1	1
9	1	3	1	2	3	4	1	2	3	4
10	1	3	2	1	4	3	2	1	4	3
11	1	3	3	4	1	2	3	4	1	2
12	1	3	4	3	2	1	4	3	2	1
13	1	4	1	2	4	3	3	4	2	1
14	1	4	2	1	3	4	4	3	1	2
15	1	4	3	4	2	1	1	2	4	3
16	1	4	4	3	1	2	2	1	3	4
17	2	1	1	4	1	4	2	3	2	3
18	2	1	2	3	2	3	1	4	1	4
19	2	1	3	2	3	2	4	1	4	1
20	2	1	4	1	4	1	3	2	3	2
21	2	2	1	4	2	3	4	1	3	2
22	2	2	2	3	1	4	3	2	4	1
23	2	2	3	2	4	1	2	3	1	4
24	2	2	4	1	3	2	1	4	2	3
25	2	3	1	3	3	1	2	4	4	2
26	2	3	2	4	4	2	1	3	3	1
27	2	3	3	1	1	3	4	2	2	4
28	2	3	4	2	2	4	3	1	1	3
29	2	4	1	3	4	2	4	2	1	3
30	2	4	2	4	3	1	3	1	2	4
31	2	4	3	1	2	4	2	4	3	1
32	2	4	4	2	1	3	1	3	4	2
	1	2		3						
Groups										

Linear Graph for L_{32} Table

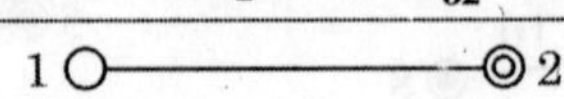

Table 9A.8

L_{16} (4^5) Series

Experiment	Factor				
	1	**2**	**3**	**4**	**5**
1	1	1	1	1	1
2	1	2	2	2	2
3	1	3	3	3	3
4	1	4	4	4	4
5	2	1	2	3	4
6	2	2	1	4	3
7	2	3	4	1	2
8	2	4	3	2	1
9	3	1	3	4	2
10	3	2	4	3	1
11	3	3	1	2	4
12	3	4	2	1	3
13	4	1	4	2	3
14	4	2	3	1	4
15	4	3	2	4	1
16	4	4	1	3	2
	1	2			
Groups					

Linear Graph for L_{16} Table

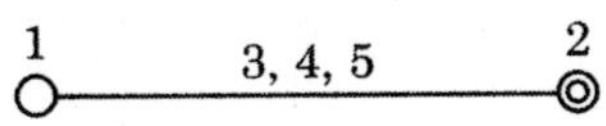

ANSWERS TO SELECTED PROBLEMS

ANSWERS TO THE SELECTED PROBLEMS

Chapter–2

2.1 Mean = 77.50
Variance = 6,963.88
Standard Deviation = 83.45

2.3 (*a*) 5 (*b*) 1 and 2 together
(*c*) 3, 5, and 6 together (*d*) 1, 3, 4, and 6 together.

2.4 (*a*) 0.05, 010 (*b*) 55%
(*c*) 0.50

2.5 (*a*) 0.2449

2.6 (*b*) 0.379 (*c*) 0.106

2.7 0.6667

2.8 *A* and *B* are mutually exclusive
B and *C* are not mutually exclusive

2.9 (*a*) 0.625 (*b*) 0.500

2.10 (*a*) 2.9% (*b*) 34.48%, 34.48%, 31.03%

2.11 (*b*) Approximate (*c*) (*i*) 0.22, (*ii*) 0.86, (*iii*) 0.59, (*iv*) 0.34

2.12 900 hours

2.13 (*a*) 0.5941 (*b*) 0.1992

2.14 (*a*) 0.9576 (*b*) 0.4571

2.15 0.0086

2.16 (*a*) 0.2240 (*b*) 0.1992

2.17 (*a*) 0.06681 (*b*) 0.86638
(*c*) 0.000214

2.18 0.3716

Chapter–5

5.1 $15,869

5.2 $37,064

5.3 $6,139 and $2,792

5.4 $113,283.20

5.5 $627

5.6 (*a*) $16,270
(*b*) $19,686.70

5.7 (*a*) $6,144.60 (*b*) $3,815.18

5.8 $123,602.89

5.9 $6,291.11

5.10 $36,363.60

5.11 (*a*) $6,904.68 (*b*) $1,294.24
(*c*) $14,800.80 (*d*) 11.83%

(e) 10.92% (f) 1.07
(g) 5 years and 6 years

5.12 (a) – $77,894.65 (b) – $17,885.19
(c) – $137,995.13

5.13 (a) – $60,102.60 (b) – $19,787.85
(c) – $94,572.61

5.14 (a) $5,625.43 (b) $1,486.45
(c) $13,011.96 (d) 24.20%
(e) 19.85% (f) 1.28

5.15 (a) $8,023.88 (b) $1,951.61
(c) $15,837.72 (d) 24.20%
(e) 18.48% (f) 1.40

5.16 $13,333.54

5.17 PW = $2,858.77, AW = $1,001.33, Endowment = $6,675.51

5.18 PW = – $1,093,738.99, AW = – $205,271.72, Endowment = – $1,710,597.65

5.19 $117,942.61

5.20 $51,961

5.21 $20,104

5.23 (a) $9,000
(b) 1st year: $15,000
2nd year: $12,000
3rd year: $9,000
4th year: $6,000
5th year: $3,000
(c) 1st year: $20,000
2nd year: $12,000
3rd year: $7,200
4th year: $4,320
5th year: $2,592

5.25 Sum of year's digit depreciation

5.26 DDB with conversion to SL

5.27 (a) $5,028.57 and (b) $14,800

5.28 $19,014.99

5.29 (a) $2,040,000 (b) 0.34
(c) $1,836,000

5.30 (a) Project should be undertaken (b) No

5.31 7.38%

5.32 8.4%

5.33 14.7%

5.34 Project B is more profitable than project A at 10% per year

5.35 0.0635

5.36 13.05%

5.37 3.53%

5.38 39,703.85

5.39 $34,705

5.40 (*a*) 2.691% (*b*) 3.489%
(*c*) 5.741% (*d*) 6.239%

5.41 (*a*) 0.801% (*b*) 4.21%
(*c*) 0.37% (*d*) 1.2%
(*e*) 2.05%

5.42 (*a*) 4.402% (*b*) 0.993%
(*c*) 5.911% (*d*) 2.079%
(*e*) 4.156%

5.43 (*a*) 2.91% (*b*) – 1.85%

5.44 (*a*) 4.56% (*b*) $304,482,000
(*c*) $16,289,787 or $16,289,000

5.45 (*a*) $37,908.00 (*b*) $37,908

5.46 The corporation should chose Company *A*'s offer

5.47 $371,063

5.48 $26.092173 millions per year or 26,092,173 per year

5.49 In years 1–5: 14.4% per year
In years 6–8: 17.7% per year
In years 9–13: 13.3% per year

5.50 Before-tax rate of return = 12%
After-tax rate of return (no inflation) = 7.1%
After-tax rate of return (5% inflation) = 17.2%

5.51 (*a*) $9,159.40 and $2,303.59 (*b*) $8,424.80 and $2,288.18

5.52 3.99 years

5.54 (*a*) Sensitive to the expected variations in sales or profits
(*b*) 8,523 volume units

5.55 20 units per hour

5.56 Above 78,739 units

5.57 (*a*) 31.564, C_T changes 9.054 times fast as X for a unit change in X and 13.741 times as fast as Y for a unit change in Y
(*b*) Standard deviation = 5.2479
(*c*) Variance = 27.5401

5.58 Subsystem *B*

5.59 Machine *A*

5.60 Alternative *A*

Chapter–6

6.1 (*a*) If $1 < x \le 40$, use m_1
If $50 < x \le 100$, use m_3
If $100 < x \le 250$, use m_2

(*b*) m_2 or m_3
(*c*) \$1200

6.2 (*a*) \$345 (*b*) \$576.90
(*c*) \$213.50

6.3 \$6,593.85

6.4 (*a*) \$15 and \$15(N) + \$25,000 + \$21,522

6.5 \$0.681, \$0.491, \$0.363, and \$0.331

6.6 (*a*) 10

6.7 (*a*) 246.67 lbs/acre (*b*) \$6,053/acre
(*c*) 388.72 lb/acre

6.8 (*b*) \$18.89 (\$ per unit)

6.9 (*a*) 261 units (*b*) \$446.84
(*c*) 9 (*d*) 185 units

6.10 (*a*) 1,400,000 units (*b*) \$100,000 and \$300,000

6.11 (*a*) \$49.84 per unit (*b*) \$71,400

6.12 1741 hours per year

6.13 (*a*) 2.9276 in^2 (*b*) \$229.94
(*c*) \$434.46 and \$314.84

6.14 40, 40, and 40

6.15 Company *A*: +\$6, and Company *B*: +\$3

6.16 Company *A*: +\$6, and Company *B*: +\$3

6.17 \$2,800

6.18 (*a*) 11.97% (*b*) 1.79 or 1.79 to 1
(*c*) 1.02 or 1.02 to 1 (*d*) 11.91 or 11.91 to 1
(*e*) 10.95 or10.95 to 1 (*f*) 80.41%
(*g*) 14.29% (*h*) D. Raghu: 11.97%, and Industry: 9.6%
(*i*) D. Raghu: 1.79%, and Industry: 2.05%
(*j*) D. Raghu: 80%, and Industry:82%

6.19 (*a*) Cost Center *A*: 27.5%
Cost Center *B*: 64.71%
Cost Center *C*: 68.75%
Cost Center *D*: 43.12%
(*b*) Cost Center *A*: 23.91%
Cost Center *B*: 38.19%
Cost Center *C*: 63.49%
Cost Center *D*: 34.05%
(*c*) Cost Center *A*: \$2.25/hr
Cost Center *B*: \$2.97/hr
Cost Center *C*: \$3.93/hr
Cost Center *D*: \$1.74/hr

6.20 Cost Center *A*: \$3.78/hr
Cost Center *B*: \$4.87/hr

Cost Center *C*: \$9.42/hr
Cost Center *D*: \$4.40/hr

6.21 (*a*) \$4,029,675 (*b*) \$8,987,103
(*c*) \$13,016,778

6.22 – 0.863, – 0.557, and – 0.377

6.23 26.672 hours, and 13.665 hours

6.24 (*a*) 131.70 minutes (*b*) 16.463 minutes
(*c*) 28.30 minutes

6.25 (*a*) 23.957 minutes (*b*) 1430.142 minutes
(*c*) 232.292 minutes

Chapter–7

7.1 $Q(t) = 0.20t - 0.01t^2$
$R(t) = 1 - 0.20t + 0.01t^2$
$$\lambda(t) = \frac{2 - 0.20t}{10 - 2t + 0.10t^2}$$

7.2 (*a*) 6.6667 years (*b*) 5 years

7.3 (*a*) 0.1813 (*b*) 0.8647

7.4 (*a*) $R(t) = \exp[-3 \times 10^{-9}t]$
(*b*) Reliability = 0.9999
MTTF = 3.333×10^9 hours
t_n = 289,819.57 hours or 33.084 years

7.5 (*a*) $f(t) = bn(1 - bt)^{n-1}$ (*b*) $Q(t) = 1 - (1 - bt)$
(*c*) $\lambda(t) = \dfrac{bn}{(1-bt)}$ for $0 \le t \le 1/b$ (*d*) $\text{MTTF} = \dfrac{1}{[b\,(n+1)]}$
(*e*) $t_{50} = \left(\dfrac{1}{b}\right)[1 - 0.5^{1/n}]$

7.6 40 days

7.7 0.1207%

7.8 $\lambda(t) = \dfrac{3t^2}{(b^3 - t^3)}$ for $0 \le t \le b$
$$\text{MTTF} = \frac{3}{4}b$$
$$\text{AFR}(t) = \frac{-\ln\left(1 - \dfrac{t^3}{b^3}\right)}{t}$$
Average failure rate up to the MTTF = $\dfrac{0.7309}{b}$

7.9 (*a*) 13.5757 years (*b*) 18.789 years

7.10 0.28352

7.11 $F(t) = \frac{2t}{a^2}\ ; 0 \le t \le a$

$R(t) = \frac{a^2 - 2t}{a^2}$

$\lambda(t) = \frac{2}{a^2 - 2t}$ (IFR)

MTTF = $a - 1$

$t_{med} = 0.25\ a^2$

$\sigma = \sqrt{(-a^2 + 2.667a - 1)}$

7.12 (*a*) 0.6277 (*b*) 33.36

7.13 0.7353

7.14 742.47 years, 514.64 years, 18,561.75 years, and 12,866.07 years

7.15 (*a*) 0.7768 (*b*) 0.5941

(*c*) Increasing failure rate (IFR) due to wear-out

7.16 $R(t) = (b^a)^{-t} = e^{-\lambda t}$

7.17 MTTF = 600, $t_{med} = 446.57$, $t_{0.95} = 93.837$, $\sigma = 500$

7.18 6811.918 hrs and 1740.82 hrs

7.19 0.8713 and 0.6936

7.20 (*a*) 0.815269 (*b*) (*i*) 2.680619, and (*ii*) 0.319381

7.21 (*i*) Design II

(*ii*) It depends on whether the $1000 increases in cost worth the very slight improvement in reliability in design I. If the response is yes, then the choice would be design I. Otherwise, the choice is design II.

7.22 76.75%

7.23 (*a*) 235 and 21.2766 (*b*) 0.76

(*c*) 191.276 hours (*d*) 0.8

7.24 99.0358 %

7.25 918,27.36 hours^2

7.26 28 months

7.27 (*a*) 0.8295 (*b*) 0.99 and 649.46 hours

(*c*) 13,284.45 hours and 6,358.65 hours (*d*) 12,698.11 hours and 11,387.7 hours

(*e*) 0.2638

7.29 (*a*) 0.9775 (*b*) 0.996625

(*c*) 5

7.30 $2e^{-\lambda t} - e^{-2\lambda t}$

7.31 $6e^{-2\lambda t} - 8e^{-3\lambda t} + 3e^{-4\lambda t}$

7.32 (*a*) 0.8306 (*b*) 0.0971

7.33 With no redundancy $R = 0.90$, With redundancy $R = 0.9508$

7.34 0.9871

7.35 (*a*) $X_1 X_4 X_2 + X_1 X_4 X_3 - X_1 X_4 X_2 X_3$

(*b*) 0.8446

(*c*) Min Path sets
1, 2, 4
1, 3, 4 4
2, 3

Min cut sets
1

(*d*) 0.8446 and 0.9373

7.36 (*b*) 1557.81 hours

7.37 $R_1 = 0.99393$
$R_2 = 0.922089$
$R_3 = 0.817550$
$R_4 = 0.720876$
$R_5 = 0.639722$

7.38 $X_1 = 14.87\%$, $X_2 = 3.27\%$, $X_3 = 0.74\%$, and $X_4 = 0.014\%$

7.39 (*a*) $[E \cup F \cup K \cup L] \cup [M \cap N \cap P] \cup [G \cup H] \cup [\{Q \cup R \cup S\} \cap J] \cup V$
(*c*) 0.1057

7.40 (*a*) $[G \cup H \cup I \cup J \cup K \cup L] \cup B \cup [M \cap N \cap P] \cup D \cup [Q \cap R \cap S \cap V] \cup F$
(*c*) 0.1530

7.41 (*a*) Sub-system *A*: 0.343
Sub-system *B*: 0.289
Sub-system *C*: 0.118
(*b*) Sub-system *A*: 0.000915
Sub-system *B*: 0.00077
Sub-system *C*: 0.000314
(*c*) Sub-system *A*: 2008.18 years
Sub-system *B*: 3452.06 years
Sub-system *C*: 545.47 years

7.42 (*a*) Sub-system *A*: 0.308
Sub-system *B*: 0.299
Sub-system *C*: 0.0506
Sub-system *D*: 0.1058
(*b*) Sub-system *A*: 0.00121
Sub-system *B*: 0.00117
Sub-system *C*: 0.000198
Sub-system *D*: 0.000415
(*c*) Sub-system *A*: 1268.87 years
Sub-system *B*: 2343.69 years
Sub-system *C*: 130.67 years
Sub-system *D*: 658.1 years

7.43 (*a*) Sub-system *A*: 0.3407
Sub-system *B*: 0.268
Sub-system *C*: 0.145
Sub-system *D*: 0.1116
Sub-system *E*: 0.164

(*b*) Sub-system *A*: 0.0033
Sub-system *B*: 0.0026
Sub-system *C*: 0.0014
Sub-system *D*: 0.00108
Sub-system *E*: 0.0016

(*c*) Sub-system *A*: 340.51 years
Sub-system *B*: 526.11 years
Sub-system *C*: 238.9 years
Sub-system *D*: 49.56 years
Sub-system *E*: 213.74

7.44 0.15%

7.45 0.0225%

7.46 0.0225%

7.47 $(A \cap B) \cup (D \cap C) \cup (D \cap E \cap F)$ and 0.98617

7.48 $(E \cup G) \cap \{F \cup [A \cup C) \cap (B \cup D)]\}$ and 0.993

7.49 $(A \cap B) \cup (D \cap C) \cup (D \cap E \cap F)$ and 0.9405

7.50 $(A \cap B \cap C) \cup [(E \cup H) \cap (D \cup F) \cap (D \cup G)]$ and 0.942

7.51 $n_1 = 2$, $n_2 = 3$, $n_3 = 3$, $n_4 = 2$, and Reliability = 0.995

7.52 $n_1 = 3$, $n_2 = 3$, $n_3 = 3$, $n_4 = 2$, and Reliability = 0.999

7.54 System failure rate = 4.831×10^{-6}
System MTTF = 207,770.62 hours

7.55 Component 1: MTTF = 3,865,481.25, R_1 system = 0.9995
Component 2: MTTF = 3,588,087.55, R_2 system = 0.9997
Component 3: MTTF = 12,835,322.81, R_3 system = 0.9998
Component 4: MTTF = 24,770,869.46, R_4 system = 0.9999
System failure rate = 6.557×10^{-7}

7.56 Component 1: MTTF = 1,231,223.84, R_1 system = 0.9961
Component 2: MTTF = 941619.59, R_2 system = 0.997
Component 3: MTTF = 653594.77, R_3 system = 0.9985
Component 4: MTTF = 1842638.66, R_4 system = 0.9989
Component 5: MTTF = 2707825.62, R_4 system = 0.9995

7.57 0.9679

7.58 0.189

7.59 0.9999

7.60 (*a*) 6.1442×10^{-6} (*b*) 6.25×10^{-9}/hour

7.61 201,699 hours

Chapter–9

9.1 For color density $(m + 4)$: \$32, and for color density $(m + 2)$: \$8

9.2 $0.25(y - 110)^2$

9.3 (*a*) Supplier *A*: \$3.6 per unit
Supplier *B*: \$9 per unit
Supplier *C*: \$18.6 per unit

(*b*) Supplier *A*: \$45,000
Supplier *B*: \$30,000
Supplier *C*: \$22,000

9.4 \$367.50

9.5 (*a*) The new production process is cost efficient.
(*b*) \$5,306,100.00

9.6 2 ± 0.005164 m

9.7 \$31.25

9.8 (*a*) The new process is cost effective
(*b*) \$6,850

9.9 2 ± 0.028284

9.10 (*a*) The new process will be cost effective to make the change
(*b*) \$138,100

9.11 (*a*) $L(y) = 11.111 \times 10^6 (y - m)^2$
(*b*) \$50.30

9.12 3.207

9.13 (*a*) Net savings due to improvement in the process capability = \$0.007 per unit of production
(*b*) \$16,800

9.14 (*a*) $\$3.67(10^9)$ (*b*) $\$3.6715(10^9)$

9.15 (*a*) $L(y) = \$140{,}625 \left[\frac{1}{y^2}\right]$
(*b*) Site 1: \$5634.32
Site 2: \$5788.94

9.16 110 ± 1.26

9.18 (*a*) 0.7071
(*b*) \$29.98 and \$15.85

9.23 (*b*) $\bar{A}_1 = (6.2 + 15.2 + 11.5)/3 = 10.966$

$\bar{A}_2 = (6.5 + 16.5 + 4.8)/3 = 9.267$

$\bar{A}_3 = (6.7 + 13.2 + 9.5)/3 = 9.8$

$\bar{B}_1 = (6.2 + 6.5 + 6.7)/3 = 6.467$

$\bar{B}_2 = (15.2 + 16.5 + 13.2)/3 = 14.967$

$\bar{B}_3 = (11.5 + 4.8 + 9.5)/3 = 8.6$

(*d*) Factor *A* at level 1, factor *B* at level 2, factor *C* at level 3, and factor *D* at level 1

9.25 (*a*) $\bar{A}_1 = (-3.2 + 6.8 + 1.9)/3 = 1.834$

$\bar{A}_2 = (-4.9 + 8.5 - 5.7)/3 = -0.7$

$\bar{A}_3 = (-5.2 + 2.7 - 2.6)/3 = -1.7$

$\bar{B}_1 = (-3.2 - 4.9 - 5.2)/3 = -4.433$

$\bar{B}_2 = (6.8 + 8.5 + 2.7)/3 = 6.0$

$\bar{B}_3 = (1.9 - 5.7 - 2.6)/3 = -2.134$

$\bar{C}_1 = (-3.2 - 5.7 + 2.7)/3 = -2.067$

$\bar{C}_2 = (6.8 - 4.9 - 2.6)/3 = -0.234$

$\bar{C}_3 = (1.9 + 8.5 - 5.2)/3 = 1.734$

$\bar{D}_1 = (-3.2 + 8.5 - 2.6)/3 = 0.9$

$\bar{D}_2 = (6.8 - 5.7 - 5.2)/3 = -1.367$

$\bar{D}_3 = (1.9 - 4.9 - 2.7)/3 = -1.9$

(*c*) Factor A at level 2, factor B at level 2, factor C at level 2, and factor D at level 2

9.26 (*a*) $72.7725 per unit

(*b*) L = $20.42 per unit, and La = $22.41 per unit

9.27 $Y_{opt} = \dfrac{A2 + B1 + C1 + D2 + E1 + F1 + G2}{8} = 159.25$

Improvement = 72.763%

9.28 (*b*) $B_1, E_2, E_xB_2, C_1, A_1, E_xC_1, D_2$

(*c*) A_1: 14.06%

B_1: 14.22%

C_1: 14.69%

E_2: 14.45%

D_2: 14.14%

(*d*) Both interactions are significant

(*e*) 270%

9.29 (*b*) 36

(*d*) For control parameter A: Optimum level is 2

For control parameter B: Optimum level is 2

For control parameter C: Optimum level is 2

For control parameter D: Optimum level is 1

SOLUTION MANUAL

CHAPTER 2

Basic Probability and Statistics

2.1. The following table gives the frequency distribution of the amounts of long distance telephone bills for the month of August 2000 for a sample of 60 families in a city.

Amount of long distance telephone bill (dollars)	Number of families
25 to less than 50	15
50 to less than 75	12
75 to less than 100	20
100 to less than 125	8
125 to less than 150	5

Calculate the mean, variance, and standard deviation.

Solution:

(*a*) The following Table P2.1 gives the frequency calculations.

Table P2.1

Amount of long distance telephone bill (dollars)	m	f	mf	m²f
25 less than 50	37.5	15	562.50	21,093.75
50 less than 75	62.5	12	750	46,875
75 less than 100	87.5	20	1750	153,125
100 less than 125	112.5	8	900	101,250
125 less than	137.5	5	87.50	94,531.25
Total		60	4050	416,875

For the given data:

$n = 60$, $\Sigma mf = 4050$, and $\Sigma m^2 f = 416{,}875$

$$\text{Mean} = \bar{X} = \frac{\sum mf}{n} = \frac{4050}{60} = 67.50$$

$$\text{Variance} = s^2 = \frac{\sum m^2 f - \left(\dfrac{\sum mf}{n}\right)^2}{n-1} = \frac{416{,}875 - 273{,}375}{60-1} = 2{,}432.20$$

Standard deviation $s = \sqrt{s^2} = \sqrt{2{,}432.20} = \$\ 49.317$

2.2. In Fig. P2.2, *E*, *T*, and *N* are the events that an automobile brought to a service garage needs an engine oil-change, tune-up, or new tires. Express in words the events represented by

(*a*) region 1

(*b*) region 3

(*c*) region 7

(*d*) regions 1 and 4 together

(*e*) regions 2 and 5 together

(*f*) regions 3, 5, 6, and 8 together.

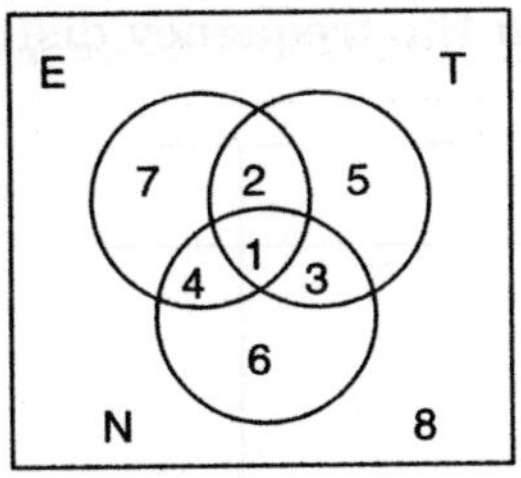

Fig. P2.2 Venn Diagram.

Solution:

(*a*) An automobile brought to the garage needs engine oil-change, tune-up, and new tires.

(*b*) An automobile brought to the garage needs tune-up, new tires but no engine oil-change.

(*c*) An automobile brought to the garage needs engine oil-change but neither tune-up nor new tires.

(*d*) An automobile brought to the garage needs engine oil-change and new tires.

(*e*) An automobile brought to the garage needs tune-up, but no new tires.

(*f*) An automobile brought to the garage does not need engine oil-change.

2.3. In Problem 2.2, List the region or combinations of regions representing the events that an automobile brought to the garage needs

(*a*) tune-up, but neither engine oil-change nor new tires

(*b*) an engine oil-change and tune-up

(*c*) tune-up or new tires, but not an engine oil-change

(*d*) new tires.

Solution:

(*a*) 5

(*b*) 1 and 2 together

(*c*) 3, 5, and 6 together

(*d*) 1, 3, 4, and 6 together.

2.4. In a particular company, 50% of the adult employees are female, 10% are divorced, and 5% are divorced females. For an employee selected at random, let

F = event the person is female

D = event the person is divorced

(*a*) find $P(F)$, $P(D)$, and $P(F \text{ and } D)$

(*b*) determine $P(F \text{ or } D)$ and interpret your answer in terms of percentages.

(*c*) find the probability that a randomly selected adult is male.

Solution:

(*a*) $P(F) = 0.50$; $P(D) = 0.10$; $P(F \text{ and } D) = 0.05$

(*b*) $P(F \text{ or } D) = 0.50 + 0.10 - 0.05 = 0.55$

55% of the employees are either female or divorced

(*c*) $P(M) = 1 - 0.50 = 0.50$

2.5 In the research and development department of a company, the number of males and female employees are shown in the frequency distribution as given below in a tabular form

Sex	Frequency
Male	30
Female	20
Total	50

Two employees selected at random from this department. The first employee obtained is not returned to the department for possible reselection; that is, the sampling is without replacement.

(*a*) determine the probability that he first employee obtained is female and the second is male

(*b*) draw the tree diagram.

Solution:

(*a*) Let

F_1 = event the first employee obtained is female

M_2 = event the second employee obtained is male

Using the general multiplication rule, we have

$$P(F_1 \text{ and } M_2) = P(F_1)\, P(M_2/F_1)$$

$$P(F_1) = \frac{f}{N} = \frac{20}{50} = 0.40$$

$P(M_2/F_1)$ is the conditional probability that the second employee is male given that the first one obtained is female. We note that, given the first employee obtained is female, there are 49 employees remaining in the department, of which 30 are male.

Hence,
$$P(M_2/F_1) = \frac{f}{N} = \frac{30}{49} = 0.6122$$

Applying the general multiplication rule, we conclude that

$$P(F_1 \text{ and } M_2) = P(F_1)\, P(M_2/F_1) = \left(\frac{20}{50}\right)\left(\frac{30}{49}\right) = 0.2449$$

When 2 employees are randomly selected from the department the probability is 0.2449 that the first employee obtained is female and the second is male.

(*b*) The tree diagram is shown in Fig. P2.5.

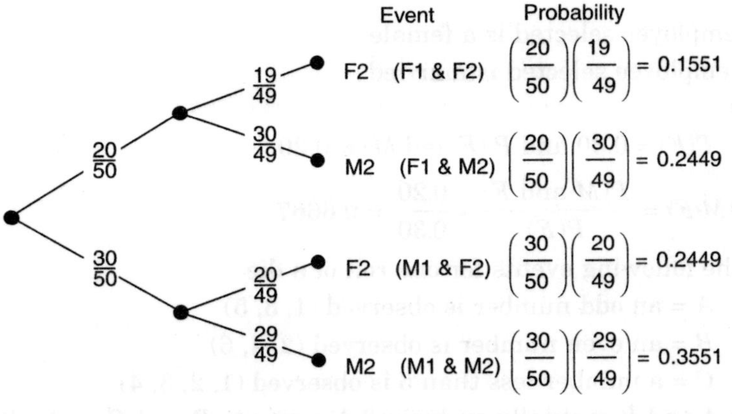

Fig. P2.5 Tree Diagram.

2.6. In a random sample of 1000 companies surveyed showed that 35 of them have no design engineer, 210 have one design engineer each, 379 have two design engineers, 270 have three design engineers, and 106 have four or more design engineers.

(*a*) prepare a frequency distribution table and calculate the relative frequencies for all categories

(*b*) if one company is selected randomly from these 1000 companies, find the probability that this company has two design engineers

(*c*) find the probability that the company selected has four or more design engineers.

Solution:

$$P \text{ (person selected is a woman)} = \frac{2}{5} = 0.4$$

$$P \text{ (Person selected is a man)} = \frac{3}{5} = 0.6$$

Yes, the sum of these probabilities is 1.0, because of the second property of probability.

(*a*)

Table P2.6. Frequency Distribution

Number of design engineers	Frequency	Relative frequency
0	35	0.035
1	210	0.210
2	379	0.379
3	270	0.270
4 or more	106	0.106
	1000	1.0

(*b*) P (Company selected has two design engineers) = 0.379

(*c*) P (Company selected has four or more design engineers) = 0.106

2.7. The probability that an employee at a university is a female is 0.3. The probability that an employee is a female and married is 0.2. Find the conditional probability that a randomly selected female employee from this university is a married person.

Solution:

Let

F = employee selected is a female

M = employee selected is married

Given that

$$P(F) = 0.30 \text{ and } P(F \text{ and } M) = 0.20$$

Hence, $$P(M/F) = \frac{P(M \text{ and } F)}{P(F)} = \frac{0.20}{0.30} = 0.6667$$

2.8. Consider the following events for one roll of a die

A = an odd number is observed (1, 3, 5)

B = an even number is observed (2, 4, 6)

C = a number less than 5 is observed (1, 2, 3, 4)

Are events A and B mutually exclusive? Are events B and C mutually exclusive?

Solution:

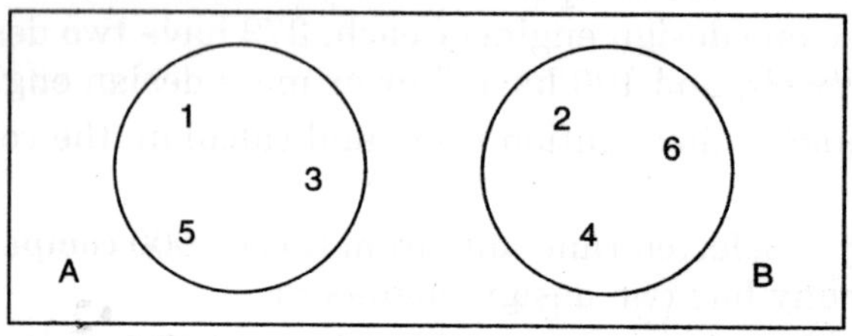

(a)

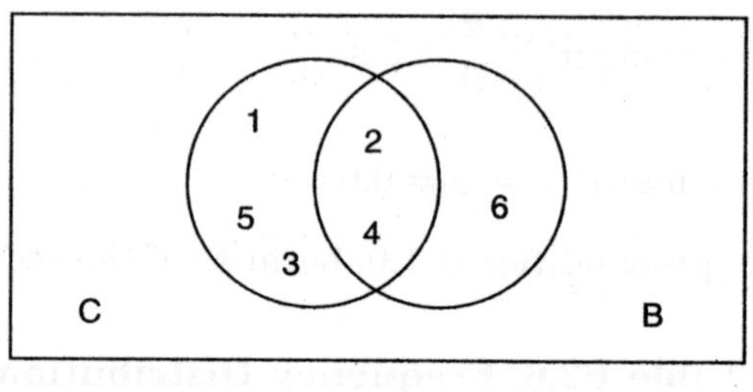

(b)

Fig. P2.8 (a) Events A and B, (b) Events B and C.

Figures P2.8 (a) and P2.8 (b) show the diagrams of events A and B and events B and C, respectively.

As we can observe from the definitions of events A and B and from Figure P2.8 (a), events A and B have no common element. For one roll of a die, only one of the two events, A and B, can happen. Hence, these are two mutually exclusive events B and C has two common outcomes: 2-spot and 4-spot. Thus, if we roll a die and obtain either a 2-spot or a 4-spot, then B and C happen at the same time. Hence, events B and C are not mutually exclusive.

2.9. Out of a collection of 100 gaskets the following are defective as indicated:

Defective type	Number of gaskets
Type I defect	10
Type II defect	8
Type III defect	3
Type I and II defect	5
Type I, II, and III defect	4
No defects	70

(*a*) if a gasket picked out of the collection has a type II defect, what is the probability that it also has a type I defect?

(*b*) if the gasket picked has a type I defect, what is the probability that it also has a type II defect?

Solution:

Using the conditional probability relationships, we have

$$(a)\ P\left(\frac{I}{II}\right) = \frac{P(I \text{ and } II)}{P(II)} = \frac{\frac{5}{100}}{\frac{8}{100}} = \frac{5}{8}$$

$$(b)\ P\left(\frac{II}{I}\right) = \frac{P(II \text{ and } I)}{P(I)} = \frac{\frac{4}{100}}{\frac{10}{100}} = 0.4$$

2.10. A mechanical system assembly plant purchases a certain component from three independent vendors; *A*, *B*, and *C*, who supply 20%, 30% and 50% respectively of the total number of components needed by the assembly plant. On an average, the percentage of defective components supplied by vendors *A*, *B*, and *C* are 5%, 3%, and 2% respectively.

(*a*) if a component is selected at random, find the probability that it is defective.

(*b*) if the selected component is defective, what is the probability that vendor *C* supplied it? By vendor *A*? By vendor *B*?

Solution:

(*a*) We have

$$\begin{aligned} P(\text{defective}) &= P\text{ (it is supplied by } A)\, P(\text{defective} \mid \text{supplied by } A) \\ &\quad + P\text{ (it is supplied by } B)\, P(\text{defective} \mid \text{supplied by } B) \\ &\quad + P\text{ (it is supplied by } C)\, P(\text{defective} \mid \text{supplied by } C) \\ &= (0.2)\,(0.05) + (0.3)\,(0.03) + (0.5)\,(0.02) \\ &= 0.01 + 0.009 + 0.01 \\ &= 0.029 \text{ or } 2.9\% \end{aligned}$$

(*b*) Applying Baye's theorem, we obtain

$$\begin{aligned} P\text{ (supplied by } C \mid \text{defective}) &= P(C \text{ supplied})\, P\text{ (defective} \mid C \text{ supplied})/ \\ &\quad [P\text{ (}A \text{ supplied)}\, P\text{ (defective} \mid A \text{ supplied)} \\ &\quad + P(B \text{ supplied})\, P(\text{defective} \mid B \text{ supplied}) \\ &\quad + P(C \text{ supplied})\, P(\text{defective} \mid C \text{ supplied}) \end{aligned}$$

$$= \frac{(0.5)(0.02)}{0.029}$$
$$= 0.3448 \text{ or } 34.48\%$$

Thus, on average vendor C supplied 34.48% of the defective components.

Proceeding in a similar fashion, we have:

$$P(\text{supplied by } A | \text{ defective}) = \frac{(0.2)(0.05)}{0.029} = 0.3448 \text{ or } 34.48\%$$

$$P(\text{supplied by } B | \text{ defective}) = \frac{(0.3)(0.03)}{0.029} = 0.3103 \text{ or } 31.03\%$$

Vendor A and C supplies the largest fraction 34.48% of the defective parts.

2.11. The frequency distribution of the number of orders received per day by a distribution company during the past 3 month (90 days) are given below:

Number of orders received per day	2	3	4	5	6
Number of days	11	20	23	22	14

(*a*) construct a probability distribution table for the number of orders received per day

(*b*) are the probabilities in (*a*) exact or approximate?

(*c*) if x denotes the number of orders received on any given day, calculate the following probabilities

(*i*) $P(x = 3)$

(*ii*) $P(x \geq 3)$

(*iii*) $P(2 \leq x \leq 4)$

(*iv*) $P(x < 4)$

Solution:

(*a*) Table P2.11 gives the probabilitv distribution

(*b*) The probabilities listed in the Table P2.11 are approximate because they are obtained from a sample of 90 days.

(*c*) (*i*) $P(x = 3) = 0.22$

(*ii*) $P(x \geq 3) = P(3) + P(4) + P(5) + P(6) = 0.22 + 0.25 + 0.24 + 0.15 = 0.86$

(*iii*) $P(2 \leq x \leq 4) = P(2) + P(3) + P(4) = 0.12 + 0.22 + 0.25 = 0.59$

(*iv*) $P(x < 4) = P(2) + P(3) = 0.12 + 0.22 = 0.34$

Table P 2.11. Probability Distribution

x	P(x)
2	11/90 = 0.12
3	20/90 = 0.22
4	23/90 = 0.25
5	22/90 = 0.24
6	14/90 = 0.15

2.12. Find the mean μ for the probability density function, ρ(x), of the life of a projector bulb, random variable x, is given as:

$p(x) = 0$ for $x < 0$

and $p(x) = 1/900\, e^{-x/900}$ for $x \geq 0$

Solution:

$$E(x) = \mu = \int_{-\infty}^{\infty} xp(x)\, dx$$

$$\int_0^{\infty} \frac{x}{900} e^{-\frac{x}{900}}\, dx = 900 \text{ hours}$$

2.13. **The daily consumption of an electric power of a certain machine (in units of power) is a random variable whose probability density is given by**

$$f(x) = \begin{cases} 1/9xe^{-x/3} & \text{for } x > 0 \\ 0 & \text{for } x \leq 0 \end{cases}$$

Determine the probabilities that on a given day

(*a*) the consumption of this machine is no more than 6 units

(*b*) the power supply is inadequate in the daily capacity if the supply is 9 units.

Solution:

$$F(x) = \frac{1}{9}\int_0^x xe^{-\frac{x}{3}}\, dx + C = \frac{1}{9}\frac{e^{-\frac{x}{3}}}{\left(\frac{1}{9}\right)}\left(-\frac{1}{3}x - 1\right) + C = C - e^{-\frac{x}{3}}\left(\frac{1}{3}x + 1\right)$$

$C = 1$

(*a*) $F(6) = 1 - 3e^{-2} = 1 - 3(0.1353) = 0.5941$

(*b*) $1 - F(9) = 4e^{-3} = 4(0.0498) = 0.1992$

2.14. **A lot contains 140 electronic components and 20 are selected without replacement for quality testing.**

(*a*) if 20 components are defective, what is the probability that at least one defective component is in the sample?

(*b*) if 5 components are defective, what is the probability that at least one defective component appears in the sample?

Solution:

Let X denotes the number of components in the sample that are defective.

(*a*) $P(X \geq 1) = 1 - P(X = 0)$

$$P(X = 0) = \frac{\binom{20}{0}\binom{120}{20}}{\binom{140}{20}} = \frac{\frac{120!}{20!\,100!}}{\frac{140!}{20!\,120!}} = \frac{120!\,120!}{100!\,140!} = 0.0424$$

$P(X \geq 1) = 1 - 0.0424 = 0.9576$

(*b*) $P(X \leq 1) = 1 - P(X = 0)$

$$P(X = 0) = \frac{\binom{20}{0}\binom{135}{20}}{\binom{140}{20}} = \frac{\frac{135!}{20!\,115!}}{\frac{140!}{20!\,120!}} = \frac{135!\,120!}{115!\,140!} = 0.4571$$

$P(X \geq 1) = 1 - 0.4571 = 0.5429$

2.15. A quality test engineer claims that 1 in 10 of certain manufactured parts is due to material defects. Using binomial distribution and rounding to four decimals, find the probability that a least 3 out of 5 of the tested parts are due to material defects.

Solution:

The probability that at least 3 out of 5 of the tested parts are due to material defects is

$$\binom{5}{3}(0.1)^3(0.9)^2 + \binom{5}{4}(0.1)^4(0.9) + \binom{5}{5}(0.1)^5(0.9)^0$$

$$= 10\,(0.001)\,(0.81) + 5\,(0.0001)\,(0.9) + (0.00001) = 0.0086$$

2.16. A photocopying machine in an office breaks down an average of three times per month. Using the Poisson probability distribution formula, find the probability that during the next month this machine will have

(*a*) exactly two breakdowns

(*b*) at most one breakdown.

Solution:

Let λ be the mean number of breakdowns per month and x be the actual number of breakdowns observed during the next month for this machine. Then,

(*a*) The probability that exactly two breakdowns will be observed during the next month is given by

$$P(X = 2) = \frac{\lambda^x e^{-\lambda}}{x!} = \frac{(3)^2 e^{-3}}{2!} = \frac{(9)(.04978)}{2} = 0.2240$$

(*b*) The probability that at most one breakdown will be observed during the next month is given by the sum of the probabilities of zero and one breakdown. Hence

$$P\text{ (at most 1 breakdown)} = P\text{ (0 or 1 breakdown)} = P(x = 0) + P(x = 1)$$

$$= \frac{(3)^0 e^{-3}}{0!} + \frac{(3)^1 e^{-3}}{1!}$$

$$= \frac{(1)(.04978)}{1} + \frac{(3)(.04978)}{1}$$

$$= 0.0498 + 0.1494 = 0.1992$$

2.17. The diameter of a machine shaft produced in a manufacturing company is normally distributed with a mean diameter of 0.001 inches and a standard deviation of 0.0002 inches.

(*a*) what is the probability that the diameter of the shaft exceeds 0.0013 inches?

(*b*) what is the probability that a diameter of the shaft is between 0.0007 and 0.0013 inches?

(*c*) what standard deviation of diameters is needed so that the probability in part (*b*) is 0.995?

Solution:

(*a*) $P(x > 0.0013) = P\left(Z > \dfrac{0.0013 - 0.001}{0.0002}\right) = P(Z > 1.5) = 0.06681$

(*b*) $P(0.0007 < x < 0.0013) = P(-1.5 < Z < 1.5) = 2P(Z < 1.5) - 1 = 0.86638$

(c) $P(0.0007 < x < 0.0013) = P\left[\frac{0.0007 - 0.001}{\sigma}\right] < Z\left[\frac{0.0013 - 0.001}{\sigma}\right]$

$$= P\left[\frac{-0.0003}{\sigma}\right] < Z < \left[\frac{-0.0003}{\sigma}\right]$$

Therefore

$$2P\left[Z < \frac{-0.0003}{\sigma}\right] - 1 = 0.995$$

and $$P\left[Z < \frac{-0.0003}{\sigma}\right] = 0.9975$$

Hence, $$\frac{0.0003}{\sigma} = 1.405$$

and $$\sigma = 0.000214$$

2.18. Out of a batch of 2800 electronic components received 25% of them were defective or non-conforming. In a sample of 50 randomly selected, the probability that between 12 and 14 are given by the binomial distribution. Use normal distribution to this binomial distribution.

Solution:

The probability that between 12 and 14 (12, 13, or 14) are defective is computed using binomial distribution as

$$P(12 \le X \le 14) = \sum_{x=12}^{14} \binom{50}{x} (0.25)^x (0.75)^{50-x} = 0.3665$$

For the normal approximation with

$$np = 50(0.25) = 12.5$$

and $$\sqrt{np(1-p)} = \sqrt{50(0.25)(0.75)} = 3.06$$

we obtain

$$P(12 \le X \le 14) = \Phi\left[\frac{14.5 - 12.5}{3.06}\right] - \Phi\left[\frac{11.5 - 12.5}{3.06}\right]$$

$$= \Phi(0.654) - \Phi(-0.327)$$

Using Table 2E in Appendix, the probabilities are found to be

$$P(12 \le X \le 14) = 0.7434 - 0.3718 = 0.3716$$

CHAPTER 3

Product and Process Design

3.1. Numerous factors affect the product development process. However, some of the main factors are as follows:

(*i*) Rate of innovation

(*ii*) Use of software tools, rapid prototyping, and virtual reality

(*iii*) Mass customization and customized production

(*iv*) Core competency, partnerships, and outsourcing

(*v*) Modern telecommunication means and internet

(*vi*) Electronic commerce (E-commerce)

(*vii*) Flexibility and agility

(*viii*) Global manufacturing

(*ix*) Automation

(*x*) Environmental issues

In order to meet the continuously changing expectations and standard of living of the users and the customers, more and more number of newer products in different models must be designed and developed within a short and limited time span by the companies. They also should do so with minimum and acceptable investment level by selecting the most appropriate technologies.

3.2. Due to globalization, products are now being collaboratively developed in many different locations all over the world. For manufacturing, this has resulted in worldwide production facilities and the use of many overseas suppliers. Critical parts that are purchased from another country must then be shipped to the various facilities. This causes many communication and logistics problems. Logistic costs that include shipping and handling costs become a significant part of the product cost. Scheduling, ordering and shipping of parts and products all over the globe also becomes very complex because of these reasons.

3.4. Stages in product development life cycle are as follows:

(*i*) Identification of customer requirements and setting up of design objectives

(*ii*) Identification of design alternatives

(*iii*) Selection of appropriate design and its detailing

(*iv*) Assessment of product design

(*v*) Manufacturing

For a large development project, there may be other stages to be considered, where a large number of groups representing several disciplines and functions may be working together and engaged to complete the entire development process.

3.5. "Collaborative design team approach" may be defined as the process of people working in teams to pursue design innovation in which information, ideas, and problem solving approaches are actively shared among the team members.

Unlike traditional approach, in "Collaborative design team approach" the team members are usually drawn from several functions, departments, or sections of a company so that the skills and the knowledge of the individual members of the team can be effectively utilized toward the common goal of developing a new product. Moreover, "Collaborative design team approach" is based on integrating product development with product life cycle issues to collectively achieve flexibility, high profit margins, and customer satisfaction.

3.6. For development of a new product, one must ensure that the product possesses a few unique features that make it an innovative one, even if a technology may seem to be innovative in nature under present condition, it may not pass the test of time. This means that many innovative products may be simply a 'novelty' item without any commercial success. The technological risk while developing a new product remains at all stages of product development.

3.7. The key steps for the requirement definition process in product development are as follows:

(*i*) Analysis of customer needs
(*ii*) Analysis of product use and users
(*iii*) Analysis and forecast of technological capability
(*iv*) Benchmarking for products and processes
(*v*) Use of tools and techniques such as prototyping, virtual reality, and house of quality

3.9. Different types of design methodologies that may be used for identifying and evaluating user needs are as follows:

(*i*) Identification of all possible design alternatives
(*ii*) Trade-off analysis
(*iii*) Transformation of product requirements into design requirements for the best alternative

At the early design or conceptual design stage, the important aspects to be considered are as follows:

(*i*) Functional task allocation
(*ii*) Producibility assessment
(*iii*) Reliability

At the detailed design stage, the types of techniques used by the design team are as follows:

(*i*) Design analysis
(*ii*) Modelling
(*iii*) Simulation

3.11. The design methodology that reduces unit production cost for the product under consideration is known as "Design-To-Cost" or DTC methodology. DTC methodology reduces unit production cost through a rigorous approach of identifying and implementing cost-reducing design and manufacturing methods and approaches.

3.12. Different considerations are as follows:

(*i*) Functional task allocation

(*ii*) Producibility assessment
(*iii*) Reliability

3.13. Trade-off-analysis is essential for product development because it examines all possible design alternatives and different parameters with the purpose of optimizing the overall performance of the product and minimizing technical risk.

The steps involved in a typical Trade-off-analysis are as follows:

(*i*) Forming a cross-functional team
(*ii*) Encouraging customer involvement and innovation
(*iii*) Defining the objectives of the trade-off analysis
(*iv*) Determining the approach and resources required
(*v*) Evaluating and selecting the preferred alternative
(*vi*) Validating the study results through experimentation and/or simulation
(*vii*) Iterating more detailed trade studies throughout the design process
(*viii*) Documenting the study and results

3.16. Functional task allocation:

(*i*) Determine performance objectives of the product
(*ii*) Determine functional requirements necessary to meet product objectives
(*iii*) Allocate these functional tasks to the user, the product, the software, and/or some combination based on analysis
(*iv*) Identify alternative design configurations
(*v*) Verify that the user and the product are able to perform the assigned functions satisfying all design requirements, for each design alternative

Producibility assessment:

(*i*) Identification of main characteristics and manufacturing resources
(*ii*) Identification of a design methodology that minimizes cost, and identification of manufacturing requirements goals
(*iii*) Trade-off analyses on alternative product and process designs including rapid prototyping
(*iv*) Development of a preliminary manufacturing plan and strategy
(*v*) Complexity analysis and simplification of product and process design

Reliability:

(*i*) Reduce the number of parts and components in the design so that the design is simple
(*ii*) Use more reliable parts, materials, and software, wherever feasible and practicable
(*iii*) Reduce stress levels for the product
(*iv*) Design for variability and increase robustness of the design so that the effect of uncontrollable noise factor on the performance of the product is minimum
(*v*) Add redundancy, self-corrective, and self maintaining features in the design of the product
(*vi*) Use standard parts and components in the assembly of the product as far as possible

3.17. The steps of concurrent engineering are as follows:

(*i*) Design: Performance, testability, manufacturability, service, cost, quality.
(*ii*) Verify

(*iii*) Review

(*iv*) Produce

(*v*) Final test

3.18. Prototyping and virtual reality in the early design phases is the use of hardware and software conceptual prototypes to study customer responses to new products and to identify areas for improvements for both requirement definition and conceptual design. It is used in product development in order to provide a communication platform for studying consumer needs and refining product use scenarios and task analysis.

3.19. In order to be sure about the quality of performance of a product when it is put to use by the user, prototypes of the products are required to be made in most cases. Physical models and software models (virtual reality) are used to gather information to reduce uncertainty in the product performance, and to optimize parameter settings of the design and to test the design. Prototyping provides valuable information regarding expected performance of the product, adequacy of the design features and parameters, types of problems and difficulties in process development and materials quality, and ways and means for improved performance of the product under consideration. It also confirms if the product design is robust.

3.21. The future of virtual reality and prototyping is to merge virtual reality software and CAD/CAE tools into a new kind of software that will allow consumers and the design team to evaluate a product in a virtual environment. The designer may shape or manipulate a new concept by hand and give it to the customers and other engineers for analysis. The customer and product designer may also experience a design concept, walk into simulation space and examine details of the conceptual design on computer screen not even visible in the real world.

3.22. In most situations, simulation and modeling can be least expensive and effective methods to gather and verify information regarding effectiveness of design and product performance. In addition, there are several advantages of using simulation as a design tool and they are as follows:

(*i*) It increases the level of knowledge regarding design issues of product-environment interface.

(*ii*) It can assess the benefits, costs, and attributes of each design requirement

(*iii*) It can perform design trade-off analyses to optimize various design elements, such as performance, producibility or manufacturability, testability and reliability.

(*iv*) It may verify the completeness of a design in relation to the design's ability to meet the requirements as specified.

For a simulation approach to be fruitful the following conditions are to be fulfilled:

(*i*) Simulation must be realistic and correct

(*ii*) Must be able to solve real problems as identified and felt by the users

(*iii*) Must involve extensive planning and management

(*iv*) Must be a cost-effective tool for design

(*v*) Must be flexible enough to accommodate changes in purpose, and should be able to adapt to different phases of product development and can accommodate new applications of the products

3.23. The basic elements of manufacturing process development are as follows:

(*i*) Identifying the company's manufacturing strategies and detailing their effects on product development process

(ii) Studying the current business environment, competitors, and new technologies with sufficient detail, and incorporating this knowledge into the product development process

(iii) Developing a detailed manufacturing plan that includes the supplier base and partners of the company

(iv) Identifying manufacturing and supplier processes that need to be developed or improved

(v) Establishing a comprehensive manufacturability program that includes design guidelines formanufacturing's capabilities, methodologies and processes and process qualification

3.24. Different manufacturing methodologies and technologies that are required to be considered for product development are as follows:

(i) Electronic commerce (e-commerce) and web purchasing

(ii) Computer-aided design and manufacturing (CAD/CAM)

(iii) Computer-aided process planning (CAPP)

(iv) Ergonomics and human factors

(v) Material requirements and enterprise resource planning (MRP/ERP)

(vi) Inventory control and just-in-time (JIT)-based manufacturing

(vii) Group technology (GT) and cellular manufacturing system (CMS)

(viii) Automation and robotics

(ix) Supplier partnerships and supply chain management

3.25. The basic elements of a typical manufacturing plan are as follows:

(i) Strategies for Manufacturing

(ii) Manufacturing or Process Planning

(iii) Process Development

(iv) Manufacturing Process Qualification and Verification

(v) Design Release and Production Readiness

(vi) Manufacturing Methodologies and Technologies

CHAPTER 4

Quality Principles in Design

1. **State a few important definitions of quality, and discuss their significance and use in product development.**

Over the years, the concept of quality has been changing. It is a dynamic issue that is adapting to customer needs and expectations continuously. Several important trends in the international business and market are also responsible for this change.

In this context, the quality of a product or a service may be defined in several ways. A particular definition of quality is dependent on the perspective or viewpoint of the user of the product or the service. Broadly speaking, there are four definitions of quality.

Quality is conformance to applicable specifications and standards. Every organization, whether profit, nonprofit, manufacturing, service, private or public, has specifications and standards for its products and services. Organizations develop these specifications and standards to measure performance and to correct deviations from expected levels of performance. For example, in a manufacturing operation, specifications detail dimensional limits or physical attributes of a quality characteristic of a part. In a service operation, standards dictate approved methods of behavior or service by the service personnel during transaction of an activity for the customer.

Quality is fitness for use. This is a market-or customer-based definition. A product or service is fit for use if it satisfies customer needs and requirements. An interesting point in this context is that a product may be fit for use in terms of satisfying the customer, but may not conform to the specifications. A surface finish specification is developed for a consumer product. The condition of the surface finish is important because it enhances the product's appearance and hence its marketability. The specification is written to include all surfaces, both external and internal. However, even if the inside product surface is blemished, it cannot be seen by a customer. So long it does not adversely influence any 'buy' decision for this product, the nonconformance is accepted. Hence, a product with a blemish may be fit for use if the blemish does not affect performance, safety, or marketability.

Quality is satisfaction of customer wants, needs, and expectations at a competitive price. The definition says that product or service quality is the producer's ability to satisfy customer needs while still being able to realize a profit. This definition has both a customer and a manufacturer orientation. While the customer is the reason for the organization's existence, the product manufacturer and service provide must still make a profit. This definition also focuses on satisfying the customer at a competitive price. Many customers will not purchase a product or service unless it is reasonably priced.

Quality is the "loss imported to the society after the product is shipped to the customers". Defined by Dr. Genichi Taguchi, an eminent authority on quality engineering, quality of a product is defined as the loss incurred due to bad or unacceptable performances of the product as a result of deviations from the target specifications or performance level. If this loss decreases, the quality of the product increases. This definition is different from all other definitions of quality. In order to measure this loss, the loss function is to be known, and, expressed in monetary terms. Taguchi proposes the concept of 'robustness' in design based on this definition.

2. Taguchi's definition of quality mentions about minimization of 'quality loss'. How is this principle applied at the design stage of a product?

   Quality defined by Dr. Genichi Taguchi, an eminent authority on quality engineering, quality of a product is defined as the loss incurred due to bad or unacceptable performances of the product as a result of deviations from the target specifications or performance level. If this loss decreases, the quality of the product increases. This definition is different from all other definitions of quality. In order to measure this loss, the loss function is to be known, and, expressed in monetary terms. Taguchi proposes the concept of 'robustness' in design based on this definition.

3. What are the important trends in quality? Discuss each.

4. Distinguish between quality control, quality assurance, and total quality management. Are they related? Discuss with suitable examples.

   Quality control: Quality defined as conformance to specification, was controlled during production instead of being inspected into products. Responsibility for quality was transferred to an independent QC department, which was now considered the "guardian" of quality. Also, the QC department was now separated from manufacturing to give its autonomy and independence. During this phase of development many organizations introduced the concept of statistical quality control (SQC) or statistical process control (SPC) in their processes. Control charting based on SQC or SPC concepts were proposed and implemented by Walter A. Stewhart in 1920s. A host of other tools and techniques, notably in the areas of process capability analysis and acceptance sampling for components and products were also made popular and worthwhile to pursue for quality control.

   Quality assurance: Quality control subsequently evolved into quality assurance (QA). The QA department focuses on assuring both process and product quality through executing operational and quality audits, importing skills-enhancement training programmes for operators, performing technical analysis, and identifying operational areas for quality improvement. The QA department, in close touch with other departments and functions, develops and implements QA programmes.

   The QC department does exist in many organizations where QA has not yet evolved. It is considered to be a functional area, which is responsible for inspecting products calibrating instruments, testing products, and inspecting incoming material.

   Total quality management: As the issue of quality becomes more prominent, QA is evolving into a company-wide quality management (CWQM) function. CWQM is also called total quality management (TQM), total quality control (TQC) or simply total quality (TQ). The quality organization is the prime facilitator and consultant in this effort. Corporate quality groups are small with more authority but less direct responsibility for quality. For example, the quality organization has authority to stop defective material from leaving the manufacturing area, while the responsibility for the control of quality rests with the manufacturing department operator.

In implementing total quality concept in an organization, its chief executive officer (CEO) often leads the quality management function. As the quality message permeates at all levels of an organization, more people become involved, and slowly a quality ethic and culture develop. The focus of the total quality programme is company-wide, customer-oriented, and competitively driven.

Quality no longer resides in one department. It is a company-wide issue essential to the organization's survival. To produce a quality product or deliver a quality service requires the attention and commitment of everyone in the organization. It is the responsibility of the person doing the work, whether it is the security guard controlling arrival of persons and transport, the manager supervising employees, the operator working on the machine, or the person delivering flowers. Every element in the organization, from the executive committee that establishes business and quality policy to the receptionist at the front desk, contributes to the quality effort. The top management usually defines a realistic policy, line management establishes operational goals and objectives, engineers design attractive, reliable, and functional products, and operators produce defect-free products in the total quality management initiative.

Another important aspect of TQM needs to be emphasized. This is customer orientation in any TQM programme. This is essential in any TQM programme because the customers' needs change and the organizations must adapt to changing needs. Designing aesthetic products, producing defect-free products, and delivering products on time, at a profit are some of the issues to be considered in adapting to this new philosophy. Most importantly, an organization must design, produce, and deliver what the customer wants, not what the organization thinks the customer wants.

5. What is 'quality loop'? Discuss each element of this loop.

The American Society for Quality Control (ASQC) has developed quality standards that are basically a set of guidelines for implementing a quality management programme for any company. This series, standards (ANSI/ASQC Q90-94), embodies most elements of the different approaches to quality management. This standard refers to the concept of 'quality loop', and identifies its elements from market identification to disposal. All these elements may not be relevant for all types of products, particularly for the so-called 'consumer' products. However, for all 'industrial' products, all the elements of the quality loop are relevant. The elements of the quality loop are shown in Figure 4.2, and briefly explained below.

(*i*) **Marketing and Market Research: Marketing's responsibility is to identify market,** identify customer requirements, develop a 'product brief', and establish feedback system for control and improvement purposes. The typical activities and responsibilities of this element are as follows:

Identify a market: Marketing takes the lead in identifying a market, its customers, and its requirements. A market can be an entirely new market, an established market, or a niche in an established market. A company may develop an entirely new product for a new market. This product is usually high priced in order to have an acceptable return on the research and development investment within a limited pay back period. This new product initially will not have any competition, so if customers want the product, they must pay its price.

For an established market, a company can produce an enhanced version of the product that is sufficiently different from competing products, or it can develop a similar product and price it lower. This is a mass marketing strategy based on lower priced products. Or a company can develop a specialized product for a niche in an established market. This product satisfies the needs of a specific target group of customers.

Identify customer requirements: Marketing identifies customer needs, wants, and expectations in both products and services. Almost any product has a service component, which may be as important as the product itself. For example, people do not simply buy an IBM computer, they buy the service that is associated with the IBM name. To identify the customer requirements, survey and research methodologies need to be used. This is the most important activity of marketing. In a constantly changing customer requirements and business scenario, keeping marketing departments up-to-date and current requirements is a challenging assignment for a marketing executive.

Develop a product brief: Once customers' needs are identified, these are communicated to the organization in terms of a set of requirements distilled into a 'product brief'. Customer requirements evolve into product and service specifications. A product brief can address many important features and aspects of product in simple and plain language that is easily understood by customers, such as performance characteristics, aesthetics, serviceability, packaging, price, and regulatory requirements.

Establish a feedback system: Finally, marketing establishes an information monitoring, and feedback system. Customer needs change, and an organization must continually accommodate to changes throughout a product's life cycle. Otherwise, products or services will age and not be able to satisfy current needs. Most products follow a life cycle consisting of four stages: introduction, growth, maturity, and decline. These four stages may show the sales pattern of a product. When a product is introduced in the marketplace, sales may be low because people are not aware of the product, or do not know what benefits will accrue from its use, or the product may be priced too high. In its growth stage, through advertising people become aware of its potential benefits, or of its ability to satisfy needs. Sales increase. In the mature stage, competitors develop enhanced products or similar lower priced products. Sales flatten out. Finally, in the decline stage, competition forces the company to develop new products or lower the price of the existing product, and sales decrease.

(*i*) **Design/Specification Engineering and Product Development:** Engineering, using the product brief, translates customer requirements into technical specifications for materials, products, and processes. It considers the following in designing products: customer requirements, cost, manufacturability, testability, and design quality.

Customer requirements: Engineering first develops a concept or an idea and the knowledge of what the customer wants and expects. Often customer requirements, whether needs, wants, or expectations, are vaguely expressed, and engineering only gets an abstract idea about the product from marketing. Marketing obtains customer information from focus groups, surveys, polls, or other sources. Marketing believes that the organization can develop a product or service to satisfy these requirements. Engineering then determines whether it can develop the product within time and budget constraints.

Cost: While always considering customer satisfaction, product and delivery costs are never far from the designer's mind. If a product is high priced and has a quality image, it will sell to a narrow, affluent market, but usually not to a mass market.

Manufacturability or Producibility: Sometimes engineers design products on a computer terminal and do not solicit inputs or advice with regard to manufacturability from manufacturing or purchasing departments. A product designed with tight tolerances is sometimes inoperable or not manufacturable. Tight tolerances might result in mating parts interfering with each other causing premature failure. A product designed with loose tolerances might result in unnecessary movement, again causing premature failure.

Testability: The ease and speed with which a product is accessible for testing is called testability. If the product is not positioned properly on a production line, it is difficult to test or even assemble. For example, in a cassette tape-recorder manufacturing facility, recorders travel on a conveyor belt where at set points, they are assembled and tested. If the production operator has to physically move the product or perform any unnecessary movements, this represents wasted effort, which adds cost to the product.

Design quality: Quality in design is essential for a final product to be defect-free, safe, and reliable. Poor design can cause a product to fail prematurely. Also, if design errors are not eliminated, they are replicated in each manufactured product.

Good design also enhances safety and health. If a nuclear power plant is not designed and constructed with safety in mind, an accident can release radioactive and other kinds of waste in the atmosphere resulting in loss of life. The Bhopal Gas tragedy, and the Chernobyl and Three Mile Island accidents are the cases of accidents in this respect.

While designing a product, one must also consider its reliability which may be defined as the long-term effects of quality or quality in the long-term. There are many instance where testability is not considered during design of the product. The resulting consequence may be very costly or catastrophic. Asbestos was originally installed in buildings because it was considered a good insulator, but its reliability was not fully understood by the researchers. Many years later it is found that asbestos is carcinogenic to those producing it, installing it, and living with it. The health risk of asbestos was not known initially when engineers developed the product. For many kinds of pharmaceutical drugs, knowing the long-term impacts on the human health and the harmful side effects is a critical issue, which must not be ignored by the designers of the products.

(*iii*) **Procurement:** Depending on the types of products and industries, many manufacturers procure as much as 40 to 90% (in terms of the manufacturing volumes) of their materials from outside suppliers. Hence, the quality of the final product or the assembly is significantly dependent on the quality of the procured components. Purchasing department plays a vital role in procuring the best quality materials and components from the 'right' suppliers. The responsibilities of the purchase department are to communicate requirements to suppliers, to select suppliers, and to measure and monitor the performance of suppliers.

A manufacturer may be thought of as being a customer to its suppliers. The customer must communicate requirements to suppliers explicitly through specifications, drawings, purchase orders, and contracts. In majority of cases, requirements include quality characteristics, service characteristics, operating environment, aesthetics and quality grade, inspection instructions, and specifications for the items to be purchased.

Supplier selection is a formal process that evaluates suppliers based on their ability to produce a defect-free product, which is competitively priced, delivered on time, and serviced properly.

Performance of the suppliers are measured and monitored continually, at periodic interval, to ensure that they are providing products that conform to specifications. Monitoring techniques include on-site assessment or audit of the supplier's quality system, product testing, and product improvement projects.

(*iv*) **Process Planning and Development:** Production processes, whether inspection, assembly, or fabrication, should be planned so that they run smoothly under controlled conditions. In a typical production system there are different kinds of fabrication and production equipment, and measuring instruments, and they are required to be monitored, calibrated, and controlled within prescribed operating limits so that defect-free products are produced. In this context, dimensional as well as process dispersion caused by operators, materials, methods, and machines should be kept at a minimum level. To achieve these, the details of the production activities, in terms of process planning are to be prescribed. The process plans are mode, and production department is to produce as per the process plan. Maintenance and measurement instructions are also included in the process plans. In modern production systems, particularly in repetitive manufacturing, where a long varieties of parts and components need to be produced with varieties of machines and equipment, computed-aided process planning (CAPP) techniques are widely used.

(*v*) **Production:** Production is responsible for creating reality out of engineering specifications and drawings. The term 'production' often used loosely. Actually, it refers to three types of activities: assembly, fabrication, and online inspection.

The responsibility for quality rests with the production supervisors and the workers. The production supervisor must communicate the importance of quality to the persons on the line. This person will only be committed if the organization is honestly dedicated to the pursuit of excellence and of never-ending improvement. If the organization's commitment to quality is not genuine, the workers may not be at all concerned about the quality they are producing. In this respect, production department's commitment to quality at all time is a pre-condition for creating an excellent system for production.

Another important issue in this context should also not be overlooked. This is the issue of process capability. The process capability is the ability of a process to manufacture a part or a component as per its given specifications. Thus, necessary tools and techniques are to be use to measure and assess the process capability on a continuous basis. Again, once a process is in operation, process should be 'in-control', and because of any reasons, the process is in 'out-of-control' state, corrective or remedial measures are to be taken to bring it back to 'in-control' state. Statistical process control techniques are used to achieve this.

(*vi*) **Inspection, Testing, and Examination:** Inspection, testing, and examination are conducted on processes and products. The level of testing or inspection depends on the product, risks to the consumer, risk to the producer, cost, and regulations. For example, products involving public health or safety have associated risks to the producers and the consumers if the product fails during its use. This type of product may be governed by government regulations which mandate extensive testing and inspection. The standards and regulatory bodies in the government govern the testing, manufacturing, and distribution of such products.

Product quality can be verified through incoming, in-process, and final inspection. SPC tools can be used at the incoming and in-process stages of a production system. Final inspection occurs just before the product is sent to inventory or to the customer. However, the degree and the frequency of inspection depend on the importance of the quality characteristic and the capability of the process. If a quality characteristic is essential to product performance, the component or the product may be 100% inspected, *i.e.*, all the units produced are to be inspected.

(*vii*) **Packing and Storage:** Quality can only be maintained if products are packaged, stored, handled, and transported properly. Proper packaging protects contents against damage due to vibration, shock, heat, abrasion, and corrosion. Some products require special storage, handling, and packaging. Electronic components may have to be packaged in antistatic containers. Perishable foods are transported in refrigerated containers.

(*viii*) **Sales and Distribution:** Once a product is sold, it must be delivered to the customer intact and in an expedient and courteous manner. If a product is damaged in transit, the problem must be handled effectively and efficiently with the customer. Also, discourteous delivery people can create a harmful impression of the product even before it is used. It is important to monitor and correct this situation because it may adversely affect future sales.

(*ix*) **Installation and Operation:** Complex industrial products may require specialized tools, equipment, methods, and trained personnel to install and to operate. The manufacturers of the complex machinery and products provide the necessary manuals in sufficient detail so that safe installation and operation procedures are followed by the customers. Installation and operating documentation includes instructions for assembly, repair, installation, and operation, spare parts lists, and product servicing information.

(*x*) **Technical Assistance and Maintenance:** Once a product is sold, the buyer of a technical or complex product might require technical assistance to maintain its operation. Technical assistance might be nothing more than answering questions on the phone. After-sales service for complex equipment might mean sending a trained person to repair, replace, or maintain a sophisticated piece of equipment. After-sales service creates product or brand loyalty, which enhances repeat product sales.

(*xi*) **Disposal after Use:** After use, a product has to be disposed of properly and safely. If the product no longer has a useful life, it is scrapped. However, a product can have its life extended if it is repaired. If it must be scrapped, this is done in such a way that it does not jeopardize safety, health, or environment. For example, environmental laws may regulate the disposal of hazardous chemicals.

After a product has gone through the quality loop, the loop will start again with another product, whether it is a new or an enhanced product. There are very few products or services that do not follow the quality loop because products must adapt and respond effectively to marketplace changes in order to survive in the long run.

6. In designing a product, discuss how the following aspects may be considered and why:

(*i*) customer requirements

(*ii*) cost

(*iii*) manufacturability or producibility

(*iv*) testability

(*v*) reliability

Engineering, using the product brief, translates customer requirements into technical specifications for materials, products, and processes. It considers the following in designing products: customer requirements, cost, manufacturability, testability, and design quality.

(*i*) **Customer Requirements:** Engineering first develops a concept or an idea and the knowledge of what the customer wants and expects. Often customer requirements, whether needs, wants, or expectations, are vaguely expressed, and engineering only gets an abstract idea about the product from marketing. Marketing obtains customer information from focus groups, surveys, polls, or other sources. Marketing believes that the organization can develop a product or service to satisfy these requirements. Engineering then determines whether it can develop the product within time and budget constraints.

(*ii*) **Manufacturability or Producibility:** Sometimes engineers design products on a computer terminal and do not solicit inputs or advice with regard to manufacturability from manufacturing or purchasing departments. A product designed with tight tolerances is sometimes inoperable or not manufacturable. Tight tolerances might result in mating parts interfering with each other causing premature failure. A product designed with loose tolerances might result in unnecessary movement, again causing premature failure.

(*iii*) **Cost:** While always considering customer satisfaction, product and delivery costs are never far from the designer's mind. If a product is high priced and has a quality image, it will sell to a narrow, affluent market, but usually not to a mass market.

(*iv*) **Testability:** The ease and speed with which a product is accessible for testing is called testability. If the product is not positioned properly on a production line, it is difficult to test or even assemble. For example, in a cassette tape-recorder manufacturing facility, recorders travel on a conveyor belt where at set points, they are assembled and tested. If the production operator has to physically move the product or perform any unnecessary movements, this represents wasted effort, which adds cost to the product.

(*v*) **Reliability:** While designing a product, one must also consider its reliability which may be defined as the long-term effects of quality or quality in the long-term. There are many instance where testability is not considered during design of the product. The resulting consequence may be very costly or catastrophic. Asbestos was originally installed in buildings because it was considered a good insulator, but its reliability was not fully understood by the researchers. Many years later it is found that asbestos is carcinogenic to those producing it, installing it, and living with it. The health risk of asbestos was not known initially when engineers developed the product. For many kinds of pharmaceutical drugs, knowing the long-term impacts on the human health and the harmful side effects is a critical issue, which must not be ignored by the designers of the products.

7. Compare and contrast inspection-based and prevention-based quality control approaches. What are the design principles you may apply for each?

Before the quality-related principles are identified prior to their application in operations and production systems, two types of quality control systems, *viz.* inspection based and prevention-based, need to be explained.

The traditional method of quality control is inspection-based. The inputs are of five types: machine, man (operator), material, methods and management. Each is a potential cause of dispersion or variation of dimensions of quality characteristics. Fabrication or assembly operations process inputs by adding value to the product in each operational step.

A process can be one step or 100 steps. At the end of the first step or the hundredth step, an operator may inspect a product to check its conformance with an acceptable specification or standard, and on acceptance, products are shipped to the customer. If rejected, products are scrapped or sent back through the process for rework. By the time product has been fully processed, value has been added to the product in terms of direct labor, raw materials, equipment, and training costs. At this point, if a product is scrapped or reworked, value has been lost. In this case, the numbers of cost-adding elements in comparison with value-adding activities are significantly more making the production process as well as the product more expensive. This type of quality control is also referred to as 'detection-based', and the guiding philosophy behind this mode of quality control has been 'production at any cost' which may not be a desirable feature of a modern production or manufacturing system where design of a best quality product with affordable price is always emphasized.

In order to achieve this objective, the quality control must be based on a philosophy which believes in the prevention of the occurrence of defects or nonconformance in the process itself. In the prevention-based quality control, the inputs are the same. Again, a process can be one step or 100 steps and value is added at each step. In the prevention mode, the process is controlled. The important difference between inspection and prevention modes is that during processing of parts, components, or raw materials, the operator continually measures the quality of these parts, and components or raw materials adjusts the process if dimensions deviate from calculated limits. Thus the operator is preventing defects form occurring by controlling the output of the operation. The objective is to send defect-free products to the customer. The prevention-based quality control is illustrated in Figure 4.4 (*b*). In this case, as the number of cost-adding elements is expected to be very less, the product as well as the production cost remains reasonably low, and hence, it is most likely that a quality product can be developed with affordable and competitive price in the market. The guiding philosophy in this quality control is 'quality is job # 1'. However, one basic pre-condition for its success is to have a workforce who can handle not only the machines, but also can operate several kinds of inspection and measuring instruments as required, and make appropriate decisions. This aspect of operations ability must not be overlooked by management.

8. Distinguish between product design and process design in the light of the specific quality principles to be applied for each. Discuss with suitable examples.

Refer to the text.

9. Explain the concept of 'extended process'. Why is this concept relevant for product development in today's business environment?

The concept of 'extended process', as proposed by W. Edwards demanding, consists of not only the processes and employees of an organization, but also other components which are directly influencing the performance of the organization. These components are outside the traditional domain and control of the organization. A typical extended process includes the components, such as customers, suppliers, shareholders or investors, and the community and the environment. A typical extended process is schematically presented in Figure 4.7. An organization consists of persons, machines, materials, methods, and capital or money. These elements are considered within the boundary of the organization. The extended process includes a number of key components that come from the system external to the organization. Among these components, customers or consumers are the most important. Satisfaction of customer is the primary goal of an organization whether it is of product or service producing type. Goals such as providing the investors with an acceptable rate of return are secondary. Achieving the primary goal—customer satisfaction—automatically causes secondary goals to be realized.

The community in which the organization operates should be considered in this extended process. This community includes consumers, employees, and anyone else who is influenced by the operations of the company, directly or indirectly.

Suppliers are another component of the extended process, and they play a very vital role. For most products, the majority of the components, parts, and raw materials required for a product need to be procured from outside suppliers. Many a time, as high as 80 percent of parts and components are to be purchased from the suppliers. Hence, the quality of the product is very much dependent on the quality of supplied parts and components, and efforts must be made to ensure that suppliers supply quality parts and components all the time.

A long-standing relationship between the suppliers and the organization is what is called for so that the organization has the maximum benefit in terms of lowered cost and improved service.

10. Describe the P-D-C-A cycle. Why and how does this cycle play a crucial role in quality improvement?

Originally called the Shewhart cycle after its originator Walter A Shewhart, or the Deming cycle when the Japanese started using the concepts of quality control and improvement on the advice and recommendation of W. Edwards Deming in 1950s, the process improvement cycle consists of four main stages: Plan(P), Do(D), Check(C), and Act(A). Hence, this cycle is also known as P-D-C-A cycle.

At the 'plan' stage, a framework is developed which describes the relationships between quality of performance as determined by the customers, and the process performance. The difference between them should be minimized, and for this, the effect of controllable process variables on process performance should be known and measured.

For example, for given product, the customers are not satisfied with the present level of rejections. The process engineer finds that the main reason of rejection is non-conformance of a given quality characteristic, say diameter, of the product with the given specification, and the process performance needs to be improved. A study is planned to determine the appropriate process parameter settings so that the variability in the process is significantly reduced. The rejection proportion is also expected to be substantially very low once the process settings are maintained.

At the 'do' stage, the theory regarding the course of action developed in the plan stage is put into action. Experiments are conducted on the process to determine the required parameter settings ensuring minimum or no variability in the measurements. Feedback is obtained from the concerned customers and from the process. At this stage, the company may test the proposed plan on a small scale.

At the next 'check' stage, the results are analyzed. It is to be verified if the difference between the customer needs and the process performance is reduced by the proposed action. The potential drawbacks relating to other quality characteristics that are important to the customer are also to be looked into at this stage. Statistical analysis is required for this purpose. The performance of the process at the recommended parameter settings should be explained in terms of mean and variance of the output quality characteristics for a representative sample size. The performance must be indicative of positive improvement in the product design.

At the 'act' stage, a decision is made regarding implementation. If the results of the analysis conducted in the check stage are positive, the proposed plan is adopted. Customer and process feedback will again be obtained after full-scale implementation. Such information will provide a true measure of the plan's success. If the results of the check stage show no significant improvement, alternative plans should be proposed and developed, and the cycle continues. In this context, it should be remembered that customers' needs are not constant. They change with time, competition, societal outlook, and other factors. Therefore, the proposed plan is continuously verified to see if it reflects or represents the current needs of the customers. If the current needs are significantly different from the anticipated, the improvement cycle starts again from the plan stage. Continuous improvement in the process performance *vis-à-vis* fulfillment of current customer needs is the main objective of working as per the Shewhart or Deming cycle of process improvement.

11. Why is 'procurement' function of an organization important for product development? How is quality of purchased parts and components maintained and improved?

Depending on the types of products and industries, many manufacturers procure as much as 40 to 90% (in terms of the manufacturing volumes) of their materials from outside suppliers. Hence, the quality of the final product or the assembly is significantly dependent on the quality of the procured components. Purchasing department plays a vital role in procuring the best quality materials and components from the 'right' suppliers. The responsibilities of the purchase department are to communicate requirements to suppliers, to select suppliers, and to measure and monitor the performance of suppliers.

A manufacturer may be thought of as being a customer to its suppliers. The customer must communicate requirements to suppliers explicitly through specifications, drawings, purchase orders, and contracts. In majority of cases, requirements include quality characteristics, service characteristics, operating environment, aesthetics and quality grade, inspection instructions, and specifications for the items to be purchased.

Supplier selection is a formal process that evaluates suppliers based on their ability to produce a defect-free product, which is competitively priced, delivered on time, and serviced properly.

Performance of the suppliers are measured and monitored continually, at periodic interval, to ensure that they are providing products that conform to specifications. Monitoring techniques include on-site assessment or audit of the supplier's quality system, product testing, and product improvement projects.

12. How are the principles of 'continuous quality improvement', and improved customer satisfaction taken care of at different stages of product development process? Explain with examples.

Originally called the Shewhart cycle after its originator Walter A Shewhart, or the Deming cycle when the Japanese started using the concepts of quality control and improvement on the advice and recommendation of W. Edwards Deming in 1950s, the process improvement cycle consists of four main stages: Plan(P), Do(D), Check(C), and Act(A). Hence, this cycle is also known as P-D-C-A cycle.

At the 'plan' stage, a framework is developed which describes the relationships between quality of performance as determined by the customers, and the process performance. The difference between them should be minimized, and for this, the effect of controllable process variables on process performance should be known and measured.

For example, for given product, the customers are not satisfied with the present level of rejections. The process engineer finds that the main reason of rejection is non-conformance of a given quality characteristic, say diameter, of the product with the given specification, and the process performance needs to be improved. A study is planned to determine the appropriate process parameter settings so that the variability in the process is significantly reduced. The rejection proportion is also expected to be substantially very low once the process settings are maintained.

At the 'do' stage, the theory regarding the course of action developed in the plan stage is put into action. Experiments are conducted on the process to determine the required parameter settings ensuring minimum or no variability in the measurements. Feedback is obtained from the concerned customers and from the process. At this stage, the company may test the proposed plan on a small scale.

At the next 'check' stage, the results are analyzed. It is to be verified if the difference between the customer needs and the process performance is reduced by the proposed action. The potential drawbacks relating to other quality characteristics that are important to the customer are also to be looked into at this stage. Statistical analysis is required for this purpose. The performance of the process at the recommended parameter settings should be explained in terms of mean and variance of the output quality characteristics for a representative sample size. The performance must be indicative of positive improvement in the product design.

At the 'act' stage, a decision is made regarding implementation. If the results of the analysis conducted in the check stage are positive, the proposed plan is adopted. Customer and process feedback will again be obtained after full-scale implementation. Such information will provide a true measure of the plan's success. If the results of the check stage show no significant improvement, alternative plans should be proposed and developed, and the cycle continues. In this context, it should be remembered that customers' needs are not constant. They change with time, competition, societal outlook, and other factors. Therefore, the proposed plan is continuously verified to see if it reflects or represents the current needs of the customers. If the current needs are significantly different from the anticipated, the improvement cycle starts again from the plan stage. Continuous improvement in the process performance *vis-à-vis* fulfillment of current customer needs is the main objective of working as per the Shewhart or Deming cycle of process improvement.

13. How are the 'extended process' concept and goal of 'continuous quality improvement' related to each other?

The concept of 'extended process', as proposed by W. Edwards demanding, consists of not only the processes and employees of an organization, but also other components which are directly influencing the performance of the organization. These components are outside the traditional domain and control of the organization. A typical extended process includes the components, such as customers, suppliers, shareholders or investors, and the community and the environment. An organization consists of persons, machines, materials, methods, and capital or money. These elements are considered within the boundary of the organization. The extended process includes a number of key components that come from the system external to the organization. Among these components, customers or consumers are the most important. Satisfaction of customer is the primary goal of an organization whether it is of product or service producing type. Goals such as providing the investors with an acceptable rate of return are secondary. Achieving the primary goal—customer satisfaction—automatically causes secondary goals to be realized.

The community in which the organization operates should be considered in this extended process. This community includes consumers, employees, and anyone else who is influenced by the operations of the company, directly or indirectly.

Suppliers are another component of the extended process, and they play a very vital role. For most products, the majority of the components, parts, and raw materials required for a product need to be procured from outside suppliers. Many a time, as high as 80 percent of parts and components are to be purchased from the suppliers. Hence, the quality of the product is very much dependent on the quality of supplied parts and components, and efforts must be made to ensure that suppliers supply quality parts and components all the time.

A long-standing relationship between the suppliers and the organization is what is called for so that the organization has the maximum benefit in terms of lowered cost and improved service.

14. Discuss, in brief, the concepts of 'quality of design', 'quality of conformance' and 'quality of performance'. How are they mapped on 'quality loop' elements?

Quality of Design:

Quality of design deals with the stringent conditions that the product or service must minimally possess to satisfy the requirements of the customer. It implies that the product or service must be designed to meet at least minimally the needs of the consumer. The objective of any design effort is to create a simple and least expensive product or service that meets the customers' expectations. This aspect should be taken into account while prescribing quality of design. Quality of design is influenced by many factors, such as the type of product, cost, demand for product, availability of parts and materials, government rules and regulations, and product safety. For example suppose the quality level of the yield strength of steel cables desired by the customer is 100 kg/cm^2. When designing such a cable, the parameters that influence the yield strength are to be selected such that the requirement of yield strength is satisfied at least minimally. In practice, the product is typically over-designed so that the product may be able to withstand excessive loads in dynamic conditions. For this, concept of safety factor, k is used. Thus, to design a product with a 50 percent stronger load characteristic over the specified weight, the value of k would 1.50, and the product is designed for yield strength of $100 \times 1.50 = 150$ kg/cm^2.

In most situations, the effect of an increase in the designed quality level is to increase the cost at an exponential rate. The value of the product, however, increases at a decreasing rate, with the rate of increase approaching zero beyond a certain designed quality level. Sometimes, it might be of interest to choose a design quality level b of a product which maximizes the difference between the value and the cost, given that the minimal customer requirements, a are met. This is done with the idea of maximizing the return on investment. It may however be observed for a particular designed equality level, the cost and the value may be equal. For any level above this the cost may exceed the value. This information is important for selection of an appropriate design level for a product or a service.

Quality of Conformance:

When the manufactured product or the service rendered meets the standards selected in the design phase it is referred to as 'quality of conformance'. With respect to the manufacturing, this phase is concerned with the degree to which quality is controlled from the procurement of raw material to the shipment of finished goods. It consists of the three main areas, *viz.*, defect prevention, defect finding, and defect analysis and rectification. As the name suggests, defect prevention deals with the means to deter the occurrence of defects, and is usually achieved using statistical process control techniques. Locating defects is conducted through inspection, test, and statistical analysis of data from the process. Finally, the causes behind the presence of defects are investigated, and remedial actions are taken.

The quality of design has an impact on the quality of conformance. Obviously, one must be able to produce what is designed. If the specification for the shaft diameter is 20 ± 0.2 mm, the question that must be addressed is how to design the tools, equipment, and operations such that the manufactured product meets the design specifications. If such a system of production is achieved, the conformance phase is capable of meeting the stringent requirements of the design phase. On the other hand, if such a production system can not be installed (for instance, if the process is only capable of producing shaft with a specification of 20 ± 0.36 mm), the design phase is affected. This feedback may suggest that the product be redesigned because the current design cannot be produced using the existing capability. Therefore, there should be a constant exchange of information between the design and the manufacturing phases so that a feasible design can be achieved.

Quality of Performance:

Quality of performance is concerned with how well the product functions or service performs when it is put to use. It measures the degree to which the product or service satisfies the customer. This is a function which is influenced and by both the quality of design and the quality of conformance. The final test of product or service acceptance always lies with the customers. Meeting their expectations is the major goal. If a product does not function well enough to meet these expectations, or if a service does not live up to customer standards, then adjustments need to be made in the design and/or conformance phase. The feedback from the performance to the design phase should help change in the design of the product or the service if so required.

15. Name a few critical organizational barriers that may slow down the product development process. How do they prevent a company from adopting the total quality philosophy?

Critical organizational barriers are as follows:

(*a*) Lack of knowledge of quality in each level/everyone of the company

(*b*) Lack of adoption power to meet customers' changing requirements

(*c*) Lack of in-time delivery of the product.

16. Discuss the responsibilities of the different departments of an organization for (*i*) product development, and (*ii*) quality control improvement. Why is 'quality of performance' a matter of concern for the designers?

(*i*) **Product Development**

(*a*) **Marketing and Market Research:** Marketing's responsibility is to identify market, identify customer requirements, develop a 'product brief', and establish feedback system for control and improvement purposes.

(*b*) **Design/Specification Engineering:** Engineering, using the product brief, translates customer requirements into technical specifications for materials, products, and processes.

(*c*) **Procurement:** Depending on the types of products and industries, many manufacturers procure as much as 40 to 90% (in terms of the manufacturing volumes) of their materials from outside suppliers. Hence, the quality of the final product or the assembly is significantly dependent on the quality of the procured components. Purchasing department plays a vital role in procuring the best quality materials and components from the 'right' suppliers. The responsibilities of the purchase department are to communicate requirements to suppliers, to select suppliers, and to measure and monitor the performance of suppliers.

(*d*) **Process Planning and Development:** Production processes, whether inspection, assembly, or fabrication, should be planned so that they run smoothly under controlled conditions. In a typical production system there are different kinds of fabrication and production equipment, and measuring instruments, and they are required to be monitored, calibrated, and controlled within prescribed operating limits so that defect-free products are produced.

(*e*) **Production:** Production is responsible for creating reality out of engineering specifications and drawings. The term 'production' often used loosely. Actually, it refers to three types of activities: assembly, fabrication, and online inspection.

(*f*) **Inspection, Testing, and Examination:** Inspection, testing, and examination are conducted on processes and products. The level of testing or inspection depends on the product, risks to the consumer, risk to the producer, cost, and regulations. For example, products involving public health or safety have associated risks to the producers and the consumers if the product fails during its use. This type of product may be governed by government regulations which mandate extensive testing and inspection. The standards and regulatory bodies in the government govern the testing, manufacturing, and distribution of such products. For example, the Nuclear Regulatory Commission (NRC) may dictate the construction, maintenance, operation, testing, and commissioning of nuclear power plants.

(*g*) **Packing and Storage:** Quality can only be maintained if products are packaged, stored, handled, and transported properly. Proper packaging protects contents against damage due to vibration, shock, heat, abrasion, and corrosion. Some products require special storage, handling, and packaging. Electronic components may have to be packaged in antistatic containers. Perishable foods are transported in refrigerated containers.

(*h*) **Sales and Distribution:** Once a product is sold, it must be delivered to the customer intact and in an expedient and courteous manner. If a product is damaged in transit, the problem must be handled effectively and efficiently with the customer. Also, discourteous delivery people can create a harmful impression of the product even before it is used. It is important to monitor and correct this situation because it may adversely affect future sales.

(*i*) **Installation and Operation:** Complex industrial products may require specialized tools, equipment, methods, and trained personnel to install and to operate. The manufacturers of the complex machinery and products provide the necessary manuals in sufficient detail so that safe installation and operation procedures are followed by the customers. Installation and operating documentation includes instructions for assembly, repair, installation, and operation, spare parts lists, and product servicing information.

(*j*) **Technical Assistance and Maintenance:** Once a product is sold, the buyer of a technical or complex product might require technical assistance to maintain its operation. Technical assistance might be nothing more than answering questions on the phone. After-sales service for complex equipment might mean sending a trained person to repair, replace, or maintain a sophisticated piece of equipment. After-sales service creates product or brand loyalty, which enhances repeat product sales.

(*k*) **Disposal after Use:** After use, a product has to be disposed of properly and safely. If the product no longer has a useful life, it is scrapped. However, a product can have its life extended if it is repaired. If it must be scrapped, this is done in such a way that it does not jeopardize safety, health, or environment. For example, environmental laws may regulate the disposal of hazardous chemicals.

(*ii*) **Quality Control Improvement**

Inspection groups evolved into quality control (QC) departments. The start of World War II required that military products be defect-free. Product quality was crucial to winning the war and could only be ensured if the inspection department could control production processes. Quality defined as conformance to specification, was controlled during production instead of being inspected into products. Responsibility for quality was transferred to an independent QC department, which was now considered the "guardian" of quality. Also, the QC department was now separated from manufacturing to give its autonomy and independence. During this phase of development many organizations introduced the concept of statistical quality control (SQC) or statistical process control (SPC) in their processes. Control charting based on SQC or SPC concepts were proposed and implemented by Walter A. Stewhart in 1920s. A host of other tools and techniques, notably in the areas of process capability analysis and acceptance sampling for components and products were also made popular and worthwhile to pursue for quality control.

Quality of Performance

Quality of performance is concerned with how well the product functions or service performs when it is put to use. It measures the degree to which the product or service satisfies the customer. This is a function which is influenced and by both the quality of design and the quality of conformance. The final test of product or service acceptance always lies with the customers. Meeting their expectations is the major goal. If a product does not function well enough to meet these expectations, or if a service does not live up to customer standards, then adjustments need to be made in the design and/or conformance phase. The feedback from the performance to the design phase should help change in the design of the product or the service if so required.

17. Define and describe the 'four absolutes' of quality management as prescribed by Crosby. Can they be maintained always in product development? Why, or why not?

In order to achieve quality goals during product development, four aspects of quality management, *viz.* definition of quality, system for achievement of quality, performance standard, and measurement should be emphasized. These four aspects are referred to as 'absolutes' of quality management by Crosby.

(*i*) **Definition of Quality:** conformance to applicable specifications

(*ii*) **System for Achievement of Quality:** the most rational approach for achievement of quality

(*iii*) **Performance Standard:** The performance standard with which the existing quality level of a product is to be judged is Zero Defects (ZD).

(*iv*) **Measurement:** The quality measurement system needs to be based on costs of quality. In order to measure the total cost of quality, four kinds of cost elements are to be considered and their measurement schemes need to be prescribed in a given situation. These costs are prevention, appraisal, internal, and external failure cost.

18. State and describe, in brief, the tasks that you need to carry out during (*i*) quality planning, (*ii*) quality control, and (*iii*) quality improvement.

The iterative process of designing and developing a new product is based on the premise that the quality levels of a product are subject to changes or modifications over time, and these changes may be dependent on changes in process design and manufacturing.

In this context, the quality improvement process assumes importance because of its impact on product development. As mentioned by Juran, the quality improvement process consists of three interacting phases, *viz.* quality planning, quality control, and quality improvement.

Introduced by Joseph Juran, the 'quality trilogy' process starts with quality planning at various levels of the organization, each of which has a distinct goal. At the upper management level, planning is termed 'strategic quality management'. Broad quality goals are established at this level. A structured approach is selected in which management chooses a plan of action and allocates resources to achieve the goals. Planning at the middle management level is termed 'operational quality management'. Departmental goals consistent with the strategic goals are established. At the workforce level, planning involves a clear assignment of work and responsibility to each worker. Each worker is made aware of how his or her individual goal contributes to departmental goals.

After the planning phase, quality control takes over. Here, the goal is to run the process effectively such that the plans are executed effectively. If there are deficiencies in the planning process, the process may operate at a high level of 'chronic waste'. Quality control will try to prevent the waste from getting worse. If unusual symptoms are sporadically detected, quality control will attempt to identify the possible causes behind this abnormal variation. Upon identifying the cause, remedial actions are taken to bring the process back to control. Figure 4.11 shows the three phases of planning, control, and improvement and the cost trends associated with poor quality. The objectives of the control phase are to eliminate the causes associated with the very high values, referred to as sporadic spikes and to bring the process output within the zone of quality control. The next phase of the trilogy process is quality improvement, which deals with the continuous improvement of the product and the process. This phase is also called the 'quality breakthrough sequence'. Such improvements usually require an action on the part of upper and middle management. They deal with actions such as creating a new design, changing methods or procedures of manufacturing, and investing in new equipment. As can be seen from Figure 4.10, quality improvement will usually cause a reduction in the cost of poor quality. The chronic waste drops to a lower level due to quality improvement. Repeating the whole cycle again result in.

A number of activities need to be carried out in a sequence during quality planning phase. These activities are: identify the customer (both internal and external), determine customer needs, develop product features that respond to customer needs, establish quality goals that meet the needs of customers and suppliers at a minimum cost, develop a process that produce the required product features, and establish process capability.

During quality improvement sequence, actions, such as installing a new machine with improved or state-of-the-art technology, reengineering the existing method of manufacturing, and investment in new processes, are required and taken by the top management. These actions result in fundamental change in the process performance. The specific steps to be followed at this stage are: justify improvement requirements identify and organize specific projects for improvement, identify the root causes of the problems, and take remedial actions. Once the remedial actions for improvement are known, their actual effectiveness should be tested. For this, actions need to be implemented, and the benefits assessed. Once the benefits in the form of process improvement are assured, there must be a control system to sustain the improved level of performance. Appropriate auditing tools may be used for this purpose. They provide a basis for further process improvement for better design of the product.

19. What are the major drawbacks of the traditional approach for product development from the viewpoint of quality improvement?

Traditional Product Development requires the following steps:

Steps:

(*i*) Customer requirements

(*ii*) Cost estimation

(*iii*) Producibility study

(*iv*) Testability measurement

(*v*) Design Quality

Quality Improvement requires the following steps:

Steps:

(*i*) Justify improvement requirements

(*ii*) Identify and organize specific projects

(*iii*) Identify root causes

(*iv*) Take remedial actions

Identify the drawbacks at each step from the literature.

20. What is the difference between chance and assignable causes? Cite three examples of each. How do you eliminate (*i*) chance causes, and (*ii*) assignable causes from a system?

There are two types of causes: common or natural and special or assignable causes for the problems that one may encounter in processes. While special causes refer to problems that arise because something unusual has occurred, the common causes refer to problems that are inherent to the system. Examples of special causes are problems due to poor quality from an unqualified vendor or use of an improper tool. With common causes, the system itself is the problem. Examples of common causes are inherent machine variability or worker capability. These problems are controllable only by management. Both Deming and Juran have claimed that about 85% of problems have common causes. Hence, only appropriate actions on the part of management can eliminate them. It is up to management to provide the necessary authority and tools to the workers so that the common causes can be removed.

There are numerous statistical techniques such as control charting that identifies special causes and common causes in processes. Variations outside the control limits are attributed to special causes. These variations are worker-controllable, and the workers are responsible for eliminating these causes. On the other hand, variations within the control limits are viewed as the result of common causes. These variations require management action. Juran says that special causes create 'sporadic problems' and common causes create 'chronic problems', and provides detailed guidelines for identifying sporadic problems.

Once the process performance is explained in terms of these causes, improvement actions may be implemented, and these tasks become the main responsibility of the design team during product development. The design team needs to work in close collaboration with quality improvement team so that benefits outweigh significantly the cost, and the product development cost is held at a minimum level.

21. What are the possible actions you may take for product development if the existing 'manufacturing' system is found to be inadequate and incapable?

Actions:

22. Management must reevaluate

23. Process must be standardized, measured, and evaluated

24. Every employee of the organization should be thoroughly trained

25. How is the quality of a product updated, and what sort of activities you may carry out for this? Discuss.

A number of fundamental principles that need to be addressed and implemented in this context are as follows:

(*i*) Everybody in an organization is required to control and improve the process(es) for which he or she is held responsible, and those closest to a process should participate in its management.

(*ii*) Natural or random variation in performance is present in all processes. Engineering design and control methods which fail to take randomness in measurements into account lead to out-of-specification products and high production costs.

(*iii*) All organizations must give top priority to fulfillment of customer requirements, and in today's market, customers demand highly reliable and low-cost quality products. Use of the state-of-the-art 'hard' technology may not result in increased market share for the product.

(*iv*) Before a new and quality product is sold to the customers, experimentation of product performance under different conditions is a must, and hence, experimental design is a valid and most sought-after tool in the process of product development.

(*v*) Use of statistical process control ensures prevention of occurrence of defects and defectives in processes, and the only way to increase market share and profits of an organization is by designing quality into all products and processes.

26. There are several dimensions of quality of a product. How are these dimensions be incorporated during product development process?

For quality assurance, at product design stage, Taguchi's three-step approach addresses the following tasks:

(*i*) ***System design:*** engineering and scientific knowledge is applied to develop a 'prototype' design which meets customer requirements. Selections of parts, components, and materials, as well as manufacturing technology are made. The emphasis here is on using the best available technology to meet customer requirements at lowest cost. In order to be a successful product designer, Taguchi recommends that time-tested and well-proven technology and low-cost parts should be used, and customer requirements need to be focused.

(*ii*) ***Parameter design:*** optimal settings of product parameters are determined at this step. The main objective is to minimize manufacturing and product lifetime costs by minimizing performance variation. This involves making the product design robust, *i.e.,* making the product performance insensitive to noise factors. A noise factor is an uncontrollable source of variation in the functional characteristics of a product. Taguchi identifies three types of noise factors: external noise, or variation in environmental conditions, such as dust, temperature, humidity, or supply voltages; internal noise, or deterioration, such as product wear, material aging, or other changes in components or materials with time or use; and unit-to-unit noise, which is differences in product units built to the same specifications caused by variability in materials, manufacturing equipment, and assembly processes. The parameter design step involves use of experimental designs to determine the impact of both controllable and uncontrollable (noise) factors on product characteristics. The goal is to set controllable factors at levels which will make the product robust with respect to noise factors. Examples of robust design include an automobile part which can withstand shock and vibration (external noise), or a food product with a long shelf life (internal noise), or a replacement part that will fit properly (unit-to-unit noise). Parameter design is a critical step in building quality engineering, and lies at the heart of Taguchi's quality engineering concept.

(*iii*) ***Tolerance design:*** tolerances around the target or nominal values of the parameters established during parameter design are determined at this step. The goal is to set tolerances as wide as possible so that the manufacturing costs are held at a minimum level, while keeping the product's functional characteristics within specified bounds.

For quality assurance at the process design stage, Taguchi's three-step approach requires completion of the following tasks:

(*i*) ***System design:*** the manufacturing process on the basis of knowledge of the product and current manufacturing technology is selected.

(*ii*) ***Parameter design:*** appropriate levels for controllable production process parameters are determined at this step. The goal here is to make the process robust, *i.e.,* to minimize the effects of uncontrollable noise factors on the performance of production process and finished product. Many types of experimental designs are usually used during this step.

(*iii*) ***Tolerance design:*** usually tolerances for the identified critical process parameters are determined. If the product or process parameter design steps are poorly done, it may be necessary here to tighten tolerances or specify higher-cost materials or better equipment. This may however increase the manufacturing costs.

During on-line quality control, Taguchi identifies three forms of production quality control methods. These are as follows:

(*i*) **Process diagnosis and adjustment:** the process is monitored at regular intervals and adjustments and corrections are made as needed.

(*ii*) **Prediction and correction:** a quantitative process parameter is measured at regular intervals. The data are used to project trends in the process. Whenever the process is found to be too far off-target, the process is adjusted to correct the situation. This method is also called 'feedback' or 'feed forward' control.

A number of activities need to be carried out during on-line customer relations building up. This may involve repair or replacement of defective products, compensation for losses, and complaint handling. Information on types of complaints and failures, and customer perceptions of products should be collected at periodic intervals and promptly fed back to relevant functions within the organization for corrective action.

CHAPTER 5

Engineering Economic Analysis

5.1. A bank loans Mr. Jones \$10,000 with an interest rate of 8% compounded annually. How much will Mr. Jones pay the bank if he pays at the end of 6 years?

Solution:

$$F = P\,(F/P, 8\%, 6)$$
$$= \$10{,}000(1.5869) = \$15{,}869$$

5.2. Linda deposits \$10,000, \$5,000 and \$20,000 at $t = 1$, 2 and 3 years, respectively. If the fund pays 8% compounded per period, what sum will be accumulated in the fund at (*a*) $t = 3$ and (*b*) $t = 5$?

Solution:

$$F_3 = \$10{,}000(F/P, 8, 2) + \$5{,}000(F/P, 8, 1) + \$20{,}000$$
$$= \$10{,}000(1.1664) + \$5{,}000(1.0800) + \$20{,}000$$
$$= \$11{,}664 + \$5{,}400 + \$20{,}000 = \$37{,}064$$

5.3. If a savings account pays 5% compounded annually, what single deposit now will accumulate \$10,000 at the end of the tenth year? If the fund pays 6% compounded annually, what single deposit is required now in order to accumulate \$5,000 at the end of the tenth year?

Solution:

(*a*) $$P = \$10{,}000(P/F, 5, 10)$$
$$= \$10{,}000(1.05)^{10} = \$10{,}000(0.6139) = \$6{,}139$$

(*b*) $$P = \$5{,}000(P/F, 6, 10)$$
$$= \$5{,}000(0.5584) = \$2{,}792$$

5.4. John decides to place his annual company bonus of \$1,000 in a fund that pays 8% compounded annually. How much money will be in the fund immediately after the 30th deposit?

Solution:

$$F = \$1{,}000(F/A, 8, 30) = \$1{,}000(113.2832) = \$113{,}283.20$$

5.5. Find the annual payments needed to accumulate \$10,000 over 10 years at a compound interest rate of 10%.

Solution:

$$A = \$10{,}000(A/F, 10, 10) = \$10{,}000(0.0627) = \$627$$

5.6. (*a*) A sum of \$10,000 is deposited into an account that pays interest at a rate of 10% compounded annually. If ten equal annual withdrawals are made from this account with the first withdrawal occurring one year after the deposit, how much money can be withdrawn each year in order to deplete the fund with the last withdrawal?

(*b*) If the first withdrawal is delayed for two years, how much money can be withdrawn each of the 10 years?

Solution:

(*a*) *A* and *P* are related by

$$A = P(A/P, i, n) = P\,(A/P, 10, 10) = \$100{,}000(0.1627) = \$16{,}270$$

(*b*) The amount in the fund at $t = 2$ equals to

$$V_2 = P(F/P, 10, 2) = \$100{,}000(1.2100) = \$121{,}000$$

Hence, the size of the equal annual withdrawals will be

$$A = V_2\,(A/P, 10, 10) = \$121{,}000(0.1627) = \$19{,}686.70$$

Therefore, delaying the first withdrawal for 2 years increases the size of each withdrawal by \$19,686.70 – \$16,270 = \$3,415.70

5.7. Mr. Jones wishes to deposit a single sum of money in a fund so that ten equal annual withdrawals of \$1,000 can be made before depleting the fund.

(*a*) if the first withdrawal is to occur 1 year after the deposit and the fund pays interest at a rate of 10% compounded annually, how much money should Mr. Jones deposit?

(*b*) if the first withdrawal will not occur until 5 years after the deposit.

Solution:

(*a*) We know that *P* and *A* are related by

$$P = A\,(P/A, 10, 10) = \$1{,}000(6.4116) = \$6{,}144.60$$

Thus, if \$6,144.60 is deposited in a fund paying 10% compounded annually, then Mr. Jones can make ten equal annual withdrawals of \$1000.

(*b*) The value of *P* to be determined occurs at $t = 0$ whereas a straightforward application of the (*P*/*A*, 10, 10) factor will yield a single sum equivalent at $t = 5$. Hence, the value obtained at $t = 5$ must be moved backward in time to $t = 0$. This operation is done easily by using the (*P*/*F*, 10, 5) factor. Hence,

$$P = A\,(P/A, 10, 10)\,(P/F, 10, 5) = \$1000(6.4116)\,(0.6209) = \$3{,}815.18$$

Therefore, deferring the first withdrawal for 5 years reduces the amount of the deposit by \$6,144.60 – \$3,815.18 = \$2,329.42

5.8. The maintenance expense for a heavy-duty earth moving machinery increases by \$1,000 per year over the 10-year life of the equipment. The initial maintenance cost was \$5,000. Determine the present worth equivalent for the maintenance costs at 7% interest rate compounded annually.

Solution:

The cash flow diagram is shown in Fig. P5.8. It can be observed that the cash flow series is made up of the sum of a uniform series of \$5,000 and a gradient series with *G* equal to \$1000.

The present worth amount for the uniform series is

$$P_1 = \$5000(P/A, 7, 10) = \$5000(7.0236) = \$35{,}118$$

The present worth amount for the gradient series is

$$P_2 = G\,(P/G, 7, 10) = \$1000(27.7156) = \$27{,}715.60$$

The present worth of maintenance cost is $P = P_1 + P_2 = \$35{,}118 + \$27{,}715.60 = \$62{,}833.60$.

Referring to Fig. P5.8, the uniform series equivalent of the cash flow series obtained by adding the base series of \$5000 and the uniform series equivalent of the gradient series. Therefore,

$$A = \$5000 + G\ (A/G, 7, 10) = \$5000 + \$1000(3.9461) = \$8,946.10$$

The conversion maintenance costs of a future worth amount are obtained by converting the present worth amount to a future worth amount.

$$F = P\ (F/P, 7, 10) = \$62,833.60(1.9672) = \$123,606.26$$

The above can be obtained by converting the uniform series amount to a future worth amount.

$$F = A\ (F/A, 7, 10) = \$8,946.10(13.8164) = \$123,602.89$$

The difference \$123,606.10 – \$123,602.89 = \$3.37 is due to the round off errors.

5.9. Seven annual deposits were made into a savings fund, which pays interest at a rate of 7% compounded annually. The first was \$900; second, third, fourth, fifth, sixth and the seventh deposits made were \$1,000, \$900, \$800, \$700, \$600, \$500, and \$400 respectively. Determine the accumulated amount in the savings fund right after the seventh deposit.

Solution:

Refer to the cash flow diagram as shown in Fig.P5.9. The cash flow series is the difference between a uniform series of \$1000 and a gradient series of \$100. Hence, the uniform series equivalent of the gradient series is given by

$$A = G\ (A/G, 7, 7) = \$100(2.7304) = \$273.04$$

Hence, a uniform series having a cash flow equal to \$1000 – \$273.04 = \$726.96, is equivalent to the original cash flow series. The future worth equivalent can be obtained from

$$F = A\ (F/A, 7, 7) = \$726.96(F/A, 7, 7) = \$726.96(8.6540) = \$6,291.11$$

5.10. The maintenance costs of an ice-breaking machine are expected to be \$2,000 at the end of the first year of the machine's life and increases 10% per year thereafter. The projected life of the machine is 20 years. If an endowment fund is set up which earns 10% per year compounded annually, determine the amount of money to be deposited initially in this endowment account.

Solution:

$$P = A_1\ (P/A_1,\ 10, 10, 20)$$

From Appendix 8B

$$= \$2000(18.1818) = \$36,363.60$$

5.11. A company invested \$100,000 in a venture with the following cash flow data:

EOY	CF	EOY	CF	EOY	CF
0	– \$100,000	3	20,000	6	21,000
1	15,000	4	25,000	7	20,000
2	20,000	5	25,000	8	15,000

If MARR is 10%, determine the following

(*a*) present worth

(*b*) annual worth

(*c*) future worth

(*d*) internal rate of return

(*e*) external rate of return

(*f*) savings/investment ratio

(*g*) payback period with and without considering the time value of money

Solution:

EOY	CF
0	– $100,000
1	$15,000
2	$20,000
3	$20,000
4	$25,000
5	$25,000
6	$21,000
7	$20,000
8	$15,000

MARR = 0.1

Payback Period

EOY	Cum CF	Cum PW
0	– $100,000	– $100,000
1	– $85,000	– $86,363.64
2	– $65,000	– $69,834.71
3	– $45,000	– $54,808.41
4	– $20,000	– $37,733.08
5	$5,000	– $22,210.05
6	$26,000	– $10,356.09

PW = $6,904.68

AW = $1,294.24

FW = $14,800.80

IRR = 11.83%

ERR = 10.92%

SIR = 1.07

PBP = 5 yrs

DPBP = 6 yrs

5.12. **An investment of $25,000 was made on a computer system that will last 5 years and a zero salvage value at the end of service life. The operating, maintenance and software costs are estimated to be $ 10,000 the first 3 years and $15,000, the last 3 years. If MARR is 10%, determine the following for this investment.**

(*a*) present worth

(*b*) annual worth

(*c*) future worth

Solution:

EOY	CF
0	– $25,000
1	– $10,000
2	–$10,000
3	– $10,000
4	– $15,000
5	– $15,000
6	– $15,000

MARR = 0.10

PW = – $77,894.65

AW = – $17,885.19

FW = – $137,995.13

5.13. A drilling machine was purchased for $50,000, kept for 5 years, and sold for $ 8,000. The annual operating and maintenance costs were $5,000. Using a MARR of 12% per year, determine for this investment the following.

(*a*) present worth

(*b*) annual worth

(*c*) future worth

Solution:

EOY	CF
0	– $50,000
1	– $5,000
2	– $5,000
3	– $5,000
4	$3,000

MARR = 0.12

PW = – $60,102.60

AW = – $19,787.85

FW = – $94,572.61

5.14. An investment of $50,000 for new earth moving machinery is being considered. The estimated salvage value of this machinery is $10,000 at the end of the estimated life of 6 years. The annual income each year for the 6 years is $10,000. Annual operating expresses are $3,000. If the money is worth 15% compounded annually, determine the following measures of investment worth in deciding whether this machinery should be purchased.

(*a*) present worth

(*b*) annual worth

(*c*) future worth

(*d*) internal rate of return

(*e*) external rate of return

(*f*) savings/investment ratio.

Solution:

EOY	CF
0	$50,000
1	$7,000
2	$7,000
3	$7,000
4	$7,000
5	$7,000
6	$17,000

MARR = 0.15

PW = $5,625.43

AW = $1,486.45

FW = $13,011.96

IRR = 24.20%

ERR = 19.85%

SIR = 1.28

5.15. A Company purchases a welding machine for $15,000; the machine is used for 4 years and is then sold for $5,000. The annual operating, maintenance costs equals $5,000/year over the 4-year period. The welding machine replaced an older system that cost $ 12,000/year. If the MARR is 12%, determine if the decision to buy this welding machine was economically sound. Use the following measures of investment worth.

(*a*) Present worth

(*b*) Annual worth

(*c*) Future worth

(*d*) Internal rate of return

(*e*) External rate of return

(*f*) Savings/investment ratio.

Solution:

EOY	CF
0	– $20,000
1	$6,200
2	$6,200
3	$6,200
4	$6,200
5	$6,200
6	$11,200

MARR = 0.12

PW = $8,023.88

AW = $1,951.61

FW = $15,837.72

IRR = 24.20%

ERR = 18.48%

SIR = 1.40

5.16. **If a fund pays 15% compounded annually, how much deposit is needed today such that $2,000 can be withdrawn every year forever?**

Solution:

$2,000/15 = $13,333.54

5.17. **If a fund pays 15% compounded annually, what deposit is required today such that $5,000 can be withdrawn every 4 years forever?**

Solution:

EOY	CF
0	$0
1	$0
2	$0
3	$0
4	$5,000

MARR = 0.15

PW = $2,858.77

AW = $1,001.33

Endowment = $6,675.51

5.18. **The maintenance costs of a numerically controlled milling machine are cyclic with the following costs occurring over a 5-year period: $5,000, $3,000, $4,000, $0, and $2,000. This sequence is expected to repeat itself every 5 years forever. Find the capitalized cost of the maintenance costs based on a time value of money of 12%.**

Solution:

EOY	CF
0	$0
1	$5,000
2	$3,000
3	$4,000
4	$0
5	$2,000

MARR = 0.12

PW = – $1,093,738.99

AW = – $205,271.72

Endowment = – $1,710,597.65

5.19. **A Company can invest in a venture, which costs $200,000 and returns $80,000/year at the end of 4 years. This investment can be renewed perpetually every 4 years. If the company's MARR is 12%, find the capitalized worth of an infinite series of these investments.**

Solution:

EOY	CF
0	– $200,000
1	$80,000
2	$80,000
3	$80,000
4	$80,000

MARR = 0.12

PW = $42,987.95

AW = $14,153

CW = $117,942.61

5.20. Determine the capital recovery cost of a high-speed computer system costing $300,000, having a life of 10 years and salvage value of $ 50,000. Use an interest rate of 12%.

Solution:

EOY	CF
0	– $300,000
1	$0
2	$0
3	$0
4	$0
5	$0
6	$0
7	$0
8	$0
9	$0
10	$50,000

CR = $51,961

MARR = 12%

5.21. Determine the capital recovery cost of a milling machine having a first cost of $70,000, a salvage value of $10,000, and a life of 10 years. Use an interest rate of 10%.

Solution:

EOY	CF
0	– $70,000
1	$0
2	$0
3	$0
4	$0
5	$10,000

CR = $20,104

5.22. A molding machine for a manufacturing company is purchased at the beginning of the fiscal year for $30,000. The estimated salvage value after 10 years is $3,000. Calculate the depreciation deduction and the resulting unrecovered investment during each year of the molding machine's life using.

(*a*) straight-line depreciation method

(*b*) double declining balance method

(*c*) sum of the year's digit depreciation method.

Solution:

(*a*)

End of year	SL deducation	Unrecovered investment
0		30,000.00
1	2,700.00	27,300.00
2	2,700.00	24,600.00
3	2,700.00	21,900.00
4	2,700.00	19,200.00
5	2,700.00	16,500.00
6	2,700.00	13,800.00
7	2,700.00	11,100.00
8	2,700.00	8,400.00
9	2,700.00	5,700.00
10	2,700.00	3,000.00

(*b*)

End of year	DDB deducation	SL deducation	DDB/SL deducation	Unrecovered investment
0				30,000.00
1	6,000.00	$2,700.00	6,000.00	24,000.00
2	4,800.00	$2,333.33	4,800.00	19,200.00
3	3,840.00	$2,025.00	3,840.00	15,360.00
4	3,072.00	$1,765.71	3,072.00	12,288.00
5	2,457.60	$1,548.00	2,457.60	9,830.40
6	1,966.08	$1,366.08	1,966.08	7,864.32
7	1,572.86	$1,216.08	1,572.86	6,291.46
8	1,258.29	$1,097.15	1,258.29	5,033.16
9	1,006.63	$1,016.58	1,016.58	4,016.58
10	805.31	$1,016.58	1,016.58	3,000.00

(*c*)

End of year	SOYD deduction	Unrecovered investment
0		30,000.00
1	4,909.09	25,090.91
2	4,418.18	20,672.73
3	3,927.27	16,745.45
4	3,436.36	13,309.09
5	2,945.45	10,363.64
6	2,454.55	7,909.09
7	1,963.64	5,945.45
8	1,472.73	4,472.73
9	981.82	3,490.91
10	490.91	3,000.00

5.23. On January first a printing press was purchased for \$50,000. The press has a 5-year useful life and a \$5,000 expected salvage value. Compute yearly depreciation using:

(*a*) straight-line depreciation

(*b*) sum of year's digit depreciation

(*c*) double declining balance depreciation.

Solution:

(*a*) Straight line depreciation

Annual depreciation charge = (\$50,000 – \$5,000)/5 = \$9,000

(*b*) Sum of years digit depreciation

Sum of years digits = (5/2) (5 + 1) = 15

1st year = 5/15(50,000 – 5,000) = \$15,000

2nd year = 4/15(45,000) = \$12,000

3rd year = 3/15(45,000) = \$9,000

4th year = 2/15(45,000) = \$6,000

5th year = 1/15(45,000) = \$3,000

Check \$45,000

(*c*) Double declining balance depreciation

1st year = 2/5(50,000) = \$20,000

2nd year = 2/5(50,000 – 20,000) = \$12,000

3rd year = 2/5(30,000 – 12,000) = \$7,200

4th year = 2/5(18,000 – 7,200) = \$4,320

5th year = 2/5(10,800 – 4,320) = \$2,592

5.24. A front-end loader cost \$80,000 and has an estimated salvage value of \$20,000 at the end of 5 years useful life. Calculate the depreciation schedule, and book value, to the end of the useful life of the front-end loader by

(*a*) straight-line depreciation

(*b*) sum of years digit depreciation

(*c*) double declining balance depreciation.

Solution:

(*a*) Straight line depreciation

$$D_c = (P - S)/N = (80,0000 - 20,000)/5 = \$12,000 \text{ per year}$$

Year	D_c	BV
1	12,000	80,000 – 12,000 = 68,000
2	12,000	80,000 – 24,000 = 56,000
3	12,000	80,000 – 36,000 = 44,000
4	12,000	80,000 – 48,000 = 32,000
5	12,000	80,000 – 60,000 = 20,000
Total	$60,000	

(*b*) Sum of years digits depreciation

$$SOYD = [N/2]\ (N + 1) = [5/2]\ (5 + 1) = 15$$

$$D_c = [N/SOYD]\ (P - S)$$

Year	D_c	BV
1	5/15(80,000 – 20,000) = 20,000	80,000 – 20,000 = 50,000
2	4/15(60,000) = 16,000	80,000 – 36,000 = 44,000
3	3/15(60,000) = 12,000	80,000 – 48,000 = 32,000
4	2/15(60,000) = 8,000	80,000 – 56,000 = 24,000
5	1/15(60,000) = 4,000	80,000 – 60,000 = 20,000
Total	$60,000	

(*c*) Double declining balance depreciation

$$DDB = [2/N]\ [P - \Sigma D_c \text{ (to date)}]$$

Year	DDB	ΣDDB	BV
1	2/5(80,000 – 0)		80,000 – 32,000 = 48,000
2	(2/5)(80,000 – 32,000) = 19,200	51,200	80,000 – 51,200 = 28,800
3	2/5(80,000 – 51,200) = 11,520	62,720	80,000 – 62,720 = 17,280
4	2/5(80,000 – 62,720) = 6,912	69,632	80,000 – 69,632 = 10,368
5	2/5(80,000 – 69,632) = 4,147	73,779	80,000 – 73,779 = 6,221

Depreciated to end of useful life means to convert to straight line.

3 Year: D_c (SL) = [BV (BOY) – S]/RUL = [28, 8000 – 20,000]/3 = $2,933.33 which is less than $11,520

4 Year: D_c (SL) = [BV (BOY) – S]/RUL= [17,280 – 20,000]/2 = – $1,360/year

Year	DDB	BV = P – ΣD_c
1	32,000	80,000 – 32,000 = 48,000
2	19,200	80,000 – 51,200 = 28,800
3	11,520	80,000 – 62,720 = 17,280
4	– 1,360	80,000 – 61,360 = 18,640
5	– 1,360	80,000 – 60,000 = 20,000
	$60,000	

5.25. Determine, which is best from the following two depreciation schedules assuming an interest rate of 6% per year.

Year	SOYD	Double digits/ Straight line
1	$16,000	19,000
2	14,000	14,000
3	12,000	10,000
4	10,000	8,000
5	8,000	6,000
6	6,000	5,000
7	4,000	5,000
8	2,000	5,000
Total	$72,000	$72,000

Solution:

Year	SOYD	DDB/SL
1	$16,000	$19,000
2	14,000	14,000
3	12,000	10,000
4	10,000	8,000
5	8,000	6,000
6	6,000	5,000
7	4,000	5,000
8	2,000	5,000
	$72,000	$72,000

The best depreciation schedule would be the one that depreciates the asset most rapidly. In other words, it is the depreciation schedule whose Present Worth is greatest.

SOYD:

$$P = 16{,}000(P/A, 6\%, 8) - 2{,}000(P/G, 6\%, 8)$$
$$= 16{,}000(621) - 2{,}000(19.842) = \$59{,}676$$

DDB/SL:

$$P = 19{,}000(P/F, 6, 1) + 14{,}000(P/F, 6, 2) + 10{,}000(P/F, 6, 3) + 8{,}000(P/F, 6, 4) + 6{,}000(P/F, 6, 5) + 5{,}000(P/F, 6, 3)\ (P/F, 6, 5)$$
$$= 19{,}000(0.9434) + 14{,}000(0.8900) + 10{,}000(0.8396) + 8{,}000(0.7921) + 6{,}000(0.7473) + 5{,}000(2.6730)\ (0.7473)$$
$$= \$59{,}588$$

Hence select the sum of year's digit depreciation.

5.26. From the two depreciation schedules for a computer laser printer given in Table P5.26, which method of depreciation should be selected assuming an interest rate of 12% per year.

Year	DDB	SOYD
1	300	286
2	200	230
3	150	180
4	100	130
5	100	85
6	100	39
	950	950

Solution:

We want to depreciate its assets as rapidly as possible. In situations where the choice between methods is not obvious, then computations are required. Select the depreciation method with the largest Present Worth of depreciation charges.

DDB with/conversion to SL

Year	DDB P/F,12%,n	
1	300 × 0.8929 = 267.87	
2	200 × 0.7972 = 159.44	
3	150 × 0.7118 = 106.77	
4	100 × 0.6355 = 63.55	
5	100 × 0.5674 = 56.74	
6	100 × 0.5066 = 50.66	
	950	705.03

SOYD

Year	DDB	P/F,12%,n
1	286 × 0.8929	= 255.37
2	230 × 0.7972	= 183.36
3	180 × 0.7118	= 128.12
4	130 × 0.6355	= 82.61
5	85 × 0.5674	= 48.23
6	39 × 0.5066	= 19.76
	950	669.22

Hence, select the DDB with conversion to SL.

5.27. A Company purchased a computer system for \$ 50,000. It is expected to last 14 years and has an estimated salvage value of \$6,000. Determine the depreciation charge on the computer system for the fourth year of its service life and the book value at the end of 8 years, using sum of digits depreciation method.

Solution:

(*a*) SOYD depreciation for 3rd year.

Sum of Years digits = $(n/2)(n+1) = (14/2)(14+1) = 105$

3rd year depreciation = (Remaining life beginning of year($P - S$))/[ΣYear digits]
= [12/105](50,000 – 6,000)
= $5,028.57

(*b*) Book value at the end of 8 years

Σ8 years of depreciation = [(14 + 13 + 12 + 11 + 10 + 9 + 8 + 7)/105](50,000 – 6,000)
= [84/105](44,000) = $35,200

Book Value = Cost – Depreciation to date = $50,000 – $35,200 = $14,800

5.28. By purchasing a machine for $30,000, a large company in the 50% income tax bracket was able to save $5,000 during year 1, $4,000 during year 2, $3,000 during year 3, $2,000 during year 4 and $1,000 during year 5. Sum of year's digits depreciation method was used to depreciate the machine over its four-year life and a zero salvage value was assumed. The company would like to sell the machine at the end of year 5. Assuming that the capital gains/losses are taxed at a rate of 28% and the machine qualities for a 10% investment tax credit, determine the resale value that would yield a 10% after-tax rate of return for the company.

Solution:

Depreciation:

SOYD = 4(5)/2 = 10

Year	SOYD depreciation
1	4/10(30,000 – 0) = 12,000
2	3/10(30,000 – 0) = 9,000
3	2/10(30,000 – 0) = 6,000
4	1/10(30,000 – 0) = 3,000

ATCF Calculations:

Year	BTCF	SOYD depr	Taxable income	50% tax	ATCF
0	–30,000	—	10% investment tax credit	+ 3,000	– 27,000
1	5,000	–12,000	–7,000	+ 3,500	8,500
2	4,000	9,000	–5,000	+ 2,500	6,500
3	3,000	6,000	–3,000	+ 1,500	4,500
4	2,000	3,000	–1,000	+ 500	2, 500
				– 500	
5	1,000	—	1,000	0.28R	500
	R	—	capital gains of R taxed at 28%		0.72 R

Solve for *R*:

$$27{,}000 = 8{,}500(P/A, 10\%, 4) - 2{,}000(P/G, 10\%, 4) + (500 + 0.72R)(P/F, 10\%, 5)$$
$$27{,}000 = 8{,}500(3.1699) - 2{,}000(4.3781) + (500 + 0.72R)(0.6209)$$
$$27{,}000 = 18{,}498.40 + 0.4471R$$

Hence $R = 8501.60/0.4471 = \$19{,}014.99$

5.29. A company has a gross income in one year of $20 million. The operating expenses are $12 million. Depreciation is $3 million. In this year there is a depreciation recapture of

$1 million realized when the company sold a machine that is no longer needed and sold for more than its book value.

(*a*) determine the company's federal income taxes

(*b*) determine the average federal tax rate

(*c*) if the state tax rate is 10 percent, determine the total income taxes paid.

Solution:

(*a*) Taxable income (TI) = Gross income – operating expenses – depreciation + depreciation recapture

$$TI = 20 - 12 - 3 + 1 = \$ 6 \text{ million}$$

Taxes = (TI range)(Marginal Rate)

Refer to Table 5.3

Taxes = (50,000)(0.15) + 25,000)(0.25) + (25,000)(0.34)
+ (235,000)(0.39) + (6 million – 335,000)(0.34)
= 7500+6250 + 8500 + 91,650 + 1,926,100 = $2,040,000

(*b*) Average federal tax rate = $2,040,000/$6,000,000 = 0.34

(*c*) Effective tax rate = 0.10 +(1 – 0.10)(0.34) = 0.10 + 0.306 = 0.406

Total income taxes = $6,000,000(0.406) = $1,836,000

5.30. A project can be summarized by the data given in Table P5.30. The company uses straight-line depreciation, pays an incremental income tax rate of 30% and requires an after-tax rate of return of 10%. Ignore any investment tax credit.

Item	Amount
Life	30 years
Investment	$100,000
Salvage	$25,000
Revenues	$26,000/year
Operating Expenses	$13,500/year

(*a*) using Net Present Worth, determine whether the project should be undertaken

(*b*) if the company used Sum-of-Years-Digits depreciation, is it possible the decision would change? (No computations needed).

Solution:

(*a*) Depreciation = [75,000 – 15,000]/30 = 2000/year.

Year	BTCF	Depreciation income	Taxable @30%	Taxes	ATCF
0	– 75,000	—	—	—	– 75,000
1 – 30	12,500	– 2,000	10,500	3,150	9,350
S	15,000	—	—	—	15,000

NPW = 9,350(*P*/*A*, 10%, 30) + 15,000(*P*/*F*, 10%, 30) – 75,000
= 9,350(9.4269) + 15,000(0.0573) – 75,000 = $14,001.02

Since NPW is > 0, we select the project.

(*b*) No. Although total depreciation is the same, SOYD is larger in the early years when it is worth more. Hence, the NPW would increase with SOYD making the project much more attractive or desirable.

5.31. A company is considering an investment in computer controlled manufacturing system to reduce its production cost. The initial cost would be $15,000 and over a 10-year period it is estimated that the company would save $2,500 annually in production costs. The salvage value of the system is estimated at $3,000. Determine the after-tax rate of return.

Solution:

Before – Tax rate of return:

Cash flow is the net profit or savings for each year. We will use straight-line depreciation to determine the depreciation charge.

Depreciation = [$15,000 – $3,000]/10 = $1,200

Table P5.31 shows the cash flow results. The before-tax rate of return is the interest rate at which the before-tax cash flow savings just equals the purchase cost of the computer controlled manufacturing system.

$$\$15{,}000 = \$2{,}500(P/A, i, 10) + \$3{,}000\ (P/F, i, 10)$$

For $i = 15\%$

$$\$15{,}000 = \$2{,}500(5.018) + \$3{,}000(0.2472) = \$12{,}547 + \$741.60 = \$13{,}288.60$$

For $i = 12\%$

$$\$15{,}000 = \$2{,}500(5.6502) + \$3{,}000(0.3220) = \$14{,}125.50 + \$966 = \$15{,}091$$

Therefore, the before-tax rate of return is slightly more than 12%.

After-Tax rate of return:

We use the after-tax cash flow in Table P5.31.

We estimate the after-tax rate of return of 6%

For $i = 6\%$

$$\begin{aligned}\$15{,}000 &= 1{,}850(P/A, i, 10) + 3{,}000(P/F, i, 10)\\ &= 1{,}850(7.3601) + 3{,}000(0.5584) = 13{,}616.19 + 1675.20\\ &= \$15{,}291 (i \text{ too low})\end{aligned}$$

For $i = 7\%$

$$\begin{aligned}\$15{,}000 &= 1{,}850(7.0236) + 3{,}000(0.5083)\\ &= 12{,}993.66 + 1{,}524.90 = \$14{,}518.56\ (i \text{ is too high})\end{aligned}$$

Table P5.31

Cash Flow Calculations

Year	Before-tax cash flow	Depreciation	Taxable income	50% Income tax	After-tax cash flow
0	–15,000				–15,000
1 to 9	2,500	1,200	1,300	– 650	1,850
10	2,500	1,200	1,300	– 650	1,850
	3,000				3,000

Therefore

$$i = 6\% + 1\%[15{,}291 - 15{,}000]/[15{,}291 - 14{,}518.56] = 6\% + 1\%(0.3767) = 7.38\%$$

5.32. A company expects to receive $60,000 each year for 15 years from the sale of a project. The initial investment will be $300,000. The expenses of manufacturing and selling the project will be $10,000 per year. Assume straight-line depreciation, a 15-year useful life and no salvage value. Use a 48% income tax rate and ignore the investment tax credit. Determine the projected after-tax rate of return.

Solution:

Straight line depreciation = $[P - F]/n$ = [300,000 – 0]/15 = $20,000 per year

Year	Before-tax cash flow	Straight line depreciation	Taxable income	48% income taxes	After-tax cash flow
0	– 300,000				– 300,000
1–15	+ 50,000	20,000	30,000	14,400	+35,600

Take the after tax cash flow and compute the rate of return at which PW of costs equals PW of benefits

$$\$300,000 = \$30,400(P/A, i, 15) \text{ or } (P/A, i, 15) = 300,000/35,600 = 8.42697$$

From interest tables and by interpolation i = 8.4%

5.33. A venture by a firm requires $2 million now, with an investment tax credit of $40,000 at the end of the first year and no salvage value. The receipts and expenses (excluding depreciation) are:

End year	Receipts, $(in thousands)	Expenses, $(in thousands)
1	3,500	2,000
2	3,800	2,800
3	2,500	1,800

A working capital of $400,000 is invested at zero time and the income tax rate is 46%. Determine the discounted cash flow rate of return assuming sum-of-the-years-digits depreciation.

Solution:

Table P5.33 shows the cash flow calculations.

Table P5.33

Cash Flow Calculations (in thousands of dollars)

	Time 0	End year 1	End year 2	End year 3
1. Receipts		3,500	3,800	2,500
2. Expenses, excluding depreciation		– 2,000	– 2,800	– 1,800
3. Receipts for tax purpose (1) + (2)		1,500	1,000	700
4. Depreciation				
(3/6)(2,000)		1,000		
(2/6)(2,000)			666	
(1/6)(2,000)				334
5. Taxable income (3) – (4)		500	334	366

6. Tax at 46%		230	153.64	168.36
7. Fixed capital	– 2,000			
8. Working capital	– 400			400
9. Investment tax credit		40		
10. Cash flow (3) – (6) + (7) + (8) + (9)	– 2,400	1,310	846.36	931.64

For the present value of the cash flow to be zero

$$P = 0 = -\$2{,}400 + 1310/(1 + r) + 846.36/(1 + r)^2 + 931.64/(1 + r)^3$$

$$P = \begin{cases} -\$8.33 \text{ for } r = 0.15 \\ +\$29.21 \text{ for } r = 0.14 \end{cases}$$

or by interpolation, $P = 0$ for $r = 14 + 1\%(29.21/(29.21 + 8.33)) = 14.78\%$.
Hence, the discounted cash flow rate of return is 14.78%.

5.34. **Compare project A with an initial investment of \$2000K, a 5-year service life, and a net present value (NPV) of \$500K, with project *B*, having an initial investment of \$3500K a 9-year service life, and a net present value of \$1500K. Both present values are at 10% per year. Also, if the initial investment is \$1K for both projects, make a correction for the difference in size and duration by changing to a premium unit cost.**

Solution:

Project A:

$$(500/2000)(A/P, 10\%, 5) = 0.25(0.2638) = 0.06595$$

Project B:

$$(1500/3500)(A/P, 10\%, 9) = 0.4286(0.1736) = 0.07440$$

From the above, we notice that project *A* has a premium of 0.06595K\$ per year per 1K\$ of original investment against 0.07440 for project *B*. Therefore, project *B* is more profitable than project *A* at 10% per year. The profitability however changes with the interest rate used.

5.35. **An initial capital investment is \$400,000 and has a 7-year life and \$100,000 salvage value. If the working capital is \$50,000 and the total net profit after taxes over 7-years is estimated at \$200,000, find the *ROI*.**

Solution:

Annual net profit = \$200,000/7 = \$28,571.43

ROI on initial capital investment = \$28,571.43/(\$400,000 + \$50,000) = 0.0635

5.36. **Find the discounted cash flow rate of return for the data in Problem 5.33.**

Solution:

Referring to Eq. (5.93), the present value of all cash flows must be zero. Hence

$$\text{NPV} = -2000/(1 + r)^0 + 700/(1 + r)^1 + 800/(1 + r)^2 + 500/(1 + r)^3 + 400/(1 + r)^4 + 300/(1 + r)^5$$

We find that

$$\text{NPV} = 43.09 \text{ for } r = 0.12$$

$$\text{NPV} = -79.77 \text{ for } r = 0.15$$

By interpolation

$$\text{NPV} = 0 \text{ for } r = 0.012 + (0.15 - 0.12)[43.09/(43.09 + 79.77)] = 0.12 + 0.0105 = 0.1305$$

Hence, the discounted cash flow rate of return is 13.05% per year.

5.37. A machine has a first cost of $20,000 and a salvage value of $5,000 after a 5-year life. Annual benefits (savings) from its use are $8,000, and the annual cost of operation is $4,000. The tax rate is 50 percent. Find the DCF rate of return.

Solution:

The annual depreciation charge using straight-line depreciation is

$$D = [P - F]/n = [20,000 - 5,000]/5 = \$3,000$$

The annual cash flow after taxes is the sum of the net receipts and depreciation.

Annual cash flow = ($8,000 – $4,000)(1 – 0.50) + $3,000(0.50) = $3,5000

Year	Cash flow
0	– 20,000
1	3,500
2	3,500
3	3,500
4	3,500
5	3,500 + 5,000(salvage value)

NPW = 0 = – 20,000 + 3500(P/A, i, 5) + 5,000(5,000(P/F, i, 5)

Try i = 5%NPW = – 20,000 + 3,500(4.3295) + 5,000(0.7835) = – $929.25

Try i = 4%NPW = – 20,000 + 3,500(4.4518) + 5,000(0.8219) = – $309.2

Try i = 3%NPW = – 20,000 + 3,500(4.5797) + 5,000(0.8626) = $341.95

Hence i = 3% + (4% – 3%)[341.95/(341.95 + 309.20)] = 3.525

DCF rate of return = 3.53%

5.38. A company in the 50% income tax bracket, is considering the purchase of a new machinery that qualifies for a 10% investment tax credit. The new machine will yield benefits of $10,000 for year 1, $15,000 for year 2, $20,000 for year 3, $20,000 for year 4, and $20,000 for year 5.

The machine is to be depreciated using the Accelerated Cost Recovery System (ACRS) over three years. The percentages 33%, 45% and 22% respectively, for years 1, 2 and 3. The company expects to sell the machinery at the end of five years of use for 25% of the original purchase price.

Determine the maximum purchase cost the company can pay if it requires a 10% after-tax rate of return? The capital gains/losses are taxed/refunded at a rate of 28%.

Solution:

Year	BTCF	ACRS depr	Taxable income	50% tax	ATCF
0	– P	—	Investment Tax credit	+ 0.10P	– 0.90P
1	10,000	0.33P	10,000 – .33P	5,000 + .165P	5,000 + .165P
2	15,000	0.45P	15,000 – .45P	7,500 + .225P	7,500 + .225P
3	20,000	0.22P	20,000 – .22P	10,000 + .110P	10,000 + .110P
4	20,000	0	20,000	10,000	10,000
5	20,000	0	20,000	10,000	10,000
	0.25P		0.25P Capital gain taxed at 28%	0.07P	0.18P

$$0.90P = 5{,}000(P/A, 10\%, 5) + 2{,}500(P/G, 10\%, 3) + 5{,}000(P/A, 10\%, 2)(P/F, 10\%, 3) + 0.18P(P/F, 10\%, 5)$$
$$= 5{,}000(3.7908) + 2{,}500(2.3291) + 5{,}000(1.7355)(0.7513) + 0.18P(0.6209)$$
$$0.90P = 18{,}954 + 5{,}822.75 + 6{,}519.41 + 0.11176P$$
$$0.90P = 31{,}296.16 + 0.11176P \quad \text{or} \quad P = 39{,}703.85$$

5.39. A company is considering building a machine using two different materials *A* & *B*. The machine could be fabricated using material *A* with an initial cost of $30,000 and will last 15 years. The maintenance expenses are $1000 per year. If material *B* is used, the annual maintenance cost would be $100 per year and it's life also 15 years. At the end of 15 years there is no salvage value for either machine. The company pays income tax at a rate of 48% and uses straight-line depreciation. There is no investment tax credit. If the minimum acceptable after-tax rate of return is 10, determine the maximum amount that should be spent on the machine fabricated using material *B*.

Solution:

Material A:

Year	BTCF	Depreciation	Tax, Inc.	Taxes@48%	ATCF
0	– 30,000	—	—	—	– 30,000
1–15	– 1,0000	2,000	3,000	– 1,440	+ 440
S	0	—	0	0	0

Material B:

Year	BTCF	Depreciation	Tax, Inc.	Taxes@48%	ATCF
0	– *P*	—	—	—	– *P*
1–15	– 100	*P*/15	– 100 – *P*/150	(– 48 – 0.032*P*)	– 52 + 0.032*P*
S	0	—	0	0	0

$NPW_A = NPW_B$ for breakeven

$$440(P/A, 10, 15) - 30{,}000 = (-52 + 0.032P)(P/A, 10, 15) - P$$
$$440(7.6061) - 30{,}000 = (-52 + 0.032P)(7.6061) - P$$
$$-26{,}653.32 = -395.52 + 0.2434P - P$$
$$P = (26{,}653.32 - 395.52)/(1 - 0.2434) = 26{,}257.80/0.7566$$
$$P = \$34{,}705$$

5.40. Calculate the average annual rate of inflation based on the CPI from the end of:

(*a*) 1990 through 1998
(*b*) 1985 through 1996
(*c*) 1970 through 1994
(*d*) 1969 through 1990

Solution:

(*a*) $391.49(1 + \bar{f}\,)^8 = 488.3$ or $0.255(1 - \bar{k}\,)^8 = 0.205$

$$\bar{f} = 0.02803 \quad \text{and} \quad \bar{k} = 0.02691 \quad \text{or} \quad 2.691\%$$

Hence, a 2.803% per-year increase in prices has the same effect as a 2.691% per-year loss in purchasing power.

(*b*) $(1 + f)^{11} = 469.9/322.2 = 1.4584$ or $f = 3.489\%$
(*c*) $(1 + f)^{24} = 444.0/116.3 = 3.8177$ or $f = 5.741\%$
(*d*) $(1 + f)^{21} = 391.4/109.8 = 3.5647$ or $f = 6.239\%$

5.41. **Using the data in Table 5.4, which gives the purchasing power of the dollar in various years, calculate the inflation rate based on consumer prices for the following years.**

(*a*) 1953
(*b*) 1982
(*c*) 1997
(*d*) 1967
(*e*) 1991

Solution:

(*a*) 1953: $(1 + f) = 1.258/1.248 = 1.0080$ or $f = 0.801\%$
(*b*) 1982: $(1 + f) = 0.371/0.356 = 1.0421$ or $f = 4.21\%$
(*c*) 1997: $(1 + f) = 0.271/0.270 = 1.0037$ or $f = 0.37\%$
(*d*) 1967: $(1 + f) = 1.012/1.0 = 1.012$ or $f = 1.2\%$
(*e*) 1991: $(1 + f) = 0.299/0.293 = 1.0205$ or $f = 2.05\%$

5.42. **For the time spans listed, find the average annual inflation rate from the producer price data given in Table 5.5.**

(*a*) end of 1961 through 1991
(*b*) end of 1958 through 1968
(*c*) end of 1971 through 1989
(*d*) end of 1986 through 1999
(*e*) end of 1945 through 1986

Solution:

(*a*) $(1 + \bar{f})^{30} = 1.067/0.293 = 3.6416$ or $\bar{f} = 4.402\%$

(*b*) $(1 + \bar{f})^{10} = 1.073/0.972 = 1.1039$ or $\bar{f} = 0.993\%$

(*c*) $(1 + \bar{f})^{18} = 0.880/0.313 = 2.8115$ or $\bar{f} = 5.911\%$

(*d*) $(1 + \bar{f})^{13} = 0.345/0.264 = 1.3068$ or $\bar{f} = 2.079\%$

(*e*) $(1 + \bar{f})^{41} = 1.832/0.345 = 5.3101$ or $\bar{f} = 4.156\%$

5.43. **Mr. Jones wants to invest some money in a bank for one year. Currently, the bank is paying a rate of 6% compounded annually. Assume that inflation is anticipated to be (*a*) 3% per year, and (*b*) 8% per year for the year Mr. Jones wants to invest. Determine *i*, *f*, and *f*′ in each case.**

Solution:

(*a*) If inflation is 3% per year:

Market rate = i
Inflation rate = f
Real interest rate = i^1

Solving for i^1, we get

$$i = i^1 + f + i^1 f$$

or $$(i - f) = i^1 + i^1 f = i^1(1 + f)$$

or $$i^1 = (i - f)/(1 + f) = (0.06 - 0.03)/(1 + 0.03) = 0.0291 \text{ or } 2.91\% \text{ per year.}$$

This means that Mr. Jones will have 2.91% more purchasing power with the dollars invested in that bank account than he had a year preciously. At the end of the year he can purchase 2.91% more goods and services than he could have at the beginning of the year.

(*b*) If inflation is 8%

As above, we would solve for i[1]

$$i^1 = (i - f)/(1 + f) = (0.06 - 0.08)/1 + 0.08) = -0.0185 \text{ or } -1.85\% \text{ per year}$$

In this case, we note that the real growth in money has decreased by 1.85%, so that Mr. Jones can now purchase 1.85% fewer goods and services with the money he had invested in the bank.

5.44. A charitable establishment is considering replacing its present building with a new facility. When the present building was constructed in 1945, the total cost was 1 million dollars. Also, at that time, a wealthy trust gifted the establishment with 1 million dollars to be used for future replacement. The administrators are now considering building a new facility in the year 2005. Assume the following data:

Inflation is 5% per year from 1945 to 2005. In 1945, the charitable establishment invested the gift at a market interest rate of 10% per year.

(*a*) determine i, i', f, $a\$$ and $r\$$ for this problem

(*b*) how many actual dollars in the year 2005 will the gift be worth?

(*c*) how much would the actual dollars in 2005 be in terms of 1945 purchasing power?

Solution:

Given:

$$F = 5\%(\text{the inflation rate})$$
$$i = 10\%(\text{the market interest rate})$$

(*a*) Therefore, $i^1 = (0.10 - 0.05)/(1 + 0.05) = 0.05/1.05 = 0.0476$ or 4.76%

One million was the cost of the building in 1945. These were the actual dollars($A\$$) spent in 1945.

(*b*) We are going from actual dollars in 1945 to actual dollars in 2005. To do this, we would use the market interest rate and discount this amount forward in time 60 years as shown in Figure P4.44(*a*).

Actual dollars in 2005 = Actual dollars in 1945(F/P,i,60)

= \$1,000,000(F/P, 10%, 60)

= \$1,000,000(304.482) = \$304,482,000

(*c*) Now we want to find the amount of real dollars based in 1945 that occur in the year 2005 which are equivalent to the \$304.482 million from part (*b*).

Method 1: We directly translate the actual dollars in the year 2005 to real 1945 based dollars in the year 2005. We can use the inflation rate to strip 60 years of inflation from the actual dollars. This is done by using the P/F factor for 60 years at the inflation rate. We are not physically moving the dollar amount in time in this case; rather we are simply removing inflation from these dollar amounts one year at a time the P/F factor will do this.

Real 1945 based dollars in 2005 [see Fig. P4.44(*b*)].

= (Actual dollars in 2005) (P/F, f, 60)

= (\$304,482,000)(P/F,5,60)

$= (\$304,482,000)(0.0535)$
$= \$16,289,787$

Method 2: In this method, we start by realizing that the actual dollars in 1945 are exactly equivalent to real 1945-based dollars that exist in 1945. That is, dollars that have today as the purchasing power base are the same actual dollars today. Refer to Fig. P4.44(*c*).

Real 1945 based dollars in 2005 = (Real 1945 based dollars in 1945)

$= (P/F, i60)$
$= (\$1,000,000)\ (F/P, 4.76\%, 60)$
$= (\$1,000,000)(16.289)$
$= \$16,289,000$

The difference between the answers in Methods 1 and 2 are due to the rounding the market interest rate off to 4.76% versus carrying it out more significant digits.

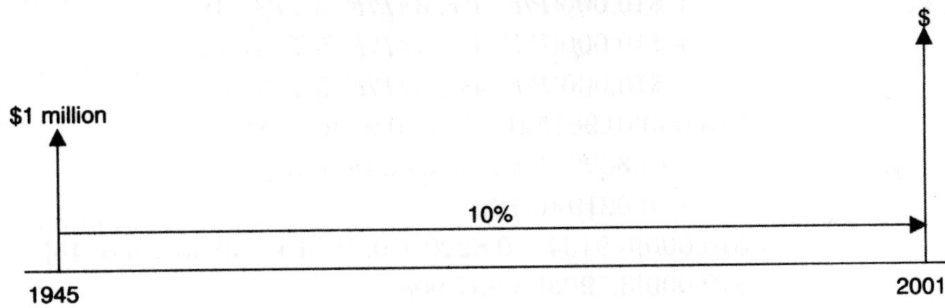

Fig. P5.44 (*a*) Discounting A$ in 1945 to A$ in 2001.

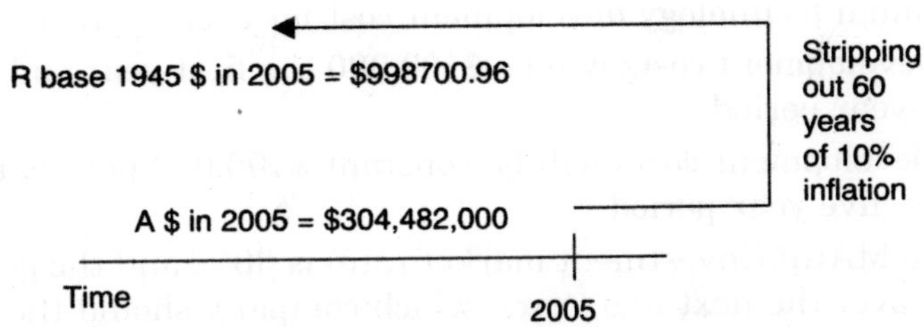

Fig. P5.44 (*b*) Translation of A$ in 2005 to R 1945 based $ in 2005.

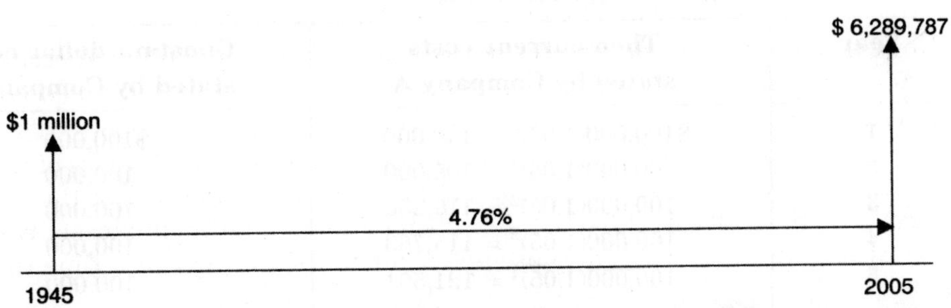

Fig. P5.44 (*c*) Translation of R 1945-based $ in 1945 to R 1945-based $ in 2005.

5.45. A series of four-end of the year payments of $10,000 has been accrued. If over the next five years the market interest rate is 10% per year and the annual inflation rate is 4%, find the present equivalent of this series using:

(*a*) actual dollar analysis

(*b*) constant dollar analysis

Solution:

(*a*) Actual dollar analysis:

$$P = \$10{,}000\ (P/A,\ 10\%,\ 5) = \$10{,}000\ (3.7908) = \$37{,}908.00$$

(*b*) Constant dollar analysis:

Inflation-free rate

$$(1 + i') = (1 + 0.10)/\ (1 + 0.04) = \frac{1.10}{1.04} = 1.0577$$

or 5.77% per year

$$\begin{aligned} P = {} & \$10{,}000(P/F, 4\%, 1)(P/F, 5.77\%, 1) \\ & + \$10{,}000(P/F, 4\%, 2)(P/F, 5.77\%, 2) \\ & + \$10{,}000(P/F, 4\%, 3)(P/F, 5.77\%, 3) \\ & + \$10{,}000(P/F, 4\%, 4)(P/F, 5.77\%, 4) \\ & + \$10{,}000(P/F, 4\%, 5)(P/F, 5.77\%, 5) \\ = {} & \$10{,}000[(0.9615)(0.95) + (0.9246)(0.89) \\ & + (0.8890)(0.85) + (0.8548)(0.80) \\ & + (0.8219)(0.76)] \\ = {} & \$10{,}000[0.9134 + 0.8229 + 0.7556 + 0.6838 + 0.6246] \\ = {} & \$10{,}000[3.7908] = \$37{,}908. \end{aligned}$$

5.46. A Corporation is interested in evaluating a new manufacturing technology. Two competitors, companies *A* and *B* offered to deliver this technology at the end of a five-year period. The annual technology development cost for each of the two companies is:

Company *A*: Development costs will be $100,000 the first year and increase at a rate of 5% over a five-year period.

Company *B*: Development costs will be constant $100,000 per year in terms of today's dollars over the five-year period.

If the corporate MARR (investment market rate) is 30%, and the general price inflation is 4% per year over the next five years, which company should the corporation chose?

Solution:

The cost for each of the two alternatives is:

Year	Then-current costs stated by Company A	Constant dollar costs stated by Company B
1	$\$100{,}000(1.05)^0 = 100{,}000$	$100,000
2	$100{,}000(1.05)^1 = 105{,}000$	100,000
3	$100{,}000(1.05)^2 = 110{,}250$	100,000
4	$100{,}000(1.05)^3 = 115{,}763$	100,000
5	$100{,}000(1.05)^4 = 121{,}551$	100,000

We inflate(or escalate)the stated yearly cost given by Company *A* by 5% per year to obtain the then current (actual) dollars each year. Company *B* costs are given in terms of today based constant dollars.

Using a constant Dollar Analysis: The then current costs stated by Company A are converted to constant today based dollars. This is done by stripping the appropriate number of years of general inflation from each year's cost using $(P/F, f, n)$ or $(1 + f)^{-n}$

Year	Constant dollar costs stated by Company A	Constant dollar costs stated by Company B
1	$\$100{,}000(1.04)^{-1} = 96{,}154$	\$100,000
2	$105{,}000(1.04)^{-2} = 97{,}078$	100,000
3	$110{,}250(1.04)^{-3} = 98{,}012$	100,000
4	$115{,}763(1.04)^{-4} = 98{,}955$	100,000
5	$121{,}551(1.04)^{-5} = 99{,}906$	100,000

$$i^1 = (i - f)/(1 + f) = (0.30 - 0.04)/1.04 = 0.25 \text{ or } 25\%$$

$$\begin{aligned}\text{PW of cost (Company } A) &= \$96{,}154(P/F, 25\%, 1)\\ &\quad + \$97{,}078(P/F, 25\%, 2)\\ &\quad + \$98{,}012(P/F, 25\%, 3)\\ &\quad + \$98{,}955(P/F, 25\%, 4)\\ &\quad + \$99{,}906(P/F, 25\%, 5)\\ &= \$96{,}154(0.80)\\ &\quad + \$97{,}078(0.64)\\ &\quad + \$98{,}012(0.512)\\ &\quad + \$98{,}955(0.4096)\\ &\quad + \$99{,}906(0.3277)\\ &= \$76{,}923 + \$62{,}130 + \$50{,}182 + \$40{,}532 + \$32{,}739\\ &= \$262{,}506\end{aligned}$$

$$\begin{aligned}\text{PW of cost (Company } B) &= \$100{,}000(P/A, 25\%, 5)\\ &= \$100{,}000(2.6893) = \$268{,}930\end{aligned}$$

Using a then-current dollar analysis: Here, we need to convert the constant dollar costs of Company B to then current dollars. This is done by adding in the appropriate number of years of general inflation into each years cost using $(F/P, f, n)$ or $(1 + f)^n$

Year	Then-current costs stated by Company A	Then-current costs stated by Company B
1	$\$100{,}000(1.05) = \$100{,}000$	$\$100{,}000(1.04)^1 = \$104{,}000$
2	$100{,}000(1.05)^1 = 105{,}000$	$100{,}000(1.04)^2 = 108{,}160$
3	$100{,}000(1.05)^2 = 110{,}250$	$100{,}000(1.04)^3 = 112{,}486$
4	$100{,}000(1.05)^3 = 115{,}763$	$100{,}000(1.04)^4 = 116{,}986$
5	$100{,}000(1.05)^4 = 121{,}551$	$100{,}000(1.04)^5 = 121{,}665$

Now we compute the present worth of costs for each alternative using the market interest rate (i).

$$\begin{aligned}\text{PW of cost (Company } A) &= \$100{,}000(P/F, 30\%, 1)\\ &\quad + \$105{,}000(P/F, 30\%, 2)\\ &\quad + \$110{,}250(P/F, 30\%, 3)\end{aligned}$$

+ $115,763(*P*/*F*, 30%, 4)
+ $121,551(*P*/*F*, 30%, 5)
= $100,000(0.7692)
+ $105,000(0.5917)
+ $110,250(0.4552)
+ $115,763(0.3501)
+ $121,551(0.2693)
= $76,920 + $62,129 + $50,186 + $40,529 + $32,734
= $262,506

PW of cost (Company *B*) = $104,000(*P*/*F*, 30%, 1)
+ $108,160(*P*/*F*, 30%, 2)
+ $112,486(*P*/*F*, 30%, 3)
+ $116,986(*P*/*F*, 30%, 4)
+ $121,665(*P*/*F*, 30%, 5)
= $104,000(0.7692)
+ $108,160(0.5917)
+ $112,486(0.4552)
+ $116,986(0.3501)
+ $121,665(0.2693)
= $79,997 + $63,998 + $51,204 + $40,957 + $32,764
= $268,930

Using either a constant dollar or then-current dollar analysis, the corporation should chose Company A's offer which has the lower worth of costs.

5.47. Determine the present worth of utility costs for the following data:
The company's minimum attractive rate of return (MARR) = 30%
The project's useful life: 10 years
The electrical costs increase at 8% per year
City water costs increase at 5% per year
Natural gas costs increase at 6% per year
The first year estimates of the utility costs (in today's dollars) are:
Electricity cost = $50,000
Water cost = $15,000
Natural gas cost = $35,000

Solution:

Here, we inflate each of the utilities independently that are used at their respective historical annual rates. We can then total them and then discount each years' total at 30% for the appropriate number of periods back to the present. The present worth of the total utility costs

= $100,000(*P*/*F*, 30%, 1) + $106,850(*P*/*F*, 30%, 2) + $114,184(*P*/*F*, 30%, 3)
+ $122,036(*P*/*F*, 30%, 4) + $130,444(*P*/*F*, 30%, 5) + $137,448(*P*/*F*, 30%, 6)
+ $149,093(*P*/*F*, 30%, 7) + $159,425(*P*/*F*, 30%, 8) + $170,494(*P*/*F*, 30%, 9)
+ $182,352(*P*/*F*, 30%, 10)

$= \$100{,}000(0.7692) + \$106{,}850(0.5917) + \$114{,}184(0.4552)$
$+ \$122{,}036(0.3501) + \$130{,}444(0.2693) + \$137{,}448(0.2072)$
$+ \$149{,}093(0.1594) + \$159{,}425(0.1226) + \$170{,}494(0.0943)$
$+ \$182{,}352(0.0725)$
$= \$76{,}920 + \$63{,}223 + \$51{,}977 + \$42{,}725 + \$35{,}129$
$+ \$28{,}479 + \$23{,}765 + \$19{,}546 + \$16{,}078 + \$13{,}221$
$= \$371{,}063$

Year	Electricity	Water	Natural gas	Total
1	$\$50{,}000(1.08)^0 = 50{,}000$	$\$15{,}000(1.05)^0 = 15{,}00$	$\$35{,}000(1.06)^0 = 35{,}000$	$100,000
2	$50{,}000(1.08)^1 = 54{,}000$	$15{,}000(1.05)^1 = 15{,}750$	$35{,}000(1.06)^1 = 37{,}100$	106,850
3	$50{,}000(1.08)^2 = 58{,}320$	$15{,}000(1.05)^2 = 16{,}538$	$35{,}000(1.06)^2 = 39{,}326$	114,184
4	$50{,}000(1.08)^3 = 62{,}986$	$15{,}000(1.05)^3 = 17{,}364$	$35{,}000(1.06)^3 = 41{,}686$	122,036
5	$50{,}000(1.08)^4 = 68{,}024$	$15{,}000(1.05)^4 = 18{,}233$	$35{,}000(1.06)^4 = 44{,}187$	130,444
6	$50{,}000(1.08)^5 = 73{,}466$	$15{,}000(1.05)^5 = 19{,}144$	$35{,}000(1.06)^5 = 46{,}838$	137,448
7	$50{,}000(1.08)^6 = 79{,}344$	$15{,}000(1.05)^6 = 20{,}101$	$35{,}000(1.06)^6 = 49{,}648$	149,093
8	$50{,}000(1.08)^7 = 85{,}691$	$15{,}000(1.05)^7 = 21{,}107$	$35{,}000(1.06)^7 = 52{,}627$	159,425
9	$50{,}000(1.08)^8 = 92{,}547$	$15{,}000(1.05)^8 = 22{,}162$	$35{,}000(1.06)^8 = 55{,}785$	170,494
10	$50{,}000(1.08)^9 = 99{,}950$	$15{,}000(1.05)^9 = 23{,}270$	$35{,}000(1.06)^9 = 59{,}132$	182,352

5.48. **The manufacturing manager of a company wants to know the equivalent annual cost of producing certain product over the five-year period. The company's before-tax MARR is 25%. The manager estimates that he will need an average of 500 lbs of each of material *A* and material *B* each day. The manufacturing company operates 7 days per week, 52 weeks per year. The current costs for material *A* are $50 per lb and $80 per lb for material *B*. Also, it was anticipated that the costs for material *A* will inflate 3% per year the next three years and then 5% the following two years. The costs for material *B* will inflate 8% per year the next two years and decrease 3% the following three years.**

Solution:

Today's cost for one year's supply of material is

Material *A*($50/lb)(500lb/day)(7 days/week)(52 weeks/year) = $9,100,000/year

Material *B*($80/lb)(500lb/day)(7 days/week)(52 weeks/year) = $14,560,000/year

Year	Material A (in millions)	Material B (in millions)	Total
0	$9.1	$14.56	
1	9.1(1.03) = 9.373	14.56(1.08) = 15.7248	$25.09780
2	9.373(1.03) = 9.65419	15.7248(1.08) = 16.982784	26.636974
3	9.65419(1.03) = 9.943815	$16.982784(1.03)^{-1} = 16.488139$	26.431955
4	9.943816(1.05) = 10.441006	$16.488139(1.03)^{-1} = 16.007903$	26.448909
5	10.441006(1.05) = 10.963057	$16.007903(1.03)^{-1} = 15.541653$	26.504710

$\text{EUAC} = [25.0978(P/F, 25\%, 1) + 26.636974(P/F, 25\%, 2) + 26.431955(P/F, 25\%, 3)$
$+ 26.448909(P/F, 25\%, 4) + 26.504710(P/F, 25\%, 5)](A/P, 25\%, 5)$
$= [25.0978(0.80) + 26.636974(0.64) + 26.431955(0.512)$
$+ 26.448909(0.4096) + 26.504710(0.3277)](0.3718)$

= [20.07824 + 17.047514 + 13.533161 + 10.833473 + 8.685593](0.3718)
= $26.092173 millions per year or 26,092,173 per year.

5.49. If the general price inflation is estimated to be 4% for the next five years, 7% for the three years after that, and 3% the following five years, at what market interest rate (i) would a company have to invest the money to maintain a real purchasing power growth rate (i^1) of 10% during those years.

Solution:

In years 1–5, the company must invest at: 0.10 + 0.040 + (0.10)(0.040)
= 0.144 = 14.4% per year

In years 6–8, the company must invest at: 0.10 + 0.070 + (0.10)(0.070)
= 0.177 = 17.7% per year

In years 9–13, the company must invest at: 0.10 + 0.030 + (0.10)(0.030)
= 0.133 = 13.3% per year

This problem illustrates the timing question: the market interest rate necessary to maintain the real purchasing power of the investment must be calculated at a time when the inflation rate is only an estimate. Most interest bearing investments generally have fixed up front rates that the investor understands going into the investment. The exception to this is the variable rate investments. Of course, inflation is not qualified and its effect on real return is not measured or assessed until the end of the year. Hence, in order to attain the conditions required in the above example problem, the company would either have to anticipate inflation and adjust the investment accordingly or accept the fact that the real investment return (i^1) may not turn out to be what was originally required.

5.50. A $10,000 investment will return annual benefits for six-years with no salvage value at the end of six years. Determine the before and after-tax rates of return assuming a straight-line depreciation and a 45% income tax rate. With no inflation, the annual benefits are $2,433 per year and with inflation equal to 6%, the benefits from the investment increase at this same rate, so that they continue to be the equivalent of $2,433 in year zero based dollars.

Solution:

The benefit schedule for the two situations is given as follows:

Year	Annual benefit for both situations, in Years–0 based dollars	No inflation, actual dollars received	6% inflation factor*	5% inflation, actual dollars received
1	$2433	$2433	$(1.06)^1$	2579
2	2433	2433	$(1.06)^2$	2734
3	2433	2433	$(1.06)^3$	2898
4	2433	2433	$(1.06)^4$	3072
5	2433	2433	$(1.06)^5$	3256
6	2433	2433	$(1.06)^6$	3451

Before-tax rate of return: Due to the fact that both situations(no inflation and 6% inflation) have an annual benefit, stated in Year-0 based dollars of 2433, they have the same before-tax rate of return.

PW of cost = PW of benefit

$10{,}000 = 2433(P/A, i, 6)$

or $(P/A, i, 6) = 10{,}000/2433 = 4.11$

From Interest Tables: Before-tax rate of return equals 12%. From interest tables in appendix-5A, Section I.

After-tax rate of return, no inflation:

Year	Before-tax cash flow	Straight line depreciation	Taxable income	45% income taxes	Actual dollars, and Year-0 based $ after-tax cash flow
0	– $10,000				– $12,000
1–6	+ 2433	$1667	$766	– $345	+ 2088

PW of cost = PW of benefit

$10{,}000 = 2088(P/A, i, 6)$ or $(P/A, i, 6) = 10{,}000/2088 = 4.79$

Hence, After-tax rate of return equals 7.1%(Appendix-5A, Section I)

After-tax rate of return, 5% inflation:

Year	Before-tax cash flow	Straight line depreciation	Taxable income	45% income taxes	Actual dollars, after-tax cash flow
0	$10,000				– $10,000
1	2579	1667	912	– 410	2169
2	2734	1667	1067	– 480	2254
3	2898	1667	1231	– 554	2344
4	3072	1667	1405	– 632	2440
5	3256	1667	1589	– 715	2541
6	3451	1667	1784	– 803	2648

Converting to Year-0 based dollars and solving for the rate of return:

Year (1)	Actual dollars, after -tax cash flow (2)	Conversion factor (3)	Year-0 based dollars, after-tax cash flow (4) (4) = (2) × (3)
0	– $10,000		– $10,000
1	2169	$(1.06)^{-1}$	2046
2	2254	$(1.06)^{-2}$	2006
3	2344	$(1.06)^{-3}$	1968
4	2440	$(1.06)^{-4}$	1933
5	2541	$(1.06)^{-5}$	1899
6	2648	$(1.06)^{-6}$	1867

After-tax rate of return = 4% = 1%[362/(362 + 25)] = 4.9%

Linear interpolation between 17% and 18%

After-tax rate of return = 17% + 1%[17/(17 + 69)] = 17.1978 or 17.2%

Present worth at 16%	Present worth at 17%	Present worth at 18%
– 10,000	– 10,000	– 10,000
1764	1749	1734
1729	1715	1700
1697	1682	1668
1666	1652	1638
1637	1623	1609
1609	1596	1582
	17	– 69

5.51. Consider a project that has the following cost series over a five-year period.

(*a*) if the rate of inflation is 3% and the market rate is 8.15%, what is the present equivalent of this series? What is the equal-annual series of payments equivalent to this series in actual dollars?

(*b*) if the rate of inflation is 5% and the market rate is 11.3%, what is the equivalent at the present of the series? What is the annual equivalent of this series in actual dollars?

Solution:

End of year	1	2	3	4	5
Estimated future cost in constant dollars	$2,000	$2,000	$2,000	$2,000	$2,000

(*a*) Inflation – free rate

$$(1 + i^1) = (1 + 0.0815)/(1 + 0.03)$$

or $$i^1 = 3\%$$

From the Table in Appendix-5A, $(P/A, 3\%, 5) = 4.5797$

$$P = \$2000(4.5797) = \$9{,}159.40$$

Also from Appendix %A, $(A/P, 8.15\%, 5) = 0.2515$

$$A = P(A/P, 8.15\%, 5) = 9{,}159.40(0.2515) = \$2{,}303.59$$

(*b*) Inflation – free rate

$$(1 + i^1) = (1 + 0.113)/(1 + 0.05)$$

or $$i^1 = 6\%$$

$$P = \$2{,}000[P/A, 6\%, 5] = \$2{,}000(4.2124) = \$8{,}424.80$$

where $(P/A, 6\%, 5) = 4.2124$ [Appendix–5A]

$$A = P[A/P, 11.3\%, 5] = \$8{,}424.80(0.2716) = \$2{,}288.18$$

where $(A/P, 11.3\%, 5) = 0.2716$ [from Appendix–5A].

5.52. The minimal acceptable return for an investment of $30,000 with a 5-year life and salvage value of $5,000 is 9%. This investment results in an annual benefit of $10,000 at an operating cost of $2,000. Determine the break-even point in terms of life at which the project is marginally justifiable if there is an uncertainty as to whether the new machinery will survive 5 years of continuous use.

Solution:

Using the annual cost method:

$$\$10{,}000 - 2000 - (30{,}000 - 5{,}000)\, F_{A/P,\,9,\,n} - 5000(0.09) = 0$$

or $$F_{A/P,\,9,\,n} = 7550/25{,}000 = 0.302$$

By interpolation from the interest tables in Appendix-5A gives n = 3.99 years. Thus, if the machine does not last 3.99 years, the investment cannot be justified.

5.53. A new plant will cost $100,000 for the first stage and $150,000 for the second stage at n years in the future. If it is built to full capacity now, its cost is $160,000. It is assumed that all facilities are expected to last 50 years and there is no salvage value. Conduct a break-even and sensitivity analysis. Assume an 8% interest rate.

Solution:

The calculations may be either annual cost or present worth. Present worth calculations are used here for simplicity.

Construct full capacity now:

PW of cost = $160,000

Two stage construction:

Compute the PW of cost for several years of n years.

PW of cost = 100,000 + 150,000(P/F, 8%, n)

n = 5 PW = 100,000 + 150,000(0.6806) = $202,090

n = 10 PW = 100,000 + 150,000(0.4632) = $169,480

n = 20 PW = 100,000 + 150,000(0.2145) = $132,175

n = 30 PW = 100,000 + 150,000(0.0994) = $1144,910

The data can be plotted. The breakeven point on the chart is the point at which both alternatives have equivalent costs ($160,000). Year 12.2 is the break-even point. The chart shows that if the second stage was needed prior to 12.2 years then one stage construction, with it's smaller PW of cost, would be preferred. If the second stage is not required until after 12.2 years, then two-stage construction is preferred.

5.54. An ice-cream company has studied the possibility of marketing a new low-calorie ice-cream in small region. The profits would depend upon the sales volume and there is some uncertainty on the sales-forecast figures. The estimated investment is $200,000 and the profits are $65,000 per year for the next 5 years. The company's MARR is 15%.

(*a*) is the investment decision sensitive to the uncertainty of the sales forecast if the estimated profits will be reduced to $40,000 per year.

(*b*) the minimum volume of sales for the investment projects to break-even, if there is a profit of $ 7 per unit volume.

Solution:

For an annual profit of $65,000

NPW = 65,000(P/A, 15%, 5) – 200,000

= 65,000(3.3522) – 200,000

= $17,893 (attractive)

(*a*) For an annual profit of $40,000

NPW = 40,000(P/A, 15%, 5) – 200,000

= 40,000(3.3522 – 200,000

= – 65,912 (not attractive)

Hence, the decision is sensitive to the expected variations in sales or profits.

(*b*) The break-even point: NPW = 0

NPW = 0 = X(P/A, 15%, 5) – 200,000

where X = minimum profit

Therefore

$$X = 200{,}000/(P/A,\ 15\%,\ 5) = 200{,}000/3.3522 = \$59{,}662.31$$

or In volume units = \$59,662.31/\$7 per unit = 8,523 volume units.

5.55. A conventional machine system will cost \$100,000 and has a service life of 10 years and an anticipated salvage value of \$10,000 at the end of life. The annual maintenance and repair cost for the machine system is estimated at \$3,000. The expected rate of return from the investment is 12%. The production rate for this machine system is 20 units per hour. The labor rate to operate this machine system is \$8.00 per hour. Assume the overhead rates on the machine system as 25% and 30% on labor.

Find the annual production output at which break-even will occur if the value added per piece of this operation is \$1.00.

Solution:

Let Q be the annual output level. The variable cost is the labor rate divided by the production rate.

Variable cost per unit = [\$8/hour/20 units/hour](1 + 30%) = \$0.52/unit.

The corresponding annual variable cost as a function of Q would be $\$0.52Q$.

The annual fixed cost is obtained on the machine system investment. Ignoring the overhead, we have

$$\begin{aligned} \text{UAC} &= 100{,}000(A/P,\ 12\%,\ 10) + 3{,}000 - 10{,}000(A/F,\ 12\%,\ 10) \\ &= 100{,}000(0.1770) + 3{,}000 - 10{,}000(0.0570) \\ &= 17{,}700 + 3{,}000 - 570 = 20{,}130 \end{aligned}$$

Adding the 25% overhead rate to this, the annual allocation of fixed cost would be \$20,130(1 + 25%) = \$25,162.50.

The sum of the fixed and variable costs would be total cost = $\$25{,}162.50 + 0.52Q$

This function can plotted.

The revenues as a function of Q would be value added times quantity or $0.70Q$.

The break-even point occurs where the two lines intersect. To obtain the break-even point, the total cost function is set equal to the revenue function. That is

$$0.70Q = 25{,}162.50 + 0.52Q$$

$$0.70Q - 0.52Q = 25{,}162.50$$

or $$Q = 139{,}792 \text{ units per year.}$$

At a production rate of 20 units per hour, this corresponds to 139,792/20 = 6989.6 hours of operation per year. Above this level of production, the company will make a profit on the investment. A loss will occur below this output level.

5.56. Refer to P5.55. Consider an alternative of an automated machine system to the machine system costing \$150,000 with an expected service life 5 years with an anticipated service life of zero is proposed whose annual maintenance costs are estimated at \$8,000. Assume the same 12% rate of return and 25% overhead rates for this system. The production rate for this system will be 50 units per hour, and one-fourth of one operator's time at \$8.00 per hour will be needed to tend the machine. The overhead rate for labor is 30%. Determine the annual production levels for which the automated system would be advantageous.

Solution:

Variable cost = 0.25[(\$8 per hour)/50 units per hour](1 + 30%)Q = 0.052Q

Fixed cost = [\$150,000(A/P, 12%, 5) + \$8,000][1 + 25%]

= [\$150,000(0.2774) + \$8,000](1.25) = \$62,012.50

The total cost is obtained by summing the fixed and variable costs.

Total cost = \$62,012.50 + 0.052Q

The total cost functions for the two production methods can be plotted. The break-even point is indicated by the intersection of the two lines. To obtain the break-even point, the two functions are set equal to each other.

$$\$25{,}162.50 + 0.52Q = \$62{,}012.50 + 0.052Q$$

$$0.468Q = \$36{,}850$$

Hence $$Q = 78{,}739$$

At a production rate of 50 units per hour on the automated machine system, this would require 1574.78 hours per year, but on the conventional machine system at 20 units per hour, 3936.95 hours per year are needed.

For an annual production levels above 78,739 units, the automated machine system would be advantageous as can be seen from a plot of the above data.

5.57. The total cost C for a project in millions of dollars is represented by

$$C = 4X^{1.3}\,Y^{-0.5} + 12Y + F$$

where X is estimated from five values: 2.6, 1.7, 1.6, 2.9, 1.5,and 2.0. Y is estimated from a high of 3 and a low of 1, F is certain with a value of 1.5. Assume normal distributions.

(a) determine the expected value for C, its sensitivity to X and Y

(b) find variation of the actual value of C with the expected value

(c) find the variance for C.

Solution:

(a) The mean value of X is (2.3 + 1.7 + 1.8 + 2.7 + 1.5 + 2.0)/6 = 2.0

The variance of X is:

$\sigma_x^2 = (2.3-2)^2 + (1.7-2)^2 + (1.8-2)^2 + (2.7-2)^2 + (1.5-2)^2 + (2.0\text{–}2.0)^2/6 - 1 = 0.192$

The mean for Y is (3 + 1)/2 = 2

The variance for Y is $\sigma_y^2 = [(3-1)/8]^2 = 0.0625$

The mean value of F is 0.6 and the variance is zero.

The expected value for C_T is obtained using the mean, or expected value of the components.

$$C_T = 4(2)^{1.3}\,(2)^{-0.5} + 12(2) + 0.6 = 31.564$$

The sensitivity of C_T to X and Y is obtained as

$$\partial C_T/X = 4(1.3)\,X^{1.3}\,Y^{-0.5} = 5.2\,X^{1.3}\,Y^{-0.5} = (5.2)(2)^{1.3}(2)^{-0.5} = 9.054$$

$$\partial C_T/Y = 4\,X^{1-3}\,(-\,0.5)Y^{-1.5} + 12 = (-\,2)\;2^{1-3}\;2^{-1.5} + 12 = 13.741$$

C_T changes 9.054 times fast as X for a unit change in X and 13.741 times as fast as Y for a unit change in Y. That is, a 1% change in X (ΔX = 0.02) would cause a change in C_T amounting to (9.054)(0.02) = 0.18108, which on a percentage basis is (0.18108)(100)/ 31.564 = 0.5737 percent. Similarly for Y, a 1 percent change in Y(at Y = 2) gives a (13.741)(0.02)(100)/31.564 = 0.8707 percent change in C_T.

The total cost C_T is more sensitive to the estimate for Y than X, since a 1 percent error in X causes a 0.5737 pe cent error C_T, but a 1 percent error in Y causes a 0.8707 percent error in C_T. This describes the sensitivity analysis.

(*b*) The variance for C_T is given by

$$\sigma^2 = (9.054)^2(0.192) + (13.741)^2 (0.0625) = 15.7392 + 11.809 = 27.5401$$

The standard deviation is then $\sigma = 5.2479$

5.58. A Company is trying to decide which of the two subsystems to install in order to reduce costs in a particular machinery system. Both subsystems cost $1500 and have the same useful life of 5 years and no salvage value. Subsystem *A* is expected to give $400 in savings annually. Subsystem *B* will provide cost savings of $500 in the first year but will decline $50 annually. With an annual interest rate of 8%, determine which subsystem the company should purchase.

Solution:

Subsystem *A*:

PW of cost = $1500

PW of benefits = 400(*P*/*A*, 8%, 5) = 400(3.9927) = $1597.08

B/*C* = *PW* of benefits/PW of costs = 1597.08/1500 = 1.0647

Subsystem *B*:

PW of cost = $1500

PW of benefits = 500(*P*/*A*, 8%, 5) – 50(*P*/*G*, 8%, 5)

= 500(3.9927) – 50(7.372) = 1996.35 – 368.60 = 1627.75

B/*C* = PW of benefits/PW of costs = 1627.75/1500 = 1.0852

To maximize the *benefit cost ratio,* select subsystem *B*.

5.59. The data for two machines *A* and *B* which are being considered for purchase by a company is as given below. Assuming 10% interest rate per year, which machines the company should buy?

	Machine A	Machine B
Initial cost	$1,000	$3,000
Uniform annual benefit	90	200
End-of-useful life salvage value	150	800
Useful life in years	6	12

Solution:

For a twelve-year analysis period, the cash flow is tabulated below:

Year	Machine A	Machine B
0	– 1000	– 3000
1–5	+ 90	+ 200
6	+ 90 – 1000 + 150	+ 200
7–11	+ 90	+ 200
12	+ 90 + 150	+ 200 + 800

Let us consider solving this problem using the *B*/*C* ratio:

B/*C* ratio = EUAB/EUAC

Also, treat the salvage value of the machines as reduction in cost, rather than increases in benefits.

Machine A:

$$EUAC = \$1000(A/P, 10\%, 6) - 150(A/F, 10\%, 6)$$
$$= 1000(0.2296) - 150(0.2296) = \$195.16$$

and $EUAB = \$90$

With the assumption that the replacement for the last 6 years has identical costs, the EUAC for the 1st 6 years equals the EUAC for all 12 years.

Machine B:

$$EUAC = 3000(A/P, 10\%, 12) - 800(A/F, 10\%, 12)$$
$$= 3000(0.1468) - 800(0.0468) = \$440.40 - \$37.44 = \$402.96$$
$$EUAB = \$200$$

Machine B – Machine A:

$$\Delta B/\Delta C = [200-90]/(402.96 - 195.16) = 110/207.80 = 0.5293 \quad (<1)$$

The *B/C* ratio is <1 and therefore represents an undesirable increment of investment. Hence, we select the lower cost alternative – machine *A*.

5.60. A section of a manufacturing company is under expansion. The initial investment and operating benefits and costs for alternatives *A*, *B*, and *C* are summarized as follows:

	Alternative A	Alternative B	Alternative C
Investment costs	\$200,000	\$150,000	\$300,000
Annual operating costs	15,000	10,000	25,000
Annual benefits	60,000	35,000	85,000

If the company's MARR is 10%, which alternative should be selected for a ten-year useful life period using benefit/cost ratio analysis?

Solution:

The *B/C* ratios for individual alternatives are

$$(B/C)_A = EUAB_A/EUAC_A = 60{,}000/[200{,}000(A/P, 10\%, 10) + 15{,}000]$$
$$= 60{,}000/[200{,}000(0.1628) + 15{,}000] = 60{,}000/47{,}560 = 1.2616$$
$$(B/C)_B = EUAB_B/EUAC_B = 35{,}000/[150{,}000(A/P, 10\%, 10) + 10{,}000]$$
$$= 35{,}000/[150{,}000(0.1628) + 10{,}000] = 35{,}000/34{,}420 = 1.0168$$
$$(B/C)_C = EUAB_C/EUAC_C = 85{,}000/[300{,}000(A/P, 10\%, 10) + 25{,}000]$$
$$= 85{,}000/[300{,}000(0.1628) + 25{,}000] = 85{,}000/73{,}840 = 1.1511$$

All of the above *B/C* rates (>1) indicate that they are economically attractive.

Incremental *B/C* analysis:

$$\text{Cost } B < \text{Cost } A < \text{Cost } C.$$

(*a*) *A* – *B*:

$$\Delta B = \$60{,}000 - \$35{,}000 = \$25{,}000$$
$$\Delta C = (\$200{,}000 - 150{,}000)(A/P, 10\%, 10) + (15{,}000 - 10{,}000) = \$13{,}140$$
$$\Delta B/\Delta C = \Delta EUAB/\Delta EUAC = 25{,}000/13{,}140 = 1.9026 \ (>1)$$

Therefore keep *A*

(*b*) *C* – *A*:

$$\Delta B = \$85{,}000 - \$60{,}000 = \$25{,}000$$
$$\Delta C = (300{,}000 - 200{,}000)(A/P, 10\%, 10) + (25{,}000 - 15{,}000) = \$26{,}280$$
$$\Delta B/\Delta C = \Delta EUAB/\Delta EUAC = 25{,}000/26{,}280 = 0.9513 \ (<1)$$

Hence keep *A* and select alternative *A*.

CHAPTER 6

Cost Evaluation

6.1. The Raghu Manufacturing Company produces metal brackets using three different machine tools m_1, m_2, or m_3. An analysis of the bracket production resulted in the following data:

Machine tool	Fixed cost/order	Variable cost/unit
m_1	\$400	\$10
m_2	\$800	\$4
m_3	\$600	\$6

(*a*) determine the most economical machine tool to use for all order sizes between 1 and 250 units

(*b*) determine the machine tool to be used for an order of size 100

(*c*) in part (*b*), determine the total production cost.

Solution:

Let x = order size, brackets/order

The total cost equations are given by

$$TC_{m1}(x) = \$400 + \$9x$$
$$TC_{m2}(x) = \$800 + \$4x$$
$$TC_{m3}(x) = \$600 + \$6x$$

The break-even points are:

For m_1 and m_2:

$$\$400 + \$9x = \$800 + \$4x$$

or $$x = 40 \text{ brackets/order}$$

For m_1 and m_3:

$$\$400 + \$9x = \$600 + \$6x$$

or $$x = 66.67$$

or 67 brackets/order

For m_2 and m_3:

$$\$800 + \$4x = \$600 + \$6x$$

or $$x = 100 \text{ brackets/order}$$

(*a*) If $1 < x \leq 40$, use m_1

If $50 < x \leq 100$, use m_3

If $100 < x \leq 250$, use m_2

(*b*) For $x = 100$, use m_2 or m_3

(*c*) $TC_{m3}(x) = \$600 + \$6x = \$600 + \$6(100) = \$1200$

$TC_{m2}(x) = \$800 + \$4x = \$800 + \$4(100) = \$1200$

6.2 **The direct labor hours for the job are estimated to be 50 hours at a rate of \$15 per hour. Direct material costs are estimated at \$1000. The overhead calculations are to be based on the following previous period cost totals:**

Total direct material dollars	**\$58,000**
Total direct labor dollars	**\$520,000**
Total direct material dollars	**\$700,000**
Total direct overhead dollars	**\$400,000**

Determine the overhead allocation based on

(*a*) direct labor hours

(*b*) direct labor dollars

(*c*) direct labor dollars plus direct material dollars.

Solution:

(*a*) Direct labor hours

Step 1: previous period direct labor hours = \$58,000
previous period overhead cost = \$400,000
Step 2: rate per direct labor hour = \$400,000/58,000 = \$6.90/hour
Step 3: estimated direct labor hours for job = 50
Step 4: estimate overhead = \$6.90/hour × 50 hours = \$345.

(*b*) Direct labor dollars

Step 1: previous period direct labor dollars = \$520,000
previous period overhead cost = \$400,000
Step 2: percent ratio per direct labor dollar = (\$400,000/\$520,000) × 100% = 76.92%
Step 3: estimated direct labor dollars for job = 50 hours × \$15/hour = \$750
Step 4: estimate overhead = 76.92% × \$750 = \$576.90.

(*c*) Direct labor dollars + direct material dollars

Step 1: previous period direct labor dollars = \$520,000
previous period direct material dollars = \$700,000
previous period overhead cost = \$400,000
Step 2: percentage ratio per (direct labor dollar + direct material dollar)
= [\$400,000/(\$520,000 + \$700,000)] × 100%
= 12.20%
Step 3: estimated direct labor dollar + direct material dollars for job
= (50 hours × \$15/hour) + \$1000 = \$1750
Step 4: estimate overhead = 12.20% × 1750 = \$213.50.

6.3. Table T6.3 summarizes the information collected for two departments of a company.

Table T6.3

	Department A	Department B	Total
Direct materials cost	$750,000	$250,000	$1,000,000
Direct labor cost	$250,000	$150,000	$400,000
Direct labor hours	25,000	15,000	40,000
Number of employees	16	9	25
First cost of equipment	$250,000	$200,000	$450,000
Annual depreciation	$25,000	$20,000	$45,000
Other factory overhead			$150,000
General overhead			$350,000

The company distributes depreciation overhead based on (1) the first cost of equipment in each department, (2) a zero salvage value of the equipment in 10 years, and (3) a constant annual (or straight-line) rate of depreciation. All overheads other than depreciation is first distributed to each department, according to the number of employees in each department and then an overhead rate per direct labor hour are computed for each department. Determine the selling price the company should quote on Job Order if the raw material costs are estimated as $1000, estimated direct labor hours required in Departments *A* and *B* are 40 hours and 100 hours respectively, and profit is to be calculated as 35% of selling price.

Solution:

For department *A*, the total overhead allocation is determined as follows:

Annual depreciation = $25,000

Other factory overhead = (16/25)($150,000) = 96,000

General overhead = (16/25)($350,000) = 224,000

Total overhead costs = $345,000

Thus, the overhead rate for department *A* per direct labor hour is

$$\text{Rate} = \frac{\$345{,}000}{25000 \text{ hours}} = \$13.80/\text{direct labor hour}$$

For department *B*, the total overhead allocation is calculated as follows:

Annual depreciation = $20,000

Other factory overhead = (9/25)($150,000) = 54,000

General overhead = (9/25)($350,000) = 126,000

Total overhead costs = $200,000

Thus, the overhead rate for department *B* per direct labor hour is

$$\text{Rate} = \frac{\$200{,}000}{15000 \text{ hours}} = \$13.34/\text{direct labor hour}$$

The estimated total cost for Job Order is then computed as

Direct material cost = $1000

Direct labor cost for Dept. *A* = ($250,000/25000)(40 hours) = 400

Overhead cost for Dept. *A* = ($13.80)(40 hours) = 552

Direct labor cost for Dept. B = (\$150,000/15000)(100 hours) = 1000
Overhead cost for Dept. B = (\$13.34)(100 hours) = 1334
Total cost = \$4286

If x = the selling price of Job Order, then

$$x = \text{total cost} + \text{profit} = \$4286 + 0.35x$$

and $$x = \frac{4286}{0.65} = \$6,593.85.$$

6.4. A machine that is used in a manufacturing process to make a certain product has an initial cost of \$60,000 and it is expected to have a zero salvage value. The maximum capacity of this machine is 8,000 units per year, and its life is estimated to be 4 years. The unit variable cost (raw material, energy, etc.) is \$15, the total fixed annual cost of supervision and taxes is \$25,000. The company uses 15% interest compounded continuously. If the machine operates at (1) 50%, (2) 60%, (3) 70%, (4) 85% of maximum capacity,
(*a*) determine the average unit cost and annual total cost
(*b*) determine the incremental production cost.

Solution:

The factor (*A/P*, 15, 4) from Tables in Appendix-5A(see Chapter 5) is 0.3587.

(*a*) Annual fixed cost = (\$60,000 – \$0)[0.3587] = \$21,522

$$\text{Average unit cost} = \frac{\text{Annual fixed Costs}}{\text{\# of units}} + \frac{\text{Variable Cost}}{\text{Unit}}$$

$$= \frac{\$25,000 + \$21,522}{N} + \$15 = \frac{\$46,522}{N} + \$15$$

Annual total production cost = \$15(N) + \$25,000 + \$21,522

(*b*)

% capacity	N(units)	Avg. unit cost (\$)	Annual total prod. cost	Inc. prod. cost
0	0	—	46522	—
50	4000	26.6305	106522	60000
60	4800	24.69208333	118522	12000
70	5600	23.3075	130522	12000
85	6800	21.84147059	148522	18000

6.5. A certain manufacturing process equipment costs \$12,000. The equipment has a life of 5 years and a salvage value of \$2,000. The maintenance, taxes, interest, and other fixed costs amount to \$500/year. The cost of power and supplies is \$5/hr, and the operator costs \$10/hr. The cycle time per unit of product is 1.2 minutes. If the MARR is 17%, determine the cost per unit of processing the product if the number of units of product made per week are 200,400,1200 and 2500. Assume the equipment operates 52 weeks per year.

Solution:

Annual cost of capital recovery with return (*A/P*, 17, 5 = 0.3126)

= (\$12,000 – \$2,000)(0.3126) + \$2,000(0.17) = \$3,466

The cost per unit for production of n units per week

$$= \frac{\$5 + \$10}{60/1.2} + \frac{\$3,466 + \$500}{n(52)}$$

n	200	400	1200	2500
Cost/unit	\$0.681	\$0.491	\$0.363	\$0.331

6.6. A large distribution company is to contain 12 million square feet of floor space. The total cost of the building (TC) is given by

$$TC = (200 + 85n + 2n^2)A$$

where n = number of floors

A = area of floor in ft²/floor.

(*a*) determine the optimal number of floors for the building based on the total cost

(*b*) Determine the optimal number of floors for the building based on the marginal cost per floor, $MC(n) = TC(n + 1) - TC(n)$.

Solution:

(*a*) $$\frac{dTC}{dn} = \frac{d}{dx}[(200 + 85n + 2n^2)(12,000,000/n)$$

$$= \frac{d}{dx}[10^7(1.2)(-200n^{-1} + 85 + 2n)] = (1.2)(10^7)[-200\, n^{-2} + 2] \quad ...(1)$$

Setting Eq.(1) to zero and solving, we get

$$0 = (1.2)(10^7)[-200\, n^{-2} + 2] - 200\, n^{-2} + 2 = 0$$

or $$n^2 = \frac{200}{2} = 100$$

or $$n = 10$$

Optimum number of floors = 10.

(*b*) Using excel spread sheet and using the definition for marginal cost, we have

$$MC(n) = TC(n + 1) - TC(n)$$

Table P6.6 summarizes the results.

Table P6.6

Floor	Factor A	TC	TC/Floor	MC
1	12,000,000	3,444,000,000	3,444,000,000	– 1,176,000,000
2	6,000,000	2,268,000,000	1,134,000,000	– 376,000,000
3	4,000,000	1,892,000,000	630,666,666.7	– 176,000,000
4	3,000,000	1,716,000,000	429,000,000	– 96,000,000
5	2,400,000	1,620,000,000	324,000,000	– 56,000,000
6	2,000,000	1,564,000,000	260,666,666.7	– 33,142,857.14
7	1,714,286	1,530,857,143	218,693,877.6	– 18,857,142.86
8	1,500,000	1,512,000,000	189,000,000	– 9,333,333.333
9	1,333,333	1,502,666,667	166,962,963	– 2,666,666.667
10	1,200,000	1,500,000,000	150,000,000	2,181,818.182
11	1,090,909	1,502,181,818	136,561,983.5	*n/a*

6.7. The production of particular type of grain is a function of fertilizer used and is given by the following relationships:

$P(t)$ = grain production, tons/acre = $0.5f - 0.001f^2$

where f = amount of fertilizer used, pounds/acre.

The selling price of the grain is \$300/ton. Assuming that 1 ton of grain weighs about 2000 lbs, the total cost of grain production is given by

$$TC(f) = \$(12200 + 2f) \text{ per acre.}$$

(*a*) determine the amount of fertilizer need to be used for this grain per acre of land to maximize the annual profit

(*b*) determine the maximum annual profit per acre of land for this grain

(*c*) find the range of fertilizer application for possible profit.

Solution:

(*a*) $\text{Profit}(f) = \$300\, P(f) - TC(f)$

$$= \$300(0.5f - 0.001f^2) - \$12200 - 2f$$

$$= -0.3f^2 + 148f - 12200$$

$$\frac{d\text{ profit}}{df} = -0.6f + 148 \quad ...(1)$$

Setting Eq.(1) to zero and solving, we obtain

$$0 = -0.6f + 148$$

or $\quad f = 246.67$ lbs/acre

(*b*) $\text{Profit}(246.667) = -0.3f^2 + 148f - 12200$

$$= -0.3(246.67)^2 + 148(246.67) - 12200$$

$$= \$6{,}053/\text{acre}$$

(*c*) The points of zero profits are obtained by solving $P(f) = 0$.

$$-0.3f^2 + 148f - 12200 = 0 \ (2)$$

Solving Eq. (2) gives

$$f = \frac{-148}{2(-0.3)} \pm \frac{\sqrt{(148)^2 - 4(-0.3)(-12200)}}{2(-0.3)}$$

$$f = 246.67 \pm 142.05$$

$$f = 388.72 \text{ and } 104.62 \text{ lb/acre}$$

Hence, the range of fertilizer application for possible profit is

$$f = 104.62 \text{ lb/acre} \quad \text{and} \quad f = 388.72 \text{ lb/acre}$$

6.8. An electronic component manufacturing company has historically operated at about 80% of line capacity with an output of \$14,000 units per month. Due to the decreasing demand for this component, the production has been lowered to 9000 units per month for the foreseeable future. Given that

Fixed cost, FC = \$80,000 per month

Variable cost, v = \$3 per unit

Revenue, r = \$10 per unit.

(*a*) determine where the reduced level of 9000 units per month will place the production relative to the linear breakeven point and the profit.

(*b*) determine the variable cost per brake, v, needed to breakeven at the 9000 units level assuming the revenue per unit, r, and fixed costs remain constant.

Solution:

From the given data, we can write, denoting Q as the quantity in units per month

$$R = rQ = 10Q \quad (\$ \text{ per month})$$
$$\text{VC} = vQ = 3Q \quad (\$ \text{ per month})$$
$$\text{TC} = \text{FC} + VC = 80{,}000 + 3Q \quad (\$ \text{ per month})$$

To find the breakeven point $Q = Q_{BE}$, we set the revenue equal to the total cost. That is

$$R = \text{TC}$$

and solve for Q_{BE}.

Therefore $\quad rQ = \text{FC} + \text{VC} = \text{FC} + vQ$

or
$$Q_{BE} = \frac{\text{FC}}{r - v} \quad \text{...(1)}$$

Solving Eq.(1), we obtain

$$Q_{BE} = \frac{80{,}000}{10 - 3} = 11{,}428.6 \text{ units per month}$$

or 11,429 units per month.

Figure P6.8 shows that at 14,000 units (80% capacity) this company has been producing just above the breakeven which occurs at about 66% of capacity.

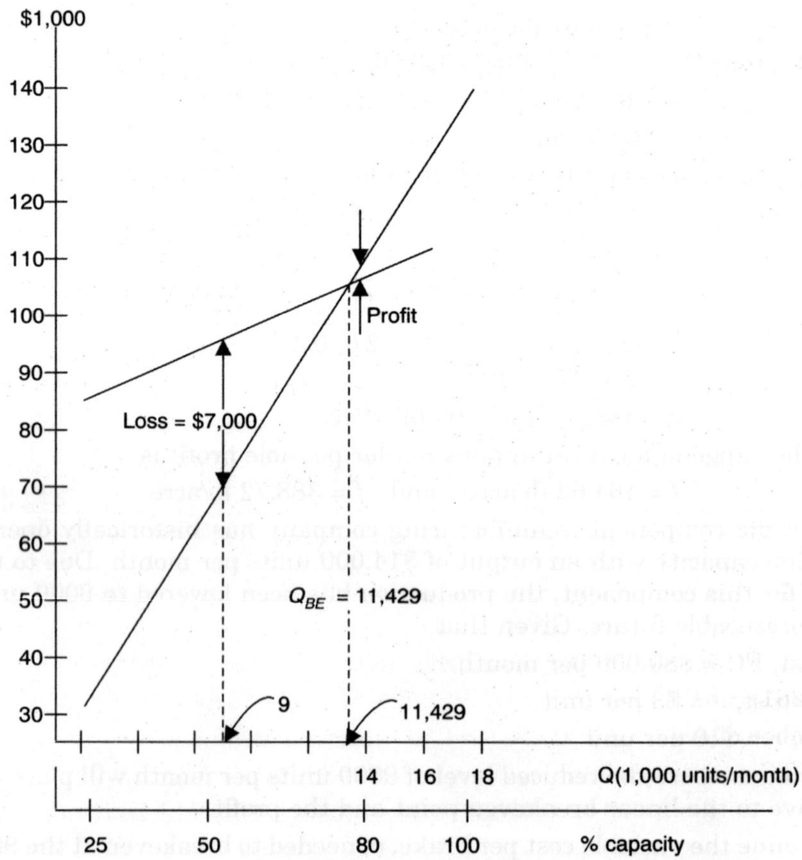

Fig. P6.8 Breakeven Graph.

In order to estimate total profit, subtract total cost from revenue for $Q = 9000$ units.

Profit $= R - \text{TC} = rQ - (\text{FC} + vQ)$

$= (r - v)Q - \text{FC}$ ...(2)

$= (10 - 3)9000 - 80{,}000 = -\$7{,}000$

There is a loss of $7000 per month. Event at the historic output level of 14,000 units per month, Eq. (2) indicates a $18000 profit.

(*b*) One way to determine v if Q_{BE} were 9000 is to state Eq. (2) at a profit of zero and $Q = 9000$. Solve for v.

$$0 = (10 - v)9000 - 80000$$

$$(10 - v) = \frac{80000}{9000}$$

$$v = \frac{80000}{9000} + 10 = \$18.89 \text{ (\$ per unit)}$$

6.9. Given that the total annual inventory cost for a particular item carried in inventory is a function of the order size Q and is given by

TC(Q) = annual cost of ordering + annual cost of carrying inventory ($/year)

$$= PC\left(\frac{A}{Q}\right) + (0.55Q + 160)$$

where

PC = cost of preparing a purchase order (constant value) = $17 per purchase order

Q = the number of items ordered each time a purchase order is placed

A = total annual demand for this item = 2200 units

A/Q = the number of orders placed each year.

(*a*) determine the value of Q that will minimize the total annual inventory cost (the economic order quantity Q^*)

(*b*) determine the corresponding total annual inventory cost, TC(Q^*)

(*c*) determine the number of orders that will be placed each year for the economic order quantity Q^*

(*d*) determine the effect on the economic order quantity if the cost of placing a purchase order is decreased by 50%.

Solution:

(*a*) $\text{TC(Q)} = \$17(2200/\text{Q}) + 0.55Q + 160 = 37400Q^{-1} + 0.55Q + 160$

$$\frac{d\text{TC}}{dQ} = -37{,}400Q^{-2} + 0.55 \quad ...(1)$$

Setting Eq. (1) to zero and solving, we obtain

$$0 = -37400Q^{-2} + 0.55$$

or $Q = 260.77$ or 261 units.

(*b*) TC(261) = $17(2200/261) + 0.55(261) + 160 = $446.84.

(*c*) Number of orders = 2200/261 = 8.43 or 9 orders.

(*d*) The TC equation can be written as

$$\text{TC}(Q) = \$8.5(2200/Q) + 0.55Q + 160 = 18{,}700Q^{-1} + 0.55Q + 160$$

$$\frac{d\text{TC}}{dQ} = -18700Q^{-2} + 0.55 \quad ...(2)$$

Setting Eq. (2) to zero and solving, we get

$$0 = -18700Q^{-2} + 0.55$$

or $$Q = 184.39 \text{ or } 185 \text{ units.}$$

6.10. The fixed costs for a particular exporting company are \$700,000 annually. Its main-line export is sold at a revenue of \$3 per unit and has \$2.50 variable costs.

(*a*) determine the annual breakeven quantity

(*b*) plot the revenue and total cost relationships and determine the annual profit if 1.6 million units are sold and if 2 million units are sold.

Solution:

(*a*) We can obtain Q_{BE} from the equation

$$Q_{BE} = \frac{FC}{r - v} = \frac{700,000}{(3 - 2.50)} = 1,400,000 \text{ units}$$

(*b*) $$R = 3Q \quad ...(1)$$

$$TC = 700,000 + 2.5Q \quad ...(2)$$

Q	R	TC
0	0	\$700,000
1 mil	\$3 mil	3.2 mil
2 mil	\$6 mil	5.7 mil

From Eqs. (1) and (2)

$$R - TC = 3Q - 700,000 - 2.5Q = 0.5Q - 700,000$$

Hence when

Q = 1.6 million units, R – TC = \$100, 000

Q = 2 million units, R – TC = \$300, 000

The plot of revenue and total cost relationships are shown in Fig. P6.10.

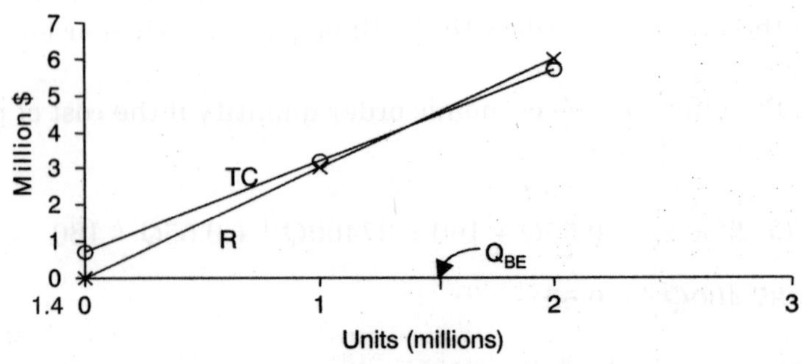

Fig. P6.10

6.11. A electronic component manufacturing company has the following fixed and variable cost components for its product over a one year period.

Fixed costs, $		Variable costs, $ per unit	
Administraive	11,000	Materials	6
Lease cost	18,000	Labor	3
Insurance	8,000	Indirect labor	5
Utilities	4,000	Other overhead	18
Taxes	11,000		
Miscellaneous operations	55,000		

(*a*) determine the revenue per unit needed to breakeven if the domestic sales volume is estimated to be 6000 units

(*b*) if international sales of 4000 units can be added to the 6000-unit domestic sales, find the total profit assuming the revenue per unit obtained in part (*a*) is realized.

Solution:

(*a*) FC = \$107,000

v = \$32/unit

Set revenue = total cost

$$R = \text{TC}$$

$$rQ = \text{FC} + vQ$$

or

$$r = \frac{\text{FC} + vQ}{Q} = [\$107{,}000 + \$32(6000)]/6{,}000$$

$$= \$49.84 \text{ per unit.}$$

(*b*) Profit $= R - \text{TC} = rQ - (\text{FC} + vQ)$

$$= (r - v)Q - \text{FC} = (49.84 - 32)(10{,}000) - 107{,}000$$

$$= \$71{,}400.$$

6.12. Two machines can be used for machining a metal part. Machine *A* costs \$900 and is expected to last 3 years. Machine *B* will cost \$1900 and last 5 years. An overhaul costing \$400 will be required after 2500 operating hours for the machine *A*, while an overhaul costing \$800 will be required for machine *B* after 10,000 hours. The operating cost of each machine is \$1 per hour. Determine the number of hours per year must the machine be required to justify the purchase of the more expensive machine. Use an interest rate of 10% per year.

Solution:

Let x = break-even hours per year.

The break-even equation is given by

$$900(A/P, 10\%, 3) + \left[\frac{400}{2500}\right]x + 1x = 1900(A/P, 10\%, 5) + \left[\frac{800}{10{,}000}\right]x + 1x$$

$$900(0.40211) + 0.16x + 1x = 1900(0.2638) + 0.08x + 1x$$

$$0.08x = 501.22 - 361.90 = 139.32$$

or

$$x = 1741 \text{ hours per year.}$$

6.13. The total cost (TC) per year of transmitting a given quantity of current over a fixed distance by a copper conductor is given by

$$\text{TC} = \$25 + \$35A + \frac{\$300}{A}$$

where A = cross-sectional area of the copper conductor.

(*a*) determine the cross-sectional area of the conductor, which will give the minimum total cost/year

(*b*) using the results in (*a*), determine the minimum total cost per year to operate the conductor system

(*c*) determine the total cost per year for 1-inch and 3-inch diameter copper conductors.

Solution:

(*a*) $$\text{TC}(A) = \$25 + \$35A + \$300A^{-1}$$

$$\frac{d\text{TC}}{dA} = 35 - 300A^{-2} \qquad \text{...(1)}$$

Setting Eq.(1) to zero and solving, we get

$$0 = 35 - 300A^{-2}$$

or $$A = 2.9276 \text{ in}^2$$

(*b*) $$\text{TC}(2.9276) = \$25 + \$35(2.9276) + \frac{\$300}{(2.9276)} = \$229.94$$

(*c*) For a 1-inch diameter

$$A = \frac{\pi}{4}(1)^2 = 0.7854$$

$$\text{TC}(0.7854) = \$25 + \$35(0.7854) + \frac{\$300}{(0.7854)} = \$434.46$$

For a 3-inch diameter

$$A = \frac{\pi}{4}(3)^2 = 7.0686 \text{ in}^2$$

$$\text{TC}(7.0686) = \$25 + \$35(7.0686) + \frac{\$300}{(7.0686)} = \$314.84$$

6.14. For a certain process in a manufacturing company the fixed costs are \$70,000, and the variable costs are \$40 per unit of production. The total cost function is given by

$$\text{TC}(x) = 70{,}000 + 40x$$

Determine the marginal cost at x = 10, 15 and 20.

Solution:

We use the method of differentiation since the total cost function is continuous. Marginal cost at 10 = first derivative of the total cost function evaluated at 10.

$$\text{MC}(x) = \frac{d}{dx}\text{TC}(x)$$

$$\text{MC}(x) = \frac{d}{dx}[70{,}000 + 40x]$$

$$\text{MC(x)} = 40$$

Hence $$\text{MC}(10) = 40$$

$$\text{MC}(15) = 40$$

and $$\text{MC}(20) = 40$$

6.15. The financial data for two manufacturing companies with a 12% cost of capital is given below (in \$ millions).

	Company A	Company B
Equity	$100	$200
Annual operating profit	$30	$40
Taxes (40%)	$12	$16
Net income	$18	$24

Determine the economic value added (EVA).

Solution:

	Company A	Company B
Capital invested	$100	$200
Operating profits	$30	$45
Taxes (40%)	$12	$18
Net operating profit after taxes	$18	$27
Cost of capital (12%)	$12	#24
EVA	+ $6	+ $3

6.16. Determine the EVA for the data in Problem 6.15 using an alternative method (based on return on assets and cost of capital).

Solution:

	Company A	Company B
Return on Assets (ROA)	18%	13.5%
Cost of capital	12%	12%
Difference	+ 6%	+ 1.5%
EVA = (Difference)(Assets)	+ $6	+ $3

6.17. The purchase price of a 200 acre agricultural farm was $200,000. The purchaser paid 30% of the purchase price in cash and the remainder was financed at a 10% annual simple interest rate with a mortgage payable to a bank over 15 years. Also purchased the seeds for $30,000, paid $12,000 in cash and for the rest gave a promissory note to the seller for the balance. The promissory note carried on 8% annual simple interest rate and was to be paid off within 5 years. The farm reported the following revenues and expenses during the first full year of operations.

Produce sold (income)	$ 17,000
Timber sold (income)	6,000
Labor expenses	2,000
Expenses for machinery hired	3,500
Professional fees	1,000
Fertilizer expenses	2,000
Property taxes	1,700
Expenses for repairs	1,000
Interest expenses	12,300
Insurance expenses	1,600
Legal fees	1,200

Expenses for miscellaneous supplies 500

Prepare an income statement for the farming operation and determine the net profit (loss) before income taxes.

Solution:

Income Statement
Year Ending 01-31-2004

Produce sold	$17,000	
Timber sold	$6,000	
Gross revenue	$23,000	
Less expenses		
Labor	$2,000	
Machinery hired	$3,500	
Professional fees	$1,000	
Fertilizer	$2,000	
Property taxes	$1,700	
Repairs	$1,000	
Interest expenses	$12,300	
Insurance expenses	$1,600	
Legal fees	$1,200	
Misc. supplies	$500	$25,800
Net profit before taxes		($2,800)

6.18. The Comparative Balance Sheet and Income Statements for D. Raghu Manufacturing for 2003 and 2004 are shown below. Based on these financial statements, determine the following ratios.

(*a*) return on assets employed

(*b*) current ratio

(*c*) acid-test ratio

(*d*) accounts receivable turnover

(*e*) inventory turnover

(*f*) debt to equity ratio

(*g*) operating income to total assets ratio

(*h*) if the industry average for return on assets employed is 9.6%, assess D. Raghu's position relative to the industry on this measure

(*i*) if the industry average for current ratio is 2.05, assess D. Raghu's position relative to the industry on this measure

(*j*) if the industry average for debt to equity ratio is 82%, assess D. Raghu's position relative to the industry on this measure.

D. Raghu Manufacturing
Comparative Balance Sheets
as of December 31, 2003 and 2004
(in thousands)

ASSETS	2004	2003
Current assets:		
Cash	$130	
Temporary investments	—	$410
Accounts receivable (net)	2,600	2,100
Inventory	2,000	2,200
Prepared expenses	70	50
Total current assets	4,800	5,060
Fixed assets net	2,200	1,300
Total assets	7,000	5,360
LIABILITIES AND CAPITAL		
Current liabilities		
Notes payable	130	60
Accounts payable	2,400	2,200
Taxes payable	140	130
Total current liabilities	2,670	2,390
Fixed liabilities		
Mortgage payable	150	90
Long term loans	300	—
Total liabilities	$3,120	$2,480
Capital		
Stock	$2,330	$1,360
Retained earnings	820	910
Earned surplus	730	610
Total capital	$3,880	$2,880
Total liabilities and capital	$7,000	$5,360

D. Raghu Manufacturing
Comparative Income Statements
For years ended December 31, 2003 and 2004
(in thousands)

Year	2,004	2,003
Net sales	$28,000	$21,000
Cost of goods sold	23,000	17,000
Gross profit	5,000	4,000
Other operating expenses	4,000	3,200
Net operating income	1,000	800
Less: Interest expenses	50	10
Less: Income taxes	210	170
Net income	$740	$620

Solution:

(*a*) Return on average assets employed

$$\frac{\text{Net income}}{\text{Average assets employed}} = \frac{\$740}{(\$5360 + \$7000)/2} = 0.1197 = 11.97\%$$

(*b*) Current ratio

$$\frac{\text{Current assets}}{\text{Current liabilities}} = \frac{\$4800}{\$2670} = 1.79 \text{ or } 1.79 \text{ to } 1$$

(*c*) Acid test ratio

$$\frac{\text{Cash receivables}}{\text{Current liabilities}} = \frac{\$130 + \$2600}{\$2670} = 1.02 \text{ or } 1.02 \text{ to } 1$$

(*d*) Accounts receivable turnover

$$\frac{\text{Net sales}}{\text{Average accounts receivable}} = \frac{\$28000}{(\$2100 + \$2600)/2} = 11.91 \text{ or } 11.91 \text{ to } 1$$

(*e*) Inventory turnover

$$\frac{\text{Cost of goods sold}}{\text{Average inventory}} = \frac{\$23,000}{(\$2000 + \$2200)/2} = 10.95 \text{ or} 10.95 \text{ to } 1$$

(*f*) Debt to equity ratio

$$\frac{\text{Total liabilities}}{\text{Total capital}} = \frac{\$3120}{\$3880} = 0.8040 = 80.41\%$$

(*g*) Operating income to total assets ratio

$$\frac{\text{Net operating income}}{\text{Total assets}} = \frac{\$1000}{\$7000} = 0.1429 = 14.29\%$$

(*h*) **D. Raghu**'s return on assets employed is 11.97% compared to the industry average of 9.6%; B&K's position is strong.

(*i*) **D. Raghu's** current ratio is 1.79 compared to the industry average of 2.05; **D. Raghu's** position is weak.

(*j*) **D. Raghu's** debt to equity ratio is 80% compared to the industry average of 82%; **D. Raghu's** position is neutral.

6.19. The soldering department of an equipment manufacturing company consists of four cost centers: *A*, *B*, *C* and *D*. The following data for the soldering department were compiled for the year 2004.

Cost center	Departmental expenses	Direct labor Hours	Direct labor Cost	Direct material
A	$22000	9800	$80000	$12,000
B	11000	3700	$17000	$11,800
C	5500	1400	$8000	$4,600
D	4700	2700	$10900	$2,900
		17600		

Determine the overhead rate (or rates) applicable using the following methods.

(*a*) percentage of direct labor cost for each year

(*b*) percentage of prime cost for each cost center

(*c*) rate per direct labor hour for each cost center.

Solution:

(*a*) Allocation based on direct labor cost

Cost center *A* = $22,000/$80,000 = 0.275 = 27.5%

Cost center *B* = $11,000/$17,000 = 0.6471 = 64.71%

Cost center *C* = $5,500/$8,000 = 0.6875 = 68.75%

Cost center *D* = $4,700/$10,900 = 0.4312 = 43.12%

(*b*) Allocation based on prime cost

Cost center *A* = $22,000/($80,000 + $12,000) = 0.2391 = 23.91%

Cost center *B* = $11,000/($17,000 + $11,800) = 0.3819 = 38.19%

Cost center *C* = $5,500/($8,000 + $4,600) = 0.6349 = 63.49%

Cost center *D* = $4,700/($10,900 + $2,900) = 0.3405 = 34.05%

(*c*) Allocation based on direct labor hours

Cost center *A* = $22,000/$9,800 = $2.25/hr

Cost center *B* = $11,000/$3,700 = $2.97/hr

Cost center *C* = $5,500/$1,400 = $3.93/hr

Cost center *D* = $4,700/$2,700 = $1.74/hr

6.20. Refer to Problem P6.19. The additional data is given below:

Cost center	Square feet occupied	Cost of machinery	No. of direct labor employees
A	1000	$11000	6
B	500	$10000	3
C	700	$12000	2
D	600	$5000	2
Total	2800	$38000	13

Expenses other than for direct labor and materials chargeable to the soldering department last year were

Maintenance	$7,000
Gas and electricity	20,000
Supervision and other indirect labor	32,000
Miscellaneous supplies	8,000
Equipment depreciation	5,000
Building depreciation	9,000

Determine an overhead rate per direct labor hour for each cost center if the soldering department expenses above are first allocated to each cost center as follows:

(*a*) maintenance expenses and equipment depreciation expenses are allocated according to the value of equipment (percent of total) in each cost center

(*b*) building depreciation expenses are allocated according to the floor space occupied (percent of total) by each cost center

(*c*) supervision and other indirect labor expenses are allocated according to the number of direct labor employees (percent of total) of each cost center

(*d*) supplies and gas and electricity expenses are allocated according to the number of direct labor hours (percent of total) for each cost center.

Solution:

(*a*) Maintenance and equipment

Cost center A = ($11,000/$38,000)($7,000 + $5,000) = $3,473.68

Cost center B = ($10,000/$38,000)($7,000 + $5,000) = $3,157.89

Cost center C = ($12,000/$38,000)($7,000 + $5,000) = $3,789.47

Cost center D = ($5,000/$38,000)($7,000 + $5,000) = $1,578.95

(*b*) Building depreciation

Cost center A = ($1,000/$2,800)($9,000) = $3,214.67

Cost center B = ($500/$2,800)($9,000) = $1,607.14

Cost center C = ($700/$2,800)($9,000) = $2,250.00

Cost center D = ($600/$2,800)($9,000) = $1,105.26

(*c*) Supervision and other indirect labor

Cost center A = (6/13)($32,000) = $14,769.23

Cost center B = (3/13)($32,000) = $7,384.62

Cost center C = (2/13)($32,000) = $4,923.07

Cost center D = (2/13)($32,000) = $4,923.07

(*d*) Supplies and gas and electricity

Cost center A = ($8,000 + $20,000)($9,800 + $17,600) = $15,590.91

Cost center B = ($8,000 + $20,000)($3,700 + $17,600) = $5,886.36

Cost center C = ($8,000 + $20,000)($1,400 + $17,600) = $2,227.27

Cost center D = ($8,000 + $20,000)($2,700 + $17,600) = $4,295.46

The overhead rate for direct labor hour

Cost center A = ($3,473.68 + $3,214.89 + $14,769.23 + $15,590.91)/9,800 hrs = $3.78/hr

Cost center B = ($3,157.89 + $1,607.14 + $7,384.62 + $5,886.36)/3,700 hrs = $4.87/hr

Cost center C = ($3,789.47 + $2,250 + $4,923.07 + $2,227.27)/1,400 hrs = $9.42/hr

Cost center D = ($1,578.95 + $1,105.26 + $4,923.07 + $4,295.46)/2,700 hrs = $4.408/hr

6.21. The costs of a plant build in the year 1985 with a two-storey building facility are as follows:

Cost of building = \$900,000

Cost of equipment = \$2,500,000

Building area = 70,000 square feet

Capacity of the plant = 8,000 tons per year

It is desired to build a two storey building in the year 2005 with the following:

Desired building area = 100,000 square feet

Desired capacity = 10,000 tons per year

Determine:

(*a*) the cost of the new building

(*b*) the cost of the new equipment

(*c*) the total estimated cost of the building and equipment.

Solution:

(*a*) The cost of the new building would be based on the construction cost index and the capacity factor for a two-storey building.

$$C(t_d, d) = C(t, t)\left[\frac{I_{ti}}{I_{tr}}\right]\left[\frac{Q_{di}}{Q_{rr}}\right]^n$$

From Table 6.15, $n = 0.75$ (multiple storey building)

$I_{ti} = 4732$ and $I_{tr} = 1381$

$$\text{Therefore}\quad C(1990, 100{,}000) = \$900{,}000\left[\frac{4732}{1381}\right]\left[\frac{100{,}000}{70{,}000}\right]^{0.75} = \$4{,}029{,}675$$

(*b*) The cost of the new equipment would be based on the manufacturing equipment index and the capacity factor for general manufacturing equipment. Refer to Table 6.14 for I_{ti} and I_{tr} which are $I_{ti} = 1038.9$ and $I_{tr} = 330.4$ and from Table 6.13, $n = 0.06$.

$$\text{Therefore, } C(1990, 10{,}000) = \$2{,}500{,}00\left[\frac{1038.9}{330.4}\right]\left[\frac{10{,}000}{8{,}000}\right]^{0.6} = \$8{,}987{,}103.$$

(*c*) The total cost estimate would be the sum of the buildings and equipment estimates, which is

$$\$4{,}029{,}675 + \$8{,}987{,}103 = \$13{,}016{,}778$$

6.22. Determine the product improvement rate curve exponents for 55%, 68% and 77% product improvement curves.

Solution:

From $P = 40\%$

$$\frac{Y_1}{Y_2} = \left(\frac{X_1}{X_2}\right)^n$$

$X_2 = 2X_1$ and for 55% product improvement curve, $Y_2 = 0.55Y_1$.

$$\frac{Y_1}{Y_2} = \frac{Y_1}{0.55Y_1} = \frac{1}{0.55}$$

$$\frac{Y_1}{Y_2} = 2.5 = \left(\frac{X_1}{2X_1}\right)^n = \left(\frac{1}{2}\right)^n = 0.5^n$$

$$n = \frac{\log(1.818)}{\log(0.5)} = -0.863$$

For $P = 68\%$

$$\frac{Y_1}{Y_2} = \frac{Y_1}{0.68Y_1} = 1.471$$

$$n = \frac{\log(1.417)}{\log(0.5)} = -0.557$$

For $P = 77\%$

$$\frac{Y_1}{Y_2} = \frac{Y_1}{0.77Y_1} = 14.299$$

$$n = \frac{\log(1.299)}{\log(0.5)} = -0.377$$

6.23. The first of a group of 100 machines takes 200 hours to build and assemble. Assuming a 70% product improvement curve, find the time needed to complete the fiftieth machine and the last hundredth machine.

Solution:

$$Y = KX^n$$

For 70% product improvement curve, $n = -0.515$, and $K = 200$ hours.

Therefore, $Y = 200X^{-0.515}$

For $X = 50$

$$Y = 200(50)^{-0.515} = 26.672 \text{ hours}$$

For $X = 100$

$$Y = 200(100)^{-0.515} = 18.665 \text{ hours}$$

6.24. The time taken for the first unit of production is 20 minutes and the total number of units produced is 8. Assuming a 95% product improvement curve, determine

(*a*) cumulative time required to produce the eight units

(*b*) the average time per unit required to manufacture a total of 8 units

(*c*) the time saved after the production of the first 8 units.

Solution:

Use Eq.(6.36)

$$K = 20 \text{ for } 90\%$$

$$X_e = 8, \ n = -0.152$$

(*a*) Cumulative time required to product the eight units

$$T = \frac{K}{n+1}\left[(X_e + 0.5)^{n+1} - 0.5^{n+1}\right]$$

$$= \frac{20}{0.848}\,[8.5^{0.848} - 0.5^{0.848}] = \frac{20}{0.848}\,[6.1397 - 0.5555]$$

$$= 131.70 \text{ min.}$$

(*b*) The average time per unit required to manufacture a total of 8 units is given by

$$T_{avg} = \frac{T}{X_e} = \frac{131.70}{8} = 16.463 \text{ min.}$$

(*c*) The time saved over the production of the first 8 units is the difference between the cumulative time for the 8 units and 8 times the first unit time.

The time saved is T_S.

$$T_S = X_e K - T = 8(20) - 131.70 = 28.30 \text{ minutes}$$

The time saved is also obtained from

$$T_S = X_e[K - T_{avg}] = 8[20 - 16.463] = 28.296 \approx 28.30 \text{ minutes.}$$

Solution:

(*a*) Here $K_1 = 20, X_1 = 10$ and $X_2 = 50$

From Table 6.16, $n_1 = -0.234$ and $n_2 = -0.074$

From Eq. (6.44)

$$K_2 = K_1 \frac{X_1^n}{X_1^{n2}} = 20.10[-0.234 - (-0.074)] = 32 \text{ minutes}$$

Now $\quad Y = K_2 X_2^{n2} = 32(50^{-0.074}) = 23.957$ minutes.

(*b*) The total time for the 50 units is

$$T_c = \left(\frac{K_1}{n_1+1}\right)[(X_1+0.5)^{n_1+1} - 0.5^{n_1+1}] + \frac{K_2}{n_2+1}[X_2+0.5)^{n_2+1} - 0.5^{n_2+1}]$$

$$= \frac{20}{(-0.234+1)}\,[10.5^{(-0.234+1)} - 0.5^{(-0.234+1)}]$$

$$+ \frac{32}{(-0.074+1)}\,[50.5^{(-0.074+1)} - 0.5^{(-0.074+1)}]$$

$$= 26.109[6.057 - 0.588] + 34.557[37.779 - 0.526]$$

$$= 142.790 + 1287.352 = 1430.142 \text{ min.}$$

(*c*) The total standard time for 50 units when each unit time is 23.927 min is 50(23.957) = 1197.85 min.

Time lost in the product improvement process for the first 50 units is

Time lost = T_c – 1197.85 = 1430.142 – 1197.85 = 232.292 min.

CHAPTER 7

Design for Reliability

7.1. The failure density function for a class of components is given by

$$f(t) = 0.20 - 0.02t$$

where t is in years. Determine the reliability functions $Q(t)$, $R(t)$, and $\lambda(t)$ and sketch them.

Solution:

The failure distribution function is

$$Q(t) = \int_0^t f(\xi)\, d\xi = 0.20t - 0.01t^2$$

The reliability function is

$$R(t) = 1 - Q(t) = 1 - 0.20t + 0.01t^2$$

The hazard function is

$$\lambda(t) = \frac{f(t)}{R(t)} = \frac{2 - 0.20t}{10 - 2t + 0.10t^2}$$

Table P7.1

t	f(t)	Q(t)	R(t)	λ(t)
0	0.2	0	1.0	0.2
2	0.192	0.36	0.64	0.3
4	0.12	0.64	0.36	0.334
6	0.08	0.84	0.16	0.5
8	0.04	0.96	0.04	1.0
10	0	1.0	0	∞

All the four functions are summarized in Table P7.1 and sketched in Fig. P7.1.

The given density function implies that the probability of a component surviving longer than 10 years is zero. In other words, if N components are put into service at $t = 0$, then all of them will fail within 10 years.

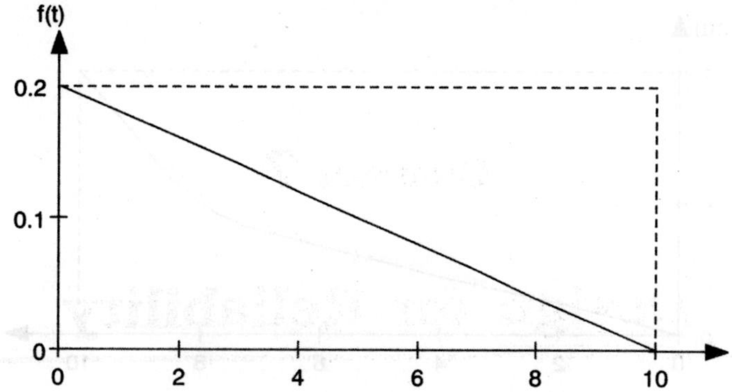

(a)

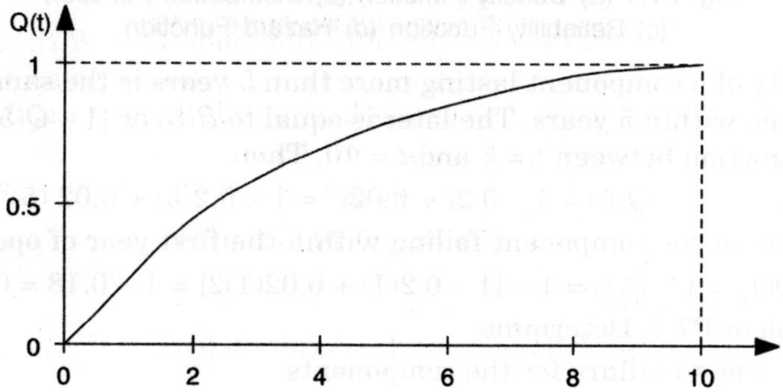

(b)

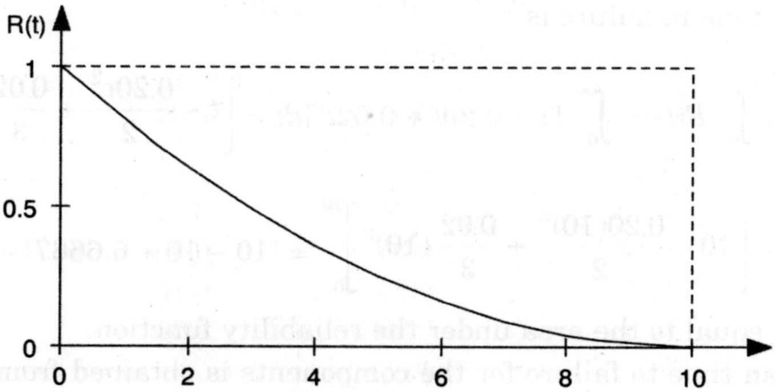

(c)

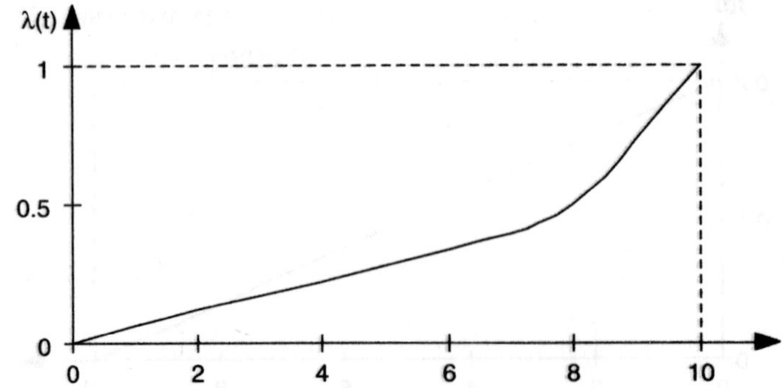

(*d*)

Fig. P7.1 (*a*) Density Function (*b*) Distribution Function (*c*) Reliability Function (*d*) Hazard Function.

The probability of a component lasting more than 5 years is the same as the probability of its not failing within 5 years. The later is equal to $R(5)$ or $[1 - Q(5)]$ or the area under the density function between $t = 5$ and $t = 10$. Then

$$R(5) = 1 - Q(5) = 1 - 0.2t + 0.02t^2 = 1 - 0.2(5) + 0.02\,(5)^2 = 0.50$$

The probability of the component failing within the first year of operation is

$$Q(1) = 1 - R(t) = 1 - [1 - 0.2(1) + 0.02(1)2] = 1 - 0.18 = 0.82$$

7.2. Refer to Problem P7.1. Determine

(*a*) the mean time to failure for the components

(*b*) the median time to failure for the components.

Solution:

(*a*) The mean time to failure is

$$\text{MTTF} = \int_0^\infty R(t) = \int_0^\infty [1 - 0.20t + 0.02t^2]dt = \left[t - \frac{0.20t^2}{2} + \frac{0.02}{3}t^3\right]_0^{10}$$

$$= \left[10 - \frac{0.20(10)^2}{2} + \frac{0.02}{3}(10)^3\right]_0^{10} = (10 - 10 + 6.6667) - 0 = 6.6667 \text{ years}$$

The MTTF is equal to the area under the reliability function.

(*b*) The median time to failure for the components is obtained from equating $Q(T_{50})$ to 0.5. Hence,

$$0.5 = 0.20\,T_{50} - (0.02)T_{50}^2$$

or $$0.02T_{50}^2 - 0.20\,T_{50} + 0.5 = 0$$

Solving for T_{50}

$$T_{50} = \frac{0.20 \pm \sqrt{(0.20)^2 - 4\,(0.02)\,(0.5)}}{2\,(0.02)} = 5$$

Since $t > 10$ has no physical relevance in this case, hence we conclude that $T_{50} = 5$ years.

7.3. The time to failure in hours of a system has the density function given as

$$f(t) = 0.002\exp[-0.002t]$$

Find the probability that the system will fail in

(*a*) less than 100 hours

(*b*) less than 1,000 hours.

Solution:

(*a*) The probability that the system will fail in less than 100 hours is

$$= \int_0^{100} f(t)dt = \int_0^{100} 0.002\, e^{-0.002t}dt$$

$$= \frac{0.002}{-0.002}\left[e^{-0.002t}\right]_0^{100} = -1[0.8187 - 1] = 0.1813$$

(*b*) The probability that the system will fail in less than 1000 hours is given by

$$\int_0^{100} f(t)dt = \int_0^{100} 0.002\, e^{-0.002t}dt = \frac{0.002}{-0.002}\left[e^{-0.002t}\right]_0^{1000}$$

$$= -1(0.1353 - 1) = 0.8647.$$

7.4. The failure rate of a certain electronic component is $\lambda = 3 \times 10^{-9}$/hour.

(*a*) determine $R(t)$ for t in hours

(*b*) find the 2-year reliability, MTTF in hours, and allowable operating time if the reliability is to be no less than 0.998.

Solution:

$$\lambda = 3 \times 10^{-9}\text{/hour}$$

$$R(t) = \exp[-3 \times 10^{-9}t]$$

$$R(2 \times 8760) = \exp[-3 \times 10^{-9} \times (2 \times 8760)] = 0.9999$$

Note: 1 year has 8760 hours.

$$\text{MTTF} = \frac{1}{3 \times 10^{-9}} = 3.333 \times 10^{9}$$

Let t_n = allowable operating time.

Then

$$\exp[-3 \times 10^{-9}t] = 0.998$$

$$e^{-3\times10^{-9}t_n} = 0.998$$

$$-3 \times 10^{-9}t_n = \text{loge}^{0.998}$$

$$t_n = \frac{-8.69458 \times 10^{-4}}{-3 \times 10^{-9}}$$

Solving for t_n, we get

$$t_n = 289{,}819.57 \text{ hours or } 33.084 \text{ years.}$$

Hence, the operating time should be limited to less than this value.

7.5. The reliability of a class of components is approximated by

$$R(t) = (1 - bt)^n \quad 0 \le t \le 1/b \text{ and } a > 0, n > 0$$

Determine

(*a*) the corresponding failure density function

(*b*) failure distribution function

(*c*) hazard function

(*d*) *MTTF*

(*e*) the median time to failure.

Solution:

Given $R(t) = (1 - bt)^n$ for $0 \le t \le 1/b$ and $b > 0$, $n > 0$.

(*a*) The failure density function is given by

$$f(t) = \frac{-d}{dt} R(t) = bn(1 - bt)^{n-1}$$

(*b*) The failure distribution function is given by

$$Q(t) = 1 - R(t) = 1 - (1 - bt)$$

(*c*) The hazard function is given by

$$\lambda(t) = f(t)/R(t) = \frac{bn}{(1 - bt)} \text{ for } 0 \le t \le 1/b$$

(*d*) The mean time to failure is given by

$$\text{MTTF} = \int_0^{1/b} (1 - bt)^n \, dt = \frac{1}{[b(n + 1)]}$$

(*e*) Median time to failure t_{50} is obtained from

$$Q(t_{50}) = 0.5$$

or

$$1 - (1 - bt_{50})^n = 0.5$$

from which

$$t_{50} = \left(\frac{1}{b}\right)[1 - 0.5^{1/n}]$$

7.6. The time to failure of a mechanical component follows the exponential distribution. The probability that the component would not survive for more than 60 days is 0.97. How often can one expect to replace this mechanical component?

Solution:

Given that

$$1 - \exp(-60\,\lambda) = 0.97$$

$$-e^{-60\lambda} = -0.03$$

$$e^{-60\lambda} = 0.03$$

$$-60\,\lambda = \log 0.03$$

Solving for λ, we get

$$\lambda = \frac{-1.522878}{-60} = 0.02538$$

$$E[\text{time-to-failure}] = \text{MTTF} = \frac{1}{\lambda} = \frac{1}{(0.02538)} = 39.40 \text{ days or say 40 days.}$$

7.7. The lifetime of an electromechanical component is estimated to be 50 hours. Find the probability that an electromechanical component of this type would survive for 2 weeks when subjected to a similar working environmental conditions.

Solution:

The failure rate $= \left(\frac{1}{50}\right)^{-1}$

2 weeks correspond to 336 hours.

Hence $R(336) = \exp(-336/50) = \frac{1}{e^{6.72}} = \frac{1}{828.82} = 0.001206$

There is a 1.206% probability that the chip would survive 2 weeks.

7.8. The reliability function for a mechanical component is given by

$$R(t) = 1 - \frac{t^3}{b^3} \quad \text{for } 0 \le t \le b$$

where b is a constant parameter. Find $\lambda(t)$, MTTF, AFR and the average failure rate up to the MTTF.

Solution:

$$R(t) = 1 - \frac{t^3}{b^3} \quad 0 \le t \le b$$

Therefore

(*a*)
$$f(t) = \frac{3t^2}{b^3}$$

(*b*)
$$\lambda(t) = \frac{\frac{3t^2}{b^3}}{\frac{b^3 - t^3}{b^3}} = \frac{3t^2}{(b^3 - t^3)} \quad \text{for } 0 \le t \le b$$

(*c*)
$$\text{MTTF} = \int_0^b \left(1 - \frac{t^3}{b^3}\right) dt = \left[t - \frac{t^4}{4b^3}\right]_0^b = b - \frac{b^4}{4b^3} = \frac{3}{4}b$$

(*d*)
$$1 - \frac{t_{med}^3}{b^3} = 0.5$$

$$t_{med} = \sqrt[3]{0.5b^3} = 0.7937\, b$$

$$\text{AFR}(t) = \frac{-\ln\left(1 - \frac{t^3}{b^3}\right)}{t}$$

Average failure rate up to the MTTF is

$$\text{AFR(MTTF)} = \text{AFR}\left(\frac{3}{4}b\right) = \frac{-\ln\left[1 - \frac{\left(\frac{3}{4}b\right)^3}{b^3}\right]}{\left(\frac{3}{4}b\right)} = \frac{0.7309}{b}$$

7.9. The hazard rate function for a system is given by

$$\lambda(t) = \frac{0.3}{700}\left(\frac{t}{700}\right)^{-0.5} \quad t \text{ in years.}$$

(*a*) calculate the design life for a reliability of 0.87

(*b*) if the burn-in period T_0 is 0.5 for a six-month burn-in period, calculate the design life for a reliability of 0.87.

Solution:

$$\lambda(t) = \frac{0.3}{700}\left(\frac{t}{700}\right)^{-0.5}$$

$$R(t) = \exp - \left(\frac{t}{700}\right)^{\frac{1}{2}} = 0.87$$

Hence, design life $t = 700(-\ln 0.87)^2 = 13.5757$ years.

For $T_0 = 0.5$, a six-month burn-in period, then

$$R\left(\frac{t}{T_0}\right) = \frac{R\,(t + 0.5)}{R\,(0.5)} = \frac{\exp - \left[\frac{(t+0.5)}{700}\right]^{0.5}}{\exp - \left(\frac{0.5}{700}\right)^{0.5}} = 0.87$$

and $$t = 700\left[-\ln 0.87 + \left(\frac{0.5}{700}\right)^{0.5}\right]^2 - 0.5$$

$$= 700[0.1393 + 0.0267]^2 - 0.5 = 18.795 \text{ years.}$$

Increase in the design life = 18.7953 – 13.5757 = 5.2196 years as a result of a six-month burn-in period.

7.10. The lifetime of a mechanical component is exponentially distributed with a mean value of 600 hours. Find the probability of this component operating successfully for at least 800 hours.

Solution:

Since the MTTF is 600 hours, the hazard rate

$$\lambda = \left(\frac{1}{600}\right) \text{hour}^{-1}$$

and $$P[\text{life time} \geq 800 \text{ hours}] = \exp[-800/600] = 0.2636.$$

If we have 3 such devices, the probability of at least one failing during the first 500 hours is equal to

$$[1 - P(\text{all 3 surviving for 500 hours})]$$

Hence, P[at least 1 fails within 500 hours]

$$= 1 - [\exp(-500/600)]^3 = 0.91793$$

Alternative method:

We use the exponential distribution to obtain the probabilities of failure and success and then to use these values in the binomial distribution. Hence, for each device,

$$p = P[\text{success in 500 hours}] = \exp(-500/600) = 0.43457$$
$$q = P[\text{failure in 500 hours}] = 1 - 0.43457 = 0.56543$$

P[at least 1 fails within 500 hours] $= 3C_1\, q\, p^2 + 3C_2\, q^2\, p + 3C_3\, q^3$
$= 0.32035 + 0.41681 + 0.18077 = 0.91793$ as before.

If we have 4 such devices, the probability of exactly two of them failing during the first 400 hours is computed as follows. For each device,

$$p = P(\text{success in 400 hours}] = \exp(-400/600) = 0.5134$$

P[exactly 2 out of 4 failing during the first 400 hours]

$$= 4C_2[1 - (0.5134)\,(0.5134)^2] = 0.37446$$

The probability of the device failing between the Tth and the $(T + 100)^{th}$ hours of operation is the "a posteriori" probability

$$Q_c(100) = 1 - \exp(-100/600) = 0.15352$$

This value does not depend upon T for the exponential distribution. It depends only on the duration of time considered, while for this example, is 100 hours.

The probability of the device failing between the Tth and the $(T + 200)^{th}$ hours of operation is the "a posteriori" probability

$$Q_c(200) = 1 - \exp(-200/600) = 0.28352.$$

7.11. The time-to-failure density function (PDF) for a system is given by

$$f(t) = \frac{2}{a^2} \text{ for } 0 \le t \le a$$

Determine $F(t)$, $R(t)$, $\lambda(t)$, MTTF, T_{med}, T_{mode}, and σ.

Solution:

$$F(t) = \int_0^t f(t')\,dt' = \int_0^t \frac{2}{a^2}\,dt' = \frac{2t}{a^2}\,; \qquad 0 \le t \le a$$

$$R(t) = 1 - F(t) = 1 - \frac{2t}{a^2} = \frac{a^2 - 2t}{a^2}$$

$$\lambda(t) = \frac{f(t)}{R(t)} = \frac{\left(\dfrac{2}{a^2}\right)}{\left(\dfrac{a^2 - 2t}{a^2}\right)} = \frac{2}{a^2 - 2t} \text{ (IFR)}$$

$$\text{MTTF} = \int_0^a R(t)\,dt = \int_0^a \left(\frac{a^2 - 2t}{a^2}\right)dt$$

$$= \frac{1}{a^2}\left[a^2 t - \frac{2t^2}{2}\right]_0^a = \frac{1}{a^2}[(a^3 - a^2) - 0] = a - 1$$

$$R(t_{median}) = \frac{a^2 - 2\,t_{med}}{a^2} = 0.5$$

or
$$1 - \frac{2\,t_{med}}{a^2} = 0.5$$

$$t_{med} = \frac{a^2}{2}(0.5) = 0.25\, a^2$$

$$\sigma^2 = \int_0^a t^2 f(t)dt - (\text{MTTF})^2 = \int_0^a t^2\left(\frac{a^2}{2}\right)dt - (a-1)^2$$

$$= \left[\frac{2t^3}{3a^2}\right]_0^a - (a-1)^2 = \frac{2a^3}{3a^2} - (a-1)^2 = \frac{2}{3}a - (a^2 + 1 - 2a)$$

$$\sigma^2 = (-a^2 + 1 + 2.667a)$$

$$\sigma = \sqrt{(-a^2 + 2.667a - 1)}$$

7.12. A vacuum cleaner is guaranteed to have more than 15 years of life. If the time-to-density function (PDF) for this vacuum cleaner is given by

$$f(t) = 0.2(1 + 0.03t)^{-3} \text{ for } t \geq 0.$$

(*a*) find the reliability for the next 15 years if it has survived a 1-year warranty period

(*b*) find the MTTF for this vacuum cleaner before the warranty period and after the warranty period.

Solution:

$$R(t) = 0.2\int_t^\infty (1 + 0.03t')^{-3} = 0.2\left[\frac{(1+0.03t')^{-2}}{-2\,(0.03)}\right]_t^\infty$$

$$= 0 - [-(1 + 0.03\,t)^{-2}]\,\frac{0.2}{0.06} = \frac{1}{3}(1 + 0.03)^{-2}$$

(*a*)
$$R\left(\frac{10}{1}\right) = \frac{R(10+1)}{R(1)} = \frac{\frac{1}{3}(1+0.03(10))^{-2}}{\frac{1}{3}(1+0.03(1))^{-2}} = \frac{(1+0.03)^2}{(1+0.30)^2} = 0.6277$$

(*b*)
$$\text{MTTF}_{before} = \int_0^\infty R(t)dt = \int_0^\infty \frac{1}{3}(1 + 0.03t)^{-2}dt$$

$$= \frac{1}{3}\left[\frac{(1+0.03t)^{-1}}{-0.03}\right]_0^\infty = 0 - \left(\frac{-1}{3\,(0.03)}\right) = 11.11$$

$$\text{MTT}_{after} = \int_{T_0}^\infty R\left(\frac{t}{T_0}\right)dt = \frac{t}{R\,(T_0)}\int_{T_0}^\infty R(T_0 + t)dt$$

$$= \frac{1}{\frac{1}{3}[1+0.03(1)^{-2}}\int_1^\infty \frac{1}{3}\,[1 + 0.03(1 + t)]^{-2}dt$$

$$= 3(1.03)^2\int_1^\infty \frac{1}{3}\,(1.03 + 0.03\,t)^{-2}dt$$

$$= \frac{3}{3}(1.03)^2 \left[\frac{(1.03 + 0.03t)^{-1}}{-(0.03)} \right]_1^\infty$$

$$= (1.03)^2 \left[\frac{1}{0.03\,[1.03 + 0.03\,(1)]} \right] = 33.36$$

7.13. Find the reliability of an electronic system for 150-hr operation, which has two independent failure modes, namely, a constant failure rate of 0.0002 and a linearly increasing failure rate given by $\lambda(t) = 3 \times 10^{-5}t$.

Solution:

$$\lambda_1(t) = 0.0002 \text{ and } \lambda_2(t) = 3 \times 10^{-5}t \text{ hr}$$

$$\lambda(t) = \sum_{i=1}^{2} \lambda_i(t) = \lambda_1(t) + \lambda_2(t) = 0.0002 + 3 \times 10^{-5}t$$

$$R(t) = e^{-\int_0^t \lambda(t')dt'} = e^{-\int_0^t (0.0002 + 3\times 10^{-5}t')dt'} = e^{-(0.0002t + \frac{t^2}{2}(3\times 10^{-5}))}$$

$$R(150) = e^{-\left[(0.0002\,(150) + \frac{150^2}{2}(3\times 10^{-5})\right]}$$

$$= e^{-[0.03 + 0.3375]} = e^{-0.3675} = 0.71353.$$

7.14. Find the system MTTF and the median time to failure if the design life is 15 years with a reliability of 0.98 is desired. Assume that the system has 25 identical components and will be replaced on failure (renewal process). Determine the component MTTF and the median time to failure if each component has a CFR.

Solution:

$$R(t_d) = e^{-\lambda(t_d)} = R(15) = e^{-\lambda(15)} = 0.98$$

The system's constant failure rate, λ is therefore given by

$$\lambda = \frac{\ln(0.98)}{15} = 1.347 \times 10^{-3}$$

$$\text{MTTF} = \frac{1}{\lambda} = \frac{15}{-\ln(0.98)} = 742.47 \text{ years}$$

and
$$t_{med} = \frac{-\ln(0.5)}{\lambda} = \frac{0.69315}{\lambda} = 0.69315(\text{MTTF})$$

$$= 0.69315(742.47) = 514.64 \text{ years}$$

Since the components are identical,

$$\lambda_i = \frac{\lambda}{n} = -\frac{\ln(0.98)}{(15)(25)}$$

$$\text{MTTF}_i = \frac{1}{\lambda_i} = \frac{(15)(25)}{-\ln(0.98)} = 18{,}561.75 \text{ years}$$

$$t_{med_i} = 0.69315(18{,}561.75) = 12{,}866.07 \text{ years.}$$

7.15. A TV remote control contains two batteries each having an MTTF of 4 operating hours (CFR).

(*a*) determine the probability of battery failure happening within three hours of operation

(*b*) find the probability of more than one failure occurring during the first 4 hours of operation, if the failed batteries are replaced instantly

(*c*) determine whether the batteries have a constant failure rate (CFR) or increasing failure rate (IFR).

Solution:

(*a*)
$$\lambda = \sum_{1}^{2} \lambda_i = \sum_{1}^{2} \frac{1}{\text{MTTF}_i} = \frac{1}{4} + \frac{1}{4} = 0.50$$

$$F(3) = 1 - R(3) = 1 - e^{-0.5(3)} = 0.7768$$

(*b*) P (more than 1 failure in first 4 hours) = 1 – P (number of failures in first 4 hours) – P(1 failure in first 4 hours)

$$= 1 - P_0(4) - P_1(4) = 1 - e^{-\lambda t} - e^{-\lambda t}(\lambda t)$$
$$= 1 - e^{-0.5(4)} - e^{-0.5(4)}(0.5)\,(4)$$
$$= 1 - 0.1353 - (0.1353)\,(0.5)\,(4)$$
$$= 1 - 0.1353 - 0.2706 = 0.5941$$

(*c*) No. Would expect batteries to have an increasing failure rate (IFR) due to wear-out.

7.16. Derive a general expression for the hazard rate function for an exponentially reliability model given by

$$R(t) = b^{-at}$$

where a and b are constant parameters.

Solution:

$$f(t) = \frac{-dR(t)}{dt} = ab^{-at}(\ln b)$$

$$\lambda(t) = \frac{f(t)}{R(t)} = a \ln b\,; \quad b \text{ is a constant}$$

Let $\lambda = a \ln b = \ln b^a$, then

$$e^{\lambda} = b^a \quad \text{and} \quad R(t) = (b^a)^{-t} = e^{-\lambda t}$$

7.17. Given the reliability function for a system given by

$$R(t) = e^{-0.002(t-100)} \quad t \geq 100$$

Find the MTTF, t_{med}, $t_{0.95}$, and the standard deviation σ.

Solution:

$$R(t) = e^{-0.002(t-100)}\,;\; t \geq 100$$

$$\text{MTTF} = 100 + \frac{1}{0.002} = 600$$

$$t_{med} = 100 + \frac{0.69315}{0.002} = 446.57$$

$$t_{0.95} = 100 - \ln\left(\frac{0.95}{0.002}\right) = 93.837$$

$$\sigma = \frac{1}{0.002} = 500$$

7.18. An electromechanical system has three components in series. Each component has a Weibull failure pattern with β equal to 2. The characteristic life times for the three components are 10,000 hr, 15,000 hr, and 20,000 hr respectively. Determine the MTTF and the design life of the electromechanical system corresponding to a reliability of 0.95.

Solution:

The electromechanical system has a Weibull distribution with $\beta = 2$ and

$$\theta = \left[\sum_{i-1}^{3}\left(\frac{1}{\theta_i}\right)^{\beta}\right]^{-\frac{1}{\beta}}$$

$$\theta = \left[\left(\frac{1}{10{,}000}\right)^2 + \left(\frac{1}{15{,}000}\right)^2 + \left(\frac{1}{20{,}000}\right)^2\right]^{-\frac{1}{2}}$$

$$= [1 \times 10^{-8} + 0.444 \times 10^{-8} + 0.25 \times 10^{-8}]^{-1/2}$$

$$= \frac{1}{1.301 \times 10^{-4}} = 7686.4 \text{ hours}$$

$$\text{MTTF} = \theta\Gamma\left(1 + \frac{1}{\beta}\right) = 7686.4(\Gamma(1.5))$$

$$= 7686.4(0.88623) = 6811.918 \text{ hrs.}$$

where, from the table in Appendix 7-B, we find $\Gamma(1.5) = 0.88623$

$$R(t_d) = e^{-\left(\frac{t_d}{\theta}\right)^{\beta}} = 0.95$$

$$t_d = \theta[-\ln(0.95)]^{1/2} = 7686.4[-\ln(0.95)]^{1/2} = 1740.82 \text{ hrs.}$$

7.19. A mechanical system has two failure modes due to external conditions and wear-out respectively. The first one has a constant failure rate of 0.06 failures per year. The second has a Weibull distribution with a characteristic life of 15 years and a shape factor of 2. Find the system reliability for a design life of 2 years. Also calculate the median time to failure.

Solution:

$$\lambda(t) = 0.06 + \frac{2}{15}\left(\frac{t}{15}\right)^{2-1}$$

or

$$R(t) = e^{-\left[0.06t + \left(\frac{t}{15}\right)^2\right]}$$

and

$$R(2) = e^{-\left[0.06(2) + \left(\frac{2}{15}\right)^2\right]} = 0.8713$$

For t_{med}, set $R(t_{med}) = e^{-\left[0.06t_{med} + \left(\frac{t_{med}}{1}\right)^2\right]} = 0.50$

And solving for t_{med} gives

$$0.06t_{med} + \left(\frac{t_{med}}{15}\right)^2 = 0.6931$$

from which

$$t_{med} = 7.45 \text{ years, and by trial and error}$$

$$t_{med} = 7.2 : 0.06(7.2) + \left(\frac{7.2}{15}\right)^2 = 0.432 + 0.2304 = 0.6624$$

$$t_{med} = 7.3 : 0.06(7.3) + \left(\frac{7.3}{15}\right)^2 = 0.438 + 0.2368 = 0.6748$$

$$t_{med} = 7.4 : 0.06(7.4) + \left(\frac{7.4}{15}\right)^2 = 0.444 + 0.2434 = 0.6873$$

$$t_{med} = 7.45 : 0.06(7.45) + \left(\frac{7.45}{15}\right)^2 = 0.447 + 0.2466 = 0.6936$$

7.20. A Physician specialist can serve four patients on any one day. The number of patients arriving has a Poisson distribution with an expected value of three per day.

(*a*) find the probability of turning away patients on any given day

(*b*) find the expected number of

(*i*) patients served per day

(*ii*) patients turned away per day.

Solution:

(*a*) P[zero patients arriving] = $3^0 e^{-3} = 0.049787$

P[1 patient arriving] = $3^1 e^{-3} = 0.149361$

P[2 patients arriving] = $3^2 e^{-3}/2! = 0.224042$

P[3 patients arriving] = $3^3 e^{-3}/3! = 0.224042$

P[4 patients arriving] = $3^4 e^{-3}/4! = 0.168037$

The sum of these five probabilities is 0.815269.

(*b*) (*i*) E(patients served per day) = 1(0.149361) + 2(0.224042) + 3(0.224042) + 4[1 – (0.815269 – 0.168031)] = 2.680619

(*ii*) P[patients turned away per day] = 1 – 0.815263 = 0.184737

E[patients turned away per day] = 3 – 2.680643 = 0.319381

7.21. Extensive testing of prototype of two designs of a mechanical component showed that the failure is a Weibull distribution in both cases. Based on the following information, which design should be selected if the component should have

(*i*) a 12-hour guaranteed life

(*ii*) a 24-hour guaranteed life

Design I : Costs \$6,000 to build and $\beta = 3$ and $\alpha = 180$ hours

Design II : Costs \$5,000 to build and $\beta = 4$ and $\alpha = 110$ hours.

Solution:

$$R(t) = \exp\left[-\left(\frac{t}{\alpha}\right)^{\beta}\right]$$

For 12 hour life

R_I : exp[– $(12/180)^3$] = exp(–0.0002963) = 0.999703

R_{II} : exp[– $(12/110)^4$] = exp(–0.000141629) = 0.999858

For 24 hour life

R_I : exp[–$(24/180)^3$] = exp(–0.0023707) = 0.9976258

R_{II} : exp[–$(24/110)^4$] = exp(–0.002266074) = 0.9977365

Cost : R_I : \$6,000

R_{II} : \$5,000

Design II is the choice, since it is slightly more reliable and costs less.

It depends on whether the \$1000 increases in cost worth the very slight improvement in reliability in design I. If the response is yes, then the choice would be design I. Otherwise, the choice is design II.

7.22. The current gain of a transistor is proportional to ln β, where β is the ratio of output to input currents. The current gains of a certain transistor are normally distributed with a mean of 1.95 and a variance of 0.01. Determine the probability that β is between 6 and 8.

Solution:

We observe that β has a log normal distribution. Also,

$$z_1 = \frac{\ln 6 - 1.95}{0.1} = \frac{1.79176 - 1.95}{0.1} = -1.5824$$

$$z_2 = \frac{\ln 8 - 1.95}{0.1} = 1.29440$$

Referring to the table of normal curve areas in Appendix-2E, we read

$$\int_{-1.5824}^{1.2944} f(z)\, dz = 0.7675$$

Hence

$$P[6 \le z \le 8] = 76.75\%.$$

7.23. The breakdown of a certain machine component is normally distributed with 90 percent of the failures occurring between 200 and 270 hours of use.

(*a*) find the MTTF and the standard deviation of failure times

(*b*) determine the reliability of the component to be used for 220 hours and then replaced

(*c*) find the design life if no more that a 2 percent probability of failure prior to replacement is to be tolerated

(*d*) determine the conditional reliability for a 20-hour use if the part has been in operation for 200 hours.

Solution:

$$(a)\ P\left\{\frac{200-\mu}{\sigma} \le z \le \frac{270-\mu}{\sigma}\right\} = 0.90 = P\{-1.645 \le z \le 1.645\}$$

$$\frac{200-\mu}{\sigma} = -1.645 \quad \text{and} \quad \frac{270-\mu}{\sigma} = 1.645$$

Therefore

$$\sigma = -\left(\frac{200-\mu}{1.645}\right) \quad \text{and} \quad \frac{270-\mu}{1.645} = -\left(\frac{200-\mu}{1.645}\right)$$

From which

$$\mu = 235 = \text{MTTF}$$

and

$$\sigma = -\left(\frac{200-235}{1.645}\right) = 21.2766$$

(*b*)

$$R(220) = 1 - \Phi\left(\frac{220-\mu}{\sigma}\right) = 1 - \Phi\left(\frac{220-235}{21.2766}\right)$$

$$= 1 - \Phi(-0.705) = 1 - 0.24 = 0.76$$

(*c*)

$$R(t_d) = 1 - \Phi\left[\frac{t_d - 235}{21.2766}\right] = 0.98$$

or

$$\Phi\left[\frac{t_d - 235}{21.2766}\right] = 0.02$$

$$\frac{t_d - 235}{21.2766} = -2.055$$

$$t_d = 191.276 \text{ hours}$$

(*d*)

$$R\frac{20}{200} = \frac{R(20+200)}{R(200)} = \frac{R(220)}{R(200)}$$

$$= \frac{0.76}{1 - \Phi\left[\frac{200-235}{21.2766}\right]} = \frac{0.76}{1 - \Phi(-1.645)}$$

$$= 0.76/0.95 = 0.8.$$

7.24. **Past experience has indicated that an average of 8% of the critical mechanical components supplied by a particular vendor is defective. In a recent purchase of a lot of 300 components, determine the probability of finding at least 35 defective components.**

Solution:

P[a component is good] = 0.92

P[a component is defective] = 0.08

With $n = 300$, the expected number of defective components is (300×0.08), or 24, and a standard deviation $\sigma = \sqrt{300\,(0.92)\,(0.08)} = \sqrt{22.08} = 4.6989$. Also, since $n = 300$, we can use the normal approximation to the binomial distribution, with $\mu = 24$ and $\sigma = 4.6989$. Hence

$$z = \frac{35-24}{4.6989} = 2.3409$$

From the table of areas under the normal curves, Appendix-2E, we obtain

$$\int_{2.3409}^{\infty} f(z)\, dz = 0.990358$$

Hence

P[finding at least 35 defective components]

$= P$[35 or more defective] = 0.990358 or 99.0358 %

The probability of finding 15 or more components defective is obtained as follows:

$$z_1 = \frac{15 - 24}{4.6989} = -1.9153$$

$$\int_{-1.9153}^{0} f(z)\, dz = \int_{0}^{-1.9153} f(z)\, dz = 0.4907$$

P[finding at least 15 or more defective components] = 0.5 + 0.4907 = 0.9907 = 99.07%

Also, P[24 or fewer components defective] = P[24 or more defective] = 0.5

7.25. A mechanical component has a normally distributed life with a mean value of 5,000 hours. The design need of such a system using these components requires that 95% of them should last for at least 4,500 hours. Find the largest variance allowable for the lifetime of the components.

Solution:

Let T be the random variable called lifetime, in hours, for the mechanical component. We need that $P[T > 4500] = 0.95$. Now, referring to the table of areas under normal curves in Appendix-2E, we observe that $z = -1.65$ for

$$\int_{z}^{\infty} f(z)\, dz = 0.95$$

Hence, $z = -1.65$ should correspond to $T = 4500$ hours, and

$$\frac{4500 - 5000}{\sigma} = -1.65$$

or $\sigma = 303.03$ hours.

Hence, Maximum allowable variance = σ^2 = 918,27.36 hours^2.

7.26. The lifetimes of a certain component manufactured by a company are found to be normally distributed with a mean of 1,000 days and a standard deviation of 100 days. Taking one month as 30 days, for how many months can the manufacturer guarantee these components so that no more than 6% of them need replacement?

Solution:

Referring to the table of areas under the normal curves, we find that

$$\int_{0}^{1.555} f(z)\, dz = 0.44$$

Denoting d_1, as the number of days by which 6% of the components fail or need replacement, we have

$$\frac{d_1 - 1000}{100} = -1.555$$

and d_1 = 844.5 days, or 844.5/30 months, or 28.15 months. Therefore, the manufacturer's guarantee should not exceed 28 months.

7.27. A mechanical system component has been found to have a Weibull distribution with a shape parameter of 2.2 and a characteristic life of 15,000 hours. Determine:

(*a*) R(7000 hr)

(*b*) the $B1$ and $B.1$ life

(*c*) the MTTF and the standard deviation

(*d*) the median and the mode

(*e*) the probability of failure in the first year of continued use.

Solution:

(*a*)
$$R(t) = e^{-\left(\frac{t}{\theta}\right)^{\beta}}$$

$$R(7000) = e^{-\left(\frac{7000}{15000}\right)^{2.2}} = 0.8295$$

(*b*)
$$B1 \to R(t_{0.99}) = 0.99$$

$$t_{0.99} = \theta\,[-\ln(0.99)]^{\frac{1}{\beta}} = 15000\,[-\ln(0.99)]^{\frac{1}{2.2}} = 1853.49 \text{ hours}$$

$$B.1 \to t_{0.999} = 0.999$$

$$t_{0.999} = 15{,}000(-\ln(0.999))^{1/2.2} = 649.46 \text{ hours.}$$

(*c*)
$$\text{MTTF} = \theta\Gamma\left(1+\frac{1}{\beta}\right) = 15{,}000\Gamma\left(1+\frac{1}{2.2}\right)$$

$$= 15{,}000\Gamma(1.4545) = 15{,}000(0.885630) = 13{,}284.45 \text{ hours}$$

$$\sigma^2 = \theta^2\left\{\Gamma\left(1+\frac{2}{\beta}\right) - \left[\Gamma\left(1+\frac{1}{\beta}\right)\right]^2\right\}$$

$$= 15{,}000^2\ \Gamma(1.909) - [\Gamma(1.4545)]^2\}$$

$$= 15{,}000^2\{0.964 - 0.7843\} = 40{,}432{,}500$$

or
$$\sigma = 6{,}358.65 \text{ hours.}$$

(*d*)
$$t_{0.50} = t_{\text{med}} = \theta[-\ln(0.5)]^{\frac{1}{\beta}} = 15{,}000[-\ln(0.5)]^{\frac{1}{2.2}}$$

$$= 12{,}698.11 \text{ hours}$$

$$t_{\text{mode}} = \theta\left[1-\frac{1}{\beta}\right]^{\frac{1}{\beta}} = 15{,}000\left[1-\frac{1}{2.2}\right]^{\frac{1}{2.2}}$$

$$= 11{,}387.7 \text{ hours}$$

(*e*)
$$F(1 \text{ year}) = F(8760) = 1 - R(8760)$$

$$= 1 - e^{-\left(\frac{8760}{15000}\right)^{2.2}} = 0.2638$$

7.28. The reliability deterioration ratio for n identical and independent components in a series structure each with a reliability of R is given by

$$\text{Deterioration ratio} = \frac{\text{Reliability of single component}}{\text{Reliability of the series structure}}$$

Compute the deterioration ratio as a function of R (say $R = 0.1$ to 0.99) and $n = 2$ to 10.

Solution:

The results are summarized in Table P7.28 and sketched in Fig.P7.28.

Table P7.28

Reliability Deterioration Ratio for n Components in Series is R/R^n or R^{1-n}

R =	0.1	0.2	0.4	0.6	0.8	0.9	0.99
$n = 2$	100	5	2.50	1.67	1.25	1.11	1.0101
$n = 4$	1000	125	15.62	4.63	1.95	1.37	1.0306
$n = 6$	100000	3125	97.65	12.86	3.05	1.69	1.0515
$n = 8$	10000000	78125	610.35	35.72	4.77	2.09	1.0729
$n = 10$	1000000000	1953125	3814.70	99.23	7.45	2.58	1.0947

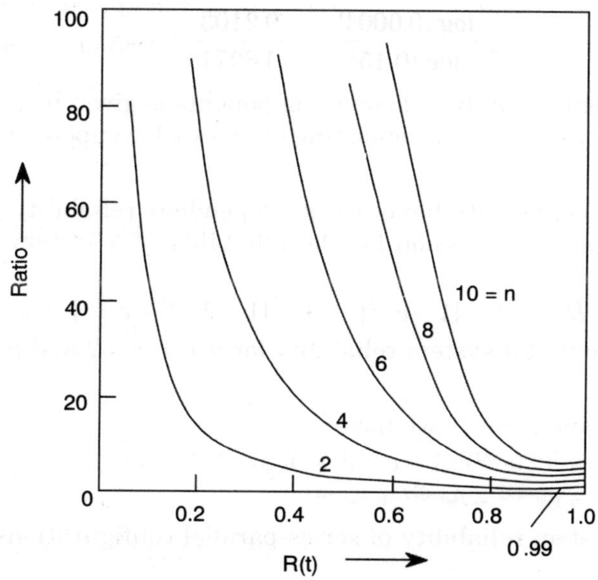

Fig. P7.28 Reliability Deterioration Ratio for the n Components.

7.29. The reliability of individual components in a parallel structure is 0.85.

(*a*) determine the reliability of the system if two such components are used

(*b*) determine the reliability of the system if three such components are used

(*c*) find the number n of components in parallel needed to achieve an overall system reliability of 0.9999.

Solution:

(*a*) The individual reliability of the component is 0.85. When two such components are in parallel,

$$R_p = 1 - (0.15)\,(0.15) = 0.9775$$

Hence, a $\left(\dfrac{0.9775 - 0.85}{0.85}\right)$ 100 or 15% improvement in reliability is accomplished over that of one component.

(*b*) When three such components are used in parallel, we have

$$R_p = 1 - (0.15)\,(0.15)\,(0.15) = 0.996625$$

or a $\left(\dfrac{0.996625 - 0.85}{0.85}\right)$ 100 or 17.25% improvement compared to the reliability of one component.

(*c*) The number of components *n* in parallel needed to accomplish an overall system reliability of 0.9999 is given by

$$0.9999 = 1 - (0.15)^n$$

which can be written as

$$(0.15)^n = 0.0001$$

from which

$$n = \frac{\log(0.0001)}{\log(0.15)} = \frac{-9.2103}{-1.8971} = 4.85 \text{ or 5 components.}$$

7.30. The time-dependent reliability of some components is given by a function of $e^{-\lambda t}$. Determine the reliability of a system consisting of two such components in parallel structure.

Solution:

Assuming some components have a time-dependent reliability given by the function $e^{-\lambda t}$, we can obtain an expression for the reliability of a system with two such components in parallel as

$$R(t) = 1 - [1 - e^{-\lambda t}]^2 = 1 - [1 - 2e^{-\lambda t} + e^{-2\lambda t}] = 2e^{-\lambda t} - e^{-2\lambda t}$$

7.31. Determine the *k*-out-of-*n* system reliability for $n = 4$, $r = 2$ and $p = e^{-\lambda t}$.

Solution:

For $n = 4$, $r = 2$ and $p = e^{-\lambda t}$, we have

$$R(t) = {}^4C_2(e^{-\lambda t})^2(1 - e^{-\lambda t})^2 + {}^4C_3(e^{-\lambda t})^3(1 - e^{-\lambda t})^1 + {}^4C_4(e^{-\lambda t})^4(1 - e^{-\lambda t})^0$$
$$= 6e^{-2\lambda t} - 8e^{-3\lambda t} + 3e^{-4\lambda t}$$

7.32. Determine the system reliability of series-parallel configurations shown in Figs. P7.32 (*a*) and (*b*).

Solution:

(*a*) $R_s = 0.95[1 - (0.1)\,(0.1)\,(0.1)]\;[1 - \{1 - 0.85[1 - (0.2)\,(0.2)]\}(1 - 0.85)]0.90$
$= (0.95)\,(0.999)\;[1 - \{1 - 0.85(0.96)\}(0.15)]0.90$
$= 0.94905[0.9724](0.90) = 0.8306$

(*b*) $R_s = 0.90[1 - (1 - 0.85)^4][1 - (1 - 0.95\{1 - (0.25)(0.25)\}\;[1 - (0.95)(0.95)(0.95)]]0.85$
$= (0.9)(0.9995)[1 - (1.8906(0.1426))]0.85$
$= 0.89955(0.1269][0.85] = 0.0971$

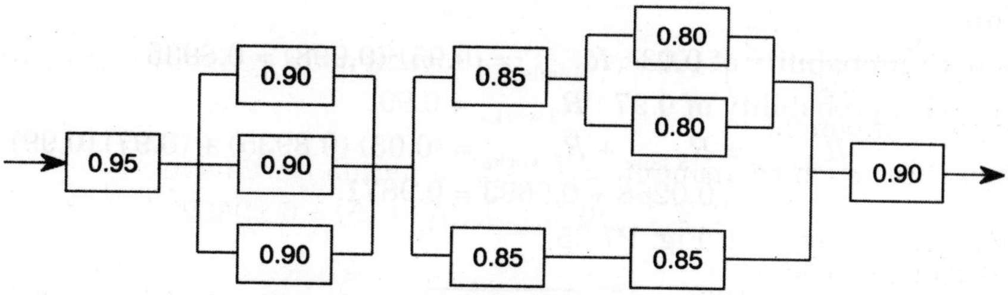

(*a*)

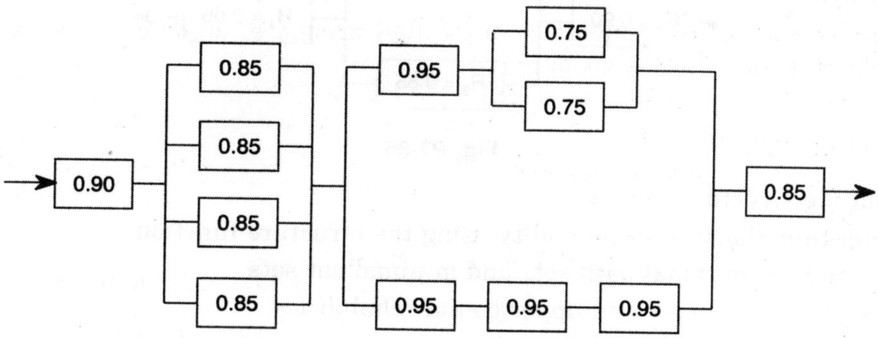

(*b*)

Fig. P7.32

7.33. Determine the system reliability if $R = 0.90$ and $R_1 = 0.98$ for the system shown in Fig. P7.34.

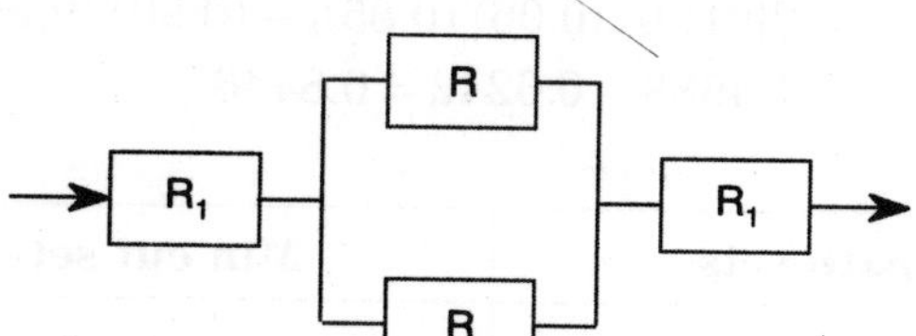

Fig. P7.33

Solution:

With no redundancy $R = 0.90$

With redundancy

$$R = R_1[1 - (1 - R)^2]R_1 = 0.98[1 - (1 - 0.90)^2]0.98 = 0.9508$$

Hence redundancy improves reliability.

7.34. Find the reliability of the system shown in Fig. P7.34.

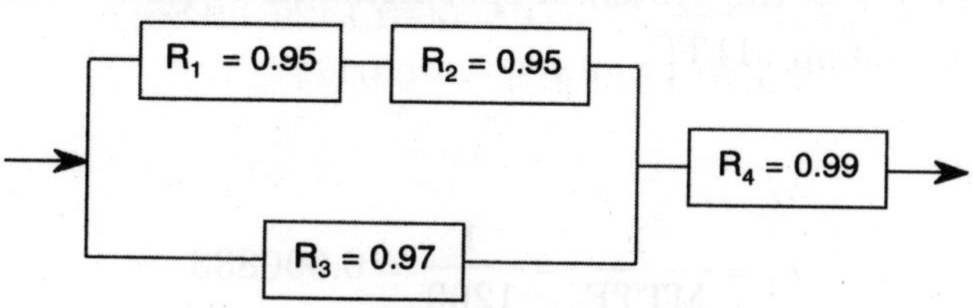

Fig. P7.34

Solution:

R_3 fails with probability of 0.03 : $R_{3\ \text{fails}} = (0.95)^2(0.998) = 0.8935$

R_3 works with probability of 0.97 : $R_{3\ \text{works}} = 0.99$

$$R_{\text{system}} = R_{3\ \text{fails}} + R_{3\ \text{works}} = (0.03)\ (0.8935) + (0.97)\ (0.99)$$
$$= 0.0268 + 0.9603 = 0.9871.$$

7.35. For the network shown in Fig. P7.35

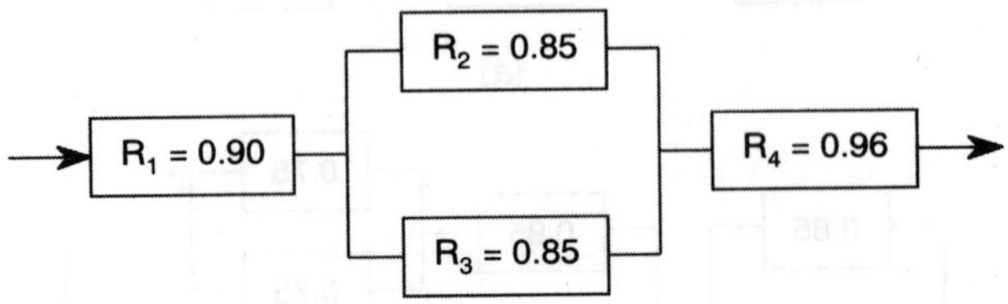

Fig. P7.35

(*a*) find the structure function

(*b*) determine the system reliability using the structure function

(*c*) identify the minimal path sets and minimal cut sets

(*d*) determine the lower and upper bound reliabilities.

Solution:

(*a*)
$$\Phi(X_1\, X_2\, X_3) = X_1[1 - (1 - X_2)\,(1 - X_3)]X_4 = X_1\, X_4(X_2 + X_3 - X_2\, X_3)$$
$$= X_1\, X_4\, X_2 + X_1\, X_4\, X_3 - X_1\, X_4\, X_2\, X_3$$

(*b*)
$$R_{\text{system}} = E[\Phi(X_1\, X_2\, X_3\, X_4)] = E[X_1\, X_4\, X_2 + X_1\, X_4\, X_3 - X_1\, X_4\, X_2\, X_3]$$
$$= R_1\, R_4\, R_2 + R_1\, R_4\, R_3 - R_1\, R_4\, R_2\, R_3$$
$$= 2[(0.90)\ (0.96)\ (0.85)] - (0.90)\ (0.96)\ (0.85)\ (0.85)$$
$$= 1.4688 - 0.6242 = 0.8446$$

(*c*)

Min path sets	Min cut sets
1, 2, 4	1
1, 3, 4	4
	2, 3

(*d*)
$$R_i = (0.90)\ (0.96)\ [1 - (1 - 0.85)^2] = 0.8446$$
$$R_u = 1 - [1 - (0.90)\ (0.85)\ (0.98)]^2 = 0.9373$$

7.36. A system consists of a primary unit and a standby unit. The MTTF of the primary unit is 1200 operating hours, and the MTTF of the standby unit is 500 hr when in operation and 3000 hr while in standby status.

(*a*) find the design life of the system if specifications require a reliability of 0.95

(b) determine the system MTTF.

Solution:

Standby system

$$\lambda_1 = \frac{1}{\text{MTTF}_i} = \frac{1}{1200} = 0.000833$$

Similarly,

$$\lambda_2 = \frac{1}{500} = 0.002 \text{ and } \lambda_3 = \frac{1}{3000} = 0.000333$$

(*a*) $$R(t) = 0.95 = e^{-\lambda_1 t} + \frac{\lambda_1}{\lambda_1 + \lambda_2 + \lambda_3}[e^{-\lambda_1 t} - e^{-(\lambda_1+\lambda_2)t}]$$

$$= e^{-0.000833t}\frac{0.000833}{(0.000833 + 0.000333 - 0.002)}[e^{-0.002t} - e^{-(0.000833+0.000333)t}]$$

$$0.95 = e^{-0.000833t} - 0.9988[e^{0.002t} - e^{-0.0005t}]$$

Hence the design life *t* is obtained by solving the above equation by trial and error method.

(*b*) $$\text{MTTF} = \frac{1}{\lambda_1} + \frac{1}{\lambda_2(\lambda_1+\lambda_2)} = \frac{1}{(0.000833)} + \frac{0.000833}{0.002\,(0.000833 + 0.000333)}$$

$$= 1200.48 + 357.33 = 1557.81 \text{ hours.}$$

7.37. **Assuming an equal allocation to each redundant subset and each redundant subset contains identical components, find the component reliability of the system shown in Fig. P7.37 if the system reliability objectives is 0.97.**

Solution:

Given

$$R_{\text{system}} = 0.97$$
$$R_{\text{system}} = 0.97 = R^5 \text{ or } R = (0.97^{1/5}) = 0.99393$$

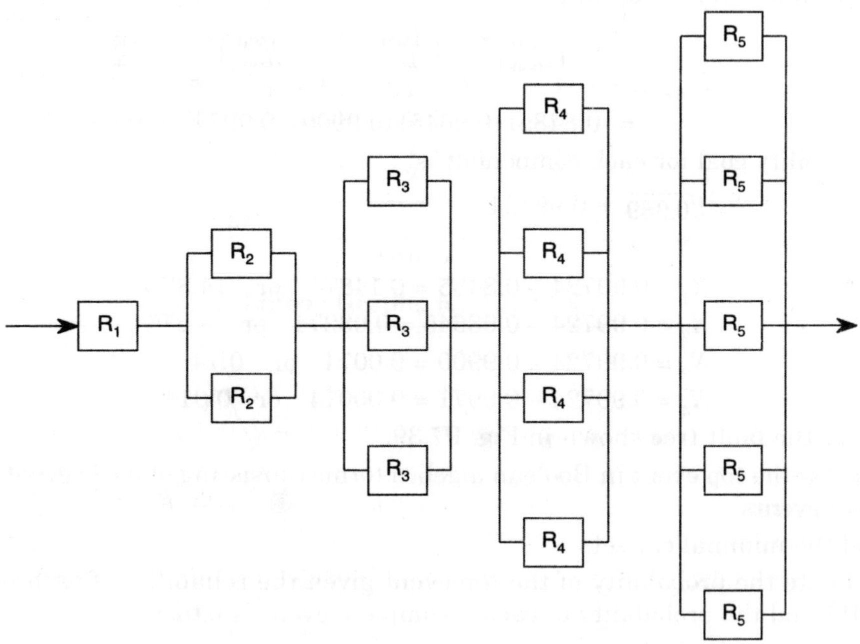

Fig. P7.37

Therefore

$$R_1 = R = 0.99393$$

$$R_2 = 1 - \sqrt[2]{(1-R)} = 1 - \sqrt[2]{(1-0.99393)} = 0.922089$$

$$R_3 = 1 - \sqrt[3]{(1-R)} = 1 - \sqrt[3]{(1-0.99393)} = 0.817550$$

$$R_4 = 1 - \sqrt[4]{(1-R)} = 1 - \sqrt[4]{(1-0.99393)} = 0.720876$$

$$R_5 = 1 - \sqrt[5]{(1-R)} = 1 - \sqrt[5]{(1-0.99393)} = 0.639722$$

7.38. The Weibull distribution parameters of a system with four components in series are given as

Component	Scale parameter, θ	Shape parameter, β
1	5,000	1.50
2	10,000	1.75
3	15,000	2.00
4	20,000	2.25

Assuming the cost of improving the current reliability is linear and equal to all the components, find the percentage of improvement needed for each component in order to achieve the system reliability goal of 0.989, at 1500 hours of operation.

Solution:

Current system reliability is

$$R_{\text{system}} = e^{-\left(\frac{1500}{5000}\right)^{1.5}} e^{-\left(\frac{1500}{10000}\right)^{1.75}} e^{-\left(\frac{1500}{15000}\right)^{2}} e^{-\left(\frac{1500}{20000}\right)^{2.25}}$$

$$= (0.8485)\,(0.9645)\,(0.9900)\,(0.9971) = 0.989$$

The reliability goal for each component is

$$\sqrt[4]{0.989} = 0.99724$$

Therefore,

$$X_1 = 0.99724 - 0.8485 = 0.14874 \quad \text{or} \quad 14.87\%$$

$$X_2 = 0.99724 - 0.93645 = 0.03274 \quad \text{or} \quad 3.27\%$$

$$X_3 = 0.99724 - 0.9900 = 0.0074 \quad \text{or} \quad 0.74\%$$

$$X_4 = 0.99724 - 0.9971 = 0.00014 \quad \text{or} \quad 0.014\%$$

7.39. Consider the fault tree shown in Fig. P7.39.

(*a*) express the top event in Boolean algebra terms consisting of basic events and incomplete events

(*b*) find the minimal cut sets

(*c*) estimate the probability of the top event given the reliability of each basic event is 0.015 and the probability of each incomplete event is 0.025.

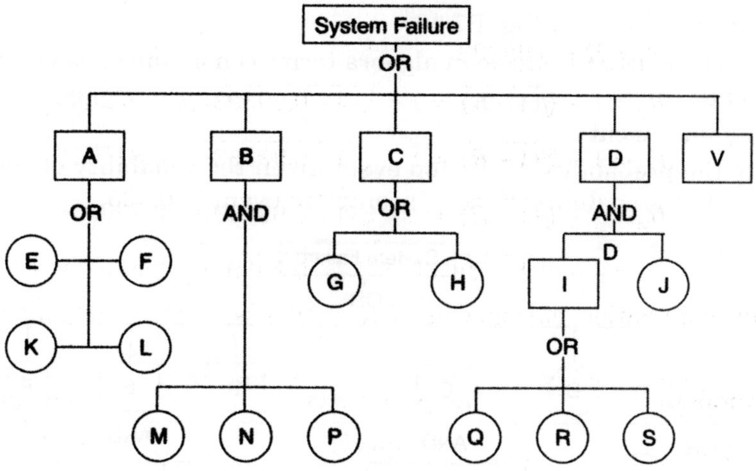

Fig. P7.39

Solution:

(*a*) Expression of the top event in Boolean algebra terms is given by

$$\text{SF} = A \cup B \cup C \cup D \cup V$$
$$= [E \cup F \cup K \cup L] \cup [M \cap N \cap P] \cup [G \cup H] \cup [I \cap J] \cup V$$
$$= [E \cup F \cup K \cup L] \cup [M \cap N \cap P] \cup [G \cup H] \cup [\{Q \cup R \cup S\} \cap J] \cup V$$

(*b*) The table of minimal cut sets is given in Table P7.39.

(*c*) The estimated probability of total system failure: The probability for each basic event is 0.015 and for incomplete events it is 0.025, but since there are no complete events we will not be using this value.

$$P(\text{SF}) \approx P(E) + P(F) + P(K) + P(L) + P(M)P(N)P(P) + P(G) + P(H) + \{P(J)[P(Q) + P(R) + P(S)]\} + P(V)$$

$$P(\text{SF}) \approx 0.015 + 0.015 + 0.015 + 0.015 + [0.015]^3 + 0.015 + 0.015 + \{0.015[0.045]\} + 0.015 = 0.1057$$

Table P7.39

Iteration	1	2	3
	A	*E*	*E*
	B	*F*	*F*
	C	*K*	*K*
	D	*L*	*L*
	V	*M, N, P*	*M, N, P*
		G	*G*
		H	*H*
		I, J	*Q, J*
		V	*R, J*
			S, J
			V

7.40 For the fault tree shown in Fig. P7.40

(*a*) express the top event in Boolean algebra terms consisting of basic events and incomplete events.

(*b*) find the minimal cut sets

(*c*) Estimate the probability of the top event given the reliability of each basic event is 0.017 and the probability of each incomplete event is 0.025.

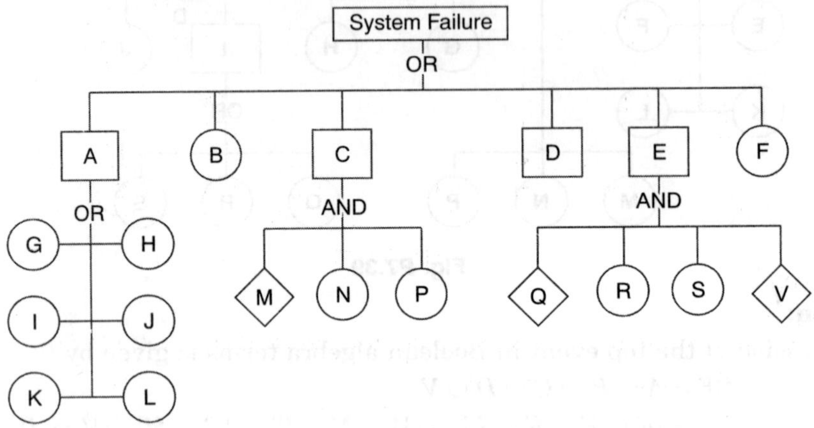

Fig. P7.40

Solution:

(*a*) Expression for the top event in Boolean algebra terms is

$$SF = A \cup B \cup C \cup D \cup E \cup F$$

$$= [G \cup H \cup I \cup J \cup K \cup L] \cup B \cup [M \cap N \cap P] \cup D \cup [Q \cap R \cap S \cap V] \cup F$$

(*b*) The table of minimal cut sets is given in Table P7.40.

Table P7.40

Iteration	1	2
	A	*G*
	B	*H*
	C	*I*
	D	*J*
	E	*K*
	F	*L*
		B
		M, N, P
		D
		Q, R, S, V
		F

(*c*) The estimated probability of total system failure: The probability for each basic event is 0.017 and for incomplete events it is 0.025.

$$P(\text{SF}) \approx P(G) + P(H) + P(I) + P(J) + P(K) + P(L) + P(B) + [P(M)P(N)P(P)] + P(D) + [P(Q)P(R)P(S)P(V)] + P(F)$$

$$P(SF) \approx 0.017 + 0.017 + 0.017 + 0.017 + 0.017 + 0.017 + 0.017 + [(0.025)(0.017)\ (0.017)] + 0.017 + [(0.025)(0.017)(0.017)(0.025)] + 0.017 = 0.1530$$

7.41. The three major subsystems of an electromechanical system each having a Weibull failure distribution with parameter values are given in Table P7.41.

Table P7.41

Subsystem	Scale parameter, θ, yr	Shape parameter, β
A	3.2	0.92
B	4.1	0.88
C	5.3	1.28

The reliability of the system should be 0.998 at the end of the first year.

(*a*) find the percentage increase in reliability for each of the subsystems needed to meet the system goal. Assume equal reliability goals for all subsystems

(*b*) apply the ARINC method to find an average failure rate goal for each subsystem assuming average failure rate for each subsystem

(*c*) determine the characteristic life goal for each subsystem assuming the reliability growth improves the characteristic life and does not improve the shape parameter.

Solution:

Subsystem	Scale parameter, θ, yr	Shape parameter, β
A	3.2	0.92
B	4.1	0.88
C	5.3	1.28

$$R_s = 0.998$$

$$\lambda_A = \frac{1}{(3.2)^{0.92}} = 0.343$$

$$\lambda_{j\,NEW} = W_j\lambda^* = W_j(-\ln(0.998)) = W_j(0.002)$$

$$\lambda_B = \frac{1}{(4.1)^{0.88}} = 0.289$$

$$W_j = \frac{\lambda_j}{0.343 + 0.289 + 0.118} = \frac{\lambda_j}{0.75}$$

$$\lambda_C = \frac{1}{(5.3)^{1.28}} = 0.118$$

$$\lambda_{A\,NEW} = \left(\frac{0.343}{0.75}\right)0.002 = 0.000915$$

$$\lambda_{B\,NEW} = \left(\frac{0.289}{0.75}\right)0.002 = 0.00077$$

$$\lambda_{C\,NEW} = \left(\frac{0.118}{0.75}\right)0.002 = 0.000314$$

$$\theta_A = \left[\frac{1}{0.000915}\right]^{\frac{1}{0.92}} = 2008.18 \text{ years}$$

$$\theta_B = \left[\frac{1}{0.00077}\right]^{\frac{1}{0.88}} = 3452.06 \text{ years}$$

$$\theta_C = \left[\frac{1}{0.000314}\right]^{\frac{1}{1.28}} = 545.47 \text{ years}$$

7.42. The four major subsystems of a mechanical system each having a Weibull failure distribution with parameter values are given in Table P7.42.

Table P7.42

Subsystem	Scale parameter, θ, yr	Shape parameter, β
A	3.5	0.94
B	4.0	0.87
C	5.5	1.75
D	6.5	1.20

The reliability of the system should be 0.997 at the end of the first year.

(*a*) find the percentage increase in reliability for each of the subsystems needed to meet the system goal. Assume equal reliability goals for all subsystems

(*b*) apply the ARINC method to find an average failure rate goal for each subsystem assuming average failure rate for each subsystem

(*c*) determine the characteristic life goal for each subsystem assuming the reliability growth improves the characteristic life and does not improve the shape parameter.

Solution:

Subsystem	Scale parameter, θ, yr	Shape parameter, β
A	3.5	0.94
B	4.0	0.87
C	5.5	1.75
D	6.5	1.20

$$R_s = 0.997$$

$$\lambda_A = \frac{1}{(3.5)^{0.94}} = 0.308$$

$$\lambda_B = \frac{1}{(4.0)^{0.87}} = 0.299$$

$$\lambda_C = \frac{1}{(5.5)^{1.75}} = 0.0506$$

$$\lambda_D = \frac{1}{(6.5)^{1.2}} = 0.1058$$

$$\lambda_{j\ NEW} = W_j\lambda^* = W_j(-\ln(0.997)) = W_j(0.003)$$

$$W_j = \frac{\lambda_j}{0.308 + 0.299 + 0.0506 + 0.1058} = \frac{\lambda_j}{0.7634}$$

$$\lambda_{A\ NEW} = \left(\frac{0.308}{0.7634}\right) 0.003 = 0.00121$$

$$\lambda_{B\ NEW} = \left(\frac{0.299}{0.7634}\right) 0.003 = 0.00117$$

$$\lambda_{C\ NEW} = \left(\frac{0.0506}{0.7634}\right) 0.003 = 0.000198$$

$$\lambda_{D\ NEW} = \left(\frac{0.1058}{0.7634}\right) 0.003 = 0.000415$$

$$\theta_{A\ NEW} = \left[\frac{1}{0.0021}\right]^{\frac{1}{0.94}} = 1268.87 \text{ yrs.}$$

$$\theta_{B\ NEW} = \left[\frac{1}{0.0017}\right]^{\frac{1}{0.87}} = 2343.69 \text{ yrs.}$$

$$\theta_{C\ NEW} = \left[\frac{1}{0.00198}\right]^{\frac{1}{1.76}} = 130.67 \text{ yrs.}$$

$$\theta_{D\ NEW} = \left[\frac{1}{0.00415}\right]^{\frac{1}{1.2}} = 658.1 \text{ yrs.}$$

7.43. The five major subsystems of an electromechanical system each having a Weibull failure distribution with parameter values are given in Table P7.43.

Table P7.43

Subsystem	Scale parameter, θ, yr	Shape parameter, β
A	3.0	0.98
B	4.0	0.95
C	5.0	1.20
D	3.5	1.75
E	4.5	1.20

The reliability of the system should be 0.99 at the end of the first year.

(*a*) find the percentage increase in reliability for each of the subsystems needed to meet the system goal. Assume equal reliability goals for all subsystems

(*b*) apply the ARINC method to find an average failure rate goal for each subsystem assuming average failure rate for each subsystem (*c*) determine the characteristic life goal for each subsystem assuming the reliability growth improves the characteristic life and does not improve the shape parameter.

Solution:

Subsystem	Scale parameter, θ, yr	Shape parameter, β
A	3.0	0.98
B	4.0	0.95
C	5.0	1.20
D	3.5	1.75
E	4.5	1.20

$$R_s = 0.99$$

$$\lambda_A = \frac{1}{(3.0)^{0.98}} = 0.3407$$

$$\lambda_B = \frac{1}{(4.0)^{0.95}} = 0.268$$

$$\lambda_C = \frac{1}{(5.0)^{1.2}} = 0.145$$

$$\lambda_D = \frac{1}{(3.5)^{1.75}} = 0.1116$$

$$\lambda_E = \frac{1}{(4.5)^{1.2}} = 0.164$$

$$\lambda_{j\,NEW} = W_j\lambda^* = W_j(-\ln(0.99)) = W_j(0.01)$$

$$W_j = \frac{\lambda_j}{0.3407 + 0.268 + 0.145 + 0.1116 + 0.164} = \frac{\lambda_j}{1.029}$$

$$\lambda_{A\,NEW} = \left(\frac{0.3407}{1.029}\right) 0.01 = 0.0033$$

$$\lambda_{B\,NEW} = \left(\frac{0.268}{1.029}\right) 0.01 = 0.0026$$

$$\lambda_{C\,NEW} = \left(\frac{0.145}{1.029}\right) 0.01 = 0.0014$$

$$\lambda_{D\,NEW} = \left(\frac{0.1116}{1.029}\right) 0.01 = 0.00108$$

$$\lambda_{E\,NEW} = \left(\frac{0.164}{1.029}\right) 0.01 = 0.0016$$

$$\theta_A = \left[\frac{1}{0.0033}\right]^{\frac{1}{0.98}} = 340.51 \text{ yrs.}$$

$$\theta_B = \left[\frac{1}{0.0026}\right]^{\frac{1}{0.95}} = 526.11 \text{ yrs.}$$

$$\theta_C = \left[\frac{1}{0.0014}\right]^{\frac{1}{1.2}} = 238.9 \text{ yrs.}$$

$$\theta_D = \left[\frac{1}{0.00108}\right]^{\frac{1}{1.75}} = 49.56 \text{ yrs.}$$

$$\theta_E = \left[\frac{1}{0.0016}\right]^{\frac{1}{1.2}} = 213.74 \text{ yrs.}$$

7.44. **For the fault tree shown in Fig. P7.44 performs a qualitative analysis for the top event in terms of non-redundant basic events using Boolean algebra if the probability of each basic event is 0.0005 and the events are independent, what is the probability of the top event?**

Solution:

Using minimal cut sets the following table can be created:

Iteration	**1**	**2**	**3**
	A *B* *C* *D* *E*	*F, G, H* *I, J* *C* *D* *E*	*F, G, H* *K, J* *L, J* *M, J* *N, J* *C* *D* *E*

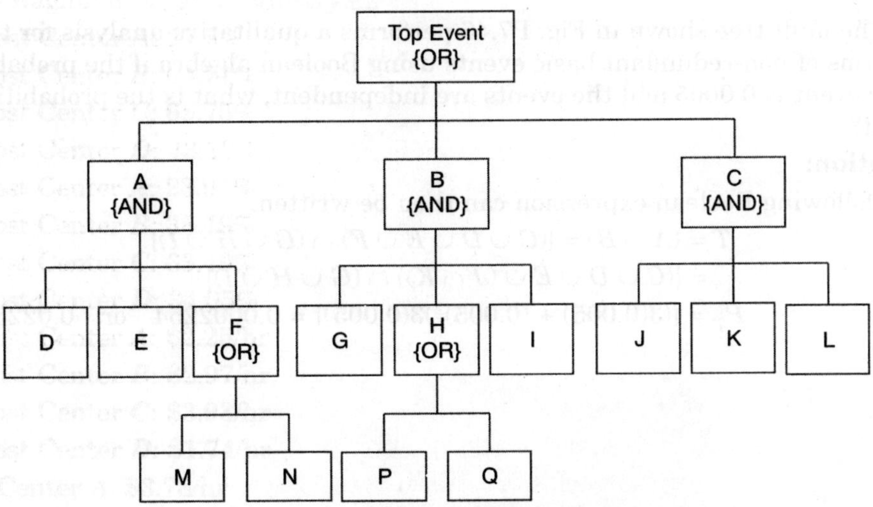

Fig. P7.44

The following Boolean expression can then be written:

$$T = [(F \cap G \cap H) \cup (K \cap J) \cup (L \cap J) \cup (M \cap J) \cup (N \cap J) \cup C \cup D \cup E]$$

or

$$T = [(F \cap G \cap H) \cup [(K \cup L \cup M \cup N) \cap J)] \cup C \cup D \cup E]$$

$$P_T = (0.0005)^3 + 4(0.0005)^2 + 3(0.0005) = 0.001501 \text{ or } 0.15\%.$$

7.45. For the fault tree shown in Fig. P7.45 performs a qualitative analysis for the top event in terms of non-redundant basic events using Boolean algebra if the probability of each basic event is 0.0005 and the events are independent, what is the probability of the top event?

Solution:

The following expression can be written:

$$T = (A \cap B) = [(C \cup D \cup E \cup F) \cap (G \cup H \cup I)]$$

$$= [(C \cup D \cup E \cup (J \cap K)) \cap (G \cup H \cup I)]$$

$$P_T = [(3(0.005) + (0.005)^2)3(0.005)] = 0.0002254 \quad \text{or} \quad 0.0225\%.$$

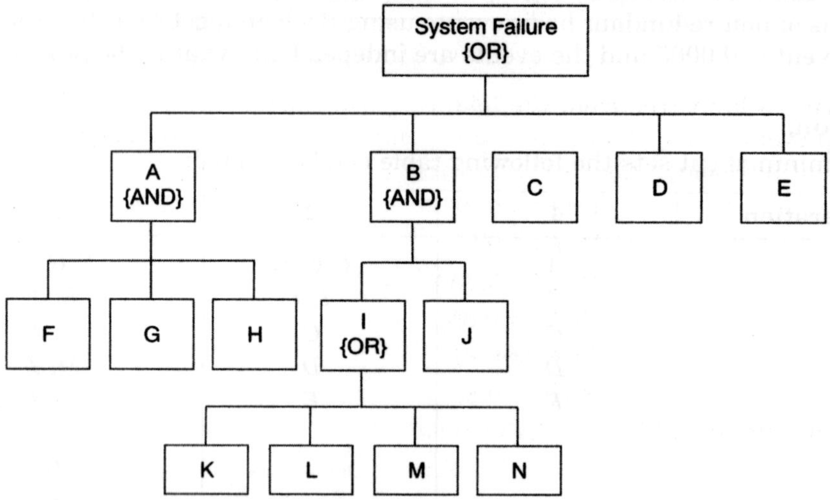

Fig. P7.45

7.46. For the fault tree shown in Fig. P7.46 performs a qualitative analysis for the top event in terms of non-redundant basic events using Boolean algebra if the probability of each basic event is 0.0005 and the events are independent, what is the probability of the top event?

Solution:

The following Boolean expression can then be written:

$$T = (A \cap B) = [(C \cup D \cup E \cup F) \cap (G \cup H \cup I)]$$

$$= [(C \cup D \cup E \cup (J \cap K)) \cap (G \cup H \cup I)]$$

$$P_T = [(3(0.005) + (0.005)^2)3(0.005)] = 0.0002254 \quad \text{or} \quad 0.0225\%.$$

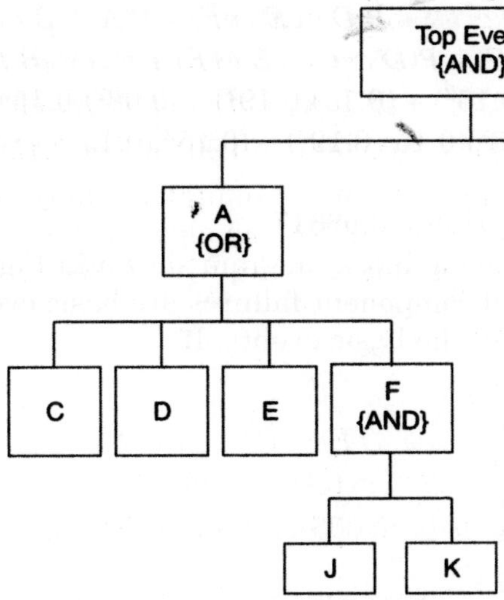

Fig. P7.46

7.47. **Construct a fault tree from the following block diagram shown in Fig.P7.47 such that the top event is a system failure and component failures are basic events. Express the top event in Boolean algebra terms of the basic events. If**

$$\Pr(A) = 0.92$$
$$\Pr(B) = 0.92$$
$$\Pr(C) = 0.85$$
$$\Pr(D) = 0.85$$
$$\Pr(E) = 0.81$$
$$\Pr(F) = 0.81$$

Compute the probability of the top event.

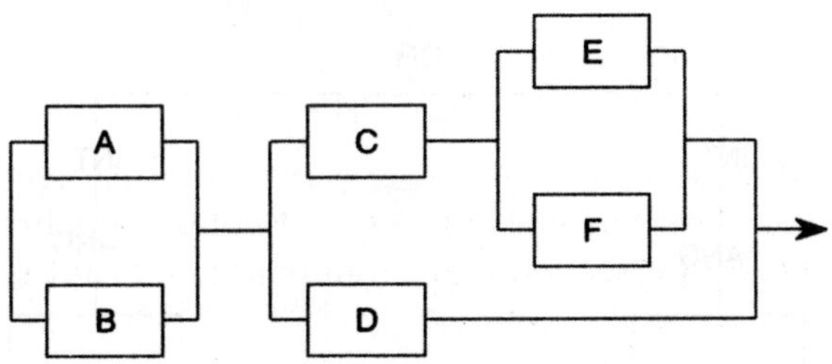

Fig. P7.47

Solution:

Refer to Fig.P7.47.

$$\Pr(A) = \Pr(B) = 0.92$$
$$\Pr(C) = \Pr(D) = 0.85$$
$$\Pr(E) = \Pr(F) = 0.81$$

$$T = (A \cap B) \cup \{D \cap [C \cup (E \cup F)]\}$$
$$= (A \cap B) \cup \{D \cap (C \cup E) \cup (C \cup F)\}$$
$$= (A \cap B) \cup \{(D \cap C) \cup (D \cup E) \cap (C \cup F)\}$$
$$= (A \cap B) \cup (D \cap C) \cup (D \cap E \cap F)$$

$$P(T) = P(A \cap B) + P(D \cap C) + P(D \cap E \cap F) - P(A \cap B \cap D \cap C)$$
$$- P(A \cap B \cap D \cap E \cap F) - P(D \cap C \cap E \cap F) + P(A \cap B \cap D \cap C \cap E \cap F)$$
$$= (0.08)(0.08) + (0.15^2) + (0.15)(0.19^2) - (0.08^2)(0.15^2)$$
$$- (0.08^2)(0.15)(0.19^2) - (0.15^2)(0.19^2) + (0.08^2)(0.15)(0.19^2)$$
$$= 0.0138$$
$$\text{Pr} = 1 - P(T) = 1 - 0.0138 = 0.98617$$

7.48. Construct a fault tree from the following block diagram shown in Fig. P7.48 such that the top event is a system failure and component failures are basic events. Express the top event in Boolean algebra terms of the basic events. If

$\Pr(A) = 0.90$
$\Pr(B) = 0.90$
$\Pr(C) = 0.90$
$\Pr(D) = 0.90$
$\Pr(E) = 0.95$
$\Pr(F) = 0.97$
$\Pr(G) = 0.95$

Compute the probability of the top event.

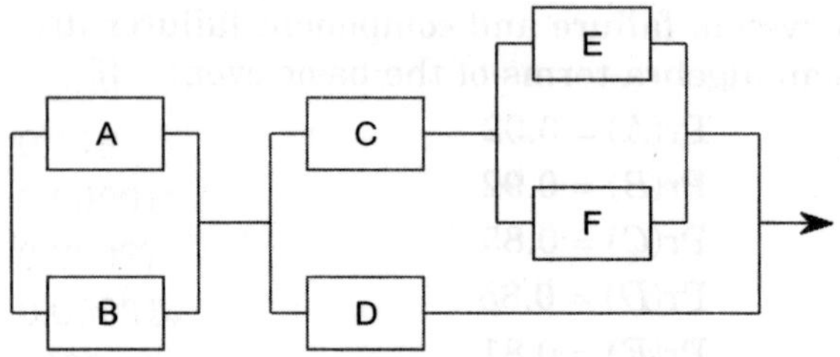

Fig. P7.48

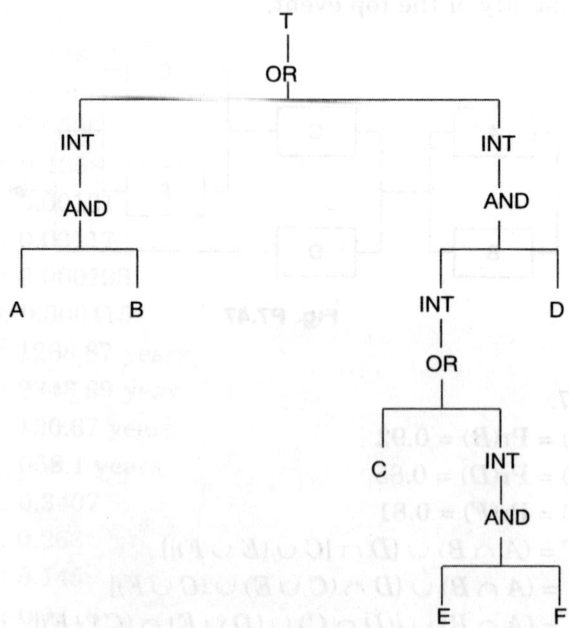

Fig. P7.48 (*a*)

Solution:

Refer to Figs. P7.48 and P7.48 (*a*)

$$\Pr(A) = \Pr(B) = 0.90$$
$$\Pr(C) = \Pr(D) = 0.90$$
$$\Pr(E) = 0.95$$
$$\Pr(F) = 0.97$$
$$\Pr(G) = 0.95$$
$$T = (E \cup G) \cap \{F \cup [A \cup C) \cap (B \cup D)]\}$$
$$P(T) = (0.10)[0.03 + (0.2)(0.2)] = 0.10[0.03 + 0.04] = 0.007$$
$$\Pr = 1 - 0.007 = 0.993$$

Alternate method:

$$R_s = 1 - \{1 - [1 - [1 - (0.9)(0.9)]^2]0.97\}\{1 - (0.95)(0.95)\}$$
$$= 1 - (0.065017)(0.0975) = 0.99366$$

7.49. Construct a fault tree from the following block diagram shown in Fig.P7.49 such that the top event is a system failure and component failures are basic events. Express the top event in Boolean algebra terms of the basic events. If

$$\Pr(A) = 0.90$$
$$\Pr(B) = 0.90$$
$$\Pr(C) = 0.80$$
$$\Pr(D) = 0.80$$
$$\Pr(E) = 0.75$$
$$\Pr(F) = 0.75$$
$$\Pr(G) = 0.75$$

Compute the probability of the top event.

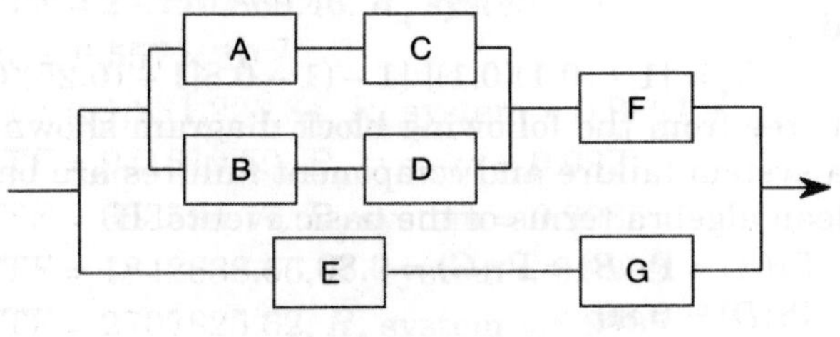

Fig. P7.49

Solution:

Refer to Figs. P7.49 and P7.49 (*a*)

$$\Pr(A) = \Pr(B) = 0.90$$
$$\Pr(C) = \Pr(D) = 0.80$$
$$\Pr(E) = \Pr(F) = 0.75$$
$$T = (A \cap B) \cup \{D \cap [C \cup (E \cup F)]\}$$
$$= (A \cap B) \cup \{D \cap (C \cup E) \cup (C \cup F)\}$$
$$= (A \cap B) \cup \{(D \cap C) \cup (D \cap E) \cap (C \cup F)\}$$
$$= (A \cap B) \cup (D \cap C) \cup (D \cap E \cap F)$$

$$P(T) = P(A \cap B) + P(D \cap C) + P(D \cap E \cap F) - P(A \cap B \cap D \cap C) - P(A \cap B \cap D \cap E \cap F) - P(D \cap C \cap E \cap F) + P(A \cap B \cap D \cap C \cap E \cap F)$$

$$= 0.01 + 0.04 + (0.2)(0.25)^2 - (0.01)(0.04) - (0.01)(0.2)(0.25)^2 - (0.2^2)(0.25)^2 + (0.01)(0.04)(0.25)^2 = 0.0595$$

$$\text{Pr} = 1 - 0.0595 = 0.9405$$

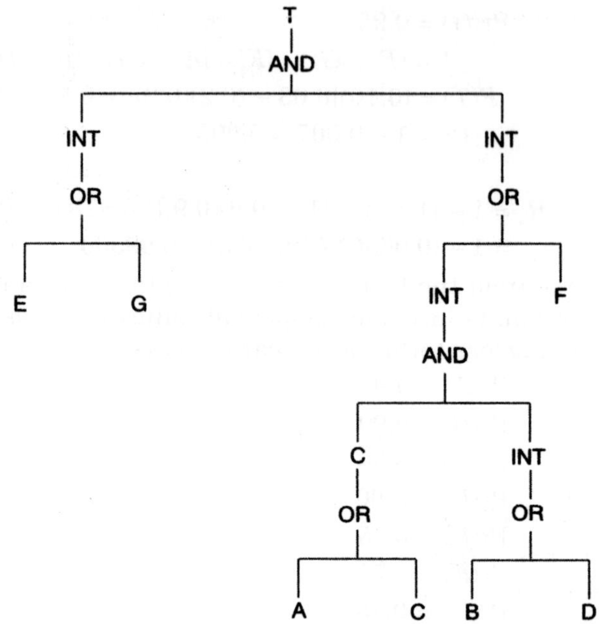

Fig. P7.49(a)

Alternate method:

$$R_s = [1 - (0.1)(0.1)]\ [1 - (1 - 0.8[1 - (0.25)(0.25)])(1 - 0.8) = 0.9405$$

7.50. Construct a fault tree from the following block diagram shown in Fig.P7.50 such that the top event is a system failure and component failures are basic events. Express the top event in Boolean algebra terms of the basic events. If

$\text{Pr}(A) = \text{Pr}(B) = \text{Pr}(C) = 0.80$

$\text{Pr}(D) = 0.80$

$\text{Pr}(E) = 0.90$

$\text{Pr}(F) = 0.70$

$\text{Pr}(G) = 0.70$

$\text{Pr}(H) = 0.90$

Compute the probability of the top event.

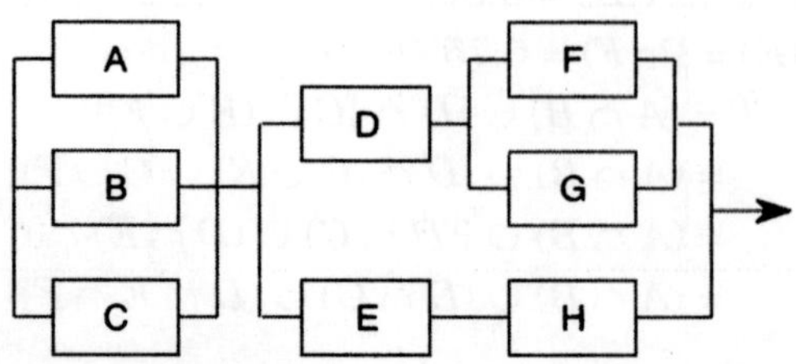

Fig. P7.50

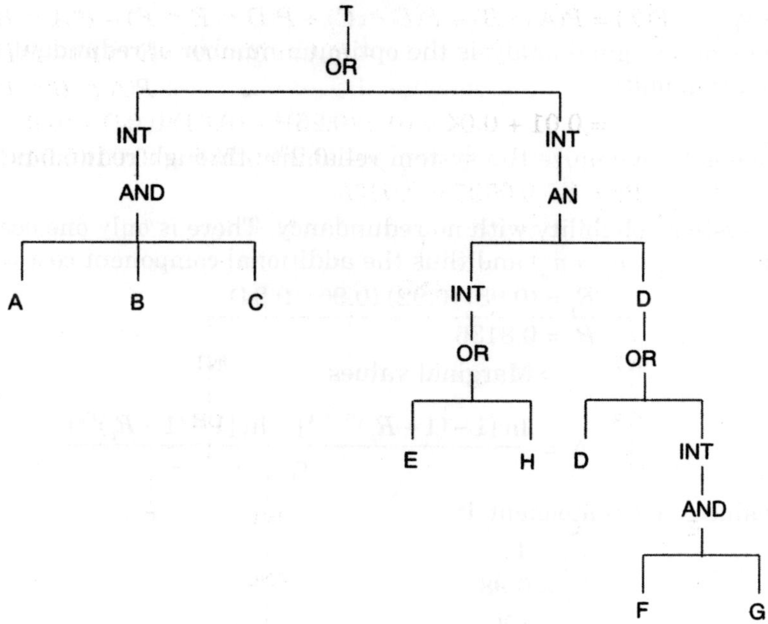

Fig. P7.50 (*a*)

Solution:

Refer to Figs. P7.50 and P7.50 (*a*)

$$Pr(A) = Pr(B) = 0.80$$
$$Pr(C) = Pr(D) = 0.80$$
$$Pr(E) = 0.90$$
$$Pr(F) = Pr(G) = 0.70$$
$$Pr(H) = 0.90$$
$$T = (A \cap B \cap C) \cup [(E \cup H) \cap (D \cup F) \cap (D \cup G)]$$
$$= (0.2)^3 + [(0.1 + 0.1)(0.2 + 0.3)(0.2 + 0.3)] = 0.058$$
$$Pr = 1 - 0.058 = 0.942$$

Alternate method:

$$R_s = [1 -(1 - 0.8)^3]\ [1 - (1 - 0.8(1 - (0.3)(0.3))(1 - (0.9)(0.9))] = 0.942$$

7.51. An engineer has a $1,000 budget (per system) to be used to increase the reliability of a four-component series system. The reliability of each component at the desired system lifetime has been established through a reliability growth-testing program with the following results:

Component	Reliability	Unit cost $
1	0.98	100
2	0.92	170
3	0.96	80
4	0.94	350

Component = k

Reliability = R_i

Unit cost = c_i

Determine using marginal analysis the optimum number of redundant units. Compute the achieved reliability.

Solution:

The objective is to maximize the system reliability through redundancy subject to the budget of $1,000.

R_i = Initial system reliability with no redundancy. There is only one occurrence of each component ($n_1 = n_2 = n_3 = n_4$) and thus the additional component cost = 0.

$$R_i = (0.98)\,(0.92)\,(0.96)\,(0.94)$$

or

$$R_i = 0.8136$$

$$\Delta_i = \text{Marginal values}$$

$$\Delta_i = \frac{\ln\left[1-(1-R_i)^{n_i+1}\right] - \ln\left[1-(1-R_i)^{n_i}\right]}{c_i}$$

Marginal value Δ_i for component 1:

$$n_1 = 1$$

$$R_1 = 0.98$$

$$c_1 = 100$$

$$\Delta_i = \frac{\ln\left[1-(1-R_i)^{n_i+1}\right] - \ln\left[1-(1-R_i)^{n_i}\right]}{c_i}$$

$$\Delta_i = 1.98 \times 10^{-4}$$

Marginal value Δ_2 for component 2:

$$n_2 = 2$$

$$R_2 = 0.92$$

$$c_2 = 170$$

$$\Delta_2 = \frac{\ln\left[1-(1-R_i)^{n_i+1}\right] - \ln\left[1-(1-R_i)^{n_i}\right]}{c_i}$$

$$\Delta_2 = 4.527 \times 10^{-4}$$

Marginal value Δ_3 for component 3:

$$n_3 = 3$$

$$R_3 = 0.96$$

$$c_3 = 80$$

$$\Delta_3 = \frac{\ln\left[1-(1-R_i)^{n_i+1}\right] - \ln\left[1-(1-R_i)^{n_i}\right]}{c_i}$$

$$\Delta_3 = 4.903 \times 10^{-4}$$

Marginal value Δ_4 for component 4:

$$n_4 = 4$$

$$R_4 = 0.94$$

$$c_4 = 350$$

$$\Delta_4 = \frac{\ln [1-(1-R_i)^{n_i+1}] - \ln [1-(1-R_i)^{n_i}]}{c_i}$$

$$\Delta_4 = 1.665 \times 10^{-4}$$

Starting with the maximum marginal value and working toward the smallest marginal value yields the results given in Table P7.51.

Table P7.51

Iteration	Max_i	k	Cumulative cost $	New_k	System reliability
1	0.0004903	3	80	0.0000192	0.846
2	0.0004527	2	250	0.0000348	0.914
3	0.0001980	1	350	0.0000039	0.932
4	0.0001665	4	700	0.0000097	0.988
5	0.0000348	2	870	0.0000028	0.994
6	0.0000192	3	950	0.0000008	0.995

The final solution for component redundancy is $n_1 = 2$, $n_2 = 3$, $n_3 = 3$, $n_4 = 2$. Meaning that working with a maximum budget of $1000, there will be two of component 1 in parallel, there will be three of component 2 in parallel, there will be three of component 3 in parallel, and there will be two of component 4 in parallel. The final system reliability is

$$R_s = \text{System reliability for } n_1 = 2, n_2 = 3, n_3 = 3, n_4 = 2$$

$$R_s = [1-(1-0.98)^{n_1}]\,[1-(1-0.92)^{n_2}]\,[1-(1-0.96)^{n_3}]$$

$$[1-(1-0.94)^{n_4}] = 0.995$$

Final calculations for the above table are as follows:

Marginal value Δ_1 for component 1

$$n_1 = 2$$
$$R_1 = 0.98$$
$$c_i = 100$$

$$\Delta_1 = \frac{\ln [1-(1-R_i)^{n_i+1}] - \ln [1-(1-R_i)^{n_i}]}{c_i}$$

$$\Delta_1 = 3.921 \times 10^{-6}$$

Marginal value Δ_2 for component 2

$$n_2 = 2$$
$$R_2 = 0.92$$
$$c_2 = 170$$

$$\Delta_2 = \frac{\ln [1-(1-R_i)^{n_i+1}] - \ln [1-(1-R_i)^{n_i}]}{c_i}$$

$$\Delta_2 = 2.772 \times 10^{-6}$$

Marginal value Δ_3 for component 3

$$n_3 = 3$$
$$R_3 = 0.96$$

$$c_3 = 80$$

$$\Delta_3 = \frac{\ln[1-(1-R_i)^{n_i+1}] - \ln[1-(1-R_i)^{n_i}]}{c_i}$$

$$\Delta_3 = 7.68 \times 10^{-7}$$

Marginal value Δ_4 for component 4

$$n_4 = 4$$
$$R_4 = 0.94$$
$$c_4 = 350$$

$$\Delta_4 = \frac{\ln[1-(1-R_i)^{n_i+1}] - \ln[1-(1-R_i)^{n_i}]}{c_i}$$

$$\Delta_4 = 9.687 \times 10^{-6}$$

7.52 Given a budget of \$1,000 and the following data on three components that must operate in series, determine using marginal analysis, the optimum number of redundant units. Compute the achieved reliability.

Component	Reliability	Unit cost $
1	0.92	250
2	0.95	150
3	0.98	100

Component = k
Reliability = R_i
Unit cost = c_i

Solution:

R_i = Initial system reliability with no redundancy. There is only one occurrence of each component ($n_1 = n_2 = n_3 = 1$) and thus the additional component cost = 0.

$$R_i = (0.92), (0.95), (0.98) \text{ or } R_i = 0.857$$

Δ_i = Marginal values

$$\Delta_i = \frac{\ln[1-(1-R_i)^{n_i+1}] - \ln[1-(1-R_i)^{n_i}]}{c_i}$$

Marginal value Δ_i for component 1:

$$n_i = 1$$
$$R_i = 0.92$$
$$c_i = 250$$

$$\Delta_1 = \frac{\ln[1-(1-R_i)^{n_i+1}] - \ln[1-(1-R_i)^{n_i}]}{c_i}$$

$$\Delta_1 = 3.078 \times 10^{-4}$$

Marginal value Δ_2 for component 2:

$$n_i = 1$$
$$R_i = 0.95$$
$$c_i = 150$$

$$\Delta_2 = \frac{\ln [1-(1-R_i)^{n_i+1}] - \ln [1-(1-R_i)^{n_i}]}{c_i}$$

$$\Delta_2 = 3.253 \times 10^{-4}$$

Marginal value Δ_3 for component 3:

$$n_i = 1$$
$$R_i = 0.98$$
$$c_i = 100$$

$$\Delta_3 = \frac{\ln [1-(1-R_i)^{n_i+1}] - \ln [1-(1-R_i)^{n_i}]}{c_i}$$

$$\Delta_3 = 1.98 \times 10^{-4}$$

Starting with the maximum marginal value and working toward the smallest marginal value yields the results given in Table P7.52.

Table P7.52

Iteration	Max_i	k	Cumulative cost $	New_k	System reliability
1	0.00032530	2	150	0.00001585	0.899
2	0.00030780	1	400	0.00002363	0.971
3	0.00019800	3	500	0.00000392	0.991
4	0.00002363	1	750	0.00000189	0.997
5	0.00001585	2	900	0.00000079	0.999
6	0.00000392	3	1,000	0.00000008	0.999

The final solution for component redundancy is $n_1 = 3$, $n_2 = 3$, $n_3 = 3$, $n_4 = 2$. Meaning that working with a maximum budget of $1000, there will be three of component 1 in parallel, there will be three of component 2 in parallel, and there will be three of component 3 in parallel. Looking at the table above it can be seen that adding a third component in parallel for each component has very little effect on improving the system reliability. A 500-dollar savings could be achieved by only having two of each component in parallel.

The final system reliability is

$$R_s = \text{System reliability}$$
$$n_1 = 3$$
$$n_2 = 3$$
$$n_3 = 3$$

$$R_s = [1-[1-0.92)^{n_1}]\,[1-(1-0.95)^{n_2}]\,[1-(1-0.98)^{n_3}] = 0.999$$

Final calculations for the above table are

Marginal value Δ_1 for component 1:

$$n_i = 3$$
$$R_i = 0.92$$
$$c_i = 250$$

$$\Delta_1 = \frac{\ln[1-(1-R_i)^{n_i+1}] - \ln[1-(1-R_i)^{n_i}]}{c_i}$$

$$\Delta_1 = 1.885 \times 10^{-6}$$

Marginal value Δ_2 for component 2:

$n_i = 3$
$R_i = 0.95$
$c_2 = 150$

$$\Delta_2 = \frac{\ln[1-(1-R_i)^{n_i+1}] - \ln[1-(1-R_i)^{n_i}]}{c_i}$$

$$\Delta_2 = 7.917 \times 10^{-7}$$

Marginal value Δ_3 for component 3:

$n_i = 3$
$R_i = 0.98$
$c_i = 100$

$$\Delta_3 = \frac{\ln[1-(1-R_i)^{n_i+1}] - \ln[1-(1-R_i)^{n_i}]}{c_i}$$

$$\Delta_3 = 7.84 \times 10^{-8}$$

7.53. A certain space probe to be launched contains a module of three experiments. Testing has resulted in a (nonredundant) reliability estimate over the duration of the mission of 0.92, 0.96 and 0.98 for the respective experiments. There is a weight limitation of 500 pounds to be allocated to the module. How many redundant experiments of each type should be included in the module if their respective weights are 180, 50 and 130 pounds? Since a minimum of one unit of each experiment is required, 360 pounds of the 500 pounds allocated are already accounted for. Maximize the overall reliability of the module.

Component	Reliability	Unit cost $
1	0.92	180
2	0.96	50
3	0.98	130

Component = k
Reliability = R_i
Unit cost = c_i

Solution:

The weight allowed for duplication of experiments is equal to a total allowed weight minus the initial 360 pounds for one of each experiment.

500 – 360 = 140 pounds.

R_i = Initial system reliability with no redundancy. There is only one occurrence of each component ($n_1 = n_2 = n_3 = 1$) and thus the additional weight would = 0 out of 140 pounds allowed.

$R_i = (0.92)(0.96)(0.98)$ or $R_i = 0.866$
Δ_i = Marginal values

$$\Delta_i = \frac{\ln [1-(1-R_i)^{n_i+1}] - \ln [1-(1-R_i)^{n_i}]}{c_i}$$

Marginal value Δ_i for component 1:

$$n_1 = 1$$
$$R_1 = 0.92$$
$$c_1 = 180$$

$$\Delta_1 = \frac{\ln [1-(1-R_i)^{n_i+1}] - \ln [1-(1-R_i)^{n_i}]}{c_i}$$

$$\Delta_1 = 4.276 \times 10^{-4}$$

Marginal value Δ_2 for component 2:

$$n_2 = 1$$
$$R_2 = 0.96$$
$$c_2 = 50$$

$$\Delta_2 = \frac{\ln [1-(1-R_i)^{n_i+1}] - \ln [1-(1-R_i)^{n_i}]}{c_i}$$

$$\Delta_2 = 7.844 \times 10^{-4}$$

Marginal value Δ_3 for component 3:

$$n_3 = 1$$
$$R_3 = 0.98$$
$$c_3 = 130$$

$$\Delta_3 = \frac{\ln [1-(1-R_i)^{n_i+1}] - \ln [1-(1-R_i)^{n_i}]}{c_i}$$

$$\Delta_3 = 1.523 \times 10^{-4}$$

Starting with the maximum marginal value and working toward the smallest marginal value yields the results given in Table P7.53.

Table P7.53

Iteration	Max_i	k	Cumulative weight	New_k	System reliability
1	If only experiment 1 is duplicated 0.00042760	1	180	0.00003282	0.935
1	0.00078440	2	50	0.00003075	0.900
2	0.00003075	2	100	0.00000123	0.902
1	If only experiment 3 is duplicated 0.00015230	3	130	0.00000302	0.883

Only allowed an additional 140 pounds. This would rule out duplicating experiment 1, as it weights 180 pounds.

The options are duplicating experiment 2 twice for an additional 100 pounds, or duplicating experiment 3 for an additional 130 pounds.

From the table above it can be seen that the highest system reliability is achieved within the 140-pound weight limit by duplicating experiment 2 twice, meaning having three of experiment 1 in parallel.

The final system reliability is

$$R_s = \text{System reliability for } n_1 = 1,\ n_2 = 3,\ n_3 = 1$$

$$R_s = [1-[1-0.92)]^{n_1}\,[1-(1-0.96)^{n_2}]\,[1-(1-0.98)^{n_3}] = 0.902$$

Final calculations for the Table P7.53 are

Marginal value Δ_1 for component 1 :

$n_1 = 32,$
$R_1 = 0.92,$
$c_1 = 180$

$$\Delta_1 = \frac{\ln[1-(1-R_i)^{n_i+1}] - \ln[1-(1-R_i)^{n_i}]}{c_i}$$

$$\Delta_1 = 3.282 \times 10^{-5}$$

Marginal value Δ_2 for component 2 :

$n_2 = 3,$
$R_2 = 0.96,$
$c_2 = 50$

$$\Delta_2 = \frac{\ln[1-(1-R_i)^{n_i+1}] - \ln[1-(1-R_i)^{n_i}]}{c_i}$$

$$\Delta_2 = 1.229 \times 10^{-6}$$

Marginal value Δ_3 for component 3 :

$n_3 = 2$
$R_3 = 0.98$
$c_3 = 130$

$$\Delta_3 = \frac{\ln[1-(1-R_i)^{n_i+1}] - \ln[1-(1-R_i)^{n_i}]}{c_i}$$

$$\Delta_3 = 3.016 \times 10^{-6}$$

7.54. Apply the AGREE allocation method to an electromechanical system containing the components given in Table P7.54.

Table P7.54

Component	Parts count, n_i	Importance index, ω_i	Operating time, hr/yr, t_i
1	177	0.96	2384
2	54	0.93	1762
3	28	0.99	2065

The warranty program requires a reliability of 0.99 over the first year of use.

Solution:

$$N = 177 + 54 + 28 = 259$$

$$24 \text{ hours} \times 365 \text{ days} = 8760 \text{ hours per year}$$

$$R^* = 0.99$$

Component 3:

$$\lambda_i = \left(\frac{-1}{2065}\right) \ln\left(1 - \frac{1 - 0.99^{\frac{28}{259}}}{0.99}\right) = (5.315)10^{-7}$$

$$\text{MTTF} = \frac{1}{(5.315)\,10^{-7}}$$

or $$\text{MTTF} = 1{,}881{,}467.54$$

$$R_{3i} = e^{(-5.315)(10^{-7})(8760)} = 0.9954$$

$$R_{3\ \text{system}} = 1 - 0.96(1 - e^{(-5.315)(10^{-7})(2065)}) = 0.9989$$

$$R_{1i}\, R_{2i}\, R_{3i} = (0.9740)(0.9889)(0.9954) = 0.9588$$

$$R_1\, R_2\, R_3 = (0.9932)(0.9978)(0.9989) = 0.99$$

System failure rate $= 3.002 \times 10^{-6} + 1.279 \times 10^{-6} + 5.315 \times 10^{-7} = 4.831 \times 10^{-6}$

System $\text{MTTF} = (4.813 \times 10^{-6})^{-1} = 207{,}770.62$ hours

Table P7.54 (*a*)

Component	Failure rate, λ_i	MTTF, $\frac{1}{\lambda_i}$ (hours)	Reliability, R_i	System reliability
1	3.002×10^{-6}	333,111.26	0.9740	0.9932
2	1.279×10^{-6}	781,860.83	0.9889	0.9978
3	5.315×10^{-7}	1,881,467.54	0.9954	0.9989
System	4.813×10^{-6}	207,770.62	0.9588	0.9900

7.55. Apply the AGREE allocation method to a system containing four components given in Table P7.55.

Table P7.55

Component	Parts count, n_i	Importance index, ω_i	Operating time, hr/yr, t_i
1	283	0.96	2000
2	154	0.97	1000
3	87	0.98	2000
4	46	1.00	1000

The warranty program requires a reliability of 0.999 over the first year of use.

Solution:

$$N = 283 + 154 + 87 + 46 = 570$$

$$24 \text{ hours} \times 365 \text{ days} = 8760 \text{ hours per year}$$

$$R^* = 0.999$$

Component 1:

$$\lambda_i = \left(\frac{-1}{2000}\right) \ln\left(1 - \frac{1 - 0.999^{\frac{283}{570}}}{0.96}\right) = (2.587)10^{-7}$$

$$\text{MTTF} = \frac{1}{(2.587)\,10^{-7}}$$

$$\text{MTTF} = 3{,}865{,}481.25$$

$$R_{1i} = e^{(-2.587)(10^{-7})(8760)} = 0.9977$$

$$R_{1\ \text{system}} = 1 - 0.96(1 - e^{(-2.587)(10^{-7})(2000)}) = 0.9995$$

Component 2:

$$\lambda_i = \left(\frac{-1}{1000}\right) \ln\left(1 - \frac{1 - 0.999^{\frac{154}{570}}}{0.97}\right) = (2.787)(10^{-7})$$

$$\text{MTTF} = \frac{1}{(2.587)\,10^{-7}}$$

$$\text{MTTF} = 3{,}588{,}087.55$$

$$R_{2i} = e^{(-2.787)(10^{-7})(8760)} = 0.9976$$

$$R_{2\ \text{system}} = 1 - 0.97(1 - e^{(-2.787)(10^{-7})(1000)}) = 0.9997$$

Component 3:

$$\lambda_i = \left(\frac{-1}{2000}\right) \ln\left(1 - \frac{1 - 0.999^{\frac{87}{570}}}{0.98}\right) = (7.791)(10^{-8})$$

$$\text{MTTF} = \frac{1}{(7.791)\,10^{-8}}$$

$$\text{MTTF} = 12{,}835{,}322.81$$

$$R_{3i} = e^{(-7.791)(10^{-8})(8760)} = 0.9993$$

$$R_{3\ \text{system}} = 1 - 0.98(1 - e^{(-7.791)(10^{-8})(2000)}) = 0.9998$$

Component 4:

$$\lambda_i = \left(\frac{-1}{2000}\right) \ln\left(1 - \frac{1 - 0.999^{\frac{46}{570}}}{1.00}\right) = (4.037)(10^{-8})$$

$$\text{MTTF} = \frac{1}{(4.037)\,10^{-8}}$$

$$\text{MTTF} = 24{,}770{,}869.46$$

$$R_{4i} = e^{(-4.037)(10^{-8})(8760)} = 0.9996$$

$$R_{4\ \text{system}} = 1 - 1.00(1 - e^{(-4.037)(10^{-8})(2000)}) = 0.9999$$

$$R_{1i}\,R_{2i}\,R_{3i}\,R_{4i} = (0.9977)(0.9976)(0.9993)(0.9996) = 0.994$$

$$R_1\,R_2\,R_3\,R_4 = (0.9995)(0.9997)(0.9998)(0.9999) = 0.999$$

System failure rate $= 2.587 \times 10^{-7} + 2.787 \times 10^{-7} + 7.791 \times 10^{-8} + 4.037 \times 10^{-8}$

$= 6.557 \times 10^{-7}$

Table P7.55(*a*)

Component	Failure rate, λ_i	MTTF, $\frac{1}{\lambda_i}$ (hours)	Reliability, R_i	System reliability
1	2.587×10^{-7}	3,865,481.25	0.9977	0.9995
2	2.787×10^{-7}	3,588,087.55	0.9976	0.9997
3	7.791×10^{-8}	12,835,322.81	0.9993	0.9998
4	4.037×10^{-8}	24,770,869.46	0.9996	0.9999
System	6.557×10^{-7}	1,525,087.69	0.9940	0.9990

7.56. Apply AGREE allocation method to a system containing five components given in Table P7.56.

Table P7.56

Component	Parts count, n_i	Importance index, ω_i	Operating time, hr/yr, t_i
1	237	0.96	5000
2	184	0.95	3000
3	93	1.00	1000
4	64	0.97	2000
5	33	0.98	1500

The warranty program requires a reliability of 0.99 over the first year of use.

Solution:

$$N = 237 + 184 + 93 + 64 + 33 = 611$$

24 hours × 365 days = 8760 hours per year

$$R^* = 0.99$$

Component 1:

$$\lambda_i = \left(\frac{-1}{5000}\right) \ln\left(1 - \frac{1 - 0.99^{\frac{237}{611}}}{0.96}\right) = (8.122)10^{-7}$$

$$\text{MTTF} = \frac{1}{(8.122)\,10^{-7}}$$

$$\text{MTTF} = 1{,}231{,}223.84$$

$$R_{1i} = e^{(-8.122)(10^{-7})(8760)} = 0.9929$$

$$R_{1\,\text{system}} = 1 - 0.96(1 - e^{(-8.122)(10^{-7})}\,(5000)) = 0.9961$$

Component 2:

$$\lambda_i = \left(\frac{-1}{3000}\right) \ln \left(1 - \frac{1 - 0.99^{\frac{184}{611}}}{0.95}\right) = (1.062)(10^{-6})$$

$$\text{MTTF} = \frac{1}{(1.062)\,10^{-6}}$$

$$\text{MTTF} = 941619.59$$

$$R_{2i} = e^{(-1.062)(10^{-6})(8760)} = 0.9907$$

$$R_{2\,\text{system}} = 1 - 0.97(1 - e^{(-1.062)(10^{-6})}\,(3000)) = 0.997$$

Component 3:

$$\lambda_i = \left(\frac{-1}{1000}\right) \ln \left(1 - \frac{1 - 0.99^{\frac{93}{611}}}{1.00}\right) = (1.53)(10^{-6})$$

$$\text{MTTF} = \frac{1}{(1.53)\,10^{-6}}$$

$$\text{MTTF} = 653594.77$$

$$R_{3i} = e^{(-1.53)(10^{-6})(8760)} = 0.9876$$

$$R_{3\,\text{system}} = 1 - 1.00(1 - e^{(-1.53)(10^{-6})}\,(1000)) = 0.9985$$

Component 4:

$$\lambda_i = \left(\frac{-1}{3000}\right) \ln \left(1 - \frac{1 - 0.99^{\frac{184}{611}}}{0.95}\right) = (1.062)(10^{-6})$$

$$\text{MTTF} = \frac{1}{(5.427)\,10^{-7}}$$

$$\text{MTTF} = 1842638.66$$

$$R_{4i} = e^{(-5.427)(10^{-7})(8760)} = 0.9953$$

$$R_{4\,\text{system}} = 1 - 0.97(1 - e^{(-5.427)(10^{-7})}\,(2000)) = 0.9989$$

Component 5:

$$\lambda_i = \left(\frac{-1}{1500}\right) \ln \left(1 - \frac{1 - 0.99^{\frac{33}{611}}}{0.98}\right) = (3.693)(10^{-7})$$

$$\text{MTTF} = \frac{1}{(3.693)\,10^{-7}}$$

$$\text{MTTF} = 2707825.62$$

$$R_{5i} = e^{(-3.693)(10^{-7})(8760)} = 0.9968$$

$$R_{5\ \text{system}} = 1 - 0.98(1 - e^{(-3.693)(10^{-7})}\ (2000)) = 0.9995$$

Table P7.56 (a)

Component	Failure rate, λ_i	MTTF, $\frac{1}{\lambda_i}$ (hours)	Reliability, R_i	System reliability
1	8.122×10^{-7}	1231223.84	0.9921	0.9961
2	1.062×10^{-6}	941619.59	0.9907	0.9970
3	1.53×10^{-6}	653594.77	0.9967	0.9985
4	5.427×10^{-7}	1842638.66	0.9953	0.9989
5	3.693×10^{-7}	2707825.62	0.9968	0.9995
System	4.316×10^{-6}	231696.01	0.9622	0.9900

7.57. For the reliability block diagram for a mechanical system shown in Fig. P7.57 determine the overall system reliability.

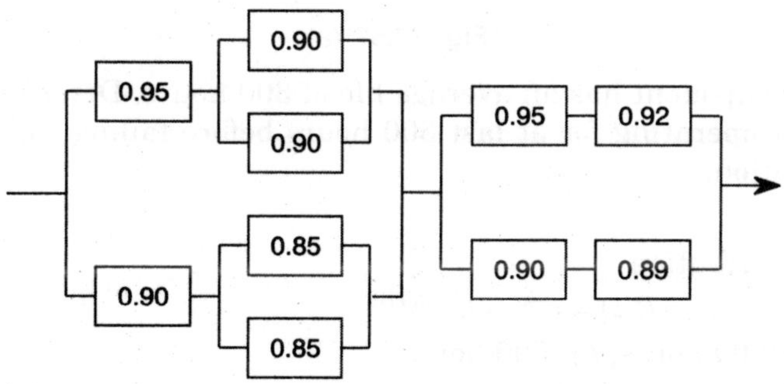

Fig. P7.57

Solution:

The reliability block diagram in Fig. P7.57 can be broken down into the following combination of series and parallel units as shown in Fig. P7.57 (*a*).

$$R_1 = 1 - (1 - 0.90)\ (1 - 0.90) = 0.990$$
$$R_2 = 1 - (1 - 0.85)\ (1 - 0.85) = 0.9775$$
$$R_3 = (0.95)\ (0.92) = 0.874$$
$$R_4 = (0.90)\ (0.89) = 0.801$$
$$R_5 = (0.95)\ (R_1) = (0.95)\ (0.990) = 0.9405$$
$$R_6 = (0.90)\ (R_2) = (0.90)\ (0.9775) = 0.8797$$
$$R_7 = 1 - (1 - R_3)\ (1 - R_4) = 1 - (1 - 0.874)\ (1 - 0.801) = 0.9749$$
$$R_8 = 1 - (1 - R_5)\ (1 - R_6) = 1 - (1 - 0.9405)\ (1 - 0.0.8797) = 0.9928$$
$$R_{\text{system}} = R_s = (R_8)\ (R_7) = (0.9928)\ (0.9749) = 0.9679$$

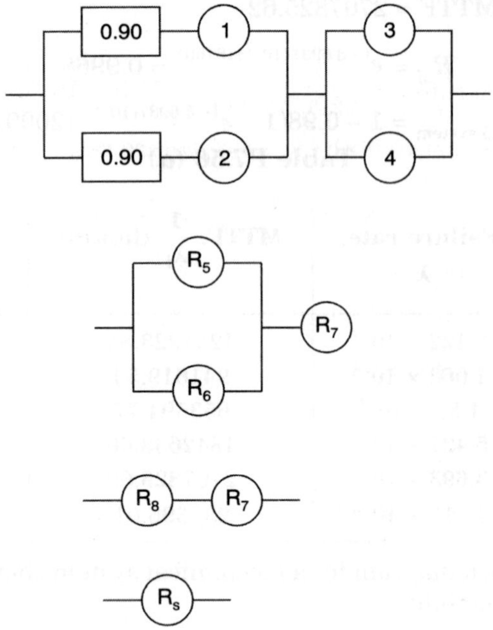

Fig. P7.57 *(a)*

7.58. An electronic component has an average life of 300 hours. Determine the probability of this component operating for at last 500 hours before failing. Assume an exponential failure distribution.

Solution:

Exponential distribution

$$R(t) = e^{-\lambda t} = e^{-t/\text{MTBF}}$$

Given MTBF = 300 hours, t = 500 hours

Therefore, the probability of this component operating for at least 500 hours before failing is given by

$$R(t) = e^{\frac{500}{300}} = e^{-1.667} = 0.189.$$

7.59. A group of components in a large electromechanical system follows a Weibull distribution. Determine the probability that one of the components will have a life of 30,000 hours. The parameters of the Weibull distribution are $\theta = 10^6$ hours, $m = 4.5$ and $t_0 = 0$.

Solution:

Weibull distribution

$$R(t) = e^{-\left[\frac{t - t_0}{\theta - t_0}\right]^m} = e^{-\left[\frac{30{,}000 - 0}{10^6 - 0}\right]^{4.5}} = 0.9999$$

7.60. A complex electromechanical system has 800 components in series configuration. Extensive testing of 200 components revealed that 3 failures occurred after 2000 hours.

(*a*) Assuming a constant failure rate, determine the reliability of the system to operate for 2000 hours

(*b*) If the desired overall system reliability is 0.99 in 2000 hours, determine the individual failure rate of the components.

Solution:

(*a*) Component failure rate $\lambda = \left(\frac{3}{200}\right)\left(\frac{1}{2000}\right) = 7.5 \times 10^{-6}$/hour

For 800 components in series with identical failure rates, we have the failure rate of the system

$$\lambda_s = (800)(7.5 \times 10^{-6}) = 6 \times 10^{-3}\text{/hour}$$

$$R(2000) = e^{-\lambda_s t} = e^{-6 \times 10^{-3}(2000)} = 6.1442 \times 10^{-6}$$

(*b*) The desired overall system reliability is 0.99 for 2000 hours.

$$R(t) = e^{-\lambda_s t}$$

$$0.99 = e^{-\lambda_s (2000)}$$

$$\ln(0.99) = -\lambda_s(2000) = -0.01$$

or $$\lambda_s = \frac{-0.01}{-2000} = 5 \times 10^{-6}\text{/hour for the system}$$

Hence $$\lambda_{\text{component}} = \frac{5 \times 10^{-6}}{800} = 6.25 \times 10^{-9}\text{/hour.}$$

7.61. The mean time to failure of a single unit and standby unit of a complex mechanical system are 50,000 hours and 30,000 hours respectively. Determine the mean time between failures of a single unit without standby that will have the same reliability as the standby system for the system to operate for 20,000 hours. Assume constant failure rate for the single unit.

Solution:

Constant failure rate

$$\lambda_1 = \frac{1}{50{,}000} = 2 \times 10^{-5}\text{/hour}$$

$$\lambda_2 = \frac{1}{30{,}000} = 3.333 \times 10^{-5}\text{/hour}$$

$$R_{(20{,}000)} = \frac{\lambda_1}{\lambda_2 - \lambda_1}\left[e^{-\lambda_1 t} - e^{-\lambda_2 t}\right] + e^{-\lambda_1 t}$$

$$= \frac{2 \times 10^{-5}}{(3.333) \times 10^{-5}}\left[e^{-2 \times 10^{-5}(2 \times 10^4)} - e^{-3.333 \times 10^{-5}(2 \times 10^4)}\right] + e^{-2 \times 10^{-5}(2 \times 10^4)}$$

$$= 1.5[0.6703 - 0.5137] + 0.6703 = 0.9056$$

For the equivalent single unit system, we have

$$0.9056 = e^{-\lambda(20{,}000)}$$

$$\ln(0.9056) = -20{,}000\,\lambda = -0.099157$$

or $$\lambda = 4.9578 \times 10^{-6}\text{/hour}$$

Therefore, the mean time between failures (MTBF) is given by

$$\text{MTBF} = \frac{1}{\lambda} = \frac{1}{4.9578 \times 10^{-6}} = 201{,}699 \text{ hours.}$$

Appendix-7A

Basic Mathematics

7A.1

$$2 \log_b t = \log_b 3 + \log_b (t + 5)$$

$$\log_b t^2 = \log_b 3(t + 5)$$

or

$$t^2 = 3(t + 5)$$

which gives

$$t^2 - 3t - 15 = 0 \text{ and}$$

$$t = \frac{3 \pm \sqrt{9 + 60}}{2} = \frac{3 \pm \sqrt{69}}{2}$$

$$t = \frac{3 + \sqrt{69}}{2} \quad \text{or} \quad \frac{3 - \sqrt{69}}{2}$$

$$t = \frac{3 - \sqrt{69}}{2} \text{ is undefined.}$$

Hence, the only root is

$$t = \frac{3 + \sqrt{69}}{2}$$

7A.2

$$\frac{1}{2}\log_3 t + \frac{1}{2}\log_3 (t - 3) = \log_3 6$$

$$\frac{1}{2}\log_3 (t^2 - 3t) = \log_3 6$$

$$\log_3 \sqrt{t^2 - 3t} = \log_3 6$$

which gives

$$t^2 - 3t = 36 \quad \text{or} \quad t^2 - 3t - 36 = 0$$

Solving for t gives

$$t = \frac{3 \pm \sqrt{9 + 144}}{2}$$

The root

$$t = \frac{3 - \sqrt{153}}{2} \text{ is undefined.}$$

Hence, the only root is

$$t = \frac{3 + \sqrt{153}}{2}$$

7A.3 (*a*) $3 \log_2 m + \frac{1}{3} \log_2 n - 3 \log_2 p = \log_2 m^3 + \log_2 \sqrt[3]{n} - \log_2 p^3$

or $\log_2(m^3 \sqrt[3]{n}) - \log_2 p^3 = \log_2\left[\dfrac{m^3 \sqrt[3]{n}}{p^3}\right]$

(b) $\dfrac{1}{3}(\log_5 c + 3 \log_5 d - \log_5 g) = \dfrac{1}{3}(\log_5 c + \log_5 d^3 - \log_5 g)$

$$= \frac{1}{3}\log_5\left(\frac{cd^3}{g}\right) = \log_5\sqrt[3]{\frac{cd^3}{g}}$$

7A.4 (a) $(27\, a^2\, b^4)^{1/3} = \left[(27^{1/3})\,(a^2)^{1/3}\,(b^4)^{1/3}\right] = 3\, a^{2/3}\, b^{4/3}$

(b) $a^{1/3}\, a^{2/5} = a^{1/3+2/5} = a^{11/15}$

7A.5 (a) $\left[\dfrac{8^{-3/2}\, x^{2/3}\, y^{-7/4}}{4^{3/2}\, x^{-1/3}\, y^{3/4}}\right]^{5/3} = \left[\dfrac{x^{2/3}\, x^{1/3}}{4^{3/2}\, 8^{3/2}\, y^{3/4}\, y^{7/4}}\right]^{5/3} = \left[\dfrac{x^{2/3+1/3}}{4^{3/2}\, 8^{3/2}\, y^{3/4+7/4}}\right]$

$$= \left[\frac{x^{(1)5/3}}{4^{(3/2)\,(5/3)}\, 8^{(3/2)\,(5/3)}\, y^{(10/4)\,(5/3)}}\right] = \frac{x^{5/3}}{4^{(5/2)}\, 8^{(5/2)}\, y^{25/6}} = \frac{x^{5/3}}{(32)^{5/2}\, y^{25/6}}$$

(b) $(4\, x^4)^{-1/2} - 5x^{-5} = \dfrac{1}{(4x^4)^{1/2}} - \dfrac{5}{x^5} = \dfrac{1}{2x^2} - \dfrac{5}{x^5} = \dfrac{x^3 - 10}{2x^5}$

7A.6 (a) $\sqrt[3]{4^3} = (\sqrt[3]{4})^3 = 4$

(b) $\sqrt[5]{2}\, \sqrt[5]{7} = \sqrt[5]{2 \times 7} = \sqrt[5]{14}$

(c) $\sqrt[3]{\sqrt{7}} = \sqrt[3\times 2]{7} = \sqrt[6]{7}$

(d) $\dfrac{\sqrt{7}}{\sqrt{2}} = \sqrt{\dfrac{7}{2}}$

7A.7 (a) $\sqrt{98} = \sqrt{(49(\,(2)} = \sqrt{49}\, \sqrt{2} = 7\, \sqrt{2}$

(b) $\sqrt{a^5\, b^4} = \sqrt{(a^4)\,(a)\,(b^2)} = \sqrt{a^4}\, \sqrt{a}\, \sqrt{b^2} = a^2 b\, \sqrt{a}$

(c) $\sqrt[3]{135} = \sqrt[3]{(27)\,(5)} = \sqrt[3]{27}\, \sqrt[3]{5} = 3\, \sqrt[3]{5}$

(d) $\sqrt[5]{32x^8\, y^{12}} = \sqrt[5]{(32)\,(x^5)\,(x^3)\,(y^{10})\,(y^2)} = \sqrt[5]{(32)\,(x^5)\,(y^{10})}\, \sqrt[5]{x^3\, y^2} = 2xy^2\, \sqrt[5]{x^3\, y^2}$

7A.8 (a) $\dfrac{\sqrt[4]{25}}{\sqrt{5}} = \dfrac{\sqrt[4]{5^2}}{\sqrt{5}} = \dfrac{5^{2/4}}{5^{1/2}} = 1$

(b) $\dfrac{\sqrt[6]{125}}{\sqrt{3}} = \dfrac{\sqrt[6]{5^3}}{\sqrt{3}} = \dfrac{5^{1/2}}{3^{1/2}} = \sqrt{\dfrac{5}{3}}$

(c) $\sqrt[9]{125x^3\, y^{16}} = \sqrt[9]{5^3\, x^3\, y^9\, y^7} = 5^{3/9}\, x^{3/9}\, y^{9/9}\, y^{7/9} = 5^{1/3}\, x^{1/3}\, y\, y^{7/9} = y\, \sqrt{5x}\, \sqrt[9]{y^7}$

7A.9 $A^{-6/8} = 0.001 = 10^{-3}$

$(A^{-6/8})^{-8/6} = (10^{-3})^{-8/6}$

$A = 10^4 = 10{,}000$

7A.10 This is a log of a quotient of two terms.

$$\log\left(\frac{x}{y+z}\right) = \log x - \log(y+z)$$

7A.11 Using logarithmic identities

$$\ln(7.3891)^{xy} = xy\ \ln(7.3891) = 2xy$$

7A.12 Apply the log identities, then use the value given for log 4

$$\log\left(\frac{1}{4}\right) = \log(4)^{-1} = -\log(4) = -0.703x$$

7A.13 Let $y = h(t)$

$y = e^u$ when $u = 2t \sin 2t$

$$\frac{dy}{du} = \frac{d}{du}(e^u) = e^u$$

Product rule:

$$\frac{du}{dt} = (2t \sin 2t)' = (2t)' (\sin 2t) + 2t(\sin 2t)'$$

$$= 2 \sin 2t + 2t[(\cos 2t)(2t)']$$

$$= 2 \sin 2t + 4t \cos 2t$$

Combining the above derivatives, we get

$$h'(t) = \left(\frac{dy}{dt}\right) = \left(\frac{dy}{du}\right)\left(\frac{du}{dt}\right) = e^u(2 \sin 2t + 4t \cos 2t)$$

$$= (2 \sin 2t + 4t \cos 2t)\, e^{2t \sin 2t}$$

7A.14 $x = \exp[y(x)]$

Differentiating both sides of the above equation w.r.t. x, we get

$$1 = \exp'[y(x)]\ y'(x) = \exp[y(x)]\ y'(x) = e^y \frac{dy}{dx}$$

Solving for $\frac{dy}{dx}$, we have

$$\frac{dy}{dx} = \frac{1}{e^y}$$

but $e^y = x$

Hence $\frac{dy}{dx} = \frac{1}{x}$

7A.15 Let $y = \ln(2t^2 - 5)$ and write the function as a composition chain

$y = \ln u,\ u = 2t^2 - 5$

Then $\frac{dy}{du} = \frac{1}{u}$ and $\frac{du}{dt} = 4t$

By chain rule, we have

$$f'(t) = \frac{dy}{dt} = \left(\frac{dy}{du}\right)\left(\frac{du}{dt}\right) = \frac{1}{u}(4t) = \frac{4t}{2t^2 - 5}$$

7A.16 Taking logarithms on both sides, we get

$$\ln y = \ln\left[\frac{(2 + x^2)\sqrt{\sin 2x}}{(3x^3 - 2x - 1)^5}\right]$$

$$= \ln[(2 + x^2)(\sin 2x)^{1/2} - \ln[(3x^3 - 2x - 1)^5]$$

Differentiating both sides, we get

$$\frac{y'}{y} = \frac{(2 + x^2)'}{(2 + x^2)} + \frac{1}{2}\frac{(\sin 2x)'}{\sin 2x} - 5\frac{(3x^3 - 2x - 1)'}{(3x^3 - 2x - 1)} = \frac{2x}{(2 + x^2)} + \frac{1}{2}\frac{2\cos 2x}{\sin 2x} - 5\frac{9x^2 - 2}{3x^3 - 2x - 1}$$

or $$y' = \left[\frac{(2 + x^2)\sqrt{\sin 2x}}{(3x^3 - 2x - 1)^5}\right]\left[\frac{4x}{(2 + x^2)} + \cot 2x - 5\frac{9x^2 - 2}{(3x^3 - 2x - 1)}\right]$$

CHAPTER 8

Quality by Experimental Design

8.1. A manufacturer is interested in investigating the surface finish of a new disc that will be used for brake system in automobiles. She suspects that change in material influences surface finish, and also knows that acceptable materials are of four different types. She decides to test four specimens of each material type. Random test sequence is necessary to prevent the effect of nuisance variables. The random run sequence and results are provided in Table P8.1. Test the hypothesis that material type influences surface finish, using $\alpha = 0.05$.

Table P8.1

Test sequence (randomized)	Specimen material type	Surface finish (micron)
11	1	3.9
19	1	3.8
20	1	4
17	1	3.9
3	1	3.5
15	2	3.1
2	2	3.0
14	2	2.9
9	2	3.2
16	2	2.9
7	3	2.8
13	3	2.6
6	3	2.9
5	3	3.0
10	3	2.8
18	4	2.3
12	4	1.9
1	4	1.8
8	4	2.0
4	4	2.1

Solution:

The random test sequence as provided in Table P8.1 implies that at the 11th experimental (test) run-specimen 1 is tested, at 19th experimental run-specimen material 1 is tested, at 20th experimental run-specimen material 1 is tested, and so on. Therefore the 1st random experimental run, as obtained from statistical package Minitab (Release 13 version), is based on specimen material type 4.

The results of single factor analysis of variance (ANOVA) are provided below:

Analysis of Variance (ANOVA) for Surface Finish

Source	DF	SS	MS	F	P
Material type	3	8.2000	2.7333	96.76	0.000
Error	16	0.4520	0.0283		
Total	19	8.6520			

The *F*-statistic shows that the material type influences the surface finish and there is a significant difference in mean, as per the analysis.

The 95% Confidence Interval (*CI*) for individual mean based on pooled standard deviation (pooled StDev) is provided below:

```
                                  Individual 95% CIs for Mean
                                  Based on Pooled Std. Deviation
Level  N   Mean     StDev         ------+----------+----------+------
1      5   3.8200   0.1924                                    (--*-)
2      5   3.0200   0.1304                               (-*--)
3      5   2.8200   0.1483                       (--*--)
4      5   2.0200   0.1924        (--*-)
                                  ------+----------+----------+------
Pooled StDev = 0.1681                  2.40       3.00       3.60
```

The different graphical plots of responses (surface finish) are provided in Figure S8.1-1:

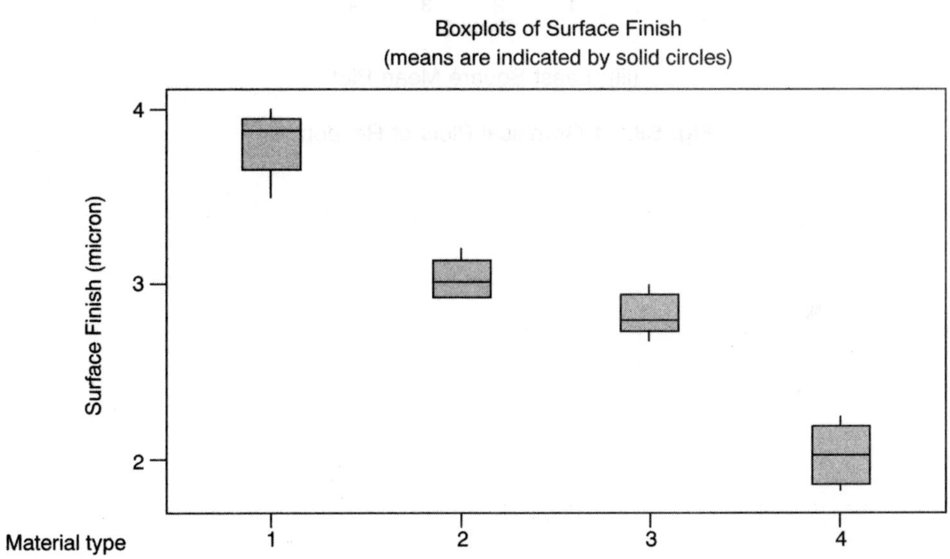

(i) Box Plot

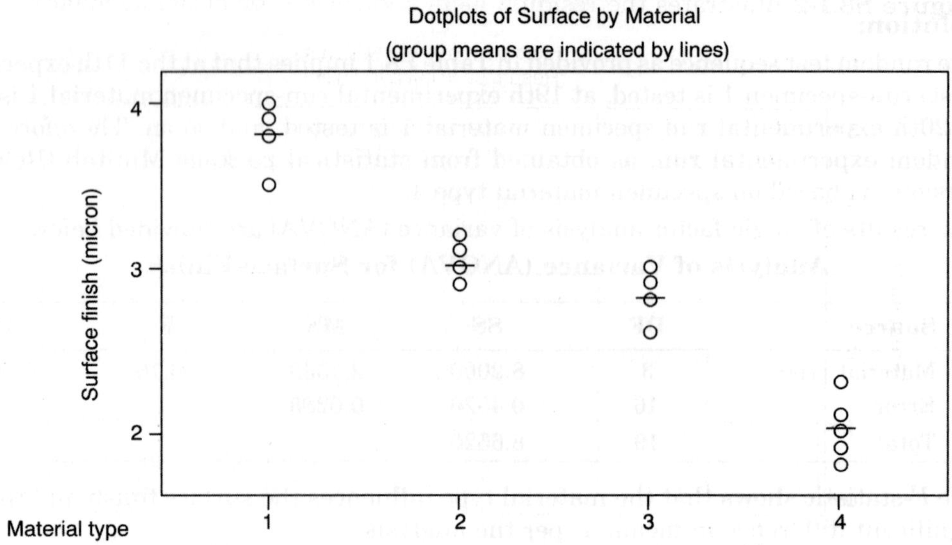

(ii) Dot Plot

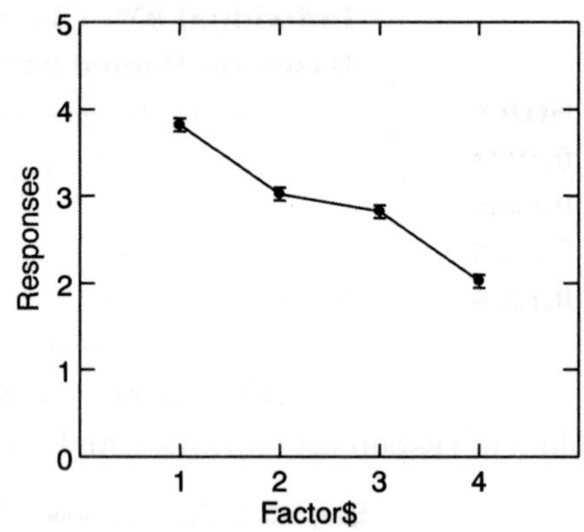

(iii) Least Square Mean Plot

Fig. S8.1-1 Graphical Plots of Responses.

Figure S8.1-2 illustrates the residual plots and checks for model adequacy

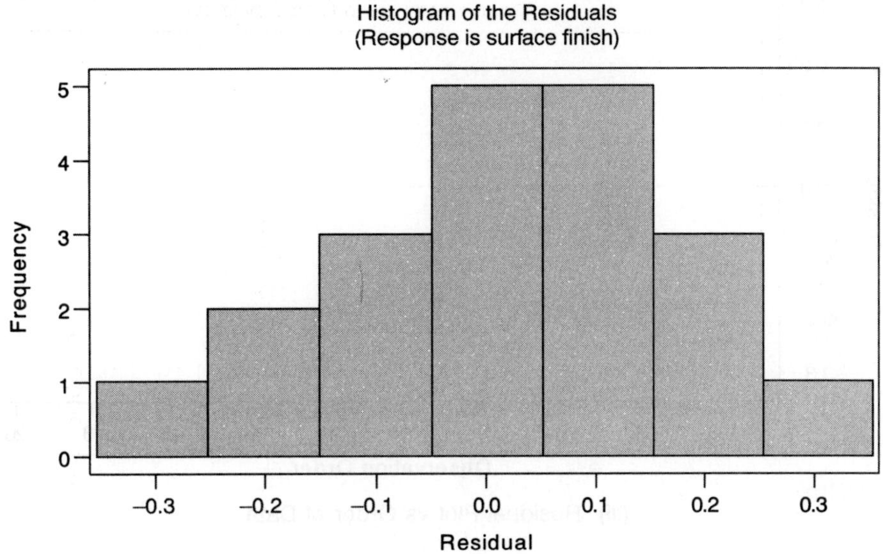

(i) Histogram of Residual

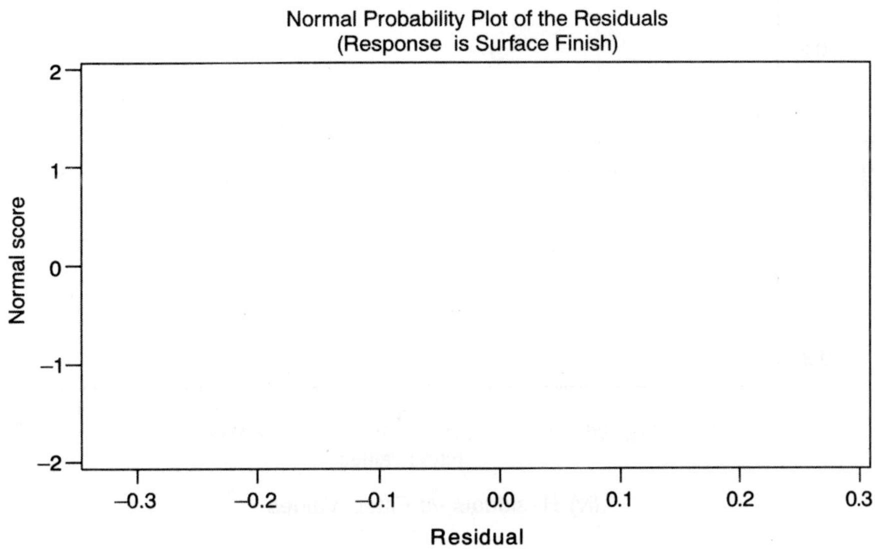

(ii) Normal Probability Plot of Residual

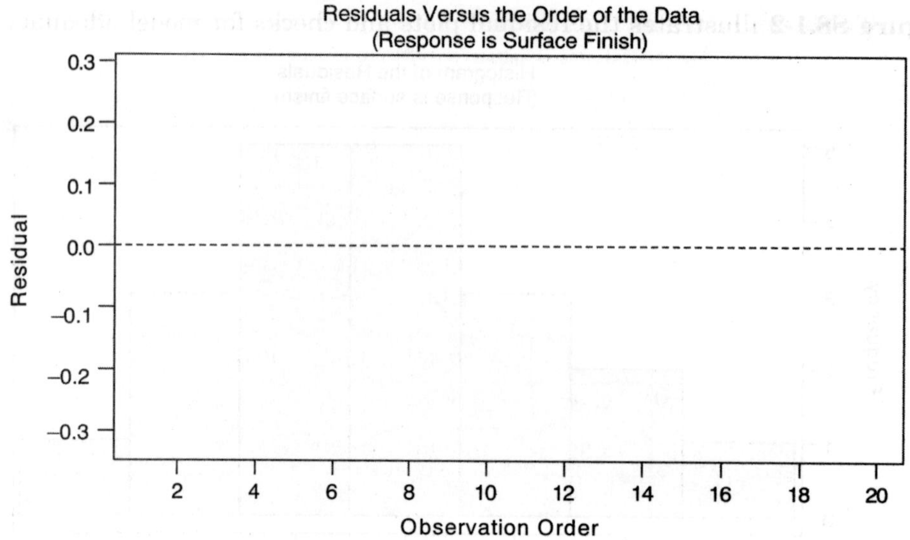

(iii) Residual Plot vs Order of Data

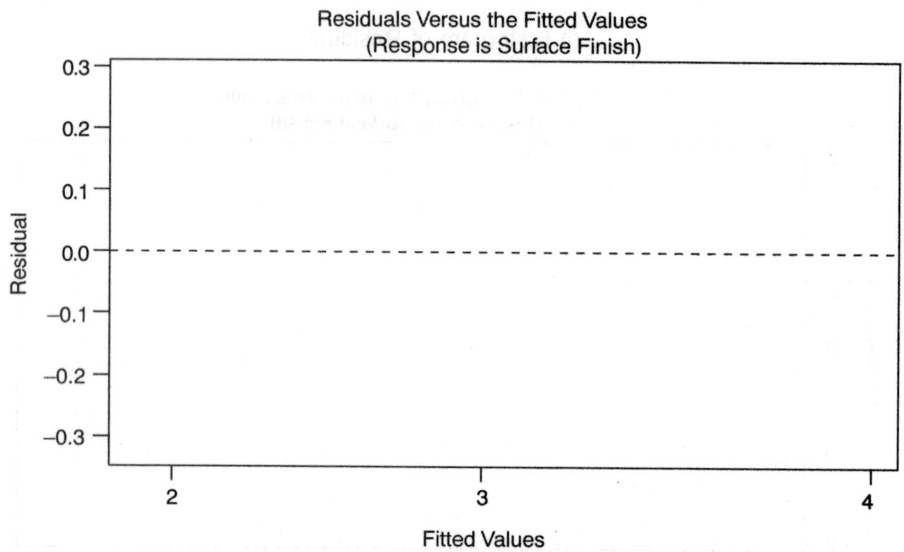

(iv) Residuals vs Fitted Values

Fig. S8.1-2 Residual Plots.

The residual plots in Figure S8.1-2 do not reveal any drift or erratic behaviour of error over time, and no unusual structure in residual is apparent.

Multiple Comparisons of mean values by Tukey's Test and Fisher LSD method in SPSS 10 package is provided in Table S8.1-1.

Table S8.1-1: Multiple Comparisons of Mean Values in Which Dependent Variable is Surface Finish

	(I) Material Type	(J) Material Type	Mean Difference (I–J)	Std. Error	Sig.	95% Confidence Interval Lower Bound	95% Confidence Interval Upper Bound
Tukey's test	1.00	2.00	.8000*	.1063	.000	.4959	1.1041
		3.00	1.0000*	.1063	.000	.6959	1.3041
		4.00	1.8000*	.1063	.000	1.4959	2.1041
	2.00	1.00	– .8000*	.1063	.000	– 1.1041	– .4959
		3.00	.2000	.1063	.274	– .1041	.5041
		4.00	1.0000*	.1063	.000	.6959	1.3041
	3.00	1.00	– 1.0000*	.1063	.000	– 1.3041	– .6959
		2.00	– .2000	.1063	.274	– .5041	.1041
		4.00	.8000*	.1063	.000	.4959	1.1041
	4.00	1.00	– 1.8000*	.1063	.000	– 2.1041	– 1.4959
		2.00	– 1.0000*	.1063	.000	– 1.3041	– .6959
		3.00	– .8000*	.1063	.000	– 1.1041	– .4959
Fisher LSD method	1.00	2.00	.8000*	.1063	.000	.5747	1.0253
		3.00	1.0000*	.1063	.000	.7747	1.2253
		4.00	1.8000*	.1063	.000	1.5747	2.0253
	2.00	1.00	– .8000*	.1063	.000	– 1.0253	– .5747
		3.00	.2000	.1063	.078	– 2.5349E – 02	.4253
		4.00	1.0000*	.1063	.000	.7747	1.2253
	3.00	1.00	– 1.0000*	.1063	.000	– 1.2253	– .7747
		2.00	– .2000	.1063	.078	– .4253	2.535E – 02
		4.00	.8000*	.1063	.000	.5747	1.0253
	4.00	1.00	– 1.8000*	.1063	.000	– 2.0253	– 1.5747
		2.00	– 1.0000*	.1063	.000	– 1.2253	–.7747
		3.00	– .8000*	.1063	.000	– 1.0253	–.5747

* The mean difference is significant at the .05 level.

8.2. An engineer is interested in investigating the tensile strength of a composite material. He suspects that change in mixing technique type influences tensile strength, and also knows that available mixing are of four different types. He decides to test four specimens at each mixing technique type. Random test sequence is necessary to prevent the effect of nuisance variables. The test sequence and corresponding tensile strength is provided in the following Table P8.2. Test the hypothesis that mixing technique type influences tensile strength, using $\alpha = 0.05$.

Table P8.2

Test sequence (randomized)	Mixing technique	Tensile strength
10	A	3120
13	A	3001
9	A	2861
12	A	2880
8	B	3100
6	B	3310
11	B	2990
16	B	3120
7	C	2785
4	C	2901
3	C	2986
2	C	3054
5	D	2605
15	D	2750
1	D	2640
14	D	2755

Solution:

The results of single factor analysis of variance (ANOVA) are provided below:

Analysis of Variance (ANOVA) for Tensile Strength

Source of variation	DF	SS	MS	F	P
Mixing technique	3	400245	133415	10.38	0.001
Error	12	154255	12855		
Total	15	554500			

The F-statistic shows that the mixing technique influences the tensile strength and there is a significant difference in mean, as per the analysis.

The 95% Confidence Interval (CI) for individual mean based on pooled standard deviation (pooled StDev) is provided below:

```
                                  Individual 95% CIs for Mean
                                     Based on Pooled StDev
Level  N   Mean    StDev   --+----------+----------+----------+----------
A      4  2965.5   120.2              (-----*-----)
B      4  3130.0   132.9                        (-----*-----)
C      4  2931.5   116.0            (-----*-----)
D      4  2687.5    76.4   (-------*-------)
                           --+----------+----------+----------+----
Pooled StDev =113.4        2600       2800       3000       3200
```

The different graphical plot of experimental tensile strength data is provided in Figure S8.2-1

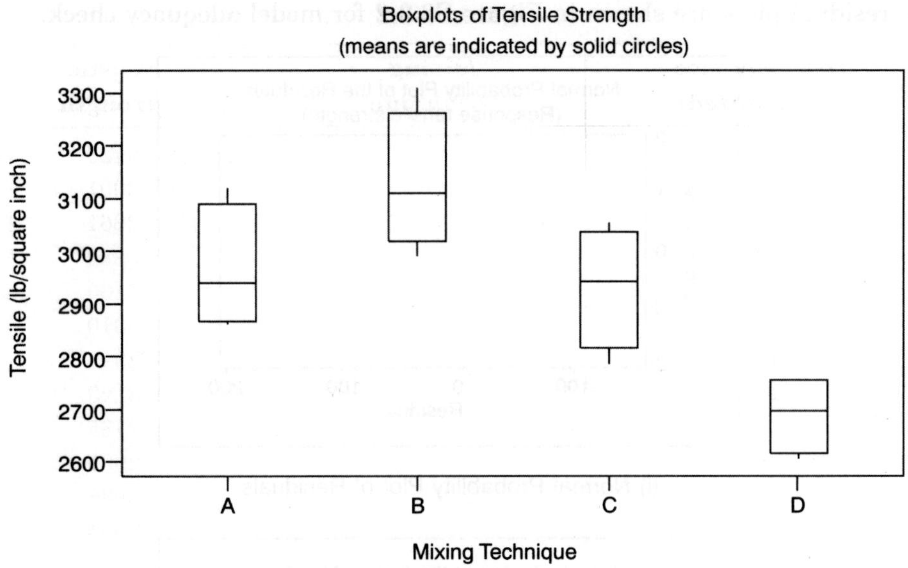

(i) Box Plots

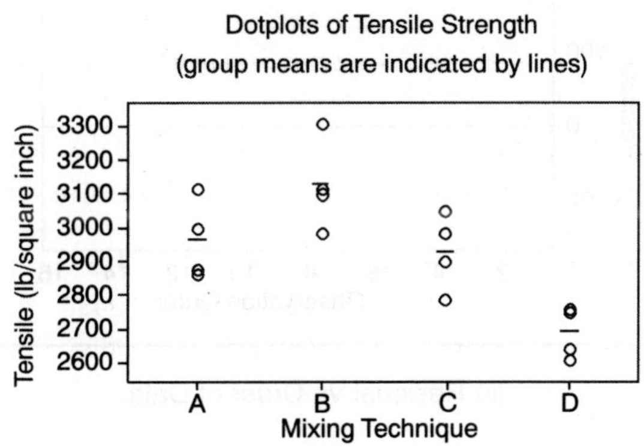

(ii) Dot Plots

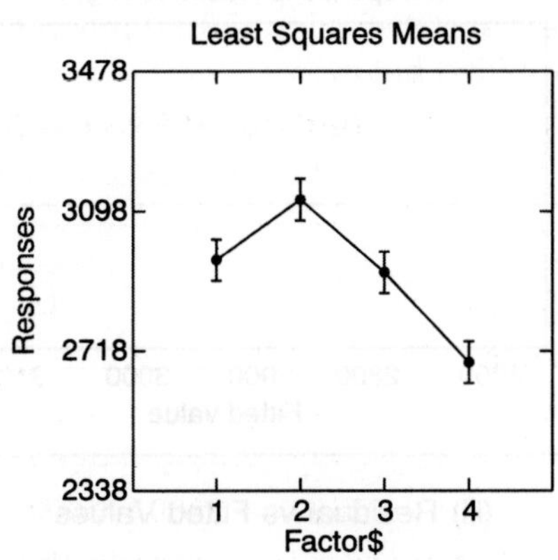

(iii) Least Mean Square

Fig. S8.2-1 Graphical Plot of Experimental Tensile Strength.

The residual plots are shown in Figure S8.2-2 for model adequacy check.

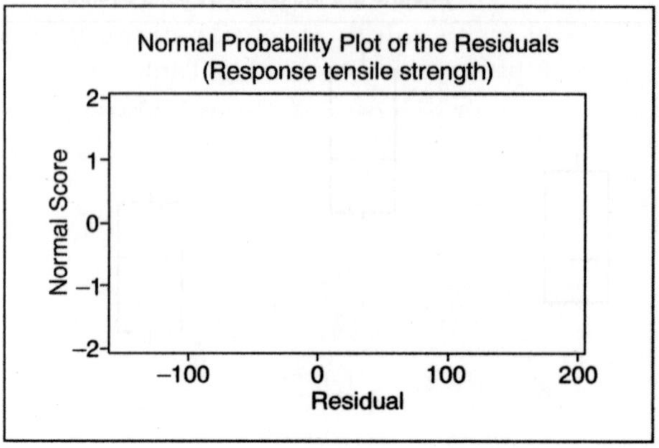

(i) Normal Probability Plot of Residuals

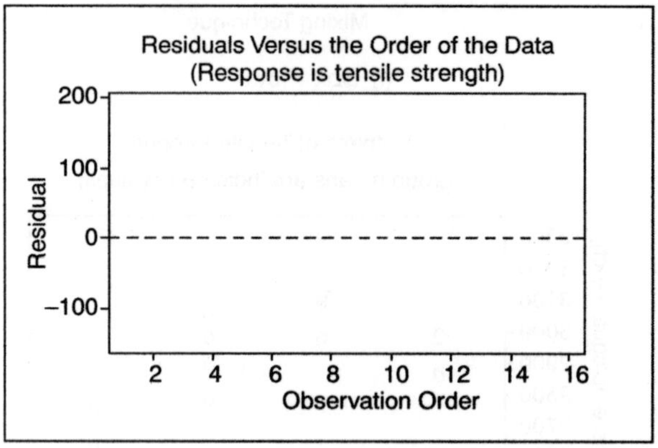

(ii) Residual vs Order of Data

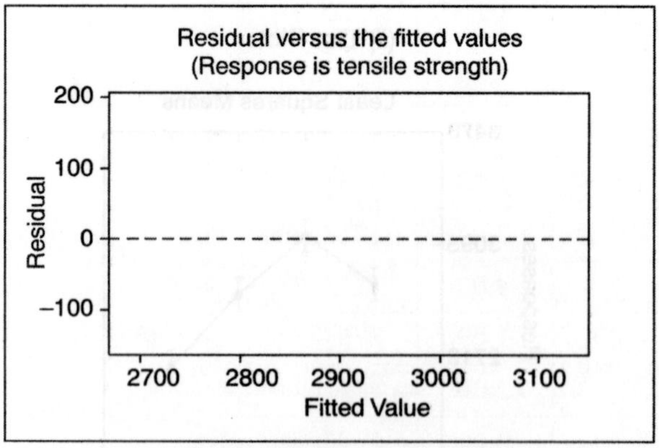

(iii) Residual vs Fitted Values

Fig. S8.2-2 Residual Plots.

The residual plots in Figure S8.2-2 do not reveal any drift or erratic behaviour of error over time, and no unusual structure in residual is apparent.

Multiple Comparisons of mean values of tensile strength by Tukey's Test and Fisher LSD method in SPSS 10 package is provided in Table S8.2-1

Table S8.2-1 Multiple Comparisons of Mean Values in which Dependent Variable is Tensile Strength

	(I) Factor Type	(J) Factor Type	Mean Difference (I–J)	Std. Error	Sig.	95% Confidence Interval	
						Lower Bound	Upper Bound
Tukey test HSD	1.00	2.00	– 164.5000	80.1704	.223	– 402.5204	73.5204
		3.00	34.0000	80.1704	.973	– 204.0204	272.0204
		4.00	278.0000*	80.1704	.021	39.9796	516.0204
	2.00	1.00	164.5000	80.1704	.223	– 73.5204	402.5204
		3.00	198.5000	80.1704	.115	– 39.5204	436.5204
		4.00	442.5000*	80.1704	.001	204.4796	680.5204
	3.00	1.00	– 34.0000	80.1704	.973	– 272.0204	204.0204
		2.00	– 198.5000	80.1704	.115	– 436.5204	39.5204
		4.00	244.0000*	80.1704	.044	5.9796	482.0204
	4.00	1.00	– 278.0000*	80.1704	.021	– 516.0204	– 39.9796
		2.00	– 442.5000*	80.1704	.001	– 680.5204	– 204.4796
		3.00	– 244.0000*	80.1704	.044	– 482.0204	– 5.9796
Fisher LSD	1.00	2.00	– 164.5000	80.1704	.063	– 339.1763	10.1763
		3.00	34.0000	80.1704	.679	– 140.6763	208.6763
		4.00	278.0000*	80.1704	.005	103.3237	452.6763
	2.00	1.00	164.5000	80.1704	.063	– 10.1763	339.1763
		3.00	198.5000*	80.1704	.029	23.8237	373.1763
		4.00	442.5000*	80.1704	.000	267.8237	617.1763
	3.00	1.00	– 34.0000	80.1704	.679	– 208.6763	140.6763
		2.00	– 198.5000*	80.1704	.029	– 373.1763	– 23.8237
		4.00	244.0000*	80.1704	.010	69.3237	418.6763
	4.00	1.00	– 278.0000*	80.1704	.005	– 452.6763	– 103.3237
		2.00	– 442.5000*	80.1704	.000	– 617.1763	– 267.8237
		3.00	– 244.0000*	80.1704	.010	– 418.6763	– 69.3237

* The mean difference is significant at the .05 level.

8.3. An industrial engineer is interested to determine whether or not four different tips produce different reading on hardness testing machine. The engineer has decided to obtain four different observations for each tip into metal test material coupons. He is also interested to remove the variability between coupons from the experimental error. Therefore he performs a randomized complete block design. The scaled value of hardness as obtained during experimentation is provided in Table P8.3. Analyze the data and interpret the results.

Table P8.3

Type of tip	Coupon (Block)	Hardness (Scaled values)
1	1	– 2.2
1	2	– 1.3
1	3	0.9
1	4	4.8
2	1	– 1.1
2	2	– 2.3
2	3	3.1
2	4	4.2
3	1	– 3.1
3	2	– 0.8
3	3	0.1
3	4	2.2
4	1	2.1
4	2	1.1
4	3	5.3
4	4	7.4

Solution:

The factors, type of factors, levels, and values assigned is provided below:

Factor	Type	Levels	Values
Type of Tip	fixed	4	1 2 3 4
Coupon	fixed	4	1 2 3 4

The ANOVA analysis is provided below:

Source	DF	Seq SS	Adj SS	Adj MS	F	P
Type of Tip	3	42.845	42.845	14.282	14.39	0.001
Material Coupon (Block)	3	89.915	89.915	29.972	30.21	0.000
Error	9	8.930	8.930	0.992		
Total	15	141.690				

It is observed from the ANOVA analysis that type of tip influences hardness reading. Also the material coupon (blocks) seems to differ significantly, as Adj SS is large relative to error.

The main effect plot is provided in Figure S8.3-1.

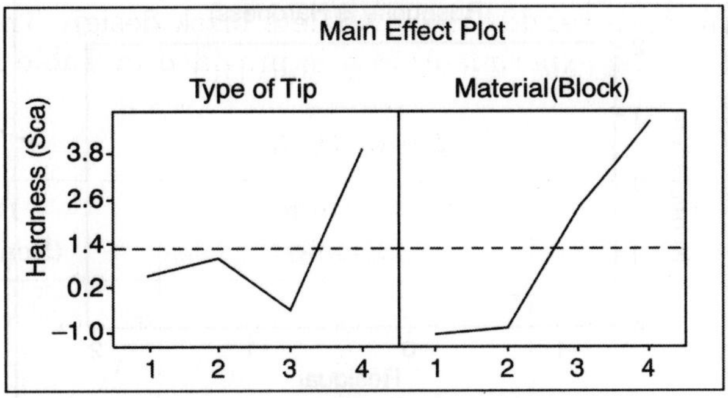

Fig. S8.3-1 Main Effects Plot.

Plots of residuals for model adequacy check are provided in Figure S8.3-2

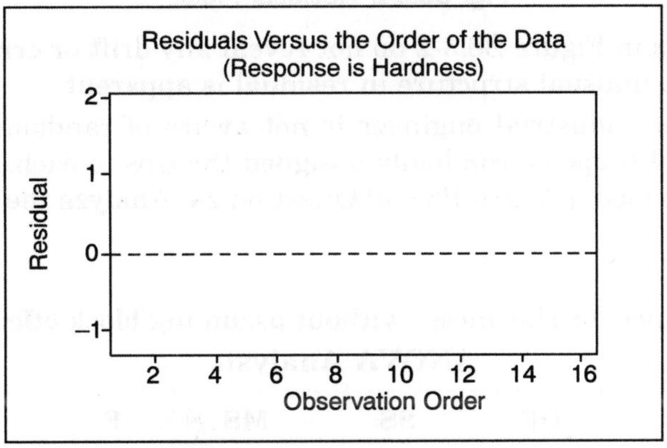

(i) Residual vs Order of Data

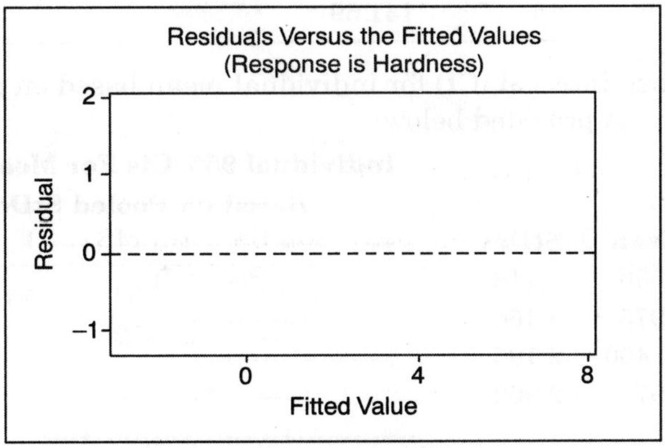

(ii) Residual vs Fitted Values

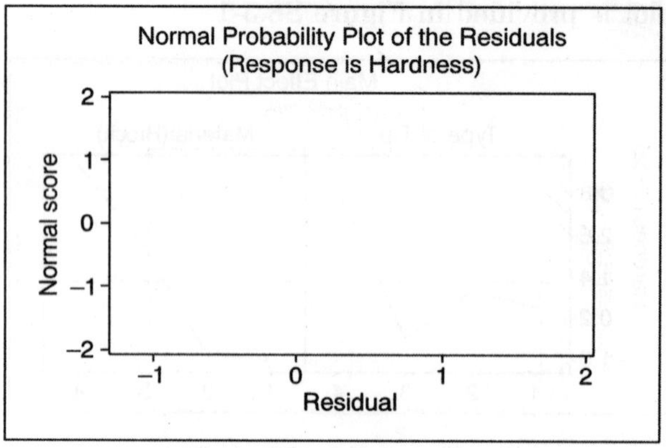

(iii) Normal Probability Plot of Residuals

Fig. S8.3-2 Residual Plots.

The residual plots in Figure S8.3-2 do not reveal any drift or erratic behaviour of error over time, and no unusual structure in residual is apparent.

8.4. Suppose the same industrial engineer is not aware of randomized block design, and uses four material coupons, randomly assigned the tips to each. The design resulted is same data as provided in Table P8.4 of Question 24. Analyze the data and interpret the results.

Solution:

Analysis of Variance for Hardness (without assuming block effect) is provided below:

ANOVA Analysis

Source	DF	SS	MS	F	P
Type of Tip	3	42.85	14.28	1.73	0.213
Error	12	98.85	8.24		
Total	15	141.69			

The 95% Confidence Interval (CI) for individual mean based on pooled standard deviation (pooled StDev) is provided below:

```
                                 Individual 95% CIs For Mean
                                     Based on Pooled StDev
Level   N    Mean     StDev    --+---------+---------+---------+----
1       4    0.550    3.118         (---------*---------)
2       4    0.975    3.160              (---------*---------)
3       4   -0.400    2.195     (---------*---------)
4       4    3.975    2.902              (---------*---------)
                               --+---------+---------+---------+----
Pooled StDev = 2.870           - 3.0      0.0       3.0      6.0
```

The ANOVA analysis shows that the hypothesis of equal hardness measurement cannot be rejected. Therefore, as the engineer fails to block when he should have, the effect may inflate the experimental error so much that important difference among the treatment means may be undetected.

The main effect plot is provided in Figure S8.4-1

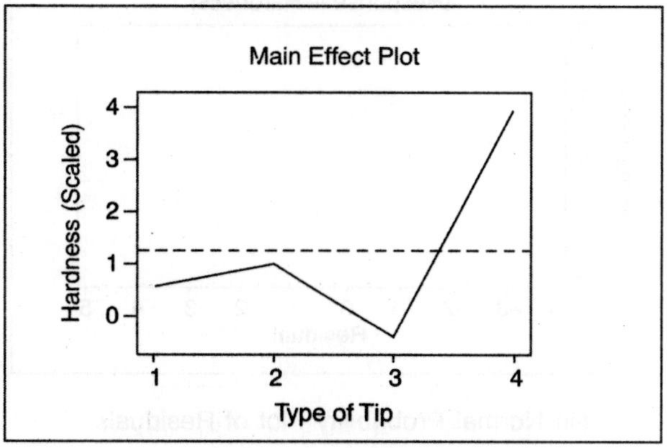

Fig. S8.4-1 Main Effect Plot.

Plots of residuals for model adequacy check are provided in Figure S8.4-2

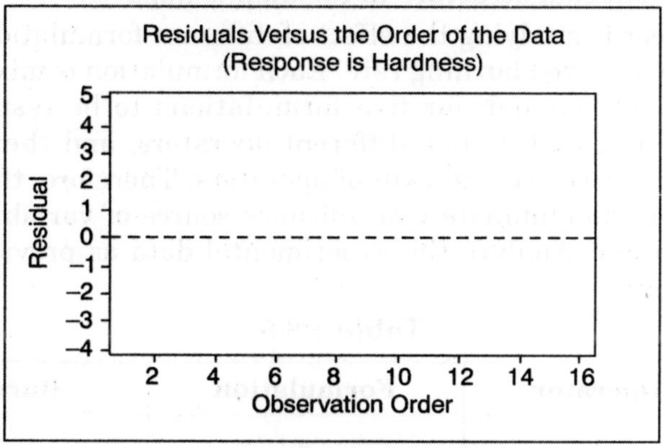

(i) Residual vs Order of Data

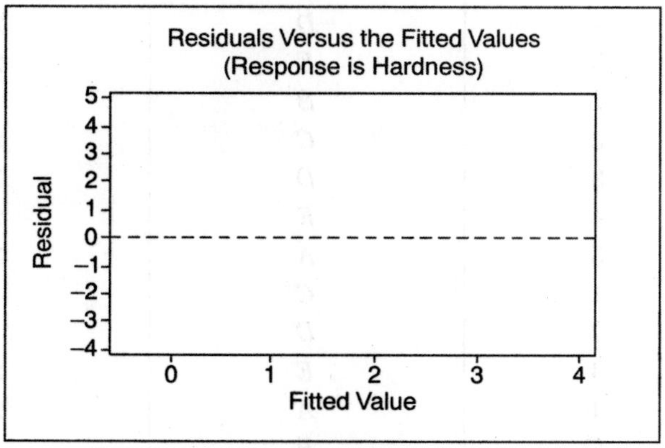

(ii) Residual vs Fitted Values

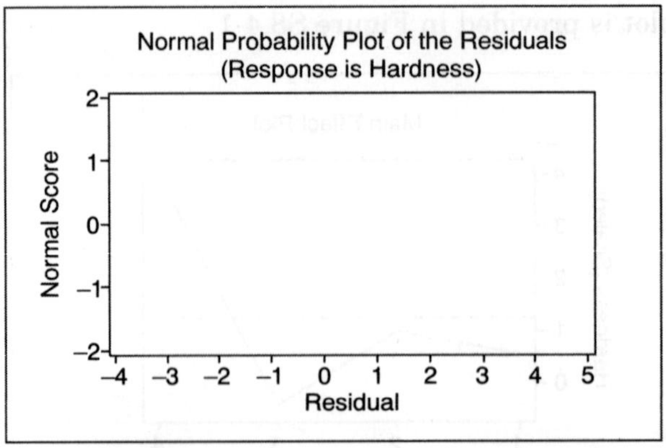

(iii) Normal Probability Plot of Residuals

Fig. S8.4-2 Residual Plots.

The residual plots in Figure S8.4-2 do not reveal any drift or erratic behaviour of error over time, and no unusual structure in residual is apparent.

8.5. A chemical engineer is studying the effect of different formulation of engine oil used in automobile on the observed burning rate. Each formulation is mixed from a batch of raw material that is only enough for five formulations to be tested. Furthermore, the formulations are prepared by five different operators, and there may be substantial differences in the experience and skill of operators. Therefore, the engineer performs a Latin square design to eliminate two nuisance source of variability, *viz.* batch of raw material and operator. Analyze the experimental data as provided in Table P8.5 and interpret the results.

Table P8.5

Batch	Operator	Formulation	Burning rate
1	1	*A*	24.3
1	2	*B*	20.2
1	3	*C*	19.2
1	4	*D*	23.9
1	5	*E*	23.8
2	1	*B*	17.3
2	2	*C*	24.1
2	3	*D*	30.2
2	4	*E*	26.9
2	5	*A*	36.3
3	1	*C*	18.4
3	2	*D*	39
3	3	*E*	26.1
3	4	*A*	27.2
3	5	*B*	21.5
4	1	*D*	26.4
4	2	*E*	30.9
4	3	*A*	25.8

Batch	Operator	Formulation	Burning rate
4	4	*B*	23.2
4	5	*C*	22.1
5	1	*E*	22.3
5	2	*A*	30.4
5	3	*B*	20.1
5	4	*C*	28.7
5	5	*D*	30.7

Solution:

The configuration of Latin-square design is 5 factors, each with 5 levels, and replication one. The Latin square design matrix is provided below:

	Operator				
Batch of raw material	1	2	3	4	5
1	*A*	*B*	*C*	*D*	*E*
2	*B*	*C*	*D*	*E*	*A*
3	*C*	*D*	*E*	*A*	*B*
4	*D*	*E*	*A*	*B*	*C*
5	*E*	*A*	*B*	*C*	*D*

In the design matrix, both batch of raw material (row) and operators (column) are orthogonal to treatments (formulations). The five formulations are denoted by latin letters, *A*, *B*, *C*, *D* and *E*, and therefore the name latin square. The row and column actually represent two restrictions on randomization.

The factors, types of factor, levels, and values assigned are provided below:

Factor	Type	Levels	Values				
Formulation	random	5	*A*	*B*	*C*	*D*	*E*
Batch	fixed	5	1	2	3	4	5
Operator	fixed	5	1	2	3	4	5

The ANOVA (Analysis of Variance) for burning rate is provided below:

Source	DF	Seq SS	Adj SS	Adj MS	F	P
Formulation	4	330.68	330.68	82.67	7.36	0.003
Batch	4	71.41	71.41	17.85	1.59	0.240
Operator	4	147.20	147.20	36.80	3.28	0.049
Error	12	134.70	134.70	11.22		
Total	24	683.98				

Therefore, there is a significant difference in mean burning rate generated by different formulation. It is also indicated that there are difference between operators, so blocking on this factor was a good precaution. Although, difference between batches is not significant, but may sometimes be useful idea.

The main effect plot is provided in Figure S8.5-1

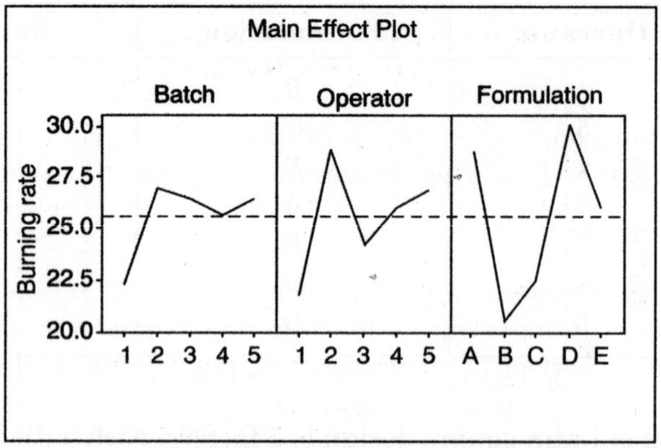

Fig. S8.5-1 Main Effects Plot.

The residual plots as provided below in Figure S8.5-2 do not reveal any drift or erratic behaviour of error over time, and no unusal structure in residual is apparent.

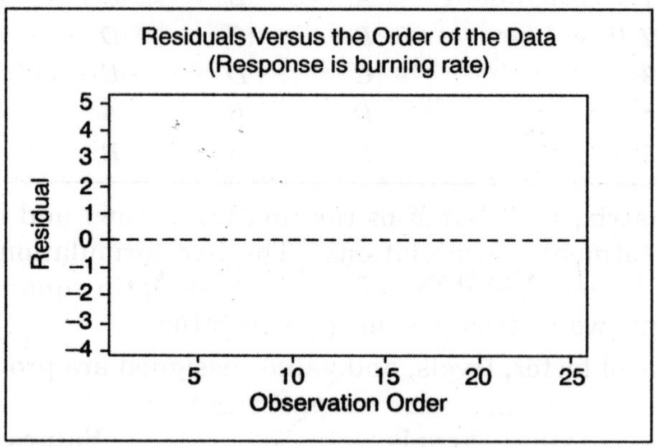

(i) Residual vs Order of Data

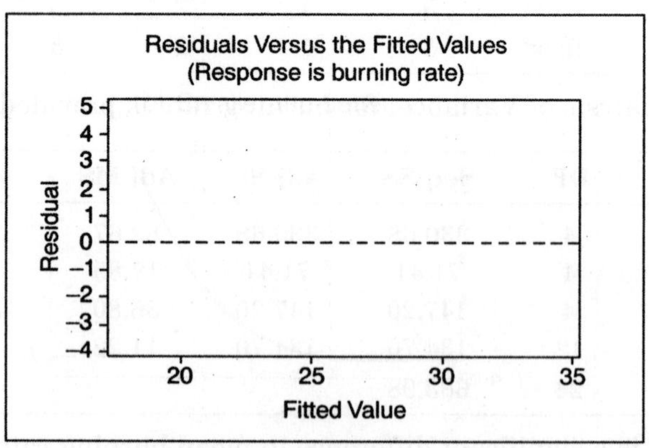

(ii) Residual vs Fitted Values

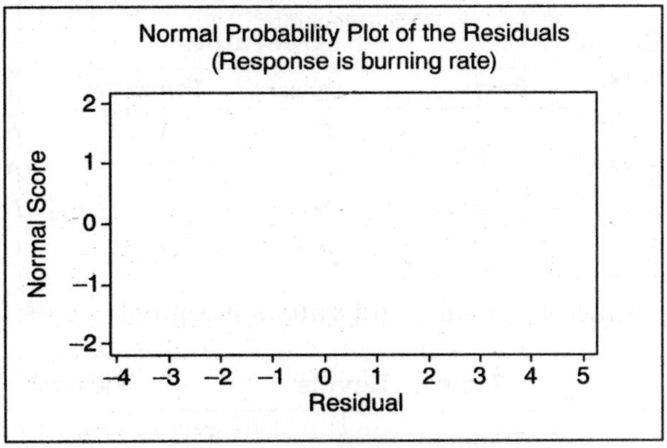

(iii) Normal Probability Plot of Residual

Fig. S8.5-2 Plot of Residuals.

8.6. **A process engineer is studying the effect of different tightening type (method) of engine cylinder head used in automobile on the observed tightening time. Four operators are selected for the study. Moreover, the tightening type can produce fatigue, regardless of the method. Therefore, the engineer performs a latin square design to eliminate two nuisance source of variability, *viz.* order of fitting and operator. Analyze the experimental data ($\alpha = 0.05$), as provided in Table P8.6, and interpret the results.**

Table P8.6

Order of tightening	Operator	Tightening type	Time
1	1	*C*	10.5
1	2	*D*	14.2
1	3	*A*	7.30
1	4	*B*	8.30
2	1	*B*	6.90
2	2	*C*	18.10
2	3	*D*	11.20
2	4	*A*	8.10
3	1	*A*	5.10
3	2	*B*	10.20
3	3	*C*	11.10
3	4	*D*	9.20
4	1	*D*	10.30
4	2	*A*	10.10
4	3	*B*	12.20
4	4	C	13.90

Solution:

The Latin-Square Design consists of 4 factors, each at 4 levels, and number of replication is one. The design matrix is provided as:

Order of fitting	Operators			
	1	2	3	4
1	*C*	*D*	*A*	*B*
2	*B*	*C*	*D*	*A*
3	*A*	*B*	*C*	*D*
4	*D*	*A*	*B*	*C*

The factors, type of factors, levels, and values assigned are as follows:

Factor	Type	Levels	Values			
Tightening	Random	4	*A*	*B*	*C*	*D*
Order of Tightening	Fixed	4	1	2	3	4
Operator	Fixed	4	1	2	3	4

Analysis of Variance is given below:

Source	DF	Seq SS	Adj SS	Adj MS	F	P
Tightening type (method)	3	72.967	72.967	24.322	13.69	0.004
Order of tightening	3	17.242	17.242	5.747	3.24	0.103
Operator	3	50.717	50.717	16.906	9.52	0.011
Error	6	10.659	10.659	1.776		
Total	15	151.584				

Therefore, the tightening type influences the time required for tightening. Although order of tightening seems to be insignificant, but there is an indication of operator difference, and therefore it was a reasonable good precaution.

The main effects plot is illustrated in Figure S8.6-1.

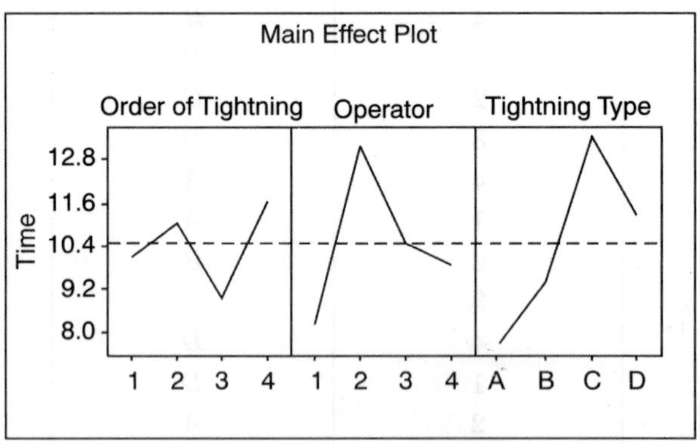

Fig. S8.6-1 Main Effect Plot.

The residual plots as provided below in Figure S8.6-2 do not reveal any drift or erratic behaviour of error over time, and no unusal structure in residual is apparent. Therefore the model assumptions are assumed to be adequate.

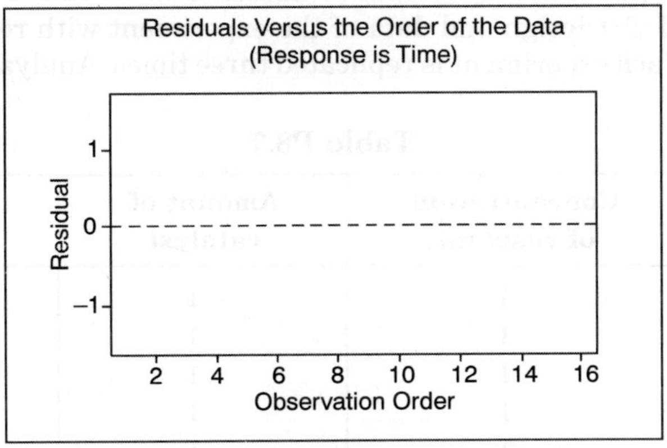

(i) Residual vs Order of Data

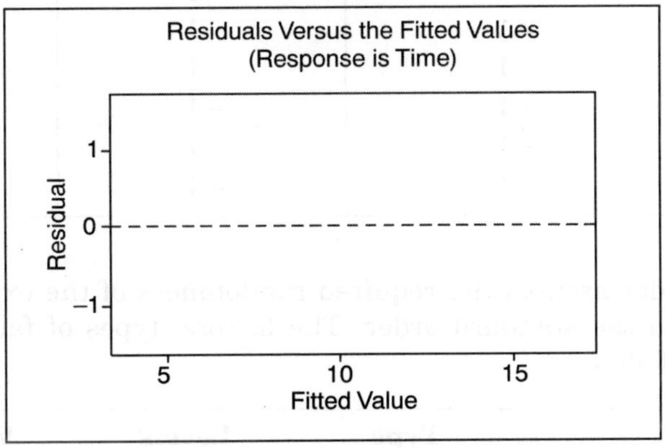

(ii) Residual vs Fitted Values

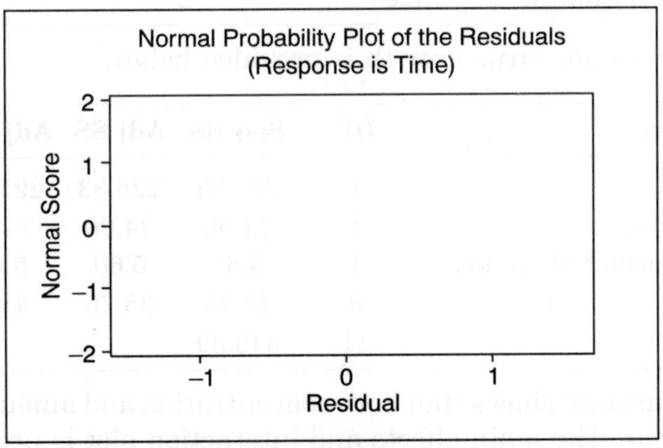

(iii) Normal Probability Plot of Residuals

Fig. 8.6-2 Residual Plots.

8.7. An engineer is interested into the change of concentration of reactant, and amount of catalyst on the growth of a typical virus. Assuming the high and low level of both the factor is represented by + 1 and – 1, respectively, a full factorial design (2^2) is performed.

The full factorial (2^2) design and data of the experiment with replication is provided in the Table P8.7. Each experiment is replicated three times. Analyze the data and interpret the results.

Table P8.7

Std. order*	Concentration of reactant	Amount of catalyst	Growth of virus
7	1	– 1	37.0
4	1	1	29.5
8	1	1	29.1
2	– 1	1	18.2
6	– 1	1	22.9
10	– 1	1	19.4
11	1	– 1	32.3
5	– 1	– 1	25.0
12	1	1	32.0
1	– 1	– 1	27.0
9	– 1	– 1	27.5
3	1	– 1	32.1

Solution:

The standard order assures the required randomness of the experiment. Experiments are run based on the standard order. The factors, types of factor, levels, and values assigned are as follows:

Factor	Type	Levels	Values
Concentration	fixed	2	– 1 1
Amount of catalyst	fixed	2	– 1 1

Analysis of Variance for virus growth is provided below:

Source	DF	Seq SS	Adj SS	Adj MS	F	P
Concentration	1	225.33	225.33	225.33	50.43	0.000
Amount of Catalyst	1	74.00	74.00	74.00	16.56	0.004
Concentration*amount of catalyst	1	5.60	5.60	5.60	1.25	0.295
Error	8	35.75	35.75	4.47		
Total	11	340.69				

The analysis of variance shows that both concentration and amount of catalyst influences the growth of virus. The main effects and interaction plot is provided in Figure S8.7-1.

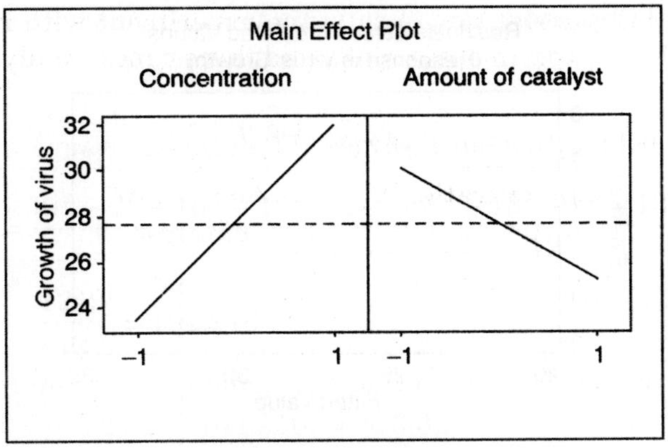

(i) Main Effects Plot

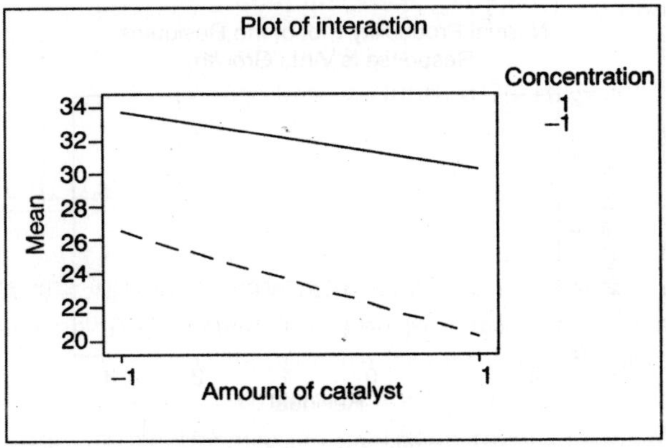

(ii) Interaction Plot

Fig. S8.7-1 Main Effect (i) and Interaction (ii) Plot.

Therefore, to improve the growth concentration at high level (+ 1) and amount of catalyst at low level (– 1) is recommended. Moreover the interaction effect is not statistically significant, as per ANOVA analysis.

The residual plot of the model is illustrated in **Figure S8.7-2.**

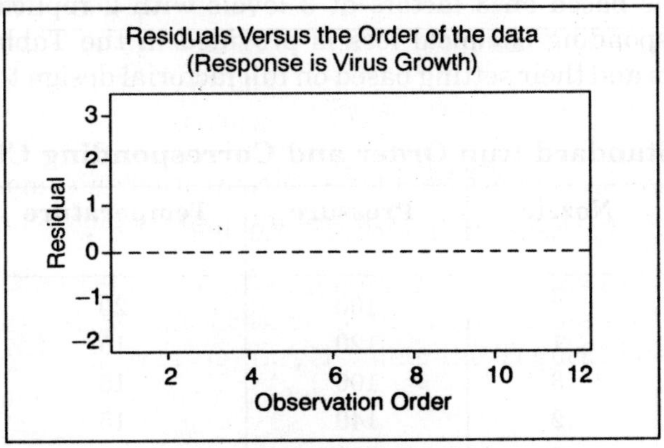

(i) Residual vs Order of Data

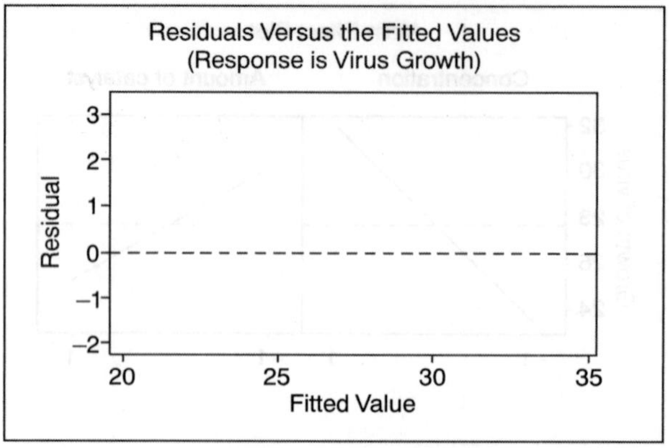

(ii) Residual vs Fitted Value

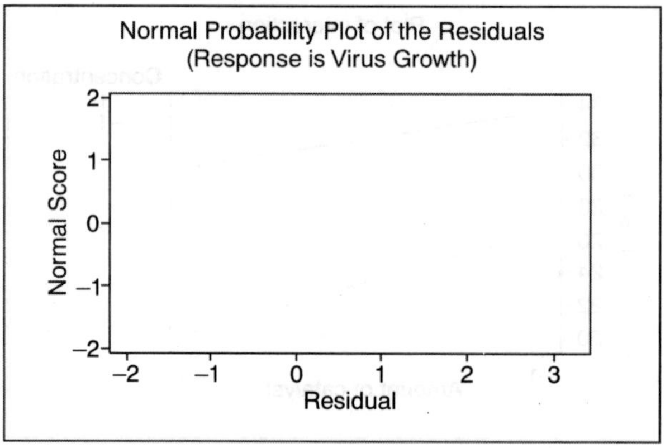

(iii) Normal Probability Plot of Residuals

Fig. S8.7-2 Residual Plots.

The residual plots as provided in Figure S8.7-2 do not reveal any drift or erratic behaviour of error over time, and no unusual structure in residual is apparent.

8.8. Three factors are thought to influence chemical loss, namely nozzle type, pressure and temperature, respectively. Each experiment is replicated twice in the design. Therefore the experiment is based on 3 factors at 3 levels with 2 replicates. The standard run order and corresponding chemical loss is provided in the Table P.8.8. Determine the influential factors and their setting based on full factorial design to minimize the chemical loss.

Table P8.8: Standard Run Order and Corresponding Chemical Loss

Std. order*	Nozzle	Pressure	Temperature	Chemical loss (Scaled data)
48	3	100	20	– 32.0
32	1	120	15	– 11.0
20	3	100	15	88.0
17	2	140	15	105.0
47	3	100	15	111.0

Std. order*	Nozzle	Pressure	Temperature	Chemical loss (Scaled data)
9	1	140	20	30.0
1	1	100	10	– 34.0
49	3	120	10	– 54.5
10	2	100	10	16.4
50	3	120	15	– 29.0
14	2	120	15	– 55.2
15	2	120	20	– 64.1
46	3	100	10	– 38.9
26	3	140	15	108.6
37	2	100	10	23.9
45	2	140	20	– 18.9
28	1	100	10	– 25.6
41	2	120	15	– 44.9
43	2	140	10	21.0
40	2	120	10	– 65.3
52	3	140	10	15.0
6	1	120	20	– 40.1
51	3	120	20	– 62.1
42	2	120	20	– 60.9
4	1	120	10	– 45.3
44	2	140	15	42.6
31	1	120	10	– 60.1
54	3	140	20	55.0
25	3	140	10	– 29.7
21	3	100	20	– 56.0
16	2	140	10	3.50
19	3	100	10	– 33.8
18	2	140	20	– 31.6
30	1	100	20	4.10
38	2	100	15	54.3
39	2	100	20	– 23.5
8	1	140	15	79.9
24	3	120	20	– 51.9
5	1	120	15	31.0
33	1	120	20	– 29.8
23	3	120	15	– 26.5
3	1	100	20	5.10
11	2	100	15	120.3
34	1	140	10	– 41.0
35	1	140	15	54.3
53	3	140	15	135.3
27	3	140	20	4.0
13	2	120	10	– 59.0

Std. order*	Nozzle	Pressure	Temperature	Chemical loss (Scaled data)
36	1	140	20	36.8
7	1	140	10	15.2
12	2	100	20	– 5.10
2	1	100	15	109.5
22	3	120	10	– 66.4
29	1	100	15	75.6

*The standard order assures the required randomness of the experiment.

Solution:

The factors and their levels are summarized below:

Factor	Type	Levels	Values		
Nozzle	fixed	3	1	2	3
Pressure	fixed	3	100	120	140
Temperature	fixed	3	10	15	20

The result of Analysis of Variance (ANOVA) for chemical loss is provided below:

Source	DF	Seq SS	Adj SS	Adj MS	F	P
Nozzle	2	1083.7	1083.7	541.8	1.26	0.298
Pressure	2	60893.6	60893.6	30446.8	71.07	0.000
Temperature	2	67744.2	67744.2	33872.1	79.07	0.000
Nozzle*Pressure	4	6508.8	6508.8	1627.2	3.80	0.014
Nozzle*Temperature	4	7427.9	7427.9	1857.0	4.33	0.008
Pressure*Temperature	4	12809.6	12809.6	3202.4	7.48	0.000
Nozzle*Pressure*Temperature	8	4569.0	4569.0	571.1	1.33	0.270
Error	27	11566.5	11566.5	428.4		
Total	53	172603.3				

The analysis shows that pressure and temperature influence the chemical loss, but nozzle type has no such influence, and interaction between pressure and temperature is significant.

The main effects and interaction plot is provided in Figure S8.8-1.

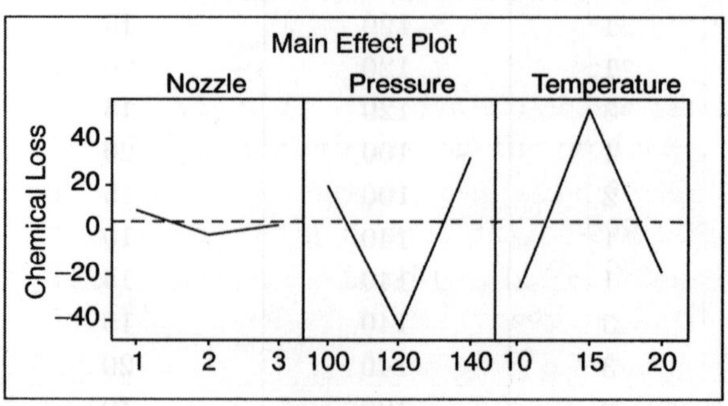

(i) Main Effects Plot

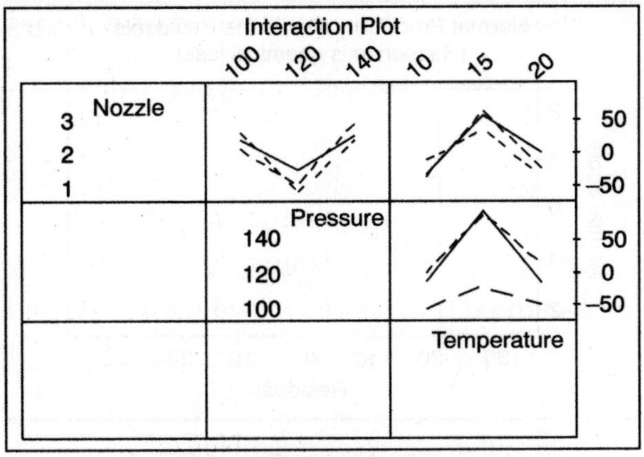

(ii) Interaction Plots

Fig. 8.8-1 The Main Effects (*i*) and Interaction (*ii*) Plot.

Suitable combinations to reduce the chemical loss may be taken as, nozzle of type 2, pressure at 120 and temperature at 10 or 20.

The residual plots as provided in Figure S8.8-2 do not reveal any drift or erratic behaviour of error over time, and no unusual structure in residual is apparent.

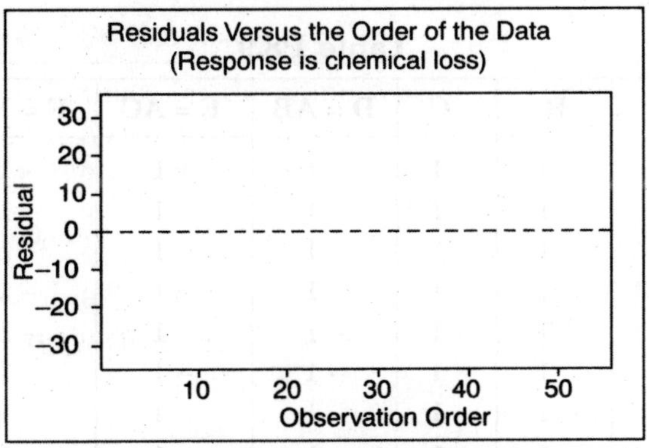

(i) Residual vs Order of Data

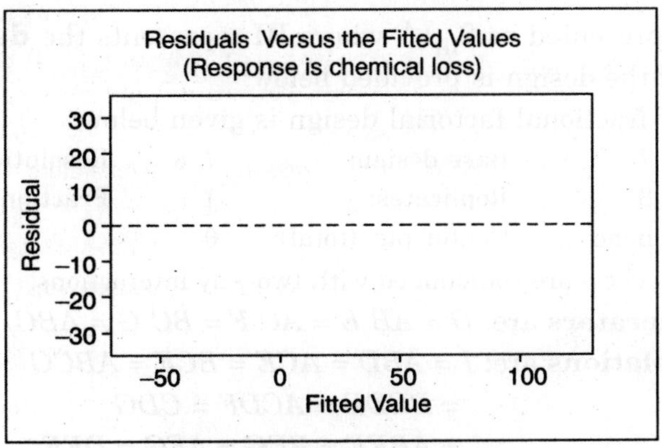

(ii) Residual vs Fitted Value

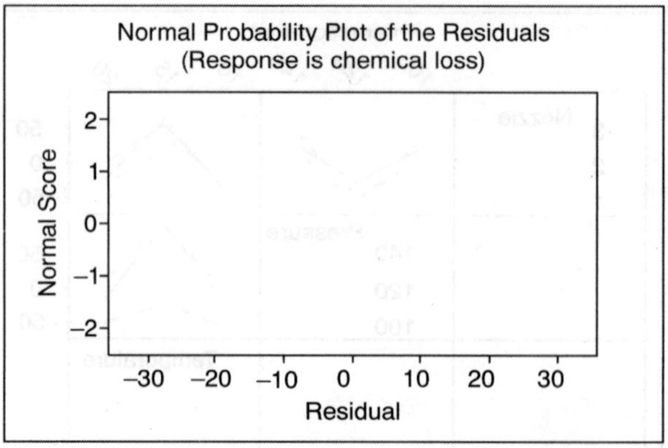

(iii) Normal Probability Plot of Residuals

Fig. S8.8-2 Residual Plots.

8.9. A process engineer wants to determine the effect of seven factors, namely *A*, *B*, *C*, *D*, *E*, *F* and *G*, on machining time. He suspects that only a few of these seven factors are of major importance. Two levels of each factor, – 1 and 1 respectively are considered. To screen these seven factors, he runs the treatment combination of 2_{III}^{7-4} design as provided in the Table P8.9, in random order. Analyze the fractional factorial design and interpret the results.

Table P8.9

Std. order	A	B	C	D = AB	E = AC	F = BC	G = ABC	Time
5	– 1	– 1	1	1	– 1	– 1	1	94.2
6	1	– 1	1	– 1	1	– 1	– 1	144
1	– 1	– 1	– 1	1	1	1	– 1	91.2
4	1	1	– 1	1	– 1	– 1	– 1	73.1
3	– 1	1	– 1	– 1	1	– 1	1	82.2
7	– 1	1	1	– 1	– 1	1	– 1	87.1
2	1	– 1	– 1	– 1	– 1	1	1	137.1
8	1	1	1	1	1	1	1	71.5

Solution:

The design is represented as 2_{III}^{7-4}, where III represents the **design resolution.**

The summary of the design is provided below:

The summary of fractional factorial design is given below:

Factors:	7	Base design:	7, 8	Resolution:	III
Runs:	8	Replicates:	1	Fraction:	1/16
Blocks:	none	Center pts. (total):	0		

Note: Some main effects are confounded with two-way interactions.

The **design generators** are: $D = AB$ $E = AC$ $F = BC$ $G = ABC$

The **defining relations** are: $I = ABD = ACE = BCF = ABCG$

$$= BCDE = ACDF = CDG$$

$$= ABEF = BEG = AFG = DEF$$

$$= ADEG = BDFG = CEFG = ABCDEFG$$

Alias Structure (up to order 2) is provided below:

I

$A + BD + CE + FG + BCG + BEF + CDF + DEG + ABCF + ABEG + ACDG + ADEF$
$+ ABCDE + ABDFG + ACEFG + BCDEFG$

$B + AD + CF + EG + ACG + AEF + CDE + DFG + ABCE + ABFG + BCDG + BDEF$
$+ ABCDF + ABDEG + BCEFG + ACDEFG$

$C + AE + BF + DG + ABG + ADF + BDE + EFG + ABCD + ACFG + BCEG + CDEF$
$+ ABCEF + ACDEG + BCDFG + ABDEFG$

$D + AB + CG + EF + ACF + AEG + BCE + BFG + ACDE + ADFG + BCDF + BDEG$
$+ ABCDG + ABDEF + CDEFG + ABCEFG$

$E + AC + BG + DF + ABF + ADG + BCD + CFG + ABDE + AEFG + BCEF + CDEG$
$+ ABCEG + ACDEF + BDEFG + ABCDFG$

$F + AG + BC + DE + ABE + ACD + BDG + CEG + ABDF + ACEF + BEFG + CDFG$
$+ ABCFG + ADEFG + BCDEF + ABCDEG$

$G + AF + BE + CD + ABC + ADE + BDF + CEF + ABDG + ACEG + BCFG + DEFG$
$+ ABEFG + ACDFG + BCDEG + ABCDEF$

$ABD + ACE + AFG + BCF + BEG + CDG + DEF + ABCG + ABEF + ACDF + ADEG$
$+ BCDE + BDFG + CEFG + ABCDEFG$

ANOVA analysis is provided below:

Analysis of Variance for Time (coded units)

Source	DF	Seq SS	Adj SS	Adj MS	F	P
Main Effects	4	5374.77	5374.77	1343.69	203.49	0.001
Residual Error	3	19.81	19.81	6.60		
Total	7	5394.58				

Estimated Effects and Coefficients for Time (coded units)

Term	Effect	Coef	SE Coef	T	P
Constant	97.55	0.9085	107.37		0.000
A	17.75	8.87	0.9085	9.77	0.002
B	– 38.15	– 19.07	0.9085	– 21.00	0.000
C	3.30	1.65	0.9085	1.82	0.167
AB	– 30.10	– 15.05	0.9085	– 16.57	0.000

The above analysis shows factor *A*, *B* and *AB* interactions are significant.
The main effects and interaction plots are shown in Figure S8.9-1.

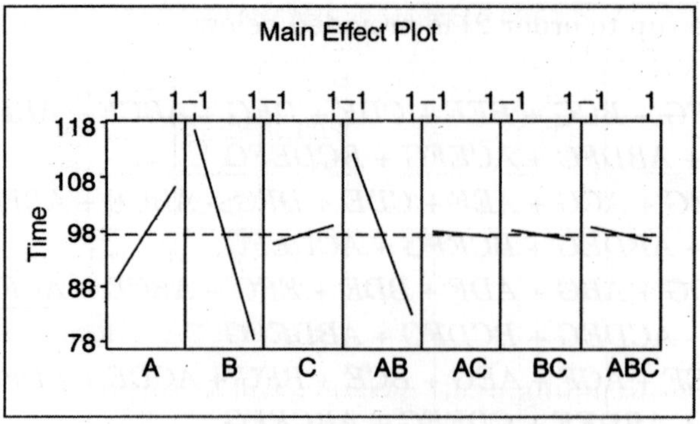

(i) Main Effects Plot

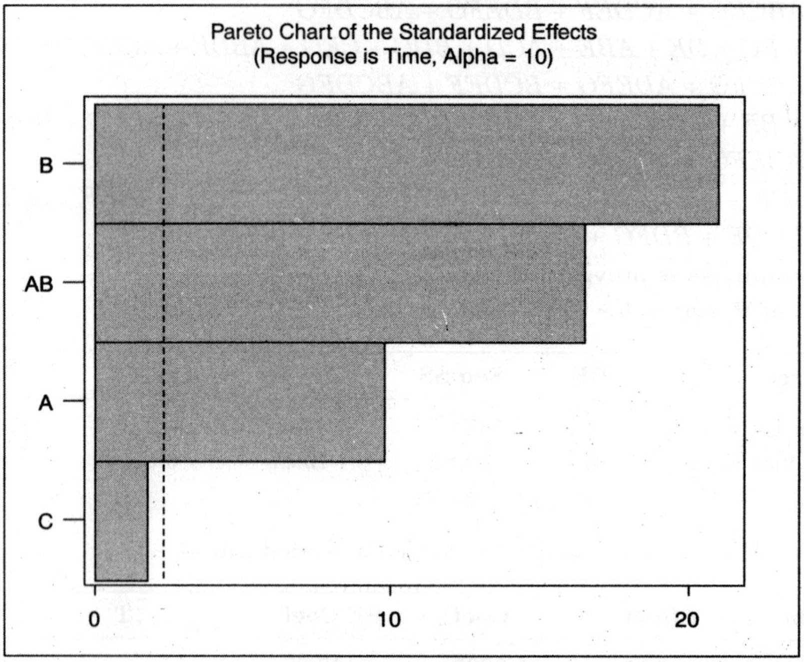

(ii) Pareto Chart of Effects

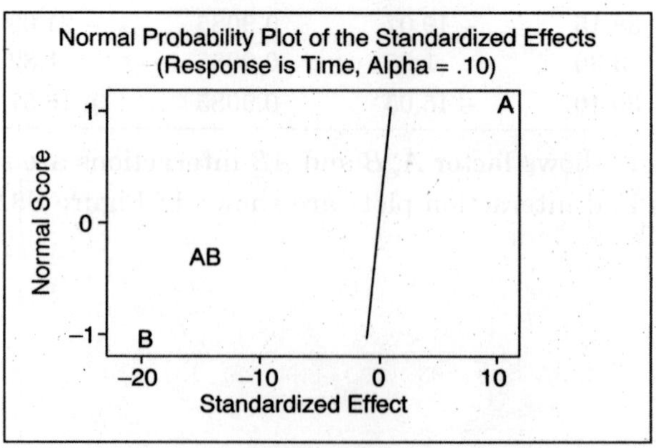

(iii) Normal Probability Plot of Effects

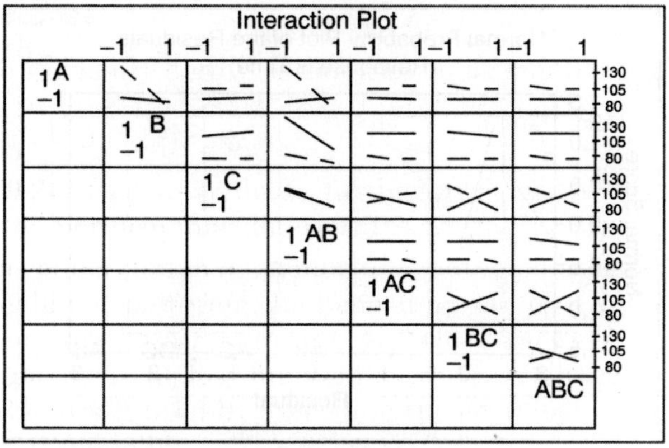

(iv) Interaction Plot

Fig. S8.9-1 Main Effects (i), (ii), (iii), and Interaction (iv) Plot.

The best possible combination of settings are *A* low (– 1), *B* high (– 1), *C* low (– 1), *D* high (+ 1), *E* high (+ 1), *F* high (+ 1), and *G* high (+ 1). Therefore, the main effect which can be considered in subsequent investigation are *A*, *B*, *C* and *D*.

The residual plots for model adequacy check are given in Figure S8.9-2

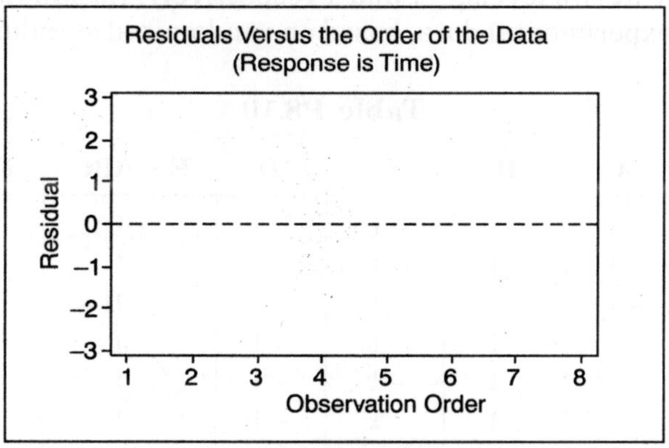

(i) Residual vs Order of Data

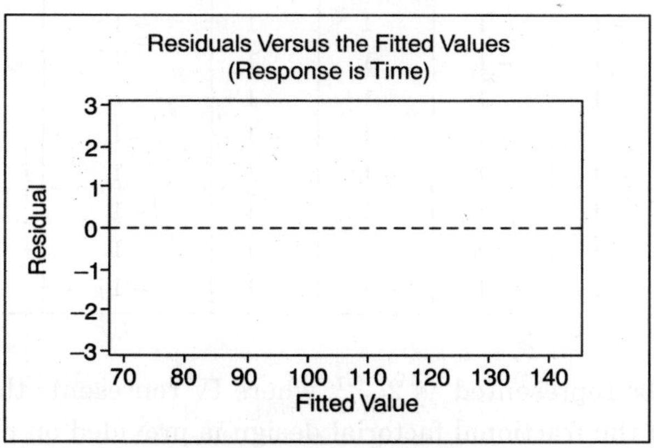

(ii) Residual vs Fitted Value

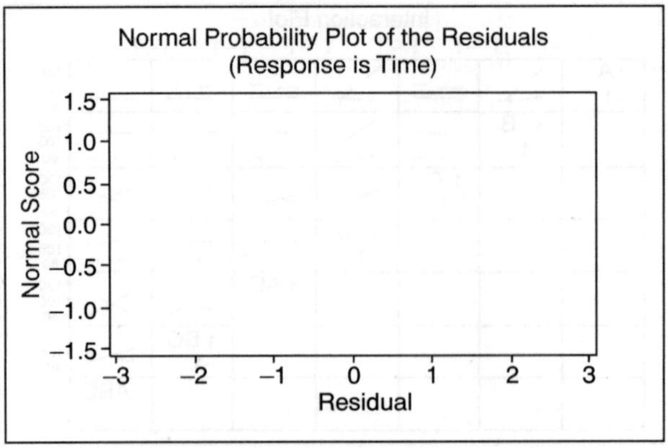

(iii) Normal Probability Plot of Residuals

Fig. S8.9-2 Residual Plots.

The residual plots as provided in Figure S8.9-2 do not reveal any drift or erratic behaviour of error over time, and no unusual structure in residual is apparent.

8.10. Determine the effect and final setting of six factors, namely *A*, *B*, *C*, *D*, *E*, and *F* to minimize the shrinkage of a molding process, based on one-half fractional factorial design. All the factors are at two levels, – 1 and 1 respectively. Assume *ABC* and *BCD* as design generators. The experimental data, based on randomized standard run order is provided in Table P8.10.

Table P8.10

Std. order	A	B	C	D	E = ABC	F = BCD	Shrinkage
1	– 1	– 1	– 1	– 1	– 1	– 1	6.1
11	– 1	1	– 1	1	1	– 1	34.2
5	– 1	– 1	1	– 1	1	1	3.9
16	1	1	1	1	1	1	51.8
4	1	1	– 1	– 1	– 1	1	59.9
6	1	– 1	1	– 1	– 1	1	15.2
8	1	1	1	– 1	1	– 1	60.2
10	1	– 1	– 1	1	1	1	12.2
15	– 1	1	1	1	– 1	1	37.1
13	– 1	– 1	1	1	1	– 1	16.3
14	1	– 1	1	1	– 1	– 1	5.1
2	1	– 1	– 1	– 1	1	– 1	10.2
9	– 1	– 1	– 1	1	– 1	1	7.8
7	– 1	1	1	– 1	– 1	– 1	25.8
3	– 1	1	– 1	– 1	1	1	32.3
12	1	1	– 1	1	– 1	– 1	59.8

Solution:

The design can be represented as 2_{IV}^{5-1}, where IV represents the **design resolution**. The summary of the fractional factorial design is provided on next page:

Factors:	6	Base Design:	6, 16	Resolution:	IV
Runs:	16	Replicates:	1	Fraction:	1/4
Blocks:	None	Center pts (total):	0		

Design Generators: $E = ABC$ $F = BCD$

Defining Relation: $I = ABCE = BCDF = ADEF$

Alias Structures are:

$I + ABCE + ADEF + BCDF$
$A + BCE + DEF + ABCDF$
$B + ACE + CDF + ABDEF$
$C + ABE + BDF + ACDEF$
$D + AEF + BCF + ABCDE$
$E + ABC + ADF + BCDEF$
$F + ADE + BCD + ABCEF$
$AB + CE + ACDF + BDEF$
$AC + BE + ABDF + CDEF$
$AD + EF + ABCF + BCDE$
$AE + BC + DF + ABCDEF$
$AF + DE + ABCD + BCEF$
$BD + CF + ABEF + ACDE$
$BF + CD + ABDE + ACEF$
$ABD + ACF + BEF + CDE$
$ABF + ACD + BDE + CEF$

Analysis of Variance for Shrinkage (in coded units) is provided below:

Source	DF	Seq SS	Adj SS	Adj MS	F	P
Main Effects	4	5830.6	5830.6	1457.66	58.06	0.000
2-way Interactions	1	548.7	548.7	548.73	21.86	0.001
Residual Error	10	251.0	251.0	25.10		
Total	15	6630.4				

Estimated Effects and Coefficients for Shrinkage (coded units)

Term	Effect	Coeff	SE Coeff	T	P
Constant		27.3688	1.253	21.85	0.000
A	13.8625	6.9313	1.253	5.53	0.000
B	35.5375	17.7688	1.253	14.19	0.000
C	– 0.8875	– 0.4438	1.253	– 0.35	0.731
D	1.3375	0.6688	1.253	0.53	0.605
$A*B$	11.7125	5.8562	1.253	4.68	0.001

From the above analysis, main effect A and B, and their interaction effect are observed to be significant.

The main effect and interaction plots are provided in Figure S8.10-1.

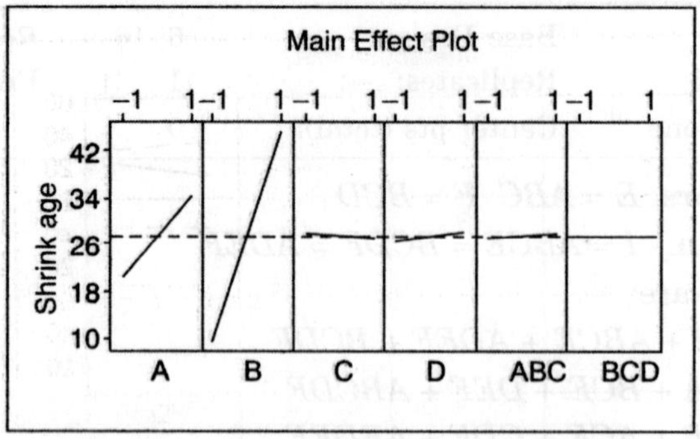

(i) Main Effects Plot

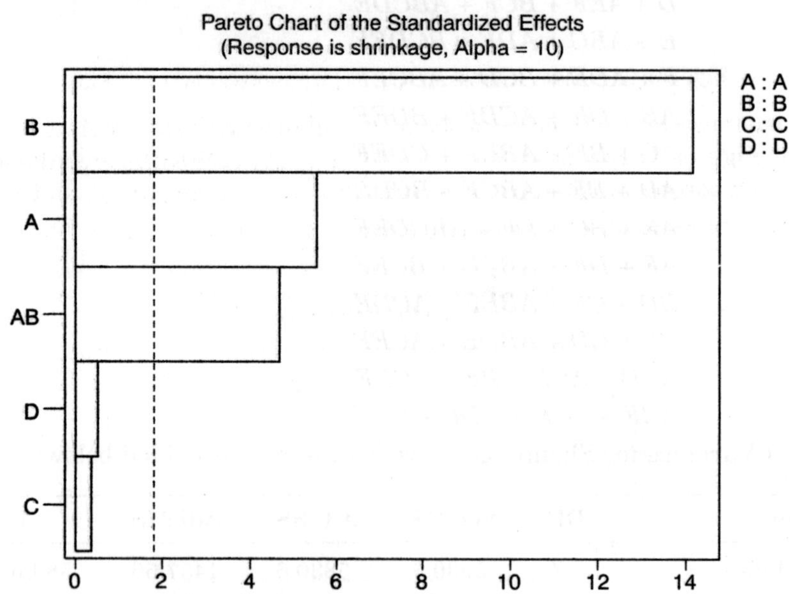

(ii) Pareto Chart of Effects

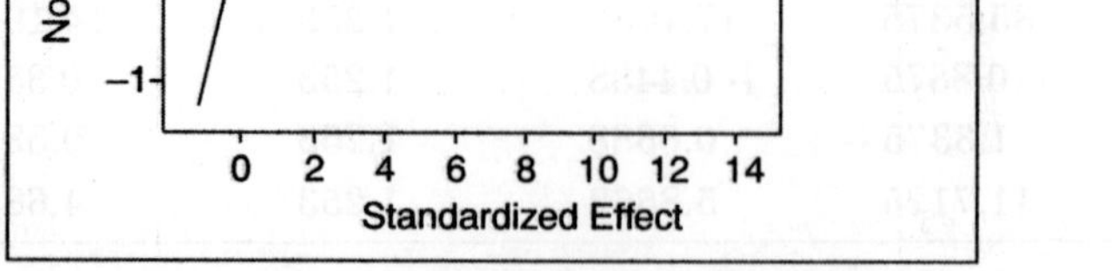

(iii) Normal Probability Plot of Effects

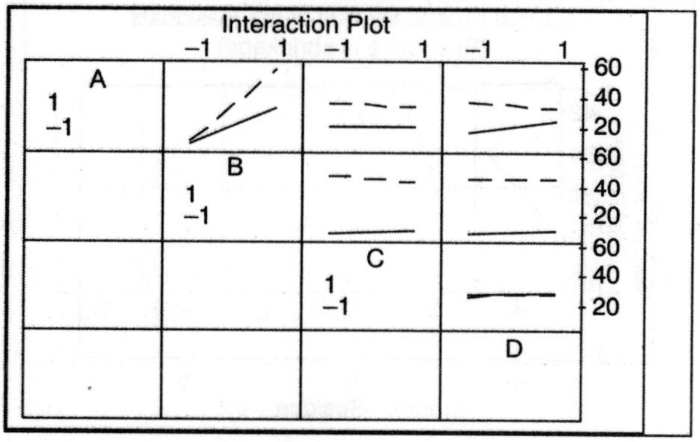

(iv) Interaction Plot

Fig. S8.10-1 Main Effects (i), (ii), (iii) and Interaction (iv) Plot.

The best combinations of factors based on the above analysis are *A* at low (– 1), *B* at low (– 1), *C* at high (+ 1) and *D* at low (– 1), so as to minimize the shrinkage. The possible factors which can be considered for subsequent experiment are *A* and *B*.

The residual plots to check model adequacy is provided in Figure S8.10-2

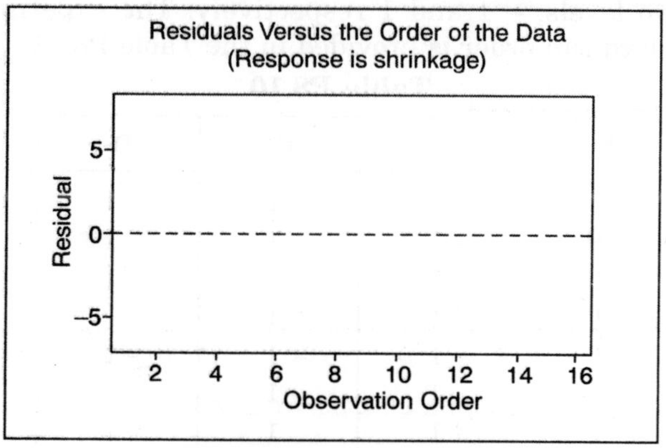

(i) Residual vs Order of Data

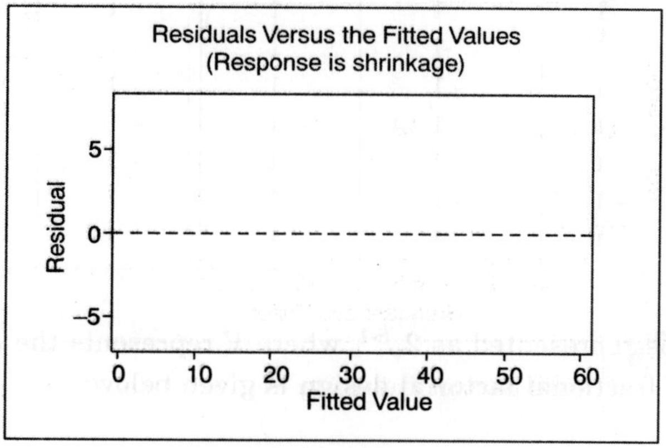

(ii) Residual vs Fitted Value

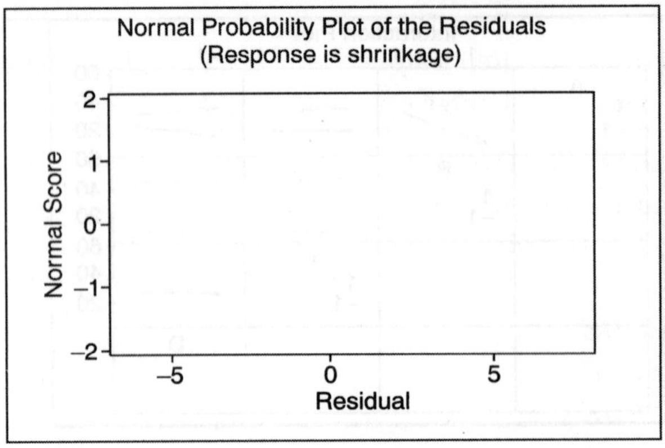

(iii) Normal Probability Plot of Residuals

Fig. 8.10-2 Residual Plots.

The residual plots as provided in Figure S8.10-2 do not reveal any drift or erratic behaviour of error over time, and no unusual structure in residual is apparent.

8.11. Determine the effect and final setting of five factors, namely A, B, C, D and E to maximize the chemical yield of a process, based on one-half fractional factorial design. All the factors are at two levels, – 1 and 1 respectively. The experimental data, based on randomized standard run order is provided in the Table P8.11.

Table P8.10

Std. order	A	B	C	D	E	Yield
11	– 1	1	– 1	1	1	30.2
14	1	– 1	1	1	– 1	20.9
10	1	– 1	– 1	1	1	10.2
15	– 1	1	1	1	– 1	43.8
3	– 1	1	– 1	– 1	– 1	34.8
12	1	1	– 1	1	– 1	50.2
6	1	– 1	1	– 1	1	21.9
1	– 1	– 1	– 1	– 1	1	8.3
7	– 1	1	1	– 1	1	45.4
8	1	1	1	– 1	– 1	61
16	1	1	1	1	1	64.1
13	– 1	– 1	1	1	1	14.9
2	1	– 1	– 1	– 1	– 1	9.1
5	– 1	– 1	1	– 1	– 1	16.3
9	– 1	– 1	– 1	1	– 1	6.1
4	1	1	– 1	– 1	1	63.4

Solution:

The design can be represented as 2_V^{5-1}, where V represents the **design resolution**. The summary of fractional factorial design is given below:

Factors:	5	Base Design:	5, 16	Resolution:	V
Runs:	16	Replicates:	1	Fraction:	1/2
Blocks:	none	Center pts (total):	0		

Design Generators: $E = ABCD$
Defining Relation: $I = ABCDE$
Alias Structure (up to order 2) is given below:

I
$A + BCDE$
$B + ACDE$
$C + ABDE$
$D + ABCE$
$E + ABCD$
$AB + CDE$
$AC + BDE$
$AD + BCE$
$AE + BCD$
$BC + ADE$
$BD + ACE$
$BE + ACD$
$CD + ABE$
$CE + ABD$
$DE + ABC$
$ABCDE$

The Analysis of Variance for Yield (coded units) is provided below:

Source	DF	Seq SS	Adj SS	Adj MS	F	P
Main Effects	5	6123.16	6123.16	1224.63	81.08	0.000
2-Way Interactions	6	323.07	323.07	53.84	3.56	0.119
Residual Error	4	60.41	60.41	15.10		
Total	15	6506.64				

From ANOVA, the main effect is only observed to be statistically significant.
Estimated Effects and Coefficients for Yield (coded units)

Term	Effect	Coeff	SE Coeff	T	P
Constant		31.288	0.9716	32.20	0.000
A	12.625	6.312	0.9716	6.50	0.003
B	35.650	17.825	0.9716	18.35	0.000
C	9.500	4.750	0.9716	4.89	0.008
D	– 2.475	– 1.238	0.9716	– 1.27	0.272
E	2.025	1.013	0.9716	1.04	0.356
$A*B$	8.500	4.250	0.9716	4.37	0.012
$A*C$	– 0.750	– 0.375	0.9716	– 0.39	0.719

$A*D$	– 0.025	– 0.012	0.9716	– 0.01	0.990
$B*C$	– 0.575	– 0.288	0.9716	– 0.30	0.782
$B*D$	– 1.600	– 0.800	0.9716	– 0.82	0.457
$C*D$	2.250	1.125	0.9716	1.16	0.311

SE coefficient of AB indicates some amount of interaction between the factors A and B. The main effect and interaction plots are provided in Figure S8.11-1.

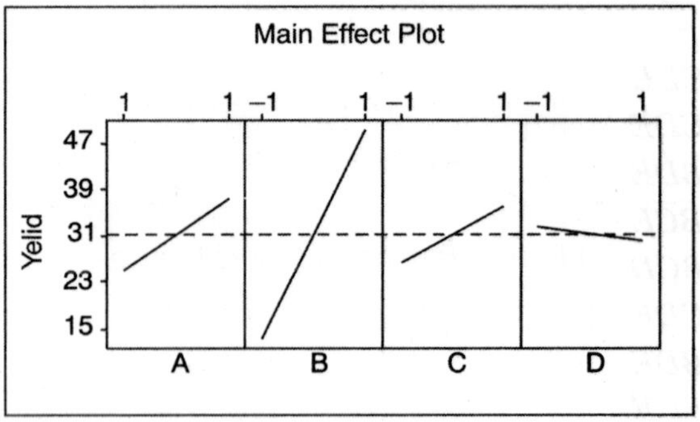

(i) Main Effects Plot

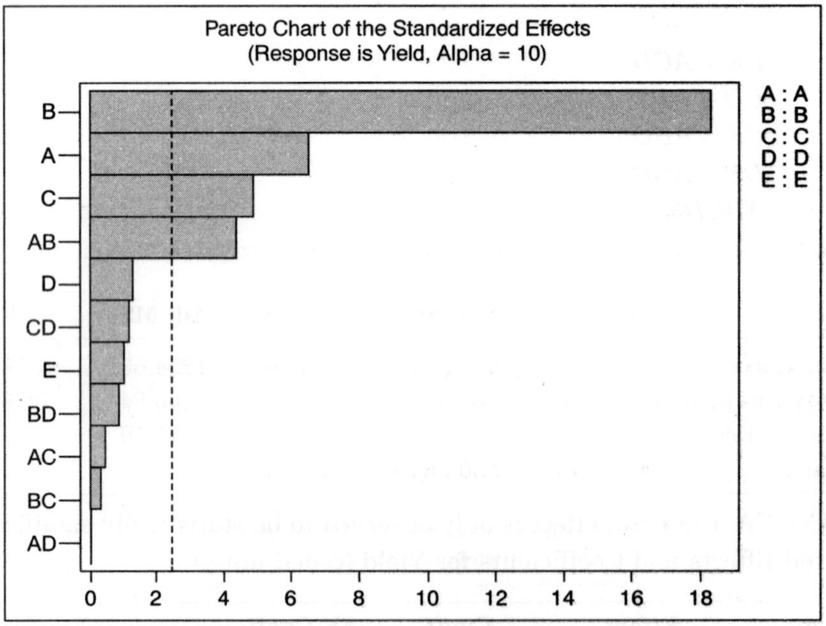

(ii) Pareto Chart of Effects

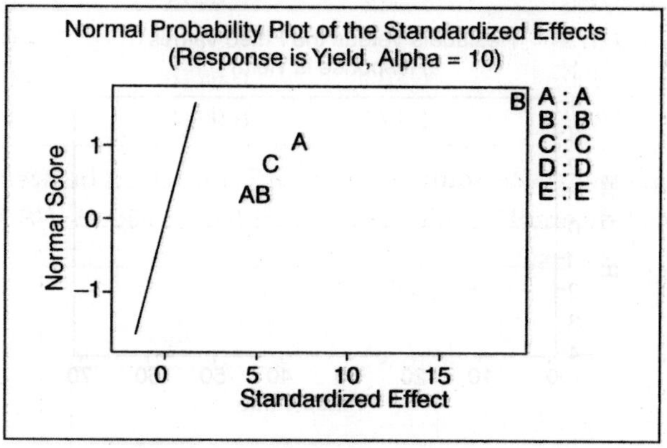

(iii) Normal Probability Plot of Effects

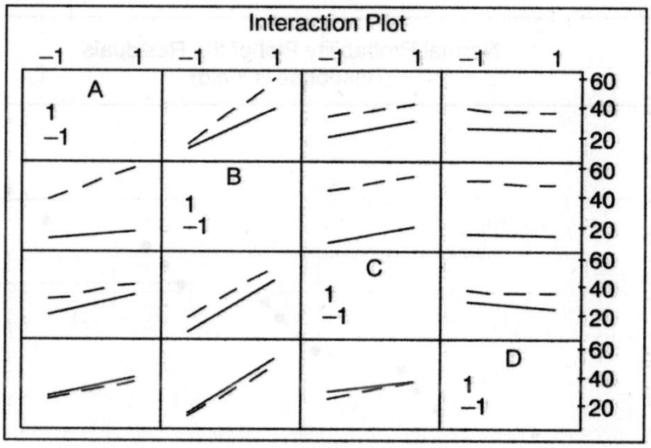

(iv) Interaction Plot

Fig. S8.11-1 Main Effects (i), (ii), (iii) and Interaction (iv) Plot.

The best combinations of process conditions are *A* at high level, *B* at high level, *C* at high level and *D* at low level, which will improve the chemical yield. The factors which needs to be considered in subsequent experimentations are *A*, *B*, and *C* for improving yield.

The model residual plots are provided in Figure S8.11-2

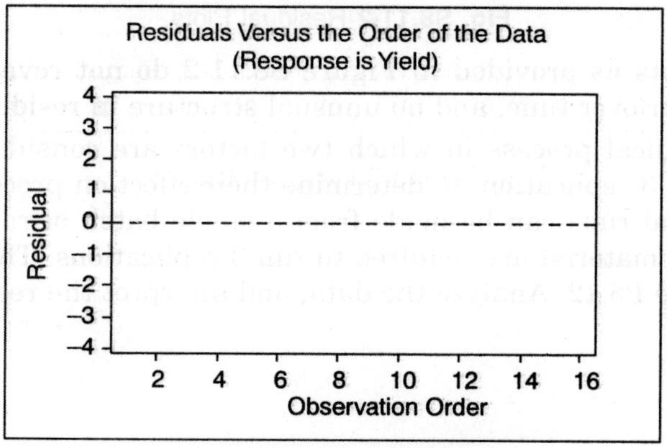

(i) Residual vs Order of Data

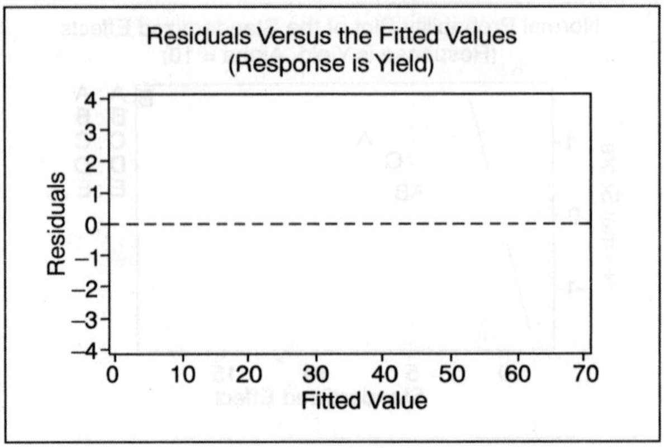

(ii) Residual vs Fitted Value

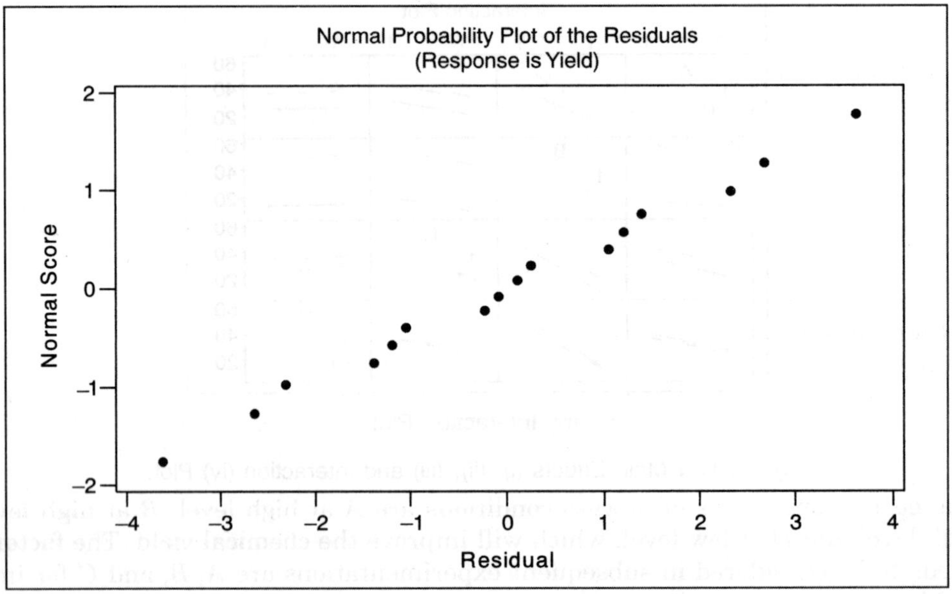

(iii) Normal Probability Plot of Residuals

Fig. S8.11-2 Residual Plots.

The residual plots as provided in Figure S8.11-2 do not reveal any drift or erratic behaviour of error over time, and no unusual structure in residual is apparent.

8.12. Consider a chemical process in which two factors are considered each at two level (– 1 and 1), with 3 replication, to determine their effect on process yield. Suppose four experimental trial runs can be made from a single batch of raw material. Therefore, 3 batches of raw material are required to run 3 replications. The experimental data is provided in Table P8.12. Analyze the data, and interpret the results.

Table P8.12

Std. order	Blocks	A	B	Yield
12	3	1	1	29.1
11	3	– 1	1	23.2
10	3	1	– 1	31.9
9	3	– 1	– 1	27.1
8	2	1	1	29.8
5	2	– 1	– 1	25.2
7	2	– 1	1	19.2
6	2	1	– 1	31.9
4	1	1	1	32
3	1	– 1	1	18.1
2	1	1	– 1	36.4
1	1	– 1	– 1	28.3

Solution:

The problem is to confound a 2^2 design, having 3 replications into 3 blocks.

The full factorial design can be summarized as given below:

Factors:	2	Base Design:	2, 4
Runs:	12	Replicates:	3
Blocks:	3	Center pts (total):	0

Block Generators: Replicates

The Analysis of Variance for Yield (coded units) is provided below:

Source	DF	Seq SS	Adj SS	Adj MS	F	P
Blocks	2	9.582	9.582	4.791	1.03	0.411
Main Effects	2	280.363	280.363	140.182	30.26	0.001
2-Way Interactions	1	9.720	9.720	9.720	2.10	0.198
Residual Error	6	27.792	27.792	4.632		
Total	11	327.457				

The above ANOVA analysis shows that block effect is relatively small, whereas the main effect is significant. The two way interactions between the variables are also not significant.

The main effect and interaction plot is provided in Figure S8.12-1

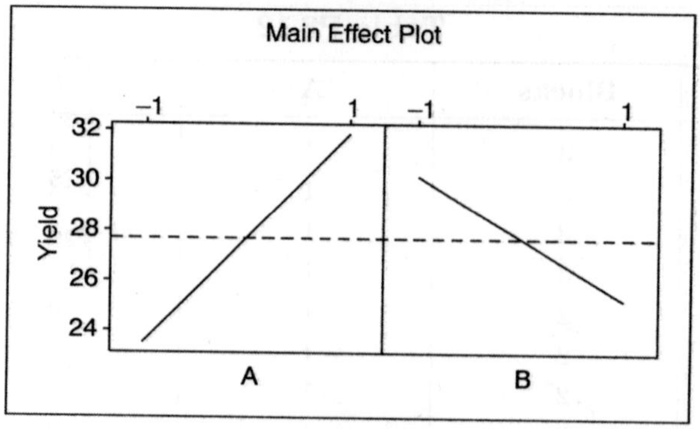

(i) Main Effects Plot

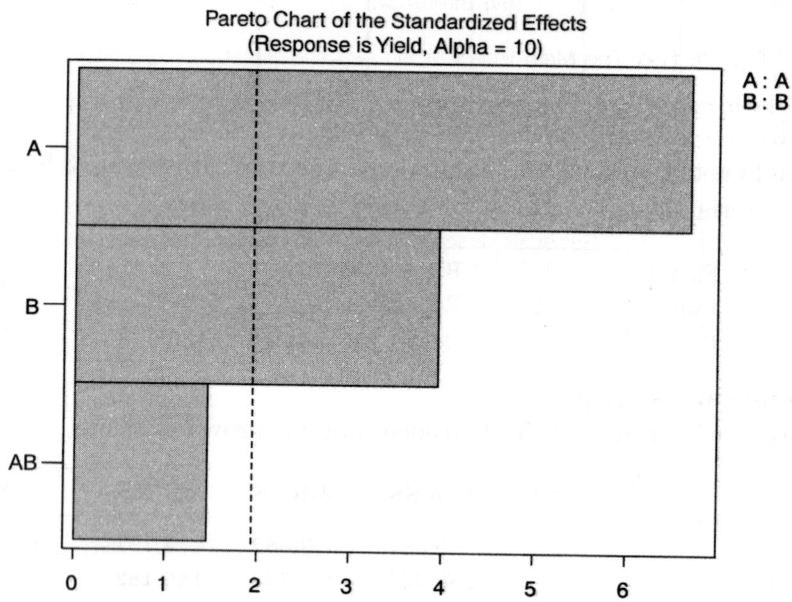

(ii) Pareto Chart of Main Effects

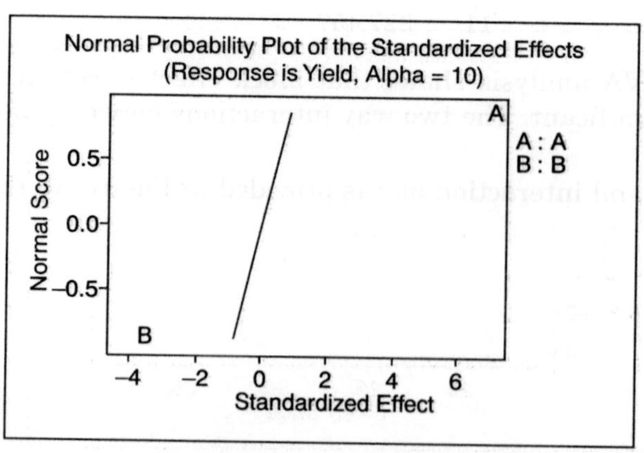

(iii) Normal Probability Plot of Effects

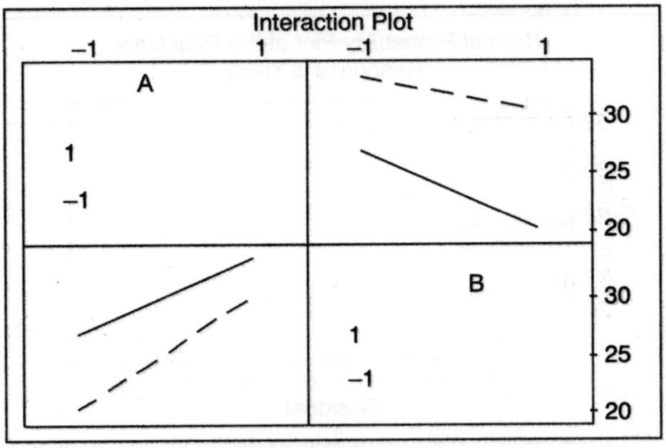

(iv) Interaction Plot

Fig. S8.12-1 The Main Effect (i), (ii), (iii) and Interaction (iv) Plots.

From the above analysis and plots, it is observed that selecting A at high (+ 1), and B at low (– 1), will improve the process yield.

The residual plots for model adequacy check is illustrated in Figure S8.12-2

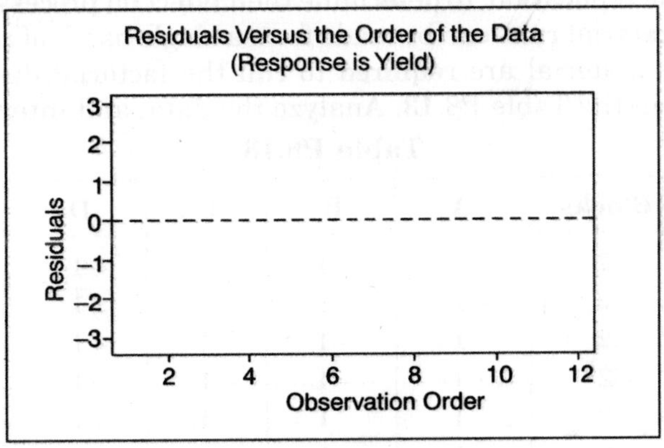

(i) Residual vs Order of Data

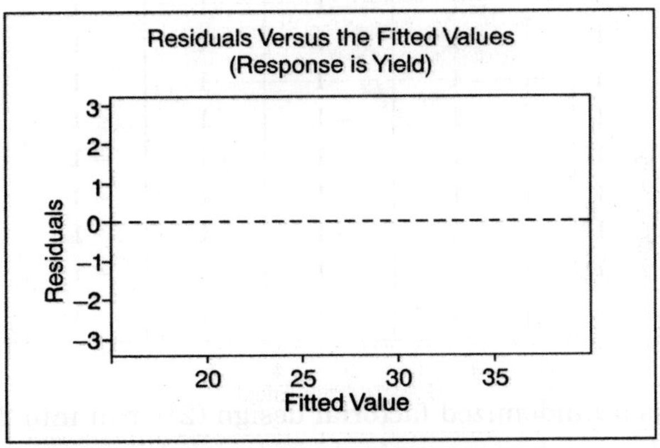

(ii) Residual vs Fitted Value

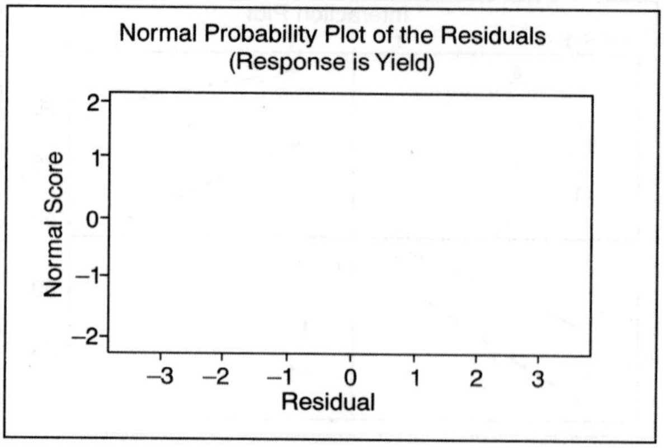

(iii) Normal Probability Plot of Residuals

Fig. S8.12-2 The Residual Plots.

The residual plots as provided in Figure S8.12-2 do not reveal any drift or erratic behaviour of error over time, and no unusual structure in residual is apparent.

8.13. Consider a chemical process in which four factors are considered each at two level (– 1 and 1), with single replication, to determine their effect on process reaction time. Suppose eight experimental trial runs can be made from a single batch of raw material. Therefore, 2 batches of raw material are required to run the factorial design. The experimental data is provided in the Table P8.13. Analyze the data, and interpret the results.

Table P8.13

Std. order	Blocks	A	B	C	D	Reaction time
12	2	– 1	1	1	– 1	60.2
14	2	– 1	1	– 1	1	24.9
11	2	1	– 1	1	– 1	39.8
13	2	1	– 1	– 1	1	80.5
16	2	1	1	1	1	76.2
10	2	1	1	– 1	– 1	45.1
9	2	– 1	– 1	– 1	– 1	24.9
15	2	– 1	– 1	1	1	55
6	1	1	1	– 1	1	104.5
8	1	– 1	1	1	1	70.2
7	1	1	– 1	1	1	85.9
1	1	1	– 1	– 1	– 1	71.1
4	1	1	1	1	– 1	65.2
5	1	– 1	– 1	– 1	1	43.3
3	1	– 1	– 1	1	– 1	68.3
2	1	– 1	1	– 1	– 1	48.1

Solution:

The experiment is a randomized factorial design (2^4), run into 2 blocks.

The factorial design can be summarized as given below:

Full Factorial Design (2^4)					
Factors:	4	Base Design:	4, 16	Resolution with blocks:	V
Runs:	16	Replicates:	1		
Blocks:	2	Center pts (total):	0		

Block Generators: *ABCD*

Alias Structure

I

Block = *ABCD*

A

B

C

D

AB

AC

AD

BC

BD

CD

ABC

ABD

ACD

BCD

Analysis of Variance for reaction time (coded units) is provided below:

Source	**DF**	**Seq SS**	**Adj SS**	**Adj MS**	**F**	**P**
Blocks	1	1406.2	1406.3	1406.25	73.75	0.000
Main Effects	4	3171.6	3171.6	792.91	41.59	0.000
2-Way Interactions	2	2465.5	2465.5	1232.74	64.65	0.000
Residual Error	8	152.5	152.5	19.07		
Total	15	7195.9				

Estimated Effects and Coefficients for Reaction time (coded units)

Term	**Effect**	**Coeff**	**SE Coeff**	**T**	**P**
Constant		60.200	1.092	55.15	0.000
Block		9.375	1.092	8.59	0.000
A	21.675	10.838	1.092	9.93	0.000
B	3.200	1.600	1.092	1.47	0.181
C	9.800	4.900	1.092	4.49	0.002
D	14.725	7.363	1.092	6.74	0.000
*A*C*	– 18.325	– 9.163	1.092	– 8.39	0.000
*A*D*	16.750	8.375	1.092	7.67	0.000

The ANOVA analysis shows that both main effect and two-way interaction are significant. In addition, the block effect, A, C, AC and AD are also found to be significant. The main effect and interaction plot is provided in Figure S8.13-1

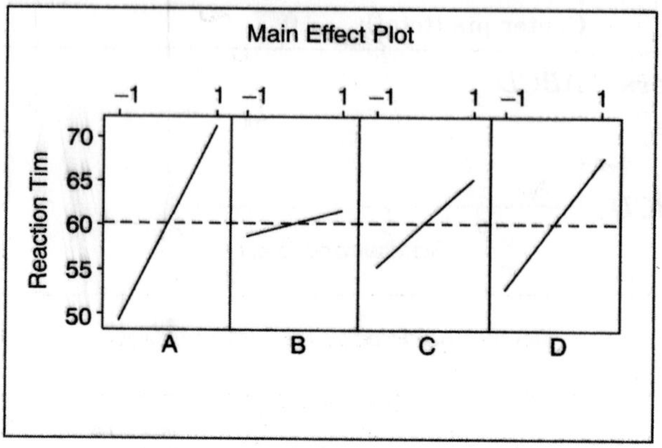

(i) Main Effects Plot

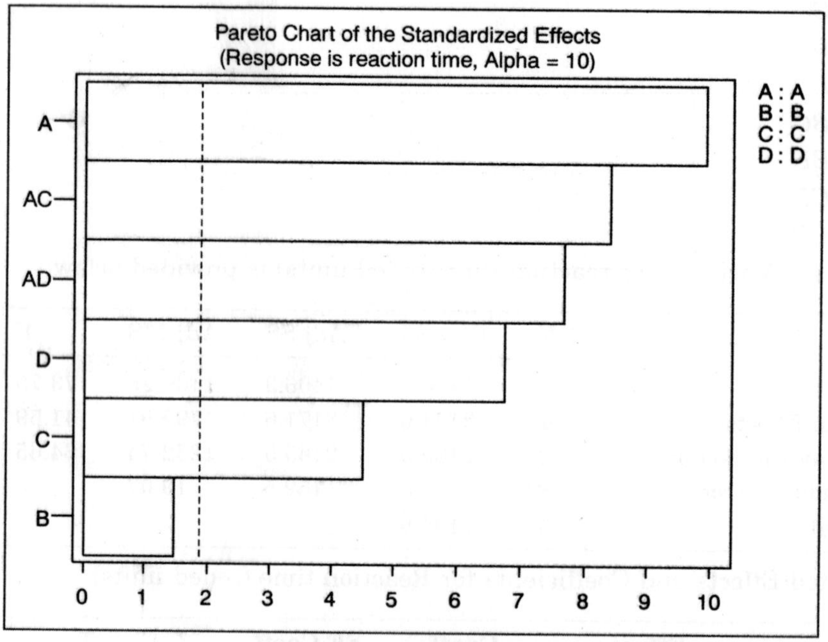

(ii) Pareto Chart of Effects

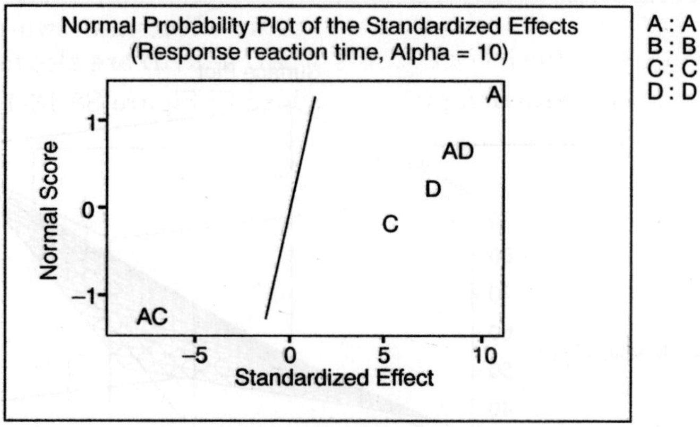

(iii) Normal Probability Plot of Effects

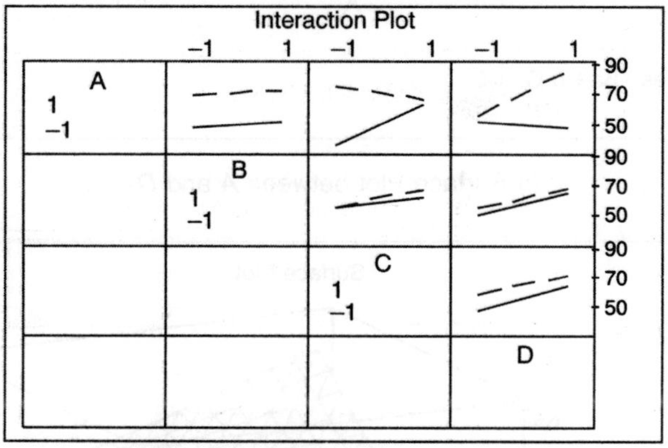

(iv) Interaction Plot

Fig. S8.13-1 The Main Effects (i), (ii), (iii) and Interaction (iv) Plots.

The surface plot between *A* and *D* and *A* and *C* is provided in Figure S8.13-2

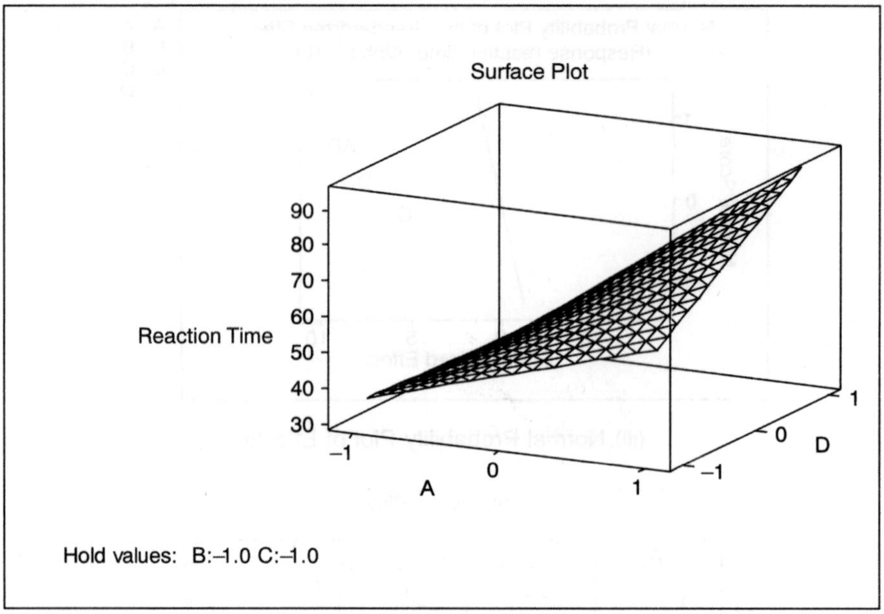

(i) Surface Plot between *A* and *D*

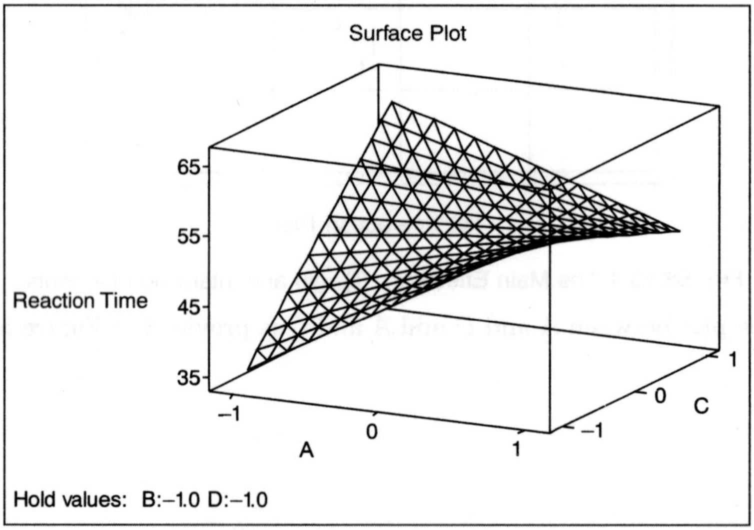

(*ii*) Surface Plot between *A* and *C*

Fig. S8.13-2 Surface Plot between (i) *A* and *D* and (ii) *A* and *C*.

From the above results and plots it is observed that to minimize the reaction time, the best possible settings are *A* low (– 1), *B* low (– 1), *C* low (– 1) and *D* low (– 1), and the batch has a significant influence on the process variability.

The residual plots as provided in Figure S8.13-3 do not reveal any drift or erratic behaviour of error over time, and no unusual structure in residual is apparent.

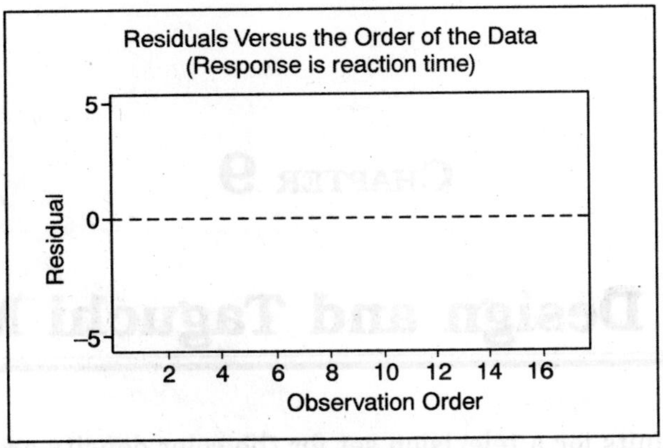

(i) Residual vs Order of Data

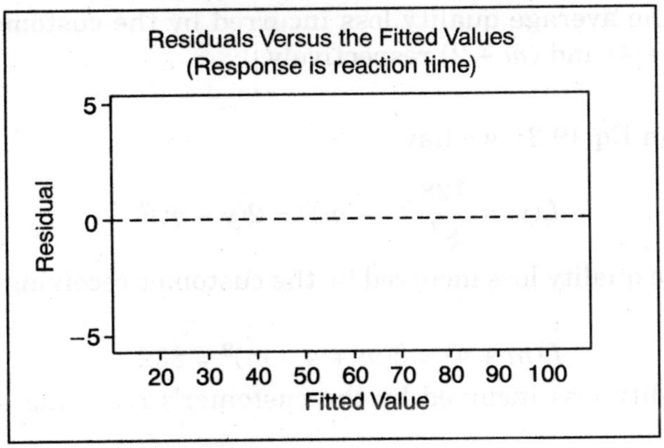

(ii) Residual vs Fitted Value

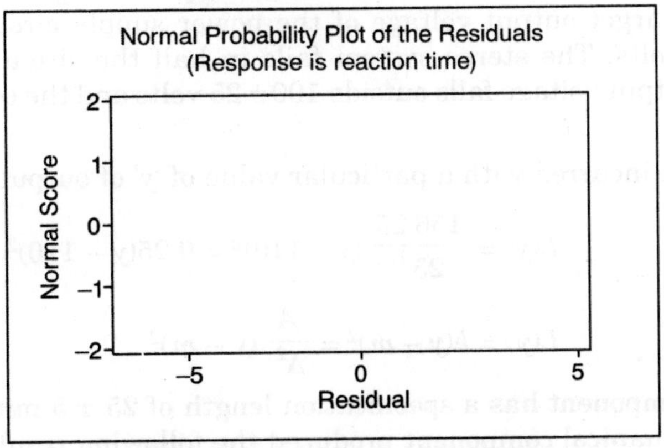

(iii) Normal Probability Plot of Residuals

Fig. S8.13-3 Residual Plots.

CHAPTER 9

Robust Design and Taguchi Method

9.1. The functional limits for a television set for the color density are $m \pm 8$, which implies that about one-half of the customers would find the television set to be defective if the color density is $m \pm 8$. If the repair of a television set in the field cost an average A = \$128. Compare the average quality loss incurred by the customers receiving sets with color density $(m + 4)$ and $(m + 2)$ respectively.

Solution:

By substituting in Eq. (9.3), we have

$$L(y) = \frac{128}{8^2}(y - m)^2 = 2(y - m)^2$$

Thus, the average quality loss incurred by the customer receiving sets with color density $(m + 4)$ is

$$L(m + 4) = 2(m + 4 - m)^2 = \$32$$

The average quality loss incurred by the customer's receiving sets with color density $(m + 2)$ is

$$L(m + 2) = 2(m + 2 - m)^2 = \$8$$

9.2. Determine the average loss associated with the output voltage value for a stereo system, given that the target output voltage of the power supply circuit used for the stereo system is 110 volts. The stereo system fails in half the situations and needs repair whenever the output voltage falls outside 100 ± 25 volts and the cost of repair is \$156.25.

Solution:

The average loss incurred with a particular value of 'y' of output voltage is given by

$$L(y) = \frac{156.25}{25^2}(y - 110)^2 = 0.25(y - 110)^2$$

$$L(y) = k(y - m)^2 = \frac{A}{\Delta^2}(y - m)^2$$

9.3. A mechanical component has a specification length of 25 ± 5 mm. Three suppliers 1, 2 and 3 of the mechanical component produced the following results.

Supplier	Mean length, $\overline{X}$	Variance, σ^2	C_{pk}
1	25	1.8	1.0
2	23	0.5	1.0
3	22	0.3	1.0

The replacement cost of component which is out of specification is \$50. The number of components out of specification in delivery of 300,000 components for the three suppliers is:

Supplier 1 = 0.30%; Supplier 2 = 0.2% and Supplier 3 = 0.15%

(*a*) compare the three suppliers based on the loss function

(*b*) compare the three suppliers based on the cost of defective units.

Solution:

From the statement of the problem, $\Delta = 5$ mm and $A = \$50$

The Taguchi loss equation is

$$L(y) = \frac{50}{(5)^2}(y - m)^2 = 2(y - m)^2$$

(*a*) $Q_A = 2[(25 - 25)^2 + 1.8] = \$3.6$ per unit

$Q_B = 2[(23 - 25)^2 + 0.5] = \$9$ per unit

$Q_C = 2[(22 - 25)^2 + 0.3] = \$18.6$ per unit

(*b*) Cost based on 300,000 parts

Supplier	$\bar{x}$	σ^2	C_{pk}	Defects %	Out-of-spec costs
A	25	1.8	1.0	0.3	\$45,000
B	23	0.5	1.0	0.2	\$30,000
C	22	0.3	1.0	0.15	\$22,500

Out-of-spec costs = (Total production) × (% defects) × (Avg. replacement cost)

Note that based on cost of replacing out-of-spec parts, supplier *C* is best. However, supplier *A* that consistently produces parts on target has the lowest cost to the customer. This is the best supplier for long-term customer relations because it is the highest quality supplier.

9.4. Customer tolerances for the length of a machine component are 2 ± 0.01 m. For a product that exceeds these limits, the cost to the customer for getting it fixed is \$75. Ten products are randomly selected and yield the following lengths (in meters): 2.03, 1.99, 2.0, 1.99, 1.98, 2.02, 2.04, 2.03, 2.01, and 2.02. Determine the average loss per product item.

Solution:

We note that target is best in this problem. The target value of m is 2 m. The loss function is given by

$$L(y) = k(y - m)^2$$

where

y = the length of the machine component

k = a proportionality constant

Now

$$k = \frac{A}{\Delta^2} = \frac{75}{(0.01)^2} = 750{,}000$$

The loss function is

$$L(y) = 750{,}000(y - 2)^2$$

The expected loss per item is

$$E[L(y)] = 750{,}000\ E(y - 2)^2$$

where $E(y - 2)^2$ are computed as follows:

$$\sum_{i=1}^{10} \frac{(y_i - 2)^2}{10} = \frac{[(2.03-2)^2 + (1.99-2)^2 + (2-2)^2 + \ldots + (2.02-2)^2]}{12} = \frac{0.0049}{10} = 0.00049$$

Therefore, the expected loss per item is

$$750{,}000(0.00049) = \$367.50$$

9.5. Refer to Prob. P9.1 regarding the length of the machine component. The manufacturing company is considering modifying the production process in order to reduce the variability in the output. The estimated additional cost for the modified production process is $30/component. The estimated annual production of these machine components is 50,000 components. Eight components are randomly selected from the new production process, resulting in the following lengths of the component: 2.01, 2.0, 1.99, 2.02, 2.02, 2.0, 1.98, and 2.01.

(*a*) Determine whether the new production process is cost effective or not

(*b*) If the new production process is cost effective, determine the annual savings.

Solution:

(*a*) The loss function from Problem P9.1 is

$$L(y) = 750{,}000(y - 2)^2$$

For the new production process, the estimated mean square deviation around the target value of 2 is given by

$$\sum_{i=1}^{8} \frac{(y_i - 2)^2}{8} = \frac{[(2.01-2)^2 + (2-2)^2 + \ldots + (2.01-2)^2]}{8} = \frac{0.0015}{8}$$

$$= 0.00018755$$

The expected loss per component for the new process is

$$750{,}000(0.0001875) = \$140.625 \approx \$140.63$$

From Problem P9.1, the expected loss per component for the original process is $367.50, which represents a savings of $367.50 – $140.63 = $226.87/component. The added cost is given as $50/component. Hence, the net savings per component using the new production process is $226.87 – $50 = $176.87. Therefore, the new production process is cost efficient.

(*b*) The net annual savings compared to the original process is therefore

$$30{,}000(\$176.87) = \$5{,}306{,}100.00.$$

9.6. Refer to Problem P9.4 regarding the length of the machine component. If the manufacturing company decides to rework the length of the component prior to shipping the component at a cost of $20 per component, determine the manufacturer's tolerance.

Solution:

The loss function from Problem P9.1 is obtained as

$$L(y) = 750{,}000(y - 2)^2$$

Assuming the manufacturer's tolerance is given by $2 \pm \delta$, then

$$\$20 = 750{,}000(\delta^2)$$

or

$$\delta^2 = \left(\frac{20}{750{,}000}\right)$$

or

$$\delta = 0.005164.$$

Hence, the manufacturer's tolerance is 2 ± 0.005164 m. In other words, for components with lengths that are equal or exceed these limits, the manufacturer should rework the components at additional cost of $20 per component in order to provide cost savings in the long-run. Otherwise, these components will incur a loss to the customer.

9.7. A manufacturer of copper wiring is concerned in reducing the variability of the diameter of the wiring. It is estimated that the loss to the customer is $20 per reel if the diameter of the wire exceeds 2 ± 0.04 mm. Each reel has 100 m of wire. A random sample of 10 gave the following diameters (in mm): 1.98, 2.01, 2.03, 1.97, 2.01, 1.99, 2.02, 2.01, 2.02, and 2.04. Determine the average loss per reel.

Solution:

The proportionality constant in the loss function, considering target is best, is given by

$$k = \frac{10}{(0.04)^2} = 6250$$

$$E(y - 2.0)^2 = [(1.98 - 2)^2 + (2.01 - 2)^2 + (2.03 - 2)^2 + (1.97 - 2)^2 + (2.01 - 2)^2 + (1.99 - 2)^2 + (2.02 - 2)^2 + (2.01 - 2)^2 + (2.02 - 2)^2 + (2.04 - 2)^2] = 50 \times 10^{-4}$$

Hence, the average loss per reel = (6250)(50 × 10^{-4}) = $31.25.

9.8. Refer to Problem P9.7 regarding the diameter of the copper wire. In order to reduce the variability in the diameter of the wire, the manufacturer is considering a new and improved production process. It is estimated that the additional cost for this improved process is $0.05 per linear meter. The annual production is 5,000 reels. Each reel has 100 m of wire. A random sample of size 10 from the new and improved production process yielded the following diameter (in millimeters):

2.01, 2.04, 1.98, 1.99, 2.02, 2.03, 2.01, 2.03, 1.97, and 1.98.

(*a*) Is the new and improved production process cost effective?

(*b*) Determine the annual savings or loss.

Solution:

(*a*) The mean square deviation around the target diameter of the wire of 2 mm for the new and improved process is estimated as

$$\Sigma(y_i - 2)^2/10 = [(2.01 - 2)^2 + (2.04 - 2)^2 + (1.98 - 2)^2 + (1.99 - 2)^2 + (2.02 - 2)^2 + (2.03 - 2)^2 + (2.01 - 2)^2 + (2.03 - 2)^2 + (1.97 - 2)^2 + (1.98 - 2)^2]/10$$

$$= \frac{58 \times 10^{-4}}{10} = 5.8 \times 10^{-4}$$

$$k = \frac{10}{(0.04)^2} = 6{,}250$$

The expected loss per reel = (6250)(5.8 × 10^{-4}) = $3.625 ≈ $3.63. The additional cost per reel for the new process = $0.05(100) = $5. Hence, the new process has an increased average cost per reel of ($5 – $3.63) = $1.37, compared to the old or the original process. The new process is therefore cost effective.

(*b*) The added expected annual savings realized using the new process

= ($1.37)(5,000) = $6850.

9.9. Refer Problem P9.7. If the manufacturer can rework the diameter of the wire prior to shipping the product at a cost of $5 per reel, determine the manufacturer's tolerance for the produced wire.

Solution:

The loss function is given by

$$L(y) = 6250(y - 2)^2$$

Assume the manufacturer's tolerance is given by $2 \pm \delta$. Then, we have

$$\$5 = 6250(\delta)^2$$

or

$$\delta = 0.028284$$

Hence, the manufacturer's tolerance should be 2 ± 0.028284.

9.10. Refer to Problem P9.7. The manufacturer is considering adopting a new process so that the average diameter of the wire produced is 2 mm, which is the target value. However, the manufacturer also estimated the standard deviation of the new process will be 0.01 mm and the cost of making this change is \$3 per reel.

(*a*) Determine if the new process will be cost effective compared to the original process

(*b*) Determine the annual savings or loss if the annual production is 5,000 reels.

Solution:

(*a*) Here, the process mean is the target value. Hence, the mean square deviation is estimated as the square of the standard deviation or

$$(0.01)^2 = 1 \times 10^{-4}$$

The expected loss per reel = $(6250)(1 \times 10^{-4})$ = \$0.625 ≈ \$0.63. Hence, the average loss per reel for the new process = \$0.63 + \$3 = \$3.63.

The original process had an average loss per reel of \$31.25.

Hence, it would be cost effective to make the change.

(*b*) The added annual savings is (5,000)(\$3.63 – \$31.25) = \$138,100.

9.11. The critical dimension of a certain turned part using a lathe machine must be 0.5 ± 0.003 inches and any part falls out of spec in the range of 0.497 and 0.503 will have to be scrapped at a cost of \$100. Fifteen samples were taken from the lathe operations of this part and their dimensions are as follows:

0.497, 0.499, 0.501, 0.502, 0.498, 0.501, 0.497, 0.500, 0.501, 0.497, 0.500, 0.498, 0.501, 0.501 and 0.497.

(*a*) write the Taguchi loss equation for the turning operation

(*b*) determine the average Taguchi loss function for the parts made using this lathe machine.

Solution:

(*a*) The Taguchi loss equation for this lathe turning operation is given by

$$L(y) = k(y - m)^2 = \frac{A}{\Delta^2}(y - m)^2$$

Given A = \$100, $\Delta = 0.003$ and $m = 0.500$. Therefore,

$$L(y) = \frac{\$100}{(0.003)^2}(y - m)^2 = 11.111 \times 10^6 (y - m)^2$$

(*b*) $Q = k[(\mu - m)^2 + \sigma^2]$

$$\mu = \bar{x} = (0.497 + 0.499 + 0.501 + 0.502 + 0.498 + 0.501 + 0.497 + 0.500 + 0.501 + 0.497 + 0.500 + 0.498 + 0.501 + 0.501 + 0.497)/15$$

$$= 7.483/15 = 0.499$$

$$\sigma = \sqrt{\left(\frac{1}{n-1}\right)\Sigma(y_1 - \mu)^2} = \sqrt{\left(\frac{1}{15-1}\right)(49 \times 10^{-6})}$$

$$= 1.8708 \times 10^{-3}$$

$$Q = 11.111 \times 10^6[(0.499 - 0.500)^2 + (1.8708 \times 10^{-3})^2]$$

$$= 11.111 \times 10^6[1 \times 10^{-6} + 3.5268 \times 10^{-6}] = \$50.30$$

i	y_i	μ	$(y_i - \mu)^2$
1	0.497	0.499	4×10^{-6}
2	0.499	0.499	0
3	0.501	0.499	4×10^{-6}
4	0.502	0.499	9×10^{-6}
5	0.498	0.499	1×10^{-6}
6	0.501	0.499	4×10^{-6}
7	0.497	0.499	4×10^{-6}
8	0.500	0.499	1×10^{-6}
9	0.501	0.499	4×10^{-6}
10	0.497	0.499	4×10^{-6}
11	0.500	0.499	1×10^{-6}
12	0.498	0.499	1×10^{-6}
13	0.501	0.499	4×10^{-6}
14	0.501	0.499	4×10^{-6}
15	0.497	0.499	4×10^{-6}
			49×10^{-6}

9.12. For the data in Problem P9.11, determine the process capability index C_p. Is the process capable of meeting the specifications?

Solution:

Refer to the solution of Prob. P9.11, we have

$$\bar{x} = 0.499 \text{ and } \sigma = 1.8708 \times 10^{-3}$$

The process capability index C_p is then given by

$$C_p = \frac{UCL - LCL}{6\sigma} = \frac{0.503 - 0.497}{6\,(1.8708 \times 10^{-3})} = 3.207$$

The value of $C_p > 1$ indicates that the process is capable of meeting the specifications.

9.13. The manufacturer of an electromechanical system wishes to improve the quality of the system by reducing deviations from the target value such that the new standard deviation will be 1.2. This improvement is technologically possible at an additional cost of \$0.10 per unit of production. Assuming that no inspection is necessary or performed. Determine

(*a*) whether the manufacturer should improve the process and

(*b*) if the production rate of this manufacturer is 200,000 units per month determine the expected savings per annum.

Solution:

(*a*) The total loss per unit of the current process is given by

$$L(y) = k\, s^2 = \$0.08\left(\frac{10}{6}\right)^2 = \$0.2222 \qquad ..(1)$$

The total loss per unit after improving the process

$$L(y) = \$0.08(1.02)^2 = \$0.1152$$

Additional cost of improvement = \$0.10 per unit.

Additional cost plus loss per unit = 0.10 + 0.1152 = \$0.2152 ...(2)

The net savings resulting from the improvement in the process capability is obtained by subtracting (2) from (1), that is \$0.2222 – \$0.215 = \$0.007 per unit of production.

(*b*) Since the production rate of this manufacturer is 200,000 units per month, the expected savings will be \$1400 per month. Hence, the expected savings per annum is \$1400(12) = \$16,800.

9.14. The tensile strength of steel cable wire of 30 samples collected for two consecutive months of production are given in Table P9.14. The lower specification limit is 10 N/m². Defect loss is \$1 per linear meter. The monthly production is 500 meter. Determine the quality level during

Month (30 work days)	Tensile strength data (N/m²)									
1	15.3	12.3	12.4	13.5	14.6	12.9	13.1	11.5	12.3	15.1
	14.7	12.5	13.9	14.1	12.6	14.1	13.7	14.1	13.1	14.7
	14.8	12.6	15.3	13.8	13.1	12.8	15.1	15.1	14.1	13.8
2	13.5	15.1	14.2	12.5	14.3	15.1	14.7	13.1	14.1	14.2
	11.6	12.7	13.7	12.1	13.2	12.2	15.1	13.6	14.6	14.6
	14.1	13.5	14.2	12.7	11.8	10.1	14.7	13.4	14.1	13.2

(*a*) the first month of production.

(*b*) the second month of production.

Solution:

This problem involves the larger-the-better tolerance. The quality levels are obtained by using Eq. (9.36).

$$L(y) = A\Delta^2\left[\frac{1}{n}\sum_{i=1}^{n}\frac{1}{y_i^2}\right]$$

Here A = \$1, Δ = 10 N/m² and n = 30.

(*a*) The quality level during the first month of production is

$$L_1 = A\Delta^2\left[\frac{1}{n}\sum_{i=1}^{n}\frac{1}{y_i^2}\right] = (1)\,(10)^2\left[\frac{1}{30}\left(\frac{1}{15.3^2}+\frac{1}{12.3^2}+\frac{1}{12.4^2}+\ldots+\frac{1}{13.8^2}\right)\right]$$

$$= \$7.3424$$

Hence

$$\text{Monthly loss} = L_1\,(500)\,(10^6) = (7.3424)\,(500)\,(10^6) = \$3.67(10^9)$$

(*b*) The quality level during the second month of production is

$$L_2 = A\Delta^2\left[\frac{1}{n}\sum_{i=1}^{n}\frac{1}{y_i^2}\right] = (\$1)\,(10)^2\left[\frac{1}{30}\left(\frac{1}{13.5^2}+\frac{1}{11.6^2}+\frac{1}{14.1^2}+\ldots+\frac{1}{14.1^2}\right)\right]$$

$$= \$7.3463$$

$$\text{Monthly loss} = (L_2)\,(500)\,(10^6) = (7.3463)\,(500)\,(10^6) = \$3.6715(10^9)$$

The quality level of the first month is better (L_1 is lower) than that of the second month of production by an amount.

$$(\text{Monthly loss of second month}) - (\text{Monthly loss of first month}) = \$1{,}936{,}867$$

The quality levels evaluated periodically would serve as an indicator of the production line and serves as an important index.

9.15. When the blower seal of a vacuum blower housing in a copier machine fails to operate, it costs $50 to replace, $25 in part costs and another $25 in installation labor costs. The machine downtime to the customer is $350 per hour. It takes approximately 30 minutes to replace the blower seal. The seal level at which the vacuum loss becomes unacceptable is 25 psi. Determine

(*a*) the larger-the-better loss function

(*b*) the expected quality loss for the vacuum blower for two separate sites. Each has ten machines operating in it. The performance data collected for the two sites is summarized in Table P9.15.

Solution:

(*a*) $\Delta = 25$ psi

$$A = \frac{\$350}{2}(\text{copier downtime}) + \$25(\text{parts cost}) + \$25(\text{installation labor cost})$$
$$= \$225$$

The larger the better loss function is given by

$$L(y) = k\left[\frac{1}{y^2}\right] = A\Delta^2\left[\frac{1}{y^2}\right] = (\$225)\,(25)^2\left[\frac{1}{y^2}\right] = \$140{,}625\left[\frac{1}{y^2}\right]$$

Table P9.15

Site	Data
1	18, 25, 30, 19, 35, 32, 26, 28, 23, 24
2	17, 24, 31, 18, 34, 31, 27, 22, 23, 28

(*b*) The expected quality loss for the larger-the-better seal adhesion for two separate sites is given by the equation

$$L(y) = k(MSD) = k\left[\frac{1}{n}\sum_{i=1}^{n}\left(\frac{1}{y_i^2}\right)\right]$$

Hence the loss function for this case is

$$L(y) = \$140{,}625\left[\frac{1}{n}\sum_{i=1}^{n}\left(\frac{1}{y_i^2}\right)^2\right]$$

After transforming the data by taking the reciprocal of the square of the measured seal strengths, we obtain the results in Table P9.15 (*a*).

Table P9.15 (*a*)

Site	Data (y_i)	L(y)
1	18, 25, 30, 19, 35, 32, 26, 28, 23, 24	$5634.32
2	17, 24, 31, 18, 34, 31, 27, 22, 23, 28	$5788.94

9.16. The output voltage of a high-voltage transformer was found to alter due to the deterioration of transistors in the power circuit. When the output voltage exceeds the tolerance limit of ± 20 *V*, the transformer is unsuitable for use and the estimated loss when the transformer exceeds the limits is $500. The manufacturer can adjust the voltage by replacing the resistor at a cost of $2 before shipping the transformer to the customer. Determine the manufacturers' specifications assuming the target voltage as 110 *V*.

Solution:

Here $L(y) = k(y - m)^2$ ...(1)

where *m* is the target value, 110 *V* and *k* is the proportionality constant. Therefore

$$k = \frac{A}{\Delta^2} = \frac{\$500}{(20)^2} = 1.25$$

Hence, the loss function is

$$L(y) = 1.25(y - 110)^2 \qquad \text{...(2)}$$

The loss or cost of adjustment to the manufacturer is given as $2. Hence, Eq. (2) becomes

$$\$2 = 1.25(y - 110)^2$$

or

$$y = 110 \pm \sqrt{\frac{2}{1.25}} = 110 \pm 1.265\ V \qquad \text{...(3)}$$

Eq. (3) gives the varying range in production and hence units whose output voltage varies from 110 *V* by more than ± 1.265 *V* are considered defective by the manufacturer as shown in Fig. P9.16. The transformer should be adjusted in the shop if the output voltage falls outside the range 110 ± 1.26. Hence, the manufacturer should adjust or repair the characteristic *y* if

$$\frac{A}{\Delta^2} = (y - m)^2 \le \frac{A_0}{\Delta_0^2}(y - m)^2 \qquad \text{...(4)}$$

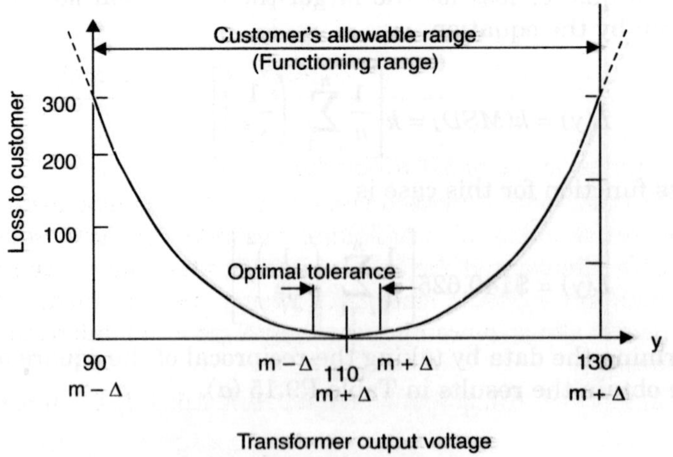

Fig. P9.16 Tolerance Determination.

The manufacturer's allowance is

$$\Delta = \sqrt{\frac{A}{A_0}}\,\Delta_0 \qquad ...(5)$$

where

2Δ = manufacturers tolerance limit

$2\Delta_0$ = customer tolerance limit or functional limit

A = manufacturer loss caused by a defective product

A_0 = customer loss caused by a defective product.

9.17. The tolerance of the deviations of the flatness of a metal component is specified as ± 0.005 inches. The observations of the derivations recorded using the existing machining system for a sample of 30 is given in Table P9.17 (*a*).

Table P9.17 (*a*)

Derivations observed on a sample of 30 (inches)				
0.002	0.002	0.005	– 0.004	– 0.003
– 0.003	0.003	0.003	– 0.003	– 0.004
– 0.004	0.001	0.004	– 0.004	– 0.005
0.003	0.005	0.005	– 0.003	0.003
– 0.003	0.002	– 0.003	0.003	0.004
0.004	– 0.002	– 0.001	0.001	0.005

A new machining system is introduced to increase the quality of the flatness. Table P9.17 (*b*) gives the derivations observed after introducing the new system for a sample size of 30. Determine

Table P9.17 (*b*)

Derivations observed on a sample of 30 (inches)				
0.001	0.002	– 0.001	– 0.002	0.003
– 0.002	0.003	– 0.001	– 0.003	0.002
– 0.003	– 0.002	0.001	– 0.001	0.001
0.003	– 0.001	0.001	0.003	– 0.001
0.000	0.000	0.002	0.002	0.001
0.001	– 0.001	– 0.002	0.001	0.000

Solution:

Follow the Examples E9.12 to E9.15 for the procedure.

9.18. The nominal thickness of the corrosion resistance coating on a metal component is found to be an important variable in establishing the wear characteristics of the component. The loss caused by the wear of the coating is \$25, when the coating thickness is ± 0.001 inch from its nominal value. A sample of twenty measurements of the coating taken from selected points along the component length gave the following deviations from the nominal thickness:

0.8, 1.0, 0, 0.4, 1.8, 1.3, 0.3, 0.7, 0.6, 1.1, 1.4, 1.5, 1.2, 1.7, 0.8, 0.9, 1.1, 1.3, 1.4, 0.5

A new coating process was found to reduce the loss by 50%. The following deviations were measured for the new process.

1.0, 1.0, 1.1, 0.2, 0.4, 0.5, 1.2, 0.9, 0.8, 0.7, 0.3, 0.2, 1.0, 1.1, 0.6, 1.0, 1.2, 0.3, 0.4, 0.5

(*a*) determine the tolerance of the coating thickness for the new process

(*b*) the loss for each unit for the old and new processes.

Solution:

The smaller the wear the better the component.

(*a*) The tolerance of the coating thickness for the new process is given by

$$\Delta_2 = \sqrt{\frac{A_2}{A_1}}\,\Delta_1 \qquad \text{...(1)}$$

where

A_1 = loss due to wear of a component produced by the old process

A_2 = loss due to wear of a component produced by the new process

Δ_1 = tolerance limit of the old process

Δ_2 = tolerance limit of the new process

Hence $$\Delta_2 = \sqrt{\frac{12.5}{25}}\,(1) = 0.7071$$

(*b*) The loss per unit for the process is

$$L_1 = \frac{A_2}{\Delta_1^2}\,[0.8^2 + 1.0^2 + 0^2 + \ldots.. + 0.5^2]/20$$

$$= \frac{25}{1^2}\,[0.8^2 + 1.0^2 + 0^2 + \ldots.. + 0.5^2]/20$$

$$= \frac{25\,(23.98)}{20} = \$29.98$$

The loss per unit for the new process is

$$L_2 = \frac{A_2}{\Delta_2^2}\,[1.0^2 + 1.0^2 + 1.1^2 + \ldots.. + 0.5^2]/20$$

$$= \frac{12.5}{(0.7071)^2}\,[1.0^2 + 1.0^2 + 1.1^2 + \ldots.. + 0.5^2]/20$$

$$= \frac{12.5}{(0.7071)^2}\,\frac{(12.68)}{20} = \$15.85$$

9.19. A high tech manufacturing company has established three factors *A*, *B*, and *C*, each at three levels that influence the satisfaction of their valued customers. The company also believes that the interaction effect $B \times C$ is important. Set up an appropriate experimental design using an orthogonal array.

Solution:

It is given that there are three factors *A*, *B*, and *C* each at 3 levels. The interaction $B \times C$ is also considered important. Hence, the $L_9(3^4)$ orthogonal array is recommended as shown in Table P9.19.

Table P9.19

Experiment number	Factor			
	B	C	B × C	A
1	1	1	1	1
2	1	2	2	2
3	1	3	3	3
4	2	1	2	3
5	2	2	3	1
6	2	3	1	2
7	3	1	3	2
8	3	2	1	3
9	3	3	2	1

9.20. A large electromechanical components manufacturing company has identified four factors, A, B, C, and D, each to be analyzed at two levels. The company also believes that the interactions $B \times C$, $B \times D$, and $C \times D$ are important. Set up an appropriate experimental design using an orthogonal array.

Solution:

The three interactions $B \times C$, $B \times D$, and $C \times D$ are treated like factors and therefore we have a total of seven factors including A, B, C, and D, each of two levels. Hence, we can select the $L_8(2^7)$ orthogonal array as shown in Table P9.20 (*a*) and the first linear graph. The assignment factors to the corresponding columns as shown in the linear graph in Table P9.20 (*b*).

Table P9.20 (*a*)

$L_8(2^7)$ Series

Experiment	Factor						
	1	2	3	4	5	6	7
1	1	1	1	1	1	1	1
2	1	1	1	2	2	2	2
3	1	2	2	1	1	2	2
4	1	2	2	2	2	1	1
5	2	1	2	1	2	1	2
6	2	1	2	2	1	2	1
7	2	2	1	1	2	2	1
8	2	2	1	2	1	1	2
	Group 1	Group 2		Group 3			

Linear Graphs for L_8 Table

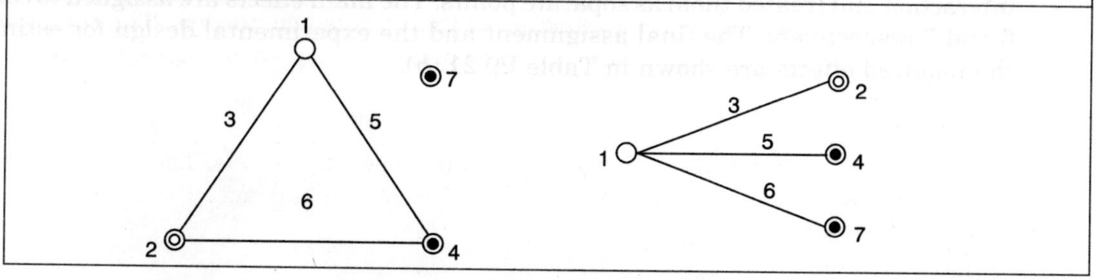

Table P9.20(*b*)

Allocation of Factors to Columns in an L_8 Orthogonal Array

Experiment	Factor						
	B	C	B × C	D	B × D	C × D	A
1	1	1	1	1	1	1	1
2	1	1	1	2	2	2	2
3	1	2	2	1	1	2	2
4	1	2	2	2	2	1	1
5	2	1	2	1	2	1	2
6	2	1	2	2	1	2	1
7	2	2	1	1	2	2	1
8	2	2	1	2	1	1	2

Linear Graph

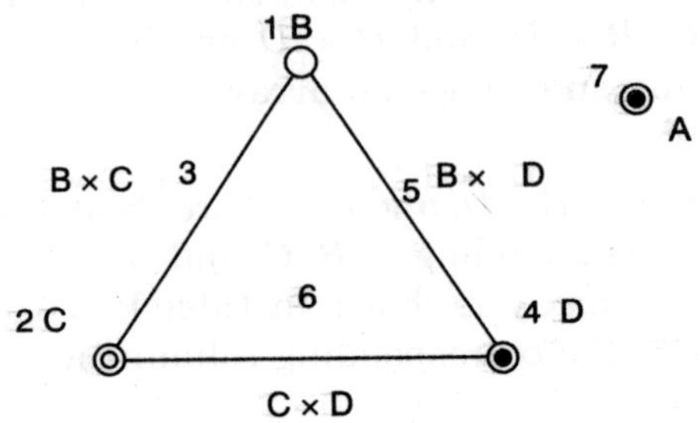

9.21. The quality engineering department of a large manufacturing company has identified five factors, *A*, *B*, *C*, *D*, and *E*, each to be investigated at two levels. The department also believes that the interactions *A* × *C* and *A* × *D* are important. Set up an appropriate experimental design for estimating the main effects and the two interactions.

Solution:

Here, we have a total of seven factors to be estimated each at two levels. Therefore, we can assign the $L_8(2^7)$ orthogonal array and its second linear graph as shown in Table P9.21 (*a*). The linear graph shows the assignment of three-interaction term. We also note that we need to estimate only two interactions. Hence, we can remove the associated two column numbers and treat them as individual points to be assigned to the main effects. The linear graph shown in Table P9.21 (*b*) illustrates this approach. We initially assign factor *A* to column 1, factor *C* to column 2, and factor *D* to column 4. The interactions *A* × *C* and *A* × *D* are assigned to columns 3 and 5 respectively. We remove columns 6 and 7 from the original graph since there is no need to estimate a third interaction and treated them as separate points. The main effects are assigned to columns 6 and 7 respectively. The final assignment and the experimental design for estimating the required effects are shown in Table P9.21 (*b*).

Table P9.21 (*a*)
$L_8(2^7)$ Series

Experiment	Factor						
	1	2	3	4	5	6	7
1	1	1	1	1	1	1	1
2	1	1	1	2	2	2	2
3	1	2	2	1	1	2	2
4	1	2	2	2	2	1	1
5	2	1	2	1	2	1	2
6	2	1	2	2	1	2	1
7	2	2	1	1	2	2	1
8	2	2	1	2	1	1	2
	Group 1	Group 2		Group 3			

Linear Graphs for L_8 Table

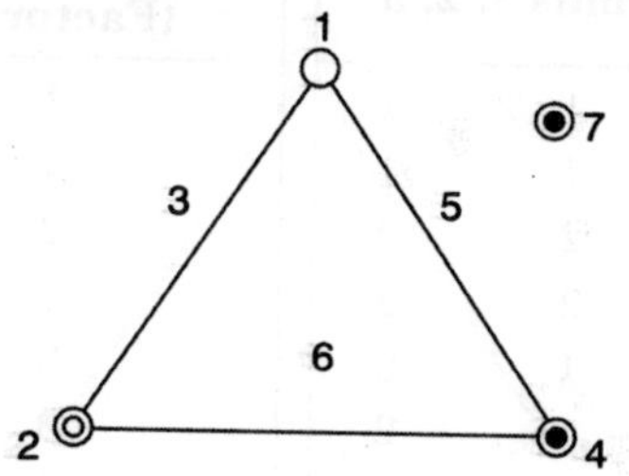

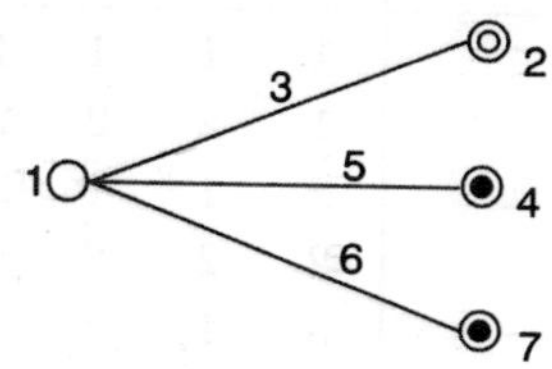

Table P9.21(*b*)
Modifying L_8's Linear Graph to Estimate Interaction Effects

Experiment	Factor						
	A	C	A × C	D	A × D	B	E
1	1	1	1	1	1	1	1
2	1	1	1	2	2	2	2
3	1	2	2	1	1	2	2
4	1	2	2	2	2	1	1
5	2	1	2	1	2	1	2
6	2	1	2	2	1	2	1
7	2	2	1	1	2	2	1
8	2	2	1	2	1	1	2

Modification of Linear Graph

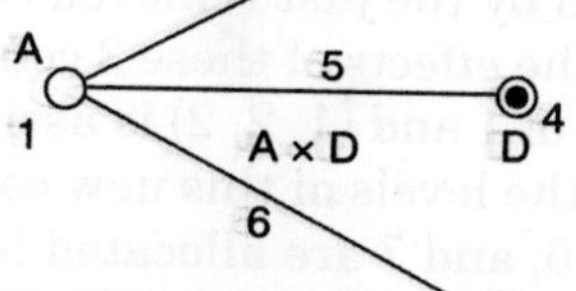

Original Graph

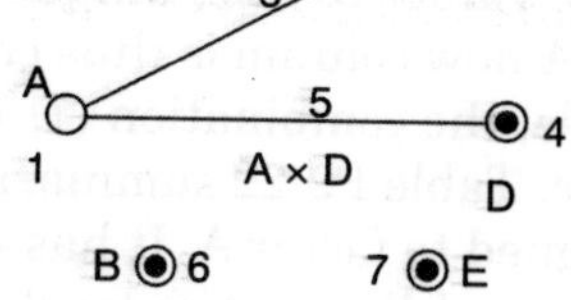

9.22. A manufacturing company has identified five factors A, B, C, D, and E that have a great impact on its volume of production. There are four levels for factor A, and each of the other four factors is to be tested at two levels. Set up an appropriate experimental design for estimating the main effects and the interactions.

Solution:

Degrees of freedom associated with the factors are calculated first. We note that the number of degrees of freedom for each factor is 1 less than the number of levels. The total number of degrees of freedom associated with an orthogonal array is 1 less than the number of experiments. The total number required in this problem is 7 (3 for factor A, and 1 for each of the factors B, C, D, and E). Also, the number of degrees of freedom associated with the L_8 orthogonal array is 7, that is, 1 less than the total number of experiments. Therefore, we can recommend the use of the L_8 array for the experimental hybrid design.

Table P9.22

Experiment	Factor A	B	C	D	E	Columns 1, 2, 3			New assignment (Factor A)
1	1	1	1	1	1	1	1	1	1
2	1	2	2	2	2	1	1	1	1
3	2	1	1	2	2	1	2	2	2
4	2	2	2	1	1	1	2	2	2
5	3	1	2	1	2	2	1	2	3
6	3	2	1	2	1	2	1	2	3
7	4	1	2	2	1	2	2	1	4
8	4	2	1	1	2	2	2	1	4

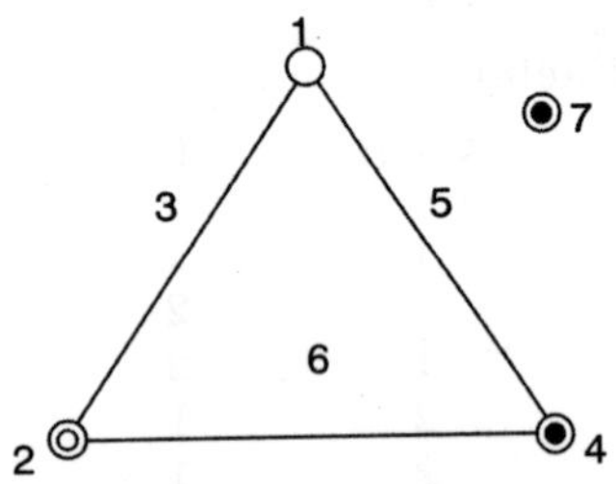

Target linear graph

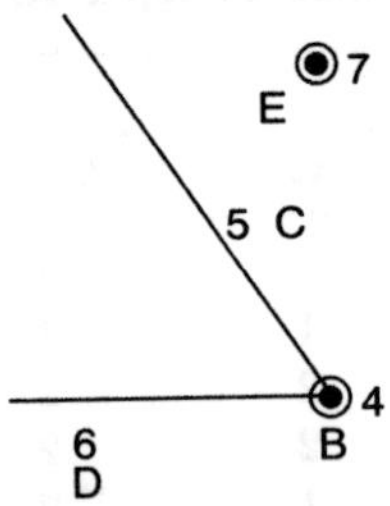

Modified graph

1,2,3 combined

It is noted here that each column in the L_8 array has 1 degree of freedom and can handle a 2-level factor. We need to study factor A at 4 levels (3 dof). Accordingly, factor A is allocated 3 columns. The remaining four columns are allocated to factors B, C, D, and E (each 1 dof). We employ linear graph to determine the columns to be allocated to factor A. If we consider the line segment connecting nodes 1 and 2 shown in Table P9.22. Notice that removing the line segment and nodes 1 and also removes node 3, we can allocate values to each unique combination generated by the just removed columns 1, 2, and 3. A new column is thus created that integrates the effects of these 3 columns. As an example, the combination (1, 1, 1) is assigned a level of 1 and (1, 2, 2) is assigned a level of 2, etc. Table P9.22 summarizes the assignment of the levels of this new column, while is assigned to factor A. It has 4 levels. Columns 4, 5, 6, and 7 are allocated to the factors B, C, D, and E have two levels each. We have ignored the interaction effects. Hence, the assignment of factor C to column 5 of the original L_8 array results in the interaction between the factors in columns 1 and 4 (or $A \times B$) is confounded with the main effect of C. In other words, the main effect of factor C can be obtained by assuming that the interaction effects as insignificant. Likewise, similar inferences can be made for the remaining factor assignments.

9.23. In a certain manufacturing operation, four factors A, B, C, and D each at three levels, is thought to be of importance in influencing the volume of parts produced. Using an L_9 orthogonal array, the factors A, B, C, and D are assigned to columns 1, 2, 3, and 4 respectively. The response variable showing the number of parts (in thousands) produced per day for each of the nine experiments is summarized below:

Experiment	1	2	3	4	5	6	7	8	9
Parts per day (in thousands)	6.2	15.2	11.5	6.5	16.5	4.8	6.7	13.2	9.5

(*a*) setup the experimental design and response variable for the corresponding experiment

(*b*) determine the main effects

(*c*) plot the average response curves

(*d*) determine the optimal settings of the design parameters.

Solution:

(*a*) The experimental design and the response variable for the corresponding experiment are summarized in Table P9.23.

Table P9.23

Experiment number	Factor				Parts per day (in thousands)
	A	B	C	D	
1	1	1	1	1	6.2
2	1	2	2	2	15.2
3	1	3	3	3	11.5
4	2	1	2	3	6.5
5	2	2	3	1	16.5
6	2	3	1	2	4.8
7	3	1	3	2	6.7
8	3	2	1	3	13.2
9	3	3	2	1	9.5

(*b*) The main effects of each factor at the three levels are found as follows:

$$\bar{A}_1 = (6.2 + 15.2 + 11.5)/3 = 10.966$$

$$\bar{A}_2 = (6.5 + 16.5 + 4.8)/3 = 9.267$$

$$\bar{A}_3 = (6.7 + 13.2 + 9.5)/3 = 9.8$$

$$\bar{B}_1 = (6.2 + 6.5 + 6.7)/3 = 6.467$$

$$\bar{B}_2 = (15.2 + 16.5 + 13.2)/3 = 14.967$$

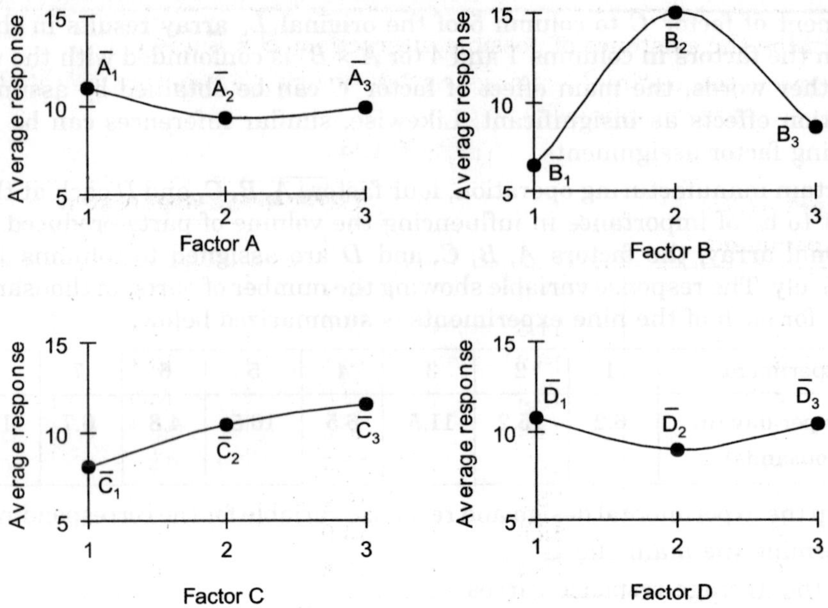

Fig. P9.23 The Average Response Curves.

$$\overline{B}_3 = (11.5 + 4.8 + 9.5)/3 = 8.6$$
$$\overline{C}_1 = (6.2 + 4.8 + 13.2)/3 = 8.067$$
$$\overline{C}_2 = (15.2 + 6.5 + 9.5)/3 = 10.4$$
$$\overline{C}_3 = (11.5 + 16.5 + 6.7)/3 = 11.567$$
$$\overline{D}_1 = (6.2 + 16.5 + 9.5)/3 = 10.734$$
$$\overline{D}_2 = (15.2 + 4.8 + 6.7)/3 = 8.9$$
$$\overline{D}_3 = (11.5 + 6.5 + 13.2)/3 = 10.4$$

(*c*) The average response curves are plotted in Fig. P9.23.

(*d*) The response variable, number of parts per day, is of the type "large-is-better". The optimum settings of the design parameters, assuming interaction effects are insignificant, are factor A at level 1, factor B at level 2, factor C at level 3, and factor D at level 1.

9.24. In a certain meat processing plant, four design parameters, A, B, C, and D, each at three levels, have been identified as having an effect on the moisture content in packed meat. Three noise factors, E, F, and G, each at two levels, are also to be investigated in the experiment. An L_9 orthogonal array is used in the inner array with the factors A, B, C, and D assigned to column 1, 2, 3, and 4, respectively. An L_4 orthogonal array is used for the outer array with the noise factors E, F, and G assigned to columns 1, 2, and 3, respectively. The moisture content in a coded scale for four replications of each of the nine experiments in the inner array is given in Table P9.24. It is highly desirable that the moisture content be around a coded value of 15.

(*a*) set up the complete parameter design layout using the Taguchi method

(*b*) determine the mean response and the appropriate signal-to-noise ratio

(*c*) plot the average *s*/*N* ratios and the average responses (*d*) determine the existence of possible interactions $B \times E$ and $C \times F$

(*e*) determine the optimal settings of the design parameters using the Taguchi's parameter design approach.

Table P9.24

Inner-array experiment	Outer-array experiment			
	1	**2**	**3**	**4**
1	13.5	16.2	15.5	14.9
2	11.8	12.1	15.8	14.5
3	16.2	16.8	15.7	14.8
4	15.3	13.5	14.8	15.8
5	11.8	16.5	16.3	16.5
6	14.1	13.5	13.4	13.5
7	15.3	16.4	12.8	14.9
8	12.5	15.0	16.1	15.4
9	15.4	14.8	14.9	14.5

Solution:

Follow the procedure given in Example E9.25 but with an L_4 orthogonal array for the outer array with the noise factors E, F, and G assigned to columns 1, 2, and 3 respectively.

9.25. The tensile strength of a steel alloy has positive and negative effects depending on the amount of alloy used. Four independent factors have been identified that have an affect on the tensile strength of the steel alloy. Each factor can be tested at three levels. An L_9 orthogonal array is used in the inner array with the factors A, B, C, and D to columns 1, 2, 3, and 4 respectively. The response variable, measuring the impact of the alloy, is recorded on a coded scale for the nine experiments. The target value is zero, with the observed coded responses being positive and negative. Table P9.25 shows the experimental design and the values of the coded response variables. For each of the settings of the design factors only one replication is conducted and the outer array of the noise factors is not shown.

(*a*) determine the main effects

(*b*) plot the average response curves

(*c*) determine the optimal settings of the design parameters.

Table P9.25

Experiment	Factor				Coded response
	A	**B**	**C**	**D**	
1	1	1	1	1	– 3.2
2	1	2	2	2	6.8
3	1	3	3	3	1.9
4	2	1	2	3	– 4.9
5	2	2	3	1	8.5
6	2	3	1	2	– 5.7
7	3	1	3	2	– 5.2
8	3	2	1	3	2.7
9	3	3	2	1	– 2.3

Solution:

We assume that the interaction effects between factors are insignificant. In our experimental design, we assign the following: factor A to column 1, B to column 2, C to column 3, and D to column 4. the interaction effect $A \times B$ will then be confounded with the main effect of C, which is assigned to column 3. Similarly, the interaction effect $A \times B$ will be confounded with the main effect of D, which is assigned to column 4.

(*a*) **Main effects:** Here we calculate the main effects of each factor at the three levels, starting with factor A where the average responses at levels 1, 2, and 3 are given by

$$\bar{A}_1 = (-3.2 + 6.8 + 1.9)/3 = 1.834$$

$$\bar{A}_2 = (-4.9 + 8.5 - 5.7)/3 = -0.7$$

$$\bar{A}_3 = (-5.2 + 2.7 - 2.6)/3 = -1.7$$

Similarly, the average responses for factors B, C and D are:

$$\bar{B}_1 = (-3.2 - 4.9 - 5.2)/3 = -4.433$$

$$\bar{B}_2 = (6.8 + 8.5 + 2.7)/3 = 6.0$$

$$\bar{B}_3 = (1.9 - 5.7 - 2.6)/3 = -2.133$$

$$\bar{C}_1 = (-3.2 - 5.7 + 2.7)/3 = -2.067$$

$$\bar{C}_2 = (6.8 - 4.9 - 2.6)/3 = -0.233$$

$$\bar{C}_3 = (1.9 + 8.5 - 5.2)/3 = 1.733$$

$$\bar{D}_1 = (-3.2 + 8.5 - 2.6)/3 = 0.9$$

$$\bar{D}_2 = (6.8 - 5.7 - 5.2)/3 = -1.367$$

$$\bar{D}_3 = (1.9 - 4.9 - 2.7)/3 = -1.9$$

(*b*) The plots of the average response for each factor against the factor levels are shown in Fig. P9.25.

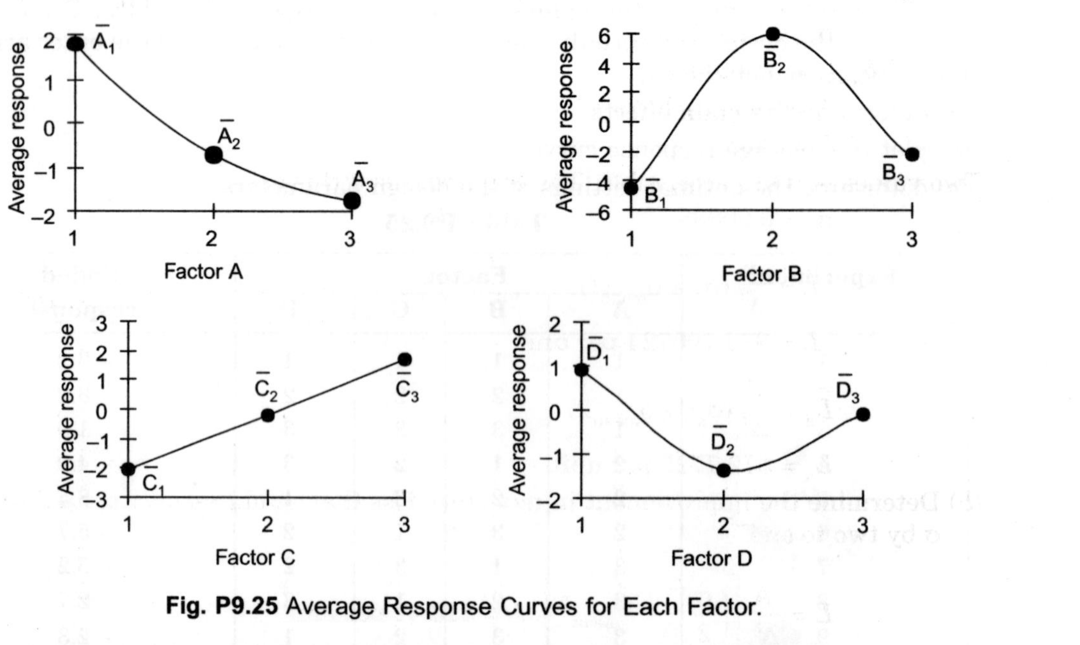

Fig. P9.25 Average Response Curves for Each Factor.

(*c*) We observe from Fig. P9.25 that factors *B* and *C* have a nonlinear affect on the average response. Hence, for minimum variability, factors *B* and *D* are set at level 2. At these levels, the response curve is flat and therefore less sensitive to variations due to noise factors.

The next step is to shift the average response as close to the target value as possible (that is the value of zero) and at the same time not to affect the variability obtained in the previous step. Factors *A* and *C* each has a linear relationship with the average response. Hence, we can set the adjustment factors *A* and *C* at level 2. Therefore, the optimum settings are factor *A* at level 2, factor *B* at level 2, factor *C* at level 2, and factor *D* at level 2.

However, confirmation experiments are recommended at this stage to validate the optimum settings proposed and/or determine more improved settings that would reduce variability considerably more.

9.26. The monetary loss induced by premature failure of a product is defined by Taguchi's loss function. For an electrical power tool, the cost of repair to the tool merchant for a specific type of failure is estimated as $465/unit. Given the deviations from the target value are given for 15 units measured at the factor are:

(0, 0.5, 0.3, 0.2, – 0.3, 0.4, 0, – 0.4, – 0.3, 0.6, 0.5, – 0.3, 0, – 0.4, 0.6)

Determine:

(*a*) the financial loss resulting from a machine

(*b*) the improvement in monetary loss that would result from a reduction of the standard deviation by two to one.

Solution:

N = 15 measurement of deviation

M = 0.25 inches

(*a*) Determine the financial loss resulting from a failure.

Given:

$$\sigma = 0.0003567$$

$$\delta_{mean} = 0.00011$$

$$\Delta = 0.001 \text{ inch}$$

$$A = \$465$$

Two values of σ are acceptable. The one above and this one:

$$\sigma_a = 0.00038$$

$$L = \frac{A}{\Delta^2}(\sigma^2 + \delta_{mean}{}^2)$$

$$L = \$64.790724 \text{ per unit}$$

$$L_a = \frac{A}{\Delta^2}(\sigma_a{}^2 + \delta_{mean}2)$$

$$L_a = \$72.7725 \text{ per unit}$$

(*b*) Determine the improvement in monetary loss that would result from a reduction in σ by two to one.

$$L = \frac{A}{\Delta^2}\left[\left(\frac{\sigma}{2}\right)^2 + \delta_{mean}{}^2\right], L = \$20.42 \text{ per unit}$$

$$L_a = \frac{A}{\Delta^2}\left[\left(\frac{\sigma_a}{2}\right)^2 + \delta_{mean}{}^2\right], L_a = \$22.41 \text{ per unit}$$

9.27. Determine the optimum combination of factor values for the orthogonal array shown below. Determine also the per cent improvement in results produced by using the optimum combination of factors. *Note:* Biggest is best.

Solution:

$A1 = 42 + 50 + 36 + 45 = 173$

$A2 = 35 + 55 + 30 + 54 = 174$

$B1 = 42 + 50 + 35 + 55 = 182$

$B2 = 36 + 45 + 30 + 54 = 165$

$C1 = 42 + 50 + 30 + 54 = 176$

$C2 = 36 + 45 + 35 + 55 = 171$

$D1 = 42 + 36 + 35 + 30 = 143$

$D2 = 50 + 45 + 55 + 54 = 204$

$E1 = 42 + 36 + 55 + 54 = 187$

$E2 = 50 + 45 + 35 + 30 = 160$

$F1 = 42 + 45 + 35 + 54 = 176$

$F2 = 50 + 36 + 55 + 30 = 171$

$G1 = 42 + 45 + 55 + 30 = 172$

$G2 = 50 + 36 + 35 + 54 = 175$

Trial No.	A	B	C	D	E	F	G	Results
1	1	1	1	1	1	1	1	42.00
2	1	1	1	2	2	2	2	50.00
3	1	2	2	1	1	2	2	36.00
4	1	2	2	2	2	1	1	45.00
5	2	1	2	1	2	1	2	35.00
6	2	1	2	2	1	2	1	55.00
7	2	2	1	1	2	2	1	30.00
8	2	2	1	2	1	1	2	54.00

Level	A	B	C	D	E	F	G
Level 1	173	182	176	143	187	176	172
Level 2	174	165	171	204	160	171	175
Totals	347	347	347	347	347	347	347

$$Y_{avg} = \frac{347}{8} = 43.375$$

Hence optimum is

$A2\ B1\ C1\ D2\ E1\ F1\ G2$ and $A2 + B1 + C1 + D2 + E1 + F1 + G2 = 1274$

$$Y_{opt} = \frac{A2 + B1 + C1 + D2 + E1 + F1 + G2}{8} = 159.25$$

$$(Y_{opt} - Y_{avg})(Y_{opt})^{-1} = 0.72763$$

Hence a 72.763% improvement is predicted.

9.28. An experiment has been designed to study the influence of five factors, which were identified during a brain storming session as influencing the torque delivered by the valve-train of a newly developed small scale engine. Two levels were chosen for the experiment. The factor and their levels are given in the following:

Factors	**Level 1**	**Level 2**
Valve guide clearance	Low	High
Valve geometry	Type 1	Type 2
Seat concentricity	Quality 1	Quality 2
Lower guide length	Location 1	Location 2
Valve face run out	Run out type 1	Run out type 2

An L_8 orthogonal array was selected and torque levels were measured under appropriate conditions for the eight experiments. Interaction is expected between the seat concentricity and the valve face run out. Also, interaction is expected between the valve geometry and the valve face run out.

The eight measured torque levels are as follows:

$$y = \begin{bmatrix} 45 \\ 34 \\ 56 \\ 45 \\ 46 \\ 34 \\ 38 \\ 43 \end{bmatrix}$$

y is measured in units of inch pounds.

(*a*) set up the experimental array in the provided L_8 format as shown in Fig. 9.28 and Table P9.28

(*b*) determine the optimum combination of factors to maximize torque.

(*c*) determine the contribution of each factor to this improvement

(*d*) are either or both interactions significant?

(*e*) estimate the overall improvement of the optimized system.

Solution:

Column / Condition	$L_8(2^7)$ 1	2	3	4	5	6	7
1	1	1	1	1	1	1	1
2	1	1	1	2	2	2	2
3	1	2	2	1	1	2	2
4	1	2	2	2	2	1	1
5	2	1	2	1	2	1	2
6	2	1	2	2	1	2	1
7	2	2	1	1	2	2	1
8	2	2	1	2	1	1	2

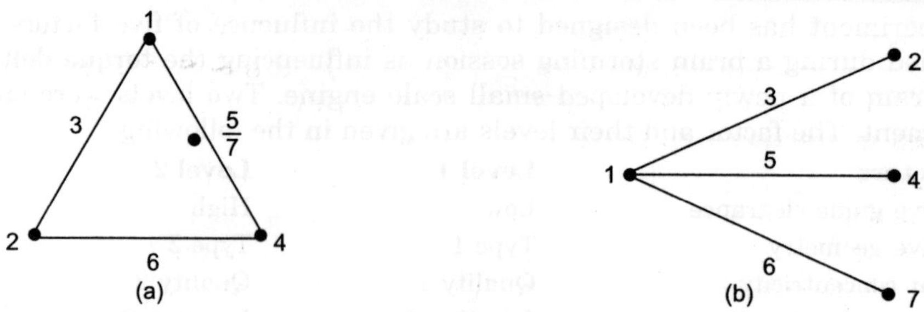

Fig. P9.28 Linear Graphs for L_8.

(*a*) **Set up the experiment array (see Fig. P9.28 and Table P9.28)**

(*b*) **Determine the optimum combination of factors to maximize torque.** ***Note:*** **Best is most.**

$B_1 = 45 + 34 + 56 + 45 = 180$ $\quad$ $E_1 = 45 + 34 + 46 + 34 = 159$

$B_2 = 46 + 34 + 39 + 43 = 162$ $\quad$ $E_2 = 56 + 45 + 39 + 43 = 183$

$E_xB_1 = 45 + 34 + 39 + 43 = 161$ $\quad$ $E_xB_2 = 56 + 45 + 46 + 34 = 181$

$C_1 = 45 + 56 + 46 + 39 = 186$ $\quad$ $C_2 = 34 + 45 + 34 + 43 = 156$

$A_1 = 45 + 56 + 34 + 43 = 178$ $\quad$ $A_2 = 34 + 45 + 46 + 39 = 164$

$E_xC_1 = 45 + 45 + 46 + 46 = 179$ $\quad$ $E_xC_2 = 34 + 56 + 34 + 39 = 163$

$D_1 = 45 + 45 + 34 + 39 = 163$ $\quad$ $D_2 = 34 + 56 + 46 + 43 = 179$

Level 1:

$B_1 = 180$ $\quad E_1 = 159$ $\quad E_xB_1 = 161$ $\quad C_1 = 186$ $\quad A_1 = 178$ $\quad E_xC_1 = 179$ $\quad D_1 = 163$

Level 2 :

$B_2 = 162$ $\quad E_2 = 183$ $\quad E_xB_2 = 181$ $\quad C_2 = 156$ $\quad A_2 = 164$ $\quad E_xC_2 = 163$ $\quad D_2 = 179$

Totals : 342 $\quad$ 342 $\quad$ 342 $\quad$ 342 $\quad$ 342 $\quad$ 342 $\quad$ 342

Sum of torque = (45 + 34 + 56 + 45 + 46 + 34 + 39 + 43) = 2739

By best is most, the highest combination is: $B_1, E_2, E_xB_2, C_1, A_1, E_xC_1, D_2$

$$B_1 + E_2 + E_xB_2 + C_1 + A_1 + E_xC_1 + D_2 = 1266$$

Note: $\quad B_1 \times E_2 = E_xB_2$ and $E_2 \times C_1 = E_xC_1$

Note: When we process mixed interactions, such as B_1 and E_2, the Taguchi method automatically accounts for the mix because it is really evaluating the interaction between columns 1 and 2 and interaction between E_2 and C_1. Here the system accounts for the interaction between columns 4 and 2, in accordance with the linear graph for an L_8 array.

(*c*) $\dfrac{B_1 \times E_2}{1.266 \times 10^3} = 0.143$ or 14.3% contribution

$\dfrac{E_2 \times C_1}{1.266 \times 10^3} = 0.1414$ or 14.1% contribution

$\dfrac{A_1}{1.266 \times 10^3} = 0.1406$ or 14.06% contribution

$\dfrac{B_1}{1.266 \times 10^3} = 0.1422$ and $\dfrac{E_2}{1.266 \times 10^3} = 0.1445$

$$\frac{C_1}{1.266 \times 10^3} = 0.1469 \quad \text{and} \quad \frac{D_2}{1.266 \times 10^3} = 0.1414$$

(*d*) Both interactions are significant.

(*e*) $\dfrac{1.266 \times 10^3 - 342}{342} = 2.7018$

Hence, an increase of 270% is achieved.

9.29. In a certain manufacturing operation, four factors *A*, *B*, *C*, and *D*, each at three levels is thought to be a great importance (control factors) in influencing the volume of the parts produced. The environmental noise factors that could affect the performance of the part are recognized as 3 noise factors at 2 levels. They are given as *E*, *F* and *G*.

Factor	Description	Level 1	Level 2
E	Time	25 hours	125 hours
F	Temperature	70 °F	160 °F
G	Humidity	30%	80%

(*a*) set up an orthogonal arrays for the inner and the outer array

(*b*) determine the number of runs required to complete the tests

(*c*) if the *s/N* ratios for the 9 experimental conditions of the control matrix are given in order as

24.03, 25.50, 25.31, 25.95, 26.92, 25.34, 25.75, 24.84 and 26.25

(*d*) discuss the results obtained from the above analysis.

Solution:

(*a*) An L_9 orthogonal array with 4 control factors, each at 3 levels can be used for the purpose as shown in Fig. P9.29. An L_4 orthogonal array for the outer array for the noise factors can be used. We can run four experiments with 3 noise factors set at one of 2 levels for each of the trials in the inner array.

(*b*) Hence, the total number of experimental runs is 9 × 4 = 36.

Experiment number	1	2	3	4
1	1	1	1	1
2	1	2	2	2
3	1	3	3	3
4	2	1	2	3
5	2	2	3	1
6	2	3	1	2
7	3	1	3	2
8	3	2	1	3
9	3	3	2	1

(*a*) $L_9(3^4)$ Orthogonal array

Experiment number	1	2	3
1	1	1	1
2	1	2	2
3	2	1	2
4	2	2	1

(*b*) $L_4(2^3)$ Orthogonal array

Fig. P9.29 Two Examples of Orthogonal Arrays for use with Taguchi Method.
(*a*) L_9 Experimental Design for 4 Control Factors, each at 3 Levels.
(*b*) L_4 Experimental Design for 3 Factors, each at 2 Levels.

The plots of each factor against the level is shown in Fig. P9.29 (*a*)

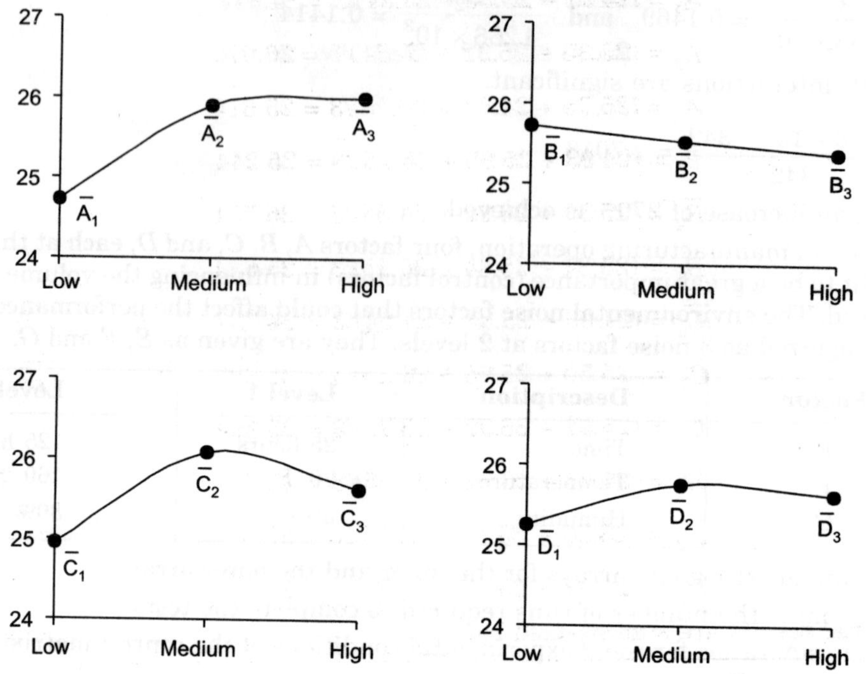

Fig. P9.29 (*a*)

(*c*) and (*d*) We obtain the optimum conditions for the control factors as follows:

	Control Array				
	A	**B**	**C**	**D**	**s/N average**
1	1	1	1	1	24.03
2	1	2	2	2	25.50
3	1	3	3	3	25.31
4	2	1	2	3	25.95
5	2	2	3	1	26.92
6	2	3	1	2	25.34
7	3	1	3	2	25.75
8	3	2	1	3	24.84
9	3	3	2	1	26.25

The average *s/N* for factor *B* at the low level (1) is

$$(24.3 + 25.95 + 25.75)/3 = 25.244$$

Similarly, the response table for S/N averages is

Level	A	B	C	D
1	24.947	25.243	24.737	25.734
2	26.070	25.753	25.897	25.530
3	25.613	25.633	25.993	25.367

$$\bar{A}_1 = (24.03 + 25.50 + 25.31)/3 = 24.947$$

$$\bar{A}_2 = (25.95 + 26.92 + 25.34)/3 = 26.070$$

$$\bar{A}_3 = (25.75 + 24.84 + 26.25)/3 = 25.614$$

$$\bar{B}_1 = (24.03 + 25.95 + 25.75)/3 = 25.244$$

$$\bar{B}_2 = (25.50 + 26.92 + 24.84)/3 = 25.754$$

$$\bar{B}_3 = (25.31 + 25.34 + 26.25)/3 = 25.634$$

$$\bar{C}_1 = (24.03 + 25.34 + 24.84)/3 = 24.737$$

$$\bar{C}_2 = (25.50 + 25.95 + 26.25)/3 = 25.897$$

$$\bar{C}_3 = (25.31 + 26.92 + 25.75)/3 = 25.994$$

$$\bar{D}_1 = (24.03 + 26.92 + 26.25)/3 = 25.734$$

$$\bar{D}_2 = (25.50 + 25.34 + 25.75)/3 = 25.530$$

$$\bar{D}_3 = (25.31 + 25.95 + 24.84)/3 = 25.366$$

The final results are summarized below:

Control parameter	Optimum level
A	2
B	2
C	2
D	1

9.30. **Determine the most robust design of paper airplanes using robust design experiment. The control parameters and noise parameters are given in Tables P9.30 (*a*) and (*b*).**

Table P9.30 (*a*)

Control Parameters

Parameter	Level 1	Level 2	Level 3
Weight of paper (*A*)	One sheet	Two sheets	Three sheets
Configuration (*B*)	Design 1	Design 2	Design 3
Width of paper (*C*)	4 in	6 in	8 in
Length of paper (*D*)	6 in	8 in	10 in

Solution:

The parameter design can be found in the article by S. Sarin, *Quality Progress,* May, 1997, pp.102–106.

Table P9.30 (*b*)

Noise Parameters

Parameter	Level 1	Level 2
Launch height (*X*)	Standing on ground	Standing on chair
Launch angle (*Y*)	Horizontal to ground	45° above horizontal
Ground surface	Concrete	Polished tile

[Adopted from S. Sarin, Quality Progress, pp.102-106, May 1997]

An elastomeric connector to a nylon tube must deliver the required pull-off performance so that it can be used in automobile engine components. Although the objective is to maximize pull-off force and minimize assembly effort, this experiment focuses on maximizing pull-off force. Researchers identified four design factors and three noise factors that appear to affect the assembly's pull-off force. Table P9.31 (*a*) shows the factor assignments and their associated levels. Each design factor was tested at three levels, the noise factors at two. The noise factors are uncontrollable during normal operations, but they were controlled over the two levels for the purpose of tests. Determine a robust design that will be insensitive to the noise factors during actual operation.

Solution:

For complete solution refer to the article: **D.M. Byrne and S. Taguchi (1987),** "The Taguchi Approach to Parameter Design", *Quality Progress,* Dec. 1987: 19-26.